Introduction to the Physics of Electron Emission

Introduction to the Physics of Electron Emission

Kevin L. Jensen
Materials Science and Technology Division,
Naval Research Laboratory, Washington, DC, USA

WILEY

This edition first published 2018

Registered Office(s)
John Wiley & Sons, Inc., 111 River Street, Hoboken, NJ 07030, USA
John Wiley & Sons Ltd, The Atrium, Southern Gate, Chichester, West Sussex, PO19 8SQ, UK

Editorial Office
The Atrium, Southern Gate, Chichester, West Sussex, PO19 8SQ, UK

For details of our global editorial offices, customer services, and more information about Wiley products visit us at www.wiley.com.

Wiley also publishes its books in a variety of electronic formats and by print-on-demand. Some content that appears in standard print versions of this book may not be available in other formats.

Library of Congress Cataloging-in-Publication Data

Names: Jensen, Kevin L., author.
Title: Introduction to the physics of electron emission / by Dr Kevin L. Jensen, Naval Research Laboratory, Washington, DC, US.
Description: First edition. | Hoboken, New Jersey : John Wiley & Sons, Inc., 2018. | Includes bibliographical references and index. |
Identifiers: LCCN 2017023351 (print) | LCCN 2017035681 (ebook) | ISBN 9781119051756 (pdf) | ISBN 9781119051763 (epub) | ISBN 9781119051893 (cloth; alk. paper) |
Subjects: LCSH: Electrons–Emission.
Classification: LCC QC793.5.E62 (ebook) | LCC QC793.5.E62 J46 2018 (print) | DDC 539.7/2112–dc23
LC record available at https://lccn.loc.gov/2017023351

Cover Design: Wiley
Cover Image: © Julia Kopacheva/shutterstock;
Back cover images: Courtesy of J. J. Petillo (Leidos) and the author

Set in 9.5/13pt, MeridienLTStd by SPi Global, Chennai, India

10 9 8 7 6 5 4 3 2 1

For Janet, David, and Cassie - truly my favorite little fermions

Contents

Part III: Exact tunneling and transmission evaluation

Part IV: The complexity of materials

Acknowledgements

Thy glass will show thee how thy beauties wear,
Thy dial how thy precious minutes waste,
These vacant leaves thy mind's imprint will bear,
And of this book, this learning mayst thou taste.
The wrinkles which thy glass will truly show,
Of mouthed graves will give thee memory,
Thou by thy dial's shady stealth mayst know,
Time's thievish progress to eternity.
Look what thy memory cannot contain,
Commit to these waste blanks, and thou shalt find
Those children nursed, delivered from thy brain,
To take a new acquaintance of thy mind.
These offices, so oft as thou wilt look,
Shall profit thee, and much enrich thy book.

– William Shakespeare, Sonnet 77

In describing the colonization of the New World by Europeans and its inauspicious impact on the native populations, Jared Diamond writes[1]

> *...for practical purposes the collision of advanced Old World and New World societies began abruptly in A.D. 1492, with Christopher Columbus's "discovery" of Caribbean islands densely populated by Native Americans.*

The quotation marks about the word *discovery* are pregnant with meaning, calling into question what it means to discover something if that something already contains many people, but the history of the event further complicates interpretation. Columbus had to pitch his proposal to many of the European princes and royalty to secure funding for his enterprise, most of whom declined until King Ferdinand and Queen Isabella of Spain agreed to outfit his three ship voyage on Columbus's second, not first, appeal. As to the alleged discovery itself, the first sighting of land (the Bahamas) was by a lookout on the *Pinta*, whose captain alerted Columbus (who captained the *Santa Maria*) by cannon-fire. To Columbus, however, went the lifetime pension granted by the King and Queen to whomsoever saw land first.

The point of the story is not to look with askance at a popular account of a wellknown event or tarnish its gold-obsessed principal investigator, but rather illuminate that although one person may claim the accolades (or pension or royalty), it is support from colleagues, collaborators, benefactors, supporters, institutions and funding agencies that is most needed for a voyage of discovery to even be possible. It was true then, no less true now, and perhaps not surprising: as put by Robert Wright[2]

> *Unlike food or spears or hides, information is shared without being actually surrendered, a fact that can make the exchange radically non-zero-sum.*

with "non-zero-sum" meaning all players can be winners.

There is sharing, though, and then there is giving. I owe much gratitude to many colleagues, having benefitted from their help, support, mentorship, collegiality, encouragement, friendship, and especially forbearance. Much of what I have written about would not have been learned without them. I am indebted the *Naval Research Laboratory* for its

[1] Jared M. Diamond, *Guns, Germs, and Steel: The Fates of Human Societies*. 1st ed., New York, NY: W. W. Norton & Company, 1997. Chat 3, p67.
[2] Robert Wright. *The Moral Animal: The New Science of Evolutionary Psychology*. New York: Pantheon Books, 1994. p195.

investments in my development, but also to the *Office of Naval Research*, the *Joint Technology Office*, the *Air Force Office of Scientific Research* and the *Department of Energy* for their crucial and well-timed support as well. Many at the national laboratories, in academia, and in industry (both in the US and abroad) shared their time and insights most generously, a testimony to the uniqueness and value of scientific research and a powerful argument for its vigorous public support. More than I can name deserve thanks; a few in particular are (*alph*) T. Akinwande, I. Ben-Zvi, V. T. Binh, C. A. Brau, I. Brodie, F. A. Buot, J. Calame, F. Charbonnier, P. Cutler, D. Dimitrov, D. H. Dowell, D. W. Feldman, D. Finkenstadt, R. G. Forbes, H. Freund, A. Ganguly, B. E. Gilchrist, D. Go, M. C. Green, J. R. Harris, C. Hernandez-Garcia, C. Holland, M. A. Hollis, C. Hunt, L. G. Il'chenko, L. Ives, B. L. Jensen, R. Kishek, M. A. Kodis, S. G. Lambrakos, Y. Y. Lau, J. L. Lebowitz, J. W. Lewellen, R. T. Longo, W. A. Mackie, C. Marrese-Reading, E. Montgomery, N. A. Moody, P. Mukhopadhyay, R. A. Murphy, R. Nemanich, K. Nguyen, G. Nusinovich, P. G. O'Shea, M. Osofsky, W. D. Palmer, Z. Pan, R. K. Parker, B. B. Pate, J. K. Percus, J. J. Petillo, L. Phillips, A. Pique, T. Rao, T. Reinecke, A. Rokhlenko, P. R. Schwoebel, A. Shabaev, J. L. Shaw, D. A. Shiffler, A. Shih, J. M. Smedley, T. Smith, C. A. Spindt, D. Temple, A. Todd, B. Vancil, D. R. Whaley, J. E. Yater, and E. G. Zaidman. Thanks to T. Antonsen and P. G. O'Shea for their interactions and hosting of a sabbatical at the University of Maryland, Institute for Research in Electronics and Applied Physics (UMD-IREAP), from which many good opportunities took root. Thanks especially to S. Chappell, E. Lessner, J. Luginsland, J. Marshall, N. A. Moody, and Q. Saulter for support in truly interesting times. I would like to pay homage to J. K. Percus and J. L. Lebowitz for exemplifying what is best in physicists.

Lastly, the constant enthusiasm of family, particularly parents, siblings, and children, invigorated this effort to its conclusion. Certainly all would have been for naught without the assistance and devotion of my wife, Janet, who was patient with my cloistered writing and manic need to share, offering encouragement rather than complaint, and ample proof that[3]

> *A good marriage is obviously a nonzero sum game, brimming with mutual cooperation.*

[3]Richard Dawkins. *The Selfish Gene.* Oxford; New York: Oxford University Press, 1989. p221.

PART I
Foundations

Tell me, Muse, of the man of many ways, who was driven far journeys ...Many are they ...whose minds he learned of, many the pains he suffered in his spirit ...

– Homer[1]

[1] Homer, *The Odyssey of Homer* (trans. Richmond Lattimore). New York: Harper & Row, 1967, p. 27.

CHAPTER 1
Prelude

Then this new kind of knowledge must have an additional quality?
What quality?
Usefulness in war.

– Plato[1]

Ever since the invention of aerial bombing, military strategists have had to assume that "the bomber will always get through". Every technological advance since then has made this more certain – with one exception.

– R.J. James [1]

In late August of 1940, during the Battle of Britain, Sir Henry Tizard led a mission to the USA at the direction of Prime Minister Winston Churchill. Tizard flew, but the rest of his mission traveled on the *Duchess of Richmond* across the Atlantic, carefully conveying a black box holding Britain's most sensitive military technical secrets [2], an account of which figures prominently in the historical review by James [1]. Tizard's mission was just shy of a year after the famed Einstein–Szilard letter to President F.D. Roosevelt urging the USA to initiate a program developing atomic weapons. Surely it is no surprise, then, what the secret documents concerned.

Radar.

The importance of radar (an acronym of **RA**dio **D**etecting **A**nd **R**anging) was (and is) great, and that importance was starkly revealed in the Battle of Britain (1940), the defeat of the Italian Navy in the Cape of Matapan (1940), the US Navy's defeat of the Japanese aircraft carriers in the Battle of Midway (1942) that marked the turning point of the war in the Pacific, and the elimination of the German U-boat threat in the Atlantic that enabled supplies to reach the British (1943), all of which testified to the pivotal and game-changing role that radar played. A summary by Skolnik [3] is to the point:

There were many factors that allowed the Allies to defeat the Axis forces in Wold War II, but the introduction of radar was one of the most important. There is no way to know whether the Allies would have lost the war if they did not have radar, but it is quite clear that there would have been more losses and a longer time needed to win the war if there were no radar.

An Allied victory was not a sure thing: both Germany and Japan in those years were formidable in military strength and scientific capability, and the edge that radar provided mattered. Ironically, radar's actual importance during World War II, compared to its public perception in the USA, is perhaps inverted from that bequeathed to atomic weapons.[2] After the war, spinoff peacetime applications and descendent technology from microwave ovens, television, cell phone networks, satellite communications, and weather imaging have caused an immense change in how people work, play, and know whether to take an umbrella.

Why is all that important here? Radio wave and microwave amplifiers are made possible by extracting power from energetic bunches of electrons. Electrons are not by nature feral: they are yanked from their host material by violently ripping them out (using scorching heat, electric fields thousands of times stronger than those associated with lightening, intense lasers brighter than the surface of the sun, and even accelerated beams of other high energy electrons). Improving the efficiency of doing so made radar better, and so it and the many technologies indebted to it served as the impetus to drive a great deal of cathode research up to the present. The physics is at its most interesting when the conditions the electron sources are subjected to are the most punishing, and so those conditions are the ones gleefully considered here.

[1] Plato, *The Dialogues of Plato: The Republic VII* (trans. Benjamin Jowett), *Great Books of the Western World*, ed. Robert Maynard Hutchins, Vol. 7, Plato. Chicago: Encyclopaedia Britannica, 1952, p. 391.

[2] For example, "The historical evidence makes clear that the popular view about the use of the bomb is a mythological construct ... there were other options available for ending the war within a reasonably short time without the bomb ... " J.S. Walker [4].

Introduction to the Physics of Electron Emission, First Edition. Kevin L. Jensen.

As to the technologies that benefit, there are many. Microwave ovens are perhaps the most obvious vacuum device, the guts being not much more than a version of the pivotal magnetron behind early radar, but in a shiny box with a fancy timer (an unkind but reasonably accurate characterization). Field emission microscopy [5], flash X-ray sources [6], and microwave devices using metallic wires [7] received intense investigation beginning in the late 1950s as possible alternates to thermionic cathodes (the ability of field emission to be to be strongly modulated being the driver). For microwave devices, using the fields that existed in cavity resonators, the field emission cathodes produced electron beams of high current density in bunches with high harmonic content from tungsten needles. Modern incarnations of high-power microwave (HPM) devices [8] use tufted carbon fiber cathodes, in which the individual fiber bundles coated with cesium iodide (CsI) salt can produce currents in excess of 1 kiloamp [9].

Vacuum electronics [10, 11], and after the 1980s vacuum microelectronics, were behind much cathode development for microwave amplifiers and "tubes" [12]. These applications encompass the class of vacuum devices that operate at microwave frequencies, for either generation or amplification. The microwave frequency range encompasses UHF (0.3–3 GHz), SHF (3–30 GHz), EHF (30–300 GHz), and, most recently, pushing into the sub millimeter (or terahertz) regime. Although "tubes" evokes images of glass-enclosed triodes and pentodes, modern tubes are rather more akin to particle accelerators, albeit small ones.

The most common microwave tubes are klystrons, traveling-wave tubes (TWTs), magnetrons, crossed field amplifiers, gyrotrons, and free electron lasers, and these are widely used in radar, communications, electronic countermeasures, directed energy devices, and particle accelerators. Depending on the frequency band, radar is used for long-range surveillance, long-range weather forecasting, airborne weather forecasting, missile tracking and guidance, marine radar, air and ballistic missile defense, high-resolution mapping and satellite altimetry, and airport surveillance, among others [3], with most modern radars at the higher frequencies. FM and television occupy VHF, microwave ovens and mobile phones occupy UHF, and radio astronomy and directed energy applications such as active denial operate in the EHF. The progress in the metric of the product of average power and frequency squared, or $P_{ave}f^2$, has been remarkably steady over decades of development, as shown in Figure 1.1 – indeed, for high values of that metric, the playing field belongs to the "tubes", with solid-state devices dominating the lower power, lower frequency regimes [13].

Many other devices reliant on electron beams have proliferated. Most people, at least those post-*Hobbit* and pre-*Harry Potter*, remember the cathode ray tubes (CRTs) that were the basis of all televisions, computer monitor displays, and oscilloscopes for decades. The heart of the CRT was a thermionic cathode that generated a beam of electrons that were accelerated towards a phosphor screen such that the beam scanned back and forth ("raster"). Before being overtaken by liquid crystal and plasma displays, flat panel field emission displays (FEDs) were developed in the 1990s using small clusters of microfabricated field emitters to individually address pixel elements on the display [14, 15]. Electron beam lithography [16], an analog of photolithography using particles instead of light, is a method of creating very small structure circuits and nanotechnology by using an electron beam to expose a resist, which can then be selectively removed. Similarly, field emission scanning electron microscopy (FESEM) images secondary electrons emitted from the surface of a sample that are created by primaries emitted from a field emission source and accelerated through a high-gradient potential (generally in the kilovolts range) [17], an improvement on the typically thermionic tungsten source that required

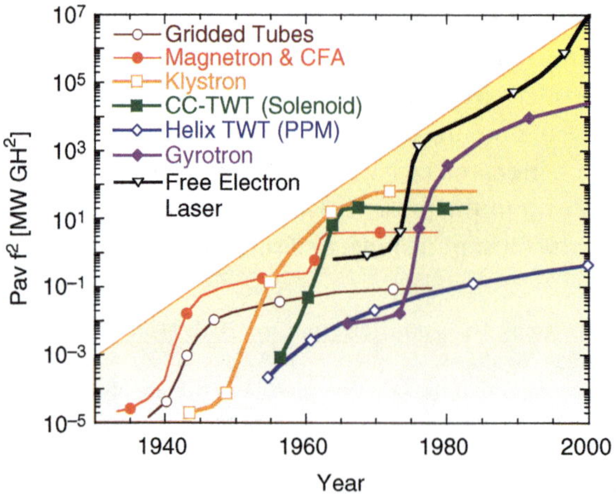

Figure 1.1 Growth in the $P_{ave}f^2$ metric. Shown for various vacuum electronic devices, based on Figure 1 in ref. [11].

long scan times. A higher brightness field emission source enables a modern field emission scanning electron microscope to achieve resolution of feature sizes on the order of 4 nm. Various cathode technologies are behind particle accelerators, radio frequency injectors [18], free electron lasers (FELs) [19, 20], energy recovery LINACs (ERL) [21–23], and X-ray FELs [24]. Other uses include charge neutralization cathodes for ion and Hall thrusters used for satellite propulsion [25], and thermionic and non-thermionic cathode alternatives for electrodynamic tethers [26] for propellentless propulsion for satellites passing through the Earth's magnetic field [27, 28].

Field and secondary emission are not always opportunities: they can cause dark current on the surface of multialkali antimonide photocathodes used in photoinjectors [29] or lead to breakdown on metal surfaces subject to high fields. Copper surfaces of particle accelerators (e.g., the Stanford linear accelerator (SLAC)) exhibit regions of melting and protrusion formation that ultimately give rise to undesirable emission off the copper cavity walls [30–32] are a precursor to seriously undesirable breakdown phenomena.

In that rather rapid summary, four emission mechanisms were identified:

1. thermal: emission by heating [33]
2. photo: emission by absorption of photons [34]
3. secondary: emission by scattering with primary electrons [35]
4. field: extraction by intense electric fields [36].

By the late 1920s, predictive equations for all of these had been developed. An introduction to the theory of the four mechanisms – and space charge – should start at a level accessible to those without prior exposure, but who nevertheless are familiar with basic methods in quantum mechanics, statistical mechanics, electricity and magnetism, mathematical methods of the kind that physicists enjoy, and a smattering of other disciplines that a well-rounded graduate curriculum would provide. Even so, this introduction will try not to presume too much of the reader, and provide opportunities to either become acquainted with or refresh one's knowledge of the physics involved in a way that is, it is hoped, at least interesting and perhaps somewhat different than the usual narratives. For those who crave more, copious citations are provided.

The ambition was to design a narrative that would be instructive, oriented to computation, and a pleasure to *read* – the kind of book that the author would like to read, which may have had unintended consequences. Sometimes this includes material to spice up otherwise mundane methods. There is a reason for that. Science and mathematics are essential to enjoy the good life, but that life also includes the world of ideas, imagination, and curiosity. Why would the two not mingle? Perhaps they should, particularly if it encourages reading for the pleasure of doing so. That is how discovery begins.

CHAPTER 2
Units and evaluation

2.1 Numerical accuracy

...que le mieux est l'ennemi du bien.

– Voltaire[1]

Mauvés ovriers ne trovera ja bon hostill.

– 13th century French proverb[2]

A mathematician and a physicist (the gender of each is not important, whether their gender is the same is not important; if of the same gender, they are distinguishable) walk into a bar (the kind of bar is not important, as long as it is not a sports bar, although if it were a wine bar it could be in Central Square, Cambridge or downtown Menlo Park). At the other end of the bar is a patron (the gender is not important, the distance is not important, the aerosol optical attenuation coefficient is less than 40 km^{-1}) that both the physicist and mathematician show an interest in talking to (why is not important, about what is not important). The mathematician sighs in despair and says "We shall have to cross half the distance remaining an infinite number of times, and so will never get there!" The physicist gets up and starts walking, saying "I'll get close enough."

The point of the anecdote may seem to be that physicists should not tell jokes.

Instead, it is a parable that something, like explaining a subject, can be done correctly in a seemingly infinite number of steps, or there is a point where one gets close enough. "Correctly" may be exact or fundamental. "Close enough" involves hand waving, approximation, analogy, reliance on simple models, and 0.064799 grams of NaCl. But close enough for what? As a practical matter, it should be "close enough to do a calculation". The most accurate theory is not always the best theory: much can be had by way of simplified models that get to the heart of the problem, lest one rues (like Hamlet) that " ...the native hue of resolution is sicklied o'er with the pale cast of thought ..."[3] The gap between understanding and doing can be large. Though number crunching (like facts) is not wisdom, wisdom nevertheless is informed by calculation. A parsimonious calculation is an art.

A good place to start is therefore with a consideration of computations and unit analyses used by electron emission and charge transport calculations, where electrons interact with atoms. For the former, the calculation of Riemann's zeta function serves as a useful example of computational challenges, given its importance and presence in many equations encountered in describing electron emission. For the latter, the prototypical hydrogen atom motivates convenient units. Those units are most easily expressed using length, energy, time, and charge scales of [nm], [eV], [fs], and [q], respectively.

Access to astonishing computational power is the birthright of the present generation, but power does not make right, so to speak. Riemann's zeta function $\zeta(p)$ (Section A2.3 and revisited in Section 6.3) illustrates a pitfall that attends all calculations, namely, the accuracy of numerical results: a computer will always give a number – whether it is what one believes it to be is another matter. The evaluation of $\zeta(3)$ and $\zeta(3/2)$ are pedagogically valuable aside from being topical in the evaluation of the electron density in Section 6.4 or the Stefan–Boltzmann constant of Eq. (3.24), and go to the heart of how much confidence should be placed in numerically calculated integrals using packaged programs. The task

[1] " ...the best is the enemy of the good." Voltaire, *La Bégueule: conte moral*, Google eBook, http://google.com/books?id=cy4HAAAAQAAJ, p. A3.
[2] "A bad workman will never find a good tool." Martin H. Manser, Rosalind Fergusson, and David Pickering, *The Facts on File Dictionary of Proverbs (Facts on File Writers Library)*, 2nd edn. New York: Checkmark Books, 2007, p. 17.
[3] Ref. [37]: *Hamlet*, III.i.84–85, p. 951.

Introduction to the Physics of Electron Emission, First Edition. Kevin L. Jensen.

is therefore to compare computed values with the known exact values[4] quoted to a ridiculous precision:

$$\zeta(3) = 1.202056903159594285399738161 51 \tag{2.1}$$

$$\zeta(3/2) = 2.612375348685488343348567567 92 \tag{2.2}$$

Begin with $\zeta(3)$: from its series definition in Eq. (A2.8) for the first 10^5 terms, a numerical evaluation gives $\zeta(3) \approx 1.2020569031097323$. Compare to a symbolic math integration program value of $\zeta(3) \approx 1.202056909910681$: though less accurate, an error of $4 \times 10^{-9}\%$ is tolerable.

However, the series expansion for $\zeta(3/2)$ does not converge as quickly and the naïve series summation is less useful and, not surprisingly, inaccurate by the third significant digit even for 10^5 terms. A symbolic math integration of $\zeta(3/2) \approx 2.6123751617794717$ is accurate to $6 \times 10^{-6}\%$ (the numerical approach of Section A2.3 is better), and therefore differences arise after the seventh digit. Gaussian quadrature methods work well (as in Section A1.2.3) if the underlying integrand function is a polynomial, which $\sqrt{x}$ (which appears in the numerator of the integrand for $\zeta(p)$ in Eq. A2.6) is not. *Caveat computantum*:[5] the numerical technique matters.

Conversely, unless one is a pedant, accuracy to six digits is often adequate and numerical evaluation of integrals and series, if done with care, is sufficient even though different numerical means give slightly different ends. Therefore, values arising from numerically calculated integrals, as $\zeta(p)$ shall be in addition to many other examples found herein, are accurate to at least six digits, an accuracy that would bring the ersatz physicist of the above parable to a small fraction of the diameter of a human hair for initial separations typical of most bars and more than "close enough".

2.2 Atomic-sized units

> *"What's in a name?" asks Juliet. "That which we call a rose by any other name would smell as sweet." What's in a name is that everyone in a language community tacitly agrees to use a particular sound to convey a particular idea ...Now any of us can convey the thought by making the noise.*
>
> – Steven Pinker[6]

> *...a word is not a relationship between a sound and an object. It is an agreement* among people *who share a joint representation of things in their world, and who share a set of conventions for communicating with each other about those things.*
>
> – Jonathan Haidt[7]

There is a tendency, as in elementary particle physics and cosmology, to economize with constants that appear frequently by setting them equal to 1, in particular, the electron mass m, the speed of light c, the unit charge q, and Plank's constant $\hbar$. While this has an advantage in making certain oft-recurring expressions easy to write, rendering electric field as dimensionless, and avoiding the problems associated with transitioning from units like centimeter-gram-seconds (cgs) to meters-kilograms-seconds-amperes (MKSA[8]), particularly with respect to the potential energy of a collection of charged particles and the forces between them, it is at the cost of a powerful ally in understanding phenomena and finding error. Finding units natural to a problem and working in them is an advantage discarded at peril. When discussing electrons, however, the choice of meters, kilograms, seconds, and amperes (MKSA) are not those units.

MKSA units are the units of technology and man-sized phenomena. Using a unit of energy equivalent to that which an electron acquires by falling through a potential of 1 volt, namely, an electron volt [eV], is a good compromise. For tunneling, distances on the length scale of atoms, or nanometers [nm], and time scales comparable to the inverse frequency of optical (visible light) transitions [fs], are also useful. Doing this, the fundamental constants acquire values that do not involve large exponents and become convenient to use, as in Table 2.1.

In particular, the fine structure constant α, which governs the strength of the electromagnetic interaction, serves as a bridge between various units, in particular cgs, where $\alpha = q^2/\hbar c$, and MKSA, where $\alpha = q^2/4\pi\epsilon_0\hbar c$ and ϵ_0 is the

---

[4] http://oeis.org/A002117 for $\zeta(3)$ and http://oeis.org/A078434 for $\zeta(3/2)$.

[5] "One who calculates, beware" (more or less).

[6] Steven Pinker, *Words and Rules: The Ingredients of Language*. New York: Basic Books, 1999, p. 2.

[7] Jonathan Haidt, *The Righteous Mind: Why Good People Are Divided By Politics and Religion*. New York: Pantheon Books, 2012, p. 240.

[8] The designation MKSA has been eclipsed by the more recent International System of Quantities (ISQ), but MKSA is used here to reinforce the contrast to the nefq units introduced below, as the acronym reinforces the units involved.

Table 2.1 Fundamental constants in units characteristic of atomic phenomena: elementary charge [q], length [nm], time [fs], energy [eV], and temperature [K].

Symbol	Definition	Value	Unit
q	Unit charge	1	q
c	Speed of light	299.792	nm/fs
m	Electron rest mass	510999	eV/c^2
k_B	Boltzmann's constant	$(11604.5)^{-1}$	eV/K
$\hbar$	Planck's constant	0.658212	eV-fs
N_A	Avogadro's number	6.02214×10^{23}	–
α	Fine structure constant	1/137.036	–

permittivity of vacuum. α has the same value (being a dimensionless constant) in all unit conventions; that it has a different defining equation depending on the units in question is a source of vexation for initiates.

There is some practicality therefore in basing a description of forces and potential energies between charges in terms of α. The fine structure constant governs the strength of the electromagnetic interaction between charged particles, and therefore it will play an integral role in descriptions of the hydrogen atom. Thus, units and conventions which push α to the fore will already be of a size and character to conveniently discuss the interaction of electrons with their environment, atomic phenomena, and tunneling phenomena.

Joining the electron charge with potential differences and electric fields gives energies and forces, respectively. As with electron charge q with potential φ giving potential energy $V \equiv q\varphi$, it is easy to speak of a force F on an electron as the product of the electron charge q and the electric field $\mathcal{E}$, or $F \equiv q\mathcal{E}$.

The hydrogen atom motivates units that are agreeable to electron emission physics as it is simple, very well understood, introduces the right scales for atomic phenomena, resolves ambiguity, and is well studied in both classical electrodynamics and quantum mechanics. It shall therefore serve below as the basis for units to be adopted.

Still, using F instead of $q\mathcal{E}$ takes patient accommodation on the part of the reader. Perhaps an analogy helps: were an inquiry as to someone's weight W to receive an answer of "784 Newtons" rather than "80 kilograms", the questioner would look askance at such a response, even though the first answer is technically correct: weight is measured in Newtons. What is being sought differs from what is being asked: it is *mass M* that is the intrinsic property but ignoring the acceleration due to gravity $g = GM_{\oplus}/R_{\oplus}^2 = 9.8$ m/s^2 in $W = Mg$ is so commonplace that in polite discourse responding to weight questions with mass answers is routine. The units of force F and electric field $\mathcal{E}$ are analogous,[9] but even more so, for in the units of [eV/nm] in the former and [GV/m] in the latter, *they are also numerically the same*. Speaking of force when one is really interested in field or vice versa is then much like speaking of weight when one is really interested in mass, but the grounds for doing so are better because the conversion factor in [nfeq] units is unity (see Tables 2.1, 2.2, and 2.3). Still, one should not lose sight of the fact that although the *difference* matters for particular units, the charged particle being an electron and the potential being volts begs for units of [eV], and also because as far as number crunching is concerned in [nefq] units, it is a difference that makes no difference.[10]

2.2.1 Bohr model of the hydrogen atom

> *Whatever the alteration in the laws of motion of the electrons may be, it seems necessary to introduce in the laws in question a quantity foreign to the classical electrodynamics, i.e., Planck's constant, or as it often is called the elementary quantum of action. By the introduction of this quantity the question of the stable configuration of the electrons in the atoms is essentially changed, as this constant is of such*

[9] A purist might object that q is analogous to M, whereas g is analogous to $\mathcal{E}$ given that the latter convey the influence of the environment on the dynamics of a massive or charged object, but that purist would thereby miss the point of the analogy in the first place.

[10] The popular adage "A difference that makes no difference is no difference" is often attributed to William James (*William James: The Essential Writings*, ed. Bruce Wilshire, Harper & Row (Harper Torchbooks, TB1616), 1971, p. xiii) but only appears in the Forward by Wilshire as "...the notion of eternally true but as yet undiscovered propositions lacks a full complement of meaning. That is, the notion is a theoretic difference which makes no difference ..."; rather, the adage is from James Blish (*Spock Must Die! [a Star Trek Novel]*, Bantam Books, 1970), which is strangely fitting: popular usage is wrong in detail even as the truth is captured.

Table 2.2 Conversion of MKSA units expressed in units of [nm], [eV], [fs], and [$q = 1$] parameters [nefq]. To go from MKSA → nfeq, multiply by the factor.

Quantity	MKSA unit	Factor	nefq unit
Length	Meter (m)	10^9	nm
Time	Second (s)	10^{15}	fs
Energy	Joule (J)	6.2415093×10^{18}	eV
Charge	Coulomb (C)	6.2415093×10^{18}	q
Force	Newton (N)	6.2415093×10^{9}	eV/nm
Potential	Volt (V)	1	eV/q
Current	Amp (A)	6241.5093	q/fs
Current density	Amp/square centimeter (A/cm^2)	$6.2415093 \times 10^{-11}$	q/(fs-nm^2)
Field	Gigavolts/meter (GV/m)	1	eV/(q-nm)
Power	Watt (W)	6241.5093	eV/fs
Capacitance	Farad (F)	6.2415093×10^{18}	q^2/eV
Resistance	Ohm (Ω)	$(6.2415093)^{-1}$	eV-fs/q^2

Table 2.3 Fundamental constants in units characteristic of atomic phenomena: elementary charge [q], length [nm], time [fs], and energy [eV]. The mixed units of the last three A–B constants reflect standard usage in commonly encountered units for the evaluation of current density. The value given for ν assumes a work function of 4.5 eV.

Definition	Formula	Value	Unit
e^- rest energy	mc^2	510999	eV
Length (a_o)	$\hbar/\alpha mc$	0.052918	nm
Time	$2\pi a_o/\alpha c$	0.1520	fs
Charge	q	1	q
Q	$\alpha\hbar c/4$	0.35999	eV nm
$2Q/a_o$	$\frac{1}{2}m(\alpha c)^2$	13.605692	eV
A_{RLD}	$(mq/2\hbar)(k_B/\pi\hbar)^2$	120.1735	A/(cm^2 K^2)
A_o	$q/(16\pi^2\hbar)$	1.521334×10^{-6}	A/eV
B_o	$(4/3\hbar)\sqrt{2m}$	6.83089	1/(eV$^{1/2}$ nm)
ν	$(8Q/9)\sqrt{2m/\Phi}$	0.772808	–

dimensions and magnitude that it, together with the mass and charge of the particles, can determine a length of the order of magnitude required.

– Niels Bohr [38]

In the Bohr model of the hydrogen atom, an electron orbits a proton but does so in quantized units of angular momentum. As the speed of light is the fundamental unit of velocity, it is convenient to speak of a particle's velocity as the product of a dimensionless constant with c. For an electron in a hydrogen atom, that constant of proportionality is α, and so the quantization condition is

$$m\alpha c r_n = n\hbar \tag{2.3}$$

The radius $r_1 \equiv a_o = \hbar/\alpha mc$ is the Bohr radius. As the velocity is non-relativistic, the kinetic energy is $KE = (1/2)m(\alpha c)^2$, which is the Rydberg energy R_y.

The centrifugal force the electron experiences is equal and opposite to the attraction between the electron and proton, and so

$$\frac{m(\alpha c)^2}{a_o} = \frac{\alpha\hbar c}{a_o^2} \tag{2.4}$$

The potential energy is $PE = -\alpha\hbar c/a_o$, from which it is seen the magnitude of KE is half that of the potential energy PE, and $PE = -2R_y$.

2.2.2 Quantum mechanical model of the hydrogen atom

The electron can no longer be conceived as a single, small granule of electricity; it must be associated with a wave and this wave is no myth; its wavelength can be measured and its interferences predicted. It has thus been possible to predict a whole group of phenomena without their actually having been discovered. And it is on this concept of the duality of waves and corpuscles in Nature, expressed in a more or less abstract form, that the whole recent development of theoretical physics has been founded and that all future development of this science will apparently have to be founded.

– Louis de Broglie [39]

The same results are obtained directly as a proper mathematical solution of the time-independent Schödinger's equation, $\hat{H}\Psi = E\Psi$. The solution for the coulomb, or $1/r$, potential will be considered at length in Section 9.1, and is moreover a centerpiece of quantum mechanics texts (40–44). The problem is considerably simplified for the ground state (angular momentum quantum number $l = 0$), for which Schödinger's equation is

$$\left\{ -\frac{\hbar^2}{2mr^2}\frac{\partial}{\partial r}\left(r^2\frac{\partial}{\partial r}\right) - \frac{\alpha\hbar c}{r} \right\}\psi = -E\psi \tag{2.5}$$

where $E \equiv \hbar^2 k^2/2m$ is positive, so that $-E$ is a bound state on the right-hand side, which allows the equation to be written as

$$\left\{ \frac{\partial^2}{\partial r^2} + \frac{2}{r}\frac{\partial}{\partial r} + \frac{[(2\alpha mc/\hbar) - k^2 r]}{r} \right\}\psi = 0 \tag{2.6}$$

The *ansatz* $\psi \propto e^{-kr}$, suggested by the large r behavior (i.e., $(\partial_r^2 - k^2)\psi \approx 0$), indicates that $[(\alpha mc/\hbar) - k]/r = 0$, or $k = \alpha mc/\hbar$ for the ground state. The expectation value

$$\frac{\int_0^\infty \left(\frac{1}{r}\right)\psi(r)^2 4\pi r^2 dr}{\int_0^\infty \psi(r)^2 4\pi r^2 dr} = k \tag{2.7}$$

is the definition of the Bohr radius $a_o^{-1} \equiv \langle 1/r \rangle$ and identifies $a_o = \hbar/\alpha mc$.

The size of α shows that the energy of interaction ($2R_y$) is a factor of α^2 smaller than the rest energy mc^2 of the electron. The combination $\hbar c$ appears very often and $\alpha\hbar c = 1.44$ eV-nm is straightforward to remember ($1.44 = 1.2^2$): it will figure prominently in the treatment of the image charge potential and the Schottky barrier factor. For the image charge, the combination appears frequently and motivates the introduction of

$$Q \equiv \frac{1}{4}\alpha\hbar c = \frac{q^2}{16\pi\varepsilon_0} \tag{2.8}$$

Because $(1.2/2)^2 = 0.36$, it follows that $4Q = 1.44$ eV-nm to a good approximation (the actual value being 4×0.3599911 eV-nm).

2.3 Units based on emission

...there is one result, though there might be a hundred anxious schemes.

– The Yi King[11]

Say not, "I have found the path of the soul." Say rather, "I have met the soul walking upon my path." For the soul walks upon all paths.

– Kahlil Gibran[12]

There are many conventions for units, by means of which the same truths should be uncovered. Still, working with physical units that reflect the magnitude of the phenomena being examined makes for an easier path in the development of intuition and order of magnitude estimates without the requirement of tracking exceptionally large exponents. The emission-based units are tailored to calculate the electron number and current density associated with emission through and over surface barriers that exist between bulk materials and vacuum, and electron transport through bulk materials, be they semiconductors or metals.

---

[11] *The Yi King, Sacred Books of the East,* Vol. 16, *The Sacred Books of China,* Vol. 2 of 6, Part II of *The Texts of Confucianism* (trans. James Legge). Oxford: Clarendon Press, 1882, Appendix note to Chapter V.31, p. 389.

[12] Kahlil Gibran, *The Prophet.* New York: Knopf, 1952, Chapter 17: On Self Knowledge.

Observe two consequences of shifting from [MKSA] to [nefq] units: mass [kg] is replaced by energy [eV] and current [A] is replaced by unit charge [$q = 1$] as the units defining the system. The first consequence is that the speed of light c can be attached to $\hbar$ and m to enable the more convenient quantities Q and mc^2 to be used in calculations: energy-based units are convenient especially when treating Fermi levels and work functions, or with tunneling phenomena where the quantity

$$B_o = \frac{4}{3\hbar}\sqrt{2m} = \frac{4}{3}\left(\frac{1}{\hbar c}\right)\sqrt{2mc^2} = \frac{4}{3}\frac{\sqrt{1.022 \times 10^6 \text{ eV}}}{197.33 \text{ eV nm}} = \frac{6.831}{\sqrt{\text{eV}}\text{nm}} \tag{2.9}$$

is easily calculated, as given in Table 2.3. The second consequence is that the electron unit charge q becomes embedded in the unit of energy [eV], charge density is numerically equivalent to number density, and current density to the number of electrons passing a unit area [nm^2] in unit time [fs]. The [nefq] units therefore simplify the treatment of Schrödinger's equation describing the wave function, and Poisson's equation relating distributions of charges to potential energy.

A third reason, but less consequential here, is that combinations of the fundamental constants associated with the hydrogen atom appear in possibly surprising places. For example, a characteristic time is defined by how long a photon takes to traverse the radius of a hydrogen atom, or c/a_o. The ratio of two energy scales, namely the potential energy of the hydrogen atom ground state $4Q/a_o$ and the rest energy of the electron mc^2, is $1/\alpha^2$. Joining them gives the Alfvén–Lawson current of Eq. (33.12). Such products of factors are simply a consequence of the fundamental units of nature and have no bearing on what is under discussion.[13]

Realizing that F is a *force* rather than a *field* takes some getting used to, but an example helps. Consider the Schottky factor $\sqrt{4QF}$, which describes how much a potential energy barrier to emission lowers when an electric field is applied to a surface: it is generally treated as lowering the work function Φ, an energy measured in units of [eV], to $\phi = \Phi - \sqrt{4QF} \equiv (1 - y)\Phi$, where

$$y \equiv \frac{\sqrt{4QF}}{\Phi} \tag{2.10}$$

The Schottky factor appears often in electron emission theory, so that familiarity with how to calculate with it is advantageous.

EXAMPLE: Consider a metal such that $\Phi = 4.5$ eV. By what fraction does the emission barrier lower for fields characteristic of thermionic emission, where the electric field is 0.1 MV/m?

SOLUTION: The electric field corresponds to $F = 0.1$ eV/μm, and so

$$y = \frac{1}{4.5 \text{ eV}}\sqrt{(1.44 \text{ eV nm})\left(0.1\frac{\text{eV}}{\mu\text{m}}\right)} = \frac{1}{375}$$

EXAMPLE: Consider a metal such that $\Phi = 4.5$ eV. By what fraction does the emission barrier lower for fields characteristic of field emission, where the electric field is 9 GV/m?

SOLUTION: The electric field corresponds to $F = 9$ eV/nm, and so

$$y = \frac{1}{4.5 \text{ eV}}\sqrt{(1.44 \text{ eV nm})\left(9\frac{\text{eV}}{\text{nm}}\right)} = \frac{4}{5}$$

[13]The consequences of this are that one can, if one is sufficiently perverse, introduce fundamental constants like $\hbar$ where they are not normally expected, as in Poisson's equation in Eq. (16.1), by using $\alpha\hbar c$ instead of $q^2/4\pi\epsilon_0 = 4Q$ because of the ubiquity of α in relating the electron velocity in a hydrogen atom to the speed of light, or $m(\alpha c)^2/2 = q^2/4\pi\epsilon_0 a_o = 4Q/a_o$, as in Section 23.4. This produces vexation and assertions that quantum mechanical units should not appear in classical equations. But one is dealing with electrons: quantum mechanical language is woven into the language of the discussion.

CHAPTER 3
Pre-quantum models

3.1 Discovery of electron emission

The question is, what is the origin of this current? How is it produced? Since we have within the globe a nearly perfect vacuum, we cannot conceive the current as flowing across vacuous space, as this is not in accordance with our pre-conceived ideas connected with higher vacua ...I have no theory to propound as to the origin of these phenomena.

– Edwin J. Houston [45]

In October, 1884, two months after the cornerstone of the Statue of Liberty was laid in New York Harbor, and two months before either the completed Washington Monument became the tallest structure in the world or the first publication of Mark Twain's satiric masterpiece *The Adventures of Huckleberry Finn* in the UK and Canada (two months again before being published in the USA), Professor E.J. Houston's presentation [45] at the first American Institute of Electrical Engineers meeting in Philadelphia described a seemingly inexplicable high-vacuum phenomena first observed by Edison.[1] Edison had invented an incandescent lamp in which a carbon fiber was enclosed in a glass vessel, the ends of which were connected to a battery. In a modified lamp, as shown in Figure 3.1, an additional strip of platinum foil was inserted and connected to a galvanometer, for which one terminal was connected to the positive terminial (+) and the other connected with the platinum strip. For ordinary current levels, light was produced with no unusual effects, but as the current levels rose so that the luminosity of the lamp increased markedly (a factor of 3 to 12), the needle of the galvanometer was strongly deflected by current passing through its coils. For the physical theories of the time, and as Houston reported to the audience, the origin of the current was a mystery. Even more mysterious, when the leads were switched so that the galvanometer was connected to the negative terminal, then the current flowed in the opposite direction but at about 1/40 of the level. Houston was baffled.

Eventually the phenomena became known as the Edison effect and the current as cathode rays. Up until the 1890s, very different conceptions about the nature of these phenomena were debated. One explanation, identified with German physicists, was that cathode rays were a process occurring in the aether because their motion in magnetic fields was in circular paths rather than straight lines. A competing explanation argued that the cathode rays were material and were the paths of particles of negative charge. In 1895, Perrin demonstrated the particle nature of the cathode rays, and then in 1897 J.J. Thomson at Cambridge University settled the matter definitively by showing the deflection of the rays passing between, essentially, a capacitor plate [47], showing that all of the particles had the same charge to mass ratio that was independent of the potential between the electrodes, and indifferent to either the residual gases within the tube or the chemical nature of the materials from which the rays were produced. For his work, Thomson received the Nobel Prize in physics in 1906.

3.2 The Drude model and Maxwell–Boltzmann statistics

...the great tragedy of Science – the slaying of a beautiful hypothesis by an ugly fact ...it is one thing to refute a proposition, and another to prove the truth of a doctrine which, implicitly or explicitly, contradicts that proposition.

– Thomas H. Huxley[2]

[1] As a result of Edison's advocacy, the Naval Research Laboratory opened in 1923 on the banks of the Potomac in Washington, DC, where pulsed radar debuted in 1934 [46]. The venerated inventor's bust greets visitors at the entrance (www.nrl.navy.mil/about-nrl/history/edison/).
[2] Thomas H. Huxley, Presidential Address at the British Association for the Advancement of Science: Biogenesis and Abiogenesis, *Collected Essays*, 8, 1870.

Introduction to the Physics of Electron Emission, First Edition. Kevin L. Jensen.

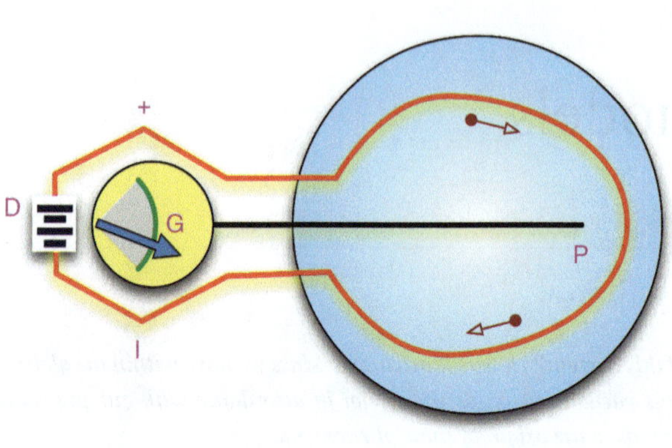

Figure 3.1 Schematic of an incandescent lamp containing a carbon fiber. The fiber (red line with current flowing in the direction of the red arrow) is connected to a battery (D) and a strip of platinum foil (P). At high current, the galvanometer (G) is deflected "violently" by a mechanism which became known as the Edison effect and was subsequently related to the thermal emission of electrons. Based on Figure 1 of ref. [45].

Drude developed a free electron theory of metals in 1900 that was instrumental in explaining electrical conductivity and its relation to thermal conductivity [48].[3] Assuming that all of the electrons in a metal had a velocity $v = \sqrt{3k_BT/m}$ much like ideal gas atoms, and that scattering with the lattice atoms randomized their direction, Drude found that the ratio between the thermal and electrical conductivities of a metal was $3(k_B/q)^2T$. Lorentz improved the model by assuming that the electron velocities were distributed according to Maxwell–Boltzmann (MB) statistics. The Drude model was a watershed event in the understanding of electrons in materials, a circumstance all the more tragic for being mistaken about the statistics that the electrons obeyed. What replaced it, however, was remarkable.

The seemingly successful use of MB statistics in the Drude model lent credence to its analogous use in the first derivation of an equation to describe thermionic emission (Section 5.7), in which electrons were thought to be boiled off a metal. Boiling increases with temperature, and therefore the temperature-dependence of thermionic emission was in stark contrast to the temperature-indifference of field emission [49], although that was because the temperatures applied were not high enough in field emission experiments [50] to reveal an effect. As a result, thermal-electrons were thought to be a different kind of particle than field-electrons, and they were dubbed "thermions" so that emitters producing them were called "thermionic" emitters (unfortunately the name stuck, even after thermal-electrons and field-electrons were recognized as being just electrons). The first formulation of the thermal emission equation came remarkably close in its form to the "modern" formulation. Its derivation shows the power of statistical reasoning even when a major feature of the problem, the indistinguishable nature of discrete fundamental particles and what that entails, was not yet realized. Understanding the physics of the early theories and the problems they faced is more than an interesting historical yarn. It helps form a deep appreciation for the bravura on the part of many who brought forth a coherent narrative of electron emission in all its forms.

First, though, the MB distribution needs to be understood if only just a little bit (for more, see [51]). Although treated at length in Section 5.4, a working understanding of MB distributions adequate to present needs is easy to come by. Let the "environment" be a small box not subject to potential gradients (like gravity or electric fields). The MB distribution can be thought of as the probability that a gas molecule in thermal equilibrium with its environment has an energy E. It is intuitive that having more energy is progressively more unlikely, and so if $E_1 < E_2$, it is expected that $f(E_2)/f(E_1) < 1$. It is also intuitive that the probability should not depend on where the zero point of the energy is measured from, and so if energy is measured from a different reference μ such that $E_n = E'_n - \mu$ then $f(E'_1)/f(E'_2) = f(E_1 + \mu)/f(E_2 + \mu)$. A different way of expressing the equality is

$$\ln\,[f(E_1 + \mu)] - \ln\,[f(E_1)] = \ln\,[f(E_2 + \mu)] - \ln\,[f(E_2)] \tag{3.1}$$

but because μ is at the moment arbitrary and therefore arbitrarily small, and because of the similarity of Eq. (3.1) to finite difference methods (see Section A1.3.2) and therefore differentiation, that in turn implies $d\ln[f(E)]/dE$ =const. $\equiv -\beta$, where the "−" is a consequence of $f(E)$ *decreasing* as E *increases*, and so $f(E) \propto \exp(-\beta E)$.

The discussion has been so general that the significance of $f(E)$ could be mistakenly overlooked. To emphasize an important point, $f(E)dE$ is the *probability* that a particle can be found with an energy E, and this will be profoundly

[3] It also treated optical properties, but wait for that until Section 31.6.

important in treating the distribution functions and statistics below. Probabilities are normalized so that their sum, or integral, is 1, and so

$$f(E) \equiv \frac{e^{-\beta E} dE}{\int_0^\infty e^{-\beta E} dE} = \beta e^{-\beta E} dE \tag{3.2}$$

The average energy is then the integral of E with its distribution, or

$$\langle E \rangle = \int_0^\infty E f(E) dE = \frac{1}{\beta} \tag{3.3}$$

If two same-sized boxes of the same kind of gas with the same number of particles are mixed, and they are characterized by β_1 and β_2, respectively, then the average energy before mixing should be equal to the average after mixing, or

$$\frac{1}{2}[\langle E_1 \rangle + \langle E_2 \rangle] = \langle E_{mix} \rangle \rightarrow \frac{1}{\beta_1} + \frac{1}{\beta_2} = \frac{2}{\beta_{mix}} \tag{3.4}$$

Because a measure of the average energy of a gas is temperature T, this shows that $\beta = 1/k_B T$, where k_B relates temperature to energy.

Still, why would anyone think of electrons as a gas? Well, as Drude showed, there some pretty good reasons, and electrical conductivity is one of them.

3.2.1 Electrical conductivity

One of the principal characteristics of materials is their ability (or lack of ability) to conduct electrical current ...if one takes the conductivity of superconductors, measured at low temperatures, into consideration, this span extends to 40 orders of magnitude ...(t)his is the largest known variation in a physical property and is only comparable to the ratio between the diameter of the universe (about 10^{26} m) and the radius of an electron (10^{-14} m).

– Rolf E. Hummel [52], p. 80

In the Drude model, electrons freely roam through the metal, their electrical charge being balanced by the ions, so the electrons act as if they are independent and therefore only respond to an external field within the material. Their motion as a consequence of the field leads to current flow. Much like a gas of particles, the electrons travel in straight lines for a time τ until they collide with something. Drude assumed that the collisions were with fixed ions, but the causes of scattering are less important than the consequences.

Particles subject to a force F accelerate, and so to prevent boundless acceleration Drude proposed that after each collision the velocity of the electrons was brought back to zero, and the acceleration begun afresh. If the number density of electrons ρ [q/cm^3] is comparable to the number density [atoms/cm^3] of the metal (or equal, if each metal atom gives rise to a single electron), then the current density J is the product of the charge q traveling at an average velocity v_{ave} with the density, or

$$J_q = -q\rho v_{ave} \tag{3.5}$$

where the subscript q is used to distinguish a current of charge from a current of heat below.

What "average velocity" means is nuanced: it is not simply the mid-point velocity between the start and just before a scattering event, and so it will not do to take $v_{ave} \approx (v_{max} + v_{min})/2$. What matters is the mean distance the electron goes and the mean time it takes to get there. "The mean time between collisions is τ" is equivalent to saying that the probability that an electron experiences a collision in time t is $P(t) \propto e^{-t/\tau}$ (see Section A3.30.1). To normalize it, take

$$P(t) = \frac{e^{-t/\tau}}{\int_0^\infty e^{-t/\tau} dt} = \frac{1}{\tau} e^{-t/\tau} \tag{3.6}$$

The average velocity is the integral of $v(t) = Ft/m$ over the probability $P(t)$ that the electron has that velocity, and so

$$v_{ave} = \frac{1}{\tau} \int_0^\infty \left(\frac{F}{m} t \right) e^{-t/\tau} dt = \frac{F}{m} \tau \tag{3.7}$$

The relationship between current flow and electric field is given by Ohm's law

$$J_q = -\sigma \mathcal{E} = -\sigma \left(\frac{F}{q} \right) = \frac{\sigma}{q} \frac{dV}{dx} \tag{3.8}$$

where σ is the conductivity of a metal. Combining Eq. (3.5) with (3.8) gives

$$\sigma = \frac{q^2\tau}{m}\rho \tag{3.9}$$

Observe that the current is therefore related to the product of a *charge* conductivity with the gradient of *potential* energy. The electron collisions are not without consequence: the kinetic energy they acquired from acceleration is ultimately transferred to the lattice (Joule heating) and the impediment to their motion by suffering those scatterings is known as "resistivity" ($1/\sigma$).

EXAMPLE: What is the conductivity of a copper-like metal for which $\tau = 25$ fs and $\rho = 8.41 \times 10^{22}$ cm^{-3}?
SOLUTION: Convert the density to $\rho = 84.1$ nm^{-3} and the electron mass $m = 5.68563$ eV [fs/nm]^2. Then, using Table 2.2,

$$\sigma = q^2 \frac{(25 \text{ fs})}{\left(5.68563 \text{ eV}\left(\frac{\text{fs}}{\text{nm}}\right)^2\right)}(84.1 \text{ nm}^{-3})$$

$$= 370\frac{q^2}{\text{eV fs nm}} = 5.93 \times 10^7 \frac{1}{\Omega \text{ m}}$$

Electrons obviously carry charge during transport from one region to another, but they also carry energy. If the temperature of the electron gas is uniform, no net transfer of energy occurs across a reference plane, but if a difference in temperature exists across that plane (a temperature gradient), then a flow of heat occurs. Metals are known to be both good current conductors *and* good heat conductors: if electrons are both the carriers of electric charge and thermal energy, then the assumption that the collision processes governing the flow of heat and charge are mediated by similar mechanisms has consequences. Therefore, consider thermal conductivity.

3.2.2 Thermal conductivity

...hot thoughts beget hot deeds ...

William Shakespeare[4]

As J_q was the flow of charge, so is J_T is the flow of heat, and is the product of a *thermal* conductivity κ with a gradient of a *temperature* energy k_BT or $J_T = -\kappa(dT/dx)$; the sign designates that energy flows from hot to cold. Electrons cross a boundary (say, the y–z plane) with a velocity v_x. Because the $+\hat{x}$ region has a different temperature than the $-\hat{x}$ region, the electron velocities from the + region will differ from those from the − region. The net flow of energy is the difference, or

$$J_T = \rho\{E_x^+ v_x^+ - E_x^- v_x^-\} \approx \rho\left(\frac{d\langle E_x v_x\rangle}{dx}\right)2l \tag{3.10}$$

$$= 2l\rho\frac{d\langle E_x v_x\rangle}{dT}\frac{dT}{dx} \tag{3.11}$$

and so $\kappa = 2l\rho(d\langle E_x v_x\rangle/dT)$, where l is the average distance the electrons travel between collisions, or the mean free path $v_{ave}\tau$. The factor of $2l$ instead of simply l occurs because, on average, the last collision an electron experiences is a distance l from the x–y plane for both sides of the plane, as in Figure 3.2.

The evaluation of the averages of a gas of particles, be they atoms of an ideal gas or a gas of electrons, requires a distribution of velocities. Drude made the reasonable assumption that the distribution $f(v)$ was MB, or the probability that a particle has an energy $E = mv^2/2$ is $f_{MB}(v) \propto \exp(-\beta E)$ where $\beta \equiv 1/k_BT$ and $E = mv^2/2$. The MB distribution is useful because

$$f_{MB}(v) = f_{MB}(v_x)f_{MB}(v_y)f_{MB}(v_z) \tag{3.12}$$

[4] Ref. [37]: *The History of Troilus and Cressida*, III.i.120, p. 996.

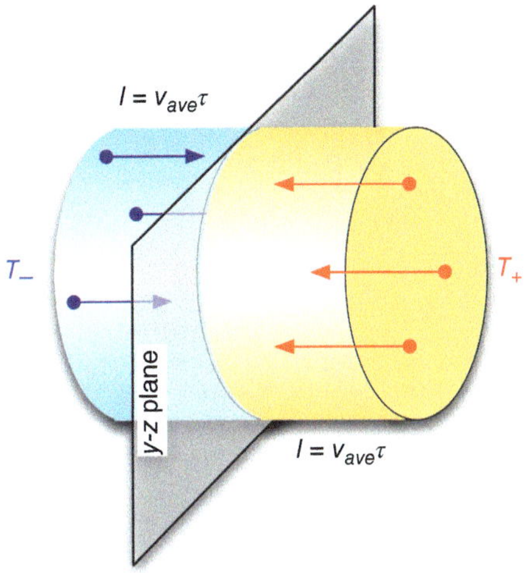

Figure 3.2 Electron flow past an x–y plane. After collisions at each of the cylinder ends that thermalize electrons, they travel a distance $l_\pm = v^\pm_{ave}\tau$ but $l_+ \approx l_-$ so that the cylinder is of length $2l$.

from which it follows that for some quantity that depends only on v_x, or $O(v_x) = O_x$, averages with respect to the normalized distribution $f_{MB}(v)$ satisfy

$$\langle O_x \rangle = \int O_x f_{MB}(v) dv_x dv_y dv_z = \int O_x f_{MB}(v_x) dv_x \tag{3.13}$$

To make the notation simpler, suppress the MB subscript on $f(v)$.

There are subtitles. $\langle v_x \rangle = 0$ as the distribution $f(v_x)$ is an even function in v_x, which makes $\langle v_x^n f(v_x) \rangle$ identically zero if n is odd. Finding a usable average velocity therefore takes one of two approaches. The first is to take $v_{ave} \approx \sqrt{\langle v_x^2 \rangle}$, that is, to approximate the average velocity by its *root mean squared* (rms) value. The second approach is to consider the average velocity in the $+\hat{x}$ direction $\langle v_x^+ \rangle$, which will be close to $\langle v_x^- \rangle$ if the temperature gradient is small.

EXAMPLE: What is the *rms* velocity of an electron thermalized with the lattice?
SOLUTION: A straightforward calculation gives

$$\langle v_x^2 \rangle = \frac{\int_{-\infty}^{\infty} v_x^2 \exp(-\beta m v_x^2/2) dv_x}{\int_{-\infty}^{\infty} \exp(-\beta m v_x^2/2) dv_x} \tag{3.14}$$

$$= \frac{2k_B T}{m} \frac{\int_{-\infty}^{\infty} s^2 e^{-s^2} ds}{\int_{-\infty}^{\infty} e^{-s^2} ds} = \frac{k_B T}{m} \tag{3.15}$$

Observe that this implies $m\langle v^2 \rangle/2 = (3/2)k_B T$ as a consequence of the equipartition theorem. At 300 K, $\sqrt{\langle v^2 \rangle} = c\sqrt{3k_B T/mc^2} = c\sqrt{(3/38.7)/(5.11 \times 10^5)} = 3.896 \times 10^{-4} c$.

EXAMPLE: What is the average positive velocity of an electron thermalized with the lattice?
SOLUTION: Including only those electrons traveling in the $+\hat{x}$ direction in the distribution,

$$\langle v_x^+ \rangle = \frac{\int_0^{\infty} v_x \exp(-\beta m v_x^2/2) dv_x}{\int_0^{\infty} \exp(-\beta m v_x^2/2) dv_x} = \sqrt{\frac{2k_B T}{\pi m}} \tag{3.16}$$

or 79.8% of $\sqrt{\langle v_x^2 \rangle}$.

A calculation similar to Eq 3.16 gives $\langle E_x^\pm v_x^\pm \rangle = (2k_B^3 T_\pm^3/\pi m)^{1/2}$ or

$$\frac{d}{dT}\langle E_x v_x \rangle = 3k_B \left(\frac{k_B T}{2\pi m}\right)^{1/2} \tag{3.17}$$

Collecting results and using $l \approx \langle v_x^{\pm} \rangle \tau$ gives

$$\kappa = \frac{6\tau k_B^2 T}{\pi m} \rho \tag{3.18}$$

3.2.3 The Lorentz number for metals

But that's not unusual/No it isn't strange
After changes upon changes/We are more or less the same.

– Paul Simon[5]

There is no *a priori* reason to take the τ of Eq. (3.9) to be that of Eq. (3.18), but if the thermal and electrical relaxation times are the same or close then the Wiedemann–Franz law predicts that the Lorentz number L is given by

$$\frac{\kappa}{\sigma T} \approx L_{crude} = \frac{12}{\pi} \left(\frac{k_B}{q} \right)^2 \tag{3.19}$$

That this might have been anticipated is well-captured by Ziman ([53], p. 232):[6]

In electrical conduction each electron carries its charge q, and is acted on by the field $q\mathcal{E}$. The current per unit field is proportional to q^2. In thermal conduction each electron carries its thermal energy $k_B T$, and is acted on by a thermal force $k\vec{\nabla}T$. The heat current per unit thermal gradient is proportional to $k_B^2 T$. The ratio of these two transport coefficients must be of the order of $k_B^2 T^2/q^2$; the factor $\pi^2/3$ comes from the fact that we are dealing only with electrons on the Fermi surface, obeying Fermi statistics.

with a well-placed teaser at the end about what is in store because of the statistics electrons obey. The Lorentz number L is therefore a ratio of fundamental constants and independent of material properties, and is approximately $L_{crude} = 2.8365 \times 10^{-8}$ [eV/q K]^2 = 2.8365×10^{-8} [A/Ω K]^2, where "crude" means that consideration of quantum effects much improves the estimate (see Eq. (3.20)). It is a striking achievement of Drude theory that for actual metals, the ratio of their thermal and electrical conductivities matches over a range of temperatures for a variety of metals, as shown in Figure 3.3.

3.2.4 Deficiencies of the Drude model

Since ...truth emerges more readily from error than confusion, we consider it useful to leave the understanding at liberty to exert itself...
– Francis Bacon[7]

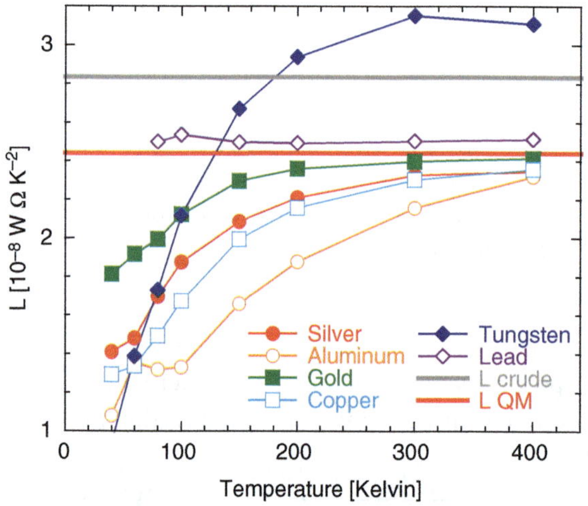

Figure 3.3 The Lorentz number for metals. Comparison of the "crude" Lorentz number (Eq. (3.19) for the gray line) with experimental values. Quantum mechanical considerations change the theoretical estimate to the red line of Eq. (3.20).

[5]Paul Simon, lyrics to *The Boxer*, 1968.
[6]Ziman uses e rather than q for unit charge, and k rather than k_B for Boltzmann's constant; to maintain consistency, the terms were rendered as used here instead of adhering to the conventions of the source material. As a general matter, Ziman is an excellent resource on transport and scattering.
[7]Francis Bacon, *Novum Organum, Great Books of the Western World*, ed. Robert Maynard Hutchins, Vol. 30, Francis Bacon. Chicago: Encyclopaedia Britannica, 1952, Aphorisms Book II, XX.

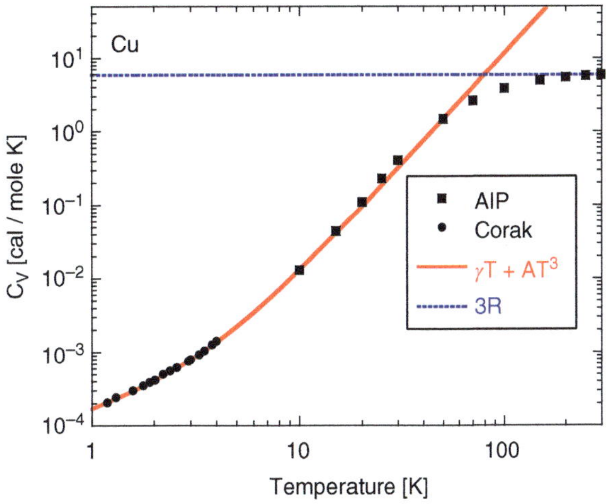

Figure 3.4 Heat capacity of copper. At low temperatures, $C_V \approx \gamma T + \tilde{A}T^3$, where $(C_e = \gamma T)$ is due to electrons and $(C_i = \tilde{A}T^3)$ to the lattice. Low-temperature data (black circles) from Figure 5 of Corak et al. [54]; high-temperature data (black squares) from the American Institute of Physics (AIP) ([55], pp. 4–106). At high temperature, the heat capacity approaches the Dulong–Petit value $(C_V \to 3N_A k_B)$ for one mole.

The limitations of Drude theory are traceable to its assumption that the electron distribution function is MB. The specific heat capacity $C_V = dE/dT$ for solids at high temperatures is known to be $3Nk_B$ (the Dulong and Petit value), as in the discussions surrounding Eq. (5.13), a finding dependent on $f(v)$ being an MB distribution. However, metals at low temperature depart from that relation, where the electron contribution to the specific heat dominates and is linear in temperature. This is shown in the low-temperature region of the heat capacity of copper in Figure 3.4, in which the red line fit to the data follows the form $C_V = 0.00016099T + 1.1921 \times 10^{-5}T^3$ in the units of cal/mole-K: C_V has been scaled by the ratio of the density with the atomic mass of copper, or $(8.96\ \text{g/cm}^3)(63.5456\ \text{g/mole})^{-1} = 0.141\ \text{mole/cm}^3$.

An early success of quantum mechanics was the explanation of heat capacity, and as a part of the new physics electrons were found to be described by Fermi–Dirac statistics, a distinction that corrects the Lorentz number to

$$L = \frac{\pi^2}{3}\left(\frac{k_B}{q}\right)^2 = 2.443 \times 10^{-8}\left[\frac{\text{A}}{\Omega\text{K}}\right]^2 \tag{3.20}$$

as shown in the treatment leading to Eq. (7.50), and produces the correct temperature dependence of $C_V = C_e + C_i = \gamma T + \tilde{A}T^3$ at temperatures well below the Debye temperature, as shall be shown when the quantum distribution functions are considered. The Sommerfeld model, which relies on the Fermi–Dirac distribution, corrects the Drude model and therefore forms the basis on which the equations of electron emission are erected.

3.3 The challenge of photoemission

...a light quantum delivers its entire energy to a single electron ...the velocity distribution of the ejected electrons will be independent of the intensity of the incident light; on the other hand the number of electrons leaving the body will, if other conditions are kept constant, be proportional to the intensity of the incident light.

– Albert Einstein[8]

Given the long run of Thomas Edison's incandescent light bulbs in dispelling darkness for much of the 20th century, it is almost intuitive that heated objects (in the light bulb, a tungsten wire filament) emit light, and the spectrum of the light they emit is related to their temperature – in the case of a tungsten wire, a temperature comparable to 2200 K produces light in the visible region of the spectrum. A warm object that radiates light is termed a "black body radiator".

In 1900, Lord Rayleigh derived the distribution in power as a function of frequency (the Rayleigh–Jeans formula). If light is imagined to be enclosed in a box and at thermal equilibrium with it then the energy density $u(\omega)$, or the amount

[8]Albert Einstein, Concerning the Emission of Cathode Rays Through Illumination of Solid Bodies, *Ann. Physik.*, **17**, 132, 1905, as translated in *American Journal of Physics*, **33**(5), May 1965, Section 8.

of energy in each phase space element, can be derived. The phase space concept will be revisited in much greater detail in Section 5.1, but for now the phase space volume is $(2/(2\pi)^3)d\vec{k}$, where the 2 in the numerator of the coefficient is for each polarization of light, a factor of 2π is associated with each component (k_x, k_y, k_z), and $\hbar\vec{k}$ is momentum. The average energy ε_ω of each mode of light (represented as little harmonic oscillators for which the kinetic and potential energy is $k_BT/2$, the sum of which is the total energy) is k_BT. Lastly, for light, the relation between energy $\hbar\omega$ and momentum is $\hbar\omega = \hbar kc$. Putting the pieces together, the Rayleigh–Jeans formula is, after integration over the angular part $(\int_0^{2\pi} d\phi \int_0^{\pi} \sin\theta d\theta = 4\pi)$,

$$\varepsilon_\omega \frac{2}{(2\pi)^3} 4\pi k^2 dk = k_B T \frac{\omega^2}{\pi^2 c^3} d\omega \equiv u_r(\omega) d\omega \tag{3.21}$$

which shows both the temperature T and angular frequency ω dependence (although the account here is beholden to modern notation and parameters). While $u_r(\omega)$ captures the long wavelength (short frequency) spectrum of black body radiators rather well, it exhibits the disquieting feature that the total energy contained in the cavity, or the integral over all frequencies, is infinite.

Shortly before, in 1896, Wien developed a relation for the high-frequency regime, thereby avoiding the so-called "ultraviolet catastrophe" that the Rayleigh–Jeans formula suffers, but departed from observations at low frequencies. Wien's formula is equivalent to

$$u_w(\omega) d\omega = \frac{\hbar\omega^3}{\pi^2 c^3} e^{-\hbar\omega/k_B T} d\omega \tag{3.22}$$

Enter Planck, who realized that the Wien and Rayleigh–Jeans formulae were high- and low-frequency limits of an equation of the form

$$u_p(\omega) d\omega = \frac{\omega^2}{\pi^2 c^3} \left(\frac{\hbar\omega}{e^{\hbar\omega/k_B T} - 1} \right) d\omega \tag{3.23}$$

Planck found that he could derive his formula if he assumed that, for whatever reason, the modes of the electromagnetic radiation appeared as integer multiples of a unit energy [43], or $\hbar\omega = n\hbar\omega_o$, transforming the integral over frequency to a sum over n. The derivation is deferred to Eq. (6.4) in Section 6.1 below (there, because photon number is not conserved, the parameter μ, or chemical potential, is 0). The total energy U, that is, the integral over $u(\omega)$ over all frequencies, thereby became finite, and the ultraviolet catastrophe averted. In fact, U is

$$U = \frac{1}{\pi^2 \hbar^3 c^3 \beta^4} \int_0^\infty \frac{x^3}{e^x - 1} dx = \frac{\Gamma(4)\zeta(4)}{\pi^2 \hbar^3 c^3 \beta^4} = \frac{\pi^2}{15 \hbar^3 c^3 \beta^4} \tag{3.24}$$

where the integral is the product of the Rieman zeta function $\zeta(4) = \pi^4/90$ (see Section A2.3) with the gamma function $\Gamma(4) = 3!$.

EXAMPLE: Calculate the coefficient of T^4 in Eq. (3.24) and compare it to the constant σ_{sb} in the Stefan–Boltzmann equation relating total power radiated per unit solid angle, or $j = \sigma T^4$.

SOLUTION: Inserting values for the fundamental constants,

$$\frac{\pi^2 k_B^4}{15(\hbar c)^3} = \frac{\pi^2}{15} \frac{(\text{eV}/11604.5\ \text{K})^4}{[(197.33\ \text{eV nm})^3]} = 4.7222 \times 10^{-24} \frac{\text{eV}}{\text{nm}^3\text{K}^4}$$

or, in MKSA, 7.5658×10^{-16} [J/K^4 m^3]. The energy $dU = Pdt$ radiated through a differential surface element is equal to the product of U with differential radial distance cdt and a factor $\cos\theta d\Omega$, where $\cos\theta$ is due to Lambert's cosine law (diffuse radiation is proportional to the cosine of the angle with the direction to the viewing location). Using $d\Omega = \sin\theta d\theta d\phi$ in polar coordinates, an integration over all ϕ and over $0 \le \theta \le \pi/2$ (beyond $\pi/2$ is inside the sphere) gives $P = (c/4)U$ and so

$$\sigma_{sb} = \frac{\pi^2 k_B^4}{60 c^2 \hbar^3} = 5.6704 \times 10^{-8}\ \text{W/K}^4\text{m}^2$$

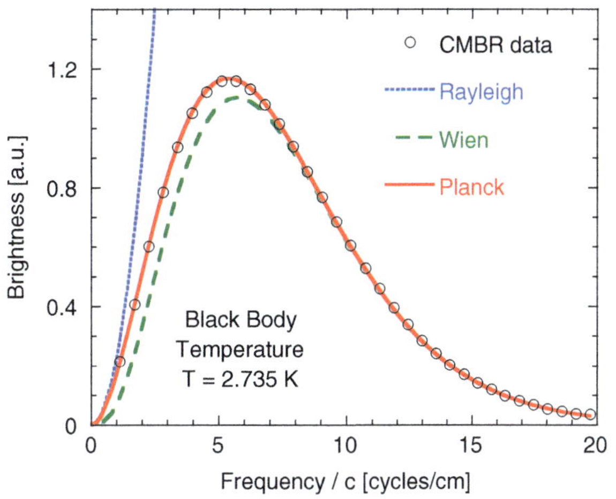

Figure 3.5 COBE satellite data: the cosmic microwave background radiation (CMBR) of the Universe as a function of frequency. Data digitally extracted from Figure 2 of ref. [56]. The three fits to the data are the Rayleigh–Jeans law of Eq. (3.21), Wien's formula of Eq. (3.22), and Planck's formula of Eq. (3.23). The x axis corresponds to $2\pi\omega/c$ in units of cycles/cm.

There is perhaps no more impressive a black body radiator in the Universe than the Universe itself: the spectrum of the cosmic background radiation, a remnant of the Horrendous Space Kablooie,[9] thermalized during the first year of the Universe's 13.7 billion year expansion. The cooled spectrum measured by the COBE (short for cosmic background explorer) satellite is reproduced in Figure 3.5 and demonstrates that the present background temperature of the Universe is $T = 2.735 \pm 0.06$ K [56]: the data is extremely well-fitted by Planck's formula assuming that $B = 2\pi\hbar/k_B T$ (the units of brightness on the y axis are indicated as arbitrary so as to not become enthralled with factors including, among other things, the aperture size of the measuring instrument and other minutia).

Einstein's genius was in taking the quantization condition that lead to Planck's formula seriously, as actually indicating something about the discrete nature of light. That was a rather bold act given the astonishing coherence of the explanation of electromagnetic radiation by Maxwell's equations. It allowed Einstein to explain at the same time photoemission [57] as the absorption of one quanta of light by an electron such that the electron had sufficient energy to surmount the barrier to emission that otherwise kept electrons inside the metal. The models of photoemission and black body radiation served as a powerful impetus in the development of quantum mechanics. Einstein was awarded the Nobel Prize in 1921 for his explanation of the photoelectric effect.

Quantization of light particles explained why the electrons were emitted immediately for photoemission from a metal surface even in low illumination conditions, rather than after a passage of time while the energy of the electron was gradually built up as a light-as-wave description would require. As to the barrier, Einstein referred to it as "the potential ...of negative electricity relative to the body" but in modern parlance it is simply referred to as "work function" and designated Φ. Einstein predicted:[10]

> *If the derived formula is correct, then Φ, when represented in Cartesian coordinates as a function of the frequency of the incident light, must be a straight line whose slope is independent of the nature of the emitting substance.*

Φ is the (positive) potential of the conductor such that an electron photo excited from it is just barely collected by a grounded conductor nearby (a technique analogous to a retarding potential analyzer in field emission studies).

How well does that prediction play out? Although the question is rhetorical (the Nobel Prize would not reward a failed prediction), it invites a reconsideration of the problem that introduces concepts of later utility. Light incident on a metal surface will excite many electrons, and so the number of electrons it liberates (a phrasing in keeping with the light-as-quanta concept) is also a measure of photoemission: by comparing the ratio of the number of excited electrons to

---

[9] "HSK" was coined in the *Calvin and Hobbes* strip of June 21, 1992; the appellation "Big Bang" is an alternate reference to the Universe's origin coined by the English astronomer Sir Fred Hoyle – ironic because he was being dismissive to the expanding universe model by giving it a snide moniker.

[10] Ref [57]: *Concerning the Emission of Cathode Rays Through Illumination of Solid Bodies*, Section 8. Therein, Einstein denoted the work function by Π, but he also denoted the speed of light by L. The quote has been rendered to use the modern notation of Φ.

the number of incident photons, or the *quantum efficiency (QE)*, another means exists for considering Einstein's prediction using a wealth of QE data.

The number of excited electrons is the ratio of charge collected over the elementary charge, or $\Delta Q/q$, and the number of incident photons is the energy deposited over the photon energy, or $\Delta E/\hbar\omega$. Therefore, QE is

$$QE = \left(\frac{\hbar\omega}{q}\right)\frac{\Delta Q}{\Delta E} \tag{3.25}$$

If the light from a laser or monochromatic source illuminates a surface for a time Δt, long by comparison to the emission time of the electron (not hard, since the latter is measured in picoseconds), and the emission area is the same as the illumination area, then $\Delta Q/\Delta E = J_q/I_\lambda$, or the ratio of the electron current over the light intensity. As in Eq. (3.5), current is the product of charge, density of electrons, and their average velocity, but if the electrons are distributed in velocity, then the current density should be a sum over velocities weighted by their distribution, or

$$J_q = \frac{q}{(2\pi)^3}\int\left(\frac{\hbar k_x}{m}\right)f(\vec{k})d\vec{k} \tag{3.26}$$

where $f(\vec{k})$ is the distribution such that $\int f(\vec{k})d\vec{k}$ is normalized. Without getting into the details, what can be inferred from J_q? Much depends on the form of $f(\vec{k})$ for the *photo excited* electrons, about which nothing so far has been said. In the absence of information, the pragmatic approach is the Kelly Johnson motto of "keep it simple, stupid"[11] (or KISS): consider it later (see Chapter 14). What is more important is how the integral in Eq. (3.26) scales. This is not as outlandish as it seems: $f(\vec{k})$ is dimensionless and therefore will not affect the scaling of J_q. Similar behavior is seen in Eq. (3.16): the MB distribution in the exponential is dimensionless, therefore the integral ratio scales as $v_x \propto \sqrt{k_B T}$.

Given that $\hbar k_x$ is the electron momentum leaving the surface, and that $d\vec{k} = k^2 \sin\theta d\theta d\phi$, one expects the integral to scale as $k^3 dk \propto (\text{velocity})^4$, and therefore Einstein's prediction suggests

$$QE \propto (\hbar\omega - \Phi)^2 \tag{3.27}$$

Consider QE data taken from copper as it is a very well characterized photocathode metal for rf photo injectors given its ruggedness and ease of use, even though its QE is relatively low, as is true for metals in general. Here, the data will be from ref. [58] for measurements of the copper photocathode used at the Linac Coherent Light Source (LCLS) at the Stanford Linear Accelerator (SLAC), although modified from its original representation of ln QE vs. λ, to use Eq. (3.27). The comparison is shown in Figure 3.6. The intercept of the graph is quite close to the commonly accepted value of the work function of copper of 4.6 eV, albeit that different crystal faces have different work functions, and the extraction field

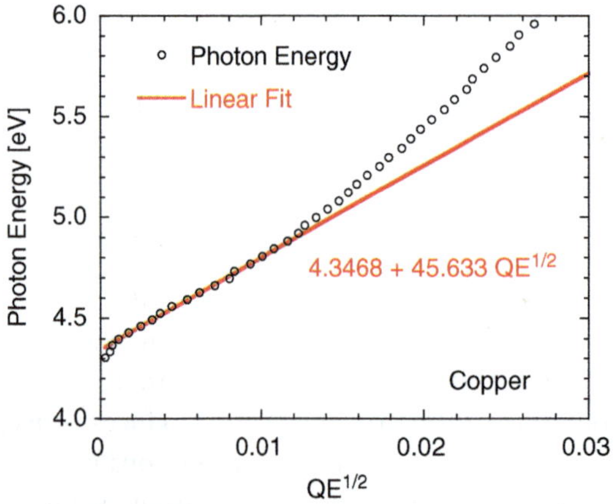

Figure 3.6 QE vs photon energy. Data of Figure 8, ref. [58] recast in the coordinates suggested by Eq. (3.27). The applied field at the surface was 50 MV/m, making the Schottky factor $\sqrt{4QF} = 0.268$ eV. The work function of copper is approximately 4.6 eV.

[11] Johnson was the designer of the world's highest and fastest aircraft, the SR-71 Blackbird, for Lockheed's Skunk Works program, so his advice is characteristically practical; see Ben Rich, *Clarence Leonard (Kelly) Johnson (1910–1990), a Biographical Memoir, Biographical Memoirs*, Vol. V67. The National Academies Press, 1995, p. 231.

used to collect the emitted charge alters the apparent work function by a small factor (as per Eq. (2.10), a 50 MV/m field reduces the work function by $\sqrt{4QF} = 0.268$). Thus, Einstein's prediction is borne out when $\hbar\omega - \Phi$ is small, although Figure 3.6 does anticipate future complications: the dipping behavior of the curve right near the intercept, as well as the non-linear behavior for higher photon energies, suggests physics yet to be encountered, but which will require an understanding of the electron distribution's statistics.

CHAPTER 4
Statistics

...intelligence...may, on the other hand, go so far as to enable a man to be moved emotionally by statistics.

– Bertrand Russell[1]

Even a small volume of metal contains an enormous number of electrons, for example 1 cm^3 of copper contains $\approx 8.411 \times 10^{22}$ electrons, comparable to the number of atoms. The physics of a large number of atoms or electrons, particularly with respect to the distribution of their energy, is the purview of statistical mechanics. The most familiar distribution may be the Maxwell–Boltzmann (MB) distribution, but understanding electrons in bulk matter and how they transport and scatter requires familiarity with the Fermi–Dirac (FD) distribution (whence "fermions"), which differs in important ways from the Bose–Einstein (BE) distribution (whence "bosons") already encountered for photons. Understanding the distributions is aided through combinatoric arguments. Practicing with distinguishable particles prepares the way and introduces randomness, probability, and the concept of entropy.

4.1 Distinguishable particles

And when I search a faceless crowd
A swirling mass of grey and
Black and white
They don't look real to me
In fact, they look so strange

– The Rolling Stones[2]

MB statistics govern non-interacting particles in thermal equilibrium and form the basis of the classical description of an ideal gas, but to the uninitiated that is hardly helpful. Analogies help make unfamiliar ideas easier to assimilate. Taking a cue then from the notion that the particles are distinguishable or individually identifiable, much like people (even in a crowd), an analogy between the role of energy for a gas atom and the role of income (money) in a person's life is actually a rather useful model on which to introduce a number of needed concepts. From two very simple hypotheses, an MB-like distribution emerges:

- In an employment hierarchy the bottom-tier positions (generally workers) are many and the top-tier positions (generally managers) are progressively fewer the higher up the management chain. Assume the tiers are labeled by j and the number of positions in each tier n_j behaves as $n_{j+1}/n_j = 1/a$, with $a > 1$ a dimensionless factor governing the number of subordinates or employees compared to bosses or employers.
- The income E_j for each tier j of the hierarchy is monotonically increasing and crudely linear.[3] Assume the variation is of the form $E_{j+1} - E_j = b$.

Instead of discussing workers and managers, it will be more convenient to generically refer to "households". An exponential relationship between the number of households n_j and an income bracket E_j exists in that the number of households n decreases exponentially with an increase in income E, specifically, $n(E) \propto \exp[-(E/b)\ln(a)]$ so that $\ln(a)/b$ takes over the role of β in an MB distribution.

[1] Bertrand Russell, *The Basic Writings of Bertrand Russell.* New York: Simon and Schuster, 1961, p. 423.
[2] Mick Jagger and Keith Richards, lyrics to *Salt of the Earth*, 1968.
[3] Perhaps very crudely, for example the 2012 GS Schedules of the Federal Government can be fitted by $E_j \approx 12.45 + 3.25j$ for $j \leq 11$ and $E_j \approx 46.29 + 13.10(j - 11)$ for $j > 11$ for the step 1 positions, with similar curves for the higher steps.

Introduction to the Physics of Electron Emission, First Edition. Kevin L. Jensen.

The simple hypothesis clearly cannot be correct over all scales: it fails at the extremes, as on the bottom ($j \to 1$) where the social safety net provides assistance to the poor, or at the top ($j \gg 1$) where income is power-law distributed (the so-called Pareto distribution in economics). The example here is meant to be more pedagogical than exact.

EXAMPLE: Using the data of Table 4.1, what are the median and mean incomes evaluated conventionally?
SOLUTION: The median value is that value for which half the population is below and half above. The number of households is 122,459,000 (sum over all n_j). If $S_k = \sum_{j=1}^{k} n_j$ then the median point lies between S_{10} and S_{11}. Interpolating gives E_{median} = \$50,694 (compare the American Community Survey (ACS) value of \$51,371). The mean value corresponds to the weighted average. For the $N_b = 42$ median values of Table 4.1, it is

$$\langle E \rangle = \frac{\sum_{j=1}^{N_b} n_j E_j}{\sum_{j=1}^{N_b} n_j} = \$71{,}274$$

A point worth observing is that a median $\langle E \rangle$ of median values E_j (which should be represented as $\langle E_j \rangle$) is the median of the population where $N = \sum_{j=1}^{N_b} n_j$, because

$$\langle E \rangle = \frac{\sum_{j=1}^{N_b} n_j \langle E_j \rangle}{\sum_{j=1}^{N_b} n_j} = \frac{1}{N} \sum_{j=1}^{N_b} \sum_{k=1}^{n_j} E_{j,k} = \frac{1}{N} \sum_{l=1}^{N} E_l \tag{4.1}$$

where $E_{j,k}$ is the kth member of the income bracket (or group) j: if all income brackets had the same number of households n, then the unique correspondence between l and (j, k) would be $l = (j-1)n + k$, but in general the number of members of the group n_j varies, be they households or energy levels.

A plot of the US Census Bureau Data 2012 distribution shown in Table 4.1 and Figure 4.1 indicates that for the middle class, the exponential distribution is more or less reasonable. The noise in the data (a consequence of grouping households into \$5000 bins) makes parameter extraction challenging, so consider instead the *summed* data shown in Figure 4.2, which represents an integrated distribution function of

$$NS_k = \sum_{j=1}^{k} n_j \to S(E) = \int_0^E f(x)dx \tag{4.2}$$

Table 4.1 US Census Bureau Data 2012 for median income E_j (in \$100,000) and number of households n_j (in millions).

E_j	n_j	E_j	n_j	E_j	n_j
0.01183	4.2040	0.71988	3.7560	1.4197	1.1190
0.07944	4.7290	0.76950	3.4140	1.4711	0.9200
0.12419	6.9820	0.81983	3.3260	1.5161	1.1430
0.17262	7.1570	0.87148	2.6430	1.5727	0.8050
0.22230	7.1310	0.91955	2.6780	1.6191	0.7310
0.27193	6.7400	0.97103	2.2230	1.6721	0.5750
0.32080	6.3540	1.0179	2.6060	1.7209	0.6160
0.37056	5.8320	1.0718	1.8380	1.7710	0.5700
0.42034	5.5470	1.1209	1.9860	1.8195	0.5020
0.47160	5.2540	1.1713	1.4640	1.8720	0.3640
0.51902	5.1020	1.2197	1.5960	1.9216	0.4320
0.57121	4.2560	1.2715	1.3270	1.9718	0.3780
0.61931	4.3560	1.3201	1.2530	2.2010	2.5490
0.67042	3.9490	1.3711	1.1400	4.3178	2.9110

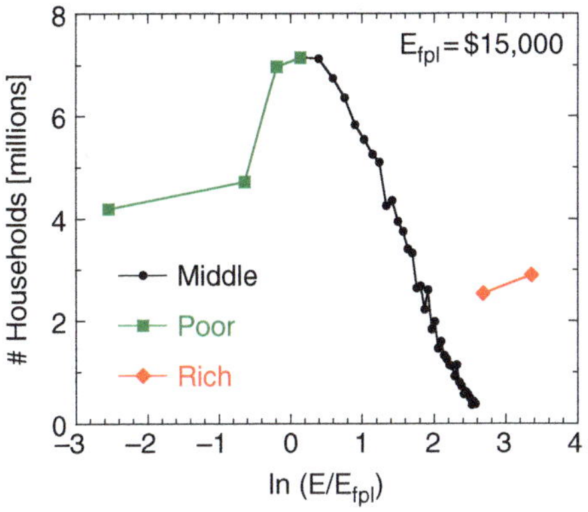

Figure 4.1 Household income vs number of households. Data is from US Census Bureau Data 2012 (Table 4.1), which gives the mean household income E_j in bins of width $5000 (high income bins were larger and therefore hold more households). E_{fpl} = $15,000 (approximately the federal poverty level for a family of two).

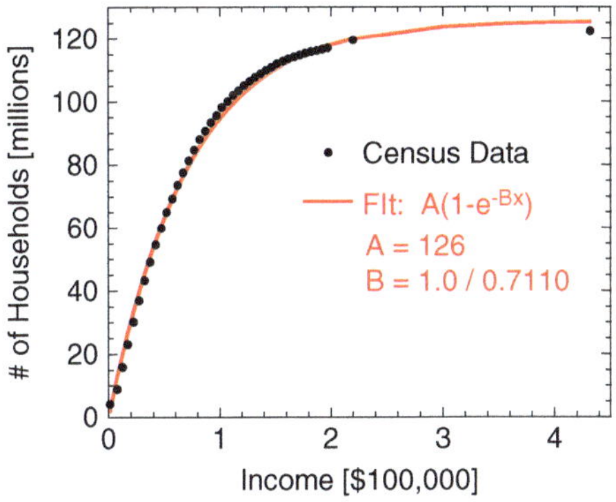

Figure 4.2 Cumulative household income vs number of households. Data is from US Census Bureau Data 2012 (Table 4.1). The y axis is the number of households below the income level E_j (x coordinate).

That data can be parametrically represented as a function of the form $A(1 - e^{-BE})$, the derivative of which with respect to E is the sought distribution function $f(E) = ABe^{-BE}$. Finding the mean income is then a trivial integration:

$$\langle E \rangle = \frac{\int_0^\infty Ef(E)dE}{\int_0^\infty f(E)dE} = \frac{1}{B} \tag{4.3}$$

or $\langle E \rangle$ = $71,100, from which $E_{median} = \langle E \rangle \ln(2)$ = $49,300, close to the numerical values found above.

It is worth pausing a moment and taking note of a number of important features. To generalize the discussion, refer to "particles" instead of "households".

1. The particle can be labeled by its position in the total population (E_l) *or* its position within its subgroup or energy level $E_{j,k}$.
2. The distribution of discrete values can take on the appearance of a distribution function, from which the statistical quantities of interest, such as mean value, are derived.
3. The distribution can be thought of as the probability that the particle l in question was chosen from a certain energy level E_j, where that probability P_j is $n_j/N \to f(E)dE$.
4. Two methods can be used to calculate the mean, and they are

$$\frac{1}{N}\sum_{l=1}^{N} E_l = \langle E \rangle = \sum_j E_j P_j \tag{4.4}$$

   where E_l refers to the energy of the l particle, E_j to the energy of the jth level containing n_j particles with $P_j = n_j/N$, and the last form is $\int Ef(E)dE$ in the continuum limit.

The income data has features that belie the simple model. Some are as follows. First, a strictly exponential distribution function would suggest that the number of households below poverty is $[1 - \exp(-BE_{flp})] = 19\%$, rather than the 13% of the first three rows in Table 4.1, indicating a "hidden" mitigating effect at one of the limiting conditions (e.g., poverty assistance programs). Second, if the "middle" census data were plotted as log(households) versus income, a discernible weak but present negative convexity at the low end and positive at the high is apparent, indicating complexity in the relation between income and number of households missed by the simple model (due to any number of causes for treating the better off even better). Third, the simple model suggests the percentage of total income held by the bottom 97.6% of the households (≤$250K) is $1-(1+2.5B)e^{-2.5B}=86.5\%$ for $1/B=0.71274$ (median income in $100K), which seems to run counter to commonly quoted statistics that put the percentage of wealth held by the bottom 99% closer to 65%. Income, although correlated with wealth, differs from wealth, so speaking of them interchangeably is incorrect (e.g., wealth includes property). Lastly, a perennial problem with logarithm plots is that the extremes do not extrapolate well from the middle, especially when the data is noisy, a case in point being the putative "top earner" characterized by $N\langle n_{top}\rangle = 1$, suggesting $E_{top} = B^{-1}\ln(N) = \$11.63\text{M}$, which is 0.3% of the income of Bill Gates in 2012.

This suggests some cautionary points worth emphasizing.

1. Data with an underlying exponential dependence always looks linear on a log plot: the non-linearity can be subtle and obfuscate important features (like subsets of emission sites that put curvature on Richardson–Laue–Dushmann (Chapter 12) and Fowler–Nordheim (Chapter 13) plots of current density as a function of temperature and field, respectively).
2. Conditions at the extremities of the log plot can signal the onset of unaccounted for physics (like space charge, which affects Miriam curves (Section 29.1) in thermionic emission and causes turnover in field emission, or even leakage current in field emission).
3. The data shown may not be the only (or even best) indication of the phenomena putatively ascribed to it, or there may be more than one cause for apparent trends (like thermal–field emission, dark current, and competing mechanisms).
4. Lastly, a relation, or *law*, about how something behaves is not the same as having a theory about why that behavior is so (a point revisited in Chapter 16).

A final feature of the income model, buried above (particularly Eq. (4.2) and the so-called important points 3 and 4), illuminates the transition from a discrete to a continuum model, and introduces concepts related to phase spaces that will be important. The transition must make reference to the size of the bin or "cell." In the census data, the bin size was $5000 and Table 4.1 gives the values of $\langle E_j\rangle$ in each of those cells. Thus, in going from a summation of a discrete set f_j to an integration over $f(x)$, the size of the cell must be respected in a manner familiar from approximating sums by integrals (as in Eq. (A1.10)):

$$\sum_j f_j \rightarrow \frac{1}{\Delta x}\int_\Omega f(x)dx \tag{4.5}$$

where Ω is the volume of space holding the population of points and Δx is the size of the bin. When finding mean values, the factor Δx appears in both numerator and denominator, and therefore factors out, so its existence was masked. If the bin were reduced to such a small region of "money" space that it contained only one income of one household, if that, then most of the cells would in fact be empty. The idea of small cells of phase space will be revisited in Section 5.1.

The analog for particles of a gas takes account of the fact that the dynamics of a particle are determined if its position and velocity are known, and so the volume element of phase space corresponds to the product $\Delta\vec{x}\Delta\vec{v}$. Actually, from both classical and quantum mechanics, it is position $\vec{x}$ and momentum $\vec{p}=\hbar\vec{k}$ that matter (e.g., Hamilton's equations in classical mechanics and conjugate variables in quantum mechanics). The product of momentum and position is related to the *action*, and the principle of least action governs classical dynamics. Thus, it is seen that $\hbar$ is the unit of action. If the smallest cell size is taken as Planck's constant $h_p = 2\pi\hbar$, then the differential volume in phase space is $d\vec{x}d\vec{p}/(2\pi\hbar)^3 = d\vec{x}d\vec{k}/(2\pi)^3$. Consequently, the phase space analog of Eq. (4.5) is

$$\sum_j f_j \rightarrow \frac{1}{(2\pi)^3}\int_\Omega f(\vec{x},\vec{k})d\vec{x}d\vec{k} \tag{4.6}$$

4.2 Probability and states

...knowledge and probability are of such contrary and disagreeing natures, that they cannot well run insensibly into each other.

– David Hume[4]

Probabilistic reasoning can be non-intuitive, but the usage of a good methodology is of great aid. Given the pervasiveness of probability in quantum mechanics, there is value in mimicking quantum mechanics methods and notation, so

[4] David Hume, *A Treatise of Human Nature*, Part 5, Section 1, *"Of Skepticism with Regard to Reason"*. New York: Clarendon Press, 1978.

the representation of probability in terms of vectors $|\psi\rangle$ and matrices $\hat{P}$ is adopted for a model to show that probabilities must sum to unity and the sub-populations, or rather a complete basis set, must all be represented in calculations lest the solution be invalid because of a neglected population.

A well-known and counterintuitive example is assessing the trustworthiness of a measurement. Measuring a feature not present is termed a "false positive." To make the example concrete, if the fraction of a population that has a trait is p, and an individual subjected to a measurement $\hat{P}$ tests positive, what is the probability that the individual has the trait? To make the example very concrete, suppose the trait is a particular low probability disease and the test is correct most, but not all, of the time.

The basis states are the presence or absence of the disease: let u designate "un-diseased" and d designate "diseased" (or, equivalently, "up" and "down" – why will be seen in a moment). Then

$$|\psi\rangle = c_u|u\rangle + c_d|d\rangle \tag{4.7}$$

with $|u\rangle$ and $|d\rangle$ a complete description of the population types (or basis states), and the coefficients c_j describing the population fraction. The basis states are "orthogonal" in that $\langle u|d\rangle = \langle d|u\rangle = 0$. The measurement $\hat{P}$ acting on $|d\rangle$ will detect the disease and its expectation value will return the fraction p of the population that has it. If $\hat{P}$ is a perfect measurement (no false positives), then $\langle\psi|\hat{P}|\psi\rangle = c_d^2\langle d|\hat{P}|d\rangle = p$, but if there are false positives, then $\langle u|\hat{P}|u\rangle = \delta$ where δ is the fraction of un-diseased (healthy) who are wrongly diagnosed as diseased. Thus, if the test is 99% accurate, $\delta = 0.01$. The question is then: given that one is diagnosed with the disease, what is the probability that one is actually diseased, or, put another way, *how confident is one of the test result?*

The notation intentionally follows Heisenberg's matrix approach to quantum mechanics (with a nod to Dirac for the bra-ket notation). So, let the basis states be given by

$$|u\rangle = \begin{pmatrix} 1 \\ 0 \end{pmatrix}, \quad |d\rangle = \begin{pmatrix} 0 \\ 1 \end{pmatrix} \tag{4.8}$$

and the coefficients corresponding to the weighting factors are

$$c_u = \sqrt{1-p}, \quad c_d = \sqrt{p} \tag{4.9}$$

One can verify that $|\langle\psi|d\rangle|^2 = p$; $|\langle\psi|u\rangle|^2 = 1 - p$, and therefore that

$$\langle\psi|\psi\rangle = \sum_j \langle\psi|j\rangle\,\langle j|\psi\rangle = 1 \tag{4.10}$$

where $|j\rangle$ corresponds to either $|u\rangle$ or $|d\rangle$. This equation is important: given that $|\psi\rangle$ is normalized, it follows that

$$\sum_j |j\rangle\langle j| = \hat{I} \tag{4.11}$$

where $\hat{I}$ is the identity matrix. Continuing, a moment's tinkering will show that

$$\hat{P} = \begin{pmatrix} \delta & 0 \\ 0 & 1-\delta' \end{pmatrix} \tag{4.12}$$

where δ accounts for false positives, and δ' for false negatives (having the disease but being told one does not).

"Confidence" C is a ratio of how many people *actually* have the disease with how many people are *diagnosed* with it and so

$$C = 100\%\frac{\langle\psi|d\rangle\,\langle d|\hat{P}|\psi\rangle}{\langle\psi|\hat{P}|\psi\rangle} \tag{4.13}$$

Conversely, "Doubt" D makes the substitution $d \to u$ in C. Therefore, in light of Eq. (4.11), $C + D = 100\%$ as there is no other outcome.

Consider a typical example. Suppose 5% of the population has the disease ($p = 0.05$) and the test for it is 99.9% accurate ($\delta = 0.001$) with no false negatives ($\delta' = 0$). Direct computation shows $C = 98.14\%$, which is fairly confident.

On the other hand, if the inaccuracy of the test is equal to the proportion of the population that does not have the disease, or $\delta = p = 0.001$, and the false negatives are the same as the false positives ($\delta' = \delta$), then one finds $C = 50\%$, which is fairly astounding: a highly accurate test is thwarted by the rareness of the disease and the deluge of false positives. In fact, working through the matrix multiplications, the general equation is

$$C_1(p) = \frac{p(1-\delta')}{p(1-\delta') + \delta(1-p)} \times 100\% \tag{4.14}$$

with the subscript 1 indicating the number of times the test is performed. Thus, the more rare the disease (smaller p), the less confidence one has in a test, especially if the disease is more rare than the test is accurate, perhaps some consolation to those who wonder why a one-in-a-million event picked *them* when they get bad test results. There is still hope, at least until the second test ($\hat{P} \to \hat{P}^2$ in Eq. (4.13), which in the present example results in $C_2(p) = 99.9\%$, the subscript denoting the power of $\hat{P}$).

The example has one more interesting perturbation to consider (a subtle, but irresistible, pun). Suppose the act of getting the test (making the measurement) has a chance of *giving one the disease*. Slightly longer tinkering will convince oneself that this will populate $\hat{P}$ with off-diagonal matrix elements. In the parlance of quantum mechanics, this can be thought of as the measurement causing one basis state to change, or "scatter", to the other basis state. Another scenario to consider: instead of disease, suppose one is talking about disgruntled employees or terrorists in situations where they constitute a tiny fraction of the population: not only do those-who-test have to worry about the many false red flags that must be investigated, they must also consider whether their testing adds to the problem by actually creating more discontents (scattering).

So much for a useful calculational method. Now what about probability itself? An ideal gas (monoatomic noninteracting point particles) is a very useful model on which to hone one's understanding of thermodynamic properties such as pressure, volume, and temperature. Some ideal gas properties anticipate those of an electron gas, but electrons, unlike people, are not distinguishable, and that makes an enormous difference in combinatorics. Knowing how to count particles is therefore necessary. Reacquaintance with probability concepts is the first step.

4.3 Probability and entropy

And on the pedestal these words appear:
"My name is Ozymandias, king of kings:
Look on my works, ye Mighty, and despair!"
Nothing beside remains. Round the decay
Of that colossal wreck, boundless and bare
The lone and level sands stretch far away.

– Percy Bysshe Shelley[5]

Early efforts to understand entropy considered the absorption of heat by gases to characterize reversible versus irreversible processes, and find the efficiencies of engines. In the reversible Carnot cycle, the isothermal (at constant temperature) and adiabatic (no exchange of heat) expansions and contractions of an ideal gas between two reservoirs on a plot of pressure P versus volume V resembles a distorted and curved parallelogram, but on an entropy S versus temperature T plot, it is a pleasing rectangle: along an isothermal, entropy changes (vertical edges), and along an adiabatic, entropy is constant (horizontal edges). In terms of ΔQ, changes in entropy ΔS are governed by

$$\Delta S = \frac{\Delta Q}{T} \tag{4.15}$$

The connection between entropy and probability, though, provides perhaps a more graspable notion of what entropy is about, particularly with respect to disorder, but that requires a mild detour into probability.

A fair coin has an equal chance of landing heads ($h = 1/2$) or tails ($t = 1/2$). Thus, the probability of a coin landing heads is $P_h = h/(h+t) = 0.5$. What about two coins? The probability of coin 1 being heads and coin 2 tails (i.e., the probability of the configuration ht where the first coin is the first position, and the second coin is the second position) is $P_{ht} = P_h P_t = 0.25$. If N coins are being tossed, then the probability of the *ad hoc* configuration *hthhtthhth* is

$$P_{hthhtthhth} = h^{n_h} t^{n_t} = (0.5)^6 (0.5)^4 = 1/1024 \tag{4.16}$$

[5] Percy Bysshe Shelley, *Ozymandias, The Norton Introduction to Literature*, 2nd edn, eds C.E. Bain, J. Beaty, and J.P. Hunter. New York, 1977, p. 733.

Extend the exercise to dice, which when thrown randomly give the numbers 1 through 6. If four distinguishable die are thrown, what is the probability of 1241 occurring? Generalizing the coin example, it follows

$$P_{1241} = \prod_{j=1}^{6} (P_j)^{n_j} = (P_1)^2(P_2)^1(P_3)^0(P_4)^1(P_5)^0(P_6)^0 = 1/1296 \tag{4.17}$$

A few points to notice: the method of calculation did not care if $P_i = P_j$, the number of times n_j an outcome j occurred was useful in grouping the results, and $n_j = 0$ indicates outcome j did not occur. Observe also that if the dice are *distinguishable* or their order matters, as would be the case if die 1 gave the first digit, die 2 the second, and so on, versus *indistinguishable* or when the order does not matter, as would be the case if only the sum of the faces was of interest, then different probabilities occur. As an example, the probability of an event that does not care about order such that face j appeared on n_j dice would be $\prod_{j=1}^{6} n_j!$ times more likely. For the 1241 example above, the probability is therefore twice as likely since there are two different ways the die with a face of 1 can be placed.

There is a very subtle concept associated with order, and it serves as a convenient introduction to the concept of entropy. To motivate the discussion, suppose a single die was thrown 288 times and the faces recorded. Question: which of the two tallies below seems more likely?

- Case A:

```
2 4 3 6 6 4 5 1 6 4 3 3 1 1 1 2 1 4 1 1 3 1 6 6 1 5 6
1 6 6 2 2 2 4 4 2 5 4 2 5 3 3 5 2 6 3 5 3 4 3 2 5 5 1
5 4 2 5 1 4 6 2 1 3 4 3 1 3 3 1 1 4 2 4 2 6 6 4 3 3 6
6 5 4 1 3 3 2 1 4 2 1 4 2 3 4 3 4 1 2 1 6 1 6 1 6 4 6
2 5 2 4 4 1 4 5 2 5 4 1 5 2 1 2 5 5 1 4 4 3 5 4 1 5 3
2 6 3 1 5 2 3 1 3 3 1 3 1 5 6 5 6 1 5 3 2 3 3 6 4 4 6
5 2 2 2 2 3 2 5 5 5 3 1 5 5 3 4 4 6 2 5 5 4 6 3 1 5 6
6 5 4 1 2 2 4 5 5 6 3 5 2 5 4 6 3 4 2 5 4 2 3 1 3 1 3
4 4 6 6 4 5 1 6 6 3 3 6 2 6 6 4 6 1 1 4 2 2 6 3 3 6 1
2 3 1 6 1 6 5 6 5 6 4 6 5 4 2 3 4 5 1 6 3 3 1 5 1 2 6
5 3 6 6 4 5 2 5 2 1 3 4 2 2 2 4 2 5
```

- Case B:

```
1 4 6 5 3 2 2 1 5 4 6 3 4 1 6 3 5 2 1 4 6 5 3 2 2 1 5
4 6 3 4 1 6 3 5 2 1 4 6 5 3 2 2 1 5 4 6 3 4 1 6 3 5 2
1 4 6 5 3 2 2 1 5 4 6 3 4 1 6 3 5 2 1 4 6 5 3 2 2 1 5
4 6 3 4 1 6 3 5 2 1 4 6 5 3 2 2 1 5 4 6 3 4 1 6 3 5 2
1 4 6 5 3 2 2 1 5 4 6 3 4 1 6 3 5 2 1 4 6 5 3 2 2 1 5
4 6 3 4 1 6 3 5 2 1 4 6 5 3 2 2 1 5 4 6 3 4 1 6 3 5 2
1 4 6 5 3 2 2 1 5 4 6 3 4 1 6 3 5 2 1 4 6 5 3 2 2 1 5
4 6 3 4 1 6 3 5 2 1 4 6 5 3 2 2 1 5 4 6 3 4 1 6 3 5 2
1 4 6 5 3 2 2 1 5 4 6 3 4 1 6 3 5 2 1 4 6 5 3 2 2 1 5
4 6 3 4 1 6 3 5 2 1 4 6 5 3 2 2 1 5 4 6 3 4 1 6 3 5 2
1 4 6 5 3 2 2 1 5 4 6 3 4 1 6 3 5 2
```

A simple numerical criteria, like the mean of the numbers, gives 3.5 for both, which is the expected mean. A close eye discerns, however, that in Case B, the sequence repeats after the 18th throw, and so Case B seems to violate what is commonly meant by *random*, even if the 18 throws are random in nature. A sequence that starts to repeat after a run of length 18 throws contains elements that are no longer probabilistically distinct. Put another way, it would be easier to describe Case B than Case A, where no order is present. To describe Case A, each entry in the entire sequence has to be given. To describe Case B, the first 18 throws are specified followed by an instruction to repeat that sequence 16 times. A measure of how much information is required to specify the sequence becomes a measure of entropy (Boltzmann followed a similar reasoning to introduce entropy into statistical mechanics [59]). Case A is therefore a plausible outcome of fair die throws, but Case B is not likely to occur because, in the idiom of gambling, the fix is in.

More starkly, if all the flips in a sequence of coin tosses were heads, a *very* ordered arrangement would be had – it would be trivial to describe ("all heads") and it represents a state of zero disorder, or zero entropy. Conversely, truly random strings of heads and tails show great disorder and entropy would therefore be deemed large. These two notions taken together lead one to expect if p is the probability of an event, that $S \propto \sum pf(p)$, where $f(p)$ is a function such that $f_s(1) = 0$ but $pf_s(p) \to 0$ as $p \to 0$.

Joint probabilities impose another requirement. In Case A, all the throws are random, so one would intuitively expect that the entropy of the long sequence would be the sum of the entropies of two shorter sequences obtained by partitioning the long sequence. In Case B, the sequence repeats and therefore looks ordered in the long view. That is, in Case A, the probability of the jth throw is independent of all others, but in Case B, the $(j+18)$th throw is same as the jth throw. Mathematically this is expressed by saying that, for uncorrelated events as in Case A, the probably that the ith and jth throws will take a particular value are p_i and p_j respectively, and their joint probability is $p_i p_j$. By comparison, Case B is correlated such that for $j = i + 18$, the outcome is determined ($p_j = 1$) and therefore the joint probability is different.

Notationwise, the conditional probability is represented as $P(i|j)$ and spoken of as the probability of outcome p_i given that outcome p_j occurred. If events are random or uncorrelated, then $P(i|j) = p_i p_j$. The entropy definition must be cognizant of that: if the probabilities are independent, then the entropy should be additive. For a random sequence like Case A, the entropy of the whole sequence is equal to the sum of the entropies of its smaller pieces if it is subdivided. Conversely, consider a volume of gas which has a partition. Removing the partition should just add the entropies because the total box is the sum of its parts (this important consideration will be revisited with consideration of the Gibbs paradox below). Mathematically, the requirement of additivity demands

$$S_{total} = \sum \sum P(p|\tilde{p})\, f_s(p, \tilde{p}) \tag{4.18}$$

$$= \sum pf_s(p) + \sum \tilde{p} f_s(\tilde{p}) = S + \tilde{S} \tag{4.19}$$

this last demand can be shown to require $f_s(p, \tilde{p}) = f_s(p) + f_s(\tilde{p})$ by insertion into the above equation and using $\sum p = \sum \tilde{p} = 1$. The only form of $f_s(p)$ which satisfies these requirements is $f_s(p) \equiv -\ln(p)$. That indicates that entropy S is defined by (known as Shannon's formula in information theory)

$$S = -\sum_{j=1}^{N} p_i \ln(p_i) \tag{4.20}$$

In physics, a state of thermodynamic equilibrium is one that maximizes S (in which case it is more common to append a factor of k_B to the definition, so that $S \to S/k_B$, see Eq. (5.22) below).

EXAMPLE: What probabilities for heads p and tails q maximize the entropy of a sequence of coin tosses?
SOLUTION: The Shannon entropy is simply $S = -p \ln p - q \ln q$, but $q = 1 - p$. Therefore,

$$\partial_p S = \partial_p\{-p \ln p - (1-p)\ln(1-p)\} = \ln\left(\frac{1-p}{p}\right) = 0$$

or $p = q = 1/2$, as can also be seen graphically by plotting $S(p)$ as a function of $0 \le p \le 1$.

EXAMPLE: What probabilities maximize the entropy of a sequence of die throws for which the probability of face j is p_j?
SOLUTION: This problem is a generalization of the previous coin example. If the die has n faces, then the requirement is that the entropy is maximized for the condition that $\sum_{j=1}^{n} p_j = 1$. Using the method of Lagrange multipliers (see Section 5.4 and ref. [60]) a constant a must be found such that

$$\partial_{p_j} S = \partial_{p_j}\left\{-\sum_{j=1}^{n} p_j \ln p_j - a\sum_{j=1}^{n} p_j\right\} = 0$$

or $-\ln p_j = a - 1$, that is, all the p_j are equal, and therefore equal to $1/n$ by virtue of the fact they sum to 1. The finding confirms the intuitive understanding that all outcomes are equally probable.

EXAMPLE: Consider N equi-spaced $x_j \equiv j\Delta$ that occur with a probability p_j, and the average of which is $x_o > 0$. Find p_j if the entropy S is maximized.

SOLUTION: There are two constraints on the p_j: first, they sum to 1, and second, $\sum_j x_j p_j = N x_o$. Using the method of Lagrange multipliers (see Section 5.4 and ref. [60]) two constants a and b must be found such that

$$\partial_{p_j}\left(S - a\sum p_j - b\sum x_j p_j\right) = 0 \tag{4.21}$$

It is found that $p_j = \exp(-a - 1 - bx_j)$ and so, using $\sum_{j=1}^{\infty} e^{-jb\Delta} = (e^{b\Delta} - 1)^{-1}$,

$$x_o = \frac{\sum_{j=1}^{N} x_j e^{-1-a-bx_j}}{\sum_{j=1}^{N} e^{-1-a-bx_j}} \approx \frac{d}{db}\ln\left(1 - e^{-b\Delta}\right)$$

which sets $b = (1/\Delta)\ln[x_o/(x_o - \Delta)]$ if $e^{-Nb\Delta} \ll 1$. Consequently, $p_i \propto \exp(-bx_i)$ and is an example of an MB distribution. For the numerical choices $N = 200$, $\Delta = 4/N$, and $b = 5$, it is found that $bx_o = 1.0508$, that is, $b \approx 1/x_o$. In other words, if N is very large, then

$$p_j \propto \exp(-x_j/x_o) \tag{4.22}$$

an outcome reminiscent of Eq. (4.3). Observe that if x_o is held fixed (and therefore b is fixed), then the consequences of increasing N are to reduce Δ: if x_j is an energy, then the spacing between the energy levels consequently decreases.

Boltzmann's equation for entropy differs subtly in appearance: his equation is referred to as the statistical definition of entropy [61] or entropy as the logarithm of the statistical weight [62]. It relates, for example, the number of ways atoms or molecules can be arranged to S, or[6]

$$S = -k_B \ln W(\langle E\rangle) \tag{4.23}$$

where $\langle E\rangle$ is the mean value of energy. Because the logarithm of the distribution function is a linear function of energy, then $\ln W(\langle E\rangle) = \langle \ln W(E)\rangle$. If the components of W are p_i, then the average of $\ln W$ is the sum over the product of $\ln p_i$ with its probabilistic weight p_i, or simply Eq. (4.20). An example of W might be the number of combinations associated with putting n objects into m places, with $m > n$. Calculating that (as will be encountered in Eqs (4.26) or (29.56)) requires familiarity with combinatorics.

4.4 Combinatorics and products of probability

How do I love thee? Let me count the ways.

– Elizabeth Barrett Browning[7]

Two boxes (volume) containing N particles (number) are equivalent if the distribution in energy of those particles are equal. If the particles all have identical characteristics (like mass), then particle a being in energy state E_1 and b in E_2 gives the same average energy as if b were in E_1 and a in E_2. So, how particles occupy energy states is a matter of counting: for a gas in equilibrium, the probability that a particle has an energy E_j is the ratio of the number of states with that energy to the total number of states.

Consider an analogy. An Afghan blanket is made of squares containing different patterns, as shown in Figure 4.3. Suppose the patterns are knitted by sequentially choosing yarns of different colors. If there are three yarns, for example red (r), green (g), and blue (b), the possible variations are trivial to write down as (rgb grb rbg gbr brg bgr) or, if the colors are labeled by numbers so rgb = 123, then

```
123 213 132 231 312 321
```

[6] The inscription on Ludwig Boltzmann's gravesite in Vienna, Austria is similar to Eq. (4.23).

[7] Elizabeth Barrett Browning, "Sonnets from the Portuguese XLIII", in M.L. Rosenthal, *Poetry in English: An Anthology*. New York: Oxford University Press, 1987, p. 630.

Figure 4.3 Color combinatorics. Afghan blanket made using six different colors such that each square region contains three different colors. No two squares are identical.

Four yarns ($n = 4$) take more patience, but give

```
1234 2134 1324 2314 3124 3214 1243 2143 1342 2341 3142
3241 1423 2413 1432 2431 3412 3421 4123 4213 4132 4231
4312 4321
```

For $n = 5$, writing the combinations without using an algorithm is tedious:

```
12345 21345 13245 23145 31245 32145 12435 21435 13425 23415 31425
32415 14235 24135 14325 24315 34125 34215 41235 42135 41325 42315
43125 43215 12354 21354 13254 23154 31254 32154 12453 21453 13452
23451 31452 32451 14253 24153 14352 24351 34152 34251 41253 42153
41352 42351 43152 43251 12534 21534 13524 23514 31524 32514 12543
21543 13542 23541 31542 32541 14523 24513 14532 24531 34512 34521
41523 42513 41532 42531 43512 43521 15234 25134 15324 25314 35124
35214 15243 25143 15342 25341 35142 35241 15423 25413 15432 25431
35412 35421 45123 45213 45132 45231 45312 45321 51234 52134 51324
52314 53124 53214 51243 52143 51342 52341 53142 53241 51423 52413
51432 52431 53412 53421 54123 54213 54132 54231 54312 54321
```

By $n = 6$ the task is tedious *and* difficult, and left to an algorithm.

EXAMPLE: Develop an algorithm to list all permutations for $n = 6$.

SOLUTION: Let C^n be an $n! \times n$ matrix containing all permutations of $[1, 2, \ldots, n]$, and let C_j^n be the jth column of that vector. Let $\bar{n}$ be a column vector of length $n - 1$ with all entries equal to n. It is obvious that $C^1 = [1]$ and it follows that

$$C^2 = \begin{bmatrix} \bar{2} & C_1^1 \\ C_1^1 & \bar{2} \end{bmatrix} \quad C^3 = \begin{bmatrix} \bar{3} & C_1^2 & C_2^2 \\ C_1^2 & \bar{3} & C_2^2 \\ C_1^2 & C_2^2 & \bar{3} \end{bmatrix} \tag{4.24}$$

which generalizes to

$$C^n = \begin{bmatrix} \bar{n} & C_1^{n-1} & C_2^{n-1} & \cdots & C_{n-1}^{n-1} \\ C_1^{n-1} & \bar{n} & C_2^{n-1} & \cdots & C_{n-1}^{n-1} \\ C_1^{n-1} & C_2^{n-1} & \bar{n} & \cdots & C_{n-1}^{n-1} \\ \vdots & \vdots & \vdots & \ddots & \vdots \\ C_1^{n-1} & C_2^{n-1} & C_3^{n-1} & \cdots & \bar{n} \end{bmatrix} \tag{4.25}$$

The n columns of C when concatenated to a single column give the permutations sought and which can then be sorted. As $n \to n+1$, the number of columns increases by 1 and the number of rows by a factor of $n+1$, showing that the number of rows grows as a factorial. An algorithm is given in Section A3.1.

Counting the number of entries for each case, the number of permutations N_p is found to be $N_p(3) = 6$, $N_p(4) = 24$ and $N_p(5) = 120$, or $N_p(n) = n!$. As a verbal instruction, it amounts to saying "There are n ways to pick the first yarn, $(n-1)$ ways to pick the second yarn…", and so on down to the last yarn, or $n \times (n-1) \times \ldots 2 \times 1 = n!$.

Now let there be more than one spool of yarn for each color, that is, let there be n_j spools of yarn for color j. Clearly, if the spools are numbered (the spools are distinguishable), the algorithm above is going to vastly overpredict the number of unique patterns because, for example, whether the red yarn came from spool 1 or 2 does not change the pattern. In other words, the yarn colors are *equivalent* (they function the same way), and as a consequence the number of distinct combinations is reduced. If P is the number of unique patterns of k colors, then the verbal description of how to calculate it might be "If P is the number of unique permutations using N spools of yarn, then it is given by the number of possible permutations $N_p(N)$ divided by the number of permutations $N_p(n_1)$ of color 1, and so on for all colors up to k." The equation is easier:

$$P = \frac{N_p(N)}{N_p(n_1)N_p(n_2)\ldots N_p(n_k)} = \frac{N!}{n_1!n_2!\ldots n_k!} \tag{4.26}$$

Revisit now the case of five yarns considered above and, for sake of argument, say three of them are red ($n_1 = 3$) and two are green ($n_2 = 2$). Then there are $5!/3!2! = 10$ combinations, and they are

```
11122 11212 11221 12112 12121 12211 21112 21121 21211
22111
```

so the equivalence of the yarn colors makes an enormous difference in the number of distinct patterns.

Probabilistic reasoning can take on peculiar turns, an iconic case being when evaluating the probability that something *occurs*, it may prove more profitable to consider when it *does not occur*. A practical example of that will be encountered in the treatment of scattering in the Monte Carlo method (Section 32.2.3; see also Section A1.2.5): here, consider a slightly more amusing and accessible variation instead that also serves the dual purpose of providing an example of how to deal with factorials such as $n!$ when n becomes large.

Let there be N objects that have n distinguishing variations (like color, or weight, or number of spiny ridges, or the contents of fields on forms labeled "DOB"). Assume $N \gg n$ and that each variation is weighted equally in the population. If all the objects are in a container and k of them are removed, what is the probability $P_n(k)$ that any two of the k objects will be of the same variation? Solving this problem straight up is challenging (try it), but because the maximum probability is 1, the question can be answered by asking instead, what is the probability $W_n(k)$ that there will *not* be a match? Clearly, $W_n(k) = 1 - P_n(k)$.

The reasoning can then proceed as follows. Think of "variations" as being unique containers or slots into which objects are placed. With the first extracted object, there are n empty slots out of n total in which to put it, and so $W_n(1) = n/n = 1$. The probability $W_n(2)$ is easy to figure out: there are $(n-1)$ remaining empty slots into which to put the second withdrawn object, which now has to be multiplied by $W_n(1)$, that is, the total probability is the product of the probability of the first event with the probability of the second (they are independent or uncorrelated). Therefore, $W_n(2) = [(n-1)/n] \times W_n(1)$. Now the reason for large N becomes apparent: if N is sufficiently large, the removal of one of the objects will not have much of an impact on the fraction of its brethren amongst the remaining $(N-1)$ objects, and so the small change in the distribution can be neglected. The reasoning can then proceed iteratively as removal of objects does not change the distributions, and so with the third selection there are $(n-2)$ slots, and by induction, with the kth selection there are $(n-k+1)$ slots in which to put it so that

$$W_n(k) = \frac{(n-k+1)}{n} W_n(k-1) = \frac{n!}{n^k(n-k)!} \tag{4.27}$$

where the second relation follows from mathematical induction.

Although Eq. (4.27) solves the problem via $P_n(k) = 1 - W_n(k)$, the solution is lacking because it requires repetitive multiplications that get cumbersome if n and k are large. Moreover, Eq. (4.27) is highly dependent on how the multiplications are performed due to the size that the numerators and denominators reach when factorials are used (see

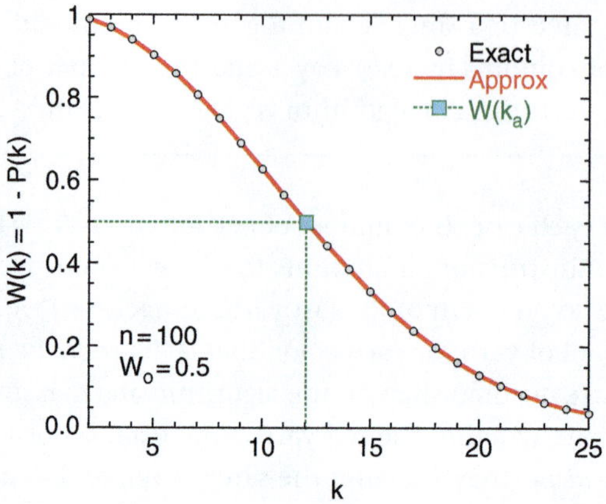

Figure 4.4 The exact evaluation of $W_n(k)$ using Eq. (4.27) compared to the approximate form of Eq. (4.28) using $n = 100$, for which $W_{100}(12) \approx 1/2$.

Section A3.2 for an appropriate algorithm). Better to now use Stirling's approximation of Eq. (A2.4) to find

$$W_n(k) \approx \left(1 - \frac{k}{n}\right)^{k-n-1/2} e^{-k} \tag{4.28}$$

where $(n-k)^p = n^p[1-(k/n)]^p$ for a power p has been used, as in Figure 4.4.

Matters become amusing when applied to real life. For the particular case $n = 365$ (the number of days in a year), the description (sometimes called the "birthday paradox") becomes the same as finding the number of people in a room that must occur so that the probability that at least two of them have the same birthday is 1/2. Using direct calculation with Eq. (4.27) or the algorithm of Section A3.2, one finds that $k_a > 22$, or about 23 people, a surprising result to the intuition of lay people.[8]

[8] Violating expectations is an astonishingly poor reason to call something a "paradox" given the accuracy of general intuition amongst lay people.

CHAPTER 5

Maxwell–Boltzmann distribution

5.1 Classical phase space

On contentious ground, I would hurry up my rear.

– Sun Tzu[1]

Position and velocity are of fundamental importance (not only in battles). To describe a box of particles requires knowledge of where they are are and how fast they are going. This takes six numbers, x, y, z, v_x, v_y, v_z, *per particle*. Figure 5.1 shows just the x coordinate of 1000 such particles for which the positions are randomly distributed. Figure 5.2 shows the corresponding v_x velocities, which are Maxwell–Boltzmann (MB) distributed. Equivalent figures could also be generated (but are not) for the $\hat{y}$ and $\hat{z}$ components. If v_x, v_y, v_z are are MB distributed, then so is $|\vec{v}|$:

$$f(v) \propto \exp\left(-\frac{v^2}{\Delta v^2}\right) = \exp\left(-\frac{v_x^2}{\Delta v^2}\right)\exp\left(-\frac{v_y^2}{\Delta v^2}\right)\exp\left(-\frac{v_z^2}{\Delta v^2}\right) \tag{5.1}$$

Understanding comes from doing, giving some impetus to see how an MB distribution can actually be generated. A simple and elegant method to do so makes use of the *central limit theorem*, which asserts that the average of a sum of n random numbers is normally distributed.[2] An algorithm to do so is given next[3] (contrast it with an alternate one utilized in Sections 33.1 and A3.33 for generating MB distributions of v_x and v_y simultaneously).

EXAMPLE: Generate $N = 1000$ random velocities v that are MB distributed.
SOLUTION: An MB distribution in velocity $f(v)$ is obtained from a normal distribution $N(0, 1) \propto e^{-x^2/2}$ by a special case of a more general algorithm in ref. [63].

Require: $\hat{R}$ is a $12 \times N$ matrix of uniform random numbers $0 \le R_{ij} \le 1$.
Require: $|v\rangle$ is a vector of length N.
  Generate $\hat{R}$
  for $j = 1$ to N **do**
    $v_j = \left(\sum_{k=1}^{12} R_{kj}\right) - 6$
  end for
  $|v\rangle$ corresponds to 1000 v values scaled by $\Delta v/\sqrt{2}$.

The phase space representation shown in Figure 5.3 is far more useful, in which v_x is plotted against x, along with an overlay showing small regions, or "cells", containing a smattering of particles – or none at all. By summing over all the cells that share a common position, a histogram of the velocities is generated in Figure 5.4. The histogram takes on the appearance of an MB distribution in velocity $f(\vec{v})$. As the number of particles becomes large and the bin size of the histogram small, a continuous representation of the histogram scaled by N results and corresponds to (note that $f(\vec{v})$ is independent of position)

$$f(v_x)\, dv_x = \frac{\int d\vec{x} \int dv_y dv_z f(\vec{v})}{\int d\vec{x} \int d\vec{v} f(\vec{v})} \tag{5.2}$$

$$f(\vec{v}) \propto \exp\left(-\beta m \left(v_x^2 + v_y^2 + v_z^2\right)/2\right) \tag{5.3}$$

[1] *Sun Tzu on The Art of War* (trans. Lionel Giles), Project Gutenberg Etext, Entry 47, p. 1682 of 2062.
[2] The designations normal distribution, Gaussian distribution, and Bell curve all designate mathematical creatures that have the same appearance. The MB distribution for $E_x = mv_x^2/2$ presently under consideration is a special case with $\langle v_x \rangle = 0$.
[3] The algorithm is based on Algorithm N5 of ref. [63] but with $n = \sigma^2 = 12$, choices which make the procedure extraordinarily easy to write and implement.

Introduction to the Physics of Electron Emission, First Edition. Kevin L. Jensen.

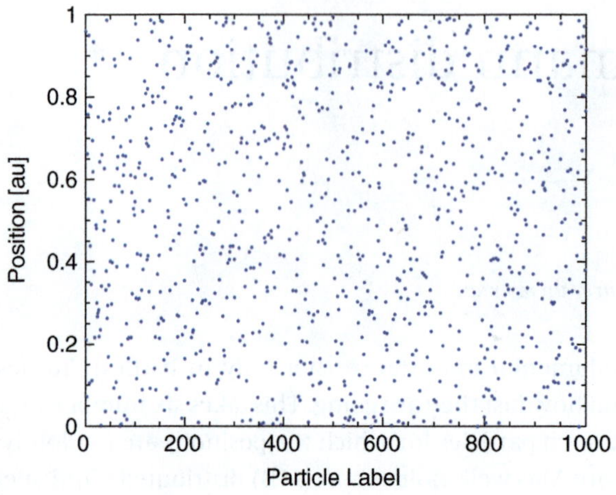

Figure 5.1 The x positions of 1000 particles labeled by particle number. The positions have been randomly chosen.

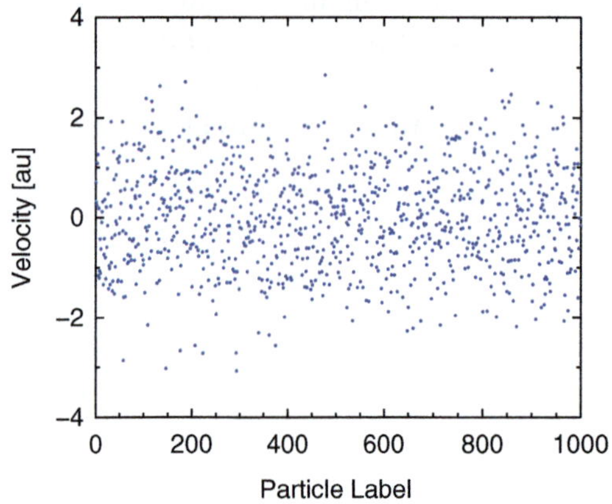

Figure 5.2 The v_x velocities of 1000 particles labeled by particle number. The velocities have been chosen according to an MB distribution.

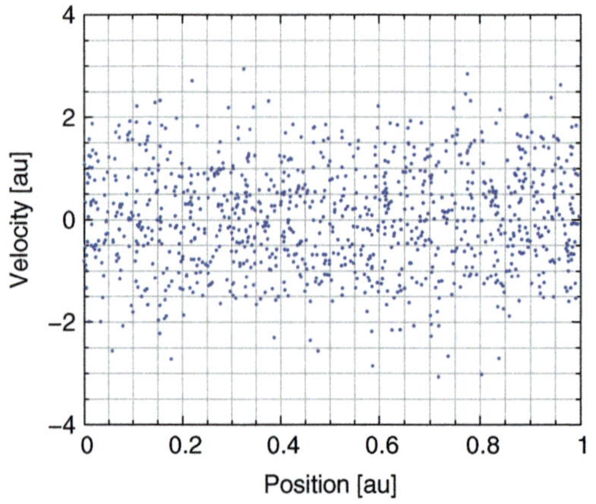

Figure 5.3 The v_x velocities of 1000 particles plotted versus their positions. Based on Figures 5.1 and 5.2. The grid overlay serves to suggest "cells" in which a particle (and sometimes many particles) finds itself: each cell is characterized by an energy ε_j where j refers to the cell, if the cells are made sufficiently small.

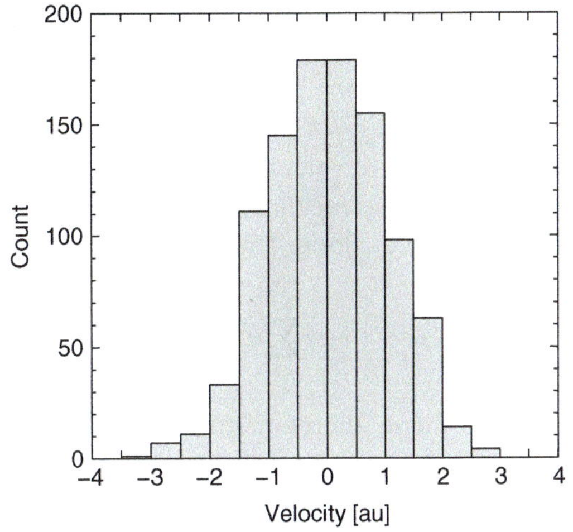

Figure 5.4 A histogram of the v_x data summed over all cells with a given v_x coordinate. In the large N limit, it would resemble $Nf(v_x)dv_x$ of Eq. (5.2).

All this talk of "velocities", though, is a bit old-school for three reasons. First, speaking so is the language of classical ideal gases, in which the particles are big, and although their speeds are respectable (e.g., a neon atom at room temperature travels about 600 m/s $\approx 2 \times 10^{-6}c$), the particles of greatest interest here are electrons that have velocities thousands of times greater in metals (e.g., $0.005c$ for an electron at the Fermi level of copper) and have wave-like properties that matter. Second, if those electrons are accelerated to large energies, as is common in particle accelerators, a phase space in which $v^2 \to c^2$ is dreary, and adopting the language friendly to the beam physics studies of Chapter 33 is handy for the transition. Third, the conjugate coordinate to position is momentum $\hbar\vec{k}$ (see Section 8.2.2). Best to just get used to it. Nevertheless, insofar as intuition can be gleaned from comparisons to ideal gases, speaking of velocity a bit longer will be occasionally useful; that usage, though, will be dropped at the earliest opportunity.

5.2 Most probable distribution

A probabilistic inference is a prediction today based on frequencies gathered yesterday. But that was then; this is now. How do you know that the world hasn't changed in the interim?

– Steven Pinker[4]

Counting may be hard, but counting with constraints is far more difficult. Working up from a simple problem to more complex arrangements is therefore good practice. But the simple problems had best be "timeless" in the sense that the conditions for how they unfold should not have changed from one day to the next: the probability that a person will do something is affected by something as trivial as weather and as consequential as mortality. Thus, while even faceless people are more interesting for understanding MB statistics, generic particles in a box are better for understanding probability.

Suppose that N particles are in said box and in thermal equilibrium, characterized by an average energy $\langle E \rangle$ such that the total energy $N\langle E \rangle \equiv U$ (the designation of the total energy of the box by U is by convention and precedent [62, 64, 65], not because it makes sense). To be clear, the box, which constitutes a "system", can have many possible energy states E_j, which are a consequence of a particular packing of the phase space cells. It is therefore useful to refer to the energy of a particle (rather than a state) by a small-case ε_j that is interpreted as the energy of the particle if it resides in the jth cell. Clearly, some of the ε_j can be equal, even if they reside in different cells: in Figure 5.3, for example, all the cells with the same vertical coordinate have the same energy because $f(v_x)$ is independent of x.

The question is, what is the most probable arrangement of the $\{n_j\}$? Simple counting says there are $N!$ ways to choose the order in which the particles are assigned to the cells (i.e., there are N ways to chose the first particle, $(N-1)$ ways the second, and so on), but if the order of the particles in each cell does not matter, then the number of arrangements

[4]Steven Pinker, *How the Mind Works.* New York: Norton, 1997, p. 351.

associated with the occupation numbers $\{n_j\}$ is instead $N!/(n_1!n_2!...)$. Overlooking the importance of this factor is unwise: when encountering Bose–Einstein (Section 6.1) and Fermi–Dirac (Section 6.2) statistics, it is this factor that changes. That is pleasingly ironic, in that even particles in a box are not quite as timeless as one would like.

Now impose the constraints $\sum_{j=1}^{\infty} n_j = N$ and $\sum_{j=1}^{\infty} n_j \varepsilon_j = U$ using the method of Lagrange multipliers [60], which allows for the maximization of the arrangements subject to constraints on the occupation numbers $\{n_j\}$ and associated energies $\{n_j \varepsilon_j\}$ (recall the treatment of Eq. (4.21)). For ease, observe that maximizing the arrangement is equivalent to maximizing the log of the arrangement, and so the later shall be done. Thus, letting A be the multiplier for $\sum n_j$ and B for $\sum n_j \varepsilon_j$, then

$$\ln\left[\frac{N!}{\prod_{j=1}^{\infty} n_j!}\right] - A\sum_{j=1}^{\infty} n_j - B\sum_{j=1}^{\infty} n_j\varepsilon_j = 0 \tag{5.4}$$

The minimization shall occur with respect to the n_j, therefore taking the derivative $\partial/\partial n_j$ of Eq. (5.4) and setting the result to 0 gives

$$-\frac{\partial}{\partial n_j}\ln\left[n_j!\right] - A - B\varepsilon_j = 0 \tag{5.5}$$

Invoking Stirling's approximation of Eq. (A2.4) for $n! = n\Gamma(n)$, then $\partial_n \ln(n!) \approx \ln(n) + (1/2n)$, which approaches $\ln(n)$ when n is large. Although n_j can be small, and as suggested by Figure 5.3 many of them can even be 0, it is reasonable to conclude that some of the n_j will be large if $N \gg 1$. Thus, the most probable arrangement is characterized by

$$n_j \approx \exp(-A - B\varepsilon_j) = \tilde{A}\exp(-B\varepsilon_j) \tag{5.6}$$

otherwise known as the *Maxwell–Boltzmann law*. For $\varepsilon_j = mv_j^2/2$ the MB distribution of Figure 5.4 and Eq. (5.2) results. Since the ratio $U/N = \sum \varepsilon_j n_j / \sum n_j$ is seen to depend only on B, the conclusion is that $B = B(T) \to \beta$. Lastly, if Z is defined by

$$Z = \sum_{j=1}^{\infty} \exp(-\beta\varepsilon_j) \tag{5.7}$$

then $\tilde{A} = N/Z$. The quantity Z is the "partition function," and is the sum over all energies that a particle has access to, so $p_j = Z^{-1}\exp(-\beta\varepsilon_j)$ is the probability that the particle has an energy ε_j. Importantly,

$$\frac{U}{N} = Z^{-1}\sum \varepsilon_j e^{-\beta\varepsilon_j} = -Z^{-1}\partial_\beta Z = -\partial_\beta(\ln Z) \tag{5.8}$$

which was foreshadowed by Eq. (5.2). If the box is not subject to any forces, then the energy is the kinetic energy $mv^2/2 = (\hbar k)^2/2m$, and the summation can be converted into a phase space integration over $\vec{k}$. Thus, with $\varepsilon = \hbar^2k^2/2m$,

$$Z \to \frac{1}{(2\pi)^3}\int e^{-\beta\varepsilon}\, d\vec{x}\, d\vec{k} = \frac{V}{2\pi^2}\int_0^{\infty} e^{-\beta\varepsilon} k^2 dk \tag{5.9}$$

Consider the simpler problem of integrating a Gaussian $e^{-\alpha x^2}$ where α is a constant. Letting $x \to r\cos\theta$ and $y \to r\sin\theta$, it is seen that

$$\left(\int_{-\infty}^{\infty} e^{-\alpha x^2}dx\right)\left(\int_{-\infty}^{\infty} e^{-\alpha y^2}dy\right) = \int_0^{\infty}\int_0^{2\pi} e^{-\alpha r^2} r\,dr\,d\theta = \frac{\pi}{\alpha} \tag{5.10}$$

Thus,

$$\int_{-\infty}^{\infty} e^{-\alpha x^2}dx = \sqrt{\frac{\pi}{\alpha}} \tag{5.11}$$

$$\int_{-\infty}^{\infty} x^2 e^{-\alpha x^2}dx = -\partial_\alpha\sqrt{\frac{\pi}{\alpha}} = \frac{1}{2\alpha}\sqrt{\frac{\pi}{\alpha}} \tag{5.12}$$

Applied to Eq. (5.9), $U/N = -\partial_\beta(\ln(\beta^{-3/2}))+$ constant $= 3/2\beta$, or

$$U = \frac{3}{2}Nk_BT \tag{5.13}$$

Thus, the average energy in each cartesian direction is $U_x = U_y = U_z = k_B T/2$, a restatement of the *principle of equipartition of energy*.

An important conclusion is thus revealed: *for an ideal gas, the average energy U/N depends only on T*. Equipartition entails that $k_B T/2$ attaches to each degree of freedom enjoyed by a particle. As a well-known example, an atom vibrating in a solid can be treated as a harmonic oscillator with three kinetic energy components (e.g., one such being $mv_x^2/2$) and three potential energy components (e.g., one such being $kx^2/2$, where k is the oscillator constant). That suggests $U = 3Nk_B T$, for which the heat capacity for one mole ($N = N_A$) is $3k_B N_A \equiv 3R = 0.2585$ meV/mole-K= 24.94J/mole-K, where R is the gas constant. The relation, experimentally confirmed in the early 1800s, is known as the *Dulong–Petit law*, as evidenced in the high-temperature thermal conductivity of metals such as copper and shown in Figure 3.4.

If the probability of particle a having an energy ε_1 is p_1, and an independent (non-interacting) particle b having an energy ε_2 is p_2, then the probability for both possessing those energies is $P_{ab} = p_a(\varepsilon_1)p_b(\varepsilon_2) = \exp(-\beta(\varepsilon_1 + \varepsilon_2))/Z^2$, and generalizing to N particles.

$$P_N = Z^{-N} \exp\left(-\beta \sum_{j=1}^{N} \varepsilon_j\right) \tag{5.14}$$

By labeling the particles, they are – for the moment – being treated as distinguishable. The factor Z^N is given by

$$Z^N = \left(\sum_{j=1}^{\infty} \exp(-\beta\varepsilon_j)\right)^N \tag{5.15}$$

If the particles were *in*distinguishable, then $Z^N \to Z^N/N!$. In what appears to be a wondrously bad idea, the power of that sum can be expanded by generalizing the binomial theorem relating the sum of two quantities u and w raised to a power n, given by

$$(u + w)^n = \sum_{k=1}^{n} \frac{n!}{k!(n-k)!} u^k w^{n-k} = \sum_{\{k+j=n\}} \frac{n!}{k!j!} u^k w^j \tag{5.16}$$

where the $\{\ldots\}$ term at the foot of the summation in the last representation is a *restriction* on the ks and js that they must be jointly considered: they have to sum to n. So, the expansion of Z^N is

$$Z^N = \sum_{\{\sum n_j = N\}} \frac{N!}{n_1!n_2!n_3!\ldots} \exp\left(-\beta \sum_{j=1}^{\infty} n_j \varepsilon_j\right) \tag{5.17}$$

The coefficient should look delightfully familiar from the earlier combinatorics exercises. Still, that is one big equation, hard to write compared to its predecessor, and just about impossible to calculate. Why writing it is a good idea is borne of experience not yet but invariably encountered.

5.3 Energy and entropy

I was moving/toward something/I thought I understood
There was an order/that was clear to me/a lawful order
Then … connections/were broken/we were sent/in a thousand directions.

– Joseph Chaikin and Sam Shepard[5]

The energy of the system can be changed by changing the cell size (e.g., by squeezing the container), changing the distribution of particles (e.g., by having them absorb energy), or changing the number of particles (e.g., by tossing more in). Whatever the origin of the thousand directions of change, they are represented by

$$dU = \sum_{j=1}^{\infty} n_j d\varepsilon_j + \sum_{j=1}^{\infty} \varepsilon_j dn_j \tag{5.18}$$

[5]Joseph Chaikin and Sam Shepard, The War in Heaven (Angel's Monologue), *Performing Arts Journal*, **9** (2/3), 1985, pp. 249–262. http://www.jstor.org/stable/3245529, p. 251.

Changing the cell size is the application of a force to the box. A force F_{ex} applied to the box face of area A that causes it to move a distance dL has changed the energy by $-F_{ex}dL = -(F_{ex}/A)AdL = -PdV$. That is, changes in cell size are classically associated with $-PdV$.

Changing the distribution $\{n_j\}$ of particles having energy ε_j is related to how much heat dQ is added, and that heat changes the entropy S by $dS = dQ/T$. For fixed particle number, dU can be written

$$dU = -PdV + TdS \tag{5.19}$$

When N is fixed, the independent variables of the free energy $F = U - TS$ are the volume V and temperature T, from which the pressure P and entropy S are defined by

$$P = -(\partial_V F)_T; \quad S = -(\partial_T F)_V \tag{5.20}$$

Its relation to the partition function Z is

$$F = -k_B T \ln(Z^N) \tag{5.21}$$

from which the thermodynamic quantities can be specified. If the particles interact, then $E(\vec{x}, \vec{k})$ contains interaction terms, making the partition function for N particles more involved than the form Z^N arising from a non-interacting ideal gas model, but such a complication need not concern the present discussion. Recalling the Shannon form of entropy in Eq. (4.20)

$$S = -k_B \sum_i p_i \ln(p_i) \tag{5.22}$$

where k_B is Boltzmann's constant, shows that $\beta = 1/k_B T$.

Particles entering and leaving the box are characterized by an average particle number $\langle N \rangle = \sum_j \langle n_j \rangle$, for which the Gibbs function $G = U + PV - TS$ is used in place of the free energy F. A modification to Eq. (5.19) to

$$dU = TdS - PdV + \mu dN = TdS - PdV + \mu \sum_{j=1}^{\infty} dn_j \tag{5.23}$$

results, and requires a third Lagrange multiplier in the generalization of Eq. (5.4) that results in $\varepsilon_j \to \varepsilon_j - \mu$ in the probability p_j and the partition function Z_N. The quantity $z = e^{\beta\mu}$ involving the chemical potential μ is often referred to as the "fugacity" [64–66].

The consideration of non-constant N allows the requirement $\sum n_j = N$ to be relaxed in Eq. (5.17), leading to an enormous simplification: a summation that was just shy of impossible has become almost enjoyable. The MB distribution will be considered first, with the Bose–Einstein and (more to the point for electrons) the Fermi–Dirac distributions treated subsequently.

5.4 The Gibbs paradox

He who confronts the paradoxical exposes himself to reality.

– Friedrich Dürrenmatt[6]

Individuality amongst particles gives rise to Gibb's paradox. Consider an ideal gas (a box of non-interacting single species particles) separated into two regions of volume V_1 and V_2 containing N_1 and N_2 particles, respectively; it does not change the argument to assume that the volumes are equal ($V_1 = V_2 = V$) and contain the same number of particles ($N_1 = N_2 = N$), but that assumption does make the exposition of the paradox unfold more rapidly. If the partition is removed, then the only real change is that, compared to one of the regions, $V \to 2V$ and $N \to 2N$, and the entropy S should be additive so that $S \to 2S$.

If the particles are distinguishable, are these expectations realized? The surprising answer is "No": removing the partition makes a difference where there should be no difference. This is seen as follows, where the partition function Z is used because the particle number is fixed (the grand partition function being employed when particle number can vary). The subscript "d" on F_d is to emphasize that the particles are considered distinguishable. It is convenient to introduce the

[6] Friedrich Dürrenmatt, *The Physicists* (trans. James Kirkup). New York: Grove Press, 1964, Appendix: "21 Points to *The Physicists*": Point 20.

"thermal wavelength" defined by

$$\lambda_T = \hbar\left(\frac{2\pi}{mk_BT}\right)^{1/2} \tag{5.24}$$

and to become reacquainted with how pressure and entropy are evaluated.

EXAMPLE: Calculate F_d for a gas of $N \gg 1$ distinguishable particles.
SOLUTION: If N is large, the energy levels are finely spaced, and the summations can be approximated using integrals. Thus, using the prescription of Eq. (4.6),

$$Z^N = \frac{1}{(2\pi)^3}\int d\vec{x}_1\cdots d\vec{x}_N \int d\vec{k}_1\cdots d\vec{k}_N e^{-\beta E}$$

For an ideal gas, $E = \sum_{j=1}^{N} \hbar^2 k_j^2/2m$ and is independent of $\vec{x}$ so that

$$Z^N = \left[\frac{V}{2\pi^2}\int e^{-\beta\hbar^2k^2/2m}k^2dk\right]^N = \left(\frac{V}{\lambda_T^3}\right)^N \tag{5.25}$$

from which $F_d = -Nk_BT\ln(V/\lambda_T^3)$.

EXAMPLE: Using Eq. (5.25), calculate the pressure P.
SOLUTION: Using Eqs (5.20) and (5.21), direct differentiation shows that

$$P = -\left(\frac{\partial F_d}{\partial V}\right)_T = \frac{Nk_BT}{V} \tag{5.26}$$

and the result is seen to be a variation of the ideal gas equation $PV = Nk_BT$.

EXAMPLE: Using Eq. (5.25), calculate the entropy S.
SOLUTION: Using Eqs (5.20) and (5.21), direct differentiation shows that

$$S = -\left(\frac{\partial F_d}{\partial T}\right)_V = Nk_B\left[\ln\left(\frac{V}{\lambda_T^3}\right) + \frac{3}{2}\right] \tag{5.27}$$

The effect of removing the partition between regions of equal size and particle number is to let $V \to 2V$ and $N \to 2N$. Does this entail $S \to 2S$ as would be required if entropy is additive? By inspection, the answer is "no" because S is augmented by an extra factor $Nk_B\ln(2)$ that arises because $\ln(2V) = \ln(V) + \ln(2)$.

If the particles are *in*distinguishable, then $Z_d^N \to Z_{in}^N/N!$ (the subscript "*in*" denoting "indistinguishable") because the ordering of the N particles is no longer unique, and the effect of $N!$ makes a subtle change in Z_{in} and S, but not P.

EXAMPLE: Calculate F_{in} for a gas of $N \gg 1$ indistinguishable particles.
SOLUTION: The procedure is as before, but using $N! \approx (N/e)^N\sqrt{2\pi N}$ (Stirling's approximation).

$$\frac{Z^N}{N!} = \frac{1}{N!}\left(\frac{V}{2\pi^2}\int e^{-\beta\hbar^2k^2/2m}k^2dk\right)^N \approx \frac{1}{\sqrt{2\pi N}}\left(\frac{eV}{N\lambda_T^3}\right)^N \tag{5.28}$$

from which $F_{in} \approx -Nk_BT\ln(eV/N\lambda_T^3)$ if $N \gg \ln(N)$.

EXAMPLE: Using Eq. (5.28), calculate the entropy S.
SOLUTION: Direct differentiation shows that

$$S = -\left(\frac{\partial F_{in}}{\partial T}\right)_V = Nk_B\left[\ln\left(\frac{V}{N\lambda_T^3}\right) + \frac{5}{2}\right] \tag{5.29}$$

where the factor $5/2 = \ln(e) + 3/2$.

Observe that $U = F + TS = (3/2)Nk_BT$ as before. Now it is clear that performing $V \to 2V$ and $N \to 2N$ results in $S \to 2S$: the problematic factor of $Nk_B \ln(2)$ does not arise because in the logarithm, the ratio V/N is found, which does not change when volume and particle number scale together (the density is constant). Consequently, there is no paradox: the removal of the partition adds the entropies of the two compartments, as intuitively it should. The same findings result if the compartments are unequal in size and particle number, so long as the density of the gas in each is the same. The factor of $N!$ in Eq. (5.28) goes by the name of *correct Boltzmann counting* and is required if the Bose–Einstein and Fermi–Dirac statistics are to merge into the MB distribution in the appropriate limits.

5.5 Ideal Gas in a potential gradient

Sors salutis/et virtutis/michi nunc contraria,
est affectus/et defectus,/semper in angaria.

– "O Fortuna", *Carmina Burana*[7]

The MB distribution for non-interacting particles has so far been considered under field-free conditions, so that $E(\vec{x}, \vec{k}) = n_k\varepsilon_k$, but that is not how life works: the pressing down of incessant forces is the fate of all. Let the potential energy $V(z)$ vary with the $\hat{z}$ coordinate such that $V(z) = -mgz$ where m is the mass of the gas atom and $g = 9.8$ m/s^2 is the acceleration due to gravity at the Earth's surface. How that modification results in the variation of pressure with height is known as the *barometric formula*.

The energy ε of an atom of gas is now $\varepsilon = \hbar^2k^2/2m - mgz$ so that Z in Eq. (5.25) becomes

$$Z = \lambda_T^{-3}\int d\vec{x}e^{-\beta mgz} = \frac{A}{mg\lambda_T^3}\left(1 - e^{-\beta mgh}\right) \tag{5.30}$$

where $A = \int dxdy$ and h is the height, or $V = Ah$. From Eq. (5.21), the relation for pressure of $P = -(\partial_V F)_T$, and the recognition that $dV = Adh$, it follows

$$P = \frac{Nmg}{A}\left(\frac{e^{-\beta mgh}}{1 - e^{-\beta mgh}}\right) \tag{5.31}$$

If $P \to P_o = \rho_o k_B T$ as $h \to 0$, where $\rho_o = N/V$ is the density of the gas at $h \to 0$ (say, sea level or some equally convenient reference point), and if the exponential argument is small such that $1 - e^{-\beta mgh} \approx \beta mgh$, then

$$P \approx P_o e^{-\beta mgh} \tag{5.32}$$

although the relation can obtained by simpler means by considering the change in pressure dP associated with a thin layer of gas $-\rho g\, dh$ and the relation $\beta P = \rho$ from the ideal gas law. The point is to obtain a familiar result from a proper analysis.

[7] "Fate is against me, in health and virtue, driven on and weighted down, always enslaved." https://en.wikipedia.org/wiki/O_Fortuna.

EXAMPLE: Find the pressure difference between the top and bottom of the Grand Canyon, assuming the depth is 1.8 km and the atmosphere is nitrogen, $N_2 = 28.135$ g/mole. Assume $T = 300$ K at the bottom.

SOLUTION: The exponential factor βmgh is

$$\frac{mgh}{k_B T} = \frac{\left(28.135 \frac{\mathrm{g}}{\mathrm{mole}}\right)\left(9.8 \frac{\mathrm{m}}{\mathrm{s}^2}\right)(1800\ \mathrm{m})}{\left(8.6173 \times 10^{-5} \frac{\mathrm{eV}}{\mathrm{K}}\right)(300\ \mathrm{K})} = 0.19897$$

Thus, $P/P_o = 0.19897/(e^{0.19897} - 1) = 0.9038$, compared to Eq. (5.32), which gives 0.8196.

The assumed constancy of the temperature in the previous example is not strictly true: an ideal gas expands adiabatically as its altitude increases such that $PV^{\gamma} =$ constant, where $\gamma \approx 5/2$ for a diatomic gas [66]. Coupled with $PV = Nk_B T$, a crude estimate is found to be $T(h)/T_o \approx (P(h)/P_o)^{2/7}$, or 8.54 K lower at the top of the Grand Canyon compared to the bottom.

5.6 The grand partition function

The one is made up of all things, and all things issue from the one.

– Bertrand Russell[8]

The grand partition function $\mathcal{Z}$, obtained from summing over all particles in $Z^N/N!$, is defined by

$$\mathcal{Z} \equiv \prod_{j=1}^{\infty} \sum_{n_j=0}^{\infty} \frac{\exp\left[\beta n_j (\mu - \varepsilon_j)\right]}{n_j!} \tag{5.33}$$

and it contains all the thermodynamic information possible about MB (as well as Fermi–Dirac and Bose–Einstein) gases. Observe that the consequence of dropping the requirement that the sum over all n_j has been to allow each n_j to appear in its own summation. By knowing that $e^x = \sum x^n/n!$, all of these sums can be performed and it follows that the grand potential Ω is

$$\Omega(T, V, \mu) \equiv -k_B T \ln \mathcal{Z} = -\frac{1}{\beta} \sum_{j=1}^{\infty} \exp\left[\beta\left(\mu - \varepsilon_j\right)\right] \tag{5.34}$$

where the summation follows from taking the logarithm of $\mathcal{Z}$. The grand potential Ω is so named because it is the basis for the calculation of the thermodynamic quantities of pressure P, density $\rho = N/V$, and entropy S via the relations [62, 65]

$$P = -(\partial\Omega/\partial V)_{T,\mu} \tag{5.35}$$

$$N = -(\partial\Omega/\partial \mu)_{T,V} \tag{5.36}$$

$$S = -(\partial\Omega/\partial T)_{V,\mu} \tag{5.37}$$

where the outside subscript indicates the terms held fixed (compare Eq. (5.20)). In the continuum limit, the (number) density ρ_{MD} becomes (compare Eq. (4.6))

$$\rho_{MD} = (2\pi)^{-3} \int d\vec{k} f_{MD}(E_k) = (2\pi)^{-3} \int d\vec{k} e^{\beta(\mu - E_k)} \tag{5.38}$$

[8] Bertrand Russell, *A History of Western Philosophy*. New York: Simon and Schuster, 1945, p. 44, Chapter IV: Heraclitus. The translation is the same as that of John Burnet, *Early Greek Philosophy*, 3rd edn. London: A.&C. Black, fragment 59, 1920. The Latin translation of the phrase is similar to *E pluribus unum*, the motto on the Great Seal of the USA.

EXAMPLE: Calculate the chemical potential μ of an MB gas (neon) at standard temperature and pressure (STP).
SOLUTION: First express the summation as an integral over phase space:

$$\sum\{...\} \rightarrow (2\pi)^{-3}4\pi V \int_0^\infty \{...\}k^2 dk \tag{5.39}$$

where $\int d\vec{x} = V$ and the integration over the angles of $\vec{k}$ give 4π. It is seen that the remaining integral is a variant of Eq. (5.12), but half as large due to the lower limit of integration on k. Performing the integration, it is found

$$\Omega = -\frac{V}{\beta}e^{\beta\mu}\left(\frac{M}{2\pi\hbar^2\beta}\right)^{3/2} \equiv -V\frac{e^{\beta\mu}}{\beta\lambda_T^3} \tag{5.40}$$

where the thermal wavelength $\lambda_T = (2\pi\hbar^2\beta/M)^{1/2}$ is used (Eq. (5.24)). The mass M of neon is 20.1797 g/mole, and standard temperature is 298.15 K ($\beta = 38.922$ eV^{-1}), so that $\lambda_T = 0.0225$ nm, or about $0.425a_o$. With $P = 101325$ Pa $= 6.3242$ eV nm^{-3}, it is found that $\mu = \beta^{-1}\ln(\beta P\lambda_T^3) = -0.3876$ eV.

Observe that the introduced length scale of the thermal wavelength λ_T is a measure of when differences in statistics become important. If the average inter-particle spacing is shorter than λ_T, then quantum effects on statistics matter. For neon, $\rho = \beta P = e^{\beta\mu}/\lambda_T^3 \equiv L^{-3}$, where $L = 152.7\lambda_T$, so that clearly MB statistics are adequate for the ideal gas example considered, and the usage of quantum statistics is overkill for gases such as neon.

EXAMPLE: Calculate the entropy S of an MB gas (neon) at standard temperature and pressure (STP).
SOLUTION: Taking the derivative of Ω with respect to T and expressing in terms of the chemical potential μ, it is found that

$$S = \frac{1}{2}k_B\beta PV(5 - 2\beta\mu) \tag{5.41}$$

or, using numbers from the prior example along with $\beta\mu = -15.086$, it is found that $S/V = 0.4329k_B$.

Eq. (5.41), which goes by the name of the Sakur–Tetrode equation, is the expression for entropy S (albeit in a different notation) that informed the derivation for the vapor pressure model of thermionic emission explored by Dushman [67] and recounted in Section 5.7.

However, for an average metal like copper where the electron number density is comparable to the atomic density, the thermal wavelength is about 4.32 nm, compared to an atomic diameter of copper of 0.256 nm. For electrons in a metal, statistics matter with a vengeance.

5.7 A nascent model of electron emission

We may then consider the evaporation of electrons from a metal as thermodynamically equivalent to the evaporation of a monatomic gas, for the electrons in the space necessarily have the same specific heat as the molecules of a monatomic gas.

– Saul Dushman [67]

The efforts to understand statistical methods and the MB distribution enable an understanding of first formulation of thermionic emission, in which electrons were modeled as evaporating off the surface of a metal. In contrast to modern derivations of the Richardson–Laue–Dushman (RLD) equation that begin by considering the tail of a Fermi–Dirac distribution, Richardson's original derivation applied a Carnot cycle to the electron gas in equilibrium with a metal. Dushman [67] reconsidered Richardson's equation and his expression for the coefficient A_{RLD} agrees with more modern treatments apart from a factor of 2 associated with electron spin (a purely quantum mechanical feature of the electron revealed by the Stern–Gerlach experiment in 1922) that Fowler [68] argued was required. It is a pleasingly different method of determining current density from thermionic emission using thermodynamic arguments without invoking the quantum mechanical ideas that form the backbone of most modern treatments.

Current density J, as in Eq. (3.10) and the expression for $\langle v_+ \rangle$ in Section 3.2.2, is the product of charge, density, and velocity. For a gas, the density $N/V = P/k_BT$, and since only half of them move in the direction of the surface, it follows

$$J = \frac{q}{2}\left(\frac{N}{V}\right)\langle v_+ \rangle = \frac{qP}{\sqrt{2\pi m k_B T}} \tag{5.42}$$

The pressure of the electron gas at the surface of a metal is, from thermodynamics, analogous to a transition wherein a liquid becomes a gas: a quantity of heat called the latent heat of transformation (L) induces the change. As known from Eq. (4.15), this means $\Delta S = L/T$. But when a gas and a liquid are in equilibrium, the change in the Gibbs energy of the gas dG_g is matched by an equal change in the Gibbs energy of the liquid dG_l, or, as in the discussion surrounding Eq. (5.23),

$$d(U - TS + PV)_{gas} = d(U - TS + PV)_{liquid} \tag{5.43}$$

which, when coupled with Eq. (5.23) itself, shows that

$$-S_g dT + V_g dP = -S_l dT + V_l dP \rightarrow \frac{\Delta S}{\Delta V} = \frac{dP}{dT} \tag{5.44}$$

and as a result it follows that

$$\frac{L}{T\Delta V} = \frac{dP}{dT} \tag{5.45}$$

Gases are far less dense than liquids, suggesting that $V_g \gg V_l$. From $\Delta V = V_g - V_l \approx V_g = Nk_BT/P$ from the ideal gas equation, the starting equation of Dushman's analysis (Eq. 1 of ref. [67]) is obtained, and is

$$\frac{L}{Nk_BT^2} = \frac{d\ln(P)}{dT} \tag{5.46}$$

a relation known as the Clausius–Clapeyron equation, and the starting point for early derivations of the thermionic emission equation (e.g., Appendix A in ref. [69]). Dushman proceeds by considering L to be the sum of a work function term $L_o = \Phi/N_A$ and terms dependent on the specific heat at constant pressure C_p for the electron gas and for the electron liquid c_p (as peculiar to modern sensibilities as that sounds). A more economical approach makes use of the Sackur–Tetrode equation (Eq. (5.41)) for the entropy of an ideal gas obeying MB statistics, from which

$$\frac{d\ln(P)}{dT} = -\frac{1}{2T}\left(5 - \frac{2\mu}{k_BT}\right) \tag{5.47}$$

$$\ln(P) = -\frac{5}{2}\ln(T) - \frac{\mu}{k_BT} + \text{constant} \tag{5.48}$$

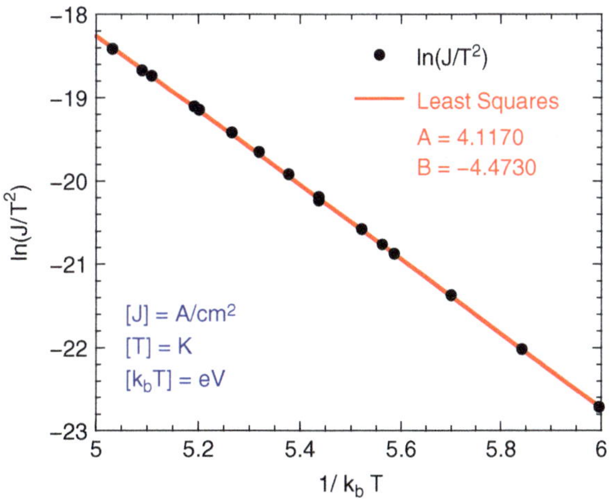

Figure 5.5 Thermionic data of Dushman. A representation of the data of Table II in Dushman (ref. [67], also Table A3.1) for thermal emission from tungsten on an RLD plot specified by $\ln(J_{RLD}/T^2)$ vs $1/k_BT$. The slope of the line $y = A + Bx$ gives the work function. The data therefore suggests $\Phi_W = -B = 4.473$ eV.

just as Dushman found. Coupled with Eq. (5.42), it has been established that $J \propto T^2 \exp(\mu/k_B T)$, where μ in this analysis is equivalent to the depth of the well the electrons find themselves within inside the metal, that is, the negative of the work function $(-\Phi)$.

The constants of integration following Eq. (5.47) are difficult to intuit from the analysis, and it is easier to bring the full power of statistical mechanics to bear, beginning with the grand potential Ω of Eq. (5.34), in the form of Eq. (5.40), from which P is

$$P = -\partial_V \Omega = \frac{k_B T}{\lambda_T^3} e^{\mu/k_B T} \tag{5.49}$$

Using the definition of λ_T from Eq. (5.24), switching over to the work function Φ notation, and inserting by hand a coefficient of 2 to account for electron spin as Fowler did gives

$$J_{RLD} = \frac{qm}{2\pi^2 \hbar^3} (k_B T)^2 \exp(-\Phi/k_B T) \tag{5.50}$$

The performance of Eq. (5.50) was quite good: data given by Dushman (Table II in ref. [67] and reproduced in Table A3.1) was remarkably linear on a plot of $\ln(J_{RLD}/T^2)$ vs $1/k_B T$, as shown in Figure 5.5. The least squares methods of Section A3.3 give the best fit parameters which can then be used to infer work function. Curiously, the fit to the data suggests $e^A = 61.375$ amp/(cm^2 K^2), which tends to obscure the contribution of electron spin of 2 given that $A_{RLD} =$ 120.1735 amp/(cm^2 K^2) (Table 2.3). A more modern perspective, faced with $e^A < A_{RLD}$, often attributes the discrepancy to crystal face variations and work function non-uniformity, more of which will be spoken of in subsequent sections.

Such is the celebrated Richardson–Laue–Dushman equation. Now begins the difficult task of achieving the same equation using the Sommerfeld model of a metal, in which the electron distribution function is not MB, as in the Drude theory, but rooted in the same Fermi–Dirac statistics required to understand field emission.

CHAPTER 6

Quantum distributions

Statistical mechanics, together with quantum mechanics, provides a foundation for modern physics which aims at the thorough understanding of physical phenomena from the microscopic viewpoint of atomic physics. Fundamental knowledge and training in statistical mechanics are therefore of vital importance not only for students studying the physical properties of matter but also for those who study nuclear physics or even astrophysics …

– Ryogo Kubo[1]

Quantum particles are indistinguishable and interchangeable, and that feature means that with regards to the combinatorics, the order the particles are put into the energy states does not matter, and the arrangement within the energy state does not matter. Whereas the (generally large) combinatoric factor of $N!/n_1!n_2!...$ arose when order mattered and particles were distinct, now each arrangement counts as one unique state, and such a factor is no longer included. This applies for Bose–Einstein (BE) as well as Fermi–Dirac (FD) statistics. How that factor plays out makes all the difference for how the distributions behave.

6.1 Bose–Einstein distribution

Brian: *You're all individuals!*
Crowd: *Yes, we're all individuals!*
Brian: *You're all different!*
Crowd: *Yes, we are all different!*
Homogenous man: *I'm not.*

– *Life of Brian*[2]

BE statistics are similar to Maxwell–Boltzmann (MB) in that the range of each n_j is unlimited. Therefore, the Grand Partition Function becomes

$$\mathcal{Z}_{BE} \equiv \prod_{j=1}^{\infty} \sum_{n_j=0}^{\infty} \exp\left[\beta n_j (\mu - \varepsilon_j)\right] \tag{6.1}$$

but that is an easy sum: making use of Eq. (A1.4) with $x^n \to 0$ as $n \to \infty$ for $x < 1$, it is found

$$-\ln(\mathcal{Z}_{BE}) = \sum_{j=0}^{\infty} \ln\left(1 - e^{\beta(\mu-\varepsilon_j)}\right) \tag{6.2}$$

and so the occupation factors $\langle n_j \rangle$ are given by

$$\langle N \rangle = \frac{1}{\beta}\partial_\mu (\ln Z)_{T,V} = \sum_{j=0}^{\infty} \langle n_j \rangle = \sum_{j=0}^{\infty} \left(e^{\beta(\varepsilon_j-\mu)} - 1\right)^{-1} \tag{6.3}$$

Observe that the quantity $\langle n_j \rangle$ is the mean number of particles in a state characterized by an energy ε_j: in the continuum limit $\varepsilon_j \to E_k$, it becomes the BE distribution $f_{BE}(E_k)$, so that

$$\rho_{BE} = (2\pi)^{-3} \int d\vec{k} f_{BE}(E_k) = (2\pi)^{-3} \int d\vec{k}\left(e^{\beta(E_k-\mu)} - 1\right)^{-1} \tag{6.4}$$

The form of E_k depends on the boson under consideration. For bosons, μ is always negative, making the term $e^{-\beta\mu}$ potentially large.

A common claim is that the high temperature limit of the BE distribution becomes the MB distribution. A more nuanced account is that at high temperature, higher energy levels can be occupied until all the occupation probabilities

[1] From the Preface to the Japanese edition of ref. [61], p. vi.
[2] Monty Python (Graham Chapman, John Cleese, Terry Gilliam, Eric Idle, Terry Jones, and Michael Palin), screenplay to *Life of Brian*, 1979.

Introduction to the Physics of Electron Emission, First Edition. Kevin L. Jensen.

$\langle n_j \rangle$ become small (< 1). When that happens, $e^{\beta(\varepsilon_j-\mu)} \gg 1$, and the MB distribution is recovered. In contrast, when the temperature is very low, $\langle n_j \rangle$ can be quite large, and BE condensation occurs [65], meaning a large crowd of bosons with the same quantum numbers can collect together in the ground state.

6.2 Fermi–Dirac distribution

...to get agreement with experiment one must assume that two electrons are never in the same state. This rule is known as Pauli's exclusion principle. *It shows us that electrons are fermions.*

– Paul A.M. Dirac [40], p. 211

FD statistics differ from MB and BE in that the range of each n_j is constrained to be either 0 or 1: fermions cannot share the same quantum numbers (of which momentum states are one). Therefore, the Grand Partition Function becomes

$$\mathcal{Z}_{FD} \equiv \prod_{j=1}^{\infty} \sum_{n_j=0}^{1} \exp\left[\beta n_j \left(\mu - \varepsilon_j\right)\right] \tag{6.5}$$

but that is an even easier sum than the BE case, and so

$$-\ln(\mathcal{Z}_{FD}) = \sum_{j=0}^{\infty} \ln\left(1 + e^{\beta(\mu-\varepsilon_j)}\right) \tag{6.6}$$

for which the occupation factors $\langle n_j \rangle$ are given by

$$\langle N \rangle = g \sum_{j=0}^{\infty} \left(e^{\beta(\varepsilon_j-\mu)} + 1\right)^{-1} \tag{6.7}$$

where a factor $g = 2$ has been inserted by hand to account for two electrons of opposite spin being allowed in the same momentum state. In the continuum limit, this becomes

$$\rho_{FD} = g(2\pi)^{-3} \int d\vec{k} f_{FD}(\varepsilon_k) = g(2\pi)^{-3} \int d\vec{k} \left(e^{\beta(\varepsilon_k-\mu)} + 1\right)^{-1} \tag{6.8}$$

For electrons, $\varepsilon = \hbar^2 k^2/2m$. Densities typical of metals entail that the chemical potential μ can become positive: for most metals, it is in the range of 2 to 11 eV, with copper in the middle ($\mu[\mathrm{Cu}] \approx 7$ eV).

The relation between number density $\rho = N/V$ and chemical potential μ is particularly important for electrons as it explains why charged particles packed extremely tightly can nevertheless be treated as "non-interacting" in the Sommerfeld model of nearly free electrons in a metal. Recall that for an MB gas, $\rho_{MB} = \exp(\mu)/\lambda_T^3$, as in Eq. (5.38). For the densities of typical gasses, the thermal wavelength λ_T is far smaller than the length scale of the interparticle separation $(1/\rho_{MB})^{1/3}$ (a requirement if quantum mechanical effects are to be neglected) and the average energy $(3/2)k_B T$ is in excess of the interaction energy between the gas atoms, which for neutral atoms is very small as per the Lenard–Jones potential (a requirement if the gas is to be treated as an ideal gas). The tight packing of electrons in a metal belies these "ideality" conditions, so why can they be treated as "free" (or nearly free)? The answer will be seen to be closely related to a key feature of fermions: from the Pauli Exclusion Principle, they do not share quantum numbers.

6.3 The Riemann zeta function

Now (1) the infinite qua infinite is unknowable, so that what is infinite in multitude or size is unknowable in quantity, and what is infinite in variety of kind is unknowable in quality. But the principles in question are infinite both in multitude and in kind. Therefore it is impossible to know things which are composed of them; for it is when we know the nature and quantity of its components that we suppose we know a complex.

– Aristotle[3]

...the more I give to thee/The more I have, for both are infinite.

– William Shakespeare[4]

[3]R.P. Hardie and R.K. Gaye, *Aristotle Physics Book I (Physica), Great Books of the Western World,* ed. Robert Maynard Hutchins, Vol. 8, Aristotle I. Chicago: Encyclopaedia Britannica, 1952, p. 262.
[4]Ref. [37]: *Romeo and Juliet,* II.ii.134–35, p. 870.

Pondering infinity has consumed philosophers since antiquity. The ζ function series that *appear* to be infinite but are not were characterized by Euler and put to use since then, as in quantum field theory [70] where Eq. (6.12) below is part of the renormalization of the effective action. While a source of amusement to mathematicians, such musings are the bane of the innumerate. The Greek philosophers struggled mightily, as the above quote by Aristotle in his disquisitions on Zeno's paradox makes plain. Zeno argued that an infinite number of bisections were necessary to close the gap between two spatially separated objects in a finite time, and therefore the concept of infinity was problematic [71], an argument lampooned in the joke opening Chapter 1.

Why does any of that matter here? The form of the FD (and BE) distribution insures that Riemann's zeta function $\zeta(p)$ appears with relentless regularity in the calculation of quantities such as the electron density ρ in its relationship to the chemical potential μ. $\zeta(p)$ is a remarkable function, and some familiarity with its calculation is necessary. The function is defined by the series relations (Eq. (A2.8)) with the form of interest here being

$$\zeta(p) = \frac{1}{1 - 2^{1-p}} \sum_{j=1}^{\infty} (-1)^{p+1} j^{-p} = \sum_{j=1}^{\infty} j^{-p} \tag{6.9}$$

where the second summation is the more common definition of $\zeta(p)$. The series for $\zeta(p)$ do not converge as quickly as the series agitating Zeno, but familiar expressions result from them all the same. For example, with $p \to 1$

$$\sum_{j=1}^{\infty} (-1)^{j+1} j^{-1} = \ln(2) \tag{6.10}$$

is well known because of the Taylor expansion of $\ln(1 + x)$ as $x \to 1$. It is the second series of Eq. (6.9) that is the more disquieting because of consequences resulting from the fact that the series is *infinite*. To see why that matters, observe the integral definition of $\zeta(p)$ with $p \to 0$ in Eq. (A2.7): the integral becomes logarithmic-infinite at the same time that $\Gamma(p)$ becomes infinite. Those infinities can be handled by isolating the divergent parts of the integrals via

$$\frac{1}{\Gamma(p)} \int_0^{\infty} \frac{dx}{x^{1-p}(e^x + 1)} = \frac{1}{2} + \frac{\int_0^{\infty} \frac{(1-e^{-x})dx}{2x^{1-p}(e^x+1)}}{\int_0^{\infty} x^{p-1} e^{-x} dx} \tag{6.11}$$

where the integral definition of $\Gamma(p)$ is used. Now, as $p \to 0$ the numerator integral is finite on the right-hand side, whereas the denominator retains its logarithmic-infinite nature, and so $\zeta(0) = -1/2$, the minus being from $(1 - 2^{1-p}) \to -1$. This is unremarkable until the series definition is considered, resulting in

$$\sum_{j=0}^{\infty} 1 = 1 + \zeta(0) = \frac{1}{2} \tag{6.12}$$

a relation found by Euler. Such relations do not sit well with those who treat ∞ as just another number, or view infinite sums as a limiting process. Their logic may be akin to the following. Consider the easier series (Eq. (A1.4))

$$\sum_{j=0}^{N-1} x^j = \frac{1 - x^N}{1 - x}$$

as N increases. As $x \to 1$ the ratio is evaluated using L'Hôpital's rule $f(x)/g(x) \to f'(x_o)/g'(x_o)$ as $x \to x_o$ if x_o is a zero of f and g, and primes denote derivatives with respect to argument. As a limiting process, therefore, the sum is N, and therefore approaches $+\infty$ as N does.

Not so fast.

First, sums such as $\sum_{j=0}^{\infty} x^j \equiv S(x)$ are different than finite sums. For one thing, it is evident that adding (or subtracting) 1 to $xS(x)$ is still $S(x)$, or

$$1 + xS(x) = S(x) \to S(x) = \frac{1}{1 - x} \tag{6.13}$$

and this is true for any x, even negative ones. Because Eq. (A1.4) contains an extra factor x^N, it differs from Eq. (6.13) because $x^N = 1$ when $x = +1$, regardless of how big N is (the case $x = -1$ shall be ignored, but it is related to the philosophically infamous light-bulb problem of predicting the final on/off state of a light bulb switched an infinite number of times [71]). Note that $S(1/2) = 2$ was encountered in Zeno's paradox.

Now consider $S(x) + S(-x) = 2S(x^2)$. It is already known that $S(-1) = 1/2$ from Eq. (6.13), and so $S(1) = 1/2 = 1 + \zeta(0)$, where the equality holds because two series are numerically equal if all of their component terms are equal.[5] Thus, $\zeta(0)$, which by its series *definition* is adding one to itself an infinite number of times, is equal to $(-1/2)$. If that is not bad enough, what about the sum of *all* positive integers? Consider the derivative of $S(x)$, or

$$S'(x) = \partial_x S(x) = \frac{1}{x}\sum_{j=1}^{\infty} jx^j = \frac{1}{(1-x)^2} \tag{6.14}$$

so that $S'(1) = 1 + 2 + 3 + \cdots = \zeta(-1)$, which is *the sum of all positive integers*. Observe that S' does not have a $j = 0$ term. Then (and being careful that $\partial_x S(-x) = -S'(-x)$) it follows that

$$S'(x) - S'(-x) = 4xS'(x^2) \tag{6.15}$$

which, for $x = 1$, gives $S'(1) - S'(-1) = 4S'(1) \rightarrow S'(1) = -1/12$. Lastly, $S'(1)$ is numerically equivalent to $\zeta(-1)$, as shown by comparing the series with $s = -1$, along with the $\Gamma(s)\Gamma(1-s)$ relation of Eq. (A2.5) and the particular value $\zeta(2) = \pi^2/6$ (as shown in Eq. (A2.7)). Either way,

$$\zeta(-1) = \sum_{j=1}^{\infty} j = -\frac{1}{12} \tag{6.16}$$

that is, the sum of all positive integers is a negative number, and a not so big one at that, an even greater insult to the intuitions of many. Properly (i.e., mathematically) speaking, $\zeta(p)$ is not defined at $p = -1$, but by analytic continuation, $\zeta(-1)$ is, leading to a conclusion that does not endear mathematical physicists to the lay public.

Therein lies the cautionary moral of Riemann's zeta function: having a formula does not mean it is understood or used correctly, or appropriate for the parameters for which it is used. More generally, using series expansion formulae requires care. Mathematicians are a bit more blunt: as remarked by Niels Abel (1828),[6] "Divergent series are the invention of the devil, and it is shameful to base on them any demonstration whatsoever ..."

6.4 Chemical potential

There is no light in earth or heaven/But the cold light of stars;

– Henry Wadsworth Longfellow[7]

Think how it wakes the seeds, –/Woke, once, the clays of a cold star.

– Wilfred Owen[8]

The allure of poetry is in the ambiguity of its language, so that using words not literally appropriate conveys understanding that would not be possible otherwise. But still, poets who speak of "cold stars" seem to have passed from ambiguous to oxymoronic. Before one speculates that either stars are emblematic of a destiny indifferent to human aspirations, or that poets literally do not know what they write about, ask if stars can indeed be understood to be "cold" in some sense that makes sense. The outer layers of the Sun are described by a hot MB distribution, but paradoxically the center (as well as stars like white dwarfs) can be better described by a cold electron gas (Section 6.6). How this is even possible, though, requires an understanding of what the chemical potential is, how it is evaluated, and what limits it takes on for the quantum distribution functions as a function of particle density.

---

[5]Be cautious: some series here include a $j = 0$ term, and some do not.

[6]As quoted in Berry and Howls, Infinity Interpreted, *Physics World*, June, 1993, p. 35.

[7]Henry Wadsworth Longfellow, "774. The Light of Stars", *English Poetry III: From Tennyson to Whitman*. Vol. XLII, The Harvard Classics. New York: P.F. Collier & Son, 1909–14. Bartleby.com, 2001. www.bartleby.com/42/.

[8]Wilfred Owen, "Futility", *A Concise Treasury of Great Poems, English and American*, ed. Loius Untermeyer. New York: Pocket Books, 1953, p. 502. Owen was killed in action on the Western Front one week before the end of World War I. His "Anthem for Doomed Youth" in the same collection as "Futility", targets the wasteful slaughter of war.

For energy parabolic in momentum, or $E_k = (\hbar k)^2/2m$, Eq. (6.8) can be rewritten as the product of a dimensioned coefficient and a dimensionless FD integral function $F_p(x)$ [72] by

$$\rho_{FD}(\mu, T) = \frac{4}{\sqrt{\pi}}\left(\frac{m}{2\pi\beta\hbar^2}\right)^{3/2} F_{1/2}(\beta\mu) \equiv \frac{N_c}{\beta^{3/2}} F_{1/2}(\beta\mu) \tag{6.17}$$

$$F_p(u) = \int_0^\infty \frac{x^p}{1+e^{x-u}}dx = \int_{-u}^\infty \frac{(x+u)^p}{1+e^x}dx \tag{6.18}$$

The first form of Eq. (6.18) is more common, but the second form is more useful when u is large and positive, as it allows for a partitioning of the integral into forms seen to be related to the Riemann zeta function.

EXAMPLE: Evaluate N_c in Eq. (6.17).
SOLUTION: Rewriting N_c gives

$$N_c = \frac{\sqrt{2m^3}}{\pi^2\hbar^3} = \frac{1}{(4\pi)^2}\left(\frac{R_y}{Q^2}\right)^{3/2} = 6.8122\ \text{eV}^{-3/2}\text{nm}^{-3} \tag{6.19}$$

where $R_y = 13.606$ eV and $Q = 0.35999$ eV-nm are used.

Three methods are considered for fermions for three regimes of $u = \beta\mu$, large positive, small positive, and negative, to illustrate three different methods that are generally useful for evaluating integrals related to density apart from a brute force numerical integration. They will be asymptotic methods, recursion techniques, and series expansions, respectively.

6.4.1 Large positive *u*

The integral for $F_p(u)$ can be partitioned into three regions, where the first region is obtained by making the replacement

$$\frac{1}{1+e^x} = 1 - \frac{1}{1+e^{-x}} \tag{6.20}$$

for $x < 0$, and collecting the remaining integrals as

$$F_{1/2}(u) = \frac{2}{3}u^{3/2} + \int_0^u \frac{\sqrt{u+x} - \sqrt{u-x}}{e^x+1}dx + \int_u^\infty \frac{\sqrt{x+u}}{e^x+1}dx \tag{6.21}$$

Asymptotically (in the large u limit), e^{-u} is negligible, allowing the last integral to be neglected. The integrand of the second can use Taylor expansions of $\sqrt{u \pm x}$ to obtain

$$\sqrt{u+x} - \sqrt{u-x} = \sqrt{u}\left\{\frac{x}{u} + \frac{x^3}{8u^3} + \frac{7x^5}{128u^5} + \dots\right\} \tag{6.22}$$

followed by the replacement of the upper limit to ∞. The middle integral of Eq. (6.21) is then well approximated by

$$F_{1/2}(u \gg 1) = \frac{2}{3}u^{3/2}\left\{1 + \frac{3}{4u^2}\zeta(2) + \frac{63}{64u^4}\zeta(4) + \frac{9765}{1024u^6}\zeta(6)\right\} \tag{6.23}$$

where numerical values of $\zeta(2)$ and higher are as in Section A2.3.

Zero temperature corresponds to the limit $u = \beta\mu \to \infty$. The value of $\mu(0) \equiv \mu_o$ is known as the *Fermi energy* (more commonly encountered as E_F), and the corresponding momentum $\hbar k_F = \sqrt{2m\mu_o}$ is called the Fermi momentum, in terms of which the density is

$$\rho_{FD}(T \to 0) = \frac{2}{3}N_c\mu_o^{3/2} = \frac{k_F^3}{3\pi^2} \tag{6.24}$$

An increase in temperature does not correspond to an increase in particle number, and so the relation $\rho_{FD}(T > 0) = \rho_{FD}(T = 0)$ can be used to develop a low-temperature expansion of $\mu(T)$ in terms of μ_o. This can be performed by replacing $\mu \to \mu_o(1 - a\delta - b\delta^2 - c\delta^3)$, where $\delta = (1/2)(\pi/2u_o)^2$, in $F_{1/2}(u) = (2/3)u_o^{3/2}$ and Taylor expanding to order δ^3. Equating

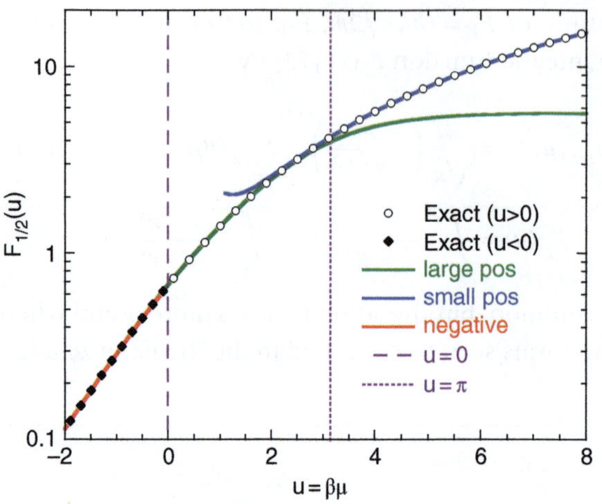

Figure 6.1 FD integral function $F_{1/2}(u)$ (Eq. (6.17). Numerical evaluation compared to the approximations of Eqs (6.23), (6.28), and (6.30) for u large positive, small positive, and negative, respectively.

the coefficients of the powers of δ separately to 0, then $a = 2/3$, $b = 4/5$, and $c = 1976/405$ is found with some effort, leading to the result

$$\frac{\mu(T)}{\mu_o} = 1 - \frac{1}{12}\left(\frac{\pi}{\beta\mu_o}\right)^2 - \frac{1}{80}\left(\frac{\pi}{\beta\mu_o}\right)^4 - \frac{247}{25920}\left(\frac{\pi}{\beta\mu_o}\right)^6 \tag{6.25}$$

For a metal like copper at room temperature, where $\mu_o = 7$ eV, $u_o \approx 270.77$, and so μ is 99.9989% of μ_o. Increasing the temperature to 3000 K ($u_o \approx 27.077$) changes that to only 99.888%, making the temperature dependence of μ for metals often ignorable even past their melting points. Clearly, the expansion is reasonable as long as $\mu_o > \pi k_B T$: at room temperature that would demand $\mu_o > 81.2$ meV, or an electron density just above 10^{20} cm^{-3}. In fact, Eq. (6.25) works reasonably well for slightly smaller μ_o, as evidenced in Figure 6.1, but then one is employing the first few terms of a divergent series for calculation (occasionally done by physicists, but distasteful to mathematicians).

A Fermi energy of 7 eV corresponds to $k_F = 13.55$ 1/nm, indicating that $\rho = 8.411 \times 10^{22}$ q/cm^3. Compare this to copper, which has 8.49×10^{22} atoms/cm^3, and it is clear that to a reasonable approximation each Cu atom gives up its outermost electron to make the electron gas of "free" electrons. Using Eq. (5.24) for room temperature conditions,

$$\rho_{FD}(\mu)\lambda_T^3 = 1024q \tag{6.26}$$

or a large number of electrons packed into a volume of sides equal to the thermal wavelength. Given the wave properties of electrons, at such densities the electrons are so overlapping that any connection to the Drude model's reliance on MB statistics of distinct, separated, and non-interacting particles is opaque, making the prediction of the Lorentz number (more or less, as in Eq. (3.19)) or Dushman's version of Richardson's equation (as in Eq. (5.50)) rather surprising.

6.4.2 Small positive *u*

In semiconductors, the number of free electrons is determined by the *dopants*. The doping density is such that even heavily doped semiconductors (e.g., with concentrations approaching 10^{19} atoms/cm^3) have electron number densities (or *carrier concentrations*[9]) that are orders of magnitude smaller than metals like copper. In fact, for such an electron density, $\rho(\mu)\lambda_T^3 = 0.143q$, or a fraction of an electron per thermal wavelength cubed, where λ_T is given by Eq. (5.24) with m_n replacing m as the electron's mass: for many semiconductors, the effective mass $m_n < m$ (e.g., the ratio of effective mass with rest mass m_n/m is 0.066 for GaAs and 0.55 for Ge, although for Si it is 1.062 [73]), making the thermal wavelength even larger. Therefore, fractions of an electron per thermal wavelength cubed require that $\mu < 0$. The boundary between small and negative is given by $u \to 0$. Comparing $F_{1/2}(0)$ to the integral definition of $\zeta(p)$,

$$F_{1/2}(0) = \left(1 - 2^{-1/2}\right)\Gamma\left(\frac{3}{2}\right)\zeta\left(\frac{3}{2}\right) = 0.678094 \tag{6.27}$$

[9] Although two types of carriers, electrons and holes, exist, the discussion of one can be bent to describe the other. If that dodge is unsatisfying, then a richer description can be found in Section 22.4.

Table 6.1 Values of g_j from Eq. (6.29) obtained via numerical integration.

j	g_j	j	g_j	j	g_j
0	0.67809	11	0.15170	22	0.085486
1	0.53608	12	0.14146	23	0.082311
2	0.43647	13	0.13257	24	0.079374
3	0.36465	14	0.12478	25	0.076649
4	0.31146	15	0.11788	26	0.074114
5	0.27105	16	0.11174	27	0.071749
6	0.23961	17	0.10624	28	0.069537
7	0.21461	18	0.10128	29	0.067464
8	0.19434	19	0.096778	30	0.065517
9	0.17762	20	0.092679	31	0.063684
10	0.16360	21	0.088930	32	0.061955

At room temperature, this would imply an electron density of $\rho = 1.1770 \times 10^{20}\ q/\text{cm}^3 = q/\lambda_T^3$. Thus, even heavily doped semiconductors have negative chemical potentials, but as will be seen, band bending near the surface of the semiconductor brought on by high applied field can render the *electro*-chemical potential small but positive. Thus, a method to find $F_{1/2}(u)$ for small u is necessary. Repeated use of Eq. (6.20) gives rise to the series

$$F_{1/2}(u) = g_0 + \sum_{j=1}^{\infty} (1 - e^{-u})^j g_j \tag{6.28}$$

$$g_j \equiv \int_0^{\infty} \frac{e^{-x}}{(1+e^{-x})^{j+1}} \sqrt{x}dx \tag{6.29}$$

Observe that $g_0 = (\pi/4)\left(2 - \sqrt{2}\right)\zeta(3/2)$. Values are tabulated in Table 6.1, but for $j > 9$ the approximation $g_j \approx 1.2502 j^{-0.86697}$ is reasonable (error $< 5\%$ by $j = 9$ or $< 1\%$ by $j = 17$). The performance of Eq. (6.28) is shown in Figure 6.1 with the series truncated at $j = 32$ and using the numerically calculated values of g_j given in Table 6.1.

6.4.3 Negative *u*

When $u < 0$, then $e^{x-u} > 1$ for all positive x in Eq. (6.18), and so the expansion of Eq. (6.13) may be employed in the form of $1/(1 + e^{x-u}) = e^{u-x} S(-e^{u-x})$. Each term in the resulting series is integrable in terms of the $\Gamma(x)$ function, and so

$$F_{1/2}(u < 0) = \frac{\sqrt{\pi}}{2} \sum_{j=1}^{M} \frac{(-1)^{j+1}}{j^{3/2}} e^{ju} \tag{6.30}$$

where M is chosen for a desired accuracy: the choice of the integer closest to $M \approx 6/|u|$ provides agreement to better than five significant digits. Note how $F_{1/2}(0)$ compares to the series definition of $\zeta(3/2)$ in Eq. (A1.8). For a very dilute gas $(-u \gg 1)$, μ decreases as

$$\mu(T) = (3/2) k_B T \ln\left[(16/9\pi)^{1/3} \mu_o / k_B T\right] \tag{6.31}$$

Consequently, $\mu(T)$ passes through zero at a temperature of

$$T(\mu = 0) = \left[\frac{3\sqrt{\pi}}{8}\left(2 - \sqrt{2}\right)\zeta(3/2)\right]^{-2/3} \left(\frac{\mu_o}{k_B}\right) = 0.99873 \frac{\mu_o}{k_B} \tag{6.32}$$

where Eq. (6.27) is used. From Eq. (6.31), it is seen that at high temperature, $\mu(T)$ varies as $-T\ln(T)$. The behavior for various μ_o cases is shown in Figure 6.2.

The regions separated in Figure 6.2 by the dashed line at $\mu = 0$ are referred to as a *degenerate* gas where $\mu > 0$ and a *non-degenerate* gas where $\mu < 0$, where non-degenerate gases obey classical statistics, but degenerate ones require quantum statistics. Instead of the imputation of moral turpitude normally associated with the word, degeneracy refers to the situation in which more than one particle can share quantum numbers: in a degenerate electron gas each set of quantum numbers (e.g., momentum) are occupied by two electrons differing only by spin. That factor of 2 increased the value of A_{RLD} in Eq. (5.50) to its modern value.

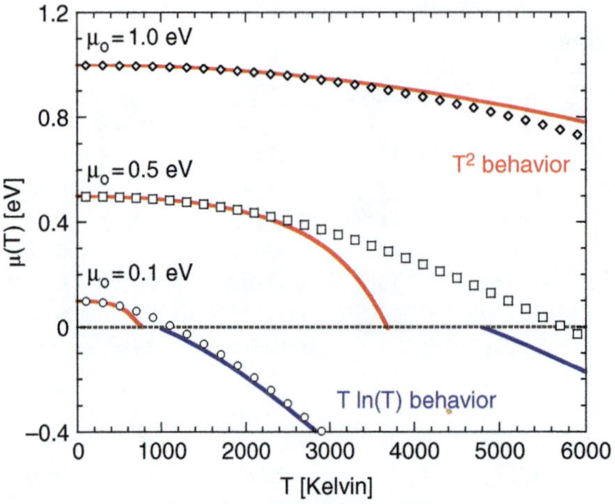

Figure 6.2 Variation of μ as a function of temperature for $\mu_o = 1.0$ eV (top), 0.5 eV (middle), and 0.1 eV (bottom). Red lines correspond to Eq. (6.25) up to quartic behavior, and blue to Eq. (6.31). The dashed gray line corresponds to $\mu(T) = 0$, and where $\mu(T)$ crosses it is specified by Eq. (6.32).

6.5 Classical to quantum statistics

If you want to find out anything from the theoretical physicists about the methods they use, I advise you to stick closely to one principle: don't listen to their words, fix your attention on their deeds.

–Albert Einstein[10]

The passage from classical (MB) to quantum (FD and BE) statistics (and vice versa) can be charted by fixing attention on the grand potential encountered in Eq. (5.34) and its transformation into Eqs (6.2) and (6.6). Close inspection reveals that all three can be obtained from

$$\Omega_s = \frac{N_c V}{2\beta} s^{-1} \int_0^\infty \sqrt{E} \ln\left[1 + s e^{\beta(\mu - E)}\right] dE \tag{6.33}$$

where N_c is from Eq. (6.19) and $s \to [-1, 0, 1]$ to obtain the BE, MD, and FD representations, respectively,[11] from which pressure P and number density $N/V = \rho$ are obtained using Eq. (5.35) to find

$$P_s = \frac{N_c}{3\beta^{5/2}} \int_0^\infty \frac{x^{3/2}}{e^{x-u} + s} dx \tag{6.34}$$

$$\rho_s = \frac{N_c}{2\beta^{3/2}} \int_0^\infty \frac{x^{1/2}}{e^{x-u} + s} dx \tag{6.35}$$

where an integration by parts has been performed on P_s to make its comparison to ρ_s more agreeable, $u \equiv \beta\mu$, and x is an integration variable. Before, the narrative focused on the accurate evaluation of μ for a given density ρ, but now consider instead the departure of FD and BE statistics from the ideal gas, or MB, behavior, so that the fixation on numerical accuracy can be relaxed a bit. Therefore, Taylor-expand the denominators of the P and ρ integrals in Eqs (6.34) and (6.35), and integrating the resulting series term by term to find

$$P_s = \frac{\sqrt{\pi}}{4} \frac{N_c}{\beta^{5/2}} \sum_{j=1}^{\infty} \frac{(-s)^{j-1}}{j^{5/2}} e^{ju} \tag{6.36}$$

$$\rho_s = \frac{\sqrt{\pi}}{4} \frac{N_c}{\beta^{3/2}} \sum_{j=1}^{\infty} \frac{(-s)^{j-1}}{j^{3/2}} e^{ju} \tag{6.37}$$

It is immediately deduced that

$$\left(\frac{\beta P}{\rho}\right)_s = \frac{\sum_{j=1}^{\infty} j^{-5/2} (-s e^u)^{j-1}}{\sum_{j=1}^{\infty} j^{-3/2} (-s e^u)^{j-1}} \tag{6.38}$$

[10]The Herbert Spencer lecture, delivered at Oxford on June 10, 1933, reproduced in Albert Einstein, *Ideas and Opinions*. New York: Crown Publishers, 1954, p. 270.

[11]Clearly, the choice of $s = 0$ should be viewed as a *limiting* procedure as $s \to 0$.

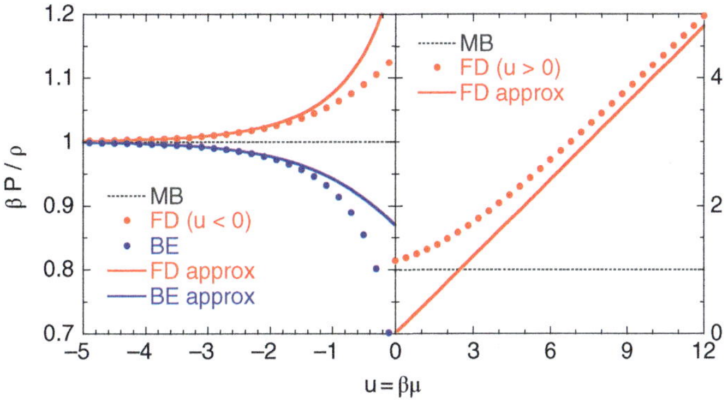

Figure 6.3 Departure of FD and BE from the ideal gas relation. Comparison of the numerically evaluated $\beta P/\rho$ of Eqs (6.34) (for P) and (6.35) (for N). Blue and red circles correspond to numerically evaluated values, and lines to Eq. (6.40). BE ($s = -1$) and FD ($s = +1$), respectively. Gray dashed line (MB) corresponds to the ideal gas relation $PV = Nk_BT$. The coordinate limits for the left ($\beta\mu < 0$) and right ($\beta\mu > 0$)) differ; otherwise the red circles and gray line would be continuous across the origin.

For MB statistics ($s \to 0$), the series are understood to truncate after the first term. Therefore, for an ideal gas described by MB statistics,

$$\left(\frac{\beta P}{\rho}\right)_{MB} = 1 \tag{6.39}$$

an alternate way of rendering the ideal gas law $PV = Nk_BT$. For BE and FD statistics when $\beta\mu = u \ll -1$, the leading order approximation to $\beta P/\rho$ is obtained by truncating the series after two terms, giving

$$\frac{\beta P_s}{\rho_s} = \frac{1 - \left(se^u/2\sqrt{8}\right)}{1 - \left(se^u/\sqrt{8}\right)} \tag{6.40}$$

and it is evident that both the BE ($s = -1$) and FD ($s = +1$) distributions, $\beta P_s/\rho_s$, approach the MB ($s = 0$) result of 1 as $u \to -\infty$, but BE does so from *below* and FD from *above* unity. Conversely, and only for FD and BE statistics ($s = \pm 1$),

$$\lim_{u\to 0} \frac{\beta P_s}{\rho_s} = \frac{1}{4}\left(s + 5 + \sqrt{2}\,(s+1)\right)\frac{\zeta(5/2)}{\zeta(3/2)} \tag{6.41}$$

or 0.51351 (BE: $s = -1$) and 1.1334 (FD: $s = +1$), which are also below and above unity, respectively. For BE statistics, there is no going further ($\mu_{BE} < 0$ always), but for a degenerate Fermi gas ($\mu > 0$), the final limit of $u \gg 1$ is obtained by

$$\lim_{u\to+\infty} \left(\frac{\beta P}{\rho}\right)_{s=+1} = \frac{2}{5}u \tag{6.42}$$

Eq. (6.42) is a rather astonishing result. Unpacking it, it says that as the temperature gets very small such that ($\beta\mu \gg 1$), the pressure goes to $P \to 2\rho\mu/5$ which is the zero temperature result: *unlike MB and BE statistics for which the pressure vanishes as the temperature vanishes, the Fermi gas has a finite pressure as* $T \to 0$. This behavior is seen in Figure 6.3, but the interpretation could easily well be that the density is becoming very large if the temperature were high. Considering such a regime seems like pompous scholasticism as the electron density even for metals, which tend to be the largest, is nowhere near such values – on Earth, anyway.

6.6 Electrons and white dwarf stars

There are more things in heaven and earth, Horatio, than are dreamt of in your philosophy.

– William Shakespeare[12]

The black dwarf material is best likened to a single gigantic molecule in its lowest quantum state. On the Fermi–Dirac statistics, its high density can be achieved in one and only one way, in virtue of a correspondingly great energy content. But this energy can no more be

[12]Ref. [37]: *Hamlet*, I.v.166–167, p. 943.

expended in radiation than the energy of a normal atom or molecule. The only difference between black dwarf matter and a normal molecule is that the molecule can exist in a free state while the black dwarf matter can only so exist under very high external pressure.

– Ralph Howard Fowler[13]

Although physics is interesting in the extremes, temperatures up to 6000 K in Figure 6.2 appear overblown for any metal or semiconductor. Yet there are natural circumstances where it does occur in stellar evolution, when a middle-weight star like the Sun nears the end of life and transitions to a white dwarf.

The Sun is a plasma sufficiently rarified and hot that the charged particles satisfy MB statistics. Near the end of life of such stars, the outward pressure opposing gravity weakens as the nuclear fuel is consumed, and the star can collapse to a point where the pressure opposing gravity is supplied by electrons: as per Eq. (6.40), an electron gas has a non-zero pressure even at zero temperature. The star continues to collapse until the electron pressure counterbalances that due to gravity. It is on the electrons to do so: the much higher mass of the ions (typically helium) means that their statistics remain well-characterized by MB statistics; moreover, He_2^{++} is characterized by BE statistics, and the pressure of a boson gas drops as the temperature drops even faster than that of an MB gas [62], a consequence of the BE condensation encountered in Section 6.1. In that case, the bosons increasingly occupy the ground state where their thermal agitation, for lack of a better phrase, is low, and insofar as thermal agitation and pressure go hand in hand, the pressure drops. Fermions, particularly electrons, cannot do so: they squeeze into whatever low energy level presents itself, with the average energy being 60% of the Fermi energy, as in Eq. (7.4). For very massive stars, the electron pressure is eventually insufficient to counterbalance gravity, causing the electrons to be forced into combination with protons to create neutrons, and the star collapses further into a neutron star, or, for the really massive, a black hole.

EXAMPLE: What is the chemical potential of the electron gas in a white dwarf star such that the e^- density is $\rho = 10^{30}$ cm^{-3} (a generally accepted value)? Assume the electron density is uniform.

SOLUTION: Using the zero temperature relation between chemical potential and density (see also Eq. (7.3) below),

$$\mu_0 = \frac{\hbar^2}{2m}(3\pi^2\rho)^{2/3} \tag{6.43}$$

$$= \frac{(197.33 \text{ eV-nm})^2}{1.022 \times 10^6 \text{ eV}}\left(\frac{3 \times 10^9}{\text{nm}^3}\pi^2\right)^{2/3} = 0.3646 \text{ MeV} \tag{6.44}$$

Observe that for a temperature of 5000 K (temperature of the photosphere of the Sun), $\beta\mu$ is ridiculously large because μ in Eq. (6.44) is large, but that even at a temperature of $\sim 1.5 \times 10^7$ K, as in the center of the Sun, $\beta\mu = 282$, which is sufficiently greater than 1 as to simplify Eq. (6.45) below: such numbers mimic a zero-temperature approximation, allowing the star to be characterized as a "cold" gas. Observe also that the chemical potential of the electrons in a white dwarf is comparable to the rest energy of the electron: this means that within such a star, the electrons are *relativistic*, which modifies the analysis.

For most stars, "more massive" goes with "larger." In contrast, white dwarf stars are smaller the more massive they are, and the explanation of that can be illuminated as follows. The largeness of $\beta\mu \gg 1$, when applied to the Grand Partition Function $\Omega_{FD} = -k_B T \ln(\mathcal{Z}_{FD})$ (Eq. (6.5)) and using the continuum distribution approach of Eq. (4.6), results in

$$\Omega_{FD} = \frac{2V}{(2\pi)^3\beta}\int_0^\infty \ln\left(1 + e^{\beta(\mu - E_k)}\right)4\pi k^2 dk \tag{6.45}$$

$$\approx \frac{4\sqrt{2}}{15\pi^2\hbar^3}m^{3/2}\mu^{5/2}V \tag{6.46}$$

From the relations for pressure $P = \partial_V \Omega$ and between electron density ρ and μ of $\mu = (\hbar^2/2m)(3\pi^2\rho)^{2/3}$ then

$$P_{el} = \frac{\hbar^2}{5m}(3\pi^2)^{2/3}\rho^{5/3} \tag{6.47}$$

[13]Quote from ref. [74]. A black dwarf is a white dwarf that has radiated away all of its energy. Fowler was addressing a seeming paradox whereby a portion of black dwarf matter, brought to free space, would not have the means to resume its gas-like (normal density) state in the absence of FD statistics.

This takes some getting used to. It was assumed that the electrons are more or less uniformly distributed throughout, and in fact, that is so; the nuclei follow suit and become fixed with respect to each other in a crystalline-like pattern termed a *quantum solid*, with densities in the center of a white dwarf close to 10^{10}–10^{11} kg/m^3, compared to the 162200 kg/m^3 for the center of the Sun [75].

The electron pressure P_{el} will be opposed by a gravitational pressure trying to collapse the white dwarf star: the evaluation of the later can be estimated crudely. On one side of an area of size $A = r^2\Delta\Omega$ a distance r from the center of the star, where $\Delta\Omega$ is a small solid angle, the force pushing out is due to the electron gas; that pushing in is due to gravitational attraction. Gravity is an inverse-square law force analogous to the interaction of charged particles, and from the latter it is known from Gauss's theorem that the force on a test charge a distance r from the center of a uniformly dense sphere of charge depends only on the amount of charge *within* a sphere of radius r, not the charge outside. So, too, with gravity (as an aside, in a problem expressly designed to torment physics graduate students, a mass dropped through a bore hole passing through the center of a non-rotating planet of mass M_o will exhibit harmonic oscillations: the gravitational force on the particle would be $F(r) = -GM(r)m/r^2$ where $G = 6.6738 \times 10^{-11}$ m^3 / kg s^2 is the gravitational constant. Since $M(r) = \rho(4\pi r^3/3) = M_o(r/R)^3$ with R the radius of the planet, then $F = -k_g r$ with $k_g = GM_o/R^3$: a force linear with displacement is the signature defining feature of a harmonic oscillator). The mass within the thin slab is $\rho\Delta\Omega r^2 dr$, so that the gravitational force attracting it to the center is

$$dF = \frac{G}{r^2}\left(m_h\rho\left(\frac{r}{R}\right)^3\right)(m_h\rho\Delta\Omega r^2)dr = \frac{4}{3}\pi G m_h^2 \Delta\Omega \rho^2 r^3 dr \tag{6.48}$$

The pressure gradient $dP = dF/A$ is therefore $dP/dr = (4\pi/3)Gm_h^2\rho^2 r$. Integrating that from the center of the star to the outer radius R gives a crude estimate of the average pressure due to gravity, or

$$P_G = \frac{3GM^2}{8\pi R^4} \tag{6.49}$$

where the average density $\rho = 3M/(4\pi R^3)$ has been used. Equating $P_{el} = P_G$ and placing all the star parameters on one side and the constants on the other gives

$$MR^3 = \frac{81\pi^2\hbar^6}{250G^3m^3m_h^5} \tag{6.50}$$

This is a weird and remarkable equation, and relates to work by Chandrasekhar to provide an upper bound on the mass of a white dwarf to $1.4M_\odot$, where the $\odot$ signifies the Sun, before gravity overcomes the electron pressure. What is weird about Eq. (6.50) is that the left-hand side refers to *stars* and the right to constants associated with gravity, quantum mechanics, chemistry, and (if the calculation were done acknowledging the near-light velocity of the electrons glimpsed in the discussion after Eq. (6.44)) relativity that are not usually expected to keep company. What is remarkable is that *the right-hand side is composed only of physical constants*, and therefore is constant itself. In other words, the mass M of the star is inversely proportional to the cube of its radius, that is, counter to general expectation, as the radius decreases, the mass increases.

EXAMPLE: Use Eq. (6.49) to estimate the pressure at the center of the Sun ($M_\odot = 1.9885 \times 10^{30}$ kg, $R_\odot = 6.96 \times 10^8$ m). Compare it to the commonly accepted value of 2.477×10^{16} Pa. Explain why the estimate appears defective.

SOLUTION: Inserting the values,

$$P_\odot = \frac{3}{8\pi}\left(6.6738 \times 10^{-11}\left[\frac{\text{m}^3}{\text{kg s}^2}\right]\right)\frac{(1.9885 \times 10^{30}\ [\text{kg}])^2}{(6.96 \times 10^8\ [\text{m}])^4} \tag{6.51}$$

$$= 1.3424 \times 10^{14}\ \text{Pa} \tag{6.52}$$

or about a factor of 185 smaller than the currently accepted value. A substantial reason for the disagreement is the approximation of uniform density: the average density of the Sun is $\rho_\odot \approx 1408$ kg/m^3 (about 40% more than that of water), whereas the center is 162200 kg/m^3, or a factor of 115 larger.

Harping on the solar density variation, but as a pedagogical exercise only, if instead the density $\rho(r)$ is taken as a power law for the form

$$\rho_s(r) = \frac{\rho_\odot}{6}\frac{(n+3)!}{n!}\left(1 - \frac{r}{R}\right)^n \tag{6.53}$$

for some n, to mimic the concentration of mass in Sun's center, where the coefficients insure that the integration over $d\vec{r}$ gives the mass of the Sun and $\rho_\odot = M_\odot / V_\odot$ is the average density as the ratio of the mass of the Sun to its volume, then the pressure $P_\odot$ would be instead

$$P_\odot = \frac{GM^2}{96\pi R^4} \frac{(n+3)^2(n+2)^2(n+1)}{(2n+1)} \tag{6.54}$$

A factor of $n \approx 8$ is therefore suggested. A power-law representation, however, is too simple a representation of the many layers involved, and much has been neglected, but the power law *ansatz* makes plausible the importance of density variation in stars where the electron gas is not degenerate.

A discourse of any length on white dwarf stars may appear a bit odd in a treatment of electron emission, but there is a subtle nod to the human dimension of the physics involved. Ralph Howard Fowler was an exceptional physicist, having made (as seen above and will be seen below) seminal contributions to the emission equations for thermal and field emission. He was wounded at Gallipoli[14] during World War I, was Lord Rutherford's son-in-law, was elected a Fellow of the Royal Academy in 1925, and had students such as Lennard-Jones and Dirac under him. He clearly possessed an excellent mastery of statistical and quantum physics, but he was equally (better?) well known as an astrophysicist and for his work on ...yes, white dwarf stars [74]. Paying homage to him by using another of his passions to illustrate a point in a tome on electron emission, a field he had much to do with creating – how poetic is that?

[14] Other physicists there were even less fortunate: Henry Moseley, who worked under Fowler's father-in-law Rutherford, and whose 1912 study of X-ray spectra from elements bombarded with high-energy electrons resulted in his demonstration that nuclear charge and not weight determined chemistry (for which he might have won the Nobel Prize), was killed at Gallipoli on August 10, 1915 [76]. Physicists were kept further from the front lines during World War II and subsequent conflicts, and thereby had much greater impact on how their countries waged war.

CHAPTER 7

A box of electrons

7.1 Scattering

> *...the Argives and their allies advancing with haste and fury, the Spartans slowly and to the music of many flute players – a standing institution in their army, that has nothing to do with religion, but is meant to make them advance evenly, stepping in time, without breaking their order, as large armies are apt to do in the moment of engaging.*
>
> – Thucydides[1]

> *...by the early sixth century, the charge of a well-trained and well-armored phalanx, its thousands of men acting as one and keeping pace by means of rhythmic shouts of the paean, proved itself all but irresistible.*
>
> – William H. McNeill[2]

At the Battle of Marathon (490 BCE), the huge invading force of 50,000, sent by the Persian king Darius I to crush the Greeks, suffered humiliating defeat at the hands of an outnumbered Athenian force (the Spartans arrived after the engagement because of their decision to conclude religious festivities first). The destruction on the Persian side, a seventh of the forces, was in stark contrast to the small losses suffered by the Greeks. The deployment of the Athenian hoplites in a phalanx had much to do with that. The phalanx, in which troops are arranged shoulder to shoulder with locked hoplons (shields), many rows deep, gave little opportunity for scattering, and as a result the seeming irrationality of being pressed onto the end of an opponent's sword or spear resulted in much fewer collective losses: being embedded in a phalanx made sense to those who viewed the risks of battle as shared equally, particularly when being routed was deadly but the urge to flee an individual matter that once acted upon could spread panic and therefore disarray [77].

Scattering has consequences ...but only if it is possible.

The electron density in a typical metal is so high that even though the electrons are cheek by jowl (i.e., their average nearest-neighbor separation is smaller than their thermal wavelength), unless they are near the Fermi level, scattering with their neighbors is not possible because there is no open space, no available state they have the energy to jump to, that is not already occupied. Sommerfeld's usage of Fermi–Dirac statistics and the newly developed quantum mechanics to describe metals resolved difficulties encountered in the Drude theory related to conductivity [78], but more significantly here paved the way for a reconsideration of the thermionic emission equation by Fowler [79] and the development of the field emission equation by Fowler and Nordheim [36]. An important application of that theory is that electrons will not scatter if their final state is occupied, and that realization fixed Drude's theory of conductivity quite nicely.

7.2 From classical to quantum mechanics

> *...there is nothing more difficult to carry out, nor more doubtful of success, nor more dangerous to handle, than to initiate a new order of things.*
>
> – Niccolò Machiavelli[3]

A key assumption in Sommerfeld's model was that the electrons behaved much like a non-interacting, or "free", gas of particles, and as quixotic as that seemed, it worked. The relation between number density $\rho = N/V$ and chemical potential μ hints why charged particles packed tightly can nevertheless be treated thusly in the Sommerfeld model of a

[1] Thucydides, *The Landmark Thucydides: A Comprehensive Guide to the Peloponnesian War*, eds Robert B. Strassler and Richard Crawley. New York: Free Press, 1996, Book 5.70, p. 343.
[2] William H. McNeill, *The Rise of the West: A History of the Human Community: With a Retrospective Essay*. Chicago: University of Chicago Press, 1991, p. 198. "Sixth century" refers to the sixth century BCE of Ancient Greece.
[3] Niccolò Machiavelli, *The Prince*. New York: New American Library, 1952, p. 49.

Introduction to the Physics of Electron Emission, First Edition. Kevin L. Jensen.

metal. Usually, an ideal or Maxwell–Boltzmann gas satisfies several conditions. First, as inferred by the density relation $\rho_{MB} = \exp(\beta\mu)/\lambda_T^3$ where the thermal wavelength λ_T is given by Eq. (5.24), λ_T must be far smaller than the length scale of the inter-particle separation $L = (1/\rho_{MB})^{1/3}$. Observe that the smallness of λ_T/L requires that the chemical potential μ be negative so that $\exp(\beta\mu) < 1$, and recall that μ can be made negative either by the gas being rarified or the temperature being so large that the gas is no longer degenerate. Moreover, the average kinetic energy $(3/2)k_BT$ must be comparable to or larger than the interaction energy between the particles if the "interaction" can be ignored.

Previously in Section 5.6, neon was taken as an example ideal gas, so for variety, consider argon ($M = 39.948$ amu = 0.039948 kg/mole) now. For it, $L/\lambda_T \approx 200$ at standard temperature and pressure (STP) conditions ($T = 273$ K, $P = 101325$ Pa). The energy of interaction between argon atoms is well described by a Lennard–Jones (LJ) 6–12 potential

$$V_{LJ}(r) = 4\varepsilon_{LJ}\left[\left(\frac{r_o}{r}\right)^{12} - \left(\frac{r_o}{r}\right)^{6}\right] \tag{7.1}$$

for which the minimum of the potential is at $2^{1/6}r_o$ and takes a value of $-\varepsilon_{LJ}$. Also, $r_o = 0.3067$ nm and $\varepsilon_{LJ} = 8.04$ meV,[4] where the average separation $r \approx L = 10.89r_o$, making the energy of interaction $V_{LJ}(L)$ very small. By comparison, the kinetic energy of an argon atom at STP is $(3/2)k_BT = 35$ meV.

So much for a gas. What about a metal? As per Eq. (6.25), $\mu(T) \approx \mu_o$ due to $(\pi/\beta\mu_o)^2 \ll 1$, so that $f_{FD}(E) \to \Theta(\mu - E_k)$ and so

$$N = \frac{2}{(2\pi)^3}\int \Theta(k_F^2 - k^2)d\vec{x}d\vec{k} = \frac{V}{2\pi^2}\int_0^{k_F} k^2 dk \tag{7.2}$$

$$\frac{N}{V} = \rho(T=0) = \frac{k_F^3}{3\pi^2} \tag{7.3}$$

where the Heaviside step function $\Theta(x) = 1$ if the argument is positive and 0 otherwise. Similarly, the average energy of an electron in a metal is

$$\langle E\rangle = \frac{1}{2\pi^2}\int_0^{k_F} \frac{\hbar^2k^2}{2m}k^2 dk = \frac{3\hbar^2k_F^2}{10m} = \frac{3}{5}\mu \tag{7.4}$$

The ratio between the average kinetic energy KE and the energy of interaction $PE = 4Q/r$ of two electrons separated by the mean spacing $r = 1/\rho^{1/3}$ with ρ from Eq. (7.3) is then

$$\frac{KE}{PE} = \frac{(9\pi)^{2/3}\hbar^2k_F}{10\alpha mc} \tag{7.5}$$

which increases as ρ (and therefore k_F) increases. At the limit of white dwarf conditions, the average kinetic energy is well beyond the potential energy of interaction of two electrons; even for more terrestrial metal-like conditions the kinetic energy is not trivial, as there, Eq. (7.5) is of order unity. Thus, fermions oddly become more "classical" as the density increases.

An "interaction" requires something to change: electrons scattering off each other swap energy and momentum, and therefore change their quantum numbers. If they are foresworn from doing so because the quantum numbers of any final state they could possibly scatter into is already taken, they cannot be said to be "interacting" in the usual sense – rather, they go on doing what they were doing and behave as though they were free. Hence, the Sommerfeld assumption is a bold but effective one.

A final distinction between classical and quantum but with regard to *statistics* resides in the consideration of the spacing of the energy levels. Recall in Section 3.3 Einstein's insight that the discrete energy spectrum used by Planck to reconcile Wien's law with the Rayleigh–Jeans formula, leading to Eq. (3.23), was fundamental to the nature of the photons. Discrete energy states are quite different from the continuum energy spectrum associated with classical physics, and that distinction is the last metric by which quantum descriptions go over into classical ones and vice versa: when the energy difference between adjacent energy levels is small by comparison to the (average) particle energy, classical statistics are good. That distinction can be viewed as a requirement on the size of the quantum numbers, for example the well-understood harmonic oscillator [44] has an energy spectrum that behaves as $\hbar\omega_n = (n + 1/2)\hbar\omega_o$, where ω_o is the ground state energy. The condition for a classical description is then that $\hbar\omega_{n+1} - \hbar\omega_n = \hbar\omega_o \ll (n + 1/2)\hbar\omega_o$, or $2n \gg 1$.

[4]See Table 11.1 of ref. [65].

7.3 Moments and distributions

Not all those who wander are lost ...

– John R.R. Tolkien[5]

Having carefully explored the landscape leading to the Sommerfeld model, the quantum refinements to Drude's model giving the correct coefficient to the Lorentz number of Eq. (3.20) can be determined (it was encountered already in the red line of Figure 3.3). The agreement is profound, and significantly strengthens the grounds by which the Sommerfeld model serves as a basis for the electron emission equations which follow. The journey shall (*i*) begin with Eq. (3.5) for the charge current density J_q but recast in the language of distribution functions $f(\vec{x}, \vec{k})$, (*ii*) recast the momentum integration to an (easier) energy integration by introducing the density of states, (*iii*) place scattering in the context of phase space (leading to Boltzmann's equation), (*iv*) incorporate the effects of occupied states on scattering, and finally (*v*) arrive at the quantum mechanical form of the Lorentz number L. The language to do so shall make use of the moments of the distribution.

Recall that the distribution of phase space points $f(\vec{x}, \vec{k})$ is normalized according to

$$N = \frac{2}{(2\pi)^3} \int d\vec{x} d\vec{k} f(\vec{x}, \vec{k}) \tag{7.6}$$

where N is the number of particles, so that the charge density is $qN/V \equiv q\rho$. Observe that the "2" in the numerator of the coefficient (previously called g) accounts for electron spin. For a constant potential, the distribution of particles will be independent of $\vec{x}$ and so

$$f(\vec{k}) = \frac{1}{V} \int d\vec{x} f(\vec{x}, \vec{k}) \tag{7.7}$$

The moments $M_n(k)$ of a distribution function $f(\vec{k})$ are often defined with respect to a mean value, but here take them to refer to integrations of powers of k with the distribution function such that

$$M_n(k_j) \equiv \frac{2}{(2\pi)^3} \int k_j^n f(\vec{k}) d\vec{k} \tag{7.8}$$

where k_j is one of the cartesian components. This form will allow for generalizations leading to the supply function in the one-dimensional emission equations. The number density is the zeroth moment, or $\rho = M_0(k_x)$. Average quantities such as $\langle k_x^n \rangle$ are the ratios of the nth moment with the zeroth, a particularly important one being current density

$$J_q = q\rho\langle v_x \rangle = \frac{2q}{(2\pi)^3} \int d\vec{k} \left(\frac{\hbar k_x}{m} \right) f(\vec{k}) \tag{7.9}$$

where the average velocity is ratio of the first moment $M_1(k_x)$ with the zeroth $M_0(k_x)$, the later of which is proportional to the number density ρ, so that

$$\rho\langle v_x \rangle = \frac{\hbar}{m} M_1(k_x) \tag{7.10}$$

Thus, Eq. (7.9) is seen to be $J_q = (q\hbar/m)M_1(k_x)$ and therefore equivalent to Eq. (3.5), except for a tiresome negative sign that indicates direction of flow but which is otherwise inconvenient to retain.[6] Eq. (7.9) is the defining equation for *charge* current density in a bulk material, and therefore the starting point to consider conductivity. Later, with modifications, it will be the starting point for the development of the one-dimensional emission equations.

Given that MB, FD, and BE statistics are more easily expressed in terms of energy E rather than wave number $\vec{k}$,[7] modifications to the integrals result and introduce the density of states (DOS) $\mathcal{D}(E)$, allowing number density to be written as

$$\rho = \int_0^\infty \mathcal{D}(E) f(E) dE \tag{7.11}$$

---

[5]J.R.R. Tolkien, *The Fellowship of the Ring, The Lord of the Rings*. Boston: Houghton Mifflin, 1987.

[6]In units of $q = 1$, the current density and the number current density are the same.

[7]Observe the change in notation to a more conventional E in the emission nomenclature, rather than the past usage of ε_k to indicate the energy of a single particle characteristic of the statistics nomenclature.

In polar coordinates, the phase space volume element is $d\vec{k} = k^2 \sin\theta dk d\theta d\phi$. A parabolic relation for energy $E = \hbar^2 k^2/2m$ depends only on the magnitude of $\vec{k}$. Finding the expression for the density of states $\mathcal{D}(E)$ is then obtained by considering the zero temperature number density or

$$\rho_{FD}(T=0) = \frac{2}{(2\pi)^3}\int_0^{k_F} 4\pi k^2 dk = \int_0^{\mu} \frac{m}{(\pi\hbar)^2}\left(\frac{2mE}{\hbar^2}\right)^{1/2} dE \tag{7.12}$$

where 4π is from integration over the angular coordinates. By introducing

$$\mathcal{D}_3(E) = \frac{m}{\pi^2\hbar^3}\sqrt{2mE} = N_c\sqrt{E} \tag{7.13}$$

as per Eq. (6.19), then $\rho = \int_0^{\mu} \mathcal{D}(E)dE$ at zero temperature, and where the subscript 3 distinguishes it from the one- and two-dimensional versions. The discussion seems to be about point-like particles, but the same relations are obtained when the states in question are manifestly quantum mechanical in origin, as shall be seen in Section 21.2 and the discussion leading to Eqs (21.10)–(21.12).

EXAMPLE: Find $\mathcal{D}(E)$ for two dimensions.
SOLUTION: In two dimensions, the area element is $kdkd\theta$ and so

$$\mathcal{D}_2(E)dE = \frac{2}{(2\pi)^2}k\left(\frac{dk}{dE}\right)dE\int_0^{2\pi} d\theta = \frac{m}{\pi\hbar^2}dE$$

or $\mathcal{D}_2(E) = m/\pi\hbar^2$ and is independent of energy.

EXAMPLE: Find $\mathcal{D}(E)$ for one dimension.
SOLUTION: In one dimension, the line element is dk and so

$$\mathcal{D}_1(E)dE = \frac{2}{(2\pi)^1}\left(\frac{dk}{dE}\right)dE = \frac{1}{2\pi\hbar}\left(\frac{2m}{E}\right)^{1/2} dE$$

Pleasingly spherical distributions render the calculation of $\mathcal{D}(E)$ no more difficult than finding the surface area of a sphere and an expression for dk/dE, but this is a consequence of the triviality of the model, not the simplicity of atomic arrangements: the potential was *taken* to be constant, whereas in the presence of atomic cores, forbidden regions open up where ranges of k values are excluded. In other words, band gaps form when atoms and the periodic arrangement of core potentials associated with them are involved. In such cases, the surface area is no longer simple, and finding all k for a given energy E is more difficult. A parabolic density of states is a model of convenience and often adequate, but not always.

Lastly, revisit the distribution of electrons given by the product $\mathcal{D}(E)f_{FD}(E)$, the integral over which gives $\rho(T)$ as in Eq. (7.11) and shown in Figure 7.1: the top plot is for a large positive Fermi level ($\mu = 7$ eV), for which not much variation is visible in the distribution; the middle plot is for conditions similar to a very heavily doped semiconductor ($\mu = 0.5$ eV), for which the temperature dependence is becoming more pronounced near $E \approx \mu$; the bottom plot is for when the chemical potential is negative ($\mu = -0.1$ eV), for which the differences are rather stark, a circumstance that anticipates the longer discourse on semiconductors in Chapter 22.

7.4 Boltzmann's transport equation

Socrates: *Heraclitus is supposed to say that all things are in motion and nothing at rest; he compares them to the stream of a river, and says that you cannot go into the same water twice.*

– Plato[8]

[8]Plato, *The Dialogues of Plato: Cratylus* (trans. Benjamin Jowett), *Great Books of the Western World*, ed. Robert Maynard Hutchins, Vol. 7, Plato. Chicago: Encyclopaedia Britannica, 1952, p. 94.

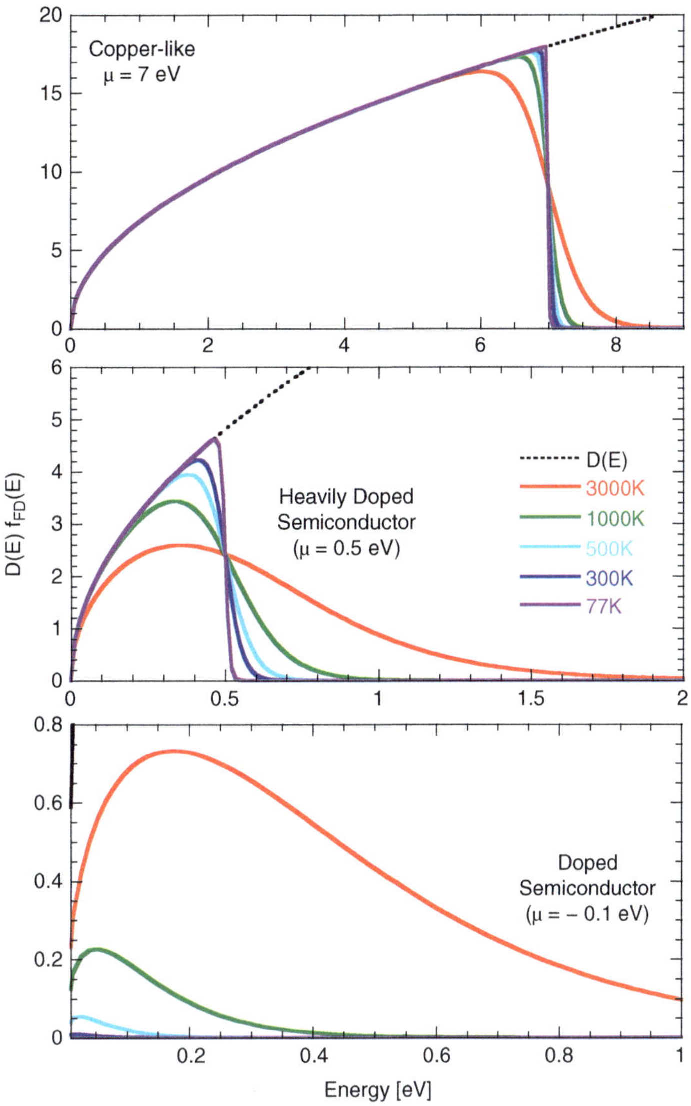

Figure 7.1 The integrand of Eq. (7.11) for various μ: for a copper-like $\mu = 7$ eV (top), for a very heavily doped semiconductor $\mu = 0.5$ eV (middle), and for a doped semiconductor $\mu = -0.1$ eV (bottom), which anticipates the electron distribution of Figure 22.7, with the difference that the zero of energy is at the bottom of the conduction band here.

Perpetual change is a better model of life than unchanging equilibrium, although if the phase space distribution is disturbed, it will relax back to the equilibrium Fermi–Dirac distribution, with the electrons thermalizing by scattering off each other (*e–e* scattering), off the atomic cores or rather groups of cores in collective motion (*e–p* scattering, where *p* refers to phonons), off individual atoms of a different type than the bulk material (impurity scattering, be they ionized or neutral), or various other mechanisms which complicate the transport of electrons away from a simple ballistic model [80, 81]. There will be a time to consider the individual processes – what matters here is that they exist, that they thermalize the energy of excited electrons to an equilibrium distribution, and that the equilibrium distribution tends to the thermal Fermi–Dirac distribution.

The waters may not be the same, but the river looks familiar: if the overall density ρ of the electrons does not change, then

$$\rho = \frac{2}{(2\pi)^3}\int f(\vec{x},\vec{k};t+dt)d\vec{x}d\vec{k} = \frac{2}{(2\pi)^3}\int f(\vec{x},\vec{k};t)d\vec{x}d\vec{k} \tag{7.14}$$

so this suggests

$$\frac{2}{(2\pi)^3}\int \left(\frac{d}{dt}f(\vec{x},\vec{k};t)\right) d\vec{x}d\vec{k} = 0 \tag{7.15}$$

which in turn indicates $df/dt = 0$ because the electrons are neither created nor destroyed and therefore have to go somewhere in phase space (the invariance of the phase space element is treated in greater detail in Eq. (33.26)). It is well known that the total time derivative is composed of three parts: the first is how f depends directly on time t, the second is due to changes in position as the particle moves with a velocity, and the third is changes in velocity as a consequence of forces, or, in short,

$$\frac{df}{dt} = \frac{\partial f}{\partial t} + \dot{\vec{x}} \cdot \vec{\nabla}_k f + \dot{\vec{k}} \cdot \vec{\nabla}_x f \tag{7.16}$$

where the odd dots over $\vec{x}$ and $\vec{k}$ indicate their time rate of change (the "over-dot", or Newton, notation). Dots, vectors, partials – in short order, the notation is maddening, and the urge to a simpler one-dimensional problem irresistible. Therefore, consider just one component of Eq. (7.16), say the $\hat{x}$ direction, and assume uniformity in the transverse directions, so that

$$\frac{df}{dt} = \frac{\partial f}{\partial t} + \dot{x}\partial_x f + \dot{k}\partial_k f \tag{7.17}$$

Temporarily ignoring the subscript x on k_x (and force F below), then $\dot{x} = \hbar k/m$ is velocity, and $\hbar\dot{k}/m = -F/m$ is acceleration (forces being associated with the negative gradients of potential energy and F being a positive quantity in the present convention). Now, if the distribution is pushed away from its equilibrium configuration f_{eq}, then df/dt is not zero, but relaxes back to the equilibrium distribution with some characteristic time, the relaxation time, represented as τ, or $df/dt = -(f - f_{eq})/\tau$, a form known as the relaxation time approximation. Put together, a simple form of Boltzmann's transport equation (BTE) results, and is

$$\frac{\partial f}{\partial t} + \frac{\hbar k}{m}\partial_x f - \frac{F}{\hbar}\partial_k f = -\frac{(f - f_{eq})}{\tau} \tag{7.18}$$

Integrating over dk on both sides of Eq. (7.16) recovers the definitions of ρ and J from Eqs (7.6)–(7.9). The same integral over the force term vanishes because $f(\pm\infty) = 0$. Finally, for the scattering term, integrals over dk for both f and f_{eq} give the same density, and so the right-hand side vanishes. Alternately, one can say that in equilibrium as many electrons scatter into a state as out of it with no net noticeable effect. The cumulative effect is to regain the continuity equation

$$\partial_t \rho + \partial_x J = 0 \tag{7.19}$$

where $\partial_x J \rightarrow \vec{\nabla} \cdot \vec{J}$ for three dimensions. With scattering, density of states, a Fermi–Dirac distribution, and a moments approach, the BTE allows for a correction to the Drude model for, in particular, conductivity and heat capacity.

7.4.1 Electrical conductivity revisited

And enterprises of great pitch and moment
with this regard their currents turn awry.

– William Shakespeare[9]

With the thermalized Fermi–Dirac distribution f_{FD} as the equilibrium condition, consider afresh the conductivity under the conditions that (*i*) the problem is time-independent ($\partial_t f = 0$), (*ii*) the current density is constant ($\partial_x f = 0$), (*iii*) the force term is only in the $\hat{x}$ direction, (*iv*) the difference between $f(k)$ and $f_{FD}(E_k)$ is small, and (*v*) $f_{FD}(E_k)$ is symmetric in k because E_k is parabolic in k, then the current flows only in the $\hat{x}$ direction. Looking closely, it is assumptions (*iv*) and (*v*) that seem simple but are not (compared to the others which seem simple and are), so examine each.

The smallness of the difference assumption means that $f \approx f_{FD} + \delta f$, where δf is a perturbation. Thus, the right-hand side of Eq. (7.18) is $-\delta f/\tau$. What causes the disturbance, the force F, is therefore small as well. In short, there are big terms like f_{FD}, small terms like F and δf, and negligible terms like products of F and δF. To leading order then (considering just the big terms), $\partial_t f_{FD} + \partial_x f_{FD} = 0$ which is clearly true given that the FD distribution is independent of time and position. The next order terms are what matters, and so neglecting $F\delta f$,

$$\frac{\hbar k}{m}\partial_x \delta f - \frac{F}{\hbar}\partial_k f_{FD} = -\frac{\delta f}{\tau} \tag{7.20}$$

Recalling Eq. (7.19), if the density is time independent, then the current density is spatially uniform ($\partial_x J = 0$), which implies $\partial_x \delta f \rightarrow 0$. Now, reconsider the current density itself: it is obtained by integrating over k_x the distribution f with

[9]Ref. [37]: *Hamlet*, III.i.86–87, p. 951.

$\hbar k_x/m$. That means that *any symmetrical function* in k_x can be subtracted from f and not change the current density because the integral of a symmetric function with an antisymmetric one vanishes. The Fermi–Dirac distribution is symmetrical in k_x, and so

$$J_q = \frac{2q}{(2\pi)^3} \int d\vec{k}\, \frac{\hbar k_x}{m}(f - f_{FD}) \tag{7.21}$$

which allows $f - f_{FD} = \delta f = (\tau F/\hbar)\partial_k f_{FD}$ to be inserted. As a result, the current density is given by

$$J_q = \frac{2q}{(2\pi)^3} \int d\vec{k} \left(\frac{\hbar k_x}{m}\right) \tau(E_k) \left(\frac{F}{\hbar}\right) \frac{\partial}{\partial k_x} f_{FD}(E_k) \tag{7.22}$$

where the likely energy dependence of the relaxation time is made explicit, and where q is appended as a subscript to J to indicate that it is a *charge* current density. A common visualization of Eq. (7.21) is that of a spherical distribution being rigidly translated a small amount in momentum space [82]. The symmetry observation also entails that the number density depends on the *symmetric* part of the distribution, but the current density depends on the *antisymmetric* part of the same distribution, a natural consequence of the moments viewpoint.

That the Fermi–Dirac distribution is the equilibrium distribution eases matters considerably, foremost because it depends only on the magnitude of k and not the angular coordinates. The quantity $k_x\partial_{k_x} f_{FD}$ is therefore easy to resolve. Recall spherical coordinates: a function $g(r)$ that depends only on the magnitude of $\vec{r}$ entails

$$\int x\partial_x g(r) d\vec{r} = \frac{1}{3}\int (\vec{r}\cdot\vec{\nabla})g\,(r)d\vec{r} = \frac{1}{3}\int r\partial_r g(r)d\vec{r} \tag{7.23}$$

Likewise, if $g(r^2)$ is the function, then $2\partial_r g(r^2)$ appears in the integral and similar relations hold for $\vec{k}$ and $\vec{\nabla}_k$. Thus, $k_x\partial_{k_x} f_{FD} = (2/3)E\partial_E f_{FD} = -(2/3)\beta E\delta_l\,(\beta(E-\mu))$ where

$$\delta_l(s) \equiv \frac{1}{(e^s+1)\,(e^{-s}+1)} \tag{7.24}$$

the notation intended to be evocative of the Dirac delta function in that $\delta_l(s)$ is sharply peaked compared to other terms in the integrand, and $\int_{-\infty}^{\infty}\delta_l(s)ds = 1$. In fact,

$$\delta_l\,(\beta(E-\mu)) = f_{FD}(E)\,(1 - f_{FD}(E)) \tag{7.25}$$

The full width at half maximum (FWHM) $2\Delta E$, determined by where $4\delta_l\,(\beta\Delta E) = 1/2$, is therefore $2\Delta E = 2k_BT\ln(2\sqrt{2}+3) = 3.5255k_BT$, or about 10 meV for $T = 329.16$ K: such a small FWHM allows terms that vary slowly by comparison to $\delta_l(s)$ to be evaluated at $E = \mu$ and taken out of the integral, to quite a good approximation.

Recall the relationship between current density J and conductivity σ of $J/F = q\sigma$: conductivity is then given by

$$\sigma = \frac{2q^2}{3m}\beta \int_0^{\infty} ED(E)\tau(E)\,\delta_l\,(\beta(E-\mu))\,dE \tag{7.26}$$

First consider this equation, then evaluate it. From Eq. (7.26), the first factor of $f_{FD}(E)$ is the probability that the energy state characterized by E is occupied, that is, there is an electron there to scatter. The second factor $(1 - f_{FD}(E))$ is the probability that it can scatter *into* a different state characterized by energy E. That is, $f_{FD}(E)$ is the probability of occupation, and $(1 - f_{FD}(E))$ is the probability of vacancy, and the probability of a scattering event occurring depends on both. But only electrons within a few k_BT of the Fermi level participate, and so all the terms other than $\delta_l(E)$ can be evaluated at $E = \mu$ and pulled outside the integral, a figurative way of speaking about how to deal with sharply peaked integrands described in Section A1.2.4.

The quality of this approximation can be assessed as follows. Change the integration variable to $s = \beta(E-\mu)$, let $u = \beta u$, and make use of properties of $\delta_l(s)$. From Eqs (7.13) and (7.24), σ becomes

$$\sigma = \frac{2q^2}{3m}\mu\tau(\mu)D(\mu)\int_{-u}^{\infty}\left(1+\frac{s}{u}\right)^{3/2}\delta_l(s)ds \tag{7.27}$$

For $u = \beta\mu \gg 1$, $\int_{-u}^{\infty}\delta_l(s)ds = 1/(1+e^{-u}) \approx 1 - e^{-u}$, which is unity to an excellent approximation. This means that the lower limit of the integral can be well approximated by $-\infty$, so that the symmetric nature of $\delta_l(s) = \delta_l(-s)$ can be exploited

(Eqs (A1.33)–(A1.35)). If $(1+(s/u))^{3/2}$ is Taylor expanded, the linear (antisymmetric) term may be discarded because it is an odd function of s and its integration with $\delta_l(s)$ vanishes. The next term is of order u^{-2} and therefore small itself. Therefore,

$$\int_{-\infty}^{\infty}\left(1+\frac{s}{u}\right)^{3/2}\delta_l(s)ds \approx 1+\frac{3\zeta(2)}{4u^2} \tag{7.28}$$

For a metal like copper ($\mu = 7$ eV) at room temperature ($T = 300$ K), the correction term is close to 1.68×10^{-5}. Consequently, the conductivity is, to a good approximation,

$$\sigma \approx \frac{2q^2}{3m}\mu\tau(\mu)D(\mu) = \frac{q^2}{m}\rho\tau(\mu) \tag{7.29}$$

and equivalent to the Drude result of Eq. (3.9) *except that the meaning of* $\tau(\mu)$ *has changed* because only the relaxation time near the Fermi level matters, that is, the same *words* are used, but the *meaning* has become profoundly different.

7.4.2 Thermal conductivity revisited

Cold indeed, and labor lost:
Then farewell heat, and welcome frost!

– William Shakespeare[10]

Thermal conductivity will be easier having gone through charge transport in such detail. Recall that for heat transport, electrons from a higher temperature region transport to a lower temperature region and thermalize through scattering events. The amount of heat they take with them is understood by reference to Eq. (5.23) for fixed volume ($dV = 0$) or, isolating the heat term dQ,

$$dQ = dE - \mu dN \tag{7.30}$$

where using E instead of U for energy reflects the parlance of the present section, and where dQ is the quantity of heat transported per electron ($dN = 1$), therefore, each electron transports a quantity of heat of $E-\mu$. Thus, instead of charge transport q in Eq. (7.21), the transport of heat (h) current makes the replacement $q \to (E-\mu)$, or

$$J_h = \frac{2}{(2\pi)^3}\int d\vec{k}\,(E-\mu)\,\frac{\hbar k_x}{m}f(\vec{k}) \tag{7.31}$$

where $f(\vec{k})$ is evaluated for a temperature at a particular location: because the temperature varies with position,

$$f_{FD}(E)|_{T+\Delta T} = f_{FD}(E)|_T + \frac{df_{FD}(E)}{dT}\frac{dT}{dx}\left(\frac{\hbar k_x}{m}\right)\tau(E) \tag{7.32}$$

where the differential element $\Delta T = (dT/dx)\Delta x$ has $\Delta x = v_x\tau = \hbar k_x\tau/m$ on the right-hand side, using arguments familiar from Section 3.2. Then, taking $J_h = \kappa(dT/dx)$, and the arguments that led to J_q in Eq. (7.21), results in

$$\kappa = \frac{2}{(2\pi)^3}\int d\vec{k}\,(E-\mu)\left(\frac{\hbar k_x}{m}\right)^2\tau(E)\frac{df_{FD}(E)}{dT} \tag{7.33}$$

Some of the same trickery from the evaluation of σ remains useful. First, the integrals can be processed according to

$$\int(\cdots)k_x^2 d\vec{k} = \frac{1}{3}\int(\cdots)k^2 d\vec{k} \tag{7.34}$$

where $(\cdots)$ sweeps up all the k^2 (or E) dependent terms. Nevertheless, show it directly by using $k_x = k\sin\theta\cos\varphi$ so that

$$\frac{\int k_x^2 d\vec{k}}{\int k^2 d\vec{k}} = \frac{\int_0^\pi \sin^3\theta d\theta\int_0^{2\pi}\cos^2\varphi d\varphi}{\int_0^\pi \sin\theta d\theta\int_0^{2\pi}d\varphi} = \frac{1}{3} \tag{7.35}$$

Second,

$$\frac{d}{dT}f_{FD}(E) = -k_B\beta^2(E-\mu)\,\delta_l(\beta(E-\mu)) \tag{7.36}$$

[10]Ref. [37]: *The Merchant of Venice*, II.vii.75–76, p. 225.

as can be shown explicitly by simply working out the derivatives. As before, terms within the integrand that vary slowly by comparison to the peakiness of $\delta_l(s)$ can be evaluated at the chemical potential and taken out of the integral, resulting in

$$\kappa \approx \frac{\rho\tau(\mu)}{m} T \int_{-\infty}^{\infty} s^2 \delta_l(s)\, ds = \frac{\pi^2}{3} \frac{\rho\tau(\mu)}{m} T \tag{7.37}$$

where, to a good approximation, $\rho \approx (2/3)\mu D(\mu)$, as shown using Eq. (7.12).

7.4.3 Heat capacity revisited

Man's capacity for justice makes democracy possible, but man's inclination to injustice makes democracy necessary.

– Reinhold Niebuhr[11]

Capacity, or the ability to hold a quantity of something desirable, is often of great interest. Heat is one such thing, and in a material it can be held by either the lattice of vibrating atoms or the electrons, although each has a different capacity for doing so. Their differences affect, for example, how fast heat can transport through a material: generally, good thermal conductors make good electrical conductors because energy is transported by free electrons. Insulators, on the other hand, which have a paucity of free electrons, have to transport heat by lattice vibrations (phonons), so that electrical insulators tend to be thermal insulators as well. The calculation of heat capacity C_V, therefore, includes a contribution due to electrons (C_e) as well as the lattice (C_i), or $C_V = C_e + C_i$: as in Figure 3.4, at low temperatures, C_e dominates, and at high temperatures, C_i dominates. The heat capacity due to the electrons behaves as $C_e = \gamma T$. At high temperatures, C_i asymptotically approaches the Dulong–Petit value of $3Nk_B$.

Focus on the electrons first. The electron heat capacity C_e is

$$C_e = \frac{d}{dT} \int E D(E) f_{FD}(E) dE \tag{7.38}$$

but to simplify the evaluation of the integral, notice that the integral can be modified to subtract $\mu\rho$ without changing much, as that product is relatively temperature-insensitive for metals, even more so at the low temperatures where C_e becomes noticeable. Replacing ρ by its integral definition, C_e can therefore be rewritten as

$$C_e = \frac{d}{dT} \int (E - \mu) D(E) f_{FD}(E) dE \tag{7.39}$$

As a result, the integration becomes familiar after the approximation $D(E) \approx D(\mu)$ is made, and then removed from the integral,

$$C_e \approx k_B^2 D(\mu) T \int_{-\infty}^{\infty} s^2 \delta_l(s) ds = \frac{\pi^2}{3} k_B^2 D(\mu) T \tag{7.40}$$

Comparison with the thermal conductivity factor κ demonstrates

$$\kappa = \frac{2\mu}{3m} \tau(\mu) C_e(T) \tag{7.41}$$

which are useful when heat diffusion phenomena associated with photo and secondary emission are considered [83–85].

EXAMPLE: Evaluate $\gamma = C_e/T$ and κ for copper-like parameters at 300 K assuming $\tau(\mu) = 25$ fs and $\mu = 7$ eV.
SOLUTION: Using $k_F = \sqrt{2m\mu}/\hbar = 13.555\ \text{nm}^{-1}$, $C_e/T = \gamma$ can be written

$$\gamma = \frac{m k_F k_B^2}{3\hbar^2} = 4.4031 \times 10^{-7} \frac{\text{eV}}{\text{K}^2\text{nm}^3} = 70.546 \frac{\text{J}}{\text{K}^2\text{m}^3}$$

from which κ is

$$\kappa = \frac{2\mu}{3m} \tau(\mu) C_e(T) = 0.0027105 \frac{\text{eV}}{\text{K fs nm}} = 434.27 \frac{\text{W}}{\text{K m}}$$

[11] Reinhold Niebuhr, *The Children of Light and the Children of Darkness*. New York: C. Scribner's Sons, 1944, p. xx.

Compare the value of γ to the value determined from Figure 3.4, found to be 0.00016099 (cal/mole K^2). Recalling that the ratio of density to atomic mass for copper is 0.141 mole/cm^3 and that 1 cal = 4.1868 J, the fitted value corresponds to 95.04 (J/m^3 K^2).

As C_e depends only on Fermi level, electron mass, and temperature, of which μ and T are either well-known by other means or specified, the discrepancy between theory and experimental values is said to reside in the mass term: an electron "thermal mass" is therefore defined by $\eta = m_{th}/m \equiv \gamma_{exp}/\gamma_{theory}$. For copper, $\eta = 1.375$, whereas for silver, it is closer to 1.0136 [52, 85]. Even so, the previous example tends to gloss over that electron scattering and relaxation times are temperature dependent and theoretically complex. Something more than approximation by a constant may be needed, even though representative values are useful.

The second contribution to the heat capacity is due to vibrations of the lattice. Without going too deeply into scattering (which will occupy Section 32.3.4), observe that the "phonons", or the lattice vibrations represented as a particle, follow Bose–Einstein statistics, and therefore a simple model is sufficient to determine the temperature dependence of heat capacity due to them. Recall the discussion of Planck's correction (Eq. (3.23)) to the Rayleigh–Jeans formula. Using those methods, each of the three directions that characterize a lattice contribute to the density of states. As with Eq. (3.21), the phase space volume couples with the relation $k = \omega/c_s$, with c_s the speed of sound, to give for the density of states factor

$$D(\omega) = \frac{4\pi V}{(2\pi)^3}\left(\frac{1}{c_l^3} + \frac{2}{c_t^3}\right)\omega^2 d\omega = \frac{3V}{2\pi^2 c_s^3}\omega^2 d\omega \tag{7.42}$$

where the longitudinal (c_l) and transverse (c_t) sound velocities are presumed to be the same and given by (c_s).

A difference with the photons of Eq. (3.23), however, is that phonons are *lattice* vibrations, and therefore the number of modes is limited by the number of atoms N, the highest mode begin when the atoms are oscillating out of phase. Therefore, an upper limit will exist such that $\omega \leq \omega_D$, where the subscript refers to Debye, in turn related to the Debye temperature T_D of a material by $\hbar\omega_D = k_B T_D$. In terms of N, then, where the factor of 3 is from the vibration directions,

$$3N = \int_0^{\omega_D} D(\omega)\, d\omega = \frac{1}{2\pi^2}\left(\frac{\omega_D}{c_s}\right)^3 \tag{7.43}$$

Thus, the Debye temperature can be specified by knowing the density of the metal and its sound velocity due to *acoustic* phonons. As specific heat is the change of energy with temperature

$$C_i = \frac{d}{dT}\left(\int_0^{\omega_D} \frac{\hbar\omega D(\omega)}{e^{\beta\hbar\omega} - 1} d\omega\right) \tag{7.44}$$

$$= \frac{3k_B^4 T^3}{2\pi^2\hbar^3 c_s^3}\int_0^{T_D/T} \frac{x^4}{(e^x - 1)(1 - e^{-x})} d\omega \tag{7.45}$$

Substituting N from Eq. (7.43),

$$C_i = 9Nk_B\left(\frac{T}{T_D}\right)^3 \int_0^{T_D/T} \frac{x^4}{(e^x - 1)(1 - e^{-x})} d\omega \tag{7.46}$$

The Debye temperature for various materials varies [55] (determined, for example, by the methods of Section 32.3.4); for copper, it is $T_D \approx 343$ K [86]. It is therefore clear that for metals that are at room temperature and higher, the denominator of the integrand can be expanded and is approximately x^2 to third order (it is an even function), so that

$$C_i(T \to \infty) \approx 9Nk_B\left(\frac{T}{T_D}\right)^3 \int_0^{T_D/T} x^2 dx = 3Nk_B \tag{7.47}$$

recognized as the high-T limit, or Dulong–Petit result, seen in Figure 3.4. Conversely, when the temperature is low, the upper limit may be extended to ∞ and $1/(1 - e^{-x}) \approx 1 - e^{-x}$. The resulting integrals are well approximated by $\zeta(p)$ and the gamma function, or

$$C_i(T \to 0) \approx 9Nk_B\left(\frac{T}{T_D}\right)^3 \int_0^{T_D/T} \left(\frac{x^4}{e^x - 1} + x^4 e^{-2x}\right) dx \tag{7.48}$$

$$= 9Nk_B\left(\frac{T}{T_D}\right)^3 \left(24\zeta(5) + \frac{3}{4}\right) \tag{7.49}$$

EXAMPLE: Evaluate $\tilde{A} = C_i/VT^3$ for copper-like parameters at 300 K assuming $T_D = 343$ K and copper values for density of 8.96 g/cm^3 and atomic mass of 63.5456 g/mole.
SOLUTION: The factor $(24\zeta(5) + (3/4) = 25.6363 \approx 282/11$. Thus, $C_i/VT^3 \approx 2538 k_B \rho_i / 11 T_D^3 = 6.7031$ J/K^4 m^3, where $\rho_i = 0.141$ mole/cm^3. Scaling to the units of Figure 3.4, it is found $\tilde{A} = 1.13545 \times 10^{-5}$ calorie/K^4 mole, reasonably close to the fit.

7.4.4 Lorentz number revisited

Because things are the way they are, things will not stay the way they are.

– Bertolt Brecht[12]

The Lorentz number is the ratio of Eq. (7.37) with Eq. (7.29). It should be constant, but because of the quantum statistics analysis, the form encountered in the Wiedemann–Franz law and crudely given by Eq. (3.19), is now corrected to be

$$\frac{\kappa}{\sigma T} = L = \frac{\pi^2}{3}\left(\frac{k_B}{q}\right)^2 \tag{7.50}$$

and shown compared to experimental data as the red line in Figure 3.3. The agreement at higher temperature is remarkable.

That tightly packed and charged electrons meandering through a lattice of vibrating ionic cores in a metal can be modeled as a non-interacting electron gas in a box without fields with the electrons subject to Fermi–Dirac statistics is an inescapable conclusion. Modeling the electrons as a gas in the metal that come to the surface only to be turned away by the surface barrier – but not always – results in the emission equations that, like the Weidman–Franz law and the specific heat evaluations, enjoy a notable ability to predictively model experimental data. The particle statistics are now on firm footing. What now remains to be done before encountering modern versions of the electron emission equations is to acquire a comfortableness with the methods of quantum mechanics.

[12] As quoted in John Cook, Steve Deger et al., *The Book of Positive Quotations*, 2nd edn. Minneapolis: Fairview Press, 2007, p. 390.

CHAPTER 8

Quantum mechanics methods

I think I can safely say that nobody understands quantum mechanics.

– Richard P. Feynman[1]

8.1 A simple model: the prisoner's dilemma

[Athenians:] ...you know as well as we do that right, as the world goes, is only in question between equals in power, while the strong do what they can and the weak suffer what they must.

–Thucydides[2]

The peoples of the world must unite, or they will perish. This war, that has ravaged so much of the earth, has written these words. The atomic bomb has spelled them out for all men to understand.

– J. Robert Oppenheimer[3]

At the conclusion of World War II in 1945, America was the sole possessor of atomic weapons. Relations with the Soviet Union deteriorated and the Cold War was underway after 1947. After the Soviets acquired their own atomic weapons in 1949, a fierce and public debate arose as to how the USA should react. The brilliant physicists and mathematicians of Los Alamos, the birthplace of the new power, were part of that discussion from the start, with J. Robert Oppenheimer (director of Los Alamos) and John von Neumann (a founder of game theory) publicly making opposing arguments. Dealings with the Soviet Union were couched in game theoretic ideas, as the utterances of Oppenheimer and von Neumann (who could quote Thucydides' excerpt above verbatim) make evident. The choices were to launch a preventive attack, as the Athenians did to the Melians and as von Neumann advocated, or take the risk of cooperation to enable mutual survival, as Oppenheimer advocated.

As shown by von Neumann, a rational course of action in a conflict (or game) between two antagonists is always possible if their aims are opposed (i.e., if one wins, the other loses). Such games are called "zero-sum", but cooperation (and survival) is a different breed. Von Neumann was hired as a consultant by the RAND corporation, where he and others applied themselves to strategic studies of intercontinental nuclear war. Two of von Neumann's new collegues at RAND, Merrill Flood and Melvin Dresher, devised a simple game for which the best outcome was not the rational one: the Prisoner's dilemma (PD) [87], which figures strongly in studies on the evolution of cooperation [88]. Here, it affords a convenient discussion of ideas that introduce a methodology illuminating to the quantum mechanical treatment of electron emission, and so the PD is used to introduce the techniques in a pleasingly simple model.

The PD [87] is concisely described thus: two criminals are charged with an offense, but held separately. If both do not confess, both will be convicted on a lesser charge and serve a year in prison. If both confess, each will serve three years in prison. But, if one turns state's evidence (confesses) and the other does not, the confessor will go free and his partner spend four years in prison. To avoid being duped, both prisoners have a powerful temptation to confess, even though collectively both would do better to trust his partner by cooperating and not confessing: in the parlance of game theory, mutual defection is the Nash equilibrium (the circumstance where no player can do better by switching his strategy unilaterally). The beauty of the problem is that it can be formulated in the language of a payoff matrix and vector states, and thereby in a simple-to-grasp and easy-to-manipulate manner allow for the exploration of recurrent ideas that will be subsequently encountered.

[1] Richard P. Feynman, *The Character of Physical Law*. Cambridge MA: MIT Press, 1989, p. 129.
[2] J. Robert Oppenheimer, Acceptance speech for Los Alamos Certificate of Appreciation, October 16, 1945, quoted in W. Poundstone [87], p. 70.
[3] "Melian Dialogue", quoted in Thucydides (trans. R. Crawley), *The Landmark Thucydides: A Comprehensive Guide to the Peloponnesian War*, ed. Robert B. Strassler. New York: Simon and Schuster, 1998, p. 352.

Introduction to the Physics of Electron Emission, First Edition. Kevin L. Jensen.

Although the PD is traditionally formulated as a parable involving the unscrupulous, it is easier to make the payoff matrix actually refer to payoffs and to speak of players and a game show host: instead of serving a sentence, let the players gain a reward depending on whether they cooperate (act in good faith) or defect (betray the other player).[4] The payoff matrix then refers to the amount the host pays out: if the players cooperate, the host pays 6 units (3 units to each player), if both players defect, the host pays 2 units (1 unit to each player), and if one player defects and the other cooperates, the host pays 4 units (all to the defector, none to the cooperator). The essence remains the same: there is an incentive to defect (or cheat) and gain a higher reward, in addition to a strong incentive not to be a dupe (one who cooperates when the other player defects) where there is no gain at all. In the bigger picture, observe that if both players cooperated, the host would have to pay out the most and the players' collective gain would be the greatest, an outcome John Stuart Mill would embrace but Ayn Rand would excoriate.

In contrast to game shows where contestants interact with each other only once, a richer game is afforded when they repeatedly interact: such a game is called an iterated PD. In a computer simulation that pitted different strategies against each other, one strategy worked surprisingly better than the others and was dubbed "tit-for-tat". This strategy cooperated in the first round, but then simply did what was done to it in subsequent rounds: it cooperated as long as it was cooperated with, but when it suffered a defection, it defected on the next turn. A strategy that performed better in the long haul was tit-for-tat with occasional forgiving of a defection. Altruism, even in this very limited sense, can perform well. Seeing why this is so is possible using the language of matrix operators and *bra-ket* vectors.

8.1.1 Matrices and vectors

A great deal of my work is just playing with equations and seeing what they give.

– Paul A.M. Dirac[5]

Using the notation introduced in Eq. (4.8), let $|u\rangle$ refer to a cooperator, and $|d\rangle$ to a defector but now both players must be considered, so that the state vector of the players $|\psi\rangle$ must take both into account. The cases where they are both cooperating or both defecting are straightforward to write:

$$|uu\rangle = \frac{1}{\sqrt{2}}\begin{pmatrix}|u\rangle\\|u\rangle\end{pmatrix} = \frac{1}{\sqrt{2}}\begin{pmatrix}1\\0\\1\\0\end{pmatrix} \tag{8.1}$$

$$|dd\rangle = \frac{1}{\sqrt{2}}\begin{pmatrix}|d\rangle\\|d\rangle\end{pmatrix} = \frac{1}{\sqrt{2}}\begin{pmatrix}0\\1\\0\\1\end{pmatrix} \tag{8.2}$$

where the factor $1/\sqrt{2}$ assures that $\langle uu|uu\rangle = \langle dd|dd\rangle = 1$, that is, the states are "normalized" and orthogonal in that $\langle uu|dd\rangle = 0$. A similar reasoning shows that when one player cooperates and one defects, two more state vectors arise depending on who does what:

$$|ud\rangle = \frac{1}{\sqrt{2}}\begin{pmatrix}1\\0\\0\\1\end{pmatrix}; |du\rangle = \frac{1}{\sqrt{2}}\begin{pmatrix}0\\1\\1\\0\end{pmatrix} \tag{8.3}$$

Here is the rub. To construct the payoff matrix, which shall be designated $\hat{H}$ to bring to mind in the future the Hamiltonian, four such vectors $|\psi_j\rangle$ for $j = 1$ to 4 are needed, with the first two, $|\psi_1\rangle = |uu\rangle$ and $|\psi_2\rangle = |dd\rangle$, already found. But $|ud\rangle$ and $|du\rangle$ are not the other two.

[4]A British daytime game show called *Golden Balls*, airing between 2007 and 2009, contained just such a PD game but with a different payout matrix and an intriguing wrinkle: the contestants could talk beforehand and agree, or pretend to agree, to cooperate, making the betrayals all the more acute.

[5]Paul A.M. Dirac, quoted in the essay "A Piece of Magic: The Dirac Equation", Frank Wilczek, contained in ref. [89], p. 102.

Therefore, find vectors $|\psi_3\rangle$ and $|\psi_4\rangle$ such that they are orthogonal to each other and those of Eqs (8.1) and (8.2). That is rather straightforward: change the sign of one of the non-zero elements, or

$$|\psi_3\rangle = \frac{1}{\sqrt{2}}\begin{pmatrix}1\\0\\-1\\0\end{pmatrix}; |\psi_4\rangle = \frac{1}{\sqrt{2}}\begin{pmatrix}0\\1\\0\\-1\end{pmatrix} \tag{8.4}$$

as the two remaining orthogonal states. $\hat{H}$ is then constructed from the orthogonal basis states by

$$\hat{H} = C_1|\psi_1\rangle\langle\psi_1| + C_2|\psi_2\rangle\langle\psi_2| + C_3|\psi_3\rangle\langle\psi_3| + C_4|\psi_4\rangle\langle\psi_4| \tag{8.5}$$

where $|\psi_j\rangle\langle\psi_j|$ are matrices, in contrast to $\langle\psi_j|\psi_j\rangle = 1$, which is a scalar. Observe that this entails

$$|ud\rangle = \frac{1}{2}(|\psi_1\rangle + |\psi_2\rangle + |\psi_3\rangle - |\psi_4\rangle) \tag{8.6}$$

$$|du\rangle = \frac{1}{2}(|\psi_1\rangle + |\psi_2\rangle - |\psi_3\rangle + |\psi_4\rangle) \tag{8.7}$$

$$|ud\rangle + |du\rangle = |\psi_1\rangle + |\psi_2\rangle = |uu\rangle + |dd\rangle \tag{8.8}$$

The matrices are straightforwardly evaluated, for example

$$|\psi_1\rangle\langle\psi_1| = \frac{1}{2}\begin{pmatrix}1\\0\\1\\0\end{pmatrix}\begin{pmatrix}1 & 0 & 1 & 0\end{pmatrix} = \begin{pmatrix}1 & 0 & 1 & 0\\0 & 0 & 0 & 0\\1 & 0 & 1 & 0\\0 & 0 & 0 & 0\end{pmatrix} \tag{8.9}$$

in which the column element of the first vector multiplies the row of the transposed second vector to give the row of the matrix. Then, by explicit expansion of Eq. (8.5),

$$\hat{H} = \frac{1}{2}\begin{pmatrix}C_1+C_3 & 0 & C_1-C_3 & 0\\0 & C_2+C_4 & 0 & C_2-C_4\\C_1-C_3 & 0 & C_1+C_3 & 0\\0 & C_2-C_4 & 0 & C_2+C_4\end{pmatrix} = \begin{pmatrix}6 & 0 & 0 & 0\\0 & 2 & 0 & 0\\0 & 0 & 6 & 0\\0 & 0 & 0 & 2\end{pmatrix} \tag{8.10}$$

if $C_3 = C_1$ and $C_4 = C_2$ are chosen and, from the payoff values, $C_1 = 6$ and $C_2 = 2$ arise because $\hat{H}|uu\rangle = 6|uu\rangle$ and $\hat{H}|dd\rangle = 2|dd\rangle$. As a result, the problem has been solved in the reverse of its usual narrative encountered when, for example, given the Hamiltonian of a dynamical system, its eigenvectors are found, in terms of which the Hamiltonian is expressed as a diagonal matrix of its eigenvalues (see, for example, the discussion of Euler's theorem and rigid body motion in Goldstein [90]). Finding eigenvectors from eigenvalues, as is more common, is considered next.

8.1.2 Eigenvectors of tit-for-tat

Wir müssen wissen/Wir werden wissen.

– David Hilbert[6]

Now extend the matrix methods to consider the time evolution of stable strategies. The power of such methods was appreciated early on by David Hilbert, whose deep and profound appreciation of matrix mechanics in particular, and mathematics in general, enabled him to advise Heisenberg and Born to seek out a differential equation with the same eigenvalues when they were unsure why matrix mechanics appeared in the physics of the atom, as in Section 9.1. Had

[6]"We must know, we will know." From David Hilbert's 1930 address to the Society of German Scientists and Physicians, reproduced and translated by the Mathematical Association of America. Available online at http://www.maa.org/book/export/html/336071 (trans. James T. Smith, San Francisco State University). It was a stinging rebuke of the maxim *ignoramus et ignorabimus* ("We do not know and will not know") that was fashionable at the time, and unfortunately still is.

Heisenberg followed Hilbert's prodding, it is speculated that he may have come across Schrödinger's equation before Schrödinger.[7]

The matrix $\hat{U}$ defines how the tit-for-tat strategy is implemented. Let two players repeatedly interact with each other over time. If a superscript designates the interaction number, for example $|\psi(t_n)\rangle \equiv |\psi^n\rangle$, where t_n is the nth interaction (or time step) of the game, then

$$|\psi^{n+1}\rangle = \hat{U}|\psi^n\rangle \tag{8.11}$$

The task is then to ferret out the mathematical meaning of "do what was done to you". A moment's reflection reveals that the matrix should be

$$\hat{U} = \begin{pmatrix} 0 & 0 & 1 & 0 \\ 0 & 0 & 0 & 1 \\ 1 & 0 & 0 & 0 \\ 0 & 1 & 0 & 0 \end{pmatrix} \tag{8.12}$$

as follows from (using the original ud notation) $\hat{U}|uu\rangle = |uu\rangle$, $\hat{U}|dd\rangle = |dd\rangle$, $\hat{U}|ud\rangle = |du\rangle$ and $\hat{U}|du\rangle = |ud\rangle$: what $\hat{U}$ does is swap the the states of the players. Clearly, there are three states where the action of $\hat{U}$ does not change the vector describing the state of the players, or $\hat{U}|\varphi\rangle = |\varphi\rangle$, and they are $|uu\rangle$, $|dd\rangle$, and $(|ud\rangle + |du\rangle)/\sqrt{2}$. Observe that $\hat{U}$ is such that

$$\hat{U}^\dagger \hat{U} = \hat{U}^2 = \hat{I} \tag{8.13}$$

where $\dagger$ is taken to mean transpose and complex conjugate (or "self-adjoint" in the appropriate parlance), although a matrix with real entries begs what is meant by complex conjugate, and $\hat{U}^{-1} = \hat{U}$ anyway: the stage is simply being set for later.

The question is then: what eigenvectors $|\varphi\rangle$ are associated with $\hat{U}$ and how do they compare with the eigenvectors $|\psi_j\rangle$ of $\hat{H}$? Finding the eigenvectors of a matrix is a standard problem in linear algebra (e.g., see refs [60, 91]). The matrix equation

$$(\hat{U} - \lambda\hat{I})|\varphi\rangle = 0 \tag{8.14}$$

only has solutions if the determinant $|\hat{U} - \lambda\hat{I}| = 0$. That is,

$$\begin{vmatrix} -\lambda & 0 & 1 & 0 \\ 0 & -\lambda & 0 & 1 \\ 1 & 0 & -\lambda & 0 \\ 0 & 1 & 0 & -\lambda \end{vmatrix} = (\lambda - 1)^2(\lambda + 1)^2 = 0 \tag{8.15}$$

Thus, for $\hat{U}$, the eigenvalues are degenerate (more than one of the same value occurs) and are equal to ± 1, or $\lambda_j = (-1)^j$ for $j = 1$ to 4. Each eigenvalue λ_j is put back into Eq. (8.14) where the components of $|\varphi_j\rangle$ are unknown and the components solved for.

EXAMPLE: Find the eigenvectors associated with each $\lambda_j = (-1)^j$.

SOLUTION: Let $\langle\varphi_j| = (a, b, c, d)$. Then from Eq. (8.14) it follows for λ_1 that

$$\begin{pmatrix} 1 & 0 & 1 & 0 \\ 0 & 1 & 0 & 1 \\ 1 & 0 & 1 & 0 \\ 0 & 1 & 0 & 1 \end{pmatrix}\begin{pmatrix} a \\ b \\ c \\ d \end{pmatrix} = \begin{pmatrix} a+c \\ b+d \\ a+c \\ b+d \end{pmatrix} = \begin{pmatrix} 0 \\ 0 \\ 0 \\ 0 \end{pmatrix}$$

[7] See Manfred R. Schroeder, Number Theory and the Real World, *The Mathematical Intelligencer* 7(4), 18, 1985. Although the often alleged claim that Hilbert also said "physics is much too hard for the physicists" is questionable (see L. Corry, *David Hilbert and the Axiomatization of Physics (1898–1918): From Grundlagen Der Geometrie to Grundlagen Der Physik (Archimedes)*, Vol. 10, softcover reprint of hardcover 1st edn, 2004, Springer, 2010, footnote 11, p. 8), he might be forgiven had he done so.

or $c = -a$ and $d = -b$. Setting $a = 1$ and $b = 0$ is the first choice, and it shall be called λ_1, with the choice $a = 0$ and $b = 1$ being the other to be called λ_3. It is simple to show that the eigenvectors associated with $\lambda_2 = +1$ correspond to $c = a$ and $d = b$. Thus,

$$|\varphi_1\rangle = \begin{pmatrix} 1 \\ 0 \\ -1 \\ 0 \end{pmatrix}, |\varphi_2\rangle = \begin{pmatrix} 1 \\ 0 \\ 1 \\ 0 \end{pmatrix}, |\varphi_3\rangle = \begin{pmatrix} 0 \\ 1 \\ 0 \\ -1 \end{pmatrix}, |\varphi_4\rangle = \begin{pmatrix} 0 \\ 1 \\ 0 \\ 1 \end{pmatrix} \tag{8.16}$$

Drawing a correspondence to the eigenvectors of $\hat{H}$, it is seen that $|\varphi_1\rangle = |\psi_3\rangle$, $|\varphi_2\rangle = |\psi_1\rangle$, $|\varphi_3\rangle = |\psi_4\rangle$, and $|\varphi_4\rangle = |\psi_2\rangle$.

It has been found that the eigenvectors of $\hat{U}$ are the same as those of $\hat{H}$, and that will have consequences. When that happens, the combination $\hat{H}\hat{U} - \hat{U}\hat{H}$ (called the commutator of $\hat{H}$ and $\hat{U}$ and designated $[\hat{H}, \hat{U}]$) vanishes. Because the eigenvectors are associated with steady-state solutions, that means the value of the payoff matrix $\hat{H}$ applied to them will not change from one interaction to the next. Formally, when the commutator vanishes,

$$\langle\psi^{n+1}|\hat{H}|\psi^{n+1}\rangle = \langle\psi^{n}|\hat{U}^{\dagger}\hat{H}\hat{U}|\psi^{n}\rangle = \langle\psi^{n}|\hat{U}^{\dagger}([\hat{H}, \hat{U}] + \hat{U}\hat{H})|\psi^{n}\rangle \tag{8.17}$$

$$= \langle\psi^{n}|\hat{U}^{\dagger}\hat{U}\hat{H}|\psi^{n}\rangle = \langle\psi^{n}|\hat{H}|\psi^{n}\rangle \tag{8.18}$$

Below, an analogy will be drawn between the Hamiltonian operator and the payoff matrix, as well as between the propagation operator and the iteration matrix, and it will be demonstrated that the time variation of an operator $\hat{O}$ will be related to its commutator with the Hamiltonian. But not yet. Here, the point is to show the mechanics of an analogous but far simpler system.

8.1.3 Quid pro quo and stable strategies

Margaret: *I cry you mercy, 'tis but quid for quo.*

– William Shakespeare[8]

Menelaos: *Zeus, lord, grant me to punish the man who first did me injury*

– Homer[9]

Ahab: *To the last I grapple with thee; from hell's heart I stab at thee; for hate's sake I spit my last breath at thee!*

– Herman Melville[10]

Quid pro quo is latin for "something for something" or, more colloquially, "tit-for-tat", and indicates an exchange of comparably valued objects or actions. But there are differences in how tit-for-tat strategies play out.

- If players believe they are to meet but once, the temptation of both to violate each other (or defect) is strong, and if acted upon then subsequent interactions, foreseen or not, continue the defections. This represents the "Ahab" initial condition of mutual defection (Ahab's harpooning of Moby Dick and the the whale's amputation of Ahab's leg).
- Were continued interactions plausible, one player might opt to initially cooperate and be soured by the other player's betrayal. This represents the "Menelaos" initial condition of one-sided cooperation (hosting a royal visit to suffer Paris making off with Menelaos's wife Helen) followed by the so-called "death spiral" of unending and alternating defections.
- Beneficent actions or those foreseen to be long term greatly and stably benefit if both actors do the right thing the first time. This represents the "Margaret" initial condition of mutual cooperation, a delivering on both sides of "mercy", and results in the most desirable long-term outcomes for both parties.

Observe the average payouts over time: mutual defection produces an average gain of (1 unit/interaction), alternating defection a gain of (2 units/interaction), and mutual cooperation a gain of (3 units/interaction). All these strategies were manifest in the Axelrod computer tournaments, and they are mathematically represented by the unchanging actions of $\hat{U}$

[8]Ref. [37]: *The First Part of King Henry the Sixth*, V.iii.109, p. 468.
[9]Homer, *The Iliad* (trans. Richmond Lattimore). Chicago: University of Chicago Press, 1962, Book III, line 351.
[10]Herman Melville, *Moby Dick*. New York: Bantam Books, 1986, Chapter 135.

on the eigenvectors $|uu\rangle$ (Margaret), $|dd\rangle$ (Ahab), and $(|ud\rangle + |du\rangle)/\sqrt{2}$ (Menelaos) that results in the same unchanging payouts over subsequent interactions. Unless something changes.

What can change is the act of forgiveness (or the act of selfishness, but it is more pleasant to focus on the positive), which would convert the never-ending betrayals of Ahab into the somewhat better alternating betrayals of Menelaos and, more importantly, converts the feuding of Menelaos into the mercy of Margaret. The phenomenon is not so academic, particularly in war-time settings, where the spontaneous withholding of fire (or directing it to where it does no damage) can be mutually reinforcing: instances abound of spontaneous benevolence between opposing forces, a poignant example being the Christmas truce of 1914 on the Western Front during World War I:

> *As darkness shrouded the wire on Christmas Eve, coloured lights appeared on the German parapet in some sectors held by Bavarian and Saxon troops. Small trees, again decorated with lights, were seen above the sandbags. A few exuberant British soldiers took potshots at the novel targets and were cheered by the enemy when a hit was scored. In other parts of the line, impromptu carol services attracted the attention of the British, who applauded the singers and responded with their own efforts ...British working parties, moving cautiously in frozen no-man's-land, were greeted with friendly shouts from the enemy trenches and ended their jobs puffing on cigars proffered from the German positions ...An English lance-corporal observed, "the Germans had a band playing ...several hymns, our National Anthem and Home Sweet Home."*[11]

Compare the end of that description to an eerily similar event almost 50 years earlier during the American Civil War:

> *By the evening of December 30 1862, the two armies confronted each other in battle formation straddling a shallow stream called Stones River, just north of town. That evening as the soldiers awaited the battle that was sure to open next morning, bands on both sides of the line took turns serenading the armies, who could clearly hear both the Rebel and the Union musicians. Finally, one of the band struck up "Home, Sweet Home" and all of the others on both sides joined in as the soldiers sang along, concluding the concert with a poignant reminder of the American culture the contending forces shared.*[12]

The face-off between soldiers of opposing armies is a PD [88], but something like forgiveness in the form of restraint can continue and expand once expressed. How would it be modeled? It would look like the insertion of a matrix with an off-diagonal element in the block representing the player that is forgiving. If it were player 1, then the matrix might look like

$$\hat{F}_1 = \begin{pmatrix} 1 & 1 & 0 & 0 \\ 0 & 0 & 0 & 0 \\ 0 & 0 & 1 & 0 \\ 0 & 0 & 0 & 1 \end{pmatrix} \tag{8.19}$$

which acts to shift player 1's state to that of cooperation regardless of the acts of player 2. It is a different kind of matrix than the payoff matrix or the interaction matrix: it does not commute with them. In fact, it kind of looks like scattering. Keep that in mind.

The discussion of the PD underscores two central matters in quantum mechanical problems. First, the initial conditions specify the subsequent dynamics of an interacting system, and although the equations can look similar, the outcomes can be unique. Second, knowing the future dynamics requires finding the complete set of commuting constants of the motion [44] for which eigenvectors, or combinations of them, are followed into future states.

8.2 Matrices and wave functions

> *For a short time it appeared that there were two new theories of the behavior of matter on the subatomic level, but it was soon discovered that Heisenberg's quantum mechanics and Schrödinger's wave mechanics were mathematically equivalent, even though they looked quite different. This seems to have pleased neither founder of quantum mechanics. Each intensely disliked the other's approach ...Heisenberg remarked that Schrödinger's theory was "loathsome" and "crap." Schrödinger, on the other hand, charged that Heisenberg's method was "disgusting."*
>
> – Richard Morris[13]

---

[11] D. Winter, Time Off From Conflict: Christmas 1914, *The RUSI Journal*, 115, 1970, 42–43.

[12] Steven E. Woodworth, *This Great Struggle: America's Civil War*. Lanham, MD: Rowman & Littlefield, 2011. p. 189.

[13] Richard Morris, *Achilles in the Quantum Universe: The Definitive History of Infinity*. New York: Henry Holt, 1997, p. 91. The tale of their competition is even stranger than suggested here: see Arthur I. Miller *Erotica, Aesthetics and Schrödinger's Wave Equation*, in ref. [89], p. 80. Physics is not without passion.

In the wave mechanics approach, the wave function $\psi(\vec{r},t)$ is acted upon by differential operators, and is therefore likely more familiar to those with exposure to differential equations. In that formulation, Schrödinger's equation takes on the familiar form

$$i\hbar\partial_t\psi(\vec{r},t) = H\psi(\vec{r},t) \tag{8.20}$$

where the Hamiltonian $H = \hat{H}_0 + \hat{V}$ is a differential operator for which the kinetic energy is $H_0 = -(\hbar^2/2m)\nabla^2$. However, as Dirac pointed out, the symbolic method using *ket* vectors like $|\psi\rangle$ and operators like $\hat{H}$ on which the discussion of the PD was based is equivalent, possesses an undeniable beauty and compactness, and seems to, as Dirac opined, "go more deeply into the nature of things" [40]. The matrix analogy makes intuitive that the order of operators matters (unless their commutator vanishes) and benefits from an appreciation of terms like eigenvector and eigenvalue. Most importantly, it gives a powerful and compact method to go from one representation or basis to another. In the symbolic notation, Schrödinger's equation becomes

$$i\hbar\partial_t|\psi(t)\rangle = \hat{H}|\psi(t)\rangle \tag{8.21}$$

Although they look similar, switching to a momentum $\vec{k}$ representation using $\psi(\vec{k},t)$ and $H_0 = \hbar^2k^2/2m$ is an ordeal using Eq. (8.20) and straightforward using Eq. (8.21). The form of Eq. (8.21) makes it clear that if $|\psi(t)\rangle$ is an eigenvector of $\hat{H}$ with eigenvalue E, then $|\psi(t)\rangle = \exp(-iEt/\hbar)|\psi(0)\rangle$, and therefore, that

$$\hat{U}(t) = e^{-i\hat{H}t/\hbar} \tag{8.22}$$

where the exponential of a matrix operator is a matrix of the exponential values of the operator's action on the wave function. Because the commutator of a function of an operator, or $f(\hat{H})$ with that operator vanishes, as can be proven by representing $f(\hat{H})$ as a Taylor expansion in terms of $\hat{H}$, it is clearly the case that $[\hat{U}(t),\hat{H}] = 0$, analogous to the PD discussion. Of greater importance here, the wave function, as per Eq. (8.21), is clearly seen to satisfy the relation

$$|\psi(t)\rangle = \hat{U}(t)|\psi(0)\rangle \tag{8.23}$$

from which several relations follow (recall $|\psi(t)\rangle$ is an eigenstate of $\hat{H}$)

$$\langle\psi(t)|\hat{H}|\psi(t)\rangle = E = \langle\psi(0)|\hat{H}|\psi(0)\rangle \tag{8.24}$$

$$|\psi(t_1+t_2)\rangle = \hat{U}(t_1)|\psi(t_2)\rangle = \hat{U}(t_1+t_2)|\psi(0)\rangle \tag{8.25}$$

$$\langle\psi(t)|\psi(t)\rangle = \langle\psi(0)|\hat{U}(t)^\dagger\hat{U}(t)|\psi(0)\rangle = \langle\psi(0)|\psi(0)\rangle \tag{8.26}$$

Observe that Eq. (8.24) implies the commutation relation $[\hat{U},\hat{H}] = 0$. The connection between the symbolic method of Eq. (8.21) with the wave mechanics approach of Eq. (8.20) is through the basis states for position $\vec{r}$ and momentum $\vec{p} = \hbar\vec{k}$, which are conjugate variables (loosely, related by Fourier transforms) such that they satisfy an uncertainty relation due to Heisenberg, for example $\Delta x\Delta p_x \geq \hbar/2$ for the $\hat{x}$ component, and another example being energy E and time t, more about which in Section 8.2.3. In d dimensions, the relations between the kets are given by

$$\langle\vec{r}|\vec{k}\rangle = (2\pi)^{-d/2}\exp(i\vec{k}\cdot\vec{r}) \tag{8.27}$$

$$\langle\vec{r}|\vec{r}\,'\rangle = \delta(\vec{r}-\vec{r}\,') \tag{8.28}$$

$$\langle\vec{k}|\vec{k}'\rangle = \delta(\vec{k}-\vec{k}') \tag{8.29}$$

The formalism is up to the task of finding the emission equations developed by the 1930s handily, but an understanding of why these relations hold has benefit in subsequent improvements of those theories. That treatment is not beholden to the three-dimensional formalism, though. In fact, the one-dimensional formalism is a lot more convenient. Thus, let x and $k_x \rightarrow k$ (i.e., suppress the x subscript on k_x) be the conjugate variables and restrict attention to one dimension. The generalization to higher dimensions is straightforward and can be stated as needed.

8.2.1 Identity and Monty Hall

Γνῶθι σαυτὸν ("Know Thyself")

– Inscription on the Temple of Apollo

In *Phaedrus,*[14] Socrates asserted "I must first know myself, as the Delphian inscription says',' referring to the engraving on the Temple of Apollo at the famed Oracle of Delphi (a structure existing in antiquity but brought to ruin for sectarian motives by 390 AD). Sage advice, even though Socrates was likely taking liberties with the meanings of the words.[15] Why is that important here? Identity operators are not as simple as multiplying by unity: they tell us something about the relation of probability estimates to the basis states.

Intuitively, the propagator $\hat{U}(t)$ is the analog of the iteration matrix $\hat{U}$ of Eq. (8.12), and that suggests an analog of the identity matrix $\hat{I}$: one might expect, based on Eq. (8.5), that $\hat{I}$ should resemble something like

$$\hat{I} = \sum_{j=1}^{4} |\psi_j\rangle\langle\psi_j| \tag{8.30}$$

What is so special about $|\psi_j\rangle$? Or, put another way, couldn't another basis set be used? The answer is of course: the normalized orthogonal states characterized by $|j\rangle$ corresponding to a column vector of zeros except for the jth entry, which is unity (e.g., $\langle 2| = (0, 1, 0, 0)/\sqrt{4}$ for $N = 4$) is a different basis, and it is trivial to explicitly show

$$\sum_{j=1}^{4} \langle\psi_k|j\rangle\langle j|\psi_l\rangle = \sum_{j=1}^{4} \langle k|\psi_j\rangle\langle\psi_j|l\rangle = \delta_{kl} \tag{8.31}$$

where the $|\psi\rangle$ are given by Eqs (8.1), (8.2), and (8.3). A modification of the basis states is still constrained by a requirement analogous to that of Eq. (8.30). The choice of basis state is dictated by other concerns, such as diagonalizing the Hamiltonian. Relations much like this one shall be encountered when the basis states refer to position and momentum (see Eq. (8.73)).

The relation between quantum mechanics and probability is such that intuition developed from tackling probability problems pays handsomely in understanding quantum mechanical notation and operations, and concepts. Apart from the pedagogical value, a perk is that the irritation of resolving probability paradoxes, where numerical calculation is often at odds with intuition, forces a deeper understanding. Consider, therefore, a seemingly routine example in which there are N containers such that $P_j = \langle\varphi_j|\varphi_j\rangle$ is the probability that a single randomly hidden object is in the jth container, with all other containers empty. Let the probability that the object was hidden in any of the containers be equally probable (uniformly distributed). This is mathematically described by saying the weighting factors w_j associated with each basis state are given by unity for all j. Thus, by construction,

$$\sum_{j=1}^{N}\sum_{k=1}^{N} w_j w_k \langle j|k\rangle = \sum_{k=1}^{N} \frac{w_j^2}{N} = \frac{1}{N}\sum_{j=1}^{N} 1 = 1 \tag{8.32}$$

Let the containers be arranged in a circle, and let one container chosen at random correspond to $|1\rangle$, with $|2\rangle$ being the next one clockwise, and so on, as schematically shown in Figure 8.1, where the figure has been rotated so the selected container is at the top. The act of opening any one of the containers at random and peering inside to find it empty is equivalent to setting its weighting factor to zero and replacing the weighting factor associated with the *unopened* containers to $1/\sqrt{N-1}$. Such a description is fairly routine, intuitive, and uncontroversial.

But here is where it gets interesting: instead of choosing from all the containers, what changes if the person who hid the object in the first place opens a container selected from all *except the first container*? That is, the randomly chosen container $|k\rangle$ with $2 \leq k \leq N$ is exposed as empty, but $|1\rangle$ is not among those selected. How then are the weighting

---

[14] Benjamin Jowett, *The Dialogues of Plato, Great Books of the Western World,* ed. Robert Maynard Hutchins, Vol. 7, Plato. Chicago: Encyclopaedia Britannica, 1952, p. 116.

[15] To sages before Plato the meaning may have been closer to "Know your measure" and therefore a caution against pride. See Eliza Gregory Wilkins, "Know Thyself", in *Greek and Latin Literature*. Chicago: The University of Chicago Libraries, 1917.

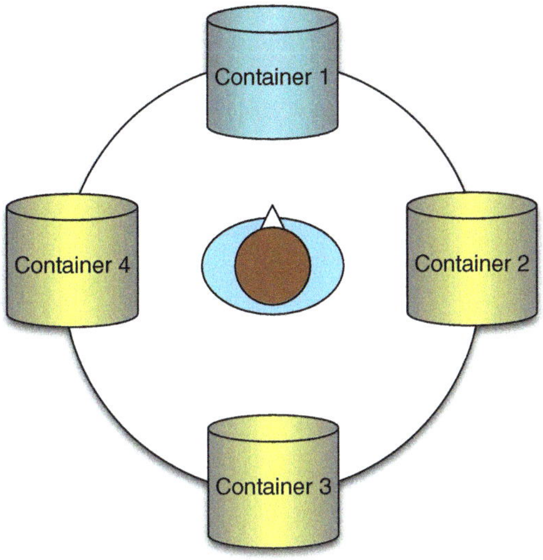

Figure 8.1 Four containers, one chosen at random to hold a prize. Containers are indexed clockwise with a container arbitrarily chosen by the contestant (in the center) rotated to the top position and labeled "1". The state vector $|\varphi\rangle$ is such that $P_j = \langle\varphi_j|\varphi_j\rangle$ is the probability that the jth container holds the hidden object.

factors redistributed over the remaining bases *given* that the kth container was empty?[16] Basis states bear a strong resemblance to vectors, and the fixing of one basis state for special treatment and eliminating the kth state is equivalent to the rotation of a coordinate system about a chosen axis until a vector $\vec{\varphi}$ no longer has a component along the $\hat{k}$ axis (alternately, a projection of $\vec{\varphi}$ onto the $\hat{k}$ axis vanishes, or $\vec{\varphi}\cdot\hat{k}=0$). In the same way, $|\varphi\rangle$ no longer has a projection along $|k\rangle$ because $w_k = 0$. The question is therefore rephrased as an inquiry into what changes when a *rotation* of the basis states with $|1\rangle$ held fixed is performed. Consideration of the case $N = 4$ and a liberal usage of Euler rotations for the three states up for rotation makes the consequences plain [60, 90]. A general rotation matrix $\mathbf{M} \equiv \mathbf{R}_z(\gamma)\cdot\mathbf{R}_y(\beta)\cdot\mathbf{R}_z(\alpha)$ is composed of[17]

$$\mathbf{R}_z(\theta) = \begin{pmatrix} 1 & 0 & 0 & 0 \\ 0 & \cos\theta & \sin\theta & 0 \\ 0 & -\sin\theta & \cos\theta & 0 \\ 0 & 0 & 0 & 1 \end{pmatrix} \tag{8.33}$$

$$\mathbf{R}_y(\theta) = \begin{pmatrix} 1 & 0 & 0 & 0 \\ 0 & \cos\theta & 0 & \sin\theta \\ 0 & 0 & 1 & 0 \\ 0 & -\sin\theta & 0 & \cos\theta \end{pmatrix} \tag{8.34}$$

Because the rotation matrices are such that $\mathbf{R}_\delta^\dagger\cdot\mathbf{R}_\delta = \hat{I}$ for $\delta = y$ or z, it follows $\mathbf{M}^\dagger\cdot\mathbf{M} = \hat{I}$. The action of $\mathbf{M}$ on $|\varphi\rangle$ can be constructed so as to change the weighting factor of one of the bases to 0, as shown in the next example.

[16]The choice of $N = 4$ here was for continuity with earlier sections, but when $N = 3$, the problem is commonly known as the Monty Hall problem. The version here is simple and classical Monty Hall rendered in quantum parlance and one door shown rather than $N - 2$ doors shown (see the second example in Section 8.2.1). Allowing the players to make quantum strategies [92] makes for a more interesting puzzle, and the notation will be familiar from the present treatment.

[17]The arrays are 4×4 because one basis state is held fixed: the remaining three correspond to the usual three-dimensional rotation of the axes in the Euler angle prescription.

EXAMPLE: Show the action of **M** on $|\varphi\rangle$ for the choices $\alpha = \pi/4, \beta = \pi/2$, and $\gamma = \pi$ is to set $w_3 \to 0$. Show that it also sets $w_4 \to 1/\sqrt{2}$.

SOLUTION: By explicit substitution and $\sin(\pi/4) = \cos(\pi/4) = \sqrt{1/2}$,

$$\mathbf{R}_z(\pi)\mathbf{R}_z\left(\frac{\pi}{2}\right)\mathbf{R}_z\left(\frac{\pi}{4}\right)|\varphi\rangle = \frac{1}{2}\begin{pmatrix} 1 & 0 & 0 & 0 \\ 0 & 0 & 0 & 1 \\ 0 & \sqrt{\frac{1}{2}} & -\sqrt{\frac{1}{2}} & 0 \\ 0 & \sqrt{\frac{1}{2}} & \sqrt{\frac{1}{2}} & 0 \end{pmatrix}\begin{pmatrix} 1 \\ 1 \\ 1 \\ 1 \end{pmatrix}$$

$$= \frac{1}{2}\begin{pmatrix} 1 \\ 1 \\ 0 \\ \sqrt{2} \end{pmatrix} \equiv |\varphi'\rangle$$

It follows $\langle\varphi'|\varphi'\rangle = \langle\varphi|\varphi\rangle = 1$, as required.

EXAMPLE: Find the rotations necessary to render $w_2 = w_3 = 0$, and find the resultant effect on w_4.

SOLUTION: The action of $\mathbf{R}_z(\pi/4)$ on $|\varphi\rangle$ gives $|\varphi'\rangle = \mathbf{R}_z(\pi/4)|\varphi\rangle = \left(\frac{1}{2}\right)\left(1 \;\; \sqrt{2} \;\; 0 \;\; 1\right)^{\dagger}$ (where the transpose $\dagger$ is used for economy). This suggests that acting on $|\varphi'\rangle$ by $\mathbf{R}_y(\arctan(\sqrt{2}))|\varphi'\rangle = |\varphi''\rangle$ results in $|\varphi''\rangle = \left(\frac{1}{2}\right)\left(1 \;\; 0 \;\; 0 \;\; \sqrt{3}\right)^{\dagger}$, as can be shown by direct computation. Therefore, $w_4 = \sqrt{3/4}$.

There remain two matters to wrap up. The first is to investigate what bearing the previous examples have on the Monty Hall problem, and the second is to specify what any of this has to do with quantum mechanics in a way that relates to emission.

The Monty Hall problem is based on a game show in which the contestant is presented with three curtains, behind one of which is a valuable prize, and behind the other two are winnings of little or no value. Once the contestant picks a curtain (say Curtain 1), the game show host reveals what is behind one of the two curtains that were not selected (say Curtain 3). The contestant is then given the opportunity to change her selection to the remaining unchosen curtain (Curtain 2), or keep her original choice (Curtain 1). The question is then, what should she do? The problem has generated a large literature, and the phrasing of the problem proves deeply important, but for the simple paraphrasing of the problem given here, the optimal and counterintuitive strategy (the one most likely to secure the valuable prize) is for the contestant to abandon her choice of Curtain 1 for Curtain 2. It is counterintuitive in the sense that when presented with the problem, many people believe the actions of the game show host do not change the probabilities of the valuable prize being behind Curtains 1 and 2, which were originally equal, and which are assumed to remain equal.

Insofar as the Monty Hall problem is an $N = 3$ version of the previously considered container problem, it is clear that the act of revealing what is behind one of the two unchosen curtains is tantamount to rotating the probability vector $|\varphi\rangle$ such that one of the weighting factors vanishes while the weighting factor for Curtain 1 is held fixed. But this is accompanied by a readjustment of the weights to make the sum over probabilities for the primed case equal unity just as for the unprimed case, or

$$\langle\varphi'|\varphi'\rangle = \langle\varphi|\mathbf{R}^{\dagger}\mathbf{R}|\varphi\rangle = \langle\varphi|\varphi\rangle = 1 \tag{8.35}$$

Why is this a paradox to many people? Simply put, it is because $P_1 = 1/3$ but P_2 now equals 2/3: after being shown an empty container, the contestant's best strategy is to switch her choice, as the probability she shall chose the prize doubles. This result is deeply unintuitive: most people (some vehemently) will maintain that the probability for each curtain is the same, or 1/2. The appreciation of why it is not is the gift the paradox provides, and it may help to reconsider the problem as a question of measurements. Before the host reveals an empty curtain, the probability that the prize is behind the curtain chosen by the contestant is 1/3 and the probability that it is not is 2/3 (a factor of 1/3 for each of the other two curtains). A measurement reveals information, and having the host reveal an empty curtain is taken as a measurement. But is it really? The contestant *knows* that one of the two curtains that the host can chose from is empty. The host shows an empty curtain. Has his doing so changed her knowledge? No, it has not. How do we know that? The initial state vector is

$$|\varphi\rangle = |1\rangle + |\bar{1}\rangle = \frac{1}{\sqrt{3}}\begin{pmatrix}1\\0\\0\end{pmatrix} + \frac{1}{\sqrt{3}}\begin{pmatrix}0\\1\\1\end{pmatrix} \tag{8.36}$$

where the bar over $|\bar{1}\rangle$ means "not $|1\rangle$", or $|\bar{1}\rangle = |2\rangle + |3\rangle$. Now the host reveals the empty curtain, that is, seems to make a measurement. If Curtain 1 has the prize, the probability is 1/2 that the host will reveal Curtain 2 and the same for Curtain 3. If Curtain 2 has the prize, the probably is unity that the host will reveal Curtain 3, and if the prize is behind Curtain 3, likewise, the probability is unity that the host will reveal Curtain 2. Therefore, the alleged "measurement" of the host is represented by a matrix **M** given by (compare Eq. (8.5))

$$\mathbf{M} = \frac{1}{2}\{|1\rangle\langle 2| + |1\rangle\langle 3|\} + |2\rangle\langle 3| + |3\rangle\langle 2| \tag{8.37}$$

and is read as "If contestant chooses 1, host shows 2 half the time and 3 half the time (the first two states); if the contestant chooses 2, host shows 3 (third state); and if the contestant chooses 3, host shows 2 (fourth state)". The advantage of Eq. (8.37) is that it is coded as in the algorithm given in Section A3.4, and can show the evolution of the winning averages over time, as in Figure 8.2. The phrasing is a consequence of the curtains being labeled, whereas in the previous formulation the labeling was fixed by the selection process. The names have changed, but the problem is the same. So, what does the measurement change? Nothing. If one works it out, $\mathbf{M}|\varphi\rangle = |\varphi\rangle$, but to show it explicitly:

$$\begin{aligned}\langle 1|\varphi'\rangle &= \langle 1|\mathbf{M}|\varphi\rangle \\ &= \frac{1}{3}\begin{pmatrix}1 & 0 & 0\end{pmatrix}\begin{pmatrix}0 & 1/2 & 1/2\\ 0 & 0 & 1\\ 0 & 1 & 0\end{pmatrix}\begin{pmatrix}1\\1\\1\end{pmatrix} = \frac{1}{3}\end{aligned} \tag{8.38}$$

$$\begin{aligned}\langle \bar{1}|\varphi'\rangle &= \langle \bar{1}|\mathbf{M}|\varphi\rangle \\ &= \frac{1}{3}\begin{pmatrix}0 & 1 & 1\end{pmatrix}\begin{pmatrix}0 & 1/2 & 1/2\\ 0 & 0 & 1\\ 0 & 1 & 0\end{pmatrix}\begin{pmatrix}1\\1\\1\end{pmatrix} = \frac{2}{3}\end{aligned} \tag{8.39}$$

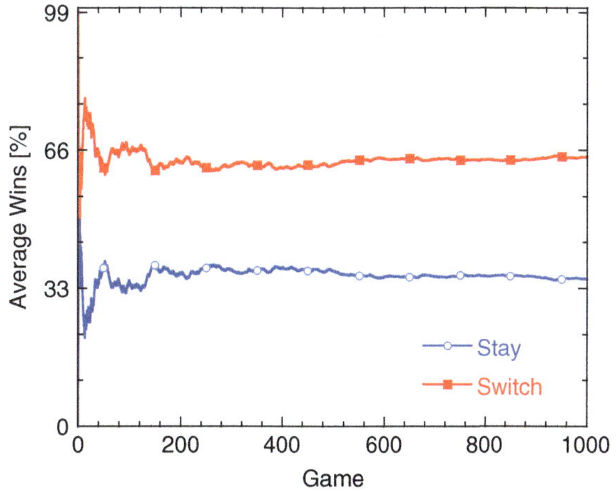

Figure 8.2 Monty Hall average winnings versus played games. The average number of winnings for the stay and switch strategies in a Monty Hall game as calculated using the algorithm of Section A3.4.

where the first is the probability the prize is behind Curtain 1, and the second is the probability that the prize is *not* behind Curtain 1, following the actions of the host. Further,

$$\langle\varphi'|\varphi'\rangle = \langle\varphi|\mathbf{M}^{\dagger}\mathbf{M}|\varphi\rangle = \langle\varphi|\varphi\rangle = 1 \tag{8.40}$$

Although $\mathbf{M}|\varphi\rangle = |\varphi\rangle$ makes $\mathbf{M}$ look like an identity operator, it is not. It is more like a rotation (something like $\hat{L}_z$ not changing $|\psi\rangle$ as far as $\mathbf{L}^2$ is concerned in Section 9.1.2), and a consequence of the fact that the $|j\rangle$ from which $|\varphi\rangle$ is built are equally weighted.

8.2.2 Conjugate variables

(the) uncertainty relation specifies the limits within which the particle picture can be applied. Any use of the words "position" and "velocity" with an accuracy exceeding that given by Eq. (1) ($\Delta x \Delta p_x \geq h$) is just as meaningless as the use of words whose sense is not defined. (Footnote: In this connection one should particularly remember that the human language permits the construction of sentences which do not involve any consequences and therefore which have no content at all ...)

– Werner Heisenberg [93], p. 15

Let $|x_n\rangle$ be a discrete basis such that the action of the position operator $\hat{X}$ on it produces the eigenvalue x_n, the upper case notation to reinforce that it is not being identified with the position operator any more than x_n is being identified with position (a rather duplicitous caveat, as they will clearly be so identified below). Let another operator $\hat{K}$ act to increment the index $n \leq N$, or

$$\hat{X}|x_n\rangle = x_n|x_n\rangle \tag{8.41}$$

$$\hat{K}|x_n\rangle = |x_{n+1}\rangle \tag{8.42}$$

If position is periodic or loops back on itself, much like the ring of curtains in Figure 8.1, so that $x_{N+j} \equiv x_j$, then it is obvious (perhaps after a small bit of pondering and the realization that $\langle x_j|x_l\rangle = \delta_{kl}$) that $\hat{K}$ is

$$\hat{K} = \sum_{n=1}^{N} |x_{n+1}\rangle\langle x_n| \tag{8.43}$$

EXAMPLE: Prove $\hat{K}^N = \hat{I}$.

SOLUTION: Using mathematical induction, first show that $\hat{K}^n = \sum_{j=1}^{N} |x_{n+j}\rangle\langle x_j|$. Clearly, it is true for $n = 1$. Assume it is true for n then prove true for $n + 1$:

$$\begin{aligned}\hat{K}^{n+1} = \hat{K}\hat{K}^n &= \sum_{j=1}^{N}\sum_{l=1}^{N} |x_{j+1}\rangle\langle x_j|x_{l+n}\rangle\langle x_l| \\ &= \sum_{l=1}^{N} |x_{l+n+1}\rangle\langle x_l|\end{aligned}$$

as required, where $\langle x_j|x_l\rangle = \delta_{jl}$ is used. Setting $n + 1 \to N$ and using

$$\hat{I} = \sum_{n=1}^{N} |x_n\rangle\langle x_n| \tag{8.44}$$

as anticipated by Eq. (8.30), completes the proof.

EXAMPLE: Prove $\hat{K}^n\hat{K}^m = \hat{K}^{n+m}$.

SOLUTION: The previous example makes clear $\hat{K}\hat{K}^n = \hat{K}^{n+1}$. Therefore, $\hat{K}^n\hat{K}^m = \hat{K}^{n-1}\hat{K}^{m+1}$. Repeating the process n times gives the desired conclusion. Alternately, using $\hat{K}^m|x_j\rangle = |x_{j+m}\rangle$, it follows $\hat{K}^n\hat{K}^m|x_j\rangle = \hat{K}^n|x_{j+m}\rangle = |x_{j+m+n}\rangle$.

The operator $\hat{K}$ has its own eigenvectors and another operator $\hat{X}$ can be constructed that increments $|k_n\rangle$ in the same way, or

$$\hat{K}|k_n\rangle = k_n|k_n\rangle \tag{8.45}$$

$$\hat{X}|k_n\rangle = |k_{n+1}\rangle \tag{8.46}$$

although the second relation must be proven below rather than baldly asserted here, but as a consequence, an equation for $\hat{X}$ similar to that of Eq. (8.43) holds as well. The relation $\hat{K}^N = \hat{I}$ has consequences: it demands that $k_n^N = 1$, which identifies

$$k_n \equiv \exp(i2\pi n/N) \tag{8.47}$$

Those familiar with Chekhov's gun[18] know full well the notational choice of k_n and $\hat{X}$ is not happenstance, but for now treat them as general. Knowing what k_n is allows for the identification of the C_j^n in the relation

$$|k_n\rangle = \sum_{j=1}^{N} C_j^n|x_n\rangle \tag{8.48}$$

where the n superscript is for now an index but, as its placement suggests, will be identified as a power. The expression $\hat{K}|k_n\rangle$ can now be expressed two ways: using the eigenvalue relation of Eq. (8.47) or using the definition of Eq. (8.43). Equating the two different forms shows that

$$e^{i2\pi n/N}\sum_{l=1}^{N} C_l^n|x_n\rangle = \sum_{j=1}^{N}\sum_{l=1}^{N} C_l^n|x_{n+1}\rangle\langle x_n|x_l\rangle$$

but knowing that $\langle x_j|x_l\rangle = \delta_{jl}$ entails $C_l^n \propto \exp(-i2\pi jn/N)$. The realization that with such a C_j^n, it follows that the kets $|k_n\rangle$ and $|x_n\rangle$ are Fourier transforms of each other, suggesting that the normalization should be the conventional $1/\sqrt{N}$. Thus,

$$|k_n\rangle = \frac{1}{\sqrt{N}}\sum_{j=1}^{N} e^{-i2\pi jn/N}|x_n\rangle \tag{8.49}$$

Identifying $x_n = na$ entails $k_n = 2\pi n/(Na)$. The $N \to \infty$, $a \to 0$ limits such that $Na = L$ is finite, plus the generalization to three dimensions, recovers Eq. (8.27). More importantly, it is seen that

$$\hat{K} = e^{ia\hat{k}} \equiv \sum_{j=0}^{\infty}\frac{1}{j!}(-ia\hat{k})^j \tag{8.50}$$

The designations x_n and k_n and the actions of $\hat{K}$ and $\hat{X}$ to increase the indices of x_n and k_n kets bring to mind position and momentum in a box of length L (and with sharp discharge, Chekhov's gun fires), but one also could have been discussing just as easily time and energy. Parameters which share a relation given by Eq. (8.49) are "conjugate variables." An eigenstate of one of them cannot simultaneously be an eigenstate of the other, and as a consequence Heisenberg's uncertainty relationship $\Delta x\Delta k \geq 1/2$ constrains the magnitude of spread in the values they can simultaneously take on.

The commutator of $\hat{x}$ with $\hat{K}$ can be used to deduce what the all-important commutator $[\hat{x}, \hat{k}]$ is. A straightforward manipulation shows that (using $x_{n+1} = x_n + a$)

$$\begin{aligned}[\hat{x}, \hat{K}]|x_n\rangle &= (\hat{x}\hat{K} - \hat{K}\hat{x})|x_n\rangle \\ &= (x_n + a)|x_n + a\rangle - x_n|x_n + a\rangle = a|x_n + a\rangle\end{aligned} \tag{8.51}$$

but for vanishingly small a, then $a|x_n + a\rangle \approx a|x_n\rangle$ and $\hat{K} \approx \hat{I} + ia\hat{k}$ using the leading order approximation to the exponential in Eq. (8.50). Taken together, this implies $[\hat{x}, \hat{K}]|x_n\rangle \approx ia[\hat{x}, \hat{k}]|x_n\rangle = a|x_n\rangle$ and so

$$[\hat{x}, \hat{k}] = i \tag{8.52}$$

[18] "Chekhov's gun" is a rule in dramatic writing that if a gun is shown on stage or mentioned in the first chapter of a literary work, it must be fired in a later scene or chapter, otherwise it should not be shown or mentioned. And, yes, it is most certainly *gun*, not *phaser*.

That is a pretty convenient relation: the commutator of the operators $\hat{x}$ and $\hat{k}$ results in a number (albeit an imaginary one), not an operator, and this can be used to establish the relationship more firmly than relying on the smallness of a by showing that if Eq. (8.52) holds, then Eq. (8.51) is recovered for arbitrary a. Using the series expansion of $\hat{K}$ directly gives

$$[\hat{x}, \hat{K}] = \sum_{j=0}^{\infty} \frac{1}{j!}(-ia)^j \left[\hat{x}, \hat{k}^j\right] \tag{8.53}$$

Applying the commutator Eq. (8.52) results in $\hat{x}\hat{k}^j = ij\hat{k}^{j-1} + \hat{k}^j\hat{x}$, easily proven using mathematical induction, which, when inserted into $[\hat{x}, \hat{k}^j]$ results in

$$[\hat{x}, \hat{k}^j] = ij\hat{k}^{j-1} \tag{8.54}$$

$$[\hat{x}, \hat{K}] = a\sum_{j=1}^{\infty} \frac{1}{(j-1)!}(-ia\hat{k})^{j-1} = a\hat{K} \tag{8.55}$$

a result seen to be identical to Eq. (8.51) because $|x_n + a\rangle = \hat{K}|x_n\rangle$, confirming again that the commutator relation of Eq. (8.52) is valid. As a side benefit, Eq. (8.54) implies that the commutator of an operator $\hat{O}(\hat{k})$ (defined in terms of its Taylor expansion in $\hat{k}$) with $\hat{x}$ is equivalent to a derivative of that operator, or

$$\left[\hat{x}, \hat{O}(\hat{k})\right] = i\partial_{\hat{k}}\hat{O}(\hat{k}) \tag{8.56}$$

a finding important to determining the current density operator. Following the procedure outlined above, it is straightforward to show also that

$$\left[\hat{x}^j, \hat{k}\right] = ij\hat{x}^{j-1} = i\partial_{\hat{x}}(\hat{x}^j) \tag{8.57}$$

$$\left[\hat{O}(\hat{x}), \hat{k}\right] = i\partial_{\hat{x}}\hat{O}(\hat{x}) \tag{8.58}$$

Lastly, from the Taylor expansion

$$|x + a\rangle = \sum_{j=0}^{\infty} \frac{1}{j!}a^j\left(\frac{\partial}{\partial x}\right)^j |x\rangle \tag{8.59}$$

then $\hat{k} \to -i\partial_x$ as seen when comparing this equation to Eq. (8.50). In the continuum limit ($N \to \infty$, $a \to 0$, $Na \to L$), this relation is important.

Before leaving, consider the identity operator of Eq. (8.44) once more, but let the "box" holding the wave functions become so large as to be considered infinite ($N \to \infty$) and consider the wave number bases, so that

$$\hat{I} = \sum_{n=1}^{\infty} |k_n\rangle\langle k_n| \tag{8.60}$$

with an equivalent version using $|x_n\rangle$. Observe that the weighting factor for each $|k_n\rangle\langle k_n|$ is unity (all the wave functions $\psi_k(x)$ are weighted equally in the wave function representation), but instead, what if the weights were chosen to reflect the probability of their actual occurrence? For example, if P_n was the probability that the state $|\varphi_n\rangle$ occurs, then the "*density operator* in the φ_n description" $\hat{\rho}$ is defined by

$$\hat{\rho} = \sum_{n=1}^{\infty} P_n|\varphi_n\rangle\langle\varphi_n| \tag{8.61}$$

The expectation value gives the number density as a function of position x, or

$$\rho(x) \equiv \langle x|\hat{\rho}|x\rangle = \sum_{n=1}^{N} P_n|\varphi_n(x)|^2 \tag{8.62}$$

such that the sum over all x_n bases gives the total particle number N. The *density matrix*, however, is more general: off-diagonal matrix elements are obtained by

$$\rho_{m,n} \equiv \langle x_m|\hat{\rho}|x_n\rangle \tag{8.63}$$

where a more matrix-friendly notation is adopted. The sandwiching between the same basis state has particular importance in treating matrices: a sum over the diagonal elements of a matrix is known as taking the "trace", or $\sum_n \rho_{n,n} \equiv Tr\hat{\rho}$. Off-diagonal elements become interesting when quantum tunneling is present, and they serve as the starting point for the development of the Wigner distribution function of Section 9.3.2. In the continuum limit, an important form of the density obtained from Eq. (8.62) that applies to electrons in a metal is

$$\rho(\vec{r}, t) = \frac{1}{(2\pi)^3} \int_{\Omega_k} d\vec{k} f_{FD}(E) |\psi_{\vec{k}}(\vec{r}, t)|^2 \tag{8.64}$$

but note that a factor of $g = 2$ for spin has *not* been included whereas electrons require it: the factor shall be part of the emission equations when they are treated.

Knowing that there is a density operator $\hat{\rho}$ makes one suspicious that there is likely a current density operator $\hat{j}$, and so there is, but to find it requires a bit more groundwork in the form of understanding the commutation relations of the conjugate operators.

8.2.3 Uncertainty

As I listen, I am uncertain – see: Heisenberg's uncertainty principle – what to do to H ...discussing math while Rome burns – if they knew what I'm thinking.

– Morris "Moe" Berg[19]

Just under five months before the surrender of Germany after the Battle of Berlin in World War II, W. Heisenberg, an extraordinary physicist and one of the fathers of quantum mechanics, doctoral advisor to Edward Teller ("father" of the hydrogen bomb), and director of the German nuclear energy program, traveled to Zurich, Switzerland to give a seminar on S-matrix theory a day or two after the onset of the Battle of the Bulge when Nazi forces pushed more than 20 kilometers into Allied-held territory. By modern sensibilities, travel for such purposes at such times by persons of such importance to projects of such consequence would already be considered unusual, but perhaps even more unusual was the former Chicago White Sox baseball icon Morris ("Moe") Berg in the audience with a pistol in his pocket. Berg, at the behest of the US Office of Strategic Services (OSS), attended the seminar and was authorized to assassinate Heisenberg then and there should Berg suspect Heisenberg was involved in efforts to build an atomic bomb for the Nazis, as it was uncertain whether Heisenberg was so engaged [94].

Uncertainty can be confusing, but coming to terms with it need not be so fraught with peril.

Heisenberg's Uncertainty Principle is a central part of quantum mechanics, and concerns the precision to which conjugate variables such as x and k (or E and t) can be measured. Arguments based on this principle afford a quick estimate of the electric fields required to extract electrons from metals via field emission. A simple means to understand the principle is afforded by the methods of the previous section.

The expectation values $\langle x \rangle \equiv \langle \psi | \hat{x} | \psi \rangle$ and $\hbar \langle k \rangle \equiv \langle \psi | \hbar \hat{k} | \psi \rangle$ give the most probable position and momentum of an electron, but there is a spread around these values: uncertainty is a constraint on how small the simultaneous spread of each is. "Spread" in statistics often makes use of the "root mean squared," or *rms*, value of a quantity, as encountered in Eq. (3.14). Applied here, the concept suggests that the square of the departure of a measurement from its mean value is the relation sought, or

$$\Delta x^2 \equiv \langle \psi | (\hat{x} - \langle x \rangle)^2 | \psi \rangle = \langle \psi | \hat{x}^2 - 2\hat{x}\langle x \rangle + \langle x \rangle^2 | \psi \rangle \equiv \langle x^2 \rangle - \langle x \rangle^2 \tag{8.65}$$

with an analogous equation holding for Δk^2. Such a formulation connects pleasingly with intuition because it accords with the usual notions concerning probability and statistics: using Eq. (8.73),

$$\langle \psi | \hat{x}^n | \psi \rangle = \int_{-\infty}^{\infty} x^n |\langle x | \psi \rangle|^2 dx = \int_{-\infty}^{\infty} x^n P(x) dx \tag{8.66}$$

for $n = 1, 2$, and $P(x) = |\psi(x)|^2$ being the probability that x obtains. Heisenberg's Uncertainty Principle places a limit on how small *both* Δx and Δk can be made.

[19] From notes scribbled by Morris Berg while attending a seminar by Heisenberg on S-matrix theory, December 18, 1944, University of Zurich, recounted in ref. [94], p. 399.

The following argument gives an intuitive idea of what that limit is. A simultaneous measurement of $\hat{x}' \equiv \hat{x} - \langle x \rangle$ and $\hat{k}' \equiv \hat{k} - \langle k \rangle$ is constrained by the fact that $|\psi\rangle$ cannot simultaneously be an eigenstate of both. Such a measurement $\hat{M}$ is a sum of $\hat{x}$ and $\hat{k}$ (not a product, as $\hat{k}$ will change the x basis states, and likewise with $\hat{x}$ for the k basis states). A general form is $\hat{M} = \hat{x}' + \lambda \hat{k}'$, where λ is complex for which the real and imaginary components remain to be specified. The question is then, how small can *both* Δx and Δk simultaneously be made? Doing this amounts to adjusting the dimensionless λ so as to minimize both of them.

By construction, $\langle \psi | \hat{M} | \psi \rangle = 0$, but $\langle \psi | \hat{M}^\dagger \hat{M} | \psi \rangle \geq 0$ because $|\varphi\rangle \equiv \hat{M}|\psi\rangle$ can be treated as a new wave function and $\langle \varphi | \varphi \rangle \geq 0$ (an intuitive observation if $|\varphi\rangle$ is thought of as a vector). Letting $\lambda \equiv \lambda_r + i\lambda_i$ then

$$\langle \varphi | \varphi \rangle = \langle \hat{x}'^2 \rangle + (\lambda_r^2 + \lambda_i^2)\langle \hat{k}'^2 \rangle + \lambda_r \{\hat{x}', \hat{k}'\} + i\lambda_i [\hat{x}', \hat{k}'] \geq 0 \tag{8.67}$$

where the anti-commutator $\{A, B\} = AB + BA$ is introduced. The real and imaginary parts of λ remain to be chosen. One constraint is that Δx and Δk be minimized simultaneously, but the absence of another means that one of the components of λ can be treated as arbitrary (the wave function can be visualized as "rotated" until this is so). It is convenient to take the real part $\lambda_r = 0$. Then, using $\Delta x^2 = \langle x'^2 \rangle$, $\Delta k^2 = \langle k'^2 \rangle$, and $[\hat{x}', \hat{k}'] = [\hat{x}, \hat{k}] = i$, Eq. (8.67) can be recast in a manner that allows λ_i to be specified:

$$\langle \varphi | \varphi \rangle = \Delta x^2 + \lambda_i^2 \Delta k^2 - \lambda_i \geq 0 \tag{8.68}$$

which is minimized when $\lambda_i = 1/2\Delta k^2$. That choice entails $\Delta x^2 \Delta k^2 \geq 1/4$, or, as more often encountered, where $p_x \equiv \hbar k$:

$$\Delta x \Delta p_x \geq \frac{\hbar}{2} \tag{8.69}$$

with an analogous equation holding for energy (ΔE) and time (Δt). The next example uses the uncertainty relation for ΔE and Δt to estimate field strengths in field emission, and is analogous to a treatment by Gomer [95] that did the same using Δp_x and Δx.

EXAMPLE: Use the energy-time version of Eq. (8.69) to estimate the fields required for field emission from a metal due to quantum mechanical tunneling processes. Assume the Fermi level and work function are those of copper ($\mu = 7$ eV, $\Phi = 4.5$ eV). Assume the energy spread of the emitted electrons is roughly 1 eV (see ref. [96]).

SOLUTION: Very crudely, an electron can pass through a forbidden region (the tunneling barrier) when product of the uncertainty in its energy with the time spent tunneling is approximately $\hbar/2$. Assume the energy uncertainty is comparable to the energy spread $\Delta E \sim 1$ eV. The uncertainty in time is comparable to the transit time $\Delta t \sim W/v_F$ through the barrier, where $W = \Phi/F$ is the width of the barrier and $v_F = \sqrt{2\mu/m}$ is the Fermi velocity. Then, from $\Delta E \Delta t \approx \hbar/2$, it follows

$$F \approx \frac{\Phi \Delta E}{\hbar}\sqrt{\frac{2m}{\mu}} = 8.7\frac{\text{eV}}{\text{nm}}$$

a number comparable to a typical field characteristic of field emission.

8.2.4 Representations

...we have interpreted the orthogonal matrix as rotating the coordinate system ... These two possibilities: (1) rotating the vector keeping the basis fixed and (2) rotating the basis (in the opposite sense) keeping the vector fixed have a direct analogy in quantum theory. Rotation (a time transformation) of the state vector gives the Schrödingier picture. Rotation of the basis keeping the state vector fixed yields the Heisenberg picture.

– George Arfken [60], p. 201–202

The expectation value of an operator $\hat{H}$ in a rotated basis $|\psi'\rangle = \hat{R}|\psi\rangle$, or $\langle \psi' | \hat{H} | \psi' \rangle$, can be thought of as the expectation value of a rotated operator $\hat{R}^\dagger \hat{H} \hat{R}$ in the unrotated basis $|\psi\rangle$, or $\langle \psi | \hat{H}' | \psi \rangle$. That is, one can put the "change" on the bases (Schrödinger) or on the operators (Heisenberg). The Heisenberg and Schrödinger representations, and the commutator relations, provide an elegant formulation of the quantum mechanical current density, and the power of the symbolic representation favored by Dirac becomes evident.

Recalling the relations of Eq. (8.24) and its companions, finding expectation values for time-dependent processes is a matter of evaluating an operator $\hat{O}$ sandwiched between a wave function $|\psi\rangle$. Thus a time-independent $\hat{O}$ between a time-dependent $|\psi(t)\rangle$ is the *Schrödinger representation*, whereas the *Heisenberg representation* would put the time dependence on the operator $\hat{O}(t)$ and use time-independent wave functions $|\psi\rangle$. Consequently, the expectation value can be given in one of two ways:

$$\langle\psi|\hat{O}|\psi\rangle = \left(\langle\psi(0)|\hat{U}^{\dagger}\right)\hat{O}\left(\hat{U}(t)|\psi(0)\rangle\right) \equiv \langle\psi(t)|\hat{O}_S|\psi(t)\rangle \tag{8.70}$$

$$= \langle\psi(0)|\left(\hat{U}^{\dagger}(t)\hat{O}_S\hat{U}(t)\right)|\psi(0)\rangle \equiv \langle\psi(0)|\hat{O}_H(t)|\psi(0)\rangle \tag{8.71}$$

Therefore, in the Schrödinger representation the wave functions are time dependent and the operators $\hat{O} = \hat{O}_S$ time independent, with the variation in time governed by Eq. (8.21). By contrast, in the Heisenberg representation, the wave functions are time independent and the operators $\hat{O}_H(t) = \hat{U}^{\dagger}(t)\hat{O}_S\hat{U}(t)$ are time dependent, with the variation in time governed by

$$\frac{\partial}{\partial t}\hat{O}_H(t) = \left(\partial_t\hat{U}^{\dagger}\right)\hat{O}_S\hat{U} + \hat{U}^{\dagger}\hat{O}_S\left(\partial_t\hat{U}\right) = (i/\hbar)\left[\hat{H}_S, \hat{O}_H(t)\right] \tag{8.72}$$

where $\hat{H}_s \equiv \hat{H}_0 + V(\hat{x})$ and the last equality is a consequence of Eqs (8.21) and (8.22) (for which the proof is pedagogically useful). This leads to the conclusion that *the time variation of an operator in the Heisenberg representation is related to its commutator with the Schrödinger Hamiltonian.*[20]

To show the connection to the wave mechanics formalism, the continuum versions of Eq. (8.44) for $|x\rangle$ but also $|k\rangle$ are useful. Recalling that conjugate variables are related by Fourier transforms, in the continuum limit, the identity operator(s) become

$$\hat{I} = \int_{-\infty}^{\infty} dx|x\rangle\langle x| = \frac{1}{2\pi}\int_{-\infty}^{\infty} dk|k\rangle\langle k| \tag{8.73}$$

$$\langle x|x'\rangle = \int_{-\infty}^{\infty}\langle x|k\rangle\langle k|x'\rangle dk$$

$$= \frac{1}{2\pi}\int_{-\infty}^{\infty} e^{ik(x-x')}dk = \delta(x-x') \tag{8.74}$$

and, similarly, $\langle k|k'\rangle = \delta(k-k')$, where $\delta(x)$ is the Dirac delta function. The identity operator allows for easy switching between x space, for which $\langle x|\psi\rangle = \psi(x)$, and k space, for which $\langle k|\psi\rangle = \psi(k)$. The commutator relation $[\hat{x}, \hat{k}]$ in the continuum case can be reestablished by noting that, first, if $|x\rangle$ is an eigenstate of $\hat{x}$ such that $\hat{x}|x\rangle = x|x\rangle$, then, using Eq. (8.73),

$$\hat{k}|x\rangle = \frac{1}{2\pi}\int_{-\infty}^{\infty}\hat{k}|k\rangle\langle k|x\rangle dk = \frac{1}{2\pi}\int_{-\infty}^{\infty} ke^{ikx}|k\rangle dk \tag{8.75}$$

$$= (-i\partial_x)\frac{1}{2\pi}\int_{-\infty}^{\infty} e^{ikx}|k\rangle dk = (-i\partial_x)|x\rangle \tag{8.76}$$

where $ke^{ikx} = -i\partial_x e^{ikx}$ is used. A vanishing of the commutator indicates that the order of the operators is not important (they can be thought of as being able to "move past" each other), but an operator like $(-i\partial_x)$ is different because it acts upon what it precedes and thereby changes it, and so the expectation is that $[\hat{x}, \hat{k}]$ will not vanish. In fact,

$$\langle x|\hat{x}\hat{k}|\psi\rangle = \int_{-\infty}^{\infty} dx'\langle x|\hat{x}|x'\rangle\langle x'|\hat{k}|\psi\rangle = x(-i\partial_x\psi(x,t)) \tag{8.77}$$

$$\langle x|\hat{k}\hat{x}|\psi\rangle = -i\partial_x(x\psi(x)) = -i\psi(x) + x(-i\partial_x\psi(x,t)) \tag{8.78}$$

Subtracting Eq. (8.78) from Eq. (8.77) results in Eq. (8.52), or $[\hat{x}, \hat{k}] = i$. Another relation of consequence is that an operator of the form $V(\hat{x})$ behaves as[21]

$$\langle\psi|V(\hat{x})|\psi\rangle = \int_{-\infty}^{\infty} dx'\int_{-\infty}^{\infty} dx\langle\psi|\hat{x}|x'\rangle\langle x'|V(\hat{x})|x\rangle\langle x|\psi\rangle \tag{8.79}$$

$$= V(x)|\psi(x,t)|^2 \tag{8.80}$$

[20] A third representation, the *interaction representation*, combines the two for Hamiltonians of the form $\hat{H} = \hat{H}_0 + \hat{H}_t$, where $\hat{H}_0$ is the time-independent part of the Hamiltonian (see Rammer [97]), but that representation is beyond present needs.
[21] Although $|\psi(t)\rangle$ instead of $|\psi\rangle$ should be used, fastidiousness would clutter, hence the more economical notation.

The relation $\langle x'|V(\hat{x})|x\rangle = V(x)\delta(x'-x)$ has been used, where $V(\hat{x})$ is an unspecified but well-behaved operator that, not surprisingly, will be identified with the potential energy, just as $\hat{H}_0 = \hbar^2\hat{k}^2/2m$ is identified with the kinetic energy, for which

$$\langle x|\hat{H}_0|\psi(t)\rangle = \frac{\hbar^2}{2m}\langle x|\hat{k}^2|\psi(t)\rangle = -\frac{\hbar^2}{2m}\left(\frac{\partial}{\partial x}\right)^2 \psi(x,t) \tag{8.81}$$

with $\partial_x^2 \to \vec{\nabla}^2$ in three dimensions. These considerations will be of great importance in finding the current density operator in Section 9.2.1.

CHAPTER 9
Quintessential problems

> *...there is one physical element which makes up the system of the bodies that move in a circle, and besides this four bodies owing their existence to the four principles ... These four bodies are fire, air, water, earth ... This world necessarily has a certain continuity with the upper motions: consequently all its power and order is derived from them. (For the originating principle of all motion is the first cause. Besides, that element is eternal and its motion has no limit in space, but is always complete; whereas all these other bodies have separate regions which limit one another.) So we must ... assign causality in the sense of the originating principle of motion to the influence of the eternally moving bodies ... men seem to have assumed that a body that was eternally in motion was also divine in nature; and, as such a body was different from any of the terrestrial elements, they determined to call it "ether".*
>
> – Aristotle[1]

The "fifth element" of Aristotle was said to fill the celestial sphere in which the perfect motion of the planets and stars occurred. Its perfection was unlike the four terrestrial essences of earth, air, fire, and water, and was therefore known as "quintessence" ("quintus" meaning "fifth" in Latin and "essence" being the constitution of something; Hamlet was therefore speaking oxymoronically and ironically when he referred to man as a "quintessence of dust"[2]). Aristotle referred to celestial sphere stuff as "aether", and as Aristotelian thought was the basis of much medieval physics, the term carried forward. Not until the famous experiments of Michelson and Morley showing definitively that the understanding of this tenet of medieval physics was mistaken did aether die out as an explanatory principle. What survived, though, is the word *quintessential* to refer to that which is the pure essence or most perfect embodiment of something, much like the perfect circles thought to describe the orbits of the planets.

Three paradigmatic problems[3] in emission physics merit the honorific of being a "quintessential problem": (*i*) the hydrogen atom, (*ii*) tunneling through and transport over a barrier, and (*iii*) the harmonic oscillator.

The reasons are easy to state. The quantum mechanical treatment of the hydrogen atom is quintessential in almost every way the word can be meant: it brings methods, concepts, and approaches (and the units of Section 2.2.2) to the physicists' lexicon, pleasingly does it by showing that there are more perfect motions than just circles, and in the process illuminating how *probability* in quantum mechanics compares to its conventional understanding. For tunneling and transport, by allowing electrons to be found where they are classically forbidden, exhibiting behavior that frustrates intuition by being just downright strange (e.g., good luck finding a consensus on what the tunneling time of an electron through a barrier is) the tunneling problem represents all that is infuriating about quantum mechanics apart from its making emission physics interesting in the first place. And lastly, the harmonic oscillator, by virtue of being so well-studied classically *and* quantum mechanically, lurks behind just about everything from mathematical techniques to particle properties and so merits the revere it so often receives. Yet so much has been written on these iconic problems that trying to upstage earlier bards is hubris.[4] Rather, it is useful to treat the quintessential problems briefly, but in a way that shows the power of the techniques they enable and the concepts they exemplify in a manner that will illuminate the remainder of the present study. As all are well covered [41–44], some familiarity with each is assumed, or at least strongly encouraged.

[1] Aristotle, *Meteorology* (trans. E.W. Webster), *Great Books of the Western World*, ed. Robert Maynard Hutchins, Vol. 8, Aristotle I. Chicago: Encyclopaedia Britannica, 1952, p. 445.
[2] Ref. [37]: *Hamlet*, II.ii.304–305, p. 947.
[3] The list is inclusive, not exhaustive.
[4] *Hubris*, or the offense of overweening pride so hated by the gods, figured prominently in Greek tragedy and more than once thrusts a hero into a bad (Odysseus) or fatal (Ajax) circumstance in the *Odyssey* of Homer. Here, it is not so dire, as the bad poetry of Christian is comical compared to that of Cyrano in the play *Cyrano de Bergerac* by Edmond Rostand, but the derisive laughter of an audience is disincentive enough to those who write.

Introduction to the Physics of Electron Emission, First Edition. Kevin L. Jensen.

9.1 The hydrogen atom

Thus water, besides the oxygen, which is one of its elements in common with many other substances, contains another element as its constituent base or radical, and for which we must find an appropriate term. None that we could think of seemed better adapted than the word hydrogen, *which signifies the* generative principle of water, *from* υδορ aqua, *and* γειvομαι gignor. *We call the combination of this element with caloric* hydrogen gas; *and the term hydrogen expresses the base of that gas, or the radical of water.*

– Antoine Lavoisier[5]

Hydrogen atoms, the unit of the gas named by the great French scientist Lavoisier, present a puzzle of such unique simplicity and complex realization that it cannot but help serve as a paradigmatic example of a physics problem: Bohr's quantized planetary model of a single hydrogen atom successfully explained the spectral line dependence of the empirical formula behind the Balmer wavelength series for hydrogen of $(1/\lambda) = R_H(2^{-2} - n^{-2})$ with n an integer and R_H the Rydberg constant for hydrogen. The Coulomb potential is spherically symmetric and so a standard account of the hydrogen atom will often start with the Laplacian in spherical coordinates, as in Section A1.3.1, use a separation of variables in Schrödinger's equation, and then slough through a tedious solution of differential equations. But the analogies between the Coulomb and gravitational potential of Newtonian physics offers pleasing similarities to identify constants of the motion that form the basis of a quantum mechanical treatment.

9.1.1 Constants of the motion

The number of commuting constants of the motion is easy to infer: three dimensions indicate three of them, but care is required because operators in quantum mechanics do not always commute and that unique feature will underlie the discrete energy spectrum of bound states. Such an outlook enables a great deal to be inferred even when it is not directly calculated, apart from its utility in providing the atomic orbitals used in Section 21.4, and it shows the power of an inspired guess. One constant of the motion is already known, energy, and so[6]

$$\hat{H}|nlm\rangle = E_{nlm}|nlm\rangle \tag{9.1}$$

Two more constants of the motion are left. Drawing inspiration from gravitational motion, when orbits (as for a satellite) are characterized by a constant radius r and velocity magnitude v, the product rv is expected to be constant. To see why, from elementary mechanics, the magnitude of the angular momentum about the axis passing through the plane of the orbit is mrv; the vector itself is $\mathbf{L} = \vec{r} \times \vec{p}$ where $\vec{p} = m\vec{v}$. For convenience, work in a coordinate system such that the symmetry axis is $\hat{z}$. Rotations in time about that axis are characterized by $\varphi = \omega t$. That is, for a satellite in a circular orbit, the radius vector is

$$\vec{r} = r(\cos\varphi\hat{x} + \sin\varphi\hat{y}) \tag{9.2}$$

whereas the velocity vector $\vec{v} = d\vec{r}/dt$ is

$$\vec{v} = v(-\sin\varphi\hat{x} + \cos\varphi\hat{y}) \tag{9.3}$$

with $v = r\omega$. Because the magnitudes r and v are constant, $\vec{r} \times \vec{v} = rv\hat{z}$ is also constant, as intuited. The physics will not change if the coordinate system is rotated itself: if rotated about $\hat{z}$ by an angle φ' (using the embedded 3×3 rotation matrix in $\mathbf{R}_z(\varphi)$ of Eq. (8.33)), the magnitude of $\mathbf{L}$ is unchanged because

$$\mathbf{R}_z(\varphi') \cdot \vec{r} = \begin{pmatrix} \cos\varphi' & \sin\varphi' & 0 \\ -\sin\varphi' & \cos\varphi' & 0 \\ 0 & 0 & 1 \end{pmatrix} \begin{pmatrix} r\sin\theta\cos\varphi \\ r\sin\theta\sin\varphi \\ r\cos\theta \end{pmatrix}$$

[5]Antoine Lavoisier (trans. Robert Kerr), *Elements of Chemistry, Great Books of the Western World,* ed. Robert Maynard Hutchins, Vol. 45, Lavoisier/Fourier/Faraday. Chicago: Encyclopaedia Britannica, 1952, p. 31. Lavoisier, the father of chemistry and member of the French Academy of Science, gave hydrogen its name and showed that hydrogen gas resulted when water was decomposed. To fund his famous laboratory, he associated with Ferme, financiers who collected taxes for the Crown. That led to his condemnation by the leaders of the French Revolution, particularly Jean-Paul Marat (a doctor by training). Lavoisier was tried and condemned, then guillotined on May 8, 1794. A magnificent portrait of him and his wife Marie-Anne Paulze is in the Metropolitan Museum of Art in New York.

[6]The notation kowtows to convention, leaving the hapless reader to know when m is the electron mass and when, like here, it is not.

$$= \begin{pmatrix} r\sin\theta\cos\left(\varphi-\varphi'\right) \\ r\sin\theta\sin\left(\varphi-\varphi'\right) \\ r\cos\theta \end{pmatrix} = \begin{pmatrix} x' \\ y' \\ z' \end{pmatrix} = \vec{r}' \tag{9.4}$$

where, although $\theta = \pi/2$ for a satellite with an equatorial orbit, θ was left in to show the general case. The rotation on $\vec{v}$ proceeds analogously, and so $\vec{r}' \times \vec{v}' = rv\hat{z}$ again. The $\hat{z}\cdot\mathbf{L} = L_z$ component of angular momentum appears to be one of the sought constants of the motion (of course it is, but the narrative phrasing is for dramatic effect). How might that be shown?

From the classical relation $\mathbf{L} = \vec{r}\times\vec{p}$, it follows that $L_z = xp_y - yp_x$. Thus, the quantum mechanical operator is $\hat{L}_z = -i\hbar\left(x\partial_y - y\partial_x\right)$ in the coordinate representation, or

$$\hat{L}_z = \hbar\left(\hat{x}\hat{k}_y - \hat{y}\hat{k}_x\right) \tag{9.5}$$

more generally. This is the "canonical procedure" for going from the Poisson brackets of classical mechanics to the commutator relations of quantum mechanics; in gauge field theories, the translation is more elegantly captured in the path-integral formalism with its alluring relations to statistical physics. Although far afield from present concerns, the deep relation between symmetries, invariances, and conservation laws deserves a brief digression.

Both the gravitational and the coulomb potentials are $1/r$ potentials, meaning that the relevant term is the magnitude of $\vec{r}$, not its direction: rotating the coordinate system will not change the potential. Said another way, $V(\vec{r}') = V(\vec{r})$ if $\vec{r}'$ and $\vec{r}$ are related by Eq. (9.4). Consider, therefore, an infinitesimal rotation about the $\hat{z}$ axis by an amount $\delta\varphi$, and using Eq. (9.4), implies

$$\begin{aligned} V(\vec{r}') &= V(x+\delta\varphi\, y, y-\delta\varphi\, x, z) \approx V(\vec{r}) + y\delta\varphi\partial_x V - x\delta\varphi\partial_y V \\ &= \left(\hat{I} + i\frac{\delta\varphi}{\hbar}\hat{L}_z\right) V(x,y,z) = \exp\left(i(\varphi/\hbar)\hat{L}_z\right) V(\hat{r}) \end{aligned} \tag{9.6}$$

The angular momentum operator is therefore seen as the *generator of rotations*, and the form of this equation looks familiar having seen the unitary operators of Eq. (8.22): just as invariance under translation results in conservation of energy and momentum, so too does invariance under rotation result in conservation of angular momentum [98]. And with that, the "canonical procedure" seems to make sense.

So, does $\hat{L}_z$ commute with the Hamiltonian $\hat{H} = \hbar^2\hat{k}^2/2m + V(\hat{r})$, where $\hat{k}^2 \equiv \hat{k}_x^2 + \hat{k}_y^2 + \hat{k}_z^2$ and similarly for $\hat{r}^2$? It should be possible to show so in any coordinate system, but taking our cue from Lear ("seek thine own ease"[7]), cartesian coordinates are particularly easy, and so they shall be used. For even greater ease, number the axes labels rather than use (xyz), that is, in the (ijk) notation, the cartesian axes are such that $(x,y,z) = (x_1,x_2,x_3)$ and similarly with k. Matters will be made still easier if some properties of commutators, introduced in Section 8.1.2, are generalized, particularly Eq. (8.52). To that end, introduce the Kronecker delta function that equals 1 when its arguments are equal $(i=j)$ but 0 otherwise, and the Levi–Civita symbol ϵ_{ijk}, defined to be +1 if (ijk) is a cyclic permutation of (123) (e.g., (312)), to be −1 if (ijk) is anti-cyclic (e.g., (321)), and 0 if two or more of the indices are equal. Defining ϵ_{ijk} in terms of its cyclic and anti-cyclic behavior allows it to be generalized to more dimensions, such as ϵ_{ijkl}, with advantages that will nevertheless not be capitalized upon here. With the new and improved notation, the relations

$$\left[x_i, p_j\right] = i\hbar\delta_{ij} \tag{9.7}$$

$$\hat{L}_k = \sum_{ij}\epsilon_{ijk}\hat{x}_i\hat{p}_j \tag{9.8}$$

are easily expressed. Also, use the $\hat{p} = \hbar\hat{k}$ notation so as to not provoke confusion between index k and wave operator $\hat{k}$. So, for example, $\hat{L}_3 = \epsilon_{312}x_1p_2 + \epsilon_{321}x_2p_1 = x_1p_2 - x_2p_1 = \hat{L}_z$ (remember that x_j and p_k commute for $j \neq k$). A relation for when products of operators appear in the commutator is needed: for operators A, B, and C, explicit evaluation shows

$$[AB, C] = A\,[B,C] + [A,C]\,B = -\,[C, AB] \tag{9.9}$$

[7] Ref. [37]: *King Lear*, III.iv.24, p. 1084.

Now to find the commutator of $\hat{H}$ with L_z. Using Eqs (9.7) and (9.9), consider the kinetic energy operator $p^2/2m$ first. Its commutator with L_z will contain

$$\begin{aligned}\sum_{ljk}\left[p_k^2, \epsilon_{ij3}x_ip_j\right] &= \left[p_1^2, x_1p_2\right] - \left[p_2^2, x_2p_1\right]\\ &= \left[p_x^2, xp_y\right] - \left[p_y^2, yp_x\right]\end{aligned} \tag{9.10}$$

with just the surviving parts being shown, where the second term on the right-hand side just has the roles of x and y reversed from the first. Making use of Eq. (8.54) and working it out, $\left[p_x^2, xp_y\right] = -2i\hbar p_xp_y$, and so the two terms on the right-hand side of Eq. ((9.10)) are the same but for their sign, and so their sum vanishes.

The harder commutator is $\left[V, L_z\right]$, but it, too, cracks quickly in the coordinate representation for which $p_j = -i\hbar\partial_j$. Taking refuge behind clever notation is not going to help now: it is better to confront what is under consideration in its full detail. Terms of the form

$$\begin{aligned}-i\hbar\left[V(r), x\partial_y\right]\psi(\vec{r}) &= -i\hbar V(r)x\left(\partial_y\psi(\vec{r})\right) + i\hbar x\partial_y\left(V(r)\psi(\vec{r})\right)\\ &= i\hbar x\left(\partial_yV(r)\right)\psi(\vec{r})\end{aligned} \tag{9.11}$$

appear, but the second term just switches the roles of x and y, and so, after the cancellations,

$$\begin{aligned}\left[V(r), L_z\right] &= i\hbar\left(x\partial_yV - y\partial_xV\right) = i\hbar\left(x\partial_yr - y\partial_xr\right)\partial_rV(r)\\ &= i\hbar\left(x(y/r) - y(x/r)\right)\partial_rV(r) = 0\end{aligned} \tag{9.12}$$

As fast as that was, it can be made faster still if the results of Section A1.3.1 are invoked, in particular, Eqs (A1.53) and (A1.54): when substituted into L_z, then in polar coordinates,

$$\hat{L}_z = -i\hbar\partial_\varphi \tag{9.13}$$

meaning that $\hat{L}_z$ depends on variations in the axial angle φ alone. Because $V(r)$ depends only on the magnitude of $|\vec{r}| = r$, the commutator of L_z with $V(r)$ must vanish, as will all commutators of operators that depend on different coordinates. Thus, $|\psi\rangle = |nlm\rangle$ is an eigenstate of $\hat{L}_z$, and so

$$\hat{L}_z|nlm\rangle \equiv m\hbar|nlm\rangle \rightarrow \langle r\theta\varphi|nlm\rangle \propto e^{im\varphi}\langle r\theta|nl\rangle \tag{9.14}$$

So is it seen that $\hat{L}_z$ commutes with the Hamiltonian $\hat{H}$ and is a constant of the motion. The values that m may take up will be considered after the remaining constant of the motion is identified.

Flush with the success of finding $\hat{L}_z$, it might be naively expected that $\hat{L}_x$ or $\hat{L}_y$ might be the last term sought. Alas, a brief calculation shows that neither of them commute with $\hat{L}_z$ because

$$\left[\hat{L}_i, \hat{L}_j\right] = i\hbar\hat{L}_k \tag{9.15}$$

with the indices a cyclic permutation of (123).

EXAMPLE: Explicitly prove Eq. (9.15) for the case $\left[\hat{L}_y, \hat{L}_z\right]$.
SOLUTION: Retain only non-zero terms, observing that $\left[x_i, p_j\right] = 0$ for $i \neq j$, and use Eq. (9.9) to find

$$\begin{aligned}\left[\hat{L}_y, \hat{L}_z\right] &= \left[-xp_z + zp_x, xp_y - yp_x\right]\\ &= \left[zp_x, xp_y\right] - \left[xp_z, yp_x\right]\\ &= z\left[p_x, x\right]p_y - y\left[x, p_x\right]p_z\\ &= i\hbar\left(zp_y - yp_z\right)\\ &= i\hbar\hat{L}_x\end{aligned}$$

But notice something interesting: if $\hat{L}_\pm$ is defined as

$$\hat{L}_\pm \equiv \hat{L}_x \pm i\hat{L}_y \tag{9.16}$$

then

$$\left[\hat{L}_\pm, \hat{L}_z\right] = \mp\hbar\hat{L}_\pm \tag{9.17}$$

but that in turn means (and one should work it out)

$$\left[\hat{L}_+\hat{L}_-, L_z\right] = 0 \tag{9.18}$$

Suddenly the quest for the final constant of the motion comes in view: if $\mathbf{L}^2 = \hat{L}_x^2 + \hat{L}_y^2 + \hat{L}_z^2$, then (work this one out, too)

$$\mathbf{L}^2 = \hat{L}_+\hat{L}_- + \hat{L}_z^2 - \hbar\hat{L}_z \tag{9.19}$$

and so $\mathbf{L}^2$ is the last commuting constant of the motion (it commutes with $\hat{L}_z$, and $\hat{L}_z$ commutes with $\hat{H}$, so therefore $\mathbf{L}^2$ commutes with $\hat{H}$). But that is great: if one has the stamina to work it out using Section A1.3.1, the Laplacian of Eq. (A1.63) is the same as

$$\nabla^2 = \frac{1}{r^2}\frac{\partial}{\partial r}\left(r^2\frac{\partial}{\partial r}\right) + \frac{\mathbf{L}^2}{r^2} \tag{9.20}$$

although one might have expected this.[8] Eq. (9.20) reveals something perhaps surprising: the eigenvalues of $\hat{H}$ do not depend on the eigenvalue m, that is, *the energy is degenerate with respect to* $\hat{L}_z$, so that states of differing m but with the same $\mathbf{L}^2$ and eigenstates for r have the same energy. In short, $\hat{H}|nlm\rangle = E_{nl}|nlm\rangle$. When spin is included in the description of the electron, then the degeneracy is removed, but that is a problem for another time.

9.1.2 Ladder operators

The operators $\hat{L}_\pm$ have a utility more than just being aesthetically tantalizing: they are known as "ladder operators" because of the effect they have on $|\psi\rangle = |nlm\rangle$, and from those effects, the nature of the eigenvalues can be inferred even without solving the differential equations which underlie the Laplacian. To see how, first observe that $\hat{L}_\pm^\dagger = \hat{L}_\mp$. Letting $\hat{L}_-|\psi\rangle \equiv |\phi\rangle$, then clearly

$$\langle\psi|\,\hat{L}_+\hat{L}_-|\psi\rangle = \left\langle\psi L_+|L_-\psi\right\rangle = \langle\phi|\phi\rangle \geqslant 0 \tag{9.21}$$

But what effect does $\hat{L}_-$ have on $|\psi\rangle$? Look at

$$\begin{aligned} L_zL_\pm|nlm\rangle &= \left(\left[L_z, L_\pm\right] + L_\pm L_z\right)|nlm\rangle \\ &= \left(\pm\hbar L_\pm + m\hbar L_\pm\right)|nlm\rangle = \hbar\,(m \pm 1)\,L_\pm|nlm\rangle \end{aligned} \tag{9.22}$$

In other words, $\hat{L}_\pm|nlm\rangle$ acts just like $|nl(m \pm 1)\rangle$. This conclusion was come to by simply considering the commutation relations, and it is the non-vanishing of the commutation relations that give quantum mechanics its particular panache. It is possible, though harder, to come to the same conclusion in the coordinate representation. From Eqs (A1.53), (A1.54), and (A1.55) and after a good amount of tinkering,

$$\hat{L}_\pm = -i(x - iy)p_z \pm iz(p_x + ip_y) \tag{9.23}$$

$$= \hbar e^{\pm i\varphi}\left(\pm\partial_\theta + i\cot\theta\,\partial_\varphi\right) \tag{9.24}$$

which, when coupled with $\langle\varphi|m\rangle = e^{im\varphi}$ shows that $e^{im\varphi} \to e^{i(m\pm1)\varphi}$ when $\hat{L}_\pm$ acts upon $|m\rangle$. The ladder operators, which act to raise or lower m by one step in $|nlm\rangle$, cannot do so forever: the eigenvalues of $\hat{L}_z^2$ are bounded in size by the eigenvalues of $\mathbf{L}^2$, after all, so there must come a point where the last application kills $|\psi\rangle$. Let $|\psi_-\rangle$ be such that $\hat{L}_-|\psi_-\rangle = 0$, and analogously, let $\hat{L}_+|\psi_+\rangle = 0$. But this means, using Eq. (9.19), that

$$\mathbf{L}^2|\psi_-\rangle = \hbar^2 m(m-1)|\psi_-\rangle \tag{9.25}$$

$$\mathbf{L}^2|\psi_+\rangle = \hbar^2 m(m+1)|\psi_+\rangle \tag{9.26}$$

[8] Actually, given the form of the Laplacian or the gravitational analog, one *should have* expected this: the angular momentum operator concerns rotations, that is, changes orthogonal to $\hat{r}$.

One sees that if $m = +l$ is the maximum value, then $m = -l$ is the minimum value. For this reason, the eigenvalues of $\mathbf{L}^2$ are taken to be the positive values $\hbar^2 l(l+1)$. Similarly, l cannot increase without bound: it, too, is bounded below by $l = 0$ (the eigenvalues of $\mathbf{L}^2$ cannot be negative) and above by the values of energy, or rather, n. The last step to finding E_{nl} is to consider the radial equation, but most of the work in dealing with the angular part of the problem *has been accomplished without having to explicitly find the form* of the eigenstates of $\mathbf{L}^2$, typically referred to as $Y_{lm}(\theta, \varphi)$,[9] simply by exploiting the consequences of symmetry, arguments about what the commuting constants of the motion have to look like (something that will be advantageous in Section 21.4), and the introduction of ladder operators.[10]

That such feats are possible is a reflection of the deep relation between symmetry groups and conservation laws, and works well beyond the rotation groups that appear from solid state to particle physics [101]. Mathematical beauty has a role to play in how physics unfolds [89]. But there is one more point that can be made before considering the last (radial) equation: angular momentum has both magnitude *and* direction, it is a vector. However, as seen in the first run-in with the hydrogen atom in Section 2.2.2, the ground state was symmetrical (rotation-invariant), and unless something breaks that symmetry, be it an electric or magnetic field or the coupling of electron spin to angular momentum, why should the energy levels care about what orientation the atom has (put another way, $\mathbf{L}^2$ has magnitude but not direction, so does direction not figure in the Hamiltonian)? In fact, they do not: the energy levels depend only on the quantum number n, or $\hat{H}|nlm\rangle = E_n|nlm\rangle$, at least in the simple model. Such will be shown in Secton 9.1.3. When the symmetry is broken or coupling to other effects taken into account, then the sub-orbitals characterized by l acquire an energy difference, and the parlance of s, p, d, and f to refer to them (taken from their historical designation of *sharp, principal, diffuse* and *fine* for $l = 0, 1, 2, 3$, respectively [41]) arises.

The ladder operator of Eq. (9.23) and the fact that there is a highest (or lowest) angular momentum state for which $\hat{L}_+|nll\rangle = 0$ allows the $Y_{lm}(\theta, \varphi)$ states to be determined, along with the normalization condition given by

$$1 = \langle lm|lm\rangle = 2\pi \int_0^{\pi} \sin\theta d\theta \left|Y_{lm}(\theta, \varphi)\right|^2 = 1 \tag{9.27}$$

where advantage has been taken of the fact that $\langle\varphi|m\rangle \propto e^{im\varphi}$. Letting $Y_{lm}(\theta, \varphi) \equiv (2\pi)^{-1/2} e^{im\varphi}\Theta_{lm}(\theta)$ then [41]

$$\partial_\theta \Theta_{ll}(\theta) = l\cot(\theta)\Theta \tag{9.28}$$

with solutions of $\Theta_{ll}(\theta) \propto (\sin\theta)^l$. Then repeated applications of the ladder operator can generate the remainder, or

$$Y_{lm} \propto \hat{L}_-^{l-m} Y_{ll} \rightarrow \Theta_{lm} \propto \frac{1}{\sin^m\theta}\left(\frac{d}{d(\cos\theta)}\right)^{l-m} (\sin\theta)^{2l} \tag{9.29}$$

but this is very close to Rodrigues's formula (Section A2.5, Eq. (A2.14)) for the generation of Legendre polynomials $P_l(\cos\theta)$. In fact, $\sqrt{2}\Theta_{l0}(\theta) = \sqrt{(2l+1)}P_l(\cos\theta)$, and so, apart from normalization,

$$\Theta_{lm} \propto \sin^m\theta \left(\frac{d}{d(\cos\theta)}\right)^m P_l(\cos\theta) \tag{9.30}$$

where the lack of concern with the normalization constants, which complicate the narrative, is because the calculation of average quantities or expectation values finds those constants in both numerator and denominator of the ratios that arise, and so cancel (apart from the fact that these normalization constants are both extensively documented or easily calculated numerically).

Standard accounts of $\langle\theta\varphi|lm\rangle = Y_{lm}(\theta, \varphi)$ hew rather closely to their representation in terms of spherical coordinates and include their proper normalizations. Yet, when probability is treated, there is advantage in their cartesian representation given the coordinate transformations of Eqs (A1.48)–(A1.50) (it aids in the visualization of chemical bonding [102]). Moreover, for the purposes of calculating averages, the normalization is far less important than the relative magnitudes. To reinforce that the *real* and *un*normalized angular functions are under discussion, refer to them as A_{lm} (in contrast to the radial functions R_{nl} below). The s sub-orbital $A_{00} = 1$ is easy: there is no angular dependence (s sub-orbitals are spherically symmetric). The normalization associated with A_{00} is $1/\sqrt{4\pi}$. The next sub-orbitals are composed of $Y_{1,0} \rightarrow A_{10}$

[9]The form of the Y_{lm}, and their normalization, are fodder for quantum mechanics [42–44] or electrodynamics [99, 100] textbooks, but the needs here are different, apart from concerns about the derisive laughter of audiences again.

[10]The term "ladder operator" will be eclipsed by the terms "creation and annihilation operators" which, apart from being a more riveting terminology, point to states being occupied and unoccupied as a result of transitions, as for the harmonic oscillator in Section 9.3.

and $Y_{1,1} \pm Y_{1,-1} \rightarrow A_{1,\pm 1}$. Keep in mind that the magnitude of the m subscript on A_{lm} relates to $\hat{L}_z$ eigenvalue, but its sign has been adapted to refer to how the Y_{lm} functions are combined. Thus,

$$\begin{aligned} A_{1,0} &= z/r \rightarrow p_z \\ A_{1,+1} &= x/r \rightarrow p_x \\ A_{1,-1} &= y/r \rightarrow p_y \end{aligned} \tag{9.31}$$

where $r^2 = x^2 + y^2 + z^2$ and the $\rightarrow$ symbol designates the association with the *spdf* notation convention favored in discussions of bonding. The normalization associated with A_{1m} is $\sqrt{3/4\pi}$. The last orbitals to be of interest here are A_{2m}, and they are also made up of combinations of the Y_{lm} to make real functions:

$$\begin{aligned} A_{2,0} &= 3(z/r)^2 - 1 \rightarrow d_{zz} \\ A_{2,+1} &= \sqrt{3}\left(2xz/r^2\right) \rightarrow d_{xz} \\ A_{2,-1} &= \sqrt{3}\,(2yz/r) \rightarrow d_{yz} \\ A_{2,+2} &= \sqrt{3}\left((x^2 - y^2)/r^2\right) \rightarrow d_{x^2-y^2} \\ A_{2,-2} &= \sqrt{3}\left(2xy/r^2\right) \rightarrow d_{xy} \end{aligned} \tag{9.32}$$

The normalization associated with A_{20} is $\sqrt{5/16\pi}$, as can be shown by numerical or analytic integration.

9.1.3 The radial equation

The contribution of angular momentum may now be included back into the radial equation for the hydrogen atom introduced in Eq. (2.6), but, with the introduction $k_o \equiv m\alpha c/\hbar = a_o^{-1}$, and the more convenient representation of

$$\psi_{nlm}(r,\theta,\varphi) = \frac{u_{nl}(r)}{r} Y_{lm}(\theta,\varphi) \tag{9.33}$$

then Schrödinger's equation becomes (where the nl subscripts are suppressed on $u(r)$ until they are needed)

$$\left\{ -\frac{\partial^2}{\partial r^2} + \frac{l(l+1)}{r^2} - \frac{2k_o}{r} \right\} u(r) = -k^2 u(r) \tag{9.34}$$

valid only for $r > 0$. The form of this equation has the look and feel of the one-dimensional Schrödinger's equations that will dominate the tunneling calculations in store, and so experience with this one is good preparation. The intent here, as with the angular part, is not to rigorously solve Eq. (9.34) (particularly because solutions to the equation have been very well treated elsewhere, see, for example, Chapter 14 in ref. [103]), but rather to highlight generally useful methods on a problem for which there is ample documentation as to its exact solution.

By rendering Eq. (9.34) to have the look and feel of a typical one-dimensional problem allows one-dimensional terminology to be employed. In particular, to prevent concerns about $r < 0$ solutions creeping in that make no sense, visualize the potential barrier as going to ∞ at $r = 0$, so the wave function is not found there. Now an effective potential of the form

$$V_{eff}(r) = \frac{\hbar^2}{2m}\left(\frac{l(l+1)}{r^2} - \frac{2k_o}{r}\right) \tag{9.35}$$

exists, where the minimum of the potential is at $r_{min} = l(l+1)a_o$, and has a depth $\hbar^2 k_o^2/(2ml(l+1))$. The effective potential is shown in Figure 9.1.

In the limit of large r, $V_{eff}(r)$ becomes negligible, and so the equation for $u(r)$ goes over into $\partial_r^2 u(r) - k^2 u(r) \approx 0$, suggesting solutions of the form $u(r) \sim e^{-kx}$, a relation already encountered in the treatment of Eq. (2.6). Pull out this part of the solution by letting $u(r) = g(r)e^{-kr}$ to obtain

$$\left\{ \partial_r^2 - 2k\partial_r + \frac{2k_o r - l(l+1)}{r^2} \right\} g(r) = 0 \tag{9.36}$$

If $g(r)$ were a polynomial, the action of ∂_r would be to reduce the power of each term in the polynomial by one, and similarly ∂_r^2 would reduce the power by two. But the factors of $(1/r)$ and $(1/r^2)$ do just that as well, and so the urge to try

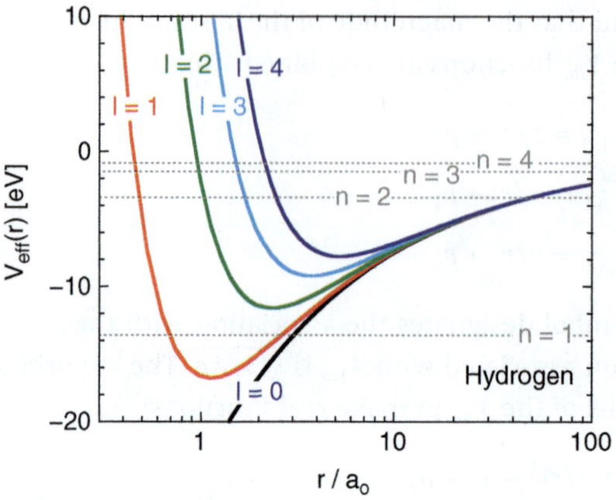

Figure 9.1 The effective potential $V_{eff}(r)$ associated with $u_{nl}(r)$ as a function of r/a_o of Eq. (9.35), where a_o is the Bohr radius (0.052918 nm) for the angular momentum states $l = 0$, 1, 2, 3, and 4. The gray lines labeled with n values correspond to $E_n = -\hbar^2/2ma_o^2n^2$.

a power series solution for $g(r)$ seems reasonable – until one appreciates that an infinite series expansion will wrongly suggest that $g(r) \sim e^{2kr}$ when r is large, as it is the solution to $\{\partial_r^2 - 2k\partial_r\}\, g(r) = 0$, and *that* grows faster than the e^{-kr} of Eq. (2.6) declines. Something therefore has to truncate the infinite series expansion, and that, in turn, will lead to particular values of energy which are allowed, that is, the bound states. This can be seen by formally inserting the power series, finding the recursion relation between the coefficients, and then finding what constraints exist on k to terminate the series. That would be the careful approach. Instead, rashly consider what happens when $g(r) \propto r^n$ (where n is an as yet unspecified power) and see what can be said about the values of n and k. Inserting $g(r) = r^n$ into Eq. (9.36) results in[11]

$$[n(n-1) - l(l+1)]\, r^{n-2} + 2(k_o - nk)r^{n-1} = 0 \tag{9.37}$$

The value of r is arbitrary, and therefore the coefficients of r^{n-2} and r^{n-1} must separately equal 0.

The second term vanishes when $nk = k_o$, and so only particular values of the energy $E(k)$ are allowed that satisfy the relation

$$k = \frac{k_o}{n} \tag{9.38}$$

which implies that the energy levels are quantized according to $E_n = \hbar^2/(2ma_o^2n^2)$, where $a_o = 1/k_o$ is the Bohr radius.

The first term on the left-hand side of Eq. (9.37) vanishes if $l = n - 1$, which additionally confirms that n is an integer because l is: the relation $l = n - 1$ specifies the maximum angular momentum state. But what about the lower angular momentum states? Setting $l = n - 2$ leaves terms behind in Eq. (9.37). Because the eigenstates of $u_{nl}(r)/r$ are orthogonal, $g(r)$ must therefore contain other powers of r, the coefficients of which can be found by enforcing the condition of orthogonality between the $u_{nl}(r)/r$. But there is a simpler way.

When l changes from $l = n - 1$ to $l = n - 2$, a term $C_{n-1}r^{n-1}$ can be added to $g(r)$ with C_{n-1} adjusted until Eq. (9.37) holds again. That is, $g(r)$ is modified by changing r^n to $r^n - C_{n-1}r^{n-1}$. Similarly, as j increases, every time $l \to n - j$, then a term $C_{n-j}r^{n-j}$ is to be appended to $g(r)$ and the resulting Eq. (9.36) set equal to zero. Setting the coefficients of the different powers of r to zero (r is arbitrary, and doing so is the only way to achieve Eq. (9.37)) will determine the C_{n-j} values. After the unnormalized $u_{nl}(r)$ are found, their normalization can be found at leisure by the orthogonality relation

$$\langle n'l'm'|nlm\rangle = \delta_{nn'}\delta_{ll'}\delta_{mm'} \to \int_0^\infty u_{n'l'}(r)u_{nl}(r)dr = \delta_{nn'}\delta_{ll'} \tag{9.39}$$

Does this prescription work? Yes, but that is better seen by example, although a reduction of Eq. (9.37) to dimensionless form in terms of $s = k_or$ proves useful. Preloading the relation $k = k_o/n$ results in

$$\left\{\frac{\partial^2}{\partial s^2} - \frac{2}{n}\frac{\partial}{\partial s} + \frac{2s - l(l+1)}{s^2}\right\} g(s) = 0 \tag{9.40}$$

---

[11]Technically, letting $g(r) = Ar^n$ is more exact, but as the wave function can be specified up to an arbitrary coefficient, it is simply easier to take $A = 1$.

Observe that this dimensionless equation can be rewritten as

$$\left\{\left(\frac{\partial^2}{\partial s^2}-\frac{l(l+1)}{s^2}\right)-\left(\frac{2}{n}\frac{\partial}{\partial s}-\frac{2}{s}\right)\right\}g(s)=0 \tag{9.41}$$

where the grouping of terms more directly suggests the relationship of k to k_o (second parentheses) and then of l to n (first parentheses) for the maximum angular momentum case for which $g(s) \propto s^n$.

EXAMPLE: Find $u_{nl}(r)$ for $n = 1$. Express in terms of $s = k_o r$.
SOLUTION: For $n = 1$, $g(s) = s^n = s$, $l = n - 1 = 0$, and the left-hand side of Eq. (9.41) identically vanishes. Therefore

$$u_{10}(r)/r = N_{10}e^{-s} \tag{9.42}$$

where N_{10} is the normalization.

EXAMPLE: Find $u_{nl}(r)$ for $n = 2$. Express in terms of $s = k_o r$.
SOLUTION:

- For $n = 2$ and $l = n - 1 = 1$, $g(s) = s^n = s^2$, and the left-hand side of Eq. (9.41) identically vanishes. Therefore
$$u_{21}(r)/r = N_{21}se^{-s} \tag{9.43}$$
- For $n = 2$ and $l = n - 2 = 0$, $g(s) = s^n - C_{n-1}s^{n-1} = s^2 - C_1 s$, and the left-hand side of Eq. (9.41) becomes $2 - C_1 = 0$. Therefore, $C_1 = 2$ and
$$u_{20}(r)/r = N_{20}(s-2)e^{-s/2} \tag{9.44}$$

EXAMPLE: Find $u_{nl}(r)$ for $n = 3$. Express in terms of $s = k_o r$.
SOLUTION:

- For $n = 3$ and $l = n - 1 = 2$, $g(s) = s^n = s^3$, and the left-hand side of Eq. (9.41) identically vanishes. Therefore
$$u_{32}(r)/r = N_{32}s^2e^{-s/3} \tag{9.45}$$
- For $n = 3$ and $l = n - 2 = 1$, $g(s) = s^n - C_{n-1}s^{n-1} = s^3 - C_2 s^2$, and Eq. (9.41) results in $6 - C_2 = 0$. Therefore, $C_2 = 6$ and
$$u_{31}(r)/r = N_{31}(s-6)e^{-s/3} \tag{9.46}$$
- For $n = 3$ and $l = n - 3 = 0$, $g(s) = s^3 - C_1 s^2 - C_2 s$, and Eq. (9.41) becomes $(9 - C_2)s + (2C_1 - 3C_2) = 0$. Requiring the coefficient of each power of s equal 0 results in $C_1 = 9$ and $C_2 = 27/2$. Therefore,
$$u_{30}(r)/r = N_{30}(s^2 - 9s + (27/2))e^{-s/3} \tag{9.47}$$

The requirement that the $u_{nl}(r)$ functions be normalized as per Eq. (9.39) is the last step to determining N_{nl}. As an example, consider N_{30}. Expressed in terms of $s = k_o r$, the normalization is determined by

$$1 \equiv \int_0^\infty u_{30}(s)^2 ds = N_{30}^2 \int_0^\infty \left(s^2 - 9s + \frac{27}{2}\right)^2 e^{-2s/3} ds \tag{9.48}$$

Employing Eq. (A2.3) shows that $N_{30} = 4\sqrt{3}/243$ (see the denominator of Eq. (9.50) in the example below for what is involved), and so for $\langle r\theta\varphi|nlm\rangle = \psi_{nlm}$ with $n = 3$ and $l = m = 0$,

$$\psi_{300}(r,\theta,\varphi) = \frac{4\sqrt{3}}{243}\left(k_o^2 r^2 - 9k_o r + \frac{28}{2}\right)e^{-2k_o r/3}\left(\frac{1}{\sqrt{4\pi}}\right) \tag{9.49}$$

where the last factor is $(1/4\pi)^{1/2} = Y_{0,0}(\theta,\varphi)$.

Fixation on the normalization factors obscures that they can often be sidestepped: there are easy ways to use the wave functions without them, and such methods will be seen to be useful in finding current density or quantum efficiency. As a straightforward example, the expectation value of $1/r$ for the $|300\rangle$ can be calculated without an explicit form for the normalization by

$$\begin{aligned}\langle 300|\,(k_o r)^{-1}|300\rangle &= \frac{\int_0^\infty s\left(2s^2-18s+27\right)^2 e^{-2s/3}ds}{\int_0^\infty s^2(2s^2-18s+27)^2 e^{-2s/3}ds}\\ &= \left(\frac{2}{3}\right)\frac{(5!)-12(4!)+48(3!)-72(2!)+36(1!)}{(6!)-12(5!)+48(4!)-72(3!)+36(2!)} = \frac{1}{9}\end{aligned} \tag{9.50}$$

where, after a change of integration variables to $s = 3w/2$, the Gamma function (Eq. (A2.4)) of Section A2.2 has been used. Should the hydrogentic wave functions $\psi_{nlm}(r,\theta,\varphi) = \langle r\theta\varphi|nlm\rangle = R_{nl}(r)Y_{lm}(\theta,\varphi)$ actually be needed, there are many tabulations of them (see, for example, refs [41, 102, 104]): it is sufficient to know that they are available (or how in principle they may be evaluated) rather than what they are.

9.1.4 Orbital shape and probability

Orbital shapes can be expressed in any convenient coordinate system, with cartesian and polar frequently encountered. In polar coordinates $\langle 300|300\rangle$ is

$$\langle 300|300\rangle = \int_0^\infty |R_{30}(r)|^2 4\pi r^2 dr \tag{9.51}$$

where the independence of the s orbitals on θ and φ allows the angular integrations to be performed and results in the factor of 4π (a factor offset by the normalization factor of $1/\sqrt{4\pi}$, being the last term in Eq. (9.49)). Apart from the r^2dr, which has units of volume, the integrand is the probability per unit volume that a measurement will reveal an electron at the position r. An indication of its behavior is given in Figure 9.2 for the first three levels of the related function $u_{nlm}(r)$ for the s states ($l = m = 0$), a function more pleasant to plot because of the absence of the $1/r$ dependence of $R_{nl}(r)$ for small r.

Visualizations of the orbitals in cartesian coordinates that portray them as solids taking the shape of spheres, weather balloons, and dumbbells show the probability density surface of a constant value (say, where $|\psi|^2 dv = 1/2$) but an image is potentially misleading if it suggests the electron is *only* inside the solid figure or *uniformly* smeared throughout it. The distributions represent the probability of finding an electron in the small volume $dv = dxdydz$ at the location (x,y,z). A plot of where repeated measurements of position found an electron will produce a fuzzy cloud that gives a better intuitive feeling of what the distribution is.

The hydrogen atom problem gives full expression to the subtle notion of probability that permeates quantum mechanics. To model the probability distribution, a uniformly distributed random number r_n can be generated and compared to $|\psi_{nlm}(x,y,z)|^2$: if r_n is smaller, an electron has been found inside the small volume (conversely, if r_n is greater, no electron was found). After many such "measurements", a graphical representation of the locations where the electron was found gives a fair picture of what the probability distribution looks like. The process of comparing a random number to a probability distribution is commonly called an "acceptance/rejection method" (alternately, Neyman's method [105] or

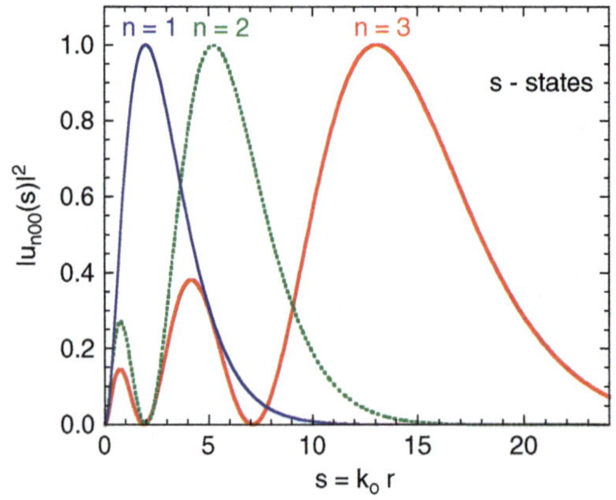

Figure 9.2 The function $u_{nlm}(s)$ for $l = m = 0$ and $n = 1, 2, 3$ (the s orbitals). The curves have been scaled by their maximum values.

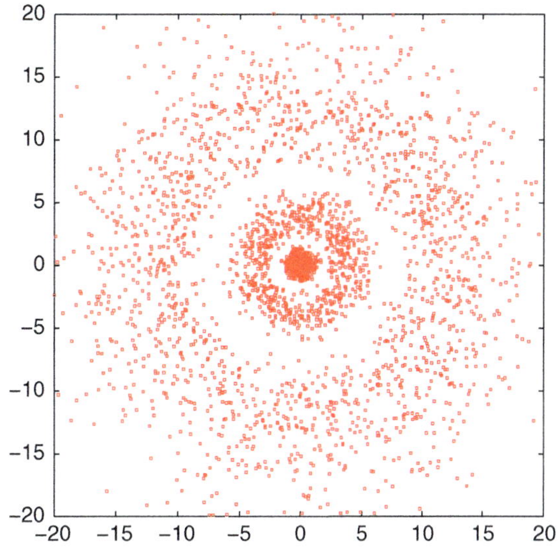

Figure 9.3 A representative distribution of measurements that find an electron for the wave function $\psi_{300}(x, y, z)$ evaluated in the x–z plane ($y = 0$) using the algorithm of Section A3.6.

"hit-or-miss" [106]) as when applied to evaluate integrals in Section A1.2.5 (as opposed to modeling carrier transport [81, 107, 108], where Monte Carlo entails a great deal more). A simple method to generate those fuzzy images is given by Section A3.6, used to generate a representative slice of sightings in the x–z plane associated with the wave function ψ_{300} (an s orbital). The results are shown in Figure 9.3. Compare those findings to the $n = 3$ line of Figure 9.2, where the gaps are anticipated by the behavior of the $u_{nl}(k_o r)$ function.

As with human anatomy, a slice looks different than the full body. By evaluating all the slices in the y direction, not just the x–z plane, and stacking them like a loaf of bread, an idea of the three-dimensional shape of the orbital emerges and resembles the probability distribution used to select the points. Visualization can still be a challenge, and the usage of projections simplifies the visualization. A projection is obtained by integrating the distribution along one of the coordinates, say $\hat{y}$, and is therefore given by

$$\rho(x, z) = \int_{-\infty}^{\infty} |\psi_{nlm}(x, y, z)|^2 dy \tag{9.52}$$

which then is a measure of the probability that an electron is found with the coordinates x and z for any y. The shape of the orbital is then related to the projection, and an example is shown in Figure 9.4 for the more visually interesting

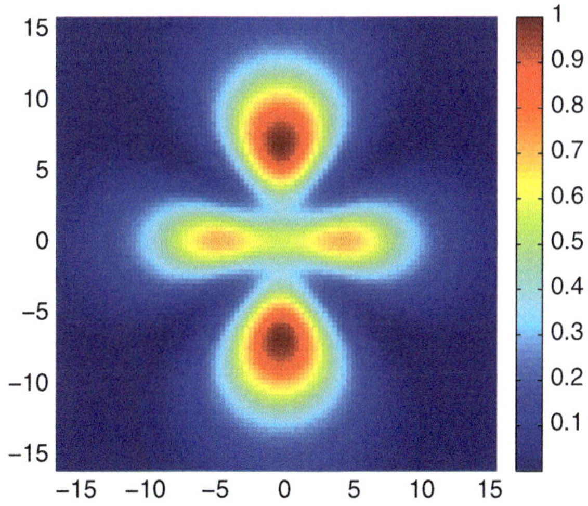

Figure 9.4 A large number of observations can be represented in a plot of how often an electron is found with particular coordinates. Shown here is $\rho_{320}(x, z)$: the sum over the y coordinate is meant to mimic Eq. (9.52). The algorithm is in Section A3.6. The orbital shown is for d_{zz} of Eq. (9.32), and $n = 3$, or $|nlm\rangle = |320\rangle$.

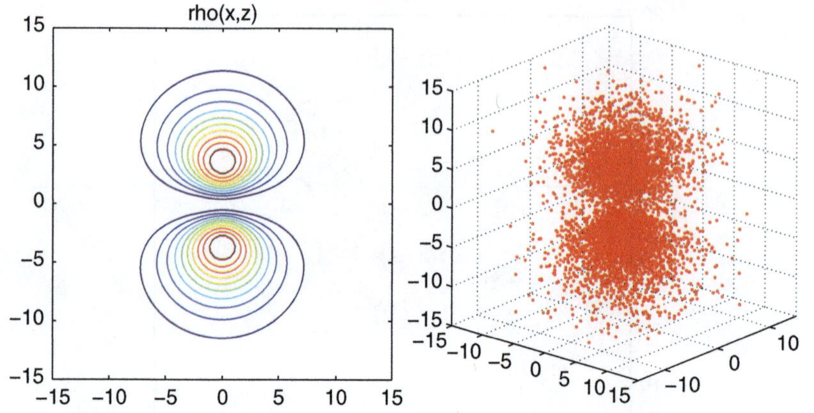

Figure 9.5 Using Eq. (9.52), a contour plot (left) of $\rho_{210}(x, z)$ compared to the "measurement" scatter plot (right).

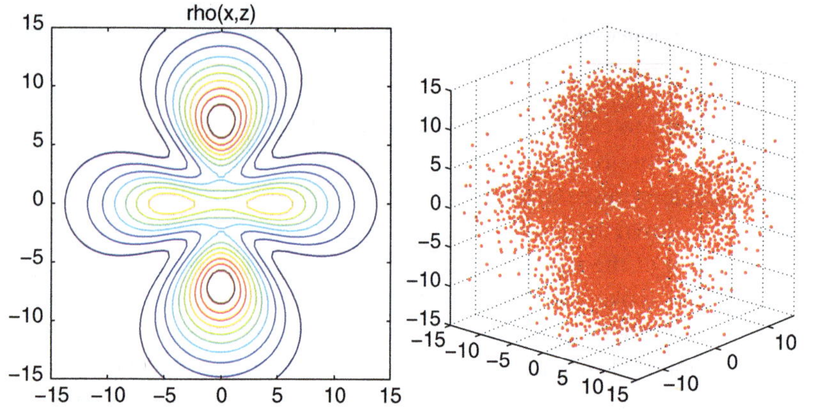

Figure 9.6 Same as Figure 9.5, but for $\rho_{320}(x, z)$.

d_{zz} orbital of $\psi_{320}(x, y, z)$, which agrees (as it should) with the more conventional plot of the contours of the integrand of $\rho_{320}(x, z)$ shown in Figure 9.6 (for chemical bonding in the larger atoms, the p orbitals are of particular importance, for which the p_z orbital is also shown in Figure 9.5).

The orbitals of many electron atoms differ from the simple orbitals of the hydrogen atom. More importantly, when atoms are brought into proximity with each other, their outermost electrons are shared and lead to chemical bonding, but *molecular orbitals* lay on even greater complexity. It is common to assume that the molecular orbitals are linear combinations of the atomic orbitals, or LCAO for short [53, 102]. In contrast to the Bloch waves favored in solid state physics, the LCAO methods are favored in treating chemical interactions [104]. They shall be encountered again in Section 21.4 when the band gap is considered.

9.2 Transport past barriers

> *If one separates the wave equation for a hydrogen atom in an homogeneous electric field in parabolic coordinates, one finds that one of the equations has a potential energy which becomes negatively infinite for infinite values of the coordinate ...There are thus no stable stationary states possible for a hydrogen atom in such a field ...The values of the field for which the dissociation becomes appreciable are ...about a tenth of the field which makes the classical Bohr orbit unstable. The effect increases very rapidly with a drop in the ionization potential of the atom, so that we should expect it to be particularly marked for certain metallic atoms at the surface of a conductor. The aeona effect, or pulling of electrons out of metal by fields of the order of 10^5 e.s.u., is probably to be accounted for in this way. This is confirmed by the fact, that when one uses the data of R.A. Millikan and C. Eyring to plot the reciprocal of the field against the logarithm of the current minus one fourth the logarithm of the field, the points so obtained lie on a straight line.*
>
> – J. Robert Oppenheimer [109]

In a pleasing parallel to Lavoisier's studies of hydrogen, consideration of the hydrogen atom in strong fields allowed Oppenheimer[12] to anticipate that tunneling phenomena, wherein an electron can pass through a region where it is classically forbidden, may explain the pulling out of electrons from a metal surface subject to intense electric fields. As Oppenheimer sagely observed, there was not a constant of the motion as normally required in the *modus operandi* of quantum mechanics: eventually, a hydrogen atom under field would lose its electron, even if it took (as Oppenheimer calculated) $10^{10^{10}}$ seconds to do so for a field of 1 Volt/cm. Therefore, the consideration of hydrogen helped motivate the next quintessential problem that will be of even greater importance to the understanding of electron emission: emission over, and *under*, a barrier.

9.2.1 Density matrix

And there's nothin' sure in this world
And there's nothin' pure in this world
Look for something left in this world
Start again

– Billy Idol[13]

Pure states are convenient to work with, but a description of probability density $\hat{\rho}$ and current density $\hat{j}$ is more complicated and requires a mixed-state representation to obtain a quantum mechanical current density $j(x,t)$, for which three essential components are required: (*i*) the continuity equation (Eq. (7.19)) relates the time-derivative of density to the spatial derivative of current density, (*ii*) the time derivative of an operator is related to the commutator of that operator with the Schrödinger representation Hamiltonian $\hat{H}_s \equiv \hat{H}_0 + V(\hat{x})$ (Eq. (8.72)), and (*iii*) the commutator of an operator with $\hat{k}$ is related to the spatial derivative of that operator (Eq. (8.58)). Once in hand, these three tools make quick work of finding a means to evaluate $j(x,t)$.

From the eigenstates defined by $|\psi_j(t)\rangle = \hat{U}(t)|\psi_j(0)\rangle$ such that $\hat{H}_s|\psi_j\rangle = E_j|\psi_j\rangle$, the density operator $\hat{\rho}(t)$ is

$$\hat{\rho}_H(t) = \sum_j f(E_j)|\psi_j(t)\rangle\langle\psi_j(t)| \tag{9.53}$$

where $f(E_j)$ is a measure of the probability that an electron has an energy E_j: in three dimensions, for example, it would be the Fermi–Dirac distribution as in Eq. (8.64), but in the one-dimensional case it is termed the "supply function'," as in Section 11.1. Letting the operator $\hat{O}(t)$ in Eq. (8.72) be the density operator, then it follows

$$i\hbar\partial_t\hat{\rho}_H(t) = \left[\hat{H}_S, \hat{\rho}_H(t)\right] \tag{9.54}$$

Observe that $\langle x|\,\hat{\rho}(t)|x\rangle = \rho(x,t)$. Because of Eq. (8.79), it follows that

$$\langle x|\,[V(\hat{x}), \hat{\rho}(t)]|x\rangle = 0 \tag{9.55}$$

and so the commutator becomes

$$\left[\hat{H}_s, \hat{\rho}_H(t)\right] = \frac{\hbar^2}{2m}\left[\hat{k}^2, \hat{\rho}_H(t)\right] = -\frac{i\hbar^2}{2m}\frac{\partial}{\partial\hat{x}}\left\{\hat{k}, \hat{\rho}_H(t)\right\} \tag{9.56}$$

where the anti-commutator notation $\left\{\hat{k}, \hat{\rho}\right\} = \hat{k}\hat{\rho} + \hat{\rho}\hat{k}$ was introduced after Eq. (8.67). If $\hat{j}$ is defined such that

$$\hat{j}_S = \frac{\hbar}{2m}\left\{\hat{\rho}_S, \hat{k}\right\} \tag{9.57}$$

$$\hat{j}_H(t) = \frac{1}{2m}\hat{U}^\dagger(t)\left\{\hat{\rho}_S, \hbar\hat{k}\right\}\hat{U}(t) \tag{9.58}$$

[12] Oppenheimer's fate, though far less grim than Lavoisier's, might be seen as symbolically parallel: On June 29, 1954 (9 years and 13 days after the Trinity test ushering in the atomic age, which Oppenheimer's leadership made possible), the Atomic Energy Commission (AEC) announced the revocation of his security clearance, deeming him a risk due to "fundamental defects in his character" and "his disregard for the obligations of security." The testimony of Edward Teller, who championed the hydrogen bomb that Oppenheimer publicly advocated against, was instrumental. P.J. McMillan summarized the incident thus: " ...by taking away the clearance of the man who had replaced Albert Einstein as the public face of scientific genius, the government told the scientists: We want your work, but we don't want you." (P.J. McMillan, *The Ruin of J. Robert Oppenheimer and the Birth of the Modern Arms Race*. New York: Viking Penguin, 2005, p. 262).

[13] Billy Idol, lyrics to *White Wedding*, 1982.

then the quantum version of the continuity equation of Eq. (7.19) is

$$\frac{\partial}{\partial t}\hat{\rho}_H(t) + \frac{\partial}{\partial \hat{x}}\hat{j}_H(t) = 0 \tag{9.59}$$

which looks much like its classical analog, but appearances are not the same as properties. Expressed in terms of its wave function representation, $\hat{j}$ is

$$\begin{aligned}\langle\psi|\hat{j}_H(\hat{x},t)|\psi\rangle &= \frac{\hbar}{2m}\langle\psi(t)|\left\{\hat{\rho}_S,\hat{k}\right\}|\psi(t)\rangle \\ &= \frac{\hbar}{2mi}\left\{\psi^\dagger(x,t)\partial_x\psi(x,t) - \psi(x,t)\partial_x\psi^\dagger(x,t)\right\}\end{aligned} \tag{9.60}$$

as is verified by expanding the anti-commutator and using Eq. (8.60) with the replacement $|x_n\rangle \to |x\rangle$.

9.2.2 Transmission probability

Insistence on clarity at all costs is based on sheer superstition as to the mode in which human intelligence functions. Our reasonings grasp at straws for premises and float on gossamers for deduction.

– Alfred North Whitehead[14]

The mathematics behind calculating a transmission probability are fairly straightforward to work out. The understanding of the implications, though, is difficult because they tend to contradict expectations. Begin with the methods first, anyway. Basis states that are chosen with the constants of the motion in mind, much like finding the stable strategies in Section 8.1.3, are particularly important. In the absence of potential variation ($V(x)$ is constant), then $\hat{H}_s = \hat{H}_0 = \hbar^2\hat{k}^2/2m$, for which the basis states are $|k\rangle$. For them, the current density operator is

$$j_k(x) = \frac{\hbar}{2mi}\left\{\psi_k^\dagger(x)\partial_x\psi_k(x) - \psi_k(x)\partial_x\psi_k^\dagger(x)\right\} \tag{9.61}$$

where $\psi_k(x) = \langle k|x\rangle = e^{ikx}$, and where the normalization for now is assumed to be unity. A simple substitution into Eq. (9.61) results in $j_k(x) = \hbar k/m$, which looks like a velocity alone, whereas the current density encountered in Eq. (3.5) was the product of charge, velocity, and number density. Remember that j_k is a *number* current density, so the *charge* current density is a product of j_k with q, and that, because the momentum was specified to a particular $|k\rangle$ rather than a distribution $\sum_k f(k)e^{ikx}$, the constant number density is subsumed in the normalization: that the number density is constant everywhere for $|k\rangle$ is a consequence of Heisenberg's uncertainty relation and the unique specification of the momentum k. Therefore, the new viewpoint does accord with the old framework.

Currents flow, and it is commonsense borne of classical mechanics that the current flowing towards the barrier from the left should be equal to the current that gets transmitted plus that which is reflected, or $j_{inc} = j_{ref} + j_{trans}$. How that intuition is preserved is edifying. Let the normalization of ψ_{inc} be unity, and choose $\psi_{trans} = t(k)e^{ik'x}$ and $\psi_{ref} = r(k)e^{-ikx}$. The values of k and k' are determined by the height (in energy) of the particle above the local "ground": if the height in Region 1 is $\hbar^2k^2/2m$ then in Region 2 it is $\hbar^2k'^2/2m + V_o$. That this is so is straightforward to show: from Schrödinger's equation in the absence of a potential,

$$-\frac{\hbar^2}{2m}\frac{\partial^2}{\partial x^2}\psi_k(x) = E_k\psi_k(x) \tag{9.62}$$

Substituting $E_k = \hbar^2k^2/2m$ and $\psi_k(x) = N_k e^{\pm ikx}$ satisfy that equation. Having a non-zero, but constant, potential V_o is tantamount to replacing $E_k \to E_k - V_o$ and $k^2 \to (k')^2 = k^2 - k_o^2$, where $k_o^2 \equiv 2mV_o/\hbar^2$. Accompanying this, one expects that the number density and the current density should be continuous at the boundary (taken to be at $x = 0$), but that requires the wave function $\psi_k(x)$ and its first derivative $\partial_x\psi_k(x)$ to be, likewise, continuous at $x = 0$. To make both requirements so, the normalization factors $r(k)$ and $t(k)$ have to satisfy

$$\psi_{inc}(0^-) + \psi_{ref}(0^-) = \psi_{trans}(0^+) \tag{9.63}$$

$$\partial_x\psi_{inc}(0^-) + \partial_x\psi_{ref}(0^-) = \partial_x\psi_{trans}(0^+) \tag{9.64}$$

where the left-hand side refers to Region 1, and the right-hand side to Region 2 of Figure 9.7, a form that looks delightfully like a matrix equation: in fact, putting in the forms for ψ_{inc}, ψ_{ref} and ψ_{trans}, and treating the coefficients as unknown

[14] Alfred North Whitehead, *Adventures of Ideas*, 1st Free Press Pbk. edn, The Free Press, 1967, p. 72.

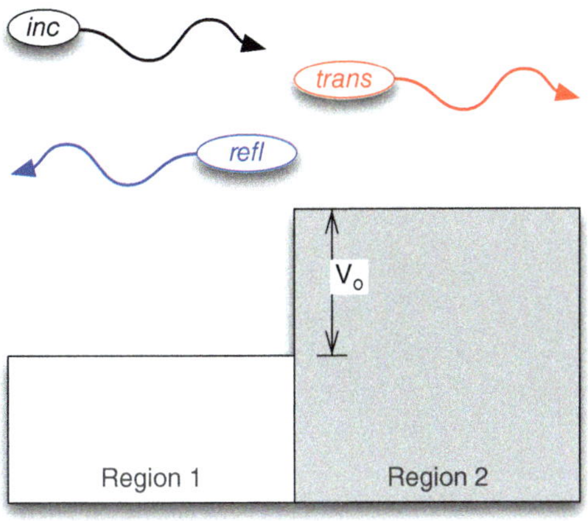

Figure 9.7 Wave function incident on a barrier. An incident wave (inc) of the form $\psi_{inc} = e^{ikx}$ hits a barrier of height V_o and separates into two waves, one reflected (ref) of the form $\psi_{ref} = r(k)e^{-ikx}$ and one transmitted (trans) of the form $\psi_{trans} = t(k)e^{ik'x}$. The relation of k' to k is $\hbar^2k^2/2m = \hbar^2k'^2/2m + V_o$.

components of the coefficients vector, but the $\exp(\pm ikx)$ as the components of the multiplying matrix (and evaluating them at $x = 0$), results in

$$\begin{pmatrix} 1 & 1 \\ ik & -ik \end{pmatrix}\begin{pmatrix} 1 \\ r(k) \end{pmatrix} = \begin{pmatrix} 1 & 1 \\ ik' & -ik' \end{pmatrix}\begin{pmatrix} t(k) \\ 0 \end{pmatrix} \tag{9.65}$$

The incident wave is normalized for convenience to unity (top component of the left-hand side coefficients vector) and no waves are incident from the right (bottom component of the right-hand side coefficients vector). In the matrices, the top row is for the wave functions, and the bottom row for their derivatives. 2×2 matrix equations are not much of a challenge: multiplying both sides by the inverse of the left-hand side matrix and expanding results in

$$\begin{pmatrix} 1 \\ r(k) \end{pmatrix} = \frac{t(k)}{2k}\begin{pmatrix} k + k' \\ k - k' \end{pmatrix} \tag{9.66}$$

Identifying the components of the column vector on the left with those on the right shows that

$$t(k) = \frac{2k}{k + k'} = \frac{2k}{k + \sqrt{k^2 - k_o^2}} \tag{9.67}$$

$$r(k) = \frac{k - k'}{k + k'} = \frac{k - \sqrt{k^2 - k_o^2}}{k + \sqrt{k^2 - k_o^2}} \tag{9.68}$$

Recall that Eq. (9.60) refers to a *probability current* and therefore, because the basis states $|k\rangle$ correspond to stationary states, the current density j_k is spatially uniform. The probability of being transmitted over the barrier is therefore a ratio of the transmitted probability to the incident probability current densities, that is, the transmission probability $D(k)$ is[15]

$$D(k) = \frac{j_{trans}(k)}{j_{inc}(k)} = \frac{4kk'}{(k + k')^2} \tag{9.69}$$

where the first relation is general and the second form is specific to step function potentials. The reflection probability is $1 - D(k)$. For $k \geq k_o$ (an ominous caveat that portends complications for energies below the barrier height), $D(k)$ increases monotonically from $D(k_o) = 0$ to $D(k \gg k_o) \to 1$, shown in Figure 9.8 for $k = k_o\sqrt{x}$, a rather odd representation, but one that will be useful below. About that $k \geq k_o$ caveat: what does happen when $k < k_o$? First, $k' \to ik'$ where now $k' \equiv \sqrt{|k^2 - k_o^2|}$. Second, $\psi_{trans} \to t(k)e^{-k'x}$ and is real, so that $j_{trans} = 0$ (all particles are reflected). Third, $r(k) \to (k - ik')/(k + ik')$ so that $|r(k)|^2 = 1$, yet another way of saying that all particles are reflected.

[15] The usage of D for the transmission probability follows the hallowed tradition established by Fowler and Nordheim [36], Murphy and Good [110], and Modinos [111], and the treatment here shall follow that precedent. Alas, the literature is not uniform, as, for example, T is used by Kemble [112] (who reserves D for the current density!) and Kane [113], among others. In short, *cave lector*, "let the reader beware" (more or less).

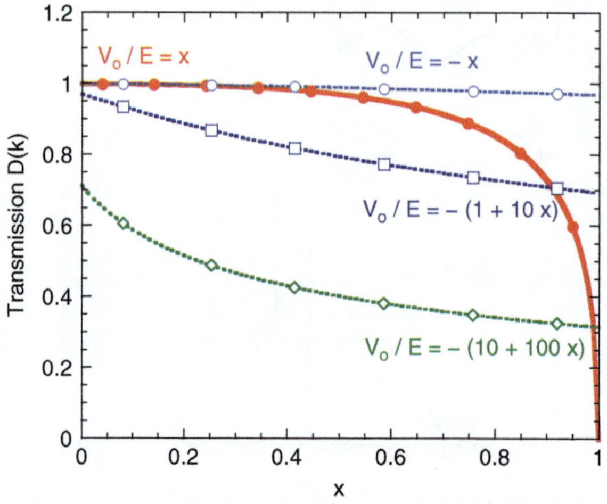

Figure 9.8 Transmission probability for a step function. The factor x is the ratio between the barrier height V_o and the kinetic energy E of the incident particle, or $x = V_o/E$. The solid circles correspond to the usual $D(k) = D(k_o/\sqrt{x})$ for a step barrier. The dashed lines correspond to transmission over a cliff for progressively higher cliffs: as the cliff height becomes infinite, the transmission probability approaches 0, meaning all particles are reflected, a paradoxical condition which has no classical analog.

The quantum result confirms the classical intuition that the incident current has to be the sum of the reflected and transmitted currents, and that step barriers reflect all particles whose energy is insufficient to surmount them. But right about here is where quantum mechanics begins getting peculiar in several ways. First, even when the kinetic energy is greater than the barrier height for a potential step, or $\hbar^2k^2/2m > V_o$, the transmission probability is less than unity: *some reflection always occurs*, and that defies a classical expectation. Second, instead of a step *barrier*, what about a step *cliff*, where V_o is negative instead of positive? Following through the treatment, one sees that this amounts to replacing $-k_o^2 \rightarrow +k_o^2$ in Eq. (9.68), and so some reflection always occurs even when going over a *cliff*. Astonishingly, the reflection becomes *greater* the *higher* the cliff is, resulting in complete reflection when the height of the cliff becomes asymptotically large. That has no classical analog. Such an outcome is coherent (follows from the mathematics) but to expectations borne of macroscopic physics it is unintelligible, and so there is little or no *clarity*. Lastly, in the case where the barrier is not a step, but rather a rectangular bump, the transmitted wave function need not be driven to zero within the barrier, such that the particle has a non-zero probability of appearing on the other side. The particle is said to have "tunneled through" the barrier, and that is where matters get very interesting.

9.2.3 Tunneling

> *I have dwelt longer upon the history of the Samians than I should otherwise have done, because they are responsible for three of the greatest building and engineering feats in the Greek world: the first is a tunnel nearly a mile long, eight feet wide and eight feet high, driven clean through the base of a hill nine hundred feet in height. The whole length of it carries a second cutting thirty feet deep and three broad, along which water from an abundant source is lead through pipes into the town. This was the work of a Megarian named Eupalinus, son of Naustrophus ...*
>
> – Herodotus[16]

Herodotus, the ancient Greek historian often referred to as the father of history, wrote a massive record of the rise of the Persian empire and its subsequent clash with the Greek city-states during the Greco–Persian wars (of which the Battle of Marathon mentioned in Section 7.1 was a part) culminating in the failed campaign of Xerxes in 480 BC to conquer the Greeks. But his *The Histories* was more than that: although it did chronicle the battles and intrigues, it also entertainingly included fantastic tales and a fairly detailed account of the ancient world, from customs and manners and beliefs to engineering accomplishments. Herodotus's identifying of the tunnel of Eupalinus as a marvel of the ancient world (a kilometer-long tunnel under a mountain to carry water to the capitol city) may have been instrumental to its discovery around 1884 on the island of Samos, birthplace of the mathematician Pythagoras. The tunnel was a

[16] Herodotus, *Herodotus: The Histories (Aubrey De Sélincourt)*. Middlesex: Penguin Books, 1954, p. 228.

masterpiece of the application of geometry and calculation to a difficult problem of excavating two holes on either side of a mountain such that they intersect in the middle. That it was executed at the behest of the tyrant Polycrates (to hide the water supply so as to prevent an invading army from denying that essential resource to the city during sieges) does not diminish the engineering accomplishment.

Tunneling calculations are difficult, and always have been. Quantum mechanics makes them complicated in a completely different way. In quantum mechanics, tunneling refers to electrons in a forbidden region, such that the probability of finding an electron where the potential energy associated with the barrier exceeds the energy of the electron is non-zero. Two exercises emphasize different aspects of the tunneling problem, and both make use of the formalism developed above: (*i*) the wave packet problem describing a single electron can be time evolved and its interaction with the step function barrier or (*ii*) the stationary problem describing the variation of electron density near the potential barrier.

9.2.3.1 Wave packet

The density associated with a single state $|k\rangle$ is constant everywhere, a consequence of Heisenberg's Uncertainty Principle. A localized electron therefore corresponds to a spread of momentum values, as in the commonly considered Gaussian wave packet, otherwise known as the minimum uncertainty wave packet, for which $\Delta x \Delta k = 1/2$. More generally, though, the wave function is given by

$$\psi(x,t) = \sum_j C_j e^{-iE_j t} \begin{cases} e^{ik_j x} + r(k_{-j}) e^{ik_{-j} x} & (x < 0) \\ t(k_j) e^{i\kappa_j x} & (x \geqslant 0) \end{cases} \tag{9.70}$$

where $\kappa_j^2 = k_j^2 - k_o^2$, $k_o = \sqrt{2mV_o}/\hbar$, and $k_{-j} = -k_j$. If $k_o > k_j$, then κ is imaginary. The negative index k_{-j} notation corresponds to states that move *away* from the barrier (to the $-\hat{x}$ direction) instead of *toward* it (to the $+\hat{x}$ direction) so that the summation need not be further divided into positive and negative momentum states. The coefficients $C_j \equiv C(k_j)$ are found from the orthonormality of the basis states $\psi_k(x)$ and so

$$C(k_j) \propto \int \psi_{k_j}(x)^\dagger \psi(x,0) dx \tag{9.71}$$

Choosing C_j to correspond to Gaussian wave packets is often treated (e.g., [44]), but their evaluation can be a challenge, especially when first encountered. To have a computationally easier time of it, consider the simpler circumstance where at $t = 0$, the wave function is given by $\psi(x,0) = \Theta(x+l)\Theta(-x)$, for which $\Theta(x - x_o) = 1$ if its argument is greater than 0, but 0 otherwise (the Heaviside step function). The C_j for such a circumstance can be evaluated easily when using plane wave basis states, but the wave packet that results is no longer a minimum uncertainty wave packet as it would be in the Gaussian case. That is a fairly minor price to pay for the ease of calculation which results.

EXAMPLE: Find the C_j for $\psi(x,0) = \Theta(x+l)\Theta(-x)$.
SOLUTION: From Eq. (9.71),

$$\begin{aligned} C_j &= C_0 \int_{-\infty}^{\infty} e^{ik_j x} \Theta(l - x) \Theta(x) dx \\ &= C_0 \int_{-l}^{0} e^{ik_j x} dx \to \frac{1}{ik_j} \left(1 - e^{-ik_j l} \right) \end{aligned} \tag{9.72}$$

where C_0 is a global normalization that is taken to be unity.

With the choice of Eq. (9.72), simulating the time evolution of a wave packet is gentle. Starting the clock a bit earlier at time $t = -t_o$ allows for the visualization of the wave packet as it comes in to collide with a barrier. It is convenient to work in a system of units such that $k_j = (j/N)k_{min}$ with N being the number of states in the packet, with k_{min} being conveniently chosen (e.g., $\pi/4$), and with lengths and energies such that $E_j = k_j^2$. Such a choice of units is friendly to calculation. For the choices of $t_o = 20$ and $N = 48$, then focusing on a region of space $-40 \leq x \leq 20$ allows the wave packet to be seen colliding with the barrier (a larger visualization region would show more than one hump in the wave packet, the location and spacing of which are a consequence of finite N and the spacing between the various momentum states chosen). A small N has the additional advantage of rendering the packet, such as it is, to look more like a wave (i.e., a smoothly varying crest) than a rectangular pulse. Simulations of that collision using the algorithm given in Section A3.5 are shown in Figure 9.9. The first square (upper left) waterfall plot in that figure represents the "no barrier" configuration: a top-down view of it would nicely show the spreading of the wave packet that often occupies discussions of wave packet evolution, and so it is shown as the left-hand image in Figure 9.10. The right-hand image shows the other peculiar case of the impact of the wave packet with the cliff potential characterized by $V_o = -1$ (lower right block of Figure 9.9), in which only a portion of the packet is transmitted.

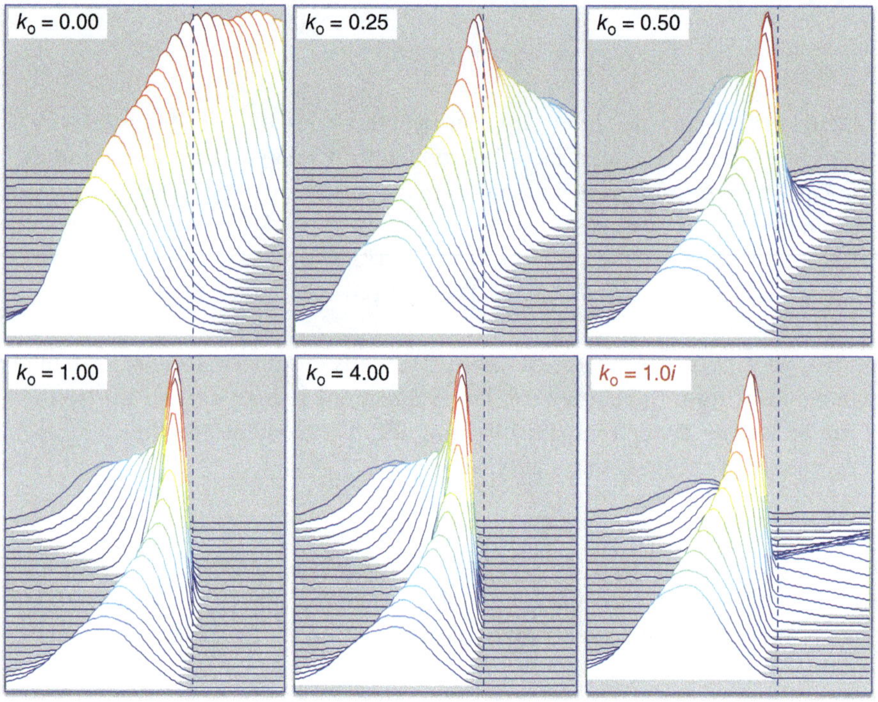

Figure 9.9 A "waterfall" plot of the time evolution of a wave packet. Each line represents the wave function at one time, with time increasing in the upward vertical direction. Units are chosen such that: $t = -20$ to $t = +20$ in 24 steps (lowest line to highest). The left boundary is at $x = -40.0$ and the right boundary is at $x = +20.0$. The potential barrier is of height $V_o = k_o^2$, and $E = k_j^2$ where $k_j = j\pi/4N$ and $N = 48$. The step potential barrier is to the right, beginning at the dashed line. The lower right-hand plot is for a negative potential $V_o = -1$ such that $k_o = i$.

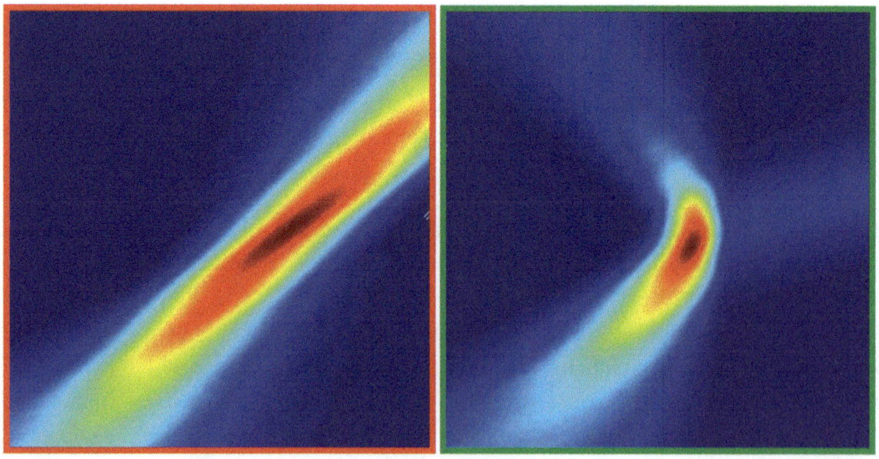

Figure 9.10 Top-down views of wave packet spreading. Left: in the absence of a barrier. Right: the wave packet encounters a cliff potential. The horizontal axis is position, the vertical axis is time. Observe in the former the elongation of the ellipse after $t = 0$ (the high point). Observe in the latter the reflection of much of the wave packet and the transmission of a portion, a behavior without a classical analog.

9.2.3.2 Density variation

Consider now not one electron with different probabilities for various energies in the packet, but rather a collection of electrons such that the probability of an energy state being filled is governed by the Fermi–Dirac distribution, or

$$|\psi(t)\rangle = e^{-i\hat{H}t/\hbar}|\psi(0)\rangle = \sum_j e^{iE_j t/\hbar} C_j |\psi_j\rangle \tag{9.73}$$

where $|C_j|^2$ is the probability that the state characterized by E_j is occupied and

$$\rho = \langle\psi(t)|\psi(t)\rangle = \langle\psi(0)|\psi(0)\rangle = \sum_j |C_j(E_j)\psi_j(x)|^2 \tag{9.74}$$

where $\psi_j(x) = \langle x|\psi(k_j)\rangle$ and is as previously defined in Eq. (9.70). The C_j, though related to $f_{FD}(E)$, are not the same as for the one-dimensional problem. Because the goal is to find a density variation that has some resemblance to the one-dimensional variation near the surface of a metal, take a look at how Eq. (3.26) is written if the distribution function corresponds to a Fermi–Dirac distribution. It becomes

$$J_q = \frac{q}{2\pi}\int \left(\frac{\hbar k_x}{m}\right) dk_x \left\{ \frac{1}{(2\pi)^2}\int f_{FD}(E_k) dk_y dk_z \right\} \tag{9.75}$$

The quantity in the brackets is the Fermi-Dirac distribution with the transverse velocity components integrated over: in the future, it will be known more generally as the supply function, but for now treat it simply as an interesting choice. For simplicity, consider the zero temperature limit, for then the Fermi–Dirac distribution function becomes a Heaviside step function itself, and the quantity in curly brackets in Eq. (9.75) becomes

$$\frac{1}{2\pi}\int_0^{\sqrt{k_F^2-k_x^2}} k_\perp dk_\perp = \frac{1}{4\pi}\left(k_F^2 - k_x^2\right) \tag{9.76}$$

An integration of Eq. (9.76) over dk_x from $-k_F$ to $+k_F$ (including the factor of $1/(2\pi)$, plus the spin of the electron $g = 2$) will, like the density integration of Eq. (6.8) before it, result in the number density of electrons at zero temperature of $\rho = k_F^3/3\pi^2$, as in Eq. (6.24).

EXAMPLE: Find the C_j that mimic Eq. (9.76).

SOLUTION: In numerically convenient units leading to Eq. (9.72), the maximum momentum k_F occurs for $j = N + 1$. Therefore, let $k_x = (jk_F)/(N + 1)$, which makes $C_{N+1} = 0$ (the supply function vanishes at $k_x = k_F$). Inserting $k_x \to (jk_F)/(N + 1)$ into Eq. (9.76) and ignoring overall factors such as $1/4\pi$ (the ratio of the density to its boundary value will eliminate common factors), it is found that

$$C_j^2 = \frac{(N+1)^2 - j^2}{(N+1)^2} \tag{9.77}$$

With Eq. (9.77), the density is given by

$$\rho(x) \propto \sum_{j=1}^{N} C_j^2 \begin{cases} \left|\left(e^{ik_j x} + r(k_{-j})e^{ik_{-j}x}\right)\right|^2 & (x < 0) \\ \left|t(k_j)e^{i\kappa_j x}\right|^2 & (x \geqslant 0) \end{cases} \tag{9.78}$$

the numerical evaluation of which, using the algorithms of Section A3.5, gives the densities normalized by their values on the left boundary (at $x = -3$), as shown in Figure 9.11. When the barrier is lower than the maximum momentum present, or $k_o < k_F$, then the density in the barrier region is non-zero ($\rho(x > 0)$) and over the barrier emission is seen to occur. However, when $k_o > k_F$, it is clear that $\rho(x \gg 1) \to 0$, but that near $x \approx 0$ the density under the barrier exponentially tapers off but is non-zero for regions close to $x = 0$. For an infinitely high barrier, the reflection probability would be unity, and the electron density would vanish at the barrier. But it is also seen that if, instead of a step function, the barrier had finite thickness, then for thin enough barriers, the density of electrons at the right-hand boundary need not vanish: if the density does not vanish, then electrons are present on the right-hand side of the barrier, and tunneling is said to have occurred.

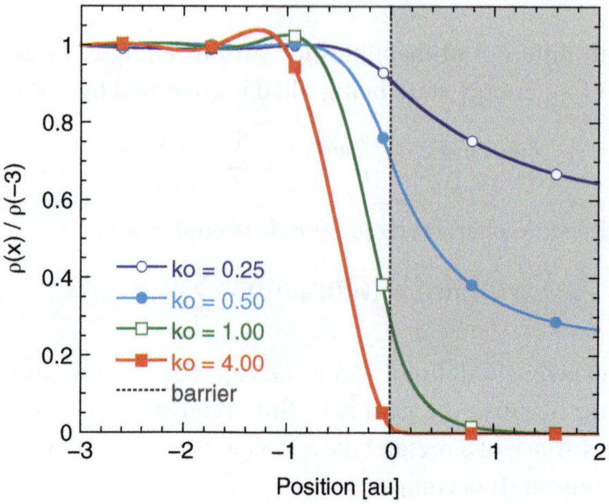

Figure 9.11 Electron density near a step function. Numerical evaluation of Eq. (9.78) for $N = 48$ using the algorithm given in Section A3.5. For $k_o \geq 1.0$, all of the wave vectors in $\rho(x)$ are smaller than k_o. The barrier is a step function starting at $x = 0$ (dashed line) and continuing to the right (gray area). Non-zero density is seen even in those cases for $x > 0$, and there is evidence of electrons quantum mechanically penetrating into forbidden region, that is, a region where they are not classically allowed.

9.3 The harmonic oscillator

What has happened will happen again, and what has been done will be done again, and there is nothing new under the sun.

– Ecclesiastes 1.9[17]

The harmonic oscillator problem is the remaining quintessential problem of emission physics, for which the quantum mechanical treatment demonstrates techniques and subtitles useful to understand. Although three-dimensional formulations can be undertaken, a one-dimensional approach is sufficient. Two approaches to solving the harmonic oscillator problem are considered: first, using a wave function approach, and second, by a less conventional but useful quantum distribution function approach.

9.3.1 Wave function approach

New things are made familiar, and familiar things are made new.

– Samuel Johnson[18]

The harmonic oscillator is characterized by the cyclic transfer between kinetic energy $KE = p^2/2m$ to potential energy $PE = Kx^2/2$ (where K is a constant). The position $x(t)$ and the momentum $p(t) = m(dx/dt)$ are therefore [90]

$$
\begin{aligned}
x(t) &= x_o \cos(\omega t + \phi) \\
p(t) &= -x_o m\omega \sin(\omega t + \phi)
\end{aligned}
\tag{9.79}
$$

where x_o is the maximum excursion and ϕ is a phase, so that the total energy $E = KE + PE = m x_o^2 \omega^2/2$ is a constant of the motion, and Schrödinger's equation is then, by virtue of the canonical procedure for transitioning between classical and quantum mechanics,

$$
\left(-\frac{\hbar^2}{2m}\frac{d^2}{dx^2} + \frac{\hbar^2 k_o^2}{2m}(k_o x)^2 \right) \psi(x) = E\psi(x) \tag{9.80}
$$

where a new term, k_o, has been introduced and defined by $K/2 = \hbar^2 k_o^4/2m$, so as to make the energy-like terms obvious and $k_o x$ dimensionless. As with the ladder operators for the hydrogen atom, Schrödinger's equation can be solved without having to endure differential equations by making use of the commutator relations between $\hat{k} = \hat{p}/\hbar$ and $\hat{x}$ given in

[17] *The New English Bible With the Apocrypha*. New York: Oxford University Press, 1971, p. 789.

[18] Samuel Johnson, *Lives of the English Poets: Prior, Congreve, Blackmore, Pope*. Cassell & Company, London, 1891. From the section on Pope.

Eq. (8.52). The first step is to write matters in a more agreeable way as

$$\left(\hat{k}^2 + k_o^4\hat{x}^2\right)|\psi\rangle = k_n^2|\psi\rangle \tag{9.81}$$

where $E = \hbar^2 k_n^2/2m$. The subscript n appended to k_n anticipates that the energy $E = \hbar^2 k_n^2/2m$ will be quantized and that the corresponding energy levels E_n will be labeled by n. The harmonic oscillator Hamiltonian offers a very direct and simple method to show the correspondence to the more familiar classical relations using Eq. (8.72): the velocity and force are (where only the commutators that survive are shown)

$$\frac{d}{dt}\hat{x} = \frac{i}{\hbar}\left[\frac{\hbar^2\hat{k}^2}{2m}, \hat{x}\right] = \frac{i\hbar}{2m}\left[\hat{k}^2, \hat{x}\right] = \frac{\hbar\hat{k}}{m} = \hat{v} \tag{9.82}$$

$$\frac{d}{dt}\hat{p} = \frac{i}{\hbar}\left[\frac{\hbar^2 k_o^4}{2m}\hat{x}^2, \hbar\hat{k}\right] = \frac{i\hbar^2 k_o^4}{2m}\left[\hat{x}^2, \hat{k}\right] = -K\hat{x} = \hat{F} \tag{9.83}$$

which, apart from the hats, have the same forms as their classical counterparts.

The left-hand side of Eq. (9.81) is suggestive: it almost begs to be represented as $(-i\hat{k} + k_o^2\hat{x})(i\hat{k} + k_o^2\hat{x})$ as it would if $\hat{k}$ and $\hat{x}$ commuted, but because they do not,

$$\begin{aligned}\left(-i\hat{k} + k_o^2\hat{x}\right)\left(i\hat{k} + k_o^2\hat{x}\right) &= \hat{k}^2 + k_o^4\hat{x}^2 + ik_o^2\left[\hat{x}, \hat{k}\right] \\ &= \hat{k}^2 + k_o^4\hat{x}^2 - k_o^2\end{aligned} \tag{9.84}$$

Therefore, analogous to the ladder operators of Section 9.1.2, introduce what will be called the creation $\hat{a}^\dagger$ and annihilation $\hat{a}$ operators

$$\hat{a}^\dagger = k_o^{-1}\left(-i\hat{k} + k_o^2\hat{x}\right) \tag{9.85}$$

$$\hat{a} = k_o^{-1}\left(+i\hat{k} + k_o^2\hat{x}\right) \tag{9.86}$$

EXAMPLE: Show that $[\hat{a}, \hat{a}^\dagger] = 2$.

SOLUTION:

$$\begin{aligned}[\hat{a}, \hat{a}^\dagger] &= k_o^{-2}\left[i\hat{k} + k_o^2\hat{x}, -i\hat{k} + k_o^2\hat{x}\right] \\ &= k_o^{-2}\left(ik_o^2[\hat{k}, \hat{x}] - ik_o^2[\hat{x}, \hat{k}]\right) \\ &= k_o^{-2}\left(2ik_o^2(-i)\right) = 2\end{aligned}$$

From the last example and Eq. (9.84), it then follows that

$$\left(\hat{a}^\dagger\hat{a} + 1\right)|\psi\rangle = \left(\frac{k_n}{k_o}\right)^2|\psi\rangle \tag{9.87}$$

As for the apocalyptic terminology, and as was done for the angular momentum ladder operators in Section 9.1.2, consider what $\hat{H} = (\hbar^2 k_o^2/2m)(\hat{a}^\dagger\hat{a} + 1)$ acting on $\hat{a}|\psi\rangle$ and $\hat{a}^\dagger|\psi\rangle$ give, and do so for the annihilation operator first, that is, find $\hat{H}\hat{a}|\psi\rangle$. Apart from a coefficient of $\hbar^2 k_o^2/2m$, it involves finding

$$\begin{aligned}\left(\hat{a}^\dagger\hat{a} + 1\right)\hat{a}|\psi\rangle &= \left(\hat{a}\hat{a}^\dagger - [\hat{a}, \hat{a}^\dagger] + 1\right)\hat{a}|\psi\rangle \\ &= \hat{a}\left(\hat{a}^\dagger\hat{a} - 1\right)|\psi\rangle = \left(\frac{k_n^2}{k_o^2} - 2\right)\hat{a}|\psi\rangle\end{aligned} \tag{9.88}$$

In other words, the state $\hat{a}|\psi\rangle$ is an eigenstate of $\hat{H}$ but with an energy $\hbar^2 k_o^2/m$ lower than the state $|\varphi\rangle \equiv \hat{a}|\psi\rangle$ (one finds, if one cares to prove it, that the action of $\hat{a}^\dagger$ increases the energy by the same amount). Observe that this amount is twice the characteristic energy $\hbar^2 k_o^2/2m$. But just as $\langle\psi|\psi\rangle \geq 0$, so to is $\langle\varphi|\varphi\rangle = \left\langle\psi\left|\hat{a}^\dagger\hat{a}\right|\psi\right\rangle = (k_n/k_o)^2 - 1 \geq 0$ and equals 0 (for the ground state) when $k_0 = k_o$ (call that case $n = 0$). Succinctly, the ground state energy is $\hbar^2 k_o^2/2m \equiv \hbar\omega/2$, and the energy E_n of the higher states, as a consequence of the arguments leading to Eq. (9.88), goes as

$$E_n \equiv (2n+1)\frac{\hbar\omega}{2} \tag{9.89}$$

and correspond to an application of $(\hat{a}^\dagger)^n$ on the ground state. Instead of referring to the eigenstates as $|\psi_n\rangle$, it is simpler to call them $|n\rangle$. And so,

$$\left\langle n' \left|\hat{a}^\dagger \hat{a}\right| n \right\rangle = 2n\delta_{nn'} \tag{9.90}$$

Normalization is imposed by demanding $\langle n|n\rangle = N_n^2 \left\langle 0\left|\hat{a}^n(\hat{a}^\dagger)^n\right|0\right\rangle = 1$. In principle, this could be accomplished by patiently commuting all the annihilation operators to the right, and the creation operators to the left. Or, one can cheat and notice

$$\langle n|\, n\rangle = \left(N_n/N_{n-1}\right)^2 \langle n-1|\, \hat{a}\hat{a}^\dagger |n-1\rangle \tag{9.91}$$

$$= \left(N_n/N_{n-1}\right)^2 \langle n-1|\, \hat{a}^\dagger \hat{a} + 2|n-1\rangle \tag{9.92}$$

$$= \left(N_n/N_{n-1}\right)^2 \langle n-1|\, 2\,(n-1) + 2|n-1\rangle \tag{9.93}$$

$$= \left(N_n/N_{n-1}\right)^2 2n = 1 \tag{9.94}$$

from which it is concluded by mathematical induction that

$$\frac{N_n}{N_0} = \frac{1}{\sqrt{2^n n!}} \tag{9.95}$$

Determination of the ground state is accomplished by observing that

$$\langle x\,|\hat{a}|\,0\rangle = \left(\partial_x + k_o^2 x\right)\psi_0(x) = 0 \tag{9.96}$$

EXAMPLE: From Eq. (9.96), find the normalized ground state wave function for the harmonic oscillator.
SOLUTION: The equation can be rewritten as $d\psi/\psi = k_o^2 x dx$, which has the solution $\psi_0(x) \propto \exp\left(-(k_o x)^2/2\right)$. Demanding that its integration over all x gives unity shows that

$$\psi_0(x) \equiv N_0 \exp\left(-\frac{1}{2}k_o^2 x^2\right) = \left(\frac{k_o}{\sqrt{\pi}}\right)^{1/2} \exp\left(-\frac{1}{2}k_o^2 x^2\right) \tag{9.97}$$

Use of the creation operators and the normalization relations allows for finding the higher energy states. The exponential factor of $\psi_0(x)$ persists in $\psi_n(x)$, so try a solution of the form

$$\langle x|n\rangle = \psi_n(x) = N_n H_n(k_o x)\exp\left(-k_o^2 x^2/2\right) \tag{9.98}$$

where $H_n(k_o x)$ is a function determined by the action of the creation and annihilation operators. From the normalization of the wave function, it follows

$$\int_{-\infty}^{\infty} e^{-(k_o x)^2} H_n(k_o x) H_{n'}(k_o x) k_o dx = \sqrt{\pi}2^n n!\delta_{nn'} \tag{9.99}$$

Changing the variables to $s = k_o x$, the actions of $a^\dagger$ and a give recursion relations

$$\hat{a}\psi_n(x) = \sqrt{2n}\psi_{n-1}(x) \rightarrow \frac{d}{ds}H_n(s) = 2nH_{n-1}(s) \tag{9.100}$$

$$\hat{a}^\dagger \psi_n(x) = \sqrt{2(n+1)}\psi_{n+1}(s) \rightarrow \frac{d}{ds}H_n(s) = 2sH_n(s) - H_{n+1}(s) \tag{9.101}$$

the combination of which shows

$$H_{n+1}(s) = 2sH_n(s) - 2nH_{n-1}(s) \tag{9.102}$$

These are the relations of the *Hermite polynomials*, one of a number of well-characterized orthogonal polynomials [60, 114]. Given the relations between the $H_n(s)$ just deduced, the question is then whether an efficient method of generating them can be found. The recursion relations of Eqs (9.100) and (9.101), and awareness of how the creation operators give rise to the higher n states, results in

$$H_n(s) = e^{s^2/2}\left(-\partial_s + s\right)^n e^{-s^2/2} \tag{9.103}$$

but a simpler relation is possible. For $y = cs^p$ for constant c and p,

$$\partial_s y + cpx^{p-1} = 2cpx^{p-1} \tag{9.104}$$

and so

$$e^y\left(-\partial_s + cpx^{p-1}\right)e^{-y} = e^{py}\left(-\partial_s\right)e^{-py} \tag{9.105}$$

With $p \to 2$ and $c \to 1/2$, then Rodrigue's formula for the Hermite polynomials results and is given by

$$H_n(s) = e^{s^2}\left(-\partial_s\right)^n e^{-s^2} \tag{9.106}$$

The first few are

$$\begin{aligned} H_0(s) &= 1 \\ H_1(s) &= 2s \\ H_2(s) &= 4s^2 - 2 \\ H_3(s) &= 8s^3 - 12s \\ H_4(s) &= 16s^4 - 48s^2 + 12 \end{aligned} \tag{9.107}$$

For the harmonic oscillator, the potential energy is symmetric in x, that is, $V(x) = Kx^2 = V(-x)$. The wave functions though come in two types: those which do not change sign as $x \to -x$ and are called the even parity solutions $\psi_{2n}(-x) = +\psi_{2n}(x)$, and those that do and are called the odd parity solutions $\psi_{2n+1}(-x) = -\psi_{2n+1}(x)$, as can be anticipated from the behavior of $(-\partial_s)^{2n}$ and $(-\partial_s)^{2n+1}$ in Eq. (9.106). Parity is one of those interesting quantum mechanical properties that is physically measurable, but which has no large-scale classical analog [41]. Represented as $\hat{P}$, it behaves as $\hat{P}|x\rangle = |-x\rangle$. Therefore, $\hat{P}^2 = \hat{I}$ is another identity operator, making identity (Section 8.2.1) more nuanced. Additionally, for potentials that depend on the relative separation of the interacting particles (as in the Coulomb potential), then parity commutes with the Hamiltonian, $[\hat{H}, \hat{P}] = 0$, so that parity is a constant of the motion, meaning that states prepared with a definite parity preserve that parity as they evolve in time.

9.3.2 Quantum phase space approach

> *In classical statistical mechanics the relative probability ...is given for statistical equilibrium by the Gibbs–Boltzmann formula ...In quantum theory there does not exist any similar simple expression for the probability, because one cannot ask for the simultaneous probability of the coordinates and the momenta.*
>
> – E. Wigner [115]

> *...the use of negative numbers as an abstract calculation permits us freedom to do our mathematical calculations in any order, simplifying the analysis enormously and permitting us to disregard inessential details. The idea of negative numbers is an exceedingly fruitful mathematical invention. Today a person who balks at making a calculation in this way is considered backward or ignorant, or to have some kind of mental block ... we have a similar strong block against negative probabilities.*
>
> – Richard P. Feynman [116], p. 236

The quantum distribution function approach updates classical phase space ideas to the quantum realm. Moments of a distribution are again related to the density, current density, and energy spread encountered in Section 7.4, but not without some effort. The Wigner function is a well-known example [117], and its consideration is immeasurably improved when it is applied to a well-understood problem like the harmonic oscillator, even though it can be applied to electron emission. For the harmonic oscillator, the quadratic nature of the harmonic potential energy $V(x) = Kx^2/2$ makes the the time evolution equation of the Wigner function particularly straightforward.

The usage of distributions to calculate number density and current density are familiar from Section 7.3, but their relation to the operators $\hat{\rho}_H$ and $\hat{j}_H$ of Section 9.2.1 requires sandwiching those operators between suitably chosen bras and kets to insure that number and current density are given by[19]

$$\rho(x,t) = \frac{1}{2\pi}\int_{-\infty}^{\infty} f(x,k,t)\,dk \tag{9.108}$$

$$J(x,t) = \frac{q}{2\pi}\int_{-\infty}^{\infty} \frac{\hbar k}{m} f(x,k,t)\,dk \tag{9.109}$$

The Wigner distribution $f(x,k,t)$ contains all the quantum mechanical effects yet satisfies relations much like Boltzmann's equation of Section 7.4. This is not as simple as it may appear. Without loosing too much, the discussion can be confined to one dimension in space and similarly for momentum. Phase space points move around with time, but the number of particles, or points, in a small region has to be conserved (a concept that will reappear in Section 33.3.1, treating Liouville's theorem), or

$$f(x,k,t)dxdk = f(x',k',t')dx'dk' \tag{9.110}$$

For infinitesimal times dt, the phase space points move as $x' = x + \hbar k dt/m$ and $k' = k + Fdt/\hbar$. The volume of phase space therefore changes: treating the probability as an incompressible fluid [65] (which entails that the number of particles in the phase space volume element does not change as it evolves), then

$$dx'dk' = \begin{vmatrix} \partial_x x' & \partial_x k' \\ \partial_k x' & \partial_k k' \end{vmatrix} dxdk = \begin{vmatrix} 1 & 0 \\ \hbar dt & 1 \end{vmatrix} dx\,dk = dx\,dk \tag{9.111}$$

and so consideration can be focused on what $f(x,k,t)$ is doing in Eq. (9.110). Before doing so, though, a small aside is instructive: observe that the linear transformations under discussion can be represented as [118]

$$\vec{\xi}' = \begin{pmatrix} x' \\ k' \\ 1 \end{pmatrix} = \begin{pmatrix} 1 & 0 & a \\ 0 & 1 & b \\ 0 & 0 & 1 \end{pmatrix}\begin{pmatrix} x \\ k \\ 1 \end{pmatrix} = \mathbf{T}(a,b)\cdot\vec{\xi} \tag{9.112}$$

where $\mathbf{T}$ is a translation operator, $a = (\hbar k dt/m)$ and $b = (Fdt/\hbar)$. Translation operators commute with each other, or $[\mathbf{T}(a',b'),\mathbf{T}(a,b)] = 0$, but they do not commute with the rotations $\mathbf{R}(\theta)$ (analogous to the lower right 3×3 matrix of Eq. (8.33) or Eq. (9.4)), that is, $[\mathbf{T}(a,b),\mathbf{R}(\theta)] \neq 0$. Non-commutation of rotations and translations is easily appreciated from experience: turning right then going forward a step lands one in a different place than the reverse order of going forward one step then turning right. The translation and rotation transformations constitute the Euclidean group and are known as canonical transformations [97, 118].

Substituting Eqs (9.108) and (9.109) into the continuity equation (Eq. (7.19)) for a particle in a linear field F results in

$$\partial_t f(x,k,t) = -\frac{\hbar k}{m}\partial_x f(x,k,t) + \frac{F}{\hbar}\partial_k f(x,k,t) \tag{9.113}$$

for which the solutions are

$$f(x,k,t+dt) = f\left(x - \frac{\hbar k}{m}dt, k - \frac{F}{\hbar}dt, t\right) \tag{9.114}$$

Compared to Eq. (7.16), $\dot{x} = \hbar k/m$ and $\dot{k} = F/\hbar$. If F is constant, then for longer times [97]

$$f(x,k,t) = f\left(x - \frac{\hbar k}{m}t + \frac{F}{2m}t^2, k - \frac{F}{\hbar}t, 0\right) \tag{9.115}$$

For a steady-state distribution ($df/dt = 0$), the paths or trajectories of physical particles must therefore follow the contour lines of a surface plot of the distribution function $f(x,k)$ in phase space. The Wigner function itself is therefore constant along paths of constant energy. That will be useful.

So far, nothing in the discussion has departed from the classical outlook. The difficulties start when the form of $f(x,k)$ in Eq. (9.108) is sought. Naively, it might be thought that $f(E)|\psi_k(x)|^2$, as suggested by Eq. (9.53), is the form, but then how Eq. (9.109) would follow from Eq. (9.56) becomes a problem. Something a good deal more involved is required.

[19] Observe that the current density J explicitly has a q in the coefficient, so that it is a *charge* current density with units of amp/cm^2.

Wigner proposed the simplest of the many possible candidates, and it is (leaving the H subscript off of the operators for convenience) [115, 117]

$$f(x,k,t) = 2\int_{-\infty}^{\infty} e^{2iky}\langle x+y|\,\hat{\rho}(t)|x-y\rangle dy \tag{9.116}$$

This is pretty good: the density $\rho(x,t)$ falls right out when an integration over dk is performed, as the integral over e^{2iky} gives a $\delta(y)$ function (see the $n=0$ form of Eq. (9.123) below). After the dy integration, the warm result of Eq. (9.108) is regained with $f(E)|\langle x|\psi(t)\rangle|^2$ as the integrand, in keeping with Eq. (9.53). The time evolution of the Wigner function, though, is a good deal more challenging. Given Eqs (9.54) and (9.56),

$$\begin{aligned}\frac{\partial}{\partial t}f(x,k,t) = -2\int_{-\infty}^{\infty} e^{2iky}dy\left\{\langle x+y|\,\frac{\hbar}{2m}\partial_{\hat{x}}\left\{\hat{\rho}(t),\hat{k}\right\}|x-y\rangle\right.\\ \left.+\langle x+y|\,\frac{i}{\hbar}\left[V(\hat{x}),\hat{\rho}(t)\right]|x-y\rangle\right\}\end{aligned} \tag{9.117}$$

where the curly brackets denote the anti-commuator $\{A,B\}=AB+BA$, compared to the commutator $[A,B]=AB-BA$. Take each expression in turn. Recalling Eq. (8.50), in general, a function $G(x)$ that can be Taylor expanded is therefore such that $G(\hat{x})|x\pm y\rangle = G(x\pm y)|x\pm y\rangle$. That indicates

$$\begin{aligned}\langle x+y|\,\partial_{\hat{x}}\left\{\hat{\rho}(t),\hat{k}\right\}|x-y\rangle &= k\left(\partial_{x+y}+\partial_{x-y}\right)\langle x+y|\,\hat{\rho}(t)|x-y\rangle\\ &= 2k\partial_x\langle x+y|\,\hat{\rho}(t)|x-y\rangle\end{aligned} \tag{9.118}$$

and likewise,

$$\langle x+y|\left[V(\hat{x}),\hat{\rho}(t)\right]|x-y\rangle = (V(x+y)-V(x-y))\langle x+y|\,\hat{\rho}(t)|x-y\rangle \tag{9.119}$$

Introducing the shorthand

$$V\left(x,k-k'\right) = -\frac{i}{\pi\hbar}\int_{-\infty}^{\infty} e^{2i(k-k')y}\{V(x+y)-V(x-y)\}\,dy \tag{9.120}$$

results in the time evolution equation for the Wigner function

$$\frac{\partial}{\partial t}f(x,k,t) = -\frac{\hbar k}{m}\frac{\partial}{\partial x}f(x,k,t)+\int_{-\infty}^{\infty} V\left(x,k-k'\right)f\left(x,k',t\right)dk' \tag{9.121}$$

The intrusion of quantum effects is now evident in the *non-local* nature of the function $V(x,k-k')$: it depends on contributions at a distance y from x.

There is much that makes a study of the Wigner function fascinating in its own right. A few examples conveys that. First, a trajectory interpretation can be had by constructing a Wigner quantum potential under which the electrons move [119, 120]. Wigner trajectories have much in common with an alternate formulation of quantum mechanics leading to Bohm trajectories [121–124] through the introduction of a quantum potential, which is introduced in Eq. (11.9). The Wigner trajectory concept sheds interesting light on the nature of tunneling in general and resonant tunneling (see Section 19.2.4) in particular [125–129]. Second, the Wigner function provides a satisfying take on the perennial problem of the spreading of the wave packet that dominates introductory treatments of quantum mechanics: the spread increases with time,[20] but to see it, the usual treatment based on finding the time-dependent wave function $\psi(x,t)$ is rather tortuous. The elegance of the Wigner function analysis for the same problem [44, 85, 97, 118] is by comparison aesthetic. In particular, and as observed by Kim and Noz [118], whereas the uncertainty product $\Delta x\Delta k$ for a Gaussian wave packet increases with time, the uncertainty associated with the Wigner representation can be formulated in a time invariant manner so that the uncertainty is represented by a sheering ellipsoid whose area remains constant over time and dictated by $\hbar$ yet is enclosed in a box of dimensions $(\Delta x)(\hbar\Delta k)$ that keeps increasing. Such concepts are tremendously interesting but, alas, not to the point here.

Instead, the concern here is with how quantum effects modify the classical description of the phase space. The answer is illuminated by Taylor expanding the terms $V(x+y)-V(x-y)$ to obtain

$$V(x+y)-V(x-y) = \sum_{n=0}^{\infty}\frac{2y^{2n+1}}{(2n+1)!}\left(\frac{\partial}{\partial x}\right)^{2n}V(x) \tag{9.122}$$

[20] See Figure 9.9, where for the high barrier case ($k_o=4$), as the wave packet moves away from the barrier, a noticeable widening of it occurs.

along with

$$\int_{-\infty}^{\infty} y^n e^{2iky}\, dy = \pi\left(-\frac{i}{2}\frac{\partial}{\partial k}\right)^n \delta(k) \tag{9.123}$$

to obtain

$$\frac{\partial}{\partial t} f(x,k,t) = -\frac{\hbar k}{m}\partial_x f + \frac{1}{\hbar}\left(\partial_x V\right)\partial_k f + \frac{1}{\hbar}\sum_{n=1}^{\infty}\frac{(-1)^n}{2^{2n}(2n+1)!}\left(\partial_x^{2n+1}V\right)\left(\partial_k^{2n+1}f\right) \tag{9.124}$$

For treating electron emission barriers, $V(x)$ is such that all of the higher order gradients must be retained [130, 131], making a simulation based on Eq. (9.121) the preferred approach. But the harmonic oscillator is different: for it, *all the higher order derivatives of the potential vanish.* This has consequences. For the time-independent harmonic oscillator distribution function, Eq. (9.124) becomes $\hat{O}f(x,k) = 0$ where

$$\hat{O} \equiv -\frac{\hbar k}{m}\frac{\partial}{\partial x} + \frac{1}{\hbar}\left(\partial_x V\right)\frac{\partial}{\partial k} \tag{9.125}$$

EXAMPLE: Show that for a smooth function $\varphi(E)$ that depends only on the total energy $E = \hbar^2k^2/2m + V(x)$, then $\hat{O}\varphi(E) = 0$ where $\hat{O}$ is defined by Eq. (9.125).
SOLUTION: Expand $\varphi(E)$ as a Taylor series in E of the form

$$\varphi(E) = \sum_{j=0}^{\infty} C_j E^j$$

With $\hat{O}E^j = jE^{j-1}\hat{O}E$, then the vanishing of $\hat{O}E$ is enough to achieve the demonstration. Explicitly,

$$\begin{aligned}\hat{O}E &= \left(-\frac{\hbar k}{m}\frac{\partial}{\partial x} + \frac{1}{\hbar}\left(\partial_x V\right)\frac{\partial}{\partial k}\right)\left(\frac{\hbar^2k^2}{2m} + V(x)\right) \\ &= -\frac{\hbar k}{m}\partial_x V + \frac{1}{\hbar}\left(\partial_x V\right)\left(\frac{\hbar^2 k}{m}\right) = 0\end{aligned}$$

as required.

EXAMPLE: Find the $n = 0$ Wigner function $f_0(x,k)$. Show that it is $f_0(\epsilon)$ along a classical trajectory defined by Eq. (9.79). Find ϵ.
SOLUTION: From Eq. (9.98),

$$\psi_0(x+y)\psi_0(x-y) = \frac{k_o}{\sqrt{\pi}}\exp\left(-k_o^2(x^2+y^2)\right)$$

Then, from Eqs (9.116) and (5.11) and using the relation

$$(k_o y)^2 - 2iky = \left(k_o y - i\frac{k}{k_o}\right)^2 - \left(\frac{k}{k_o}\right)^2$$

it is found that

$$\begin{aligned}f_0(x,k) &= \frac{2k_o}{\sqrt{\pi}}e^{-k_o^2x^2}\int_{-\infty}^{\infty} e^{-k_o^2x^2+2iky}dy \\ &= 2\exp\left(-k_o^2x^2 - \left(\frac{k}{k_o}\right)^2\right) \equiv 2e^{-\epsilon} \to f_0(\epsilon)\end{aligned} \tag{9.126}$$

where $\epsilon \equiv 2E/\hbar\omega = k_o^2x^2 + (k/k_o)^2$.

EXAMPLE: Find the $n = 1$ Wigner function $f_1(x, k)$. Show that it is $f_1(\epsilon)$ along a classical trajectory defined by Eq. (9.79). Find ϵ (again).

SOLUTION: From Eq. (9.98),

$$\psi_1(x+y)\psi_1(x-y) = \frac{k_o}{\sqrt{\pi}} H_1(k_o(x+y))H_1(k_o(x-y))e^{-k_o^2(x^2+y^2)}$$
$$= \frac{k_o^3}{\sqrt{\pi}}\left(x^2 - y^2\right) e^{-k_o^2(x^2+y^2)}$$

Then, using the same relations as the previous example, and using Eqs (5.11) and (5.12),

$$f_1(x, k) = 2(2\epsilon - 1)e^{-\epsilon} \to f_1(\epsilon) \tag{9.127}$$

where $\epsilon \equiv 2E/\hbar\omega = k_o^2 x^2 + (k/k_o)^2$.

The last two examples explicitly show the harmonic oscillator, $f(x, k)$ depends only on the combination $\epsilon = k_o^2 x^2 + (k/k_o)^2$, a rather interesting result that generalizes to higher n. Along the classical trajectories,

$$f\left(k_o^{-1}\cos\theta, k_o \sin\theta\right) = f\left(k_o^{-1}, 0\right) = f(0, k_o) \tag{9.128}$$

for $2\pi \geq \theta \geq 0$. The second example shows further that for x in units of $1/k_o$ and k in units of k_o, the Wigner function for the harmonic oscillator appears as rotationally symmetric in phase space, reinforcing that the classical trajectories in those particular units are circles. But here, it means more: it means there is an easier way to find the Wigner function for larger n than 0 or 1 because $f_n(x, k)$ need only be evaluated at $k = 0$, after which $\epsilon = k_o^2 x^2 \to k_o^2 x^2 + (k/k_o)^2$ with $x \to k_o^{-1}\cos(\theta)$ and $k \to k_o \sin(\theta)$.

This means that there exists a (simpler) way of finding $f_n(x, k)$ by simply finding $f_n(x, 0)$ first, and then replacing $k_o^2 x^2 \to \epsilon$ at the end. As practice, one can show this easily for $f_0(x, k)$ and $f_1(x, k)$. For general n, a bit more is required. Let

$$f_n(x, 0) = 2N_n^2 \int_{-\infty}^{\infty} H_n(u+s)H_n(u-s)e^{-u^2-s^2}ds \tag{9.129}$$

where $s = k_o y$ and $u = k_o x$. The functions $H_n(x)$ are *polynomials*, meaning that $H_n(u \pm s)$ can be Taylor expanded about s (not u, because the use of the orthogonality relation of Eq. (9.99) will be needed) with a finite number of terms, and because of the astonishingly convenient relation of Eq. (9.100), it is found

$$H_n(u \pm s) = (\pm 1)^n \sum_{j=0}^{n} \frac{n!}{j!(n-j)!}(\pm 2u)^j H_{n-j}(s) \tag{9.130}$$

This form allows the orthogonality relations to be used, and it follows that

$$f_n(x, 0) = 2\int_{-\infty}^{\infty} \psi_n(x+y)\psi_n(x-y)dy$$
$$= 2N_n^2 \int_{-\infty}^{\infty} H_n(u+s)H_n(u-s)ds$$
$$= 2(-1)^n e^{-u^2} \sum_{j=0}^{n} \frac{\left(-2u^2\right)^j}{(j!)^2\,(n-j)!} \tag{9.131}$$

which, after the replacement $u^2 \to \epsilon$, gives the final form of $f_n(\epsilon)$. One can check that $f_0(\epsilon)$ and $f_1(\epsilon)$ are covered by this equation, but also that the next few are

$$f_2(\epsilon) = 2\left(1 - 4\epsilon + 2\epsilon^2\right)e^{-\epsilon} \tag{9.132}$$

$$f_3(\epsilon) = \frac{2}{3}\left(-3 + 18\epsilon - 18\epsilon^2 + 4\epsilon^3\right)e^{-\epsilon} \tag{9.133}$$

$$f_4(\epsilon) = \frac{2}{3}\left(3 - 24\epsilon + 36\epsilon^2 - 16\epsilon^3 + 2\epsilon^4\right)e^{-\epsilon} \tag{9.134}$$

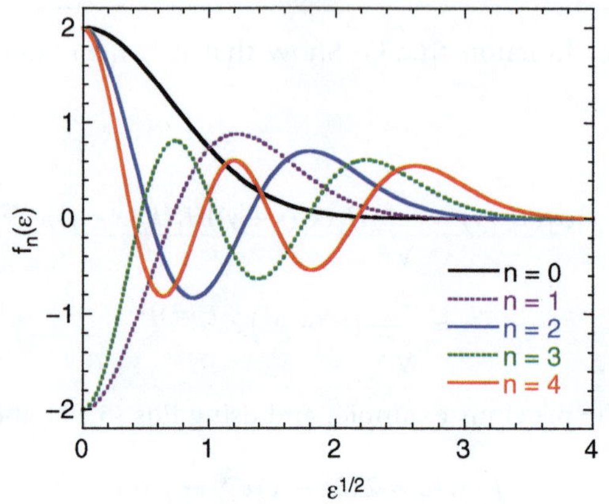

Figure 9.12 The Wigner function $f_n(\epsilon)$, where $\epsilon = k_o^2x^2 + (k/k_o)^2$ for the harmonic oscillator (see Eqs (9.132)–(9.134)). The energy $E = (\hbar^2 k_o^2/2m)\epsilon$. Observe that $f_{2n+1}(\epsilon \to 0) = -1$ and $f_{2n}(\epsilon \to 0) = +1$. Observe also that only $f_0(\epsilon)$ is non-negative.

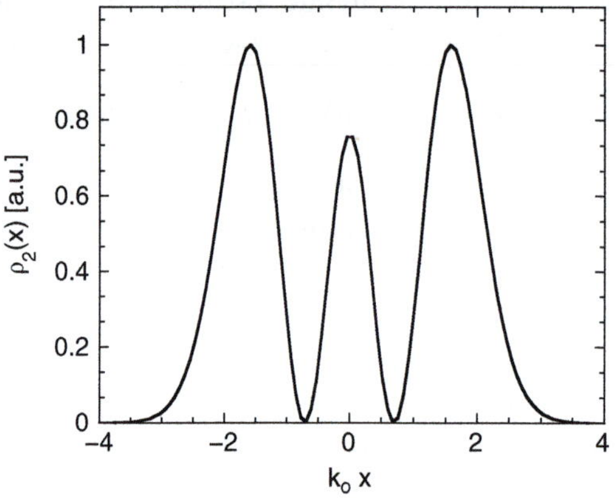

Figure 9.13 The density $\rho_n(x)$ for $n = 2$ (Eq. (9.135)), normalized to its maximum value. Observe that $\rho_2(x) \geq 0$.

and are shown in Figure 9.12: the lines rotated about the vertical axis would give a surface plot of the phase space $f_n(x, k)$. Lastly, observe that although $f_n(x, k)$ can be less than 0, the density $\rho_n(x)$ given by Eq. (9.13) is not. As an example, the case $\rho_2(x)$ is shown in Figure 9.13, for which

$$\rho_2(x) = \frac{1}{2\pi}\int_{-\infty}^{\infty} f_2(x, k)dk = \frac{k_o}{2\sqrt{\pi}}\left\{4x^4 - 4x^2 + 1\right\} e^{-x^2} \tag{9.135}$$

as can be shown by substitution and straightforward integration after the replacement of $\epsilon \to (k_o x)^2 + (k/k_o)^2$ in Eq. (9.132).

PART II
The canonical equations

And he that calls on thee, let him bring forth
Eternal numbers to outlive long date.

– William Shakespeare[1]

The word *canonical*, particularly in mathematics and the natural sciences, refers to an authoritative or orthodox formula or rule. The "canonical equations" here are the four equations of electron emission (thermal, field, photo, and secondary), plus one other imposing limits on that emission (space charge), all of which were developed by the 1930s even if subsequent developments generalized or modified the equations. The canonical forms are pervasively used in the literature: before examining how they have changed, understanding from whence they arose is profitable.

The enduring popularity of the canonical equations is in no small measure derived from their reasonably good account of the phenomena they treat, the simplicity of the equations, and the distillation of the phenomena into a small number of empirical (albeit not always well-understood) parameters. But reverence is not the same as uncritical acceptance, so the equations will be brought forth in a manner not necessarily the same as the way they were first developed: as Thomas Jefferson advised, "...laws...must go hand in hand with the progress of the human mind. As that becomes more developed, more enlightened, as new discoveries are made, new truths disclosed, and manners and opinions change with the change of circumstances, institutions must advance also, and keep pace with the times."[2]

[1]Ref. [37]: Sonnet 38, p. 1459.

[2]Jefferson to H. Tompkinson (*aka* Samuel Kercheval), July 12, 1816, in *The Writings of Thomas Jefferson*, ed. P. L. Ford. New York: G.P. Putnam's Sons, 1892. The quote appears on the Jefferson Memorial in Washington, DC.

CHAPTER 10

A brief history

Det er ganske sandt, hvad Philosophien siger, at Livet maa forstås baglænds.
Men derover glemmer man den anden Sætning, at det maa leves forlænds.

– Søren Kierkegaard[1]

The theoretical equations schematically illustrated in Figure 10.1 are often designated by the names of those who first developed them. Visually, the emission mechanisms are distinguished by where the emitted electrons originate and by how those electrons are able to move past a finite thickness barrier. Most metals have a reasonably hefty barrier, characterized by the work function Φ that is several electronvolts above their Fermi energy μ: for copper, $\Phi = 4.5$ eV, whereas for cesium, $\Phi = 2.14$ eV. For emission over the barrier, regardless of thickness, the electrons have to be provided with sufficient energy so that they may overcome it, which can be accomplished by heating, absorption of photons, or acquiring energy via collisions with other electrons. For emission under the barrier, the barrier must be made thin enough that electrons have some probability of tunneling through it. Most tunneling electrons originate near the Fermi level, and the insensitivity of μ to the temperature T is reflected in the insensitivity of the field emission current to T as well for metals. Following the numbering established in Figure 10.1, the canonical equations are described by the following terms (along with references to the articles introducing them):

- **field emission** (Chapter 13) for electron emission by the application of high electric fields that enable quantum mechanical tunneling, leading to the Fowler–Nordheim equation [36]
- **thermal emission** (Chapter 12) for electron emission by heating the cathode so that a thermal tail of the distribution function has sufficient energy to overcome the surface barrier, leading to the Richardson–Laue–Dushman equation [33]
- **photoemission** (Chapter 14) for electron emission by the absorption of photons such that the electrons have sufficient energy to pass over the barrier, leading to the Fowler–DuBridge equation [34]
- **secondary emission** (Chapter 15) for electron emission by bombardment with high-energy primary electrons that transfer energy to secondary electrons such that they have sufficient energy to pass over the surface barrier, leading to the Baroody equation [35]
- **space-charge** limited emission (Chapter 16) for the limiting of current due to charge in the anode–cathode (AK) gap that acts to suppress the field at the surface of the cathode or limit the amount of charge crossing the gap [132].

10.1 Thermal emission

The development of an equation describing thermal emission by Richardson in 1901 predicted that $\ln(J/\sqrt{T})$ was a linear function of $1/T$. Some 12 years later after an experimental demonstration that (*i*) the effect was not the consequence of chemical processes and (*ii*) that electric current in metals was carried by electrons, Richardson modified that to $\ln(J/T^2)$ being linear [133], saying

> *Taken together these experiments prove that the emission of electrons does not arise from any interaction between the hot filament and surrounding gases or vapors nor from any process involving consumption of the material of the filament. It thus follows that the emission of electrons from hot tungsten, which there is no reason for not regarding as exhibiting this phenomenon in a typical form, is not a chemical but a physical process.*

Motivated by the difficulties faced by that theory in the explanation of conductivity and specific heat, Richardson (in 1912) and von Laue (in 1918) ascertained the correct T^2 dependence of the coefficient. In 1923, Dushman's quantum

[1] "It is quite true what Philosophy says, that life must be understood backwards. But do not forget the other Phrase, that it must be lived forwards." Søren Kierkegaard, *Journalen JJ*, 167 (1843), Søren Kierkegaards Skrifter, Søren Kierkegaard Research Center, Copenhagen, 1997, vol. 18, p. 306.

Introduction to the Physics of Electron Emission, First Edition. Kevin L. Jensen.

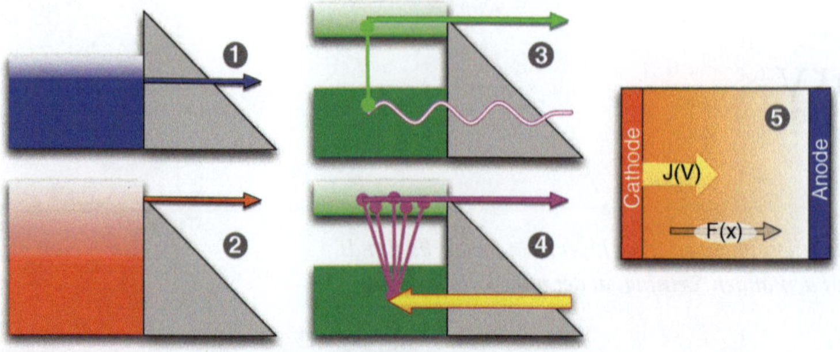

Figure 10.1 Electron emission schema: (1–4) material is to the left, the vacuum region is to the right; (5) cathode to the left, anode to the right, vacuum in the anode–cathode (AK) gap. Field (1) and thermal (2) shown for metals (no band gap); photo (3) and secondary (4) for semiconductors (band gap). Field emission relies on tunneling *through* the surface barrier, the others on emission *over* the barrier. Space charge (5) is not an *emission* mechanism but rather a current *limitation* mechanism.

mechanical analysis related A_{RLD} to fundamental constants. Fowler in 1929 modified A_{RLD} to include a factor of 2 for electron spin. By the 1930s, the so-called "Edison effect" was dubbed "thermionic emission" and, given its preeminence in vacuum tubes used in radio communications, experiments were directed to exploring low work function coatings such as barium, cesium, and thorium to improve electron emission at lower temperatures. It rapidly became clear that the surface work functions were not uniform nor characterized by a single work function Φ: rather, different crystal faces of polycrystalline refractory metal surfaces (often tungsten) exhibited different values of Φ. In fact, most early thermionic cathodes were tungsten because they were dependable, easy to clean by heating, and hard to kill. By comparison, more exotic cathodes that relied on other materials were poisoned by unknown processes. The high work function, however, was a limitation to achieving high current density, and so thorium (which was unfortunately radioactive) and submonolayer coverages of alkali (e.g., cesium) and alkali earth (e.g., barium) metals were investigated and found to substantially reduce the work function and greatly improve thermal emission current.

In one of the more emblematic instances of the impact of good luck on the fortunes of science, Arthur Wehnelt in 1904 in Germany was working on cathode ray tubes using thermionic emitters made of platinum which, like other metals, had a fairly high work function. Vacuum technique being what it was in those days, stopcock grease was used in the vacuum systems, and some of it had contaminated the cathode surface, curiously leading to much higher emitted current. Wehnelt, who was fluent in the technology behind vacuum tubes, electron guns, and emission from metal oxides, surmised that the metal oxides were behind the superior emission, and thus undertook an exploration of what the best metal oxide was for thermal emission. He concluded that it was barium oxide (BaO). Thus was born the metal oxide thermionic cathode, used in radio, radar, TV, displays, microwave ovens, and all manner of electron beam devices. As Sommer observed [134][2]

> *The BaO cathode is a truly amazing case of a researcher interpreting an accidental observation and putting it to practical use. It is ironic that this observation could not have been made in a modern high vacuum system!*

The need for radar in World War II drove the development of semiconductor, oxide, and matrix emitter thermionic cathodes. A pivotal improvement was the invention of dispenser cathodes that continually rejuvenated the low work function barium coatings on the surface by diffusion through a porous sintered tungsten matrix, to replace the barium lost to evaporation [135]. Increasing the temperature of a dispenser cathode increases the current density but does so at the cost of increasing evaporation rates (Section 29.3). As barium exhaustion signals the lifetime of the dispenser cathode has been reached, keeping the current density low by using beam compression gets the needed current densities in the beam tunnel (Section 33.2).

10.2 Field emission

Subjecting surfaces to very strong electric fields results in the emission of electrons, as first observed by Wood in 1897 [135]. Later in Germany, Lilienfield (University of Leipzig) was laboring to improve his X-ray tubes in the early

[2] Lest fans of other emission mechanisms knowingly smirk, the article containing the quote is entitled *The Element Of Luck In Research – Photocathodes 1930 to 1980*, humbling for its implications.

1900s, when he found that the measured current between two metals in vacuum was larger than he had anticipated, with a fluorescence that was observed in the "anti-cathode" region of his tube. Lilienfeld described the phenomena thus [136]:

> *The tubes at present in general use depend for the production of the electrons upon one of two effects, known for many decades: Either the ionization of residual gases (gas-filled tubes) or the heating of the cathode to a bright glow (thermionic or Edison–Richardson effect). The tube to be described, however, utilizes the electrons released by a phenomenon which I have only recently discovered. The main feature of the phenomenon is the releasing of the electrons on an unheated and cold cathode in the absence of any kind of ionization, either the ionization of the residual gases or an ionization through a hot spark discharge (electric arc) which pulls molecules from the electrodes and converts them into ions.*

and coined the phrase "Autoelektronische Entladung" (or "auto-electron discharge" which became "field emission"), attributing it to electrons being liberated from his metal tips [137]. In 1923, Schottky attempted to give a non-quantum mechanical account of both thermal and field emission [138] by proposing a lowering of the emission barrier whereby $\phi = \Phi - \sqrt{4QF}$, which also contained his clairvoyant conjecture about field enhancement effects for small bumps on larger bumps being multiplicative (more about this in Section 30.4). Yet it was the successful prediction of the behavior of current density as a function of field that gave an early success to the new field of quantum mechanics. In January of 1928, Oppenheimer wrote an elegant manuscript (see Section 9.2) in which, using a method hauntingly similar to those of Eqs (11.7) and (11.8), he predicted that atoms subjected to fields would, in a process unknown to classical physics, eventually lose their electron. He conjectured that the "pulling of electrons out of a metal by fields on the order of 10^5 e.s.u."[3] explained the data of Millikan and Eyring [49]. Oppenheimer predicted that the current J observed by Millikan and Eyring was such that $\ln(J/F^{1/4})$ was linear in $1/F$. On May 1 of the same year, Fowler and Nordheim [36] argued instead that $\ln(J/F^2)$ was linear in $1/F$: as a practical matter, Millikan's data supported both formulations fairly well (the first cautionary moral in Section 4.1). Fowler and Nordheim's theory was amended by Nordheim afterwards to include the Schottky factor for barrier lowering that led to $\Phi \rightarrow \phi = \Phi - \sqrt{4QF}$ in the exponential argument.

10.3 Photoemission

In 1887, Heinrich Hertz was conducting experiments to demonstrate the existence of electromagnetic waves, in which he used a Tesla transformer to produce sparks in a primary gap that caused sparking in a secondary gap a distance away. His investigations led him to conclude that premature sparking at the secondary gap was a consequence of the ultraviolet light from the primary spark [139]. Hertz's assistant, Philipp Lenard, continued the investigations and observed that light striking certain metals induced the emission of electrons, for which Lenard received the Nobel Prize in 1905. The difficulties in explaining both the promptness of photoemission and the insensitivity of the energy of photoemitted electrons to the intensity of light (particularly, red light of any intensity did not lead to photoemission) paved the way for Einstein's explanation using Planck's corpuscular theory of light (Section 3.3), in which Einstein boldly proposed [57]:

> *The usual conception that the energy of light is continuously distributed over the space through which it propagates, encounters very serious difficulties when one attempts to explain the photoelectric phenomena, as has been pointed out in Herr Lenard's pioneering paper. According to the concept that the incident light consists of energy quanta ...one can conceive of the ejection of electrons by light in the following way. Energy quanta penetrate into the surface layer of the body, and their energy is transformed, at least in part, into kinetic energy of electrons. The simplest way to imagine this is that a light quantum delivers its entire energy to a single electron: we shall assume that this is what happens.*

By 1928, Fowler [68] was able to exploit Sommerfeld's electron theory of metals in much the same way as he did in his treatment of field emission to show that the photoelectric threshold frequency was sharp (at a particular frequency ω_o), and that the corresponding energy was equal to the thermionic work function, or $\hbar\omega_o = \Phi$. DuBridge [140] modified Fowler's formulation and consequently accounted for the effects of temperature on the energy distribution of the electrons, as well as the total yield, the metric of which is the quantum efficiency. Photoemission from semiconductors, particularly from compound photoemitters, such as the I-V material ("I" referring to a column 1 element of the Periodic Table and "V" to a column 5 element) cesium antimonide (Cs_3Sb), was important in the development of semiconductor band structure theory. High quantum efficiency materials are important, likewise, for photodetectors.

[3] In CGS units, an esu is related to MKSA units by $10^6 c^{-1}$ statV/cm = 1 V/m.

10.4 Secondary emission

Secondary electron emission was discovered in 1902 by Austin and Starke, but its theoretical description was complicated by the fact that the energy loss mechanism of the primaries and the generation of the secondaries demanded a quantum mechanical treatment that was difficult [141], one of the first being by Wooldridge [142]. Baroody [35] lamented this fact, observing

> *There would seem to be qualitative value, however, in a much simpler theory in which the primary is considered to interact in a classical manner with a free electron gas, the possibility of momentum transfer to the lattice being introduced by assuming a finite mean free path for elastic collision ...Although the use of classical principles greatly simplifies the calculations, the problem remains complex and it is not clear that all the mathematical difficulties have been adequately met. The final results are therefore probably more limited in significance than would have been judged from the starting assumptions. It is partly because of this circumstance, and partly because it is felt that more attention should be given the Sommerfeld model, that the present paper is offered.*

and so unlike the other canonical equations, Baroody's formulation came after the madness of World War II rather than before it. His formulation makes use of concepts concerning penetration into the metal, generation of the excited electrons, and their transport back to and emission from the surface, for which analogs exist in Spicer's semi-empirical three-step model of photoemission. Secondary emission is utilized in photomultipliers, magnetrons, crossed-field amplifiers [12], and diamond current amplifiers [143, 144]), but secondary emission processes also figure in damage and degradation, as when some of a high-energy beam in a particle accelerator strikes cavity walls: a rough distinction between the secondaries utilized by electron sources and those contributing to wall damage is that the secondary yield of cathodes is desired to be high, but the secondary yield of walls and cavities is purposely suppressed by exploiting low-yield materials to suppress emission.

10.5 Space-charge limited emission

Space-charge limited emission theory is often associated with thermal emission: Langmuir's ground-breaking treatment observed that an understanding of space charge was crucial in showing that current due to thermal emission was not, as argued in Germany at the time, due to chemical reactions. Because current from a thermal cathode rose as the temperature rose, departures were observed from Richardson's equation for a given anode voltage. Langmuir wrote [132]

> *This apparent discrepancy between the results of calculation by Richardson's equation and the facts observed with a tungsten lamp seemed at first to confirm the growing opinion that in a very high vacuum the thermionic current is very small, if not entirely absent. Experiments on the Edision [sic] effect in tungsten lamps, made some time ago by the writer, throw a great deal of light on the cause of the apparent failure of Richardson's equation at high temperatures. The observations therefore seem of sufficient interest to warrant their publication.*

Langmuir mentioned in a footnote that C.D. Child found the same relation for the maximum current pulled across a planar gap for positive ions in 1911. For electrons, the relation therefore goes by the name of the Child–Langmuir equation (or often more compactly as *Child's law*), and it has since become " ...one of the best-known limits in the fields of non-neutral plasma physics, accelerator physics, sheath physics, vacuum electronics, and high power microwaves ..." [145] Space charge is inextricably linked with geometry: 10 years after Langmuir's first paper, he and Blodgett [146] found space-charge limited current between coaxial cylinders, and a year later between concentric spheres [147]. Later refinements include initial velocities, voltage differences of such a magnitude that electrons traversing the gap became relativistic [148], and, ultimately, quantum mechanical effects for very small AK gaps [149]. Space-charge effects additionally appear for electron transport in semiconductors: when scattering dominates, the Child–Langmuir transforms into the Mott–Gurney law [81]. Within thin films, space-charge effects during transport can lead to self-quenching behavior, as shown in simulations by Mumford and Cahay of Metal/CdS/LaS cold cathodes [150].

10.6 Resources and further reading

> *Those who cannot remember the past are condemned to repeat it.*
>
> – George Santayana[4]

[4]George Santayana, *Reason in Common Sense, The Life of Reason*, Vol. 1. New York: Dover Publications, 1980.

Learning from others is second nature to humans: we do it more readily and precisely than any other animal.

– Franz de Waal[5]

A rich history is available for the various electron emission mechanisms. A sampling of useful resources is suggested here, grouped according to kind: full-length books, book chapters, review articles, and general articles that have emerged as standard references. Of the books, a few describe more than one emission mechanism in depth, while others focus on one emission mechanism. A quasi-chronological account is:

- De Boer, *Electron Emission and Adsorption Phenomena* [151]: Extensive and historically useful overview of emission physics, phenomena, and literature up to the time of publication (1935).
- Birdsall and Bridges, *Electron Dynamics Of Diode Regions* [148]: Standard resource on space-charge limited flow and the Child–Langmuir law.
- van der Ziel, *Solid State Physical Electronics* [152]: Concise chapters on all emission mechanisms, solid state physics, and semiconductor device concepts.
- Dobretsov and Gomounova, *Emission Electronics* [153]: Comprehensive and economical treatment of all emission mechanisms and space charge.
- Modinos, *Field, Thermionic, and Secondary Electron Emission Spectroscopy* [111]: An excellent and thorough resource.
- Gomer, *Field Emission and Field Ionization* [95]: A classic treatment of both theoretical and experimental matters with a historical background.
- Fransen *et al.*, *On the Electron-Optical Properties of the ZrO/W Schottky Electron Emitter* [154]: Extensive analysis and data of thermal-field emission (or the Schottky emitter), particularly useful for treating the transition from thermal emission to field emission.

The series *Advances in Imaging and Electron Physics* (ed. P.W. Hawkes) and the earlier *Advances in Electronics and Electron Physics* series contain many invaluable books and book chapters, in particular

- Dyke and Dolan, *Field Emission* [155]
- Swanson, and Bell, *Recent Advances in Field Electron Microscopy of Metals* [156]
- Murata and Kyser, *Monte-Carlo Methods and Microlithography Simulation for Electron and X-ray-beams* [157]
- Hawkes, *Electron Optics* [158]
- Binh, Garcia, and Purcell, *Electron Field Emission from Atom-Sources: Fabrication, Properties, and Applications of Nanotips* [159]
- Fransen, Van Rooy, Tiemeijer, Overwijk, Faber, and Kruit [160], *On the electron-optical properties of the ZrO/W Schottky electron emitter* [160]
- Binh and Semet, *Electron Field Emission from Atom-Sources: Fabrication, Properties, and Applications of Nanotips* [161]
- Jensen, *Electron Emission Physics* [85]
- Swanson and Schwind, *A Review of the Cold-Field Electron Cathode* [162].

Book chapters and reviews that focus on one of the canonical mechanisms abound, with particularly good treatments being the following:

- Thermal emission
  1. Dushman, *Thermionic Emission* [33]
  2. Becker, *Thermionic Electron Emission and Adsorption Part I. Thermionic Emission* [163]
  3. Herring and Nichols, *Thermionic Emission* [164]
  4. Haas and Thomas, *Thermionic Emission and Work Function*, Chapter 2 [135]
- Field emission
  1. Dyke and Dolan, *Field Emission* [155]
  2. Gadzuk and Plummer, *Field Emission Energy Distribution* [165]
  3. Zhu, *Vacuum Microelectronics* (all chapters) [166]
  4. Swanson and Schwind, *A Review of the Cold-Field Electron Cathode* [162]
- Photoemission
  1. Görlich, *Recent Advances in Photoemission* [167]
  2. Sommer and Spicer, *Photoelectric Emission* [168]
  3. Rougeot and Baud, *Negative Electron Affinity Photoemitters* [169]
- Secondary emission
  1. McKay, *Secondary Electron Emission* [170]

[5]Franz B.M. de Waal, *The Ape and the Sushi Master: Cultural Reflections By a Primatologist*, 1st edn. New York: Basic Books, 2001, p. 20.

2. Dekker, *Secondary Electron Emission* [141]
3. Hachenberg and Brauer, *Secondary Electron Emission From Solids* [171]

Reviews or extensive treatments have been listed in the bibliography, but there are particularly notable treatments for thermal emission [110, 172, 173], field emission [79, 96, 110, 173–177], photoemission [134, 178–181], secondary emission [35, 157], and space charge [145, 182].

CHAPTER 11
Anatomy of current density

"Beauty is truth, truth beauty," – that is all
Ye know on earth, and all ye need to know.

– John Keats[1]

***Truth**, n. An ingenious compound of desirability and appearance.*

– Ambrose Bierce[2]

Current density as encountered in Eq. (3.5) is complicated by several considerations: (*i*) particles have a distribution of energies, leading to the distribution approach of Eq. (7.9), (*ii*) current is directed into a barrier such that the transverse components of the distribution can be summed over, leading to the notion of a "supply function" suggested by Eq. (9.75), and (*iii*) a transmission probability governs the distribution of particles transported *past* the barrier. A formulation of the current density which incorporates these wrinkles and bears a resemblance to the current density of the distribution function is

$$J(F,T) = \frac{q}{2\pi}\int_0^{\infty}\frac{\hbar k}{m}D(k)f(k)dk \tag{11.1}$$

where $\hbar k/m$ is the velocity and $D(k)f(k)$ is the distribution of particles making their way past the barrier. Eq. (11.1) looks and feels right at a gut level, even though a compelling derivation of it is lacking so far: it has *truthiness*.[3] It also requires unpacking:

- The equation is one-dimensional, and so $k_x \to k$.
- The lower limit of integration, normally $-\infty$, is 0 because no electrons are incident from the vacuum boundary (right) side.
- The supply function $f(k)$ is the integral of the distribution over the transverse momentum as in the brackets $\{\cdots\}$ of Eq. (9.75), and includes spin.
- The factor $D(k)f(k)$ is the distribution of *emitted* particles.
- The arguments of $J(F,T)$ reflect that the supply function is temperature dependent (T) and the transmission probability is field dependent (F).

Were the vacuum to be replaced by a material and the barrier by an interface barrier characteristic of metal–semiconductor (and metal–insulator–semiconductor, MIS) boundaries as typical of Schottky barriers, then electrons would be incident from the right, the integration region extended to $-\infty$, and the distribution function $f(k)$ would have to be replaced by something accounting for the rightmost boundary, most simply by including a right-hand side supply function. Indeed, Eq. (11.1) would generalize to the Tsu–Esaki formula [183]

$$J_{TE}(V_b) = \frac{q}{2\pi}\int_0^{\infty}\frac{\hbar k}{m}D(k)\left(f(k)-f(k')\right)dk \tag{11.2}$$

where $E_{k'} = E_k + qV_b$ and V_b is the voltage bias across the barrier, as will be used in treating resonant tunneling diodes (RTDs; two barriers) of Section 19.2.4 to understand *negative differential resistance*. For his work with RTDs, Esaki shared the Nobel Prize in 1973.

The genesis of all four of the emission equations can be traced back to limiting forms of Eq. (11.1) after (*i*) the integral is recast in terms of energy, (*ii*) the supply function is specified, and (*iii*) the transmission probability is approximated

[1] John Keats, "Ode on a Grecian Urn", in *Poetry: An Introductory Anthology*, ed. Hazard Adams. Boston: Little, Brown and Company, 1968, p. 166.
[2] Ambrose Bierce, *The Devil's Dictionary*. New York: Dover Publications, 1993, p. 136.
[3] The word *truthiness* to suggest something held to be true not because of evidence but because of gut feeling was coined by the comedian Stephen Colbert in 2005 and, owing to its popularity, has made it into some dictionaries a decade later, for example http://dictionary.reference.com.

Introduction to the Physics of Electron Emission, First Edition. Kevin L. Jensen.

based on an image charge potential. Although far simpler than their original genesis, such an approach reproduces the canonical emission equations quickly and compactly. Performing the first step, switch from a momentum integration over dk to an energy integration over dE under the assumption that energy is parabolic in momentum, that is, using $\hbar^{-1}dE = (\hbar k/m)dk$ so that

$$J(F,T) = \frac{q}{2\pi\hbar}\int_0^\infty D(E)f(E)dE \tag{11.3}$$

Using this equation as the starting point, consideration of the supply function $f(E)$ and the transmission probability $D(E)$ can be undertaken.

11.1 Supply function

...two social and economic forces ...do indeed play a fundamental role in determining wage inequality, even in more sophisticated theories: the supply and demand of skills. In practice, the supply of skills depends on, among other things, the state of the educational system ...The demand for skills depends on, among other things, the state of the technologies available to produce goods and services ...At a minimum, they influence the relative power of different social groups.

– Thomas Piketty[4]

The relation of demand to supply is a general and powerful notion appearing across disciplines (albeit that physics benefits from a less convoluted relation than economics). The demand, to push the analogy, is current density in the form of electrons making their way past the barrier, and it is dependent upon the (temperature dependent) supply of electrons from the bulk material. The supply function is a boundary condition, specifying the coefficients attached to the incoming wave functions in Eq. (9.78), and was implicit in Eq. (9.75). To make $f(k)$ explicit,

$$f(k_x) = \frac{g}{(2\pi)^2}\int_0^\infty \frac{2\pi k_\perp dk_\perp}{1+\exp\{\beta\left(E_x + E_\perp - \mu\right)\}} \tag{11.4}$$

where $\beta \equiv 1/k_BT$, $E_x \equiv \hbar^2k_x^2/2m$ and $E_\perp = \hbar^2k_\perp^2/2m$ are used, and the x suffix has been temporarily reinstated to reduce ambiguity. Pleasantly enough, the integral is straightforward: letting $u = \beta(E_\perp + E_x - \mu)$, then

$$f(E_x) = \frac{m}{\pi\beta\hbar^2}\int_{\beta(E_x-\mu)}^\infty \frac{du}{1+e^u} = \frac{m}{\pi\beta\hbar^2}\ln\left(1+e^{\beta(\mu-E_x)}\right) \tag{11.5}$$

The dropping of the x subscript can now be done because all vestiges of three-dimensionality have been subsumed in Eq. (11.5), and so let $E_x \to E$ below. To emphasize an important point, past discussions of $f(k)$ will now be replaced by present discussions using $f(E)$, as both refer to $f(E(k))$. Mathematicians would look askance, but it is convenient to do so.

EXAMPLE: Verify that an integration over the zero-temperature supply function reproduces the density $\rho = k_F^3/3\pi^2$. Let $E_x \to E = \hbar^2k^2/2m$.
SOLUTION: For $\beta = 1/k_BT \to \infty$, Eq. (11.5) is non-zero only for $E < \mu$, for which $f(E) = (m/\pi\hbar^2)(\mu - E) \to \left(k_F^2 - k^2\right)/2\pi$. Performing the integration over k from $-k_F$ to k_F (the range over which the integrand is non-zero) results in

$$\rho = \frac{1}{2\pi}\int_{-k_F}^{k_F}\frac{k_F^2 - k^2}{2\pi}dk = \frac{k_F^3}{3\pi^2}$$

11.2 Gamow factor

Cry woe, destruction, ruin, and decay
The worst is death, and death will have his day.

– William Shakespeare[5]

[4]Thomas Piketty, *Capital in the Twenty-First Century* (trans. Arthur Goldhammer). Cambridge, MA: Belknap Press of Harvard University Press, 2014, p. 305.
[5]Ref. [37]: *King Richard II*, III.ii.102–103, p. 653.

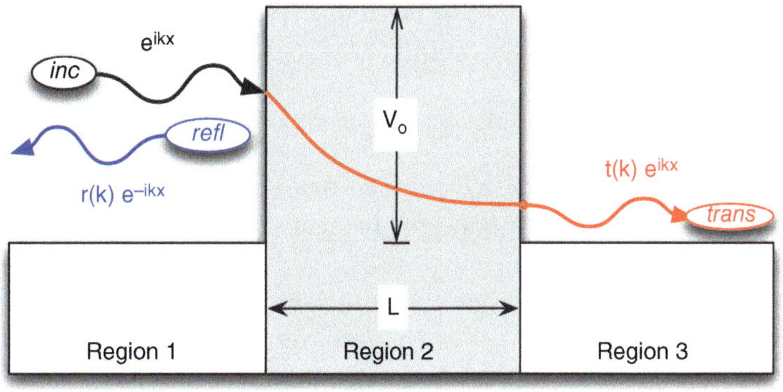

Figure 11.1 A barrier of finite thickness L and the decay of the transmitted wave function passing through it when $E < V_o$.

When an electron of energy E encounters a one-dimemsional step function potential of height V_o such that $E < V_o$, the wave function in the barrier region is no longer oscillatory as with the incident wave function, but decays as in Section 9.2.3. Setting $\hbar^2\kappa_o^2/2m = |E - V_o|$, then the wave function dies out as $\psi_{trans}(x) = t(k)\exp\left(-\kappa_o x\right)$.[6] If instead of a step function potential, a potential barrier of width L were encountered, then the wave function would have decayed to $e^{-\kappa_o L}$ before emerging from the other side to continue its oscillatory motion, as in Figure 11.1. The transmission probably should then go as something like $D(E) \approx \exp\left(-2\kappa_o L\right)$. Why? A hand-waving explanation is because the current density relation contains terms of the form $(\hbar/m)\psi^\dagger\partial_x\psi \sim (\hbar\kappa_o/m)\rho$ which resemble the product of velocity and density, and as transmission probabilities are ratios of transmitted current to incident current, the expectation of how $D(E)$ varies is anticipated to be $D(E) \sim e^{-2\kappa_o L}$. The Gamow factor is contained in the argument of the exponent, and for the rectangular barrier is $\theta_{rec}(E) = 2\kappa_o(E)L$, the subscript designating "rectangular". Such a hand-waving argument can be put on firmer footing.

Gamow introduced $\theta(E)$ to describe the decay of the emission probability of an alpha particle attempting to escape the nucleus of an atom [184]. In dealing with potentials that arise from the consideration of real surfaces from which electrons long to escape, the decay of a nucleus via alpha particle emission serves as a good template of how to proceed. Normally, the evaluation of the decay factor would require a wave function matching method commonly known as WKB [185], although the acronym JWKB, referring to Jeffries–Wentzel–Kramers–Brillouin, would be more accurate. The JWKB method is an approach to match wave functions at points where the wave function passes from above to below, or vice versa, when encountering a smoothly varying potential. The Gamow factor is often used in place of a proper JWKB treatment because it captures the most important physics with minimal effort. How the transmission probability $D(E)$ is obtained from $\theta(E)$ can be done several ways [186], and will become justified when the full mechanics of treating general barrier problems is undertaken in Chapter 18.

A simple argument can render the Gamow factor plausible here (similar to an approach by Miller and Good [187]). Start with Schrödinger's equation

$$i\hbar\partial_t\psi(x,t) + \left(\frac{\hbar^2}{2m}\partial_x^2 - V(x)\right)\psi(x,t) = 0 \tag{11.6}$$

and assume ψ takes the form

$$\psi(x) = R\exp(iS) \tag{11.7}$$

for two real functions $R(x,t)$ and $S(x,t)$. Clearly, $R(x,t)^2 = \rho(x,t)$ is the number density. When Eq. (11.7) is inserted into Eq. (11.6), a pot pourri of real and imaginary terms spill out. Collecting the real terms together, and doing the same with the imaginary terms, both have to be separately set equal to zero, and so

$$\begin{aligned} 0 &= \frac{\hbar^2}{2m}\left\{(\partial_x S)^2 - \frac{1}{R}\partial_x^2 R\right\} + V(x) + \hbar\partial_t S \\ 0 &= i\left\{\frac{\hbar}{R}\partial_t R + \frac{\hbar^2}{2mR}\left[2\left(\partial_x R\right)\left(\partial_x S\right) + R\partial_x^2 S\right]\right\} \end{aligned} \tag{11.8}$$

[6] The term κ_o is used instead of the simpler κ because for general and awkwardly shaped potentials a spatially dependent $\kappa(x)$ will useful, as in Eq. (11.21). Hence, using κ_o reinforces that it is constant.

Comparing to the Hamilton–Jacobi equation of classical mechanics [90, 122], the term $(\hbar\partial_x S)^2/2m$ resembles kinetic energy. Similarly, the *quantum potential* introduced by Bohm [121] is

$$\Omega(x) \equiv -\frac{\hbar^2}{2mR}\partial_x^2 R(x) \tag{11.9}$$

and is added to $V(x)$ for Bohm's prescription to avoid a conflict between macroscopic and microscopic realism in quantum mechanics through the introduction of a trajectory concept analogous to that in Section 9.3.2. If the electron density is smoothly varying, then $\Omega(x)$ will be small and ignorable.

The point to focus on is that the kinetic energy term identifies $\hbar\partial_x S/m$ as a velocity: letting the (number) current density J be the product of a velocity and a density, as in Eq. (3.5), results in

$$J(x,t) = R(x,t)^2\left(\frac{\hbar}{m}\partial_x S(x,t)\right) \tag{11.10}$$

The same form is obtained using Eq. (11.7) in Eq. (9.60) (try it). Inserting $J(x,t)$ into Eq. (11.8) results in the imaginary equation becoming (try this too)

$$\frac{i\hbar}{2\rho}\left(\partial_t \rho + \partial_x J\right) = 0 \tag{11.11}$$

for which the quantity in parentheses is the familiar continuity equation of Eq. (7.19). But there is more: $J(x)$ inside Eq. (11.8) in place of $\partial_x S$ gives

$$\partial_x J(x) = i\left[\frac{2\rho}{\hbar}\left(E - V(x) - \Omega(x)\right) - \frac{m}{\hbar\rho}J(x)^2\right] \tag{11.12}$$

For steady-state problems ($\partial_t \rho = 0$), the current density is spatially uniform ($\partial_x J = 0$) and so Eq. (11.10) shows that the phase $S(x)$ evolves according to

$$\partial_x S = i\sqrt{\frac{2m}{\hbar^2}\left[E - V(x) - \Omega(x)\right]} \tag{11.13}$$

Therefore, for E larger than the maximum of the potential and $\Omega(x)$ small, then $\partial_x S$ behaves much like κ_o, but it is when $E < V_{max}$ that $\kappa_o \to i\kappa_o$ and exponential decay results because $e^{i\kappa_o x} \to e^{-\kappa_o x}$. Thus, a reasonable expectation is that the prototypical Gamow factor of $2\kappa_o L$ generalizes for an arbitrary potential to the form

$$\theta(E) \equiv 2\frac{\sqrt{2m}}{\hbar}\int_{x_-}^{x_+}\sqrt{V(x) - E}\,dx \tag{11.14}$$

where $V(x_\pm) = E$ so that the integrand vanishes at the integration limits, and the factor of 2 out front is because the density behaves as $|\psi|^2$. Eq. (11.14) is the usual formulation of the Gamow factor.

EXAMPLE: Find the Gamow factor $\theta_{rec}(E)$ for a rectangular barrier of height $V_o = \hbar^2 k_o^2/2m$ and width L_o, that is, for $V(x) = V_o\Theta(x)\Theta(L_o - x)$, using Eq. (11.14).

SOLUTION: Define $L_{rec} \equiv L_o$. Then,

$$\theta_{rec}(E) = 2\int_0^{L_o} dx\sqrt{k_o^2 - k^2} = 2\sqrt{k_o^2 - k^2}L_o \tag{11.15}$$

$$\equiv 2\kappa_o(E)L_{rec} \tag{11.16}$$

where $E = \hbar^2k^2/2m$ and $V_o = \hbar^2k_o^2/2m$ in Eq. (11.14).

The quantity $\theta_{rec} = 2\kappa_o L$ is the prototypical or constant barrier Gamow factor containing three important factors [188]: a numerical coefficient characteristic of the overall barrier *shape* (2), a wave number characteristic of the barrier *height* (κ_o), and a length characteristic of the barrier *width* (L). Other simple barriers can be treated analogously.

EXAMPLE: Find the Gamow factor $\theta_{tri}(E)$ for a triangular barrier $V(x > 0) = V_o - Fx$ using Eq. (11.14).
SOLUTION: Define the length $L_{tri}(E) \equiv \left(V_o - E\right)/F$. Then,

$$\theta_{tri}(E) = 2\frac{\sqrt{2m}}{\hbar}\int_0^{L_{tri}(E)} dx\sqrt{V_o - E - Fx} \tag{11.17}$$

$$= \frac{4}{3}\kappa_o(E)L_{tri}(E) \tag{11.18}$$

EXAMPLE: Find the Gamow factor $\theta_{quad}(E)$ for a parabolic barrier $V(x) = \phi\left(1 - (2x/L_o)^2\right) + \mu$, where L_o is the width of the barrier at $E = \mu$, using Eq. (11.14). Observe that $V_o = V(0) = \mu + \phi$.
SOLUTION: Define the length $L_{qua}(E) = L_o\sqrt{(\mu + \phi - E)/\phi}$ as the positive root of the equation $V(L_{qua}/2) - E = 0$. Then

$$\theta_{qua}(E) = 2\frac{\sqrt{2m}}{\hbar}\int_{-L_{qua}(E)/2}^{L_{qua}(E)/2} dx\sqrt{\frac{\phi}{L_o^2}\left(L_o^2 - 4x^2\right) + \mu - E} \tag{11.19}$$

$$= \frac{\pi}{2}\kappa_o(E)L_{qua}(E) \tag{11.20}$$

In all three cases, the factor $\kappa_o(E)$ is as defined in Eq. (11.15), or $\left(\hbar\kappa_o(E)\right)^2/2m = V_o - E$. Observe that the shape factors (2 for rectangular, 4/3 for triangular, and $\pi/2$ for quadratic) are dimensionless, different for each case, and in proportion to how much of the area of height $\kappa(E)$ and width $L(E)$ is taken up by the integrand (completely in the case of the rectangular barrier). The shape factor approach will be the focus of a more general discussion in Section 11.3, and make a brief appearance in Section 25.2. Returning to the wave function $\psi(x)$, for general potentials $V(x)$ it is useful to define a generalization of κ_o that is spatially dependent, or

$$\kappa(x) \equiv \frac{\sqrt{2m}}{\hbar}\sqrt{V(x) - E} \tag{11.21}$$

and the zeros $x_\pm$ such that $\kappa(x_\pm) = 0$. With the neglect of the quantum potential $\Omega(x)$, then from Eq. (11.13) $\partial_x S = \kappa(x)$, and for time-independent situations where the current density J is constant, $R(x)^2 \propto 1/\kappa(x)$. Thus, the leading order approximation is (said this way because the same result is obtained using a perturbation-like approach as done by Miller and Good [187] and Schiff [42])

$$\psi(x) \propto \frac{1}{\sqrt{\kappa(x)}}\exp\left(-\int_{x_-}^{x_+}\kappa(x')dx'\right) \tag{11.22}$$

when the energy of the incident electron is below the barrier maximum. As density is proportional to $|\psi(x)|^2$ it follows that $D(E) \propto \exp(-2\kappa L)$ to this order. A careful reader might have a disquieting feeling that something is being obfuscated (it is), that shifting back and forth between a real $\kappa(x)$ and imaginary $i\kappa(x)$ in the definition of $\partial_x S$ is cavalier (it is), and that a proper account of the barrier might reveal weirdness that the present analysis hides (it will), but $\theta(x)$ is adequate to the needs of developing the canonical equations. Those discussions will wait.

11.3 Image charge potential

… I feel it,/I know my image now … I start the fire I suffer.

–Ovid[7]

[7] R. Humphries, *Ovid Metamorphoses, Narcissus*. Indiana University Press, 1983, lines 462–465.

The method of images is a remarkably convenient tool to solve boundary value problems. The method is based on the simple observation that if the boundary conditions imposed by conductor surfaces can be mimicked by a distribution of charges (used with abandon in Section 30.3), the replacement of the metal surface by those charges is also a solution to the boundary value problem, and by the uniqueness theorem (if two solutions agree at the boundaries, they are the same solution) the resulting potential is the only solution [99]. For a flat metal plane over which a charge q is suspended, the metal surface is held at zero potential and all the electric field lines are normal to the surface,[8] but the same boundary conditions can be had if a charge of equal and opposite sign is equidistant from the surface but *inside* the metal. As every point on the metal surface is equidistant between the charge and its image, the potential vanishes, and so the electrostatic problem is solved. Representing a metal surface by an image charge is preferable to the horror of trying to find the surface charge density directly.

The form of the image charge potential is revealed by considering the force between the external charge and its image: the charges are equal and opposite in magnitude, and equidistant from the surface boundary. If x is the distance of the external charge to the surface, then the distance between the charge and its image is $2x$. Consequently, the force on a charge due to its image is

$$F(x) = -\frac{q^2}{4\pi\varepsilon(2x)^2} \equiv -\frac{Q}{x^2} \tag{11.23}$$

The image charge potential is then determined by integrating over the force, or

$$V_{image}(x) = \int_x^{\infty}\left(-\frac{q^2}{4\pi\varepsilon_o(2x)^2}\right)dx = -\frac{q^2}{16\pi\varepsilon_o x} = -\frac{Q}{x} \tag{11.24}$$

For an electron emitted from a metal surface, the consequence is that in addition to a barrier described by V_o subject to an electric field giving rise to $-Fx$, there is a Coulomb potential of the form $-Q/x$ (see Eq. (2.8)) due to the image charge inside the metal and serving in its stead. The barrier for metals is characterized by its height above the Fermi level μ, or $V_o = \mu + \Phi$, where Φ is the work function, so that

$$V_{image}(x) \equiv \mu + \Phi - Fx - \frac{Q}{x} \tag{11.25}$$

for $x > x_{min}$ where x_{min} is the smaller root of the equation $V_{image}(x) = 0$. The potential $V_{image}(x)$ is shown alongside the triangular barrier (Eq. (11.25)) with $Q \to 0$) in Figure 11.2. There is much to notice. First, the image charge potential does not encounter the $x = 0$ origin, and the location of the maximum of the barrier will clearly move around with field. Second, the barrier maximum is reduced by comparison to the triangular barrier by an amount commonly called the

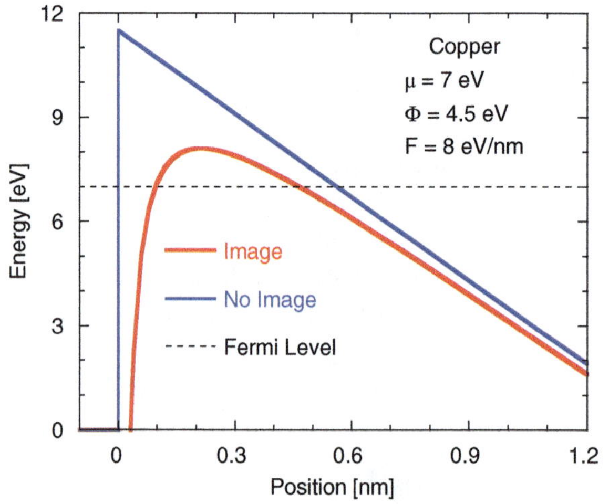

Figure 11.2 Triangular vs image charge barrier. The image charge barrier (Eq. (11.25)) compared to the triangular barrier ($Q \to 0$) for copper-like parameters. The dashed horizontal line corresponds to the Fermi energy μ. $V(x)$ for small x is truncated at 0.

[8]Were the electric field lines not normal to the surface, electrons at the surface would move in response until a distribution of charge was established such that the field lines became normal.

Schottky barrier-lowering factor. Third, the zeros of the $\theta(E)$ integrand, or $x_{\pm}(E)$, change depending on the location of E, but arguably the most important instance is at $E = \mu$. Consider each in turn.

The combination $\mu + \Phi - E$ appears with such regularity that it is convenient to define $\Delta(E) \equiv \mu + \Phi - E$. The zeros of the $\theta(E)$ integrand are then solutions to $-Fx_{\pm}^2 + \Delta x_{\pm} - Q = 0$, whereas the location of the barrier maximum is the solution to $\partial_x V(x_o) = 0$, or

$$x_{\pm}(E) \equiv \frac{1}{2F}\left\{\Delta \pm \sqrt{\Delta^2 - 4QF}\right\} \tag{11.26}$$

$$x_o^2 \equiv x_+ x_- = \frac{Q}{F} \tag{11.27}$$

$$L(E) \equiv x_+ - x_- = \frac{1}{F}\sqrt{\Delta^2 - 4QF} \tag{11.28}$$

$$V_{max} = \mu + \Phi - \sqrt{4QF} \tag{11.29}$$

Observe in Eq. (11.29) that the height of the barrier above the Fermi level is $\Phi - \sqrt{4QF}$, where the Schottky lowering factor $\sqrt{4QF}$ is the amount by which the barrier is lowered due to the image charge compared to the triangular barrier. Giving the image-reduced work function its own symbol is desirable: since the barrier is *reduced*, use ϕ, or

$$\phi \equiv \Phi - \sqrt{4QF} \equiv (1 - y)\,\Phi \tag{11.30}$$

where $y^2 = 4QF/\Phi^2$ represents the fractional height of the image charge barrier compared to the image-free barrier. In terms of ϕ and x_o, an alternate method of expressing the image charge potential of Eq. (11.25) is

$$V_{image}(x) = \mu + \phi - \frac{F}{x}\left(x - x_o\right)^2 \tag{11.31}$$

which is suggestive in that near $x \approx x_o$ the potential resembles a parabola, yet for large x and small F, when x_o is negligible, the potential more closely resembles a triangle. Observe that ϕ *is* the height of the barrier to an electron *at* the Fermi level $E = \mu$, for which $\Delta(\mu) = \Phi$.

EXAMPLE: For an electron at the Fermi level $E = \mu$ (alternately, $\Delta = \Phi$) find the width of the barrier, the zeros $x_{\pm}(\mu)$, the location x_o and y^2. Assume copper parameters of $\mu = 7$ eV and $\Phi = 4.5$ eV. Let $F = 8$ eV/nm.

SOLUTION:

$$L(\mu) = \frac{1}{F}\sqrt{\Phi^2 - 4QF} = 0.36934 \text{ nm}$$

$$x_+(\mu) = \frac{1}{2F}(\Phi + LF) = 4.6592 \text{ nm}$$

$$x_-(\mu) = \frac{1}{2F}(\Phi - LF) = 0.96581 \text{ nm}$$

$$x_{min} = x_-(0) = 0.03202 \text{ nm}$$

$$x_o = \frac{Q}{F} = 0.21213 \text{ nm}$$

$$y^2 = 0.56887$$

CHAPTER 12
Richardson–Laue–Dushman equation

If the theory ...is correct and if all of the usual disturbing factors are eliminated in the design, the resulting retarding-potential current should have the same exponential dependence on the voltage characteristic of the temperature extending to the point of saturation ...Equally significant is the flatness of the saturated region. This is rarely seen in retarding-potential experiments, since the accelerating voltage usually manages to collect more current by influencing trajectories, reducing space charge, altering the average potential outside of patchy surfaces, and by other devious means.

– Haywood Shelton [189]

12.1 Approximations

...in science, we can make successive approximations to the truth, in which each new stage results from an improvement, not a rejection, of what has gone before.

– Bertrand Russell[1]

The Richardson–Laue–Dushman (RLD) equation predicts current density due to thermal emission, generally from metals that, through the application of coatings such as barium to a porous tungsten base, have a low work function (e.g., dispenser cathodes, Figure 12.1). Typical operational parameters are $T = 1173$ K, $F = 0.1$ eV/μm, and $\Phi = 2.0$ eV. Standard accounts of the RLD equation are given by Guth and Mullin [190] and Murphy and Good [110].

For the transmission probability, observe that fields for typical thermal emission measurements are comparatively weak,[2] so weak in fact that electrons even a wee bit below the barrier maximum $\mu + \phi$ are prevented from escaping. Next observe that even at temperatures that thermal cathodes endure, only electrons in the thermal, or Maxwell–Boltzmann, tail of the supply function $f(E)$ contribute to the current density. Using the approximation $\ln(1+x) \approx x$ in Eq. (11.5) pulls out the Maxwell–Boltzmann term. Therefore, and using the convenient term $\beta \equiv 1/k_B T$,

$$D_{RLD}(E) \approx \Theta[E - (\mu + \phi)] \tag{12.1}$$

$$f_{RLD}(E) \approx \frac{m}{\pi \hbar^2 \beta} \exp[-\beta(\mu - E)] \tag{12.2}$$

When these approximations are inserted into Eq. (11.3) then

$$\begin{aligned} J_{RLD}(T) &= \left(\frac{qm}{2\pi^2 \hbar^3 \beta}\right) \int_{\mu+\phi}^{\infty} e^{-\beta(\mu-E)} dE \\ &= A_{RLD} T^2 \exp\left(-\frac{\phi}{k_B T}\right) \end{aligned} \tag{12.3}$$

where the coefficient $A_{RLD} \equiv (qmk_B^2/2\pi^2\hbar^3)$ is as encountered in Eq. (5.50) and, importantly, ϕ is the Schottky reduced work function of Eq. (11.30). When $\phi \to \Phi - \sqrt{4QF}$ in Eq. (12.3), the equation is often referred to as the Schottky emission equation (*sans* a coefficient of $(1-r)$ that accounts for the small reduction due to quantum mechanical reflection, and will be encountered again in Section 19.2.3).

[1] Bertrand Russell, *A History of Western Philosophy, and Its Connection With Political and Social Circumstances From the Earliest Times to the Present Day.* New York: Simon and Schuster, 1945, p. 836.

[2] Calling such fields "weak" is most certainly a relative statement: compared to typical macroscopic fields, they are rather strong, for example the fields of 0.1 MV/m characteristic of gradients leading to lightening strikes are similar to a dispenser cathode grid at 100 V and a cathode–anode gap of 250 μm for a field of 0.4 MV/m [191].

Introduction to the Physics of Electron Emission, First Edition. Kevin L. Jensen.

Figure 12.1 Dispenser cathode. A small thermionic dispenser cathode compared to a US dime (diameter 17.9 mm). The white arrow points to the emission surface. *Image courtesy of Bernard Vancil (e-Beam).*

12.2 Analysis of thermal emission data

Using Eq. (12.3) to analyze the data of Shelton (Figure 6 of ref. [189] and Figure 12.2, a resource worthwhile for its explanation of the complexities in thermal emission measurements) is undertaken with the data given in Table 12.1. Retarding potential (RET) and saturation current (SAT) measurements are two methods of measuring the current density from a hot source. They are different in that the RET measurements assess the temperature-independent thermal constants, but the SAT experiments give an A coefficient that must be reduced to account for changes in the work function due to a temperature variation given by (see Eq. (28.1))

$$\Phi(T) = \Phi_o + aT \tag{12.4}$$

Both data sets are fitted by $y(x) = A\exp(-Bx)$, but because the RET measurements give the actual thermal constants, $A_{rec} \equiv A_{RLD}$. A fit to the logarithmic data gives $B_{rec} = 53.0$ for which $\Phi_o = B/1000k_B = 4.57$ eV. For the SAT measurements, $A_{sat} = 974.37$ A/cm^2 K^2 and $B_{sat} = 55.14$. Because of the temperature dependence of Φ in Eq. (12.4), $A_{sat} = A_{RLD}\exp(a/k_B)$,

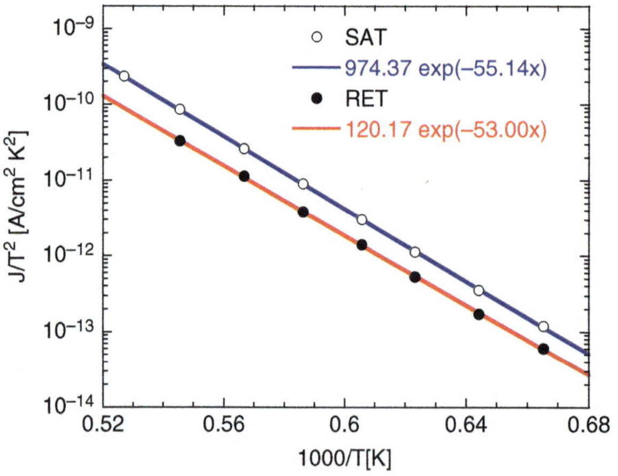

Figure 12.2 Thermal emission data taken from two [211] tantalum crystals using a saturation current (SAT) method and a retarding-potential (RET) method (based on Figure 6, ref. [189], data from Table 12.1). The fits to $y(x) = A\exp(-Bx)$ give $A = A_{RLD}$ for RET; A and B were determined by fitting for SAT. The difference $B_{sat} - B_{ret}$ exhibits the linear temperature dependence of $\Phi(T) = \Phi_o + \alpha T$.

Table 12.1 Thermal emission data of Shelton: the retarding potential and saturation current methods of measuring thermal current density as a function of temperature. The columns labeled RET (retarded potential) and SAT (saturation current') correspond to J/T^2 and are therefore in units of A/cm^2 K^2. Data digitally extracted from Figure 6 of ref. [189].

1000/*T* (K)	RET	SAT
0.66531	6.0567×10^{-14}	1.1969×10^{-13}
0.64416	1.7273×10^{-13}	3.5677×10^{-13}
0.62314	5.3231×10^{-13}	1.1295×10^{-12}
0.60566	1.4199×10^{-12}	3.0421×10^{-12}
0.58625	3.8574×10^{-12}	9.0186×10^{-12}
0.56664	1.1364×10^{-11}	2.6164×10^{-11}
0.54546	3.3281×10^{-11}	8.5922×10^{-11}
0.52692	–	2.3619×10^{-10}

so that

$$a_{Ta} = k_B \ln\left(\frac{A_{sat}}{A_{RLD}}\right) = 2.061 k_B = 0.1776 \text{ eV/K} \tag{12.5}$$

for tantalum, the material under investigation.

CHAPTER 13

Fowler–Nordheim equation

The main features of the phenomenon of the extraction of electrons from cold metals by intense electric fields are well known, and an approximate theory of the effect was first developed by Schottky. More recently the experimental data have been much improved, notably by Millikan and Eyring, and Millikan and Lauritsen. The theory has been considered afresh by O.W. Richardson and by Houston working with Sommerfeld. It seems to us, however, that there is still room for improvement in the theoretical exposition and its correlation with the experiments. Neither O.W. Richardson nor Houston really treat the theory in the simple straightforward way which is now possible in the new mechanics, using the revived electron theory of metals which we owe to Sommerfeld. Again, while Millikan and Lauritsen seem to have established quite definitely the laws of dependence of the emission on the field strength F, they speak of the implications of their result in a way which is hard to justify and might in certain circumstances prove to be definitely misleading.

– Ralph H. Fowler and Lothar W. Nordheim [36][1]

The Fowler–Nordheim (FN) equation predicts current density due to field emission, generally from metals that, due to closely spaced extraction grids or anodes, experience very high surface fields over small emission areas (e.g., Spindt-type field emitter arrays in Figure 13.1). Typical operational parameters are $T = 300$ K, $F = 8$ eV/nm, and $\Phi = 4.5$ eV, as characteristic of tungsten wires [182] or conical molybdenum field emitters [175]. Standard accounts are given by Dyke and Dolan [192] and Murphy and Good [110], with an updated reformulation by Forbes and Deane [193].

For the transmission probability, typical field emission gradients are quite strong, so strong in fact that those electrons with a longitudinal energy close to the Fermi level are most likely to tunnel, even though the supply function is small there. The supply function itself is well approximated by its zero temperature limit. Therefore,

$$D_{FN}(E) \approx P(E)\exp\left(-\frac{B}{F} - \frac{C}{F}(\mu - E)\right) \tag{13.1}$$

$$f_{FN}(E) \approx \frac{m}{\pi\hbar^2}(\mu - E)\Theta(\mu - E) \tag{13.2}$$

where the argument of the exponential in $D(E)$ is approximated by the linear expansion in energy E about the Fermi level μ, the supply function has made use of the approximation $\ln(1+x) \approx \ln(x)$, and a pre-factor $P(E)$ has been attached to $D(E)$ to accommodate the consequences of the coefficient of $\psi(x)$ in Eq. (11.22), its exact form will be determined using the Airy function approach in Eq. (18.59). Because the transmission probability $D(E)$ increases exponentially with energy but the supply function only linearly below the Fermi energy, their product will be peaked near $E \sim \mu$: other functions (such as $P(E)$) vary slowly with energy, and can therefore be evaluated at $E = \mu$ and withdrawn from the integral. Let $(C/F)(\mu - E) = x$, then

$$J_{FN}(F) = \frac{q}{2\pi\hbar}\left(\frac{m}{\pi\hbar^2 C}\right)P(\mu)e^{-B/F}\int_0^{C\mu} xe^{-x}dx \tag{13.3}$$

For large $C\mu$, the upper limit of the integration may be taken to $+\infty$ and so the integral is equal to unity (the form when $C\mu$ is not large, as for semiconductors, will be considered in Section 24.3.2). As a result the non-F-dependent terms can be grouped, and J_{FN} written as

$$J_{FN}(F) \approx AF^2\exp\left(-\frac{B}{F}\right) \tag{13.4}$$

with A defined by comparison with Eq. (13.3). An immediate consequence of the largeness or smallness of $C\mu$ is thereby revealed: in the case of metals, $C\mu \gg 1$, and the resulting $J_{FN}(F)$ is *independent of the chemical potential* μ.

Finding the forms of A, B, and C takes additional work, and depends on the nature of the Gamow factor for the determination of B and C via $\theta(E) \approx -(B/F) - (C/F)(\mu - E)$. The influential derivation of Murphy and Good [110] upgraded an earlier treatment by Nordheim [194] in which elliptical integral functions were recast as $v(y)$ and $t(y)$ to account for the barrier being given by an image charge potential. The present approach is intended to be more intuitive, and will be undertaken in three steps: (i) find A, B, and C in Eq. (13.3) using the simple triangular barrier approximation that

[1] The quote is the abstract in whole of Fowler and Nordheim's seminal 1928 manuscript.

Introduction to the Physics of Electron Emission, First Edition. Kevin L. Jensen.

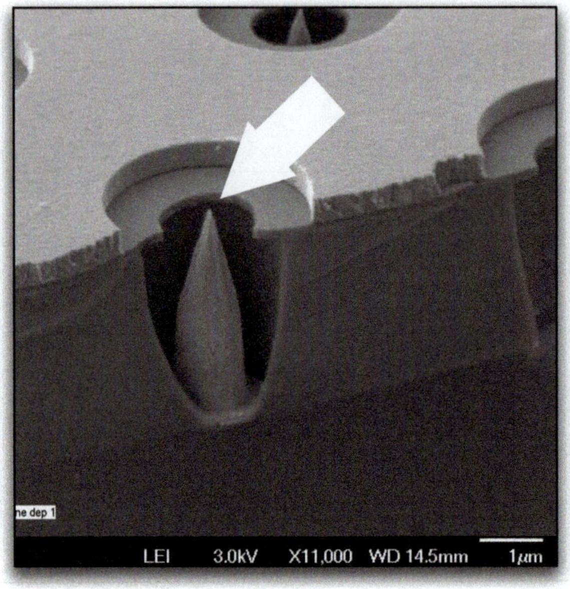

Figure 13.1 Close-up of a single Spindt-type gated field emitter (in an array) that is 3.5 μm in height; gate is 3 μm in diameter with a 1.6 μm silicon nitride shield. White arrow points to the emitter apex, where field emission occurs. Pulsed emission has yielded > 3 mA from individual tips. *Image courtesy of Christopher E. Holland, Capp A. Spindt (SRI).*

constituted the original FN equation, (*ii*) invoke scaling arguments based on how the integrand in the Gamow factor behaves in changing its appearance from triangular-barrier-like for small fields to quadratic-barrier-like for large fields, and (*iii*) consider the image charge potential directly and show that the behavior anticipated from the scaling arguments is what in fact occurs.

13.1 Triangular barrier approximation

What we are doing here, that is the question. And we are blessed in this, that we happen to know the answer.

– Samuel Beckett[2]

The evaluation of $\theta(E)$ using the triangular barrier $V_{tri}(x) = \mu + \Phi - Fx$ in $\theta(E)$ is known from Eq. (11.17) and is given by

$$\theta_{tri}(E) = \frac{4}{3}\kappa(E)L_{tri}(E) = \frac{4\sqrt{2m}}{3\hbar F}(\mu + \Phi - E)^{3/2} \tag{13.5}$$

where the second relation follows from the definitions $\kappa(E) = \sqrt{2m\Delta(E)}/\hbar$ and $L(E) = \Delta(E)/F$ with $\Delta(E) = \mu + \Phi - E$ as in Eq. (11.26). Taylor expanding $\theta(E)$ to first order in $(\mu - E)$ is sufficient to evaluate B and C. Doing so gives

$$B = \theta(\mu)F = \frac{4}{3\hbar}\sqrt{2m\Phi^3} \tag{13.6}$$

$$C = \theta'(\mu)F = \frac{2}{\hbar}\sqrt{2m\Phi} \tag{13.7}$$

where the prime indicates derivative with respect to argument $\theta'(E) = \partial_E\theta(E)$, and $\theta'(\mu)$ is $\theta'(E)$ evaluated at $E = \mu$. Both B and C as defined do not depend on field (but only work function Φ) and so are constant.[3] Comparing Eq. (13.3) to Eq. (13.4) reveals that

$$A = \frac{q}{16\pi^2\Phi\hbar}P(\mu) \tag{13.8}$$

[2]Samuel Beckett, *Waiting for Godot; Tragicomedy in 2 Acts*. New York: Grove Press, 1954, Act II, p. 52.
[3]Including Φ inside B and C is temporary: by Eq. (13.25) the Φ dependence will be pulled out and shown explicitly.

Nothing in the analysis so far reveals the nature of $P(\mu)$, but in Section 18.2.4 it will be evaluated and shown to be given by

$$P(\mu) = \frac{4\sqrt{\mu\Phi}}{\mu + \Phi} \tag{13.9}$$

and thus is on the order of unity (e.g., 1.9522, if $\mu = 7$ eV and $\Phi = 4.5$ eV). All together, the oft-cited[4] FN equation introduced in ref. [36] is

$$J_{FN}(F) = \frac{q}{16\pi^2\Phi\hbar}\left(\frac{4\sqrt{\mu\Phi}}{\mu+\Phi}\right)F^2 \exp\left(-\frac{4}{3\hbar F}\sqrt{2m\Phi^3}\right) \tag{13.10}$$

The triangular barrier model contains enough to anticipate through scaling arguments what the effects of the image charge potential should be. From Eq. (13.5) for $\theta(E)$, $\kappa(E)$ is height of the integrand and $L(E)$ is the width at the energy E. The effect of taking the derivative of $\theta(E)$, or $d\theta/dE$, is to *bring $\kappa(E)$ to the denominator of the integrand without changing the length of the integration region*. Therefore, $\theta(\mu)$ will scale as $\kappa(\mu)L(\mu)$ but $\theta'(\mu)$ will scale as $L(\mu)/\kappa(\mu)$. What effect will that have?

For low fields, the image charge barrier remains more or less triangular in appearance, with the height of the barrier above μ being $\Phi \to \Phi - \sqrt{4QF} = \phi$. Introduce $y = \sqrt{4QF/\Phi^2}$, so that $\Phi \to (1-y)\Phi$. If $v(y)$ is introduced such that $B \to Bv(y)$ when $\Phi \to (1-y)\Phi$, then the question is how does $v(y)$ vary with y based on how $\theta(\mu)$ scales? As $\kappa(\mu) \propto \Phi^{1/2}$, then $\kappa \to \kappa\sqrt{1-y}$ for $\Phi \to (1-y)\Phi$. Similarly, $L(\mu) \propto \sqrt{\Phi^2 - 4QF} = \Phi\sqrt{1-y^2}$, and so $v(y)$ is a decreasing function of y that behaves more or less as

$$v(y) \sim (1-y)(1-y^2)^{1/2}$$

the performance of which, if compared to its accurate numerical evaluation via Eq. (13.16) below, would be seen to be in the right direction but not particularly praiseworthy: the fault lies in the neglect of how the shape factor coefficient scales. Nevertheless, scaling arguments suggest that $v(y)$ is a decreasing function of y and therefore F, showing that the image charge current density improves over the simple triangular barrier form of Eq. (13.10). Such a conclusion is intuitively obvious due to the barrier being both lower and thinner when the image charge is included. In a similar manner, for $C \to Ct(y)$ as $\Phi \to (1-y)\Phi$ means $\theta'(\mu)$ scales as $L(\mu)/\kappa(\mu)$ or

$$t(y) \sim \frac{(1-y^2)^{1/2}}{(1-y)^{1/2}} = \sqrt{1+y}$$

which, like the expression above for $v(y)$, is indicative of the behavior of $t(y)$ (increases as field increases) without being a particularly good approximate form of it if compared to Eq. (13.19) below.

13.2 Image charge approximation

...if thou gaze long into an abyss, the abyss will also gaze into thee.

– Friedrich Nietzsche[5]

The Murphy and Good (MG) equation for $J_{MG}(F)$ must take on the accurate evaluation of the elliptical integral functions[6] $v(y)$ and $t(y)$, so-called here because they can be expressed in terms of elliptical integrals such as Eq. (30.104). In an earlier era, tabulating functions was the *modus operandi* of dealing with them, and so Abramowitz and Stegun [103] was never far away. More recently, though, very good approximations to $v(y)$ and $t(y)$ have become available [193], accurate to within 0.3% over the range $0 \le y \le 1$. Even so, for those of a more computational disposition, evaluating the one-dimensional integral for $R(w)$ below used to construct $v(y)$ and $t(y)$ is no longer taxing and can be performed accurately and rapidly.

The image charge represents the metal looking back (as it were) at a charge outside its surface. The degree to which it looks back is governed by the elliptical integral functions $v(y)$ and $t(y)$, so-called here because they can be expressed in

[4]Be forewarned: at times in the literature the Murphy and Good formulation of $J_{FN}(F)$ is given and ref. [36] incorrectly cited. The equation of Fowler and Nordheim in their 1928 article is Eq. (13.10) and is based on a triangular barrier without a Schottky factor.

[5]Friedrich Wilhelm Nietzsche, *Beyond Good and Evil* (trans. Helen Zimmern), Kindle edition, 2011, Chapter IV: Apophthegms and Interludes, Aphorism 146.

[6]Alternately, R. Forbes has advocated calling them *Schottky–Nordheim barrier functions*.

terms of elliptical integrals such as Eq. (30.104). Remember that $v(y)$ modifies $\theta(\mu)$ and $t(y)$ modifies $\theta'(\mu)$, that is, if $\theta(E)$ is approximated by a linear function in energy about $E = \mu$, then the relation

$$\theta(E) \approx \frac{4}{3\hbar F}\sqrt{2m\Phi^3}\, v(y) + \frac{2}{\hbar F}\sqrt{2m\Phi}\, t(y)(\mu - E) \tag{13.11}$$

is the defining equation for $v(y)$ and $t(y)$. The form of Eq. (13.11) will reappear slightly rearranged in Eq. (17.10). Whether one heroically tackles numerically evaluating the elliptical integral functions or sheepishly uses an approximation, the MG approximation to the FN equation takes the form

$$J_{MG}(F) = \frac{qF^2}{16\pi^2\hbar\Phi t(y)^2}\exp\left(-\frac{4}{3\hbar F}\sqrt{2m\Phi^3}v(y)\right) \tag{13.12}$$

One notices right away that $P(\mu)$ has disappeared: that is not so bad as it would otherwise introduce yet another parameter (the Fermi energy μ) in a coefficient $P(\mu)$ that is of order unity in an equation that already faces many challenges when applied to experimental data. The need to retain a level of accuracy that is hidden in the uncertainty of other parameters themselves or experimental error in general[7] to predict field emission from metals reflects more on the compulsion of theorists than the needs of experimentalists. Therefore, for decades Eq. (13.12) was treated as "the" FN equation even though many approximations to the $v(y)$ and $t(y)$ functions (and a host of other factors [186]) customize the formula and therefore its predictions in numerous small (and occasionally large) ways.

Still, uncertainty is no excuse to evaluate the elliptical integral functions poorly (or ignore them) when ways to obtain good values are available. In an earlier era, tabulating functions was the *modus operandi* of dealing with them, and so Abramowitz and Stegun [103] was generally within reach for the theorists. The functions $v(y)$ and $t(y)$ have been tabulated [195], series expanded [196, 197], interpolated for spreadsheets [198], and put into a recursion relation [193]. For those of a more computational disposition, though, evaluating $v(y)$ and $t(y)$ numerically, accurately, and rapidly is possible. Therefore, use may be made of Eq. (11.26) and its companions, particularly $L(\mu)$ and $x_{\_}(\mu)$. Replace the integration variable $x \rightarrow= x_{\_}(\mu) + L(\mu)\sin^2(u)$ in the Gamow factor integral of Eq. (11.14) and evaluate it at $E = \mu$. All the factors with the exception of $w = x_{\_}/L$ may be pulled out of the integrand, leaving [197, 199]

$$\theta(\mu) = \frac{4}{\hbar}\sqrt{2mFL(\mu)^3}R(w) \tag{13.13}$$

$$R(w) \equiv \int_0^{\pi/2}\frac{(\cos u)^2(\sin u)^2}{\sqrt{w + \sin^2 u}}du \tag{13.14}$$

$$w(y) = \frac{x_{\_}(\mu)}{L(\mu)} = \frac{1}{2}\left(\frac{1}{\sqrt{1 - y^2}} - 1\right) \tag{13.15}$$

in terms of which the sought-for $v(y)$ function becomes

$$v(y) = 3(1 - y^2)^{3/4}R(w) \tag{13.16}$$

The asymptotic limits of $R(w)$ can easily be found.

EXAMPLE: Find $R(w)$ in the limit of small field.

SOLUTION: Small field corresponds to small y and therefore to $w \rightarrow 0$. The integral defining $R(w)$ becomes

$$R(0) \rightarrow \int_0^{\pi/2}\cos^2 u\ \sin u\ du = \int_0^1 x^2\ dx = \frac{1}{3}$$

As F is coupled to Q, small field also corresponds to the same limit as $Q \rightarrow 0$, meaning the barrier appears increasingly triangular. For the triangular barrier, $L(\mu) = \Phi/F$, and Eq. (13.6) is recovered.

[7] This says nothing about the level of accuracy of the FN equation itself, which is already an approximation because of how the transmission probability is treated (Chapter 17 and Section 18.2.4).

EXAMPLE: Find $R(w)$ in the limit of large field.
SOLUTION: Large field corresponds to $y \to 1$ and therefore to $w \to \infty$. The integral defining $R(w)$ becomes

$$R(w \gg 1) \to \frac{1}{\sqrt{w}} \int_0^{\pi/2} \cos^2 u \sin^2 u \, du = \frac{\pi}{16\sqrt{w}} \tag{13.17}$$

For large F, the barrier for $E > \mu$ appears parabolic.

In a premonition of things to come, Eq. (13.11) is not the only starting point. The strength of the field just as the top of the image charge barrier is dragged down to the Fermi level is given by $F_{max} = \Phi^2/4Q$, obtained by setting $\phi = 0$ in Eq. (11.30). In that limit, $L \to 0$ so that $x\_(\mu)/L(\mu) = (\Phi/2FL(\mu)) - (1/2) \to \infty$. Consequently,

$$\lim_{F \to F_{max}} \theta(\mu) \approx \frac{\pi}{2} \frac{\sqrt{m\Phi^3}}{\hbar F} \left(1 - \frac{4QF}{\Phi^2}\right) \tag{13.18}$$

as could be inferred from Eq. (11.20) if one is careful about what the relationship is between the quadratic barrier and the image charge parameters when F approaches F_{max} from below.[8] Eq. (13.18), however, is a high field limit, and the evaluation of the field emission current comes from the low field side.

Returning to the elliptical integral functions, the ease with which $v(y)$ may be numerically calculated using $R(w)$ means the more painful analysis using elliptical integrals and tabulated values has been dodged, and so attention turns to $t(y)$. It can be shown that $t(y)$ is defined by

$$t(y) = \left(1 - \frac{2y}{3}\frac{\partial}{\partial y}\right) v(y) = \frac{3(2w+1)R(w) - 4w(w+1)R'(w)}{\sqrt{2w+1}} \tag{13.19}$$

where $R'(w) = dR/dw$ or

$$R'(w) = -\frac{1}{2} \int_0^{\pi/2} \frac{\cos^2 u \sin^2 u}{(w + \sin^2 u)^{3/2}} du \tag{13.20}$$

EXAMPLE: Find $v(1)$ and $t(1)$ exactly.
SOLUTION: Begin with $v(y)$ as defined by Eq. (13.16) and use $w(y)$ as defined by Eq. (13.15). Then in the limit $w(y \to 1) \to \infty$, Eq. (13.17) is useful, giving

$$v(y \to 1) = \frac{3R(w)}{(2w+1)^{3/2}} \to \frac{3\pi\sqrt{2}}{64w^2} = 0 \tag{13.21}$$

The evaluation of $t(y \to 1)$ is more difficult, but the same analysis giving Eq. (13.17) when applied to Eq (13.20) shows that $R'(w) \to -\pi/32w^{3/2}$ for large w, and so when used in Eq. (13.19)

$$t(y \to 1) = \frac{\pi(8w+5)}{16[w(2w+1)]^{1/2}} \to \frac{\pi\sqrt{2}}{4} = 1.11072 \tag{13.22}$$

Such formulae, however, can still be off-putting. As a result, the "quadratic" approximations of Spindt *et al.* [175]

$$v_{spindt}(y) \approx v_o - y^2$$
$$t_{spindt}(y) \approx t_o$$

where v_o and t_o are given by $v_o = 0.95$ and $t_o^2 = 1.1$ are frequently employed for their reasonable accuracy and huge simplicity.

[8] For the image charge potential, $\theta(\mu)$ differs from $\theta_{qua}(\mu)$ by a factor of $\sqrt{2/(1+y)}$, which approaches 1 as $y \to 1$, a pleasing little derivation for those who enjoy such things.

And so matters sat for years until Forbes found a stunningly simple approximation of fairly good accuracy (better than 0.33% over the whole range of y) by numerical experimentation using MAPLE. In what began as a happy accident [200] but which was put on a better mathematical footing [193, 201] Forbes and Deane found

$$v(y) = 1 - \frac{y^2}{3}(3 - \ln(y)) \tag{13.23}$$

$$t(y) = 1 + \frac{y^2}{9}(1 - \ln(y)) \tag{13.24}$$

for which it is easy to see $v(y)$ varies from 1 to 0 and $t(y)$ from 1 to 1.1111 as in Figure 13.2 (the latter behavior justifying the widespread approximation of $t(y)$ by a constant, but observe that the exact value of $t(1)$ is not 10/9 but rather $\pi\sqrt{2}/4$ from Eq. (13.22)). What is now clearly seen is that each contains a logarithmic term, the presence of which frustrates polynomial and other kinds of expansions near $y = 0$ and $y = 1$ (the logarithmic term was anticipated in ref. [197], but the form there lacked the elegant simplicity and therefore computational utility of the Forbes–Deane version).

The form of Eq. (13.23) can be used to "correct" the Spindt approximation by demanding that $v(y)$ be linear in y^2 with the coefficient of y^2 set to unity, which gives $y_o = 1/\sqrt{e} = 0.606531$, $v_o = 1 - (1/6e) = 0.9387$ and $t(y_o) = t_o = 1 + (1/6e) = 1.06131$ [85, 202]. What of it? Although the usage of t_o makes sense (similar to the reasons for abandoning $P(\mu)$), use of the quadratic form of $v(y) \approx v_o - y^2$ is ill-advised as in Figure 13.3, particularly when fields are high such that Eq. (13.23) is correct but the quadratic form is not, and this will matter when emission near the barrier maximum is considered such

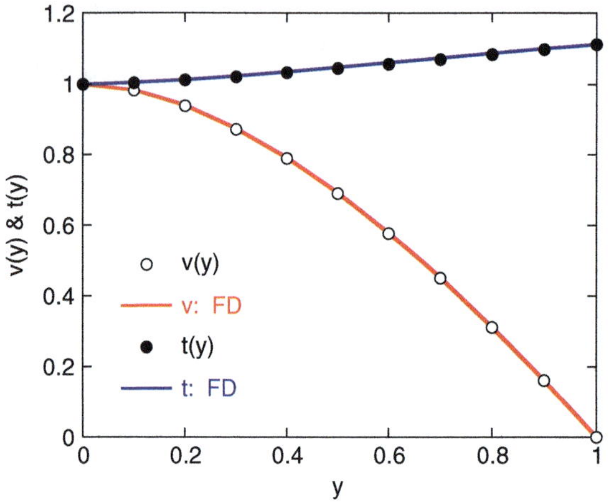

Figure 13.2 The Nordheim functions. A comparison of the numerically calculated $v(y)$ and $t(y)$ functions with the approximations of Eq. (13.23): the differences are generally less than 0.33%, as shown in Figure 13.3.

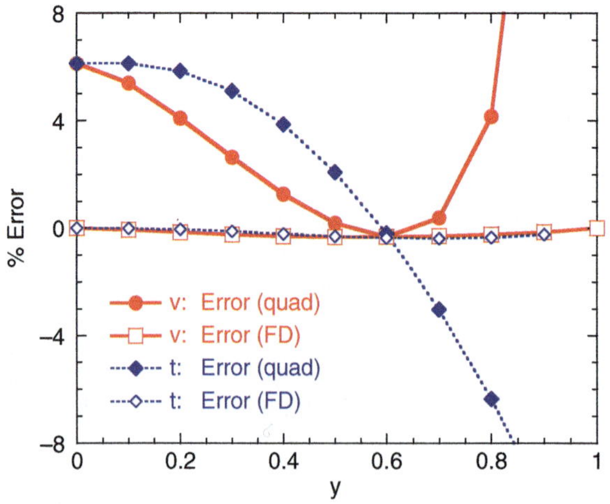

Figure 13.3 Nordheim function approximation error. The error, defined as $100\%(f - f_a)/f$, where f_a is the approximation to the exact value f, of the functions $v(y)$ and $t(y)$. Quad refers to $v_{quad}(y) = 0.9387 - y^2$ and $t_{quad}(y) = 0.9387 + y^2/3$. FD refers to Eqs (13.23) and (13.24).

that Eq. (13.18) becomes a better starting point in the thermal-field equations of Chapter 17. If one craves simplicity at the cost of a small bit of accuracy (and for *modeling* electron emission in the beam codes of Chapter 33, one does), then the *recommended* approximations are to use Eq. (13.23) and let $t(y) \approx t_o$ if one must, and therefore render $J_{MG}(F)$ as [188]

$$J_{FN}(F) \equiv \frac{A_o}{\Phi t_o^2}\left(\frac{\Phi^2 e^6}{4Q}\right)^{\nu} F^{2-\nu} \exp\left(-B_o \frac{\Phi^{3/2}}{F}\right) \tag{13.25}$$

$$A_o \equiv \frac{q}{16\pi^2\hbar} = 1.5414 \times 10^{-6} \left[\frac{\text{A}}{\text{eV}}\right] \tag{13.26}$$

$$B_o = \frac{4}{3\hbar}\sqrt{2m} = 6.8309 \left[\frac{1}{\text{nm}\sqrt{\text{eV}}}\right] \tag{13.27}$$

$$\nu = \frac{2B_o Q}{3\sqrt{\Phi}} = \frac{8Q}{9}\left(\frac{2m}{\Phi}\right)^{1/2} = \left(\frac{2.68754\text{ eV}}{\Phi}\right)^{1/2} \tag{13.28}$$

where A_o, B_o and ν are in Table 2.3. The point of introducing A_o and B_o is because they are collections of constants both fundamental and mathematical ("universal constants" in the terminology of Forbes [198]), so that giving these constants special status is desirable.

The form of Eq. (13.25) has illuminating consequences. Observe that with the Forbes–Deane approximation, the *form* of the MG Eq. (13.12) and the original Fowler–Nordheim Eq. (13.10) become hauntingly close, albeit that $P(\mu)$ is neglected and that A_o and B_o are different than A and B by pulling out the work function Φ explicitly. The *close* family similarity between Eq. (13.25) and Eq. (13.10) explains (but does not justify) why the triangular barrier formula is occasionally (and wrongly) used for metals: the slope of the current density in so-called Fowler–Nordheim coordinates $\ln(J_{FN}/F^2)$ vs $1/F$ seems to be the same for both, and often it is the slope that is of interest. This is because the Schottky barrier lowering factor manifests itself in the new coefficient of $(\Phi^2 e^6/4Q)^{\nu}$ and a factor of $F^{-\nu}$ in the coefficient, both of which do not affect the slope (much) even though they *do* affect the magnitude of J_{FN} by quite a bit, so that including it is a must. Still, the neglect of $P(\mu)$ by setting it equal to unity is desirable to avoid the proliferation of the quantity of parameters needed to evaluate J_{FN} (for now, that is: when field emission from semiconductors is considered, the behavior of $P(E)$ will come back into consideration, and it will not be ignored so blithely, see Section 24.3.3.).

Eq. (13.25) is understood to be the MG form utilizing the Forbes–Deane approximation for the evaluation of the Fowler–Nordheim current density. Because a subscript acronym of FDMGFN is unwieldy (although it would reflect genealogy) it has been contracted back to FN. When J_{FN} is mentioned hereafter, Eq. (13.25) is what is in mind.

13.3 Analysis of field emission data

Priestley: *I thought details mattered to you, sir.*
Lavoisier: *Only if they are relevant.*

–Carl Djerassi and Roald Hoffmann[9]

Turning attention now to the ability of Eq. (13.25) to account for features evident in experimental data, consider first the data of Millikan and Eyring [49], invoked by both Oppenheimer [109] and Fowler and Nordheim [36] to give credence to their equations of tunneling and emission. The data given in Table 13.1 and shown in Figure 13.4 were designed to show the small but nevertheless non-zero impact of temperature on field emission: it was the insensitivity of field emission current to temperature, recall, that led to the initial speculation that thermal emission electrons ("thermions") were a different breed than field emission electrons. Millikan and Eyring's work demonstrated that conduction band electrons were the carriers of emission current, and that conclusion led them to speculate on how the thermal and field emission equations could be unified, a subject returned to in Section 13.4.

The data in Figure 13.4 shows an exponential behavior over the range of anode potentials measured, a common feature of field emission. Considering the 300 K line (both increasing and decreasing data) and performing a best fit of the data using the forms of field emission hypothesized by Oppenheimer ($AF^{1/4}\exp(-B/F)$) to that hypothesized by Fowler and Nordheim ($AF^2\exp(-B/F)$) (recall that, by Eq. (13.25), the parameterization by these forms is *reasonable*) it is seen the data are a bit too uncertain to conclusively choose between the two forms (*Cautionary Moral* 1). It is common to represent

[9] Carl Djerassi and Roald Hoffmann, *Oxygen: A Play in Two Acts*. Weinheim, New York: Wiley-VCH, 2001, Scene 9, p. 89.

Table 13.1 Field emission data of Millikan and Eyring. The values are equivalent to those of Table VII in ref. [49]. The data refer to current vs anode potential for two temperatures, 300 K and 1100 K (the third column of Table VII is not reproduced given its closeness to the 300 K values shown). The data are plotted in Figure 13.4.

V_a (V)	I(300 K) (μA)	I(1100 K) (μA)
3050	0.12	0.14
3500	0.60	0.75
4000	4.50	5.40
4400	21.00	24.00
4750	54.00	59.00
5000	100.00	110.00
4800	63.00	69.00
4200	12.00	15.00
3750	2.10	2.70
3200	0.24	0.36
3000	0.09	0.15

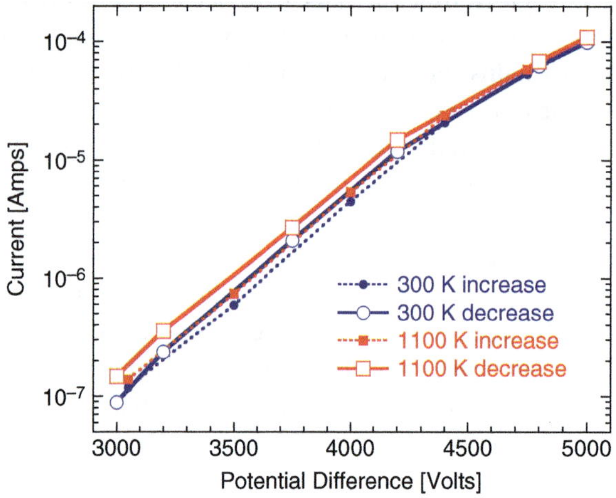

Figure 13.4 Graphical representation of data of Millikan and Eyring ([49], Table VII) given in Table 13.1.

$I(V)$ data on a so-called Fowler–Nordheim plot wherein $\ln(I/V^2)$ vs $1/V$ is considered, based on the presumed relationship between current I and current density J via $I = AJ$, where A is emission area, and likewise the relationship between anode potential V and field F via $F = \beta_g V$, the subscript emphasizing that the β_g field enhancement factor, when relating voltage to field, has dimensions of *inverse length*, as is common for presenting experimental data (e.g., Barbour *et al.* [182]). Such discussions are premature before considering what the area factor A could be (see Chapter 30), but a best-fit analysis, shown in Figure 13.5, shows that the hypotheses of Oppenheimer, and separately Fowler and Nordheim, are borne out: the agreement with each of their hypothetical equations is reasonable given the accuracy of the data. Lastly, the data is recast on a Fowler–Nordheim plot, or representation, in Figure 13.6.

13.4 The Millikan–Lauritsen hypothesis

It was all very well to say "Drink Me," but the wise little Alice was not going to do that in a hurry. "No, I'll look first," she said, "and see whether it's marked 'poison' or not"; for she had read several nice little histories about children who had got burnt, and eaten up by wild beasts and other unpleasant things, all because they would not remember the simple rules their friends had taught them ... and she had never forgotten that, if you drink much from a bottle marked 'poison,' it is almost certain to disagree with you ...

– Lewis Carroll[10]

[10]Lewis Carroll, *Alice's Adventures in Wonderland*. Racine, WI: Whitman Publishing Company, 1945.

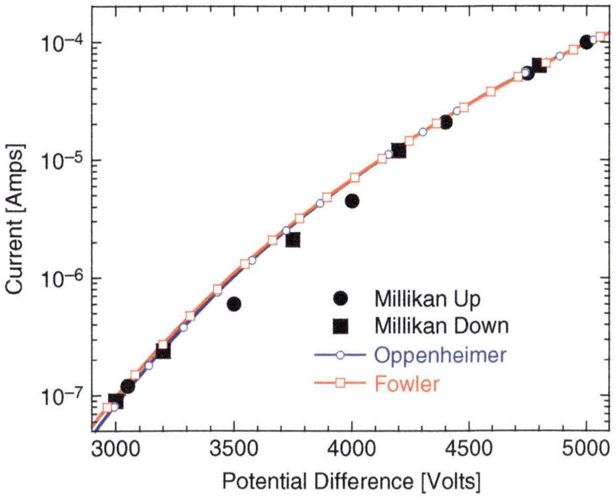

Figure 13.5 Millikan–Eyring data and theory ([49], Table VII). The 300 K data of Figure 13.4 compared to the predictions of Oppenheimer [109] and Fowler and Nordheim [36]: observe the uncertainty in the data.

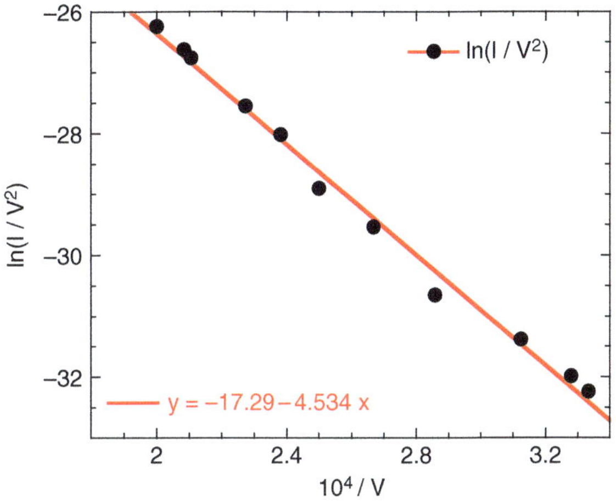

Figure 13.6 Millikan–Eyring data on a Fowler–Nordheim plot. The 300 K data of Figure 13.4 represented on a typical Fowler–Nordheim plot. No distinction is made between the increasing vs decreasing directions of the data, as in Table 13.1.

The only rational way of educating is to be an example – if one can't help it, a warning example.

– Albert Einstein[11]

Cautionary tales, in which incautious children meet grisly fates for heedlessly failing to follow sage precepts from those with experience greater than their own, are efficacious ways to warn of perils. Any tale centered on the origins of a clever hypothesis, and conclusions drawn from it, that proves to be not quite right deserves attention for the object lessons it provides.

Millikan, who measured the electron charge using his iconic oil drop experiment (a feat for which he received the Nobel Prize in 1923; he also measured the value of Planck's constant, but that is another story), was an accomplished and discriminating experimentalist, whose careful research on the difference between current emitted from heating metals compared to that pulled from the metals by high field were important to both Oppenheimer and Fowler and Nordheim in the development of their theories of field emission.

Millikan and Lauritsen attempted to distinguish between "thermionic currents" and "field currents", and their careful measurements showed that temperature effects on the field currents did not manifest themselves until astonishingly high temperatures, from which came "the conclusion that the 'field-electrons' at ordinary or low temperatures have nothing

[11] Albert Einstein, *Ideas and opinions. Based on Mein Weltbild*, Essay, "Education and Educators". New York: Crown Publishers, 1954, p. 57. (A letter to a young girl, published in *Mein Weltbild*. Amsterdam: Querido Verlag, 1934).

to do with thermions, i.e., that they must have a static, not a dynamic, interpretation" [50]. In other words, the emission of thermions was thought to be due to a process by which they shared in the "thermal agitation" of the atoms, in contrast to the notion that "...field currents are due to conduction electrons pulled from minute peaks on the surface" [49], as argued in their earlier work.[12] Though the origins of the current were by different *mechanisms*, one associated with the behavior of the transmission probability (field) in contrast to the other associated with the thermal tail of the distribution (thermal), current was nevertheless carried by the same *particle*: without the conceptions introduced by Sommerfeld and seized upon by Fowler and Nordheim, speaking of the processes coherently was difficult.

It had already been established that empirical data for the total current from thermal emission from an area A at a temperature T produced a linear relation when $\ln(I/V^2)$ was plotted versus $1/T$, or

$$I(V) = AT^2 \exp\left(-\frac{B}{T}\right) \tag{13.29}$$

with A and B related to the slope and intercept of the aforementioned linear relation. Similarly, when Millikan and Lauritsen examined $\ln(I/V^2)$ in their field emission studies, they found that it, too, satisfied a linear relation with $1/V$, or

$$I(V) = A'V^2 \exp\left(-\frac{B'}{V}\right) \tag{13.30}$$

The astonishing similarity between these two equations, in which the T of the former has the same role as the F of the later,[13] led Millikan and Lauritsen to hypothesize that the Richardson–Laue–Dushman (RLD) equation was the low field limit of a "general" formula given by

$$J_{ML}(F,T) = A(T+cF)^2 \exp\left(-\frac{b}{T+cF}\right) \tag{13.31}$$

(their Eq. (5)) from which they concluded [50] "...the application of an external field *is equivalent to increasing the temperature* of the electrons within the metal" (their emphasis). Seems reasonable.

However, as Hamlet's mother was rightly admonished,[14] "seems" and "is" are different: appearances can suggest relations which do not survive in the light of a coherent theory. As attractive and perhaps intuitive as Eq. (13.31) seems to be, it is errant, as are the conclusions based on it. The arguments of Fowler and Nordheim shortly afterwards (their 1928 abstract of ref. [36] starts Chapter 13), and the energy distribution measurements of Gadzuk and Plummmer [96] several decades later would drive home how different the processes were; even later, a general thermal field equation, developed in Chapter 17, would show the actual equation even *looks* different, in that $J(F,T)$ is the *sum* of a thermal part and a field part, not one expression with differing limits. Herein lies several cautionary tales: that similarity is not equivalence, that equations with fitting parameters can model data just fine and still motivate misguided conclusions, that an equation which fits the data is not the same as a theoretically justified equation which fits the data (although a theoretical equation that *doesn't* fit the data is worse). But it is also true that Hamlet's hesitation itself is cautionary in that understanding need not be required for action and that delay can be fatal to ambition.

The behavior of the transmission probability $D(E)$ and the supply function $f(E)$ give an explanation as to why Eq. (13.31) cannot be the general equation of thermal and field emission. In thermal emission, $D(E)$ is a step function which acts to mask all but the thermal tail of $f(E)$ so that the product of the two (the current density integrand) is peaked about the barrier maximum $\mu + \Phi$; in field emission, $D(E)$ exponentially increases but slower than the thermal tail decreases, so that the maximum of the current density integrand is peaked about the Fermi level μ (Figure 8 of ref. [110]; Figure 3 of ref. [155]) and shown by measurements [96]. Field effects appear in thermal emission via the Schottky barrier lowering factor. Thermal effects do not appear in field emission because the dominant part of the current density integrand is where the supply function $f(E)$ in Eq. (11.5) behaves as $f(E \sim \mu) = (m/\pi\hbar^2)(\mu - E)$ until conditions are such that the states in the thermal tail ($E > \mu$) of the supply function begin to contribute appreciably. Such a way of looking at emission came after Millikan's hunch, though, so he is entitled to some leeway by posterity.

The cautionary tale about empirical formulae, however, stays.

[12] The passage goes on to say "...the fatigue effects of both current treatment and heat treatment being due to the rounding off of these peaks by positive ion bombardment or by temperature ..." [49].

[13] This assertion presumes field F and anode potential V are linearly related, as well as I and J; that assumption will come under great scrutiny in Chapter 30.

[14] "Seems, madam? Nay, it is, I know not 'seems."', ref. [37]: *Hamlet* I.ii.77, p. 936.

CHAPTER 14

Fowler–Dubridge equation

The precise hypothesis which succeeds so well in correlating the observed effect near the threshold is that the photoelectric sensitivity or number of electrons emitted per quantum of light absorbed is to a first approximation proportional to the number of electrons per unit volume of the metal whose kinetic energy normal to the surface augmented by hν is sufficient to overcome the potential step at the surface.

– Ralph H. Fowler [203]

The Fowler–Dubridge (FD)[1] equation predicts quantum efficiency generally for metals due to optical and UV light incident on the surface. Typical operational parameters are $T = 300$ K, $F = 50$ eV/μm, and $\Phi = 4.5$ eV for metals (work function) or $\chi = 1.6$ eV for cesiated semiconductors, as characteristic of cesium antimonide or multialkali antimonide surfaces (electron affinity). A standard account is given by DuBridge [204] and Bechtel [205]; the latter gives a generalization to multiphoton effects.

14.1 Approximations

Heisenberg: *It starts with Einstein.*
Bohr: *It starts with Einstein ...the universe only exists as a series of approximations. Only within the limits determined by our relationship with it. Only through the understanding lodged inside the human head.*

– Michael Frayn[2]

Beginning with Einstein's assertion that a single electron absorbed the whole of the incident photon's energy (Section 3.3), Fowler anticipated that the consequences could be accounted for not by altering $f(E_x)$, but rather by modifing $D(E_x)$. Appending the entire photon energy $\hbar\omega$ to E_x may not seem proper: after absorbing a photon, the electron should be heading off in any direction, only some of which are towards the surface, but Fowler pointed out that appending the photon energy to the transmission probability term via $D(E_x) \to D(E_x + \hbar\omega)$ worked surprisingly well (moving beyond that approximation is considered in Section 31.4). Therefore, in Eq. (11.3) make the following substitutions

$$D_{FD}(E) \approx \Theta[E + \hbar\omega - (\mu + \phi)] \tag{14.1}$$

$$f_{FD}(E) = \frac{m}{\pi\beta\hbar^2} \ln\{1 + \exp[\beta(\mu - E)]\} \tag{14.2}$$

where the x subscript on E_x is hidden. That the supply function $f(E)$ is not being approximated by any of its limits (as done for both thermal and field emission previously) might cause apprehension, as it should: the integration is going to have contributions from the Maxwell–Boltzmann tail characteristic of thermal emission, but also from the linear behavior in E of field emission, and both can contribute depending on the size of $E + \hbar\omega$ compared to $\mu + \Phi$.

The ratio of the number of electrons *emitted* $\Delta Q/q$ with the number of photons *absorbed* $\Delta E/\hbar\omega$ defines the quantum efficiency QE, or

$$QE \sim \frac{\Delta Q/q}{\Delta E/\hbar\omega} = \left(\frac{\hbar\omega}{q}\right)\frac{\Delta Q}{\Delta E} \tag{14.3}$$

where ΔQ is the total emitted charge and ΔE is the total absorbed energy. Assume next that the duration over which the metal surface is illuminated is long compared to time scales associated with photon absorption or electron transport within the metal to the surface. Consequently, electrons are assumed to be emitted as soon as the surface is illuminated, and cease to be emitted as soon as the illumination is extinguished. Further, assume that the light intensity is not so

[1] Observe that in the present section the subscript FD refers to Fowler–Dubridge, not Fermi–Dirac.
[2] Michael Frayn, *Copenhagen*. New York: Anchor Books, 2000, Act 2, p. 73.

Introduction to the Physics of Electron Emission, First Edition. Kevin L. Jensen.

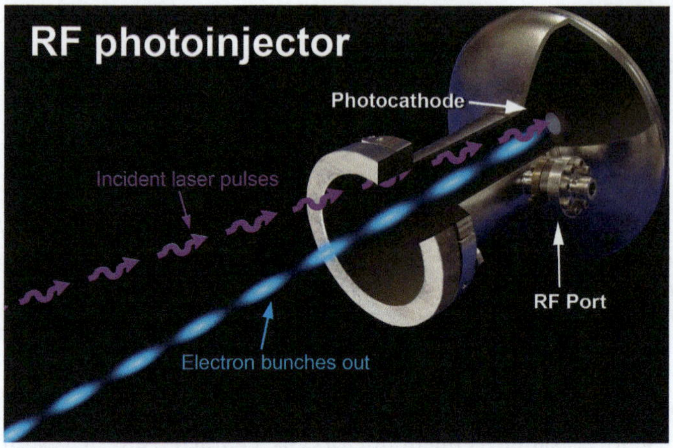

Figure 14.1 Representation of a photocathode in an rf injector. Image courtesy of C. Hernandez-Garcia, *Thomas Jefferson National Accelerator Facility* (JLAB). Reproduced from C. Hernandez-Garcia, P.G. O'Shea, M.L. Stutzman, *Physics Today* **61**, 44 (2008) with the permission of the American Institute of Physics.

severe that the metal appreciably heats up (as would happen with powerful lasers characteristic of photo-injectors as in Figure 14.1), and therefore no temperature-dependent effects occur (e.g., thermal emission from a heated metal surface). If Δt is the duration of the illumination and I_λ is the illumination intensity, then $\Delta Q/\Delta E \approx (J(\hbar\omega)\Delta t)/(I_\lambda \Delta t) = J(\hbar\omega)/I_\lambda$. In fact, if the electron emission tracks the time dependence of the illumination, then

$$J(\hbar\omega) \propto \left(\frac{q}{\hbar\omega}\right) I_\lambda \tag{14.4}$$

A number of factors affect Eq. (14.4). A factor of $(1 - R(\omega))$ describes the fraction of photons absorbed, where $R(\omega)$ is the reflectivity of the surface. Not all accounts of quantum efficiency include this factor, with some treatments defining it as the ratio of the emitted current density to absorbed (not incident) light intensity, so *cave, lector.*[3] Next, not all of the electrons that are excited make their way to the surface unhindered: they can scatter with other electrons and/or vibrational modes of atoms (*aka* phonons) or are lost by other means such as recombination so that only a fraction F_λ survive. At the surface, the remaining electrons encounter a surface barrier, for which only a faction $P(\hbar\omega)$ of the electrons with an energy above the barrier experience transmission into vacuum, and so

$$J(\hbar\omega) = \left(\frac{q}{\hbar\omega}\right) [1 - R(\omega)]\, F_\lambda I_\lambda P(\hbar\omega) \tag{14.5}$$

describes the current density. If QE is defined as the ratio of the number of electrons out $(\pi r^2 J\Delta t/q)$ to the number of incident photons $(\pi r^2 I_\lambda \Delta t/\hbar\omega)$ for a given area πr^2 and pulse length Δt, then

$$QE \equiv \left(\frac{\pi r^2 J\Delta t}{q}\right)\left(\frac{\hbar\omega}{\pi r^2 I_\lambda \Delta t}\right) = [1 - R(\omega)]\, F_\lambda P(\hbar\omega) \tag{14.6}$$

EXAMPLE: Expressing J in units of A/cm^2, intensity I_λ in units of W/cm^2 and using wavelength λ in nm, find the constant of proportionality relating QE to $J/\lambda I_\lambda$.
SOLUTION: from $QE = (\hbar\omega/q)(J/I_\lambda)$ and $\omega = 2\pi c/\lambda$, it is seen that

$$QE = \left(\frac{2\pi\hbar c}{q}\right)\frac{J}{\lambda I_\lambda} \to \left(\frac{2\pi\hbar c}{q}\right) = 1239.8\ \frac{\text{eV}}{q}\text{nm} = 1239.8\ \frac{\text{W}}{\text{A}}\text{nm}$$

Until one knows how to account for the impact of scattering processes, F_λ is mysterious (at least until Eq. (31.30)). Therefore, the factor of most concern at present is $P(\hbar\omega)$, and how it accounts for the number of electrons above the

[3] Latin for "beware, reader." More or less.

barrier compared to all the electrons incident on the barrier. In the language of transmission and supply, this is familiar, and is

$$P(\hbar\omega) = \frac{\int_0^\infty D(E)f(E)dE}{\int_0^\infty f(E)dE} = \frac{\int_0^\infty D(E)\ln\left(1+e^{\beta(\mu-E)}\right)dE}{\int_0^\infty \ln\left(1+e^{\beta(\mu-E)}\right)dE} \tag{14.7}$$

but because tunneling is not appreciable, $D(E)$ acts as a step function, and so

$$P(\hbar\omega) \to \frac{\int_{(\mu-(\hbar\omega-\phi))}^\infty \ln\left(1+e^{\beta(\mu-E)}\right)dE}{\int_0^\infty \ln\left(1+e^{\beta(\mu-E)}\right)dE} \tag{14.8}$$

Let $z = \beta(\mu - E)$ and introduce the Fowler functions $U(x)$ defined by [204]

$$U(x) = \int_{-\infty}^{x} \ln(1+e^z)\,dz \tag{14.9}$$

such that $U(0) = \zeta(2)/2$ is already familiar from Eq. (A2.6), to find

$$P(\hbar\omega) = \frac{U[\beta(\hbar\omega-\phi)]}{U(\beta\mu)} \tag{14.10}$$

EXAMPLE: Find $U(x) + U(-x)$.

SOLUTION: Separating off the negative region by using $U(0)$, it follows that the remaining integral in Eq. (14.9) can be combined (where $x \to -x$ as the integration term in the integral contained in $U(-x)$) to give

$$U(x) + U(-x) = \frac{\pi^2}{6} + \int_0^x \ln\left(\frac{1+e^z}{1+e^{-z}}\right)dz \tag{14.11}$$

$$= \frac{\pi^2}{6} + \int_0^x z\,dz = \frac{\pi^2}{6} + \frac{x^2}{2} \tag{14.12}$$

When the argument of U is negative, then $e^z < 1$ over the entire integration region in Eq. (14.9), and so the logarithm may be Taylor expanded in a procedure familiar from Section A1.2.2. It follows that

$$U(-x) = \sum_{j=1}^{\infty} \frac{(-1)^{j+1}}{j} \int_x^\infty e^{-jz}\,dz = \sum_{j=1}^{\infty} \frac{(-1)^j}{j^2} e^{-jx} \tag{14.13}$$

$$U(x) = \frac{x^2}{2} + \frac{\pi^2}{6} - U(-x) \tag{14.14}$$

where x is treated in both as a positive quantity. The last form of Eq. (14.13) can be treated as a *polylogarithm* or Jonquière's function $Li_2(e^{-x})$. These equations are equivalent to the findings of DuBridge (see Eqs (2) and (3) of ref. [204]). If numerical integration methods are not available, a reasonable approximation is [206]

$$U(x) \approx \begin{cases} e^x(1-be^{ax}) & x < 0 \\ \dfrac{x^2}{2} + \dfrac{\pi^2}{6} - e^{-x}(1-be^{-ax}) & x \geqslant 0 \end{cases} \tag{14.15}$$

with $b = 1 - (\pi^2/12) = 0.17753$ and $a = (1 - b - \ln(2))/b = 0.72843$. The performance of the approximation is shown in Figure 14.2 for negative argument (the positive argument being a trivial addition). For photoemission, wavelengths such that $\hbar\omega > \phi$ are chosen to get appreciable emission, therefore near room temperature conditions, where $\beta \approx 38.682$ eV^{-1}, the arguments of $U(x)$ in Eq. (14.10) are large, and so

$$P(\hbar\omega) \approx \frac{3(\hbar\omega-\phi)^2 + (\pi k_B T)^2}{3\mu^2 + (\pi k_B T)^2} \tag{14.16}$$

or more succinctly, $P(\hbar\omega) \propto (\hbar\omega - \phi)^2$ if the temperature dependence is taken as small.

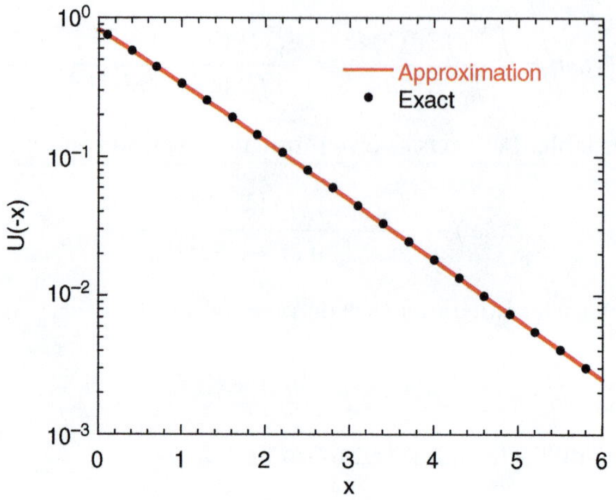

Figure 14.2 Fowler function $U(x)$. The exact (Eq. (14.13)) vs the approximate (Eq. (14.15)) values of the Fowler function $U(-x)$ (for negative argument). For positive argument, use Eq. (14.14) (that is, add $(3x^2 + \pi^2)/6)$).

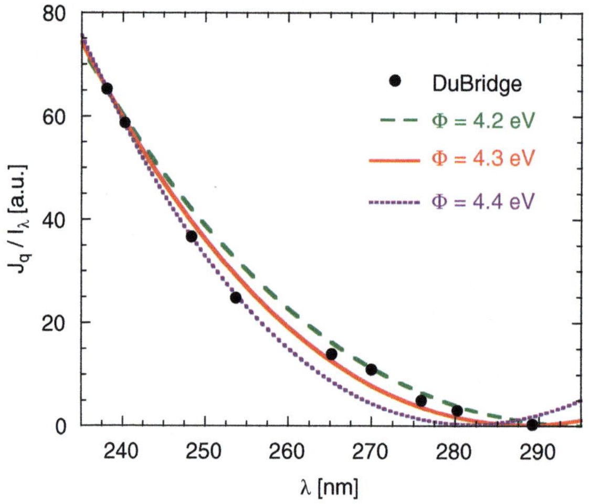

Figure 14.3 DuBridge photoemission data. The ratio of photoemitter current density with incident light intensity, based on Figure 1 of DuBridge and Roehr [34] under the assumption of no field and room temperature. The theoretical lines were calculated according to Eq. (14.16), but normalized so that the theory at $\lambda = 238$ nm passes through the experimental data point ($y = 65.3$). The correspondence is closest to $\Phi = 4.3$ eV.

Table 14.1 Photoemission data. Data numerically extracted from DuBridge and Roehr [34], Figure 1.

λ (nm)	J_q/I_λ (au)
238	65.3
240	58.8
248	36.7
254	24.9
265	14.0
270	11.0
276	5.01
280	3.09
289	0.341

14.2 Analysis of photoemission data

...the variation of the photocurrent with frequency and temperature predicted by Fowler's theory are due directly to the characteristics of the Fermi distribution of velocities among the electrons, and to nothing else. The success of the theory thus constitutes one of the most direct and most convincing tests yet obtained of the applicability of the Fermi–Dirac statistics to electrons in metals.

– Lee A. DuBridge [204]

When compared to experimental data, the agreement afforded by Eq. (14.16) is striking. We use the data behind Figure 1 of DuBridge and Roehr [34] showing the ratio of the electron current to the incident intensity as a function of wavelength for room temperature conditions from molybdenum, reproduced here in Figure 14.3 and in Table 14.1. Figure 14.3 is not the quantum efficiency because the wavelength has not been inserted (as in the example after Eq. (14.6)), that is, the y axis is J_q/I_λ, but $QE \propto J_q/(\lambda I_\lambda)$. Several curves based on Eq. (14.16) with different values of Schottky-reduced work function $\phi = \Phi - \sqrt{4QF}$ are overlaid on top of the data, showing that the work function can be approximated from this data even though the wavelength dependence of the reflectivity is ignored.

CHAPTER 15
Baroody equation

The calculations are almost entirely classical and are intended to give insight into the mechanism of emission and to emphasize properly the influence of electron scattering and absorption ...Although the development gives some insight into a variety of secondary emission phenomena, it is most successful in showing how the Sommerfeld model can clarify the role of work function, and account in a qualitative way for the velocity distribution of the secondaries.

– Eugene M. Baroody [35]

Baroody's equation gives a rough account of the approximate yield (number of secondary electrons emitted versus the number of incident primaries) of emitted electrons after a metal or other material, such as the diamond thin film shown in Figure 15.1, is struck by high-energy primary electrons. Typical operational parameters are primary electron energies E_o of hundreds of electronvolts to tens of kiloelectronvolts. Many metals have maximum secondary yields of 1–2 (e.g., Pt has a yield of 1.8, semiconductors like Si or Sb have yields of 1.1–13, salts like NaCl have yields of 6.8, oxides like MgO have yields of 24; some good photoemitters like Cs_3Sb also have high yields, and pure diamond has a very high yield of over 100 [207]). Good accounts of secondary emission are given by van der Ziel [152], Lye and Dekker [208], Dekker [141], and Hacehnberg and Brauer [171], although the present account is after ref. [209]. A history of the early theory efforts is provided by McKay [170].

15.1 Approximations

...it is Zeus himself, the Olympian, who gives people good fortune, /to each single man, to the good and the bad, just as he wishes; /and since he must have given you yours, you must even endure it.

– Homer[1]

Whereas thermal and field emission were surface effects (the barrier to emission being the largest determinant to the current density), photoemission[2] and secondary emission are concerned with a three-step process beginning with the fortunes of the excited electrons. For secondaries it unfolds as first, the penetration of highly energetic primaries that burrow their way deeply into bulk, leaving secondary electrons in their wake as a consequence, second, the propagation of the secondary electrons that do not suffer recombination or scattering loss events back to the surface, and third, the ability of a secondary electron to surmount the surface barrier and be emitted. Such a description is imprecise for secondary emission because three kinds of emitted electrons are a consequence of the fortunes they endure after a material is struck with a high-energy primary beam:

1. **back-scattered primaries:** primary electrons that have suffered a collision and changed their direction to the surface, to be emitted with approximately half their incident energy
2. **intermediate energy primaries:** either primaries that have given up much of their initial energy yet still have a respectable amount left, or secondaries that have acquired energy from a collision with a primary
3. **low-energy secondaries:** electrons, many in number, and generally with a low energy (< 20 eV) that have been excited by the primary electrons.

The first group causes a high-energy peak at about half the incident beam energy. The second is poorly represented in the emitted population. It is the last group, sometimes called the *true* secondaries, that are generated by primaries whose large initial energy is bled off through repeated collisions with the bulk atoms: these secondaries diffuse through the bulk material until they either suffer recombination or get to the surface where, if their energy is favorable for it

[1] *The Odyssey of Homer* (trans. Richmond Lattimore). New York: Harper & Row, 1967, Book VI, lines 188–190.
[2] The three-step model of photoemission of Spicer follows an analogous development in Section 31.3.

Introduction to the Physics of Electron Emission, First Edition. Kevin L. Jensen.

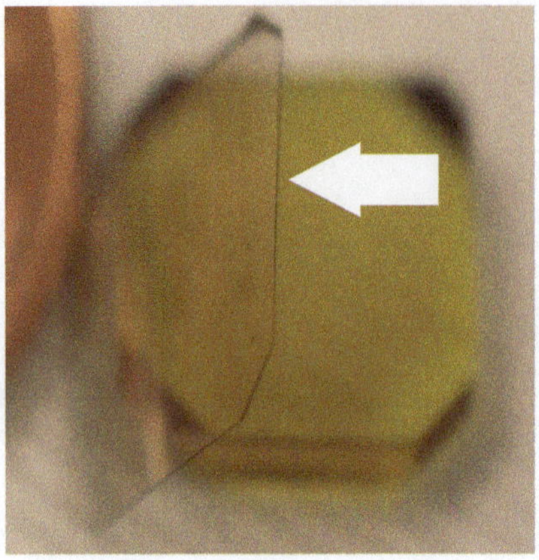

Figure 15.1 Diamond secondary emitter. A thin diamond film developed for a diamond amplifier based on secondary emission. The white arrow points to a diamond film tens of microns thick where the high-energy primary electrons create many secondaries. The yellow square below the film is less than 1 inch across. The orange disk to the left is a US penny. Image courtesy of J.E. Butler, B.B. Pate, and J. Yater (NRL); see also Figure 4 of ref. [210].

(larger than the work function), they are emitted. Such "true" secondaries – the *good* ones for electron sources that have endured much before their wanderings have brought them to the surface – are the ones under discussion.

Naming the secondary emission equation after Baroody gives short shrift to the contributions of van der Ziel, Lye and Dekker, Jonker, Young [211], and others. Moreover, Baroody's equation will not even be the last word. However, Baroody was the first observe that if δ_m is the *maximum* secondary emission yield of a material being struck by primary electrons of energy E_m (with E and δ being the primary energy and the resultant yield), then the yield ratio $\delta(E)/\delta(E_m)$ as a function of E/E_m tends to follow a universal curve for both metals (Figure 15.2) and insulators (Figure 15.3).

The relationship between photoemission yield and Eq. (11.1) is easier to visualize when it is "one photon to one electron," but in secondary emission, high-energy primary electrons generate *many* secondaries through collisions with atoms, after which the escape of the secondaries has much in common with the photoemission processes. Assume that each secondary is created with more or less the same energy ΔE, corresponding to the amount of energy needed to liberate a core electron from an atom. Primary electrons, with their considerable energy, plow straight through the material, shedding secondaries as they go. If the energy lost between the positions x and $x + \delta x$ is δE, and if $\delta E/\delta x \to dE/dx$, then the number of emitted secondaries whose origins are at x is $d\delta \equiv BdE/dx$.[3] Next, *contra* Tolkien (Section 7.3) some who wander do get lost, through recombination or a reduction of their energy to below escape conditions. Their loss will depend on how many are present $N(x)$ and the fraction of them that get lost per unit length γ, or $dN/dx = -\gamma N$.

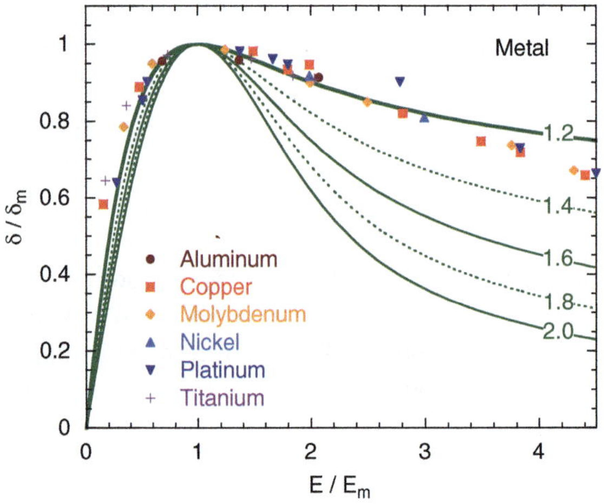

Figure 15.2 Baroody secondary emission data for various metals compared to Eq. (15.8) for various values of n, for which the values $1.2 \leq n \leq 1.3$ work best. Data digitally extracted from Figure 1 of Baroody [35].

[3]The usage of B follows convention: it should not be mistaken for the the same B used in Chapter 13 when discussing field emission.

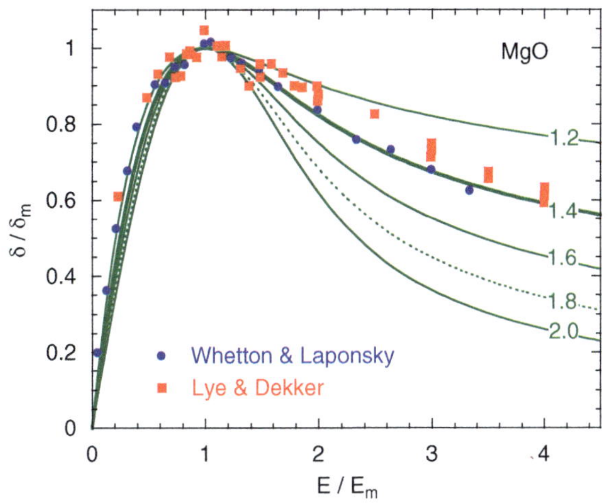

Figure 15.3 Lye–Dekker secondary emission data for the insulator MgO compared to Eq. (15.8) for various values of n, for which the values $1.35 \leq n \leq 1.4$ work best. Data digitally extracted from Figure 5 of Lye and Dekker [208].

Solving, $N(x) = N_o \exp(-\gamma x)$. But N and BdE/dx refer to the same secondaries, and so, summing over all generated secondaries and accounting for their losses incurred in racing to the surface,

$$\delta(E) = -B \int_0^{R(E)} \left(\frac{dE}{dx} \right) e^{-\gamma x} dx \tag{15.1}$$

where $R(E)$ is the range of the primary electrons of beam energy E, that is, how far the primaries penetrate into the bulk material before stopping. There are reasons (to be treated at greater length in the Bethe model of Section 32.1.1) behind the intuition that slower electrons will shed their energy faster than higher energy ones, and so to surmise that dE/dx decreases with increasing E as[4]

$$\frac{dE}{dx} \approx -\frac{A}{E^{n-1}} \tag{15.2}$$

The case $n = 2$ is Whiddington's law, from which Baroody deduced the universal relation between δ/δ_m and E/E_m. A general power law relation better accounts for the penetration depth studies of Young [208, 211] in addition to being of greater flexibility in matching Figures 15.2 and 15.3. Recovering Baroody's deduction without specifying n (yet) is possible.

The form of Eq. (15.2) makes trivial the finding of the range R, where the primary electron finally sputters to a stop in the bulk material, or

$$\int_0^x dx' = \int_{E(0)}^{E(x)} \left(\frac{dx}{dE} \right) dE = -\frac{1}{A} \int_{E_o}^{E} E^{n-1} dE \tag{15.3}$$

where E_o now represents the initial energy (i.e., the energy of the primaries in the incident beam of electrons) so that it is visually distinct from $E(x)$; it and the range $R(E_o)$ are then given by

$$E(x) = (E_o^n - nAx)^{1/n} \tag{15.4}$$

$$R(E_o) = \frac{E_o^n}{nA} \tag{15.5}$$

such that $E[R(E_o)] = 0$, which *seems* circular, but is not. Whiddington's law recommends $n = 2$ so that the range is parabolic in energy, but measurements by Young for aluminum [211], reproduced in Figure 15.4, show that $n = 1.35$ is a better approximation. Eq. (15.4) makes possible finding the solution to $\delta(E_o)$ in Eq. (15.1): inserting the former into the latter

[4]The power $(n - 1)$ is used here; Lye and Dekker [208] use n, so *cave, lector iterum*.

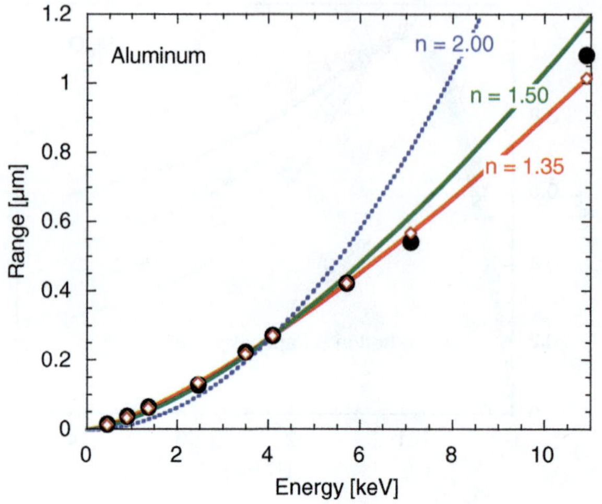

Figure 15.4 Primary electron range vs primary beam energy for aluminum, based on Figure 5 of ref. [211]. The lines were all calculated according to R_n [μm] = $0.271(E_o/4.098\text{ keV})^n$. Young's data digitally extracted (black dots). The best match occurs for $n = 1.35$ (red open diamonds on solid red line).

and introducing the dimensionless term $\lambda \equiv \gamma R(E_o)$ gives

$$\delta(E_o) = B\left(\frac{E_o}{n}\right)\int_0^1 \frac{e^{-\lambda t}}{(1-t)^{1/n}}dt \equiv BE_oG_n(\lambda) \tag{15.6}$$

$$G_n(\lambda) \equiv e^{-\lambda}\int_0^1 \exp\left(\lambda s^n\right) ds \tag{15.7}$$

where, in going from the first integral representation to the second, the substitution $t = s^n - 1$ is made. Baroody's analysis made use of the notation F (Lye and Dekker used G); also Baroody sets up the integral such that the upper bound is a variable so as to make use of tabulated functions (a desirable arrangement in his day). Here, having a fixed upper bound of 1 allows for a bit more transparency in the argument to be made. Clearly (either immediately or after some reflection), λ depends on the primary energy E_o, and as $\delta \to \delta_m$ as $E_o \to E_m$, then so too $\lambda \to \lambda_m$. Introduce the dimensionless parameter $p = E_o/E_m$ and observe $\lambda = \lambda_m p^n$ by its definition, so that

$$\frac{\delta}{\delta_m} = p\frac{G_n(\lambda_m p^n)}{G_n(\lambda_m)} \tag{15.8}$$

The case of $n = 2$ gives Baroody's result, and so Eq. (15.8) is referred to here rather generously as *Baroody's equation*[5] although Baroody presented it differently: his Eq. (24) of ref. [35]) appears as

$$\left.\frac{\delta}{\delta_m}\right|_{n=2} = 1.848e^{-(0.9241p)^2}\int_0^{0.9241p} e^{x^2}dx \tag{15.9}$$

which a change of integration variables shows is equivalent to Eq. (15.8) for $\lambda_m(n = 2) = (0.9241)^2$. But to the point: δ/δ_m depends only on the variable p (and the dimensionless constant λ_m which, recall, will depend on the value of n chosen). That is, there is a universal relation that the yield (in relation to its maximum value) satisfies, and although Baroody came to his conclusion using $n = 2$, Figures 15.2 and 15.3 show that other values of n perform better.

EXAMPLE: Find the limiting behavior of $G_n(\lambda)$ as $\lambda \to 0$ and $\lambda \to \infty$.
SOLUTION: For small argument, use the expansion $e^x \approx 1 + x + (x^2/2)$ in the integrand of Eq. (15.6) to obtain

$$G_n(\lambda \ll 1) \approx 1 + \frac{\lambda n}{n+1} + \frac{(\lambda n)^2}{(n+1)(2n+1)} \tag{15.10}$$

[5]The present treatment has more in common with Lye and Dekker [208]: their Eq. (11) is equivalent to the present Eq. (15.8).

For large argument, use the expansion $s^n = 1 - n(1 - s)$ in the argument of the exponential in the integrand of Eq. (15.6) to obtain

$$G_n(\lambda \gg 1) \approx \frac{1 - e^{-\lambda n}}{\lambda n} \tag{15.11}$$

The limiting forms are compared to a numerical evaluation in Figure 15.5.

EXAMPLE: Find an approximation to $\delta(E_o)$ and determine its accuracy.
SOLUTION: Using Eq. (15.11) to approximate $G_n(\lambda)$ along with the relations $p = E_o/E_m$ and $\lambda = \lambda_m p^n$ in Eq. (15.6) results in

$$\frac{\delta}{\delta_m} \approx \frac{1}{p^{n-1}} \left(\frac{1 - e^{-\lambda_m n p^n}}{1 - e^{-\lambda_m n}} \right) \tag{15.12}$$

The approximation is compared to a numerical evaluation in Figure 15.6.

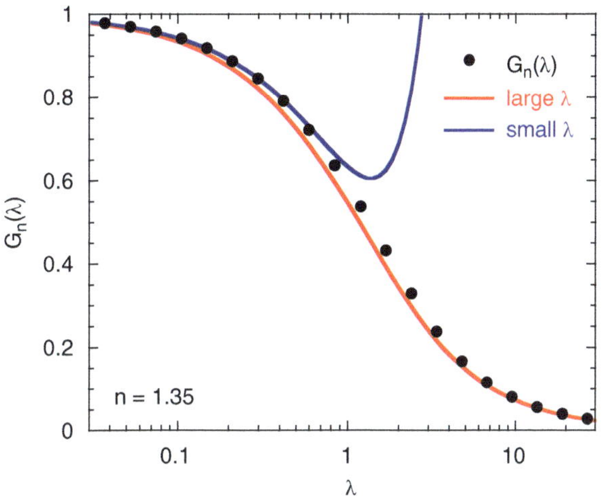

Figure 15.5 The function $G_n(\lambda)$. Comparison of $G_n(\lambda)$ to its asymptotic limits of Eqs (15.10) and (15.11) for the case $n = 1.35$. Although Eq. (15.11) was derived under the assumption of $\lambda \gg 1$, it is seen to perform reasonably well over the entire range of λ.

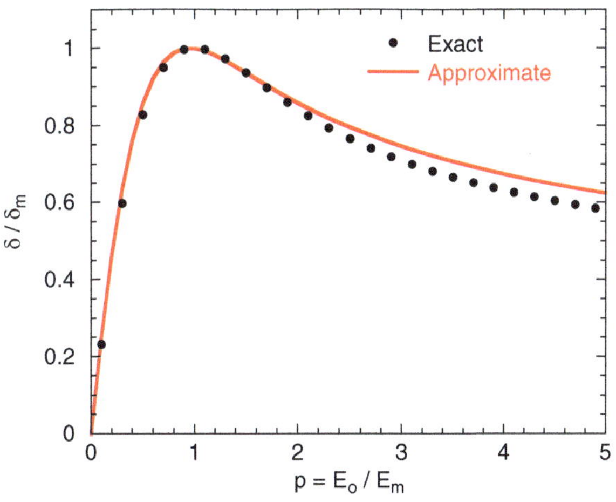

Figure 15.6 δ/δ_m vs E_o/E_m. Comparison of the exact (Eq. (15.6)) to the approximate (Eq. (15.12)) forms of δ/δ_m for $n = 1.35$, for which $\lambda_m = 1.7944$.

15.2 Analysis of secondary emission data

The mob rushes in where individuals fear to tread.

– Burrhus Frederic Skinner[6]

Much depends on the parameter A in Eq. (15.4) for $E(x)$, and the parameter B in Eq. (15.6), without which the analysis of data is hamstrung. Approximations will be considered in Section 32.1.1, so for now Baroody's equation will be used to explain the genesis of the n curves in Figures 15.2 and 15.3, and show how the value of n is determined from the dimensionless form of δ/δ_m given in Eq. (15.8), compared to the approximate form of Eq. (15.11). The maximum is characterized by $\partial_p(\delta/\delta_m) = 0$ at $p = 1$, and so differentiation of Eq. (15.8) and the evaluation of the result at $p = 1$ gives a function $H_n(\lambda)$ of the form (apart from multiplicative factors that can be ignored as ratios are taken)

$$H_n(\lambda) = \int_0^1 (1 + n\lambda(s^n - 1)) \exp(\lambda(s^n - 1))\, ds \tag{15.13}$$

such that $H_n(\lambda_m) \equiv 0$. The zeros are indicated in Figure 15.7. As n decreases, the zero-crossing moves to higher vales of λ. Analytically, $H_1(\lambda) = e^{-\lambda}$, so that $\lambda_m(n = 1) = \infty$. As for the other zeros, $1/\lambda_m(n) \approx 0.952n - 0.730$, as shown in Figure 15.8.

Although the zeros of $H_n(\lambda)$ can be graphically determined from Figure 15.7, a numerical method known as bisection works better in practice for finding roots. In bisection, a maximum and minimum value of λ is chosen such that $H_n(\lambda_{min}) > 0$ ($\lambda_{min} = 0$ works well) and $H_n(\lambda_{max}) < 0$ (a large value like ($\lambda_{max} = 10$ works well). Then the logic unfolds according to

Require: Initially set $\lambda_{min} = 0$.
Require: Initially set $\lambda_{max} = 10$.
  for $j = 1$ to N **do**
    Generate $\lambda_h = (\lambda_{min} + \lambda_{max})/2$
    If $H_n(\lambda_h) > 0$ then $\lambda_{max} \to \lambda_h$
    If $H_n(\lambda_h) < 0$ then $\lambda_{min} \to \lambda_h$
  end for
  $\lambda_m = (\lambda_{min} + \lambda_{max})/2$.

The number of times N that the calculation of λ_h and its replacement of λ_{max} or λ_{min} is performed is governed by the desired level of accuracy: if on the first iteration $\lambda_{max} - \lambda_{min} = 1$ (for example), then after N iterations, $\lambda_{max} - \lambda_{min} = 1/2^N$. Thus, every 10 iterations improves the accuracy (constrains the region in which λ_m is found) by 3 orders of magnitude because $2^{10} = 1024$ and λ_m lies at their midpoint by construction. An example code implementing the bisection method

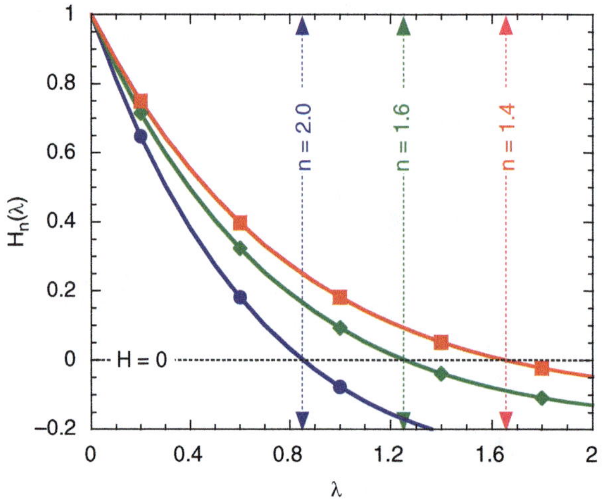

Figure 15.7 $H_n(\lambda)$ function. Behavior of $H_n(\lambda)$, defined in Eq. (15.13), for various n such that $H(\lambda = \lambda_m) \equiv 0$.

[6] Burrhus Frederic Skinner, *Walden Two (Hackett Classics)*. Indianapolis: Hackett Publishing Company, Inc., 2005, p. 37.

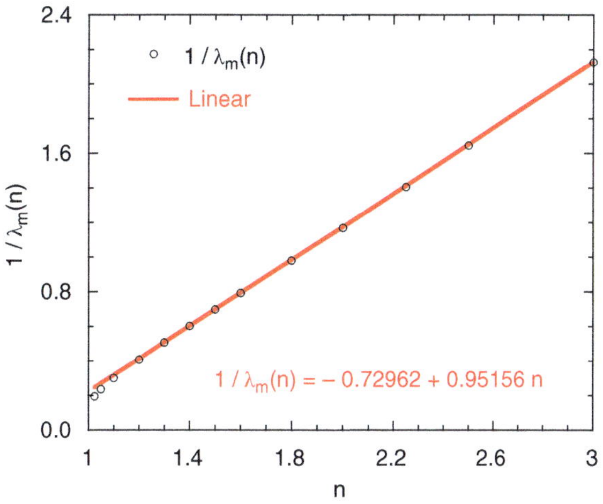

Figure 15.8 Zeros of the function $H_n(\lambda)$, designated $\lambda_m(n)$. To a good approximation, $1/\lambda_m(n)$ is linear in n, the best fit line in red, with the equation for $\lambda_m(n) \to y(n)$ likewise in red, although the fit wrongly suggests that $\lambda_m(1)$ is finite.

as applied to the determination of the zeros of $H_n(\lambda)$ is given in Section A3.7.1. N can be kept small ($N = 16$) by making the final step use linear interpolation between $H_1 = H_n(\lambda_{max})$ and $H_0 = H_n(\lambda_{min})$ to obtain λ_m.

15.3 Subsequent approximations

It is observed that in spite of the great difference in yields the data follow rather closely a single reduced yield curve, which implies that, as with metals, the individual crystal characteristics modify the yield curves principally in the role of scale factors and do not influence markedly their functional form.

– Robert G. Lye and A. J. Dekker [208]

Transmission electron measurements of primary electrons in the energy range $0.3\text{ keV} \le E_o \le 7\text{ keV}$ through aluminum oxide films led Young [211] to conclude that $n = 1.35$ is a better value to match the data, in contrast to the Whiddington law estimate of $n = 2$. In fact, $n = 1.35$ was found to apply for primaries up to 7 MeV. Lye and Dekker proposed the "power law" formula for the range seen in Eq. (15.2), but as they note, Young's experiments cast some doubt on that as well: the fraction of energy lost by the primaries behaves like

$$\left(\frac{dE}{dx}\right)_{eff} \approx \frac{E_o}{R(E_o)} \tag{15.14}$$

where the subscript eff is for "effective." This is a rather different result: in contrast to the power law relation which posits that the range of all primary electrons is the same, Young argued that the data revealed an almost linear decrease in the number of electrons with depth. Usage of Eq. (15.14) in Eq. (15.1) makes for an easy integration: recalling that $\lambda = kR(E_o)$, then

$$\delta = B\left(\frac{E_o}{R(E_o)}\right)\int_0^{R(E_o)} e^{-kx}dx = \frac{BE_o}{\lambda}(1 - e^{-\lambda}) \tag{15.15}$$

resulting in, in the terminology of Lye and Dekker (Eq. (17) of ref. [208]), a "universal reduced yield equation" given by

$$\frac{\delta}{\delta_m} = \frac{1}{p^{n-1}}\left(\frac{1 - \exp(-\lambda_m p^n)}{1 - \exp(-\lambda_m)}\right) \tag{15.16}$$

where, as before, $p = E_o/E_m$. Such an equation is hauntingly familiar from Eq. (15.12), but slightly different in that the factor of n in the exponential terms is absent. That is not much of a difference (the A factors could be rescaled, for instance), but Young's formulation becomes convenient when dE/dx is evaluated using the Bethe model of Section 32.1.1, and there it shall wait.

CHAPTER 16

Child–Langmuir law

Creon: *You knew the order not to do this thing?*

Antigone *I knew, of course I knew. The word was plain.*

Creon: *And still you dared to overstep these laws?*

Antigone *For me it was not Zeus who made that order. /Nor did that Justice who lives with the gods below/mark out such laws to hold among mankind. /Nor did I think your orders were so strong/that you, a mortal man, could over-run the gods' unwritten and unfailing laws.*

– Sophocles[1]

Antigone, the daughter of King Oedipus and his wife/mother Queen Jocasta of Thebes, defied the alleged tyrant Creon, Antigone's uncle, by giving her brother Polyneices a proper burial, after Creon ordered Polyneices' body to be left on the battlefield for the carrion-eaters after Polyneices' attempt to take Thebes by force. For that defiance, Antigone was sentenced by Creon to death. The great eponymous play about her fate, written by Sophocles around 441 BCE, explored the differences between the laws of the state and moral (or even natural) laws. Given Antigone's family tree, her positioning herself as a defender of the divine moral order may seem tenuous, but the larger philosophical point implicit in her argument, that there is a difference between man's law (injunctions for his benefit) and laws of nature (descriptions of how things are), parallels similar contemplations on what a "law" in science is.

More than 2000 years after *Antigone*, the experiments of Langmuir on the Edison effect (*aka* thermal emission) [132] showed that the amount of current I drawn between a planar diode is related to the voltage V on the anode and the anode–cathode (AK) separation D by $I \propto V^{3/2}/D^2$ (Child having found the same relation for positively charged ions). Langmuir himself considered how the relation was modified for current flow between coaxial cylinders [146] and concentric spheres [147]. Interest extended to triodes, tetrodes and pentodes, given that space charge controlled their performance, and then on to klystrons and traveling wave tubes (TWTs). The relation kept adapting to deal with each new complication to the physics, such as temporal variation of the electric field, initial velocity of the electrons from the cathode, a statistical distribution of initial velocities, relativistic effects, and so on. The problems are incredibly rich, and the physics deceptively subtle, but the caveats and special conditions even for the simple diode problem led van der Ziel (p. 155 of ref. [152]) to advise about Child's so-called law that "This expression is usually not very accurate ..." (he gave a more accurate version, as will we). Such a history of a law has more of Creon in it than Zeus, but referring to the limits placed on electron emission by cathode geometry and anode potential as the "Child–Langmuir limiting relation for particular circumstances" lacks both panache and gravitas compared to referring to it as a "law".

But even that flippant quip is a bit off the mark. Observe that the other canonical equations are referred to as *equations*, but the Child–Langmuir relation is often referred to as "Child's *law*". Why the difference? Simply, because a "law" is an *empirical* relation that works even when there is not an underlying theory to explain it (a usage of "law" different than the lay understanding of it). Aye, there's the rub. In elegant phrasing, Thomas Kuhn,[2] in an almost Jeffersonian sentiment, argued

> *...empirical laws ...can be confronted directly with observation or experiment ... they fill an apparent gap, supplying information that was previously lacking. As science develops, they may be refined, but the original versions remain approximations to their successors, and their force is therefore either obvious or readily recaptured. Laws, in short, to the extent that they are purely empirical, enter science as net additions to knowledge and are never thereafter entirely displaced.*

[1] Sophocles, *Antigone*, in *The Complete Greek Tragedies: Sophocles I* (trans. E. Wyckoff). Chicago: University of Chicago Press, 1954, lines 447–455.
[2] Thomas S. Kuhn, *The Essential Tension: Selected Studies in Scientific Tradition and Change*. Chicago: University of Chicago Press, 1977, p. 19. Compare Jefferson's quote in Section II.

Introduction to the Physics of Electron Emission, First Edition. Kevin L. Jensen.

Read closely, such a description applies rather well the evolution of Child's law over decades, and points to its consequential importance. Still, although the theoretical foundation has much improved, the nickname has not been displaced either.

16.1 Constant density approximation

Die Tragdie besteht darin, daß sich der Baum nicht biegt, sondern bricht.

– Ludwig Wittgenstein[3]

The first space-charge behavior to be considered, that associated with charged particles in fixed locations, represents a solvable instance of Poisson's equation. It has, however, more to do with band bending in semiconductors than it does the space-charge limited *flow* to be encountered in Section 16.2. Understanding it is necessary to the explanation of why band-to-band tunneling, a kind of "breakage" in holding off current flow, occurs in Section 24.4 (see in particular Figure 24.13), where the bending becomes extreme enough to allow tunneling.

For continuity, though, the treatment shall be conducted in the language of AK gaps and capacitors. The consideration of two planar conductors, the anode being at V_a and the cathode being at zero potential energy, a distance D apart with the intervening gap filled by a constant charge density ρ_o is a model of space charge in semiconductors caused by doping (impurity atoms that give up or take up charge, being "donors" and "acceptors," respectively). It is also a good simple model to introduce methods and notation. The potential energy $V(x)$ between the parallel plates is found by solving Poisson's equation, which, in one dimension, is

$$\frac{d^2}{dx^2}V(x) = -\frac{q^2}{\varepsilon_0}\rho(x) \tag{16.1}$$

Note a few things about this form compared to usual accounts in electrostatics [99, 100]: first, because $V(x)$ is a potential *energy* and ρ_o is a *number* density, a factor of q^2 on the right-hand side is explicitly present; second, because *electrons* are the particles under consideration, the anode potential *energy* is $-V_a$. Therefore, although Eq. (16.1) is equivalent to more engineering-friendly accounts, there are differences to heed. Had not Benjamin Franklin, as clever as he was, randomly established the convention for electricity such that "negative" charge became associated with the electron [212], there would be a friendlier symmetry between the equations for potential energy and potential.

A constant density $\rho(x) = \rho_o$ makes the first integration of Eq. (16.1) trivial: with the identification $F(x) \equiv -dV(x)/dx$, then

$$F(x) - F_c = \frac{q^2\rho_o}{\varepsilon_o}x = 16\pi Q\rho_o x \tag{16.2}$$

where $-F_c$ is the field at the surface of the cathode (insofar as possible, terms are taken as positive and their signs explicitly evident), and the second form makes use of Eq. (2.8). The second integration is trivial, and it follows that for the anode potential energy at V_a and for an AK gap D, then, if the cathode is said to be grounded (held at $V(0) = 0$), a second integration yields

$$\frac{V_a}{D} = F_c + 8\pi Q\rho_o D \tag{16.3}$$

In the absence of charge in the AK gap ($\rho_o = 0$) then the usual parallel plate capacitor equation $F_c = V_a/D$ results, but other circumstances are of greater consequence. The field at the surface of the cathode can be made to vanish if $\rho_o D \equiv \sigma_o$ is such that $V_a/D = 16\pi Q\sigma_o = \sigma_o/\varepsilon_0$: such an equation is well-known in electrostatics,[4] as a one-dimensional *sheet* of charge gives rise to a field on either side of the sheet that does not taper off with (i.e., is independent of) distance. Consequently, a stack of sheets gives rise to the same field as if all the charge density were in a single sheet. Thinking of the sheets as having a thickness dx such that $dF = 16\pi Q\rho_o dx$ regains Eq. (16.2).

The relation of depletion layers to the doping of semiconductors will figure prominently in Section 24.4 and be related to Eq. (16.3). An alternate designation is the "space-charge region" where band bending occurs as near the surface of a semiconductor, or near semiconductor/metal interfaces [213]. For now, a simple treatment replaces the vacuum

---

[3] "You get tragedy where the tree, instead of bending, breaks," Ludwig Wittgenstein, *Culture and Value* (trans. Peter Winch). Chicago: University of Chicago Press, 1984, p. 1.

[4] See, for example, Section 3.6 (Surface Charges and Double Layers) of ref. [99].

between the parallel plates in the capacitor model with an insulator (diamond or silicon dioxide SiO_2, for example) with a dielectric constant K_s, which is accounted for by replacing $\varepsilon_0 \rightarrow K_s \varepsilon_0$ in Poisson's equation. The dielectric properties of the insulator decrease the internal field compared to what would exist for a vacuum. In the absence of charges in the gap, the (constant) field is $F = V_a/(K_s D)$ for a dielectric-filled capacitor.

If the dielectric material contains charged impurities introduced by *doping*, then ρ_o will be related to the doping density N_d. Of particular importance is the length over which the field is completely shielded out, known as the *depletion width* or w. Shielding of the field is taken to mean $F_c \rightarrow 0$ in Eq. (16.3) or, equivalently, $F(w) = 0$ in Eq. (16.2). Finding the depletion width w is straightforward. Because N_d is constant[5] with typical values of 10^{15}–10^{19} atoms/cm^3, a second integration of Eq. (16.2) with $F(0) = 0$ as a boundary condition gives

$$V_a = \frac{q^2 N_d}{2K_s \varepsilon_o} w^2 = 8\pi Q N_d \frac{w^2}{K_s} \tag{16.4}$$

Observe the units: V_a has units of eV, Q of eV-nm, and w of nm. Therefore, because N_d has units of #/nm^3, both sides of the equation are in units of eV.

EXAMPLE: A diamond thin film is exposed to an external field of $F_{vac} = 10$ eV/μm (corresponding to an electric field of 10 MV/m). Assuming a doping density of 10^{15} boron atoms per cm^3 and that the donors are singly charged, find the depletion width w. Why is the dielectric constant of diamond ($K_s = 5.7$) not needed in this calculation? By how much do the bands bend?

SOLUTION: The field just *inside* the diamond is $F(0) = F_{vac}/K_s$, so that, using Eq. (16.2) with $F(w) = 0$, it is found that

$$\frac{F_{vac}}{K_s} = \frac{q^2 \rho_o}{K_s \varepsilon_o} w = 16\pi Q \rho_o \frac{w}{K_s} \tag{16.5}$$

or $w = F_{vac}/(16\pi Q \rho_o) = 552.64$ nm (f/n), where f is the field in eV/μm and n is the doping density in 10^{14} atoms/cm^3. With $f = 10$ and $n = 10$, then $w = 0.52264$ μm. The independence of K_s is a consequence of specifying the vacuum field rather than the internal field at the surface boundary. Then, from Eq. (16.4), $V_a = 0.4848$ eV.

16.2 Constant current approximation

Plus l'obstacle est puissant, plus on reçoit de gloire …

– Molière[6]

Lest van der Ziel's limited praise of Eq. (16.18) seem confusing, return to the odd bit about excess current being returned to the cathode when F_m changes from accelerating electrons away from the cathode to pushing them back: that is, a potential barrier is created that must be surmounted for current to cross the gap, and its neglect undercuts the ability of J_{CL} to predict the current. The origin and location of the barrier may be found via the constant current approximation, but will take some development.

If the current density $J_o \approx q^2 \rho(x) v(x)$ is taken as constant,[7] that implies $\rho(x) = J_o/q^2 v(x)$, and so the number density varies with x. From conservation of energy $(1/2)mv(x)^2 = V_m - V(x)$, and so Eq. (16.1) becomes

$$dF(x) = \frac{8\pi Q}{q} J_o \sqrt{\frac{2m}{V_m - V(x)}}\, dx \tag{16.6}$$

[5] N_d is constant as far as present needs are concerned here. In fact, $N_d(T)$ is a temperature dependent quantity and the donors ionize only when conditions are favorable for doing so.

[6] "The more powerful the obstacle, the more glory is received", *L'étourdi, ou les Contre-Temps* (*The Blunderer*), EBook #6563, Project Gutenberg, http://www.gutenberg.org, 2004, v.xi.

[7] Care is required here: J is taken as a *charge* current density with units of A/cm^2, yet ρ is being treated as a *number* density with units of #/cm^3, in sympathy with usual treatments of doping density which do likewise, rather than a charge density with units of q/cm^3, as was the case in Eq. (3.5). To respect that, an additional q must be appended to the coefficient.

where V_m is a reference potential, either the surface of the cathode for now, or where the potential energy is at a maximum for later, so that letting $U(x) \equiv V_m - V(x)$ is convenient. Multiplying both sides by $F = dU/dx$ results in

$$d(F^2) = \frac{16\pi Q}{q} J_o \sqrt{\frac{2m}{U}} dU \equiv 2aU^{-1/2} dU \tag{16.7}$$

$$a \equiv \frac{8\pi Q J_o}{q} \sqrt{2m} \tag{16.8}$$

Each side is now easily integrated, giving

$$F^2 - F_m^2 = 4a\sqrt{U} \tag{16.9}$$

where $F(0) = F_m$ is the field at the cathode surface (or potential energy maximum), where $U(0) = 0$. Then, remembering that $F = dU/dx$,

$$\int_0^{U_a} \frac{dU}{(4aU^{1/2} + F_m^2)^{1/2}} = \int_0^D dx \tag{16.10}$$

for which the right-hand side is straightforward but the left-hand side is not. On the left, there is a change of variables to $U^{1/2} = (F_m^2/4a)u$ and the integral is

$$\int \frac{udu}{\sqrt{u+1}} = \frac{2}{3}(u-2)\sqrt{1+u} \tag{16.11}$$

After rearranging a bit, recalling $U(D) = -V_a$ if $V_m = 0$ where D is the AK gap separation and V_a refers to the anode, then

$$\left(2aV_a^{1/2} - F_m^2\right)\sqrt{4aV_a^{1/2} + F_m^2} = 6a^2D - F_m^3 \tag{16.12}$$

In a manipulation more tedious than tantalizing, that equation can be recast as

$$F_m^3 - F_m^2\frac{Va}{D} + \frac{a}{3D}\left(4V_a^{3/2} - 9aD^2\right) = 0 \tag{16.13}$$

The limiting cases of this equation are instructive. First, observe that a is proportional to J_o (Eq. (16.8)), so that what is said about one applies to the other. The first limit of interest, then, corresponds to no current flowing across the AK gap, or $a \to 0$, for which Eq. (16.13) is trivial ($F_mD - V_a = 0$) and reproduces the usual relation between anode potential, gap distance, and field: the potential energy decreases linearly across the gap. If the current density is just very small rather than zero, then to leading order in a,

$$F_m \approx \frac{V_a}{D} - \frac{64\pi}{3q} QDJ_o\sqrt{\frac{m}{2V_a}} \tag{16.14}$$

which indicates that as more current is injected into the gap (i.e., as J_o becomes larger), the field at the surface F_m decreases, and the potential energy acquires a curved characteristic. As ever more current is injected, at some point the surface field itself will vanish and the potential energy across the gap will have a parabolic shape such that the maximum of the curve is at the surface of the cathode and the surface field $F_m = 0$. Under that special condition, J_o is called J_{CL}, and from Eq. (16.13) it corresponds to $a = 4V_a^{3/2}/9D^2$, or

$$J_{CL}(V_a, D) = \frac{q}{18\pi Q\sqrt{2m}}\left(\frac{V_a^{3/2}}{D^2}\right) \tag{16.15}$$

which allows Eq. (16.14) to be recast as

$$F_m = \frac{V_a}{D}\left(1 - \frac{16J_o}{27J_{CL}}\right) \tag{16.16}$$

In fact, if dimensionless terms f and j are introduced such that $F_m = fV_a/D$ and $j = J_o/J_{CL}$, then Eq. (16.13) takes on a distinctly more appealing form of [214]

$$f^2(1-f) = \frac{16}{27}j(1-j) \tag{16.17}$$

such that $j = 0$ corresponds to $f = 1$ and $j = 1$ to $f = 0$. Observe that the limiting cases $f^2 \to 1$ and $j^2 \to 0$ give rise to Eq. (16.14).

The representation of Eq. (16.15) for J_{CL} is not in a form commonly encountered. To connect with its common representation, let $V_a \to q\varphi_a$, where φ_a is the anode potential in volts, and use the definition of Q in Eq. (2.8) to obtain the conventional and venerated form of Child's law representing the *maximum* current that can be pulled from the cathode before a non-zero fraction of emitted electrons are returned to the cathode, or

$$J_{CL}(\varphi_a) = \frac{4\varepsilon_0}{9D^2}\left(\frac{2q}{m}\right)^{1/2}\varphi_a^{3/2} \tag{16.18}$$

where the coefficient of $\varphi^{3/2}$ is often referred to as the *perveance* and designated K.

EXAMPLE: What is the perveance K for $D = 1$ cm?

SOLUTION: Inserting the values of $\varepsilon_0 = 8.8542 \times 10^{-12}$ C/V-m and, for the electron, $q = 1.6022 \times 10^{-19}$ C and $m = 9.1094 \times 10^{-31}$ kg in MKSA units, one finds

$$K = 2.334 \frac{\mu\text{A}}{\text{V}^{3/2}\text{cm}^2}$$

an enjoyable calculation for those who are inordinately fond of exponential notation. One may alternately start from $J_{CL} = K'V_a^{3/2}$ using Eq. (16.15). As $Q = 3.6$ eV-nm and $mc^2 = 5.11 \times 10^6$ eV then $K'D^2 = 0.01457\ q/\text{eV}^{3/2}$ fs. Observing that (1 q)/(1 fs) = 160.22 μA gives $K'D^2 = 2.334\ \mu\text{A/eV}^{3/2}$. Using $K = K'q^{3/2}$ with $q = $ eV/V returns the previous result.

Let the position of the barrier maximum be at x_m in front of the cathode surface and at a potential V_m higher than that of the cathode (held at $V = 0$), as in Figure 16.1. The yellow shaded region would give rise to Eq. (16.13) if x_m were at the surface of the cathode and V_m were at the cathode potential. Thus, one expects that the modifications to Eq. (16.13) of $V_a \to V_a + V_m$ and $D \to D - x_m$ correct the equation, or

$$J_o \to \frac{q}{18\pi Q\sqrt{2m}} \frac{(V_a + V_m)^{3/2}}{(D - x_m)^2} \tag{16.19}$$

But determining the values of V_m and x_m is not straightforward: as methods for their determination are found elsewhere for thermionic emitters [148, 152], an abbreviated account suffices. If the cathode emits a current J_o such that the emitted electrons encounter a potential hump that only a fraction r of them surmount, then rJ_o is the current that reaches the anode, and $(1 - r)J_o$ is reflected back to the cathode. If the AK gap is divided into two regions, Region 1 being before x_m and Region 2 being after it, then $\rho_2(x) = rJ_o/v_2(x)$, but in Region 1 the reflected electrons also contribute to the density,

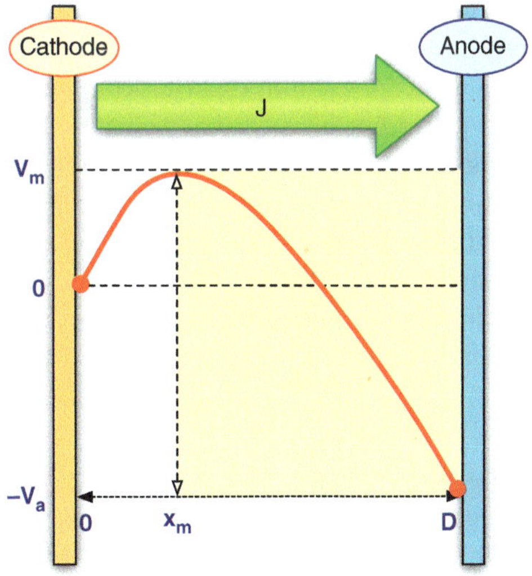

Figure 16.1 Potential terms for Child's law. V_m and x_m refer to the potential energy maximum. Cathode (orange) to the left, anode (blue) to the right. Electrons in the yellow shaded region (to the right of x_m) only travel to the right; current in excess of J_{CL} injected into the unshaded region (to the left of x_m) is reflected.

and so summing the *there* and *back* components is needed, or $\rho_1(x) = rJ_o/v_1(x) + (1-r)J_o/v_1(x)$ or $(2-r)J_o/v_1(x)$. This entails the solution of two equations,

$$\frac{d^2}{dx^2}U_1(x) = -\frac{q^2}{\varepsilon_0}\frac{rJ_o}{v_1(x)} \tag{16.20}$$

$$\frac{d^2}{dx^2}U_2(x) = -\frac{q^2}{\varepsilon_0}\frac{(2-r)J_o}{v_2(x)} \tag{16.21}$$

subject to the boundary conditions that $U_1(0) = 0$, $U_1(x_m) = U_2(x_m) = V_m$, and $U_2(D) = -V_a$, from which x_m and V_m can be determined. Now, the limiting current density is $rJ_o = J_{CL}(V_a + V_m, D - x_m)$, where J_o is the current density the cathode *generates* under particular conditions (like temperature for thermionic cathodes) regardless of how much current is *returned*, and van der Ziel's comment is now properly understood. Moreover, current density has been treated as constant, but oscillations and complex behavior attend particular conditions such that interesting physics continues to bedevil "space-charge limited" flow [215–218].

A reasonable question would be, why run cathodes so hard that a potential energy maximum forms *outside* the cathode surface? If temperature is required for thermal emission, this means more power is expended heating the cathode than needed, and for many systems (particularly space-based ones) power is at a premium. The short answer is because cathode surfaces are typically non-uniform, to have a positive F_m at the cathode pulls electrons away from the surface as soon as they are liberated, and so whatever non-uniformity of emission exists at the cathode surface is replicated in the beam produced. Often, this is undesirable (Section 33.4). The creation of a virtual cathode into which electrons are injected and distributed thereby has considerable advantage in creating a smooth, uniformly emitting beam [215, 219, 220]. An additional advantage is that flicker noise is reduced, which is a problem when space-charge smoothing cannot be employed [221]. The transition from temperature-limited emission to space-charge limited emission reappears in the introduction of the Miram curve (Section 29.1).

16.3 Transit time approximation

A time of love, a time of hate/A time of peace ...
And a time for every purpose under heaven.

– Pete Seeger[8]

Focusing on current density and space charge obscures the importance of another parameter. Because the *charge* in the gap determines the space-charge reduction in the field at the surface of the cathode, that other parameter must be *time* or, more specifically, how long the electrons take to get from cathode to anode, otherwise known as the *transit time*. Charge per unit area σ creates a field F from a sheet of charge (an isolated sheet gives $F = \sigma/2\varepsilon_0$: recall the discussion following Eq. (16.4)). For constant current density, $\sigma = J_o\tau$, where τ is the transit time. Seen another way from Eq. (16.3), the space-charge field goes as $q^2\rho_o D$, but from $\rho_o = J_o/q\langle v\rangle$ and $\langle v\rangle\tau = D$, the views are equivalent.

The transit time model originates from a model of vacuum capacitance [222], and can be extended to treat relativistic and quantum mechanical conditions [149] (but not here). It has been used to infer scaling laws for cylindrical and spherical diode geometries [223]. It is useful when (as for photoemission) the pulse duration of the beam bunch can be shorter than the transit time across the gap [224], as then not only can there be more current than allowed by J_{CL}, but an oscillatory behavior in the THz frequency can be seen via simulation for a train of electron bunches [225]. But it is the pedagogical utility of the transit time model as a means to understand space-charge behavior that counts now.

Under equilibrium conditions for constant current density, the transit times of all electrons across the gap are the same in the planar diode if the initial velocities are negligible. Are they? Recall that the mean velocity of the emitted electrons for thermal emission is approximately

$$\langle v_x(0)\rangle \approx \frac{\int_0^\infty (\hbar k_x/m)e^{-\beta E_x}dk_x}{\int_0^\infty e^{-\beta E_x}dk_x} = \sqrt{\frac{2k_BT}{\pi m}} \tag{16.22}$$

where $E_x = (\hbar k_x)^2/2m$. The approximation changes a bit if the emission mechanism is tunneling: for field emission, as the electron emerges from the tunneling barrier, its kinetic energy is comparable to the Fermi energy μ. Compare such

[8]Pete Seeger, lyrics to "Turn, Turn Turn", 1954.

energies to those a short distance away from the cathode surface, where the emitted electron has an energy comparable to the potential drop near the extraction grid (held at kilovolts for gridded dispenser cathodes) or gate (held at hundreds of volts for Spindt-type field emitters), values which tend to dominate the few electronvolts or less of the emitted electron kinetic energy. Conditions for photoemission and secondary emission, in which the cathode surface is subject to fields of several megavolts per meter (corresponding to $F \approx 10$ eV/μm) are comparable. In all these cases, the initial kinetic energy (and therefore velocity) only mildly affects the transit time. A simple account can therefore treat the initial velocity as negligible.

If the transit time is known, Child's law can be almost trivially inferred from Eq. (16.3) (although this will be put on better footing subsequently)

$$\frac{V_a}{D} \approx F_m + \frac{qJ_o\tau}{\varepsilon_0} \tag{16.23}$$

with $J_o \rightarrow J_{CL}$ when $F_m \rightarrow 0$, but a difficulty with that is that the definition of "transit time" is not unique in that it can be approximated different ways: it can refer to how long one electron by itself takes to cross the gap (the "ballistic" or space-charge-free transit time), the actual transit time (which begs the question of what J_{CL} is), or a transit time associated with emission occurring all at once so that an infinitesimally thin sheet crosses the gap (the "pulse" transit time), as may occur for rapid photoemission pulses. For all that, the transit time does not vary as strongly with the anode potential as does the current density, so that a reasonable estimate of τ, if one can be had, is good enough to anticipate when space-charge effects begin to matter. Stacked next to the exact approach of Eq. (16.17), the virtue of "good enough" is a rather anemic endorsement, but that is not the point: the transit time concept becomes useful when visualizing space charge and discussing how matters are different for multidimensional surfaces, even if the approximation is inexact.

The ballistic or space-charge-free transit time τ_o is obtained using simple kinematical equations, and therefore easiest. From $x(t) = x_o + v_o t + (a/2)t^2$ with v_o being the (neglected) initial velocity and $ma = F_o \equiv V_a/D$ from Newton's equation, then $x(\tau) = D$ gives

$$\tau_o = \sqrt{\frac{2mD}{F_o}} \tag{16.24}$$

Such a transit time is a measure of how long a particle takes to fall down a potential energy hill, as on the left side of Figure 16.2.

The second transit time τ is for when space charge is present, as on the right side of Figure 16.2, where the fall of the particle is impeded by charge in the gap. It is defined by

$$\tau \equiv \int_{x_+}^{D} \frac{dx}{v(x)} \approx \int_0^D \frac{dx}{\langle v \rangle} = \frac{q\sigma(D)}{J} \tag{16.25}$$

where $x_+(E)$ is the location where the electron of energy E emerges through or over the barrier characterized by the image charge potential, but generally $x_+ \ll D$ and so it is negligible. The sheet charge density varies with position, so let $\sigma(x)$ be defined by

$$\sigma(x) = \int_0^x \rho(x')dx' \tag{16.26}$$

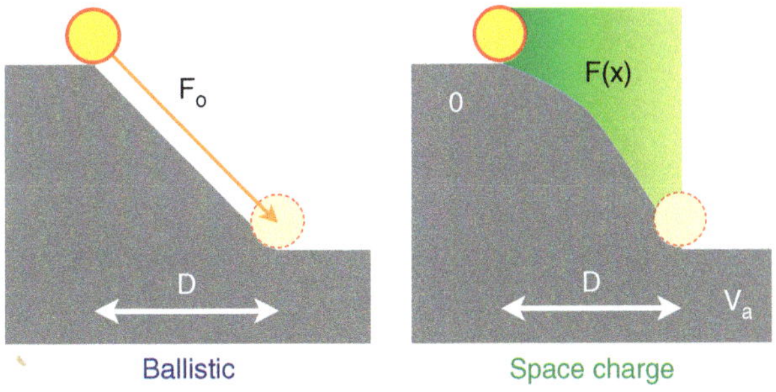

Figure 16.2 Ballistic transit time (left) corresponds to how long a particle takes to fall down a potential energy hill when the field is linear, in contrast to (right) when charge is present and the descent is thereby slowed.

The potential energy $V(x)$ can be calculated from Poisson's equation subject to the conditions $V(0) = 0$, $V(D) = V_a$, and $\partial_x V(0) = F$, where F is the field at the surface, and so

$$V_a = FD + \frac{q^2}{\varepsilon_0}\int_0^D \sigma(x)dx \tag{16.27}$$

$$\equiv FD + \frac{q^2 D}{\varepsilon_0} N_\tau \sigma(D) \tag{16.28}$$

where N_τ is defined by this equation, but which is seen to be $N_\tau \equiv \int_0^D \sigma(x)dx/(D\sigma(D))$. Compare this to a different way of solving Poisson's equation, relying on the relatively common trick of multiplying both sides by $2F = 2dV/dx$ to get

$$d[F^2] = 2\frac{q^2}{\varepsilon_0}\rho dV = \frac{2qJ}{\varepsilon_0 v}dV \tag{16.29}$$

which, using the velocity $v(x) = [2(V(x) + E_o)/m]^{1/2}$ where $mv_o^2/2 = E_o$ is the initial kinetic energy (KE), allows for a simple integration of both sides and gives (after a wee effort)

$$F(x) = \left\{\frac{2q\sqrt{2m}}{\varepsilon_0}J\left[\sqrt{V(x) + E_o} - \sqrt{E_o}\right] + F^2\right\}^{1/2} \tag{16.30}$$

where F without an argument is $F \equiv F(0)$ or the field at the surface. In an exact approach, a second integration is required to obtain a self-consistent relation (see, for example refs [214, 226]), but the transit time approach proceeds differently. Assume the initial kinetic energy $E_o = 0$, so that

$$F(x) = \left(F^2 + \frac{2q}{\varepsilon_0}J\sqrt{2mV(x)}\right)^{1/2} \tag{16.31}$$

Then, from $\sigma(D) = (\varepsilon_o/q^2)((V_a/D) - F)$ plus the right-hand side of Eq. (16.25) it follows

$$\tau = \frac{\sqrt{8mV_a}}{F}\left[1 + \left(1 + \frac{qJ}{\varepsilon_o F^2}\sqrt{8mV_a}\right)^{1/2}\right]^{-1} \tag{16.32}$$

Thus τ decreases as F increases. From both Eqs (16.25) and (16.28)

$$\frac{V_a}{D} \equiv F_o = F + \frac{q}{\varepsilon_o}N_\tau J\tau \tag{16.33}$$

As $N_\tau J\tau$ is the average charge per unit area within the gap, the form of Eq. (16.33) is as expected, and it is the defining equation relating transit time to current density. Observe that the derivation depends on the one-dimensional relation between σ and F: for two- and three-dimensional charge distributions field varies with distance, and the definition of τ in such cases is still open.

Space-charge limited current arises when $F \to 0$, for which $J \to J_{CL}$, and the value of N_τ is affected by the choice of τ. For $\tau \to \tau_o$, as in Eq. (16.24), then $N_\tau = 9/8$, a finding first reported in ref. [222]. In contrast, if Eq. (16.32) is used for τ, then $N_\tau = 3/4$ when $F \to 0$. Showing that these limits are so is edifying. Using τ_o underestimates space-charge effects, but the $F = 0$ limit of Eq. (16.32) overestimates them. The usage of both in Eq. (16.33) easily identifies when space-charge effects are significant: field emission experimental data, for example, lie between the two choices [209], but showing that now is a bit premature until the geometry models are introduced.

The two transit times considered so far, namely, the ballistic transit time τ_o of Eq. (16.24) and the averaged charge density transit time $\tau(F)$ of Eq. (16.32), presume that the current density is unchanging. When it is not, a third transit time appears, and its consideration is instructive. Imagine what happens if charge is emitted in an instant, say by a short-duration laser pulse causing photoemission. The resultant sheet will take time to cross the AK gap, and the time it takes represents the third transit time. The density can be modeled as $\rho(x) = (\sigma/D)\delta(x - \bar{x}(t))$, where $x(t)$ is the location of the charged sheet and σ is the charge per unit area of the sheet. An important limiting case is when $q^2\sigma/\varepsilon_0$ of the sheet is enough to eliminate the field on the cathode: when that happens, let $\sigma \to \sigma_o$ and $q^2\sigma_o = J\delta t = \epsilon_o F_o$, where $\delta t \ll \tau$ is the very short duration of the laser pulse injecting the charge. Excess emitted charge density is pushed back to the cathode. The fields on either side of the sheet, shown in Figure 16.3, are related by [209, 227])

$$V_a = F_+(D - \bar{x}) + F_-\bar{x} \tag{16.34}$$

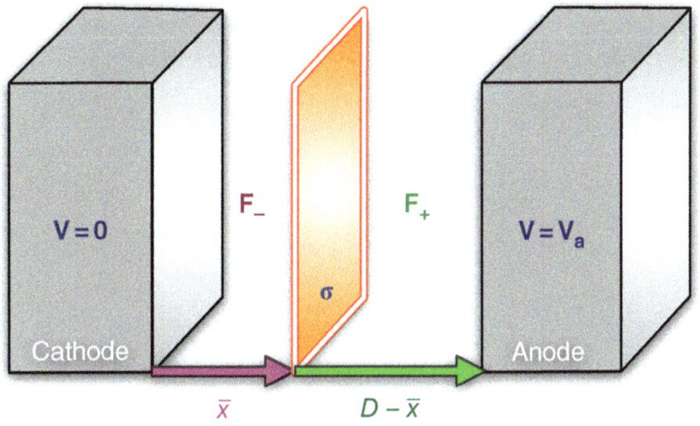

Figure 16.3 When a sheet of charge is injected into the AK gap, the fields $F_{\pm}$ on each side of the charge differ by an amount proportional to the area charge density σ of the sheet. The position of the sheet is $\bar{x}(t)$.

where F_- is the field between the cathode surface and the sheet, and F_+ between the sheet and anode: the two generally are not equal, but can be related further by applying Gauss's law to the sheet (*aka* the Guassian pillbox argument [99]). For an arbitrary sheet charge density σ that argument results in

$$\frac{q^2}{\varepsilon_0}\sigma = F_+ - F_- \tag{16.35}$$

The signs are because the fields due to the sheet point in opposite directions and so on the – side the force and the normal are antiparallel. Solving Eqs (16.34) and (16.35) simultaneously gives

$$F_+(\bar{x}(t)) = F_o + \frac{q^2\sigma}{\varepsilon_o D}\bar{x}(t) \rightarrow F_o\left(1 + \frac{\bar{x}(t)}{D}\right) \tag{16.36}$$

$$F_-(\bar{x}(t)) = F_o - \frac{q^2\sigma}{\varepsilon_o}\left(1 - \frac{\bar{x}(t)}{D}\right) \rightarrow F_o\frac{\bar{x}(t)}{D} \tag{16.37}$$

where the arrows correspond to $\sigma \rightarrow \sigma_o$. A tricky point is that the force under which the sheet moves (i.e., under which $\bar{x}$ evolves) is the average of F_+ and F_- (not their sum, as perhaps thought: in the limit of vanishing σ the fields on each side are equal to F and the field on the sheet is likewise F), and so

$$\frac{d^2}{dt^2}\bar{x}(t) = \frac{F_o}{2m} + \frac{F_o}{mD}\bar{x}(t) \tag{16.38}$$

Just as equations of the form $\partial_t^2 x = -\omega^2 x$ bring to mind harmonic oscillators and sines and cosines, so too do equations of the form $\partial_t^2 x = \omega^2 x$ bring to mind hyperbolic sines and cosines. One therefore anticipates solutions of the form $x(t) = A\cosh(at) + B\sinh(at) + C$. When $x(0) = 0$, then $C = -A$. Because the initial velocity is assumed to be zero, $B = 0$. And lastly, so that Eq. (16.38) holds, $A = D/2$ and $a^2 = F_o/mD = \sqrt{2}/\tau_o$. The general solution is therefore given by

$$\bar{x}(t) = \frac{D}{2}\left(\cosh\left(\sqrt{2}\frac{t}{\tau_o}\right) - 1\right) \tag{16.39}$$

The transit time τ is then found by $\bar{x}(\tau) = D$, or

$$\tau = \frac{\tau_o}{\sqrt{2}}\cosh^{-1}(3) = 1.2465\tau_o \tag{16.40}$$

That is a fairly astonishing result: the difference in transit times between an electron crossing the AK gap on its own (τ_o) versus traveling with so many friends that an infinitesimally thin sheet density reduces the field at the cathode to zero is a difference of about 25%. The uncertainty in about every other aspect of electron emission, from approximations to the Gamow factor and uncertainty about the electron trajectories, to work function and crystal face variation, and then on to statistical variation in field enhancement factors, makes the overall uncertainty in predicting the current density of physical electron sources such that 25% seems a bargain. It is therefore not too far from reasonable to take the transit time as something simple like τ_o and devote one's effort to better approximating current density.

CHAPTER 17

A General thermal–field–photoemission equation

In nova fert animus mutatas dicere formas
corpora; di, coeptis (nam vos mutastis et illas)
adspirate meis …

– Ovid[1]

In Ovid's *Metamorphosis*, or the *Book of Transformations*, the Roman poet fashioned from the mythologies of earlier times and writers something at once familiar and yet new. In the same way, the emission equations can be refashioned in a manner that – in the appropriate limits – transforms into the canonical equations of thermal (Eq. (12.3)), field (Eq. (13.25)), and photoemission (Eq. (14.16)). A single equation that becomes the various emission equations in the relevant limits is made possible by a reconsideration of the transmission probability $D(E)$ and how it behaves in conjunction with the supply function $f(E)$, from which the boundary between the thermal and field regimes naturally emerges.

Is such a single equation really useful? After all, the fields on a dispenser cathode are three or four orders of magnitude lower than for field emission cathodes, and thermal or field emission is undesirable for a laser-switched photocathode's operation, so are these equations really seen operating at the same time? Consider two instances.

Measurements of a low work function Ba-dispenser cathode (a thermionic emitter) with different metal and metal-oxide top layers were made by Geittner *et al.* [228] and are reproduced in Figure 17.1, in which emission from a tungsten (W-I) cathode is compared to that from a cathode containing scandium (Sc/Re-I). The current as a function of anode potential shows a change of behavior in keeping with a switch from a thermal characteristic ($\ln(I) \propto \sqrt{F}$) to a field emission characteristic ($\ln(I) \propto 1/F$). Even better, and shown to guide the eye, the dashed (thermal) and solid (field) curves are based on the Richardson–Laue–Dushman (RLD) and Fowler–Nordheim (FN) equations, respectively, using the following methods, required because the measurements are in terms of an anode potential (U) rather than a macroscopic field ($F_o = U/D$). Experimentally, field enhancement factors relate F_o to U, and are often called beta (β) factors (Section 30.1). The conventional reference to field enhancement factors by β is another cruel joke of nomenclature by employing an already overworked symbol. Even worse, experimental beta factors (which convert potential differences to fields) have units of length whereas theoretical beta factors (which compare microscopic to macroscopic fields) are dimensionless. For the moment, let the beta factors be of the experimental kind, and subscripted by whether they are to be used in the thermal (β_r) or field emission (β_f) equation.[2] The thermal (dashed) lines of Figure 17.1 are therefore constructed based on Eq. (13.25) to be

$$I_r(U) = I(U_o)\exp\left[\gamma(\sqrt{U} - \sqrt{U_o})\right] \tag{17.1}$$

where $\gamma = \sqrt{4Q\beta_r}/(k_B T)$, as intuited from Eq. (12.3). For the W line, $U_o = 3$ kV and for the Sc/Re line, $U_o = 1.5$ kV, with $\gamma = 4\ \text{kV}^{-1/2}$ for both. The choice of γ entails that $\beta_r U_o = 12.4$ eV/μm for W and 17.4 eV/μm for Sc/Re if the temperature is 300 K. The field (solid) lines are constructed using

$$I_f(U) = I(U_o)\left(\frac{U}{U_o}\right)^{2-\nu} \exp\left[B\left(U_o^{-1} - U^{-1}\right)\right] \tag{17.2}$$

where $B = (4/3\hbar\beta_f)\sqrt{2m\Phi^3}$, as intuited from Eq. (13.25) for current density.[3] For the W line, $U_o = 5$ kV, $B = 65.2$ kV, and $2 - \nu = 1.227$ ($\Phi = 4.5$ eV and $\beta_f = 1\ \text{nm}^{-1}$), and for the Sc/Re line, $U_o = 3.5$ kV, $B = 33.8$ kV, and $2 - \nu = 0.6615$

[1] "My intention is to tell of bodies changed/To different forms; the gods, who made the changes, /Will help me – or so I hope …" Ovid, *Metamorphoses* (trans. R. Humphries). Bloomington: Indiana University Press, 1955, p. 3.
[2] Later, the distinction between experimental and theoretical beta factors will be denoted by referring to the former with a subscript, as in β_g.
[3] This approach presumes that the emission area is constant, which is not true; the impact of emission area is considered in Section 30.3.

Introduction to the Physics of Electron Emission, First Edition. Kevin L. Jensen.

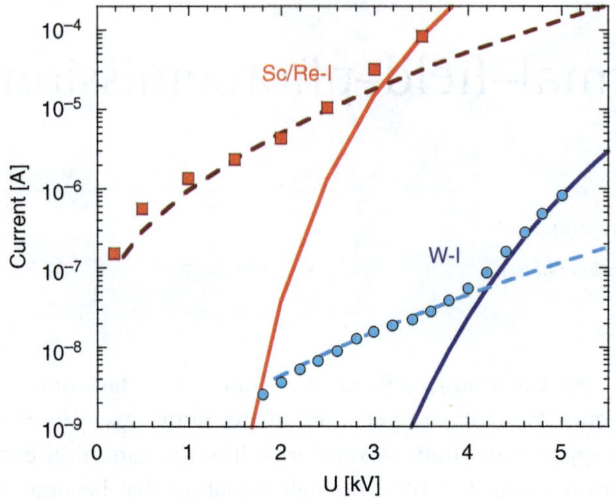

Figure 17.1 Emission data from Ba-dispenser cathodes; the Sc/Re-I cathode contains scandium. Solid lines correspond to a Fowler–Nordheim-like equation; dashed lines correspond to a Richardson-like equation. Experimental data digitally extracted from Figure 2 of ref. [228].

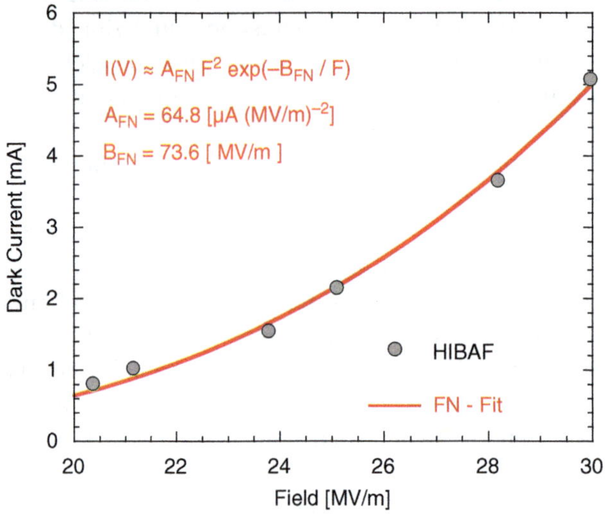

Figure 17.2 Dark current measured from the CsK_2Sb photocathode in the Los Alamos high-brightness accelerator FEL (HIBAF). The solid curve is based on a conventional FN-fit. Based on Figure 5 of ref. [230]; numerical data courtesy of P.G. O'Shea (UMD).

($\Phi = 1.5$ eV and $\beta_f = 0.37$ nm^{-1}). Clearly, both thermal and field emission current components contribute to the total current from a low work function dispenser cathode if the anode potential is sufficiently high.

Measurements of a photoemission cathode were undertaken in 1991 on the Los Alamos High Brightness Free Electron Laser (HIBAF), which employed a photoinjector to replace the thermionic cathode. The switch from a thermal to a photoemitter was motivated by the desire to achieve a lower emittance and brighter beam (Chapter 33), and thereby enable the production of micropulses containing 10 nC of charge per bunch. Dark current (current that existed when the laser was off) was found, and argued to arise from field emission from the CsK_2Sb photocathode [29, 229]. A comparison of that data, shown in Figure 17.2 and compared to a microscopic model of field emission based on the point charge model of Section 30.3 applied to that photocathode material, shows that hypothesis to be warranted [230]. A slightly faster approach is to show that the data lies along a Fowler–Nordheim fit, as the solid line on Figure 17.2 shows, the parameters of which are reasonable for the work function, the presumed size of the protrusions leading to the field emission current, and the fact that a fine polishing of the photocathode surface eliminated the field emission current.

So indeed, field emission, photoemission, and thermal emission effects can appear simultaneously. Having one equation that simultaneously treats them is practical beyond its economy.

17.1 Experimental thermal–field energy distributions

When we mean to build,
We first survey the plot, then draw the model...

– William Shakespeare[4]

The consideration of field emission from hot metal bodies is a good place to explore the transition from *cold* to *hot* emission. Refractory metals like tungsten and molybdenum can take punishing temperatures that melt other metals, and so making field emitters out of them is prudent: a heat treatment is often required to prepare them for high field operation, during which contaminants like carbon and absorbed gasses are removed before measurements are made [175]. These metals therefore are useful to examine the energy spread of the emitted electrons as the anode potential is progressively raised for a range of temperature conditions. The detailed analysis of experimental data performed by Gadzuk and Plummer [96] on the total energy distribution curves for a tungsten emitter held at 1570 K and subjected to progressively higher anode potentials is considered here, although ref. [154] gives a thorough account of Schottky emitters (field emission from hot emitters) and ref. [160] examines the energy distributions of carbon nanotube emitters. In the transition from thermal to field emission, the migration of the hump in the energy distribution peak moves from near the top of the emission barrier $\mu + \phi$ for thermal emission down to the Fermi level μ for field emission in a manner suggested by Figure 2 of Dolan and Dyke [192] and in Section 13.4. In particular, the work of Gadzuk and Plummer is synopsized in Figures 17.3 and 17.4.

Relating those findings to the treatment of current density here, however, is hampered by the distinction between total and normal energy. The x axes of Figures 17.3 and 17.4 are of total energy E, and therefore differ from the normal energy E_x into the barrier (called W by Gadzuk and Plummer, following the treatment of Young [231], but updated for modern sensibilities by Forbes [232]) that forms the basis of the emission equations approaches developed herein as in the supply function of Eq. (11.4) and the Gamow factor $\theta(E_x)$ in the transmission probability $D(E_x)$ in the integrand of Eq. (11.3). E and E_x are related by $E = E_x + E_\perp$, where $E_x = \hbar^2 k_x^2/2m$ and $E_\perp = \hbar^2(k_y^2 + k_z^2)/2m$. In terms of the total energy E, the current density can be written as

$$J(F,T) = \int_0^\infty j'(E)dE \tag{17.3}$$

Thus, $j'(E)$ is what is plotted in Figures 17.3 and 17.4. The total energy distributions bear a resemblance to the normal energy distributions that lay behind Eq. 11.3. Such a viewpoint is useful for understanding effects related to band structure [96, 165], so understanding the distinction between total energy versus normal energy formulations is important.

The total energy distribution is best approached from the viewpoint afforded by the precursor to Eq. (11.1), rewritten as (including electron spin)

$$J(F,T) = \frac{2q}{(2\pi)^3} \int \left(\frac{\hbar k_x}{m}\right) D(E_x) f_{FD}(E) d\vec{k} \tag{17.4}$$

where $E = E_x + E_\perp = \hbar^2 k^2/2m$. Instead of using k_x and $k_\perp$, consider instead the magnitude k and the polar angles, so that

$$J(F,T) = \frac{2q}{(2\pi)^2} \int k^2 dk \left(\frac{\hbar k}{m}\right) f_{FD}(E) D_t(E) \tag{17.5}$$

$$D_t(E) = \int_0^{\pi/2} D(E\cos^2\theta)\cos\theta\sin\theta d\theta \tag{17.6}$$

where integration over the azimuthal angle $d\varphi$ has been performed, the "total energy" transmission probability $D_t(E)$ has been introduced, and the $\cos\theta$ from $k_x = k\cos\theta$ is moved to the D_t integral. Observe that in contrast to the normal energy distribution, where the Fermi–Dirac distribution was modified to obtain the supply function (i.e., $f_{FD}(E) \to f(E_x)$), in the total energy distribution the modifications flow to the transmission probability (i.e., $D(E_x) \to D_t(E)$). Making the substitution $s = E\cos^2\theta$, it follows

$$D_t(E) = \frac{1}{2E} \int_0^E D(s)ds \tag{17.7}$$

[4]Ref. [37]: *The Second Part of King Henry the Fourth*, I.iii.41–42, p. 713.

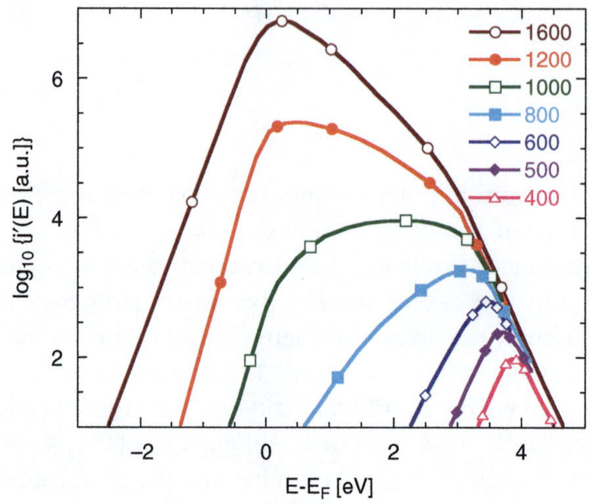

Figure 17.3 The migration of the total energy distributions of thermal emission (peak near $E - \mu \approx 4$ eV) as they migrate to field emission (peak near $E - \mu \approx 0$ eV) for a tungsten field emitter held at a steady temperature of 1570 K. Gadzuk and Plummer offered $\mu = 8$ eV, $\Phi = 4.8$ eV as the values giving the best fit to their theoretical analysis. Plot numbers refer to the anode potential (in volts). Based on Figure 2a of ref. [96].

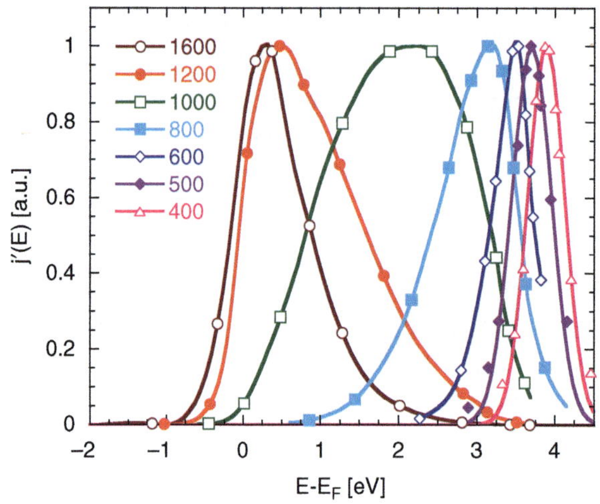

Figure 17.4 Same as Figure 17.3 but normalized and not on a log scale.

From Eq. (17.4) the integrand $j'(E)$ behind Figures 17.3 and 17.4 is

$$j'(E) = \left(\frac{mq}{\pi^2 \hbar^3}\right) \frac{ED_t(E)}{e^{\beta(\mu - E)} + 1} \tag{17.8}$$

Just as the supply function $f(E)$ could be analytically found, so too can $D_t(E)$, depending on the form $D(E_x)$ takes, and its derivation shall be undertaken next.

17.2 Theoretical thermal–field energy distributions

Again, to the intermediate some extremes show a certain likeness, as that of rashness to courage and that of prodigality to liberality; but the extremes show the greatest unlikeness to each other; now contraries are defined as the things that are furthest from each other, so that things that are further apart are more contrary.

– Aristotle[5]

[5]R.P. Hardie and R.K. Gaye, *Aristotle Nicomachean Ethics Book II, Great Books of the Western World*, ed. Robert Maynard Hutchins, Vol. 9, Aristotle II. Chicago: Encyclopaedia Britannica, 1952, p. 354 (Book II, Chapter 8).

A description of the so-called "thermal-field" conditions and their evolution was theoretically challenging (e.g., Cutler and Nagy [233]). Gadzuk and Plummer based their analysis, as did Murphy and Good [110] before them, on the use of the Kemble form of $D(E_x)$ given by [112]

$$D(E_x) = \frac{1}{1 + \exp[\theta(E_x)]} \tag{17.9}$$

where $\theta(E_x)$ is the Gamow factor (Section 11.2). Near the Fermi level $E_x \approx \mu$, θ is large, and the Fowler–Nordheim equation follows. Conversely, near $E_x \approx \mu + \phi$, θ is small, and $D(E_x)$ is comparable to unity so that the Richardson equation follows. The Kemble approximation is useful in that an approximation to it based on its *extremes* yields a convenient representation of its *middle*, the extremes being the form of the transmission probability $D(E)$ in the thermal and field limits. Note the guarded phrase "convenient representation": the approximation that shall be used here to capture the total energy distribution behavior of the previous section will not in fact be good, but it will suggest a path to get to an approximation that is. The first step to doing so, familiar from the derivation of the Fowler–Nordhiem equation, is to consider a linear form of the Gamow factor expressed in an unusual way by

$$\theta(E) \approx \theta(E_m) + \theta'(E_m)(E - E_m) \equiv \beta_F(E_c - E) \tag{17.10}$$

where β_F and E_c are defined by this equation: as per best practice for numerical approximation, the Taylor expansion point E_m is where the integrand is maximum. Two new terms β_F and E_c are introduced. E_c functions as the point where $D(E_c) = 1/2$ in the Kemble approximation if $\theta(E)$ were linear in energy,[6] and β_F acts as an energy slope factor, much like $\beta_T \equiv 1/k_B T$: to emphasize the distinction, it is given an F subscript. Thus, β_F acts as the energy slope factor for the transmission probability $D(E)$ much like β_T acts as an energy slope factor for the supply function $f(E)$ accounting for temperature effects. When Eq. (17.10) is used for the Gamow factor $\theta(E)$ in the Kemble approximation, it shall be referred to as the *linearized* Kemble approximation. A complication that will eventually command attention is that the integrand maximum changes location along the E axis as conditions change, therefore these "constants" are not constant, but depend on field and temperature, and as a result shift with conditions.

For field-like conditions, the factors β_F and E_c are as follows. The behavior of the Gamow factor $\theta(E)$ for the triangular barrier in Eq. (13.1) is modified by the image charge and leads to Eq. (13.11) but suggestively rewritten as

$$\theta(E) \approx \frac{C}{F} t(y) \left(\mu + \frac{Bv(y)}{Ct(y)} - E \right) \tag{17.11}$$

Therefore, it is easy to identify (using the definitions of B and C from Eqs (13.6) and (13.7)) that the linearized Gamow factor $\theta(E)$ for $E \approx \mu$

$$\beta_F \approx \frac{2}{\hbar F}\sqrt{2m\Phi}\, t(y) \tag{17.12}$$

$$E_c \approx \mu + \frac{2v(y)}{3t(y)}\Phi \tag{17.13}$$

It is worth emphasizing that E_c *is not the same as the location of the maximum* of the current density integrand, but off-set. For thermal-like conditions, the factors β_F and E_c change, but so as not to confuse the narrative, that discussion will wait until Section 17.3.

The important point is that as E approaches and then passes from below the barrier ($E < \mu + \phi$) to above it ($E > \mu + \phi$), $\theta(E)$ likewise appears to be driven to zero, but how it behaves for E above the barrier maximum has not yet been adequately dealt with. However, by demanding that $\theta(E)$ and its first derivative be continuous across $E = \mu + \phi$, the linear form $\beta_F(E_c - E)$ results, and as E increases, it becomes more negative. As a result, the linearized Kemble form of $D(E) \to 1$ as it should when the energy significantly exceeds the barrier height. Such a treatment may seem unsettling, but it is not without grounds: although the Gamow factor vanishes, the Kemble form is an approximation to the solutions to Schrödinger's equation that do behave in a manner mimicked by the linearized Kemble form. To see why requires the numerical solutions of Section 18.1, but until such time, Eq. (17.11) using Eqs (17.12) and (17.13) even for energies above the barrier maximum will have to be the *comic premise*,[7] although they change in the transition region and especially

[6] ...which it is not, and more on that below.

[7] In theater and humor, the comic premise is a paradoxical or unrealistic setting that must be accepted for the drama or joke to unfold, for example in the gag "A priest, a rabbi, and a duck walk into a bar, and the bartender says 'What is this, some kind of joke?'" requires suspension of disbelief that characters not normally together frequent establishments they do not normally patronize (even with a duck).

in the thermal regime. Usage of Eq. (17.11) in the linearized Kemble approximation allows $D_t(E)$ to be found, perhaps with effort, to be

$$D_t(E) = \frac{1}{2\beta_F E} \ln\left(\frac{1 + e^{-\beta_F(E_c - E)}}{1 + e^{-\beta_F E_c}}\right) \tag{17.14}$$

When $\beta_F E_c \gg 1$, the denominator of the fraction within the logarithm is approximately unity, and so

$$j'(E) = \left(\frac{mq}{2\beta_F \pi^2 \hbar^3}\right) \frac{\ln\left(1 + e^{-\beta_F(E_c - E)}\right)}{e^{\beta_T(E-\mu)} + 1} \tag{17.15}$$

A trained eye will see Figure 17.3 emerge, but to assist the imagination, examples help.

EXAMPLE: Find $j'(E)$ in units of eV-fs-nm-q for contemporary tungsten-like parameters of $\mu = 11$ eV and $\Phi = 4.5$ eV, and for a fixed temperature of $T = 1570$ K with $E = x\mu$. Assume β_F and E_c values for field emission conditions and use the Forbes–Deane approximation for $v(y)$, but for simplicity let $t(y) = t(y_o) = 1.06131$. Consider only the cases $F = 1.0$ eV/nm, 3.5 eV/nm, and 4.5 eV/nm.

SOLUTION: Evaluation of Eq. (17.12) reveals $\beta_F = 32.068/F$ and $E_c = 11 + 2.8267v(y)$ with $y = 0.26666\sqrt{F}$. Thus, for 1 eV/nm,

$$j'(x\mu) = 0.0456 \ln\left(1 + e^{243(x-1.24)}\right) / \left(1 + e^{81.3(x-1)}\right)$$

for 3.5 eV/nm,

$$j'(x\mu) = 0.155 \ln\left(1 + e^{71.5(x-1.18)}\right) / \left(1 + e^{81.3(x-1)}\right)$$

and for 4.5 eV/nm,

$$j'(x\mu) = 0.198 \ln\left(1 + e^{56.1(x-1.16)}\right) / \left(1 + e^{81.3(x-1)}\right)$$

A plot of the three equations in the above example (plus one more for good measure) gives a feel of how the total thermal–field energy distribution evolves as a function of field in Figure 17.5. Note in particular that when $F = 3.07$ eV/nm then $\beta_F \approx \beta_T$.

The difference between the total energy distribution and the normal energy distribution is then manifest in the differences in the integrands before the final integration (over E in the total energy distribution and over E_x in the normal energy distribution). The integrands are

$$j'_{norm}(E_x) = \left(\frac{qm}{2\pi^2\hbar^3}\right) \frac{\ln\left(1 + e^{\beta_T(\mu - E_x)}\right)}{\beta_T\left(1 + e^{\beta_F(E_c - E_x)}\right)} \tag{17.16}$$

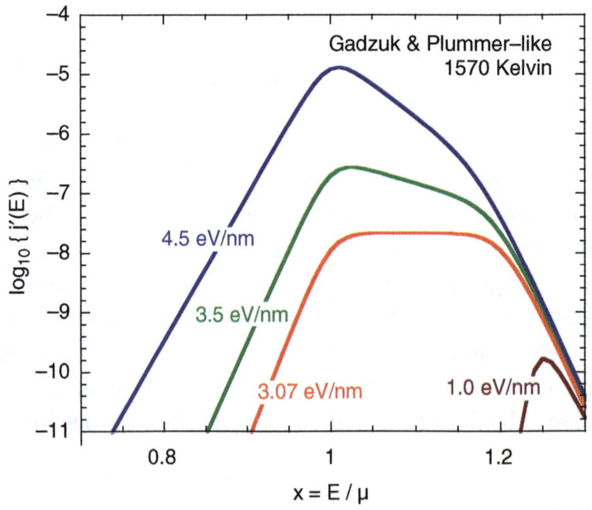

Figure 17.5 Evolution of the total energy distribution $j'_{tot}(E)$ for various fields using contemporary tungsten-like parameters $\mu = 11$ eV and $\Phi = 4.5$ eV at a temperature of $T = 1570$ K in the linearized Kemble approximation. Note the similarity to Figure 17.3. The line labeled "3.07 eV/nm" corresponds to $\beta_T = \beta_F$.

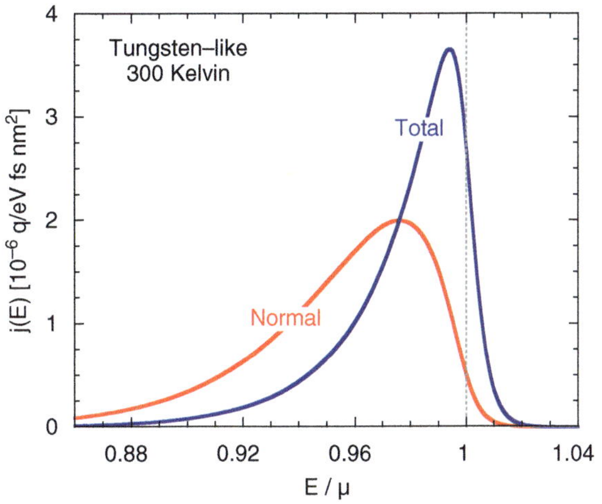

Figure 17.6 The energy distribution integrands for the normal (Eq. (17.16)) and total (Eq. (17.17)) distributions for the values of Gadzuk and Plummer of $\mu = 8$ eV, $\Phi = 4.8$ eV, and room temperature (300 K) conditions.

$$j'_{tot}(E) = \left(\frac{qm}{2\pi^2\hbar^3}\right)\frac{\ln\left(1+e^{\beta_F(E-E_c)}\right)}{\beta_F\left(1+e^{\beta_T(E-\mu)}\right)} \tag{17.17}$$

where the first equation is for the normal energy distribution and the second is for the total energy distribution: observe the subtle symmetries between the two. A comparison of their behavior as a function of their argument is shown in Figure 17.6 for room temperature conditions: as the temperature drops, the total distribution becomes higher and sharper near $E \approx \mu$ compared to the normal distribution (which peaks at a lower energy), giving a small advantage to the more gentle normal distribution for numerical evaluations. Having shown what we have set out to, we can return to considerations of the normal energy distribution approach that forms the basis of the earlier sections.

Understanding or visualizing the migratory behavior of the normal distribution integrand $D(E_x)f(E_x)$ is necessary before formulating a general thermal–field equation. An algorithm to do so is developed in Section A3.8. Although the temperature is that used by Gadzuk and Plumer, the analysis will relax back to the preferred copper-like values of $\mu = 7$ eV and $\Phi = 4.5$ eV. A waterfall plot for 6 eV $\leq E \leq$ 10 eV for values of F equispaced between 2.4 eV/nm and 3.6 eV/nm in steps of 0.2 eV/nm is created by the algorithm and shown in Figure 17.7. The transition region between thermal and field emission is evidently represented by the broad line near $F = 3$ eV/nm with a width almost spanning the energy range $\mu \leq E \leq \mu + \phi$.

With the Section A3.8 algorithm and a free afternoon, different circumstances can be investigated. At room temperature, the transition region drops to a much lower field and the curves sharpen, a characteristic of "cold" field emission. As the temperature increases, the transition region rises to higher values of F and the distribution becomes less abrupt, developing softer declines away from the peak (visually, the yellow-green bands narrow). But it is the middle condition

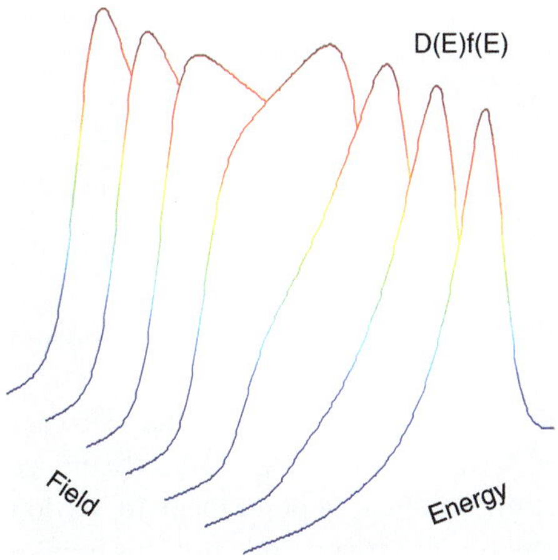

Figure 17.7 The migration of the normal energy distribution of thermal emission to field emission for parameters mimicking Figure 17.3: a metal-like emitter with $\mu = 7$ eV and $\Phi = 4.5$ eV is held at 1570 K. A line for each $D(E_x)f(E_x)$ is shown for 6 eV $\leq E_x \leq$ 10 eV with values of F equispaced between 2.4 eV/nm and 3.6 eV/nm in steps of 0.2 eV/nm, with the lower field values closer and the higher values to the rear. It is seen the behavior of Figure 17.4 is captured.

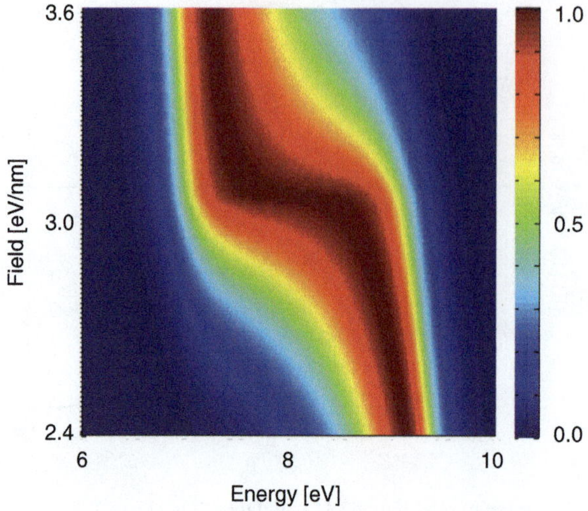

Figure 17.8 Same as Figure 17.7 but densely spaced and viewed top-down. The color calibration bar to the right is as used for the previous plot. Migration from thermal to field conditions is demarcated by the bar of red near $F \approx 3$ eV/nm: the peak initially centered near $E \approx \mu + \phi$ (right-hand side for $F \approx 2.4$ eV/nm) broadens and then coalesces again near $E \approx \mu$ (left-hand side for $F \approx 3.6$ eV) as the field increases.

when $\beta_F \approx \beta_T$ that corresponds to the red bar in the energy range $\mu \leq E \leq \mu + \phi$ (where $\mu = 7$ eV and $\phi \approx 2.4$ eV) and $F \approx 3$ eV/nm in Figure 17.8.

The transition region is nicely parameterized by comparing the beta factors governing how the integrand in $D(E_x)f(E_x)$ falls off: the energy factor β_T captures the fall-off of $f(E_x)$, while β_F captures the fall off (as E_x decreases) of $D(E_x)$. They can be compared. Consider the ratio

$$n \equiv \frac{\beta_T}{\beta_F} \tag{17.18}$$

Taking β_F as from Eq. (17.12) (i.e., $n \approx \hbar F/2k_B T\sqrt{2m\Phi}t(y)$), and letting $t(y) \approx t(y_o) = 1.06131$, then $n \approx 0.320412F$ if F is measured in eV/nm. In that case, the red bar in Figure 17.8 denoting the transition between thermal and field emission conditions, seen to occur near $F = 3$ eV/nm, *is characterized by the condition* $n \approx 1$. That is significant: when the energy slope factors are comparable, the transition from one regime to the other is initiated.

The β_F up to now is as given by Eq. (17.12), corresponding to a Taylor expansion of $\theta(E_x)$ to first order about $E_x = \mu$, that is, the linearized $\theta(E_x)$ is the tangent line at $E_x = \mu$. For field emission, this is close to where the maximum of $j'_{norm}(E_x)$ occurs. However, as $n \to 1$, the maximum occurs between μ and $\mu + \phi$, and failure to account for that reduces the accuracy of the approximation to $J(F, T)$. When emission becomes thermal-like, β_F should be found for $\theta(E_x)$ linearized near the barrier maximum $\mu + \phi$. Near there the image charge barrier $V(x)$ is approximately parabolic, and $\theta(E)$ can be evaluated in a procedure similar to that leading to Eq. (11.20). For a potential of the form $V_q(x) = V_o - a(x - x_o)^2$ such that x_o is the maximum of the image charge potential in Eq. (11.25) and a is such that $\partial_x^2 V_q(x_o) = \partial_x^2 V_{image}(x_o)$, and so (recalling that $\phi = \Phi - \sqrt{4QF}$)

$$\theta_{quad}(E) = \frac{\pi Q^{1/4}}{\hbar F^{3/4}}\sqrt{2m(\mu + \phi - E)} \tag{17.19}$$

from which the β_F and E_c factors are easily seen to be

$$\beta_F \approx \frac{\pi Q^{1/4}}{\hbar F^{3/4}}\sqrt{2m} \tag{17.20}$$

$$E_c \approx \mu + \phi \tag{17.21}$$

for thermal emission conditions. This has consequences for the values n and s (and even u) take on in the general emission equation. Representative values will be considered in Section 17.3 below.

Before taking leave of the energy slope factors, return to the passing observation above that the image charge barrier looks parabolic at high field: it is more than that. When the energy E is near the barrier maximum $\mu + \phi$ for the image charge potential, the barrier apex resembles the quadratic barrier generally, regardless of field (a point made in Section 13.2 when the functions $v(y)$ and $t(y)$ were introduced, and which anticipated the present treatment). But this implies that β_F for the image charge barrier and the quadratic barrier should be related in the limit that $E_x \to \mu + \phi$. This can be shown analytically: in fact, the ratio Eqs (17.12) and (17.20) is

$$\frac{\beta_F(image)}{\beta_F(quad)} = \frac{\pi}{2t(y)}\left(\frac{QF}{\Phi^2}\right)^{1/4} \tag{17.22}$$

In the limit that $E \to \mu + \phi$, then y in $t(y)$ approaches unity, or $\sqrt{4QF} \to \Phi$. Substituting in the limiting field $F = \Phi^2/4Q$ results in

$$t(1) = \frac{\sqrt{2}}{4}\pi = 1.1107 \tag{17.23}$$

which is known to be true as the limiting case [196] (and even well approximated by the Forbes–Deane formula of Eq. (13.24), which gives $10/9 = 1.1111$), as the adventurous in spirit can show using Eq. (13.19). It can also be shown numerically, as in Section 17.4 below.

17.3 The $N(n, s, u)$ function

There are three species of government: republican, monarchical, and despotic ...we must now inquire into those laws which directly conform to this nature, and consequently are the fundamental institutions.

– Montesquieu[8]

That the thermal, field, and photoemission canonical equations of Chapter 11 were all obtainable from a master equation suggests that the three kinds of emission represent three domains of an underlying function which also merit inquiry. The master equation is

$$J(F,T,\omega) = \frac{q}{2\pi}\int_0^\infty \frac{\hbar k_x}{m} D(E_x + E_\omega) f(E_x)) dk_x \tag{17.24}$$

$$= \frac{q}{2\pi\hbar}\int_0^\infty D(E_x + E_\omega) f(E_x) dE_x \tag{17.25}$$

where the domains are specified by the forms that $D(E_x)$ and $f(E_x)$ take in either high or low T and F limits, and whether or not there was a photon term present ($E_\omega = 0$ when there is not). Using the Kemble form of the transmission probability $D(E_x)$ (Eq. (17.9)) and the supply function $f(E_x)$ (Eq. (11.5)), Eq. (17.24) can be more compactly written as [85, 202]

$$J(F,T,\omega) = A_{RLD} T^2 N\left[\frac{\beta_T}{\beta_F}, \beta_F(E_o - \mu), \beta_F E_o\right] \tag{17.26}$$

where E_o is the generalization of E_c in Eq. (17.13) to accommodate photons, more about which in Section 17.3.2. With the introduction of the important quantities

$$n = \beta_T/\beta_F \tag{17.27}$$

$$s = \beta_F(E_o - \mu) \tag{17.28}$$

$$u = \beta_F E_o \tag{17.29}$$

where $E_o = E_o(E_m)$ is understood to be evaluated *at the integrand maximum* E_m, then the N function $N(n, s, u)$ is an integral of the form

$$N(n,s,u) \equiv n\int_{-\infty}^{u} \frac{\ln[1 + e^{n(z-s)}]}{1 + e^z} dz \tag{17.30}$$

where the integration variable z arises from $\beta_F(E - \mu)$, the denominator in the integrand arises from the transmission probability, and the numerator arises from the supply function. The three domains are specified by whether s is positive or negative and, if positive, whether n is greater than or less than 1.

[8] Charles de Secondat, Baron de Montesquieu, *The Spirit of Laws* (trans. Thomas Nugent, rev. J. V. Prichard), Kindle eBook, 2011, Book II, Section 1. Montesquieu went on to advocate three branches of republican government (executive, legislative, and judicial) in Book XI.6. Many things of profound interest come in threes.

There have been heroic efforts to carefully approximate similar integrals, for example by Guth and Mullin [190] (who also study the Fowler–DuBridge function of Eq. (14.9)). Any time spent crafting analytical approximations to Eq. (17.30) reveals it to be rather ghastly: its integration range is such that the usual recourse to Taylor expansions of $\ln(1+x)$ and $1/(1+x)$ (Section A1.2.1) abuts the divergent series, causing the crimes against intuition treated in Section 6.3: the case $n = 1$ is frustrating as it is there that the divergent series reveal their "devilish" nature, but that is the very point of greatest interest in investigating the transition, and coincidentally the region that brought ruin to the Millikan–Lauritsen hypothesis of Eq. (13.31). One must proceed cautiously. The first step is to appreciate the variation of parameters involved as described by Eqs (17.12) and (17.20).

Each of the parameters n, s, and u depend on β_F, and as there are different limiting cases of β_F, it is evident that something unusual is going on between those limiting cases. Consider the parameter n from Eq. (17.18) again as it approaches unity starting from a high-field, low-temperature condition compared to starting from a low-field, high-temperature condition: the ratio of the field-like n (Eq. (17.12)) to the thermal-like n (Eq. (17.20)) is

$$\frac{n(\text{field})}{n(\text{thermal})} = \frac{\pi\sqrt{2y}}{4t(y)} \tag{17.31}$$

As with Eq. (17.23), as $y \to 1$, the two become the same,[9] as evident in the pinching of the contour lines in Figure 17.9. That a difference exists between the n limits is a wrinkle that will need revisiting below, but for the development of a general thermal–field equation, what it means here is that before approximations can be invoked, an understanding of the values n, s, and u may take on is needed. The limiting cases are field and thermal emission conditions, and so representative conditions are taken for each:

- **For field conditions:** If the forms of β_F (Eq. (17.12)), E_o (Eq. (17.13)), and B and C are from Eqs (13.7) and (13.6)), respectively, then

$$n = \frac{F}{Ck_BTt(y)} = \frac{\hbar F}{2\sqrt{2m\Phi}k_BTt(y)} \tag{17.32}$$

$$s = \frac{B}{F}v(y) = \frac{4\sqrt{2m\Phi^3}}{3\hbar F}v(y) \tag{17.33}$$

$$u = \frac{B}{F}v(y) + \frac{C}{F}\mu t(y) \tag{17.34}$$

where as usual $y = \sqrt{4QF}/\Phi$ in the Nordheim functions $v(y)$ and $t(y)$.

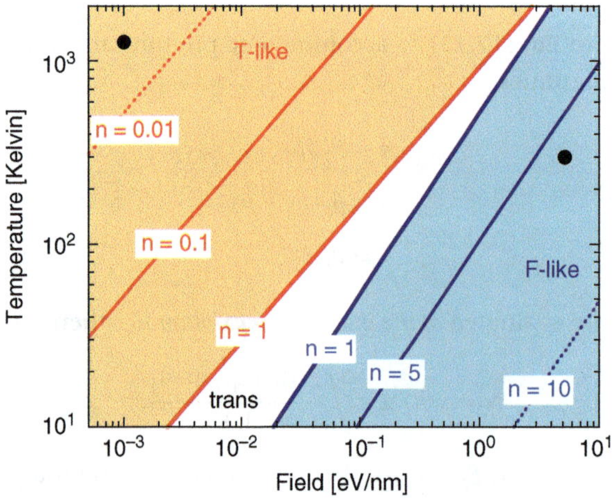

Figure 17.9 Boundaries defined by $n = 1$ using field-like (F-like; Eq. (17.12); bluish region) and thermal-like (T-like; Eq. (17.20); reddish region) energy slope factors. The intermediate regime (trans; white region between two $n = 1$ lines) corresponds to where $n = 1$. The two black dots correspond to typical operating temperatures for thermal (top left: 1273 K, 1 eV/μm) and field (middle right: 300 K, 5.1734 eV/nm) conditions treated in the two examples below.

[9] To prove this requires Eq. (13.22), as Eq. (13.24) is only approximately correct at $y = 1$.

- **For thermal conditions:** From the forms of β_F (Eq. (17.20)), E_o (Eq. (17.21)), and the terms B and C, then

$$n = \frac{\hbar}{2\pi k_B T}\left(\frac{4F^3}{Qm^2}\right)^{1/4} \tag{17.35}$$

$$s = \beta_F(\Phi - \sqrt{4QF}) = \beta_F \phi \tag{17.36}$$

$$u = \beta_F(\mu + \phi) \tag{17.37}$$

Under these approximations, the typical values are found in the subsequent examples using the following limiting case formulae.

EXAMPLE: *Field-like conditions:*
Consider copper-like parameters ($\mu = 7$ eV, $\Phi = 4.5$ eV) at room temperature ($T = 300$ K) such that $y = y_o = e^{-1/2}$. This chooses a value of field ($F = \Phi^2/4eQ = 5.1734$ eV/nm) typical of field emission yet simplifies $v(y)$ and $t(y)$ when using Eqs (13.23) and (13.24).
SOLUTION: For the chosen value of F, $C = 4.2014$ eV^{-1}, $B = 65.207$ eV/nm, $t(y) = 1 + (1/6e) = 1.0613$, $v(y) = 1 - (7/6e) = 0.57081$ and therefore $\beta_F = Ct(y) = 4.4590$ eV^{-1}. For room temperature, $\beta_T = 38.682$ eV^{-1}. Substituting, it is found $n = 8.6749$, $s = 7.1946$, $ns = 62.413$, and $u = 38.408$.

EXAMPLE: *Thermal-like conditions:*
Consider parameters typical of a barium dispenser cathode ($\mu = 7$ eV, $\Phi = 2.0$ eV, $T = 1273$ K) for a large field (by thermionic cathode standards) of $F = 1$ eV/μm, corresponding to an electric field of 1 MV/m.
SOLUTION: For the given temperature and field, $\beta_T = 9.1159$ eV^{-1}, $\beta_F = 2217.0$ eV^{-1}, $\phi = \Phi - \sqrt{4QF} = 1.9621$ eV, and therefore $n = 3.7388 \times 10^{-3}$, $s = 4349.8$, $ns = \beta_T\phi = 17.886$, and $u = 19869$.

What is seen from these examples is that thermal conditions are characterized by very small n and field conditions by large n, but also that s is large and ns very large for field conditions, whereas ns is large and s very large for thermal conditions. These observations will decide which terms to keep in an expansion of the integrand terms in $N(n, s, u)$ when discussing thermal and field emission.

What about photoemission? In the treatment leading to the one-dimensional Fowler–DuBridge theory of Chapter 14, the approach was to augment the energy E_x in the transmission probability by $E_\omega = \hbar\omega$, or $D(E_x) \rightarrow D(E_x + E_\omega)$. Careful attention to how s is defined reveals the presence of photons causes $\mu \rightarrow \mu + E_\omega$, or $s(\text{photo}) \rightarrow \beta_F(E_o - \mu - E_\omega)$); coupled with $E_o < \phi$ and $E_\omega > \phi$, photoemission entails that *s is negative*. Therefore, consider the positive and negative regimes of s separately.

17.3.1 The $s > 0$ regime

The scientific approach to the examination of phenomena is a defense against the pure emotion of fear. Keep tight hold and continue while there's time.

– Tom Stoppard[10]

To tame difficult integrals, it is often useful to separate them into manageable parts by either subtracting out a dominant (and analytic) contribution or partitioning the integration region into sections where Taylor expansions can and do lead to convergent series. Both techniques are useful here [202]. Below, this will amount to

$$N = N_1 + N_2 + N_3 + N_4 \tag{17.38}$$

[10] Tom Stoppard, *Rosencrantz & Guildenstern are Dead*, New York: Grove Press, 1967, Act 1, p. 17.

where the N_j shall be defined below, and which rely on tricks such as $\ln(1+e^x) = x + \ln(1+e^{-x})$. Given how s and ns behave, matters shall be set so that N_1 and N_2 dominate when field emission is ascendant, and N_3 and N_4 when thermal emission is ascendant, therefore refer to the integrals that will arise as either "field-dominated" or "thermal-dominated".

The field-dominated regions matter most when $n > 1$ and are characterized by $z > s$. They are

$$N_1(n,s,u) = n\int_s^u \frac{n(z-s)}{e^z+1}dz \tag{17.39}$$

$$N_2(n,s,u) = n\int_s^u \frac{\ln\left[1+e^{-n(z-s)}\right]}{e^z+1}dz \tag{17.40}$$

Similarly, the thermal-dominated integrals matter most when $n < 1$ and are characterized by $z < s$. They are

$$N_3(n,s,u) = n\int_{-\infty}^0 \frac{\ln\left[1+e^{n(z-s)}\right]}{e^z+1}dz \tag{17.41}$$

$$N_4(n,s,u) = n\int_0^u \frac{\ln\left[1+e^{n(z-s)}\right]}{e^z+1}dz \tag{17.42}$$

The choice of the integrand and the integration region is to a definite purpose each time. In the case of N_1, the integral is analytic. In the case of the others, use can be made of familiar series expansions (Section A1.2.1) of the terms behaving as $\ln(1+x)$ or $1/(1+x)$, where x stands in for an exponential term (e.g., e^z and $e^{\pm n(z-s)}$) (an analogous approach formed the basis of Guth and Mullin's treatment [234]). The integration regions then allow for a term-by-term integration of the series expansions over regions where the series are convergent.

An important approximation concerns u: when $u \gg 1$ then the upper limit of the integral can be extended to ∞, thereby reducing the number of terms that survive in the series expansions that arise and making for much easier formulae. Practically, it means that $N(n,s,u) \approx N(n,s,\infty) \equiv N(n,s)$ and likewise with the N_j, and that approximation shall be assumed henceforth.

Two integrals enable finding $N_1(n,s)$, and they are

$$\int_s^\infty \frac{dx}{e^x+1} = \ln(1+e^{-s}) \tag{17.43}$$

$$\int_s^\infty \frac{xdx}{e^x+1} = U(-s) + s\ln(1+e^{-s}) \tag{17.44}$$

where $U(s)$ is the Fowler–DuBridge function from Eq. (14.9). To show it, make the substitution $z = -\ln(x)$ in the first integral, and use integration by parts in the second. It is then found that

$$N_1(n,s) = n^2U(-s) \tag{17.45}$$

a compact result, but given that the other three integrals will be given in terms of a series representation, it is more useful to express N_1 as a series itself. The next example concerns finding that series representation.

EXAMPLE: Find the series expansion of $N_1(n,s)$.

SOLUTION: From Eq. (A1.5) followed by an integration of each term in the series,

$$N_1(n,s) = n^2\int_{-\infty}^{-s} dx \sum_{j=1}^{\infty}(-1)^{j+1}\frac{e^{jx}}{j} \tag{17.46}$$

$$= n^2\sum_{j=1}^{\infty}(-1)^{j+1}\frac{e^{-js}}{j^2} \tag{17.47}$$

Observe that $N_1(n,0) = n^2\zeta(2)/2$. The result is reminiscent of the polylogarithm function discussed in conjunction with Eq. (14.13).

The other N_j entail double summations, one for each expansion using Eqs (A1.3) and (A1.5) for $1/(1+x)$ and $\ln(1+x)$, respectively, after which the integration over the exponentials may be performed using

$$\int_s^\infty e^{-ax}dx = \frac{1}{a}e^{-as} \tag{17.48}$$

In an exercise requiring attention to bookkeeping, a straightforward integration and simplification for each of the N_j result in

$$N_2(n,s) = n\sum_{k=1}^\infty\sum_{j=1}^\infty \frac{(-1)^{k+j}e^{-ks}}{j(jn+k)} \tag{17.49}$$

$$N_3(n,s) = -n\sum_{k=0}^\infty\sum_{j=1}^\infty \frac{(-1)^{k+j}e^{-jns}}{j(jn+k)} \tag{17.50}$$

$$N_4(n,s) = n\sum_{k=1}^\infty\sum_{j=1}^\infty \frac{(-1)^{k+j}}{j(jn-k)}(e^{-ks}-e^{-jns}) \tag{17.51}$$

where the limits of the k summation for N_2 differ from the summations for the other N_j: this is a consequence of having to use $1/(e^z+1) = e^{-z}/(1+e^{-z})$ when $z > 0$ in N_2 and N_4, but not N_3. It is more useful to separate out the $k = 0$ part of the summation so that when one combines N_1 through N_4 it is found that

$$\begin{aligned} N(n,s) &= \sum_{j=1}^\infty \frac{(-1)^{j+1}}{j^2}(n^2e^{-js}+e^{-jns}) \\ &+ 2n^2\sum_{j=1}^\infty\sum_{k=1}^\infty \frac{(-1)^{j+k}}{n^2j^2-k^2}(e^{-ks}-e^{-jns}) \end{aligned} \tag{17.52}$$

Before continuing, notice a feature of importance: the denominator $(n^2j^2 - k^2)$ vanishes whenever $n = k/j$, which appears to reveal a singular (infinite) term in the summation. Note, though, that at the same value of n, then by L'Hôpital's rule governing the ratio of two differentiable functions that vanish at the same point, when $n \to j/k$ then

$$\frac{e^{-ks}-e^{-jns}}{n^2j^2-k^2} \to -\frac{s(2+s(k+jn)}{2(k+jn)} \to \frac{s}{2k} - \frac{s^2}{2} \tag{17.53}$$

There are three important points to be made:

1. In the quest for a computationally useful approximation to $N(n,s)$, the appearance of infinities for particular values of n signals that they are an artifact of the approximation, not of $N(n,s)$ itself.
2. In the thermal regime ns is large but s is larger so that $e^{-s} \ll e^{-ns}$, whereas in the field regime, s is large but ns is larger so that $e^{-s} \gg e^{-ns}$. Thus, away from $n \approx 1$ either the thermal term dominates or the field does, but one of them is negligible as n moves away from unity.
3. $N(n,s)$ is well-behaved as $n \to 0$ and $n \to \infty$, as it must be, with the leading order surviving terms giving rise to the RLD and FN equations.

The last two points figure strongly in the treatment of an approximation to the function $\Sigma(x)$ in Eq. (17.65).

Eq. (17.52) contains a great many terms, but they are of two types: N_1 and N_2 involve e^{-ks} whereas N_3 and N_4 involve e^{-jns}. Remember now that s and ns are both large, but for field conditions ns is very large and for thermal conditions s is very large. More importantly, only the $k = 1$ term need be kept in the field-dominant series, and only the $j = 1$ term in the thermal-dominant series. Knowing that, and keeping in mind that j and k are dummy labels, so that when they are swapped, a $-$ sign is introduced because of the denominator, it is found that

$$N(n,s) = e^{-s}n^2 2\sum_{j=1}^\infty \frac{(-1)^{j+1}}{(jn)^2-1} + e^{-ns}2\sum_{j=1}^\infty \frac{(-1)^{j+1}}{(j/n)^2-1} \tag{17.54}$$

In fact, this can be made even more compact: if the function $\Sigma(x)$ is introduced (where x is used to indicate a general argument, n having been reserved for β_T/β_F in particular) such that

$$\Sigma(x) = 1 + 2\sum_{j=1}^\infty \frac{(-1)^{j+1}}{(j/x)^2-1} \tag{17.55}$$

then the elegant result

$$N(n,s) = e^{-s} n^2 \Sigma\left(\frac{1}{n}\right) + e^{-ns}\Sigma(n) \tag{17.56}$$

is obtained, with the first term being the *field*-dominated term (which will lead to J_F), and the second the *thermal*-dominated term (which will lead to J_T). Recall from Eq. (17.26) that current density is the product of N with $A_{RLD}T^2$. Therefore, the "field" and "thermal" J's are defined by

$$J_F = A_{RLD}(k_B\beta_T)^{-2} n^2 \Sigma\left(\frac{1}{n}\right) e^{-s} \tag{17.57}$$

$$J_T = A_{RLD}(k_B\beta_T)^{-2}\Sigma(n)e^{-ns} \tag{17.58}$$

The symmetry between the two forms is far better appreciated, however, by putting in the definition of n and rewriting:

$$J_F = A_{RLD}(k_B\beta_F)^{-2}\Sigma\left(\frac{\beta_F}{\beta_T}\right)\exp[-\beta_F(E_o - \mu)] \tag{17.59}$$

$$J_F = A_{RLD}(k_B\beta_T)^{-2}\Sigma\left(\frac{\beta_T}{\beta_F}\right)\exp[-\beta_T(E_o - \mu)] \tag{17.60}$$

Now the pleasing symmetry between T and F that Millikan intuited must be there is readily apparent. The compact form of the general thermal–field equation for current density, or $J_{GTF}(F,T)$, is given by

$$J_{GTF}(F,T) = \begin{cases} n^{-2}J_F + J_T & (n < 1) \\ J_F + n^2 J_T & (n > 1) \end{cases} \tag{17.61}$$

with the helpful mnemonic that n^2 attaches itself to whichever J is smaller in such a way as to increase its contribution. Importantly, in the large n limit ($T \to 0$), $J_F \to J_{FN}$, and in the small n limit ($F \to 0$), $J_T \to J_{RLD}$. Eq. (17.61) *in its entirety* is the means by which thermal terms are introduced into field emission, and field into thermal emission, situations which matter when $n \approx 1$.

A certain aesthetic attractiveness attends Eqs (17.59) and (17.60) until they are put into practice. The problem is because $\Sigma(n)$ and $\Sigma(1/n)$ are calculated from an infinite series given by Eq. (17.55), there are values of n for which the denominator can vanish, whereas as seen before in the full definition of $N(n,s)$ there are no such singularities. This is an aforesaid artifact of the approximation. The solution, therefore, is to find a low-order polynomial approximating $\Sigma(x \approx 1)$, as for commonly encountered parameters, either e^{-s} or e^{-ns} quickly overpowers $\Sigma(1/n)$ or $\Sigma(n)$, respectively. The first term in $\Sigma(x)$ is singular at $x = 1$, the very region where all four N_j are retained. No "low-order polynomial" approximation exists in the vicinity of that singularity. For that case, the summations are regrouped until the problemmatic $x = 1$ singular part is isolated, and that part is separated from the polynomial part. Using the relation

$$\frac{1}{(j/x)^2 - 1} = \sum_{k=1}^{\infty}\frac{x^{2k}}{j^{2k}} \tag{17.62}$$

and the definition of the Riemann $\zeta(j)$ function series of Eq. (A1.8) it is found that

$$\Sigma(x) = 1 + \sum_{j=1}^{\infty}(1 - 2^{1-2j})\zeta(2j)x^{2j} \tag{17.63}$$

which hardly seems an improvement as $\zeta(2j) \to 1$ as $j \to \infty$: the replacement of a series with a singularity in it with a divergent series at $x = 1$ (bringing to mind Section 6.3) is not obviously wise: as Virgil said, *aegrescit medendo*[11]. The reason

[11] Latin for "the remedy is worse than the disease", the interpretation given to "grown worse with healing" (trans. James Rhoades), *The Poems of Virgil, Great Books of the Western World,* ed. Robert Maynard Hutchins, Vol. 13, Virgil. Chicago: Encyclopaedia Britannica, 1952, p. 355, *Aeneid,* Book XII, line 48. The Romans of Virgil's time distrusted their physicians.

to do so is because the offending part may be isolated by

$$\Sigma(x) = 1 + 2\sum_{j=1}^{\infty} x^{2j} + 2\sum_{j=1}^{\infty}\left\{\left(1 - 2^{1-2j}\right)\zeta(2j) - 1\right\}x^{2j}$$

$$= \frac{1+x^2}{1-x^2} + 2\sum_{j=1}^{\infty}\left\{\left(1 - 2^{1-2j}\right)\zeta(2j) - 1\right\}x^{2j} \tag{17.64}$$

$$\approx \frac{1+x^2}{1-x^2} + (\zeta(2) - 2)x^2 + \frac{1}{4}(7\zeta(4) - 8)x^4 \tag{17.65}$$

where now the singular term is isolated and the other terms are small by comparison when $x < 1$ (or even x a little larger than 1, as will be needed). Thus, at $n = 1$, the singularity of $\Sigma(n)$ cancels that of $\Sigma(1/n)$ via

$$\lim_{n\to 1}\left\{\Sigma(n) + \Sigma\left(\frac{1}{n}\right)\right\} = \frac{7}{2}\zeta(4) + 2\zeta(2) - 8 = -0.9220 \tag{17.66}$$

It is, however, the behavior of Eq. (17.56) as $n \to 1$ that matters. In the series summation definition of $\Sigma(n)$ in Eq. (17.55), there are two kinds of terms: those that are problematic as $n \to 1$ (the first term of the summation) and those that are not (the rest of the summation), and so for $n \approx 1$,

$$\Sigma(n \approx 1) = 1 + \frac{2n^2}{1-n^2} + 2\sum_{j=2}^{\infty}\frac{(-1)^{j+1}}{j^2 - 1} \tag{17.67}$$

where in the second summation, n is set to 1 as doing so does not encounter a zero in the denominator. The equation continues to hold for $n \to 1/n$. The summation term is easily evaluated

$$2\sum_{j=2}^{\infty}\frac{(-1)^{j+1}}{j^2-1} = \sum_{j=2}^{\infty}\left\{\frac{(-1)^{j+1}}{j+1} - \frac{(-1)^{j+1}}{j-1}\right\}$$

$$= \sum_{j=1}^{2}\frac{(-1)^{j+2}}{j} = -\frac{1}{2} \tag{17.68}$$

and so with the approximation $e^{-ns} \approx e^{-s}(1 - (n-1)s)$,

$$e^{-s}\frac{(n+1)(n^2+1) + s(3n^2+1)}{2(n+1)} \to (s+1)e^{-s} = N(1,s) \tag{17.69}$$

The behavior of $N(n,s)$ for $n \approx 1$ is the most important for J_{GTF}, but how the other limits of $n \to 0$ and $n \to \infty$ fare, as in the discussion following Eq. (17.52), is important to reassess. How the behavior of J_{GTF} is obtained in the $n \ll 1$ and $n \gg 1$ limit given that $\Sigma(x)$ from Eq. (17.64) is a converging series only for $x < 1$ is the point: J_{GTF} asks for arguments in $\Sigma(x)$ for $x > 1$ (i.e., if $n < 1$ then $1/n > 1$ and same with $n > 1$ entailing $1/n < 1$). Any untoward behavior as a function of n that occurs is a consequence of the approximation to $\Sigma(x)$. The resolution is understood by being specific about how n approaches the limits of 0 and ∞ when using Eq. (17.65):

- In the thermal regime, $n \to 0$ behavior is obtained by driving $F \to 0$. The polynomial part of the $\Sigma(1/n)$ approximation then diverges as $1/F^4$ but it is attached to an exponential part that declines as $\exp(-a/F^p)$ where p is on the order of unity and a sweeps up all other factors.
- In the field regime, $n \to \infty$ behavior is obtained by driving $T \to 0$. The polynomial part of the $\Sigma(n)$ approximation then diverges as $1/T^4$ but it is attached to an exponential part that declines as $\exp(-a/T)$ where a sweeps up all other factors.

Consequently, the nature of the polynomial part of the approximation to $\Sigma(x)$ to become large (as x^4) when its range is extended to $x > 1$ is overshadowed by the exponential terms to which it is attached, and so the numerical values given by the $\Sigma(x)$ approximation continue to work well into the traditional regimes of the canonical equations. The canonical equations are thereby recovered with acceptable numerical accuracy. This is additionally demonstrated in Figure 17.10, where the field part is always less than the thermal part in the thermal regime, whereas the thermal part is always less than the field part in the field regime, by several orders of magnitude. The quantities are evaluated in terms of an argument x such that $x = F/0.05 \to 0$ corresponds to $n \to 0$ and $x = T/50 \to 0$ corresponds to $n \to \infty$, using parameters typical of dispenser cathodes ($T = 1200$ K and $\Phi = 2.0$ eV) or field emitters ($F = 8$ eV/nm and $\Phi = 4.5$ eV). To reiterate, the thermal and field parts of $N(n,s)$ do not become comparable until within the $n = 1$ regime.

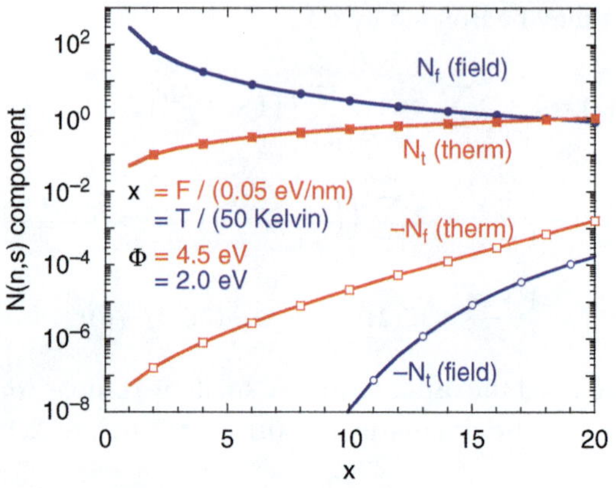

Figure 17.10 Behavior of the field-dominated and thermal-dominated parts of $N(n,s)$ (first and second terms of Eq. (17.56), respectively) as a function of $T = 50x$ K (field) and $F = 0.05x$ eV/nm (thermal), where $n(F,T)$ and $s(F)$ are given by Eqs (17.32) and (17.33) in the field regime, and Eqs (17.35) and (17.36) in the thermal regime.

Speaking of n and s as parameters themselves is, however, misguided. Although β_F has been treated *as though* it can only take on one of two forms (from $\theta(E \approx \mu)$ for field emission leading to Eq. (17.12)), or $\theta(E \approx \mu + \phi)$ for thermal emission leading to Eq. (17.20)), in reality, β_F, as well as n and s, change with F and T in smooth and continuous ways. $N(n,s)$ is good to approximate J_{GTF} and understand its behavior, but more computationally friendly forms await in Section 17.5, and a proper numerical evaluation in Section 17.6.

17.3.2 The $s < 0$ regime

Whatever may be the plays on words and the acrobatics of logic, to understand is, above all, to unify.

– Albert Camus[12]

Now comes taming the difficult integrals associated with $N(n,-s)$. Proceed in a manner much like that of Section 17.3.1 for $+s$. The notation bears emphasis because s is treated as a *positive quantity;* the situations when a negative argument arises are explicitly indicated by $-s$ so that the meanings of the integration limits will always be transparent. As before, the integral of $N(n,-s)$ shall be partitioned into a sum of other integrals such that convergent expansions of $\ln(1+e^{n(z+s)})$ (related below to summations over j) and $1/(1+e^z)$ (related below to summations over k) are obtained in Eq. (17.30) for $s \to -s$, except that now the limit of $u \to \infty$ is assumed from the start. Although the notation shall be similar albeit with the introduction of a prime where $N'(n,s) \equiv N(n,-s)$, the terms N'_l will not share the same integration region as N_l for index l. Expansions and careful regrouping of terms, the painstaking details of which are left as an exercise, give the following

$$N'_1(n,s) = n\int_{-\infty}^{-s} \frac{\ln(1+e^{n(z+s)})}{1+e^z}dz \tag{17.70}$$

$$= n\sum_{j=1}^{\infty}\sum_{k=0}^{\infty}(-1)^{j+k+1}\frac{e^{-ks}}{j(k+jn)} \tag{17.71}$$

Make note of whether the summations begin at 1 or 0, reflecting whether the exponential term is smaller or larger than 1. Continuing,

$$N'_2(n,s) = n\int_{-s}^{0} \frac{\ln(1+e^{n(z+s)})}{1+e^z}dz \tag{17.72}$$

$$= n(U(s) - U(0) - s\ln(2))$$

$$+ n\sum_{j=1}^{\infty}\sum_{k=0}^{\infty}(-1)^{j+k+1}\frac{(e^{-ks}-e^{-jsn})}{j(-k+jn)} \tag{17.73}$$

[12] Albert Camus (trans. J. O'Brien), "An Absurd Reasoning," *The Myth of Sisyphus and Other Essays*. New York: Vintage Books, 1991, p. 13.

and lastly,

$$N_3'(n,s) = n \int_0^\infty \frac{\ln(1+e^{n(z+s)})}{1+e^z} dz \tag{17.74}$$

$$= n(U(0) + s\ln(2)) + n \sum_{j=1}^{\infty} \sum_{k=1}^{\infty} (-1)^{j+k} \frac{e^{-jsn}}{j(k+jn)} \tag{17.75}$$

As before, let both e^{-s} and e^{-ns} be taken as small, so that the leading order terms need be retained. Regrouping,

$$N'(n,s) = 2U(0) + n^2(U(s) + e^{-s}) - \Sigma(n)e^{-ns} - n^2\Sigma\left(\frac{1}{n}\right)e^{-s} \tag{17.76}$$

which should look unsettlingly familiar: the terms involving $\Sigma(n^{\pm 1})$ are, in fact, $N(n,s)$, and so (using the leading order expansion for $U(s) \approx 2U(0) - e^{-s} + (s^2/2)$ in Eqs (14.9) and (14.14)) it is found [202] (where $N'(n,s) \to N(n,-s)$) that

$$N(n,s) + N(n,-s) = \frac{1}{2}n^2s^2 + \zeta(2)(n^2+1) \tag{17.77}$$

which is an equation with a certain aesthetic allure. In the case where $N(n,s)$ is small, as occurs when the thermal–field terms are small, then the photoemission term $N(n,-s)$ is, to leading order, the Fowler–DuBridge relation of Eq. (14.14) under the conditions that n is very small but ns is large. Because photo excited electrons pass over the barrier $\mu + \phi$ having been excited from near the Fermi level μ, and because the energy E_x is augmented by the photon energy $\hbar\omega$ in $D(E_x + E_\omega)$, then (recalling the convention about $-s$) it follows that $-s = \beta_F(\mu + \phi - (\mu + \hbar\omega))$ or $s = \beta_F(\hbar\omega - \phi)$ for photoemission, and so $ns = \beta_T(\hbar\omega - \phi)$. In both cases of the next example, as with thermionic emission, ns is large, but s is very large. Therefore, to leading order, the Fowler–DuBridge result is recovered.

EXAMPLE: Numerically evaluate n, s, and ns for parameters characteristic of photoemission. Use copper-like values ($\mu = 7$ eV, $\Phi = 4.5$ eV) at room temperature ($T = 300$ K) for laboratory fields (2 kV/cm → 0.0002 eV/nm) and for fields comparable to an rf photo injector (25 MV/m → 0.025 eV/nm). Assume $\lambda = 266$ nm ($\hbar\omega = 4.661$ eV).

SOLUTION:

- For laboratory conditions, $\phi = \Phi - \sqrt{4QF} = 4.483$ eV. From Eq. (17.20), $\beta_F = 7413$ eV^{-1}. $\beta_T = 1/k_BT = 38.68$ eV^{-1}. Therefore, $n = 0.005218$, $s = 1320$, and $ns = 6.886$.
- For rf photo injector conditions, $\phi = 4.31$ eV, $\beta_F = 198.3$ eV^{-1}, β_T is unchanged, $n = 0.1951$, $s = 69.56$ eV^{-1}, and $ns = 13.57$.

17.4 Brute force evaluation

W-e-e-elll, a satirical piece in The Times *is one thing, but bricks and baseball bats really get right to the point ...*

– Woody Allen[13]

The behavior of β_F when it is evaluated near the barrier maximum compared to how it behaves near the Ferm level (the difference between Eqs (17.12) and (17.20)) must now be dealt with explicitly. This is necessary because numerically evaluating the integral behind the current density $J(F,T)$ requires knowing E_m, that is, knowing where the integrand is at a maximum and evaluating the energy slope factor β_F *there*. Finding the location of the maximum E_m and therefore $\beta_F(E_m)$ is analytically difficult in the transition region, and so the behavior will be investigated using the bricks and baseball bats approach of numerical evaluation. Subtlety can wait.

The numerical evaluation of the Gamow factor $\theta(E_x)$ is the first step. Recall that the canonical equations assumed $\theta(E_x)$ was linear in energy, but strictly, it is not. Even the triangular barrier Gamow factor of Eq. (13.5) shows a more complicated non-linear behavior, and that trend persists in the variation of β_F between the image charge potential and the quadratic barrier when E approaches the barrier maximum. Still, the variation of $\theta(E_x)$ is *smooth*, and that suggests a polynomial is a good approximation if one can be found.

[13]'Manhattan',' *Four Films of Woody Allen*. New York: Random House, 1982, p. 203.

The numerical evaluation of $\theta(E_x)$ will make use of a generalization of $\theta(\mu)$ (Eq. (13.13)) and $R(w)$ (Eq. (13.14)) for arbitrary energy E instead of just the case $E = \mu$, given that the integrand of $R(w)$ has desirable features that make numerical integration techniques in Section A1.2 profitable (the slope of the integrand at the boundaries vanishes so that Eq. (A1.13) renders the trapezoidal series approximation better than expected). Therefore,

$$\theta(E_x) = \frac{4\sqrt{2mF}}{\hbar} L(E_x)^{3/2} R\left(\frac{x_-(E_x)}{L(E_x)}\right) \tag{17.78}$$

$$x_-(E_x) = \frac{2Q}{FL(E_x) + [4QF + (FL(E_x))^2]^{1/2}} \tag{17.79}$$

$$L(E_x) = \frac{1}{F}[(\mu + \Phi - E_x)^2 - 4QF]^{1/2} \tag{17.80}$$

A numerical integration is economized by the introduction of $z = (\mu + \Phi - E_x)/\Phi$ and the usual $y = \sqrt{4QF}/\Phi$, in terms of which

$$x_-(E_x) = \frac{\Phi}{2F}\left(\frac{y^2}{z + \sqrt{z^2 - y^2}}\right) \tag{17.81}$$

$$L(E_x) = \frac{\Phi}{F}\sqrt{z^2 - y^2} \tag{17.82}$$

Consider three cases for which $R(w(E_x))$ is obtained via numerical integration to show the behavior of $\theta(E_x)$ evaluated using the numerical algorithm of Section A3.9: they are shown below for copper-like parameters at various fields of 1 eV/nm (Figure 17.11), 4 eV/nm (Figure 17.12), and 8 eV/nm (Figure 17.13).

Four features are evident: first, $\theta(E_x)$ is *mostly* linear; second, the linear approximation becomes better for high field, but is increasingly distressed for lower fields characteristic of thermal–field emission; third, the slope of $\theta(E_x)$ at μ and $\mu + \phi$ is well-matched by the Fowler–Nordheim (Eq. (17.12)) and quadratic (Eq. (17.20)) β_F equations, respectively; and lastly the mild convexity of $\theta(E_x)$ indicates that there is a single E_x for which $n = 1$ or where the integrand is at a maximum, in contrast to the behavior indicated in Figure 17.9, in other words finding E_m is a moving target.

A purely numerical treatment for finding E_m would be faced with many evaluations of $\theta(E_x)$, each requiring an integration. For that matter, finding $J(F, T)$ would likewise be faced with two integrations in the calculation of $J(F, T)$: one for the evaluation of $\theta(E_x)$ and one for the integration over E_x. To transition from current density J to total current I (as in Section 30.2) requires yet another integration over surface. As is generally true, numerical methods are very good for the evaluation of one-dimensional integrals, but become increasingly less accurate as more dimensions are added. In fact, multidimensional integrals are better handled using Monte Carlo integration methods (Section A1.2.5) when the dimensionality d becomes too large. Although "brute" is a synonym of *forceful*, it need not be one of *stupid*. The smooth

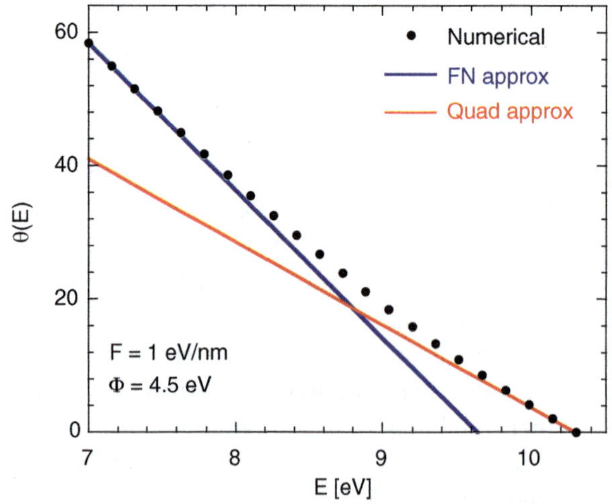

Figure 17.11 $\theta(E_x)$ evaluated numerically from Eq. (17.78) for copper-like parameters and a low field of 1 eV/nm. The blue line is the slope to $\theta(E_x)$ evaluated at $E_x = \mu$ (i.e., the Fowler–Nordheim approximation). The red line is the slope evaluated at $E_x = \mu + \phi$ (i.e., the quadratic barrier approximation).

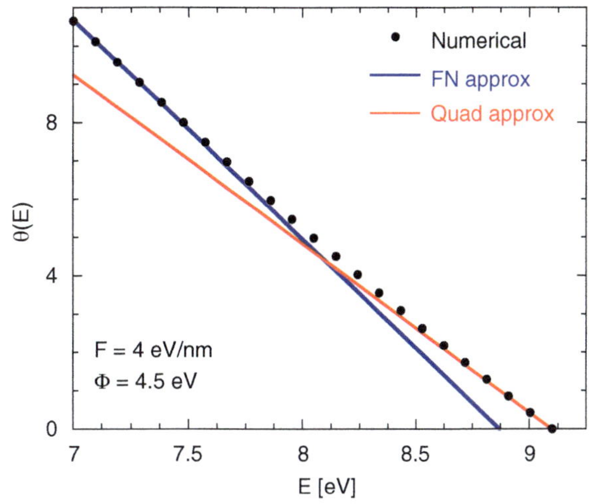

Figure 17.12 Same as Figure 17.11 but for an intermediate field of 4 eV/nm.

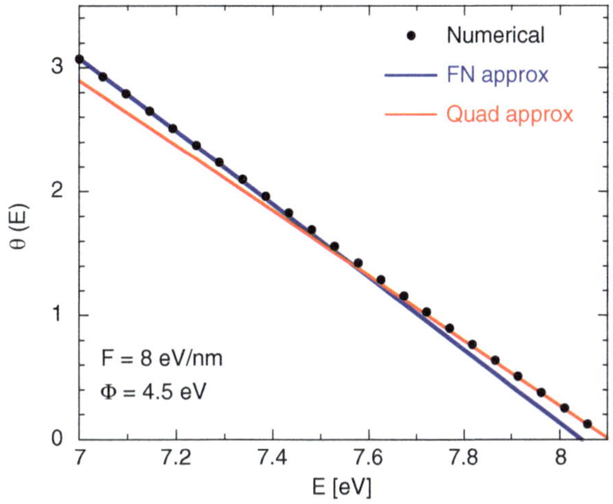

Figure 17.13 Same as Figure 17.11 but for a high field of 8 eV/nm. Observe that, as predicted by Eq. (17.22), $\beta_F(\mu)$ is closer to $\beta_F(\mu+\phi)$.

behavior of $\theta(E_x)$ and its known behavior at the boundaries of the transition region suggests a very good way to approximate $J(F,T)$ in the transition region via a numerical integration by replacing $\theta(E_x)$ with a polynomial expansion, and so a little cleverness can significantly ease the computational burden, if a reliable method to obtain the polynomial expansion can be found. It can.

Recall from Section A1.2 that the trapezoidal integral approximation becomes quite accurate when the slopes of the integrand at the limits are included, leading to Eq. (A1.17). The reason? Inclusion of the slopes is tantamount to finding a cubic polynomial, which has higher order accuracy for smooth functions. Therefore, knowing $\theta(E_x)$ and $\theta'(E_x) \equiv \partial\theta/\partial E_x = -\beta_F(E_x)$ at both $E_x = \mu$ and $E_x = \mu + \phi$, for a total of four conditions, allows for the specification of a cubic polynomial for $\theta(E_x)$, and given its reasonable behavior such an approximation is very good. A cubic in E_x, however, lacks elegance. Instead, introduce the dimensionless $p = (E_x - \mu)/\phi$ as the values $p = 0, 1$ for the limits $E_x = \mu$ and $E_x = \mu + \phi$, respectively. In terms of p, the cubic polynomial is

$$\theta(E_x(p)) = \sum_{j=0}^{3} C_j p^j \tag{17.83}$$

The boundary conditions can be put into a compact matrix notation enabling a rapid inversion and solution. The 4×4 matrix multiplying the coefficients vector $\vec{C}$ corresponds to the values that p takes on with the (i,j)th component corresponding to p^j for $i = 1$ and $i = 3$ (the θ rows), and to jp^{j-1} for $i = 2$ and $i = 4$ (the θ' rows). Observing that

$\phi(\partial/\partial E_x) = \partial_p$, the matrix equation is

$$\begin{pmatrix} 1 & 0 & 0 & 0 \\ 0 & 1 & 0 & 0 \\ 1 & 1 & 1 & 1 \\ 0 & 1 & 2 & 3 \end{pmatrix} \begin{pmatrix} C_0 \\ C_1 \\ C_2 \\ C_3 \end{pmatrix} = \begin{pmatrix} \theta(\mu) \\ -\phi\beta_F(\mu) \\ 0 \\ -\phi\beta_F(\mu+\phi) \end{pmatrix} \tag{17.84}$$

where on the right-hand side the third entry is 0 because $\theta(\mu + \phi) = 0$ at the barrier maximum. Inversion of the matrix equation gives the C_j, regrouping gives

$$\begin{aligned} \theta(\mu + p\phi) =& (1-p)^2(2p+1)\theta(\mu) \\ &- \phi p(1-p)\{(1-p)\beta_F(\mu) - p\beta_F(\mu+\phi)\} \end{aligned} \tag{17.85}$$

Notice the general structure behind the polynomial: factors of $p(1-p)$ prevent those terms from appearing in $\theta(E_x)$ at the boundaries so that at $p = 0, 1$, only the correct boundary value survives. Such factors routinely appear in Lagrangian interpolation coefficients [103] for a similar reason. Lastly, observe that in keeping with the comic premise of the linearized θ factor and the associated linearized Kemble approximation (see discussion prior to Eq. (17.14)), the cubic form of Eq. (17.85) becomes negative for $p > 1$ as in Figure 17.14, which allows the Kemble form of the transmission probability to gracefully approach unity above the barrier maximum such that $\theta(E_x)$ and $\theta'(E_x)$ are continuous across $E_x = \mu + \phi$.

Use of the cubic approximation to $\theta(E)$ given by Eq. (17.85) results in the sinuous current integrand shown in Figure 17.15: clearly, $j(E_x)$ is more sharply peaked than suggested by the linear $\theta(E)$ used in Figure 17.9 due to the higher

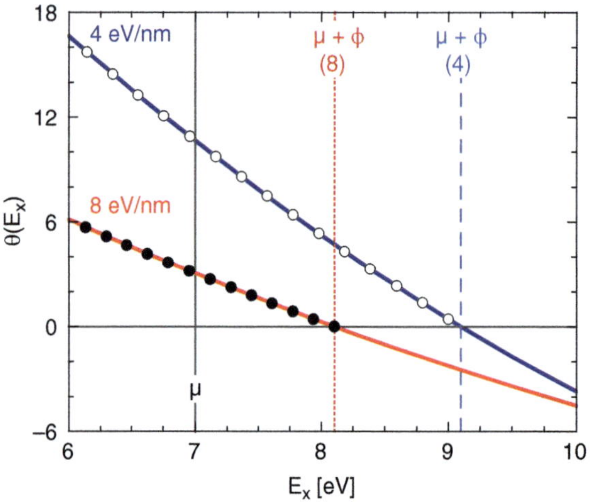

Figure 17.14 A comparison of the exact (numerically evaluated) $\theta(E)$ compared to the polynomial approximation of Eq. (17.85) for the fields of $F = 8$ eV/nm (red) and $F = 4$ eV/nm (blue). The approximate θ has been extended into the negative regime, whereas the numerical θ there would become imaginary. Dashed lines indicate the locations of the barrier maximum for each field for copper-like conditions ($\mu = 7$ eV, $\Phi = 4.5$ eV).

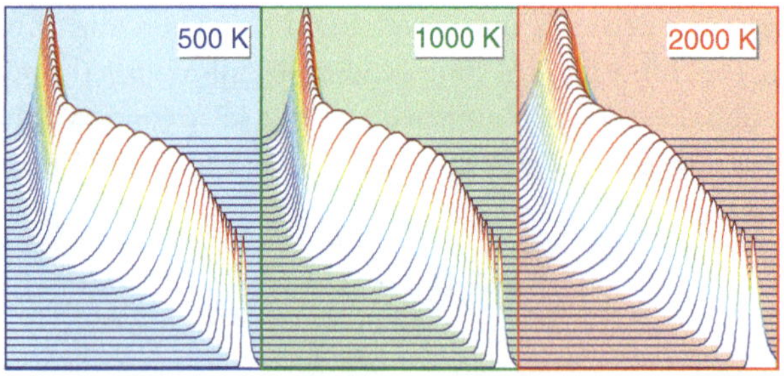

Figure 17.15 The integrand $j(E_x)$ for the temperatures 500 K (left/blue), 1000 K (middle/green) and 2000 K (right/red). The 32 equispaced lines are for increasing field in steps of 0.05 eV/nm (500 K), 0.1 eV/nm (1000 K), and 0.2 eV/nm (i.e., $F_j = j\Delta F$). In all cases, 6 eV $\le E_x \le$ 11.5 eV, $\mu = 7$ eV, and $\Phi = 4.5$ eV.

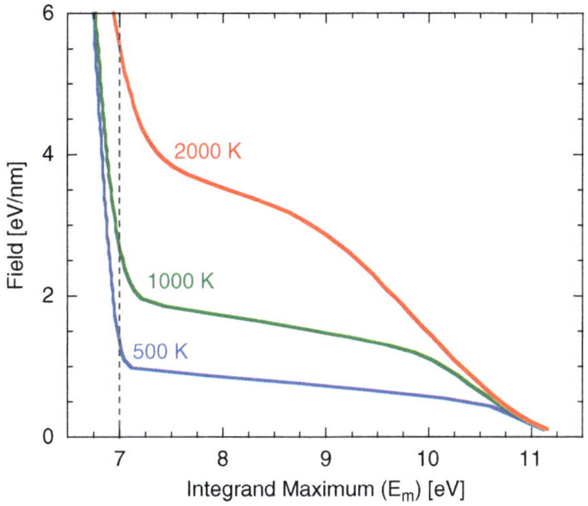

Figure 17.16 The location E_m of the maximum of $j(E_x)$ for the temperatures considered in Figure 17.15.

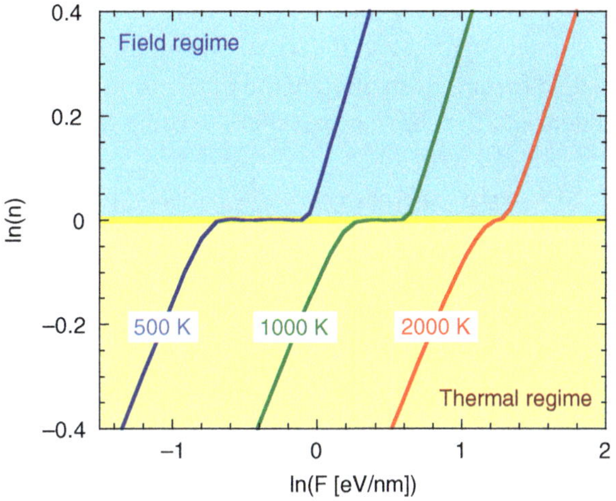

Figure 17.17 The values of $n(F,T) = \beta_T/\beta_F(E_m)$ along the lines shown in Figure 17.16.

order terms that cause faster fall-off on each side of peak located at $E_x = E_m$. The locations of E_m as a function of field are shown in Figure 17.16. But the cubic form of $\theta(E)$ does more: it allows for the calculation of $\beta_F(E_m) = -\theta'(E_m)$ along the lines of Figure 17.16, and so the value of $n(F,T)$ may be evaluated along the spine that snakes from right to left as the field increases. The results of doing this are shown in Figure 17.17 and are perhaps surprising: *in the transition region, the value of $\beta_F(E_m)$ is defined by the relation $n = 1$ to a good approximation!* And with that observation, a very convenient analytical model becomes possible.

17.5 A computationally kind model

> ***Kindness**, n. A brief preface to ten volumes of exaction.*
>
> – Ambrose Bierce[14]

Just as the linear form of $\theta(E_x)$ made analytic forms of $J(F,T)$ possible using the Kemble form of the transmission probability $D(E_x)$, the cubic $\theta(E_x)$ revealed a structure of $n(F,T)$ in Eq. (17.17) that can be exploited. Such an ability to avoid evaluation by numerical integration altogether makes the formulation useful for when general thermal–field

[14]Ambrose Bierce, *The Devil's Dictionary*. New York: Dover Publications, 1993, p. 72.

evaluations are required, for example in computing the current from a hot needle-like structure (e.g., Zro/W Schottky emitters [154]) subject to high fields, as thermal emission dominates along the sides, field emission dominates at the apex, and in between the thermal and field emissions are comparable. Similar complications arise for field-enhanced photoemission [235].

17.5.1 Intermediate $n = 1$ regime

...make me happy in your unity.

– William Shakespeare[15]

As suggested by Figures 17.11 thru 17.13, when parameters are such that n is either well below or well above 1, then n and s are obtainable from Eqs (17.35) plus (17.36) (thermal) or Eqs (17.32) plus (17.33) (field), respectively, and total current density J_{GTF} from Eq. (17.26) (with $\beta_F E_o \to \infty$), which is a rather gentle means of obtaining the thermal or field current. But when $n = 1$, that requires work before J_{GTF} can use Eq. (17.69) because although n has become simple, s and ns (which become equal) have acquired complexity as a result of the dependence of $s = \beta_F(E_m)(E_o(E_m) - \mu)$ on E_o. Knowledge of how E_m changes in the transition regime is required, a point the conclusions of Section 17.4 sought to make clear. The first step is to bound the problem.

Consider, therefore, the difference in behavior of $n(F, T)$ exhibited by Eqs (17.32) and (17.35): the former behaves as $n \propto F/Tt(y)$ and the latter as $n \propto F^{3/4}/T$. That is, each exhibits a different power that F is raised to, which is the cause of the convergence of the T-like and F-like lines in Figure 17.9. If υ is that power, then it may be found by $\upsilon = F\partial_F \ln(n)$. For the thermal regime, $\upsilon = 3/4$, but for the field regime, evaluation of the power is more involved: approximating $t(y) \approx 1.0613$, sets υ to 1, but the weak field dependence of $t(y)$ changes that slightly. Using Eq. (13.24) gives

$$\upsilon = 1 - F\partial_F \ln(t(y)) = \frac{1}{t(y)}\left(1 + \frac{y^2}{18}\right) \tag{17.86}$$

The variation of y between 0 (for $F = 0$) and 1 (for $F = 4Q/\Phi^2$), results in a variation of υ between 1 and $19/20 = 0.95$, with the latter value being more common when the fields are high. For copper-like parameters and a field of 8 eV/nm, $\upsilon = 0.95427$.

As seen from Figure 17.17, the intersection of the power law with the line $n = 1$ is different for each regime. In the thermal regime, $\ln(n)$ as a function of $\ln(F)$ intersects the $n = 1$ boundary when $T_{max} = 1/k_B\beta_F(\mu + \phi)$ (from Eq. (17.20)), whereas in the field regime, it intersects at $T_{min} = 1/k_B\beta_F(\mu)$ (from Eq. (17.12)). Explicitly,

$$T_{min} = \frac{\hbar F}{k_B\sqrt{m\Phi}}\left(\frac{\sqrt{2}}{4t(y)}\right) \tag{17.87}$$

$$T_{max} = \frac{\hbar F}{k_B\sqrt{m\Phi}}\left(\frac{1}{\pi\sqrt{y}}\right) \tag{17.88}$$

Both T_{min} and T_{max} are proportional to the product of $\hbar F/k_B\sqrt{m\Phi}$ with a factor containing y, and that by virtue of the field dependence of $y = \sqrt{4QF}/\Phi$, the power υ is off-set from unity.

EXAMPLE: Find T_{max} and T_{min} for Cu parameters and $F = 4$ eV/nm.
SOLUTION: The quantity $\hbar c/\sqrt{mc^2\Phi} = 0.13013$ nm. For the chosen field, $y = 0.53333$ and so $t(y) = 1.0515$. Therefore

$$T_{min} = 2031 \text{ K}$$

$$T_{max} = 2633 \text{ K}$$

[15] Ref. [37]: *The Tragedy of King Richard the Third*, II.1.31.

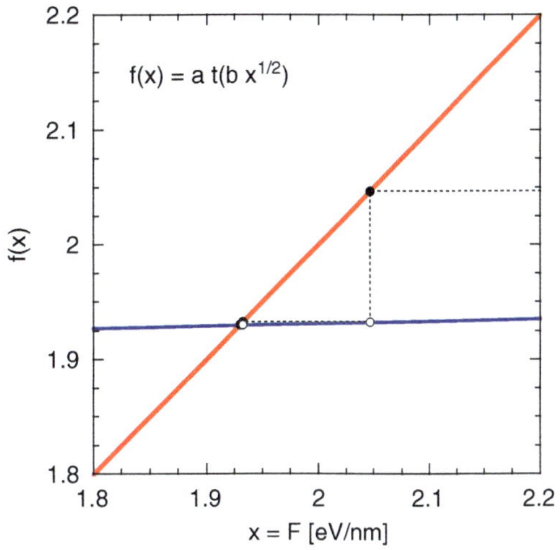

Figure 17.18 A graphical representation (based on Figures A3.1 and A3.2) of the process used to find a solution to Eq. (17.90) using iteration of $x = at(b\sqrt{x})$ with $a = 0.53389$ and $b = 0.26666$ for a temperature of $T = 1000$ K. Open circles represent $f(x_{j-1})$ and black circles represent x_j. a scales linearly with T, and so to leading order, F_{max} will as well.

An equivalent calculation can be used to find the maximum and minimum fields for a given temperature, although the more complicated relationship of $t(y)$ on F makes their determination more involved. For a given temperature T, and using Eq. (17.20), the thermal regime extends for all fields up to

$$F_{min} = \left[4Qm^2\left(\frac{\pi k_B T}{\hbar}\right)^4\right]^{1/3} \tag{17.89}$$

Conversely, the field regime is for all fields greater than

$$F_{max} = \frac{2k_B T t(y_{max})}{\hbar}\sqrt{2m\Phi} \tag{17.90}$$

where the max and min subscripts refer to the fields at the upper and lower $n = 1$ regime, and where $y_{max} = \sqrt{4QF_{max}}/\Phi$. Although Eq. (17.89) can be evaluated directly, Eq. (17.90) must be solved by other means if the weak dependence of $t(y)$ on field is respected. The means may include the following:

- The method of bisection, as in Section A3.7.1 is a rugged and straightforward approach: extreme values of F_{max} and F_{min} are chosen and the midpoint F_{mid} is used in calculating $F - 2t(y(F))\sqrt{2m\Phi}/\beta_T\hbar$. If the answer is greater than 0, F_{min} is set to the value of F_{mid}, and if not, then F_{max} is. With each step, the uncertainty in the location of the correct value of F is reduced by a factor of 2. Though rugged, the method is not fast by comparison to iteration.
- The method of iteration, as in Section A3.7.2, is a more elegant method for solutions to equations of the form $x = f(x)$ as long as $f(x) < 1$. Eq. (17.90) is such an equation. Let $x_{j+1} = f(x_j)$ given an initial guess x_1, then after several iterations a reasonable approximation to x_∞ results. With $x \to F$ and $x_\infty \to F_{max}$, the derivative of the right-hand side of Eq. (17.90) is rather small: if it were zero, then iteration would give the correct answer in one step. But small is good: as seen in Figure 17.18, even for a poorly chosen initial guess of $F = 10$ eV/nm, a very small number of iterations determine $F_{\rm max}$ as a consequence of the weak variation of the right-hand side of Eq. (17.90) with F_{max}.

17.5.2 Evaluation of the general thermal–field current density

Some say the world will end in fire,
Some say in ice.
From what I've tasted of desire
I hold with those who favor fire.
But if it had to perish twice,
I think I know enough of hate

To say that for destruction ice
Is also great
And would suffice.

– Robert Frost[16]

The domination of current by thermal or cold (*aka* field) emission is related to the position of E_m. Although E_m is easily approximated by μ in the field regime or $\mu + \phi$ in the thermal regime, in the intermediate ($n = 1$ regime), it changes subject to the constraint

$$\beta_F(E_m)|_{n=1} = \beta_T \tag{17.91}$$

The behavior of $\beta_F(E)$ is obtained from Eq. (17.85) and $\partial_E = \phi^{-1}\partial_p$ to give

$$\begin{aligned}\beta_F(\mu + p\phi) = &\frac{6\theta(\mu)}{\phi} p(1-p) + (1-p)(1-3p)\beta_F(\mu)\\ &+ p(3p-2)\beta_F(\mu+\phi)\end{aligned} \tag{17.92}$$

an equation which is quadratic in p. Therefore, Eq. (17.91) in Eq. (17.92) defines E_m, from which $s|_{n=1} = \beta_T(E_o(E_m) - \mu)$ may be calculated from[17]

$$E_o = E_m + \frac{\theta(E_m)}{\beta_F(E_m)} = E_m + \frac{\theta(E_m)}{\beta_T} \tag{17.93}$$

The use of s so calculated in $N(1, s)$ leads to the current density via

$$\begin{aligned}J_{GTF}(F,T)|_{n=1} &= A_{RLD}T^2 N(1,s)\\ &\approx A_{RLD}T^2(s(E_m)+1)e^{-s(E_m)}\end{aligned} \tag{17.94}$$

Each term in Eq. (17.94) takes some unpacking, and so consider an example worked out in detail to appreciate what is involved.

EXAMPLE: Find J_{GTF} for $\mu = 7$ eV, $\Phi = 4.5$ eV at a field of $F = (F_{max} + F_{min})/2$ and a temperature of $T = 1000$ K.
SOLUTION: Proceed in steps.

1. From Eq. (17.89) the value of F_{min} is 1.100 eV/nm.
2. The first three iterations of Eq. (17.90) using an initial value of 4 eV/nm give 4.000, 1.970, and 1.930, after which the last significant digit does not change in the third decimal, and so $F_{max} = 1.930$ eV/nm. Therefore, $F = 1.515$ eV/nm.
3. $\phi = \Phi - \sqrt{4QF} = 3.023$ eV.
4. $\beta_T = 1/k_B T = 11.60$ eV^{-1}.
5. Using Eq. (17.78), $\theta(\mu) = 36.58$.
6. Using Eq. (17.20), $\beta_F(\mu + \phi) = 9.129$ eV^{-1}.
7. Using Eq. (17.12), $\beta_F(\mu) = 14.71$ eV^{-1}.
8. Eq. (17.92) is $11.60 = 72.61p(1-p) + 9.129p(3p-2) - 14.71(1-p)(3p-1)$, which has the roots $p = 0.6034$ and $p = -4.689$, for which the positive root is the correct one.
9. From p, $E_m = \mu + p\phi = 8.823$ eV.
10. From Eq. (17.93) and $\beta_F = \beta_T$, $E_o = 9.900$ eV.
11. From Eq. (17.26), $s(E_m) = 33.66$.
12. From Eq. (17.94), $J_{GTF}(F,T) = 1.006 \times 10^{-5}$ A/cm^2.

How does $J_{GTF}(F,T)$ as calculated using the computationally kind approach compare to its competitors? It is expected that J_{GTF} should exceed J_{FN} because β_F in the transition region is smaller than near μ, and should exceed J_{RLD} because

[16] Robert Frost, *Fire and Ice*, 1920, https://www.poetryfoundation.org.
[17] This form is possible because even though s is defined by $s(E_m) = \beta_F(E_m)(E_o(E_m) - \mu)$, the $n = 1$ regime allows β_F to be replaced by β_T, which is easier to evaluate. When $n \neq 1$, then the usage of β_T is *not* correct.

of the influence of tunneling electrons. Calculation confirms this. Using the parameters of the previous example and Eq. (13.25), then $J_{FN}/J_{GTF} = 0.00081674$. Using Eq. (12.3)), then $J_{RLD}/J_{GTF} = 0.0069563$. The thermal–field, or $n = 1$ regime, exhibits behavior not accounted for by either the canonical thermal or field emission equations, and therefore the conditions which signify its presence, detailed in Eqs (17.87), (17.88), (17.89), and (17.90), require attention for general thermal–field emission conditions.

Lastly, there is the question of accuracy: use of $N(1, s)$ in the thermal–field, or intermediate $n = 1$, region is only approximately correct by retaining just the leading order part of a factor $(1 + \exp(\beta_F(E_o - E))/(1 + \exp(\theta(E)))$ that plays the role of $f(x)$ in Section A1.2.4. Because this function is peaked itself at $E = E_m$ (and equal to unity at E_m), $N(1, s)$ is an *overestimate* of the actual integrand involving $\theta(E)$ by a factor on the order of 2, as indicated by comparing the integrands in Figure 17.19. On a log plot, such differences are easily overlooked. They are not altogether insignificant, though: factors of 2 are preferable to orders of magnitude differences that characterize the usage of the FN and RLD equations in regimes where they are not warranted, as in Figures 17.20 and 17.21.

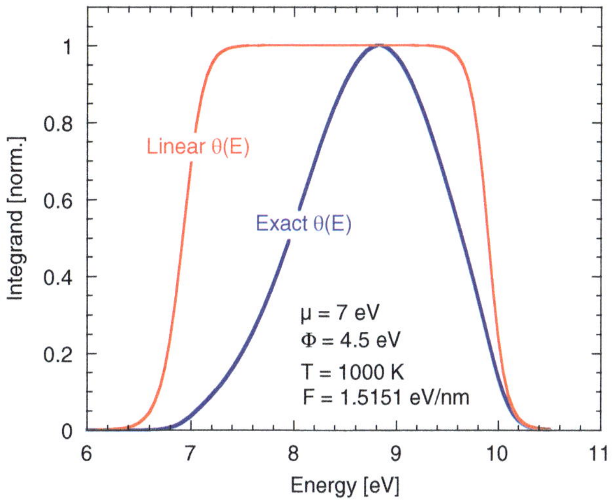

Figure 17.19 Comparison of the integrands $j(E_x)$ using $\theta(E)$ (exact) and $\beta_F(E_o - E)$ (linear). The red line is the integrand of $N(1, s)$. Conditions are such that $n = 1$, and F is equidistant from its maximum and minimum values (Eqs (17.90) and (17.89)) for $T = 1000$ K and copper-like parameters.

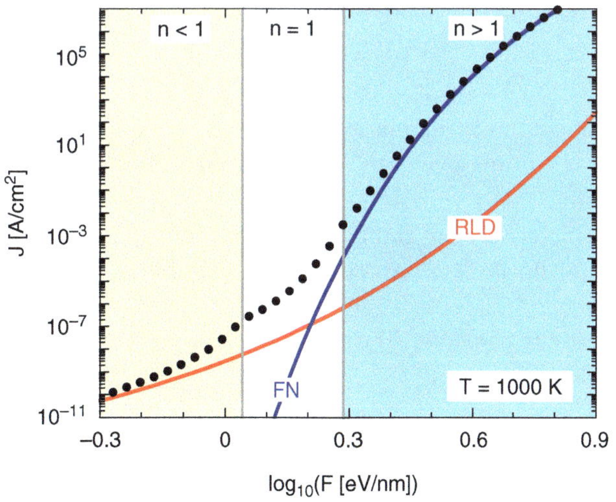

Figure 17.20 General thermal–field emission from a metal ($\mu = 7$ eV, $\Phi = 4.5$ eV) held at a temperature of 1000 K. The thermal region is labeled $n < 1$, the field region is labeled $n > 1$, and the intermediate region is labeled $n = 1$. The gray lines at the boundaries mark the fields specified by Eqs (17.89) and (17.90). RLD refers to Eq. (12.3) and FN to Eq. (13.25).

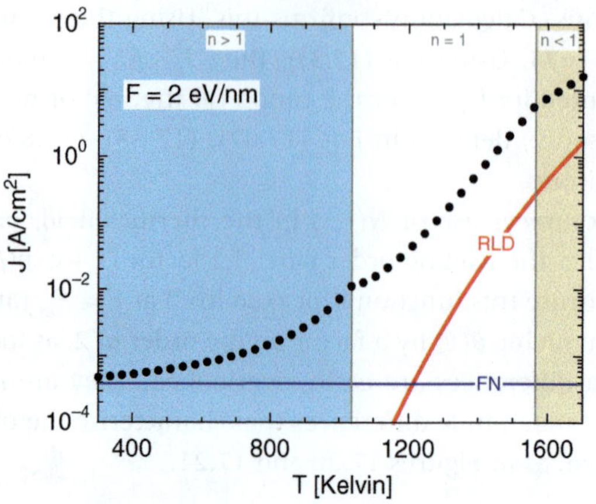

Figure 17.21 General thermal–field emission from a metal ($\mu = 7$ eV, $\Phi = 4.5$ eV) held at a field of 2 eV/nm. The thermal region is labeled $n < 1$, the field region is labeled $n > 1$, and the intermediate region is labeled $n = 1$. The gray lines at the boundaries mark the temperatures specified by Eqs (17.87) and (17.88). RLD refers to Eq. (12.3) and FN to Eq. (13.25).

17.6 General thermal–field emission code

> ...*Wisdom comes alone through suffering.*
>
> – Aeschylus[18]

> ...*a person that started in to carry a cat home by the tail was gitting knowledge that was always going to be useful to him, and warn't ever going to grow dim or doubtful.*
>
> – Mark Twain[19]

Doing is learning. The final and arduous task in the thermal–field emission theory treatment is to describe the means for its numerical implementation using a common programming language, as the tasks to be performed for even one pair of field and temperature values is already more involved than for either the thermal or field canonical equations. The code developed in the present section is complete (it will run as written). The algorithm is presented in Fortran using F95 conventions as Fortran is the quintessential scientific programming language: even the name connotes turning scientific equations into code (**FOR**mula **TRAN**slation) [236, 237], and it is particularly adept at handling vectors and matrices (although that capability is not fully exploited in the program). The reformulation into other languages should be straightforward.

The code is broken into blocks (subroutines and formula) for easier implementation. For simplicity, no optimizations are made in the form of vectorizations or the replacement of formulae used repeatedly by their numerical equivalent. The components are in the forms of *modules, subroutines,* and *function calls,* as follows:

1. Modules
   - **Fundamentals**: Fundamental constants used in main program and subroutines, such as $\pi, m, k_B, \hbar, \alpha, c, \varepsilon_0$, and Q (called `Qo`).[20] Two additional terms are given: `Arld` is encountered in Eq. (12.3); `Ampcm2` converts A/cm^2 from nm-eV-fs-q units to MKSA (Table 2.1).
   - **MaterialEnvironment**: Parameters used to control the output sent to written data files: they can be altered after the first run by editing the name list file `Jfrin.nml`. Default parameters are used in the output below.
2. Subroutines
   - **NameDropping**: Reading in of the name list file `Jfrin.nml`: if the file does not exist, a sample one is created and populated with the default values.
3. Function calls
   - **RJfno**: Fowler–Nordheim equation as given by Eq. (13.25).
   - **RJrldo**: Richardson–Laue–Dushman equation as given by Eq. (12.3).

[18] Aeschylus, *Agamemnon* (trans. R. Lattimore), *The Complete Greek Tragedies: Aeschylus I*. Chicago: University of Chicago Press, 1953, lines 77–78.
[19] Mark Twain, *Tom Saywer Abroad*, A Public Domain Book, Kindle eBook, 2013, Chapter 10.
[20] Values derived from NIST: http://physics.nist.gov/constants.

- **RJgtf**: General thermal–field emission equation based on Eq. (17.56).
- **RNns**: The function $N(n, s)$ based on Eq. (17.65).
- **ThetaE**: Evaluation of $\theta(E)$ based on Eq. (17.85).
- **BetaE**: Evaluation of $\beta_F(E)$ based on Eq. (17.92).

4. Output
   - `JTemp.txt` contains four columns for the vectors T, J_{GTF}, J_{RLD} and J_{FN} for a fixed field. Because J_{FN} is independent of temperature, its values are constant
   - `Jfield.txt` contains four columns for the vectors F, J_{GTF}, J_{RLD} and J_{FN} for a fixed temperature, where J_{RLD} varies because of the field dependence of $\phi = \Phi - \sqrt{4QF}$.

17.6.1 Simulation code

```
!-------------------------------------------------------------
module Fundamentals
!Source: http://physics.nist.gov/constants
!        (converted from MKSA)
!Units:
!  electron volts, nanometers,
!  femtoseconds, charge q = 1
!  terms with different units are combinations of Units
!  eg: [rmo] = [eV]/c^2, [c] = nm/fs
!--------------------------------------------------------------
  real, parameter :: pi = 3.14159265359, &
&     rmo = 5.685629853, rkb = 1.0/11604.5192, &
&     rhbar = 0.658211899, &
&     alpha = 7.29735257E-3, c = 299.7924580, &
&     epso  =  1.0/(4*pi*alpha*rhbar*c), &
&     Qo = alpha*rhbar*c/4, Ampcm2 = 1.602176565E10, &
&     Arld = (rmo*rkb**2)/(2*rhbar*(pi*rhbar)**2)
!--------------------------------------------------------------
end module Fundamentals
!--------------------------------------------------------------
!--------------------------------------------------------------
module MaterialEnvironment
!--------------------------------------------------------------
  real     :: Chem,Phi,Fmax,Fmin,Field,Tmax,Tmin,Temp
  integer :: Nx
  character*1:: tb = char(9)
  data Chem,  Phi, Nx / 7.0, 4.5, 41 /
  data Fmax, Fmin, Field, Tmax,  Tmin,    Temp &
  &    / 8.0,   1.0,   2.0, 1700.0, 300.0, 1000.0 /
!--------------------------------------------------------------
end module MaterialEnvironment
!--------------------------------------------------------------

!=============================================================
PROGRAM GenThermField
!=============================================================
!.....Calculation of Generalized Thermal
!           Field Current Density
!.....Units:  Field = eV/nm; Temp = Kelvin;
```

```
!              J = Amp/cm^2
!.....When not specified, units are eV-nm-fs-q
!---------------------------------------------------
use MaterialEnvironment  !parameters used in subroutines
real                  :: Tmp, Fld
!....Read in needed parameters
call NameDropping
!---------------------------------------------------
!....Create temperature variation output
!....Let ln(F(j)) be evenly spaced
!---------------------------------------------------
write(*,*) 'Creating_file_Jfield.txt:__field_variation'
write(*,*) '__Temp_taken_to_be_T_=_',Temp,'_Kelvin'
open(unit=11,file='Jfield.txt')
write(11,*) 'F_[eV/nm]',tb,'JGTF_[A/cm^2]',tb, &
&           'JRLD',tb,'JFN'
do i = 1, Nx
  Fld = Fmin*(Fmax/Fmin)**(float(i-1)/(Nx-1))
  write(11,*) Fld,tb,RJgtf(Phi,Chem,Fld,Temp), tb, &
 &            RJrldo(Phi,Fld,Temp),tb,RJfno(Phi,Fld)
end do
close(unit=11)
!---------------------------------------------------
!....Create field variation output
!....Let T(j) be evenly spaced
!---------------------------------------------------
write(*,*) 'Creating_file_Jtemp.txt:_thermal_variation'
write(*,*) '__Field_taken_to_be_F_=_',Field,'_eV/nm'
open(unit=12,file='JTemp.txt')
write(12,*) 'T_[K]',tb,'JGTF_[A/cm^2]',tb,&
&           'JRLD',tb,'JFN'
do i = 1, Nx
  Tmp = Tmin + (Tmax-Tmin)*float(i-1)/(Nx-1)
  write(12,*) Tmp,tb,RJgtf(Phi,Chem,Field,Tmp), tb, &
 &            RJrldo(Phi,Field,Tmp),tb,RJfno(Phi,Field)
end do
close(unit=12)
!---------------------------------------------------
END PROGRAM GenThermField
!===================================================

!===================================================
subroutine NameDropping
!READING/CREATING OF DATA INPUT FILE
!---------------------------------------------------
use MaterialEnvironment
!---------------------------------------------------
implicit none
integer               :: istat, idebug
namelist/JftParam/ Chem, Phi, Fmax, Fmin, Field, &
  &                Tmax, Tmin, Temp, Nx
write(*,*) 'PROGRAM_GenFNRLD'
```

```
write(*,*) '__<Jfrin.nml>_being_opened...'
open(unit=11,file='Jfrin.nml',status='old',iostat=istat)
close(unit=11)
if(istat == 0) then  !A namelist file exists; read it
  open(unit=11,file='Jfrin.nml')
  read(11,JftParam)
  close(unit=11)
  if(idebug>0) write(*,JftParam)
else !A namelist file does not exist; create it
  write(*,*) '__input_file_Jfrin.NML_not_found.'
  write(*,*) '__Creating_input_data_file_now...'
  open(unit=11,file='Jfrin.nml')
  write(11,*) 'Symbol____Definition___________Unit'
  write(11,*) 'Chem______Chemical_Potential___[eV]'
  write(11,*) 'Phi_______Work_Function________[eV]'
  write(11,*) 'Fmax______Max_field_for_Jfield__[eV/nm]'
  write(11,*) 'Fmin______Min_field_for_Jfield__[eV/nm]'
  write(11,*) 'Field_____Field_for_for_Jtemp___[eV/nm]'
  write(11,*) 'Tmax______Max_temp__for_Jtemp___[Kelvin]'
  write(11,*) 'Tmin______Min_temp__for_Jtemp___[Kelvin]'
  write(11,*) 'Tmin______Temp_for__for_Jfield__[Kelvin]'
  write(11,*) 'Nx________Number_of_plot_points'
  write(11,JftParam)
  close(unit=11)
  stop
end if
return
!-------------------------------------------------------------
! END subroutine NameDropping
! READING/CREATING OF DATA INPUT
!=============================================================
end

!=============================================================
function RJfno(Phi,Field)
!-------------------------------------------------------------
!Internal Units:  eV-nm-fs-q
!Implicit use is made of the Forbes-Deane
!         Approximation to v(y)
!The Nordheim paraemeter t(y) is approximated by tyo
!Output Units:  Amps / cm^2
!-------------------------------------------------------------
use Fundamentals
implicit none
real, parameter :: Ao = 1.0/(16*rhbar*pi**2), &
  &                Bo = 4*sqrt(2*rmo)/(3*rhbar), &
  &                tyo = 1.0613132
real               :: Phi, Field, RJfno, nu
nu = 2*Bo*Qo/(3*sqrt(Phi))
RJfno = (Ao/(Phi*tyo**2))*(((Phi**2)*exp(6.0)/&
 &      (4*Qo))**nu)*(Field**(2.0-nu))* &
 &      exp(-Bo*(Phi**1.5)/Field)*Ampcm2
```

```
return
!---------------------------------------------------------
end
!=========================================================

!=========================================================
function RJrldo(Phi,Field,Temp)
!---------------------------------------------------------
!Internal Units:  eV-nm-fs-q
!Output Units:  Amps / cm^2
!---------------------------------------------------------
use Fundamentals
implicit none
real            :: Phi, phix, Field, Temp, RJrldo
phix   = Phi - sqrt(4*Qo*Field)
RJrldo = Arld*(Temp**2)*exp(-phix/(rkb*Temp))*Ampcm2
return
!---------------------------------------------------------
end
!=========================================================

!=========================================================
function RJgtf(Phi,Chem,Fld,Tmp)
!Internal Units:  eV-nm-fs-q
!Output Units:  Amps / cm^2
!---------------------------------------------------------
use Fundamentals
implicit none
real :: y, phix, vy, ty, Qmu, Bfmu, Bfmphi, Tmin, &
 &    Tmax, Tmp, Em, s, rn, RJgtf, Phi, Chem, &
 &    Fld, BetaT, &
 &    RNns, Ap, Bp, Cp, p, ThetaE, BetaE
! Create terms needed for the evaluations
betaT = 1.0/(rkb*Tmp)
y  = sqrt(4*Qo*Fld)/Phi
phix = Phi - sqrt(4*Qo*Fld)
vy = 1.0 - (3.0 - alog(y))*(y**2)/3
ty = 1.0 + (1.0 - alog(y))*(y**2)/9
Qmu    =  4*sqrt(2*rmo*Phi**3)*vy/(3*rhbar*Fld)
Bfmu   =  2*sqrt(2*rmo*Phi)*ty/(  rhbar*Fld)
Bfmphi = pi*sqrt(y*rmo*Phi)   /(  rhbar*Fld)
Tmin   = (rhbar*Fld/(4*ty*rkb))*sqrt(2.0/(rmo*Phi))
Tmax   = (rhbar*Fld/(pi*rkb))  *sqrt(1.0/(y*rmo*Phi))
! Three regimes:
if(Tmp.le.Tmin) then          !Field Emission Regime
  s  = Qmu
  rn = betaT/Bfmu
else if(Tmp.gt.Tmax) then     !Thermal Emission Regime
  s  = Bfmphi*Phix
  rn = betaT/BFmphi
else                          !Intermediate Regime
  rn = 1.0
```

```
  Ap = 3*(phix*(Bfmu + Bfmphi) - 2*Qmu)
  Bp = 2*(3*Qmu - phix*(2*Bfmu + Bfmphi))
  Cp = phix*(Bfmu - BetaT)
  p = (-Bp-sqrt(Bp**2 - 4*Ap*Cp))/(2*Ap)
  Em = Chem + p*phix
  s  = betaT*(Em - Chem + &
 &   (ThetaE(Phi,Chem,Fld,Em)/BetaE(Phi,Chem,Fld,Em)))
end if
RJgtf = (Arld/(rkb*betaT)**2)*RNns(rn,s)*Ampcm2
return
!-----------------------------------------------------------
end
!-----------------------------------------------------------

!-----------------------------------------------------------
function RNns(n,s)
! (7*pi^4 - 720)/720 = -0.10593434
! (  pi^2 -  12)/6   = -0.35506593
!-----------------------------------------------------------
implicit none
real   :: n, s, RNns, sn, sd, x, y, z, sng
x = n**2
y = 1.0/x
sn = -x*(0.10593434*x + 0.35506593)
sd = -y*(0.10593434*y + 0.35506593)
z = (n-1.0)*s
if(abs(z).gt.0.001) then
  sng = (n**2 + 1.0)*((n**2)*exp(-s) - &
  &     exp(-n*s))/(n**2 - 1.0)
else
  sng = (0.5*(n**2 + 1.0)*exp(-s)/(n+1.0))* &
      & ((1.0-n)*(s**2) + 2*(1.0+n) + 2*s)
end if
RNns = sng + sn*exp(-n*s) + x*sd*exp(-s)
return
!-----------------------------------------------------------
end
!===========================================================

!===========================================================
function ThetaE(Phi,Chem,Field,E)
! Cubic in p = (E - mu)/phix
!-----------------------------------------------------------
use Fundamentals
implicit none
real  :: Qmu, Bfmu, Bfmphi, p, ThetaE, Phi, Chem, &
 &      Field, E, vy, ty, y, phix
y  = sqrt(4*Qo*Field)/Phi
phix = Phi - sqrt(4*Qo*Field)
vy = 1.0 - (3.0 - alog(y))*(y**2)/3
ty = 1.0 + (1.0 - alog(y))*(y**2)/9
Qmu    =  4*sqrt(2*rmo*Phi**3)*vy/(3*rhbar*Field)
```

```
Bfmu    =  2*sqrt(2*rmo*Phi)*ty/(  rhbar*Field)
Bfmphi = pi*sqrt(y*rmo*Phi)    /(  rhbar*Field)
p = (E - Chem)/phix
ThetaE = Qmu*(2*p+1.0)*(1.0 - p)**2 -  &
 &       phix*Bfmu*p*(1.0 - p)**2 + &
 &       phix*Bfmphi*(1.0 - p)*p**2
return
!----------------------------------------------
end
!==============================================


!==============================================
function BetaE(Phi,Chem,Field,E)
! Quadratic in p = (E - mu)/phix
!----------------------------------------------
use Fundamentals
implicit none
real   :: Qmu, Bfmu, Bfmphi, p, BetaE, Phi, Chem, &
 &                Field, E, vy, ty, y, phix
y  = sqrt(4*Qo*Field)/Phi
phix = Phi - sqrt(4*Qo*Field)
vy = 1.0 - (3.0 - alog(y))*(y**2)/3
ty = 1.0 + (1.0 - alog(y))*(y**2)/9
Qmu    =  4*sqrt(2*rmo*Phi**3)*vy/(3*rhbar*Field)
Bfmu    =  2*sqrt(2*rmo*Phi)*ty/(  rhbar*Field)
Bfmphi = pi*sqrt(y*rmo*Phi)    /(  rhbar*Field)
p = (E - Chem)/phix
BetaE = (6*Qmu/phix)*p*(1.0 - p) - &
  &     Bfmu*(1.0 - p)*(3*p - 1.0) + &
  &     Bfmphi*p*(3*p - 2.0)
return
!----------------------------------------------
end
!==============================================
```

17.6.2 Input/Output

Input file Jfrin.nml (containing default values):

```
 Symbol    Definition             Unit
 Chem      Chemical Potential     [eV]
 Phi       Work Function          [eV]
 Fmax      Max field for Jfield   [eV/nm]
 Fmin      Min field for Jfield   [eV/nm]
 Field     Field for for Jtemp    [eV/nm]
 Tmax      Max temp  for Jtemp    [Kelvin]
 Tmin      Min temp  for Jtemp    [Kelvin]
 Tmin      Temp for  for Jfield   [Kelvin]
 Nx        Number of plot points
 & JFTPARAM
 CHEM =    7.00000,
 PHI =   4.50000,
```

```
FMAX =   8.00000,
FMIN =   1.00000,
FIELD =   2.00000,
TMAX =   1700.00,
TMIN =   300.000,
TEMP =   1000.00,
NX =   41,
/
```

Output file `JTemp.txt` in tab-delimited format:

```
T [K]   JGTF [A/cm^2]    JRLD      JFN
  300.000        4.792653E-04     0.000000         3.925210E-04
  335.000        4.970490E-04     0.000000         3.925210E-04
  370.000        5.179189E-04     1.088887E-31     3.925210E-04
  405.000        5.423434E-04     2.599633E-28     3.925210E-04
  440.000        5.709133E-04     1.825727E-25     3.925210E-04
  475.000        6.043782E-04     4.938201E-23     3.925210E-04
  510.000        6.436993E-04     6.255702E-21     3.925210E-04
  545.000        6.901248E-04     4.292822E-19     3.925210E-04
  580.000        7.452986E-04     1.782111E-17     3.925210E-04
  615.000        8.114256E-04     4.874463E-16     3.925210E-04
  650.000        8.915255E-04     9.393571E-15     3.925210E-04
  685.000        9.898394E-04     1.345306E-13     3.925210E-04
  720.000        1.112504E-03     1.494791E-12     3.925210E-04
  755.000        1.268712E-03     1.334567E-11     3.925210E-04
  790.000        1.472803E-03     9.854653E-11     3.925210E-04
  825.000        1.748101E-03     6.164562E-10     3.925210E-04
  860.000        2.134147E-03     3.333098E-09     3.925210E-04
  895.000        2.700365E-03     1.584359E-08     3.925210E-04
  930.000        3.571666E-03     6.717056E-08     3.925210E-04
  965.000        4.975671E-03     2.571505E-07     3.925210E-04
  1000.00        7.327671E-03     8.984302E-07     3.925210E-04
  1035.00        1.137798E-02     2.891130E-06     3.925210E-04
  1070.00        1.294884E-02     8.637821E-06     3.925210E-04
  1105.00        1.615842E-02     2.412849E-05     3.925210E-04
  1140.00        2.172754E-02     6.340260E-05     3.925210E-04
  1175.00        3.093601E-02     1.575728E-04     3.925210E-04
  1210.00        4.600899E-02     3.721621E-04     3.925210E-04
  1245.00        7.071256E-02     8.388857E-04     3.925210E-04
  1280.00        0.111365         1.811501E-03     3.925210E-04
  1315.00        0.178509         3.760179E-03     3.925210E-04
  1350.00        0.289663         7.525453E-03     3.925210E-04
  1385.00        0.473769         1.456118E-02     3.925210E-04
  1420.00        0.778335         2.730675E-02     3.925210E-04
  1455.00        1.28079          4.974182E-02     3.925210E-04
  1490.00        2.10631          8.819163E-02     3.925210E-04
  1525.00        3.45551          0.152470         3.925210E-04
  1560.00        5.64702          0.257467         3.925210E-04
  1595.00        7.51378          0.425305         3.925210E-04
  1630.00        9.70862          0.688221         3.925210E-04
  1665.00        12.6294          1.09234          3.925210E-04
  1700.00        16.5061          1.70257          3.925210E-04
```

Output file `Jfield.txt` in tab-delimited format:

F [eV/nm]	JGTF [A/cm^2]	JRLD	JFN
1.00000	3.610247E-08	2.808417E-09	1.161073E-18
1.05336	8.879720E-08	4.052475E-09	3.366324E-17
1.10957	2.282641E-07	5.904351E-09	8.256117E-16
1.16878	3.464692E-07	8.688134E-09	1.727458E-14
1.23114	5.675248E-07	1.291519E-08	3.108433E-13
1.29684	1.009689E-06	1.940028E-08	4.847257E-12
1.36604	1.965366E-06	2.945578E-08	6.598127E-11
1.43893	4.220286E-06	4.521744E-08	7.894216E-10
1.51572	1.009293E-05	7.020131E-08	8.355946E-09
1.59660	2.718130E-05	1.102586E-07	7.873680E-08
1.68179	8.351229E-05	1.752450E-07	6.643776E-07
1.77154	2.973305E-04	2.819540E-07	5.048154E-06
1.86607	1.250405E-03	4.593549E-07	3.472470E-05
1.96564	5.079909E-03	7.580564E-07	2.173307E-04
2.07053	1.580514E-02	1.267604E-06	1.243542E-03
2.18102	5.320935E-02	2.148536E-06	6.534721E-03
2.29740	0.185017	3.692663E-06	3.167354E-02
2.41999	0.643032	6.437693E-06	0.142182
2.54912	2.18956	1.138879E-05	0.593417
2.68514	7.22165	2.045251E-05	2.31123
2.82843	22.9292	3.730041E-05	8.42985
2.97935	69.8631	6.911160E-05	28.8888
3.13834	204.007	1.301498E-04	93.3138
3.30580	570.791	2.492153E-04	284.951
3.48220	1530.86	4.854400E-04	824.970
3.66802	3939.07	9.623223E-04	2270.50
3.86375	9735.08	1.942366E-03	5955.74
4.06992	23138.0	3.993702E-03	14925.8
4.28709	52960.2	8.368768E-03	35820.3
4.51586	116904.	1.788174E-02	82502.4
4.75683	249222.	3.897999E-02	182747.
5.01066	513861.	8.673330E-02	390070.
5.27803	1.026172E+06	0.197095	803816.
5.55967	1.987526E+06	0.457671	1.602017E+06
5.85634	3.738639E+06	1.08660	3.093205E+06
6.16884	6.838934E+06	2.63918	5.795330E+06
6.49802	1.218124E+07	6.56176	1.055209E+07
6.84476	2.115210E+07	16.7105	1.869902E+07
7.21000	3.584957E+07	43.6160	3.229345E+07
7.59474	5.937064E+07	116.755	5.442438E+07
8.00000	9.617879E+07	320.750	8.961740E+07

17.6.3 Remarks

The output data files `JTemp.txt` and `Jfield.txt` have been graphically presented in Figures 17.21 and 17.20, respectively. In the later, the dovetailing of the general thermal–field form with the canonical FN and RLD equations is good, apart from the mild bumps in the curve of $J_{GTF}(F, T)$ near the left boundary of the $n = 1$ region, discussed in Figure 17.19. Although greater accuracy can be obtained by using the polynomial $\theta(E)$ of Eq. (17.85) or, better, the exact form in Eq. (17.78), in a numerical integration of $j(E_x)$ exact approaches begin to be attractive.

PART III
Exact tunneling and transmission evaluation

> *"Do you mean that you think you can find out the answer to it?" said the March Hare.*
>
> *"Exactly so," said Alice.*
>
> *"Then you should say what you mean," the March Hare went on.*
>
> *"I do," Alice hastily replied; "at least – at least I mean what I say – that's the same thing, you know."*
>
> *"Not the same thing a bit!" said the Hatter.*
>
> *"You might just as well say that 'I see what I eat' is the same thing as 'I eat what I see'!"*
>
> – Lewis Carroll[1]

Exact solutions to models are indispensable for identifying when approximations go awry, and are "exactly so" in two ways. The first type are exact (not approximate) solutions to simple models, of which the rectangular and triangular barriers are iconic. The second type are numerical methods for exact (accurate) solutions of more general barriers. Both types illuminate how to modify the Kemble approximation and the underlying Gamow factor and identify where they are appropriate – or when they are not.

The treatment of the exact methods will first analyze the simplest of the barriers, namely, the rectangular and triangular barriers, and so introduce numerical methods for dealing with them. Those methods shall then be generalized to develop a transfer matrix approach (TMA) first using plane waves and then using the more accurate Airy function method. Lastly, because the TMA methods can analyze resonances, the treatment will close with a consideration of several configurations where tunneling, resonance, and emission appear alongside one another.

[1] Lewis Carroll, *Alice's Adventures in Wonderland*. Racine, WI: Whitman Publishing Company, 1945, Chapter VII: A Mad Tea-Party.

CHAPTER 18

Simple barriers

Cora: *...But the decision I've come to is I'd rather make the gravest of mistakes than surrender my own judgment.*

– *Last of the Mohicans*[1]

The development of the general thermal–field–photoemission equation (Eq. (17.26) using Eq. (17.30)) implicitly used a feature of the Kemble approximation of Eq. (17.9) that when $\theta(E) = 0$, then $D(E) = 1/2$. Even from elementary considerations, this is not quite acceptable: recall the step function potential barrier of Section 9.2.2, where $D(E(k))$, as given in Eq. (9.69), vanishes at $E = V_o$. The step function barrier is the limiting cases of a rectangular barrier ($L \to \infty$) or a triangular barrier ($F \to 0$) for the Gamow factors of Section 11.2, and for these cases $D(E = V_o)$ does not in general equal 1/2. Nevertheless, Eq. (11.1) for $J(F, T)$ was remarked to have had a kind of *truthiness* in its ability to anticipate the behavior of the energy distributions of thermal–field emission: the Gamow factor (Section 11.2) provided a means of estimating the tunneling current, the Kemble approximation of Eq. (17.9) enabled treating the integrand near the barrier maximum, the polynomial representation of the Gamow factor $\theta(E)$ for $E > \mu + \phi$ that was extended into the region $E_x > \mu + \varphi$ enabled the development of the general thermal–field current density J_{GTF} in terms of $N(n, s)$ of Eq. (17.56), and enabled the recovery of the canonical Fowler–Nordheim (Eq. (13.25)) and Richardson (Eq. (12.3)) equations, as well as the Fowler Dubridge (Eq. (14.3)) equation. These canonical equations were shown to have good correspondence with experimental data. Lastly, there is the weighty authority of those whose names became synonymous with the canonical equations, and whose pronouncements command a certain deference.

All that may be true, but it does not settle whether the Kemble form is reasonable for general barriers, or whether the Gamow factor is reasonable in general. Moreover, the justification for extending $\theta(E)$ into situations where $E > \mu + \varphi$ because that prescription seems to work is rather weak praise, particularly when $\theta(E > V_o)$ should vanish. How the extension of the wave function formalism of Section 9.2.2 justifies the canonical equations beyond the step potential models of Section 9.2.3.1 is intriguing, although the Bohm approach of Section 11.2 lends a certain plausibility. Lastly, deference to the authority of past masters is often – but not always – valid.

A better approach is not to surrender one's judgment and undertake assessing accuracy using an exact methodology to identify when the Kemble form is good, what approximation to $\theta(E_x)$ is good, and how $D(E_x)$ really behaves near the barrier maximum. To emphasize the point: exact methods can treat barriers that induce resonant behavior, but the methodologies covered till now cannot. The first step to understanding what changes is a reconsideration of the simple barriers, not from the perspective of the Gamow factor, but from the perspective of solving Schrödinger's equation properly.

18.1 Rectangular barrier

Thus there exists an essential truth that must be disengaged from the outward appearance of the objects to be represented. This is the only truth that matters.

– Henri Matisse[2]

Consider the simple rectangular barrier, a staple of all quantum mechanics textbooks. Although straightforward, it portends a great deal that appears with other and more complex potentials, in addition to being exquisitely suited to

[1] Michael Mann and Christopher Crowe, screenplay of *Last of the Mohicans*, based on the novel by James Fenimore Cooper, 1992.
[2] Henri Matisse, "Exactitude is Not Truth", *Matisse on Art*, ed. Jack D. Flam. Berkeley: University of California Press, 1995, Essay 40, p. 179. The title of the essay refers to a quote by Eugène Delacroix (*Journal de Eugène Delacroix*, entry for January 25, 1857, p. 236: "La froide exactitude n'est pas l'art ...", translated as "Cold exactitude (precision) is not art ..." In another of life's little ironies, Matisse did not preserve the quote exactly.

Introduction to the Physics of Electron Emission, First Edition. Kevin L. Jensen.

demonstrate the methodology that will lead to the transfer matrix approach of Chapter 19. Rather than cleaving to the traditional narrative, consider it instead in a way that sheds light on both the Kemble approximation and the Gamow factor.

A rectangular barrier is a modification to the step barrier by the addition of a second step, therefore methods used in the analysis of the step potential of Sections 9.2.2 and 9.2.3 can be generalized. Now, however, instead of treating two regions, as in Figure 9.7, there are three, as in Figure 11.1, as considered when treating the Gamow factor in Section 11.2: each side of the barrier (regions 1 and 3) and the barrier itself (region 2). Using the convention that $\hbar k_n$, t_n, and r_n are the momentum, transmission, and reflection coefficients, respectively, in region n, and that the boundary between regions n and $n+1$ occurs at x_n, then for a barrier the single matrix equation of Eq. (9.65) (for $n = 1$) becomes two matrix equations. Writing them out is an ordeal, so analogous to the methods described in Section 8.2, introduce the simplifying notation for a 2×2 matrix $\hat{M}_n(x)$ and a two-component coefficient vector $|\xi\rangle$ defined by

$$\hat{M}_n(x) \equiv \begin{pmatrix} e^{ik_n x} & e^{-ik_n x} \\ ik_n e^{ik_n x} & -ik_n e^{-ik_n x} \end{pmatrix}; \quad |\xi_n\rangle \equiv \begin{pmatrix} t_n \\ r_n \end{pmatrix} \tag{18.1}$$

in terms of which the two matrix equations are delightfully simple, one for each side of the barrier:

$$\begin{aligned} \hat{M}_1(x_1)|\xi_1\rangle &= \hat{M}_2(x_1)|\xi_2\rangle \\ \hat{M}_2(x_2)|\xi_2\rangle &= \hat{M}_3(x_2)|\xi_3\rangle \end{aligned} \tag{18.2}$$

The elimination of $|\xi_2\rangle$ simplifies matters even more, giving

$$\begin{aligned} |\xi_1\rangle &= \left\{\hat{M}_1(x_1)^{-1} \cdot \hat{M}_2(x_1)\right\} \cdot \left\{\hat{M}_2(x_2)^{-1} \cdot \hat{M}_3(x_2)\right\} |\xi_3\rangle \\ &\equiv \hat{S}_1(x_1) \cdot \hat{S}_2(x_2)|\xi_3\rangle \end{aligned} \tag{18.3}$$

where the matrix $\hat{S}_n(x) \equiv \hat{M}_n(x)^{-1} \cdot \hat{M}_{n+1}(x)$ is defined. Thus, a factor of $\hat{S}_n(x_n)$ is associated with each interface between adjacent regions where $V(x)$ changes: the change in $V(x)$ can be a change in *slope*, but also a discontinuous change in *height*, or *both*, but for now consider just discontinuous changes in height. The generalization to more than one barrier should be rather easy when the time comes.

Unlike the step potential, two cases are now in play. Letting the k associated with the barrier height V_o be defined by $V_o \equiv \hbar^2 k_o^2/2m$, then over the barrier transport is defined by $k_1 > k_o$. Conversely, tunneling occurs when $k_1 < k_o$. The following assumptions and conventions are used

- The problem is one dimensional, and so the total energy is $E = \hbar^2 k^2/2m$. The case presently under consideration (Figure 11.1) is $k_1 = k_3 = k$ because in regions 1 and 3, $V(x) = 0$ whereas in region 2, $V(x) = V_o$ (but this need not be so generally).
- The boundary between region 1 and 2 occurs at $x_1 = 0$; the boundary between 2 and 3 occurs at $x_2 = L$.
- The quantity $\kappa \equiv \sqrt{|k_o^2 - k^2|}$ is so constructed that *it is always real,*[3] so that when $E > V_o$, $k_2 = \kappa$, but when $E < V_o$, then $k_2 = i\kappa$.

Whether k_2 is imaginary ($i\kappa$) or real (κ) determines whether the transport is over ($E > V_o$) or under ($E < V_o$) the barrier, respectively. Consider each in turn.

For over-the-barrier transport, using the simple inverse of a 2×2 matrix followed by matrix multiplication, the expression for $S_1(0)$ is easily shown to be

$$\hat{S}_1(0) = \frac{1}{2k}\begin{pmatrix} k+\kappa & k-\kappa \\ k-\kappa & k+\kappa \end{pmatrix} \tag{18.4}$$

Were the problem to end here, the step potential findings of Eq. (9.65) result, but the problem does not end here: there is the matter of $S_2(L)$, given by

$$\hat{S}_2(L) = \frac{1}{2\kappa}\begin{pmatrix} (k+\kappa)e^{i(k-\kappa)L} & (k-\kappa)e^{-i(k+\kappa)L} \\ (k-\kappa)e^{i(k+\kappa)L} & (k+\kappa)e^{-i(k-\kappa)L} \end{pmatrix} \tag{18.5}$$

[3] An attempt to preserve this convention when linear fields are introduced is the motivation for the Zi functions in Section 19.2, along with the desire that Zi account for outgoing waves.

One might think the next step is to consider the product $\hat{S}_1 \cdot \hat{S}_2$, but that is a torturous path for the writing of equations: it is easier to make use of the coefficient vectors, where

$$|\xi_1\rangle = \begin{pmatrix} 1 \\ r(k) \end{pmatrix}; \quad |\xi_3\rangle = \begin{pmatrix} t(k) \\ 0 \end{pmatrix} \tag{18.6}$$

(i.e., $t_1 = 1$, $r_1 = r(k)$, $t_3 = t(k)$ and $r_3 = 0$) because of the ease of dealing with two-component columns rather than four-component matrices, especially when multiplications by 0 are involved, to wit

$$\begin{pmatrix} 1 \\ r(k) \end{pmatrix} = \frac{t(k)}{4k\kappa} e^{ikL} \begin{pmatrix} 4k\kappa \cos(\kappa L) - 2i(k^2 + \kappa^2)\sin(\kappa L) \\ -2i(k^2 - \kappa^2)\sin(\kappa L) \end{pmatrix} \tag{18.7}$$

where $e^{\pm i\kappa L} = \cos(\kappa L) \pm i \sin(\kappa L)$ renders the components a bit more compact. Equating the components of the column vectors results in expressions for $r(k)$ and $t(k)$, or

$$r(k) = \frac{-i(k^2 - \kappa^2)\sin(\kappa L)}{2k\kappa \cos(\kappa L) - i(k^2 + \kappa^2)\sin(\kappa L)} \tag{18.8}$$

$$t(k) = \frac{2k\kappa}{2k\kappa \cos(\kappa L) - i(k^2 + \kappa^2)\sin(\kappa L)} e^{-ikL} \tag{18.9}$$

From the relations $j_{inc} = \hbar k/m$ and $j_{trans} = (\hbar k/m)|t|^2$, and Eq. (9.69), it follows that the transmission probability for a rectangular barrier $D_{rec}(k)$ is

$$D_{rec}(k > k_o) = \left\{ 1 + \frac{1}{4}\left(\frac{k}{\kappa} - \frac{\kappa}{k}\right)^2 \sin^2(\kappa L) \right\}^{-1} \tag{18.10}$$

For under-the-barrier transport, the problem could be started afresh by setting up the matrix relations at each interface again, but then it is far easer to realize that doing so means repeating all the same steps with the replacement $\kappa \to i\kappa$. The result is simply to make the same changes in Eq. (18.10) and use the relation $\sin(i\kappa L) = i \sinh(\kappa L)$ to obtain

$$D_{rec}(k < k_o) = \left\{ 1 + \frac{1}{4}\left(\frac{k}{\kappa} + \frac{\kappa}{k}\right)^2 \sinh^2(\kappa L) \right\}^{-1} \tag{18.11}$$

Therefore, the transition from under-the-barrier to over-the-barrier is occasioned by $\kappa \to i\kappa$, but *this is equivalent to* $(k < k_o) \to (k > k_o)$, or, in other words, for all k, $D_{rec}(k)$ is given by

$$D_{rec}(k) \equiv \left\{ 1 + \left[\left(\frac{k_o^2}{2k}\right) \frac{\sinh\left(L\sqrt{k_o^2 - k^2}\right)}{\sqrt{k_o^2 - k^2}} \right]^2 \right\}^{-1} \tag{18.12}$$

This form can be shoe-horned into the Kemble form of $D(k) = 1/(1 + e^{\theta})$ if $\theta(k)$ is identified as

$$\theta(k) = 2 \ln\left(\frac{k_o^2 \sinh\left(L\sqrt{k_o^2 - k^2}\right)}{2k\sqrt{k_o^2 - k^2}} \right) \tag{18.13}$$

with the realization that when $k > k_o$, the argument of the sinh –function becomes imaginary and the sin function results so that $e^{\theta(k)}$ goes from exponentially increasing to acquiring an oscillatory nature.

It is difficult to see how the Gamow factor hides in Eq. (18.13), but it is there. Recall that the Gamow factor for the rectangular barrier, from Eq. (11.15), is given by $2\kappa L = 2L\sqrt{k_o^2 - k^2} \leftrightarrow \theta_{rec}(k)$, therefore as the argument of the sinh-function becomes large, $\ln(\sinh(2\kappa L)) \to 2\kappa L$, and there it is. However, other factors complicate matters when $k \approx k_o$ or when $k \to 0$. The case $k_o L = 1$ corresponds to a barrier that is as wide as $1/k_o$ is tall, and the case $k/k_o = 1$ corresponds to an energy $\hbar^2 k^2/2m$ equal to the barrier height V_o. Then $\theta_{rec}(k)/(2k_o L) = \sqrt{1 - (k/k_o)^2}$, regardless of the magnitude of $k_o L$, against which $\theta(k)/(2k_o L)$ may be compared, as in Figure 18.1. Thus, it is seen that when $k_o L$ is large, the Gamow factor is a reasonable approximation, but when $k_o L \to 1$, it is less so. At the boundaries of the figure, $\theta(k \to 0)/(2k_o L) \to -2 \ln(2x)$

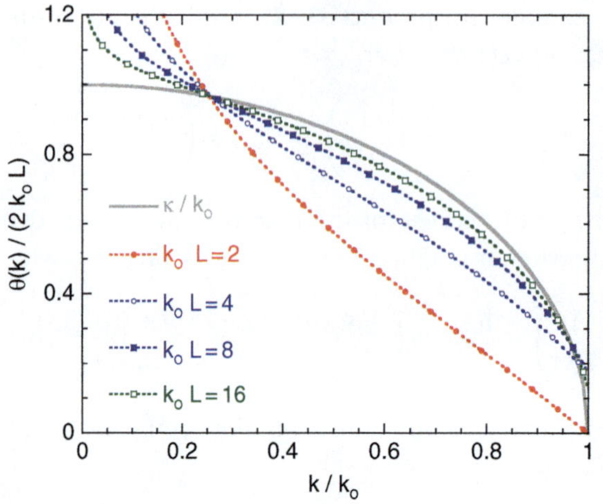

Figure 18.1 Comparison of $\theta_{rec}(k)/(2k_oL)$ (using Eq. (18.13)) to $\kappa/k_o = \sqrt{1-(k/k_o)^2}$.

becomes large whereas θ_{rec} approaches unity, and that $\theta(k \to k_o) \to 2\ln(k_oL/2)$ whereas $\theta_{rec} \to 0$. Therefore, the Gamow factor is best in the intermediate regime $k/k_o \sim 1/\sqrt{2}$ for high and thin barriers ($k_oL \gg 1$).

The relation of $D_{rec}(k)$ to the Kemble approximation is even more nuanced. Recall that the Kemble approximation is $D(k) \approx 1/(1+\exp(\theta(k)))$, which suggests that $D(k_o) = 1/2$ because $\theta(k_o) = 0$. In contrast, Eq. (18.12) can be used to show

$$D_{rec}(k_o) = \left\{ 1 + \left(\frac{k_oL}{2}\right)^2 \right\}^{-1} \tag{18.14}$$

which gives the Kemble result $D(k_o) = 1/2$ only for $k_oL = 2$.

For $k \geq k_o$, more interesting behavior is revealed: not only is $\theta(k)$ and its first derivative continuous across $E_k = V_o$ (as assumed by the general thermal–field treatment and shown in Figure 17.4), but also Eq. (18.13) can be rewritten, equivalently, as

$$\theta_{rec}(k > k_o) = 2\ln\left(\frac{k_o^2 \sin\left(L\sqrt{k^2 - k_o^2}\right)}{2k\sqrt{k^2 - k_o^2}} \right) \tag{18.15}$$

Now it is apparent that when $\kappa L = j\pi$ for integer j, then $D \to 1$. That is, the transmission probability is oscillating, with wider barriers experiencing more oscillations, as shown in Figure 18.2. Notice something else, though: the step potential barrier result of Eq. (9.69) (where k' there is κ here) is lurking in the background, as shown in Figure 18.3, where the step function $D(k)$ of Eq. (9.69) is compared to the $n = 16$ case as a function of κ/k_o, with oscillations developing

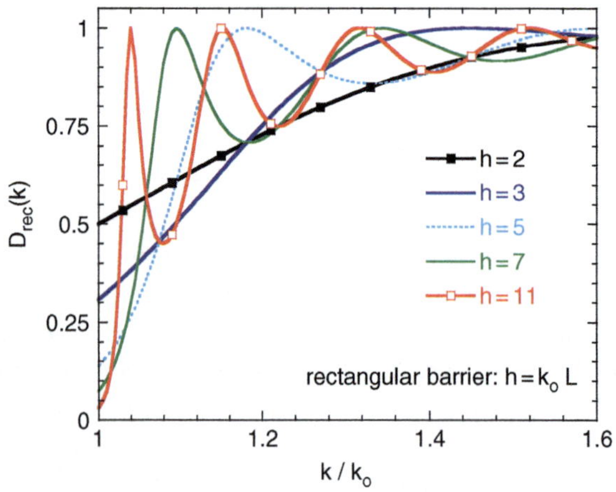

Figure 18.2 The transmission probability $D_{rec}(k)$ for $k > k_o$ for the values $k_oL = h = 2, 3, 5, 7$, and 11, respectively. Observe the continuity of $D(k)$ and $\partial_E D(k)$ across the $k/k_o = 1$ boundary.

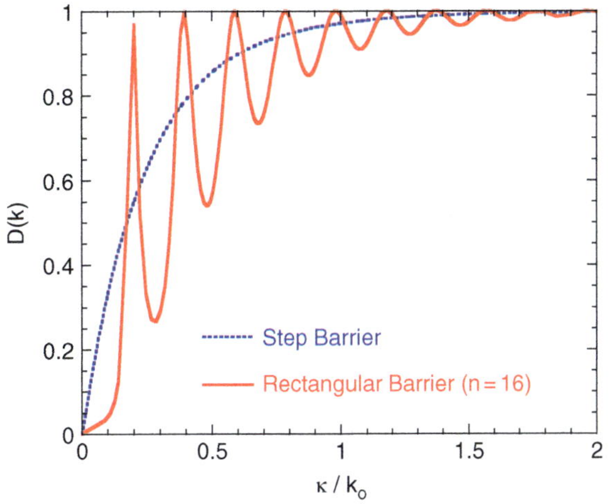

Figure 18.3 The transmission probability $D_{rec}(k)$ for $k > k_o$ for the values $k_o L = h = 16$ compared to the step function transmission probability of Eq. (9.69).

in the rectangular barrier case because of constructive and destructive interference of the wave function components that manifest themselves across a single barrier (a premonition of the truly bizarre behavior in store when the resonant tunneling diode (RTD) is considered in Section 19.2.4 below). As seen, the overall trend is anticipated by the step function results.

Right about now, it appears that the previous embrace of the linearized Kemble approximation of Section 17.2, used by $J_{GTF}(F,T)$ and discussed near Eq. (17.10), is running afoul of features of the simple square barrier model. For the purposes of discussion, let $h = k_o L$. Then (*i*) although $D(k)$ is continuous across $k = k_o$, the condition $D(k_o) = 1/2$ only occurs for comparatively short barriers such that $h = 2$, (*ii*) $D(k) = 1/2$ for $k > k_o$ when h is larger, and (*iii*) for higher values of $h = k/k_o$, an oscillatory behavior is distinctly visible but which has no analog in the linearized Kemble form. But the linearized Kemble approximation *also* assumed that $\partial_E \theta(E(k)) \equiv \theta'(E)$ was continuous through $E = V_o$ (or $k = k_o$), which is the case for Eq. (18.13) but not so for $\theta_{rec}(E) \propto \sqrt{E - V_o}$ for which $\theta'_{rec}(E) \propto -1/\sqrt{E - V_o}$ and therefore singular at $E = V_o$.

The triangular barrier (as well as the quadratic barrier) is not singular at $E = V_o$ for the Gamow factors by comparison. It remains to be seen if that will make a difference. But it is rather clear already that one moral of the story is that simple models (such as those based on the Kemble approximation) are wonderful for their pedagogical insights, but as with any toy model, the differences with models that are more faithful to reality (more "essential" as Matisse might have meant it) matter.

18.2 Triangular barrier: general method

Joseph: *There you are. It's clever. It's German. It's quality work. And there are simply too many notes. Do you see?*
Mozart: *There are just as many notes, Majesty, neither more nor less, as are required.*

– Peter Shaffer[4]

The triangular barrier model, which formed the basis of Fowler and Nordheim's celebrated equation [36] given in Eq. (13.10), simplifies the rectangular barrier model. It does so by returning to two regions instead of three, as for the rectangular barrier. But that visual simplicity comes at the cost of greater mathematical complexity by making the wave functions of region 2 more difficult to describe: the potential is no longer constant and therefore plane waves can no longer be used. That is, whereas Schrödinger's equation in the rectangular barrier region

$$\left(-\frac{\hbar^2}{2m}\frac{\partial^2}{\partial x^2} + V_o \right) \psi_k(x) = E_k \psi_k(x) \tag{18.16}$$

[4]Peter Shaffer, *Amadeus*. New York: Harper Perennial, 2001, Act 1, Scene 8, p. 39.

implies $\psi_k(x)$ is a combination of $\exp(\pm i\sqrt{k^2-k_o^2}x)$ with $E_k = \hbar^2k^2/2m$ and $V_o = \hbar^2k_o^2/2m$ (using the notation of the last section), the triangular barrier potential of

$$\left(-\frac{\hbar^2}{2m}\frac{\partial^2}{\partial x^2} + V_o - Fx\right)\psi_k(x) = E_k\psi_k(x) \tag{18.17}$$

makes $\psi_k(x)$ something a good deal more challenging. To see what it is, Eq. (18.17) should be simplified because in its present form it has "too many notes", and the functions that will arise because of that will only add to the cacophony.

The first step is to prune Eq. (18.17) until Airy's differential equation is clearly visible [85, 173, 238]. Introduce the term $f \equiv 2mF/\hbar^2$ (which has units of $f = 1/\text{nm}^2$) so that Eq. (18.17) can be written as

$$-\partial_x^2\psi_k(x) + (k_o^2 - k^2 - fx)\psi_k(x) = 0 \tag{18.18}$$

and then introduce the term w such that $w(x) \equiv f^{-2/3}(k_o^2 - k^2 - fx)$: with that notational simplification, the cacophony of Eq. (18.17) is reduced to

$$\partial_w^2\psi - w\psi = 0 \tag{18.19}$$

or Airy's differential equation, which is a far simpler tune. Consequently, ψ will be linear combinations of the functions $\text{Ai}(w)$ and $\text{Bi}(w)$.

The sign of various terms creates issues. The first issue is that the field gradient has been taken so far as negative (i.e., F appears with an explicit $-$ sign as in $V(x) = V_o - Fx$). When more complex potentials are encountered, occasions arise where both positive and negative fields are encountered. There is advantage in keeping f itself a positive quantity, so introduce a factor s to account for the sign of the gradient of the field: $s = -1$ will correspond to the usual potential term $(-Fx)$ but when the potential is rising, as in $(+Fx)$, then $s = +1$. Thus, $V(x) = V_o - E_k - Fx$ becomes, in this notation, $\kappa(x)^2 = |k_o^2 - k^2 + sfx|$ with $s = -1$, where κ has been generalized to include a dependence on position x. Because s appears in Eq. (18.19) as $s^2 = 1$ (one should convince oneself of this), its consideration need not trouble the discussion till later.

The second issue is that $k_o^2 - k^2 + sfx$ can change sign, so careful bookkeeping for following the consequences of that sign through the calculations is important. Giving the sign associated with $k_o^2 - k^2 + sfx$ its own name does that: call it c here.[5] In the asymptotically small f limit, sines and cosines, as well as exponentially increasing and decreasing functions, that arise in the step potential problem of Section 9.2.2 should be recovered, suggesting that c corresponds to the four roots of the equation $c^4 = 1$, or $c = \pm 1, \pm i$. Rather than dealing with w, which can be positive or negative, deal instead with z such that it is only positive and defined by $w \equiv c^2 z$, with

$$z(x) = f^{-2/3}|k_o^2 - k^2 + sfx| \equiv \frac{\kappa(x)^2}{f^{2/3}} \tag{18.20}$$

where $c^2 = 1$ if $(k_o^2 - k^2 + sfx) > 0$ and $c^2 = -1$ if $(k_o^2 - k^2 + sfx) < 0$. Now terms such as f and κ are always, by definition, positive, so that the value of c denotes the Airy function under consideration, of which there are four: $\text{Ai}(c^2z) = \text{Ai}(\pm z)$ and $\text{Bi}(c^2z) = \text{Bi}(\pm z)$.

The literature on Airy functions is far too extensive and thorough to be improved upon. But it is not the behavior of Airy functions which is of interest. Instead, it is the behavior of decaying, growing, or propagating wave function solutions which command attention, and to force attention back to that key point, speak instead of a function $\text{Zi}(c,z)$ which is a solution to the equation

$$\partial_z^2\,\text{Zi}(c,z) - c^2z\,\text{Zi}(c,z) = 0 \tag{18.21}$$

How does $\text{Zi}(c,z)$ behave? There are two kinds of behavior that are sought: exponential and oscillatory behavior. Therefore,

$$\begin{aligned}
\text{Zi}(1,z) &= \text{Bi}(z)\\
\text{Zi}(-1,z) &= \text{Ai}(z)\\
\text{Zi}(i,z) &= \text{Ai}(-z) - i\,\text{Bi}(-z)\\
\text{Zi}(-i,z) &= \text{Ai}(-z) + i\,\text{Bi}(-z)
\end{aligned} \tag{18.22}$$

[5] c here should not be confused with the speed of light.

because the first grows exponentially, the second decays exponentially, and the last two correspond to traveling waves to the left or right. All can be expressed in a single equation as

$$\begin{aligned}\mathrm{Zi}(c,z) = &-\frac{1}{4}(c-1)(c^2+2c+3)\,\mathrm{Ai}(c^2 z)\\ &+\frac{1}{4}(c+1)(3c^2-2c+1)\,\mathrm{Bi}(c^2 z)\end{aligned} \tag{18.23}$$

Alas, such a response just begs the question[6] by failing to illuminate how any of the functions behave by defining them in terms of each other. Before one reflexively reaches for Watson [239], which any other time would be an excellent idea, make use of the fact that the Gamow factor for the triangular barrier is already known: $\theta_{tri} \propto \kappa^3/F$. One therefore has good reason to suspect that $\mathrm{Zi}(c,z) \sim \exp(az^{3/2})$. In the large z limit, insertion of that trial form into Eq. (18.21) suggests letting $9a^2 - 4c^2 = 0$, so take $a = 2c/3$. Making use of Eq. (11.22) to hone intuition, a more realistic ansatz for $\mathrm{Zi}(c,x)$ is

$$\mathrm{Zi}(c,z) = \frac{G_c(z)}{2\sqrt{\pi}} z^{-1/4} \exp\left(\frac{2}{3}cz^{3/2}\right) \tag{18.24}$$

where the coefficient $z^{-1/4}$ comes from the factor $\kappa^{-1/2}$ in Eq. (11.22), and where $G_c(z)$ is (likely) a smooth function of z. Inserted into Eq. (18.21), the result is

$$\left\{16z^2\left(\frac{\partial}{\partial z}\right)^2 + 8z(4cz^{3/2}-1)\frac{\partial}{\partial z} + 5\right\} G_c(z) = 0 \tag{18.25}$$

and is not evidently an improvement, but switching variables to $z = u^{-2/3}$ and defining $G_c(z(u)) = H_c(u)$ gives

$$\{36u^2\partial_u^2 + 24(3u-2c)\partial_u + 5\}H_c(u) = 0 \tag{18.26}$$

which is in the direction of simplicity.

The convention used here, given that $G_c(z)$ and $H_c(u)$ are the same function, is that $G_c(z)$ will be used when $z \leq 1$ and $H_c(u)$ will be used when $u \leq 1$. The convention is therefore to facilitate the consideration of the limiting (small argument) case of $H_c(u)$: consideration of the function $G_c(z)$ is eclipsed by the availability of good polynomial expansions of $\mathrm{Ai}(x)$ and $\mathrm{Bi}(x)$ when $|x| < 1$, the forms of which will be taken up in Section 18.3.

Observe that $H_c(0)$ (which is just $G_c(\infty)$) can be determined from the asymptotic forms of the Airy functions themselves [103]. it is found that

$$\begin{aligned} H_1(0) &= 2; & H_{-1}(0) &= 1\\ H_i(0) &= \sqrt{2}(1-i); & H_{-i}(0) &= \sqrt{2}(1+i)\end{aligned} \tag{18.27}$$

The substitution of $z(u) = u^{-2/3}$ renders Eq. (18.24) as

$$\mathrm{Zi}(c,z(u)) = \frac{H_c(u)}{2\sqrt{\pi}} u^{1/6} \exp\left(\frac{2c}{3u}\right) \tag{18.28}$$

The introduction of the $\mathrm{Zi}(c,z)$ functions enables investigation of the smooth passage to the transmitted (e^{ikx}) and reflected (e^{-ikx}) waves when the field vanishes and the electron energy is above the barrier ($c^2 = -1$). That is, if $\mathrm{Zi}(c,z)$ is known, say at $z = z_o$ where $x = 0$, then how does it change a short distance away? It does so by

$$\begin{aligned}\frac{\mathrm{Zi}(i,z)}{\mathrm{Zi}(i,z_o)} &\approx \exp\left\{\frac{2i}{3f}[(k^2-k_o^2-sfx)^{3/2}-(k^2-k_o^2)^{3/2}]\right\}\\ &\approx \exp[-is(k^2-k_o^2)^{1/2}x]\end{aligned} \tag{18.29}$$

[6] "Begging the question" is a logical fallacy that is commonly, but incorrectly, thought to be failing to give an answer that pertains to the question. In the words of D.N. Walton, "To commit the fallacy of begging the question, two requirements must be met, in a given case: (i) there must be a circular sequence of reasoning, where the conclusion to be established is either identical to one of the prmises, or the premise in question depends on the conclusion, and (ii) the circular sequence of reasoning must be used illicitly in a context of dialogue (conversation) to escape the proper fulfillment of a legitimate burden of proof in that context." (see Chapter 16 of Hans V. Hansen and Robert C. Pinto, *Fallacies: Classical and Contemporary Readings*. University Park, PA: Pennsylvania State University Press, 1995, p. 230), which is what was done here.

as $f \to 0$, but that is the equation for a transmitted wave over a step barrier. Therefore, $\mathrm{Zi}(\pm i, z)$ *are the analogs of* $e^{\pm ikx}$ *when a field exists*, making their usage, rather than the real valued Ai and Bi, a bit more useful for tunneling calculations in spite of the overhead required to understand them.

The behavior of $H_c(u)$ for non-zero u is needed to progress. Differential equations with constant coefficients (what Eq. (18.26) resembles in the $u \to 0$ limit) motivates trying a solution of the form $H_c(u) \approx H_c(0)e^{-\varphi}$ because terms of the form $H^{-1}\partial_u H$ suggest its utility. The leading order behavior of φ is then determined from the resulting equation

$$36u^2((\partial_u \varphi)^2 - \partial_u^2 \varphi) + 24(2c - 3u)\partial_u \varphi + 5 = 0 \tag{18.30}$$

The leading order behavior is obtained by neglecting terms of order u^2 and higher, giving $\varphi \approx (5/72)\ln[1 - (3u/2c)]$, but keeping only the linear order term in u from a Taylor expansion of φ does not work (try it). The next approximation just shy of a quadratic function is based on a Padé approximation, where φ is a ratio of polynomials in u, or, more to the point here, a function of the form $\varphi(u) = Au/(Bu + 1)$ (there is no constant in the numerator because of the expansion associated with the *ln* term). Doing so, and neglecting terms of order u^2, gives $\varphi = 5u/12(4c - 3u)$, or

$$H_c(u) \approx H_c(0) \exp\left\{\frac{5u}{12(4c - 3u)}\right\} \tag{18.31}$$

which is adequate for u small.

Recall, though, that the definition of current in Eq. (9.60) involved more than just $\psi(x,t)$, it also involves $\partial_x \psi(x,t)$. Therefore, in addition to $\mathrm{Zi}(c,z)$, $\partial_x \mathrm{Zi}(c,z) \equiv \mathrm{Zi}'(c,z)$[7] is also required. The plane wave asymptotic forms $e^{\pm ikx}$, where $\partial_x e^{\pm ikx} = \pm ike^{ikx}$, suggest the analogy

$$\mathrm{Zi}'(c,z) \equiv \partial_x \mathrm{Zi}(c,z) = \mathrm{Di}(c,z)\mathrm{Zi}(c,z) \tag{18.32}$$

such that in the limit $f \to 0$, $\mathrm{Di}(c,z) \to \pm\kappa$ (for $c = \pm 1$) or $\pm ik$ (for $c = \pm i$). See how this is so. Observe that $\mathrm{Zi}(\pm 1, z)$ is real because $\mathrm{Ai}(z)$ and $\mathrm{Bi}(z)$ are, and that $\mathrm{Zi}(-i,z) = \mathrm{Zi}(i,z)^\dagger$ because $\mathrm{Zi}(-i,z)\mathrm{Zi}(i,z) = \mathrm{Ai}(-z)^2 + \mathrm{Bi}(-z)^2$ is also real. Lastly, because the Wronskian equation (Eq. (A2.27)) is real, $\mathrm{Di}(-i,z) = \mathrm{Di}(i,z)^\dagger$. These properties are useful. From the definition of $z(x)$ in Eq. (18.20), it follows

$$\frac{\partial}{\partial x} = \left(\frac{\partial z}{\partial x}\right)\frac{\partial}{\partial z} = sf^{1/3}\frac{\partial}{\partial z} \tag{18.33}$$

Then

$$\begin{aligned}\mathrm{Di}(c,z) &= sf^{1/3}\left(-\frac{1}{4z} + c\sqrt{z} + \frac{\partial_z G_c(z)}{G_c(z)}\right)\\ &= \frac{s}{4}\left(\frac{f}{u}\right)^{1/3}(4c - u - 6u^2\partial_u \ln\{H_c(u)\})\end{aligned} \tag{18.34}$$

Because terms of order u^2 are neglected, the factor containing $\partial_u \ln(H_c(u))$ can be neglected because of the dependence of $\varphi(u)$, resulting in the approximation

$$\mathrm{Di}(c,z) = \frac{sf^{1/3}}{4u^{1/3}}(-u + 4c) \tag{18.35}$$

Because the $\mathrm{Di}(c,z)$ functions can also be expressed directly in terms of the Airy functions, this means the following approximations, valid for $u \ll 1$, to combinations of the Airy functions results, the first two for $c^2 = 1$, the second for $c^2 = -1$:

$$\frac{4-u}{4u^{1/3}} \approx \frac{\mathrm{Bi}'(z)}{\mathrm{Bi}(z)} \tag{18.36}$$

$$\frac{-4-u}{4u^{1/3}} \approx \frac{\mathrm{Ai}'(z)}{\mathrm{Ai}(z)} \tag{18.37}$$

$$-\frac{1}{2}u^{2/3} \approx -2\frac{\mathrm{Ai}(-z)\mathrm{Ai}'(-z) + \mathrm{Bi}(-z)\mathrm{Bi}'(-z)}{\mathrm{Ai}(-z)^2 + \mathrm{Bi}(-z)^2} \tag{18.38}$$

$$\frac{1}{u^{1/3}} \approx \frac{(1/\pi)}{\mathrm{Ai}(-z)^2 + \mathrm{Bi}(-z)^2} \tag{18.39}$$

[7] Although such a convention invites confusion with respect to the similar Airy functions $\mathrm{Ai}'(z)$ and $\mathrm{Bi}'(z)$ where the prime denoted differentiation with respect to argument, whereas for Zi, it denotes differentiation with respect to position x, the convention was established earlier [85, 202] and is retained for continuity.

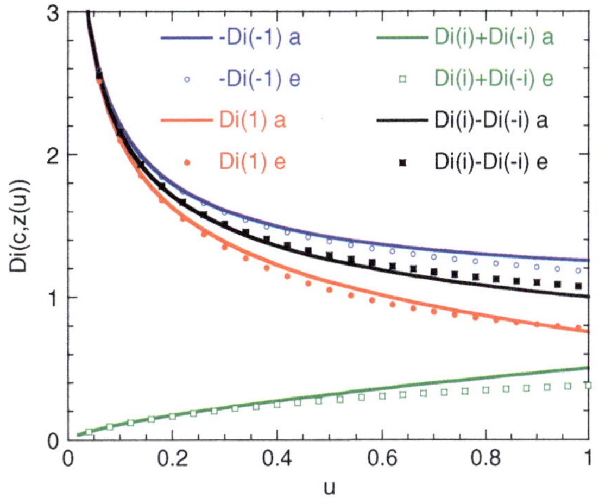

Figure 18.4 Comparison of the ratio of the approximation to the Di functions, given by Eq. (18.35), with $sf^{1/3}$ with their exact evaluation in terms of Airy functions. For $c = -1$, the negative of the curves are shown.

In these four equations, the left-hand side is proportional to the ratio with $sf^{1/3}$ of $\mathrm{Di}(1, z)$ and $\mathrm{Di}(-1, z)$ (for $c^2 = 1$), and $\mathrm{Di}(i, z) + \mathrm{Di}(-i, z)$ and $\mathrm{Di}(i, z) - \mathrm{Di}(-i, z)$ (for $c^2 = -1$), expressed using Eq. (18.35). In all, the right-hand side is the exact relation in terms of Airy functions. The limit that $u \to 0$ for $c = i$ is an important case, giving

$$\lim_{z\to\infty} \mathrm{Di}(i, z) = \frac{sf^{2/3}}{iu^{3/2}} = i\sqrt{k^2 - k_o^2} \tag{18.40}$$

(where $s = -1$) or ik' in the parlance of Eq. (9.65), or $i\kappa(0)$ here. The leading order approximations are compared to the exact Airy function relations in Figure 18.4, and the agreement is reasonably good.

18.2.1 Over the barrier: $D(k > k_o)$

Although Eq. (9.66) almost immediately suggests the form of $t(k)$ for $f > 0$, walking through the evaluation explicitly is useful. The generalization of Eq. (9.65) becomes

$$\begin{pmatrix} 1 & 1 \\ ik & -ik \end{pmatrix} \begin{pmatrix} 1 \\ r \end{pmatrix} = \begin{pmatrix} \mathrm{Zi}(i, z_o) & \mathrm{Zi}(-i, z_o) \\ \mathrm{Zi}'(i, z_o) & \mathrm{Zi}'(-i, z_o) \end{pmatrix} \begin{pmatrix} t \\ 0 \end{pmatrix} \tag{18.41}$$

where the onset of the barrier is at $z_o = \kappa(0)^2/f^{2/3}$. The transmission coefficient t is therefore

$$t(k) = \frac{2k}{(k - i\mathrm{Di}(i, z_o))\,\mathrm{Zi}(i, z_o)} \tag{18.42}$$

The remaining piece of the effort to find the transmission probability $D(k)$ is the evaluation of the outgoing current j_{trans}, as needed in Eq. (9.69). As the outgoing wave $\psi_k(x)$ corresponds to $t(k)\mathrm{Zi}(i, z)$, this means that the quantity $\mathrm{Zi}(-i, z)\mathrm{Zi}'(i, z) - \mathrm{Zi}(i, z)\mathrm{Zi}'(-i, z)$ must be found (recall that for the Zi, the prime indicates ∂_x, *not* derivative with respect to argument). Invoking Eq. (18.22), and bypassing even thinking of $\mathrm{Di}(c, z)$ by using the fortuitous Wronskian relation of Airy functions of Eq. (A2.27), it is found that

$$\frac{(\hbar/2mi)\,(\mathrm{Zi}(-i, z)\,\mathrm{Zi}'(i, z) - \mathrm{Zi}(i, z)\,\mathrm{Zi}'(-i, z))}{\hbar k/m} = -\frac{sf^{1/3}}{\pi k} \tag{18.43}$$

where $s = -1$ because a barrier that is *increasing* to the right has no *transmitted* waves. This elegant result, along with $u = f/\kappa^3$, enables finding $D(k) = (f^{1/3}/\pi k)|t(k)|^2$ to be

$$D(k > k_o) = \frac{16k\kappa}{(|\mathrm{Di}(i, z)|^2 + k^2)|H_i(u)|^2 + 8k\kappa} \equiv \frac{16k\kappa}{\Pi_i(z) - 8sk\kappa} \tag{18.44}$$

$$\Pi_c(u) = \{k^2 + |\mathrm{Di}(c, z(u))|^2\}|H_c(u)|^2 \tag{18.45}$$

for transmission over the triangular barrier of Fowler and Nordheim. The introduction of the real function $\Pi_c(u)$ should be noted: although apparently just a convenience here for $D(k > k_o)$, it returns when considering tunneling in $D(k < k_o)$ in an interesting way.

EXAMPLE: In the limit that the field $F \to 0$, show that Eq. (9.69) is recovered from Eq. (18.44).

SOLUTION: A triangular tunneling barrier entails $s = -1$. Vanishing field implies $f \to 0$. Because z is evaluated at $x = 0$, $z = (k^2 - k_o^2)/f^{2/3} = \kappa^2/f^{2/3}$. Because $u = z^{-3/2}$, $u = f/\kappa^3 \to 0$. As a result, $\mathrm{Di}(\pm i, z) \to \pm i\kappa$ and $|H_i(0)|^2 = 4$. Inserting the components,

$$\lim_{F\to 0} D(k > k_o) = \frac{16k\kappa}{4(k^2+\kappa^2)+8k\kappa} = \frac{4k\kappa}{(k+\kappa)^2} \tag{18.46}$$

18.2.2 Under the barrier: $D(k < k_o)$

The treatment of tunneling involves moving from above the barrier $(k > k_o)$ to below the barrier $k < k_o$, but that is not as simple as changing c in the previous analysis, although it would amount to simply changing z to $-z$ in the arguments of the Airy functions if they were used as the basis states. In fact, usage of Ai and Bi for the particular case of the triangular barrier is a bit simpler [240]. The Zi approach will not reveal its utility until more general potentials that suffer small values of f (and therefore large arguments of Airy functions), as well as shifting locations where $\psi_k(x)$ emerges from the barrier, are considered. The Zi approach, however, requires accommodation that the Airy approach does not when a wave function passes from under the potential to over it, or vice versa. The reason is because $\mathrm{Zi}(i, z)$ corresponds, like e^{iix}, to a traveling wave and therefore contains both Ai and Bi components, the coefficients of which have to be matched. That is, for the wave function going from under the barrier to over the barrier, the wave function changes from exponential growth/decay to oscillatory (and the reverse when going from over the barrier to under it). The transition between the two behaviors occurs where $z = 0$.

For the triangular barrier $(s = -1)$ the relationship between the coefficients to the left (under) to those on the right (over) can be found using the same methods of Eqs (18.2) and (18.3) but with the Zi functions instead of the $\exp(\pm ikx)$ functions:

$$\hat{M}_1(x_o)|\xi_1\rangle = \hat{M}_i(x_o)|\xi_i\rangle \tag{18.47}$$

where the subscript refers to the value c takes on, and because x_o is such that $k^2 - k_o^2 - sfx_o = 0$, then

$$\hat{M}_c(x_o) = \begin{pmatrix} \mathrm{Zi}(c,0) & \mathrm{Zi}(-c,0) \\ \mathrm{Zi}'(c,0) & \mathrm{Zi}'(-c,0) \end{pmatrix}; \quad |\xi_c\rangle = \begin{pmatrix} t_c \\ r_c \end{pmatrix} \tag{18.48}$$

where the circumstance shown in Eq. (18.47) is for under the barrier to over (the wave function emerges from the usual triangular barrier), as in Figure 18.5. Were a wave function to *enter* a triangular barrier $(s = +1)$, then the 1 and i subscripts would swap places. For the simple triangular barrier, $t_1 = 1$, $r_1 = r(k)$, $t_i = t(k)$ and $r_i = 0$, analogous to Eq. (18.6).

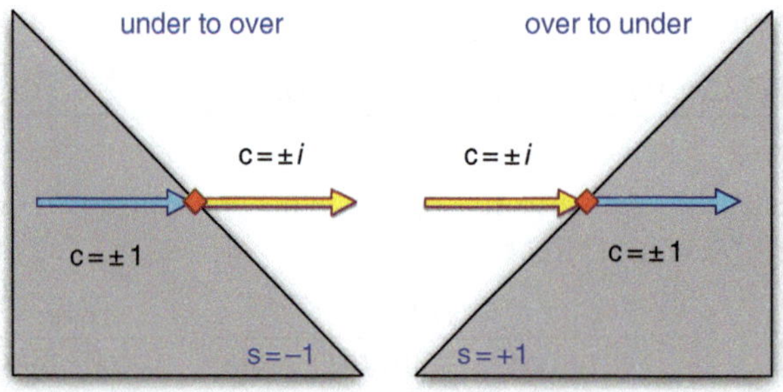

Figure 18.5 Schematic of wave function emerging from under a barrier to over it (left) and over a barrier to under it (right). The change in c occurs at the red diamond.

For the under-to-over case ($s = -1$, $c = 1$ on the left, and $c = i$ on the right), shown in the leftmost diagram of Figure 18.5, and in the notation introduced in Eq. (18.3), $\hat{S}(x_o) = \hat{M}_1(x_o)^{-1}\hat{M}_i(x_o)$, so then

$$\hat{S}(x_o) = \begin{pmatrix} \mathrm{Zi}(1,0) & \mathrm{Zi}(-1,0) \\ \mathrm{Zi}'(1,0) & \mathrm{Zi}'(-1,0) \end{pmatrix}^{-1} \begin{pmatrix} \mathrm{Zi}(i,0) & \mathrm{Zi}(-i,0) \\ \mathrm{Zi}'(i,0) & \mathrm{Zi}'(-i,0) \end{pmatrix} \tag{18.49}$$

This apparently entails dreadful matrix inversions and multiplications, but there are several simplifications behind it. First, the inverse of the first matrix contains a Wronskian and can use Eq. (A2.27). Second, although matrix multiplication results in superficially difficult factors (e.g., $[\hat{S}(x_o)]_{1,1} = \mathrm{Zi}(i,0)\mathrm{Zi}'(-1,0) - \mathrm{Zi}(-1,0)\mathrm{Zi}'(i,0)$) the arguments of the Zi and Zi′ are 0 so that when converted to Airy functions using Eq. (18.22) there is a good deal of cancelation and what remains is a Wronskian itself! This means that all the elements of $\hat{S}(x_o)$ are either ± 1 or $\pm i$. In fact, direct computation shows that the transition matrices that must be introduced whenever $c = 1 \to i$ (under to over) or $c = i \to 1$ (over to under) are

$$\hat{S}(x_o) = \begin{cases} \begin{pmatrix} -i & i \\ 1 & 1 \end{pmatrix} & (\textit{under} \to \textit{over}) \\ \frac{1}{2}\begin{pmatrix} i & 1 \\ -i & 1 \end{pmatrix} & (\textit{over} \to \textit{under}) \end{cases} \tag{18.50}$$

and yes, the factor of (1/2) is correctly attached to the second form (work it out). Thus, whenever a change in c occurs in the Zi functions (either $1 \to i$ or $i \to 1$), a transition matrix $\hat{S}(x_o)$ must be inserted, the form dictated by whether the wave function goes into a tunneling region or emerges from it.

Incorporating the transition matrix $\hat{S}(x_o)$ of Eq. (18.50) into Eq. (18.41) so as to treat the case when $k < k_o$ gives

$$\begin{pmatrix} 1 & 1 \\ ik & -ik \end{pmatrix}\begin{pmatrix} 1 \\ r \end{pmatrix} = \begin{pmatrix} \mathrm{Zi}(1,z_o) & \mathrm{Zi}(-1,z_o) \\ \mathrm{Zi}'(1,z_o) & \mathrm{Zi}'(-1,z_o) \end{pmatrix} \times \begin{pmatrix} -i & i \\ 1 & 1 \end{pmatrix}\begin{pmatrix} t \\ 0 \end{pmatrix}$$

Although the manipulations proceed analogously to the case $k > k_o$, there is a bit more bookkeeping involved because $\mathrm{Zi}(\pm 1, z_o)$ is real. Not surprisingly, however, the final result, using Eq. (18.45) for $\Pi_{\pm 1}(u)$, is

$$D(k < k_o) = \frac{16k\kappa}{\Pi_1(u)e^{4/3u} + \Pi_{-1}(u)e^{-4/3u} - 8sk\kappa} \tag{18.51}$$

where, remember, $u = f/\kappa^3$. Notice how $\Pi_c(u)$ changes when $c = i \to c = \pm 1$: the arguments of the exponentials in the underlying Zi functions become real, resulting in the $e^{\pm 4/3u}$ terms.

For the simple triangular barrier, it may be that the Zi approach feels unnecessarily busy, but when dealing with potentials having several regions, particularly those for which the arguments of the Airy functions become uncomfortably large (e.g., as occurs when the field is small), the use of the Zi functions makes for easier calculations, as shall be seen when potentials composed of several linear segments are considered. Still, we will explore the one case where Airy functions themselves are simpler next.

18.2.3 Airy function method

Nothing is more simple than greatness; indeed, to be simple is to be great.

– Ralph Waldo Emerson[8]

Using the Airy function definitions behind the Di and Zi functions gives an equivalent form to Eq. (18.45) for the case of the triangular barrier where $s = -1$, which has the virtue of being simple and therefore compact [196, 240] in

[8] Ralph Waldo Emerson, An Oration delivered before the Literary Societies of Dartmouth College, July 24, 1838, http://www.emersoncentral.com/litethics.htm (09/03/2009 18:36:13).

that there are only two regions and one site of matching the wave function and its derivative, so that $\pm z$ is enough to determine which Airy function is used. The straight Airy function version makes use of

$$|\mathrm{Zi}(i,z)|^2 = \mathrm{Ai}(-z)^2 + \mathrm{Bi}(-z)^2 \tag{18.52}$$

$$|\mathrm{Zi}'(i,z)|^2 = f^{2/3}(\mathrm{Ai}'(-z)^2 + \mathrm{Bi}'(-z)^2) \tag{18.53}$$

$$\mathrm{Zi}(1,z)^2 + \mathrm{Zi}(-1,z)^2 = \mathrm{Ai}(-z)^2 + \mathrm{Bi}(-z)^2 \tag{18.54}$$

$$\mathrm{Zi}'(1,z)^2 + \mathrm{Zi}'(-1,z)^2 = f^{2/3}(\mathrm{Ai}'(-z)^2 + \mathrm{Bi}'(-z)^2) \tag{18.55}$$

so that

$$D(k) = \frac{4kf^{1/3}}{\pi[k^2(\mathrm{Ai}^2 + \mathrm{Bi}^2) + f^{2/3}(\mathrm{Ai}'^2 + \mathrm{Bi}'^2)] + 2kf^{1/3}} \tag{18.57}$$

where, for compactness, the arguments of $\mathrm{Ai}(\pm z)$ and $\mathrm{Bi}(\pm z)$ and their derivatives are suppressed, but positive arguments are associated with under the barrier ($k < k_o$) and negative with over ($k > k_o$). Now the continuity at the barrier maximum, where $k = k_o$ and $z_o = 0$, is fairly obvious: from the values of the Airy functions for zero argument (Section A2.6),

$$D(k = k_o) = \frac{4k_o f^{1/3}}{1.5839k_o^2 + 2f^{1/3}k_o + 0.84179f^{2/3}} \tag{18.58}$$

With Eq. (18.58), the Kemble approximation can be appreciated for the simplest imaginable barrier: the behavior of $D(k_o)$ is shown in Figure 18.6 for the usual metal parameters of $\mu = 7$ eV and $\Phi = 4.5$ eV. The thin gray line therein corresponds to $D = 1/2$: in the range of field typical of field emission, $D(k_o)$ conveniently hovers around the Kemble assumption ($D(k_o) \approx 0.5$) for the triangular barrier treated by Fowler and Nordheim [36]. Its doing so bodes well for the general thermal–field equation of Eq. (17.61) in the $Q \to 0$, or triangular barrier, limit.

The reasonableness of the Kemble approximation for the triangular barrier D_{tri} is consequently of a different character than for the rectangular barrier D_{rec}: at the barrier maximum, the transmission probability $D_{tri}(E)$ smoothly passes from $E < \mu + \Phi$ to $E > \mu + \Phi$ because the underlying Airy functions in the combinations[9] $\mathrm{Ai}^2 + \mathrm{Bi}^2$ and $(\mathrm{Ai}')^2 + (\mathrm{Bi}')^2$ are well-behaved, smooth, and finite as their argument passes from negative to positive values, as apparent in Figure 18.7 for metal-like parameters at several fields, generated using the algorithm of Section A3.10. In the tunneling regime,

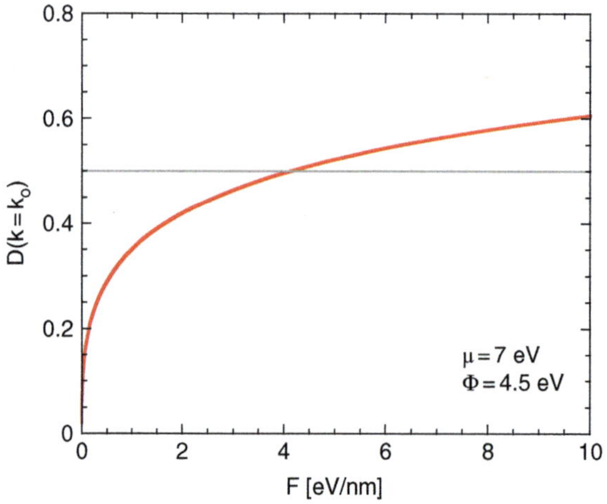

Figure 18.6 The transmission probability $D(k)$ evaluated at $k = k_o$ = the barrier height for a triangular barrier using Eq. (18.58). Metal-like parameters of $\mu = 7$ eV and $\Phi = 4.5$ eV are presumed. The thin gray line corresponds to $D = 1/2$, as suggested by the Kemble approximation.

[9] The argument x is being used for the general case, as is done in Section A2.6, to allow for positive and negative values, in contrast to z, which, in the present section, is only positive.

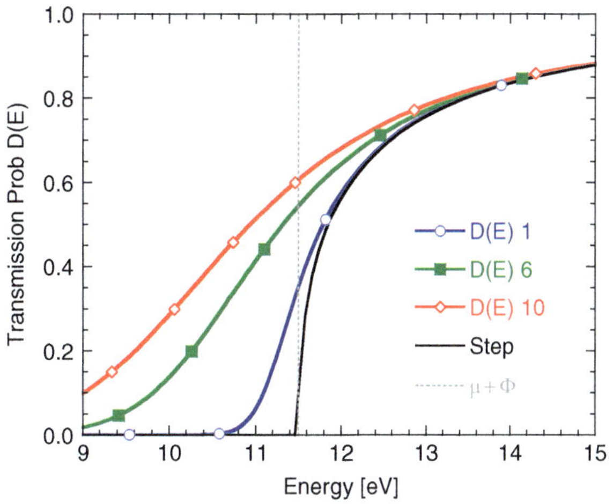

Figure 18.7 The transmission probability $D(E)$ for a triangular barrier: the numbers in the legend refer to F in eV/nm. Metal-like parameters of $\mu = 7$ eV and $\Phi = 4.5$ eV are presumed. The thin dashed gray line corresponds to $E = \mu + \Phi$. The black line labeled Step corresponds to Eq. (18.46).

the Gamow-like exponents of $\pm(2/3)z^{3/2}$ likewise approach the origin $z = 0$ in a smooth way, so that the linearization of $\theta_{tri}(E)$ there is well-behaved. The approximations made in the development of the general thermal–field equation of Chapter 17 if applied to a triangular barrier would therefore be reasonable, as they should be given that the triangular barrier is the $Q \to 0$ limit of the image charge barrier.

18.2.4 A Fowler–Nordheim refrain

The day returns, but nevermore
Returns the traveller to the shore,
And the tide rises, the tide falls.

– Henry Wadsworth Longfellow[10]

A *refrain* is a poetic device of repeating a line or phrase at the end of stanzas (as "and the tide rises, the tide falls" is in Longfellow's poem of the same name) in order to draw attention to an idea or notion through its repetition, sometimes to emphasize *differences* between sections of the poem, or to make the content of the refrain take on more meaning. The technique is likewise powerful in the development of methodologies to explore particularly useful problems in physics by tackling the *same* problem with *different* approaches, that is, through repetition with variation. The original Fowler–Nordheim equation for the triangular barrier, first encountered in Section 13.1, is a grand example of the same problem that can be considered in multiple ways. Treating it now will be through limiting forms of the components of $D(k < k_o)$ because of the magnitude of the quantities involved, in order to ascertain the form of the coefficient of $J_{FN}(F)$ that eludes identification using only the Kemble approximation and Gamow factor.

EXAMPLE: Determine the magnitude of the parameters k, k_o, f, κ, z and u for metal-like conditions ($\mu = 7$ eV and $\Phi = 4.5$ eV) under the assumption that the incident electron is at the Fermi level, the slope of the triangular barrier corresponds to 6 eV/nm, and x_o occurs at the left edge of the barrier for which $x_o = 0$.
SOLUTION: Use $2m/\hbar^2 = 26.247$ (1/eV nm^2). Then $f = 2mF/\hbar^2 = 157.48$ nm^{-3}, $k_F^2 = 2m\mu/\hbar^2 = 183.73$ nm^{-2}, $k_o^2 = 2m(\mu + \Phi)/\hbar^2 = 301.84$ nm^{-2}, and $\kappa^2 = k_o^2 - k_F^2 = 118.11$ nm^{-2}. From $z = \kappa^2/f^{2/3} = 4.0501$, it follows $u = z^{-3/2} = f/\kappa^3 = 0.12269$.

[10] Henry Wadsworth Longfellow, "The Tide Rises, the Tide Falls" in M.L. Rosenthal, *Poetry in English: An Anthology*. New York: Oxford University Press, 1987, p. 1880.

The smallness of u (equivalent to the largeness of z) entails that $H(\pm 1, u) \approx 2$ and $\text{Zi}(1, z) \gg \text{Zi}(-1, z)$. The leading order approximation of $\text{Di}(1, z) \approx -(f/u)^{1/3} = -\kappa$, as obtained from Eq. (18.35). And so, for Cu-like parameters,

$$D(k < k_o) \approx \frac{4kf^{1/3}}{\pi(k^2 + \kappa^2)Z(1, z)^2} = \frac{4k\kappa}{(k^2 + \kappa^2)} \exp\left(-\frac{4\kappa^3}{3f}\right) \tag{18.59}$$

But $\kappa^2 = k_o^2 - k_F^2$ when $k = k_F$, and so, using $k_o^2 = (2m/\hbar^2)(\mu + \Phi)$, $k_F^2 = 2m\mu/\hbar^2$, and $f = 2mF/\hbar^2$, it follows, as anticipated in Eq. (13.9), that

$$D_{FN}(E = \mu) = \frac{4\sqrt{\mu\Phi}}{(\mu + \Phi)} \exp\left(-\frac{4\sqrt{2m\Phi^3}}{3\hbar F}\right) \tag{18.60}$$

where the argument of $D(E)$ has returned to energy ($E = \hbar^2 k^2/2m$) in a one-dimensional problem, to facilitate the correspondence to Section 13.10. Recall that in the development of the triangular barrier Fowler–Nordheim equation (Eq. (13.10)), $-\ln(D(E))$ is taken as linear in E, so the weak variation of the coefficient of Eq. (18.59) is neglected, leaving only the exponential of $\theta_{tri}(E = \mu)$ of Eq. (13.5).

18.3 Triangular barrier: numerical

What a fine creed that is! So far as I can see, your religion consists of arithmetic. Men, I must say, can get some weird notions in their heads, and people who've studied a lot aren't always a lot wiser.

– Molière[11]

Generating $D(E)$ for a triangular barrier using Eq. (18.57) and the code in Section A3.10 relies on the numerical evaluation of the Airy functions $\text{Ai}(x)$ and $\text{Bi}(x)$ and their derivatives. These function calls are generally a part of programming languages such as MATLAB or the IMSL libraries used by FORTRAN. Two problems arise. First, one may not have access to either the programming languages or libraries that possess such capabilities, and, second, even if one does, the arguments of the Airy functions required (as can be the case for the more complex potential barriers composed of multiple segments) are such that numerical sensitivity or errors are introduced that can be overcome by reliance on the Zi functions, or rather $H_c(u)$ and $\text{Di}(c, z)$ if they were available. There are ways to obtain these functions, and there is advantage to doing so: raw computational evaluation using Airy functions themselves will require exponentially large (or small) terms, but clever recasting using H_c and Di allows for a more intuitive understanding using functions that do not become unmanageable, and finding them becomes a good exercise in how to deal with functions that are rather good at hiding their behavior.

18.3.1 Large argument evaluation

Life is what the least of us make most of us feel the least of us make the most of.

– Willard Quine[12]

The form of Eq. (18.26) indicates that $H_c(u)$ may be represented as a power series in u and that the coefficients of the power series can be related to each other, a technique known as the method of Frobenius [60]. That is, substituting

$$H_c(u) = \sum_{j=0}^{\infty} b_j u^j \tag{18.61}$$

into the differential equation and collecting terms associated with various powers of u requires that each coefficient of each different power of u separately be zero because the value of u is arbitrary. Of course, that is easier said than done.

---

[11] Molière, *Don Juan* (trans. Richard Wilbur). Harcourt: San Diego, 2001, Act 3, Scene 1, p. 73.

[12] Willard Quine, as quoted in John D. Barrow, *The Constants of Nature: From Alpha to Omega – the Numbers That Encode the Deepest Secrets of the Universe*, 1st American edn. New York: Pantheon Books, 2002, p. 141.

Proceeding slowly, insert the power series of Eq. (18.61) into Eq. (18.26) and consider all terms containing a particular b_j: it will be found that

$$36j(j-1)\,u^j b_j + 24j\,(3u-2c)\,u^{j-1} b_j + 5u^j b_j = 0 \tag{18.62}$$

Each one of these terms is in a summation over j, and so each summation can be re-indexed so that the powers of u are the same rather than the index of b_j. Doing so, one finds

$$(36j^2 - 36j + 5)b_{j-1}u^{j-1} = 48cjb_j u^{j-1} \tag{18.63}$$

an equation from which b_j can be found if b_{j-1} is known [85, 241]:

$$b_j = \frac{(6j-5)(6j-1)}{48cj} b_{j-1} \tag{18.64}$$

or, using the asymptotic forms [103] to find b_0 as done in Eq. (18.27),

$$b_j = \frac{(-c)^j}{\sqrt{\pi}} \frac{1}{3^{2j}(2j)!} \Gamma\left(3j + \frac{1}{2}\right) b_0 \tag{18.65}$$

$$b_0 = H_c(0) = \frac{1}{4}\left\{(c+3)(c^2+1) + 2\sqrt{2}(c-1)(c^2-1)\right\} \tag{18.66}$$

which corresponds to Eq. (A2.34). This is fairly nice until one thinks about it: for large j, $b_j \sim (3j/4)b_{j-1}$, which means b_j for j large behaves as $(3/4)^j j!$. That's not good. Eq. (18.61) is a devilish divergent series, reminiscent of those that Niels Abel inveighed against in Section 6.3. It does not mean that a Taylor series expansion about $u = 0$ with only the first few terms kept is precluded, but rather it means that the infinite series for finite u to find $H_c(u)$ should not be done because it will not converge.

If the truncated Taylor expansion leaves one ill at ease, there are other ways. The values of $H_c(0)$ and $H_c(1)$ are known (the latter from the next section) and so it is possible to use finite difference methods [242] to make headway in the solution of the differential equation for $H_c(u)$ [85]. In the general case, let $f(x)$ be defined on N equi-spaced points $x_j = j\Delta x$ such that $f(x_j) \equiv f_j$. Further, let the boundary conditions f_0 and f_{N+1} be known. The differential operators d/dx and d^2/dx^2 are then approximated in a second-order central differencing scheme discussed in Section A1.3.2, by

$$\frac{d}{dx} f(x) \approx \frac{1}{2\Delta x}\{f_{j-1} - f_{j+1}\} \tag{18.67}$$

$$\frac{d^2}{dx^2} f(x) \approx \frac{1}{\Delta x^2}\{f_{j-1} - 2f_j + f_{j+1}\} \tag{18.68}$$

Proceed now in two steps. First, use the finite difference representation of d/du and d^2/du^2 in Eq. (18.26). Second, group terms according to which term of $H_c(u_{j-1})$, $H_c(u_j)$, or $H_c(u_{j+1})$ that they multiply. Third, since $0 \le u \le 1$, therefore $\Delta u = 1/(N+1)$. An equation of the form

$$D_j H_c(x_{j-1}) + C_j H_c(x_j) + U_j H_c(x_{j+1}) = 0 \tag{18.69}$$

results, where the notation D, C, and U refer to "down", "center", and "up", in anticipation of the diagonals they will occupy in the matrix equation to be developed. Equations of the sort that Eq. (18.69) is can be put into a matrix equation where the two known values, $H_c(x_0)$ and $H_c(x_{N+1})$ serve as the boundary vector terms. The equation will resemble

$$\hat{O}|H\rangle = |B\rangle \tag{18.70}$$

where

$$\hat{O} = \begin{pmatrix} C_1 & U_1 & 0 & \cdots & 0 \\ D_2 & C_2 & U_2 & \cdots & 0 \\ \cdots & D_j & C_j & U_j & \cdots \\ 0 & \cdots & D_{N-1} & C_{N-1} & U_{N-1} \\ 0 & \cdots & 0 & D_N & C_N \end{pmatrix} \tag{18.71}$$

and

$$|H\rangle = \begin{pmatrix} H_c(u_1) \\ \vdots \\ H_c(u_j) \\ \vdots \\ H_c(u_N) \end{pmatrix}; \quad |B\rangle = \begin{pmatrix} -D_1 H_c(0) \\ \vdots \\ 0 \\ \vdots \\ -C_N H_c(1) \end{pmatrix} \tag{18.72}$$

where $u_0 = 0$ and $u_{N+1} = 1$ are used in the boundary value vector $|B\rangle$. The following relations hold for D_j, C_j and U_j:

$$U_j = 12\{-2c(N+1) + 3j(j+1)\} \tag{18.73}$$

$$C_j = -72j^2 + 5 \tag{18.74}$$

$$D_j = 12\{2c(N+1) + 3j(j-1)\} \tag{18.75}$$

where any constant factor (such as a denominator $(N+1)$) can be eliminated from both sides of the matrix equation given by Eq. (18.70).

Efficient algorithms exist to solve tridiagonal matrix equations, that is, to find $|H\rangle = \hat{O}^{-1}|B\rangle$ by using an algorithm akin to Gauss–Jordan elimination [236], a procedure well known from elementary linear algebra [91]. The process can be made even more elegant by noting that because there is no integer j for which C_j vanishes, numerical checks designed to prevent a divide by zero situation, which would otherwise arise if one (or more) of the C_j were to equal zero, do not have to be included. The streamlined tridiagonal matrix algorithm unfolds in the following sequence of steps:

Downward

Require: C, U, D and B are vectors of length N. X is a scalar.

$B_j = 0$; $B_1 = -D_1 H_c(0)$; $B_N = -U_N H_c(1)$
for $j = 2$ to $N-1$ **do**
  $X = C_j - D_j U_{j-1}$
  $C_j = C_j / X$
  $B_j = -D_j B_{j-1} / X$
  $B_N = (B_N - D_N B_{N-1})/(C_N - D_N C_{N-1})$
end for
$B_N = (B_N - D_N B_{N-1})/(C_N - D_N C_{N-1})$

Upward

$H_N = B_N$
for $j = 2$ to $N-1$ **do**
  $H_j = B_j - C_j H_{j+1}$
end for

where, first, in the downward elimination, the instruction $B_j = -D_j B_{j-1}/X$ is used instead of the general $B_j = (B_j - D_j B_{j-1})/X$ because all B_j are initially zero except at $j = 1$ or N, and second, the dummy term X need not be saved after it is used to update B_j. The labeling of "downward" and "upward" reinforces that the lower diagonal D is being eliminated in a Gauss–Jordan procedure followed by the upward elimination of the modified upper diagonal U: performing the same operations on the boundary vector B is what solves the matrix equation. The code is implemented in Section A3.11. The behavior is graphically shown in Figure 18.8.

As a consequence of the above numerical algorithms, $H_c(u)$ may be obtained for $u \leq 1$. Values of $H_c(u)$ for u between the u_j points, an example of which is in Table 18.1, are obtainable through interpolation [243], and finite difference methods may be employed to calculate derivatives of $H_c(u)$ so as to obtain $\mathrm{Di}(c, z(u))$ from Eq. (18.34). The details of doing so are where the pain is [85, 197, 199]. If the u_j are evenly spaced ($u_{j+1} - u_j = \Delta u$), and $h_j = H_c(u_j)$, then, for example, a simple

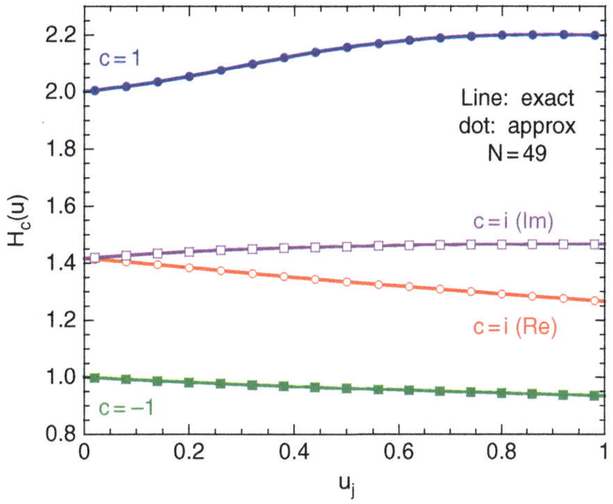

Figure 18.8 The functions $H_1(u)$, $H_{-1}(u)$, $\Re(H_{-i}(u))$, and $\Im(H_{-i}(u))$, where $\Re$ and $\Im$ denote real and imaginary parts.

Table 18.1 Values of $H_c(u)$, where $\Re$ and $\Im$ denote real and imaginary parts.

u_j	$H_1(u)$	$H_{-1}(u)$	$\Re[H_i(u)]$	$\Im[H_i(u)]$
0.00	2.0000	1.00000	1.4142	1.4142
0.10	2.0229	0.99031	1.3985	1.4276
0.20	2.0524	0.98178	1.3817	1.4382
0.30	2.0895	0.97411	1.3648	1.4463
0.40	2.1260	0.96711	1.3482	1.4523
0.50	2.1555	0.96067	1.3324	1.4567
0.60	2.1767	0.95468	1.3172	1.4597
0.70	2.1901	0.94907	1.3029	1.4618
0.80	2.1971	0.94381	1.2892	1.4630
0.90	2.1992	0.93884	1.2763	1.4636
1.00	2.1975	0.93413	1.2640	1.4637

quadratic interpolation can make use of[13]

$$h(u_j + p\Delta u) = -\frac{1}{2}p(1-p)h_{j-1} + (1-p^2)h_j + \frac{1}{2}p(1+p)h_{j+1} \tag{18.76}$$

which has the required behavior that for $p = -1, 0, 1$, then $u \to u_{j-1}, u_j, u_{j+1}$ and $h(u) \to h_{j-1}, h_j, h_{j+1}$, as they should.

18.3.2 Small argument evaluation

> **Logic**, *n. The art of thinking and reasoning in strict accordance with the limitations and incapacities of the human misunderstanding.*
>
> – Ambrose Bierce[14]

One might suspect that a comparable ordeal to that of evaluating $H_c(u)$ awaits for the calculation of $G_c(z)$, but instead existing small argument series expansions of Airy functions [103] that are suitably truncated can be used [85]. The algorithms here preserve the accuracy required to perform numerical tunneling calculations when the time comes. For

[13]See the Lagrange three-point interpolation formula in Davis and Polonsky, Chapter 25 in *Numerical Interpolation, Differentiation, and Integration* [103]. Although more complex formulae may be used with different orthogonal polynomials, the least tiring way to achieve greater accuracy is simply to decrease Δu by increasing the number of u_j.

[14]Ambrose Bierce, *The Devil's Dictionary.* New York: Dover Publications, 1993, p. 79.

small argument such that $-1 \leq x \leq 1$ (where $x = \pm z$)

$$\mathrm{Ai}(x) = \frac{P(x)}{3^{2/3}\Gamma(2/3)} - \frac{S(x)}{3^{1/3}\Gamma(1/3)} \tag{18.77}$$

$$\mathrm{Bi}(x) = \frac{P(x)}{3^{1/6}\Gamma(2/3)} + \frac{S(x)}{3^{-1/6}\Gamma(1/3)} \tag{18.78}$$

The truncated polynomials $P(x)$ and $S(x)$ are then given by

$$P(x) = 1 + \sum_{k=1}^{N} \frac{x^{3j}}{(3k)!}\left(\prod_{j=1}^{k}(3j-2)\right) \tag{18.79}$$

$$S(x) = x\left[1 + \sum_{k=1}^{N} \frac{x^{3j}}{(3k+1)!}\left(\prod_{j=1}^{k}(3j-1)\right)\right] \tag{18.80}$$

In both cases, five-digit accuracy is obtained by setting $N = 4$. Derivatives of the Airy functions are then obtained by finding $P'(x) = \partial_x P(x)$ and $S'(x) = \partial_x S(x)$, which in turn enable the evaluation of $\mathrm{Di}(c, z)$, or the functions $G_c(z)$ of Figure 18.9 and Table 18.2.

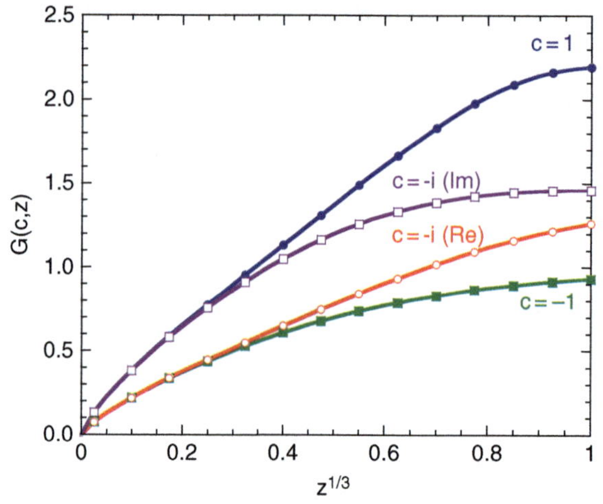

Figure 18.9 The functions $G_1(z)$, $G_{-1}(z)$, $\Re(G_{-i}(z))$, and $\Im(G_{-i}(z))$. The value of the curves as they exit the right-hand side of the graph correspond to the values of the curves exiting the right-hand side of Figure 18.8.

Table 18.2 Values of $G_c(z)$.

$z^{1/3}$	$G(1,z)$	$G(-1,z)$	$\Re(G(-i,z))$	$\Im(G(-i,z))$
0.00	0.00000	0.00000	0.00000	0.00000
0.10	0.38791	0.22365	0.22396	0.38736
0.20	0.65542	0.37438	0.37828	0.64831
0.30	0.89836	0.50160	0.51764	0.86777
0.40	1.1353	0.61006	0.65120	1.0523
0.50	1.3736	0.70072	0.78123	1.2012
0.60	1.6113	0.77434	0.90574	1.3133
0.70	1.8350	0.83231	1.0202	1.3899
0.80	2.0220	0.87667	1.1199	1.4357
0.90	2.1471	0.90981	1.2014	1.4577
1.00	2.1975	0.93413	1.2640	1.4637

CHAPTER 19

Transfer matrix approach

"This earthquake is nothing new" replied Pangloss, "the town of Lima in America experienced the same shocks last year. The same causes produce the same effects. There is certainly a vein of sulphur running under the earth from Lima to Lisbon." "Nothing is more likely," said Candide, "but oil and wine, for pity's sake!" "Likely!" exclaimed the philosopher. "I maintain it's proved."

– Voltaire[1]

The 8.5–9.0 magnitude earthquake on All Saints' Day (November 1, 1755) some 200 km southwest of the coast of Portugal, and the subsequent tsunami it generated, all but obliterated Lisbon and caused the death of several tens of thousands. Voltaire used the occasion to savagely lampoon the adage of the German polymath Gottfried Leibniz (the co-inventor with Newton of calculus), that this was "the best of all possible worlds" by making Dr. Pangloss, who pontificated about such hokum, an object of scorn. A fatal defect of Pangloss (aside from his insufferable optimism) was a tendency to draw conclusions on flimsy grounds. That is something to be avoided in the claims about how $D(k)$ behaves, particularly for rounded barriers.

Specifically, it may seem reasonable to expect that the qualitative behavior of the transmission probability for a parabolic barrier lies between the rectangular barrier (Figure 18.2) and the triangular barrier (Figure 18.7) results, because as $E(k)$ approaches the barrier maximum the parabolic Gamow factor θ decreases faster than for the square barrier but slower than for the triangular barrier. It may then seem likely that the parabolic barrier transmission probability is described by the Kemble approximation. But one should not end there with such a Panglossian claim, even though the relation between the parabolic barrier and the Kemble approximation was behind the general thermal–field equation of Eq. (17.61), largely because of the similarity of the image charge potential (at its apex) to a parabolic potential. Although Fröman and Fröman [244] show that the parabolic potential transmission probability *is* given by the Kemble form (their Eq. 9.12b), the mechanics to do so may leave one crying for more wine than oil. What can be done without suffering through an analysis more onerous than that deployed for the Airy functions?

Numerically, there are some simple ways to proceed, and they will be referred to collectively as the transfer matrix approach (TMA) methods. There are two flavors: the more palatable approach [183, 245–248] uses a generalization of the plane wave basis behind Eq. (18.3), and is suggested by Figure 19.1. The Airy function generalization [85, 196, 197, 249–252] is, however, better for general potentials, particularly those for emission into vacuum under an applied field, and require far less discretization of the potential for regions where the potential can be approximated by a linear function. Thus, the Airy method is more powerful and accurate, but takes more work to set up. The numerical results of these methods will add gravitas to the otherwise Panglossian "proof" based on the rectangular and triangular barrier results that the Kemble approximation for a parabolic barrier is good. It will become even more useful and even obligatory when more complex ion and resonant potentials are undertaken.

19.1 Plane wave transfer matrix

Paradox though it may seem – and paradoxes are always dangerous things – it is none the less true that Life imitates art far more than Art imitates life.

– Oscar Wilde[2]

Representing a potential region as closely packed rectangular barriers of various sizes enables the wave function in those regions to be represented as plane waves, which are easy to manipulate, allowing for straightforward reasoning about the nature of tunneling and resonances. Proceed systematically. As in Figure 19.1, let a general potential $V(x)$ be defined at the points x_j (not necessarily equi-spaced) such that $V_j = V(x_j)$. Let the potential between the points x_j and x_{j+1} be constant and equal to V_j. Schrödinger's equation then has solutions of the form $\psi_k(x) = \exp(\pm i\kappa(k)x)$ or $\exp(\pm\kappa(k)x)$,

[1] Voltaire (*trans.* John Butt), *Candide*. Penguin Books, Middlesex, England, 1947. page 34.

[2] Oscar Wilde, *The Complete Collection (the Picture of Dorian Gray, 14 Short Stories, 9 Plays, All Poems, Selected Essays and Letters)*, Kindle eBook, 2013, Essay: *The Decay of Lying (1889)*.

Introduction to the Physics of Electron Emission, First Edition. Kevin L. Jensen.

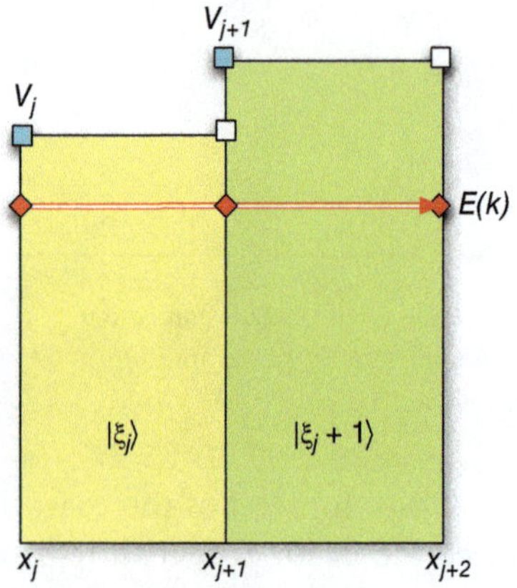

Figure 19.1 Labeling associated with a two-region potential barrier with constant potential segments. The $|\xi_j\rangle$ correspond to the two-component vectors containing the transmission (t_j) and reflection (r_j) coefficients. The blue squares identify the potential points $V_j = V(x_j)$. The red diamonds represent where the matrix equations $\hat{M}|\xi\rangle$ are matched by matrix equations analogous to Eq. (18.2). Compare to Figure 19.9.

for electrons with energy $E > V_j$ or $E < V_j$, respectively, and the wave function will be a combination of them such that the wave function and its first derivative are elegantly expressed in the matrix equation of Eq. (18.1), with the coefficients inside $|\xi_n\rangle$ being such that the $\psi(x)$ and $\partial_x\psi(x)$ are continuous at $x = x_j$. Although plane waves are simple, that is not the only choice: two other linearly independent solutions are sines and cosines [253], which entail different advantages but are otherwise employed in much the same way.

If there are n flat segments that approximate $V(x)$, then the generalization of Eq. (18.3) will be (using a more attentive notation)

$$|\xi_0\rangle = \left\{ \prod_{j=0}^{n} \hat{S}(\kappa_j, \kappa_{j+1}; x_{j+1}) \right\} |\xi_{n+1}\rangle \equiv \mathbb{S}_n |\xi_{n+1}\rangle \tag{19.1}$$

where x_{j+1} is the location where the matching of the wave function and its first derivative is occurring, κ_{j+1} refers to κ to the right of it, and κ_j to the left, where

$$\hbar\kappa_j = \sqrt{2m\left(E - V\left(x_{\leq}x < x_{j+1}\right)\right)} \equiv \sqrt{2m\left(E - V_j\right)} \tag{19.2}$$

When $E < V_j$, then $\kappa_j \rightarrow i\kappa_j$. Although the notation $c = \pm 1, \pm i$ might be introduced as for the *Zi* functions, for the plane wave approach, such extra bookkeeping turns out to be unnecessary because $\kappa(x)$ does not switch from real to imaginary (or vice versa) between x_{j+1} and x_j when there is no field term. Consequently, $\hat{S}$ can be expressed in terms of the center and difference wave numbers $K = (k_a + k_b)/2$ and $\Delta k = (k_a - k_b)/2$, in terms of which

$$\hat{S}(k_a, k_b; x) \equiv \frac{1}{2k_a} \begin{pmatrix} Ke^{-i\Delta kx} & \Delta k\, e^{-iKx} \\ \Delta k\, e^{iKx} & Ke^{i\Delta kx} \end{pmatrix} \tag{19.3}$$

The calculation of the matrix elements is straightforward (an example of the numerical algorithm is in Section A3.12.1). Once the product of matrices that defines $\mathbb{S}_n$ is computed, then [196]

$$D(k) = |t(k)|^2 \frac{k_{trans}}{k_{inc}} \rightarrow \frac{1}{[\mathbb{S}_n]_{1,1}} \frac{k_{trans}}{k_{inc}} \tag{19.4}$$

where the notation $[\mathbf{M}]_{i,j}$ denotes the ith row, jth column of the matrix $\mathbf{M}$, k_{inc} is the incident wave number (left-hand side) and k_{trans} is the transmitted wave number (right-hand side) if both left and right are at constant potential. It is convenient to choose the potential of the left- and right-hand sides to be the same ($V(\pm\infty) = 0$), so that $k_{trans} = k_{inc} = k$, and so $D(k) = |t(k)|^2$. If the potential is under bias (as it would be, for example, for a resonant tunneling diode (RTD) through which current is flowing), then the factor k_{trans}/k_{inc} would have to be reinserted.

To evaluate the matrices at x_j, multiply them in the correct order, and use the appropriate boundary conditions expressed in the $|\xi\rangle$ vectors (an example of the numerical algorithm is in Section A3.12.2). Applied to the rectangular barrier leading to Eqs (18.10) and (18.11), a single barrier is specified by two points (where it begins and where it ends) and one potential (how high it is). The numerical evaluation of $D(k)$ then requires three x_j and a corresponding number of V_j, as in Figure 19.2. Using Eq. (19.4) and the algorithm of Section A3.12.2 results in Figure 19.3, showing that the rectangular barrier results are recovered.

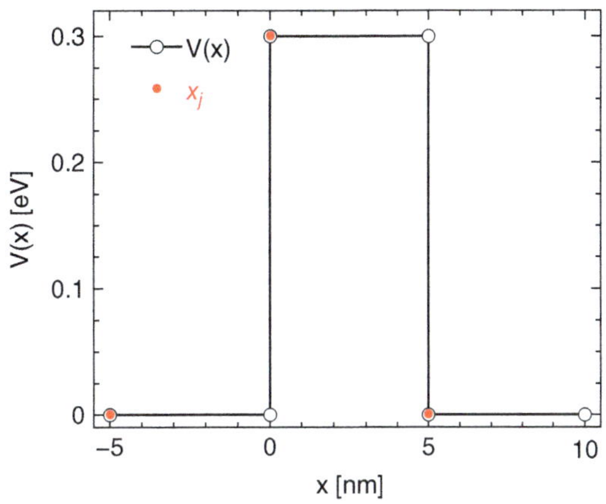

Figure 19.2 The potential points required for a rectangular barrier. Black circles with red centers denote the (leftmost) positions and potentials. Open circles designate the other (rightmost) positions and potentials of each constant potential segment. A point prior to the onset of the barrier at a flexible location is needed, and the potential after the barrier is chosen to be 0, although it need not be.

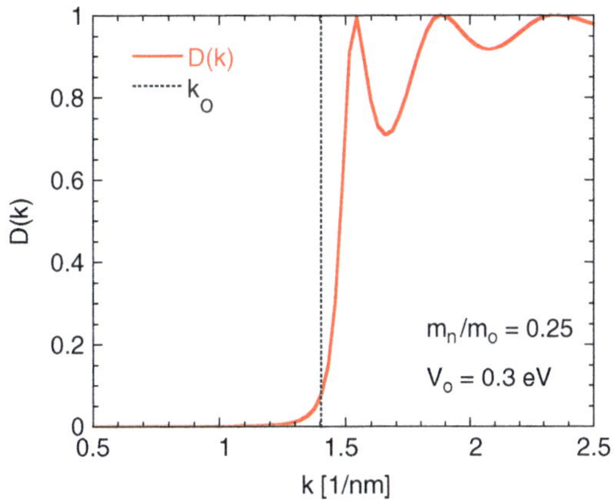

Figure 19.3 The transmission probability $D(k)$ associated with the potential profile of Figure 19.2 for $m_n/m = 0.25$ (where m_n is the effective mass) and $V_o = 0.3$ eV.

19.1.1 Kemble approximation revisited

Let your reason serve/To make the truth appear where it seems hid…

– William Shakespeare[3]

The single barrier can now be altered by putting several barriers alongside each other and tailoring their heights for some desired profile. When their heights vary quadratically with how far they are from the center of the barrier, the parabolic barrier behind the Kemble approximation is modeled. Thus, let the barrier envelope be a parabola given by

$$V(x) = V_o\left\{1 - \left(\frac{x}{L}\right)^2\right\} \tag{19.5}$$

for a barrier of width $2L$ at the base. Assuming that 16 barriers are placed adjacently, the left boundary of each barrier being at x_j and the height of the barrier being $V_j \equiv V(x_j)$, the barrier constructed of flat segments will resemble Figure 19.4. Of course more segments could be used, and therefore greater fidelity to the parabolic barrier achieved, but 16 segments is adequate for present purposes.

The $D(k)$ evaluated using the TMA is to be compared to the Kemble approximation using the quadratic Gamow factor $\theta_{qua}(E)$ of Eq. (11.19). The present formulation differs slightly (Eq. (11.19) measured the barrier height from μ and took the base width as L_o, whereas now the barrier height is V_o and the base width is $2L$). Those small changes result in $\theta_{qua}(E)$

[3]Ref. [37]: *Measure for Measure*, V.1.65–66.

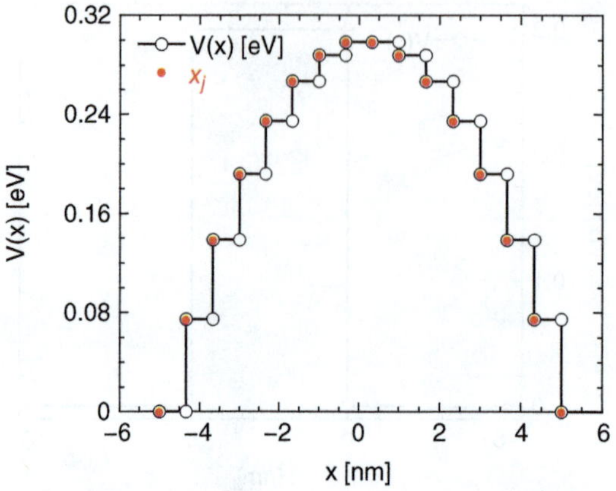

Figure 19.4 A multiple segment barrier where each segment is of the same width and the heights follow the parabolic profile of Eq. (19.5). As with Figure 19.2, open circles identify changes in $V(x)$, red dots identify only the left boundary of a segment.

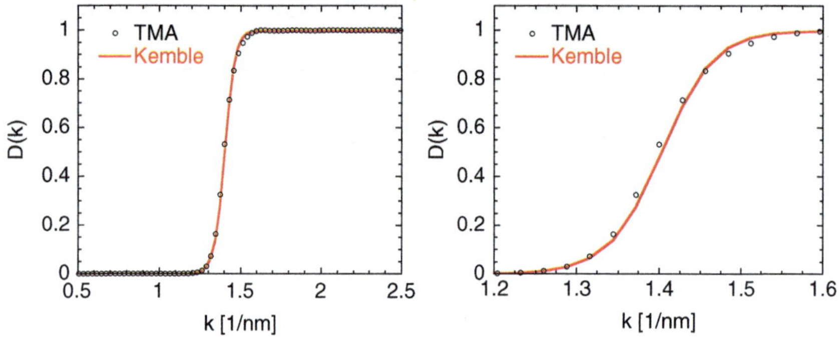

Figure 19.5 Comparison of $D(k)$ calculated using the TMA applied to $V(x_j)$ of Figure 19.4 and the Kemble approximation of Eq. (17.9) using Eq. (19.6) for θ_{qua}, for $r = m_n/m = 0.25$, $L = 5$ nm, $V_o = 0.3$ eV. Left, wide range of k; right, focus on $k \approx k_o$.

becoming

$$\theta_{qua}(E) = \frac{\pi L}{\hbar}\left(\frac{2m}{V_o}\right)^{1/2}(V_o - E) \tag{19.6}$$

where the electron mass $m \to m_n$ is an effective electron mass, if transport is occurring in bulk material rather than in vacuum. Evaluating $D(k)$ using the TMA and comparing it to the Kemble approximation $1/(1 + \exp(\theta_{qua}(E)))$ results in Figure 19.5. The agreement is fairly good, even for the lumpy, ill-tempered potential shown in Figure 19.4, which gives fairly reasonable agreement with the Kemble form. Increasing the number of rectangular barriers that make up $V(x_j)$ improves the correspondence. Therefore, without having actually proven the conjecture, the grounds for believing that the Kemble approximation works well for parabolic potentials are substantially strengthened.

One matter that creates apprehensions is why, after long using metal work functions of several eV and a free space electron rest mass, would one switch to smaller effective masses and shorter barriers? Superficially, such changes help spread out the $D(k)$ that results, making it more appealingly rounded near $k \approx k_o$ and less step-like as one crosses from under the boundary to over it. But there is another reason. The TMA, as used here, owes much to the very interesting study of superlattices by Tsu and Esaki [183], the two-barrier version of which is the RTD.[4] RTDs typically have barriers and wells composed of, for example, GaAs/AlGaAs layers, for which m_n/m, V_o and L are small. Generic values of those parameters for semiconductors are used here. The interest in RTDs is because they exhibit *negative differential resistance*, wherein an increase in voltage across the barriers results in a seemingly paradoxical decrease in current through it: the increase is caused by energy levels in the well region for which $D(k)$ approaches unity for $k < k_o$ [126, 130, 255, 256]. A two-barrier potential is shown in Figure 19.6, created using the methods of Section A3.12.4. A barrier such that $V_o = 0.3$ eV and a rest mass of $m_n/m = 0.25$ corresponds to $k_o = 1.403$ 1/nm, so the sharp peak ($D(k) \to 1$) at $k = 1.3609$ 1/nm

[4] Esaki shared the Nobel Prize in 1973 for his work on tunneling and resonance behavior. Tellingly, his Nobel lecture [254] begins with the triangular barrier of the Fowler–Nordheim equation.

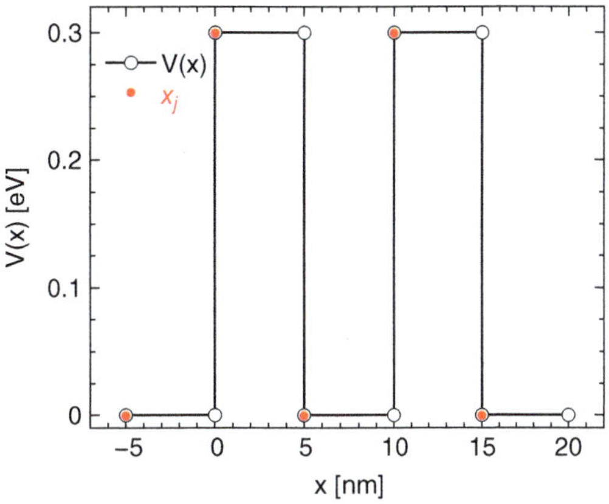

Figure 19.6 An RTD at zero bias. As with Figure 19.2, open circles identify changes in $V(x)$, red dots identify only the left boundary of a segment.

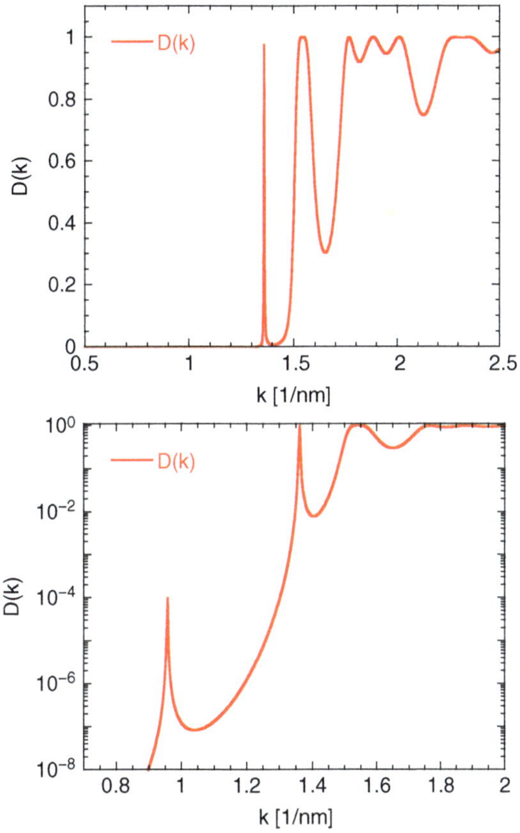

Figure 19.7 $D(k)$ calculated using the TMA applied to $V(x_j)$ of Figure 19.6 for $r = m_n/m = 0.25$, $L = 5$ nm, $V_o = 0.3$ eV. Top, linear y-axis scale to show the oscillatory behavior for $k > k_o$ and the suggestion of bands, or regions, where $D(k)$ drops precipitously; bottom, log-scale showing the resonance states for $k < k_o$, where narrow spikes in the transmission probability appear.

indicates a resonance level in Figure 19.7. Others are visible in the second (semi-log plot) figure: the resonances appear as narrow Lorentzians (Section A2.7). The peaks at the lower k resonances are generally larger than shown, a consequence of the coarseness of the spacing of points chosen in the figures. The behavior as parameters are varied exhibits a fascinating evolution, particularly when the number of barriers, the heights, and the right-hand boundary value are changed.

So far, $D(k)$ has been investigated under no bias conditions (the right boundary potential is at the same level as the left boundary potential). An RTD under zero bias does not have a net current flowing through it. In contrast, emission is enhanced when an adsorbed atom or incident ion affects a field emission site by providing a resonance level and leading to increased tunneling effects [257–259]. The RTDs differ from field emitters in that there is a supply function from both

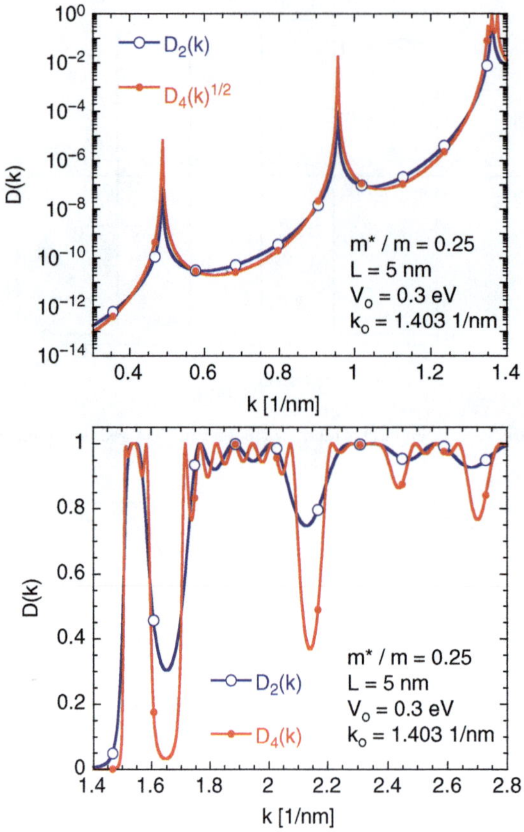

Figure 19.8 A comparison of the two-barrier and four-barrier superlattice potentials. Top, for $k < k_o$ observe that $D_4(k)^{1/2} \approx D_2(k)$; bottom, for $k > k_o$ observe that regions, or bands, appear to be forming, an effect that becomes more pronounced as the number of barriers increases.

sides of an RTD (recall the Tsu–Esaki formula of Eq. (11.2)), and so at zero bias, the same current flows in each direction ($k > 0$ flows to the right, $k < 0$ to the left). Under bias, the right-hand side is brought to a lower potential energy level, and less current flows through the double barrier from right to left. As long as electrons from the left-hand side are at an energy such that they can resonantly tunnel, they will contribute significantly to the current. When the resonant level drops below the left-hand side conduction band minimum, the contribution of the resonant level declines and so the current versus voltage declines, that is, the differential resistance (change in current with applied bias) becomes negative rather than positive as it normally is.

More unusual behavior can be brought about. For a three-barrier potential, if the wells are such that the resonant levels align, then the transmission through the resonant level will continue unabated, but the addition of more barriers means that the sequential tunneling (first through one barrier, then the next, and so on) will be increasingly suppressed. Doubling then quadrupling the barriers and watching the behavior of $D(k)$ demonstrates the effect nicely [85], but a simple case catches the effect. Taking the barrier of Figure 19.6 and doubling it (accomplished by setting $N_b = 4$ in Algorithm A3.12.4), it is expected that $D_4(k)$ (the subscript indicating the number of barriers) would be comparable in magnitude to $D_2(k)^2$ or, equivalently, $D_4(k)^{1/2} \approx D_2(k)$ for $k < k_o$ and away from the resonance values, as seen in Figure 19.8.

But there is more. Above the barriers ($k > k_o$), greater structure is put on $D(k)$ as the number of barriers increases. Transmitted and reflected waves interfere with each other so that for some values of k, $D(k)$ drops, as shown in the bottom plot of Figure 19.8. When many barriers are present, the size of the drop increases, and regions, or "bands", open up where transmission occurs between regions where it is suppressed, and anticipates the Kronig–Penny model of Section 21.3.

19.1.2 Gamow factor revisited

...there is a pleasure in recognizing old things from a new point of view.

– Richard P. Feynman [260]

Bizarre quantum effects should not overshadow the relationship between the behavior of $D(k)$ revealed by the TMA compared to the fluid-like description of Section 11.2 introducing the Gamow factor. It may not be clear how they are related, but there is a simple way to see it. Reconsider the TMA for the single rectangular barrier. Let the left side of the barrier at $x = 0$ be labeled by a 1 index, and the right at $x = L$ by a 2 index. Then, writing out the matrix equation in its entirety without the aesthetic simplifications of Eqs (19.1) and (19.3):

$$\begin{pmatrix} 1 \\ r \end{pmatrix} = \begin{pmatrix} 1 & 1 \\ ik & -ik \end{pmatrix}^{-1} \begin{pmatrix} 1 & 1 \\ \kappa & -\kappa \end{pmatrix} \times \begin{pmatrix} e^{\kappa L} & e^{-\kappa L} \\ \kappa e^{\kappa L} & -\kappa e^{-\kappa L} \end{pmatrix}^{-1} \begin{pmatrix} e^{ikL} & e^{-ikL} \\ ike^{ikL} & -ike^{-ikL} \end{pmatrix} \begin{pmatrix} t \\ 0 \end{pmatrix} \tag{19.7}$$

but instead of grouping the first two matrices into $\hat{S}_1$ and the second two into $\hat{S}_2$, consider the middle two matrices and call them $\mathbf{B}$, or

$$\mathbf{B} = \begin{pmatrix} 1 & 1 \\ \kappa & -\kappa \end{pmatrix} \begin{pmatrix} e^{\kappa L} & e^{-\kappa L} \\ \kappa e^{\kappa L} & -\kappa e^{-\kappa L} \end{pmatrix}^{-1} = \begin{pmatrix} \cosh(\kappa L) & -\kappa^{-1}\sinh(\kappa L) \\ -\kappa \sinh(\kappa L) & \cosh(\kappa L) \end{pmatrix} \tag{19.8}$$

When $\kappa L \gg 1$, then $\cosh(\kappa L) \approx \sinh(\kappa L) \approx e^{\kappa L}/2$, and so

$$\begin{pmatrix} \cosh(\kappa L) & -\kappa^{-1}\sinh(\kappa L) \\ -\kappa \sinh(\kappa L) & \cosh(\kappa L) \end{pmatrix} \rightarrow \frac{e^{\kappa K}}{2\kappa} \begin{pmatrix} \kappa & -1 \\ -\kappa^2 & \kappa \end{pmatrix} \tag{19.9}$$

Consequently, the term $e^{\kappa L}$ appears as an overall multiplier, and so, apart from the k and κ odds and ends from the other matrices, $D_1(k) = |t(k)|^2 \propto e^{-2\kappa L}$, where the index on D is for the number of rectangular barriers from which it is composed. But then, a barrier represented by two adjacent rectangular barriers would have a similar factor for each, and so $D_2(k) \propto e^{-2\kappa_1 L_1 - 2\kappa_2 L_2}$. Generalizing to N rectangular barriers,

$$D_N(k) \propto \exp\left(-2\sum_{j=1}^{N} \kappa_j L_j\right) \tag{19.10}$$

In the limit that all the $L_j \rightarrow dx$ and the summation moves over to an integral as per Eq. (A1.10), then the argument of the exponent simply becomes the familiar Gamow factor of Eq. (11.14). Thus, the Gamow factor naturally emerges from the TMA, where it was hidden all along.

19.2 Airy function transfer matrix

...mathematicians may be completely repelled by the liberties taken here...just as a poet often has license from the rules of grammar and pronunciation, we should like to ask for "physicists' license" from the rules of mathematics in order to express what we wish to say in as simple a manner as possible.

– Richard P. Feynman [261]

The Airy TMA of Eqs (18.41) and (18.51) is an improvement on the plane waves used in the S matrix in Eq. (19.3). This is because not only can trapezoidal segments better represent a general potential barrier than rectangular segments, but also the presence of a field at the right-hand boundary can be accounted for. The cost of that fidelity is that attention must be paid to when passage from under the barrier to over it occurs, or vice versa, as in Figure 19.9. When using the *Zi* functions this necessitates the introduction of the transition matrix of Eq. (18.50). The transition matrices and their bookkeeping, coupled with the greater effort required to numerically evaluate $Zi(c, z)$, seems to tilt preferences to the plane wave TMA for the work-adverse. Those with greater fortitude realize that the plane wave approach requires many potential segments to obtain only reasonable correspondence for curved potentials, but with the Airy approach, a few segments are generally enough to achieve good accuracy. As a trivial example, the number of segments for an RTD under bias for the plane wave TMA requires many potential segments as it approximates linear regions as a staircase potential, but the Airy approach need not require any more segments than the unbiased case and remains exact because all the segments of the potential are linear in field.

The task is therefore to find the matrix replacement using Airy functions to take over for $\mathbf{S}$ in Eq. (19.3). The approach will make full use of the *Zi* functions of Eq. (18.24). $\mathbf{S}_a$ will be the product of two matrices, or $\mathbf{S}_a = \mathbf{M}_l \cdot \mathbf{M}_r$, where l (for left) indicates the linear region before x_j (the $j-1$ region) and r (for right) the linear region after x_j (the j region).

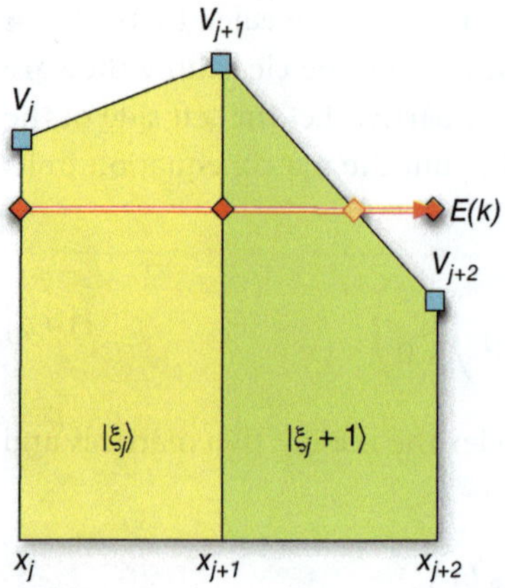

Figure 19.9 Labeling associated with a two-region potential barrier with linear potential segments. The $|\xi_j\rangle$ correspond to the two-component vectors containing the transmission (t_j) and reflection (r_j) coefficients. The blue squares identify the potential points that define k_o, f, c, and s for each region. The red diamonds represent where the matrix equations $\hat{M}|\xi\rangle$ are matched by matrix equations analogous to Eq. (18.2). The orange diamond represents where a transition matrix (Eq. (18.50), in this case for under-to-over) must be inserted. Compare with Figure 19.1.

Consider the components for the j region (for the $(j-1)$ region, the indices are simply altered by $j \to (j-1)$). Introducing the nomenclature (designed to facilitate coding):

- $f_j = 2mF_j/\hbar^2$; as a practical matter, if $F_j < 0.005$ eV/nm, then $f_j \to 0$ and the plane wave representation is used
- $c_j = 1$ if $(k_j^2 + s_j f_j(x_{j+1} - x_j)) < k^2$ and i otherwise.
- $q_j = s_j f_j^{1/3}$
- $z_j = |k_j^2 + s_j f_j(x_{j+1} - x_j)|/f_j^{2/3}$
- $R_j = \exp[(4/3)c_j z_j^{3/2}]$
- $C_j = 2\pi/(c_j s_j f_j^{1/3}(1 - 3c_j^2))$

in terms of which

$$\mathbf{M}_l = C_{j-1}\exp\left(-\frac{2}{3}c_{j-1}z_{j-1}^{3/2}\right) \times \begin{pmatrix} q_{j-1}Gi'(-c_{j-1}, z_{j-1}) & -Gi(-c_{j-1}, z_{j-1}) \\ -q_{j-1}R_{j-1}Gi'(c_{j-1}, z_{j-1}) & R_{j-1}Gi(c_{j-1}, z_{j-1}) \end{pmatrix} \tag{19.11}$$

$$\mathbf{M}_r = \exp\left(\frac{2}{3}c_j z_j^{3/2}\right)\begin{pmatrix} Gi(c_j, z_j) & R_j Gi(-c_j, z_j) \\ q_j Gi'(c_j, z_j) & q_j R_j Gi'(-c_j, z_j) \end{pmatrix} \tag{19.12}$$

where the "reduced" $Gi(c, z)$ functions are defined by

$$Gi(c, z) \equiv Zi(c, z)\exp\left(-\frac{2}{3}cz^{3/2}\right) \tag{19.13}$$

Lastly, should $c = 1 \to i$ (under to over) or $c = i \to 1$ (over to under), then $\mathbf{M}_r \to \mathbf{M}_r \cdot \mathbf{S}(x_o)$, where $\mathbf{S}(x_o)$ is the matrix of Eq. (18.50). As a consequence of using the "reduced" $Gi(c, z)$ functions, an overall multiplicative exponential term appears in front of each matrix, the form of which is instrumental in leading to the Gamow factor in the approach of Sections 19.1.2 and 19.2.2 below.

Assuming that the matrix evaluations are performed and the matrices multiplied with care to include instances where over to under (or vice versa) crossings are accounted for, then an N-component potential barrier will give rise to a transmission coefficient $t_N(k)$. Now, though, a field f_N exists on the right-hand boundary. The evaluation of $D(k)$ thereby becomes more involved compared to the plane wave version of Eq. (19.4). Evaluating j_{trans} using Eq. (9.61) and comparing it to j_{inc} for the case where $\psi_N(k)$ is a sum of Airy functions corresponding to an outgoing wave as in Eq. (18.22), the transmission probability is now [199]

$$D(k) = \left(\frac{f_N^{1/3}}{\pi k}\right)|t_N(k)|^2 \tag{19.14}$$

When applied to the parabolic potential or, more importantly, the image charge potential, to take full advantage of the economy of the Airy approach requires an artful placement of where the segments are and how wide they should be made. And it is an Art. A few general guidelines encapsulate the aesthetics involved: (*i*) the x_j points should be more densely placed when the potential varies strongly (e.g., near the origin for the image charge barrier), (*ii*) placed sparsely when the potential is either constant or linear, and (*iii*) placed carefully where the potential undergoes a maximum or a minimum (ideally on the maximum or minimum). Although the $\{x_j, V_j\}$ could be placed mechanically, densely, and uniformly without an unacceptable burden on modern desktop computers, it sacrifices elegance and economy of execution.

In the case of the image charge potential of Eq. (11.25), the zeros and maximum are analytically known from Eq. (11.26) and its cohorts. For a more general potential, finding the zeros of a potential or field can make use of the methods of Section A3.7. A method employed below for the image charge is to find the $\{x_j\}$ such that the potential is reasonably linear between each x_j and x_{j+1}. If the artistry of the placement is well considered, then even a small number of points is enough for a good account of $D(k)$ and therefore $J(F, T)$.

Consider the Airy method applied to the field emission problem represented by Figure 19.10: for metal-like and typical environmental parameters, a calculation of $D(k)$ and its numerical integration with the supply function $f(E_k)$ of Eq. (11.5) results in Figure 19.11. As far as the big picture is concerned, the Fowler–Nordheim (FN) equation of Eq. (13.25) gives reasonably good correspondence, but a comparison of the numerical values reveals that the FN equation is consistently smaller in the range of 20–50% (performing better at higher F). This is significant for an odd reason: much has been written about the Nordheim functions $v(y)$ and $t(y)$ and how to obtain them to great precision, but such efforts are

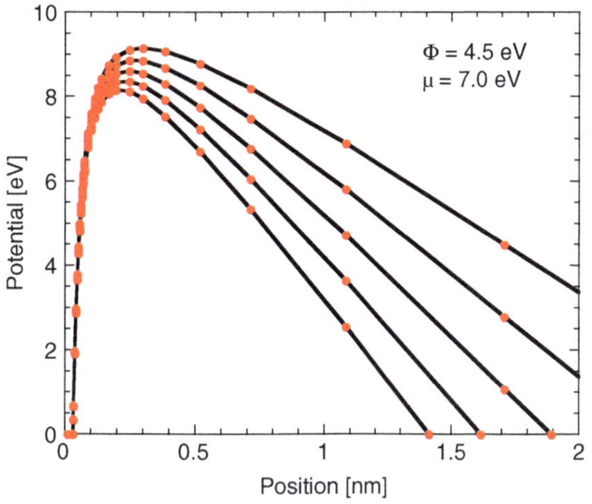

Figure 19.10 The image charge barrier rendered as linear segments between points chosen such that the potential between them is reasonably linear.

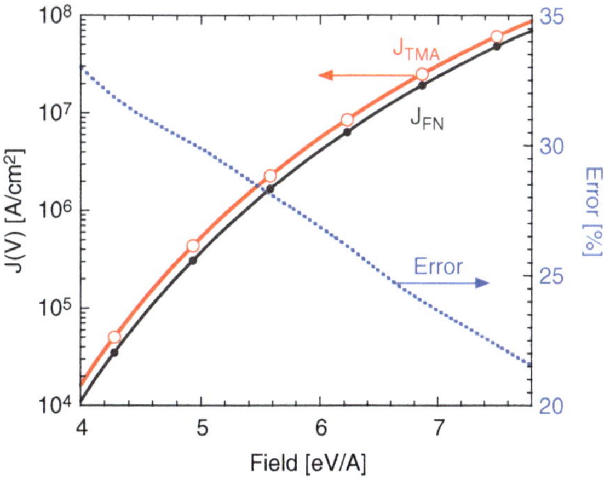

Figure 19.11 The current density for the potential of Figure 19.10 for the general parameters of $\mu = 7.0$ eV, $\Phi = 4.5$ eV, and $T = 300$. The gray line is the FN relation of Eq. (13.25), which tends to be 20–40% smaller than the Airy function TMA, shown by the blue dashed line and the right-hand side axis.

misguided when the discrepancy between the numerical solution of the field emission current density using TMA differs as much as it does from the FN equation itself. The very good approximations of Eqs (13.23) and (13.24), accurate to better than 0.4% and trivial to use in the guise of Eq. (13.25), are more than adequate for most considerations of field emission from metals. Therefore, efforts to "improve" the FN equation are better spent on what impact the pre-factors have [262], or identifying how the equation changes as emission near the barrier maximum occurs in thermal–field conditions (Chapter 17).

19.2.1 Kemble approximation revisited, again

Help me then to draw out the conclusion which follows from our admissions; for it is good to repeat and review what is good twice and thrice over…

– Plato[5]

The ease of the plane wave TMA formulation stands in contrast to the more involved Airy function TMA so that affection for the Lear approach ("seek thine own ease", Section 9.1.1) might encourage exclusive usage of the former. Yet one must be careful in doing so, as there are times when the planar TMA causes unphysical artifacts to appear if and when the segment widths approach the inverse wave numbers $1/k$. If that is opaque, an example clears up the nature of the problem, and revisiting the Kemble approximation is a good laboratory to explore what that problem is, particularly because it is already by now so familiar.

Therefore, consider the parabolic potential again, where the form of $V(x)$ as seen in both the planar and Airy TMA is shown in Figure 19.12, and where the x axis is normalized to the base width of the parabolic potential. The potential is partitioned into 10 equally thick slabs: because of the manner in which the partitioning is numerically executed, in which $x_j = L(j-1)/(N-1)$ and $V_j = V(x_j)$, the first partition for the planar model has zero height. For $m_n = m$ and $L = 2$ nm both the planar and Airy TMA are in reasonable agreement, as shown in the left plot of Figure 19.13. However, when *only* L is changed to 10 nm, then the right plot of Figure 19.13 results. Wave function interference now has the opportunity to occur because the width of the partition is too large, and the consequences on the evaluation of $D(k)$ are stark. In contrast, the Airy TMA approach performs well, in that the results do not change appreciably when the number of partitions is increased ($N = 20$) whereas the planar TMA, although much improved, shows interference (rippling) for $k > k_o$ even though double the number of partitions is used.

19.2.2 Gamow factor revisited, again

This act is as an ancient tale new told
And, in the last repeating, troublesome …

– William Shakespeare[6]

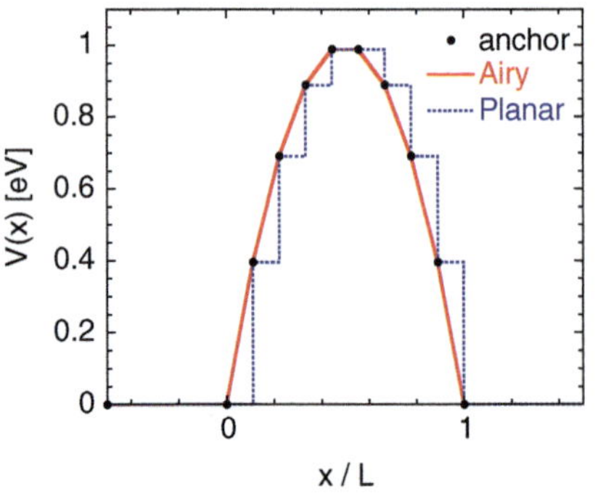

Figure 19.12 The planar (dashed blue line) and Airy (solid red line) representation of the quadratic potential barrier. The x_j are indicated by the solid black dots. When the x axis is scaled by L, then the $L = 2$ nm and 10 nm representations coincide.

[5] Plato, *The Dialogues of Plato: Gorgias* (trans. Benjamin Jowett), *Great Books of the Western World*, ed. Robert Maynard Hutchins, Vol. 7, Plato. Chicago: Encyclopaedia Britannica, 1952, p. 280.
[6] Ref. [37]: *The Life and Death of King John*, IV.ii.18–19.

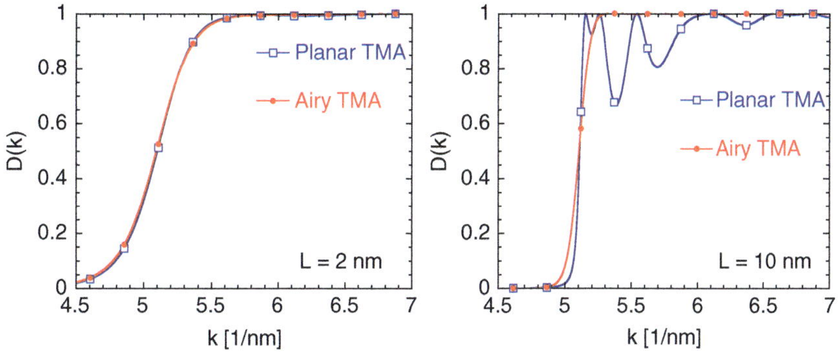

Figure 19.13 Left, when $L = 2$ nm, both the planar and Airy TMA are close in their predictions. Right, when $L = 10$ nm, the width of the partitions is sufficiently large that interference effects become manifest and undermine the accuracy of the planar TMA method for a low number of partitions, but not the more robust Airy method, for which the results do not change appreciably as the number of partitions increases.

Reconsider the Gamow factor using the *Zi* functions. As in Eq. (19.7), the center two matrices describe the barrier itself, and so when the potential $V(x)$ varies linearly between the left- (where $z = z_o$) and right- (where $z = z_L$) hand sides, and the energy of the electron is well below the barrier height, the matrix **B** is given by

$$\mathbf{B} = \begin{pmatrix} Zi(1,z_o) & Zi(-1,z_o) \\ Zi'(1,z_o) & Zi'(-1,z_o) \end{pmatrix} \begin{pmatrix} Zi(1,z_L) & Zi(-1,z_L) \\ Zi'(1,z_L) & Zi'(-1,z_L) \end{pmatrix}^{-1} \tag{19.15}$$

where no transition matrix is needed if the energy is such that the electron does not pass from under to over the barrier in the section. If the left-hand side is at the origin and the right-hand side is at L, then $z_o = (k_o^2 - k^2)/f^{2/3}$ and $z_L = (k_o^2 - k^2 + sfL)/f^{2/3}$. The asymptotic approximations $H(1,u) \to 2$, $H(-1,u) \to 1$, and $Di(c, z(u)) = csf^{1/3}\sqrt{u}$ can be made. Effort shows

$$\mathbf{B} \approx -\frac{\cosh\left[\frac{2}{3}(z_L^{3/2} - z_o^{3/2})\right]}{\pi(z_L z_o)^{1/4}} \begin{pmatrix} sf^{1/3}\sqrt{z_L} & 1 \\ f^{2/3}\sqrt{z_L z_o} & sf^{1/3}\sqrt{z_o} \end{pmatrix} \tag{19.16}$$

where, for large argument, $\cosh(x) \approx \sinh(x)$, as was done in Eq. (19.9). Notice what has happened (again): the cosh term, being large, emerges as a coefficient to the matrix, so that for more general potentials constructed of trapezoidal slabs sandwiched together these pre-factors are going to cluster together in the form of a product. Moreover, the argument of the cosh term is seen to be

$$\frac{2}{3}\left(z_L^{3/2} - z_o^{3/2}\right) = s\int_0^L \sqrt{k_o^2 - k^2 + sf\,x}dx \tag{19.17}$$

The s on the right-hand side seems to be a problem, but is not: the potential is either increasing ($s = 1$) or decreasing ($s = -1$), making $z_L > z_o$ or $z_L < z_o$, respectively, and therefore making the argument of the cosh term positive or negative, respectively.

When the argument of $\cosh(x)$ is largish, then $\cosh(x) \to e^x/2$, and so, as with the plane wave approach, the product of exponentials results in the exponential of a summation, which will be of the form

$$\sum_{j=0}^{N-1} \int_{x_j}^{x_{j+1}} \sqrt{k_j^2 - k^2 + s_j f_j (x - x_j)}dx \to \int_{x_0}^{x_N} \kappa(x)dx \tag{19.18}$$

where $\kappa(x)$ is the smooth curve that passes through all the $k_j^2 - k^2$, $\hbar k_j = \sqrt{2mV_j}$, and the property of integrals, namely that

$$\int_a^b \cdots dx + \int_b^c \cdots dx = \int_a^c \cdots dx \tag{19.19}$$

has been used. With that last step, the Gamow factor is recovered as twice the expression given in Eq. (19.17), a necessary but not surprising result.

The behavior of multiplicative transmission probabilities can finally be seen in the next most simple barrier possible, namely, a rectangular (r) barrier pushed up against a triangular (t) barrier, or an rt barrier for short, which can also be (more or less) evaluated analytically. If the rectangular region is of width L, then it can be shown [188] that to a good approximation,

$$D_{rt}(k) \approx D_r(k)D_t(k) = e^{-\kappa(k)L} D_{FN}(k) \tag{19.20}$$

with $\kappa(k)^2 = k_o^2 - k^2$, which recapitulates Eq. (19.18) rather well.

19.2.3 Schottky deviations

Next...come the philosophers...they still boast that they can see...things which are all so insubstantial...they bring out their triangles, quadrilaterals, circles and similar mathematical diagrams, piled on top of each other and intertwined like a maze, and then letters of the alphabet which they marshal in line and deploy hither and thither.

– Erasmus[7]

An interesting usage of the Gamow factor analysis is to shed light on the deviations visible in a Schottky plot for thermal emission, wherein a plot of $\ln(J(F,T))$ vs $F^{1/2}$ is mostly linear. The simple model motivated by Eq. (19.20) will exaggerate (by quite a bit) the magnitude of the departures from linearity, whereas the actual deviations are so minute that Erasmus would have lampooned their investigation. A plot of experimental data reveals them [135, 164], an example of which is shown in Figure 19.14 (based on Figure 6 of Haas and Thomas [135]) to give a sense of their behavior and magnitude: the periodic nature visible is a consequence of the variations in the average reflection coefficient term $(1 - r)$ that normally attends the RLD equation, first encountered in the discussion following Eq. (12.3).

The theoretical derivation of the deviations is involved, with the work of Cutler and Gibbons [263] being notable, but rather than reproduce it, a simple model and Eq. (19.20) are enough to justify why those deviations are plausible. The image charge potential can be very crudely treated as a rectangular barrier pushed up against a triangular barrier (a "slab + triangular barrier" [188], shown in Figure 19.15), on which Eq. (19.20) is applied. The trick is to determine the height and width of the slab, which, as Figure 19.15 demonstrates, is field dependent.

Some artistry is involved in setting the slab parameters. Choose the smaller zero of $V(x) - \mu = 0$, as the image charge barrier methods for calculating $x_{\pm}(\mu)$ via Eq. (11.26) are available. Thermal emission is characterized by comparatively small fields, and so the small field limit is $x_{min} \approx Q/\Phi$. As for the right-hand side of the slab, its height should be the same as the height of the image charge barrier, and its location should change as the location of the barrier maximum

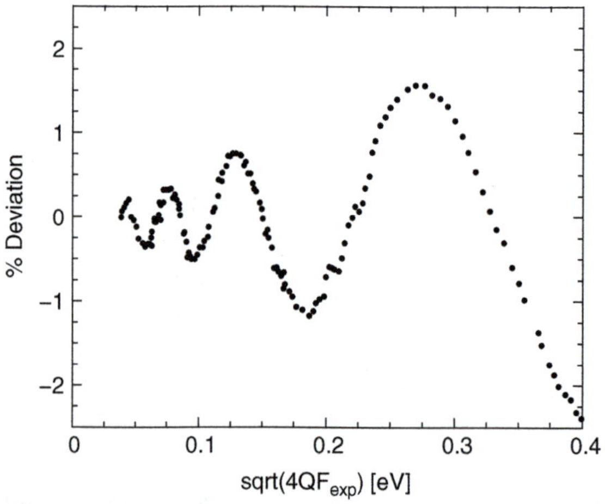

Figure 19.14 Deviations from the Schottky relation for a molybdenum filament at a temperature of 1950 K. F_{exp} refers to the experimental field (the field at the surface is larger if field enhancement is present). The deviations are revealed by removing the linear portion of a plot of $\ln(J)$ as a function of $\sqrt{F}$. The data was digitally extracted from Figure 6 of Haas and Thomas [135].

[7]Desiderius Erasmus (trans. Betty Radice), *Praise of Folly; and, Letter to Maarten Van Dorp, 1515*. London, New York: Penguin Books, 1993, passage [52], p. 151–152).

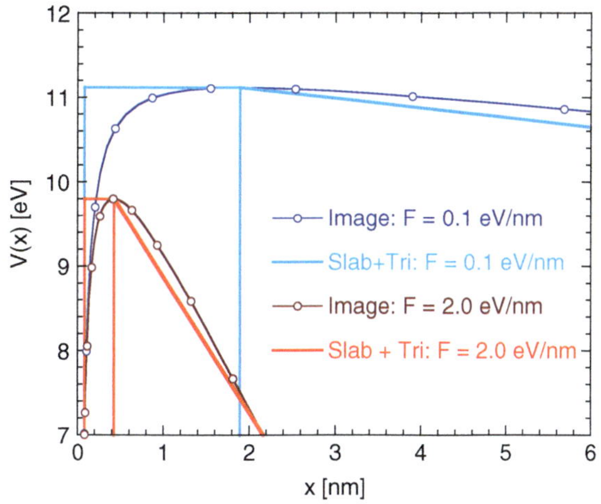

Figure 19.15 The image charge barrier and the approximate slab + triangular barrier representation, shown for fields of 0.1 eV/nm (blue) and 2.0 eV/nm (red) for $\mu = 7$ eV and $\Phi = 4.5$ eV. The rectangular barrier spans from $x_{min} = Q/\Phi$ to $x_c = \sqrt{Q/F}$. The triangular barrier is such that it coincides with $V_{image}(x_{max}) = \mu$. The slab is the rectangular barrier to the left of the triangular region.

$x_o = \sqrt{Q/F}$ changes. A convenient choice is x_o itself. Between the two, the width of the barrier is approximately

$$W = x_o - x_{min} \approx \sqrt{\frac{Q}{F}} - \frac{Q}{\Phi} \tag{19.21}$$

where the symbol W for the width of the slab is because L is already equated to the width of the image charge barrier itself. The slab therefore mimics the thin portion of the potential with dimensions that vary in response to the applied field.

Approximating a smoothly varying potential region by an abrupt rectangular slab comes at a cost. The price is anticipated by the smooth behavior of $D(E)$ for the quadratic potential (Figure 19.5) compared to the rectangular barrier potential (Figure 19.5), that is, oscillatory behavior for the rectangular barrier is much larger than for the quadratic barrier. Insofar as the thickness variation of the rectangular (or slab) barrier region affects the current density, it does so to leading order as $J \approx D(k_m)J_{RLD}$, where k_m is the value of k that maximizes the current integrand $(\hbar k/m)D(k)f(k)$. The expectation is that k_m will satisfy something like $\hbar^2 k_m^2/2m = \mu + \phi + nk_BT$, where n is of order unity, that is, the integrand peak will be about a thermal energy above the barrier maximum, which is intuitively reasonable.

The deviation is obtained by removing the part of $\ln(J)$ that is linearly dependent on $\sqrt{4QF}$. That is difficult, but because $\ln(J_{RLD})$ varies linearly with $\sqrt{4QF}$ as shown by Eq. (12.3) (remember that $\phi = \Phi - \sqrt{4QF}$), and because the triangular barrier part captures most of J_{RLD}, consider the ratio $\ln(J_{RLD}/J)$ as a reasonable proxy. From Eq. (18.10) it is

$$-\ln(D_{rec}(k_m)) \approx \ln\left\{1 + \frac{1}{4}\left(\frac{k_m}{\kappa} - \frac{\kappa}{k_m}\right)^2 \sin^2(\kappa W)\right\} \tag{19.22}$$

with $\hbar^2 k_m^2/2m = \mu + \phi + \hbar^2\kappa^2/2m$, where $\hbar^2\kappa^2/2m = nk_BT$ is a small energy, and $\hbar^2 k_m^2/2m$ is about the average energy of an electron just emitted over the barrier. The smallness of $k_BT \approx 0.168$ eV at $T = 1950$ K compared to μ and ϕ means, then, that to leading order

$$\frac{1}{4}\left(\frac{k_m}{\kappa} - \frac{\kappa}{k_m}\right)^2 \approx \frac{\mu + \phi}{4nk_BT} \tag{19.23}$$

which itself varies weakly for fields of the magnitude shown in Figure 19.14 (assuming $n \approx 1$). That means the deviations can be traced to the sin term of Eq. (19.22), which causes similar oscillations to those observed in the experimental data, albeit on a far grander scale. In fact, assuming that $n = 1$, and that field enhancement due to surface roughness and/or curvature is about a factor of 4 (that is, the field at the emission site is 4 times greater than the experimental field $F = 4F_{exp}$, the later of which is used in Figure 19.14), then for $\mu = 7$ eV, $\Phi = 4.5$ eV, and $T = 1950$ K, calculation shows that Eq. (19.22) is as shown in Figure 19.16. Consequently, the oscillations in the deviations of thermal emission data from a linear Schottky relation are plausibly related to the wave nature of the electron.

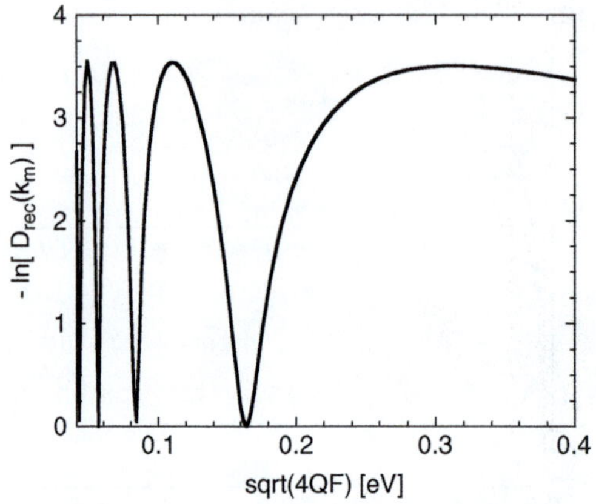

Figure 19.16 Behavior of Eq. (19.22) for $T = 1950$ K, $\hbar^2\kappa^2/2m = k_BT$, $\mu = 7$ eV, and $\Phi = 4.5$ eV, and a field enhancement factor of 4.

19.2.4 Resonant transmission

> *...I would like to point out that many high barriers exist in this world: Barriers between nations, races and creeds. Unfortunately, some barriers are thick and strong. But I hope, with determination, we will find a way to tunnel through these barriers easily and freely, to bring the world together...*
>
> – Leo Esaki [254]

The RTD barrier of Figure 19.6 is shown at zero bias, that is, $V(x \to -\infty) = V(x \to +\infty)$. When a bias is applied, then the right-hand side of the barrier is brought down to a level $-V_b$. The simplest approximation is that the bias causes a linear drop across the potential barrier, as shown in Figure 19.17, starting a tad before the first barrier and stopping a tad after the second. The barriers of 0.3 eV are characteristic of an RTD with AlGaAs barriers sandwiched by regions of GaAs, for which relevant parameters are $m_n = 0.0667m$, a doping density of 10^{18} cm^{-3} ($\mu \approx 0.08$ eV), $k_F \approx 0.374$ nm^{-1}, $T = 300$ K, and $k_BT = 0.0259$ eV [264, 265], although the barrier and well widths in the present example are chosen more for dramatic effect than for technological relevance. Lastly, effective mass changes from one heterostructure layer to another are an important modification [266, 267] and make changes to the matrix equations defining the transmission probability, but for simplicity now, treat the effective mass as the same throughout.

The transmission probability $D(k)$ associated with the potential profile shown in Figure 19.17 is shown in Figure 19.18, where the supply function $f_o(k)$ for the aforementioned GaAs parameters is shown by the gray dashed line overlay, normalized to 1 at $k = 0$, and labeled by the biases associated with Figure 19.17. A resonance level (sharp peak) is accessible by the supply function (i.e., they overlap where $f(k)$ is not negligible), but as the bias increases, the peak

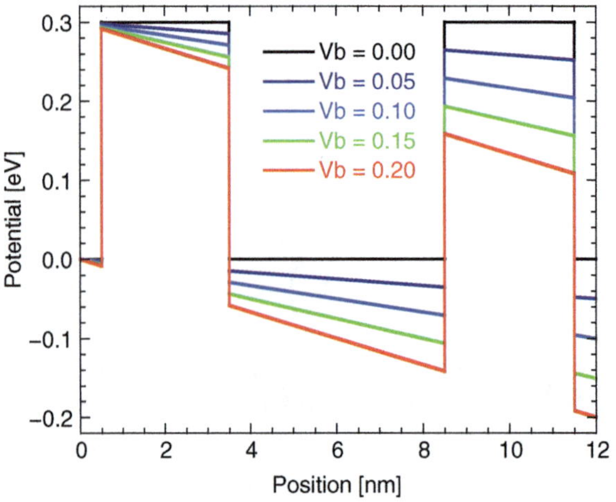

Figure 19.17 Potential barrier profile for an RTD under bias. Black line ($V_b = 0.0$) corresponds to an unbiased state (Figure 19.6). The beginning and end of the linear slope lie a short distance before and after the barriers, respectively, just for the fun of it (although band bending would cause an analogous behavior).

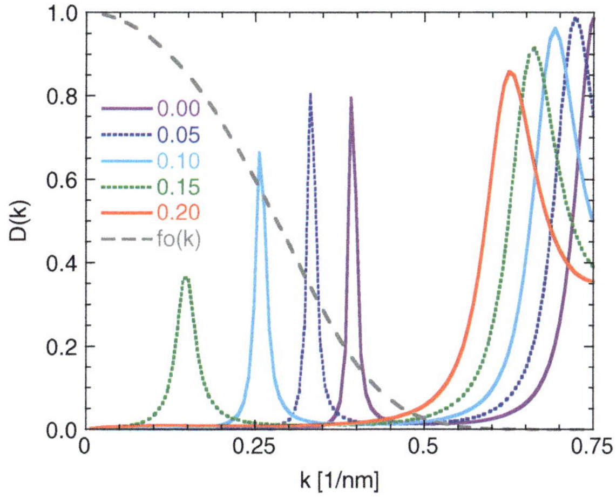

Figure 19.18 The transmission probability for each of the biases shown in Figure 19.17 for a barrier modeled after an AlGaAs/GaAs RTD. The resonant peaks occur at lower and lower k values until by $V_b = 0.2$ eV the resonant peak disappears altogether. Gray line is the normalized supply function for comparison.

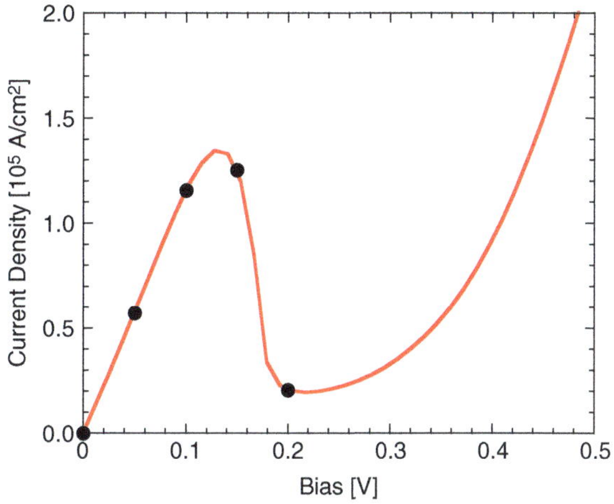

Figure 19.19 The current density versus applied bias for an RTD. The black dots identify the special bias cases shown in Figures 19.17 and 19.18. When V_b is between 0.15 and 0.20 eV, the current declines with increasing bias, a condition known as negative differential resistance.

shifts to lower k until it vanishes, at which point there is no resonant peak accessible until the bias becomes substantially higher. As a result, the current density goes through a maximum but then declines as voltage increases, as in Figure 19.19, creating a region where an *increase* in voltage results in a *decrease* in current, that is, *negative differential resistance* occurs.

The bias across the potential barriers was linear for simplicity, but in real RTDs the well region can fill up or deplete depending on conditions, and band bending (Section 24.2) occurs before and after the barriers. As the presence of charge density leads to space-charge contributions, that well density has a consequence. Numerical simulations to investigate it reveal that the charge in the well region (and in the regions just before and after the barriers) is enough to bring the resonant level into and out of the range where resonant tunneling can occur. As a result, oscillations in the current density arise with a time scale (in the THz) dictated by how long it takes the well to repopulate and deplete. The $I(V)$ characteristic that results therefore develops a hysteresis, that is, ramping the voltage up gives rise to a different $I(V)$ curve than ramping the voltage down in the negative differential resistance region, although revealing the hysteresis requires more formidable approaches than TMA methods [127, 268, 269]. Such odd circumstances lead to unexpected intrinsic bistability and current oscillations, and bear on the thorny problem of quantum mechanical tunneling times [255].

Now consider what seems to be a minor modification to a well-examined potential barrier: the triangular barrier of height $\mu + \Phi = 11.5$ eV and subject to a field of $F = 5$ eV/nm with a modification suggested by Eq. (19.18), namely, remove one of the slabs from it. Visually, what is being done is shown in Figure 19.20: a trapezoidal piece of width $L = 0.6$ nm centered at $x_c = 0.7$ nm is excised from the triangular region, making the width of the first trapezoidal barrier

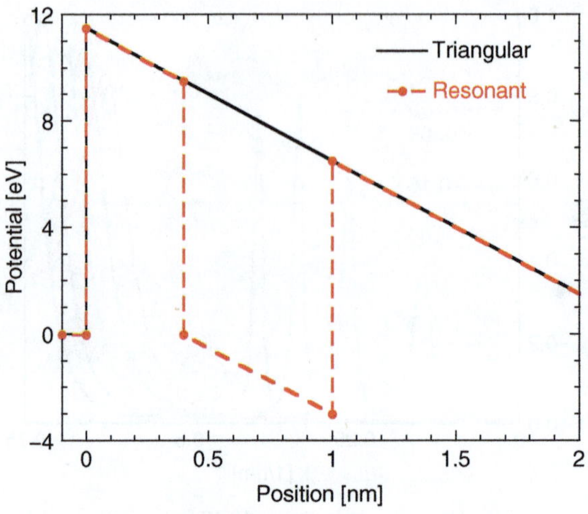

Figure 19.20 The triangular barrier (black solid line) with a trapezoidal region of width 0.6 nm excised (center of well at 0.7 nm), leaving the potential with the red dashed outline. The depth of the well is 9.5 eV below the triangular barrier. The field is 5 eV/nm for both the barrier and well regions.

$L_o = x_c - L/2 = 0.4$ nm. The potential well is made of a depth such that the right edge of the first barrier is at $V = 0$, so the well is of a depth of $\mu + \Phi - FL_o = 9.5$ eV. The height of the triangular barrier at x_c is $V_c = 8$ eV. If the slab were not excised from the triangular barrier, then $D(k)$ would be given by Eq. (18.57): call this solution $D_{tri}(k)$ to serve as a reminder that it is the triangular barrier transmission probability. Now the slab is excised. From Eq. (19.18) it is therefore expected that the transmission probability becomes

$$D(k) \approx \exp\left(2\int_{-L/2}^{L/2} \sqrt{\kappa_c^2 - \frac{fs}{2}} ds\right) D_{tri}(k)$$

where $\kappa_c(E)^2 = (2m/\hbar^2)(\mu + \Phi - E - Fx_c)$. Taylor expanding the integrand, the linear term vanishes because only even powers of s (i.e., s^{2j} and not s^{2j+1}) survive over the integration region. To leading order, therefore,

$$D(k) \approx \exp(2\kappa_c(E_k)L)D_{tri}(k) \tag{19.24}$$

The current density J_{seq} due to the sequential tunneling terms is then seen to be (more or less) comparable to

$$J_{sec}(F) \approx \exp(2\kappa_c(\mu)L)J_{FN}(F) \tag{19.25}$$

where $J_{FN}(F)$ is familiar from Eq. (13.10). The approximation is better than the off-hand treatment here, as can be shown more rigorously [199], but accuracy is not the goal, rather, seeing how the sequentially tunneling component stacks up to the resonant tunneling term is more interesting.

The form of Eq. (19.24) is enough to suggest that if $\exp(-2\kappa(E)L)D(k)/D_{FN}(k)$ is plotted, any unusual deviation from a line of order unity should be indicative of a resonant energy level within the well. Sure enough, Figure 19.21 reveals two of them. In general, near a resonance, $D(E_k)$ behaves as [255]

$$D(E_k \approx E_{res}) \approx \frac{D_o}{1 + \{(E_k - E_{res})/\Delta E\}^2} \tag{19.26}$$

where E_{res} is the energy of, and ΔE is the width of, the the resonant level.[8] At the resonances, the peaked behavior evokes the Lorentzian functions of Section A2.7, and the temptation to treat $\delta_n(E)$ (where $n = 1/\Delta E$) as a Dirac delta function is irresistible. One finds (in the zero temperature limit)

$$J_{res} \approx \left(\frac{q}{2\pi\hbar}\right)\int_0^{\mu}\left(\frac{m}{\pi\hbar^2}\right)(\mu - E_{res})D_o\pi\Delta E\delta_n(E - E_{res})dE \tag{19.27}$$

$$= \frac{qmD_o\Delta E}{2\pi\hbar^3}(\mu - E_{res}) \tag{19.28}$$

[8]The notation is being handled a bit cavalierly: $D(E)$ is shorthand for $D(E_k) = D(k)$, where k in the latter corresponds to $\sqrt{2mE_k}/\hbar$ in the former, as introduced in Section 11.2.

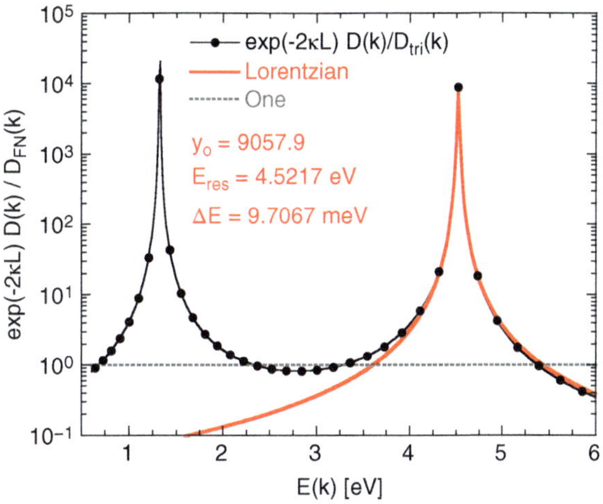

Figure 19.21 The ratio $\exp(-2\kappa(E)L)D(k)/D_{tri}(k)$ for the dashed line potential of Figure 19.20. The red line is a best fit Lorenztian term for the higher energy resonant level associated with the barrier.

For $\mu = 7$ eV, $\Phi = 4.5$ eV, and the parameters of Figure 19.21, direct numerical evaluation shows that J_{res} is about 28× larger than J_{seq}, and only the larger E_{res} was being considered in the comparison. Resonance terms can be substantial.

An account of why the form of the resonance is like a Lorentzian relates to the concept of an "attempt frequency" f in a tunneling formalism by Price [270, 271], although the present treatment owes more to the groundwork laid out in Section 11.2. In the Bohm approach, for which $\psi = Re^{iS}$ as in Eq. (11.7), the quantity $(\hbar/m)\partial_x S$ was a velocity.[9] A bound state for an infinite potential well [44] entails that $\psi(x)$ is real, so that $S = 0$ and no current flows because Eq. (9.60) is identically zero. For practice, consider how that changes for transport past a potential barrier by introducing the quantity $\gamma(x) = \psi(x)^{-1}\partial_x\psi(x)$. Using the Bohm form, γ is

$$\gamma(x) = \partial_x \ln(R) + i\partial_x S \tag{19.29}$$

Assuming that $\ln(R)$ varies slowly, the second term becomes the more important one, and is known from Eq. (11.13) (where the quantum potential $\Omega \approx 0$ in the same approximation). Clearly, γ is a function of E as well, and so can be expanded around the resonance energy E_o according to

$$\gamma(E) \approx (E - E_o)\gamma'(E_o) \tag{19.30}$$

where the prime indicates derivative with respect to argument and the absence of a first term $\gamma(E_o)$ is a reflection of the vanishing of γ near a "quasi-level" (which acts like the resonance level). If γ retains a semblance of $\partial_x S$, then γ' will resemble

$$\gamma'(E_o) \sim \frac{\sqrt{2m}}{2\hbar}(E_o - V)^{-1/2} \sim \frac{1}{\hbar v} \tag{19.31}$$

where $v^2 = 2(E_o - V)/m$ acts like a velocity in the form of $\delta x/\delta t$. If $\delta t = 2/f$ is a transit time (where f is an "attempt frequency"), then $\delta x = 1/k$ is a characteristic length for which the wave number k is appropriate, and so the expectation is that Eq. (19.30) should be something like[10]

$$\gamma'(E_o) \sim \frac{2k}{\hbar f} \tag{19.32}$$

which means that

$$\gamma(E) \approx (E - E_o)\frac{2k}{\hbar f} \tag{19.33}$$

Compare this to transport problems, where the incident and reflected waves are combined into ψ as

$$\psi = N\left(e^{ikx} + re^{-ikx}\right) \tag{19.34}$$

[9] This is trivially seen for a constant potential region, for which $\partial_x S = k$ and $v = \hbar k/m$.
[10] Compare Eq. (6) of ref. [270]: Price's treatment is more rooted in Gamow's tunneling theory, and therefore may have greater appeal for the more fastidious reader.

where r is the reflection coefficient. If that ψ is used in the definition of $\gamma = \psi^{-1}\partial_x\psi$ and evaluated at $x = 0$ it results in

$$\gamma = ik\frac{1-r}{1+r} \tag{19.35}$$

the inversion of which, plus Eq. (19.33), results in

$$r = \frac{\hbar f - 2i(E - E_o)}{\hbar f + 2i(E - E_o)} = \frac{(\hbar f - 2i(E - E_o))^2}{(\hbar f)^2 + 4(E - E_o)^2} \tag{19.36}$$

the denominator of which has the structure of a Lorentzian

$$r \propto \left\{1 + \left[\frac{2}{\hbar f}(E - E_o)\right]^2\right\}^{-1} \tag{19.37}$$

where the width of the Lorentzian is governed by the tunneling rate factor f.

CHAPTER 20

Ion enhanced emission and breakdown

...the Chimaira...a thing of immortal make, not human, lion-fronted and snake behind, a goat in the middle, and snorting out the breath of the terrible flame of bright fire...

– Homer[1]

Emission mechanisms due to thermal, field, photoemission, secondary emission, and now resonant tunneling have been up till now as treated as distinct processes, not often seen operating together in the explanation of a phenomenon. Their segregation is seemingly natural, but wildly different parts grafted together need not be horridly monstrous. One circumstance where all of the emission mechanisms play a part makes for an interesting bit of physics [272]. This happens in the gaseous breakdown across gaps [273, 274], otherwise known as discharge. Even more interesting here is when the gaps are micro-scale and so-called "micro-plasmas" are generated. Conditions can be pushed to the point where the discharge becomes self-sustaining, and an avalanche breakdown occurs. Interest, initially motivated by concerns over spark discharges and erosion of material in microelectromechanical systems, has since grown due to possible applications for gas sensing, lighting, energy conversion devices, and micro-plasma physics [275], but ambitions here are a bit more circumscribed, being more towards how the emission mechanisms manifest themselves.

20.1 Paschen's curve

Our doubts are traitors
And makes us lose the good we oft might win,
By fearing to attempt.

– William Shakespeare[2]

Electrical breakdown or arcing is so complex, dependent on the gases involved, the current levels present, the voltages that exist, whether the surface of the cathode is rough or smooth, and what the origin of the initial emission mechanism is, that the prospects of fashioning a reasonable account of it may seem doubtful. Electrons in the anode–cathode (AK) gap generally always exist, having been created by cosmic rays, radioactivity, or photoemission, as suggested in Figure 20.1. Likewise, different regimes exist in the current versus voltage relations: as the voltage increases, the current increases but plateaus, continuing to raise the voltage and at some point the current starts to increase again, until it takes off and becomes self-sustaining (even if the initial electron generation mechanisms are suppressed). When that happens, *breakdown* occurs [12]. The conditions for the onset of breakdown depend on pressure P, AK gap separation D, and the anode voltage at breakdown $q\varphi_b = V_b$, and the curves that result depend on the nature of the gas involved. The characteristic V_b versus D (or the product of gap distance and background pressure PD) curves for each gas are called Paschen characteristics, the shape of the curve being dictated by Paschen's law. The relation for some typical gases is shown in Figure 20.2.

A bit of reflection shows, however, that the behavior of the curves is qualitatively explicable: to the left of the curve minima, the breakdown voltage V_b increases as the product of pressure P and AK gap separation D decreases because the separation between the gas molecules becomes larger, making the probability of another ionizing collision between generated electrons and neutrals less likely. Conversely, to the right of the minimum, V_b increases with the PD product, and reflects the fact that when sufficient opportunity arises for electrons to have collisions with neutrals, a higher voltage

[1] Homer, *The Illiad* (trans. Richmond Lattimore). Chicago: University of Chicago Press, 1951, Book 6, lines 179–183. Hesiod, in *Theogony* (lines 319–325), gives a similar description of the beast.
[2] Ref. [37]: *Measure for Measure*, V.4.78–80.

Introduction to the Physics of Electron Emission, First Edition. Kevin L. Jensen.

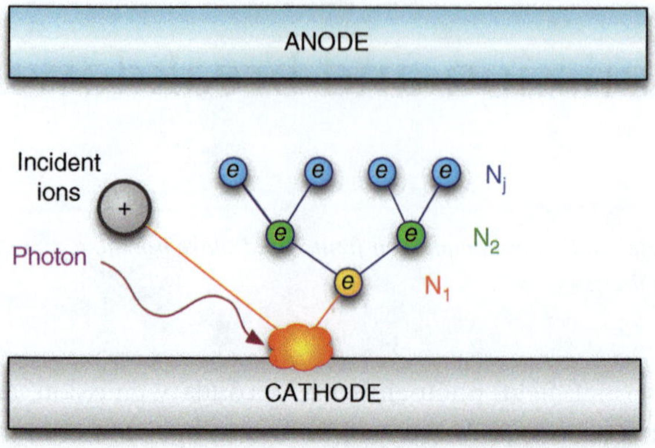

Figure 20.1 Steps leading to breakdown: incident photons, cosmic rays, or ionized gas molecules strike the cathode surface and liberate electrons. Electrons accelerate in field $F = V_a/D$ and when their energy is sufficient, they can collide with other gas molecules and ionize them. Ions are drawn to cathode, and liberated electrons are further accelerated, both of which lead to more electrons.

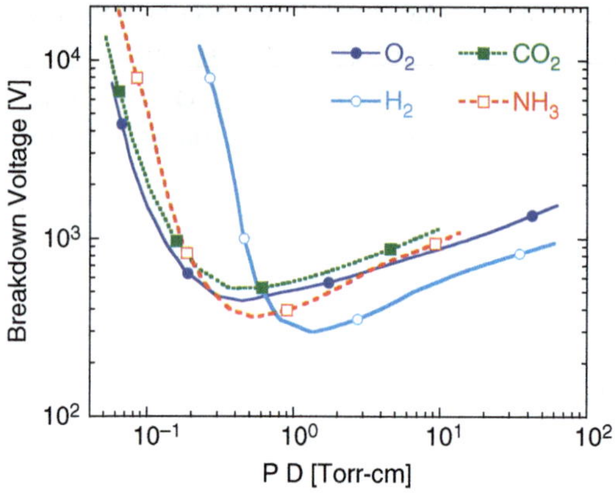

Figure 20.2 The Paschen relation for breakdown voltage versus the product of the background pressure P and the AK gap separation D for the gases hydrogen, oxygen, ammonia, and carbon dioxide. The noble gases have a similar curve (see Figures 7 and 8 of ref. [274]).

is required so that the electrons have a chance of being accelerated to a high enough energy to ionize neutrals when a collision occurs.

The mechanism behind the exponential growth in generated electrons (Figure 20.1) begins with the creation of a free electron. When that accelerated electron acquires enough energy and collides with a neutral, it liberates n electrons ($n = 2$ in the figure), and each of those electrons typically accelerates again until they too have enough energy, $E \approx V_a l/D$, to liberate electrons from the neutrals they strike. The growth is exponential because if there are N_1 electrons after the first collision, there will be $N_2 = nN_1$ after the second, and by iteration, $N_j = \alpha^j N_1$. If the electrons travel a mean free path l, pointed mostly towards the anode, then the iteration index goes as the distance from the cathode. This suggests that $dN/dx = \alpha N(x)$ (in preference to $N_{j+1} = nN_j$), which has the solution

$$N(x) = N_o \exp(\alpha x) \tag{20.1}$$

The number of electrons generated per unit time, or the current density of electrons, follows the same behavior, or

$$J_e(x) = J_e(0) \exp(\alpha x) \tag{20.2}$$

with the total number of generated electrons being

$$J_e(D) - J_e(0) = J_e(0)(e^{\alpha D} - 1) \tag{20.3}$$

The generation of the ions $J_i(x)$ goes in the opposite direction, increasing towards the cathode. The total number of generated ions is the same as the number of generated electrons, or

$$J_i(0) - J_i(D) = J_e(D) - J_e(0) \tag{20.4}$$

where no ions are generated at the anode, or $J_i(D) = 0$. When the ions impact on the cathode, each ion creates secondary electrons, the number of secondaries per ion being γ, or $J_e(0) = \gamma J_i(0)$. By joining Eqs (20.3) and (20.4), and using $J_i(D) = 0$, it follows that when the current of electrons and ions is self-sustaining then

$$J_i(0) = \gamma J_i(0)(e^{\alpha D} - 1) \rightarrow \alpha = \frac{1}{D} \ln (1 + \gamma^{-1}) \tag{20.5}$$

Recall, though, that α is the number of ionizations per unit length, that is, α should be something like $\alpha \sim \Delta N_e / l$, where ΔN_e is the number of created electrons, and l is the mean free path, or how far the electrons travel on average before they endure their next collision. But ΔN_e should go as the number of electrons N_e present and the probability of their collision P_{coll}, and so

$$\alpha \sim \frac{\Delta N_e}{l} = -\frac{N_e}{l} \left\{ \frac{N_i A_i}{(V/l)} \right\} = -\frac{N_e N_i A_i}{V} \tag{20.6}$$

where N_i is the number of neutrals that will become ions after a collision, A_i is the cross-section of the neutrals, $V/l = A_o$ is the unit cross-sectional area, and the probability of collision behaves as $P_{coll} \approx N_i(A_i/A_o)$.[3]

From the ideal gas law, the density $\rho = N_i/V$ of the gas of neutrals is simply $P/k_B T = \beta P = \rho$, which, when coupled with Eq. (20.6), results in

$$N_e(x) = N_o \exp(-\beta P A_i x) \tag{20.7}$$

but that implies

$$\alpha = \Delta N_e / l \approx \alpha_o \beta P A_i \exp(-\beta P A_i l) \tag{20.8}$$

Because l is the distance over which the electron is accelerated in a field characterized by $F = q\mathcal{E}$ until it has sufficient energy to cause an ionizing collision, it can be related to the ionization energy E_i of the neutral atoms by $E_i = Fl$, and so

$$\begin{aligned} \alpha &\approx \alpha_o \beta P A_i \exp(-\beta P A_i E_i / F) \\ &\equiv AP \exp(-BP/F) \end{aligned} \tag{20.9}$$

where the constants A and B are specific to the species of gas atoms or molecules which make up the background gas in the gap. The field in the gap is simply $F = V_b / D$ at breakdown, and that, coupled with Eq. (20.5), finally allows for the breakdown voltage V_b to be related to the gap separation D (or the product of pressure and gap separation PD, as shown in Figure 20.2) by

$$V_b = \frac{BPD}{\ln [APD / \ln (1 + \gamma^{-1})]} \tag{20.10}$$

where from the derivation it is clear that other factors matter as well, such as temperature and the dimensions and ionization energies of the neutral gas molecules involved.

The form of the breakdown voltage in Eq. (20.10) can be rendered more generally as $y(x) = x / \ln(ax)$, and from it one sees the general contours of Figure 20.2 emerge: for large values of x (or PD), y (or V_b) appears to be increasing roughly linearly with x, but when x approaches $(1/a)$, y rapidly increases due to the singular behavior of $1/\ln(ax)$. In between, $y(x)$ goes through a minimum at $y = x = e/a$, as shown in Figure 20.3. Augment the curves by a factor to change scale, and obtaining agreement with experimental data becomes quite easy.

20.2 Modified Paschen's curve

She knows there's no success like failure
And that failure's no success at all.

– Bob Dylan[4]

[3]Be alert to the perils of notation: V is a volume, but $V_b = q\varphi_b$ is a breakdown *energy* as φ_b is a breakdown *voltage*. The factors A_i and A_o are areas here, but in Eq. (20.9) A is a factor in the Paschen relation. Context determines content.

[4]Bob Dylan, lyrics to *Love Minus Zero, No Limit*, 1965.

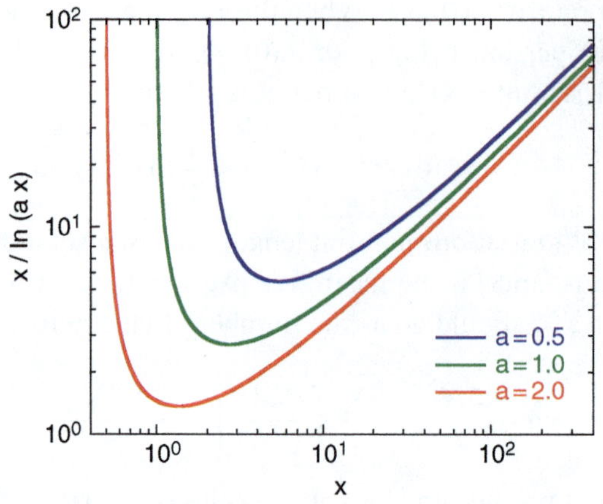

Figure 20.3 The Paschen relation $y = x/\ln(ax)$, based on Eq. (20.10), for various values of a. The minimum occurs at $y_m = x_m = e/a$ and a singularity occurs for $x = 1/a$.

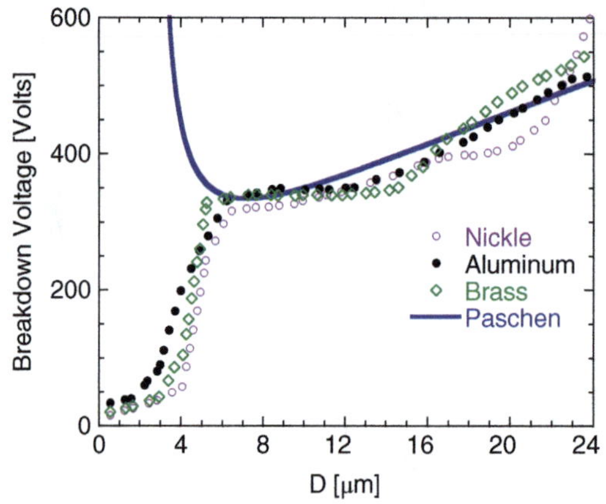

Figure 20.4 Based on Figure 4 of Torres and Dhariwal [276] (data digitally extracted from a line). The Paschen curve is given by $V_b = 46.73D/\ln(0.38D)$ for V in volts and D in microns.

The Paschen relation works well until the gaps narrow to about 15 μm or smaller, at which point Eq. (20.10) fails to account for the experimental breakdown voltage and another model is needed to successfully account for the failure mechanism. Consider the data from Figure 4 of Torres and Dhariwal [276], reproduced in Figure 20.4, where the AK gap separation is in microns and the voltage across it is in the hundreds of volts. Where there should be singular behavior, there is instead a plummeting of the breakdown voltage towards the origin. At the minimum of the Paschen curve (about $D = 7.2$μm), the breakdown voltage is 334 V, corresponding to $V_b/D = 46$ MV/m (or about two orders of magnitude lower than standard metal-like fields leading to field emission). As Torres and Dhaliwal ruefully note, the plummeting V_b as D approaches microscopic spacings bodes poorly for the fabrication of micro-devices, and so its causes are more than just a challenging physics problem.

Hypotheses on the origins of the modified Paschen curve of Figure 20.4 speculated as to the role ions play as they approach the surface. At different distances from the surface, the manner in which the ion potential modifies the emission barrier changes. Field emission (assisted by field enhancement effects due to surface roughness) [275, 277, 278], thermo-field emission [279], and resonant mechanisms [272] (similar in behavior to the resonant states of a single atom on a field emitter tip [258]) can all contribute. Numerical simulations that include such emission physics match the small-gap behavior [280]. Therefore, before applying the transfer matrix approach (TMA; Chapter 19), consider an approximate treatment first.

A natural place to start is reconsidering the factor γ in Eq. (20.5) by allowing field emission to affect it [275]. That is, let

$$\gamma \rightarrow \gamma_o + \gamma_f \tag{20.11}$$

where γ_o is analogous to the large-gap, high-voltage term of Eq. (20.10), and γ_f accounts for field emitted electrons and therefore will be dependent upon the field that exists at the surface at the point of breakdown. Therein lies the first problem: machined metal surfaces always have some level of roughness [281, 282], surfaces which endure arcing or discharge can become pitted [283], the presence of particulates or micro particles can induce field emission sites, and local heating giving rises to protrusions [31, 284]. The field on the apex of a protrusion can be treated using a field enhancement factor β_g, as was encountered in Section 13.3 and will be considered in Section 30.1. For now, simply extrapolate from elementary results. It is well-known from elementary electrostatics that the field enhancement factor of a hemisphere causes the field at the apex to be three times greater than the background field (Eq. (30.10)). Couple that with the conjecture of Schottky [138], and others [285, 286], that field enhancement factors are multiplicative. If so, a bump on a bump on a bump (a reasonable approximation of a surface emission site) would endure a field at the apex comparable to $3 \times 3 \times 3 = 27$ times larger than the background field. These simple hand-waving numbers will be improved upon in Chapter 30, but the conjectured magnitude is about right. Notice that the discussion has been about the field at the emission site compared to its background value $F_o = V_b/D$, where field enhancement and beta factor refer to the dimensionless ratio of the apex field to the background field.

The field dependence of the probability of emission is dominated by the exponential factor in the Fowler–Nordheim equation of Eq. (13.25). A reasonable expectation is then that γ_f will manifest a similar behavior, or

$$\gamma_f(F_o) \sim C \exp\left(-\frac{B_o \Phi^{3/2}}{\beta F_o}\right) \equiv Ce^{-b/F_o} \tag{20.12}$$

where C is a constant to be determined and, for convenience, $b \equiv 4\sqrt{2m\Phi^3}/(3\hbar\beta)$. Because departures from Paschen's curve in Figure 20.4 occur at about a gap spacing of 7.2 μm and a breakdown voltage of 335 V, there is where γ_o is likely comparable to γ_f, and so choose $C = \gamma_o \exp(b/F_c)$ so that

$$\gamma_f(F_o) = \gamma_o \exp\left\{ b\left(\frac{1}{F_c} - \frac{1}{F_o}\right)\right\} \tag{20.13}$$

where $F_c = (335/7.2)$ eV/μm = 0.05317 eV/nm (corresponding to an electric field of about 53 MV/m). Given the rapid decline of field emission current density when the field declines, for $F_o < F_c$ the influence of field emission will fade rapidly and so Eq. (20.10) holds with γ replaced by γ_o. Conversely, once F_o surpasses F_c, the same observation means field emission will rapidly dominate, or $\gamma_f \gg \gamma_o$. Since $1/\gamma_f$ is then small, the approximation of $\ln(1+x) \approx x$ for x small means that

$$V_b(V_a > F_c D) \approx \frac{BPD}{\ln(\gamma_f APD)} = \frac{BPD}{\{\ln(\gamma_f) + \ln(APD)\}} \tag{20.14}$$

Inserting γ_f from Eq. (20.13),

$$V_b = \frac{BPD}{\ln(\gamma_o APD) + (b/F_c) - (b/F_o)} \tag{20.15}$$

After recalling that $F_o = V_b/D$, then, surprisingly, isolating V_b results in an equation much like the original Paschen equation, or

$$V_b(F_o > F_c) = \frac{(BP+b)D}{\ln(CAPD)} = \frac{(BP+b)D}{\ln(\gamma_o APD) + (b/F_c)} \tag{20.16}$$

with the first form related to Eq. (10) of Radmilović-Radjenović and Radjenović [287] (but see also Rumbach and Go [278]), who observe that the coupling of PD in the standard Paschen relation of Eq. (20.10) has been altered: pressure P and gap separation D now appear separately. The factor $b = (4/3\hbar\beta)\sqrt{2m\Phi^3}$ tends to dominate all the other terms with which it appears, therefore, to leading order, $V_b \approx F_c D$, although the neglected terms in Eq. (20.16) cause a mild visible curvature in Figure 20.5.

And yet, a bit of mystery remains: for small values of D, the breakdown voltage in Figures 20.4 and (20.16) falls even faster than governed by $V_b \approx F_c D$, which indicates additional physics. Rumbach and Go [278] demonstrate that at very small gap spacings the ion concentration near the cathode can increase by orders of magnitude as the anode potential V_a increases. The manner in which they change the emission barrier needs investigation. As ions approach the surface, the Coulomb field they have adds to the surface field and therefore the Schottky barrier reduction (contributing to thermal emission), and when the ion is close enough, tunneling can be increased (contributing to field emission). When adjacent to the surface, resonant tunnelling becomes a possibility. All intermingle as the ion approaches.

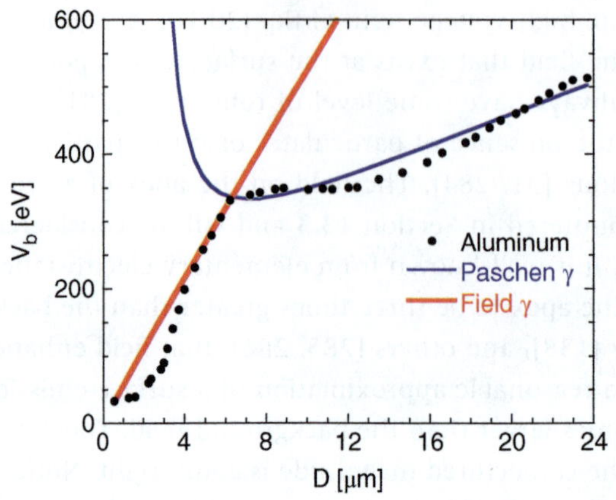

Figure 20.5 The aluminum data of Figure 20.4, for which $\Phi = 4.08$ eV, compared to Eqs (20.10) (Paschen γ) and (20.16) (Field γ), where the latter approximately goes as $V_b \approx F_c D$.

20.3 Ions and the emission barrier

Leave your crisp channels, and on this green land
Answer your summons ...

– William Shakespeare[5]

How the ions induce a path by which electrons answer the call of freedom can be investigated using the Airy function TMA. Much like electrons and the image charge potential, ions too induce an image charge, but the emitted electrons need not be on the axis defined by the ion and its image, but rather are generally off-side by an amount ρ (in cylindrical (ρ, z) coordinates). That is, the potential an electron emitted from a metal experiences that is modified by the presence of an ion of charge $+Zq$ resembles (after [288], but see also Eq. (4) of ref. [279])

$$V(z) = \mu + \Phi - Fz - \frac{Q}{z} - V_{ion}(z) \tag{20.17}$$

$$V_{ion}(z) = 4ZQ\left\{\frac{1}{r_+(z)} - \frac{1}{r_-(z)}\right\} \tag{20.18}$$

For an ion located at the cylindrical coordinates (ρ_i, z_i), then $r_\pm(z)$ are defined by

$$r_\pm(z) \equiv \left\{(z \pm z_i)^2 + \rho_i^2\right\}^{1/2} \tag{20.19}$$

where ρ_i is the distance of the electron from the axis defined by the ion and its image. That this is an *approximation* is not obvious, but is important to realize: for general three-dimensional potentials, the tunneling probability must be calculated along lines of steepest gradient [289], and an electron plowing through the ion potential along a line parallel to the ion-image axis is not following that path. Nevertheless, the one-dimensional potential given by Eq. (20.17) is both reasonable and unambiguous, and so it is used.

How $V_{ion}(z)$ alters the usual image charge potential depends on just how far from the surface z_i that the ion is, and what value of ρ_i is being used. Consider as a representative the image charge plus ion potential as the location of z_{ion} moves progressively closer to the surface, as in Figure 20.6. As the ion approaches, first the barrier is lowered (thermal emission increases), then the barrier thins (field emission increases), after which well energy levels can facilitate transmission for energy levels well below the Fermi level (resonant transmission), and lastly when the ion is at the surface, the overall barrier is greatly reduced such that both over and under barrier emission occurs (thermal–field emission) whether a resonance level contributes or not.

EXAMPLE: If an argon atom at rest is singly ionized 5 nm from the surface, crudely estimate the number of electrons emitted if the average current density across an area equal to twice the cross-section of the atom is

[5] Ref. [37]: *The Tempest*, IV.1.130–131.

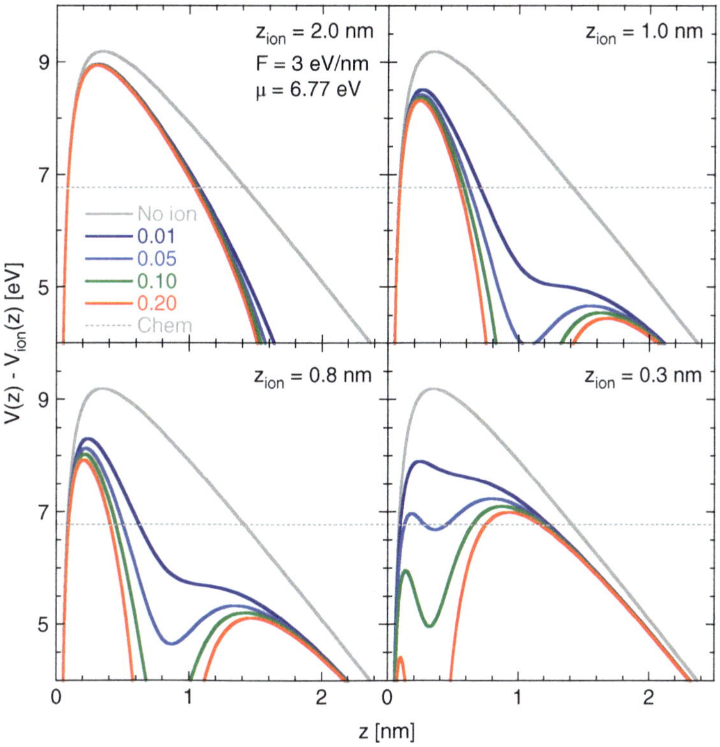

Figure 20.6 The image + ion potential given by Eq. (20.17) for z_i = 2, 1, 0.8, and 0.3 nm for values of ρ_i = 0.01, 0.05, 0.10 and 0.20 nm. The values of μ = 6.77 eV and F = 3 eV/nm were used. The gray line in background is the image charge potential without the ion potential.

$J = 2.2 \times 10^8$ A/cm^2 during the time the ion approaches the surface. Let the surface F be 3 eV/nm. Such numbers are for illustration only.

SOLUTION: The mass of the argon atom is 39.948 g/mole = 3.72×10^{10} eV/c^2. The acceleration is therefore F/M. The ion will travel a distance l = 5 nm in $t = \sqrt{2lM/F}$ = 1.18 ps. If the radius is R = 0.07 nm, then the cross-sectional area is $\pi(2R)^2$ = 0.061575 nm^2. The crude estimate for the number of emitted electrons is then $J\pi(2R)^2 t/q \approx 1$.

In the example above, the choice of 5 nm was not entirely random: the first ionization energy of argon is 1520.4 kJ/mole = 15.758 eV. An electron accelerated in a field of 3 GV/m would have to travel 5 nm or more to acquire enough energy to ionize an argon atom after a collision. Applied to the estimate of γ, the emission of electrons as the ion approaches the surface represents a more credible description of how γ_f contributes to γ, as discussed by Tirumala and Go [277], where emission was integrated over time and area during the ion's approach. Their resulting analytical model compared well with simulation and experiments dealing with modified Paschen's curve behavior using only a field emission contribution.

The potential barriers of Figure 20.6 change dramatically for small changes in the position of the ion, and therefore those potentials do not enable a convenient analytical solution for either a Gamow factor or a current density equation. However, the vales of $D(k)$ can be evaluated using the TMA of Eq. (19.14). Assume that the potentials have been artfully represented with linear segments densely placed in regions of rapid variation, further apart where long linear segments suffice, and having made sure that over/under and under/over matrices are properly placed, a program to evaluate $D(k)$ returns an array of values $D(k_j)$. The supply function can likewise be specified at the same k_j values to give $f(k_j)$, and from these two the current density through and over the potential barrier is found as a summation using Eq. (A1.17). If k_{min} is chosen such that $D(k_{min}) \ll 1$ and k_{max} so that $f(k_{max}) \ll f(k_F)$, then the corrections due to the boundaries are negligible, and so, with $k_j = k_{min} + (k_{max} - k_{min})(j-1)/(N-1)$ for j from 1 to N,

$$J(F,T) = \frac{\Delta k}{2\pi} \sum_{j=1}^{N} \left(\frac{\hbar k_j}{m} \right) D(k_j) f(k_j) \tag{20.20}$$

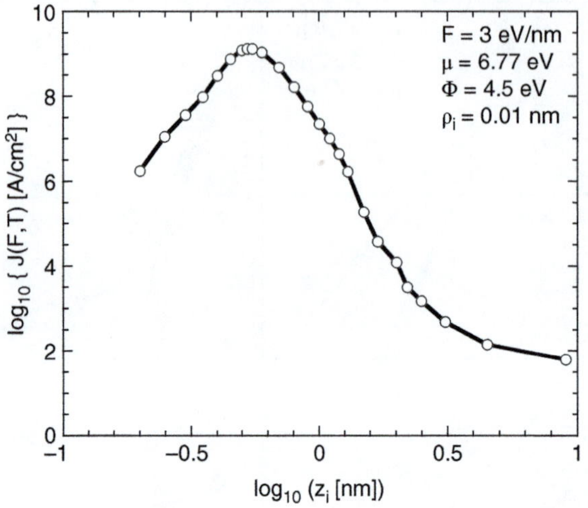

Figure 20.7 The current density $J(F, T)$ (in A/cm^2) as a function of the distance z_i (in nm) an ion ($Z = 1$) is away from the surface. Logarithms are in base 10. The current density peaks at $z_i = 0.525$ nm.

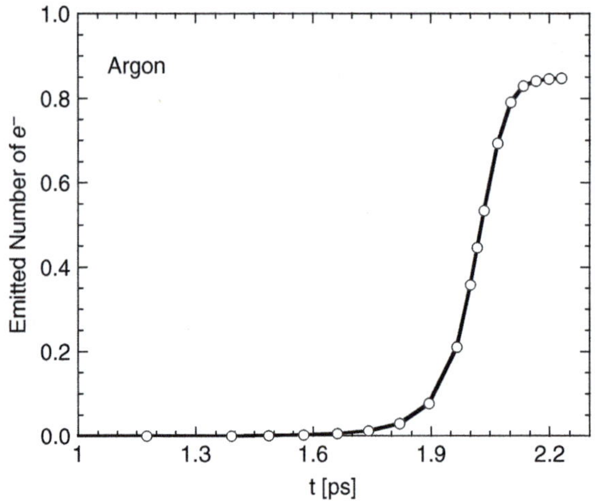

Figure 20.8 The current density $J(F, T)$ (in A/cm^2 as a function of the distance z_i (in nm) an ion ($Z = 1$) is away from the surface. Logarithms are in base 10. The current density peaks at $z_i = 0.525$ nm.

where $\Delta k = (k_{max} - k_{min})/(N - 1)$. For ease, a uniform spacing was assumed, but that can be modified, for example spacing k quadratically results in a summation equivalent to an energy integration.

If applied to the potentials shown in Figure 20.5 and letting $\rho_i = 0.01$ nm (so that the incident wave is approaching the ion core potential almost dead center), repeated evaluations of Eq. (20.20) as z_i moves from far away to in close give the J values shown in Figure 20.7. Position can be converted to time by the relation $z(t) = l - (F/2m)t^2$. As a result, the total emitted charge, or the sum over the product of $J(t_j)$ with the time increment $t_{j+1} - t_j = \Delta t_j$, makes for an estimate of the number of field emitted electrons as a function of time, as shown in Figure 20.8, which is related to the value of γ_f. The field $F = 3$ eV/nm was intentionally chosen low for the purposes of this example: a small increase in field, charge state Z, or dwell time of the argon atom on the surface will substantially increase the current density, and thereby the emitted charge.

If the origin of the current density shown in Figure 20.7 were due to standard field emission processes from a copper-like metal with the effect of the ion to simply reduce the effective work function, then the Fowler–Nordheim equation itself could be used to back out what the presumptive work function would be. From Eq. (13.25), a reasonable quadratic fit gives $\Phi(J)$ [eV] $\approx 4.9193 - 0.1169\ln(J) - 0.010228(\ln(J))^2$ for J in units of A/cm^2. Consulting Figure 20.7, the maximum J is 1.3342×10^9 A/cm^2, for which $\Phi(J) = 2.0115$ eV, and therein lurks a problem. The Schottky barrier-lowering factor for $F = 3$ eV/nm is $\sqrt{4QF} = 2.0784$ eV: the barrier maximum has been dragged below the Fermi

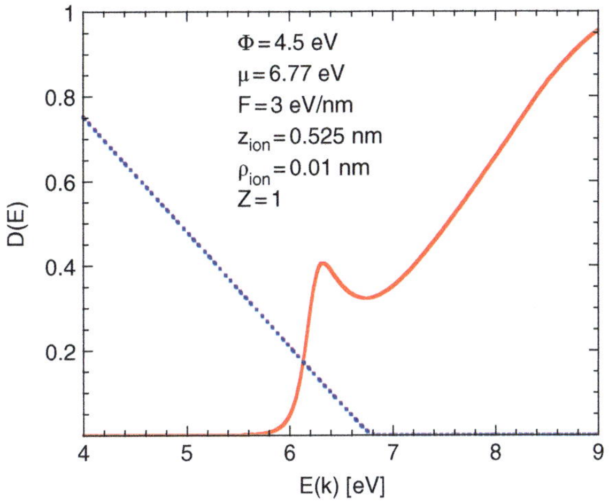

Figure 20.9 The transmission probability $D(E_k)$ (solid line) alongside the supply function ratio $f(E_k)/f(0)$ (dashed line). The resonance level is just below the Fermi level μ.

level μ, and the Fowler–Nordheim equation under such circumstances is not an adequate representation. Something else is involved, and that something else might be resonances.

To see if the resonances hypothesis is reasonable, return to the Airy TMA values of $D(k)$ for when the ion is 0.525 nm away from the putative surface. Graphing $D(E_k)$ alongside the supply function $f(E_k)$, as in Figure 20.9, shows that a resonance level has been brought to just below the Fermi level (at room temperature, $f(E)$ appears to be linear and cross the axis at $E = \mu$, after which it vanishes quickly). Creating similar graphs for z_i further out, or closer in, shows that the resonance becomes less favorable to emission. Therein lies the power of the Airy TMA, as the resonances would not otherwise be revealed. The modified Paschen relation is therefore interesting because a variety of emission mechanisms act in tandem and therefore the physics is rich.

PART IV

The complexity of materials

The underlying physical laws necessary for the mathematical theory of a large part of physics and the whole of chemistry are thus completely known, and the difficulty is only that the exact application of these laws leads to equations much too complicated to be soluble. It therefore becomes desirable that approximate practical methods of applying quantum mechanics should be developed, which can lead to an explanation of the main features of complex atomic systems without too much computation.

– Paul A.M. Dirac [290]

CHAPTER 21

Metals

The individuals see the good they reject; the public wills the good it does not see. All stand equally in need of guidance.

– Jean Jacques Rousseau[1]

...men that are free, well-born, well-bred, and conversant in honest companies, have...honor. Those same men, when by base subjection and constraint they are brought under and kept down, turn aside from that noble disposition...

– Francois Rabelais[2]

Many treatments of energy bands in metals and semiconductors start with how electrons move through the bulk material [82, 86], but an equally good place to start is how the atoms behave as they are brought together to form that bulk [104]. Each illuminates in a unique way the different behavior of electrons when free or bound, how these behaviors are manifest as collections of atoms are brought together to form the bulk material, and how electrons are then induced to differentiate between where they can go and where they cannot: whereas a physics approach focuses on how the Bloch waves which roam over the (perfect) lattice behave, a materials approach considers the *linear combination of atomic orbitals* (LCAO) and the chemical bonds they enable.

Plane waves are useful basis states, given their simple relationship between $\vec{k}$ and energy $E(k)$, and especially because of their utility in treating tunneling and emission. Nevertheless, inside the metal (or semiconductor), freely roaming electrons are from atoms that have lost their outermost electron, and what is left behind are *ions* that give rise to long-range Coulomb potentials. The free electrons that migrate long distances over the bumps and lumps of the underlying ions gives an understanding of how the band gaps behind the Kronig–Penny model arise.

The opposite focus on the atoms that are brought into close proximity considers using basis states characteristic of the Coulomb potential, such as those behind the hydrogen atom of Section 9.1. Those potentials interact, and how they do so gives another way of appreciating how band gaps originate, which is the viewpoint behind the LCAO approach.

21.1 Density of states, again

I have told thee often, and I retell thee again and again...

– William Shakespeare[3]

A potential well with infinite walls located at $\pm L/2$ constitutes a straightforward and simple quantum mechanical problem in the determination of its energy levels. Within the well, Schrödinger's equation relates energy to momentum as $E = \hbar^2 k^2/2m$, so the most general state of energy E is of the form (using the notation of Section (8.1.1)) where particles travel to the right ($+k$) or left ($-k$), or $|\psi\rangle = a_+|k\rangle + a_-|-k\rangle$, where $a_\pm$ are coefficients to be determined. As in Section 9.2.2, the normal procedure is to match the wave function and its first derivative at boundaries, but for infinite walls there is nothing beyond them. This makes the problem not one of finding the coefficients, but of finding restrictions that are placed on the values of k. Because the wave function vanishes at both boundaries, the two equations for the two unknown a values are

$$\begin{pmatrix} e^{ikL/2} & e^{-ikL/2} \\ e^{-ikL/2} & e^{ikL/2} \end{pmatrix} \begin{pmatrix} a_+ \\ a_- \end{pmatrix} = \begin{pmatrix} 0 \\ 0 \end{pmatrix} \tag{21.1}$$

[1]Jean Jacques Rousseau (trans. G.D.H. Cole), *Montesquieu/Rousseau, Great Books of the Western World*, ed. Robert Maynard Hutchins, Vol. 38. Chicago: Encyclopaedia Britannica, 1952, "The Social Contract", Book II, Chapter 6, p. 400.
[2]Francois Rabelais (trans. Sir Thomas Urquhart and Peter Motteux), *Gargantua and Pantagruel, Great Books of the Western World*, ed. Robert Maynard Hutchins, Vol. 24, Rabelais. Chicago: Encyclopaedia Britannica, 1952, Book 1, Chapter 57.
[3]Ref. [37]: *Othello* I.3.361–362.

Introduction to the Physics of Electron Emission, First Edition. Kevin L. Jensen.

where the top row holds for $x = L/2$ and the bottom for $x = -L/2$. From linear algebra theory, if the determinant of the matrix for a homogeneous equation (one of the form $\mathbb{M}|a\rangle = 0$) vanishes, then non-trivial solutions are possible (a trivial solution is where all the elements of $|a\rangle$ are zero). Setting the determinant of the matrix in Eq. (21.1) to zero (i.e., $det(\mathbb{M}) = 0$) therefore requires

$$e^{ikL} - e^{-ikL} = 2i\sin(kL) = 0 \tag{21.2}$$

which is satisfied for $k \to k_j = j\pi/L$ where j is an integer. Putting that result into the equation from the top row of Eq. (21.1) (or the bottom, just pick one) shows that $a_+ = -a_- \equiv a_j$, and so the general solution is

$$\psi_j(x) = a_j \sin\left(\frac{j\pi x'}{L}\right) \tag{21.3}$$

where $x' = x - (L/2)$. The energy levels for the one-dimensional well with infinite sides is consequently

$$E_j = \frac{\hbar^2}{2m}\left(\frac{j\pi}{L}\right)^2 \tag{21.4}$$

Elegant, but so what? Well, remember that in developing the canonical equations of electron emission in Part II, it was useful to switch over from an integration over dk to one over dE, as done in Section 7.3. When dealing with electron transport with scattering in bulk materials, energy integrations are advantageous. Consequently, if there is a restriction on what values k can take, better to know how the distribution of energy states, or rather the density of states (DOS) function $\mathcal{D}_1(E)$, is affected,[4] where, yes, the subscript refers to the number of dimensions of the well and, yes, higher dimensions entail other difficulties.

The DOS function has already been encountered for the continuum case in Section 7.3, but the circumstance now is how discrete phase space points are handled. That makes the energy levels discrete, and counting the number of levels within a given energy range ΔE is the task. Graphically, there is a very simple way to do that by using a histogram, exactly as was done with incomes in Section 4.1 or for velocities in Section 5.1. Because E_j goes as j^2, the number of levels is expected to decrease in each bin of size ΔE as E itself increases (at least in one dimension). The calculation is simple: let the highest filled energy level in the well be μ (the Fermi level at zero temperature, where there are no unoccupied levels below it as per Fermi–Dirac statistics (see Section 6.2)). Calculate the values E_j and put them into bins of size ΔE so that the energy of the last electron in just fits into the last bin for the one-dimensional case. For N_b bins, then $\Delta E = E_{20}/N_b$, for example. "Filling a bin" is taken to mean bin n holds all the energy levels for which $(n-1)\Delta E \leq E_j < n\Delta E$. With the filled bins, consider the resulting histogram that counts the number of energy levels E_j in each bin. If the number of bins is 20, then the number of levels in each bin looks like Figure 21.1 if $N = 2000$ and L is cleverly chosen so that $E_j = j^2$. The data seem to fall on a familiar curve, and indeed comparing it to a power law shows that the curve is well represented by the form $y(E) \propto 1/\sqrt{E}$. The suggested curve gets better as bins for larger n are considered.[5]

EXAMPLE: For an electron in an infinite well with a ground state energy of $E_o = 1$ meV (so that $E_j = j^2$), find the width L of the well.

SOLUTION: Let $E_o = E(j = 1) = \pi^2\hbar^2/2mL^2$. Using $m = 510999$ eV/c^2 and $\hbar c = 197.327$ eV-nm, then $L =$ 19.3915 nm. When L is large, the energy levels are very finely spaced.

What happens if the number of dimensions is now 2 and the well is of width L in both dimensions? Now the wave function is [43, 44]

$$\psi_{nj}(x,y) = a_{nj}\sin\left(\frac{n\pi x'}{L}\right)\sin\left(\frac{j\pi y'}{L}\right) \tag{21.5}$$

(here, ignore the prime on the $x' = x - (L/2)$ and $y' = y - (L/2)$: the discussion is more or less the same) and so $E_{nj} = (\hbar^2/2m)(\pi/L)^2(n^2 + j^2)$ with the restriction being that the sum $n^2 + j^2$ must be such that $E_{nj} \leq \mu$. By straightforward extension, in three dimensions, this would become $E_{njl} = (\hbar^2/2m)(\pi/L)^2(n^2 + j^2 + l^2)$ such that $E_{njl} \leq \mu$. When the

[4] Calling the DOS $D(E)$ like Kittel [86] does would have been pleasant, but that name has already been taken by the transmission probability. Ziman [82] favored $\mathcal{N}(E)$, but that looks like a number function. The differing notation of different authors is irksome. So $\mathcal{D}(E)$ is chosen, following the convention of Section 7.3.

[5] Such findings are not a surprise for those familiar with Eq. (7.13) and the examples which follow it.

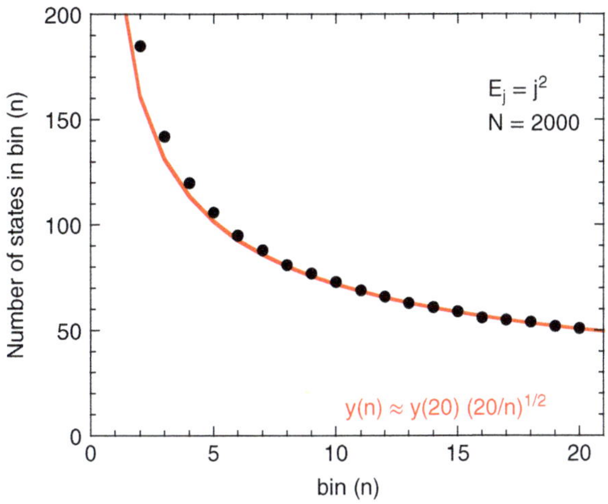

Figure 21.1 A histogram with 20 bins of the one-dimensional energy states for $E_j = (\hbar^2/2m)(j\pi/L)^2$ for the infinite well with a width of $L = \pi\hbar/\sqrt{2mE_o}$ with $E_1 \equiv E_o$ being the ground state energy. The black dots represent the number of energy levels E_j in each bin. A curve of the form $y(20)\sqrt{20/n}$ is also included to guide the eye, where $y(20) = 51$.

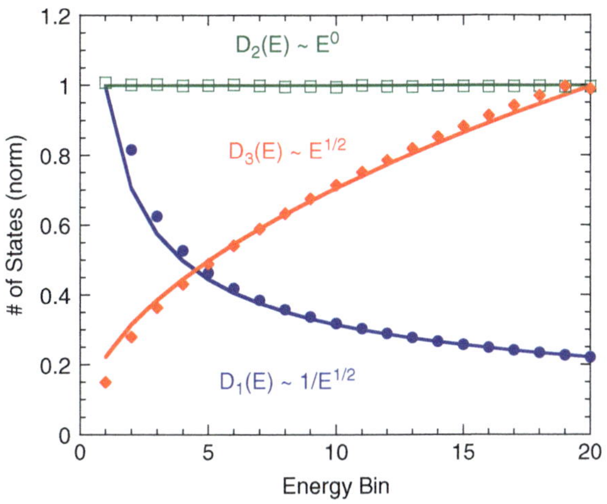

Figure 21.2 A histogram with 20 bins of the one-, two-, and three-dimensional energy states $E_j \propto j^2$, $E_{jk} \propto n^2 + j^2$ and $E_{njl} \propto n^2 + j^2 + l^2$, respectively. The curves to guide the eye behave as $y_1(x) \propto x^{-1/2}$, $y_2(x) \propto x^0$ and $y_3(x) \propto x^{1/2}$, respectively.

histograms for two and three dimensions are plotted alongside the one-dimensional case (with all of the curves normalized so that they can appear on the same graph) and compared to best-fit power laws, it is found the two-dimensional case is well-fitted by $y_2(E) =$ constant, and $y_3(E) \propto \sqrt{E}$, as shown in Figure 21.2.

21.2 Spheres in *d* dimensions

> *The World is Spherical...whether because this figure is the most perfect of all, as it is an integral whole and needs no joints; or because this figure is the one having the greatest volume and thus is especially suitable for that which is going to comprehend and conserve all things; or even because the separate parts of the world i.e., the sun, moon and stars are viewed under such a form; or because everything in the world tends to be delimited by this form...And so no one would hesitate to say that this form belongs to the heavenly bodies.*
>
> – Nicholaus Copernicus[6]

[6] Nicholaus Copernicus (trans. Charles Glenn Wallis), *On the Revolutions of the Heavenly Spheres Great Books of the Western World*, ed. Robert Maynard Hutchins, Vol. 16, Ptolemy/Copernicus/Kepler. Chicago: Encyclopaedia Britannica, 1952, p. 511. The "World" referred to not only the Earth, but also the heavens. To the modern ear, attributing divine status to geometric figures sounds strange, but the paradigms of that time were different, for example 76 years after Copernicus, Johannes Kepler related the orbits of the planets to musical scales in his work *Harmonices Mundi* (*Harmony of the World*), which also sounds (no pun) strange. Concurrently, Kepler also had to defend his mother against serious accusations of witchcraft. His times had peculiar notions, as perhaps all times do.

Clearly something deeper is going on, and that something is geometry: the number of discrete states in a d-dimensional phase space that lie in a particular energy bin, where energy is proportional to the square of the magnitude of a d-dimensional momentum vector $\vec{k}$, is now a counting problem, but if the phase space points are equally spaced on a grid, then the count will be proportional to the volume of phase space given by product of the surface area of a hollow sphere and the thickness of its shell. Knowledge of this was presumed in Section 7.3, but consider the problem in greater detail. Borrowing the "bin" analogy of above, the bin holds all the energies that lie in the "element" $d\Omega$. In one dimension, the differential element $d\Omega$ is a line, in two dimensions it is an area, and in three dimensions a volume: each case corresponds to a d-sided object. The construction of the d-sided differential element $d\Omega_d = dk_1 dk_2 \cdots dk_d$ proceeds by iteratively finding the differential element dk_d that is orthogonal to all the other $d-1$ elements. The process symbolically represented in Figure 21.3 shows how this is done for up to three dimensions, and the method can be generalized to higher dimensions: algorithmically, $d\Omega_2 = dk_2 d\Omega_1 = (kd\theta)(dk) = kdkd\theta$, and similarly, $d\Omega_3 = dk_3 d\Omega_2 = (k\sin\theta d\varphi)(kdkd\theta) = k^2 \sin\theta dkd\theta d\varphi$. For d dimensions, then,

$$d\Omega_d = k^{d-1} dk \prod_{j=0}^{d-2} (\sin\theta_j)^j d\theta_j \tag{21.6}$$

The problems of interest constrain $d \leq 3$ where $d\theta_1 = \theta$ and $\theta_0 = \varphi$ are the usual polar and azimuthal angles, respectively, in spherical coordinates.

The important point is that $E(k)$ depends only on the magnitude $\vec{k}$ in the form of k^2. Therefore, all the angular integrations can be performed to give:

$$d\Omega_1 \rightarrow dk \tag{21.7}$$

$$d\Omega_2 \rightarrow kdk \int_0^{2\pi} d\varphi = 2\pi kdk \tag{21.8}$$

$$d\Omega_3 \rightarrow k^2 dk \int_0^{\pi} \sin\theta d\theta \int_0^{2\pi} d\varphi = 4\pi k^2 dk \tag{21.9}$$

The final step defines $d\Omega_d \equiv (d\Omega_d/dE)dE$. From $E = \hbar^2 k^2/2m$ and $d\Omega_d \equiv D_d(E)dE$, it is then straightforward to show (observe the factor of $2/(2\pi)^d$ in each equation)

$$D_1(E) = \frac{2}{(2\pi)^1}\frac{1}{2\hbar}\left(\frac{2m}{E}\right)^{1/2} = \frac{1}{2\pi\hbar}\sqrt{\frac{2m}{E}} \tag{21.10}$$

$$D_2(E) = \frac{2}{(2\pi)^2}\frac{2\pi m}{\hbar^2} = \frac{m}{\pi\hbar^2} \tag{21.11}$$

$$D_3(E) = \frac{2}{(2\pi)^3}\frac{4\pi m}{\hbar^2}\left(\frac{2mE}{\hbar^2}\right)^{1/2} = \frac{m}{\pi^2\hbar^3}\sqrt{2mE} \tag{21.12}$$

from which the behavior of Figure 21.2 makes perfect sense.

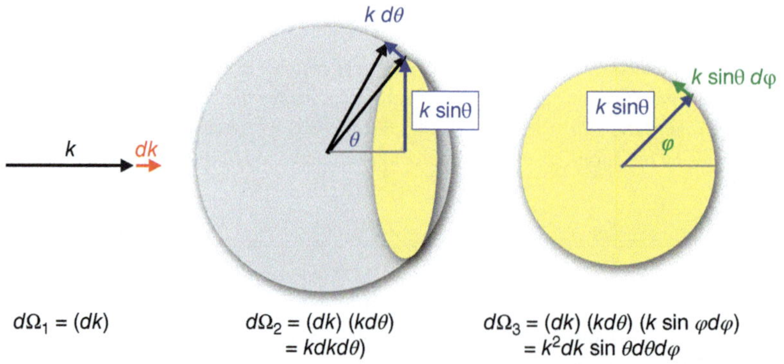

Figure 21.3 Spheres in d dimensions. Left, a one-dimensional "sphere" is a line k, and its element is dk. Middle, a two-dimensional "sphere" is a circle, and its "volume" element is $kdkd\varphi$. Right, a three-dimensional sphere really is a sphere, and its volume element is $k^2 dk \sin\theta d\theta d\varphi$.

21.3 The Kronig Penny model

The principal idea of Bloch's theory is the assumption that the interaction of a given electron with the other particles of the lattice may be replaced in first approximation by a periodic field of potential. With this model an interpretation of the specific heat, the electrical and thermal conductivity, the magnetic susceptibility, the Hall effect, and the optical properties of metals could be obtained.

– R.D. Kronig and W.G. Penney [291]

A very wide well bounded by infinite walls makes for simple expressions for $\mathcal{D}(E)$, but whether such an idealized potential describes a bulk material needs consideration. Matter is complicated. In preparation for how atoms cause that complexity, consider the next simplest potential well with infinite walls, for which half of the well is at a higher potential than the other half. That is, instead of $V(x) = 0$ for $-L/2 \le x \le L/2$ (in one dimension), let the potential for $x < 0$ remain unchanged, but for $x > 0$, let $V(x > 0) = V_o \equiv \hbar^2 k_o^2/2m$, akin to a step barrier stuck into a well with infinite walls, as schematically suggested in Figure 21.4. Now what happens to the energy levels?

How the energy levels change depends on whether k is smaller or larger than k_o. Consider the case $k > k_o$ first, as in the limit $k_o \to 0$, the findings of Section 21.1 should be reclaimed. Let the real κ be defined by $\kappa = \sqrt{k^2 - k_o^2}$. Finally, let region 1 (deep well) be for $x < 0$ and region 2 (shallow well) be for $x > 0$. The wave functions are specified by

$$\psi_1(x) = Ae^{ikx} + Be^{-ikx} \tag{21.13}$$

$$\psi_2(x) = Ce^{i\kappa x} + De^{-i\kappa x} \tag{21.14}$$

where the coefficients a_j and b_j are to be determined by the boundary conditions. As is now routine, one equation comes from $\psi_1(0) = \psi_2(0)$, and one from the derivatives $\psi_1'(0) = \psi_2'(0)$. The third equation is from the left wall, $\psi_1(-L/2) = 0$, and the fourth from the right wall, $\psi_2(L/2) = 0$. The conditions on the coefficients can now be put in a matrix equation $\mathbb{M}\,|u\rangle$ where the matrix $\mathbb{M}$ arises from the boundary conditions and the vector $|u\rangle$ arises from the coefficients, or

$$\begin{pmatrix} 1 & 1 & 1 & 1 \\ ik & -ik & -i\kappa & i\kappa \\ e^{-ikL/2} & e^{ikL/2} & 0 & 0 \\ 0 & 0 & e^{i\kappa L/2} & e^{-i\kappa L/2} \end{pmatrix} \begin{pmatrix} A \\ B \\ C \\ D \end{pmatrix} = 0 \tag{21.15}$$

where each row encodes one of the aforementioned equations. To have non-trivial solutions of the equations, the determinant of the 4×4 matrix must equal zero. An evaluation of the determinant followed by collecting terms gives

$$k\left(e^{ikL} + 1\right)\left(e^{i\kappa L} - 1\right) + \kappa\left(e^{ikL} - 1\right)\left(e^{i\kappa L} + 1\right) = 0 \tag{21.16}$$

for which a clever use of the half-angle formulae for sines and cosines results in

$$(k + \kappa)\sin\left[(k + \kappa)L/2\right] = (k - \kappa)\sin\left[(k - \kappa)L/2\right] \tag{21.17}$$

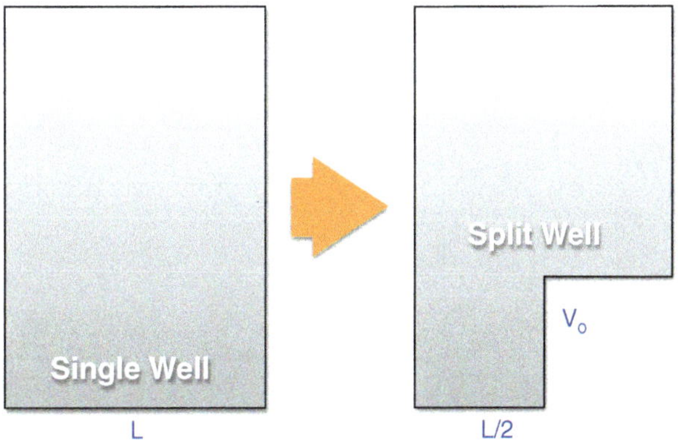

Figure 21.4 Representation of how a well with infinite side walls is modified to make a split level well such that $V(-L/2 \le x \le 0) = 0$ and $V(0 \le x \le L/2) = V_o$.

In the trivial limit that $k_o/k \rightarrow 0$ (i.e., $\kappa \rightarrow k$) the right-hand side vanishes and the usual relation $k = j\pi/L$ results, as it did in Section 21.1. The values of allowed k may be obtained graphically, solutions to the problem being where the rapidly oscillating function on the left-hand side crosses the much more slowly varying function on the right.

On the other hand, the case $k_o > k$ results in the substitution $\kappa \rightarrow i\kappa$ in the above analysis. Not using the half-angle formulae this time results in

$$\frac{k}{\kappa} = \tan(kL/2)\coth(\kappa L/2) \tag{21.18}$$

Now consider the limit when $k_o \rightarrow \infty$, which drags κ with it, and makes $1/\kappa \rightarrow 0$ and $\tanh(\kappa L/2) \rightarrow 1$. In that limit, solutions exist for $k = 2j\pi/L$, that is, only every other energy level survives as one would expect for a well half the size. For general k_o, the number of energy levels is between the two limits and decreases as k_o increases.

When the infinite walls are removed and the deep well/shallow well is repeated over and over, something akin to the one-dimensional potential shown in Figure 21.5 results (the labeling of "superlattice" to draw attention to the similarity of the potentials of Section 19.1, and because "ion cores" associated with the bulk material atoms have not yet been introduced). The general problem, in which the well regions and barrier widths are of different size, is known as the Kronig–Penny model [52, 86, 291]. A flavor of what is to come has already been hinted at in Figure 19.8: not only will discrete energy levels in the wells ($k < k_o$) arise, but *regions where there are no allowed solutions* for ($k > k_o$) arise as well, and the latter will come to be called "bands."

Imagine the number of wells significantly expanded to N wells: these will come in time to be associated Coulomb potentials of the ion cores but first, rectangular wells look nothing like that, and second, the problem at hand is one-dimensional and a lattice of atoms in a bulk material is very much three-dimensional. Nevertheless, keeping the destination in mind makes the journey more comprehensible. If N is large, the first well and the Nth well have very little influence on the center wells, and most of the wells can be visualized as existing in similar circumstances, but to really remove any distinguishing feature between them, impose periodic boundary conditions so that the $N + 1$ well is taken as the same as the first well. Now all the wells are equivalent with respect to the physics of interest, such as energy levels.

A translation (or rotation, if periodicity is obtained by attaching the Nth well to the first well to form a ring of N links) that does not change the energy levels means that the wave function $\Psi(x)$ describing the ring is the same as for the rotated ring up to an overall phase factor. If the distance from center to center of two adjacent wells is a distance $l = a + b$, where a is the well width and b is the barrier width, and $Nl = L$, then it is hopefully apparent that $V(x + l) = V(x)$. But this relation also has to be reflected in the wave function itself, with the only difference being that if the wave function is shifted a distance a, that is only reflected in an overall phase factor. Therefore, the behavior of $V(x)$ demands

$$\Psi(x + a) = e^{i\varphi}\Psi(x) \tag{21.19}$$

Periodicity of the boundary conditions (i.e., the identification of the $N + 1$ well with the first well), however, means that

$$\Psi(x + Nl) = \Psi(x) \tag{21.20}$$

but that simply means that $\exp(iN\varphi)$ is one of the N roots of 1, or

$$\varphi = \frac{2\pi j}{N} \tag{21.21}$$

Coupled with Eq. (21.19), that entails

$$\Psi(x) = e^{i2\pi jx/Nl}u_k(x) \tag{21.22}$$

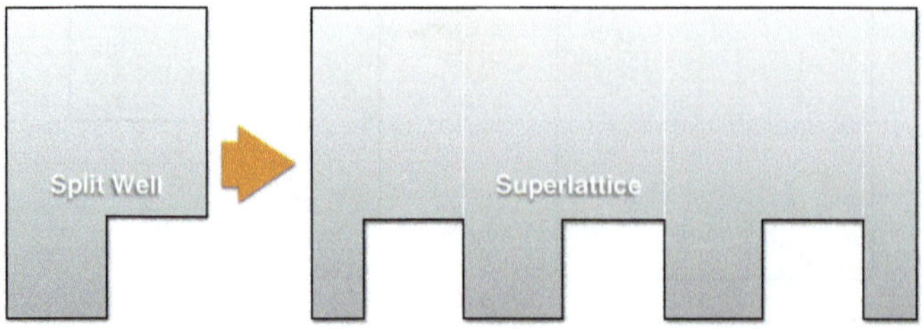

Figure 21.5 Representation of how the split level well of Figure 21.4 is repeated to make a superlattice (the infinite side walls are now removed; the faint white lines at the barrier edges are only to guide the eye). The wells and barriers need not be equally wide.

where $u_k(x)$ is a function with the periodicity of the lattice, or $u_k(x+l) = u_k(l)$. Identifying $k_j = 2\pi j/Nl = 2\pi j/L$, where $L \gg l$ is now the dimensions of the box containing all the wells, results in

$$\Psi(x) = e^{ik_j x} u(x) \tag{21.23}$$

which is the one-dimensional version of the Bloch theorem: when the potential is periodic, the eigenfunctions are the product of the plane wave with a function $u(x)$ that has the periodicity of the lattice [82, 86, 104]. As usual, let $V_o = \hbar^2 k_o^2/2m$ and $\kappa^2 \equiv |k_o^2 - k^2|$. When Eq. (21.23) is inserted into Schrödinger's equation, two equations result: the first (call it $u_a(x)$) corresponds to a well and the second (call it $u_b(x)$) corresponds to a barrier. The equations they satisfy are, for $k_o < k$,

$$\left\{\partial_x^2 + 2ik_j\partial_x + k^2 - k_j^2\right\} u_a(x) = 0$$

$$\left\{\partial_x^2 + 2ik_j\partial_x - \kappa^2 - k_j^2\right\} u_b(x) = 0 \tag{21.24}$$

Solutions are found by trying $u = e^{iwx}$ in each region and finding what w is. For each region, it is found

$$w_a^2 + 2k_j w_a + k_j^2 - k^2 = 0 \tag{21.25}$$

$$w_b^2 + 2k_j w_b + k_j^2 - \kappa^2 = 0 \tag{21.26}$$

where it is seen that the role k plays for w_a is the role that $i\kappa$ plays for w_b, much like for the rectangular barrier of Section 18.1. From the quadratic equation, two solutions exist, and it is straightforward to show that they are $w_a^{\pm} = k_j \pm k$ and $w_b^{\pm} = k_j \pm i\kappa$, with the most general wave function being a combination, or

$$u_a(x) = A\exp\left[-i(k_j - k)x\right] + B\exp\left[-i(k_j + k)x\right] \tag{21.27}$$

$$u_b(x) = C\exp\left[(\kappa - ik_j)x\right] + D\exp\left[-(\kappa + ik_j)x\right] \tag{21.28}$$

The matching of the wave function and its first derivative at $x = 0$ where the first well meets the first barrier specifies two equations. Two more are specified by the periodicity conditions that relate the u_a and u_b wave functions and their derivatives. Because these equations are set to 0, they can be multiplied by any convenient factor that will make taking the determinant of the matrix $\mathbb{M}$ simpler. Therefore, consider the four equations given by

$$u_a(0) - u_b(0) = 0 \tag{21.29}$$

$$\partial_x\left\{u_a(x) - u_b(x)\right\}_{x=0} = 0 \tag{21.30}$$

$$e^{ik_j a}\left\{u_a(a) - u_b(-b)\right\} = 0 \tag{21.31}$$

$$e^{ik_j a}\left(\partial_x\left\{u_a(x)\right\}_{x=a} - \partial_x\left\{u_b(x)\right\}_{x=-b}\right) = 0 \tag{21.32}$$

In matrix evaluations, there is no substitute for cleverness, and in creating $\mathbb{M}$ from the above four equations, the combinations $A \pm B$ and $C \pm D$ appear, so that rewriting the equations in terms of $A' = A + B$, $B' = A - B$, $C' = C + D$, and $D' = C - D$ simplifies the resulting matrix to

$$\mathbb{M} = \begin{pmatrix} 1 & 0 & -1 & 0 \\ -ik_j & ik & ik_j & -\kappa \\ \cos(ka) & i\sin(ka) & -\varphi\cosh(\kappa b) & \varphi\sinh(\kappa b) \\ h_1 & h_2 & \varphi h_3 & -\varphi h_4 \end{pmatrix} \tag{21.33}$$

where, for compactness, the following abbreviations are used: $\varphi = e^{ik_j a}$ and

$$h_1 = -ik_j\cos(ka) - k\sin(ka)$$

$$h_2 = k_j\sin(ka) + ik\cos(ka)$$

$$h_3 = ik_j\cosh(\kappa b) + \kappa\sinh(\kappa b)$$

$$h_4 = ik_j\sinh(\kappa b) + \kappa\cosh(\kappa b) \tag{21.34}$$

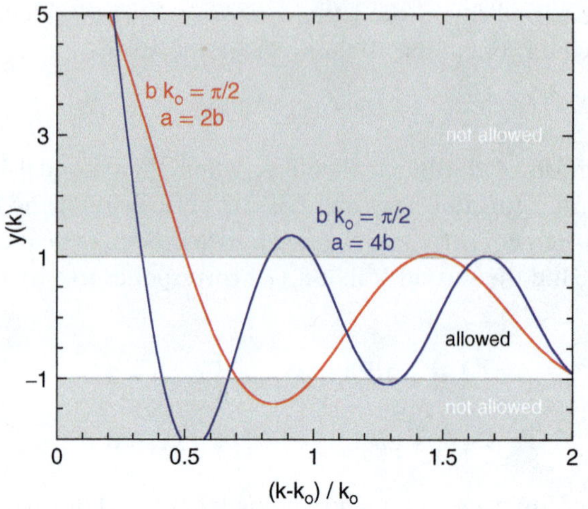

Figure 21.6 The behavior of the right-hand side of Eq. (21.36) (called $y(k)$) for parameters such that $bk_o = \pi/2$ and $a = 2b$ (red), and $a = 4b$ (blue). Because the left-hand side of Eq. (21.36) is ≤ 1, the greyed regions in the figure correspond to values of k that are forbidden. As a result, there are gaps, or bands, between allowed values of k, and therefore $E(k)$.

The advantage of the relabeling of the coefficients is to render the top row of Eq. (21.33) easier with respect to the evaluation of determinants by the method of cofactor expansion [91]. Even with the simplifications, it takes a bit of effort to show that setting the determinant of $\mathbb{M}$ to 0 results in the relation for $k < k_o$:

$$\cos\left(k_j l\right) = \left(\frac{\kappa^2 - k^2}{2k\kappa}\right)\sin(ka)\sinh(\kappa b) + \cos(ka)\cosh(\kappa b) \tag{21.35}$$

where, recall, $l = a + b$. When $k > k_o$, then $\kappa \to i\kappa$ and so

$$\cos\left(k_j l\right) = -\left(\frac{\kappa^2 + k^2}{2k\kappa}\right)\sin(ka)\sin(\kappa b) + \cos(ka)\cos(\kappa b) \tag{21.36}$$

The limits of these equations are illuminating. For the case where the barriers between the wells are removed, the energy should be given by that of the plane wave states, and indeed as $k_o \to 0$, then $\kappa \to k$ in Eq. (21.35), or

$$\cos(k_j l) = \cos(ka)\cos(kb) - \sin(ka)\sin(kb) = \cos(kl) \tag{21.37}$$

from which $E(k = k_j) = \hbar^2 k_j^2/2m$, as expected. In the opposite limit, where the height of the barriers becomes large, or $k_o \to \infty$, then

$$\frac{\sin(ka)}{\cos(ka)} = -\frac{2k}{k_o} \to 0 \tag{21.38}$$

or $k = (j\pi/a)$ for integer j (compare the implications of Eq. (21.2)), that is, the energy levels of the single well with infinite walls of Section 21.1 are recovered. Such levels are very narrow, but when k_o is of finite size, some wiggle room is recovered and the levels broaden, particularly for energies near the barrier height V_o.

For the general $k > k_o$ case, though, the left-hand side of Eq. (21.36) oscillates between +1 and −1, but the right-hand side can take on values outside those limits. Those values of k that are associated with the right-hand side being greater than 1 or less than −1 are therefore disallowed: in other words, *bands occur* as anticipated by Figure 19.8, and correspond to gaps at finite intervals in the continuous energy spectrum of the electron as it moves through the lattice. Graphically, this is seen in Figure 21.6, where the right-hand side of Eq. (21.36) is shown as a function of $(k - k_o)/k_o$. The regions for which $|y(k)| \leq 1$ (within the band of white) represent where allowed k values exist. By contrast, the lines in the gray region correspond to regions where no k value satisfies the relation of Eq. (21.36). A close-up of the band regions is shown in Figure 21.7.

21.4 Atomic orbitals

"Modern materials science is at least as concerned with defects in crystals and noncrystalline materials as it is with the properties of perfect crystals. (In a phrase often attributed to F.C. Frank "crystals are like people – it is the defects in them that makes them interesting.")

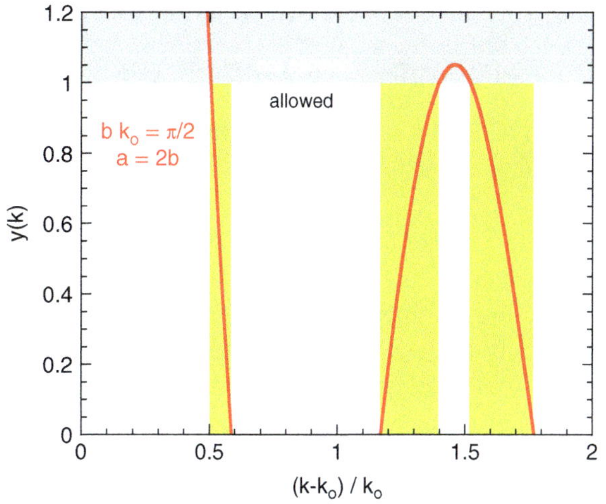

Figure 21.7 Close-up of Figure 21.6, showing the region of bands (pale green) for the $a = 2b$ case.

Band theory and free electron theory have very little to say about such things, because they rely on perfect translational symmetry and Bloch's theorem. To understand the properties of defects...we have to take an altogether different viewpoint and set the problem up in real space."

– Adrian P. Sutton [104], Chapter 1.

A different outlook to envisioning free electrons roaming over the entire bulk material is to consider their behavior near two interacting atoms. At some point, the differing views will need reconciliation, but for now consider two hydrogen atoms (the simplest type) and how they interact. Quantum mechanically, a state vector $|\Psi\rangle$ will describe them both, and the Hamiltonian $\hat{H}$ acting on the system of both atoms will give the total energy $\hat{H}\,|\Psi\rangle = E\,|\Psi\rangle$, or specifically,

$$\hat{H}\left\{C_1\,|\psi_1\rangle + C_2\,|\psi_2\rangle\right\} = E\left\{C_1\,|\psi_1\rangle + C_2\,|\psi_2\rangle\right\} \tag{21.39}$$

If the hydrogen atoms are far apart from each other, then both $\psi_1(\vec{r}) = \langle\vec{r}|\psi_1\rangle$ and $\psi_2(\vec{r}) = \langle\vec{r}|\psi_2\rangle$ are going to look very much like the hydrogenic wave functions of Eq. (9.33), and $\langle\psi_1|\psi_2\rangle \approx 0$. It is convenient to retain this approximation and take $\psi_j(\vec{r})$ to be those hydrogenic wave functions to leading order and to take their overlap as $\langle\psi_1|\psi_2\rangle = 0$. In actuality, their overlap is not zero, but it does fall off exponentially with r, and it can therefore be regarded as negligible, simplifying the problem considerably. For both atoms,

$$\hat{H}_j\left|\psi_j\right\rangle = \left\{\frac{\hbar^2\hat{k}^2}{2m} + V_j(\hat{r})\right\}\left|\psi_j\right\rangle = E_o\left|\psi_j\right\rangle \tag{21.40}$$

Compare that to the full Schrödinger's equation, which is

$$\hat{H}\,|\Psi\rangle = \left\{\frac{\hbar^2\hat{k}^2}{2m} + V_1(\hat{r}) + V_2(\hat{r})\right\}|\Psi\rangle = E\,|\Psi\rangle \tag{21.41}$$

That means there are two equations for the two unknown C_j in Eq. (21.39) to be determined by $\langle\psi_j|\hat{H}|\Psi\rangle = EC_j$ for $j = (1, 2)$. For convenience, take the hydrogenic wave functions to be in their ground state $(1s)$, so that the form of $V(\hat{r})$ is the same for both. In writing out the equations for C_j, terms of the form $\langle\psi_1|\,V(\hat{r})\,|\psi_2\rangle$ arise and represent the influence of one nucleus on the other. Because the hydrogenic wave functions are real for s states,

$$\langle\psi_1|\,V(\hat{r})\,|\psi_2\rangle = \langle\psi_2|\,V(\hat{r})\,|\psi_1\rangle \equiv -\delta E \tag{21.42}$$

and so $E = E_o - \delta E$, the sign convention chosen so that δE itself is positive (the potential is attractive and the energy change therefore negative). The two equations for the C_j encapsulated in $\langle\psi_j|\hat{H} - E|\Psi\rangle = 0$ can therefore be put into a matrix equation of the form

$$\begin{pmatrix} E_o - E & -\delta E \\ -\delta E & E_o - E \end{pmatrix}\begin{pmatrix} C_1 \\ C_2 \end{pmatrix} = 0 \tag{21.43}$$

For solutions to exist, the determinant of the matrix of coefficients must equal zero, and so

$$\left(E_o - E\right)^2 - \delta E^2 = 0 \rightarrow E = E_o \pm \delta E \tag{21.44}$$

As in Section 8.1.2, the eigenvectors can be found depending on whether the + or − sign is chosen for the eigenvalue. The two eigenstates are determined to be $C_2 = \pm C_1$ for $E = E_o \pm \delta E$, and so the normalized states are

$$\left|\psi_\pm\right\rangle = \frac{1}{\sqrt{2}}\left\{\left|\psi_1\right\rangle \pm \left|\psi_2\right\rangle\right\} \tag{21.45}$$

corresponding to a symmetrical (+) and anti-symmetrical (−) state when $\left|\psi_1\right\rangle$ and $\left|\psi_2\right\rangle$ are interchanged. In terms of charge density, the antisymmetrical density $\langle\psi_-|\psi_-\rangle$ vanishes at the point equidistant between the two nuclei (think of the antisymmetry of $\sin x$ about $x = 0$, for example). In contrast, $\left\langle\psi_+|\psi_+\right\rangle$ does not. Physically, the antisymmetric state pushes electron charge out of the center region and is therefore identified as an antibonding state, compared to the symmetrical case where charge is piled up between the nuclei and pulls them together, identifying it as a bonding state [104]. The point here, however, is that a consideration of the local interaction between the atoms has resulted in a discrete energy shift between one state and the other in a way that is a prelude to how bonding and antibonding between dissimilar atoms introduces the notion of "electronegativity" (how much different atoms seem to crave an electron they share) in Section 21.5, and anticipates an analogous difference in the behavior of the electron density for energy states just above and below what will be called the "band gap" for a sinusoidal potential of Section 21.6.

Regarding the bonding and anti-bonding states, these eigenstates are the analogs of the "stable strategies" discussed in Section (8.1.3): a system set up in such a state persists and its density $\rho_\pm = \left\langle\psi_\pm|\psi_\pm\right\rangle$ is time independent. On the other hand, what if the electron were confined initially to one of the atoms (say, atom 1, or $C_1(0) = 1, C_2(0) = 0$). How then would the density associated with atom 1, or $|C_1(t)|^2$, evolve over time? Such an initial state is a superposition of the symmetric and antisymmetric states in equal measure. From the time-dependent Schrödinger's equation $i\hbar\partial_t\,|\psi(t)\rangle = \hat{H}\,|\psi(t)\rangle$, it follows that

$$|\psi(t)\rangle = \frac{1}{\sqrt{2}}\left(e^{iE_+t/\hbar}\left|\psi_+\right\rangle + e^{iE_-t/\hbar}\left|\psi_-\right\rangle\right) \tag{21.46}$$

Then, knowing $C_1(t) = \left\langle\psi_1|\psi(t)\right\rangle$ leads to the conclusion

$$C_1(t) = \frac{1}{2}\left(e^{i\omega_+t} + e^{i\omega_-t}\right) = e^{iE_ot/\hbar}\cos\left(\frac{\delta E}{\hbar}t\right) \tag{21.47}$$

The same logic shows that $C_2(t)$ is the same form with cos → sin, so that the probability density of the electron on atoms 1 and 2 is

$$|C_1(t)|^2 = \cos^2\left(\frac{\delta E}{\hbar}t\right); \qquad |C_2(t)|^2 = \sin^2\left(\frac{\delta E}{\hbar}t\right) \tag{21.48}$$

respectively, and $|C_1(t)|^2 + |C_2(t)|^2 = 1$, as it must. The location of the electron oscillates back and forth between atoms 1 and 2 over time by tunneling through the potential barrier between them. When the atoms are too far apart, then the overlap factor δE is small and is an indication that the barrier between the atoms to the electron's migration is large, making the tunneling probability small; conversely, when the atoms are close, the overlap is significant and the passage back and forth more likely. The language of discourse is comfortably recognized from the discussion of the tunneling probabilities associated with a barrier in Section 9.2.3.

21.5 Electronegativity

I will pick up the hook.
You will see something new.
Two things. And I call them
Thing One and Thing Two.

– Theodor Geisel[7]

[7]Theodor Geisel (*aka* Dr. Seuss), *The Cat in the Hat*. New York: Random House, 1957, p. 33.

The most obvious generalization to the two hydrogen atoms problem of the previous section is two dissimilar atoms, called 1 and 2 in deference to convention. When the atoms are isolated, $\hat{H}_o |\psi_j\rangle = E_j |\psi_j\rangle$ for each atom. This changes matters because now the ground state energy of each atom is different. The modification to Eq. (21.43) is now

$$\begin{pmatrix} E_1 - E & -\delta E \\ -\delta E & E_2 - E \end{pmatrix} \begin{pmatrix} C_1 \\ C_2 \end{pmatrix} = 0 \tag{21.49}$$

Although visually not much of a change, it results in the eigenstate energies for the bonding and antibonding states acquiring an interesting twist. Setting the determinant of the multiplying matrix to zero as before gives

$$(E_1 - E)(E_2 - E) - \delta E^2 = 0 \tag{21.50}$$

With the introduction of the average energy $E_o = (E_1 + E_2)/2$ and the difference energy $\Delta E = (E_1 - E_2)/2$, E is then (compare Eq. (21.44))

$$E_\pm = E_o \pm \sqrt{\delta E^2 + \Delta E^2} \tag{21.51}$$

As before, the eigenstates are obtained by putting each eigenenergy $E_\pm$ into the matrix equation of Eq. (21.49) and finding the relationship between C_1 and C_2. Those relationships are

$$\frac{C_2(E_\pm)}{C_1(E_\pm)} = \frac{\Delta E}{\delta E} \mp \sqrt{1 + \left(\frac{\Delta E}{\delta E}\right)^2} \equiv \frac{1}{\sqrt{1 + r^2} \pm r} \tag{21.52}$$

with the (+) sign on $E_\pm$ associated with the bonding state $|\psi_+\rangle$ and the (−) sign associated with the antibonding state $|\psi_-\rangle$. The factor $r \equiv \Delta E/\delta E$ is the ratio of the difference in energies $(E_1 - E_2)/2$ factor to the overlap factor δE. The behavior is shown in Figure 21.8: when $r \to 0$, both atoms are identical and the values of the coefficients are equal in magnitude ($C_2 = \pm C_1$) whereas $r \gg 1$ signifies one atom has a much deeper (in terms of well depth) ground state, and the atom with the higher electronegativity starts to dominate the time the electron spends with it. In the limit of a purely ionic bond r will be very large: when it is infinite, the "shared" electron ceases to spend any time in the vicinity of the lower electronegative atom. In short, one atom attracts the shared electron more strongly than the other. Pauling empirically described the *electronegativity* E_P of an atom as a measure of its ability to pull electrons to itself from its neighbors: atoms with a greater electronegativity therefore hold on to the electron more of the time. In Pauling units,[8] fluorine is 3.98 and cesium is 0.79. The idea of an electronegativity of an individual atom is unusual because generally atoms are in molecules and electronegativity refers to how atoms behave there. An alternate definition due to Mulliken defined electronegativity E_M in terms of the energy required to rip an electron off an atom (I_A, or ionization potential) and the energy added to an

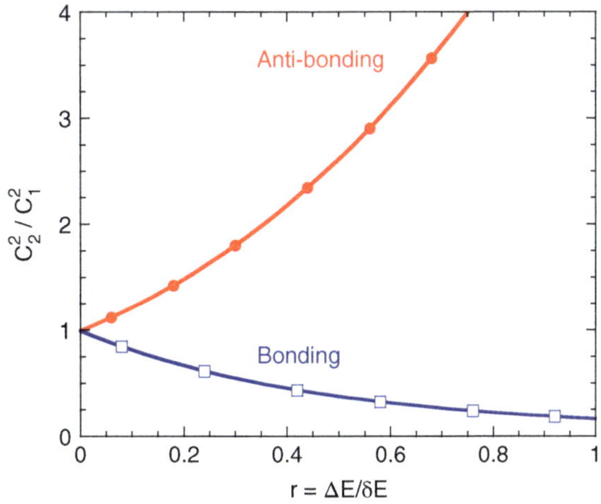

Figure 21.8 Ratio of the coefficients of the states |1⟩ and |2⟩ for dissimilar atoms, as per Eq. (21.52), for the antibonding (solid circles; E_-) and bonding (open squares; E_+) states of Eq. (21.51).

[8]The units have been revised since Pauling's introduction of them: Pauling had set fluorine to be 4.0.

atom when an electron joins it (χ, or electron affinity): it is therefore not predicated on an arbitrary scale and is defined by

$$E_M \equiv \frac{I_A + \chi}{2} \tag{21.53}$$

with units of eV (or kJ/mole, depending on preference). The relationship between E_P and E_M is reasonably linear and given by

$$E_P = 0.336\left(E_M - 0.615\right) \tag{21.54}$$

The aforementioned values for fluorine and cesium follow the general trend that moving up and to the right on the Periodic Table corresponds to an increase in electronegativity, but moving down and to the left corresponds to a decrease, trends seen in Figures 21.9 and 21.10. Intuitively, this makes sense: the alkalis (first column of the Periodic Table) have low electronegativities because they readily abandon their outermost electron to other atoms. Cesium being so large would have the lowest electronegativity, and the alkali earth metals such as barium would not be far behind.

Unexpectedly, tungsten has a strong electronegativity compared to other metals near it on the Periodic Table. The passing comment that early work on thermionic emission, discussed in Section 10.1, investigated the effects on emission due to coating tungsten with cesium and barium in particular, takes on a whole new significance. Tungsten covets the electron that cesium or barium contributes more than they do, resulting in a more substantial lowering of the work function barrier. A simple argument shows why, one that will be significantly expanded when Gyftopoulos–Levine theory is introduced (Section 31.8). A flat surface of tungsten upon which a layer of cesium rests will have the top layer of tungsten atoms dominate the shared electrons that the cesium top layer offers. A layer of cesium atoms will therefore acquire a positive sheet charge density and the topmost layer of tungsten atoms a negative sheet charge density on average, much like a capacitor (a visualization is shown in Figure 31.29). The potential drop across a capacitor of plate separation D is given by $\Delta V = q\sigma D/\varepsilon_0$, where σ is the (number) charge density on one of the capacitor plates. Assuming that the charge density is a fraction f of an electron (reflecting the reduced amount of time the electron spends with the

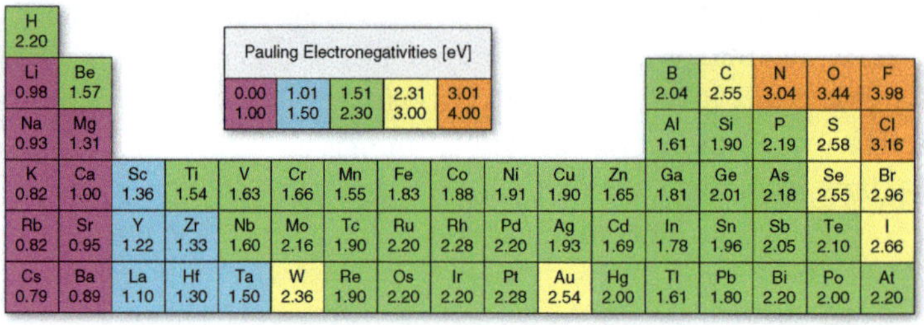

Figure 21.9 Pauling electronegativities of the elements, excluding the noble gases and the lanthanum and actinium series.

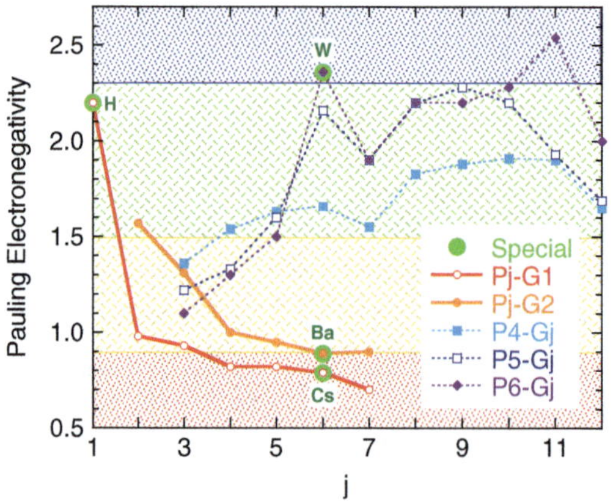

Figure 21.10 A graph of the electronegativities of the alkali and alkali earth elements (Groups 1 and 2, period indexed by j), and transition metals (Periods 4, 5, and 6, group indexed by j). The background color corresponds to the color scheme used in Figure 21.9 to denote E_p magnitude. Light green halos identify particular elements of interest, namely hydrogen (P1-G1), tungsten (P6-G6), barium (P6-G2), and cesium (P6-G1).

cesium atom) per unit area (taken to be $(2a_c)^2$, where a_c is the radius of the cesium atom, or 0.225 nm). Let $D \approx 2a_c$. Then the potential energy drop across the fictitious capacitor standing in for the dipole layers is

$$\frac{\Delta\Phi}{D} \approx \frac{q^2\sigma}{\varepsilon_0} \to \Delta\Phi = 8\pi Q\left(\frac{\theta f}{a_c}\right) \tag{21.55}$$

EXAMPLE: Assume the radius of a cesium (Cs) atom in a surface monolayer is its covalent radius of $a_c = 0.253$ nm (a covalent bond exists between atoms with an electronegativity difference smaller than 1.7). Assume one Cs atom occupies a square area of $(2a_c)^2$ at full monolayer coverage. Assume that the average fractional charge on the Cs atom is fq. Using the empirical value of $\Delta\Phi = 2.9$ eV (tungsten (W) has a work function of 4.5 eV; the minimum work function for W partially coated with Cs is 1.6 eV.), and knowing that the minimum work function occurs for a partial monolayer of 60%, find f.

SOLUTION: Let $d = 2a_c$. Taking the number density of the Cs monolayer to be $\sigma = fq\theta/(2a_c)^2$ then from Eq. (21.55),

$$f = \frac{a_c\Delta\Phi}{8\pi Q\theta} \approx 0.1352 \tag{21.56}$$

EXAMPLE: Although the capacitor model of Eq. (21.55) is crude, use Eq. (21.56) of the previous example to provide a guestimate of r.

SOLUTION: When $C_1 = \pm C_2$ the magnitude of both is 1/2. Therefore, let f be the amount in excess of 1/2. The two equations to be solved are therefore

$$C_1^2 + C_2^2 = 1$$
$$C_1^2 - C_2^2 = 2f$$

which is solved by $C_1 = \sqrt{(1/2)+f}$ and $C_2 = \sqrt{(1/2)-f}$. Using Eq. (21.52), it follows $r = 2f/\sqrt{1-4f^2} \approx 0.281$.

21.6 Sinusoidal potential and band gap

I am death, shatterer of worlds,
annihilating all things.

– Bhagavad Gita[9]

Although a ring of atoms can be used to illustrate further how band gaps and changes to the electron density arise as a consequence of the overlap integrals in the LCAO framework (as in ref. [104]), consider instead a potential closer in spirit to the Kronig–Penny model, in which $V(x) = -V_o\cos(k_l x)$ with $k_l = 2\pi l/L$ and L is the width of the one-dimensional box. The negative sign insures that $x = 0$ is the location of a minimum. The problem is best handled using time-independent perturbation methods [41, 43, 44]. The present treatment (based on ref. [85]) side-steps much of a discussion that would be entailed were a general potential $V(x)$ being considered by considering instead a cos variation, which simplifies the treatment by allowing the utilization of methods and notations introduced in Section 8.2.2, along with the

[9] Stephen Mitchell, *Bhagavad Gita: a new translation*. New York: Harmony Books, 2000, verse 11.32, p. 138. The passage is spoken by Krishna to Prince Arjuna when Krishna reveals his divine form. Robert Oppenheimer recalled the passage when the first atomic weapon test was conducted at Alamogordo (July 16, 1945): another passage describing Krishna (verse 11.12) had an uncanny resemblance to the Trinity test itself, and its wording was used in the title *Brighter than a Thousand Suns: A Personal History of the Atomic Scientists* by Robert Jungk and James Cleugh. Krishna, though, is not associated only with *annihilation*, but also with *creation*.

creation and *annihilation* operators of Section 9.1.2. The notation gets difficult quickly, so to simplify it, instead of referring to momentum eigenstates as $|k_n\rangle$, represent the momentum states instead as $|n\rangle$. Thus, for example, Eq. (8.46) is generalized to

$$\hat{X}_{\pm}|n\rangle = |n \pm 1\rangle \tag{21.57}$$

where $\hat{X} \leftrightarrow \hat{X}_-$ and $\hat{X}^{\dagger} \leftrightarrow \hat{X}_+$, which mimics the creation/annihilation operators of the harmonic oscillator in Eqs (9.85) and (9.86). Switching to a ($\pm$) subscript is convenient to avoid entanglements with powers of the operators, and because a dagger notation is intrusive. The potential $V(x)$ can be cast in terms of $\hat{X}_{\pm}$ by

$$V(x) = -\frac{1}{2}V_o\left(e^{ik_l x} + e^{-ik_l x}\right) = -\frac{1}{2}V_o\langle x|\hat{X}_+^l + \hat{X}_-^l|0\rangle \tag{21.58}$$

The utility of the notation is powerful when coupled with the usual formulation of the perturbation technique. Because $\langle n|l\rangle = \delta_{nl}$, then $\langle n|\hat{X}_{\pm}^l|0\rangle = \delta_{n,\pm l}$, and Kronecker delta functions make summations rather easy. The perturbation methodology expresses the potential matrix elements, and changes to the wave function, in terms of the unperturbed Hamiltonian eigenstates. The "unperturbed" Hamiltonian is taken to be $\hat{H}_0$ such that

$$\hat{H}_0|n\rangle = \frac{\hbar^2\hat{k}^2}{2m}|n\rangle = \frac{\hbar^2 k_n^2}{2m}|n\rangle \tag{21.59}$$

The question then becomes, what are the energies and eigenstates of $(\hat{H}_0 + \hat{V})|\psi\rangle = E|\psi\rangle$? It would be nice to express the perturbed parts of the eigenvectors using a notation much like $|n\rangle$, but doing so is not so simple. A workable but not altogether satisfactory notation is to introduce numerical subscripts to n which denote the "order" of the perturbation. The perturbation order can be understood as follows. Suppose the potential operator is turned on mathematically by increasing a factor λ from 0 to 1, where $\hat{H}_\lambda \equiv \hat{H}_0 + \lambda\hat{V}$ and $|n_\lambda\rangle = |n_0\rangle + \lambda|n_1\rangle$. It is reasonable to then expect that the modifications to the wave function are turned on in a similar way, or

$$\left(\hat{H}_0 + \lambda\hat{V}\right)\left(|n_0\rangle + \lambda|n_1\rangle\right) = E_0|n_0\rangle + \lambda E_1|n_1\rangle \tag{21.60}$$

so that now the $|n_0\rangle$ are the eigenstates of $\hat{H}_0$. But this seems odd – there are factors on the left-hand side that go as λ^2, without a corresponding term on the right. This is fixed by adding on higher order perturbation terms to both the energy and the wave function, but the process is unending, and so

$$\left(\hat{H}_0 + \lambda\hat{V}\right)\sum_{j=0}^{\infty}\lambda^j|n_j\rangle = \sum_{j=0}^{\infty}\lambda^j E_j|n_j\rangle \tag{21.61}$$

To reiterate, the n in $|n_j\rangle$ refers to k_n, the j subscript to the order of λ^j. The point is, the value of λ is arbitrary, and so the terms associated with each λ^j on the left must correspond to the same λ^j on the right. The first few, for λ^0, λ^1, and λ^2, are

$$\hat{H}_0|n_0\rangle = E_0|n_0\rangle \tag{21.62}$$

$$\hat{H}_0|n_1\rangle + \hat{V}|n_0\rangle = E_0|n_1\rangle + E_1|n_0\rangle \tag{21.63}$$

$$\hat{H}_0|n_2\rangle + \hat{V}|n_1\rangle = E_0|n_2\rangle + E_1|n_1\rangle + E_2|n_0\rangle \tag{21.64}$$

where $E_0 \equiv \hbar^2 k_n^2/2m$. A second requirement is that the wave function be normalized, or $\langle n|n\rangle = 1$, and this, too, creates a set of conditions for each power of λ^j encountered because $|n\rangle = \sum\lambda^j|n_j\rangle$. The first three for λ^0, λ^1 and λ^2 are

$$1 = \langle n_0|n_0\rangle \tag{21.65}$$

$$0 = \langle n_0|n_1\rangle + \langle n_1|n_0\rangle \tag{21.66}$$

$$0 = \langle n_0|n_2\rangle + \langle n_1|n_1\rangle + \langle n_2|n_0\rangle \tag{21.67}$$

and so on for the higher order terms, but of the batch the first two are particularly important. The first of the equations entails that the unperturbed wave functions corresponding to $|n_0\rangle$ are already normalized. The second equation entails that $\langle n_0|n_1\rangle = 0$, that is, to order λ^1 the perturbed wave functions are orthogonal to the unperturbed wave functions.[10] However, that means if Eq. (21.63) is multiplied on the left by $\langle n_0|$ then

$$\langle n_0|\hat{V}|n_0\rangle = E_1\langle n_0|n_0\rangle = E_1 \tag{21.68}$$

which is a standard result. But when Eq. (21.58) is taken into account, it is seen that the left-hand side is 0 for the sinusoidal potential, and so *the first order energy change for the sinusoidal potential perturbation vanishes.* That is rather important. It means that to leading order, the energy is still given by the friendly relation $E = \hbar^2 k_n^2/2m$.

Having found that the energy does not change to first order, the question turns to the wave function, and from it the density profile. To zeroth order, the density is constant, as

$$\sum f_{FD}\left(E(k_n)\right)\langle n_0|n_0\rangle = \sum f_{FD}\left(E_0(k_n)\right) \tag{21.69}$$

is position independent. To find the first-order perturbation $|n_1\rangle$, return to Eq. (21.63) and rewrite it in light of $E_1 = 0$ for the sinusoidal potential, to obtain

$$\left(E_0(k_n) - \hat{H}_0\right)|n_1\rangle = \hat{V}|n_0\rangle \tag{21.70}$$

where the explicit k_j dependence of the energy terms is shown. Multiplying all this by the identity matrix $\hat{I}$ does not change anything, but let $\hat{I}$ be given by

$$\hat{I} = \sum_l |j_0\rangle\langle j_0| \tag{21.71}$$

(compare Eq. (8.30)). Observe that the terms for $j = n$ on both sides vanish, and so the summation can be constrained by the restriction $j \neq n$. Equating the coefficients of the remaining $|j_0\rangle$ then gives

$$|n_1\rangle = \sum_{j\neq n}|j_0\rangle\frac{\langle j_0|\hat{V}|n_0\rangle}{E_0(n) - E_0(j)} \tag{21.72}$$

where for convenience the notation is modified so that $E_0(k_n) \to E_0(n)$ (freeing up k to be used as a summation index below) and so

$$\rho_n(x) = |\langle x|n\rangle|^2 = 1 + \Sigma_1 + \Sigma_2 \tag{21.73}$$

$$\Sigma_1 = \sum_{j\neq n}\frac{\langle x|j_0\rangle\langle j_0|\hat{V}|n_0\rangle\langle n_0|x\rangle + c.c.}{E_0(n) - E_0(j)} \tag{21.74}$$

$$\Sigma_2 = \sum_{j\neq n}\sum_{k\neq n}\frac{\langle x|k_0\rangle\langle k_0|\hat{V}|n_0\rangle\langle n_0|\hat{V}|j_0\rangle\langle j_0|x\rangle}{\left(E_0(n) - E_0(j)\right)\left(E_0(n) - E_0(k)\right)} \tag{21.75}$$

where $|\langle x|n_0\rangle|^2 = 1$ and the notation *c.c.* means *complex conjugation.*[11] Now the consequential form of $\hat{V}$ in Eq. (21.58) can be used to great advantage because

$$\langle j_0|\hat{X}_\pm^l|n_0\rangle = \delta_{j,n\pm l} \tag{21.76}$$

which makes the summations much easier. The undaunted shall find that

$$\Sigma_1 = \frac{4V_o\cos\left(k_l x\right)}{E_0(l) - E_0(2n)} \tag{21.77}$$

$$\Sigma_2 = \frac{2V_o^2}{E_0(l)\left(E_0(l) - E_0(2n)\right)}\left\{\cos(2k_l x) + \frac{E_0(l) + E_0(2n)}{E_0(l) - E_0(2n)}\right\} \tag{21.78}$$

---

[10] Well, not quite: the second equation simply says the real part of $\langle n_0|n_1\rangle$ vanishes, but because the wave functions are specified up to an overall phase factor, the wave function can be rotated until only the real part remains, and so the statement is essentially correct (see ref. [41]).

[11] This equation is *less* complex than the more general perturbation equation because properties of the sinusoidal potential $V(x)$ have been invoked.

Analogous to $E_0(j) = (2\pi^2\hbar^2/mL^2)j^2$, let $V_o \equiv (2\pi^2\hbar^2/mL^2)v^2$, which allows Σ_1 and Σ_2 to be more compactly written as

$$\Sigma_1 = v^2 \left\{ \frac{4\cos(k_l x)}{l^2 - 4n^2} \right\} \tag{21.79}$$

$$\Sigma_2 = v^4 \left\{ \frac{\cos(2k_l x)}{l^2(l^2 - 4n^2)} + \frac{(l^2 + 4n^2)}{l^2(l^2 - 4n^2)^2} \right\} \tag{21.80}$$

where the denominators are the consequence of simplifications like $E_0(n \pm l) - E_0(n) \propto l(l \pm 2n)$ when processing Eq. (21.72) and the combination of the fractions in which they reside.

When their denominators are looked at closely, Eqs (21.80) are rather remarkable, given that they give rise to a sinusoidal perturbation in the electron density that matches the period of the potential. Remember that the lattice periodicity matches that associated with the ionic cores that exist in a real bulk material, thereby making l a rather important index. As a result, both Σ_1 and Σ_2, going as $1/l^2$ and $1/l^4$, respectively, will be small *except when* $2n = l \pm 1$ ($2n \neq l$ because of the restriction imposed by Eq. (21.70)). As a result, when the n of k_n jumps the value $n = l/2$, the whole behavior of the density term $\rho_n(x) = |\langle x|k_n\rangle|^2$ switches from being *more* piled up over the ion cores corresponding to the troughs of the sinusoidal potential to being *less* piled up over those cores, in analogy with the bonding/antibonding character of the two hydrogen atoms (see discussion following Eq. (21.45)), and as would be seen for a ring of atoms in an LCAO model [104]. This follows because as $n = (l/2) - 1$ goes to $n = (l/2) + 1$, the denominators $l^2 - 4n^2$ change from $4(l-1)$ to $-4(l+1)$, and so the sign of the denominators changes: what was a trough in the density now becomes a peak.

More realistic potentials are not sinusoidal, but they do have periodicity to them, and so the potentials can be represented by their Fourier transforms $V(x) = \sum_l V_l \cos(k_l x)$. As a result, as with the Kronig–Penny model, band gaps open up in the $E(k)$ relations. When that happens, $E(k_n)$ is not parabolic with k_n near those edges where $n \approx l/2$, but away from there the density $\rho(x)$ is reasonably constant and the relationship between $E(k)$ and k is reasonably parabolic.

21.7 Ion potentials and screening

> *...the effects of [electron interactions] would be mollified by electron correlation, tending to keep electrons apart, and surrounding any given electron by a virtual hole in the charge cloud. This hole has the effect of a positive charge, which screens out the electric potential of the electron at all but very short distances. Two electrons could then only interact if they came very close. Such a* screened potential *occurs naturally in any system where a cloud of electrons is available to spread itself out about a Coulomb source.*
>
> – John M. Ziman [82], p. 159.

The extension to three dimensions complicates the description of materials and crystals, requiring the introduction of Bloch states, Brillioun zones, and reciprocal lattice spaces [53, 78, 86, 292]. It shows that bands can overlap: for a metal, the valence and conduction bands do so, so that instead of the valence band filling up completely, the conduction band becomes partially filled. For an intrinsic semiconductor (a semiconductor that is without any dopant species), a band gap between the conduction and valence band exists with the (negative) Fermi level equidistant between them, and electrons in the conduction band are a consequence of thermal excitations from the valence band. Doped semiconductors can have positive Fermi levels measured from the bottom of the conduction band, but the carrier density is still well below that of metals. This will have consequences for how external fields are shielded compared to metals, as taken up in Chapter 22, but more immediately it affects how the Coulomb potentials associated with the ionic cores are shielded.

Mobile electrons in the conduction band will migrate so as to surround the bare ionic charges of the lattice atoms and shield their influence. Changes in the density of the electrons are reflected in changes in the chemical potential μ, or $\delta\mu = \delta\phi$, and governed by (see also Section 25)

$$\frac{\delta\rho}{\delta\phi} = \frac{2}{(2\pi)^3} \int \frac{\delta f_{FD}(E(k))}{\delta\mu} d\vec{k} \tag{21.81}$$

21.7.1 Low temperature/high density

How frozen and faint I then became, ask it not, Reader, for I do not write it, because all speech would be little.

– Dante Alighieri[12]

For conditions such that $\beta\mu \gg 0$, such as are characteristic of high carrier concentration and/or low temperature (see Section 6.4 for $\mu > 0$), then

$$\frac{\delta\rho}{\delta\phi} = \beta \int_0^\infty \frac{\mathcal{D}(E)e^{\beta(E-\mu)}}{\left(1 + e^{\beta(E-\mu)}\right)^2} dE \approx \mathcal{D}(\mu) \tag{21.82}$$

as a consequence of Eq. (7.24), where $\mathcal{D}(E) = \mathcal{D}_3(E)$ is the three-dimensional density of states of Eq. (21.12) (the 3 index is dropped for simplicity), and the relation holds for $\beta\mu \gg 1$. Thus,

$$\delta\rho\left(\vec{r}\right) \approx \mathcal{D}(\mu)\delta\phi\left(\vec{r}\right) \tag{21.83}$$

The Coulomb potential is rotationally symmetric, so from Poisson's equation,

$$\frac{1}{r^2}\frac{\partial}{\partial r}\left(r^2\frac{\partial}{\partial r}\delta\phi\right) = \frac{q^2}{\varepsilon_0}\delta\rho = 16\pi Q\mathcal{D}(\mu)\delta\phi \tag{21.84}$$

The right-hand side is of the form $k^2\delta\phi$. Equivalently, one can show that a solution of the form $\phi = (4Q/r)\exp(-kr)$ is a solution,[13] but whichever path one favors, it follows that

$$k = (16\pi D(\mu)\,Q)^{1/2} = \sqrt{\frac{4k_F}{\pi a_o}} \leftrightarrow k_{TF} \tag{21.85}$$

where a_o is the Bohr radius, and k_{TF} is the Thomas–Fermi screening constant.

EXAMPLE: Find k_{TF} for copper, where $\mu = 7$ eV.
SOLUTION: Evaluate $k_F = \sqrt{2m\mu}/\hbar = 13.5546$ nm^{-1}. Then, in units of eV, nm, fs, and q,

$$k_{TF} = \left(\frac{4 \cdot 13.5546}{\pi \cdot 0.0529177}\right)^{1/2} = 18.0592\text{nm}^{-1}$$

21.7.2 High temperature/low density

Till swoll'n with cunning, of a self-conceit,
His waxen wings did mount above his reach
And melting, heavens conspired his overthrow!

– Christopher Marlowe[14]

For conditions such that $-\beta\mu \gg 1$, which are characteristic of low carrier concentration and/or high temperature (Section 6.4 for $\mu < 0$), then the leading order electron density is given by the first term in Eq. (6.37). Take the background

---

[12] *The Divine Comedy of Dante Alighieri* (trans. Charles Eliot Norton), *Great Books of the Western World*, ed. Robert Maynard Hutchins, Vol. 21, Dante. Chicago: Encyclopaedia Britannica, 1952, Canto XXXIV.16. In Dante's *Inferno*, the ninth circle was a frozen, hardened wasteland.

[13] If one is unhappy with the *ansatz* approach here, the problem is revisited in Section 26.6, where it is put on a better footing. The point here is to show how the equations introduce length scales that identify relevant terms like k_{TF} without having to solve equations like Eq. (21.84) explicitly.

[14] Christopher Marlowe and S. Barnet, *Doctor Faustus*. New York: New American Library, 1969, Prologue, p. 23. The passage refers to the Greek myth of Icarus, who flew on contrived wings in the rarefied upper atmosphere where the heat of the sun melted the wax holding the feathers in place, causing Icarus to plunge to his death. The story has long been a metaphor of hubris.

positive charge as fixed in position. In a Jellium-like model, where the background charge is ρ_o, then the electron density near the charge will be $\rho = \rho_o \exp(\beta\phi)$, so that Poisson's equation becomes

$$\frac{1}{r^2}\frac{\partial}{\partial r}\left(r^2\frac{\partial}{\partial r}\phi\right) = \frac{q^2}{\varepsilon_0}\rho_0\left(e^{\beta\phi} - 1\right) \tag{21.86}$$

If the electrons are not large in number so that shielding is weak, and if the temperature is sufficiently high, $\beta\phi$ will not be large, then with the approximations $e^{\beta\phi} \approx 1 + \beta\phi$ and $\phi = (4Q/r)\exp(-kr)$,

$$\frac{1}{r^2}\frac{\partial}{\partial r}\left(r^2\frac{\partial}{\partial r}\phi\right) = \beta\frac{q^2}{\varepsilon_0}\rho_o\phi = \frac{16\pi Q\rho_o}{k_B T}\phi \tag{21.87}$$

$$k_D = \left(\frac{16\pi Q\rho_o}{k_B T}\right)^{1/2} \tag{21.88}$$

where k_D is Debye's screening constant.[15] A modification for semiconductors is that the dielectric constant K_s of the medium (e.g., $K_s = 11.9$ for silicon) affects the factor Q by making the replacement $Q \rightarrow (K_s - 1)Q_o/(K_s + 1)$ where now $Q_o = \alpha\hbar c/4 = 0.359991$ eV-nm [99].

EXAMPLE: Find the Debye screening constant for a semiconductor doped such that the density of electrons in the conduction band is 10^{18} cm^{-3}. Assume the temperature is 300 K and that $K_s = 11.9$.

SOLUTION: Converting, $\rho_o = 0.001$ nm^{-3} and $k_B T = 0.025852$ eV. Therefore, in units of eV, nm, fs, and q,

$$k_D = \left(\frac{16 \cdot 0.001 \cdot \pi \cdot 0.359991 \cdot 0.844961}{0.025852}\right)^{1/2} = 0.769046 \text{ nm}^{-1}$$

21.7.3 Many cores

What are the core values that we, as Americans, hold in common? That's not how we usually frame the issue, of course; our political culture fixates on where our values clash.

– Barack Obama[16]

The presence of multiple shielded cores in a lattice can modify the apparent shape of the shielded Coulomb potential. A suggestion of how is afforded by a screened Coulomb potential (called the Yukawa potential) for which

$$\phi(sR) = \left(\frac{4Q}{R}\right)\frac{\exp(-\kappa Rs)}{s} \tag{21.89}$$

where s is a dimensionless measure of the radial distance from the core, and R is a measure of the distance of the atom in a bulk material to its neighbor. For simplicity in the present example, assume a cubic lattice and let R be twice the covalent radius. Then, apart from the overall energy scale $4Q/R$, the screened Coulomb potential has a similar form governed only by the factor κR. Use the previous two example calculations, and take $R_{Cu} = 0.276$ nm and $R_{Si} = 0.222$ nm, so that $(\kappa R)_{Cu} = 4.98434$ and $(\kappa R)_{Si} = 0.170728$. A comparison of the dimensionless potential $\phi/(4Q_o/R)$ is shown in Figure 21.11. Contrast that with ion potentials that are close packed on a cubic lattice, and let s be measured along the $\hat{x}$ axis. Then the cumulative potential is approximated by summing over an array of screened potentials on a cubic lattice by

$$\phi(s)\left(\frac{4Q_o}{R}\right)^{-1} = \sum_{jkl=-N}^{N}\frac{e^{-\kappa R s_{jkl}}}{s_{jkl}} \tag{21.90}$$

where $s_{jkl}^2 = (j - s)^2 + k^2 + l^2$ and the summation is over each index spanning the range from $-N$ to N. The potential is specified up to an additive energy constant, and so subtracting off a value so that the midpoint between the middle atom and its nearest neighbor is approximately zero gives the behavior shown in Figure 21.12. The potentials were generated using the algorithm in Section A3.12.5.

[15] As observed by Kubo [61], Mott first derived the relation, and so the appellation "Mott's screening constant" may be a better designation.
[16] Barack Obama, *The Audacity of Hope: Thoughts on Reclaiming the American Dream*. New York: Crown Publishers, 2006, p. 52.

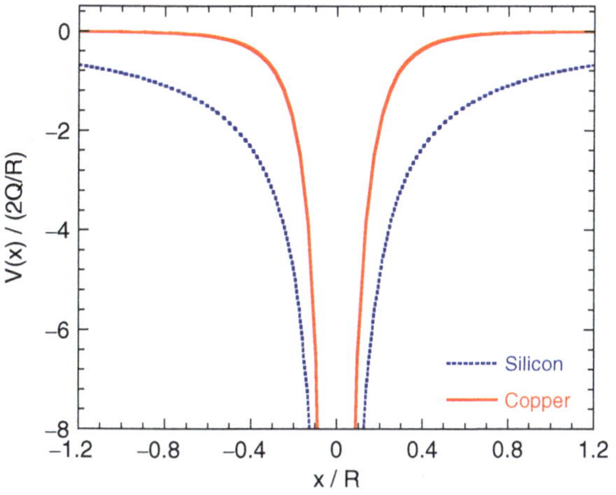

Figure 21.11 Screened, or Yukawa, potential for copper-like and silicon-like parameters using the algorithm of Section A3.12.5.

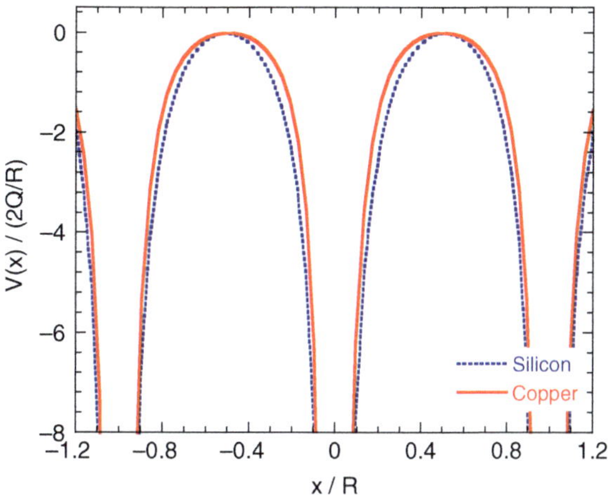

Figure 21.12 Screened, or Yukawa, potential for copper-like and silicon-like parameters taking into account contributions of neighbors as in Eq. (21.90) using the algorithm of Section A3.12.5.

The ability of electrons to migrate in response to the presence of a charge anticipates that they will also migrate so as to shield the inside bulk of a material from an externally applied field. Therefore, electron emission from semiconductors will be complicated by accounting for how the conduction bands bend in response to the magnitude of those external fields.

CHAPTER 22
Semiconductors

22.1 Resistivity

The effect of heat in increasing the conducting power of many substances ...is well known. I have lately met with an extraordinary case of this kind ...which is in direct contrast with the influence of heat upon metallic bodies ... The substance presenting this effect is sulphuret of silver ... On applying a lamp under the sulphuret between the poles, the conducting power rose rapidly with the heat ...and the susulpheret was found conducting in a manner of a metal. On removing the lamp and allowing the heat to fall, the effects were reversed ....

– Michael Faraday[1]

Faraday is credited with having first measured (c. 1833) a characteristic feature of semiconductors [294]: their conductivity increases with temperature. More accurately, Faraday had measured an *intrinsic* semiconductor: *doping* the semiconductor to make it extrinsic complicates matters (Section 22.4). Metal conductivity generally decreases: resistivity (the inverse of conductivity) for metals commonly behaves as

$$\rho(T)|_{metal} \approx \rho(T_o) + \alpha(T - T_o) \tag{22.1}$$

whereas for intrinsic semiconductors (like silicon), the behavior is closer to

$$\rho(T)|_{semi} \approx \rho(T_o)\exp\{-\alpha(T - T_o)\} \tag{22.2}$$

For example, intrinsic (high-purity) silicon data [295] is compared to data [296] for the metals palladium (Pd), gold (Au), and silver (Ag) in Figure 22.1, with the equivalent parameters shown in Table 22.1 for $T_o = 273.15$ K.

Why the difference? Consider a simpler question first: what distinguishes an insulator and a metal (because a semiconductor is betwixt them)? A distinguishing feature of metals is that the flow of electrons is comparatively easy; that of insulators is where the flow is restricted. But as Figure 22.1 shows, that distinction is not helpful when pure silicon obstructs current compared to a metal at room temperature yet has comparable resistivity when the temperature is only 100 K higher. A more precise definition [297] bases the distinction on identifying the behavior of the conductivity:

- The conductivity of a *metal* is finite as $T \to 0$ K.
- The conductivity of an *insulator* goes to 0 as $T \to 0$ K.

At $T = 0$, the valence band of an insulator is therefore completely full, and the conduction band is empty. Although band theory [53] provides a much more nuanced account, an adequate simple-minded treatment is to take a semiconductor to be an insulator with an appreciable number of electrons thermally excited into the conduction band: the difference between a great insulator like diamond and one that conducts at room temperature like silicon is related to the size of the band gap.

As introduced in Section 7.4.1, current flow is characterized by a drift velocity $v_d \sim l/\tau$ where l is comparable to the mean free path and τ to the average time between collisions of the electron with its fellow electrons and background atoms: a common measure of this is electron mobility [52], obtained by drawing a relationship between current and the electric field as in $J = \sigma\mathcal{E} = \sigma(F/q)$, and current as given by Eq. (3.5), or $J = qv_d\rho$ so that

$$\sigma = q\rho\mu_e \tag{22.3}$$

where the conductivity μ_e, with units of cm^2/V s, is the ratio between the drift velocity and the electric field. Therein lies a reasonable way to look at it: a mean time between collisions and a scattering rate presupposes scattering events made

[1] Michael Faraday, *Experimental Researches in Electricity*, ref. [293], Sections 432–434. "Sulphuret of silver" as known to Faraday is known now as "silver sulfide", Ag_2S, a semiconductor with a band gap of 0.9 eV, more commonly experienced as the irritating black tarnish on silverware and jewelry.

Introduction to the Physics of Electron Emission, First Edition. Kevin L. Jensen.

Table 22.1 Resistivity parameters for metals compared to intrinsic silicon in terms of Eq. (22.1) for metals and Eq. (22.2) for semiconductors. Data based on refs [296] and [295], and shown in Figure 22.1.

Material	Eq.	ρ_o (Ω-cm)	α (1/K)
Palladium (Pd)	22.1	955.74	3.9563
Gold (Au)	22.1	206.61	0.82405
Silver (Ag)	22.1	14.21	0.61115
Silicon (Si)	22.2	1.1369×10^6	0.057158

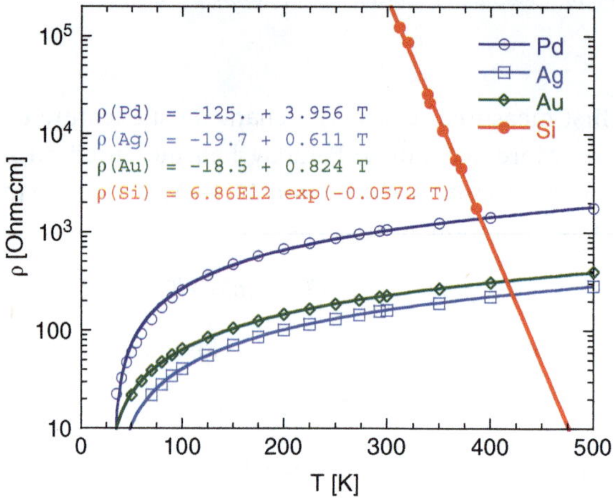

Figure 22.1 Behavior of the resistivity of intrinsic silicon (data digitally extracted from line 6 of Figure 2 of ref. [295]) to the resistivity of some metals (specifically, palladium (Pd) gold (Au), and silver (Ag), from data in the tables of ref. [296]). The lines based on the equations show the linear fit to the metals (appearing curved because of the log plot) and the exponential fit to the semiconductor.

possible by an electron picking up energy from fields existing in the bulk material. The only reason that they would not scatter is if *they had no place to go* (bringing to mind the analogy of the phalanx in Section 7.1) because the final states they would like to scatter to are occupied. Accelerating electrons increase in energy. Insulators have a filled band, as in Figure 22.2, with no empty states available in the gap that an accelerated electron can access, so current is prevented from flowing. Gaps are generally of such a magnitude that electrons do not often acquire the energy (from thermal agitation or long wavelength photons) to cross it. Tellingly, diamond field emitters, for example, show good transport: the problem is getting electrons into the conduction band [298]. As shown in Table 22.2, good insulators, like diamond, have a large band gap whereas the more commonly used semiconductors, like silicon, germanium, and gallium arsenide, have smaller

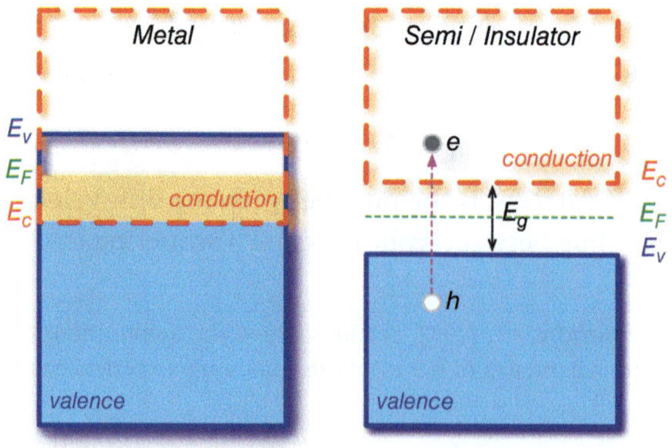

Figure 22.2 Left, a metal in which the bottom of the conduction band E_c is below the top of the valence band E_v. Right, a semiconductor/insulator: an electron e that is promoted (via thermal excitation, photo-excitation, and so on) from the valence band into the conduction band leaves behind a hole h.

Table 22.2 Parameters for common semiconductors based on Figure 22.4 and Eq. (22.14). K_s is the static dielectric constant. χ is the electron affinity (difference between the vacuum level and bottom of the conduction band) for semiconductors. The number density is $\rho(T)$ (#/cm^3). E notation denotes that nEp is understood as $n \times 10^p$, for example with $n = 4$ and $p = 5$, then 4E5 is understood as 4×10^5. χ for diamond (C) is for a clean diamond surface; hydrogen-terminated surface of diamond has negative electron affinity (NEA) properties where the vacuum level is below E_c. χ is measured from E_c. The factors a and b are from Figure 22.4.

Parameter	Units	C	Si	GaAs	GaP
$E_g(0)$	eV	5.4068	1.1693	1.5179	2.3236
a	eV	0.5565	0.0850	0.1708	0.3616
b	eV	1290.3	320.70	329.74	614.89
S	eV	2.5025	1.5379	3.0055	3.4122
$\langle\hbar\omega\rangle$	eV	0.1112	0.0276	0.0284	0.0530
θ	–	1290.3	320.7	329.7	614.9
$E_g(300)$	eV	5.3992	1.1249	1.4326	2.2701
χ	eV	0.35	4.05	4.07	4.3
r_e	–	0.57	1.1	0.068	0.35
r_h	–	0.8	0.56	0.5	0.5
K_s	–	5.7	11.8	13.2	11.1
$\rho(300)$	#/cm^3	6.2079E-27	6.21430E09	1.84140E06	5.80013E-01

ones. Metals, on the other hand, are characterized by electrons always having a level to scatter to: the conduction band is partially filled (the valence and conduction bands overlap) even as $T \to 0$.

Metals and semiconductors differ in how full and how close the bands are. If a unit cell is the smallest part of a crystal that reflects the arrangement of atoms in the bulk (in the proper language, the number of unit cells corresponds to the number of Brillouin zones), then there are two electron states (one for each spin direction) per unit cell per band. Then, the number of electrons in the unit cell can be related to whether the solid is a metal or not. Paraphrasing Ziman [53],

- Alkali metals (column 1 in the Periodic Table, e.g. Cs, K, and Na) are monovalent (single electron in the unit cell). Metals in column 11 (Cu, Ag, Au) likewise have a half-filled band. They are metals. An odd number of electrons in a unit cell is a metal, as in column 13 (Al, Ga, etc.).
- Materials with half-filled bands in column 15 (As, Sb, Bi) have five electrons per atom but arrange themselves so that there are two atoms per unit cell. Thus, their outermost band is almost full, but because of partial overlap with the higher band, there is some current. These are *semi-metals*.
- Materials with an even number of electrons in the unit cell would seem to be insulators, unless there is band overlap.

This can continue with increasingly more complicated examples of the divalent and transition elements (where the d shells make their influence known). The tetravalent elements are particularly interesting. Ziman writes:[2]

> *There is an interesting progression in Group IV of the Periodic Table, from C, which, as diamond, forms a semiconductor with such a large energy gap that in practice it is an insulator, to Si and Ge which are semiconductors, to Sn, which is both metal and semiconductor, to Pb which is a metal. The energy band structure of all these elements can be represented by N.F.E.*[3]*models …*

The tour has come full circle: from relating the origin of the band gap to both the behavior of the wave functions over a corrugated potential barrier and to the overlap integrals of the atomic orbitals basis states, a nearly free electron model is encountered again.[4] From the NFE vantage point, what semiconductors bring to the emission equations can be considered once the number of electrons in the conduction band is found.

22.2 Electrons and holes

> *O, cursèd be the hand that made these holes!*
>
> – William Shakespeare[5]

[2]Experiencing Section 4.1 of ref. [53] is worth the effort.
[3]NFE stands for "nearly free electron".
[4]There is *much* more to be said on the relation between the "band" and "bond" approaches. Sample Ziman [53], Ashcroft and Mermin [292], Sutton [104], and Quéré [78].
[5]Ref. [37]: *Richard III*, I.ii.14.

When an electron does acquire enough energy to skip across the band gap E_g through thermal agitation (or photo-absorption or collisions with a high-energy primary electron, but "thermal" shall be a proxy for all), its absence is noticed. An electron in the conduction band can roam the bulk, but its vacancy is a missing electron on the atom it leaves behind, and that vacancy can be filled by robbing an adjacent atom of one of its electrons. In that way, the vacancies move, and can do so with velocities similar to those of the electrons. It is very, very convenient to let those vacancies be represented as positively charged particles and endow them with mass and velocity; when that is done, the particles are called *holes*, the flow of which constitutes a current in the opposite direction of electrons. More generally, both holes and electrons flow, although one overshadows the other in doped materials (Section 22.4) and so the phrase "dominant charge carrier" need not refer to electrons exclusively.

The hole distribution can be made to follow the appearance of the electron distribution in Eq. (6.17). If all the states in the valence band up to E_v are filled, then the total number of electrons N in a volume V or, better, $\rho = N/V$, is simply (compare Eq. (7.11))

$$\rho = \int_0^{E_v} \mathcal{D}(E)dE \tag{22.4}$$

which is analogous to Eq. (7.12), but as the temperature increases some of those electrons are thermally excited into the conduction band, and so

$$\rho = \int_0^{E_v} \mathcal{D}(E)f_{FD}(E)dE + \int_{E_c}^{\infty} \mathcal{D}(E)f_{FD}(E)dE \tag{22.5}$$

The total number of electrons does not change as the temperature increases, and if the volume does not either[6] then Eq. (22.4) = Eq. (22.5) or, after collecting terms,

$$\int_0^{E_v} \mathcal{D}(E)(1 - f_{FD}(E))dE = \int_{E_c}^{\infty} \mathcal{D}(E)f_{FD}(E)dE \tag{22.6}$$

Actually, a bit more is required: the holes do not have to have the same mass as the electrons in the conduction band, often they will not. It is common to append subscripts to $\mathcal{D}(E)$, as in $\mathcal{D}_e(E)$ and $\mathcal{D}_h(E)$, and to speak of electron and hole masses m_e and m_p explicitly.

The function $\bar{f}(s) = 1/(e^s + 1)$ is very convenient. It is straightforward to show $1 - \bar{f}(s) = \bar{f}(-s)$. The form of the Fermi–Dirac distribution is analogous to $f(s)$, and therefore the side of Eq. (22.6) that pertains to holes is easily manipulated and recasts Eq. (6.17) as

$$\begin{aligned} \rho_e(T) &= \frac{\sqrt{2}}{\pi^2\hbar^3}(m_e k_B T)^{3/2} F_{1/2}[\beta(\mu - E_c)] \\ &\equiv \frac{N_c}{\beta^{3/2}} F_{1/2}[\beta(\mu - E_c)] \end{aligned} \tag{22.7}$$

$$\begin{aligned} \rho_h(T) &= \frac{\sqrt{2}}{\pi^2\hbar^3}(m_p k_B T)^{3/2} F_{1/2}[\beta(E_v - \mu)] \\ &\equiv \frac{N_v}{\beta^{3/2}} F_{1/2}[\beta(E_v - \mu)] \end{aligned} \tag{22.8}$$

For intrinsic semiconductors, not many electrons are in the conduction band and fewer still at lower temperatures. Whereas metals have about 10^{23} electrons/cm^3, intrinsic semiconductors have many orders of magnitude less, for example room temperature intrinsic silicon has a number density on the order of 10^{10} cm^{-3}, or about three freely roaming electrons in 100 nm^3. Given that $N_c/\beta^{3/2} \approx 2.8 \times 10^{19}$ cm^{-3} (Eq. (6.19)), μ is going to be quite negative, meaning that it resides below the conduction band minimum E_c and in the band gap. If so, then Eq. (6.30) can simplify matters considerably because the expansion can be truncated after the first term. Letting $m_e = r_e m$ and $m_h = r_h m$, where r_e and r_h are the ratios of the effective mass to the electron rest mass for electrons and holes, respectively (Section 23), then $\rho_e(T) = \rho_h(T)$ results in

$$r_e^{3/2} \exp[\beta(E_v - \mu)] = r_h^{3/2} \exp[\beta(\mu - E_c)] \tag{22.9}$$

[6] It does. Penance for deceptively ignoring it will be atoned for in Section 22.3.

Isolating μ then shows

$$\mu(T) = \frac{1}{2}(E_c + E_v) - \frac{3}{4}k_B T \ln\left(\frac{r_h}{r_e}\right) \tag{22.10}$$

At room temperature, $k_B T$ weighs in at a relatively pathetic 0.025852 eV, and since the electron and hole masses are expected to be comparable ($r_e \sim r_h$), the temperature-dependent term is going to be small indeed. Therefore, although the Fermi level does depend on T for intrinsic semiconductors, for the most part it simply sits midway between the conduction and valence band edges to a very good approximation. Additionally, the leading order terms in an expansion of the product of the electron density ρ_e with the hole density ρ_h are starightforward to obtain: keeping only the first terms in the expansion of $F_{1/2}(x)$ and recalling that the band gap is defined by

$$E_g \equiv E_c - E_v \tag{22.11}$$

then

$$\rho_e \rho_h \approx \left(\frac{m\sqrt{r_e r_h}}{4\pi\beta\hbar^2}\right)^3 e^{-\beta E_g} = (r_e r_i)^{3/2}\left(\frac{2e^{-\beta E_g/2}}{\lambda_T^3}\right)^2 \equiv \rho_i^2 \tag{22.12}$$

where the thermal wavelength λ_T is from Eq. (5.24). When the electron and hole masses are equal, then $\rho_e = \rho_h = \rho_i$, where ρ_i is the intrinsic carrier density.

22.3 Band gap and temperature

A man who doesn't know the truth is just an idiot, but a man who knows the truth and calls it a lie is a crook.

– Bertolt Brecht[7]

In developing Eq. (22.6), the volume of a crystal was implied to not change (or not change much) with temperature, an assumption that contradicts the tendency of solids to expand when heated. If the volume changes with temperature, then the eigenstates change, and so the band gap must change as well. It is well known that the variation of band gap with pressure on a semiconductor can be understood from the effects of pressure on the atomic orbitals [299], in that increasing the pressure moves the cores closer together. Similarly, an increase in temperature changes the magnitude of the atomic vibrations, pushing the cores apart and so it, too, must lead to a change in the band gap. Claiming the change is small does not mitigate the offense, being more wishful thinking than accurate. Penance is due.

A commonly used empirical relation, known as *Varshni's equation*, is

$$E_g(T) \approx E_g(0) - \frac{AT^2}{T + B} \tag{22.13}$$

but the following form [300] gives a closer correspondence to data[8]

$$E_g(T) \approx E_g(0) - \frac{2Sk_B\theta}{e^{\theta/T} - 1} \equiv E_g(0) - 2Sk_B T g(\theta/T) \tag{22.14}$$

where $k_B\theta = \langle\hbar\omega\rangle$ is an average phonon energy and S is a dimensionless "coupling constant". The behavior of $g(\theta/T)$ is shown in Figure 22.3. Eq. (22.14) gives a satisfactory fit to experimental relations for a number of semiconductors, as shown in Figure 22.4 and Table 22.2. Comparing the asymptotes of Eqs (22.13) and (22.14) for $T \to 0$ and $T \to \infty$ shows that $B \leftrightarrow \theta/2$ and $A \leftrightarrow 2Sk_B$.

22.4 Doping of semiconductors

I cannot praise a fugitive and cloistered virtue, unexercised and unbreathed ...Assuredly we bring not innocence into the world, we bring impurity much rather; that which purifies us is trial, and trial is by what is contrary.

– John Milton[9]

---

[7] Bertolt Brecht (trans. C. Laughton), *Galileo*. New York: Grove Press, 1966. The quote is from Scene 9 in the German original, but Laughton's translation regrettably omitted it, as observed in the Introduction by E. Bentley (p. 26).

[8] Ref. [300] prefers $E_g(T) \approx E_g(0) - S\langle\hbar\omega\rangle(\cosh(\langle\hbar\omega\rangle/2k_B T) - 1)$ but they are equivalent.

[9] John Milton, *Areopagitica, Great Books of the Western World*, ed. Robert Maynard Hutchins, Vol. 32. Chicago: Encyclopaedia Britannica, 1952, p. 390–391.

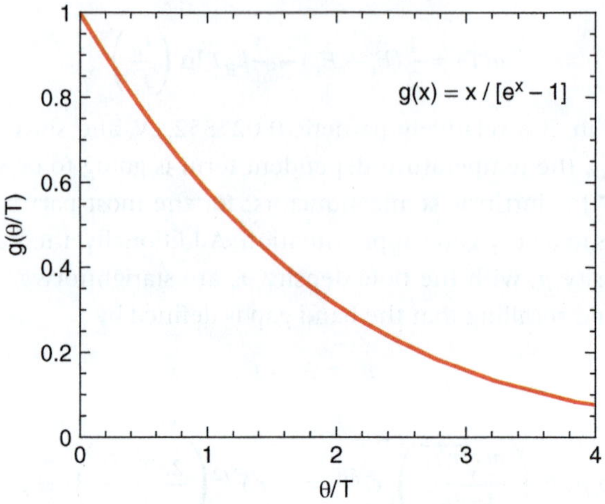

Figure 22.3 The function $g(\theta/T)$ defined by Eq. (22.14). Observe that large temperature corresponds to small values of θ/T.

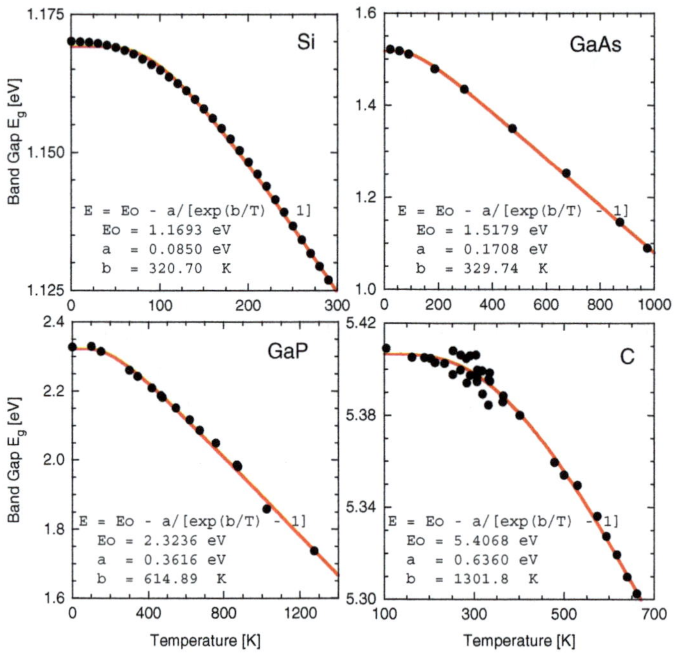

Figure 22.4 Temperature dependence of the band gap for several common semiconductors. Data for silicon from ref. [301]. Data for GaAs and GaP from ref. [302]. Diamond (carbon) data digitally extracted from Figure 8 of rf. [303]. Fits are based on Eq. (22.14).

In an impassioned polemic against censorship written in 1644 during the first English Civil War, John Milton (author of *Paradise Lost*) argued that Good is not a consequence of averting consideration of Bad. Rather, exposure to "impurities" and the ensuing struggle to determine what is right or best or worthy is essential to vet such value judgments: to know from experience and forego indulgence is praiseworthy and of greater moral character than not indulging because of a lack of comprehension. Impurities, in a sense, make possible that which is valued.

In a way, so too with semiconductors. Much of their technological advantage arises because the intentional inclusion of different atoms in an otherwise pure semiconductor can give rise to mobile electrons (such atoms are *donors*) or take up electrons and thereby give rise to mobile holes (such atoms are *acceptors*). For example, in an impressively good insulator like diamond, boron is an acceptor as it lies to the left of carbon in the Periodic Table; in contrast, nitrogen is a donor as it lies to the right of carbon. Silicon, lying one row below carbon, is affected by those so-called dopants in much the same way and a variety of donor (e.g., P, As, and Sb) and acceptor (e.g., B, Ga, In, Al) atoms are used. Semiconductors doped by donor impurities are called n-type, and those doped by acceptor impurities are called p-type. Semiconductors that include dopants are called *extrinsic* semiconductors, in contrast to the pure and undoped intrinsic semiconductors discussed in Section 22.1.

Impurity atoms ionize, and it is the number of ionized donors (or acceptors) that dictate the electron (or hole) density. Because it takes a bit of energy to ionize a donor atom and release its electron into the conduction band, the electron concentration due to donors is a temperature-dependent quantity. A similar argument applies to holes: there, it takes a bit of energy for an electron to jump to the acceptor site, leaving a hole behind. The energy level of a donor with respect to the bottom of the conduction band is referred to as the binding energy of the donor, and similarly with acceptors. As the binding energies of the more useful "shallow" donors and acceptors tend to be tens of millielectronvolts, at room temperature conditions, most are ionized for common semiconductor materials and conditions. In equilibrium, a semiconductor with no net charge entails

$$\rho_h(T) - \rho_e(T) + N_D^+ - N_A^- = 0 \tag{22.15}$$

About those irritating $\pm$ superscripts: they are needed because not all the donors or acceptors are ionized, but rather a fraction of them. At $T = 0$, none of the impurities are charged (electrons in the conduction band fall into the donor level within the band gap, and the electrons at the acceptor levels rejoin with holes in the valence band). Conversely, at high enough temperature, electrons from the donor sites can easily jump to the conduction band, and electrons in the valence band can easily jump to the acceptor sites, leaving behind a hole. This temperature dependence can be accounted for. The distribution of charged donors and acceptors is evaluated in much the same way that the Fermi–Dirac distribution was calculated in Section 6.2 (or see ref. [304]), but the degeneracy of the number of charges that donor or accepter states allow must be factored in. If the energy level of the donor is E_D and the acceptor E_A, then accounting for the degeneracy amounts to (remember that $\beta = 1/k_BT$)

$$N_D^+(T) = \frac{N_D}{1 + g_D e^{\beta(\mu - E_D)}} \tag{22.16}$$

$$N_A^-(T) = \frac{N_A}{1 + g_A e^{\beta(E_A - \mu)}} \tag{22.17}$$

where g_D is the donor degeneracy factor (often 2), and g_A is the acceptor degeneracy factor (often 4). If the modified energy levels

$$\beta E'_D \equiv \beta E_D - \ln(g_D) \tag{22.18}$$

$$\beta E'_A \equiv \beta E_A + \ln(g_A) \tag{22.19}$$

are used then the distributions are more easily expressed in terms of the Fermi–Dirac distribution function f_{FD} encountered in Eq. (6.8) by

$$N_X^\pm(T) = N_X f_{FD}(E'_X) \tag{22.20}$$

where X denotes either D or A. Thus, Eq. (22.15) is easily written as

$$\frac{N_v}{\beta^{3/2}} F_{1/2}(z_v) - \frac{N_c}{\beta^{3/2}} F_{1/2}(-z_c) + N_D f(z_D) - N_A f(z_A) = 0 \tag{22.21}$$

where $z_D = \beta(E'_D - \mu)$, $z_A = \beta(E'_A - \mu)$, $z_v = \beta(E_v - \mu)$ and $z_c = \beta(E_c - \mu)$, and where $f(z) \equiv 1/(1 + e^z)$. This equation is important: all of the z factors depend on the chemical potential μ, and so this equation is *a means of determining what μ is* as a function of the valence and conduction band edges E_v and E_g, the doping of acceptors N_A and donors N_D, and the temperature T.

The evaluation of μ is not as difficult as it may appear. Doping levels generally vary between 10^{15} and 10^{18} atoms/cm^3, or a bit smaller than $N_c F_{1/2}(\beta\mu)/\beta^{3/2} = 1.92 \times 10^{19}$ nm^{-3} at room temperature for $\mu = 0$. Thus, for typical doping levels, μ is negative, such that below room temperature the leading order term of Eq. (6.30), or $F_{1/2}(z) \approx (\sqrt{\pi}/2)e^z$ for $z < 0$), is often sufficient. Such an approximation is often conveniently assumed from the start [52, 73, 213, 304] given the pedagogical utility of doing so, but if the doping level is higher than 5×10^{18} cm^{-3} then more terms of Eq. (6.30) should be retained or the integral definition of $F_{1/2}(z)$ used. Both are compared here to show that the approximation is good, but less so at higher temperatures.

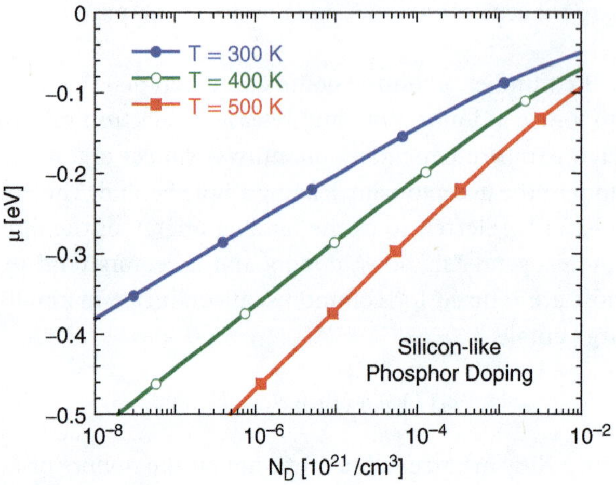

Figure 22.5 Relation between the doping concentration N_D and the chemical potential μ for various temperatures. Remember that the number of *ionized* dopants is governed by $N_D^+ = N_D f_{FD}(E_D')$ as per Eq. (22.16).

To keep the discussion specific, consider phosphorous-doped silicon ($E_g = 1.17$ eV, $E_D = -0.045$ eV below the conduction band). To keep the discussion simple (and reflecting a bias for electrons), let $N_A = 0$. Lastly let the electron and hole masses be the same as the rest mass of the electron ($m_e \approx m_h \approx m$): that approximation is not in general true (electron and hole masses differ, and then there are also heavy and light holes) but the assumption lets $N_v \sim N_c$, does not alter the qualitative features of the figures below, and has the virtue of allowing terms to be collected in an aesthetically pleasing way. Using the leading order term of Eq. (6.30) and rearranging Eq. (22.21) results in

$$N_D \approx \frac{\sqrt{\pi} N_c}{2\beta^{3/2}} \left(e^{\beta\mu} - e^{-\beta(\mu+E_g)} \right) \left(1 + 2e^{\beta(\mu-E_D)} \right) \tag{22.22}$$

as shown in Figure 22.5. As a consequence, N_D vanishes when $\mu = -E_g/2$, corresponding to an intrinsic semiconductor when $m_e = m_h$. The general trend is clear: for small doping levels ($\ll 10^{15}$ cm^{-3}) the semiconductor Fermi level is near the middle of the band gap and the semiconductor is *intrinsic*; as the doping level increases ($\gg 10^{15}$ cm^{-3}) the doping dictates the carrier concentration and the semiconductor is *extrinsic*, as shown in Figure 22.6. For very high doping, μ approaches the conduction band minimum, for which another terminology is introduced: when the energy separation between E_c and μ is larger than $k_B T$ (commonly, when $\mu - E_c < 3k_B T$) the semiconductor is said to be *non-degenerate*, and when $\mu - E_c > 3k_B T$, then it is said to be *degenerate*. Alternately, and colloquially, when the semiconductor starts looking more like a metal (lots of electrons in the conduction band), then it is said to be degenerate (although such a way of

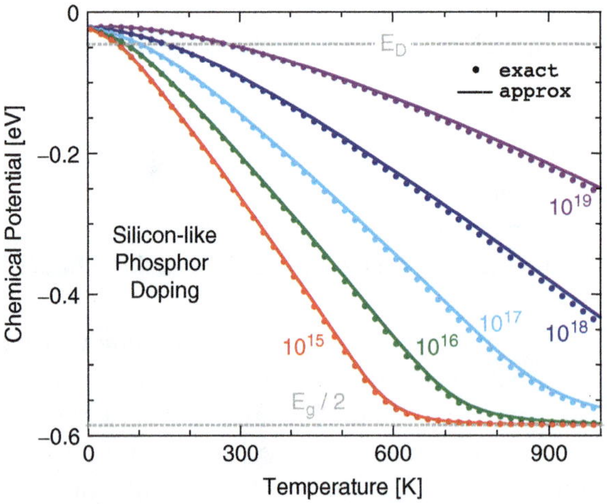

Figure 22.6 Chemical potential $\mu(T)$ as a function of temperature. The exact relations (closed circles) are given by solutions to Eq. (22.21). The approximate relations (lines) are given by solutions to Eq. (22.22). The algorithm for the evaluation is in Section A3.13.

speaking blurs the significant differences between metals and semiconductors). This is readily visible by looking at the distribution of holes versus electrons as μ climbs from the center of the gap (designated E_i) to the conduction band minimum E_c: the closer it gets, the more the holes (dashed line) diminish and the electrons (solid line) increase, as shown in Figure 22.7.

Lastly, consider the very curious Figure 22.8, which compares the electron carrier concentration for the silicon-like example to the doping level and which is calculated using the algorithm of Section A3.13. At low temperature, $\rho_e \approx (\sqrt{\pi}/2)N_c\beta^{-3/2}e^{\beta\mu}$. The $T^{3/2}$ and negative $\beta\mu$ dependence drives ρ_e to zero as $T \to 0$: the carrier concentration is said to "freeze out". Conversely, in the high temperature limit μ is increasingly negative, but asymptotically becomes $-E_g/2$ as in the high-temperature behavior of Figure 22.6: the chemical potential is then characteristic of an intrinsic semiconductor. Lastly, at typical temperatures all of the dopants are ionized, as was assumed in Section 16.1 in the calculation of the depletion layer w of Eq. (16.4). All the curves of Figure 22.8 bear the same family resemblance of fall off at low T and exponential growth at high enough T, a behavior reasonably captured in the range shown.

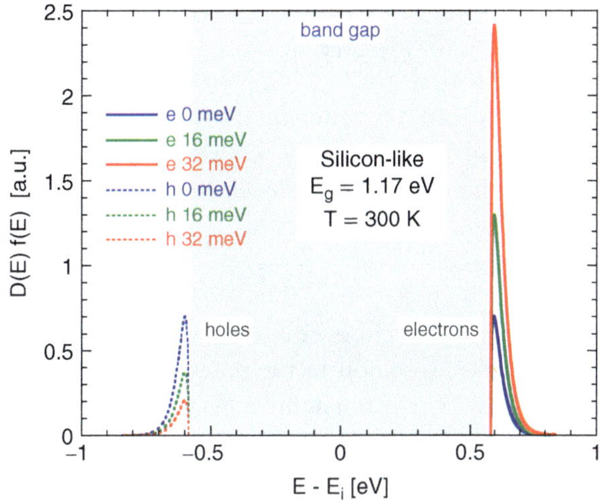

Figure 22.7 The hole ρ_h and electron ρ_e distributions (labeled h and e, respectively) as a function of energy for different values of the chemical potential μ measured in millielectronvolts from the center of the band gap. Compare Figure 7.1.

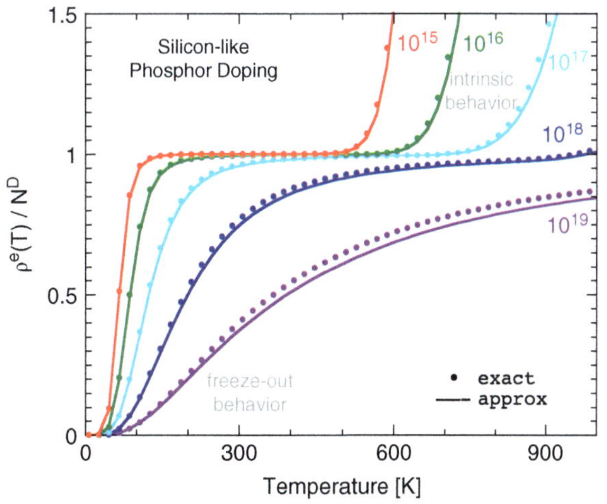

Figure 22.8 Ratio of electron concentration $\rho_e(T)$ to doping concentration N_D as a function of temperature. Freeze-out occurs when $\mu \sim E_D$. Intrinsic occurs when the temperature is so high that $\mu \sim -E_g/2$. The exact relations (closed circles) are given by solutions to Eq. (22.21). The approximate relations (lines) are given by solutions to Eq. (22.22). The algorithm for the evaluation is in Section A3.13. The approximation is reasonably good, although it degrades at higher doping densities.

22.5 Semiconductor image charge potential

Prometheus: *Do you think I will crouch before your Gods – so new – and tremble? I am far from that …*
Hermes: *Just such the obstinancy that brought you here, to this self-willed calamitous anchorage.*
Prometheus: *Be sure of this: when I set my misfortune against your slavery, I would not change.*

– Aeschylus[10]

The electrons in a metal seem free, but they must move in response to an electric field: they therefore pile up at the surface until the bulk of the metal is shielded, that is, held at an equipotential. As a consequence, electric fields terminate at the surface and are normal to it: any transverse surface field would start a flow of electrons in response. Such is the fate of those "nearly free" electrons (just as Hermes must travel the world at the whim of the god of lightning). In an insulator, however, the electrons are bound to their atoms like Prometheus to his rock, the effect of the electric field only causing the charge distribution of the atoms to distort. Dielectrics can be treated as an arrangement of little dipoles orienting in response to an electric field, but not moving charge over a distance because of it.

The little dipoles in a dielectric material placed in an electric field therefore incompletely shield the internal or material region (call it Region 2) from the field in the external or vacuum region (call it Region 1). $\vec{F}_2$ in the material is related to $\vec{F}_1$ in the vacuum by the *dielectric constant* ε of the material according to [99]

$$\left(\vec{F}_1 - \varepsilon\vec{F}_2\right) \cdot \hat{n} = 0 \tag{22.23}$$

where $\hat{n}$ is the normal to the surface. The tangential component of the fields is continuous across the boundary; alternately the potential is continuous across the surface.[11] A charge q_s embedded in the dielectric will likewise cause the dipoles to reorient themselves in response, so that the potential at a distance r from q_s is then

$$\varphi(r) = \frac{V(r)}{q} = \frac{q_s}{4\pi K_s \varepsilon_0 r} = \frac{4Q}{r}\left(\frac{q_s}{q}\right) \tag{22.24}$$

where the notation of the dimensionless K_s is preferred over $\varepsilon \equiv K_s\varepsilon_0$. Such a seemingly simple observation has consequences for the nature of the image charge contribution to the potential energy barrier. Outside a metal, a departing electron experiences a full image of itself across the plane defined by the metal surface, but if the metal is replaced by a semiconductor, what the electron experiences is somewhat lacking by comparison.

To anticipate the dependence of the image charge potential on K_s, think back to the metallic image charge potential (Section 11.3). In response to an external charge, the distribution of charge at the metal surface rearranges such that it produces a potential in vacuum equivalent to a mirror charge at the same distance from the surface but inside the metal, equal in magnitude and opposite in sign, to the external charge. That is the vacuum region perspective. Inside the metal, though, no field exists, so what perspective accounts for that? It is as though an image charge *overlaps* the real external charge, again opposite in sign, but of the same magnitude and distance from the surface, making the sum of the external charge and the image identically vanish *as seen from the inside of the metal.* The two regions therefore differ in their outlook. When the metal is replaced by a semiconductor with a dielectric constant of K_s, however, the magnitude of the overlapping image charge changes from q_s to $q_i < q_s$.

To find the image charge q_i, consider a more general problem. Let a charge q_s sit a distance a outside the surface of a semiconductor of dielectric constant $\varepsilon = K_s\varepsilon_0$. In cylindrical coordinates, the distance from q_s to an observation point at the location (ρ, z) is r_-, where $r_\pm^2 = \rho^2 + (z \pm a)^2$, so that $r_+ = r_-$ when $z = 0$. Outside the semiconductor, the potential energy is

$$V(\rho, z > 0) = \frac{qq_s}{4\pi\varepsilon_0 r_-} - \frac{qq_i}{4\pi\varepsilon_0 r_+} \tag{22.25}$$

where q_i is the image charge magnitude, whereas inside the semiconductor

$$V(\rho, z < 0) = \frac{qq_s}{4\pi\varepsilon_0 r_-} - \frac{qq_i}{4\pi\varepsilon_0 r_-} \tag{22.26}$$

[10] Aeschylus (trans. David Grene), *Aeschylus II: Prometheus Bound, The Complete Greek Tragedies,* ed. David Grene, Richmond Lattimore. Chicago: The University of Chicago Press, 1956, lines 960–967.

[11] Remember the convention here: fields $\mathcal{E}$ and potentials φ are multiplied by unit charge q to convert them to forces F and potential energies V, but in units of $q = 1$, the discussion can use either terminology, and so the more familiar one is used.

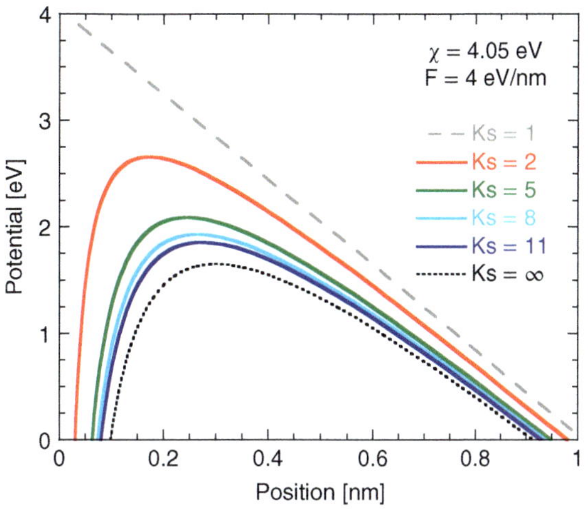

Figure 22.9 Semiconductor image charge potential barrier based on Eq. (22.29) with $\chi = 4.05$ eV (after silicon) and $F = 4$ eV/nm.

where r_- is in the denominator of the second term because q_i is in the vacuum. Now impose Eq. (22.23), which can be written as

$$\varepsilon_0 \partial_z V(\rho, 0_-) = K_s \varepsilon_0 \partial_z V(\rho, 0_+) \tag{22.27}$$

Using $r_+ = r_-$ for $z = 0$ and eliminating all the common factors, which turns out to be most of them, all that is left is

$$q_s - q_i = K_s(q_s + q_i) \rightarrow q_i = \frac{K_s - 1}{K_s + 1} q_s \tag{22.28}$$

Applying this to the image charge potential for an emitted electron, the modification to Eq. (11.25) when a semiconductor replaces the metal is [305]

$$V_{image}(x) \equiv \chi - Fx - \left(\frac{K_s - 1}{K_s + 1} \right) \frac{Q}{x} \tag{22.29}$$

It is seen that a metal corresponds to the limit $K_s \rightarrow \infty$ and the replacement of the electron affinity χ by the sum of chemical potential μ and work function Φ. The specification of the barrier height by χ for semiconductors is because, as seen in Sections 22.3 and 22.4, doping levels and fields can affect the relationship of μ to Φ, whereas χ remains a fixed value above the conduction band minimum. As shown in Figure 22.9, the inclusion of K_s makes the image charge potential barrier somewhat higher than for metals.

Although Eq. (22.29) describes the image charge modification for field emission from bulk semiconductors, it is nevertheless the case that often semiconductors are *coatings* on metals that can be rather thin, even so thin as to (in the case of adsorbates) constitute a monolayer. In such cases there are image charges of image charges to enforce boundary conditions. A single layer of atoms resting on a metal surface can be treated as a thin film of thickness l and dielectric constant K_s, so that the method of images applied to that configuration results in [111]

$$V_{image}(x) \rightarrow \chi - Fx - \left(\frac{K_s - 1}{K_s + 1} \right) \frac{Q}{x} - \left(\frac{2}{K_s + 1} \right) \frac{Q}{x + l} \tag{22.30}$$

when a layer is present. For more than a few atomic layers, however, the last term becomes small as $l \rightarrow \infty$, and its neglect is common. Field emission from adsorbate-covered surfaces (an in-depth treatment found in Chapter 6 of ref. [111]) bears a resemblance to work function reduction due to adsorbates behind the Gyftopoulos–Levine theory in Section 31.8, and so will be revisited there.

22.6 Dielectric constant and screening

> *…self-deception is the handmaiden of deceit: in hiding the truth from ourselves, we are able to hide it more fully from others.*
>
> – David Livingston Smith[12]

[12]David Livingston Smith, *Why We Lie: The Evolutionary Roots of Deception and the Unconscious Mind.* New York: St. Martin's Press, 2004, p. 3.

Much like hiding a lie, a charged inclusion can be hidden inside a material if the material participates in shielding it. A charged inclusion inside a dielectric pulls charges of the opposite sign towards it, and the more effectively the material allows the inclusion to do so, the more effectively the effects of that inclusion's charge are hidden. The material's *dielectric constant* $\varepsilon \equiv K_s\varepsilon_0$ is a measure for how well a material can do so. A K_s of unity signals that the material is no better than vacuum for hiding or shielding the charge from the outside, that is, the charged inclusion is evident to all. A K_s that is very large (as occurs for metals) means that charge separation is increasingly effortless, so that the inclusion's existence can be hidden very well. Expressing the dielectric constant as $\varepsilon = K_s\varepsilon_0$ makes revisiting the Yukawa potential of Eq. (21.89) conducive to understanding semiconductors, particularly with respect to screening [86, 292] and band bending at interfaces [53].

If a positive charged particle, represented by ρ_i, is in a material[13] then Poisson's equation is in the form

$$-\nabla^2\Phi(\vec{r}) = \frac{q^2}{\varepsilon_0}\rho_i(\vec{r}) = 16\pi Q\rho_i(\vec{r}) \tag{22.31}$$

Then, just as with the free and bound charges treatment of the image charge in Section 22.5, let the potential $\phi(\vec{r})$ be related to the charge distribution, including both the bare charge and the charges which screen it (denoted by the s subscript), or

$$-\nabla^2\phi(\vec{r}) = 16\pi Q(\rho_i(\vec{r}) - \rho_s(\vec{r})) \tag{22.32}$$

Now draw an analogy with bulk dielectrics. Recall that a parallel plate capacitor with a vacuum in the anode–cathode (AK) gap has a field between the two plates of $F = V_a/D = \sigma/\varepsilon_0$, where σ is the charge density $\sigma = Q/A$ and A is the area of the capacitor plate (in this way, one derives that the capacitance is $C = Q/V_a = \varepsilon_0 A/D$). If, instead of vacuum, the AK gap is filled with an insulator of dielectric constant K_s, then the field in the gap becomes $F \to F/K_s$, which means that $\sigma/\varepsilon_0 \to \sigma/K_s\varepsilon_0 \equiv \sigma/\varepsilon$. Because of the position-dependent nature of the ion charge potential, the more general equation is now

$$\phi_i(\vec{r}) = \int_\Omega \varepsilon(\vec{r} - \vec{r}\,')\phi(\vec{r}\,')d\vec{r}\,' \tag{22.33}$$

Equations of this form can use the convolution theorem [60]: if tildes denote the Fourier transform of a function (e.g., $\tilde{g}(\vec{k})$ is the Fourier transform of $g(\vec{r})$), then

$$\begin{aligned}\int \varepsilon(\vec{r} - \vec{r}\,')\phi(\vec{r}\,')d\vec{r}\,' &= (2\pi)^{-3}\iiint d\vec{r}\,'\, d\vec{k}\, d\vec{k}'\, \tilde{\varepsilon}(\vec{k})e^{i\vec{k}\cdot(\vec{r}-\vec{r}\,')}\tilde{\phi}(\vec{k}')e^{i\vec{k}'\cdot\vec{r}\,'}\\ &= \iint d\vec{k}d\vec{k}'\, \tilde{\varepsilon}(\vec{k})\, \tilde{\phi}(\vec{k}')e^{i\vec{k}\cdot\vec{r}}\delta(\vec{k} - \vec{k}')\\ &= \int d\vec{k}\, \tilde{\varepsilon}(\vec{k})\, \tilde{\phi}(\vec{k})e^{i\vec{k}\cdot\vec{r}}\end{aligned} \tag{22.34}$$

This is convenient because the Fourier transform of Eq. (22.33) can be written as

$$\tilde{\phi}_i(\vec{k}) = \tilde{\varepsilon}(\vec{k})\tilde{\phi}(\vec{k}) \to \tilde{\phi}(\vec{k}) = \frac{\tilde{\phi}_i(\vec{k})}{\tilde{\varepsilon}(\vec{k})} \tag{22.35}$$

For the Coulomb potential of a point charge, $\tilde{\phi}_i(\vec{k})$ is interestingly tricky because although all of the integrals involved are elementary, the limits of integration are not. The integrals are

$$\tilde{\phi}_i(\vec{k}) = Q\int \frac{e^{i\vec{k}\cdot\vec{r}}}{r}d\vec{r} = 2\pi Q\int_0^\infty rdr\int_{-1}^{1} e^{ikr\cos\theta}\sin\theta\, d\theta \tag{22.36}$$

A common technique is to append a factor $e^{-\delta r}$ to the integrand, and at the end of the calculation take the limit $\delta \to 0$ (or $1/\delta \to \infty$).[14] The intermediate integral step is then

$$\tilde{\phi}_i(\vec{k}) = \frac{2\pi iQ}{k}\int_0^\infty \left(e^{-ikr} - e^{ikr}\right)e^{-\delta r}dr = \frac{4\pi Q}{k^2 + \delta^2} \tag{22.37}$$

[13] A positive charge puts the sign back into Poisson's equation as in standard accounts, and recall that ρ is a *number* density. The sign will be important to remember in Eq. (22.40).

[14] Memories of the troubles one can get into by failing to appreciate which infinite limit to take first may arise, that of $1/\delta$ or the upper integration limit for r, as in Section 6.3, but here it is not so critical.

Now the limit $\delta \to 0$ is well-behaved, and so

$$\tilde{\phi}_i(\vec{k}) = \frac{4\pi Q}{k^2} \tag{22.38}$$

Replacing $\phi_i(\vec{r})$ by its Fourier transform in Eq. (22.31), the action of the Laplacian operator amounts to replacing $\nabla^2 \to -k^2$ because

$$-\nabla^2 \int d\vec{k} e^{i\vec{k}\cdot\vec{r}}\{\cdots\} = \int d\vec{k} e^{i\vec{k}\cdot\vec{r}} k^2 \{\cdots\} \tag{22.39}$$

as $\{\cdots\}$ contains no $\vec{r}$-dependent terms. Now assemble the pieces. From Eq. (22.35) and Eq. (21.83)

$$\begin{aligned}\tilde{\varepsilon}(\vec{k}) = \frac{\tilde{\phi}_i(\vec{k})}{\tilde{\phi}(\vec{k})} &= 1 - \frac{(\tilde{\rho}(\vec{k}) - \tilde{\rho}_i(\vec{k}))}{\tilde{\rho}(\vec{k})} \\ &= 1 + \frac{16\pi Q D(\mu)\delta\phi}{k^2 \delta\phi} = 1 + \frac{k_{TF}^2}{k^2}\end{aligned} \tag{22.40}$$

To get $\phi(\vec{r})$ requires taking the Fourier transform, and so (where $s = \cos\theta$)

$$\begin{aligned}\phi(\vec{r}) &= \frac{16\pi Q}{(2\pi)^3} \int_0^\infty \left(\frac{k^2}{k^2 + k_{TF}^2} \right) dk \int_{-1}^{1} e^{ikrs} ds \int_0^{2\pi} d\varphi \\ &= \frac{4Q}{r} \exp(-k_{TF} r)\end{aligned} \tag{22.41}$$

which is, of course, the *ansatz* offered right after Eq. (21.84).

CHAPTER 23

Effective mass

...the dynamics of an electron wave packet (or, as we shall now say, of an "electron in a state **k***") may be very different from those of an elementary free classical particle. However attractive may be the analogy between* $\hbar\mathbf{k}$ *and momentum, we must not be led into believing that the velocity of an electron in a solid is anything like this quantity divided by the mass.*

– John M. Ziman [82], p. 92

"You keep using that word!" the Spaniard snapped. "I don't think it means what you think it does."

– William Goldman[1]

Treating electrons as particles creates an expectation that the the mass of the electron is related to the energy E by $E = p^2/2m$, where $\vec{p}$ is the momentum. The relation can be written as

$$\frac{1}{m} = \frac{1}{p}\frac{\partial}{\partial p}E \tag{23.1}$$

A difficulty with transposing that understanding to a quantum mechanical treatment arises because electrons are *not* point particles, but rather wave packets composed of a distribution of $|k\rangle$ with different velocities such that $p = \hbar k$ for each $|k\rangle$. The same relation will be shown to hold between m and E, but the meaning of the mass term changes.

23.1 Dispersion relations

Glory is like a circle in the water,
Which never ceaseth to enlarge itself,
Till by broad spreading it disperse to naught.

– William Shakespeare[2]

The concept of *dispersion* relates frequency ω to wave number k. It is the origin of distinctions between phase velocity and group velocity, as will be important in treating lattice waves (e.g., phonons), but it arises for wave phenomena beyond the examples considered here. Simply, dispersion arises because the different waves that make up a wave packet have different wave numbers and therefore different velocities, causing the wave packets to spread out over time (as in the positive time regions of Figures 9.9 and 9.10 for the no barrier case) and "disperse to naught", as it were. A few examples give the flavor of how $\omega(k)$ varies with k.

Photons are a gentle case, given that $\omega/k = c$ is the phase velocity of the photon and $\hbar\omega$ is its energy, so that $\omega = ck$ denotes a linear dispersion relation. The *group velocity* $v_g = d\omega/dk$ is then the same as the phase velocity c, a consequence of all photons traveling at the same speed. When they do not, as when light passes through a high index of refraction material such as a prism, the various wavelengths *disperse* and, for white light, a familiar rainbow pattern emerges.

The dispersion relations of matter waves are revealed by Schrödinger's time-dependent equation: inserting $\psi \propto e^{i(kx-\omega t)}$ into $-i\hbar\partial_t\psi = -(\hbar^2/2m)\partial_x^2\psi$ indicates that $\omega = \hbar k^2/2m$ and $d\omega/dk = \hbar k/m$. This is commensurate with the non-relativistic $E = \hbar^2k^2/2m$, so that $\omega = \hbar k^2/2m$ and $d\omega/dk = \hbar k/m$.

The group velocity v_g will be the concept from which a better understanding of effective mass shall emerge. It is common to introduce v_g in the context of the evolution of wave packet by [41]

$$\Psi(x,t) = \int_{-\infty}^{\infty} A(k)\psi_k(x,t)dk \tag{23.2}$$

[1] William Goldman, *The Princess Bride*. New York: Ballantine Books, 1998, p. 102.
[2] Ref. [37]: *Henry the Sixth, Part 1*, I.ii.134–138.

Introduction to the Physics of Electron Emission, First Edition. Kevin L. Jensen.

where $\psi_k(x,t) = e^{i(kx-\omega t)}$ and $A(k)$ is a distribution in k. It is then shown that $\Psi(x,t)$ can be put in the form of plane waves encased in an envelope function traveling at the group velocity $v_g = d\omega/dk$, and that a correspondence with a classical particle then demands that $d\omega/dk = p/m$. The derivation is elegant, but not quite simple.

A more appealing approach uses the Wigner function of Section 9.3.2 to arrive at the same result,[3] which is possible because $f(x,k,t)$ contains all the information in it that $\psi_k(x,t)$ does, but hews to phase space intuitions. The group velocity of the wave packet is

$$\frac{\hbar}{m}\langle k\rangle = \int_{-\infty}^{\infty}\int_{-\infty}^{\infty}\frac{\hbar k}{m}f(x,k,t)dxdk \tag{23.3}$$

All that shall be required is that $f(x,k,t)$ is initially centered at (x_o, k_o), is symmetrical about k_o, declines sufficiently fast (say, at least exponentially fast) as $x \to \pm\infty$ and same with k, and let the potential that the wave function propagates in be at worst linear in position (e.g., $V(x) = V_o - Fx$). Because the Wigner trajectories follow the classical particle trajectories as in Eq. (9.115), and using Eq. (9.111),

$$\begin{aligned}\frac{\hbar}{m}\langle k\rangle &= \int_{-\infty}^{\infty}\int_{-\infty}^{\infty}\frac{\hbar k}{m}f\left(x-\frac{\hbar k}{m}t+\frac{F}{2m}t^2, k-\frac{F}{\hbar}t, 0\right)dxdk \\ &= \int_{-\infty}^{\infty}\int_{-\infty}^{\infty}\frac{\hbar(k'+(Ft/m))}{m}f(x',k',0)dx'dk'\end{aligned} \tag{23.4}$$

Now use the symmetry of $f(x,k,0)$: symmetric (or even) functions of $(k-k_o)$ (such as constants or $(k-k_o)^2$) integrated over all space with it will be non-zero, but anti-symmetric (or odd) functions, such as $(k-k_o)$ itself, vanish because the integral of an even and odd function is identically zero. Therefore, by writing $k'+(Ft/m)$ as $(k'-k_o)+(k_o+(Ft/m))$, the term in the first parentheses is odd, and the second even. Therefore,

$$v_g = \frac{\hbar k_g(t)}{m} \tag{23.5}$$

where $k_g(t) = k_o + (Ft/m)$ is the velocity of the center of the wave packet if $f(x,k,0)$ is a Gaussian in both $x - x_o$ and $k - k_o$. Now Eq. (23.5) gives the velocity of the "particle" as desired, and shows that it is the group velocity.

The other term, namely mass $m \to m_n$, requires comparable scrutiny, in recognition that the "effective mass" of the electron in an energy band in the first Brillouin zone of a material can be something different than its vacuum value.[4] The mass term r_e is a consequence of electron transport in materials – it does not *a priori* have to be constant. If it is not, though, consequences accrue, the first being the relationship between the force on the electron and its change in momentum, or

$$F = \frac{dp}{dt} = \frac{d}{dt}(m_n v) = m\left(r_e + v\frac{dr_e}{dv}\right)\frac{dv}{dt} \tag{23.6}$$

Because $r_e = \hbar k/mv$ and the chain rule, then

$$r_e + v\frac{dr_e}{dv} = r_e - v\frac{\hbar k}{mv^2} + \frac{\hbar}{m}\frac{dk}{dv} = \frac{\hbar}{m}\frac{dk}{dv} \tag{23.7}$$

because first two terms are equal and opposite. The final step is recalling that $v = d\omega/dk = \hbar^{-1}dE/dk$ so that $dv/dk = \hbar^{-1}d^2E/dk^2$ and so replacing dv/dt in Eq. (23.6) gives

$$F = \frac{r_e m}{\hbar}\frac{d^2E}{dk^2}\frac{dk}{dt} \tag{23.8}$$

Therefore, matters have been engineered so that the form of the old familiar equations like Eq. (23.1) can be retained to an extent:

$$r_e m v = m_n v \equiv \hbar k \tag{23.9}$$

$$\frac{1}{r_e m} = \frac{1}{m_n} \equiv \frac{1}{\hbar^2 k}\partial_k E \tag{23.10}$$

---

[3]This is not to say one should deprive oneself of the treatment (see Ziman, Section 2.9 of ref.[82]), which dives right into the complications that attend describing packet dynamics with classical-like equations and effective masses: such a treatment is edifying, but beyond needs here.

[4]An asterisk (*) on $m^* = r_e m$ is another common way to denote effective mass, as is letting m be the effective mass and m_o be the electron rest mass (confusing), or (as in Eq. (22.9) and earlier) introducing a coefficient r_e so that $m_n = r_e m$. For consistency, the coefficient r_e method is retained here, so that m continues to refer to the electron rest mass and is constant.

as long as it is recognized that the definitions hinge on what is meant by E and its relation to k. That will take some work, but at least now it is understood what the words mean: m_n is not the mass of a particle, but rather the curvature of the $E(k)$ relation, and therefore an effective mass.

23.2 The $k \cdot p$ method

Travaillons sans raisonner ...c'est le seul moyen de rendre la vie supportable.

– Voltaire[5]

The Bloch states $u(k)$ of Eq. (21.23) that arise in the Kronig–Penny model (Section 21.3) signal the first hint of where complications are to come in the revised Schrödiger-like equations of Eq. (21.24) (albeit for one dimension): a new term $2ik_j\partial_x$ appears and $k^2 \to k^2 - k_j^2$. An argument can be made about how to solve such an equation, but making use of the work already developed in solving Schrödinger's equation suggests a perturbation approach would make the experience easier. For three dimensions, let k_j go to a continuum and $\vec{p} = -i\hbar\vec{\nabla}$, resulting in $2ik_j\partial_x \to (2/\hbar)\vec{k} \cdot \vec{p}$, or

$$\left\{ -\frac{\hbar^2}{2m_n}\nabla^2 + V(\vec{r}) - i\frac{\hbar^2}{2m_n}\vec{k} \cdot \vec{\nabla} \right\} u(\vec{k}, \vec{r}) = E'(\vec{k})u(\vec{k}, \vec{r}) \tag{23.11}$$

where

$$E' \equiv E - \frac{\hbar^2 k^2}{2m_n} \tag{23.12}$$

which looks very much like the familiar Schrödinger's equation with the addition of a perturbation defined by

$$\hat{H}_{kp} = \frac{\hbar}{2m_n}\vec{k} \cdot \hat{p} \tag{23.13}$$

What kind of difference does $\hat{H}_{kp}$ make? To answer, start by considering a seemingly asinine choice for the Kronig–Penny model: let the height of the barrier vanish, or $V_o \to 0$, but otherwise keep the periodicity of the lattice as R (i.e., the wave function satisfies $\psi(\vec{k}, \vec{r} + \vec{R}) = \psi(\vec{k}, \vec{r})$). Thus, the wave function can be written

$$\psi(\vec{K}, \vec{r}) \propto \exp(i\vec{k} \cdot \vec{r}) \exp\left(i(\vec{K} - \vec{k}) \cdot \vec{r}\right) \tag{23.14}$$

where $\vec{k}$ is the reduced wave vector (in the first Brillouin zone), and $\vec{K} - \vec{k} \equiv \vec{G}$ is a reciprocal lattice vector such that $\vec{G} \cdot \vec{R} = 2\pi n$ for n an integer. Thus, for the special case $V_o = 0$, then $u(\vec{k}, \vec{r}) \propto \exp(i\vec{G} \cdot \vec{r})$. When $\vec{K}$ is in another Brillouin zone, it can be brought back into the first Brillouin zone by a translation of $\vec{G}$. This is commonly demonstrated by diagrams of the form shown in Figure 23.1. By such a construction, k is restricted to the first zone ($-\pi < ka < \pi$ in the figure, where a is the one-dimensional lattice constant in the shameful units of $(a = 1)$).

When V_o is then finite, the $\vec{k} \cdot \vec{p}$ term acts as a perturbation. Bands open up where the crossings occurred, and near the bottom of the band the usual parabolic relations emerge, but as k increases, the effects of the $\vec{k} \cdot \vec{p}$ makes itself known. Here is where the problem becomes rapidly complicated and difficult, requiring as it does an understanding of the electron orbitals introduced in Section 21.4. The Kane model [294, 306] introduces appropriate concepts, which are moreover useful for the alpha semiconductor model of Section 23.4, to treat the non-parabolic nature of $E(k)$ in semiconductors. A new wrinkle is an inclusion of the electron spin of $\pm 1/2$ for which the spin operator is $\vec{\sigma}$. Spin coupling introduces another term H_σ into the Hamiltonian already shown in Eq. (23.11) of the form

$$\hat{H}_\sigma = \frac{\hbar}{4m^2c^2}\left(\vec{\nabla}V \times \vec{p}\right) \cdot \vec{\sigma} \tag{23.15}$$

The actual form of $\hat{H}_\sigma$, however, is less important than what it does by introducing a small splitting in the energy depending on whether the spin is "up" or "down". The impact of both $\hat{H}_\sigma$ and $\hat{H}_{kp}$ will be determined using the unperturbed basis states $|u_i\rangle$ that are the solution of

$$\left(\frac{\hat{p}^2}{2m} + V(\hat{r})\right)|u_i\rangle = E_i|u_i\rangle \tag{23.16}$$

[5] "Let us work without disputing ...it is the only way to render life tolerable." Voltaire (trans. John Butt), *Candide*. Middlesex: Penguin Books, 1947, p. 34.

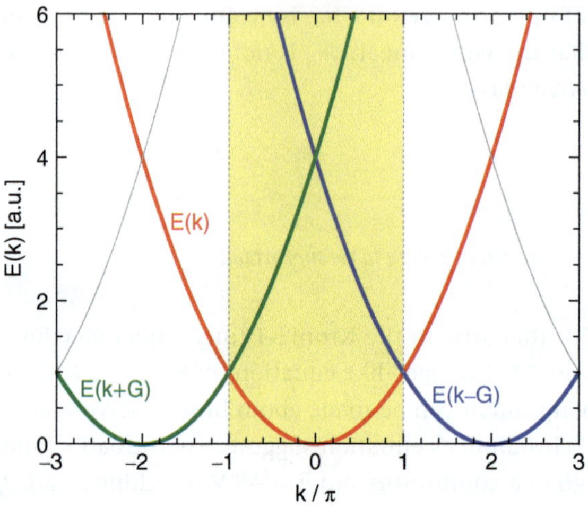

Figure 23.1 Band structure in the first Brillouin zone (yellow region) for $V_o = 0$ in a one-dimensional model with a lattice constant of $a = 1$. When $V_o > 0$ band gaps are introduced at the cross-over points where $E(k \pm G)$ is degenerate with $E(k)$.

where $\hat{p}$ is the momentum operator $-i\hbar\nabla$ as before. The basis states in the conduction band are s-like in the language of Section 21.4, but now they are additionally cognizant of spin, in recognition of which let $|u_c\rangle = |s\pm\rangle$. The valence band states are made up of the p-like states, but so as not to confuse the momentum operator and the state, designate them by $|u_v\rangle = |X\pm\rangle, |Y\pm\rangle$, and $|Z\pm\rangle$. Introduce two energies $\mathcal{P}$ and Δ: assuming the spin is along the $\hat{z}$ axis, they are defined by

$$\mathcal{P} = -\frac{i\hbar}{m}\langle S|P_z|Z\rangle \tag{23.17}$$

$$\Delta = \frac{3i\hbar}{(2mc)^2}\langle X|P_y(\partial_x V) - P_x(\partial_y V)|Y\rangle \tag{23.18}$$

where Δ is the spin-orbit energy splitting of the valence band from the $\hat{H}_\sigma$ term, and $\mathcal{P}$ arises from $\langle S|\hat{H}_{kp}|P\rangle$. Using the s- and p-orbital wave functions and the methods of ref. [306], these quantities can be found, but there is no imperative to do so: what matters here is that they exist and with cleverness approximations to them or limits for them will prove useful (e.g., $\mathcal{P} \propto \hbar c/246 = 2.228Q$ generally holds for III–V compound semiconductors as observed by ref. [306], although $\mathcal{P}$ is given in Eq. (23.24) below). For the four chosen basis states, the roots obtained by diagonalizing the Hamiltonian result in

$$\begin{vmatrix} E_s & 0 & k\mathcal{P} & 0 \\ 0 & \left(E_p - \frac{1}{3}\Delta\right) & \frac{\sqrt{2}}{3}\Delta & 0 \\ k\mathcal{P} & \frac{\sqrt{2}}{3}\Delta & E_p & 0 \\ 0 & 0 & 0 & \left(E_p + \frac{1}{3}\Delta\right) \end{vmatrix} = 0 \tag{23.19}$$

where E_s indicates the conduction band and E_p the valence band. Explicit expansion then collection of like terms results in the determinant equation being given by

$$\left(3E_p + \Delta\right)\left\{E_s(3E_p + \Delta)(3E_p - 2\Delta) - 3k^2\mathcal{P}^2(3E_p - \Delta)\right\} = 0 \tag{23.20}$$

The fourth row/column corresponds to the heavy hole band, and it sets the choice $E_p = -\Delta/3$. E_s is at the bottom of the conduction band, so the band gap appears to be $E_s - E_p = E_g + \Delta/3$ before the perturbation, but a diagonalization after the perturbation restores $E_s - E_p = E_g$ as the band gap energy for $k = 0$, the details of which are less useful here than the result. The ratio Δ/E_g is an important parameter and identifies how much the band gap decreases as a result of the spin-orbit interaction. In the limit that it is large, E_k in the conduction band[6] takes on the form

$$E_k = \frac{\hbar^2k^2}{2m} + \frac{1}{2}\left(E_g + \sqrt{E_g^2 + \frac{8}{3}k^2\mathcal{P}^2}\right) \tag{23.21}$$

[6] E_k can also be found in the valence band, as in ref. [306], Eqs (58–60), but conduction band *electron* emission is the *casus belli* for the present assault.

For small values of k^2, using the expansion $\sqrt{1+x} \approx (x/2)$ shows that E_k is indeed parabolic in k, or

$$E_k - E_g \approx \frac{\hbar^2 k^2}{2m}\left[1 + \frac{2m\mathcal{P}^2}{3\hbar^2}\left(\frac{2}{E_g} + \frac{1}{E_g + \Delta}\right)\right] \tag{23.22}$$

This is pretty heady stuff: it says that the effective mass of the electron in the conduction band is dependent on the free electron mass m, the band gap E_g, the $k \cdot p$ factor $\mathcal{P}$, and the spin-orbit energy factor Δ, or

$$\frac{m}{m_n} = \frac{1}{r_e} = 1 + \frac{2m\mathcal{P}^2}{3\hbar^2}\left(\frac{2}{E_g} + \frac{1}{\Delta + E_g}\right) \tag{23.23}$$

It is worth pausing for a moment and considering the implications. That the dynamics of an electron transporting *through a material* can be fairly well accounted for by changing the value of the *electron's mass* in the equations of motion is rather remarkable. Solving Eq. (23.23) for $\mathcal{P}$ then gives

$$\mathcal{P}^2 = \frac{3\hbar^2}{2m}\left(\frac{1}{r_e} - 1\right)\left(\frac{2}{E_g} + \frac{1}{E_g + \Delta}\right)^{-1} \tag{23.24}$$

which approaches $\hbar^2 E_g/2m_n$ when r_e is small and $\Delta/E_g \to 0$.

EXAMPLE: Using the approximation of $\mathcal{P} \to \hbar c/246$ and a band-gap of $E_g = 1.424$ eV for GaAs, estimate $r_e = m_n/m$ in the two limits $\Delta/E_g \to \infty$ and $\Delta/E_g \to 0$. Compare to the commonly accepted value for GaAs of $r_e = 0.067$.
SOLUTION: From Eq. (23.23), in the limit $\Delta/E_g \to \infty$, then

$$r_e = \left(1 + \frac{4mc^2}{3 \times 246^2 E_g}\right)^{-1} = 0.1123$$

In the limit $\Delta/E_g \to 0$, then

$$r_e = \left(1 + \frac{2mc^2}{256^2 E_g}\right)^{-1} = 0.07776$$

Having found what E_s and E_p must be, and then choosing the energy reference to be the top of the valence band, attention returns to Schrödinger's equation in Eq. (23.11) for the general relationship between E and k. It can be found to be[7]

$$E'(E' - E_g)(E' + \Delta) - k^2\mathcal{P}^2\left(E' + \frac{2}{3}\Delta\right) = 0 \tag{23.25}$$

but now Eq. (23.24) can be used to eliminate $\mathcal{P}^2$ in favor of more familiar terms

$$\frac{\hbar^2 k^2}{2m}\left(\frac{1}{r_e} - 1\right) = \frac{E'(E' - E_g)}{E_g}\left[\frac{(E' + \Delta)(3E_g + 2\Delta)}{(E_g + \Delta)(3E' + 2\Delta)}\right] \tag{23.26}$$

which hardly seems an improvement until some interesting features are noticed. First, for a number of semiconductors, $1/r_e$ is big (for GaAs, it is about 15), so that the left-hand side, to a reasonable approximation, is $\hbar^2k^2/2m_n$, where the effective mass m_n takes over the role previously held by m. Now look at the limits of small ($\Delta/E_g \to 0$) vs large ($\Delta/E_g \to \infty$) spin-orbit coupling. One finds *the same limit in both cases*, because of the term in the square brackets, to be

$$\lim_{\Delta/E_g \to 0}\left(\frac{\hbar^2 k^2}{2m_n}\right) = \lim_{\Delta/E_g \to \infty}\left(\frac{\hbar^2 k^2}{2m_n}\right) = \frac{E'(E' - E_g)}{E_g} \tag{23.27}$$

This is a most agreeable simplification, and it works fairly well in practice. It is generally called a *hyperbolic dispersion relationship*, which sounds unusual but is actually more familiar than one might expect: such relations are found in treatments of relativity.

[7] See Eq. (53) of ref. [306] (alternately, Eq. (2.66) of ref. [294]) for a bit of discussion regarding it.

23.3 Hyperbolic relations

Confusion never stops
Closing walls and ticking clocks

– Coldplay[8]

All motion is relative. Perhaps it is you who have moved away – by standing still.

– Jerome Lawrence and Robert E. Lee[9]

If a ball is thrown against a wall, bounces off, and returns to its launch point, as in Figure 23.2, the time of flight $2\Delta t$ can be related to the distance traveled $2L$ and the velocity of the ball v_b by $2\Delta t = 2L/v_b$. If, on the other hand, the same process is repeated from a platform moving at a velocity v_p in a direction parallel to the surface of the wall but with velocity of the ball was unchanged, a person in the moving frame would say the transit time $\Delta t'$ is longer because the velocity in his frame is smaller, or

$$2\Delta t' = \frac{2L}{\sqrt{v_b^2 - v_p^2}} \tag{23.28}$$

For an observer in the rest frame, however, the velocity is still v_b but the path length has increased, so

$$2\Delta t = 2\frac{\sqrt{L^2 + (v_p \Delta t)^2}}{v_b} \tag{23.29}$$

Solving for Δt and comparing to Eq. (23.28) shows $\Delta t = \Delta t'$. Time passes the same for both the rest frame and the moving frame.

Put that understanding in a useful mathematical representation. Designate the moving platform with primed coordinates and the stationary platform with unprimed coordinates (as is usual). Relating the physics behind the differing views on the platforms is then accomplished by a *Galilean* transformation: if the unprimed coordinates in space and time are (x, y, z, t), then the primed coordinates satisfy [307]

$$\begin{pmatrix} t' \\ x' \\ y' \\ z' \end{pmatrix} = \begin{pmatrix} t \\ x - vt \\ y \\ z \end{pmatrix} = \begin{pmatrix} 1 & 0 & 0 & 0 \\ -vt & 1 & 0 & 0 \\ 0 & 0 & 1 & 0 \\ 0 & 0 & 0 & 1 \end{pmatrix} \begin{pmatrix} t \\ x \\ y \\ z \end{pmatrix} \tag{23.30}$$

where it is convenient to drop the p subscript on the platform velocity v_p and simply call it v. Because force F involves the second derivative with respect to time, then Newton's equations under a Galilean transformation stay the same, or $F = m(d^2x/dt^2) = m(d^2x'/dt^2)$, where dt and dt' are the same. But when such Galilean transforms are applied to the

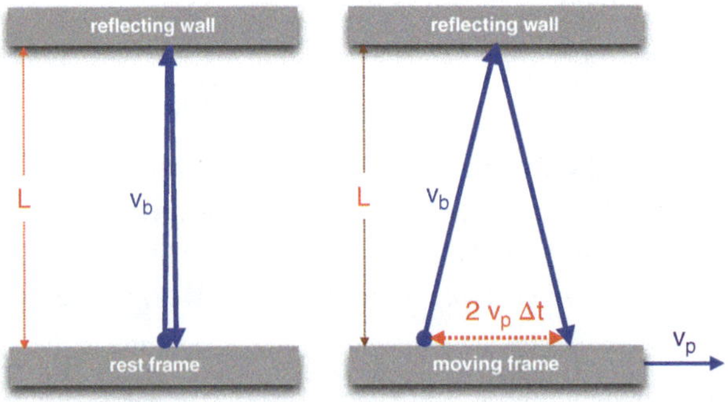

Figure 23.2 Left, a ball traveling at a velocity v_b with respect to a stationary platform will travel to a wall a distance L and back in a time $2\Delta t = 2L/v_b$. Right, if the platform is moving at a velocity v_p, then the transit time increases because the path length increases; an observer in the moving frame would see the ball, however, as traveling at a velocity $\sqrt{v_b^2 - v_p^2}$ over a distance $2L$.

[8]C. Martin, J. Buckland, G. Berryman, and W. Champion, lyrics to *Clocks*, 2002.
[9]J. Lawrence and R.E. Lee, *Inherit the Wind*. New York: Bantam Books, 1960, Act 2, Scene 2, p. 60.

equations of electrodynamics, the form of the wave equations for electric and magnetic field associated with light alter, showing that the equations are not "invariant" under a Galilean transformation.

Enter Einstein, who asserted that if the ball were replaced by a beam of light, then both an observer in the rest frame and one in the moving frame would measure the speed of light as the same. But if each observer experiences light as traveling at a velocity c, then the equivalence of the transit times no longer holds. This means that something is peculiar with the passage of time: an observer in the rest frame would interpret the moving frame results as being due to light traveling a path length of $2\sqrt{L^2 + (\beta ct)^2}/c$ and therefore argue that more time passed than the observer in the moving frame, who would say the transit time is $2\Delta t$. The passage of time in the moving frame has slowed: writing c as the path length over the transit time, first in the rest frame and then in the moving frame, and using the very common notation that $\beta \equiv v/c$, then

$$\frac{2L}{2\Delta t} = c = \frac{2\sqrt{(c\Delta t)^2 + (\beta c\Delta t')^2}}{2\Delta t'} \tag{23.31}$$

where the left-hand side is for the rest frame and the right-hand side is for the moving frame with $L \to c\Delta t$. Solving the rightmost equation for $\Delta t'$ gives

$$\Delta t' = \frac{\Delta t}{\sqrt{1-\beta^2}} \equiv \gamma \Delta t \tag{23.32}$$

where the second of the common symbols of relativity, namely $\gamma = 1/\sqrt{1-\beta^2}$, is introduced. If the passage of time has slowed, then the length L of a rod with its axis in the direction of motion shortens: determining the length of a rod by measuring the transit time of a light pulse from one end of it to another requires $L'/\Delta t' = c$, so that if $\Delta t' = \gamma \Delta t$, then so too must $L' = \gamma L$.

The invariant in going from one coordinate system to another is the speed of light: a point on the spherical wave front of an expanding pulse of light from a point source, measured in both the unprimed (rest) and primed (moving) frames is the same, or

$$c = \frac{d}{dt}|\vec{r}| = \frac{d}{dt'}\left|\vec{r'}\right| \tag{23.33}$$

To keep the discussion simple, assume that the coordinate axes in both frames are parallel such that $y' = y$ and $z' = z$. Introduce the differential length $dr = \sqrt{dx^2 + dy^2 + dz^2}$, so that Eq. (23.33) can alternately be written

$$c^2 dt^2 - dx^2 - dy^2 - dz^2 = c^2 dt'^2 - dx'^2 - dy'^2 - dz'^2 \tag{23.34}$$

which can be expressed as $ds^2 = ds'^2$ where ds is the invariant quantity in inertial frames.[10] How, then, must the Galilean transformation matrix of Eq. (23.30) be altered to respect the constancy of the speed of light in any inertial frame plus retain the equal validity of all inertial frames in describing the physics? Ignoring the rows and columns associated with the $\hat{y}$ and $\hat{z}$ axes (they are unaffected by uniform motion in $\hat{x}$), then the simpler 2×2 matrix for (x, t) need only be considered. The quantity $ds^2 = ds'^2$ may then be expressed as

$$c^2 dt^2 - dx^2 = \begin{pmatrix} cdt & dx \end{pmatrix} \begin{pmatrix} 1 & 0 \\ 0 & -1 \end{pmatrix} \begin{pmatrix} cdt \\ x \end{pmatrix} \equiv \langle \zeta | \mathbb{L} | \zeta \rangle \tag{23.35}$$

where $\mathbb{L}$ is the Lorentz metric (it should be a 4×4 matrix with only diagonal elements non-zero, such that $L_{11} = 1$ and $L_{jj} = -1$ for j from 2 to 4, but as the $\hat{y}$ and $\hat{z}$ axes are the same in the rest and moving frames, they can be ignored). The unusual usage of *bra-ket* notation is only because it captures the matrix operations and vector properties that are relevant[11] and so is convenient. The invariance of ds^2 therefore entails that in a moving frame $|\zeta'\rangle = \mathbb{U}|\zeta\rangle$ for which ds^2 is unchanged can be represented by introducing a coordinate transformation matrix $\mathbb{U}$ where

$$\mathbb{U} = \begin{pmatrix} a & b \\ f & g \end{pmatrix} \tag{23.36}$$

[10] If $ds^2 = 0$, it is said to be "light-like", if positive, then "time-like", and if negative, "space-like".

[11] No imputation that $|\zeta\rangle$ is a wave function or basis state is implied, because it is not: $|\zeta\rangle$ is nothing more than a two-component real vector here, and $\mathbb{L}$ and $\mathbb{U}$ are just matrices, as emphasized by the symbol representing them.

The quantities (a, b, f, g) are real quantities to be determined such that

$$\mathbb{U}^{\dagger}\mathbb{L}\mathbb{U} = \mathbb{L} \rightarrow \begin{pmatrix} a^2 - f^2 & ab - fg \\ ab - fg & b^2 - g^2 \end{pmatrix} = \begin{pmatrix} 1 & 0 \\ 0 & -1 \end{pmatrix} \tag{23.37}$$

where † denotes "transpose" as the matrix elements of $\mathbb{U}$ are real. These represent *three* equations for four unknowns because the off-diagonal entries are the same. One more is needed, but it has already been encountered: Eq. (23.32) shows that $a = \gamma$, and therefore, $g = \gamma$ and $b = f = \beta\gamma$. Thus,

$$\begin{pmatrix} ct' \\ x' \\ y' \\ z' \end{pmatrix} = \begin{pmatrix} \gamma(ct + \beta x) \\ \gamma(\beta ct + x) \\ y \\ z \end{pmatrix} \leftrightarrow \begin{pmatrix} ct \\ x \\ y \\ z \end{pmatrix} = \begin{pmatrix} \gamma(ct' - \beta x' \\ \gamma(x' - \beta ct') \\ y' \\ z' \end{pmatrix} \tag{23.38}$$

where the other coordinates have been returned to their rightful place.

Now that the coordinates have been considered, turn to energy and momentum. The equivalence of any reference frame for the description of dynamics means that saying $F = dp/dt$, where F is force and p momentum, has consequences as well. In a reference frame moving just shy of the speed of light, the application of force is not changing the particle velocity (which is just shy of c) that much, so something else must be affected, and that something is mass. The way dt' is related to dt suggests that $F = d(mv)/dt \rightarrow d(\gamma\beta m_o c)/dt$, where m_o indicates the mass of the particle in its rest frame and $m \equiv \gamma m_o$ is the mass in a moving frame.[12] A more rigorous exploration confirms this expectation [99]. Analogous to the invariance of ds^2, the invariant quantity in the energy-momentum relation is the rest mass m_o, and so

$$m_o^2 = m^2 - (p/c)^2 \tag{23.39}$$

A more agreeable representation is to define $mc^2 \equiv E$. Therefore, in terms of the energy-momentum terms (E, cp_x, cp_y, cp_z), the invariance relationship becomes

$$E^2 = (m_o c^2)^2 + (cp)^2 = (\gamma m_o c^2)^2 \tag{23.40}$$

In the non-relativistic limit $(\beta \ll 1)$, γ can be Taylor expanded to show that $E \approx m_o c^2 + (1/2)m_o(\beta c)^2 = m_o c^2 + (1/2)m_o v^2$, the second term easily being seen to correspond to kinetic energy. As tempting as it is to head off in that direction, in actuality Eq. (23.39) is the equation sought. If $E = \gamma m_o c^2$ is the total energy, and $m_o c^2$ is the rest energy, then because $\gamma^2 + \beta^2\gamma^2 = 1$,

$$(pc)^2 = (\gamma^2 - 1)(m_o c^2)^2 = (\beta\gamma m_o c^2)^2 = (\gamma m_o c^2)^2(\beta c)^2 \tag{23.41}$$

which recovers $p = mv$ but with $m = \gamma m_o$ and $v = \beta c$. Putting momentum back into the form of $p = \hbar k$ then results in

$$(\hbar ck)^2 = E^2 - (m_o c^2)^2 \tag{23.42}$$

which is in the form of a *hyperbolic dispersion relationship*, where the designation indicates that k and E can be expressed in terms of hyperbolic trigonometric functions cosh and sinh. That is, by dividing throughout by the rest mass term $(m_o c^2)^2$ and juggling,

$$1 = \cosh^2\Theta - \sinh^2\Theta \tag{23.43}$$

$$\tanh\Theta = \frac{\hbar kc}{E} \tag{23.44}$$

where the dispersion relationship is now seen to be expressible in hyperbolic functions.

[12]Alas, relativity has forced a distinction in the mass of an electron in the rest frame m_o compared to the moving frame m, breaking with the earlier convention of calling m the mass of the electron in vacuum. Context will have to determine content, again.

23.4 The alpha semiconductor model

The electron dispersion relation thus has the form of a relativistic particle of rest mass m_n and a limiting upper velocity $\alpha_o c$. The quantity $\alpha_o c$ is roughly constant for the III-V compounds.[13]

– Barbara L. Jensen [306]

Invoking relativity to provide an example of a hyperbolic dispersion relation seems more ostentatious than serviceable, but there is a reason for it. The limitation represented by the speed of light for inertial frames has direct consequences. An analogous limitation on the electron velocity in a crystal gives rise to a method of estimating the effective electron mass m_n. Actually, it does more than that: it provides a means of estimating optical quantities, acts as a bridge between the (quasiclassical) Drude theory and a quantum mechanical formulation, and provides some of the scattering terms needed for photoemission predictions from multialkali antimonides [308] and a number of compound semiconductors. So as not to get distracted, consider just the dispersion relation and its implications for effective mass and a hypothetical limiting velocity though. What is *particularly* interesting is that for a certain class of semiconductors, namely, the II-VI and III-V semiconductors, the variation of band gap E_g with effective mass $r_e = m_n/m$ is very intriguing, as shown in Figure 23.3.

Return to the dispersion relation of Eq. (23.27), but now, instead of measuring energy from the top of the valence band, measure it from the center of the band gap, or $E' \to W + E_g/2$, and use the $\Delta/E_g \to \infty$ form of the equation to write

$$\frac{\hbar^2 k^2}{2m_n} = \frac{(W + E_g/2)(W - E_g/2)}{E_g} = \frac{W^2 - E_g^2/4}{E_g} \tag{23.45}$$

or

$$W^2 = \frac{E_g^2}{4} + \frac{E_g \hbar^2 k^2}{2m_n} \tag{23.46}$$

This is suggestive of the relativistic dispersion relation, but for it to *be* like those dispersion relations, then W would have to act like E, therefore investigate the consequences of $W = \gamma m_n (\alpha_o c)^2$, where γ is to be determined[14] and where $\alpha_o c$ takes over the role of a limiting velocity. With this substitution, then

$$\gamma^2 = \left\{ \frac{E_g}{2m_n(\alpha_o c)^2} \right\}^2 + \frac{E_g \hbar^2 k^2}{2m_n^3 (\alpha_o c)^4} \tag{23.47}$$

The first (or constant) term on the right-hand side is unity if

$$E_g \equiv 2m_n(\alpha_o c)^2 \tag{23.48}$$

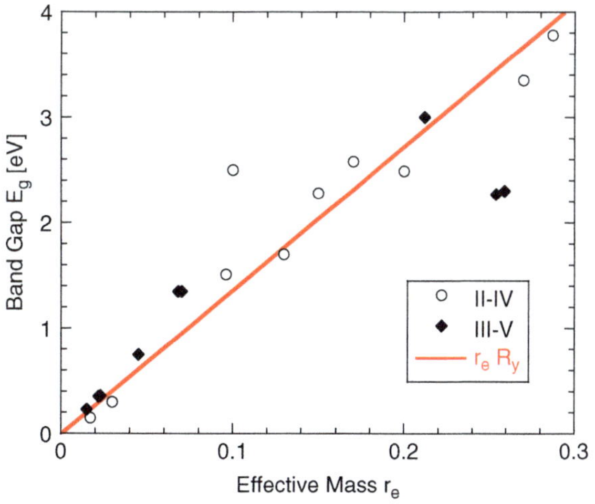

Figure 23.3 Band gap E_g vs effective mass $r_e = m_n/m$ for selected II-VI and III-V semiconductors. Values used are based in Table 23.1.

[13]In ref. [306], $\alpha_o = (E_g/2m_n c^2)^{1/2} \leftrightarrow \alpha/2$, that is, it is half the value of the fine structure constant α. Because α is reserved here for the fine structure constant, this version of the quote modifies α_o by appending a o subscript, as done in ref. [308].

[14]No one is fooled: the intention is to bend the meanings of the terms until γ functions in the same way that the relativistic γ does.

Table 23.1 Semiconductor band gaps and effective electron masses. Select II-VI and III-V semiconductor parameters of E_g and $m_n = r_e m$. "Ratio" refers to the quantity $R = (E_g/R_y)(m/m_n)$, where $R_y = m(\alpha c)^2/2 = 13.6057$ eV is the Rydberg energy: the alpha semiconductor model assumes $R = 1$. The data are graphically represented in Figure 23.3.

Semiconductor	Type	$r_e = m_n/m$	E_g [eV]	Ratio	Ref.
CdO	II-VI	0.1000	2.5000	1.8375	[309]
CdS	II-VI	0.2050	2.4100	0.8641	[310]
CdSe	II-VI	0.1300	1.7000	0.9611	[311]
CdTe	II-VI	0.0960	1.5100	1.1561	[310]
CdTe	II-VI	0.1100	1.5000	1.0023	[311]
HgSe	II-VI	0.0300	0.3000	0.7350	[309]
HgTe	II-VI	0.0170	0.1500	0.6485	[309]
ZnO	II-VI	0.2700	3.3500	0.9119	[311]
ZnS(W)	II-VI	0.2870	3.7800	0.9680	[311]
ZnSe	II-VI	0.1700	2.5800	1.1155	[311]
ZnTe	II-VI	0.1500	2.2800	1.1172	[311]
AlP	III-V	0.2120	3.0000	1.0401	[310]
AlSb	III-V	0.1200	1.5800	0.9677	[311]
GaAs	III-V	0.0680	1.3500	1.4592	[309]
GaP	III-V	0.1300	2.2500	1.2721	[311]
GaSb	III-V	0.0450	0.7500	1.2250	[310]
InAs	III-V	0.0219	0.3540	1.1881	[310]
InP	III-V	0.0700	1.3500	1.4175	[311]
InSb	III-V	0.0150	0.1700	0.8330	[311]

and if the second term is represented as $(\beta\gamma)^2$, then

$$\hbar k \equiv \gamma m_n(\beta\alpha_o c) \tag{23.49}$$

This is perfect: $\beta\alpha_o c$ acts like the velocity, γm_n acts like the relativistic mass, and E_g acts like the rest energy. But there is more: the ratio of E_o with the Rydberg energy $R_y = (1/2)m(\alpha c)^2$ shows that if

$$\alpha_o \equiv \frac{\alpha}{2} \tag{23.50}$$

then

$$\frac{E_g}{R_y} = \frac{2\alpha_o}{\alpha}\left(\frac{m_n}{m}\right) = \frac{m_n}{m} \tag{23.51}$$

just as shown in Figure 23.3.

In short, for semiconductors that are describable by the alpha semiconductor approximation, then if the band gap is known, the effective mass is known, and a convenient simple model is available to calculate parameters and optical constants associated with electron transport and emission. That is, if one is interested in how some physical feature (such as photoemission, for example [308]) scales with band gap, then a means of taking into account the changes wrought on the effective mass as a consequence are easily incorporated. The basis behind this approximation is that the dispersion relation *introduces a factor acting like the limiting velocity in an alpha semiconductor of* $\alpha_o c$ (a form reminiscent of the electron velocity in the Bohr atom in Section 2.2.1 being αc), and the velocity of the electron being defined by $\beta\alpha_o c$. so that Eq. (23.46) becomes

$$\gamma^2 = 1 + \beta^2\gamma^2 \leftrightarrow \gamma = \frac{1}{\sqrt{1-\beta^2}} \tag{23.52}$$

Finally, some of the scattering terms in electron transport yet to be encountered are dependent on the energy of the electron measured from the bottom of the conduction band, but W is thereby an odd choice, as it is measured with respect to the *center* of the band gap (the roots of W^2, being $\pm W$, therefore act like (positive energy) electrons ($W > 0$) and (negative energy) anti-electrons). In such calculations, finding a term that goes as the actual energy in the conduction band has utility. A candidate is the dimensionless parameter

$$u = \frac{W - (E_g/2)}{E_g} \tag{23.53}$$

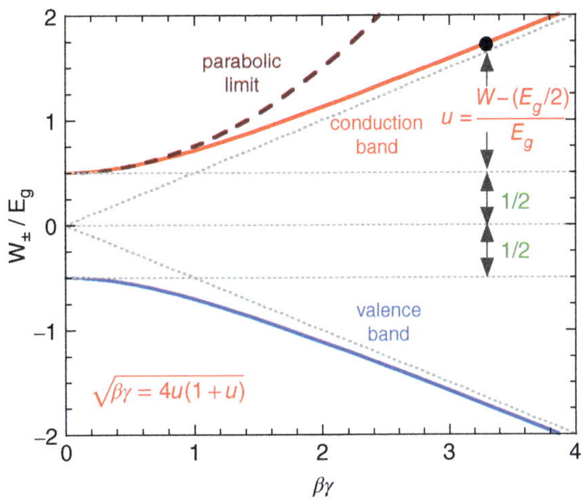

Figure 23.4 A band gap diagram for the hyperbolic dispersion relationship of W in Eq. (23.46) using the alpha semiconductor approximation. Observe that $\beta\gamma = \hbar k/m_n \alpha_o c$.

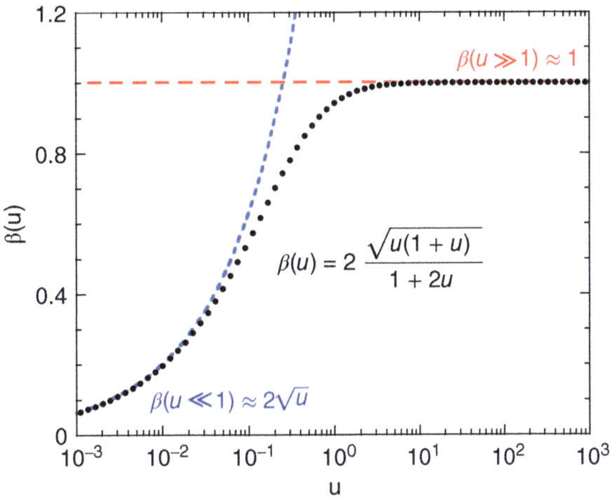

Figure 23.5 Behavior of $\beta(u)$ defined by Eq. (23.54).

which is the ratio of the excess energy above the conduction band minimum with the band gap energy

$$\gamma = 1 + 2u \tag{23.54}$$

In terms of u, the limiting behavior of W is readily visible in Figure 23.4. Further, the behavior of $\beta(u)$ as u increases mimics the relativistic relationship between kinetic energy and velocity $v = \beta c$ of Section 23.3, and shown in Figure 23.5, where

$$\beta(u) = \frac{\sqrt{4u(1+u)}}{1+2u} \tag{23.55}$$

23.5 Current and effective mass

"Keep ancient lands, your storied pomp!" cries she
With silent lips. "Give me your tired, your poor,
Your huddled masses yearning to breathe free,
The wretched refuse of your teeming shore.
Send these, the homeless, tempest-tost to me,
I lift my lamp beside the golden door!"

– Emma Lazarus[15]

It is a perennial and contentious question: if a barrier to freedom exists, who gets through, or, put another way, do the weighty ones have an advantage? For metals, the effective mass of the electrons in the bulk metal was close enough to the free electron mass that such questions did not arise, but with semiconductors, the effective masses are sufficiently different that some effect must arise that will affect emission.

The transverse momenta $\hbar\vec{k}_\perp$ are conserved across the interface, so attention can be gratifyingly restricted to the one-dimensional problem. Although the effect of a changing electron mass from bulk to barrier can be ascertained using a formalism like that leading Eq. (17.4) (examples being the treatments of Stratton [305, 312]), there is a way to intuit the behavior based on reconsidering the rectangular barrier of Section 18.1. The matrix equations leading to Eq. (18.10) can be reformulated to take into account that the incident wave function is characterized by an effective mass $m_n = r_e m$. Of the two matricies that are involved, Eqs (18.4) and (18.5), only S_1 will change as a consequence of $m \to m_n$ to become

$$\hat{S}_1(0) \to \frac{1}{2k'}\begin{pmatrix} k'+\kappa & k'-\kappa \\ k'-\kappa & k'+\kappa \end{pmatrix} \tag{23.56}$$

where $k' = \sqrt{2m_n E}/\hbar = k\sqrt{r_e}$ and $\kappa = \sqrt{k_o^2 - k^2}$ is the same as before. The effect on the transmission coefficient $t(k)$, originally considered in Eq. (18.9) for transmission over the barrier, is then[16]

$$t(k) = -\frac{4ik'\kappa e^{-ikL}}{2(\kappa^2 - kk')\sinh(L\kappa) + 2i\kappa(k+k')\cosh(L\kappa)} \tag{23.57}$$

and so for large κL,

$$D(k') = 16\left(\frac{m_n}{m}\right)\frac{4k'k\kappa^2}{4\kappa^2(k+k')^2 + 4(\kappa^2+k'^2)(\kappa^2+k^2)\sinh^2(L\kappa)} \tag{23.58}$$

If $k \to k'$, Eq. (18.11) is recovered. If $D(k)$ represents $k' \to k$ in $D(k')$, then in the limit $\exp(L\kappa) \gg 1$,

$$\frac{D(k')}{D(k)} \approx \frac{m_n k'(\kappa^2+k^2)}{mk(\kappa^2+k'^2)} \approx r_e^{3/2}\frac{k_o^2}{k_o^2 - (1-r_e)k^2} \tag{23.59}$$

similar to the findings of Stratton [305] but by different means. The dependence of the ratio of the effective mass transmission probability $D(k')$ with the free electron mass transmission probability $D(k)$ on $r_e^{3/2} = (m_n/m)^{3/2}$ may have been anticipated in Eq. (11.4) by observing the changes in going from an integration over $d\vec{k}$ to $E^{1/2}dE\,d\Omega$ when changing from momentum to energy, where a $\sqrt{m_n}$ is associated with each momentum component.[17] In any event, Eq. (23.58) confirms that, for field emission from semiconductors and in a manner of speaking, the weighty ones do indeed have an advantage.

[15] E. Lazarus, *The New Colossus*, http://www.libertystatepark.com/emma.htm. A poem written in 1883 and mounted on a plaque at the Statue of Liberty in 1903.

[16] The notation leaves something to be desired: it would appear that $t(k)$ is a function of k', but keep in mind that $k' = k\sqrt{r_e}$.

[17] Effective masses do not have to be the same along each axis (e.g., silicon), in which case $m_n^{3/2} \to (m_x m_y m_z)^{1/2}$. For silicon, $m_y = m_x$, so the effective mass is commonly written as $(m_l m_t^2)^{1/2}$, where the subscripts refer to longitudinal and transverse [73].

CHAPTER 24

Interfaces

All lies and jest
Still, a man hears what he wants to hear
And disregards the rest.

– Paul Simon and Art Garfunkel[1]

Differing materials abutting one another with or without a potential drop across or between them exhibit complex behavior dependent upon the nature of the metals and semiconductors involved, the doping, and the manner in which they are in contact [52, 53, 213, 304, 312]. The description of emission from semiconductors (field emission in particular) is significantly complicated by the numerous differences between semiconductors and metals, beyond the effective mass changes encountered in Section 23.5, to include field-dependent electrochemical potentials and temperature-dependent band gaps, the orders of magnitude smaller electron density values near the surface in the conduction band *or* the nature of field emission from the valence band itself, changes to the image charge potential affecting the Schottky barrier reduction factor, the non-linear potentials that attend doping profiles, and on and on. Most vexing is the differences that arise depending on whether the doping is p-type or n-type [111, 172, 305, 312, 313]. Not all of that complexity relates to what is needed to understand emission. Why listen to a laborious disquisition on it all?[2] Focus instead on the processes which change the emission barrier or the supply of electrons abutting it. The shape of the band bending affecting the supply of electrons, like that of the barrier impeding their transport, can be understood by assessing how charge is distributed near the interface.

24.1 Metal–insulator–metal current density

When two metallic electrodes are separated by an insulating film, the equilibrium conditions require that the top of the energy gap of the insulator be positioned above the Fermi level of the electrodes. Thus, the action of the insulating film is to introduce a potential barrier between the electrodes which impedes the flow of electrons between the electrodes. The electronic current can flow through the insulating region between the two electrodes if: (a) The electrons in the electrodes have enough thermal energy to surmount the potential barrier and flow in the conduction band. (b) The barrier is thin enough to permit its penetration by the electric tunnel effect.

– John G. Simmons [314]

Electron *emission* is into vacuum, but analogous equations to transport through interface barriers arise when the vacuum is replaced by an insulator. When the insulator is sandwiched between two metal layers, as in Figure 24.1, by the addition of a right contact to replace the vacuum, then both contacts can provide electrons into the insulator region, and the net current will go as their difference. The bottom of the conduction band of the insulator takes over the role of the vacuum level in the emission equations, and modifications to account for a differing dielectric constant and effective mass of the electron in the insulator introduce changes. When a bias is applied (the right metal contact is held at a lower potential energy $V_b = q\varphi_b$ than the left contact), then a difference in the left and right current through the structure occurs, so that total current flow, being the difference between them, depends on whether the rightmost edge of the insulator conduction band is above or below the Fermi level of the left metal contact. If the right barrier is lower, then the total current should follow an equation like (compare Eq. (11.2))

$$J = \frac{q}{2\pi\hbar} \int_0^\infty D(E)(f_{\text{left}}(E) - f_{\text{right}}(E))dE \tag{24.1}$$

[1]Paul Simon and Art Garfunkel, lyrics to *The Boxer*, 1968.
[2]Apart from the obvious reason that interfaces are technologically quite important to electronics, that is.

Introduction to the Physics of Electron Emission, First Edition. Kevin L. Jensen.

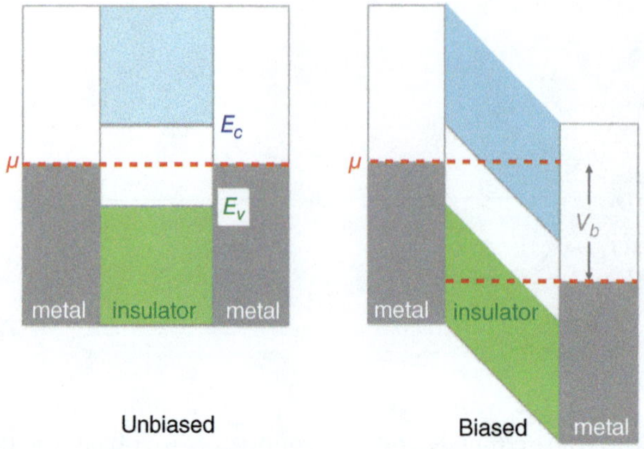

Figure 24.1 Left, a model of an unbiased metal–insulator–metal interface (compare the middle figure of Figure 24.12); observe that the Fermi levels of the metal contacts are aligned. Right, the same interface under bias, in which the right contact is held at a lower potential $V_b = q\varphi_b$ than the left contact. Image charge effects round the rectangular barrier [314] but are not shown.

where the x subscript on E_x is suppressed and $f(E)$ is the supply function of Section 11.1 for the left and right metal contacts. If the metal layers were replaced by semiconductors, the description would be more complex, requiring the introduction of doping levels and the specification of whether the semiconductors were p- or n-type but decidedly non-trivial [213, 315]. How $D(E)$ behaves and how the equations for J that result compare to the canonical Richardson–Laue–Dushman (RLD) and Fowler–Nordheim (FN) equations of Part II is more to the point here.

Before doing this, a small historical aside is useful. The treatment of transport across a metal–insulator–metal (MIM) region[3] evolved differently than Eq. (24.1) would suggest, focusing as it did on electron transport *from* doped semiconductors *into* a metal contact (metal–semiconductor transport). The earlier approaches therefore used drift-diffusion models oft used in semiconductor analyses. In contrast, Bethe hypothesized that the current limiting mechanism was the barrier between the metal and semiconductor: when the bias was such that flow was from metal into semiconductor, the model is akin to the description of thermionic emission in Chapter 12, so that Bethe's theory is perhaps not surprisingly known as the thermionic emission model. The Bethe approach has more in common with the present treatment, and although both the drift-diffusion and thermionic emission models give comparable equations, the Bethe approach is less problematic in its interpretation [316]. Actual transport from semiconductor to metal, however, contains elements of both. A final simplification is that being concerned with electron transport means the usual complications of hole transport and recombination effects do not trouble the discussion.

For the MIM configuration, three cases of interest dominate. The first is that the insulator region is so wide, and the bias so small, that only electrons in the thermal tail of the distribution functions have adequate energy to surmount the barrier: the resulting current density should be analogous to the RLD equation of Section 12. The second is that (as shown in the rightmost figure of Figure 24.1) if the bias is large enough, and perhaps the conduction band of the insulator low enough, tunneling from the leftmost metal into the conduction band of the insulator is possible: the resulting current density should be analogous to the FN equation of Section 13.1. Lastly, the insulator region may be sufficiently thick that tunneling through the insulator is possible, so that the transmission probability $D(E)$ has features of the rectangular barrier of Section 18.1. Consider each in turn.

24.1.1 Thermal model

A significant advance in our understanding of metal–semiconductor contacts came during the Second World War as a result of the use of silicon and germanium point-contact rectifiers in microwave radar. This advance was considerably helped by developments in semiconductor physics. Perhaps the most important contribution during this period was Bethe's thermionic-emission theory (1942), according to which the current is determined by the process of emission of electrons into the metal, rather than by drift and diffusion in the semiconductor as was supposed by Mott and Schottky.

– E. H. Rhoderick [316], p. 2

[3] The designation "metal–oxide–metal" diode is also encountered.

When the bias is small, then only the electrons in the population with a forward energy E that exceeds the interface barrier height are able to cross the insulator from metal contact to metal contact. Depending on what is being emphasized, the process is called Schottky emission, thermionic emission (into a semiconductor rather than vacuum), or the Bethe model, but all include over-the-barrier transport. The transmission probability is then a step function $D(E) \approx \Theta(E - \mu - \Phi_b)$, where Φ_b plays the role of the work function in thermal emission theory, but is now the energy difference between the conduction band in the insulator and the Fermi level in the metal. If the bias across the insulator is V_b, corresponding to the potential energy difference between the left and right metal contacts as labeled in the right schematic of Figure 24.1, then clearly electrons from the right contact have to surmount a barrier of height $\Phi_b + V_b$ to get to the left contact electrode because their "zero energy", as it were, is V_b lower than their brethren in the left contact. Thus, Eq. (24.1) becomes

$$J \approx \frac{q}{2\pi\hbar} \int_{\mu+\Phi_b}^{\infty} (f(E) - f(E + V_b))dE \tag{24.2}$$

If the barrier height $\Phi_b > 4k_BT$, which is often the case, then $f(E)$ can be replaced by its Maxwell–Boltzmann limiting case because the integration region is restricted to those energies for which that approximation is reasonable, resulting in

$$f(E) = \frac{m_n}{\pi\beta\hbar^2} \ln\left(1 + e^{\beta(\mu-E)}\right) \approx \frac{m_n}{\pi\beta\hbar^2} e^{\beta(\mu-E)} \tag{24.3}$$

and so Eq. (24.2) becomes the simple integration over exponentials, and is

$$J \approx \frac{qm_n}{\pi\beta^2\hbar^3} e^{-\beta\Phi_b}\left(1 - e^{-\beta V_b}\right) \tag{24.4}$$

Making use of A_{RLD} as defined in Eq. (12.3),

$$J \approx r_e A_{RLD} T^2 e^{-\beta\Phi_b}\left(1 - e^{-\beta V_b}\right) \tag{24.5}$$

for which all the factors outside the parentheses can be scooped up into an RLD-like equation J_{RLD} but with $\Phi \to \Phi_b$. If βV_b is large, then the second term in the parentheses is negligible, making what remains much like the RLD thermal emission equation amended by a factor r_e denoting the effective mass of the electron inside the insulator.

But that is not all – just as the application of a field lowered the barrier to thermionic emission by the Schottky factor $\sqrt{4QF}$, so too does the field existing across the insulator lower Φ_b, albeit with a few modifications. First, the Q in the Schottky factor has to be amended by $Q \to Q/K_s$ where K_s is the dielectric constant of the insulator[4]. Thus,

$$\Phi_b \to \phi_b = \Phi_b - \sqrt{4QF/K_s} \tag{24.6}$$

is the Schottky factor lowered barrier height above the left metal Fermi level, and the field across the insulator is approximated as $F = V_b/L$ as would be the case if the insulator is narrow and free from doping. Being free from all doping is rather optimistic, as Section 22.4 makes apparent, and so F is instead the field that exists at the metal–semiconductor interface on the semiconductor side.[5]

An equation like Eq. (24.4) is important because it leads to *current rectification*, which is a means of transforming alternating current into direct current. However, the more common device type associated with this is the metal–semiconductor diode (also called rectifiers and Schottky-barrier diodes), for which current from the semiconductor to the metal is described by an equation analogous to Eq. (24.4), given that the electron distribution in the semiconductor is really a Maxwell–Boltzmann distribution. The barrier shape has more in common with the depletion layers encountered in Section 24.5, but for the thermal emission-type behavior the shape is not as important as the height of the barrier, and so a discussion of the metal–semiconductor barrier is delayed until then. The process of rectification is straightforward: if the bias potential on the semiconductor side sinusoidally oscillates so that the barrier oscillates between small and large, the current is positive (read: from semiconductor to metal) and vanishes for the part of the cycle where the barrier is

---

[4] Those who recall Section 22.5 may feel disoriented, thinking that Eq. (22.29) describes the image charge when semiconductors are involved, but now the insulator has *replaced* the vacuum and the charge in the insulator is *outside a metal surface*, not a semiconductor one, so Q/K_s is correct.

[5] The barrier shape would also become quadratic (sloping-like) instead of linear (triangular-like). Patience – that discussion will come in Section 24.5.

large; current does not flow in the reverse direction as the field changes sign. This is just an inkling of what such devices do [152] and the physics they share with *p–n* junctions [316].

24.1.2 Tunneling model

> *By evaporating an aluminum strip on a glass plate, oxidizing the aluminum surface, and then evaporating another strip over it at about right angles, one obtains a metal–oxide–metal diode consisting of a thin oxide layer (20–200 angstroms thick) sandwiched between two metal electrodes. If a potential difference is maintained between the electrodes, current will flow by field emission (tunneling through the barrier). The theory of (field emission) is applicable to this problem with some slight modification…*
>
> – Aldert van der Zeil [152], p. 148

The same three considerations that affected a thermal emission-like equation in Section 24.1.1 also apply to developing a field emission-like equation here for the treatment of tunneling through a MIM structure: the effective mass, the barrier height, and a Schottky lowering factor. When the field across the insulator is sufficiently strong, then the tunneling electrons from the left-hand contact experience a barrier analogous to the image charge affected triangular barrier of Fowler and Nordheim, as suggested in the right side of Figure 24.1, but now the incentive to consider the effects of the right-hand contact are severely attenuated, given the great sensitivity of tunneling current to the Gamow factor $\theta(E)$. Given Eq. (13.25), the MIM version of the tunneling equation not surprisingly involves the same replacements of (*i*) $m \rightarrow m_n = r_e m$, (*ii*) $\Phi \rightarrow \Phi_b$, and (*iii*) $Q \rightarrow Q/K_s$ as before. But actually, that is not quite true. The tunneling process is very rapid, taking approximately $L/v_F < 1$ fs for copper-like parameters and typical field emission fields and work functions. Consequently, the passage of an electron through a dielectric medium should make reference to the *optical* (or high frequency) dielectric constant $\varepsilon_\omega = K_\omega \varepsilon_0$, rather than the static $K_s\varepsilon_0$, when tunneling is through a material like an insulator [152]. The point is, though, that the image charge term Q is modified by a dielectric constant term, reflecting the fact that the tunneling is occurring in a material, not in the vacuum.

Such is the qualitative understanding. The quantitative understanding begins with observing that values of the dielectric constant are such that simply focusing on the triangular barrier model and ignoring the image charge lowering makes for a good and simple first approximation. Then, use the shape factor approach of Section 11.2: recall that the factor $\theta(\mu)$ most affects the form of the tunneling current $J(F) \propto \exp(-\theta(\mu))$ (see Eq. (13.10)), so focus on it. Electrons traveling from the left contact to the right experience the usual triangular barrier Gamow factor $\theta_l(\mu)$ (Eq. (11.17)), but those from the right experience $\theta_r(\mu)$ (red and blue shaded regions of Figure 24.2, respectively). In terms of a shape factor (4/3), κ factor of $\sqrt{2m\Phi_b}/\hbar$, and a length factor of $L\Phi_b/V_b$ (the field across the dielectric is $F = V_b/L$) then

$$\theta_l(\mu) = \frac{4L}{3\hbar V_b}\sqrt{2m_n\Phi_b^3} \tag{24.7}$$

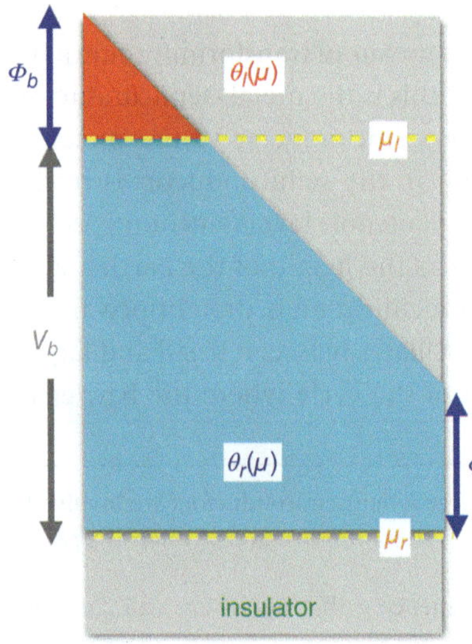

Figure 24.2 Close-up of the insulator region of Figure 24.1 showing the regions used in the calculation of the left and right $\theta(\mu)$ factors used to define the left and right transmission probabilities (red and blue, respectively). The width of the blue region is always L (the thickness of the insulator); the width of the red region $L' = L(\Phi_b/V_b)$ depends on the bias and changes.

The blue region of $\theta_r(\mu)$ is not so simple. First, $V_b \geq \Phi_b$ or the electrons entering from the left would not be experiencing a triangular barrier. Second, electrons from the right experience a barrier that increases rather than decreases, and so

$$\theta_r(\mu) = 2\frac{\sqrt{2m_n}}{\hbar}\int_0^L \sqrt{\Phi_b + Fx}dx = \left[\left(1+\frac{V_b}{\Phi_b}\right)^{3/2} - 1\right]\theta_l(\mu) \tag{24.8}$$

Therefore, for the minimum condition $V_b = \Phi_b$, the right $\theta_r(\mu)$ is larger by a factor of $2\sqrt{2} - 1 = 1.83$ than $\theta_l(\mu)$, which substantially reduces the current from right to left as the bias increases: by the time $V_b = 2\Phi_b$, then that factor increases to 4.2. Ignoring the right current, then, is very reasonable, and so the tunneling current density through an MIM structure is

$$J_{mim}(F) = \frac{q}{16\pi^2\Phi_b\hbar}\left(\frac{4\sqrt{\mu\Phi_b}}{\mu+\Phi_b}\right)\left(\frac{V_b}{L}\right)^2 \exp\left(-\frac{4L}{3\hbar V_b}\sqrt{2m_n\Phi_b^3}\right) \tag{24.9}$$

recognizable as the modified form of the triangular barrier Fowler–Nordheim equation of Eq. (13.10). Incorporating an image charge effect amounts to introducing the elliptical integral functions $v(y_s)$ and $t(y_s)$, where the functions are defined by Eqs (13.23) and (13.24), and where $y_s = \sqrt{4QV_b/LK_\omega}$ accounts for Schottky barrier lowering in a dielectric material. An equation much like it will be encountered again in Section 25.1, but for which the field is a bit different.

24.1.3 Numerical model

For truth is rightly named the daughter of time, not of authority. It is not wonderful, therefore, if the bonds of antiquity, authority, and unanimity, have so enchained the power of man, that he is unable (as if bewitched) to become familiar with things themselves.

– Francis Bacon[6]

The musings of Bethe and Fowler and Nordheim (not to mention Rhoderick or van der Ziel) about tunneling junctions and barriers command deference for good reason, and if they maintain that the canonical thermal and field emission equations can be adapted to treating an MIM diode, it must be so. And yet, that second interface on the right-hand side of Figure 24.2 surely does *something* to $D(k)$, especially if the wave nature of the electron manifests itself in other tunneling and transmission barriers of greater complexity (Section 19.2.3). If the barrier Φ_b and width L are "small", so that $D(k) \approx \exp(-\theta(k))$ becomes lacking, then that, too, must matter. One should perhaps honor the elders, but undertake the calculation anyway: respect but verify.

Fortunately, the problem represents a minor modification to an existing and well-understood problem: the triangular barrier of Fowler and Nordheim as treated using the techniques of Sections 18.2 and 19 makes the transfer matrix solution to $D(k)$ the proper way to judge the adequacy of the thermal and tunneling approximations to an MIM diode, leading only to minor changes to Eq. (19.15) in Section 19.2.2. For the sheer perversity of it, use the Airy functions $Ai(w)$ and $Bi(w)$ directly, with the caution that the argument $w(x)$ can be both positive and negative (in contrast to the usage of z in the $Zi(c, z)$ functions, where it is positive only). In terms of $k_o \equiv \sqrt{2m_n\Phi_b}/\hbar$ and $f = 2m_nF/\hbar^2$, then

$$w(x) = \frac{k_o^2 - k^2 - fx}{f^{2/3}} \tag{24.10}$$

The values of $w(x)$ at the interface are what matters, so let $w_o = w(0)$ and $w_b = w(L)$. Because the right-hand side of the diode is characterized by a potential drop V_b rather than a field, then $F = V_b/L$ and introduce $k_b \equiv \sqrt{2m_nV_b}/\hbar$. The wave number in the right-hand side metal is then

$$k' \equiv \sqrt{k^2 + k_b^2} \tag{24.11}$$

[6]Francis Bacon, *Novum Organum, Great Books of the Western World,* ed. Robert Maynard Hutchins, Vol. 30, Francis Bacon. Chicago: Encyclopaedia Britannica, 1952, p. 121, Aphorism 94. The paraphrase "Truth is the daughter of Time, not of Authority" was uttered by Galileo in the Bertolt Brecht play of that name when he was unsuccessfully entreating the Prince and his courtiers to look through a telescope at the moons of Jupiter, which would have upended the authority of Aristotle.

All that modifies the Fowler–Nordheim triangular barrier analysis is the insertion of the matrices for the interface at $x = L$, or

$$\begin{pmatrix} 1 \\ r \end{pmatrix} = \hat{S}_0 \cdot \hat{S}_L \cdot \begin{pmatrix} t \\ 0 \end{pmatrix} \tag{24.12}$$

$$\hat{S}_0 = \begin{pmatrix} 1 & 1 \\ ik & -ik \end{pmatrix}^{-1} \begin{pmatrix} \mathrm{Bi}(w_o) & \mathrm{Ai}(w_o) \\ -f^{1/3}\mathrm{Bi}'(w_o) & -f^{1/3}\mathrm{Ai}'(w_o) \end{pmatrix}$$

$$= \frac{1}{2k} \begin{pmatrix} k & -i \\ k & i \end{pmatrix} \begin{pmatrix} \mathrm{Bi}(w_o) & \mathrm{Ai}(w_o) \\ -f^{1/3}\mathrm{Bi}'(w_o) & -f^{1/3}\mathrm{Ai}'(w_o) \end{pmatrix} \tag{24.13}$$

$$\hat{S}_L = \begin{pmatrix} \mathrm{Bi}(w_b) & \mathrm{Ai}(w_b) \\ -f^{1/3}\mathrm{Bi}'(w_b) & -f^{1/3}\mathrm{Ai}'(w_b) \end{pmatrix}^{-1} \begin{pmatrix} 1 & 1 \\ ik' & -ik' \end{pmatrix}$$

$$= \frac{\pi}{f^{1/3}} \begin{pmatrix} -f^{1/3}\mathrm{Ai}'(w_b) & -\mathrm{Ai}(w_b) \\ f^{1/3}\mathrm{Bi}'(w_b) & \mathrm{Bi}(w_b) \end{pmatrix} \begin{pmatrix} 1 & 1 \\ ik' & -ik' \end{pmatrix} \tag{24.14}$$

where it is evident that if $\hat{S}_L$ were absent, the formulation would give the usual Fowler–Nordheim relation for a triangular barrier (Eqs (18.41) and (18.51)). The rendering of Eq. (24.12) into a numerical algorithm for the evaluation of $D(k)$ is given in Section A3.16 and the output of that calculation is shown in Figure 24.3 for generic parameters. The grey vertical line in these figures denotes the location of k_o, which corresponds to the barrier height Φ_b by $\Phi_b = \hbar^2 k_o^2/2m_n$. The thermionic emission model would simply assert that $D_{therm}(k)$ is unity for $k > k_o$ but vanishes otherwise. More interesting is the leading order behavior of the tunneling term in Eq. (18.59) and shown as the dashed lines, calculated according to

$$D_{FN}(k) = \frac{4k}{k_o} \exp\left(-\frac{4}{3f}(k_o^2 - k^2)^{3/2}\right) \tag{24.15}$$

Clearly, the second interface makes a non-trivial difference, although it is equally clear that the thermionic prescription of Bethe and the triangular barrier model of Fowler and Nordheim do capture the general behavior. In practical barriers, where a rounding of the sharp edges is present in the simple MIM model, a softening of the oscillations that are evident occurs, as is generally true for rounded or smeared out potential regions (remember the parabolic barrier of Section 19.1.1). This makes the qualitative prescriptions better, but does not completely erase the impact of the wave nature of the electron and the interference effects which arise because of it. The moral, that the wave nature matters, is going to become more evident as other modifications to the simple potential barriers are considered.

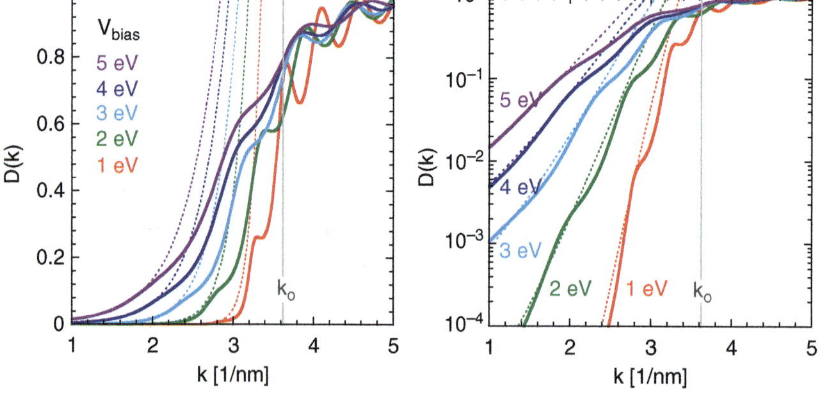

Figure 24.3 Transmission probability as a function of k for an MIM diode based on Eq. (24.12), using the algorithm of Section A3.16. The following parameters are used: the height of the insulator conduction band is $V_o = 1$ eV above the left metal Fermi level μ_l, the effective mass is $m_n = 0.5m$, and the width of the insulator region is $L = 5$ nm. The lines are labeled by the value of the bias voltage V_b [eV]. The grey line labeled $k_o = \sqrt{2m_n\Phi_b}/\hbar$ shows the location of the barrier maximum (as in Figure 24.2). The dashed lines correspond to $D_{FN}(k)$ of Eq. (24.15) (see also Eq. (18.59)).

24.1.4 Interface roughness

The ultimate measure of a man is not where he stands in moments of comfort and convenience, but where he stands at times of challenge and controversy.

– Martin Luther King, Jr.[7]

The more challenging the physical circumstances, the more illuminating the insight. The idealized barriers and smooth interfaces that form the one-dimensional models do not give a full account (although they are certainly suggestive) because they are silent about what happens when circumstances are *rough*. Roughness, in this case, refers to the undulations and irregularities of actual interfaces, for which the idealized models do not fully account for the complexity of the physics involved. Specifically, the electron momentum $\hbar\vec{k}_\perp$ (which is parallel to the plane of the interface) is taken to be conserved in passing from one side of the interface to the other (specular transmission), an approximation that ignores how structure complicates wave function matching in crossing the region over which roughness is expressed (diffuse transmission) [317].

In a planar, or one-dimensional, model, a one-to-one matching of the wave function on one side of an interface with a wave function on the other side exists, and so conservation of transverse momentum components holds. But when the roughness characterizing the interface is confined to a region between x_a and x_b, then the wave function to the left of x_a with an energy E_a is

$$\left|a(\vec{k}'_a)\right\rangle = \sum_{\vec{k}_a} A_l(\vec{k}'_a, \vec{k}_a)\left|\vec{k}_a\right\rangle \quad (x < x_a) \tag{24.16}$$

$$= \sum_{\vec{k}_a} A_r(\vec{k}'_a, \vec{k}_a)\left|\vec{k}_a\right\rangle \quad (x > x_a) \tag{24.17}$$

whereas to the right of x_b with an energy E_b it is

$$\left|b(\vec{k}'_b)\right\rangle = \sum_{\vec{k}_b} B_l(\vec{k}'_b, \vec{k}_b)\left|\vec{k}_b\right\rangle \quad (x < x_b) \tag{24.18}$$

$$= \sum_{\vec{k}_b} B_r(\vec{k}'_b, \vec{k}_b)\left|\vec{k}_b\right\rangle \quad (x > x_b) \tag{24.19}$$

In specular transmission the left–right coefficients have to be equal so that $\vec{k}$ is conserved, but in diffusive transmission a mash-up of the wave functions occurs and the coefficients A_r and A_l (as well as the Bs) no longer have to be the same. Current can then be evaluated by treating the passage through the interface region as a transition from the a state to the b state (a method credited to Bardeen), with all the mixing that entails. The diffuse case can be shown to then yield the specular case in the appropriate limits.

Dry mathematical discourse hides ominous consequences, however, which take time to appreciate. As Stratton [317] puts it,

The basic assumption underlying our technique for treating tunneling through a barrier with diffuse boundaries is that the tangential momentum of an electron passing through the boundary is not conserved. (emphasis added)

That is going to make matters difficult. What applies to interfaces also applies to surfaces, and such calculations as thermal and photoemission emittance (emittance being a metric of great concern for electron beams) is going to be altered in ways profoundly difficult to anticipate (possibly even reflected in the observation that experimental emittance often is in excess of theoretical estimates based on one-dimensional theories [181]). The models are therefore indicative of what is happening, even if they underestimate a complete measure of it. With this caveat in mind, therefore, consider how material considerations affect the transport equations, resulting in forms which often bear similarity to those of Part II, but which subtly change what the terms signify.

[7]Martin Luther King Jr., *Strength to Love*, Gift ed. Fortress Press, 2010, p. 26.

24.2 Band bending

If it takes my whole life I won't break I won't bend
It'll all be worth it worth it in the end

– Sarah McLachlan[8]

A reed got into an argument with an oak tree. The oak tree marvelled at her own strength, boasting that she could stand her own in a battle against the winds. Meanwhile, she condemned the reed for being weak, since he was naturally inclined to yield to every breeze. The wind then began to blow very fiercely. The oak tree was torn up by her roots and toppled over, while the reed was left bent but unharmed.

– Aesop[9]

The conduction band minimum for metals will not bend. In contrast, the much lower carrier density in semiconductors results in an inability to resist the penetration of an external field, and so their conduction bands do. In dealing with screening and the dielectric constant, a subtle concept behind Eq. (21.81) may have been missed, but for identifying what happens to the emission barrier at the surface of a semiconductor (or at the interface between a semiconductor and a metal), it is rather important. That concept concerns how the chemical potential μ depends on the potential ϕ, conjoined because of their relation to the density. Because charge piles up at the surface to screen out an applied field, speaking of the *electrochemical* potential $\mu(x)$ is useful: it is measured with respect to the conduction band, but graphically it is common to represent the electrochemical potential as a straight line in a band diagram, and let the conduction band deform in response to changes in density occurring because of an applied electric field. In other words, the semiconductor bands (like the reed) *bend* in response to a force.

In the absence of a force and at equilibrium, the chemical potential μ is not going to change from place to place because if it did, there would be regions where the average electron energy would be higher (or lower), and flow would occur until conditions balanced. But when a force is present that caused the density to change, then if the variation in the potential characterizing charge buildup is not too rapid, the local density might appear somewhat flat, and therefore speaking of a local chemical potential does not seem too audacious. But that local chemical potential will be the chemical potential raised or lowered by the local $\phi(x)$, the gradient of which gives $F(x)$ inside the material, and so in one dimension

$$\mu(x) \equiv \mu + \phi(x) \tag{24.20}$$

When $\mu(x)$ appears with an argument henceforth, then it is the *electrochemical potential*, but when μ is alone, it is the *chemical potential* of the bulk.[10] The electrochemical potential $\mu(x)$ allows the density $\rho(x)$ to be evaluated from Eq. (6.8) with $\mu \to \mu(x)$. Mostly, attention shall be devoted to material in the negative x region (i.e., $x < 0$), with that material being metal or semiconductor, so that the region for $x > 0$ is vacuum (or a differing semiconductor).

When treating the image charge for semiconductors, the continuity of the field across the semiconductor–vacuum interface, or Eq. (22.23), had to be respected, but the fields are related to the variation in $\phi(x)$. Therefore, band bending can be understood by relating the FD distribution, the density, and the potential through Poisson's equation expressed as

$$\frac{\partial^2}{\partial x^2}\phi(x) = \frac{4Q}{K_s}(\rho(x) - \rho_o) \tag{24.21}$$

where ρ_o is the (constant) background charge in a jellium model (the positive cores are represented by a uniform distribution of charge). By introducing $F = \partial_x \phi$ then the left-hand side becomes

$$\frac{\partial^2}{\partial x^2}\phi = \frac{\partial}{\partial x}F = \left(\frac{\partial}{\partial x}\phi\right)\frac{\partial}{\partial \phi}F = \frac{1}{2}\frac{\partial}{\partial \phi}F^2 \tag{24.22}$$

The right-hand side can be rendered, using Eq. (6.8) in the form of Eq. (6.17), as[11]

$$\frac{1}{2}\frac{\partial}{\partial \phi}F^2 = \frac{4\pi Q}{K_s}\left(\frac{N_c}{\beta^{3/2}}\right)\left\{F_{1/2}(\beta(\mu+\phi)) - F_{1/2}(\beta\mu)\right\} \tag{24.23}$$

---

[8]Sarah McLachlan, lyrics to *Answer*, 2004.

[9]Aesop, "The Oak Tree and the Reed" (trans. Laura Gibbs), *Aesop's Fables*. Oxford, New York: Oxford University Press, 2008, Fable 202, p. 102. The Gibbs translation provides the *endomythium* fables (the moral is inside the story) rather than the *epimythium* fables (the moral is an italicized sentence after the story) as in the perhaps better known (1894) Jacobs translation (which, truth be told, often seems to miss the point).

[10]It would be nice to refer to the bulk chemical potential as μ_o, but then that would run foul of Section 6.4, where μ_o referred to the $T \to 0$ limit; the presence-of-an-argument notation will have to do.

[11]Keep in mind that $F(x)$ is a *field* but $F_p(u)$ is a dimensionless *function*.

Up till now, no mention was made of the *sign* of $\phi(x)$ but now there is cause to. When $\phi > 0$, then $\mu(x) > \mu$ and the electron density increases because $F_{1/2}(\beta\mu(x))$ increases: such a region is known as an *accumulation region*. Conversely, when $\phi < 0$ then $F_{1/2}(\beta\mu(x))$ and so the electron density decreases, connoting an absence of electrons: such a region is known as a *depletion region*. Each region affects how Poisson's equation is solved, theoretically or numerically.

24.3 Accumulation layers

It becomes indispensable to accumulate, to acquire property, in order to retain one's good name.

– Thorstein Veblen[12]

A semiconductor such as silicon, when exposed to a vacuum field, has its conduction band pulled down, as in Figure 24.4. Letting the surface be at $x = 0$, and taking the vacuum field to be F_{vac}, then at the surface $F(x = 0) = F_{vac}/K_s$ and $\phi(0) = \phi_s$ whereas deep into the bulk the field and potential both vanish. An integration of both sides of Eq. (24.23) from $\phi = 0$ to ϕ then gives a relationship between F and ϕ that can be used to find the dependence of each on position x. To do so, however, requires knowing the integral of $F_p(u)$. Using Eq. (6.18) it is

$$G_p(u) \equiv \int_0^u F_p(s)ds = \int_0^\infty s^p \ln\left\{\frac{1+e^{u-s}}{1+e^{-s}}\right\} ds \tag{24.24}$$

and so the integration of Eq. (24.23) gives

$$F(\phi)^2 = \frac{8\pi Q}{K_s}\left(\frac{N_c}{\beta^{5/2}}\right)\{G_{1/2}(u) - G_{1/2}(u_o) - \beta\phi F_{1/2}(u_o)\} \tag{24.25}$$

where $u = \beta(\mu + \phi)$ and $u_o = \beta\mu$.

> **EXAMPLE:** Express the coefficient $\sqrt{8\pi QN_c/K_s\beta^{5/2}}$ as a field F_β and determine its magnitude if the temperature is $T = 300$ K and $r_e = 1$.
>
> **SOLUTION:** From Eq. (6.19), $N_c = 6.8122\ \text{eV}^{-3/2}\ \text{nm}^{-3}$. Then, using $Q = 0.35999$ eV nm and $\beta = 38.682\ \text{eV}^{-1}$, it is found that $F_\beta = 0.023691$ eV/nm $\rightarrow \mathcal{E} = 23.691$ MV/m.

If conditions are such that $\beta\mu(x) < -4.6$ then $e^{\beta\mu(x)} \ll 1$. Then,

$$G_{1/2}(u) \approx F_{1/2}(u) \approx \int_0^u s^{1/2}e^{s-u}ds = \frac{\sqrt{\pi}}{2}e^{-u} \tag{24.26}$$

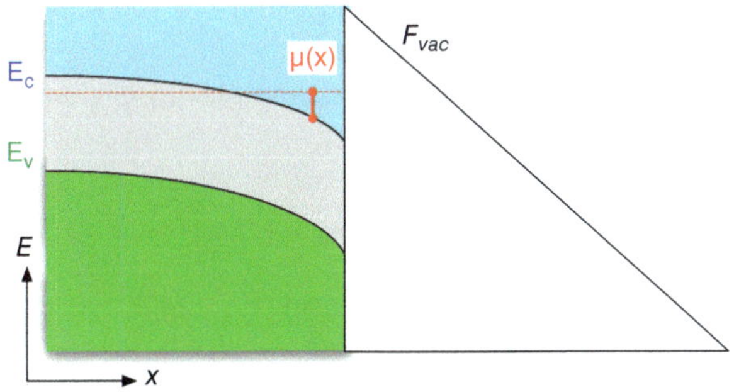

Figure 24.4 A semiconductor surface subject to high field, for which band bending results in an accumulation region near the surface. The semiconductor occupies the region for which $x < 0$. For the configuration shown, in the bulk ($x \rightarrow -\infty$), the electron density is non-degenerate, but near the surface, it becomes degenerate.

[12]Thorstein Veblen, *The Theory of the Leisure Class: An Economic Study of Institutions*, Introduction by C. Wright Mills, *New American Library*, 1953, p. 37.

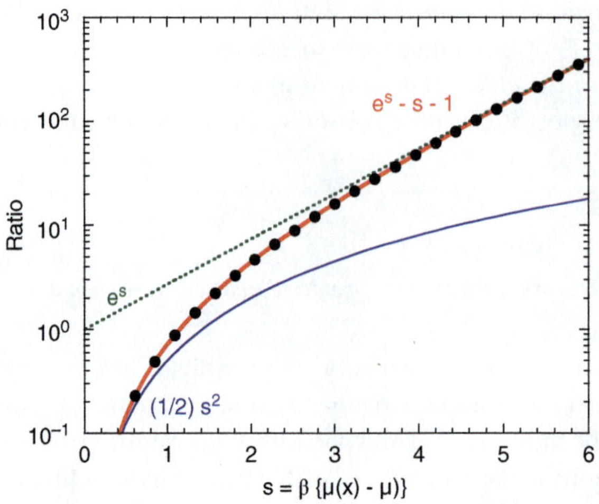

Figure 24.5 The ratio of $F(\phi)^2$ with $(8\pi Q N_c \sqrt{\pi} e^{u_o}/2K_s\beta^{5/2})$ (black dots) compared to the approximation of Eq. (24.27) (red line) and its asymptotic forms (dashed green for s large, solid blue for s small) for $T = 300$ K, $\mu = -0.1$ eV, and $\beta\mu = -3.8682$.

and so[13]

$$F(\phi)^2 \approx \frac{8\pi Q}{K_s}\left(\frac{N_c}{\beta^{3/2}}\right)\frac{\sqrt{\pi}}{2\beta}e^{u_o}\{e^{u-u_o} - 1 - (u - u_o)\} \tag{24.27}$$

The ratio of $F(\phi)^2$ with the coefficient of the bracketed term $\{\cdots\}$ in Eq. (24.25), compared to the approximation in Eq. (24.27), is shown in Figure 24.5 for silicon-like parameters. The approximation is seen to be rather good. When conditions are such $u - u_o = \beta(\mu(x) - \mu) = \beta\phi$ is small, then the exponential can be Taylor expanded to second order, and the relation simplifies to

$$F(\phi)^2 = \frac{8\pi Q}{K_s}\left(\frac{N_c}{\beta^{5/2}}\right)\left\{\frac{\sqrt{\pi}}{4}(\beta\phi)^2 e^{\beta\mu}\right\} = \frac{k_D^2}{4K_s}\phi^2 \tag{24.28}$$

where $\rho \approx N_c\sqrt{\pi}e^{\beta\mu}/\beta^{3/2}$ from Eq. (6.37), and the Debye screening constant k_D from Eq. (21.88) has been used. Eq. (24.28) is very familiar: it is of the form

$$F = \frac{\partial}{\partial x}\phi = \frac{\phi}{\lambda} \rightarrow \phi(x) = \phi_s \exp(x/\lambda) \tag{24.29}$$

and so $\mu(x) - \mu$ decays exponentially (remember that $x < 0$) into the bulk with a length constant of $\lambda = 2\sqrt{K_s}/k_D$.

EXAMPLE: For silicon-like parameters ($\mu = -0.1$ eV, $K_s = 11.8$ and $T = 300$ K, but let $r_e = 1$ for simplicity) estimate what vacuum field is required to give $\phi(x = 0) \equiv \phi_s = 40$ meV.

SOLUTION: Use $F_\beta = 0.023691$ eV/nm from the previous example. To evaluate the remaining terms, first find $\beta\phi = 38.682 \times 0.04 = 1.5473$ and then $e^{\beta\mu} = 0.020897$. Lastly

$$\frac{\sqrt{\pi}}{4}(e^{\beta\phi} - \beta\phi - 1) = 1.9066$$

The product of these factors with their appropriate powers, using Eq. (24.27) as a guide, results in

$$0.023691(1.9066 \times 0.020897)^{1/2} = 0.0047288$$

which is the field F_s (in [eV/nm]) just inside the semiconductor. Outside the semiconductor, the vacuum field is $F_{vac} = K_sF_s = 0.05580$ eV/nm, corresponding to an electric field of about 56 MV/m.

[13] Compare Eq. (2.28) in Mönch [213]. Also, keep in mind $F(x) = F(\phi(x)) \leftrightarrow F(\phi)$.

Conversely, when the electron density is large, then either $u_o = \beta\mu$ can become large and positive (as for metals) or ϕ can become sufficiently large that $\beta\mu(x) = \beta(\mu + \phi)$ is large and positive (as for field emission conditions), that is, the electron density right near the barrier is *degenerate* [312], as in the discussion following Eq. (22.22). When $u \approx u_o \gg 1$,

$$F(\phi)^2 \approx \frac{8\pi Q}{K_s}\left(\frac{N_c}{\beta^{3/2}}\right)\frac{2u_o^{5/2}}{15\beta}\left\{2\left(\frac{u}{u_o}\right)^{5/2} - 5\frac{u}{u_o} + 3\right\} \tag{24.30}$$

The dominant term in the $\{\cdots\}$ brackets, equivalent to Eqs (24) and (25) of ref. [312], are familiar in the treatment of field emission from semiconductors. If ϕ/μ is small, a Taylor expansion of $(u/u_o)^{5/2} = (1 + (\phi/\mu))^{5/2}$ allows Eq. (24.30) to become

$$F(\phi)^2 \approx \frac{8\pi Q}{K_s}\left(\frac{N_c}{\beta^{3/2}}\right)\left\{\frac{\sqrt{\mu}}{2}\beta^{3/2}\phi^2\left(1 + \frac{\phi}{6\mu}\right)\right\} \approx \frac{k_{TF}^2}{4K_s}\phi^2 \tag{24.31}$$

where k_{TF} is defined by Eqs (21.85) and (21.12). Again, $\phi(x) = \phi_s \exp(x/\lambda)$ decays exponentially into the bulk, but now with $\lambda = 2\sqrt{K_s}/k_{TF}$.

Because $F(\phi)$ in Eq. (24.27) is only a function of ϕ, a rapid method of numerically finding $\phi(x)$ is available. Using $F_\beta = \sqrt{8\pi Q N_c/K_s\beta^{5/2}}$ shows that the equation is the same as

$$F(\phi) \approx F_\beta\left(\frac{\pi}{4}\right)^{1/4} e^{\beta\mu/2}\{e^{\beta\phi} - 1 - \beta\phi\}^{1/2} \tag{24.32}$$

which is sufficiently accurate for finding $\phi(x)$ compared to Eq. (24.25) for semiconductors, but far easier to numerically evaluate.

Let $\phi_j \equiv \phi(x_j)$ be evenly distributed from $\phi = 0$ to $\phi = \phi_s$, where ϕ_s is the value at the surface ($x = 0$). Similarly, let $F_j \equiv (d\phi(x)/dx)|_{x=x_j}$. Finite difference methods, discussed in Section A1.3.2, show that a second-order accurate approximation to the derivative results in

$$\frac{\phi_{j+1} - \phi_j}{x_{j+1} - x_j} \approx \frac{F_{j+1} + F_j}{2} \tag{24.33}$$

which can be reworked to find x_{j+1} in terms of the other factors, or

$$x_{j+1} = x_j + 2\left(\frac{\phi_{j+1} - \phi_j}{F_{j+1} + F_j}\right) \tag{24.34}$$

Starting with $x_1 = 0$ and $\phi_j = \phi_s(N - j + 1)/N$, the values of x_j are built up sequentially as j progresses from 2 to N. Because $\phi_{j+1} < \phi_j$, as j increases, x_j becomes more negative. Observe that $\phi_{N+1} = 0$, which should occur at $x_{N+1} = \infty$ (i.e., far into the bulk), but the algorithm fails at that boundary because a finite difference is a poor approximation for an infinite step. But then, the calculation need not be run to $j = N$: it can just as readily truncate when x_j passes a threshold value x_{max}, a flexibility that can be put to good computational advantage. Two examples are shown. Figure 24.6 (left) shows typical silicon-like parameters with a moderate doping level, for which Eq. (24.29) is seen to be a reasonable approximation. Open circles ($\phi(x)$) refer to a numerical solution using Eq. (24.34) and implemented in Section A3.14;

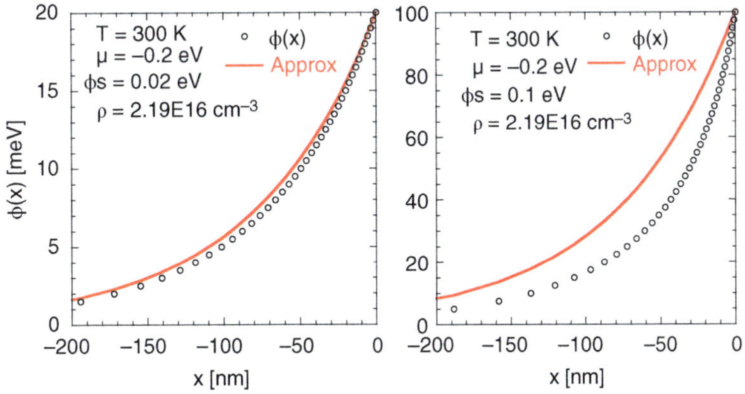

Figure 24.6 Left, silicon-like parameters at a temperature of $T = 300$ K with small $\phi_s = 0.02$ eV. Right, same as left, but for $\phi_s = 0.1$ eV, so that $\beta\phi_s$ is no longer small.

the red line is the approximate solution of Eq. (24.29). The vacuum conditions correspond to $K_s = 11.9$, $T = 300$ K (corresponding to $\beta = 38.68$ 1/eV), $F_{vac} = K_s F(0) = 36.2$ MV/m, $\lambda_D = 78.78$ nm, $\phi_s = 0.02$ eV, and $\rho = 2.19 \times 10^{16}$ cm^{-3}. Figure 24.6 (right) shows the same parameters but for $F_{vac} = K_s F(0) = 3.47$ MV/m and $\phi_s = 0.1$ eV, for which Eq. (24.29) is no longer adequate given that $\exp(\beta\phi_s)$ is no longer sufficiently small.

24.3.1 Factors affecting emission equations

Strange, is it not? that of the myriads who
Before us pass'd the door of Darkness through,
Not one returns to tell us of the Road,
Which to discover we must travel too

– Omar Khayyám[14]

Accumulation layers drag down the conduction band minimum and thereby changes the barrier height, affecting how electrons strive to get over or through the barriers and into the darkness beyond. Even though the doping may be such that the electron statistics are non-degenerate at zero field (Maxwell–Boltzmann), as the field increases, electrons can accumulate until the statistics become degenerate. Therefore, field and thermal emission regimes will show markedly different behavior [312, 318]. Nevertheless, restricting attention to the field emission regime forgoes examining some of the more profound differences that field emission from semiconductors exhibit [305, 313]) and the assumption of degenerate statistics sidesteps, so as to use β_F, s, and J_F evaluated by Eqs (17.12), (17.33) and (17.59). The parameters they use, though, are modified as follows:

1. The semiconductor image charge potential of Eq. (22.29) shows that $Q \to \eta Q$ with[15] $\eta \equiv (K_s - 1)Q/(K_s + 1)$ for every term dependent upon the Schottky factor $\sqrt{4QF}$, although Q in the coefficient of Eq. (24.27) does not thereby acquire the factor η because it is not associated with a Schottky factor (the K_s attending ε_0 in Poisson's equation has already been accounted for).
2. The height of the barrier above the Fermi level μ now changes with field, so that the work function Φ is replaced by $\chi - \phi$, with ϕ defined in Eq. (24.27).
3. The factor $\phi(F)$ must be determined for each field.

Observe that item 3 creates a particular problem because the inversion of Eq. (24.27) is not simple. An easier approach is to specify ϕ instead, and calculate the associated $F(\phi)$. The algorithm to do so behind the findings here is described in Section A3.17, which first finds the maximum and minimum ϕ to be considered based on the vacuum field range to be considered, and then implements the three prescriptions above in terms of the electron affinity of the semiconductor χ, the vacuum field $F_{vac}(\phi)$, the dielectric constant K_s, and the effective mass $m_n = r_e m$. Although Eq. (13.25) is convenient when the Forbes–Deane approximations to $v(y)$ and $t(y)$ (Eqs (13.23) and (13.24)) are used, the Murphy and Good form of Eq. (13.12) shows where the modifications take root:

$$\eta = \frac{K_s - 1}{K_s + 1}; \quad y = \frac{\sqrt{4\eta Q F(\phi)}}{\Phi_b} \tag{24.35}$$

$$A = \frac{r_e^{3/2}}{16\pi^2 \hbar \tilde{\Phi} t(y)^2}; \quad B = \frac{4\sqrt{2 r_e m \Phi_b^3}}{3\hbar} v(y) \tag{24.36}$$

$$n = \frac{\hbar F(\phi)}{2 k_B T \sqrt{2 m \Phi_b} t(y)} \tag{24.37}$$

where $\Phi_b = \chi - \phi$, then if the J_T contribution is negligible compared to J_F,

$$J_{semi}(F(\phi), T) \to A F(\phi)^2 \exp\left(-\frac{B}{F(\phi)}\right) \Sigma\left(\frac{1}{n}\right) \tag{24.38}$$

[14] Omar Khayyám, *The Rubáiyát of Omar Khayyám* (trans. Edward FitzGerald). Garden City, NY: Garden City Books, 1952, p. 167, Quatrain LXIV.
[15] Observe that Stratton [305] uses the notation ν in place of η as used here, the connection being $\nu = \sqrt{\eta}$, but this would cause confusion with ν as used in Eq. (13.25), and so Stratton's convention is not used.

where Φ_b acts as a work function (height of barrier above Fermi level) but means the difference between the electron affinity (χ) and the changing band bending at the surface (ϕ). For rather generic semiconductor parameters of χ = 4 eV, $r_e = 0.2$, $K_s = 8$, and $\mu_o = -0.1$ eV, then the results of the algorithms in Section A3.17 are shown in Figure 24.7 for $F(\phi)$ and Figure 24.8 for $J(F, T)$ for two temperatures of 300 K and 600 K. Although the J curves are rather close (and possess a slightly visible non-linearity), they are distinct, whereas the traditional copper-like curves shown alongside for the same temperatures overlap. As expected, temperature changes have a greater consequence for semiconductors.

24.3.2 Small barriers

...like so many unfortunate events in life, just because you don't understand doesn't mean it isn't so.

– Lemony Snicket[16]

Something becoming "smaller" is often synonymous with "easier",- but there are times where the surprise must be subdued and the consequences dealt with. In the development of Eq. (13.3), the size of the upper limit of the current

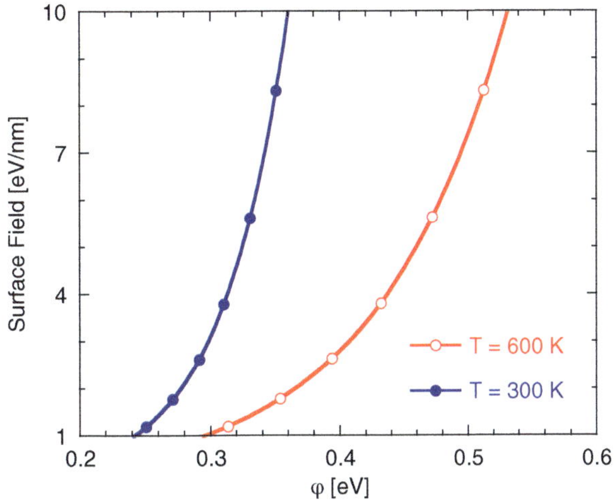

Figure 24.7 An evaluation of the surface band bending associated with an applied field, based on Eq. (24.27), using the algorithms of Section A3.17.

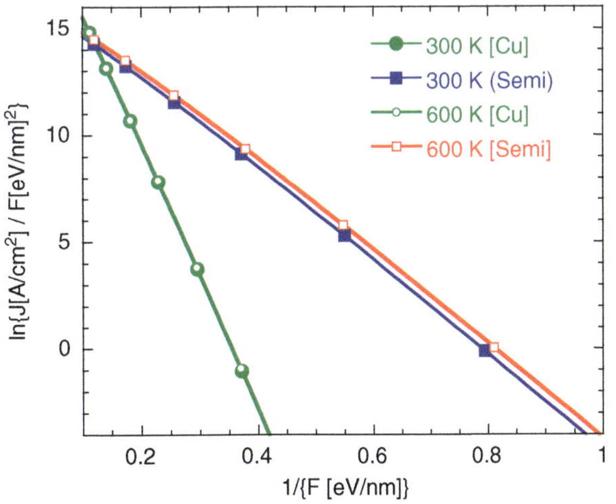

Figure 24.8 Evaluation of the field emission current density from a generic semiconductor for two representative temperatures, with an effective mass $m_n = 0.2m$, a dielectric constant $K_s = 8$, a zero field $\mu = -0.1$ eV, and an electron affinity of $\chi = 4$ eV, using the algorithms of Section A3.17. For plotting the maximum F_{max} was set to 10 eV/nm and the minimum to 1 eV/nm.

[16]Lemony Snicket and Brett Helquist, *A Series of Unfortunate Events: The Bad Beginning*. New York: HarperCollins Publishers, 1999, p. 162.

density integral, designated $C\mu$, was assumed large, allowing it to be taken to infinity without much error, and resulting in a *simpler* equation: the current density became independent of μ. Semiconductors are a different case, for example a typical metal-like barrier has a height of $\mu + \Phi = 11.5$ eV (copper), whereas a typical semiconductor-like barrier has $\chi = 4.05$ eV (silicon). No longer taking the limit $C\mu \to \infty$ in Eq. (13.3) will introduce a μ-dependent term to the current density. That term can be found by

$$\int_0^{C\mu} xe^{-x}dx = \left(-\frac{\partial}{\partial_\lambda}\right)\int_0^{C\mu} e^{-\lambda x}dx = \left(-\frac{\partial}{\partial_\lambda}\right)\left(\frac{1-e^{-\lambda C\mu}}{\lambda}\right) \tag{24.39}$$

where $\lambda \to 1$ at the end. Performing the differentiation then setting $\lambda \to 1$ gives

$$J_{sFN}(T) = AF^2 \exp\left(-\frac{B}{F}\right)\left[\Sigma\left(\frac{1}{n}\right) - (C\mu + 1)e^{-C\mu}\right] \tag{24.40}$$

$$C = \frac{2}{\hbar F(\phi)}\sqrt{2r_e m\Phi_b}\, t(y) \tag{24.41}$$

a form comparable to Eq. (24) of Stratton [305], hence the small s subscript, to treat electrons emitted from the conduction band of semiconductors, but there are a few differences. First, and like Murphy and Good (Eq. (55) of ref. [110]) before him, Stratton uses an analytical integration of a limiting form of the current integral to obtain the temperature dependence, giving rise to

$$\int_0^{\infty} s^{-p-1}\ln(1+s)ds = \frac{\pi}{p\sin(\pi p)} \tag{24.42}$$

and resulting in a factor $\pi Ck_BT/\sin(\pi Ck_BT)$ in place of the $\Sigma(1/n)$-based terms. For small argument, both thermal modifications are about the same, but the singularity in $\Sigma(1/n)$ as $n \to 1$ is balanced by that in $\Sigma(n)$, as discussed in the treatment of Eq. (17.65))), a behavior in contrast to $\pi Ck_BT/\sin(\pi Ck_BT)$ when $Ck_BT \to 1$.

EXAMPLE: Find the temperature at which the factor $\pi Ck_BT/\sin(\pi Ck_BT)$ faces difficulties. Assume a field of $F = 1$ eV/nm, a work function term of $\Phi_b = 3$ eV, and an effective mass ratio of $r_e = 0.5$.
SOLUTION: "Difficulties" can be taken to correspond to $Ck_BT \approx 1/2$, for after that point, the denominator ceases to increase and begins to decline. Because $t(y)$ is already close to (but larger than) unity, approximate it by unity. Then, expressing all factors in units of [eV], [nm], [fs], and [q],

$$T = \frac{\hbar}{4k_B}\frac{F}{\sqrt{2r_e m\Phi}} = \frac{11605 \cdot 0.65821 \cdot 1}{4\sqrt{2 \cdot 0.5 \cdot 5.6856 \cdot 3}} = 462.36K$$

EXAMPLE: Compare the size of the correction term in Eq. (24.40) for metal-like, then semiconductor-like, values of relevant parameters at room temperature, assuming the surface field is approximately 7 eV/nm.
SOLUTION: Take metal-like to mean $\mu = 7$ eV and $\Phi = 4.5$ eV, for which $C\mu = 3.3367$ and

$$(C\mu + 1)e^{-C\mu} = 1.7486893 \times 10^{-9}$$

For semiconductor-like, from Figure 24.7, $\phi \approx 0.342$ eV when $F = 7$ eV/nm, and let $\chi = 4.05$ eV, $r_e = 0.2$, $\mu = -0.1$ eV, and $K_s = 8$, for which $C(\mu + \phi) = 0.3298$ (the electrochemical potential $\mu(\phi) = \mu + \phi$ from Eq. (24.20) must be used) and

$$(C\mu(\phi) + 1)e^{-C\mu(\phi)} = 0.9562$$

24.3.3 Pre-factors of transmission probability $D(E)$

Teaching about FN tunnelling does not use FN's full quantum-mechanical approach, because the simple-JWKB (Jeffreys–Wentzel–Kramers–Brillouin) approximate method is easier to understand. When applied to FN's exact triangular barrier, the simple-JWKB method generates an approximate formula without the tunnelling pre-factor.

– Richard G. Forbes [319]

A wary reader may be unsettled that the electrochemical potential can be negative as well as positive. A negative $\mu(\phi)$ means the barrier is substantially larger at lower fields, and therefore a possibly stark change in the current-field plots should be expected. Stratton and others, such as Baskin *et al.* [313], consider such a circumstance and find behavior very much different than that of metals: the path to demonstrate this is arduous, but, although the results are aesthetically interesting, the operational regime for field emission from semiconductors is for circumstances different from where the atypical behavior occurs. Here, instead, consider the more relevant modifications to the transmission probability coefficient.

Where the smallness of the electrochemical potential $\mu(\phi)$ makes a difference is in the oft-neglected coefficient $P(E)$ of the transmission probability $D(E)$, called the "tunneling pre-factor" by Forbes [319]: its effects can be substantial, especially for drawing inferences from experimental data. Its neglect is perhaps due to the expectations created by metal parameters. That is, for the triangular barrier, $P(E \approx \mu)$ was given by Eq. (13.9): for copper parameters, $P(\mu) = 1.9522$. But metals are not all there is: the pre-factor matters for more than just the *magnitude* of the overall current density, rather it affects the *behavior* of the current density integrand. As a consequence, the distribution in energy of the emitted electrons (the normal distribution, anyway, recall the distinction between normal and total energy distributions in Figure 17.6) is also affected. For metals, the integrand was concentrated about μ, so $P(E)$ was reasonably evaluated at $E = \mu$ and $P(\mu)$ could be pulled out of the integral. Such an approximation is reasonably good, where the presence or absence of $P(\mu)$ changes the height but not so much the general shape. For semiconductors, the electrochemical potential $\mu(\phi) = \mu + \phi$ can be small, and so the rapid decline of the integrand evinced in metals due to the behavior of the transmission probability $D(E)$ is not so strong. Both cases are shown in Figure 24.9. As a result, with semiconductors the approximation of $P(E)$ by its evaluation at μ and its extraction from the integrand (and often, the subsequent overlooking it endures) is poor because the low-energy behavior of $P(E)$ now affects the integrand.

The rising importance of $P(E)$ threatens the utility of the otherwise simple Fowler–Nordheim equation, but, curiously, there is recourse. Recall from Eq. (18.59) that $P(E)$ behaves, to leading order, as

$$P(E(k)) \approx 4k/k_o \tag{24.43}$$

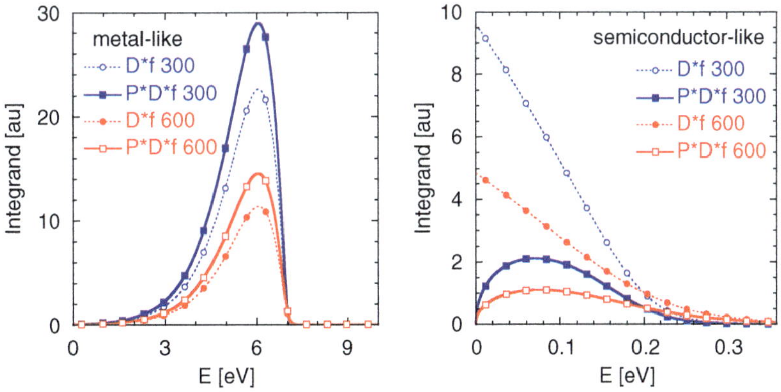

Figure 24.9 A comparison of the current density integrands $dJ(E)$ for metal parameters ($m_n = m$, $\mu = 7$ eV) (left) and semiconductor-like parameters ($m_n = 0.2m$ and $\mu + \phi = 0.2$ eV) (right). Parameters common to both: $T = 300$ (blue) and 600 K (red), $F = 5$ eV/nm. For convenience, $\Phi = \Phi_b = 4.5$ eV and $K_s = 8$ for each.

for $E = \hbar^2 k^2/2m_n$, a result for the triangular (no image charge) barrier. But observe that, first, $V(x)$ for semiconductors is not as blunted as for metals as in Figure 22.9 because of K_s and, second, that what remains is fairly triangular itself. This has led to the suggestion that k_o be replaced in $P(E)$ by a term that goes as the actual barrier height but otherwise retaining the image charge form of the Gamow factor $\theta(E)$, or

$$k_o^2 \to \frac{2m_n}{\hbar^2}\left(V_o - \sqrt{4\eta QF}\right) \tag{24.44}$$

where V_o is the height of the barrier above the conduction band, or $\chi - \mu(\phi)$, where $\mu(\phi) = \mu + \phi$ is the electrochemical potential. Such an approximation works quite well [318], giving the current integrand behavior up to a scale factor close to unity, as compared to an Airy function transfer matrix solution of $D(k)$, as shown in Figure 24.10 for semiconductor-like parameters (based on Figures 1 and 5 of ref. [318].).

Alas the calculation of $J(F)$ in Figure 24.10 is by the numerical evaluation of an integral rather than the relative simplicity of a formula no more complex than the J_{FN} in Eq. (13.4): the cause is $P(E)$. The modifications to the current density formulation are most transparently revealed by comparing the current density $J(F)$ evaluated with $P(E)$ compared to that evaluated with $P(E) \to P(\mu)$. The latter shall be called $J_\mu(F)$ and leads to Eq. (13.12) if $P(\mu) \to 1$. The ratio of the two current densities is therefore

$$\frac{J(F)}{J_\mu(F)} = \int_0^{C\mu} se^{-s}\sqrt{1 - \frac{s}{C\mu}}ds \tag{24.45}$$

where $C = \sqrt{2m_n\Phi_b}t(y_b)/\hbar F$, the b subscripts denoting that these quantities are different than their metal analogs by $\Phi_b = \chi - \mu(\phi)$ and $y_b = \sqrt{4\eta QF}/\Phi_b$. The integral is a special case of the Kummer function (alternately known as the confluent hypergeometric function) for all the good that does: its evaluation will, for most who need it, be accomplished numerically anyway. Knowing the limits is helpful, particularly when $C\mu$ is large (as for metals) or small (as can be the case for semiconductors). For small $C\mu \equiv p$, Taylor expand e^{-s} to find

$$\frac{J(F)}{J_\mu(F)} \approx \frac{4}{105}p^2(7 - 4p) \tag{24.46}$$

Conversely, for large $C\mu \equiv p$, Taylor expand $(1 - (s/p))^{1/2}$ to find

$$\frac{J(F)}{J_\mu(F)} \approx \frac{p-1}{p} - \frac{p^2-2}{2p}e^{-p} \tag{24.47}$$

where both expansions are only to first order. A comparison of these expansions to a numerical evaluation is shown in Figure 24.11. Whether using the exact evaluation or the approximations for the Kummer function, as $C\mu$ decreases the impact on $J(F)$ can be substantial, thereby boding poorly for the neglect of $P(E)$ in the $J(F)$ relations.

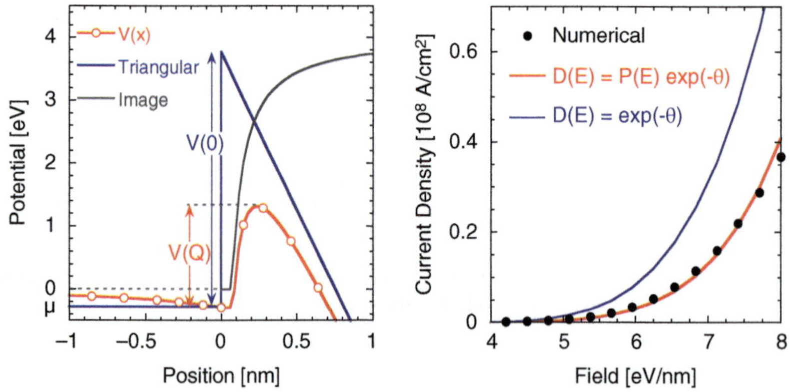

Figure 24.10 Potential (left) and associated current density (right) without and with the $P(E)$ contribution for heavily doped GaAs-like semiconductor parameters ($T = 300$ K, $\rho = 10^{19}$ cm^{-3}, $\mu = 0.28$ eV, $K_s = 9$, $m_n = 0.06m$, $\chi = 3.6$ eV). $V(Q)$ refers to $V_o - \sqrt{4\eta QF}$, so that $V(0) = V_o$. On the left, $V(x)$ shows band bending in bulk, so that ϕ is reasonably large at the suface. On the right, "Numerical" refers to the J evaluated using an Airy transfer matrix approach to finding $J(F)$. Based on Figures 1 and 5 of ref. [318].

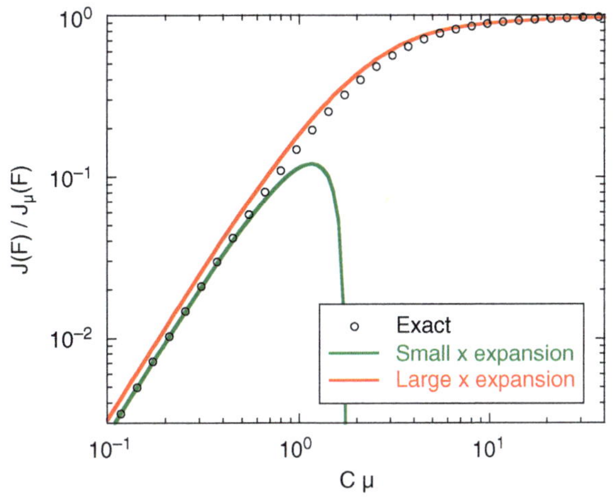

Figure 24.11 The ratio of Eq. (24.45) compared to its small and large x approximations (Eqs (24.46) and (24.47), respectively). Although the large x expansion appears sufficient throughout, it is a factor of $\approx 5/4$ larger than the exact case for small argument.

24.4 Depletion layers

Don't it always seem to go
That you don't know what you've got
Till it's gone...

– Joni Mitchell[17]

The opposite of electrons piling up (accumulating) at an interface and thereby causing the bands to bend downward is the pushing away (depletion) of electrons and the consequent bending of the bands upward, such that ϕ in $\mu(\phi) = \mu + \phi$ becomes negative. Although a common situation in semiconductor devices, where the doping is so low that $F_{1/2}(\beta\mu(x))$ may be approximated by a Maxwell–Boltzmann distribution and Poisson's equation readily solved, consider instead an interestingly different situation in which the electrons in the conduction band (and holes in the valence band) are negligible compared to the presence of *ionized dopants*: they, too, result in band bending. Several cases are typical and of interest to electron emission:

- For field emission, the ionization of acceptor atoms causes the shielding of the bulk material from an externally applied field, as occurs with boron doping in diamond.
- At the interface between diamond and a metal, intentional nitrogen dopants become charged and lead to an internal field emission barrier [298, 320, 321].
- A nitrogen layer can keep generated secondary electrons from being absorbed by a metal back contact [322].

The creation of a depletion layer that attends the nitrogen doping cases (introduced in Section 16.1) is suggested by the sequence shown in Figure 24.12. The scheme presented is a simplified representation, sometimes characterized as an abrupt junction with no trapped charge states between a metal and an n-type semiconductor (a classical Schottky barrier): the existence of trapped charge and the presence of dielectric phases due to the intermixing of the metal and semiconductor makes the behavior of the current flow between metal and semiconductor more complex [213, 323], and such complications change the barrier parameter (and introduce new ones), but such complexity can be built upon the simple model of the classical Schottky barrier.

The greater understanding of semiconductor features, and the intention to demonstrate a numerical method of finding the band bending profiles associated with more exotic doping profiles, is dealt with by, first, exploring more fully the depletion width associated with boron charged impurities, followed by considering nitrogen *and* boron effects next. Such a model will be useful in Chapter 32.1 when considering the transport of generated secondary electrons in a thin diamond film.

Were diamond a perfect insulator, then the potential across a thin layer of it deposited upon a metal surface and subject to an electric field would be $\phi(x) = (F_{vac}/K_s)x$ for x measured from the metal–diamond interface. But because charge in

[17] Joni Mitchell, lyrics to *Big Yellow Taxi*, 1970.

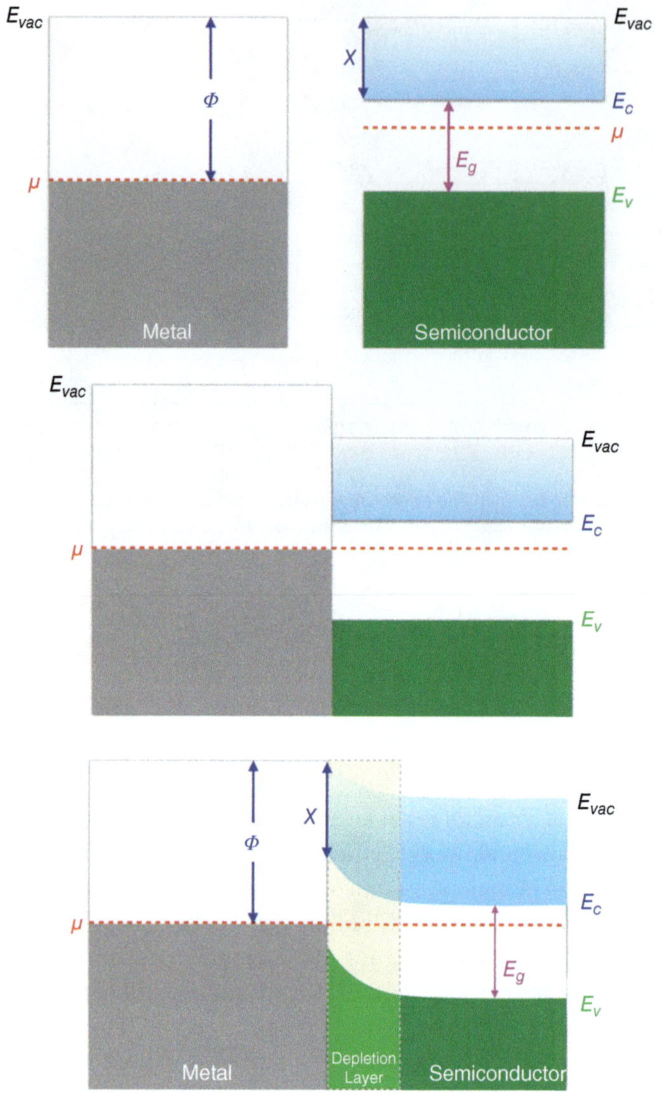

Figure 24.12 Top, a metal (left) and an n-type semiconductor (right) before contact; Middle, after contact but before charge migration (a discontinuity in E_{vac} is unphysical); Bottom, after contact and charge migration, which creates a depletion layer.

the form of thermally excited electrons, holes, and charged donors and acceptors exists, then the potential must instead be determined by the solution to Poisson's equation in the form [213, 324]

$$\partial_x^2 \phi(x) = \frac{16\pi Q}{K_s} \left(n(\phi) - p(\phi) + N_a^+(\phi) - N_d^-(\phi) \right) \tag{24.48}$$

where n and p are the number densities of the electrons in the conduction and valence bands, respectively, and N_a^+ and N_d^- are the number densities of the ionized acceptor and donor atoms, respectively. The rationale for making them functions of ϕ are because it determines the density of electrons via $\mu(x)$ for electrons (and analogously for holes) and because it is the bending of the bands that determines whether charge can tunnel away from donors or to acceptors [298, 320]: a nitrogen atom with an energy level 1.7 eV below the conduction band sees something looking much like a tunneling barrier when the diamond layer is under a field, and so the tunneling of its electron (or alternately the injection of its hole) will leave the nitrogen atom charged. Moreover, trace amounts of boron are always present even in high-quality diamond, so that doping levels of even 10^{14} cm^{-3} boron are considered quite pure (about 1 boron atom every 215 nm). The different processes that can be involved are schematically illustrated in Figure 24.13 for diamond on a semiconductor.[18] If conditions are favorable to the ionization of a boron impurity (in contrast to the nitrogen ionization

[18]For diamond on a metal, see Figure 6.8 of *Novel Cold Cathode Materials* [166].

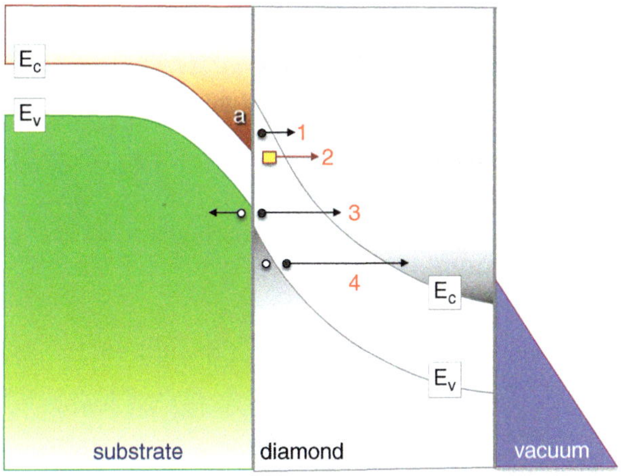

Figure 24.13 Processes in band-to-band transport/internal field emission for a semiconductor/insulator interface (an accumulation region (a) occurs if the substrate is a semiconductor, although the substrate could be a metal). The processes are: 1, electron injection from the accumulation region; 2, tunneling from donor or impurity; 3, hole injection; 4, Zener tunneling. The diamond surface shown is a positive electron affinity (PEA) surface; a hydrogen-terminated surface would be a negative electron affinity (NEA) surface. Levels are not to scale. Based on Figures 2 and 3 of Cutler *et al.* [298].

of process (2) in Figure 24.13, boron atoms close to the surface can accept stray electrons and also become charged), then a layer of negative charge can build up and screen the applied field from the bulk. The charging of the boron creates a different kind of depletion layer.

The nature of the depletion layer for two kinds of dopants in an insulator provides a rather good model to introduce a second numerical method to compliment the one introduced in Section 24.3. Take the thickness of the diamond layer on the substrate to be L, and the vacuum field applied to its surface to be F_{vac}, so that in the absence of charged impurities[19] the field across the layer would be $F_o = F_{vac}/K_s$, where $K_s = 5.7$ in Table 24.1. If L is sufficiently large, then it will exceed the depletion width $w = \varepsilon_0 F_{vac}/N_a q^2 = F_{vac}/16\pi Q N_a$ (Eq. (16.4)). Even for low dopings, this width is small, for example for $N_a = 10^{15}$ atoms/cm^3, then $w = 55.26$ nm for a field of $F_{vac} = 1$ eV/μm (equivalent to an electric field of 1 MV/m). Now consider Poisson's equation applied to ϕ when two dopants are present:

$$\frac{d^2}{dx^2}\phi(x) = -\frac{16\pi Q}{K_s}\begin{cases}(N_b - N_n) & x < x_n \\ N_b & x \geq x_n\end{cases} \tag{24.49}$$

where the subscripts refer to boron (b) and nitrogen (n) for the doping densities N, and the charge of the doping is associated with the sign of the term. In contrast to the boron, however, let the nitrogen doping levels be rather high, and

Table 24.1 Parameters used in the numerical solution of Poisson's equation in diamond.

Symbol	Definition	Value	Unit
m	Mass of electron (vacuum)	0.511	eV/c^2
m^*	Mass of electron (diamond)	0.57m	eV/c^2
E_G	Diamond band gap	5.5	eV
ρ_o	Reference density	10^{16}	#/cm^3
n_b	Boron factor (ρ_b/ρ_o)	0 or 0.023	ρ_o
n_n	Nitrogen factor (ρ_n/ρ_o)	0 or 10	ρ_o
K_s	Static dielectric constant	5.7	[–]
L	Diamond flake thickness	8	μm
x_n	Nitrogen layer thickness	0.3	μm
ϕ_b	Bias potential $\phi(L)$	[–]	eV

[19]By virtue of its large band gap, diamond is already relatively free of roaming electrons and holes.

the diamond grown so that only the immediate region $0 < x < x_n$ near the back contact has significant doping, where x_n is the boundary past which no nitrogen is present. Now what does the potential look like?

Re-expressing the equation in dimensionless terms is a good precursor to preparing the problem to be solved numerically. Set the back contact at $x = 0$ and the surface at $x = L$. Scaling all lengths then by L, let $s = x/L$ and $s_n = x_n/L$. A good energy scale emerges naturally from Poisson's equation. Let $\phi = y\phi_o$, where ϕ_o is to be determined. Further, let $N_a \equiv n_b\rho_o$ and $N_d \equiv n_n\rho_o$ so that n_b corresponds to boron doping and n_n to nitrogen doping. Also, ρ_o is a reference density which can be rather arbitrarily set: $\rho_o = 10^{16}$ atoms/cm^3 is an agreeable choice, being a mid-range doping density. Then Poisson's equation is, in terms of the dimensionless quantities s and y for $s < s_n$

$$\frac{d^2}{ds^2}y(s) = -\frac{16\pi Q\rho_o L^2}{K_s\phi_o}(n_b - n_n) \equiv -(n_b - n_n) \tag{24.50}$$

with an analogous equation for $s \geq s_n$. This serves to identify ϕ_o as

$$\phi_o = \frac{q^2}{K_s\varepsilon_0}L^2\rho_o = \frac{16\pi Q}{K_s}\rho_o L^2 \tag{24.51}$$

A dimensionless replacement for the field is also sought: from $\partial_x\phi(x) = -F_{vac}/K_s$, then $\partial_s y = -R$ demands $R = F_{vac}L/K_s\phi_o$.

Solving Eq. (24.49) is as close to trivial as ordinary differential equations get: as the right-hand side is a constant, $y(s)$ must be a quadratic function in s, and so solutions are of the form

$$y(s < s_n) = A_2s^2 + A_1s + A_0 \tag{24.52}$$

$$y(s \geq s_n) = B_2s^2 + B_1s + B_0 \tag{24.53}$$

where the constants A_j and B_j must be determined. As there are six unknown coefficients, six conditions or equations are required to specify them, found by demanding

1. from Poisson's equation, $\partial_x^2 y = -(n_b - n_n)$ for $s < s_n$
2. from Poisson's equation, $\partial_x^2 y = -n_b$ for $s \geq s_n$
3. the potential vanishes at the origin, or $y(0) = 0$
4. the field at the surface demands $\partial_s y = -R$
5. continuity of $\partial_x\phi(x)$ at x_n demands $\partial_s y(s_n^-) = \partial_s y(s_n^+)$
6. continuity of $\phi(x)$ at x_n demands $y(s_n^-) = y(s_n^+)$

where the $\pm$ on the $s_n^{\pm}$ indicates infinitesimally larger or smaller than s_n. These equations may be solved by tediously substituting one equation into the next, or rapidly and elegantly by translating the six equations into one large matrix equation, like this

$$\begin{pmatrix} 0 & 0 & 2 & 0 & 0 & 0 \\ 0 & 0 & 0 & 0 & 0 & 2 \\ 1 & 0 & 0 & 0 & 0 & 0 \\ 0 & 0 & 0 & 0 & 1 & 2 \\ 0 & 1 & 2s_n & 0 & -1 & -2s_n \\ 1 & s_n & s_n^2 & -1 & -s_n & -s_n^2 \end{pmatrix} \begin{pmatrix} A_0 \\ A_1 \\ A_2 \\ B_0 \\ B_1 \\ B_2 \end{pmatrix} = \begin{pmatrix} n_n - n_b \\ -n_b \\ 0 \\ -R \\ 0 \\ 0 \end{pmatrix} \tag{24.54}$$

the inversion and solution of which (perhaps by Gauss–Jordan elimination if one must, perhaps by symbolic mathematical packages if one can) results in finding (*i*) $A_0 = 0$, (*ii*) $A_1 = -R + n_b - n_n s_n$, (*iii*) $A_2 = -(n_n - n_b)/2$, (*iv*) $B_0 = -n_n s_n^2/2$, (*v*) $B_1 = -R + n_b$, and lastly, (*vi*) $B_2 = -n_b/2$, and so

$$y(s < s_n) = \frac{1}{2}(n_n - n_b)s^2 + (n_b - n_n s_n - R)s$$

$$y(s \geq s_n) = -\frac{1}{2}n_b s^2 + (n_b - R)s - \frac{1}{2}n_n s_n^2 \tag{24.55}$$

The depletion width w is recovered by setting $n_n = 0$ and evaluating $\partial_s y(s = 0) = 0$ to find $R = n_b$, which shows $w = F_{vac}/16\rho_o\pi Q n_b$, but which is equivalent to the definition of w in Eq. (16.4).

EXAMPLE: Find the characteristic energy ϕ_o if the diamond layer L is the same as the depletion width w for a boron doping density of 10^{15} cm^{-3} $= n_b\rho_o$ (i.e., $n_b = 0.1$), and when the surface field is $F_{vac} = 10$ eV/μm (equivalent to an electric field of 10 MV/m). Assume all the boron atoms are ionized, and the dielectric constant of diamond to be $K_s = 5.7$. Then find the ratio $R = F_{vac}L/\phi_o$, again assuming $L = w$. Why does R not change when the surface field changes for these conditions?

SOLUTION: From the definition of w in Eq. (16.4) inserted into Eq. (24.51), and using $N_a = n_b\rho_o = 10^{16}$ cm^{-3} $= 10N_a$, then, in [nfeq] units,

$$\phi_o = \frac{1}{16\pi K_s Q\rho_o}\left(\frac{F_{vac}}{n_b}\right)^2 = \frac{1}{0.0010314}\left(\frac{0.01}{0.1}\right)^2 = 9.6954 \text{ eV}$$

Observe that for these parameters, $w = 0.5526$ μm. Substituting,

$$R(L = w) = \frac{F_{vac}w}{K_s\phi_o} = n_b = 0.1$$

The independence here of R on F_{vac} is a consequence of the special case $L = w$, as it is not in general true.

Eq. (24.55) represents the rather special case of regions of constant doping and the comparative absence of carriers (electrons or holes), so that a polynomial expansion of $\phi(x)$ provides an easy solution. This is not the general circumstance: were electrons to be present in the conduction band in any appreciable number, then the difficulties encountered in the treatment of space charge, as in Section 10.5, come back in force, and would result in a *varying* charge density, making numerical solutions perhaps a preferred means of finding $\phi(x)$. To show correspondence, however, the examples below shall apply to the piecewise constant doping density case so as to compare the numerical accuracy against the analytical model. This will be done even though the doping may be graded and additional charge present, as in the right-hand side of Eq. (24.48).

The numerical methods of Section A1.4 are quick and quite adaptable, and therefore possibly preferable to more sophisticated methods that require a higher investment of coding, memory, or both [236]. When both the left and right boundary conditions $\phi(0)$ and $\phi(L)$ are known, then the very fast and low-memory algorithm of Section A1.4.2 is attractively simple.[20]

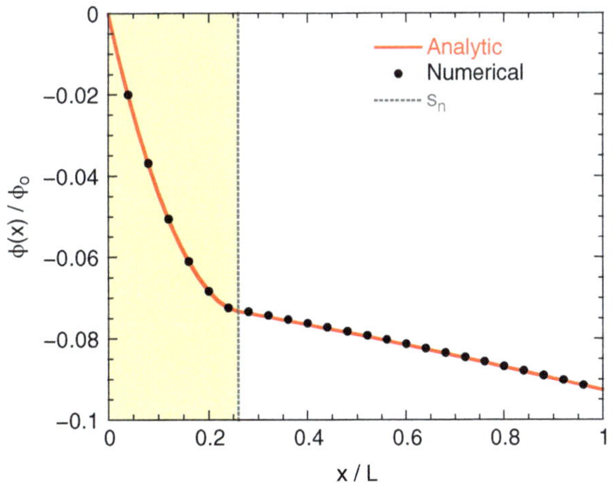

Figure 24.14 A comparison of the analytic solution of Eq. (24.55) to its numerical evaluation using `DopePhi` of Section A3.15.2 using the default parameters of that program ($N = 24$, $n_b = 0.01$, $n_n = 2$, $R = 0.03$, and $s_n = 0.26$). The yellow region is where nitrogen doping occurs; boron doping occurs throughout.

[20] Were the field at the rightmost boundary F_{vac}/K_s known instead of $\phi(L)$, then one could use the Section A1.4.2 algorithm in conduction with a bisection search algorithm. There are occasions where this is a reasonable approach, instead of the more conventional finite difference solutions to differential equations [242] that are always available.

The algorithm of DopePhi of Section A3.15.2 gives an example of how such a calculation might proceed, the results of which are shown in Figure 24.14. By using a dimensionless formulation, the values of L, ϕ_o, K_s and F_{vac} need not be explicitly given, as the solution only relies on the factors s_n, R, n_b and n_n. The factor s_n is somewhat unusual: *it is in between two of the grid points* at s_{nc} and s_{nc+1} (where $nc \leftrightarrow$ nc is an index term in the code). This is commonly the case: a discrete grid results in ambiguity as to where an exact point for an abrupt transition should be situated. Were the doping profile smooth and continuous, deciding where s_n is placed and how the abrupt transition is numerically represented would not be an issue. For all that, even a rather coarse discretization as the one shown provides good correspondence to the analytical solution of Eq. (24.55). Consequently, the algorithm can be profitably used in analyzing the impact of electron density variation in tunneling structures [325] (it will also be useful when $\partial_x^2 u = v$ appears elsewhere, such as heat diffusion in the coupled thermal equations associated with the laser impinging on a photocathode [85]).

24.5 Modifications due to non-linear potential barriers

...the fundamental goal of any emancipatory movement must be to demystify and democratize the production of scientific knowledge, to break down the artificial barriers that separate "scientists" from "the public".

– Alan D. Sokal[21]

The misuse of scientific jargon and sloppy pseudo-theories common to postmodern analyses was sharply satirized through the publication of a fraudulent article successfully pushed through a peer-review process to publication by A.D. Sokal in 1996. Lest pranking scientific journals seem an odd way of driving this point home, consider that Hans Bethe and friends G. Beck and W. Riezler did the very same when they were post-docs in getting a fake derivation of the fine structure constant published in *Naturwissenschaften* meant to satirize questionable numerological efforts fashionable at the time.[22] Although taking aim at the idea that "truth" is a social convention, Sokal's hoax (as it is called) brushed rather closely in places with a concern here: that in the efforts to describe and understand barriers, inordinate care with the mathematics and precision with the ideas is required lest the results border on rubbish.

Finding the two roots of $V(x) - E$ for an image charge barrier plus linear field is straightforward for the evaluation of the Gamow factor $\theta(E)$ needed to develop a field emission equation or find the Schottky reduced barrier for the other emission mechanisms. Inserting a non-linear component makes for more of a challenge: although the root equations may even be possible to write down, actually finding the roots takes more care than amateurs generally bring to the problem, largely because of the image charge term. Material interfaces provide a good non-linear potential candidate to practice upon. When two dissimilar materials are joined, charge migration occurs so that their electrochemical potentials are aligned, as in Figure 24.12. The band diagram that results depends on the doping levels, whether the semiconductors are n- or p-type, and how big the energy band gaps and electron affinities of each are. The intensity of the treatments [111, 213, 312, 315, 323, 326, 327] is indicative of the complexity of the problem, and while many of these works are concerned with *emission*, hole transport [323, 328] and transport between material layers can be treated with the same methodology as the linear potential barrier becomes non-linear in a way that will be useful for field emission barriers associated with field enhancement in Chapter 30.

What is meant by "quadratic" is somewhat ambiguous. The periodic Schottky deviations of Section 19.2.3 encountered a potential barrier treated by Cutler and Gibbons [263] and subsequently applied to field emission by Cutler and Nagy [233]. Those barriers are "quadratic" in the sense that

$$V(x) = \mu + \Phi - Fx - \frac{Q}{x} \rightarrow \mu + \Phi - Fx - \frac{Q}{x} + \frac{\gamma}{x^2} \tag{24.56}$$

or the potential is quadratic in $1/x$. It is flexible in modeling departures from the standard image charge potential while retaining the linear field asymptotically, but complicates analysis near the origin. In contrast, depletion barriers are better described by

$$V(x) = \mu + \Phi - Fx - \frac{Q}{x} + \gamma x^2 \tag{24.57}$$

---

[21] Alan D. Sokal, Transgressing the Boundaries: Toward a Transformative Hermeneutics of Quantum Gravity, *Social Text*, **46/47**, spring/summer 1996, p. 217–252.

[22] The episode is recounted in S.S. Schweber, *In the Shadow of the Bomb: Bethe, Oppenheimer, and the Moral Responsibility of the Scientist*, Princeton Series in Physics. Princeton, NJ: Princeton University Press, 2000, p. 87.

Apart from the Q/x term, the remaining terms of Eq. (24.57) are quadratic in x, with γ having units of [eV/nm^2].

The unpleasant lack of a closed form solution of the quadratic potential Gamow factor is bracing after the success of Fowler and Nordheim's solution for the image charge potential. With an almost Dylan Thomas-esque defiance,[23] there have been efforts to smuggle the Fowler–Nordheim equation back into the quadratic barrier [329–331], treating the non-linear potential barrier at interfaces by introducing a sheet charge that causes a discontinuous change in the electric field and putting a kink in the triangular barrier by causing an abrupt change in slope, as a consequence of charged traps. Doing so allows the current density to be recast as a product of Fowler–Nordheim-like terms so that much of the triangular barrier intuition does not have to be jettisoned, at the price that the position of the sheet charge and its charge per unit area require specification. The model is often used to treat metal–insulator–semiconductor layers for its ease.

By contrast, a disinclination to rely on parametric values gleaned from data requires coming to terms with the actual barrier, and for the Gamow factor that means coming to terms with finding the three roots of $V(x) - E = 0$ [235]. It is one thing to speak of the roots of Eq. (24.57), and another matter to actually calculate them: iterative numerical methods can perform poorly if the initial guesses are poor; bisection methods are straightforward and always work once the roots are bracketed, but bracketing those roots (which, depending on the size of γ, can be closely spaced) takes artistry. When many $J(F,\gamma)$ evaluations are needed, fancy artistry does not compete with a fast method using at most a few iterations.

Some observations simplify the development of an appropriate method. First, in the limit that $\gamma \to 0$, two of the roots have to coalesce into the standard image charge roots of Eq. (11.26) with the last root approaching ∞. Second, a relationship between the roots and the usual factors V_o, F, Φ, and now γ is established by demanding that $V(x) - E$ from Eq. (24.57) be matched to

$$V(x) - E \equiv \frac{\gamma(x - x_1)(x - x_2)(x - x_3)}{x} \tag{24.58}$$

where x_j are the roots with $j = (1, 2, 3)$ such that x_1 and x_2 converge to $x_\pm$ when $\gamma \to 0$, and x_3 is the largest of the roots which lies in a region where $V(x)$ crosses back over the x axis beyond $x_2 > x_+$, and which is therefore a parameter specifying $V(x) - E$ rather than a physical root, as in Figure 24.15. Equating these forms, and then equating the coefficients of powers of x gives [235]

$$F = \gamma(x_1 + x_2 + x_3) \tag{24.59}$$

$$V_o = \gamma(x_1 x_2 + x_1 x_3 + x_2 x_3) \tag{24.60}$$

$$Q_s = \gamma(x_1 x_2 x_3) \tag{24.61}$$

where $V_o \equiv \mu + \Phi - E$. If $K_s > 1$, then $Q \to Q_s = Q/K_s$.

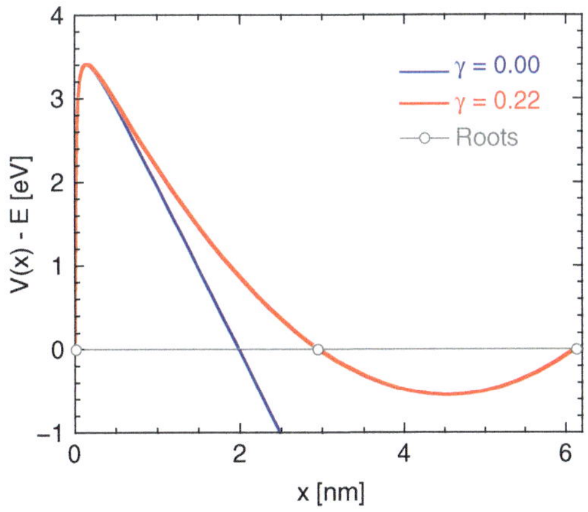

Figure 24.15 A comparison of the quadratic image charge potential $V(x) - E$ for the parameters used in Table 24.2. The standard image charge corresponds to $\gamma = 0$. In practice, in the evaluation of $\theta(E)$, the potential would be truncated at its lowest point and set to a constant thereafter if the bias across the semiconductor were set equal to 0.

[23] Dylan Thomas's popular poem beginning, "Do not go gentle into that good night, /Old age should burn and rave at close of day; /Rage, rage against the dying of the light." (*The Poems of Dylan Thomas*, New Directions Publishing Corp., New York, 1971, p. 239) was a futile but valiant protest of that which no man can overcome.

The third and last observation makes possible an iterative approach: for any of the roots x_j for which $V(x_j) - E = 0$, a grouping of terms shows that it can be related to itself via

$$x_j = \frac{Q_s + Fx_j^2}{V_o + \gamma x_j^2} \tag{24.62}$$

Such a form is ripe for using the iteration methods of Section A3.7.2: in fact, those iteration methods converge fast for a good first guess. In the case of the smallest root x_1, a great first guess is x_- from the standard image charge result of Eq. (11.26). Skip over x_2 for a moment, as a trick will be useful for it. That means consider x_3 next: when γ is small, x_3 is large, and so the x_3^2 terms on the right hand side will dominate the neighboring terms in the numerator and denominator of Eq. (24.62): this is because $\gamma x_3^2 = x_3(Q_s/x_1 x_2) \approx x_3(Q_s/x_+ x_-) = Fx_3$, which is large. Therefore, $x_3 \approx F/\gamma$ is the asymptotic approximation. Let that approximation give the starting approximation to x_3 in Eq. (24.62). Now return to x_2: rather than using x_+, it is better to use the relation for Q in Eq. (24.61). The first approximations to the three roots are then given by

$$x_1 \approx \frac{Q_s}{V_o + \sqrt{V_o^2 - 4Q_s F}} \tag{24.63}$$

$$x_3 \approx \frac{F^3 + Q_s \gamma^2}{\gamma(F^2 + \gamma V_o)} \tag{24.64}$$

$$x_2 \approx \frac{Q_s}{\gamma x_1 x_3} \tag{24.65}$$

where the approximation to x_2 is evaluated last. How well does it work? Using the algorithm of Section A3.18 with the choices $E = \mu$, $\Phi = 4.0$ eV, $F = 2$ eV/nm, $\gamma = 0.22$ eV/nm^2, and $K_s = 8$ (in the algorithm, $Q_s = Q/K_s$), the first six iterations are in Table 24.2, and the convergence is rapid indeed, being adequate by $n = 4$ even though γ is rather large, as suggested in Figure 24.15. Notice also in Table 24.2 that rapid convergence of x_2 is a consequence of the rapid convergence of x_1 and x_3, and that x_1 is closely approximated by x_-.

With a method to obtain the roots, the quadratic Gamow factor $\theta(E)$ in Eq. (11.14) can be approximated. Remembering that $V_o = \mu + \Phi - E$,

$$\theta(E) = 2\frac{\sqrt{2m_n\gamma}}{\hbar}\int_{x_1}^{x_2}\left\{\frac{(x-x_1)(x_2-x)(x_3-x)}{x}\right\}^{1/2} dx \tag{24.66}$$

The length scale that matters most is that between the first two zeros, so let $x = x_1 + L\sin^2(s)$ where $L \equiv x_2 - x_1$. The substitution results in [235]

$$\theta(E) = 4L^2\frac{\sqrt{2m_n\gamma}}{\hbar}\int_0^{\pi/2}\left[\frac{L\cos^2\varphi + x_3 - x_2}{L\sin^2\varphi + x_1}\right]^{1/2}\cos^2\varphi\,\sin^2\varphi\, d\varphi \tag{24.67}$$

Table 24.2 Quadratic image barrier roots. Iterative solution for x_j as given by Eq. (24.62) indexed by n using the algorithm of Section A3.18 with its default parameters ($V_o = 4.0$ eV, $F = 2$ eV/nm, $\gamma = 0.22$ eV/nm^2, $K_s = 8$). The values for $n = 1$ are from Eq. (24.63).

n	x_1	x_2	x_3
1	0.0113137	2.42554	7.45359
2	0.0113136	2.88460	6.43117
3	0.0113136	2.94774	6.15451
4	0.0113136	2.94899	6.13078
5	0.0113136	2.94899	6.13061
6	0.0113136	2.94899	6.13061

Although x_1/L is small, $(x_3 - x_2)/L$ is generally large for smallish γ, and so the radical of the numerator in the integrand may be expanded as a series in $\cos^2\varphi$. If the function $R_n(p)$ is defined by

$$R_n(p) \equiv \int_0^{\pi/2} \frac{\cos^{2+n}\varphi \sin^2\varphi}{\sqrt{p+\sin^2\varphi}} d\varphi \tag{24.68}$$

then

$$\theta(E) = \frac{4}{\hbar}\sqrt{2\gamma m_n L^3(x_3 - x_2)}\left\{ R_2\left(\frac{x_1}{L}\right) + \frac{L}{2(x_3 - x_2)} R_4\left(\frac{x_1}{L}\right)\right\} \tag{24.69}$$

For small p, the $R_n(p)$ functions can be replaced by $R_2(0) = 1/3$ and $R_4(0) = 1/5$, as can be shown using elementary trigonometric integrations. As a result,

$$\theta(E) = \frac{2}{15\hbar}\sqrt{2\gamma m_n L^3}\left\{ \frac{10x_3 - 7x_2 - 3x_1}{\sqrt{x_3 - x_2}} \right\} \tag{24.70}$$

where the dependence on E is through the dependence of the x_j on $V_o = \mu + \Phi - E$.

A scaling argument can be obtained. In the limit that $x_3 \to \infty$, $\gamma \to 0$, and $Q_s \to 0$, the triangular barrier θ_{tri} of Eq. (13.5) is recovered: small γ is synonymous with large $x_3 \approx F/\gamma$ (as can be shown by substituting $x_\pm$ in for x_2 and x_1 in Eq. (24.65)), and the neglect of x_2 and x_1 by comparison to x_3. Thus

$$\lim_{\gamma \to 0} \frac{\theta}{\theta_{tri}} = \frac{1}{v(y_s)}\left(\frac{FL}{V_o}\right)^{3/2} \tag{24.71}$$

where $y_s = \sqrt{4Q_s F}$. If both Q_s and γ vanish, then $L \to L_o = V_o/F$ and $v(0) = 1$, showing that $\theta/\theta_{tri} \to 1$. For small γ, therefore, $\theta(E)$ increases as $(L/L_o)^{3/2}$, where $L_o = V_o/F$ is the triangular barrier width.

CHAPTER 25

Contacts, conduction, and current

You see a number of great objects, and when you look at them there seems to you to be one and the same idea (or nature) in them all ...
– Plato[1]

25.1 Zener breakdown

In 1934, the development of the energy-band theory of solids prompted Zener to propose interband tunneling, or internal field emission, as an explanation for dielectric breakdown. ...he showed that an energy gap could be treated in the manner of a potential barrier.
– Leo Esaki [254]

If the band bending at the interface is sufficiently strong, and the band gap sufficiently small, then electrons can tunnel directly through the band gap from the valence band, leaving a hole behind, a circumstance shown in Figure 24.13. The semiconductor is assumed to be n-type – were it p-type, then the charge carrier would be holes. Such a process of tunneling current through the interface is more common with p–n junctions rather than metal–semiconductor interfaces. The strong similarity of the tunneling barrier to Fowler–Nordheim field emission means that the tunneling current can be almost immediately written down after three modifications: (*i*) the height of the barrier Φ is now the band gap, (*ii*) the mass m is now the effective mass m_n, and (*iii*) the field is the slope of the band bending at the interface. The energy difference is just the band gap energy $\Phi \to E_g$. The field at the interface is, from Eq. (24.49), $F = 16\pi Qw/K_s$, where from Eq. (16.4), $w = K_s V_d / 8\pi Q N_d$, and where V_d is the amount $V(x)$ changes over the depletion width. The tunneling current equation for Zener breakdown is then most easily given by modifying Eq. (13.12) to get [152]

$$J_{zener}(F) = \frac{qF_o^2}{16\pi^2 \hbar E_g t(y_z)^2} \exp\left(-\frac{4}{3\hbar F_o} \sqrt{2m_n E_g^3}\, v(y_z) \right) \tag{25.1}$$

where $F_o = (32\pi Q N_d V_d / K_s)^{1/2}$, $y_z^2 = 4QF_o/K_s E_g$ and m_n instead of m is used. Lastly, when there is no bias then V_d is the difference between the left side (metal) and right side (semiconductor) where the "work function" is measured with respect to the Fermi level (not the bottom of the conduction band). When a positive or negative bias V_b is applied to the semiconductor, the location of the conduction band minimum changes, but this is not so for the metal. Hence, under bias, $V_d \to V_d + V_b$ in F_o. If $V_b = V_o \cos(\omega t)$ such that $V_d > V_o$, then the current J_{zener} would act as a rectifier by only appreciably passing current in one direction.

25.2 Poole–Frenkel transport

The fact that the illumination of an electronic semiconductor results in an additional increase of the conductivity independent of $\mathcal{E}$ shows that the increase of electrical conductivity in intense fields is due to the increase of the number of free electrons, and not of their mobility ...In an external field $\mathcal{E}$ this energy is further decreased by a mechanism similar to that of the Schottky effect in the thermoelectronic emission from metals.
– Jakov Ilich Frenkel [332]

Charged inclusions into the crystal structure of semiconductors like silicon [333], $Al_xGa_{1-x}As$:Te, and Ge:Hg [334], or undopted chemical vapor deposition (CVD) diamond [335] can mimic the ionic potentials considered in Section 20 on

[1] Plato, *Parmenides* (trans. Benjamin Jowett). *Great Books of the Western World*, ed. Robert Maynard Hutchins, Vol. 7, Plato. Chicago: Encyclopaedia Britannica, 1952, p. 489.

Introduction to the Physics of Electron Emission, First Edition. Kevin L. Jensen.

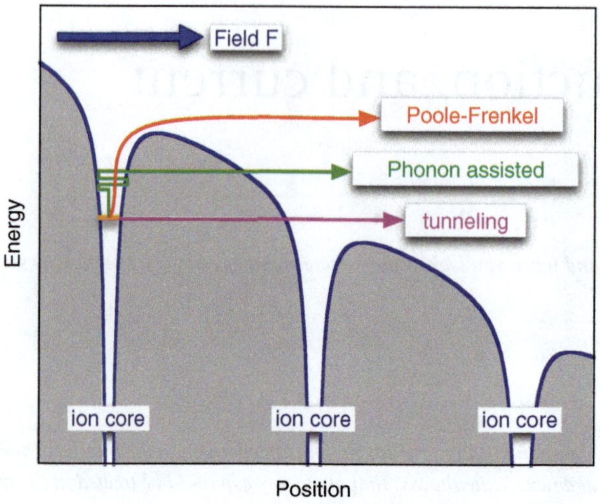

Figure 25.1 Various mechanisms by which an electron attached to a charged impurity can escape by "hopping" from core to core: thermal-like emission over the barrier (Poole–Frenkel), tunneling through the barrier (direct tunneling), and interactions with lattice vibrations until tunneling occurs (phonon-assisted tunneling) (based on Figure 1 of ref. [334]). The line is modeled after "32" of Figure 25.2.

breakdown. When a sufficient number of charged inclusions[2] are present, then electrons can "hop" from one ion core potential to another by a variety of means suggested in Figure 25.1: thermally assisted electron hopping (Poole–Frenkel emission), direct tunneling, or interactions with lattice vibrations (phonon-assisted tunneling) which temporarily excite the electron energy above the rim of the ion potential barrier.

The type of mechanism responsible for the hopping leaves its dirty little fingerprints on the behavior of the current–voltage (IV) relationships that arise. Start by considering emission from the CVD diamond films grown and examined by Göhl *et al.* [335]: if the metal–insulator barrier (perhaps somewhat like that shown in Figure 24.14) dictated the behavior of the electron emission from thin (3.2–27.5 μm) films deposited onto a metal, then the current as a function of bias voltage should have behaved as $\ln(I/V^2)$ and $1/V$. That was not observed. Rather, the current was better explained by a Poole–Frenkel conduction mechanism. It is common (but not necessarily correct or even a good idea) to represent the ion cores as being equi-spaced so that the potential along the $\hat{z}$ coordinate normal to the interface (or surface) is modeled by (in cylindrical coordinates (ρ, z))

$$V(\rho, z) = V_o - Fz - \frac{4Q}{K_s} \sum_{j=-N}^{N} \frac{e^{-kr_j}}{r_j} \tag{25.2}$$

$$r_j = \{\rho^2 + (ja - z)^2\}^{1/2} \tag{25.3}$$

where k is the screening constant familiar from Section 21.7, and a is the average separation between the trap centers (e.g., the representative trap density of $\rho_i = 7.1 \times 10^{17}$ cm^{-3} $= a^{-3}$ suggested in ref. [335] for CVD diamond corresponds to $a = 5.2$ nm). Choosing the $\hat{z}$ origin in the center of the layer is convenient for some of the numerics below, so that there are N traps to either side of the origin. And lastly, about that cylindrical coordinate factor of ρ: it would be zero if the $\hat{z}$ axis passed directly through the trap core centers, but for graphing purposes it does not hurt to let ρ be arbitrarily small but finite so that singularities are not encountered.

If the energy levels of the trap centers are such that the barriers that electrons have to cross are high and wide, then hopping will have a thermionic-like (over the barrier) characteristic, and current will then be the difference between left-hopping and right-hopping electrons. In the absence of a field, these two processes are equal, but when a field is present, as in Figure 25.2, then the barriers to the right are slightly lower than those to the left, so that a net current is possible. Assuming the occupation of the energy levels conforms to a Maxwell–Boltzmann distribution, and that the activation energy (amount of energy the electron must acquire to escape the well when the field F is zero) is E_a, then the energy difference between the right and left barriers confining the electron to the core can be calculated. Let the

[2] Charged inclusions make the discussion easier by using terminology familiar to those who have endured earlier sections and "ion cores", but effects can also occur for neutral inclusions.

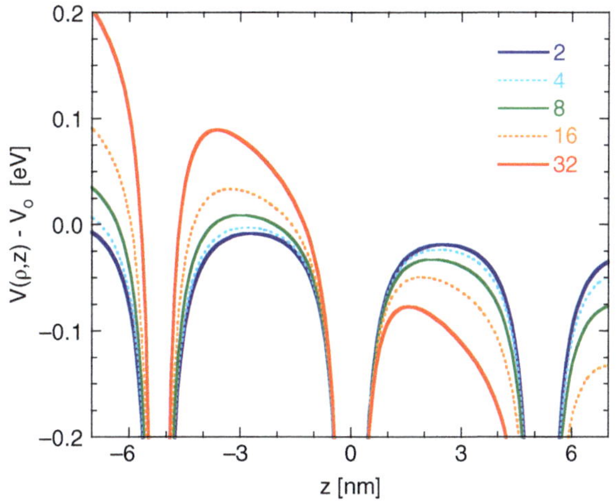

Figure 25.2 The ion core potentials for diamond-like parameters suggested by ref. [335] ($K_s = 5.7$, core-to-core separation = 5.2 nm, corresponding to a trap density of about 7.1×10^{18} cm^{-3}, and screening $k = 0.769$ nm^{-1}) for various representative (albeit low) fields in units of [eV/μm]) calculated according to Eq. (25.3).

location of the barrier maximum be a distance l to the right of the core center (location of the hump in Figure 25.2): the presumed simplicity of Eq. (25.3) then indicates that the energy difference is

$$\Delta\Phi = V(\rho, -a + l) - V(\rho, l) = Fa + \frac{2Q}{K_s}\left(\frac{e^{-kr_L}}{r_L} - \frac{e^{-kr_R}}{r_R}\right) \tag{25.4}$$

where $r_R^2 = \rho^2 + (Na + l)^2$ and $r_L^2 = \rho^2 + ((N+1)a - l)^2$. In either the limit is that N becomes large or the field is weak so that $l \to a/2$, then $r_L \approx r_R$, leading to $\Delta\Phi \to Fa$. Although N does not need to be that large (Figure 25.2 does not visually change much if N is changed from 600 to 6 for the parameters of the figure) saying it is has other uses: if the number of trap centers to the left and right differ, then N_L does not need to even equal N_R if both are large for the correction term to Fa to be small. The point is, for an electron in the center trap, the barrier to the left appears to be $E_a + \Delta\Phi/2$ and the barrier to the right appears to be $E_a - \Delta\Phi/2$: the difference between the left and right barrier heights gives rise to a net hopping current. Of course, such an aesthetic model presumes too much by pretending the trap centers are uniformly spaced, but it allows for a simple understanding and a means of contrasting Poole–Frenkel (or thermal) hopping here with tunneling hopping next.

Borrowing from methods familiar from derivation of the RLD equation in Section 12, the occupation of the trap states is taken to be Maxwell–Boltzmann distributed, and the current, such as it is,[3] is most easily approached in cylindrical coordinates (k_ρ, k_z), so that

$$J_{pf} = \frac{2q}{(2\pi)^3} N_c \int_0^\infty 2\pi k_\rho dk_\rho e^{-\beta E_\rho} \int_0^\infty \frac{\hbar k_z}{m} dk_z D(E_z) e^{-\beta E_\rho} \tag{25.5}$$

where the total energy $E = E_\rho + E_z = (\hbar^2/2m)(k_\rho^2 + k_z^2)$, N_c is a normalization constant that sweeps up any straggling material-dependent parameters (it can, for example, be found by specifying the number density, an integral resembling J_{pf} apart from a factor of q and $\hbar k_z/m$), and the integrals in cylindrical coordinates fall into easily separated terms that can be integrated parts by virtue of the Maxwell–Boltzmann distribution. The first integral is straightforward:

$$\int_0^\infty 2\pi k_\rho dk_\rho e^{-\beta E_\rho} = \frac{2\pi m}{\beta \hbar^2} \tag{25.6}$$

When the transmission probability $D(E)$ is a step function, the second integral is almost as easy. But now the current traveling to the left must be subtracted from that traveling to the right, and so

$$\int_0^\infty \frac{\hbar k_z}{m} dk_z \{D_R(E_z) - D_L(E_z)\} e^{-\beta E_\rho} = \frac{2\pi m}{\beta \hbar^2}\left(e^{-\beta\Phi_r} - e^{-\beta\Phi_l}\right) \tag{25.7}$$

[3]It is not current in the usual sense: rather, the electrons occupy various energy levels within the traps due to their interactions with the thermal agitations of the host lattice. It would therefore be better to speak of a collision frequency and an escape probability for a charge q in a well, but if one does so, the concept of flux, and therefore current, is not far behind anyway.

where the shorthand $\Phi_{l,r} = E_a \pm \Delta\Phi/2$ is used. Invoking the definition of the sinh function and combining all terms,

$$J_{pf} = \frac{2qm^2}{\pi\hbar^3} N_c (k_B T)^2 e^{-\beta E_a} \sinh(\beta F a/2) \tag{25.8}$$

Observe that the quantity $(J_{pf}/T^2)e^{\beta E_a}$ is a function of βF and *therefore should have the same behavior for different temperatures,*[4] that is, a plot of $(J_{pf}/T^2)e^{\beta E_a}$ vs βE_a for different temperatures will lie along the same line.

Finding how such a prediction plays out is pleasing to behold: consider, therefore, the measurements of Göhl *et al.*, who examined undoped very thin diamond films (3.2–27.5 μm) grown epitaxially on silicon and measured using a field emission scanning microscope (doing so allowed the examination of μm-scale areas). High (354.2 K) and low (300.1 K) temperature measurements of the current, for which data extracted from Figure 4 of ref. [335] are shown in Figure 25.3 (intermediate measurements were also presented, but the argument is adequately made with the limiting cases). For the $\hat{x}$ axis, convert the field in the diamond to units of [eV/nm] by multiplication of 0.001 to bring [1/μm] to [1/nm]. For the $y(x)$ axis, construct $y = \ln(J_{pf}/T^2) + \beta E_a$, where E_a is chosen so as to make the lines overlap as in Figure 25.4, (the value found, $E_a = 0.43$ eV, is the same as that found in ref. [335]). The scaled data then overlaps as suggested

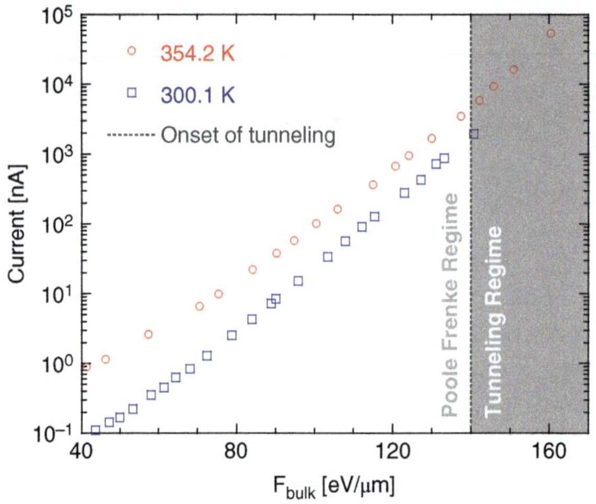

Figure 25.3 Current vs field for a 6.7 μm diamond sample for T = 354.2 K (red circles) and 300.1 K (blue squares) based on Figure 4 of ref. [335].

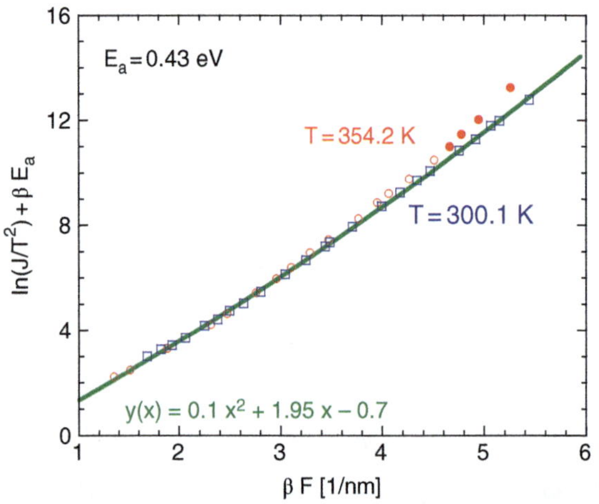

Figure 25.4 The data of Figure 25.3 recast by subtracting βE_a from $\ln(J/T^2)$ and adjusting E_a until lines overlap, based on Figure 5 of ref. [335]. A value of E_a = 0.43 eV is found to give good overlap. Units of the $\hat{x}$ axis are [1/nm] for F in [eV/nm]. Solid red circles are those points in Figure 25.3 in the gray shaded area signaling the onset of tunneling.

[4]Compare Eq. (2) of ref. [335] and the discussion surrounding it.

by Eq. (25.8). The deviation of the higher points of the 354.2 K data set are attributed to the onset of a tunneling contribution, where the potential barrier has become sufficiently thin that electrons can tunnel out of the traps in addition to jumping over them. This allowed Göhl *et al.* to conclude that field emission from thick undoped CVD diamond films was mediated by the Poole–Frenkel mechanism of conduction rather than a Fowler–Nordheim-like current through an interface barrier (between the substrate and diamond or between the diamond and vacuum), as had been speculated prior to their demonstration [335].

25.3 Tunneling conduction

After much effort I succeeded in "proving" this theorem on the basis of the similarity of triangles ...

– Albert Einstein[5]

The high field data of Figure 25.3 showed evidence of a tunneling contribution when the internal field is sufficiently strong, wherein the electrons may start passing through the barrier in addition to hopping over it. Tunneling transport has a different signature than over the barrier (or Poole–Frenkel) transport, and therefore monitoring the dependence of the emission rate on F shows it. The average distance of 5.2 nm between the trap centers for the diamond studies of Göhl *et al.*, was fairly small. Compare that to Ganichev *et al.* [334], who examined the emission rate associated with several deep-defects in semiconductors, for which charged mercury atoms in germanium (or Ge:Hg) at a concentration of $10^{14} - 10^{15}$ cm^{-3} is closer to an average separation of 100–200 nm. Under such conditions, the Coulomb potential $2Qe^{-kr}/K_s r$ of a trap is similar to the familiar Q/x potential associated with the image charge barrier, in the presence of which a field acts to reduce the barrier by a factor $\sqrt{4QF}$. Consequently, the ionization probability $w(F)$ of the trap goes as the degree to which the barrier is reduced, or for small fields,

$$\ln\left\{\frac{w(F)}{w(0)}\right\} \propto F^{1/2} \tag{25.9}$$

as shown in Figure 25.5. Frenkel [332] estimated the coefficient of proportionality using a method similar to that which gives the Schottky factor $\sqrt{4QF}$ for the image charge barrier encountered in Eq. (11.30). Paraphrased, it suggests the potential near the ion behaves something like

$$V(r,\theta) = V_o - Fr\cos\theta - \frac{2Q}{K_s r}\exp(-kr) \tag{25.10}$$

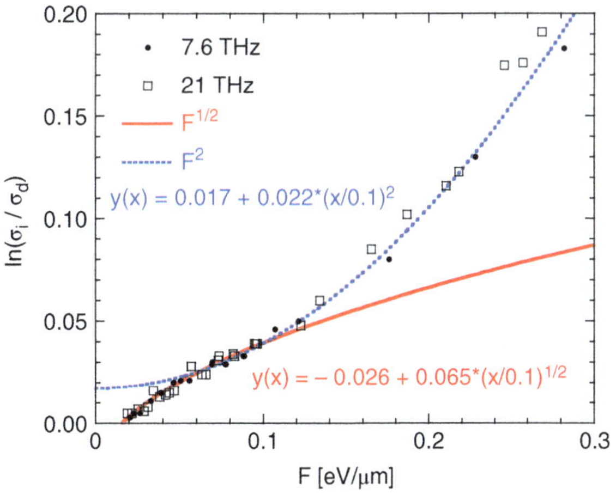

Figure 25.5 Variation of the log of the irradiated and dark conductivities (which are proportional to the emission rates $w(F)$) measured at two frequencies of the electric field. The low-field behavior goes as $\sqrt{F}$ and is indicative of Poole–Frenkel transport. The high-field behavior goes as F^2 and is indicative of phonon-assisted tunneling. Based on Figure 4 of ref. [334].

[5]Einstein was speaking in reference to proving the Pythagorean theorem when he was 12 years old, as quoted in W. Isaacson, *Einstein: His Life and Universe*, Simon & Schuster, New York, 2008, p. 17. The popular canard that Einstein failed math as a young student is hokum: he always received top grades in the subject, and before he was 15 he had mastered calculus.

where K_s is the dielectric constant, k is the screening factor, and V_o is a constant. The maximum of the barrier radially away from the core center is then found by $\partial_r V = 0$: it is smallest in the forward direction $\theta \to 0$ and, if the screening factor is small, then the maximum lies at $r_o = (2Q/K_s F)^{1/2}$. At that point, therefore, the potential has dropped by an amount

$$\Delta V = 2\left(\frac{2QF}{K_s}\right)^{1/2} \tag{25.11}$$

The number of free electrons due to thermal ionization in the absence of a field goes as $\exp(-\beta V_o/2)$, and is associated with a conductivity of σ_o. If the barrier lowers by ΔV, then the conductivity increases to $\sigma = \sigma_o \exp(\beta \Delta V/2)$, and so Eq. (25.9) is suggested to really be

$$\ln\left\{\frac{w(F)}{w(0)}\right\} \approx \beta\left(\frac{2QF}{K_s}\right)^{1/2} \tag{25.12}$$

The relation is about a factor of 2 off, a consequence of the multidimensional nature of the emission of the carriers from the traps [334]. The power (1/2) that F is raised to, though, is seen in the low field regime of Figure 25.5.

The high-field behavior of the same figure is another matter: the change of the power from $p = 1/2$ to something bigger is a signal that over-the-barrier transport is yielding to phonon-assisted tunneling in Figure 25.1. Why that "something bigger" power is somewhere between $1 \leq p < 2$ can be understood by an argument similar to Albert Einstein's youthful and innovative proof of the Pythagorean theorem when he was entering his teen years. Consider the right triangles shown in Figure 25.6. The area of a right triangle is proportional to the square of its hypotenuse (if the hypotenuse doubles, then the area increases fourfold, for example), or $C = \alpha c^2$, where α is the constant of proportionality.[6] But the triangles A and B have the same constant of proportionality, or $A = \alpha a^2$ and $B = \alpha b^2$. Therefore, because $A + B = C$, it follows $a^2 + b^2 = c^2$. With a little bit of contemplation, one sees that if each of the legs a and b scale with respect to a third factor x, then the area of C scales with x^2, but a can be called the "height" of C, and b its "length". In short, scaling arguments and powers can be related.

Now return to Figure 25.2 with an eye towards understanding how the Gamow factors associated with them will behave: compared to the almost field-free case of the dark blue line (2), the higher field lines look increasingly triangular, such as the red line (32). If tunneling is occurring from one well region to the next, then the Gamow factor θ associated with the tunneling is going to mimic that behavior. Scaling, as in Section 11.2, suggests θ goes as the product of a shape factor, a height κ_o (where the integrand tops out), and a length L (the distance between the integration limits of the Gamow factor), the latter two of which are field dependent. Therefore, the difference of the Gamow factors $\theta_o - \theta$ (where θ_o is the field-free case) will behave approximately as F when the field is weaker and $L \approx a$. The question then becomes, how does it change when the field is strong enough so that L reduces as F increases?

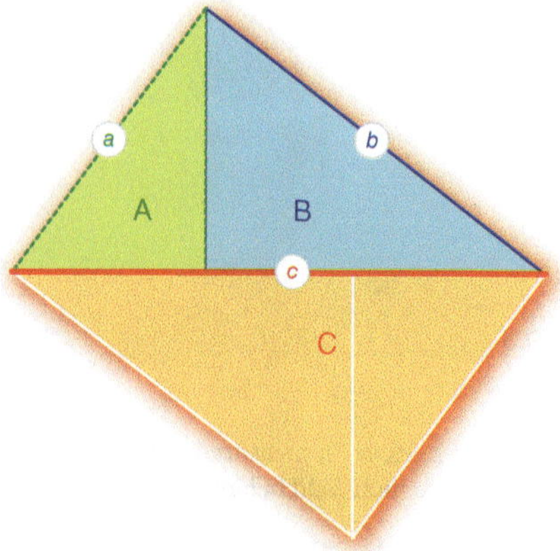

Figure 25.6 The triangles A, B, and C are right triangles, with hypotenuses a, b, and c, respectively. The area of triangle C is clearly the sum of the areas $A + B$ (being the lower half of a rectangle with a diagonal of c).

[6] If ϕ is the smallest interior angle of the right triangle, then $\alpha = \sin(2\phi)/2$, as can be shown by elementary means, but the value of α is not important for the argument. The same factor is found for the triangles A and B, because the triangles are similar.

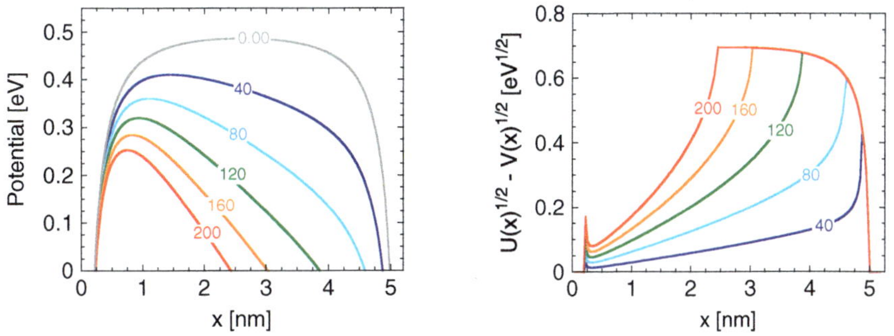

Figure 25.7 Behavior of the integrands of the Gamow factor $\theta(E)$ and $\theta_o(E)$. Left, gray (0.00) is $U(x)$; color is for $V(x)$ for increasing fields (in jumps of 40 eV/μm). Right, the difference $\sqrt{U(x)} - \sqrt{V(x)}$ for the color lines of the left. For "40" small increases in F increases the height, but by "80", increases in F affect both height and length.

Using arguments similar to those leading to Eq. (25.4) via Eq. (25.3), let $U(\vec{r})$ be the same as $V(\vec{r})$ in Eq. (25.3) but with $F = 0$, or

$$U(\rho, z) = V_o - \frac{4Q}{K_s} \sum_{j=-N}^{N} \frac{e^{-kr_j}}{r_j} \tag{25.13}$$

$$r_j = \left\{\rho^2 + (ja - z)^2\right\}^{1/2} \tag{25.14}$$

Thus, it will be that θ_o is evaluated using $U(\vec{r})$, and θ using $V(\vec{r})$: the integrands of θ_o and θ are shown on the left of Figure 25.7, with the gray line (labeled "0.00") being the line associated with U and the color lines being those associated with V at ever higher fields (the labeling being in [eV/μm]). For circumstances in which the transmission probability behaves as $D(E) \approx \exp(-\theta(E))$, then the increased rate of jumping will go as the ratio of the transmission probabilities (recall Eq. (25.9))

$$\frac{w(F)}{w(0)} \sim \frac{D(E)}{D_o(E)} \approx D_o(E) \exp\{\theta_o(E) - \theta(E)\} \tag{25.15}$$

where E is understood as the energy of the level in the Coulomb potential region such that the activation energy (distance from E to the top of the barrier) is equivalent to $E_a = V_o$ in the present treatment. Assuming the cores line up just so, so that $\rho \to 0$ makes the potential along z look like Figure 25.2, then the problem has again the look and feel of a one-dimensional potential barrier problem, and so the indolent urge to go back to speaking of $U(x)$ and $V(x)$ can be indulged. The argument of the exponent becomes

$$\theta_o - \theta = 2\frac{\sqrt{2m}}{\hbar} \int_{x_-}^{x_+} \left(\sqrt{U(x)} - \sqrt{V(x)}\right) dx \tag{25.16}$$

Rewrite the integral as

$$\int_{x_-}^{x_+} \left(\sqrt{U(x)} - \sqrt{V(x)}\right) dx = FL \int_0^1 \frac{s}{\sqrt{U[x(s)]} + \sqrt{V[x(s)]}} ds \tag{25.17}$$

by virtue of the definitions of U and V, where $x = x_- + sL$ and $L = x_+ - x_-$ is the length of the integration region over which the integrand is positive, an approach suggested by the shape factor approach of Section 11.2. If $L \sim F^p$, then $\theta_o - \theta \sim F^{p+1}$, but given the quirky behavior demonstrated in Figure 25.7, a numerical approach teases out the behavior more clearly.

Consider parameters, therefore, that are (nominally) motivated by the experimental data of Figure 25.5 in the case of mercury (Hg) in germanium (Ge). Evaluate Eq. (25.17) for the values $a = 100$ nm and $K_s = 16.2$ (note that such a large value of a means that N in Eq. (25.14) can be kept small[7]). Also let $V_o = 0.3$ eV and $k = 0.769$ 1/nm, For low fields, something like 0.1 eV/μm, then the difference between θ_o and θ (as seen in Figure 25.8) will be slight: $L \approx a$ and the height of the integrand will go as $V_o - \sqrt{16QF/K_s}$, making $\Delta\theta = \theta_0 - \theta \sim F^{1/2}$. At higher fields, something like 3 eV/μm, both the height and length of the integrand exhibit field dependence, and so a quadratic nature to $\Delta\theta$ becomes evident.

[7] $N \to 0$ is not bad, although $N = 5$ is used for the figures.

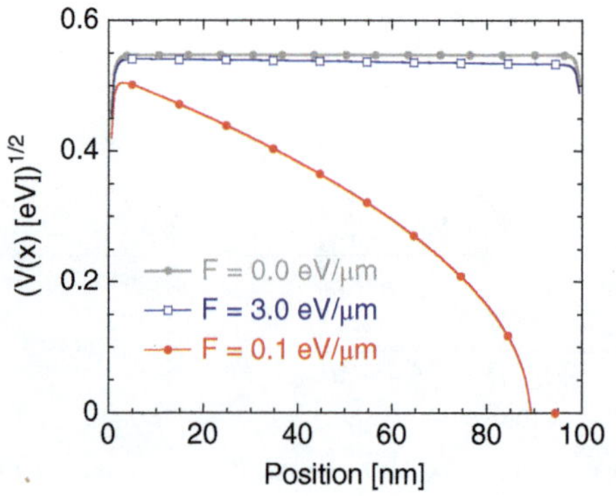

Figure 25.8 The terms of the integrand of θ (ignoring factors of m_n and $\hbar$) for zero, low, and high fields for $a = 100$ nm, $K_s = 16.2$ (after germanium), $V_o = \chi - E = 0.3$ eV, and screening factor $k = 0.769$ 1/nm

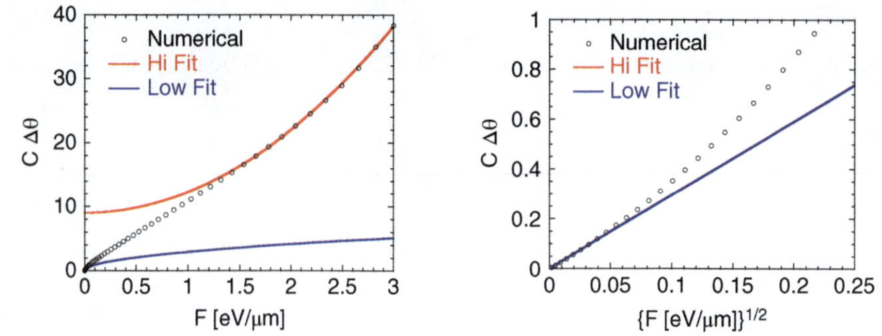

Figure 25.9 The integrand of $\Delta\theta = \theta_o - \theta$ as a function of field for the parameters of Figure 25.8 for $C = \hbar/2\sqrt{2m_n}$ to remove the dependence on effective mass and $\hbar$ (the result is simply the integration of $\sqrt{U} - \sqrt{V}$). The right-hand figure shows the low-field behavior vs $\sqrt{F}$. The "Hi Fit" is given by $9.034 + 29.18F^2$; the "Low Fit" is given by $3.165F^{1/2}$.

These expectations are clearly seen in the numerical evaluation shown in Figure 25.9 when the numerical evaluations of Eq. (25.17) are compared to $9.034 + 29.18F^2$ for the high-field fit, and $3.165F^{1/2}$ for the low-field fit (the fits being found numerically). Observe how rapidly the low fit diverges from the numerical values as the field dependence of L comes into play as suggested by the triangle arguments near Figure 25.6. Similarly, the quadratic nature of the high-field behavior does not continue unabated: it only holds until F increases to the point that θ becomes small.

Naturally, the problem is even more vigorously complicated by the behavior of the energy levels in the wells, and their interactions with the lattice vibrations that are always present – ref. [333] gives a flavor of how rapidly the complexity escalates when physics is added. But simple models motivated by methods familiar from the general thermal–field models of Chapter 17 lead to expectations that are approximately and pleasantly realized.

25.4 Resonant tunneling in field emission

…the stupider one is, the closer one is to reality. The stupider one is, the clearer one is. Stupidity is brief and artless, while intelligence wriggles and hides itself.

– Fyodor Dostoyevsky[8]

For metal to semiconductor interfaces, following the thermal hopping of Poole–Frenkel and the tunneling of hopping conduction, one last configuration invites examination: resonant tunneling as through a SiO_2 layer atop a metal (or semiconductor) or, even more interesting, an SiO_2–Si–SiO_2 sequence of layers that creates a well [259] as schematically shown in Figure 25.10. There are similarities to the resonant-like and hopping-like effects of atoms atop field emitters [258] or charged centers in the oxides which cover them [336]. Such territory is familiar from Chapter 20 on the Paschen curve, or Section 19.2.4 on resonant tunneling diodes (RTDs): insofar as the rightmost configuration of Figure 25.10 resembles

[8]Fyodor Dostoyevsky, *The Brothers Karamazov* (trans. Constance Garnett), *Great Books of the Western World*, ed. Robert Maynard Hutchins, Vol. 52, Dostoyevsky. Chicago: Encyclopaedia Britannica, 1952, p. 121.

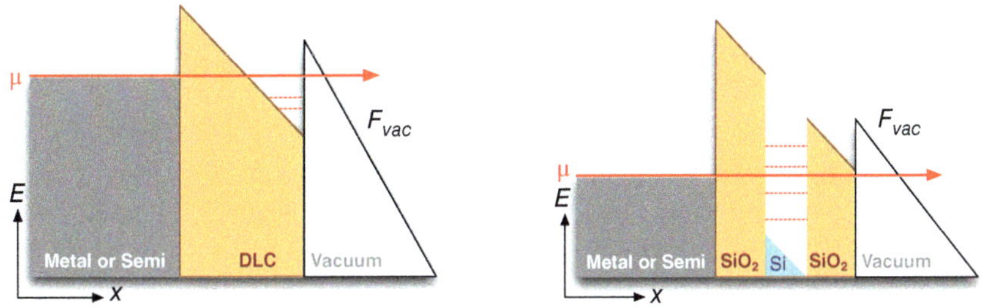

Figure 25.10 Two types of resonant effects in transmission through a thin diamond-like carbon film (DLC, left) or through a resonant tunneling δ-doped layer (SiO_2–Si-SiO_2–, right), also called a multilayer cathode (MLC). Dashed red lines indicate resonant levels. Based on Figures 1 and 2 of ref. [259].

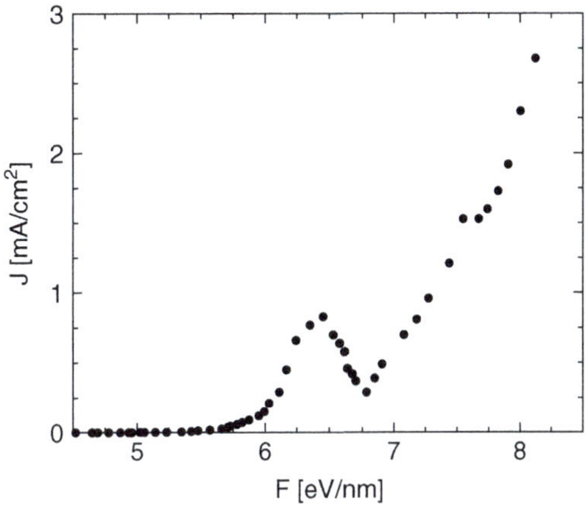

Figure 25.11 Measured current versus field through a multilayer cathode (MLC) structure (see Figure 25.10), based on Figure 14a of ref. [259], for which the hump was attributed to resonant tunneling effects.

an RTD, Litovchenko *et al.* refer it as a *multilayer cathode* (MLC), for which the $J(F)$ relation shown in Figure 25.11 shows an odd bump in an otherwise exponentially increasing current density as the field increases. But there are differences between the resonant field emission problem and what has come before: the right-hand boundary is vacuum and subject to a constant field, and there are no incident electrons from it.

Familiar or not, a proper account of the problem [259] is challenged to tease out the influences of the resonant levels (suggested by the dashed red lines in Figure 25.10) on the emission current, complicated by the replacement of the metal by silicon because of the development of an accumulation region with levels itself. The resonant levels in both the accumulation region and the quantum well are related to zeros of the Airy function, the determination of which builds character but not necessarily enlightenment: the location of the energy levels matters less than how those energy levels shift as a field is applied until enhanced transport is possible. In fact, dogged calculation using the methods of Chapter 19 and fidelity to the actual potential obscures what should be important here.

Consider, therefore, a purposefully "foolish" model[9]. In that sense, "stupidity" as apparent foolishness is instructive in the sense of simpleminded that exposes the consequences of resonance on field emission current density through layered structures or thin films. The methods also apply to the other emission mechanisms if one wants to be shrewd about the supply function, but as experimental data from Figure 10 of ref. [259] will be considered, follow its lead. Though simple, the model becomes powerful because it is easily analyzed by exploiting much what has come before, enabling the behavior of interest to be teased out easily.

[9] (Germaine Greer, *Shakespeare*, Oxford University Press. Oxford. 1986. p91). The epithet "foolish" is nowadays treated as a synonym of "stupid" but (at least in the time of Shakespeare or Erasmus) it was not. Fools spoke truth, especially to power, without being tortured or executed in the many innovative ways of the time that the more intelligent were for speaking impoliticly: King Lear's fool was not particularly humorous, but he retained the King's favor after speaking truthfully whereas Lear's daughter Cordelia did not. Rather, a fool is a "natural", *simple* as we say, and by extension, still in a state of nature. Our word *silly* is descended from an Old English word meaning "blessed", surviving still in our idea of blissful idiots.

Place, therefore, a barrier before the triangular barrier of Fowler and Nordheim such that it has the same height, and let the bottom of the well be at the bottom of the conduction band, that is, at the zero of the energy scale. To keep matters *really* simple, ignore effective mass variations ($m_n = m$). Teasing out the resonances is going to be difficult unless some artifice is employed. As the resonances are what is under investigation, extract their behavior from $D(k)$ as calculated using the transfer matrix approach of Section 19.2 by first extracting the behavior of the triangular barrier by considering the ration $D(k)/D_{\triangle}(k)$ (where the denominator is given by Eq. (18.52)), and then remove the effects of the leading barrier by multiplying the ratio by $e^{\kappa x_b}$ following the methods of Section 19.2.4 (see also ref. [199]), or

$$G(E(k)) \equiv \exp[\kappa(k)x_b]\frac{D(k)}{D_{\triangle}(k)} \tag{25.18}$$

Defined thus, $G(E)$ rather effectively identifies the resonances of the quantum well: for parameters (similar to those of Litovchenko *et al.*) of a barrier width of $x_b = 0.5$ nm and a well width of $x_w = 1$ nm, as shown in to the left of Figure 25.12, increasing the field does remarkably little to budge the location of the maxima of $D(k)$, which correspond to the resonant levels in the well, as shown to the right of Figure 25.12. Therefore, the bumps and blemishes on the I-V curves that Litovchenko calculates and measures have to be due to some other effect, and that effect is the lowering of the well in response to field, just as one would (or should) expect from Figure 25.10.

In the simplest of modifications, then, let the well drop by a small amount in proportion to the field: a reasonable approximation is to take that amount to be how much the center of the well drops in response to the field if the field F/K_s across the surface layer is constant, where F is the vacuum field and the bulk material is metal. This drop is

$$V_w(F) = -\frac{F}{K_s}\left(x_b + \frac{x_w}{2}\right) \tag{25.19}$$

if the well depth is at 0 for $F = 0$. For the *ad hoc* choice of $K_s = 10$, then the modifications to $G(k)$ are shown in the left part of Figure 25.13. The energy levels for a square (flat bottom) well depend only on the depth of the well, and so the expectation is naturally that the resonant peaks will move in proportion to the field, as they do in the right part of Figure 25.13.

The resonant peaks follow the same Lorentzian behavior of Eq. (19.26) that RTDs do, naturally, so that the Lorentzian is modeled near the jth resonant peak as

$$G(E \approx E_j) \approx \frac{1}{\gamma^2(E - E_j)^2 + \varepsilon^2} \equiv \frac{\pi}{\varepsilon\gamma}\delta_{\varepsilon}[\gamma(E - E_j)] \tag{25.20}$$

Because the resonances slightly differ in height, a better correspondence with the simulation is achieved by allowing the γ to vary according to $\gamma_j = n_j/(aE_j)$, or

$$G(E) = \sum_{j=1}^{N} \frac{1}{\gamma_j^2(E - E_j)^2 + \varepsilon^2} \tag{25.21}$$

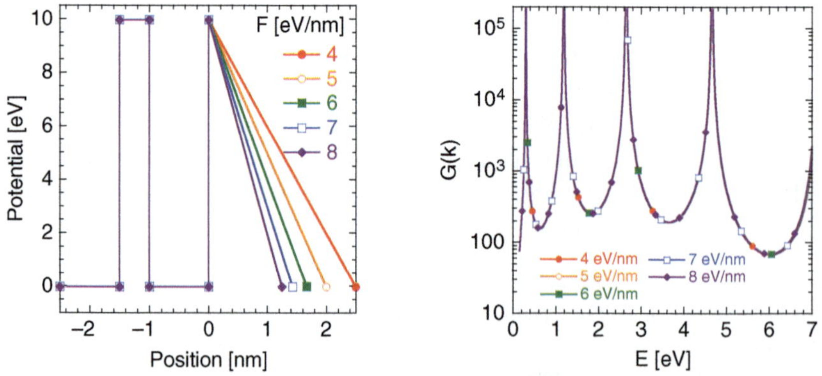

Figure 25.12 A simple model of resonance in field emission, in which the quantum well (left) is held at a fixed depth for a barrier of fixed height before a Fowler–Nordheim-like triangular barrier. All of the resonances (right) overlap, where $G(k)$ is defined in Eq. (25.18). The parameters of $V_b = V_o = 10$ eV, $x_w = 2x_b = 1$ nm, and $m_n = m$ are used, similar to but not the same as those in ref. [259]. Lines are labeled by the vacuum field F in [eV/nm].

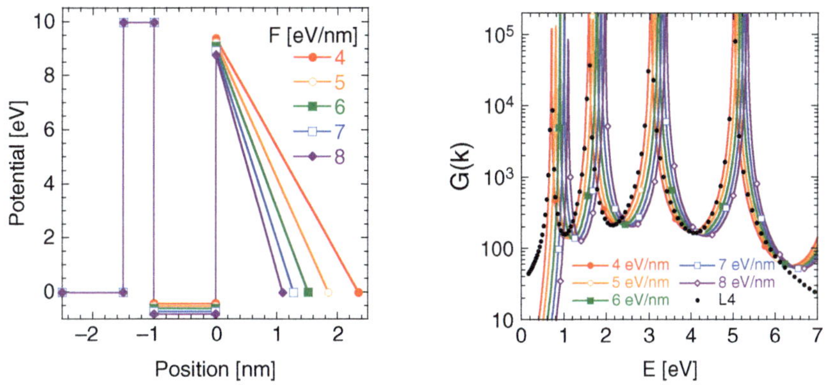

Figure 25.13 Same as Figure 25.12, but now the well depth is given by Eq. (25.19) for $K_s = 10$. Observe that the triangular barrier drops by the same amount. Now, the resonances are shifted by an amount that scales with the field. Lines are (still) labeled by the vacuum field F in [eV/nm]. Black dots represent the analytical model of Eq. (25.21).

where $N = 4$, $\{n_j\} = [1, 1, 1, 2]$, $\{E_j\}[\text{eV}] = [0.6949, 1.5749, 3.0238, 5.0042]$, $a = 3.2$, and $\varepsilon = 10^{-8}$, parameters which all depend on the specifics of the potential heights and well widths and where the values were found by appropriate means.[10]

The form of the Lorentzian of the resonant level of Eq. (25.20) gives a surprisingly good match to the numerical evaluation of $G(E)$ defined by Eq. (25.18), as suggested by the discussion in Section 19.2.4. Hence, the excellent fitting of the peaks of Figure 25.13 by Lorentzians is not entirely unexpected. The utility of Lorentzians $\delta_\varepsilon(x)$ is due to their behaving much like Dirac delta functions $\delta(x)$ when ε is small (Section A2.7). This allows the current density contributions from the resonant peaks to be almost immediately written down as

$$J_{res} = \frac{q}{\pi^2\hbar^3}\sum_{j=1}^{N}\left(\frac{\pi}{\varepsilon_j\gamma_j}\right) D(E_j)(\mu - E_j) \tag{25.22}$$

whereas the non-resonant part of $G(E)$ does not vary rapidly and so

$$J_{non-res} \approx G_o e^{-\kappa x_b} J_{FN}(F) \tag{25.23}$$

an equation anticipated by Eq. (19.20), with $G_o \sim 10 - 100$ as implied in Figure 25.13, and $\kappa = (2/\hbar)\sqrt{2m_n V_b}$. The total current is then the sum of the resonant and non-resonant parts.

The simplicity of the model constrains its predictions to being qualitative, but that is enough. The resonant current contributes substantially when a resonant level is accessible by the supply of electrons from the metal (or semiconductor accumulation layer) but when *as a consequence of increasing field* the resonant level moves outside the range of energies that overlap with the supply function, the current density consequently drops. The total current is then a combination of a steadily increasing non-resonant part with bumps occasioned by the contribution of the resonant parts as they move into and out of range, an understanding that matches Figure 25.11.

[10] In the present circumstance, "appropriate means" was visually or by hand. It does not have to be done thus, but since only a few E_j were required, doing so was simple. Sometimes eyeballing graphs is close enough.

CHAPTER 26

Electron density near barriers

Wiseman John: *As thou removest the Book from its cradle, you must recite these words. Clatoo, Verata, Nicto.*

Ash: *Clatto Verata Nicto. Okay.*

Wiseman John: *Repeat them ...*

Ash: *I got it. I got it. I know your damn words. All right?*

– The Army of Darkness[1]

Using words without understanding their significance invites trouble. Accepting that quantum mechanics describes electrons, but then ignoring their wave nature in treating the accumulation layer (as in Eq. (24.25)), causes the density at the barrier to be overestimated, and misses the effects of electron penetration into the barrier. As a result, errors include overpredicting the degree of band bending [326] or not quite accounting for the magnitude of the dipole caused by electron penetration into the barrier that modifies the work function [337, 338]. Admittedly, calling such consequences "trouble" is hyperbole, but sloppy talk misses effects. Fortunately, a few simple models capture the effects of density variation and dipole contributions to the barrier height quite well.

26.1 An infinite barrier

Compared with the classical case the electrons are repelled from the surface when quantum size-effects occur. A correct description of the spatial variations of the potential and the electron density in accumulation and inversion layers thus needs self-consistent solutions of Schrödinger's and Poisson's equations.

– Winfried Mönch [213], p. 30.

For atomically flat surfaces, an important concept is the electrical surface, which is "the plane where the field appears to start". In basic models the electrical surface coincides with the classical image-plane, but in principle the concepts are different.

– Richard Forbes [339]

That density variations that occur near a barrier are caused by the wave nature of the electron should not be a surprise: the variations were already evident in the simple models of Section 9.2.3. What is new is that the accumulation region is a well (recall Figure 24.13) with energy levels, and the wave function of those levels determines the behavior of density near the interface. The simplest model to compliment the step potential model of Section 9.2.3 is therefore to let the well region for $0 \le x \le L$ be at zero potential, and the walls to be infinitely high so that $\psi_k(0) = \psi_k(L) = 0$, so as to first appreciate the consequences of the contribution of discrete states.

The well with infinite barriers is one of the most familiar problems of quantum mechanics [43]. Let the sides of the well be such that the left side is at $x = -L$ and the right at $x = 0$. The usual $\psi_k(x) = e^{ikx} + r(k)e^{-ikx}$ can only vanish at $x = 0$ (i.e., $\psi(0) = 0$) if $r(k) = -1$ for all k. Similarly, $\psi_k(-L) = 0$ only if k is quantized such that $k_j = j\pi/L$ for integer j, and so the eigenstates are[2]

$$\psi_j(x) = \langle j|x\rangle = N \sin\left(\frac{j\pi x}{L}\right) \tag{26.1}$$

[1]Sam Raimi and Ivan Raimi, screenplay to *The Army of Darkness*, shooting script February 26, 1991. The phrase Ash speaks is a homage to the cautionary science fiction classic *The Day the Earth Stood Still* (1951) made at the dawn of the Cold War era. Although Ash's version is altered from the original "Klaatu barada nikto" and a good deal less serious, in both films getting the words *right* mattered.

[2]The usage of $\langle j|$ may seem odd, but the intent is not to create confusion with $\langle k_j|$. It is seen that $\langle j| = \langle k_j| - \langle -k_j|$.

Introduction to the Physics of Electron Emission, First Edition. Kevin L. Jensen.

where N is the normalization constant from the requirement that

$$1 = \int_{-L}^{0} |\langle j|x\rangle|^2 dx = N^2 \int_{-L}^{0} \sin^2\left(\frac{j\pi x}{L}\right) dx = \frac{1}{2} LN^2 \tag{26.2}$$

or $N = \sqrt{2/L}$. If C_j is the weight for the state $\psi_j(x)$, then density is

$$\rho = \sum_{j=1}^{N} C_j^2 |\langle j|x\rangle|^2 \tag{26.3}$$

written so because the C_j can be chosen to be real.

That this reproduces the results of Section 9.2.3 is easy to show by starting from the continuum formulation introduced there and continued in Section 11.1. There, for a metal surface at zero temperature, the density is found by summing over all momenta up to the maximum of k_F, and, for the moment, the wave nature of the electron is ignored, so that (recall Eq. (6.24))

$$\rho_o = \frac{2}{2\pi} \int_0^{k_F} f(k) dk = \frac{1}{2\pi^2} \int_0^{k_F} \left(k_F^2 - k^2\right) dk = \frac{k_F^3}{3\pi^2} \tag{26.4}$$

which is nothing more than Eq. (7.3) for a Fermi distribution. Next, the effects of inserting an infinite barrier at $x = 0$ result in setting the reflection coefficient to be $r = -1$ (no current is possible), and so $\psi_k(x) = 2^{-1/2}\left(e^{ikx} - e^{-ikx}\right) = i\sqrt{2}\sin(kx)$, which satisfies $\psi_k(0) = 0$. Now respecting the wave nature makes the replacement $f(k) \rightarrow f(k)|\psi_k(x)|^2$. That, plus using the identity $2\sin^2(kx) = 1 - \cos(2kx)$, results in the density $\rho(x)$ being given by

$$\rho_e = \frac{2}{2\pi^2} \int_0^{k_F} \left(k_F^2 - k^2\right)(1 - \cos(2kx))\, dk \equiv \rho_o\, \xi(2k_F x) \tag{26.5}$$

where the function $\xi(s)$ has been introduced and is [188]

$$\xi(s) \equiv \left\{1 + 3\frac{\cos s}{s^2} - 3\frac{\sin s}{s^3}\right\} \tag{26.6}$$

Because $s \equiv 2k_F x$ is dimensionless, many different density configurations give rise to the same oscillatory behavior near the walls, where $\rho/\rho_o \approx s^2/10$ for $s \ll 1$. The oscillations near the origin, seen in Figure 26.1 and all too familiar from metal surface studies [340, 341], are known as Friedel oscillations [85, 342, 343], and such oscillations also arise in how electron density screens charged inclusions [97] and vacancies [78].

Net charge, or the difference between the electron $\rho_e(x)$ and the background positive charge $\rho_i = \rho_o$, where the variation is only in x, corresponds to *sheets* of charge, and those sheets give rise to *fields*, as per Eq. (16.3). A foreboding that the undulations are going to complicate the determination of a self-consistent potential or change the surface barrier looms: right from the start, the origin for the electrons (where $x = 0$) cannot be where the background positive charge ends and vacuum begins if overall charge neutrality is required. That is, the absence of electrons right next to the barrier itself would lead to quite a slab of unbalanced positive charge there. Therefore, the edge of the background positive charge

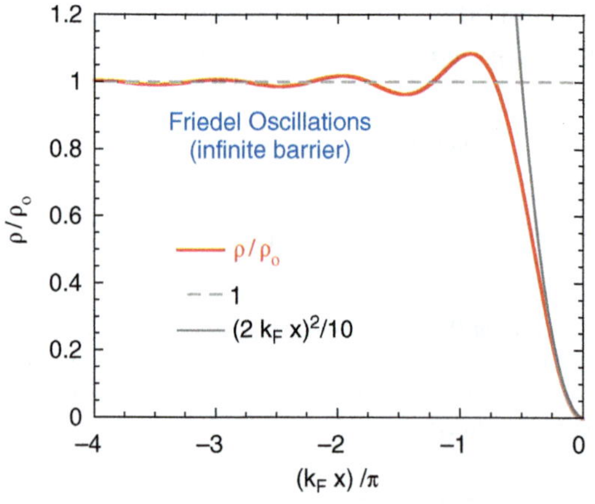

Figure 26.1 Friedel oscillations in the electron density ρ near an infinite barrier for $x > 0$, where $\rho_o \equiv k_F^3/3\pi^2$ and Eq. (26.5) is used where $s = 2k_F x$.

ends at $x = x_i < 0$. Where the electric field starts, and where background positive charge ends, need not be at the same spot, a simple statement masking a complicated circumstance [339, 344, 345], and affected by the degree of penetration of electron wave functions into the barrier.

A metal surface not exposed to a field will be uncharged, regardless of whether the electron density wiggles and the ion density does not. This can be expressed as

$$\int_{-\infty}^{0} \{\rho_e(x) - \rho_o \Theta(x - x_i)\} dx = 0 \tag{26.7}$$

where the Heaviside step function Θ term is attached to the background positive charge term. Invoking Eq. (26.6), the relation is alternately expressed as

$$\int_{-\infty}^{0} (\xi(s) - 1)\, ds = 2k_F x_i \tag{26.8}$$

As is typical, solving the integral on the left-hand side can be done a hard way or an easy way. The hard way is edifying for techniques, useful later, that it used, so consider it first and use series expansion methods [188]. The function $\xi(s)$ is equivalent to

$$\xi(s) = 1 + \frac{3}{s}\frac{d}{ds}\left\{\frac{\sin(s)}{s}\right\} \tag{26.9}$$

That is fairly convenient: it entails, for example, that for $s_j \equiv -2\pi j$, then[3]

$$\int_{s_j}^{s_{j+1}} s\,(\xi(s) - 1)\, ds = 0 \tag{26.10}$$

Such a relation suggests that the infinite integral be replaced by an infinite sum over integrals over a region where s varies by only 2π, or

$$\int_{-\infty}^{0} (\xi(s) - 1)\, ds = \sum_{j=0}^{\infty} \int_{0}^{-2\pi} \{\xi(s - 2\pi j) - 1\} ds \tag{26.11}$$

The $j = 0$ term is primary, being adjacent to the infinite barrier: call it N_P. The remainder will be called N_R. A quick integration by parts on each, followed by Taylor expanding the trigonometric functions, results in

$$\int_{-\infty}^{0} (\xi(s) - 1)\, ds = 3\left(N_P + N_R\right) \tag{26.12}$$

where (the form of the integration makes use of $\xi(s)$ being an even function)

$$N_P = \int_{0}^{2\pi} \left\{\frac{s\cos(s) - \sin(s)}{s^3}\right\} ds = \frac{1}{2\pi}\sum_{j=1}^{\infty} \frac{(-2\pi)^j}{(2j-1)!\left[4j^2 - 1\right]} \tag{26.13}$$

and

$$\begin{aligned} N_R &= -\sum_{j=1}^{\infty} \int_{-\pi}^{\pi} \frac{\sin(s)}{\left[s + (2j+1)\pi\right]^3} ds \\ &\approx \frac{\pi^4 - 96}{16\pi^3} + \frac{\left(\pi^6 - 960\right)\left(\pi^2 - 6\right)}{48\pi^5} + \dots \end{aligned} \tag{26.14}$$

where the denominator in N_R is Taylor expanded and Eqs (A2.8) and(A1.8) are used to show (mind the lower limit!)

$$\sum_{j=0}^{\infty} (2j+1)^{-p} = (1 - 2^{-p})\zeta(p) \tag{26.15}$$

The series are summable numerically to find $N_P + N_R = -0.7854$, a wholly uninspiring result until one realizes in a flash of insight that that number is the same as $-\pi/4$. Clearly something is going on.

[3]If one balks at the tediousness of carrying the – sign about, one can ignore it by considering positive s and observing that $\xi(-s) = \xi(s)$ is an even function.

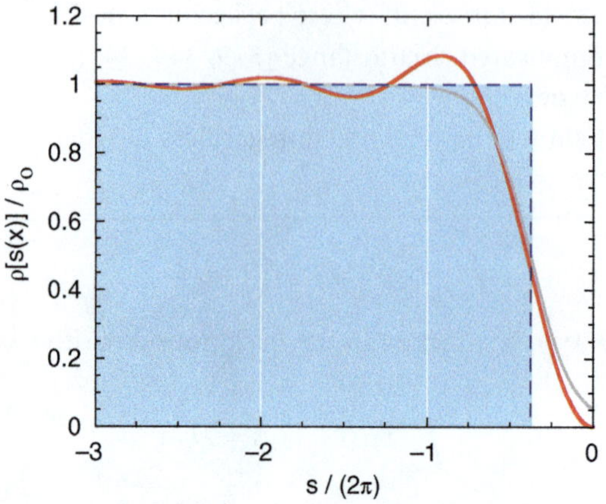

Figure 26.2 Electron density variation (red line) overlaid on the background positive charge (blue region) for an infinite barrier at $x = 0$; the blue region begins at $x = x_i$. Thin white lines show $s_j = -2\pi j$. Also included is a grey line showing the hyperbolic tangent approximation of Eq. (26.52). Based on Figure 4 of ref. [188].

The easy approach has piquancy by comparison. Approach the integral using the calculus of residues [60]. Observe first that the integrand in Eq. (26.8) is even, so the limits can be extended to $\pm\infty$ by including a factor of (1/2). Next, observe

$$S_3 \equiv \frac{1}{2}\int_{-\infty}^{\infty}\left(\frac{\cos s}{s^2} - \frac{\sin s}{s^3}\right) ds = \frac{1}{2}\mathrm{Re}\left\{\int_{-\infty}^{\infty}\frac{e^{is}(i+s)}{s^3}ds\right\} \tag{26.16}$$

The pole $s = 0$ lies on the real axis and is of order $n = 3$ (hence the 3 on S_n). Therefore, the sum of residues elegantly results in

$$S_3 = \frac{1}{2}\mathrm{Re}\left\{\frac{\pi i}{(n-1)!}\left(\frac{d}{ds}\right)^{n-1}\left[e^{is}(s+i)\right]\right\}\Bigg|_{s=0,\, n=3} = -\frac{\pi}{4} \tag{26.17}$$

Using either method, the location of the onset of the background positive charge necessary to balance the electron charge is found to be

$$x_i = -\frac{3\pi}{8k_F} = -1.1781 k_F^{-1} \tag{26.18}$$

for a single infinite barrier at $x = 0$. The constant background positive charge is overlaid on the electron density variation in Figure 26.2.

26.2 Two infinite barriers

> *...for there is nothing either good or bad but thinking makes it so...I could be bounded in a nutshell and count myself a king of infinite space...*
>
> – William Shakespeare[4]

For what is to come, having a discrete model is useful. The discreteness is imposed by the second boundary condition that $\psi_k(-L) = 0$ (considering the well to the left of the origin continues the convention of having the metal to the left of the metal/vacuum interface) regains Eq. (26.1), with $C_n^2 \leftrightarrow (n^2 - j^2)$ in Eq. (26.3). The next example explores its connection with the usual continuum model.

EXAMPLE: Find the discrete approximation to ρ_o using the composite modified trapezoidal rule of Section A1.2.1. Compare it to the usual $\rho_o = k_F^3/3\pi^2$.

SOLUTION: Let $k \to k_j = j\pi/L$ for a length scale L. At zero temperature, the Fermi level k_F corresponds to the maximum wave number, and so let $k_F \equiv n\pi/L$, which serves to define n. Thus, $dk \to (\pi/L)\Delta j$, but $\Delta j = 1$.

[4]Ref. [37]: *Hamlet*, II.2.247–253.

Putting the pieces together,

$$\rho_o \approx \frac{\pi}{2L^3}\left(\sum_{j=0}^{n}\left(n^2-j^2\right)-\frac{n^2}{2}\right)=\frac{\pi}{2L^3}\left(\frac{2}{3}n^3-\frac{1}{6}n\right)$$

where the summation uses Eq. (A1.2) (mind the limits) and where the subtraction of the factor $n^2/2$ in the second line is a consequence of the f_0 term in Eq. (A1.10). Observe that $k_F^3/(3\pi^2) = \pi n^3/3L^3$ after the change of terms, so the approximation is good as n becomes large.

The quantization condition imposed by $\psi_k(x = -L) = 0$ therefore parallels the equi-spaced k_j in transitioning to a series approximation of the continuum integral. Without further ado, therefore, the quantized level form of ρ becomes,[5]

$$\frac{\rho}{\rho_o} \approx \frac{3}{2n^3}\sum_{j=1}^{n}\left(n^2-j^2\right)\sin^2(\lambda_n j) \tag{26.19}$$

where $\lambda_n(x) \equiv \pi nx/NL$ and $n \equiv k_F L/\pi$. The expectations borne of the continuum example and the appreciation of the quantization conditions lead to two interesting consequences. The first is that some vestige of Eq. (26.5) will persist near the wall's edge and, second, that some scaling relation for the discrete case will render the behavior of the densities visibly similar even as they are different in magnitude. This will be seen even if n is not large, that is, *even if there are only a small number of energy levels involved.*

Consider an L so small that there are only a few energy levels that can be squeezed into the well before k_F is exceeded. The density profile for several of them is shown in Figure 26.3 based on the numerical evaluaton of Eq. (26.19): in the center region, the rippled nature is increasingly suppressed, even as the oscillations near the well walls seem to be confined to an ever smaller region near the walls; to those familiar with the Gibbs phenomenon and Fourier transforms [60], this is expected. Analogous to the dependence the density variations on s in Eq. (26.5), the densities for small n are similar when the x axis is expressed in terms of $k_F x \propto nx/L$, as in Figure 26.4. Thus, n does not have to be very large before conditions near the barrier at $x = 0$ look suspiciously close to the $L \to -\infty$ limit.

Why does this matter? Simply, because insofar as accumulation regions near a barrier can be (very) crudely modeled as a well region, the electrochemical potential model surrounding Eq. (24.20) indicates that band bending is not going quite the way that Figure 24.4 might suggest, namely, with an exponential shape. Rather, because the density drops to zero rather than increasing right up to the barrier, the quantum mechanical nature of the electron causes the potential drop to not be as great as if the density were increasing right up to the barrier's edge. To say more requires a triangular well model.

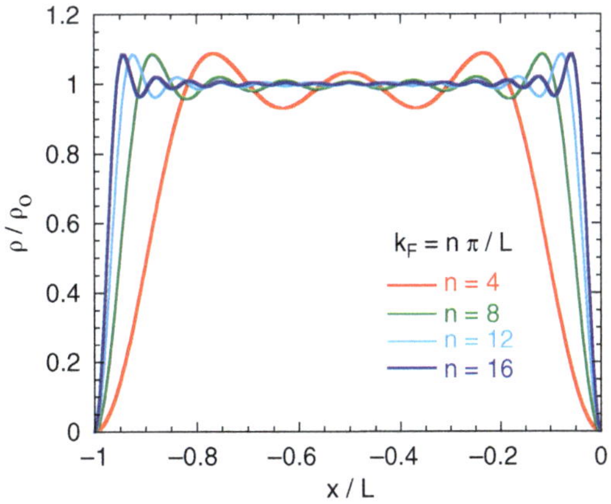

Figure 26.3 The density of Eq. (26.19) for several $k_F = n\pi/L$ characterized by low values of n. Observe that larger n flattens the middle but gives rise to behavior at the well walls that seems increasingly compressed into a region that becomes smaller as n becomes larger.

[5]The same result is obtained if the incantations of quantum mechanics are properly expressed, but keep in mind that the quick end-run here was a direct consequence of the assumption of infinite barriers: were the barriers finite, the k_j levels would *not* be equi-spaced, and the analogy would fail. Getting the "words" right (so to speak) does matter, however convenient the analogy otherwise appears.

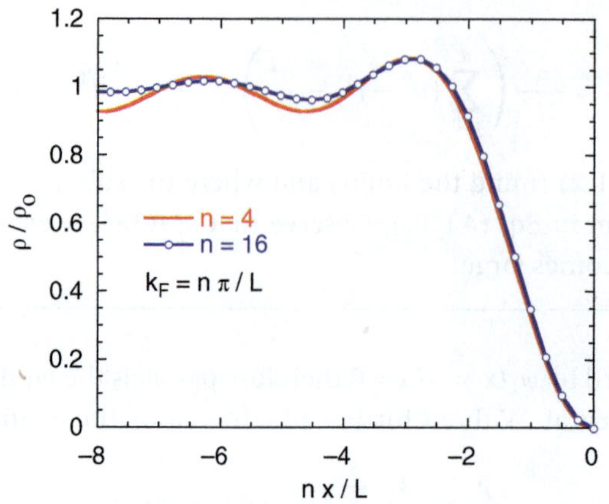

Figure 26.4 Same as Figure 26.3 but only for the largest and smallest n shown there, and for the x axis scaled by n. Observe how rapidly the oscillations adopt the behavior of Figure 26.1.

26.3 A triangular well

The simple analysis based on square (flat bottom) wells with infinite walls is suggestive but limited because accumulation layers do not resemble flat *square* wells so much as *triangular* ones. The energy levels are going to be very different. Fortunately, though, the methodology is much the same: find Airy wave functions such that $\psi_k(0)$ still vanishes. The zeros of the Airy function then determine the spacing of the energy levels within the triangular well region. The potential is defined by

$$V(x) = \begin{cases} -Fx & (x < 0) \\ \infty & (x \geq 0) \end{cases} \tag{26.20}$$

where there is no harm in emphasizing that $V(x)$ increases as x becomes more negative, because $F = \hbar^2 f/2m$ is taken as a positive number.

But which Airy function, $\mathrm{Ai}(w)$ or $\mathrm{Bi}(w)$ or a combination? Consider the argument w. In Section 18.2, $w(x)$ is obtained by setting the barrier height to 0 at $x = 0^-$, thereby setting $k_o \to 0$ in Eq. (18.20). The argument of the Airy function is[6]

$$w(x) = -f^{-2/3}\left(k^2 + fx\right) \tag{26.21}$$

Thus, near the origin, $w(x)$ is negative and the Airy function oscillatory, but as x becomes more negative, $w(x)$ becomes positive and increasingly large, for which $\mathrm{Bi}(w)$ exponentially grows as $\mathrm{Ai}(w)$ decays. Therefore, the boundary condition that $\psi_k(x) \to 0$ as $x \to -\infty$ dictates that $\mathrm{Ai}[w(x)]$ is the solution, but not every $\mathrm{Ai}(w)$, just the ones for which $w(0) = w_n$ such that $\mathrm{Ai}(w_n) = 0$.

There are a few ways to find the w_n zeros, the most straightforward being to look them up in ref. [103] if one is content to know only the first ten of them. The more stimulating approach is to fiddle with Eqs (18.22) and (18.24) to convince oneself that a crude approximation is $w_n \approx -[3\pi(4n-1)/8]^{2/3}$. With effort, a better approximation is

$$\lambda_n = \frac{3\pi}{8}(4n-1)$$
$$w_n = -\lambda_n^{2/3}\left(1 + \frac{5}{48\lambda_n^2}\right) \tag{26.22}$$

Finally, if accuracy is an obsession, then start with either the crude approximation or the better one; a few rounds of Newtonian iteration, by which

$$w_n \leftarrow w_n - \frac{\mathrm{Ai}(w_n)}{\mathrm{Ai}'(w_n)} \tag{26.23}$$

(where, recall, the prime indicates derivative with respect to argument) is sufficient to produce a very accurate approximation to w_n. An algorithm to produce the w_n is given in Section A3.19, the results of which are shown in Table 26.1.

[6]The present treatment is not a tunneling problem, and so using $w(x)$ instead of $c^2 z(x)$ is easier as $w(x)$ can change sign, simplifying the discussion. Nevertheless, the notational conventions of Section 18.2 are retained, such as the definition of f.

Table 26.1 Zeros of the Airy function $\text{Ai}(w)$: simple approximation to w_n (Eq. (26.22)) and exact w_n (application of Eq. (26.23)). As n increases, agreement improves.

n	Simple	Exact	n	Simple	Exact
1	−2.338107347	−2.339599848	11	−13.69148922	−13.69148922
2	−4.087949276	−4.088062286	12	−14.52783012	−14.52783012
3	−5.520559788	−5.520586014	13	−15.34075546	−15.34075546
4	−6.786707878	−6.786717415	14	−16.13268471	−16.13268471
5	−7.944133759	−7.944138050	15	−16.90563393	−16.90563393
6	−9.022650719	−9.022653580	16	−17.66130066	−17.66130066
7	−10.04017448	−10.04017544	17	−18.40113258	−18.40113258
8	−11.00852394	−11.00852489	18	−19.12638092	−19.12638092
9	−11.93601513	−11.93601608	19	−19.83812904	−19.83812904
10	−12.82877636	−12.82877731	20	−20.53733253	−20.53733253

Clearly, then, the well wave functions are $\psi_n(x) = N_n \,\text{Ai}[w_n(x)]$, where $w_n(x)$ is given by Eq. (26.21) for $k \to k_n$ such that $w_n(0)$ is now the zero of the Airy function, showing that $k_n^2 = -f^{2/3} w_n(0)$ (it is positive because $w_n(0)$ is negative). Finding the normalization N_n is found by demanding

$$1 = N_n^2 \int_{-\infty}^{0} \text{Ai}[w_n(x)]^2 dx = N_n^2 f^{-1/3} \int_{w_n}^{\infty} \text{Ai}(w)^2 dw \tag{26.24}$$

where w_n is shorthand for $w_n(0)$. As bad as the integral looks, a bit of cleverness induces it to reveal itself. Perform an integration by parts to obtain[7]

$$\int_{w_n}^{\infty} \text{Ai}(w)^2 dw = w\,\text{Ai}(w)^2 \Big|_{w_n}^{\infty} - 2\int_{w_n}^{\infty} w\,\text{Ai}\,(w)\text{Ai}'\,(w) dw \tag{26.25}$$

The first term vanishes by definition of a "zero", and $\text{Ai}(\infty) = 0$. The second term benefits from the relation $w\,\text{Ai}(w) = \text{Ai}''(w)$ by virtue of Airy's differential equation, and so

$$-2\int_{w_n}^{\infty} w\,\text{Ai}\,\text{Ai}' dw = -\int_{w_n}^{\infty} \left(\frac{d}{dw}(\text{Ai}')^2\right) dw = \text{Ai}'(w_n)^2 \tag{26.26}$$

where for compactness the argument was left off Ai′ and Ai in the integrals, and where $\text{Ai}'(\infty) = 0$ as well. Consequently, the normalized eigenfunctions are

$$\psi_n(x) = f^{1/6} \frac{\text{Ai}(w_n(x))}{|\text{Ai}'(w_n(0))|} \tag{26.27}$$

an elegantly simple result. The denouement, though, is how the density behaves. Let

$$\rho_n(x) \propto \mathbf{S} \equiv \sum_{j=1}^{n} \left(n^2 - j^2\right) \left[\frac{Ai\left(w_n - f^{-1/3}x\right)}{Ai'(w_n)}\right]^2 \tag{26.28}$$

The behavior of **S** is shown in Figure 26.5. Why is **S** scaled by a factor of n^2? In three dimensions, density goes as n^3, as in Eq. (26.19). The density for the triangular barrier is more akin to a two-dimensional gas, and therefore it is presumed to go as n^2. By taking the ratio of **S** with n^2, the density heights should be comparable, which, numerically, they are. As for the horizontal coordinate, the dimensionless measure of position in $w_n(x)$ is $f^{1/3}x$, and as with the square barrier case the expectation is that the $\mathbf{S}/n^2$ lines will be comparable if plotted as a function of $f^{1/3}x/n$, an expectation also realized numerically.

If electrons were point particles, the density would monotonically increase right up to the barrier edge, but because of the wave nature of the electrons, the density instead drops. This has definite consequences when the barrier is made finite in height (the wave functions can then penetrate the barrier as in Figure 9.11), and Poisson's equation is used to relate density to electrochemical potential $\mu(x) = \mu + \phi$ because of the variation of ρ with $\mu(x)$ (Section 24.3 and Eq. (24.21)).

[7] For brevity, if w_n appears without an argument, it is taken to be the Airy function zero $w_n(0)$.

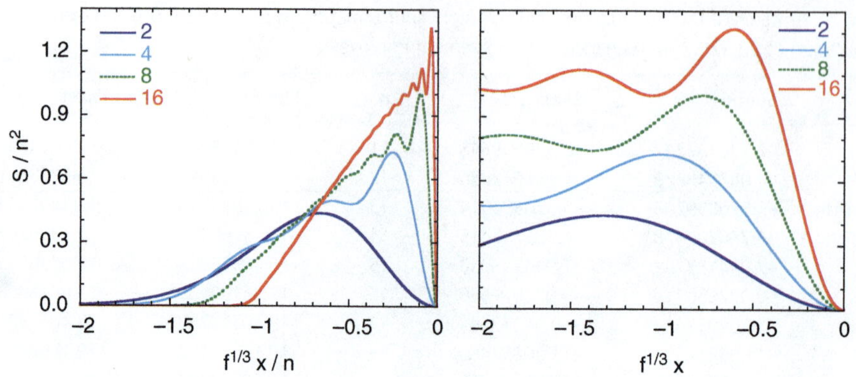

Figure 26.5 Lest, behavior of the scaled density (calculated by Eq. (26.28) but with $f^{1/3}x$ scaled by n. Right, behavior of the scaled density with unscaled position $f^{1/3}x$. Observe that close to the barrier at $x = 0$ the density variation visually resembles the square well case and that the quantity $f^{1/3}x$ is dimensionless.

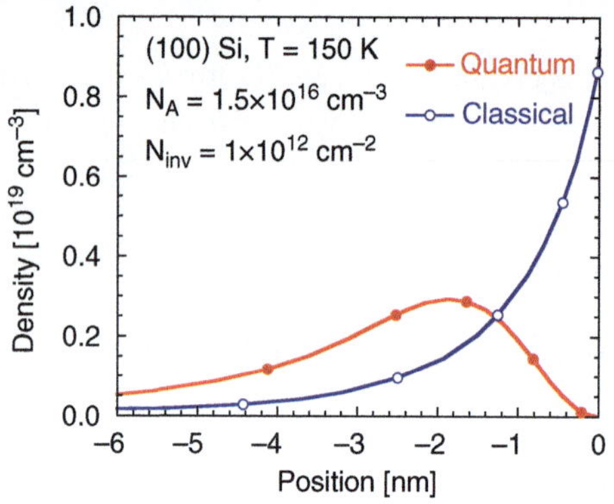

Figure 26.6 Comparison of theoretical classical and quantum predictions of electron density near an interface for silicon, based on Figure 6 of ref. [326], except that here the x axis has been shifted to the negative side to match the orientation of prior figures.

In fact, such a calculation has been performed by Stern [326], who made a much more detailed analysis of wells at interfaces using properties of semiconductors that for simplicity are ignored here. Stern's findings for (100) silicon are reproduced in Figure 26.6 (based on Figure 6 of ref. [326]; see also Chapter 2 of Mönch [213]). Although the physical potential well is only approximately triangular for the depletion layer considered by Stern, the qualitative shape is in keeping with behavior evinced in the idealized triangular well model of the present analysis.

26.4 Density and dipole component

When the barrier is not infinite, then wave function penetration into the barrier region can occur. To find the effect of electron density, consider the rectangular barrier as its transmission behavior is well understood: the reflection coefficient is known analytically from Eq. (18.9) with the change $\kappa \to i\kappa$ for transmission when $k < k_o$. Thus, for under the barrier,

$$r(k < k_o) = \frac{(k^2 + \kappa^2)\sinh(\kappa L)}{2ik\kappa\cosh(\kappa L) + (k^2 - \kappa^2)\sinh(\kappa L)} \tag{26.29}$$

where $\kappa^2 = 2m(V_o - E)/\hbar^2 = k_o^2 - k^2$ and $k^2 = 2mE/\hbar^2$. Intuitively, if the wave function penetrates the barrier, then it pushes a bit more into the barrier before it is turned back, and so a phase change between the incident and reflected waves is introduced. This can be seen by representing $r(k)$ as

$$r(k) \equiv -R(k)\exp(2i\varphi(k)) \tag{26.30}$$

Recalling that $\kappa^2 = k_o^2 - k^2$ for $k < k_o$, it is straightforward to show that

$$R(k) = \left\{1 + \left(\frac{2\kappa k}{k_o^2\sinh(\kappa L)}\right)\right\}^{-1/2} \tag{26.31}$$

$$\tan(2\varphi) = -\frac{2\kappa k\cosh(\kappa L)}{(\kappa^2 - k^2)\sinh(\kappa L)} = -\frac{2\kappa k}{k_o^2\tanh(\kappa L)} \tag{26.32}$$

where $R(k) = |r(k)|$ and φ are real functions. Observe that the usual limits obtain: for $k_o \to \infty$, then $R(k) \to 1$ and $\varphi \to 0$; for $L \to \infty$, then $R(k) \to 1$ and $r \to (\kappa + ik)/(\kappa - ik)$ (recall Eq. (9.68) with $k' \to i\kappa$). For finite tunneling barriers with a width of about a nanometer and barrier of a few electronvolts, then k_oL is quite large (explicitly, a 1 eV barrier that is 1 nm thick entails $k_oL \approx 26$ for $m_n = m$), so that to a very good approximation $\cosh(\kappa L) \approx \sinh(\kappa L)$. As a result, $\tan(\kappa L) \to 1$ and so φ is independent of the barrier width L to a good approximation. To leading order, therefore,

$$r(k) \approx -\frac{\kappa + ik}{\kappa - ik} \tag{26.33}$$

$$R(k) = |r(k)| \approx 1 \tag{26.34}$$

$$\varphi(k) = -\frac{1}{2}\arcsin\left(\frac{2k\kappa}{k_o^2}\right) \approx -\frac{k}{k_o}\left(1 + \frac{k^2}{6k_o^2}\right) \tag{26.35}$$

where the form of $r(k)$ is familiar from Eq. (9.68) (a wide barrier for $k < k_o$ is indistinguishable from a step potential). For tall barriers, the onset of the barrier is what matters, not its extent (or shape), as the same conclusion can be drawn for barriers of different shape that are also high [188, 338, 346]. For the calculation of density, the wave functions with smaller k are weighted more, particularly if the weighting factor for the wave functions behaves like the supply function $f(k)$, as in Eq. (26.5). Consequently, for the purposes of finding electron density variation near an interface, $\varphi \approx -kx_o$ where $x_o \equiv \kappa/k_o^2 \approx 1/k_o$ to a good approximation. As a result,

$$|\langle n|x\rangle|^2 \propto 1 - \cos\left(2k_F(x - x_o)\right) \tag{26.36}$$

It is as though *the infinite barrier density result were rigidly moved over a distance* x_o (not exactly, but pretty close) that is more or less constant at $1/k_o$, that is, Eq. (26.5) is simply modified by

$$\rho \to \rho_o\xi[2k_F(x - x_o)] \tag{26.37}$$

So far, temperature and field effects on the density have been ignored by considering the zero-temperature limit of the supply function $f(k)$ and the infinite barrier approximation. How good these approximations are may be estimated by evaluating $\delta\rho(x)$, where

$$\rho(x) = \rho_o\xi(2k_Fx) + \delta\rho(x) \tag{26.38}$$

and where $\rho_o\xi(2k_Fx) \equiv \rho_o(x)$ so that when $\rho_o(x)$ appears with an argument, it is to be understood as the infinite barrier, zero-temperature solution of Eq. (26.5). Clearly, $\delta\rho$ must depend on both temperature and field simultaneously, but it is sufficient to treat each effect separately with respect to how $\delta\rho(x)$ modifies $\rho_o(x)$.

Consider temperature first. The temperature effects treated here are a consequence of the presence of higher momentum components in the supply function (another temperature effect which gives rise to changes in the work function Φ on the order of k_BT is due to the thermal expansion of the bulk material and are described in ref. [164], but are not considered further here). Because of the thermal tail in the supply function, higher momentum $\hbar k$ states are present and increase the density near $x = 0$, the maximum of which is pushed away from the barrier by a length π/k_F, as in Figure 26.1. The shape of the density near the origin therefore has a thermal component which can be approximated as follows. Subtracting the infinite barrier, zero temperature integral of Eq. (26.5) from the temperature-dependent integral defining total density, and using the identity

$$s - \ln(1 + e^s) = \ln(1 + e^{-s}) \tag{26.39}$$

for $s = s(k) = \beta\hbar^2(k_F^2 - k^2)/2m$ results in

$$\delta\rho \approx \frac{2m}{\pi^2\hbar^2\beta}\sin^2(k_Fx) \times \left\{\int_0^{k_F} \ln\left(1 + e^{-s(k)}\right) dk - \int_{k_F}^{\infty} \ln\left(1 + e^{-s(k)}\right) dk\right\} \tag{26.40}$$

where $\sin^2(kx) \to \sin^2(k_Fx)$ is evaluated at the maximum of the integrand and pulled out of the integration because it is slowly varying by comparison to the ln terms. Do a change of integration variables to s and $ds/dk = \beta\hbar^2k/m \to \beta\hbar^2k_F/m$ because it, too, is weakly varying by comparison to the ln term. What is left is therefore

$$\delta\rho \approx \frac{2m}{\pi^2\hbar^2\beta}\sin^2(k_Fx)\left(\frac{m}{\beta\hbar^2k_F}\right)\int_0^{\infty} \ln(1 + e^{-s})\, ds \tag{26.41}$$

The integral is half of the familiar Riemann zeta function $\zeta(2)/2 = \pi^2/12$, and so putting the parts together,

$$\frac{\rho(x)}{\rho_o} = \xi(2k_F x) + \left(\frac{\pi}{2\beta\mu}\right)^2 \sin^2(k_F x)$$
$$= \xi(2k_F x) + 2\left(\frac{\pi}{4\beta\mu}\right)^2 (1 - \cos(2k_F x)) \tag{26.42}$$

For metals, the size of $\beta\mu$ makes the correction term very small, so that the finite temperature density $\rho(x)$ is not visually distinguishable from the zero-temperature result for typical temperatures and metal-like μ (unless μ is small, T is high, or both), yet another reminder that the electron gas in a metal is well-approximated by its zero temperature limit. As far as excitement goes, that was meager.

Consider field next, where the prospects for exciting behavior are better [338, 347]. Instead of a rectangular barrier of thickness L and height V_o, now consider a triangular barrier of height V_o and field F. If the barrier is sufficiently high, then $\mathrm{Bi}(z) \gg \mathrm{Ai}(z)$ when $z = (k_o^2 - k^2)/f^{2/3} = \kappa^2/f^{2/3}$ (Section 18.2), so that for under the barrier, the analog to Eq. (26.29) for triangular barriers is

$$r(k < k_o) \approx -\frac{-ik\,\mathrm{Bi}(z) + f^{1/3}\mathrm{Bi}'(z)}{ik\,\mathrm{Bi}(z) + f^{1/3}\mathrm{Bi}'(z)} \tag{26.43}$$

Because $\mathrm{Bi}'(z)/\mathrm{Bi}(z) \approx -f^{-1/3}\mathrm{Di}(1, z)$ (remember that $s = -1$ and $c = 1$ for the triangular barrier conditions under consideration) and

$$\mathrm{Di}(1, z) \approx f^{1/3}\left(\frac{1 - 4z^{3/2}}{4z}\right) = \frac{f}{4\kappa^2} - \kappa \tag{26.44}$$

it follows that

$$r(k < k_o) \approx \frac{4\kappa^2(ik + \kappa) - f}{4\kappa^2(ik - \kappa) + f} \tag{26.45}$$

so that $|r| = 1$ for these conditions as well, and as a result the effect of the barrier is to introduce a modified phase factor $2\varphi'$. Watch closely: if we let

$$\kappa' = \kappa - \frac{f}{4\kappa^2} \tag{26.46}$$

then substitution into r gives the hauntingly familiar result for the triangular barrier

$$r(k) = -\frac{\kappa' - ik}{\kappa' + ik} \tag{26.47}$$

which is just Eq. (26.33) with $\kappa \to \kappa'$, meaning all the prior findings can be carried over by making that replacement. In particular, whereas before for the wide and high rectangular barrier $\varphi \approx k\kappa/(\kappa^2 + k^2) = k\kappa/k_o^2$ so that $x_o \approx \kappa/k_o^2 \approx 1/k_o$, now for the high triangular barrier with field $f = 2mF/\hbar^2$,

$$x_o' \approx -\frac{k\kappa'}{\kappa'^2 + k^2} = -\frac{4\kappa^2(4\kappa^3 - f)}{(4\kappa^3 - f)^2 + 16k^2\kappa^4} \tag{26.48}$$

The variation of both $k_o x_o$ and $k_o x_o'$ with k is shown in Figure 26.7. That variation suggests replacing x_o and x_o' by a constant value, specifically $1/k_o$, as a reasonably good (although not great) approximation, primarily for its considerable ease of use.

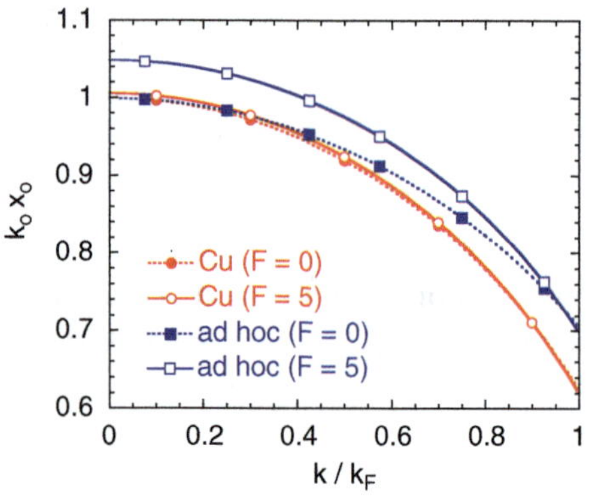

Figure 26.7 Variation of $k_o x_o$ (for $F = 0$ eV/nm) and $k_o x_o'$ (for $F = 5$ eV/nm) for copper-like ($\mu = 7$ eV, $\Phi = 4.5$ eV) and *ad hoc* ($\mu = \Phi = 1.5$ eV) parameters, where $x_o = \kappa/k_o^2$ and x_o' is given by Eq. (26.48).

An integration over the force $F(x)$ from $x = -\infty$ to the surface at $x = 0$ gives the change in potential due to the density oscillations of Eq. (26.5). Using Poisson's equation after integration by parts [85, 348, 349],

$$\begin{aligned}\Delta\Phi &= -\int_{-\infty}^{0} F(x)dx = \int_{-\infty}^{0} x\partial_x F(x)dx \\ &= \int_{-\infty}^{0} x\partial_x^2 \phi(x)dx = 4\pi Q \int_{-\infty}^{0} x\{\rho_e(x) - \rho_i(x)\}dx\end{aligned} \tag{26.49}$$

In light of Eqs (26.5) and (26.10), this is an excellent result: using $\rho_o = k_F^3/3\pi^2$ and Eq. (26.18), it is the same as

$$\begin{aligned}\Delta\Phi &= 4\pi Q\rho_o \left\{ \frac{1}{4k_F^2}\left(\int_{-\infty}^{0} 3\frac{d}{ds}\left(\frac{\sin s}{s}\right) ds \right) - \frac{1}{2}x_i^2 \right\} \\ &= \frac{Qk_F}{32\pi}\left(32 - 3\pi^2\right)\end{aligned} \tag{26.50}$$

EXAMPLE: Find $\Delta\Phi$ for a bulk electron density characteristic of copper.
SOLUTION: For copper, $\mu = 7$ eV, so that $k_F = \sqrt{2m\mu}/\hbar = 13.5546$ nm^{-1} (or $\rho_o = k_F^3/3\pi^2 = 84.1088$ nm^{-3}). Therefore,

$$\Delta\Phi = \frac{32 - 3\pi^2}{32\mu}(0.359991 \text{ eV nm})(13.5546 \text{ nm}^{-1}) = 0.116063 \text{ eV}$$

Even before the satisfaction of the previous calculation fades, apprehensions arise: temperature *must* make a difference because the higher momentum states are weighted more as $\beta = 1/k_B T$ decreases in the supply function $f(k)$. Because incident waves with higher k penetrate deeper into a barrier of height $V_o = \hbar^2 k_o^2/2m$, they cause x_i to shift from its $T = 0$ location, and even if this only by a small amount, that will change $\Delta\Phi$. Moreover, application of an external field induces a shift in the electron density and affects where x_i (the origin of the background positive charge) is said to be. That shift gives rise to a slab of electrons not matched by the background positive charge and which therefore contributes to the barrier height. If that shift in the origin of the background positive charge compared to where it is at F or $T = 0$ is δx_i, then the induced change in barrier height $\Delta\Phi$ is

$$\Delta\Phi = \frac{\rho_o}{\varepsilon_0}\int_0^{\delta x_i}\left(\int_0^{x'} dx''\right)dx' = \frac{8}{3\pi}Qk_F^3(\delta x_i)^2 \tag{26.51}$$

Investigating these effects using a simultaneous solution of the wave functions for each k and the corresponding phase factors, then integrating to obtain the density variations from deep in the bulk to the surface takes some doing [188, 338, 347]. A simpler model on which to base the analysis is to be welcomed even if it is less accurate.

The hyperbolic tangent model (or "tanh model" for short) [173, 342, 350] is already useful in relating the work function Φ to the electron density variation via the exchange-correlation potential [340, 351], but it can also be used to explore the sensitivity of the electron density to temperature effects. Ignoring the oscillations in $\rho(x)$ in Figure 26.4, a squinty look at the curve just makes out a hyperbolic tangent, giving the approximation its name. In practice, though, the relation is easier to represent as[8]

$$\rho_a(x) = \frac{\rho_o}{1 + \exp\left[\lambda k_F(x - x_i)\right]} \tag{26.52}$$

where λ is a parameter to be determined. The infinite barrier approximation affords a good place to start, and for it, first let $\lambda = \lambda_o$ and, second, determine λ_o by demanding that $\partial_x\rho(x) = \partial_x\rho_a(x)$ at $x = x_i$, where $\rho(x)$ is from Eq. (26.5). They are tedious but straightforward derivatives, and when the dust settles,

$$\lambda_o = \frac{72}{\Delta^3}\cos(\Delta) + \frac{24(\Delta^2 - 3)}{\Delta^4}\sin(\Delta) \tag{26.53}$$

[8]The connection is because $(1/2)\left[1 - \tanh(x)\right] = 1/(1 + e^{2x})$, as easily shown.

where $\Delta = 2k_F x_i$. In light of Eq. (26.18), however, $2k_F x_i = -(3\pi/4)$, and so

$$\lambda_o = \frac{64\sqrt{2}}{9\pi^4}\left(-3\pi^2 + 12\pi + 16\right) = 2.48711 \tag{26.54}$$

How Eq. (26.52) behaves using λ_o from Eq. (26.54) was shown in Figure 26.2 by the gray line, indicating that the approximation captures (more or less) the behavior in the region near the barrier. The reason for its consideration can now be revealed: it allows for a simple approximation to $\Delta\Phi$ via Eq. (26.49) as

$$\begin{aligned}\Delta\Phi &= 4\pi Q \int_{-\infty}^{\infty} x\{\rho_a(x) - \rho_i(x)\}dx \\ &= 4\pi Q \int_{-\infty}^{\infty} (x - x_i)\{\rho_a(x) - \rho_i(x)\}dx\end{aligned} \tag{26.55}$$

where the second form follows by the assumption that no net charge difference exists so that the total integral over ρ_a is the same as that over ρ_i. As a result, the integrations become straightforward after switching to $s = \lambda(x - x_i)$ to obtain

$$\Delta\Phi = \frac{4\pi Q}{\lambda^2 k_F^2}\rho_o \left\{ \int_{-\infty}^{0} s\left(\frac{1}{1+e^s} - 1\right) ds + \int_0^{\infty} \frac{s}{1+e^s} ds \right\} \tag{26.56}$$

Those integrals are old friends (Eq. (A2.7)): the first integral is the same as the second but with $s \to -s$, and with $\rho_o = k_F^3/3\pi^2$, all simplifies to

$$\Delta\Phi = \frac{8Qk_F}{3\pi\lambda^2}\int_0^{\infty}\frac{s}{1+e^s}ds = \frac{2\pi Q k_F}{9\lambda^2} \tag{26.57}$$

EXAMPLE: Find λ for $r_s a_o = 0.2, 0.3, 0.5$ [nm] as in Table 1 of Appelbaum and Hamann [351], who use the hyperbolic tangent model to evaluate the exchange-correlation terms to explore the image charge potential.
SOLUTION: Observing that $k_F = (9\pi/4)^{1/3}/(r_s a_o)$, then k_F [1/nm] = 9.5958, 6.3972, and 3.8383. From Appelbaum and Hamann's values of $(1/\lambda k_F)$ [nm] = 0.0626, 0.0626, and 0.0610, respectively, it follows that λ = 1.6647, 2.4971, and 4.2710.

The manner in which the hyperbolic tangent approximation is *used* affects the values of λ that are thereby useful. Why is this useful? It is because Eq. (26.57) captures how $\Delta\Phi$ *scales* with f and L. When a barrier thins (L decreases) and/or a field is stronger (f increases), then greater penetration of the barrier occurs, and as a consequence, λ decreases in magnitude, resulting in an increase in $\Delta\Phi$. To say more or to quantify the account, the density must be found from a thermal distribution of wave functions properly matched at the barrier's edge, and the barrier need not be abrupt (as rectangular and triangular barriers are). Likewise, the self-consistently matched image charge potential [340, 341] or its more rudimentary and routine form of Eq. (11.25) are relatively abrupt as well. When wave function penetration into the barrier is then accounted for, the shifting position of x_i in response to the changing density is accounted for, and then the small changes to the work function barrier can be evaluated numerically [188, 347].

CHAPTER 27

Many-body effects and image charge

All many-body effects are contained in the exchange and correlation contributions to $\bar{\mu}$ and in their effect on the barrier potential $\Delta\phi$. In particular, the image-force effect on Φ may be regarded as contained in the disappearance of part of the correlation energy when one electron is moved away from the metal surface.

– Norton D. Lang and Walter Kohn [348]

The electron density and the potential under which electrons move is clearly related. In the bulk, electrons that seemingly move around in a quasi-free manner kept the physics simple, but when those electrons get to the surface, that quasi-free narrative must be reworked: electrons interact with each other and with the ion cores around which they move, and the interaction of all the electron plus ion components affects the dynamics.

To describe a group of N electrons (e) interacting with ions (b), the Hamiltonian of their interaction must be specified, meaning that the various contributions to the total energy must be accounted for: the kinetic and potential energy of the electrons and their interactions, the interaction between the electrons and the ions, and even the energy of the ions themselves. Assuming the ions are represented as a smeared-out distribution of positive charge characterized by $\rho_+(\vec{r})$, these three components can be expressed as

$$H^N = KE + V_{e-e} + V_{e-b} + V_b \tag{27.1}$$

where the e and b subscripts denote "electron" and "background", respectively. In the jellium model, where the background positive charge is uniformly distributed, the last two terms do not figure prominently in the story line.[1] The ones that do are

$$KE = \sum_{i=1}^{N} \frac{(\hbar k_i)^2}{2m} \tag{27.2}$$

$$V_{e-e} = 4Q \sum_{i<j=1}^{N} \frac{e^{-\epsilon|\vec{r}_i - \vec{r}_j|}}{|\vec{r}_i - \vec{r}_j|} \tag{27.3}$$

where ϵ is a small convergence factor which eventually is taken to zero ($\epsilon \to 0$) at the end of the calculation, and where the zero temperature limit is implicitly assumed so that the occupation probability from the Fermi–Dirac distribution is unity for all electrons in the sum – a good approximation that significantly eases all the integrals and summations. A proper account of the so-called exchange-correlation potential is involved (see refs [337, 348, 352]; the correlation part is a rather thorny problem in many body statistical physics, and relies on the summation of a large class of Feynman diagrams [64, 85]. The present treatment aims to be more empathetic to those not acquainted with the formalism. The best place to start, of course, is with what is known already, namely, the kinetic energy term.

27.1 Kinetic energy

Sense sure you have,
Else could you not have motion ...

– William Shakespeare[2]

[1]The present treatment is purposely simplified from the operator notation from whence it comes and the contributions not important to the present narrative have been hidden. See [85] and references therein.
[2]Ref. [37]: *Hamlet*, III.iv.72–73.

Introduction to the Physics of Electron Emission, First Edition. Kevin L. Jensen.

The kinetic energy of Eq. (27.2) has been glimpsed in Eq. (7.4), and so tackling it is straightforward. The approximation of the summation by an integration causes KE to be approximated by

$$KE \rightarrow \frac{2}{(2\pi)^3} \int_0^{k_F} \frac{\hbar^2 k^2}{2m} 4\pi k^2 dk = \frac{\hbar^2 k_F^5}{10\pi^2 m} \tag{27.4}$$

The integration is over all of k space, and so KE is the kinetic energy per unit volume. It is agreeable to represent it that way by making use of $\rho_o = k_F^3/3\pi^2$, so that

$$KE = \rho_o \varepsilon_k \tag{27.5}$$

where ε_k refers to the average kinetic energy of the N electrons. Using $\mu = \hbar^2 k_F^2/2m$ and $\rho_o = k_F^3/3\pi^2$, it is seen that Eq. (27.4) is the same as

$$KE = \frac{3}{5}\mu\rho_o \tag{27.6}$$

and so $\varepsilon_k = 3\mu/5$. The other terms will unfold in a similar manner: the integrals treating the potential energy terms will be the product of density with an average energy, with the average energy part labeled ε with a subscript denoting its origin, for examlple ε_k for kinetic energy.

27.2 Exchange energy

If a system in atomic physics contains a number of particles of the same kind, e.g. a number of electrons, the particles are absolutely indistinguishable one from another. No observable change is made when two of them are interchanged. This circumstance gives rise to some curious phenomena in quantum mechanics having no analogue in the classical theory ...

– Paul A.M. Dirac [40], p. 207

The interactions between the N electrons contained in Eq. (27.3) is simpler in k space: letting $\vec{R} = \vec{r}_i - \vec{r}_j$, then a continuum version of it is

$$\int \frac{e^{-\epsilon R}}{R} e^{i\vec{k}\cdot\vec{R}} d\vec{R} = 2\pi \int_0^{\infty} R e^{-\epsilon R} dR \int_{-1}^{1} e^{ikRs} ds = \frac{4\pi}{k^2 + \epsilon^2} \tag{27.7}$$

where $s = \cos\theta$. Well, almost – it is a bit more complicated than that. Because $\vec{R}$ is a distance between two particles, the operator associated with V_{ex} (call it $\hat{V}_{ex}$) acts on the ket $|r_i, r_j\rangle$ to get to the left-hand side of Eq. (27.7) with $\vec{R} = \vec{r}_i - \vec{r}_j$. Transforming to the momentum representation, terms of the form $\langle k_1, k_2 | \hat{V} | k_3, k_4 \rangle$ will have to be evaluated. A quantum mechanical field operator formalism [64, 78, 97, 353] leads to the Feynman diagram approach to evaluating the contributions, but the outcome can be appreciated without as much trouble [85]. The problem with the ket vector $|k_1, k_2\rangle$ is that it seems to treat the electrons as distinguishable, and yet they are not. Wave functions for multi-particle statistics are constructed differently than simply gluing kets together, as in $|k_1\rangle |k_2\rangle = |k_1, k_2\rangle$. Specifically, many particle wave functions must show the proper symmetry for fermions and bosons in that if any two particle labels in the ket (e.g., k_i and k_j) are interchanged, then the wave function must be either symmetric (for bosons) or antisymmetric (for fermions) [65]. That is, $\hat{P}_{ij}$ is a permutation operator that swaps k_i and k_j in an N-body ket such that

$$\hat{P}_{ij} \left| k_1 \cdots k_i \cdots k_j \cdots k_N \right\rangle^{(\pm)} = \pm \left| k_1 \cdots k_j \cdots k_i \cdots k_N \right\rangle^{(\pm)} \tag{27.8}$$

with the + associated with bosons and the – associated with fermions. A rapid method of making the proper fermion kets is to make use of so-called Slater determinants, which, for fermions, are defined by (in the case of two particles)

$$\left| k_1 k_2 \right\rangle^{(-)} = \frac{1}{\sqrt{2}} \det \begin{bmatrix} |k_1\rangle & |k_2\rangle \\ |k_1\rangle & |k_2\rangle \end{bmatrix} = \frac{1}{\sqrt{2}} \left(|k_1 k_2\rangle - |k_2 k_1\rangle \right) \tag{27.9}$$

and where the N-particle version has instead N rows and columns, and a coefficient of $1/\sqrt{N}$. A particle exchange is then tantamount to an interchange of columns in the determinant, the consequences of which are that the determinant changes sign. $\hat{P}_{12}$ acting on $|k_1 k_2\rangle^{(-)}$ clearly now gives $\hat{P}_{12} |k_1 k_2\rangle^{(-)} = -|k_1 k_2\rangle^{(-)}$.

As a consequence, an *exchange* term associated with the energy of interaction amongst the electrons arises. It will be the $\epsilon \to 0$ limit of

$$V_{ex} = -4Q\frac{1}{(2\pi)^6}\int d\vec{k}_1 \int d\vec{k}_2 \frac{4\pi}{|\vec{k}_1 - \vec{k}_2|^2 + \epsilon^2} f_{FD}(E_1) f_{FD}(E_2) \quad (27.10)$$

In the zero-temperature limit, $f_{FD}(E) \to \Theta(\mu - E)$. Introducing the difference and average wave vectors $\vec{k} = \vec{k}_1 - \vec{k}_2$ and $\vec{k}' = (\vec{k}_1 + \vec{k}_2)/2$, respectively, the integration over $\vec{k}'$ can be written as

$$\begin{aligned} \delta O(k) &\equiv \int d\vec{k}' \Theta(\mu - E_1)\Theta(\mu - E_2) \\ &= \int d\vec{k}' \Theta\left(k_F - \left|\vec{k}' + \frac{\vec{k}}{2}\right|\right)\Theta\left(k_F - \left|\vec{k}' - \frac{\vec{k}}{2}\right|\right) \end{aligned} \quad (27.11)$$

since the denominator of Eq. (27.10) is independent of $\vec{k}'$. This integral takes some getting used to. Were only one Θ factor present, then the interpretation is easy: $\int d\vec{r}\Theta(r - |\vec{x} - \vec{y}|)$ is the volume of a sphere of radius r that has been displaced from the origin by a distance $\vec{y}$. But because two Θ functions are present, their joint action denotes where two such displaced spheres actually overlap. Hence the problem is one of finding the overlap region of two spheres, as indicated in Figure 27.1. Finding the cap section of a sphere (the shape of the piece that results when a bit of a sphere is sliced off, as it were) is a routine problem in calculus, and $\delta O(k)$ is simply twice that. Therefore,

$$\delta O(k) = 2\pi k_F^3 \Theta(2k_F - k)\int_0^{\theta_m} \sin^3\theta \, d\theta \quad (27.12)$$

where $\theta_m(k) = \arccos(k/2k_F)$, and the $\Theta(2k_F - k)$ factor insures that $k/2$ does not exceed k_F, as insisted upon by the original Θ functions of Eq. (27.11). The integral gives $(2 - 3\cos\theta_m + \cos^3\theta_m)$, or, using the definition of θ_m,

$$\delta O(k) = \frac{\pi}{12}\left(4k_F + k\right)\left(2k_F - k\right)^2 \Theta(2k_F - k) \quad (27.13)$$

Observe that the volume of a sphere, $4\pi k_F^3/3$, results when $k = 0$, as it should. The remaining k integration is straightforward, made easier by taking the $\epsilon \to 0$ limit *before* performing the integral, or

$$V_{ex} = -4Q\frac{1}{(2\pi)^6}\int_0^{\infty} \frac{4\pi}{k^2 + \epsilon^2}\delta O(k) 4\pi k^2 dk = -\frac{Q}{\pi^3}k_F^4 \quad (27.14)$$

Now, as with Eq. (27.5), it is desired to represent V_{ex} as $\rho_o \varepsilon_{ex}$ and as can easily be shown,

$$V_{ex} = -\frac{Q}{\pi^3}k_F^4 = -\frac{m}{\pi^3 \hbar^2 a_o}\mu^2 \quad (27.15)$$

where a_o is the Bohr radius (Table 2.3).

A more commonly encountered nomenclature involves defining the quantity $r_s a_o$ as the radius of said sphere that holds one electron, so that the electron density can be expressed as [64, 78, 354]

$$\frac{1}{\rho_o} = \frac{4}{3}\pi(r_s a_o)^3 \leftrightarrow r_s = \left(\frac{9\pi}{4}\right)^{1/3}\frac{1}{k_F a_o} \quad (27.16)$$

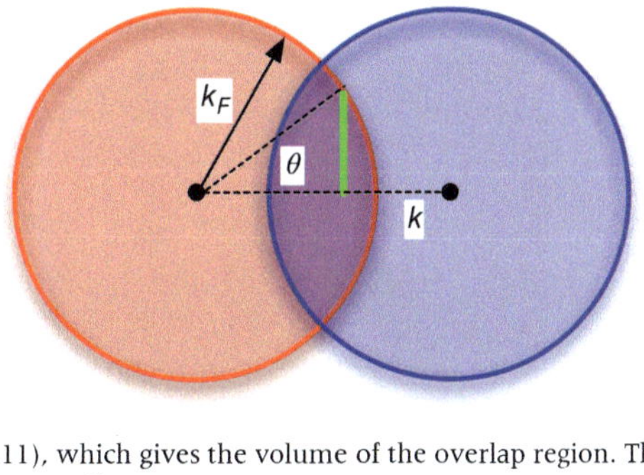

Figure 27.1 Overlap of two spheres, as in Eq. (27.11), which gives the volume of the overlap region. The green line of thickness dk' rotated about the $\hat{k}_x$ axis is the volume element (a disk) of the integral over $d\vec{k}'$ in Eq. (27.11), with $\cos\theta = k/2k_F$. The radius of the green element is simply $k_F \sin\theta$. Therefore, the volume element is $\pi k_F^3 \sin^3\theta \, d\theta$.

(r_s is dimensionless) to obtain

$$KE + V_{ex} = R_y\rho_o\left\{\left(\frac{3}{2\pi}\right)^{1/3}\frac{9\pi}{10r_s^2} - \left(\frac{3}{2\pi}\right)^{5/3}\frac{\pi}{r_s}\right\}$$

$$= R_y\rho_o\left\{\frac{2.2099}{r_s^2} - \frac{0.91633}{r_s}\right\} = \rho_o(\varepsilon_k + \varepsilon_{ex}) \qquad (27.17)$$

where $R_y = 2Q/a_o = 13.606$ eV and the remaining appalling mess of factors have been distilled into their numerical equivalents in the parentheses.

27.3 Correlation term

...a given electron in a dense electron gas acts to polarize its immediate surroundings: it pushes other electrons out of its way until its associated screening cloud possesses a charge nearly equal (and opposite) to its own. The quasi particles (electrons plus screening clouds) interact via an effective short-range interaction of the order of the interparticle spacing, in contrast to the original long-range Coulomb interaction.

– David Pines [355], p. 11

The electron is sometimes described as carrying a positive hole around with it ...A part of the potential due to this type of interaction can be seen in the classical image force that acts on an electron as it approaches the surface of a conductor ... this interaction, of course, cannot be determined by classical methods when the electron is within a few angstroms of the surface.

– G.A Haas and R.E. Thomas [135], p. 124

The exchange potential energy is not the last of it: there are other Feynman diagrams associated with Eq. (27.3). A small r_s expansion to augment Eq. (27.17) is good for a high-density electron gas (see, for example, Feynman [64]), but a typical metal like copper has $r_s = 2.6756$ when $\mu = 7$ eV, and therefore is not small in the required way. As a result, many other Feynman diagrams need summing. In the asymptotic limit when r_s becomes large, the electrons crystallize into a lattice [353]: Wigner proposed an early form of the so-called correlation energy V_{cor} (perhaps Feynman's tongue-in-cheek appellation of "stupidity energy" [64] to designate all the rest that is hard to know is superior) but a best fit to the data of Table 1 of Ceperley and Alder [356], reproduced here in Table (27.1) and based on simulation, is given by

$$\varepsilon_{cor} = -R_y\left\{\frac{c_0}{r_s + c_1\sqrt{r_s} + c_2}\right\} \qquad (27.18)$$

where $c_0 = 0.8740, c_1 = 3.3618$, and $c_2 = 2.946$ correspond to a least squares fit,[3] as shown in Figure 27.2.

Table 27.1 First two columns: ground state energy of electrons as a function of r_s, where $\varepsilon_{tot} = (KE + V_{ex} + V_{cor})/(R_y\rho_o)$ given in Table 1 of Ceperley and Alder [356]. Second two columns: various metals and their associated r_s from Table 1 of Kiejna [349].

r_s	ε_{tot}	Metal	r_s
1	1.1740	Al	2.07
2	0.0041000	Zn	2.30
5	−0.15120	Pb	2.30
10	−0.10675	Mg	2.65
20	−0.063290	Cu	2.67
50	−0.028840	Au	3.01
100	−0.015321	Ag	3.02
		Li	3.28
		Na	3.99
		K	4.96
		Rb	5.23
		Cs	5.63

[3] Compare to the values of $c_0 = 0.862849, c_1 = 3.22016$, and $c_2 = 3.03546$ by Kiejna and Wojciechowski [352], which result in visually very similar relations.

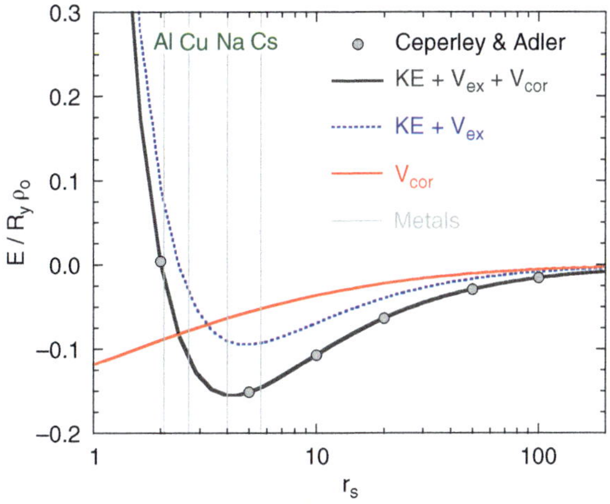

Figure 27.2 Numerical data from Table 27.1 compared to $KE + V_{ex}$, V_{cor}, and their sum, as calculated using Eqs (27.17) and (27.18).

EXAMPLE: Find r_s for copper.
SOLUTION: For copper, $\mu = 7$ eV, so that $k_F = \sqrt{2m\mu}/\hbar = 13.5546\ \text{nm}^{-1}$. Therefore, from Eq. (27.16)

$$r_s = \left(\frac{9\pi}{4}\right)^{1/3} \frac{1}{k_F a_o} = \frac{1.9192}{(13.555\ \text{nm}^{-1})(0.052918\ \text{nm})} = 2.6756$$

Compare to Table 27.1.

27.4 Core term

The assumption of a uniform positive background charge distribution to represent the lattice of positive ions that compliment the smeared distribution of electrons has the virtue of simplicity, but not accuracy. A more faithful approximation is to consider the *sphere* of radius r_s to be composed of an ionic core (the nucleus and the electrons of the lower orbits) and an electron that is smeared more or less uniformly throughout it [164]; a metal is then a lattice of such spheres. Such an approximation is quite good as long as the so-called Wigner–Seitz sphere [357] is surrounded by other such spheres making up the bulk of the material (near the surface, the charge distribution within the sphere deforms and a dipole contributes that becomes a part of the work function). The configuration will give rise to a self-energy of the spherical electron cloud, and an energy associated with the interaction of the electron cloud with the core. Extended further, an assembly of non-overlapping atomic potentials centered on each atom, otherwise known as "muffin-tin potentials" (an array of depressions in an otherwise flat surface) can, for example, explain features of the field emission total energy distribution that the free electron model cannot [358]. The present model, though conceptually similar, is simpler in its application and therefore more of an introduction.

Therefore, for an electron in the sphere of radius $r_s a_o$ add to Eq. (27.17) a self-interaction term ε_{ee} and an interaction with the ion at the core ε_{ei}. These energies contribute to Σ_ε. Let the electron in the Wigner–Seitz sphere be a uniform density of charge, and the core be a bare ion. The electron self-interaction term can be evaluated using familiar methods [99]: inside a uniform gas of charge, a test charge at a radius r feels only the net charge inside that radius, as the cumulative effect of all charge outside that radius cancels. As a result, a differential shell of electron gas of volume $4\pi r^2 dr$ at a distance r from the origin experiences a charge of magnitude $q\rho_o(r/r_s a_o)^3$, where r is the (dimensioned) radial distance to the shell from the origin. Remembering Eq. (27.16) for ρ_o, the self-interaction term is then

$$\varepsilon_{ee} = \int_0^{r_s a_o} \frac{4Q}{r} \rho_o \left(\frac{r}{r_s a_o}\right)^3 4\pi r^2 dr = \frac{6}{5r_s} R_y \tag{27.19}$$

The interaction with the ion core is different: first, the interaction energy is negative, and second, *all* the ion charge is at the center, so

$$\varepsilon_{ei} = -\int_0^{r_s a_o} \frac{4Q}{r} \rho_o 4\pi r^2 dr = -\frac{3}{r_s} R_y \tag{27.20}$$

Table 27.2 Values of sphere radius r_s and ionic radius r_i used in Figure 27.4 from ref. [86] ($r_i = a_i/a_o$ and similarly for r_s, where $a_o = 0.052918$ nm is the Bohr radius and a_i is the physical radius). Li–Cs are labeled "alkali" and Cu–Au are labled "other" in the figure.

	Li	Na	K	Rb	Cs	Cu	Ag	Au
r_i	1.29	1.83	2.51	2.80	3.16	2.55	2.38	2.59
r_s	3.25	3.93	4.86	5.20	5.63	2.67	3.02	3.01

Thus, the sum of the energy components (except for ε_{cor} to save space, and because it is not affected by including ε_{ee} and ε_{ei}) is now[4]

$$\varepsilon_k + \varepsilon_{ex} + \varepsilon_{ee} + \varepsilon_{ei} = R_y \left\{ \frac{2.2099}{r_s^2} - \frac{0.91633}{r_s} + \frac{6}{5r_s} - \frac{3}{r_s} \right\} \tag{27.21}$$

Reinserting ε_{cor} and defining $\Sigma_\varepsilon = \varepsilon_k + \varepsilon_{ex} + \varepsilon_{cor} + \varepsilon_{ee} + \varepsilon_{ei}$ gives

$$\frac{\Sigma_\varepsilon}{R_y} = \frac{2.2099}{r_s^2} - \frac{2.7163}{r_s} - \frac{0.8740}{r_s + 3.3618\sqrt{r_s} + 2.946} \tag{27.22}$$

for which the kinetic repulsion is balanced by the potential attraction. Stability occurs at the minimum, which is at $r_s = 1.588$ or $r_s a_o = 0.08403$ nm. Although this number is comparable to the radius of the monovalent metals as in Table 27.2, it is *smaller* and independent of the metal species, a bad sign given that metal atoms of course have different radii. Something has been overlooked, and that something is the ionic core.[5]

A modification nees to be made to the calculation by letting the core be of a finite size specified by $r_i a_o$ (the i subscript is for "ion"). Of all the ways to account for the core, assuming it to be impenetrable is easiest in that no part of the electron cloud gets past the boundary at r_i, a useful albeit draconian "effective potential" [78]. Now the integrals are considered afresh, after it is carefully observed that *the density ρ_o has changed*: the exclusion of the electron from the core region means the core volume must be excised from the volume that the electron occupies. Therefore, let $1/\rho'_o = (4\pi/3)(r_s^3 - r_i^3)$. Now the integrals are[6]

$$\varepsilon_{ee} = \int_{r_i a_o}^{r_s a_o} \frac{4Q}{r} \rho'_o \left(\frac{r}{r_s a_o} \right)^3 4\pi r^2 dr = \frac{6R_y}{5r_s^3} \left(\frac{r_s^5 - r_i^5}{r_s^3 - r_i^3} \right) \tag{27.23}$$

$$\varepsilon_{ei} = -\int_{r_i a_o}^{r_s a_o} \frac{4Q}{r} \rho'_o 4\pi r^2 dr = -3R_y \left(\frac{r_s^2 - r_i^2}{r_s^3 - r_i^3} \right) \tag{27.24}$$

To leading order in (r_i/r_s), their combination is

$$\varepsilon_{ee} + \varepsilon_{ei} \approx -\frac{9}{5r_s} + \frac{3r_i^2}{r_s^3} \tag{27.25}$$

Observe that the second term *acts to increase the kinetic energy* by tightening up the space that the electron can occupy. Σ_ε becomes

$$\Sigma_\varepsilon = \varepsilon_k + \varepsilon_{ex} + \varepsilon_{cor} + \varepsilon_{ee} + \varepsilon_{ei} \tag{27.26}$$

then Σ_ε/R_y is shown in Figures 27.3 and 27.4 for various values of r_i (the algorithm being given in Section A3.20), where the inclusion and exclusion of Eq. (27.25) is governed by C_i being equal to 1 or 0, respectively. The points labeled "Data" in the later figure are for various metals.[7] It may seem that the total exclusion of electrons from the core is too Draconian, so that softening the value of C_i may mimic that – although the correspondence is improved, the grounds for doing so are *ad hoc*: a softening of the core potential requires a better model, the one given next being a suitable template.

[4]It bears repeating because it is easy to overlook: r has dimensions of length, but $r_s = a_s/a_o$ and $r_i = a_i/a_o$ below are dimensionless.
[5]Calling the core an *ion* is not the best use of words: what is meant is the atom inside the Wigner–Seitz cell apart from the last orbiting electron under consideration. It is a terminology of convenience to suggest that an atom stripped of its last electron has a certain size, even if it is stuck in a lattice of similarly afflicted atoms.
[6]At the risk of irritating the reader (again), remember that r has the dimensions of [nm], but that r_i and r_s are dimensionless.
[7]Data is from Table 9 (ionic radii, p. 78) and Table 1 (Fermi parameters, p. 150) of ref. [86].

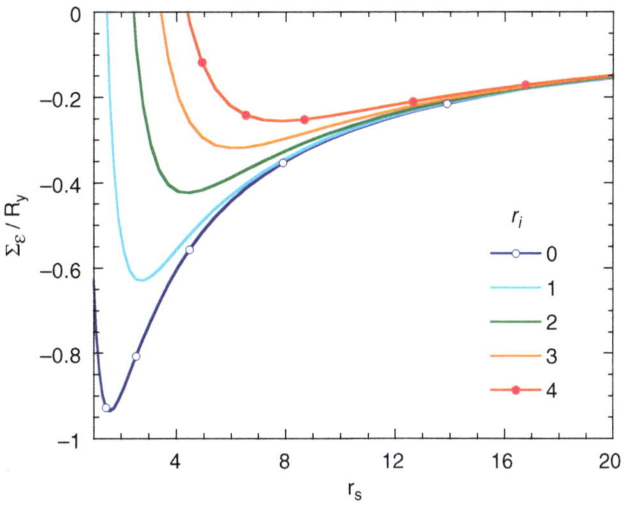

Figure 27.3 The variation of Σ_ε/R_y as a function of r_s for various values of r_i after ε_{ee} and ε_{ei} (Eq. (27.25)) are included in Σ_ε. A minimum as a function of r_i is evident, and the variation of its location with r_i is shown in Figure 27.4.

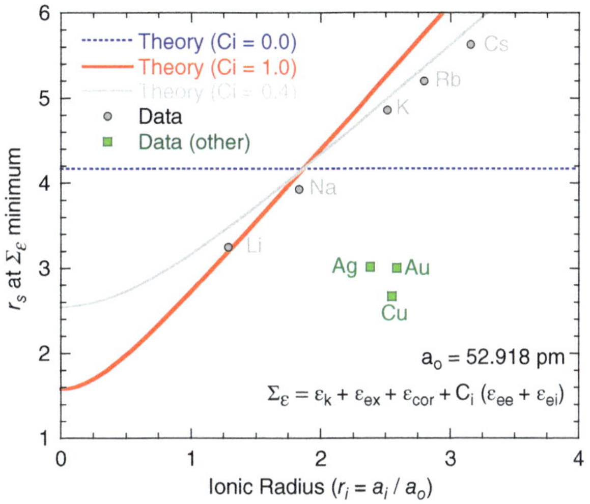

Figure 27.4 Value of r_s that minimizes Σ_ε/R_y (Eq. (27.26)) as a function of r_i. "$C_i = 0$" neglects $\varepsilon_{ee} + \varepsilon_{ei}$; "$C_i = 1$" includes it. Other single valency ($z = 1$) metals (e.g., copper, silver, and gold), require additional considerations [78]. Data from ref. [86]. Treating C_i as adjustable ($C_i = 0.4$ line) improves correspondence, but requires justification.

Reconsider the core model afresh by adopting the methods of scattering theory [135, 357–359]. Given the symmetry of the Wigner–Seitz sphere, the wave function inside, $u(r)\exp(i\vec{k}\cdot\vec{r})$, must be matched to outside, $\exp(i\vec{k}\cdot\vec{r})$. The Bloch functions $u(r)$ (recall Section 23.2) are given by spherical Bessel functions $j_l(x)$ and $n_l(x)$ with $l = 0$, where $j_0(x) = \sin(x)/x$ and $n_0(x) = -\cos(x)/x$: they are solutions to Helmholtz's equation and describe radial waves. If the core were of zero size ($r_i \to 0$) then $u(r)$ would only be given by $j_l(r)$ as it alone vanishes for $r \to 0$, but for finite size the $n_l(kr)$ solutions contribute so that

$$u(r) = j_0(kr) + ka\, n_0(kr) \tag{27.27}$$

where k and a will be determined by boundary conditions. The first boundary condition is that the derivative of the wave function be continuous across the sphere boundary, and that in turn demands that at $r = r_s a_o$,

$$j_0'(kr_s a_o) + ka\, n_0'(kr_s a_o) = 0 \to ka = -\frac{j_0'(kr_s a_o)}{n_0'(kr_s a_o)} \tag{27.28}$$

For the ground state, the small kr limits of the spherical functions give

$$ka = \frac{kr\cos(kr) - \sin(kr)}{kr\sin(kr) + \cos(kr)} \approx -\frac{1}{3}k^3 r^3 \tag{27.29}$$

or, using $r = r_s a_o$, then

$$k^2 = -\frac{3a}{(r_s a_o)^3} \tag{27.30}$$

where the introduced length a can be related to phase shift δ_0 in scattering theory [41]. Using the *first Born approximation* [359], the phase shifts are given by

$$\tan\delta_l = -\frac{2m}{\hbar^2}k\int_0^\infty j_l(kr)^2 V(r) r^2 dr \tag{27.31}$$

so that $a = k^{-1}\tan\delta_0$ for $k \to 0$, and so

$$a = -\left(\frac{2m}{\hbar^2}\right)\int_{r_i a_o}^{r_s a_o}\left(\frac{\sin(kr)}{kr}\right)^2\left(\frac{4Q}{r}\right) r^2 dr \tag{27.32}$$

The small k limit can be taken immediately, and the result is $a = a_o(r_s^2 - r_i^2)$, which, when coupled with k^2 in Eq. (27.30), results in

$$\frac{\hbar^2 k^2}{2m} = -\frac{3}{r_s}R_y + \frac{3r_i^2}{r_s^3}R_y \tag{27.33}$$

This is simply the leading order terms from Eq. (27.21). An adjustable C_i is then seen to be analogous to reducing the ability of the ion core to keeping the electron out, thereby causing the kinetic energy correction term to be smaller.

Two questions are left. First, what about polyvalent atoms? The theory can be modified by replacing $r_s \to r_s z^{-1/3}$ in the derivations [78], where z is the valency number.[8] Second, why does the model work so well for the alkali and not for other (valency = 1) metals like copper, silver, and gold (Cs, Ag, Au)? It is because of the (implicit) assumption that there is no overlap between the ion cores,[9] whereas for Cu, Ag, and Au, the d shell electrons do overlap on adjacent ions; moreover, in polyvalent metals, the electrons occupy more than one Brillouin zone, among other defects of the model [135].

27.5 Exchange-correlation and a barrier model

We should like to point out that it is possible, formally, to replace the many-electron problem by an exactly equivalent set of self-consistent one-electron equations.

– Walter Kohn and Lu Jeu Sham [360]

The passage of the electron from metal to vacuum affects the approximation that the potential is modified by an image charge term near the surface. It is useful to imagine the opposite problem of an electron getting *out* by imagining what goes on when it tries to get *in* the surface of a metal [104]: as the electron approaches the surface, the correlation term reduces the density of charge about the incoming electron, essentially creating a correlation hole that, by the time the electron enters the metal, has a charge about equal and opposite to that of the electron. Consequently, the plummeting of the image charge potential $-Q/x$ to $-\infty$ is sidestepped. The indistinguishability of the incoming electron to those in the metal also gives rise to an exchange contribution that, when coupled with the correlation interaction, matches up with the image charge potential further than a few angstroms from the surface of the metal.

To find the variation of the effective potential, the variation in electron density serves to specify it for a single electron. That this is so, as shown by Kohn and Sham [360], is because the exact ground state energy is a functional of density and can be written as

$$E_G = KE[\rho] + \int V(\vec{r})\rho(\vec{r})d\vec{r} + 2Q\int\int\frac{\rho(\vec{r})\rho(\vec{r'})}{|\vec{r}-\vec{r'}|}d\vec{r}d\vec{r'} + E_{xc}[\rho] \tag{27.34}$$

If the exchange-correlation potential V_{xc} is defined as

$$V_{xc}(\rho) = \frac{d}{d\rho}E_{xc} = \frac{d}{d\rho}\left(\rho\epsilon_{xc}(\rho)\right) \tag{27.35}$$

---

[8] The key phrase is "in the derivations" because making replacements at the end misses the subtle changes.

[9] Explicitly, the assumptions are that the electrons can be treated as plane waves over most of the Wigner–Seitz sphere, and that the ion cores do not have appreciable overlap for adjacent atoms.

then the density ρ which satisfies these equations is that given by the one-particle Schrödinger's equation

$$\left\{-\frac{\hbar^2}{2m}\nabla^2 + V(\vec{r}) + V_{xc}(r)\right\}\psi_{\vec{k}}(\vec{r}) = \varepsilon_{\vec{k}}\psi_k(\vec{r}) \tag{27.36}$$

where $V(\vec{r})$ contains the familiar components of barriers and electrostatic contributions from Poisson's equation. Solving these equations is the artistry of *density functional theory* and doing so is very effective in modeling the behavior of surface barriers [361].

V_{xc} accounts for (most of) the changes in passing from the high electron density of a metal to the emptiness of a vacuum ($r_s \to \infty$). Seeing how that unfolds is edifying, with a good example being sodium (Na). Sodium is unique in this regard, as r_s(Na) is close to the value which which minimizes the energy in Figure 27.2 or right near the cross-over of Figure 27.4 where the ion core effects are minimized (thereby enabling their neglect) and the potential energy can therefore be treated without the ε_{ee} and ε_{ei} corrections [362]. It is expected that

$$V_{xc} = -\mu - (\Phi - \Delta\Phi) \tag{27.37}$$

where $\Delta\Phi$ is an electrostatic contribution from Eq. (26.50). The zero of energy is chosen so that $V_{xc}(r_s \to 0) = 0$. First switch from density ρ to r_s by

$$\frac{d}{d\rho} = \left(\frac{d\rho}{dr_s}\right)^{-1}\frac{d}{dr_s} = -\frac{4}{3}\pi a_o^3 r_s^4 \frac{d}{dr_s} \tag{27.38}$$

and then apply it to V_{xc} (where momentarily the s subscript is suspended for simplicity) to get

$$\frac{V_{xc}}{R_y} = -\left(\frac{18}{\pi^2}\right)^{1/3}\frac{1}{r} - 1.1653\frac{r + 2.9416\sqrt{r} + 2.2095}{(r + 3.3618\sqrt{r} + 2.9460)^2} \tag{27.39}$$

EXAMPLE: Estimate Φ for sodium (Na) assuming $r_s = 3.93$ (corresponding to $\mu = 3.2446$ eV). Compare to the accepted value of $\Phi(\text{Na}) = 2.29$ eV.

SOLUTION: Using Eqs (27.39) and (26.50) for V_{xc} and $\Delta\Phi$, respectively,

$$\begin{aligned}\Phi &= V_{xc} - \mu + \Delta\Phi \\ &= (5.2650 - 3.2446 + 0.0790)\text{ eV} = 2.0995\text{ eV}\end{aligned}$$

which good to within 10%.

That the exchange-correlation contribution accounts for the image charge behavior can be seen by a simple calculation for sodium. Using the algorithm of Section A3.21, the results of which are shown in Figure 27.5, already shows

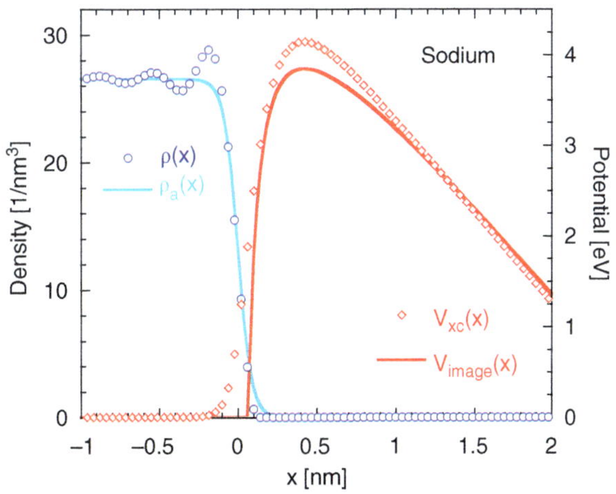

Figure 27.5 Simple exchange-correlation potential for sodium parameters compared to the image charge approximation using the algorithm and output of Section A3.21.

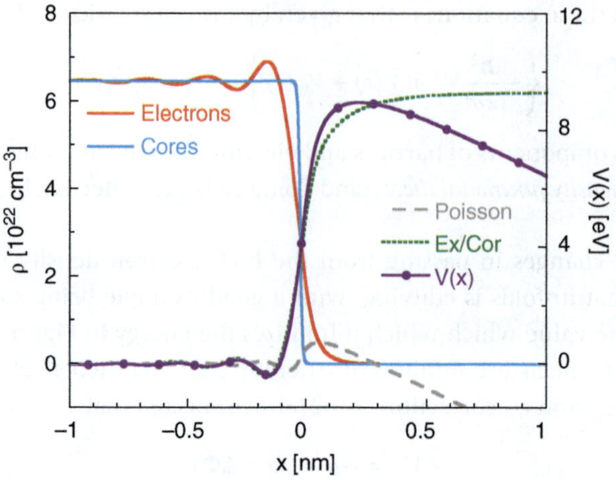

Figure 27.6 Sum of the exchange-correlation potential and the potential from Poisson's equation. The origin is set at the edge of the background positive charge and is determined by the insistence on global charge neutrality. Based on Figure 2 of ref. [347] for molybdenum parameters.

fairly good agreement between the exchange-correlation potential and the image charge approximation for a field of $F = 2$ eV/nm. Numerical evaluations of the exchange-correlation potential coupled with Poisson's equation, and the determination of the location of x_i (the edge of the background positive charge due to the positive cores), followed by summing the various components, yields a representation much like Figure 27.6 [347, 363]), showing the contribution of the various components.

Such acrobatics drives home rather forcefully that the work function is not a fixed quantity: it depends on how the electron density varies near the surface, on the nature of coatings like cesium that coat the surface, of which the crystal face is present, and even on features (such as surface corrugation [85, 135, 364, 365]) that have not as yet been treated. Therefore, look askance at the idea that the work function is an *unchanging* input factor when approaching the emission equations. That seems to cast doubt on how much confidence to put into those equations, but that doubt is premature and the next section considers an approach.

CHAPTER 28

An analytic image charge potential

> **Garnet:** *Don't answer the question they're asking. If a dishonest man has formed the question, there will be no honest answer. Answer the question beneath the question. The* equivalent *question. Answer the question really asked. And answer it with your life.*
>
> – Bill Cain[1]

Questions of utility broached at the end of both Sections 26.4 and 27.5 remain regarding how to account for the variation of $V(x)$ near the surface. The question "What does the emission barrier look like?" is not easily answerable: the image charge approximation so widely used is bereft of effects associated with wave function penetration of the barrier and thermal expansion of the lattice, and so is silent about changes to the work function that are a consequence. The evaluation of those effects is numerically achievable but not practically easy, and are increasingly complicated the more the potential is faithful to the variations in electron density and how it changes in response to field and temperature, not to mention coatings, crystal face, and non-planar structure. To model experimental data becomes difficult.

In practice, though, what the barrier looks like is not really the question being asked. Given the utility of the canonical equations in modeling experimental data, the question behind the question is, can those equations be retained with, at worst, some "effective" parameters used within an *analytic* image charge barrier? That is an unambiguous question, and one worth answering. The answer is "Yes."

28.1 Work function and temperature

> **Einstein:** *You say that as if it were nothing.*
> **Möbius:** *How else should I say it?*
>
> – Friedrich Dürrenmatt[2]

The temperature variation of the work function Φ is empirically known to vary with temperature, although the variation is not large. Over a good temperature range, the variation is more or less linear, or

$$\Phi(T) = \Phi_o + \tilde{\alpha} T \tag{28.1}$$

where $\tilde{\alpha}$ on the order of k_B [135, 366] (the tilde on it is to help prevent its being confused with the fine structure constant). That behavior is manifested in numerical simulations [347] in which the exchange-correlation potential and Poisson's equation for the proper treatment of the dipole term are self-consistently solved, as in Figure 28.1. Clearly over the range typical of thermionic dispenser cathodes (800–1200 K), the linear approximation is reasonable to account for changes on the dipole term wrought by increased penetration of the barrier by higher energy electrons. This is not the entire story, however. Thermal expansion of the bulk material, different surfaces associated with different crystal faces [364], and other physical effects bring in their own contributions [164], so that $\tilde{\alpha}$ can even be negative, as shown for some materials in Table 28.1. For example, for copper, the measured value of $\tilde{\alpha}/k_B$ was (–5.69, –5.60, –4.74) [100], (–4.48) [110], (–3.45) [221], and (–2.76, –3.19) [111], where the crystal face is in square brackets, and each number refers to a different measurement [367], the data for which are shown in Figure 28.2.

28.2 Work function and field

> *No one can be a great thinker who does not recognize that as a thinker it is his first duty to follow his intellect to whatever conclusions it may lead.*
>
> – John Stuart Mill[3]

[1] Bill Cain, *Equivocation* Dramatists Play Service, 2014, Act II, Scene 3.
[2] Friedrich Dürrenmatt, *The Physicists* (trans. James Kirkup). New York: Grove Press, 1964, Act II, p. 74.
[3] John Stuart Mill, *On Liberty*. New York: Liberal Arts Press, 1956, p. 41.

Introduction to the Physics of Electron Emission, First Edition. Kevin L. Jensen.

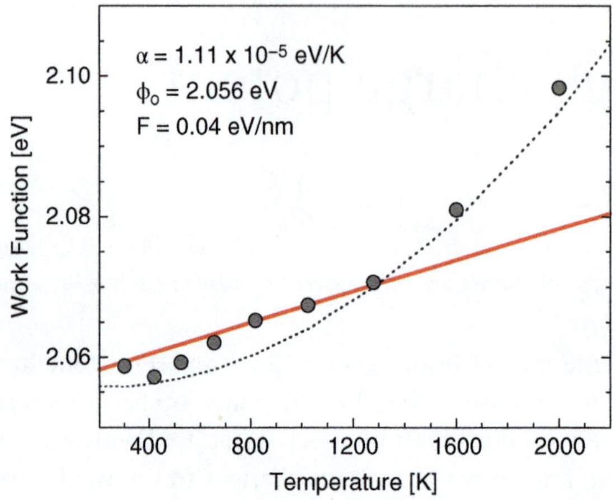

Figure 28.1 Work function corresponding to the exchange-correlation potential and dipole contribution for cesium-like parameters. The applied field is set at 40 MV/m, based on Figure 5 of ref. [347]. The slope of the red line corresponds to $\tilde{\alpha} = 0.1288k_B$. The dashed grey line to guide the eye corresponds to $\Phi(T)[\text{eV}] = 2.056 - 0.051k_BT + 1.6(k_BT)^2$.

Table 28.1 Work function and temperature: the value of $\tilde{\alpha}$ in Eq. (28.1) for various metals. Adapted from ref. [135] for representative cases except for Cu [367]. Values of $\tilde{\alpha}$ can vary depending on crystal face and method of measurement. For a number of elements, the quoted value is representative, not unique.

Name	*Atomic number*	*Symbol*	Φ_o	$\tilde{\alpha}/k_B$
Barium	4	Ba	2.3	4.31
Carbon	6	C	4.39	1.46
Silicon	14	Si	4.02	2.24
Chromium	24	Cr	4.58	0.52
Manganese	25	Mn	3.83	0.95
Iron	26	Fe	4.31	0.52
Cobalt	27	Co	4.4	0.78
Nickel	28	Ni	4.89	−1.46
Copper	29	Cu	4.4	−4.74
Strontium	38	Sr	2.3	0.43
Yttrium	39	Y	2.95	0.17
Niobium	41	Nb	3.95	0.26
Molybdenum	42	Mo	4.55	−0.34
Silver	47	Ag	4.31	0.09
Hafnium	72	Hf	3.6	1.21
Tantalum	73	Ta	4.33	0.22
Tungsten	74	W	4.5	0.13
Rhenium	75	Re	4.93	0.34
Iridium	77	Ir	5.4	−0.26
Platinum	78	Pt	5.03	1.72
Gold	79	Au	4.25	0.13
Thorium	90	Th	3.38	0.39
Uranium	92	U	3	2.33

The data points of Figure 28.1 were evaluated using an assumed field of 40 MV/m. When the data points are evaluated using a higher field of 100 MV/m, the fitting lines as a whole are shifted to a different Φ_o. In other words, the field F makes a difference in the fitted value of the zero-temperature work function Φ_o. Alternately, holding temperature constant for evaluations analogous to that of Figure 28.1 results in Figure 28.3, using the value $x_o = 1/k_o = \hbar/\sqrt{2mV_o}$ with V_o corresponding to the maximum of $V(x)$ (for the image charge potential, it would be $\mu + \phi$, but in the simulation $V(x)$ differs from $V_{img}(x)$). This dependence is not entirely unexpected: recall from Section 26.1 that changes in barrier height correspond to a rigid translation of the density by an amount corresponding to x_o as in Eq. (26.37), to a good

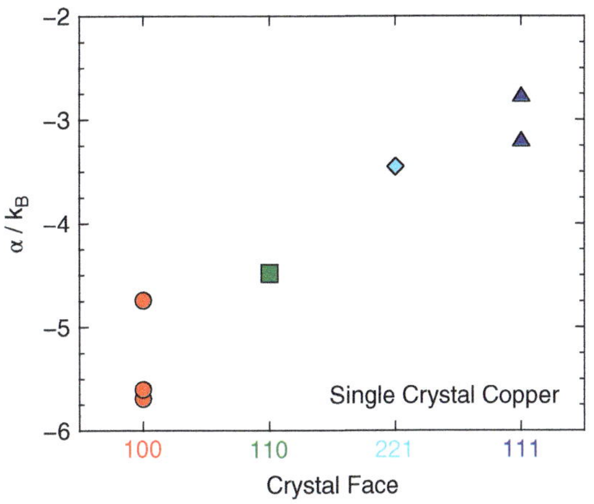

Figure 28.2 Variation of $\tilde{\alpha}/k_B$ for crystal faces of single crystal copper (adapted from Table 2 in ref. [367]).

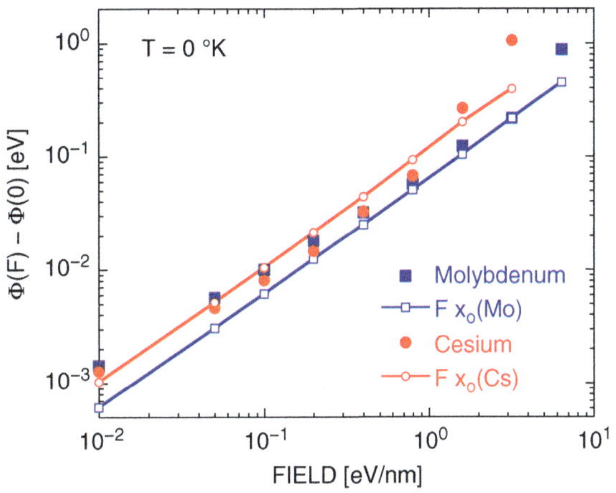

Figure 28.3 Work function corresponding to the exchange-correlation potential and dipole contribution for molybdenum-like parameters, based on Figure 5 of ref. [347]. Observe that x_o is directly evaluated from the numerically determined barrier maximum.

approximation. As the image charge potential and the onset of the external field are with respect to the surface associated with the positive background charge in the jellium approximation, both shall be affected, and so there are good reasons to expect $Q/x \to Q/(x+x_o)$, an expectation that is put on a good foundation from a more complex analysis of the image charge potential using Thomas–Fermi theory [368]. But Figure 28.3 is something different: the present result suggests that $\Phi(F) \to \Phi(0) + Fx_o$.

Grouping all the field and temperature effects (the term in Eq. (26.51) is included implicitly) together in one equation, a reasonable expectation is that an analytical potential barrier might look like [85, 347]

$$V(x) \sim \mu(T) + \Phi_o + \tilde{\alpha}T - F(x - x_o) - \frac{Q}{x + x_o} \tag{28.2}$$

but if an *effective* work function of the form

$$\Phi_{eff} \equiv \Phi_o + 2Fx_o + \tilde{\alpha}T \tag{28.3}$$

is proposed, then

$$V(x) \to \mu(T) + \Phi_{eff} - F(x + x_o) - \frac{Q}{x + x_o} \tag{28.4}$$

This is a fairly convenient outcome: it means that *the canonical equations can still be used* with just a minor change in the parameter associated with the work function, because the form of the image charge potential is retained through a redefinition of position via $x \to x + x_o$, which affects neither the barrier height for a *given* work function Φ_{eff} nor the calculation of the Gamow factor for the tunneling probability. That is, although $J(F, T)$ will change, the means to calculate

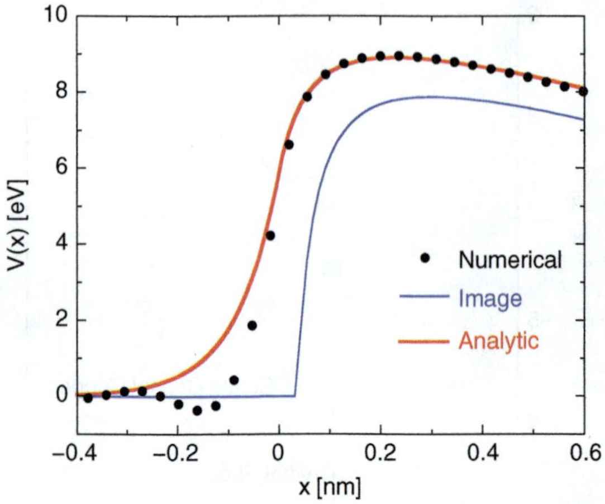

Figure 28.4 Comparison of the numerically evaluated $V(x)$ including exchange-correlation and dipole effects for a field of 4 eV/nm, compared to the equivalent analytical image charge potential for molybdenum parameters. Based on Figure 4 of ref. [347].

it using, for example, the general thermal field equation, will proceed no differently than before. However, x_o must still be determined, although a reasonable expectation is that it is close to the value given by the analytic image charge potential. In fact, a comparison of Eq. (28.4) to the numerically found potential profile is shown in Figure 28.4 and the agreement is fairly good in terms of barrier height and shape for energies near to the Fermi level or the barrier maximum. Observe, though, that the analytic image charge barrier is more unfavorable to emission than the standard image charge barrier.

28.3 Changes to current density

Non sum qualis eram

–Horace[4]

The changes to the estimation of current density using the canonical equations can be traced to the behavior of Φ_{eff} of Eq. (28.3). The impact is relatively straightforward to ascertain. Let Φ_o be the fixed work function and $\Phi = \Phi_{eff}$ be the effective work function modified by temperature or field effects. Consider the effect of each one applied separately to the limiting cases of thermal emission (via the Richardson–Laue–Dushman (RLD) equation of Eq. (12.3)) and field emission (via the Fowler–Nordheim (FN) equation of Eq. (13.25)).

- **Changes to RLD current density** The dependence of the Richardson–Laue–Dushman on work function is simple, and so a comparison of $J_{RLD}(T, \Phi)$ to $J_{RLD}(T, \Phi_o)$ is most easily considered by examining their ratio, or

$$\frac{J_{RLD}(T, \Phi_o + \tilde{\alpha} T)}{J_{RLD}(T, \Phi_o)} = \exp\left(-\frac{\tilde{\alpha}}{k_B}\right) \tag{28.5}$$

  and therefore a constant change in the current density is obtained, for example for tungsten the current is 88% of what it would be if the work function had no temperature variation.[5] Because the change is constant, the effect is subsumed in the value of the work function.

- **Changes to FN current density** For field emission, the work function is more pervasive throughout the equation, appearing in the coefficient, in the factor $v(\Phi)$, and in the exponent in the factors of Eq. (13.25). Additionally, x_o is dependent on the height of the barrier: to leading order it may be evaluated from the simple image charge barrier by $x_o \approx \hbar/\sqrt{2m(\mu + \phi)}$ with $\phi = \Phi_o - \sqrt{4QF}$. Even so, the evaluation of $J_{FN}(F; \Phi)$ is nuanced, as in the algorithms of Section A3.22 used to generate Figure 28.5. The primary impact on $\ln(J/F^2)$ vs $(1/F)$ will be a shift in the intercept plus a small change to the slope.

The consequences of a temperature- and field-dependent effective Φ_{eff} therefore are masked in the traditional representations of thermal and field current density. Where the changes become more consequential is near the thermal–field

[4] "I am not what I used to be", Horace, *Horatus Flaccus, Carmina: Liber IV*, http://www.thelatinlibrary.com/horace/carm4.shtml (*The Fourth Book of the Odes of Horace: Ode I. To Venus*).

[5] The temperature change of $\mu(T)$ for tungsten is too insignificant to treat for its effect on $J(T)$.

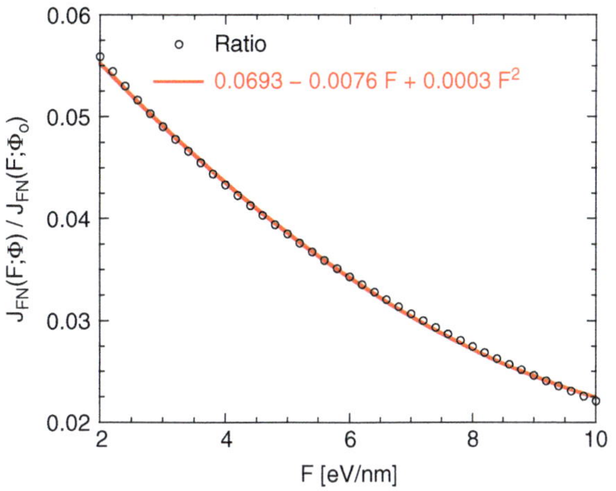

Figure 28.5 Consequences of a field-dependent work function $\Phi(F) = \Phi_o + 2Fx_o(F)$ with $x_o = \hbar/(2m(\mu + \phi))^{1/2}$ and the approximation $\phi = \Phi_o - \sqrt{4QF}$, as calculated using the algorithm of Section A3.22, for copper-like parameters.

transition as a consequence of the changes wrought on the $\theta(E)$ factor in the Kemble approximation. Away from that region, the analytical image charge potential allows for a more or less standard usage of the canonical equations with at worst a willingness to be flexible with what Φ is, with Φ_{eff} being a field and temperature dependent quantity. With Φ_{eff} in hand, though, use may be made of the usual emission formulations.

PART V

Application physics

We shall not cease from exploration
And the end of all our exploring
Will be to arrive where we started
And know the place for the first time.

– T.S. Eliot[1]

[1] T.S. Eliot, *Little Gidding*, No. 4 of *Four Quartets*. Orlando, FL: Harcourt Brace, 1943, Poem V.

CHAPTER 29
Dispenser cathodes

The field of thermionic electron emission is often considered a "mature" technology in which only incremental progress, if any, can be made. However, in recent years with the advent of modern surface analytical capabilities that enable one to better understand the physics and chemistry of emitter surfaces significant improvements have been made in thermionic electron sources...

– Richard E. Thomas, John W. Gibson, George A. Haas, and Richard H. Abrams Jr. [369]

The dispenser cathode developed by Phillips Laboratory [366], engineered to continually dispense low work function coatings such as barium[1] throughout its uncommonly long life, was a groundbreaking achievement. No wonder, then, that the physical processes behind its operation received great attention [228, 366, 369–380]. Although developed as a *thermionic* source, its potential use as a photocathode [381–386] came to be appreciated based on its ability to rejuvenate the low work function coating on which its performance depends.

The surface of a dispenser cathode is shown in Figure 29.1, revealing an assortment of tungsten grains pressed together with pores visible between the grains. Impregnants containing barium are such that when heated, barium diffuses up and out through the pores in a process generally attributed to Knudsen flow [371]. The grain sizes vary, but if the area of a representative number of them are converted to equivalent circular diameters via $A = (\pi/4)D^2$, the distribution of diameters can be used to infer a mean size, as in Figure 29.2. Similarly, the separation between the nearest-neighbor pores is well represented by a log-normal distribution (similar to a normal or Gaussian distribution with $x \rightarrow \ln(x)$ in the argument [387]) given by [382, 388]

$$L(x; \mu, \sigma^2) \equiv \frac{1}{\sqrt{2\pi}\sigma x} \exp\left\{-\frac{1}{2\sigma^2}\left[(\ln(x/\mu))^2\right]\right\} \tag{29.1}$$

The analysis of a typical sintered random pore dispenser cathode gives a grain size of approximately 5–6 μm on average, and a pore-to-pore separation of 2.5 μm. The surface of the dispenser cathode, though machined flat, nevertheless has microscale surface roughness to it. Therefore, the apparent work function of the cathode depends on how coatings come out of the pores, how those coatings spread across the surface, how they evaporate away, and how they depend on the way in which the cathode evolves with usage.

29.1 Miram curves and the Longo equation

The data are presented as "Miram plots" which are normalized plots of emission versus temperature at various current densities...Such a plot is basically a "fingerprint" for any given cathode; it allows meaningful comparisons between cathodes as well as of the change of emission of a given cathode with time.

– Morris Feinleib and Michael C. Green [389]

For a dispenser cathode, higher current density operation comes at the cost of a shorter lifetime, so that knowledge of where the optimal operation point is for dispenser cathodes is exceedingly important, particularly for applications such as space-based traveling-wave tubes (TWTs) [378], where the prospects of replacing a failed cathode are non-existent. As in Figure 29.3, changing the current density from 0.75 A/cm^2 to 3 A/cm^2 (a factor of 4×) reduces the lifetime of an M-type cathode from 30 years to 1 year: as a rule of thumb, each 40 °C increase in temperature approximately halves the cathode's lifetime. As a result, even though thermionic cathodes can be driven harder (e.g., 460 A/cm^2 has been achieved by scandate dispenser cathodes [375]), the preference is often to hold them to a more modest 1 A/cm^2 or less.

Thermionic cathodes are generally run space-charge limited (SCL; apart from being the highest current deliverable across a gap, SCL operation smooths the noisiness of the cathode [221], making other benefits accrue), but the CL

[1] Why the work function is reduced is the topic of Section 31.8.

Introduction to the Physics of Electron Emission, First Edition. Kevin L. Jensen.

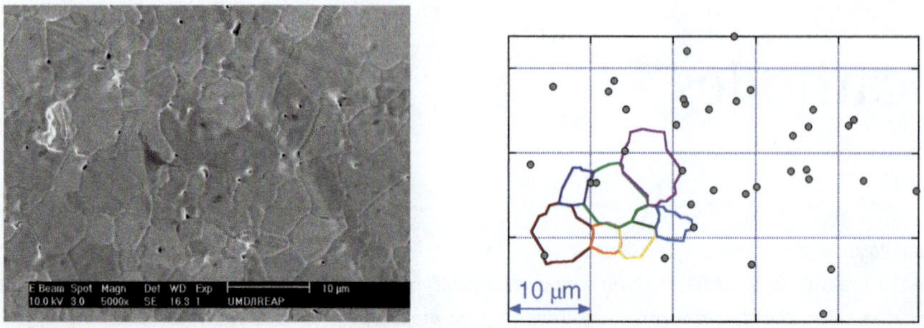

Figure 29.1 Left, the surface of a dispenser cathode (image courtesy of N.A. Moody (UMD / LANL)). Right, grain outlines and pore locations for analysis in Figure 29.2. A figure based on the left image appears as Figure 12 in N.A. Moody, K.L. Jensen, D.W. Feldman, E.J. Montgomery, and P.G. O'Shea, *J. Appl. Phys.*, **102**, 104901 (2007), and is reproduced with permission of AIP Publishing.

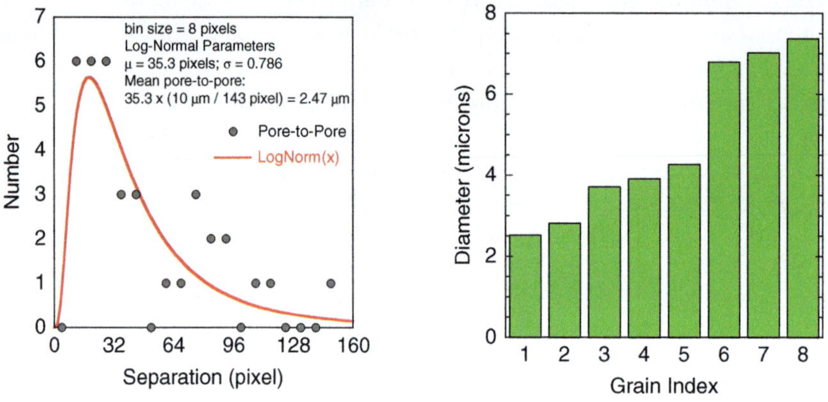

Figure 29.2 Left, determination of mean distance between pores of Figure 29.1 as fitted to the log-normal distribution of Eq. (29.1). Right, area of the grains of Figure 29.1 (right) represented as an equivalent circular diameter in microns, where the labeling is from smallest area to largest.

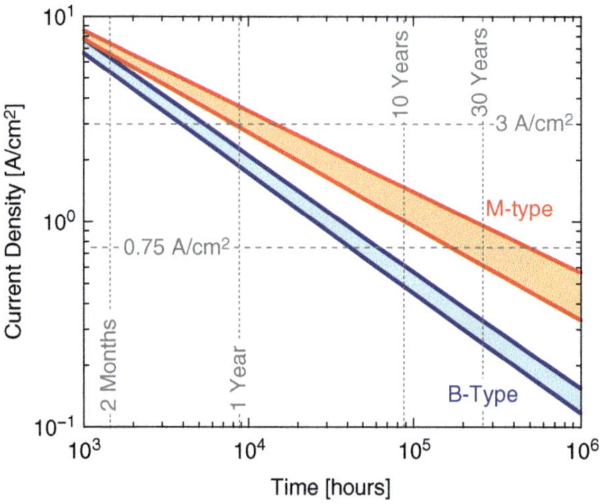

Figure 29.3 Lifetime versus operating current density for typical B- and M-type dispenser cathodes. Based on Figure 5.27 of A.S. Gilmour, *Microwave Tubes*, Artech House, Dedham, MA, 1986, in turn based on a graphic by M.C. Green.

relation of Eq. (16.15) is temperature independent. Increasing the temperature beyond that just required to reach J_{CL} is wasted effort: no more current can be drawn across the gap, and if it is generated anyway, the excess is forced back to the cathode. A convenient method to see where *source-limited current* such as J_{RLD} transitions over to *space-charge-limited current* is given by a Miram plot, in which the ratio of the current density with J_{CL} is given, so that when the relation approaches 100%, no further increase in temperature improves J. The preferred operation point of the dispenser cathode is then at the "knee" of the curve. Knowing how the location of that knee progresses and what causes the sharpness of its bend over time is desirable.

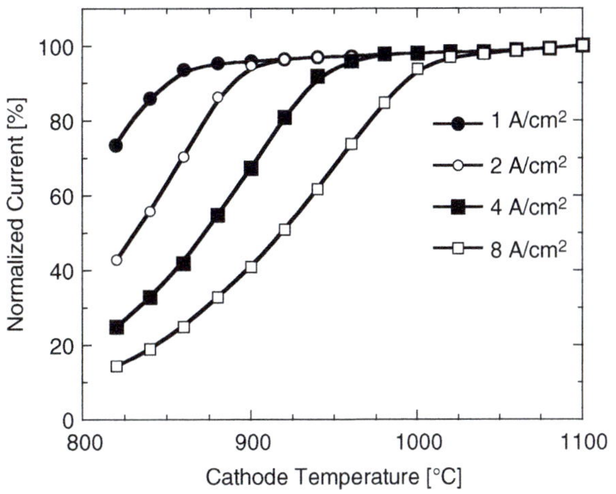

Figure 29.4 A typical Miram plot for experimental data from a dispenser cathode (data courtesy of M.C. Green (Varian Medical Systems) based on M. Feinleib and M.C. Green, *Technical Digest of the International Electron Devices Meeting*, **30**, 314, 1984.

A quintessential Miram plot[2] is shown in Figure 29.4. Such plots are reminiscent of an empirical formula relating the current density J of a thermionic source to the two relevant limiting forms of thermal (J_{RLD}) and SCL (J_{CL}) current density. The Longo equation, often used in life-test models of barium dispenser cathodes, relates them by

$$\frac{1}{J^{\alpha}} = \frac{1}{(J_{RLD})^{\alpha}} + \frac{1}{(J_{CL})^{\alpha}} \tag{29.2}$$

Consequently, if $y = J/J_{CL}$ and $x = J_{RLD}/J_{CL}$, then a Miram plot represents

$$y[x(T)] = \frac{x}{(1+x^{\alpha})^{1/\alpha}} \tag{29.3}$$

The original empirical Longo equation set $\alpha \to 1$; in the later incarnation that is Eq. (29.2), α is referred to as a "shape factor". If T_o is the temperature for which $J_{RLD}(T_o) = J_{CL}$, then a convenient representation becomes

$$x(T) = \left(\frac{T}{T_o}\right)^2 \exp\left(B\frac{T-T_o}{T}\right) \tag{29.4}$$

where $B = \phi/k_B T_o = (\Phi - \sqrt{4QF})/(k_B T_o)$. For example, if $T_o = 1273$ K and $\phi = 2$ eV, then $B = 18.762$. The insertion of Eq. (29.4) into Eq. (29.3) forms the basis of Figure 29.5.

What then if the work function evolves over time, as it does for aging dispenser cathodes? Physically, "aging" is noticed by the requirement of a higher temperature T_o' to achieve J_{CL}, and it is straightforward to infer what change in ϕ' causes this. Setting $J_{RLD}(\Phi, T_o) = J_{RLD}(\Phi', T_o')$ (both are equal to J_{CL}) is solved for ϕ' in terms of ϕ, T_o and T_o' and using Eq. (12.3) as[3]

$$\phi' = \frac{T}{T_o}\left\{\phi + 2k_B T_o \ln\left(\frac{T_o'}{T_o}\right)\right\} \tag{29.5}$$

Small changes in Φ over time therefore do not signficantly change the qualitative shape of the Miram curve, as Figure 29.6 shows (although the knee does become a bit sharper). But change is there: the value of T_o' has increased, which accelerates the aging of the dispenser cathode if the same level of current density is demanded, because the loss of barium is accelerated. On a standard Miram curve like Figure 29.4 (rather than the scaled Figure 29.6), this is manifested as a shift of the knee to the right, and therefore a monitoring of the knee position is a means of monitoring the health of the cathode.

---

[2]This figure by M.C. Green and K.L. Jensen provided the image used on the plaque of the first *International Vacuum Electron Sources Conference Miram Award* (2012) received by Professor Yiman Wang (Beijing University of Technology) in recognition of her meticulous work on scandate cathodes.

[3]Remember that $\phi = \Phi - \sqrt{4QF}$; there is no point distinguishing what the field F is, as it is not changing, so that the change in ϕ is the same as the change in Φ.

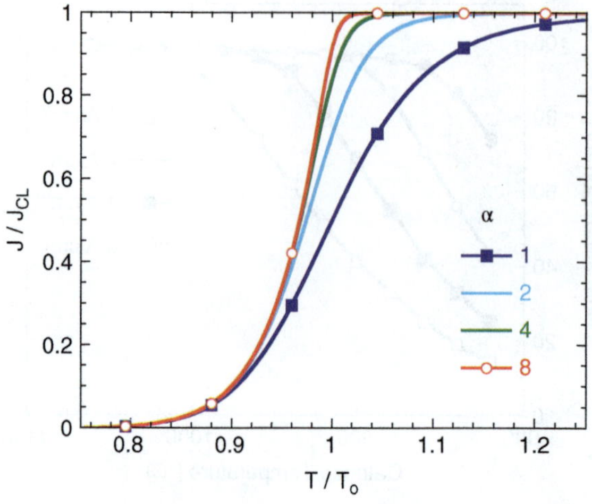

Figure 29.5 A Miram plot (Eq. (29.3)) using the Longo equation (Eq. (29.4)) for various values of α, with $T_o = 1273$ K and $\phi = 2$ eV.

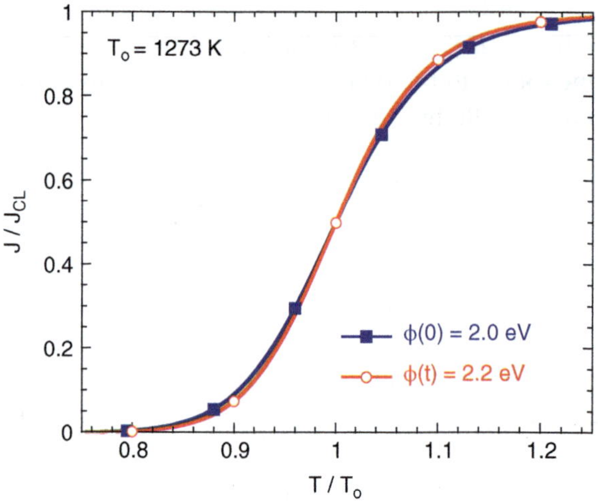

Figure 29.6 Change in the Miram curve for a change in ϕ of 10% (from 2.0 eV to 2.2 eV) for an initial $T_o = 1273$ K and $\alpha = 1$. Observe that were the x axis a temperature scale, the red curve would be undesirably shifted approximately 114 K to the right ($T_o' = 1387$ K) by Eq (29.5).

EXAMPLE: Let T_o' be 10% larger than $T_o = 1273$ K for $\phi = 2$ eV. What has the work function ϕ' changed to as a percentage of ϕ?

SOLUTION: Using $T_o' = 1.1T_o$, then

$$\phi' = 1.1\phi + 2.2k_B T_o \ln(1.1) = (2.2 + 0.023)\text{eV}$$

therefore, ϕ' is 11.15% larger than ϕ. Generally the second term is smaller than the first, so that, crudely, the percentage change in T_o corresponds to the percentage change in Φ.

How quickly J transitions from J_{RLD} to J_{CL} is governed by α, which is affected by work function evolution due to coverage. Causes of its variation are therefore complex, so much so that Vaughan [390] argued that there is no theoretical justification for α (which he called n), entailing that it is simply a fitting parameter for which "good" cathodes are characterized by $6 \leq \alpha \leq 10$, with larger α causing the "knee" of the transition to sharpen. A number of factors typically affect the knee [391], but three are particularly important for the relationship between the Longo equation, the Miram curve, and the transit time model of Section 16.3:

- cathode evolution changes the degree of coverage and therefore the work function
- emitted electrons have a thermal spread in launch velocity (initial velocities are neglected in Eq. 16.32)

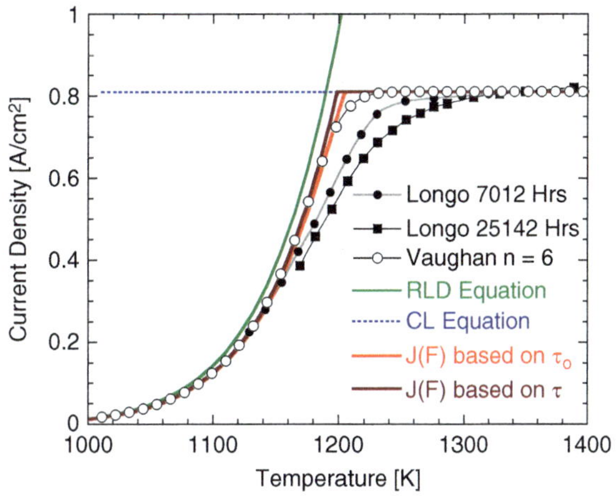

Figure 29.7 Current density from a Ba dispenser cathode after different operational lengths is compared to Vaughan's "good cathode" curve to J_{RLD}, J_{CL}, and the predictions of a transit time model using the ballistic τ_o and full τ in solving Eq. (16.33). Based on Figure 2 of K.L. Jensen, J. Lebowitz, Y.Y. Lau, and J. Luginsland, *J. Appl. Phys.*, **111**, 054917 (2012) using data digitally extracted from Figure 4 of R.T. Longo, *J. Appl. Phys.*, **94**, 6966 (2003).

- dispenser cathodes are made by "sintering" (pressing together under temperature) tungsten grains, so that many crystal faces are exposed to the surface [392], each of which when covered by barium exhibits a different work function.

Such factors introduce velocity spread into the beam, which in turn causes a spread in transit times, by creating changes in the accelerating field near the surface of the cathode that can be affected by either surface structure (protrusions) or different crystal faces (work function variation). The scale of variations in Φ limits the length scale associated with field variation near the cathode, and the Φ variations are associated with the average pore-to-pore separation, which depends on the type of dispenser cathode: a conventional dispenser cathode has randomly spaced pores [372], whereas uniformly spaced pores characterize controlled porosity dispenser cathodes [392]. There are also length scales determined by the variations due to machining [281] and crystal grain size [84]. Variations in velocity after electrons are emitted are due to undulations in field, so the length over which those changes can be induced is related to the feature sizes, as further away the field homogenizes (becomes more uniform). Even though feature sizes are hundreds to thousands of times smaller than the anode–cathode (AK) gap (distance from anode to cathode), non-uniformity contributes to the form of the knee, but in ways difficult to quantify outside of simulation.

A softening of the knee accompanies aging for dispenser cathodes. Longo made measurements of cathodes at different times in their life, generating the curves of Figure 29.7. The two data sets shown, the first corresponding to 7012 hours (0.8 years) with $\alpha \approx 1.8$ and the second corresponding to 25142 hours (2.9 years) with $\alpha \approx 2.5$, show a rather pronounced rounding of the knee that accompanies aging. To relate this to transit time, non-uniformity due to randomly spaced pores and patchy emission [219], and their relation to the impact of space charge in the anode cathode gap, recall that charge in the gap suppresses the surface field F at the cathode, thereby affecting the Schottky barrier lowering factor (the surface field is governed by Eq. (16.33)). When the ballistic transit time τ_o of Eq. (16.24) is used to find F, the bright red line results; if the transit time is given by Eq. (16.32), then the darker red line results. The lower temperature region of theory is thereby brought into correspondence with data, but near the knee other factors contribute. The theory curve is rather close to Vaughan's "good cathode" implying that "goodness" is correlated with a uniform emission surface (an implicit assumption in the one-dimensional models).

29.2 Diffusion of coatings

Theoretical treatments of the diffusion of single adatoms on clean metal surfaces are difficult in that it appears to be necessary to include both the relaxation between substrate atoms and the adatom, and also the variation that can occur in the bond strength per bond when the number of nearest neighbour atoms is varied, if agreement between theory and experiment is to be obtained.

– Martin Prutton [343], p. 149

Because thermionic dispenser cathodes are operated at high temperature, mobile barium atoms migrate across the surface rapidly (as would cesium atoms) and give rise to a low work function coating. The physics of diffusion and evaporation of Ba and Cs coatings on metals at high temperature [373, 393–396] is therefore a central feature of the operation of dispenser cathodes (such processes also affect the operation of field emission [397, 398] and photoemission [380–382] cathodes). Atoms giving rise to surface coatings are provided through *randomly* spaced pores at a temperature-dependent rate, whereas uniform emission implies uniform coatings (surface density in atoms per unit area), which in turn reflects a balance between how material diffuses away from the pores and how fast material evaporates from the surface. The combination of evaporation, diffusion, and flow dictate the "lifetime" of a dispenser cathode. Thermionic dispenser cathodes die by *barium exhaustion*,[4] by which time the barium impregnates have run out of their chemical reactions and the amount of barium that can make passage to the surface becomes insufficient to pull a desired current from the cathode at a preferred temperature [366, 378].

Dispenser cathodes will therefore be the occasion to explore diffusion, evaporation, and flow, but the related problem of work function reduction due to degree of coverage of the coatings themselves shall instead be taken up in Section 31.8. This creates a small problem, as it is useful to know *why* coverage and work function reduction are related at all. The very simple model behind Eq. (21.55) is adequate for now, and it suggests that the coating of cesium or barium causes a work function reduction in proportion to its fractional coverage factor θ, with $\theta = 1$ corresponding to one atom per unit area. Herein is an ambiguity that occasionally surfaces: what is the unit area for one atom on the surface? Analogous to setting the volume per atom (of, say, cesium) $V = (2r)^3$ to correspond to $M\rho$, where $M = 132.905$ grams/mole and $\rho = 1.87$ grams/cm^3 to obtain $r = 0.245$ nm (close to the covalent radius 0.253 nm of cesium), the bond length of Cs–Cs, or $2r = 0.5309$ nm [84], serves as a good choice to determine the unit area because on the surface Cs is not packed into the same lattice structure it experiences in bulk, but conforms in part to the potential peaks and valleys of the underlying substrate and to the proximity of its nearest neighbors in a manner suggested by Figure 29.8. The bond length-derived value comports well with experiment, and so the surface density σ and the coverage factor θ are chosen to be related by

$$\sigma \equiv \frac{\theta}{4r_c^2} \tag{29.6}$$

where r_c is half the bond length, although taking it as the covalent radius is more convenient.

29.2.1 Point source model

...the main concern of physics is not with drunks who stagger home from lampposts ...Instead of talking in terms of a drunk taking steps, let us revert to the less alcoholic vocabulary of physics and think of a particle performing successive steps....

– Frederick Reif [51], pp. 6–7.

Two methods to treat diffusion offer a nice contrast to each other: an analytic approach by assembling solutions to the diffusion equation, and a numerical approach to mimic the actual migration, or random walk, of the atoms across

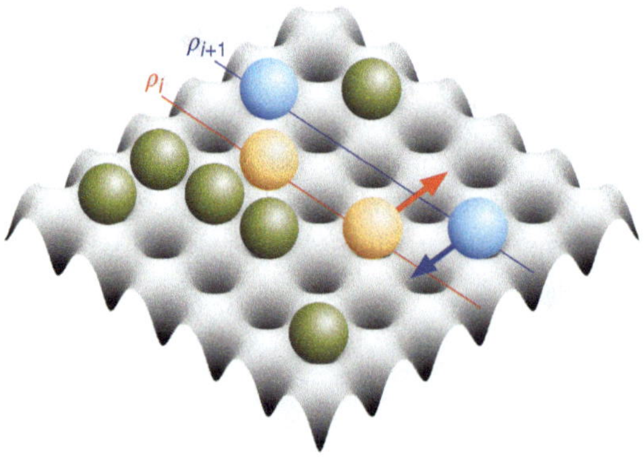

Figure 29.8 Schematic representation of spherical adatoms (cesium or barium) resting on the corrugated potential surface of a bulk material like tungsten. Two fictitious lines, one red and one blue, provide a reference point for atoms to hop off and on to (they can also move along the line, but that does not change the number density), as suggested by the red and blue arrows, for the purposes of investigating the relationship between hopping and diffusion, codified in the rate of change of the number densities ρ_n and ρ_{n+1}.

[4]Cesium dispenser photocathodes end their useful life analogously, but for brevity one or the other is explicitly mentioned and the other implicitly understood.

the surface of Figure 29.8. The analytic model speaks for itself: find the characteristic solutions to the diffusion equation and assemble them to represent a distribution, then start the timer and watch. The analytic approach is well-suited to incorporating parametric models of evaporation and sources. The numerical model treats the atoms as a gas that can migrate, and tracks them as they hop over potential barriers between adjacent sites (diffuse) or hop off the surface altogether (evaporate), with the energies associated with the later typically a factor of 5× or so larger than the former; by way of example, consider BaO, for which the diffusion activation energy is 0.7 eV, but the binding energy is about 2.1 eV [366]. The inter-atomic potentials affect the barrier heights over which atoms jump and the well depths that keep them from escaping [398], features that become important in the lattice gas model of Section 29.2.2. Jump probabilities then are synonymous with random steps.

As the surface atom (taken to be either barium or cesium, but referred to generically as an *adatom*) vibrates back and forth, it strikes the boundaries that confine it. Those boundaries due to the substrate atoms (taken to be tungsten) themselves vibrate, and so the energy of the adatom varies with each collision or interaction. If all the surface atoms along a line are imaginatively strung together like pearls on a chain, visually suggested by the red and blue lines marked ρ_i and ρ_{i+1} respectively in Figure 29.8, with the subscript indexing the chain, then the number of atoms on each chain can be kept track of by the value of ρ at a time $t = t_n$; let it be indicated by a superscript $\rho(t_n) \equiv \rho^n$. In the amount of time taken for the adatom to typically strike the barrier and bounce back, or $\Delta t = t_{n+1} - t_n$, let the probability that the adatom instead hops the barrier to the adjacent site be given by p. The future value of the number of adatom atoms is kept track of by monitoring those that stay put, adding the ones that hop on from the adjacent chains, and subtracting the ones that hop off. In short,

$$\rho_i^{n+1} = \rho_i^n + \frac{p}{2}(\rho_{i+1}^n - \rho_i^n) + \frac{p}{2}(\rho_{i-1}^n - \rho_i^n) \tag{29.7}$$

where each difference indicates the net loss or gain from a changing of atoms between two adjacent chains, and where a factor of 1/2 appears with p because half the atoms are on average headed to the $(i+1)$ line and half to the $(i-1)$ line. A regrouping of terms transforms that equation into

$$\rho_i^{n+1} - \rho_i^n = \frac{p}{2}(\rho_{i+1}^n - 2\rho_i^n + \rho_{i-1}^n) \tag{29.8}$$

which is exactly the form of the Euler or the upwind differencing scheme (UDS) for the approximation of a derivative in time (left-hand side) and the central difference scheme (CDS) for the second derivative in space (right-hand side) using the finite difference methods of Section A1.3.2. In fact, if the spacial separation between the lines marked ρ_i and ρ_{i+1} in Figure 29.8 is Δx and a diffusion constant

$$D \equiv p\frac{\Delta x^2}{2\Delta t} \tag{29.9}$$

is defined, then the continuum form of Eq. (29.8) is

$$\frac{\partial}{\partial t}\rho(x,t) = D\frac{\partial^2}{\partial x^2}\rho(x,t) \tag{29.10}$$

If the "number of atoms" term is scaled by the length (or unit length) of the chain, then ρ_i is a *line* number density, but the treatment is equally applicable to an *area* number density for the diffusion of two-dimensional sheets through a three-dimensional volume. As will be investigated shortly, extensions to higher dimensions is straightforward, amounting to $\partial_x^2 \to \nabla^2$, the resulting equation dubbed Fick's (second) law and the associated D called the Fick's law diffusion coefficient. More importantly, this D is the *chemical diffusion coefficient D* to distinguish it from the *tracer diffusion coefficient* treated in Section 29.2.2 and designated D^* [399].

The diffusion equation of Eq. (29.10) has much in common with the time-dependent Schrödinger's equation and the heat conduction equation. No surprise, then, that methods to deal with these equations have common elements (e.g., Green's function approaches [400]). The *point source solution* has an advantageous viewpoint that recurs in point charge models for the field emission of Section 30.3. It involves finding the solution to

$$\frac{\partial}{\partial t}u(x,t) = D\frac{\partial^2}{\partial x^2}u(x,t) \tag{29.11}$$

Use of the Fourier transform representation

$$u(x,t) = \frac{1}{2\pi}\int_{-\infty}^{\infty} e^{ikx}u(k,t)\,dk \tag{29.12}$$

results in[5]

$$\partial_t u(k,t) = -Dk^2 u(k,t) \tag{29.13}$$

for which the solution is

$$u(k,t) = \exp(-Dk^2 t)u(k,0) \tag{29.14}$$

The reason for calling the method under discussion the "point source solution" is now revealed by the initial conditions: assume that whatever is diffusing is initially all piled up at the origin $x = 0$, that is, assume $u(x,0) \propto \delta(x)$. Then from Eq. (8.74),

$$u(k,0) \propto e^{ikx} \tag{29.15}$$

Inserting $u(k,0)$ into Eq. (29.14), that in turn into Eq. (29.14), and lastly demanding that $u(x,t)$ be *normalized* so that its integral over all x is unity (more on which near Eq. (29.20)) results in

$$u(x,t) = \frac{1}{\sqrt{4\pi Dt}} \exp\left(-\frac{x^2}{4Dt}\right) \tag{29.16}$$

From these, a general ρ that satisfies Eq. (29.10) can be approximated by

$$\rho(x,t) \approx \sum_{j=1}^{N} w_j \, u(x - x_j, t) \tag{29.17}$$

where w_j are weights and x_j are suitably chosen positions.

Adding dimensions is straightforward. In Cartesian coordinates, more dimensions simply adds to the Cartesian Laplacian (see Section A1.3.1), so that in two dimensions $\partial_x^2 \to \partial_x^2 + \partial_y^2$, and so on for three dimensions. The method of separation of variables then posits solutions to be

$$u_2(x,y,t) = u(x,t)\, u(y,t) \tag{29.18}$$

$$u_3(x,y,z,t) = u(x,t)\, u(y,t)\, u(z,t) \tag{29.19}$$

for two and three dimensions, respectively. Consider two dimensions (it is easier, and the generalization to three dimensions is straightforward) and return to the normalization question that should have been taken on in Eq. (29.16). There are two ways to express the normalization of u_2:

$$\iint u_2(x,y,t)\, dxdy = \left(\int_{-\infty}^{\infty} u(x,t)dx\right)^2 = 2\pi \int_0^{\infty} u(r,t) r dr \tag{29.20}$$

where the first form is made possible because the integrals over x and y are identical, and the second because $r^2 = x^2 + y^2$ (a trick familiar from Eq. (5.11)). The second form, though, is easier because of the transformation

$$\int_0^{\infty} e^{-ar^2} r dr = \frac{1}{2}\int_0^{\infty} e^{-as} ds = \frac{1}{2a} \tag{29.21}$$

where $a = 1/(4Dt)$, and so the normalization of $u(x,t)$ in Eq. (29.16) follows straightforwardly.

Solutions to configurations modeled by Eq. (29.17) can apply to experiments designed to infer the approximate magnitude of D for adatoms diffusing over the surface of tungsten, needed to quantify what temperatures are most desirable to balance the speed of restoring the surface to a uniform condition with the losses to evaporation that result when higher temperatures are used. Dispenser cathodes used as thermionic sources do not generally have such a problem: the high temperature of operation insures that barium or cesium spreads rapidly, but for dispenser photocathodes that are operated at room temperature or so [381] the concern is more relevant, as the reduction of the "rejuvenation time" of a cesium dispenser photocathode is desirable even as the evaporation losses that attend heating are not.

The migration of coatings can be observed,[6] as shown in Figure 29.9, by evaporating cesium through a square mesh onto the surface of 70% dense sintered tungsten and monitoring its evolution using a photoemission electron microscope. Insofar as work function $\Phi(\theta)$ depends on coverage θ and the quantum efficiency of emission (which is measured

[5] There are assumptions required to go from Eq. (29.12) to Eq. (29.13), such as that ∂_t can be brought inside the integral, and D is independent of position [60, 400].

[6] Images courtesy of E.J. Montgomery and D.W. Feldman (UMD), and J.L. Shaw (NRL).

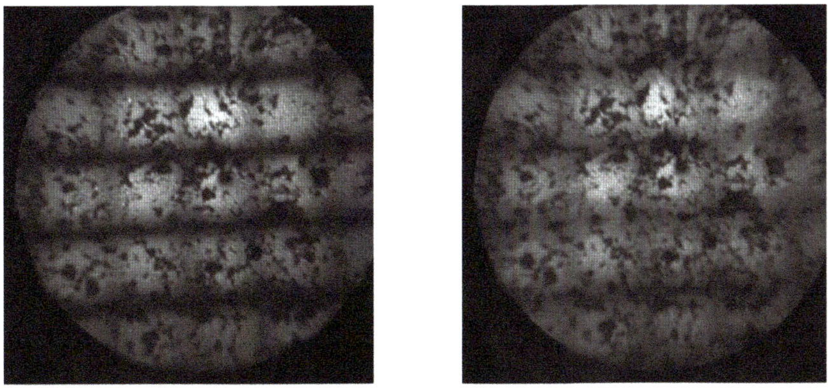

Figure 29.9 Images taken using NRL's photoemission electron microscope (PEEM) by E.J. Montgomery, D.W. Feldman, and J.L. Shaw of cesium deposited through a mesh grid onto the surface of sintered tungsten. The approximate time between images is 8 minutes. Each image is 512 pixels across, or approximately 125 μm. Images courtesy of E. Montgomery (IREAP, UMD, January 2012).

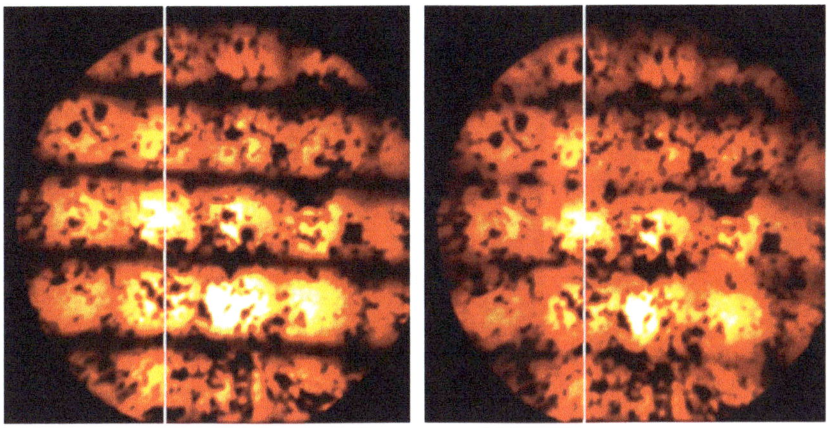

Figure 29.10 Images of Figure 29.9 processed by Algorithm A3.23. The thin white line at horizontal pixel 205 in each marks the pixel column represented in Figure 29.13.

by the brightness of the image) on Φ, the brightness is a proxy measure of the coverage, and so its evolution over time reveals the diffusion process. The images are approximately 125 μm across and the grid regions about 25 μm on a side. The temperature was held below 100 °C to reduce loses due to evaporation. The dark areas between the squares or, even better, between adjacent rows show clear evidence of brightening over the 8 minute duration between the two images, corresponding to Cs diffusion across the 10 μm scale in minutes (the bright regions dim as well). Digitally processing the images shows the effect better in the color renditions of Figure 29.10, where the brighter intensity regions are more visible. The intensity as a function of position for a vertical line drawn through the images can be extracted and represented: the lines chosen corresponded to the $x = 205$ pixels, faintly visible as a vertical white line. The image itself is approximately 512 pixels across, with each pixel corresponding to roughly 0.25 μm. The intensity along the line at two different measurements separated by 8 minutes is shown in Figure 29.11. The regions of greater intensity diminish, and those of weaker intensity increase, conforming to how the cesium is expected to diffuse. The script taking the images and rendering them as well as extracting the data along the line is given by the code in Section A3.23. The rapidity with which the peaks diminish and the valleys fill serves to quantify the value of D using the point source solutions.

Such behavior can be crudely modeled by preparing an array of point sources that mimic the experimental arrangement and then assessing how the magnitude decreases as time progresses. For example, consider the configuration defined by

$$\rho(x, y) \propto \sum_{k=-3}^{3} u(x + k\Delta, t)\{u(y - \Delta, t) + u(y + \Delta, t)\} \tag{29.22}$$

a configuration which creates two parallel columns of seven sources a distance 2Δ apart, where the units of Δ and time are chosen so that $4D = 1$. Figure 29.12 shows the evolution of Eq. (29.22) for the time steps 0.4, 0.8, and 1.2. A slice

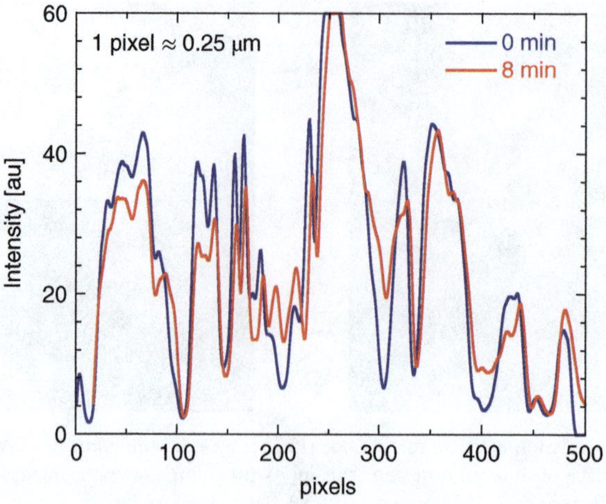

Figure 29.11 Intensity along the white lines of Figure 29.10, in which the numerical values of the intensity along the 205 pixel column are shown as a function of their pixel position. Because the images do not overlap, the pixel coordinate of the 8 minute data was shifted until there was overlap at the leftmost minimum.

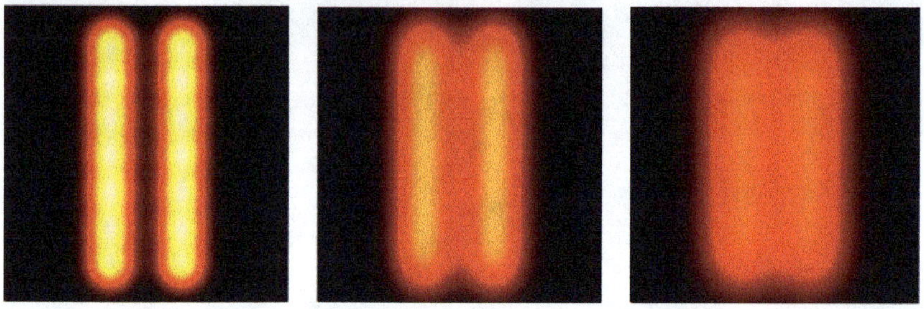

Figure 29.12 Analytic model of Eq. (29.22), where the separation of the centers of the point sources in two parallel columns is 2Δ. From left to right, the times are 0.4, 0.8, and 1.2, in units such that $4D = 1$.

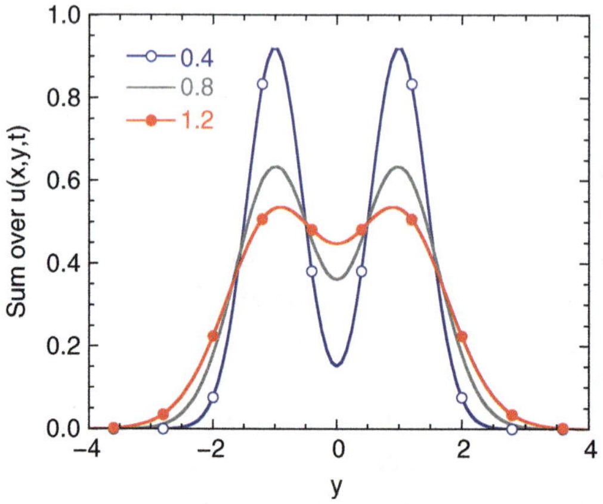

Figure 29.13 The numerical values of ρ along a horizontal slice through the center of the images is represented in Figure 29.12.

through the horizontal center shows the diffusion in an alternate manner as a cross-section: it is seen how rapidly the sharp peaks spread, and the center valley literally fills in, as in Figure 29.13. Consider the behavior near the 425 pixel mark of Figure 29.11: $\Delta \approx 50$ pixels = 12.5 μm, chosen for the low concentration and reasonably clear signature. Thus, for cesium on tungsten for $T < 373$ K and for the 8 minutes = 480 seconds of the experiment, then a crude estimate for D is

$$D \sim \frac{(12.5\ \text{cm})^2}{4(480\ \text{s})} = 8 \times 10^{-10} \frac{\text{cm}^2}{\text{s}} \tag{29.23}$$

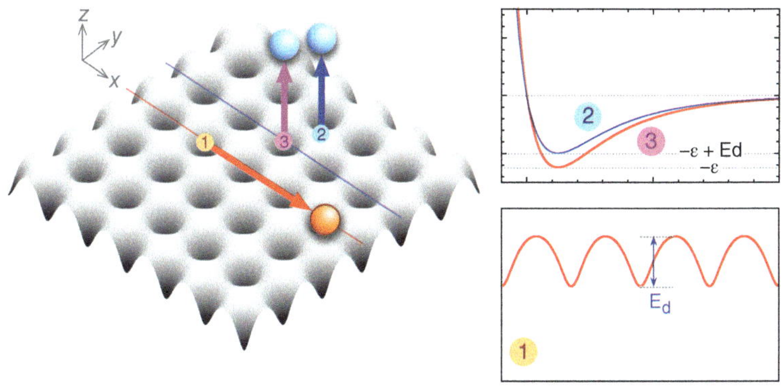

Figure 29.14 Left, motion along the plane of the surface (1) corresponding to diffusion and normal to the surface from the high (2) and low (3) potential points corresponding to evaporation. Right, based on Figure 1 of Gomer [399]) a Lennard–Jones potential normal to the surface on top and the surface undulations along the surface on bottom. E_d refers to the activation energy to get over a barrier or to a higher potential position.

How good is the approximation? An empirical Arrhenius relation due to Langmuir and Taylor [393] for the value of D for cesium diffusion on tungsten gives[7]

$$D_{LT} = D_o e^{-\beta E_d} \tag{29.24}$$

where $\beta = 1/k_B T$, $D_o = 0.1995$ cm^2/s and $E_d = 0.6072$ eV (a value that compares favorably to that of $E_d = 0.586$ eV of ref. [402]). The grounds for such a form will be taken up in Section 29.2.2. With these values, $D_{LT}(T) = 8 \times 10^{-10}$ cm^2/s if $T = 364.4$ K. Such numbers are compatible with the temperature of the tungsten in the experiment behind Figure 29.11, specified to be less than 100 °C = 373 K. The calculation, though, is more suggestive than accurate, given the immense subtlety that attends inferring diffusion parameters from experimental measurements (see Gomer [399]), where surface imperfections and movement of clusters of atoms complicate the analysis.

29.2.2 Lattice gas

> *Surface diffusion in adsorbed layers, a subject slighted in the past, has recently undergone a renaissance motivated both by the need for more information about transport processes on heterogeneous catalysts and by a concern with the dynamics of atomic jumps on solids. In principle at least, an adsorbed atom moving over a surface can be viewed as a convenient probe, both of interactions with the substrate and with other adatoms in its vicinity.*
>
> – David A. Reed and Gert Ehrlich [403]

When an Arrhenius relation such as Eq. (29.24) arises, the parameter E_d is often referred to as an *activation energy*. For atoms hopping around the surface of a metal, its intuitively appealing interpretation relates to the size of the potential barrier that an atom must hop over to get to an adjacent site (or, as in Section 29.3, the barrier that must be overcome for the atom to be evaporated), as schematically suggested in Figure 29.14. Diffusion over the surface of a material can then be modeled by allowing atoms to migrate by simulating their jumps over a lattice, giving rise to *lattice gas models* [398, 399, 403–408] as a means of investigating surface diffusion phenomena. Actually, as Figure 29.14 suggests, there are two kinds of jumps possible: those that carry the adatom to an adjacent site (diffusion) and those that allow the atom to jump completely off the surface of the metal (evaporation). These atoms wiggle in a well often modeled by a Lennard–Jones potential of the form

$$V(r) = 4\varepsilon \left\{ \left(\frac{r_m}{r}\right)^{12} - \left(\frac{r_m}{r}\right)^{6} \right\} \tag{29.25}$$

where $r = r_m$ is the closest approach for an atom with a zero total energy and $-\varepsilon$ is the depth of the well, as schematically shown in the top right plot of Figure 29.14.

EXAMPLE: From ref. [402] for cesium, $r_o = 0.5336$ nm at the potential minimum (their parameter $\varepsilon = 0.586$ eV at 350 K is not needed here, but is comparable to the value of Langmuir and Taylor in Eq. (29.24)). Find r_m.

---

[7]Nothing like Eq. (29.24) is found in ref. [393]. What is found (their Eq. (1)) is $\log_{10} D = -0.70 - (3060/T)$, which is equivalent (see an excellent review by Antczak and Erlich [401]). The point here is to give their form in a way that introduces E_d.

SOLUTION: Setting $r_m \partial_r V = 0$ at $r = r_o$ using Eq. (29.25) results in

$$-12\left(\frac{r_m}{r_o}\right)^{13} + 6\left(\frac{r_m}{r_o}\right)^{7} = 0$$

Isolating r_m results in

$$r_m = 2^{-1/6} r_o \tag{29.26}$$

from which $r_m = 0.4754$ nm.

29.2.3 A simple model of the diffusion coefficient

Into this Universe, and Why *not knowing*
Nor Whence, *like Water willy-nilly flowing;*
And out of it, as Wind along the Waste,
I know not Wither, *willy-nilly blowing.*

– Omar Khayyàm[8]

What, then, should the diffusion term D look like? Because adatoms have energy, they have thermal agitation, causing them to rattle about in their confining wells, and the atoms of the underlying substrate do the same. Because these atoms are constantly bouncing off each other and the metal atoms, they are constantly exchanging energy. The probability that the adatom finds itself with an energy E_d sufficient to jump the barrier to an adjacent empty site is therefore proportional to $\exp(-\beta E_d)$. Then, from the discussion after Eq. (29.8), and letting $\Delta x \to r_o$ be the length of a lattice hop (from well region to well region), D can be written as[9]

$$D = \frac{1}{4} r_o^2 \nu \exp(-\beta E_d) \tag{29.27}$$

where $2\Delta t \to 1/\nu$ is the round-trip time and ν is therefore the frequency of the particle making that round trip. The coefficient of (1/4) reinforces that the adatom has one of four locations to jump to on a two-dimensional surface, as in Figure 29.8. Near $r \approx r_o$, the Lennard–Jones potential is approximately parabolic, indicating that it is a harmonic oscillator, as in Section 9.3, for which a spring constant K can be defined [409]. Taylor expanding $V(r)$ in Eq. (29.25) about $r \approx r_o$ gives

$$V(r) \approx -\varepsilon + 36\frac{\varepsilon}{r_o^2}(r - r_o)^2 \leftrightarrow -\varepsilon + \frac{1}{2}K(r - r_o)^2 \tag{29.28}$$

Equating the last two forms shows that $K = 72\varepsilon/r_o^2$, from which the ground state angular frequency is $\omega_o = \sqrt{K/M}$, where M is the mass. The oscillating adatom is therefore in a higher energy level $\hbar\omega = (2n + 1)\hbar\omega_o/2$, and the frequency ν is $\nu = \omega/2\pi$.

EXAMPLE: If cesium is treated as a harmonic oscillator in the Lennard–Jones potential well, find the energy spacing $\hbar\omega_o$ and estimate n. Use the cesium parameters $r_o = 0.5336$ nm and $E_d = 0.6072$ eV (after Eq. (29.24)), and take the mass M of a cesium atom to be 0.132905 kg/mole.

SOLUTION: The energy spacing is

$$\hbar\omega_o = \hbar\sqrt{\frac{K}{M}} = \frac{6\hbar}{r_o}\sqrt{\frac{2\varepsilon}{M}} = 6.827 \text{ meV}$$

Next, equating the two forms of $D(T)$ of Eq. (29.24) and (29.27), and letting $\hbar\nu/2\pi = (2n + 1)\hbar\omega_o/2$, then

$$n = \frac{8\pi D_o}{r_o^2 \omega_o} - \frac{1}{2} \sim 169$$

from which $\nu = 280$ THz.

[8] Omar Khayyàm (trans. Edward FitzGerald), *The Rubaiyat of Omar Khayyàm*. Garden City, NY: Garden City Books, 1952, p. 149, Quatrain XXIX.

[9] When the hopping D is encountered again in Section 29.2.5, it will go by the name of D^* (following Gomer [399]) and is called the *tracer* diffusion coefficient.

The notion of hopping over a barrier provides an intuitively appealing means of understanding the Arrhenius relation of Eq. (29.24). In a simple view, every time the adatom bangs against the edges of the Lennard–Jones potential, the thermal agitations of the lattice cause its characteristic harmonic number n to change (it switches, essentially randomly, to a new energy level E_n), so that the probability that the adatom will surmount the barrier to the empty adjacent site is a reflection of the probability that its energy level exceeds the barrier height. As was repeatedly used in Chapter I, the probability that a particle is in an energy state $E_n = (2n+1)\hbar\omega_o/2$ is proportional to $\exp(-\beta E_n)$. Therefore, the probability of a jump is the probability that E_n of the adatom exceeds the barrier height E_d, where $P_n \propto \exp(-\beta E_n)$. The sums are analytic and easy, resulting in

$$P(E_n > E_d) = \frac{\sum_{n=n_o}^{\infty} e^{-\beta E_n}}{\sum_{n=1}^{\infty} e^{-\beta E_n}} = \exp(-\beta n_o \hbar\omega_o) \tag{29.29}$$

where n_o is such that $[n_o + (1/2)]\hbar\omega = E_d$. Given the size of n_o, this is about the same as $\exp(-\beta E_d)$, just as the Arrhenius relation would have it.

29.2.4 Modeling diffusion using a shuffle algorithm

> *...accuracy for us means the absence of bias.* Inaccuracy *is then measured by the magnitude of the bias.*
>
> – John Mandel [410], p. 105

> *It is simply a question of sacrificing ideals to expediency. In Monte Carlo work justice will have been done if the final verdict is fair, i.e., the answers come out approximately right.*
>
> – J.M. Hammersley and D.C. Handscomb [106], p. 26[10]

In light of Eq. (29.29), $D(T)$ of Eq. (29.27) can be written as a product of a jump length squared, an attempt frequency, and the probability of a successful jump. A simple numerical model focuses on the last factor: a surface is populated with adatoms positioned on random locations, and then at each time step the probability that it will hop to a random adjacent site is calculated. If chosen, it jumps. Were the adatoms far apart such that rarely did an adatom have an adjacent neighbor, such a description would suffice, but as soon as neighbors in adjacent sites becomes probable, a subtle problem in the simulation of the diffusing atoms can arise because of *bias*. Diffusion then becomes a convenient vehicle to discuss mitigation of that bias.

Bias, by which is meant an unwarranted predisposition to an outcome, can skew results in remarkably indirect ways. It arises here when a method by which adatoms are chosen for jumping and updating show a preference to a particular direction. This can be seen using a very simple model of a linear chain of atoms rather than a two-dimensional surface. If all the atoms were far apart, then regardless of the order the atoms are chosen for updating, their options (move to the left, move to the right, or stay put) are the same. On the other hand, if all the atoms started in a line at the center of the string (so that only the leftmost and rightmost atoms can initially jump to an unoccupied site, all the rest having neighbors on both sides), the order they are picked matters: if the updating always occurs from leftmost to rightmost atom, a drift to the left occurs because jump sites preferentially open up to the leftmost atoms first.[11] The problem is, *that which does not make a difference should not make a difference*: physically, the order that the atoms are chosen should not matter to the outcome, so if it does something is wrong and a bias is shown.

An obvious improvement would be to randomly choose the order that the adatoms were updated. This would provide a much better outcome because it would forestall a net drift, but a problem with randomly choosing the adatoms is that not all of them would be chosen with the same frequency: because they are chosen randomly, some would be chosen more often and some less often. That is, if N_x atoms were on a chain and N_t is the number of time steps, then the number of updatings to be performed is $N \equiv N_t N_x$. For a randomly chosen order, though, the mean number of times that the atom j is updated would be $\langle N_j \rangle = N_t$, but there would be fluctuations about the mean number N_t, and those fluctuations would become more pronounced the smaller N is.

---

[10]Although the quote was in reference to the generation of pseudorandom numbers, it reflects a viewpoint that applies more generally.

[11]If this is hard to see, consider the problem *in extremis*: suppose the jump probability was 100% and the atom had to go to an available site if the opportunity permitted. Then, if the atoms were chosen sequentially from left to right, every one of them would have no choice but to jump left, and so the block of atoms would move left *en masse*, which is unphysical.

A way to insure that *all* the atoms are updated the same number of times N_t, but that they are chosen in an order without bias, is to use a *shuffling* algorithm [411]. A proper way to implement it bears little similarity to how a deck of cards is shuffled: cards experience a "riffle" shuffle, a method of randomization that, in the words of Knuth, is "miserably inadequate" as a technique of randomization and which further leaves a trace of the ordering, or a *bias*, in the cards that bridge players have learned to exploit when they "finesse". The card analogy, though, makes the discussion easier: the atoms correspond to cards, and the arrangement of a deck of cards corresponds to a particular order in which the atoms are selected. The shuffle algorithm proceeds by starting with the topmost card, then swaps its position with a randomly chosen card *below* it; the process is iterated with each card in the deck down to the bottom card. A version of the algorithm is part of Section A3.24.

A simple algorithm for hopping using the shuffle algorithm constitutes the remainder of Section A3.24. It assumes a fixed jump probability p_o and no interaction energy between the adatoms: if the jump is allowed, a direction is chosen, and if the site in that direction is empty, the adatom jumps. Periodic boundary conditions are assumed, so that an adatom that leaves one boundary re-enters the simulation region at the opposing boundary. A group of atoms, initially confined to the center of the simulation region, will spread their way out and fill the available space. The visual spread of the adatoms, by comparing the initial and final state positions, is shown in Figure 29.15.

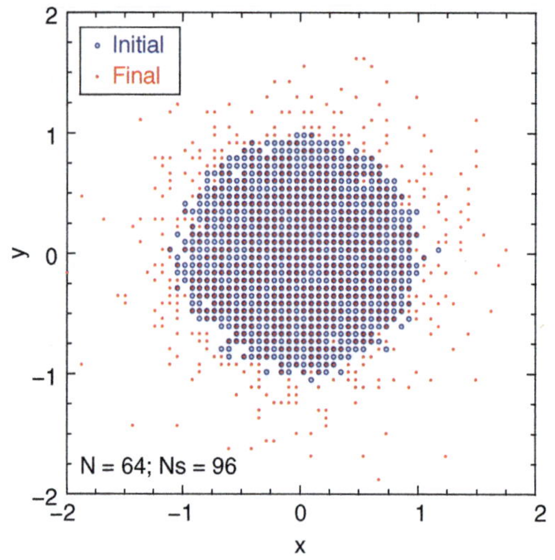

Figure 29.15 A two-dimensional area with 64 × 64 sites is initially populated with adatoms preferentially positioned near the center and marked as blue circles. At each time step the adatoms are randomly selected using a shuffling algorithm to determine if they make a hop, with the probability of a hop being $p_o = 0.2$ At the end of the simulation, the adatoms are in the red dot final state positions. The simulation used Algorithm A3.24 with $N = 64$ and $N_s = 96$, so that the number of time steps was $6 \times N_s$, with 6 corresponding to the intermediate time steps shown in Figure 29.16.

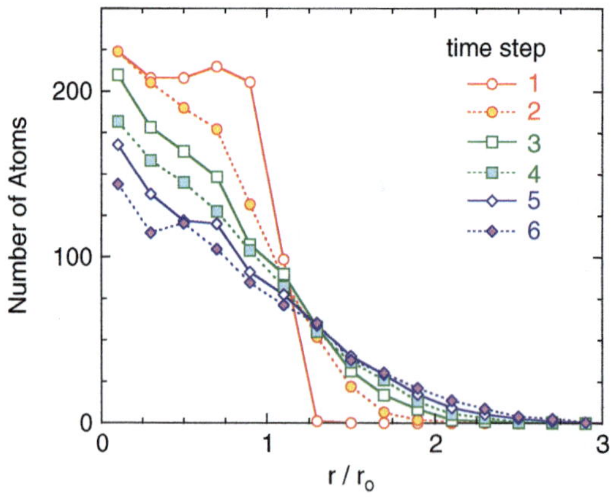

Figure 29.16 The distribution of radial distances from the center for the adatoms of Figure 29.15 for the time increments $j \times N_s$ with $j \in \{1, 6\}$ and $N_s = 96$, showing a spreading pattern compatible with diffusion.

29.2.5 Adatom motion and diffusion coefficients

...explanations in engineering freely use vernacular language, no matter how flaky the explanation may seem to the logician or theoretical physicist. When the words are chosen carefully, they can insightfully convey what is going on in a human contrivance without equations or computer simulations, at least at the executive-summary level.

– Steven Pinker[12]

In the hopping simulation of the previous section, neither the unit of the time step nor the site-to-site separation is explicitly given, both in some sense related to the magnitude of the jump probability p_o. A constant p_o implies that the conditions for all the adatoms are the same, but a moment's thought reveals just how tenuous this assumption is: adatoms in adjacent sites must surely affect the jump probability by altering the nature of the potential barrier between the adatoms, that is, the jump probability really should be coverage dependent, just as the work function will be found to be coverage dependent in Section 31.8.

Appreciating that the concentration of the adatoms on the surface affects how the dynamics of their motion unfolds, revisit Eq. (29.7) with a more critical eye, based as it was on a dilute approximation: the jump probability p_o reflects, as it were, the desire of the adatom to hop, but says nothing about whether it actually can. Were the site the adatom wished to hop to occupied, its ability to hop would be frustrated. The correction of this oversight identifies a key feature of a Langmuir gas, and brings a better derivation of Fick's law with it. Instead of using p_o or ν or the like, speak of a jump rate as a function of coverage $\Gamma(\theta)$. Clearly, what has been discussed so far is $\Gamma(0)$, or the jump rate in the absence of meddlesome companions occupying jump destinations. The probability of a jump $\Gamma(\theta)$ has to be tempered by the availability of spaces to jump to. If all the spaces were occupied, then the rate would have to be 0, or $\Gamma(1) = 0$, and if none were, then $\Gamma = \Gamma(0)$ as before. In short, the jump probability is related to the proportion of available sites. The simplest relation of the "effective jump rate" $\Gamma(\theta)$ is therefore [399, 403]

$$\Gamma(\theta) = \Gamma(0)(1 - \theta) \tag{29.30}$$

Next, the number of atoms n_1 along line 1 is a product of the width of the strip from $-\lambda/2$ to $\lambda/2$ (where λ is the distance from the initial jump site to the target jump site and is assumed to be the same on both sides), the number of sites N_s per unit area, and the fraction θ of the sites that are filled, or $n_1 = \lambda\theta_1 N_s$, so that the rate of change of the number of atoms along that line is

$$\frac{d}{dt}n_1 = \Gamma(\theta_2)\lambda N_s\theta_1 \tag{29.31}$$

The equation for the second line n_2 follows exactly the same kind of logic, and so is given by the replacement $1 \leftrightarrow 2$ in Eq. (29.31). The net flux J of adatoms is the difference between the change in n_1 and the change in n_2, or

$$\begin{aligned} J &= \frac{dn_1}{dt} - \frac{dn_2}{dt} = \Gamma(0)N_s\lambda[(1-\theta_2)\theta_1 - (1-\theta_1)\theta_2] \\ &= -\Gamma(0)N_s\lambda(\theta_2 - \theta_1) \approx -\Gamma(0)\lambda^2(N_s\partial_x\theta) \end{aligned} \tag{29.32}$$

where $N_s\partial_x\theta = \partial_x C$ is the "concentration" gradient so that $D = \Gamma(0)\lambda^2$, and so, happily, observing the concentration evolve gives information about the isolated jump probability $\Gamma(0)$, assuming the jump distance λ is known. When Eq. (29.32) is joined with the continuity equation $\partial_t C + \partial_x J = 0$, the result is

$$\partial_t C = D\partial_x^2 C \tag{29.33}$$

which hearkens back to Eq. (29.10). Well, almost. Reed and Ehrlich [403] draw attention to the transport of adatoms being driven by changes in the chemical potential μ rather than the concentration C, but even before that, observe that when the hopping of an adatom was the focus, diffusion was in the language of the random walk of a single one, but when the discussion is about concentration (or chemical potential) gradients, then diffusion is in the language of a flow of adatoms. This suggests the diffusion coefficient for each way of looking differs in some fundamental way. Following Gomer, let D in Eq. (29.27) now be called D^*, and reserve D for the term in Eq. (29.32). How does D differ from D^*?

Recall what D^* represents: it follows an adatom around on a "random walk" over the surface. The random walk topic is a quintessential problem in statistical physics [51]. Just to be quick about it, consider the problem in one dimension,

[12] Steven Pinker, *The Stuff of Thought*. New York: Viking, 2007, p. 226.

as the generalization to two dimensions is straightforward. Let a particle[13] have a probability p of taking a step $\Delta x = v_o \Delta t$ to to the right, and therefore a probability of $q = 1 - p$ to the left at each time step. Of course, $p = q$, but the problem unfolds much better by not making that explicit until the end. The random walk problem is therefore concerned with two questions: first, how far away from the origin on average is the adatom at a later (usually long) time and, second, if the adatom does the walk over and over or, equivalently, many non-interacting adatoms all start out at the same time,[14] what is the *dispersion* or their average distance (in particular, their *root mean square* distance) from the origin?

A simple way to answer both questions uses the methods familiar from Section 4.4. The probability of a step left or right is, of course, unity $(1 = p + q)$. So too is the probability of taking any combination of two steps left and right, or $1 = (p + q)^2 = pp + pq + qp + qq$, where each combination has been spelled out. For n steps, of course,

$$1 = \sum_{k=0}^{n} \frac{n!}{k!(n-k)!} p^k q^{n-k} \equiv \sum_{k=0}^{n} W_{k,n-k}\, p^k q^{n-k} \tag{29.34}$$

by the binomial theorem, where each term in the sum represents a particular order of events with $W_{k,j} = (k+j)!/k!j!$ being the probability of a total of k steps to the left and j to the right, regardless of the order taken (so, for example, for $n = 2$, the probability W_{11} is $1/2$ that there will be one step to the left and one to the right, as seen by $pq + qp$ accounting for two of the four possible combinations). Now consider

$$\begin{aligned} n = n\sum_{k=0}^{n} W_{k,n-k}\, p^k q^{n-k} &= \sum_{j,k=0}^{n} (k+j) W_{k,j}\, p^k q^j \delta_{n,j+k} \\ &= \sum_{j,k=0}^{n} k W_{k,j}\, p^k q^j \delta_{n,j+k} + \sum_{j,k=0}^{n} j W_{k,j}\, p^k q^j \delta_{n,j+k} \end{aligned} \tag{29.35}$$

where $\delta_{i,j}$ is the Kronecker delta (unity if its arguments are the same, zero if its arguments are not). This appears to be an unnecessary complication, but consider what it represents: on the second line, the first term is the average $\langle k \rangle$ number of steps to the right, and the second is the average number $\langle j \rangle$ to the left. But if $p = q$, the two sums are in fact identical because there is a one-to-one correspondence in their terms due to the symmetry of $W_{k,j}$ and $p^k q^j = p^{k+j}$, meaning that the average number of steps to the left is the same as the average number to the right, or $\langle k \rangle \equiv k_o = n/2$, as intuitively understood (if the probability of going to the right on each step is p, then it is np for n steps). As a result, *the mean displacement is zero* because $\langle k \rangle - \langle j \rangle = 0$. Therefore, the adatom tends to hover around the origin on average.

The dispersion is a measure of the deviation from an average, or $\langle \Delta k^2 \rangle \equiv \langle (k - k_o)^2 \rangle$. Because

$$\langle (k - k_o)^2 \rangle = \langle k^2 \rangle - 2k_o \langle k \rangle + k_o^2 = \langle k^2 \rangle - k_o^2 = \langle \Delta k^2 \rangle \tag{29.36}$$

the task at hand is to find $\langle k^2 \rangle$. Using the convenient trick[15] that $k^2 p^k = (p\partial_p)^2 p^k$ (and treating q as independent of p for now),

$$\begin{aligned} \langle k^2 \rangle &= (p\partial_p)^2 \sum_{k=0}^{n} W_{k,n-k}\, p^k q^{n-k} \\ &= (p\partial_p)^2 (p+q)^n = np(np+q)(p+q)^{n-2} = np(np+q) \end{aligned} \tag{29.37}$$

where the last step follows from $p + q = 1$. Now setting $p = q = 1/2$, then

$$\langle k^2 \rangle = \frac{1}{2} n(n+1) \to \langle \Delta k^2 \rangle = \langle k^2 \rangle - \langle k \rangle^2 = npq = \frac{n}{4} \tag{29.38}$$

The *root mean squared* displacement is therefore

$$\Delta k_{rms} = \sqrt{\langle \Delta k^2 \rangle} = \sqrt{n/4} \tag{29.39}$$

[13]The particle is a drunkard and the origin is a lamppost in the usual formulation, but such details invite many more that are better simply avoided.

[14]This is one of the details best avoided by not talking about drunkards: there is no need to consider why so many alcoholics would aggregate at the same lamppost, nor what failure of society led to so many to begin with.

[15]This trick can also be used to evaluate $\langle k \rangle = np$, but variety in technique is good pedagogy.

The relation of the random walk to the tracer diffusion D^* is by the two-dimensional generalization [399]

$$\Delta r_{rms} = \langle |\vec{r}(t) - \vec{r}(0)|^2 \rangle = (n_x + n_y)\lambda^2 \tag{29.40}$$

where $r(t)^2 = x(t)^2 + y(t)^2$ and $x(t) = n_x\lambda$ with a similar equation for $y(t)$. The number of jumps $n_x + n_y$ is the product of the time t with the effective jump frequency ν_{eff}, and the effective jump frequency is simply the "try frequency" ν_o with the probability of success $\exp(-\beta E_d)$, so that

$$\Delta r_{rms}^2 = \lambda^2 t\nu_o \exp(-\beta E_d) \equiv 4D^* t \tag{29.41}$$

The relationship between position $r(t)$ and velocity $v(t)$ allows for expressing D^* in terms of *velocity correlation functions*. Using

$$\begin{aligned}\langle |\vec{r}(t) - \vec{r}(0)|^2 \rangle &= \left\langle \int_0^t \vec{v}(t')dt' \cdot \int_0^t \vec{v}(t'')dt'' \right\rangle \\ &= \int_0^t dt' \int_0^t dt'' \langle \vec{v}(t') \cdot \vec{v}(t'') \rangle \end{aligned} \tag{29.42}$$

Recall that $\langle \cdots \rangle$ is an average, and in the absence of external forces for an equilibrium condition of a particle engaged in a random walk the averages depend on the time *differences*, not the times themselves, and so

$$K(s) \equiv \langle \vec{v}(t) \cdot \vec{v}(t+s) \rangle = \langle \vec{v}(0) \cdot \vec{v}(s) \rangle = K(-s) \tag{29.43}$$

$$\int_0^t dt' \int_0^t dt'' K(t' - t'') = 2\int_0^t ds(t-s)K(s) \tag{29.44}$$

Although Eq. (29.44) can be shown more formally [51], it can be shown quickly if correlations in velocity are assumed to die out exponentially fast with a decay constant τ, or $K(s) \approx K_o \exp(-|s|/\tau)$. With that ansatz, direct integration of both sides of Eq. (29.44) shows equivalence. As a result

$$D^* = \lim_{t\to\infty} \frac{2}{4t}\int_0^t (t-s)K(s)ds = \frac{1}{2}\int_0^\infty K(s)ds \tag{29.45}$$

for one adatom. For N of them,

$$D^* = \frac{1}{2N}\sum_{j=1}^{N}\int_0^\infty \langle \vec{v}_j(0) \cdot \vec{v}_j(t) \rangle dt \tag{29.46}$$

Now recall what D represents: it follows currents of adatoms over the surface. As a result [399]

$$D = \frac{1}{2(\Delta N^2)}\int_0^\infty \langle \vec{J}(0) \cdot \vec{J}(t) \rangle dt \tag{29.47}$$

$$= \frac{1}{2(\Delta N^2)}\int_0^\infty \left\langle \sum_{j=1}^{N} \vec{v}_j(0) \cdot \sum_{j=1}^{N} \vec{v}_j(t) \right\rangle dt \tag{29.48}$$

with ΔN^2 arising as a result of fluctuations of particles in the finite region under consideration, which is embedded in an infinite open system. If the particle velocities are correlated (the adatoms are affected by their neighbors), then further reduction of the equation is difficult and one is deep into the methods of Kubo–Green relations. Alas, as shall be seen, adatoms are in fact subject to the influence of their neighbors, but suppose that the interaction is small enough that treating the velocities as uncorrelated is a reasonable first approximation. The consequence is that sums of the form

$$\sum_{j=1}^{N} v_j(0) \cdot \sum_{j=1}^{N} v_j(t) = \sum_{j=1}^{N} v_j(0) \cdot v_j(t) + \sum_{j\neq k}^{N} v_j(0) \cdot v_k(t) \tag{29.49}$$

where the second summation is understood as a double sum but where the indices are unequal. Uncorrelated means that integrals over time involving the second summation are zero: compared to the reference velocity $\vec{v}_j(0)$, when t is sufficiently large that all memory dies out, $\vec{v}_k(t)$ appears random by comparison, and sums over random quantities are on average zero. As a result,

$$\frac{D}{D^*} = \frac{N}{\Delta N^2} \rightarrow \frac{\langle N \rangle}{\langle \Delta N^2 \rangle} \tag{29.50}$$

where the second form is a consequence of fluctuations in the finite region that is part of the infinite open system, so that average numbers of particles and their dispersion is what matters.

29.2.6 The Darken equation

That's just the way it is/Some things will never change
That's just the way it is/Ah, but don't you believe them.

– Bruce Hornsby[16]

Furthering an understanding about the way things are and how they change requires the methods of Chapter 5 concerning energy and entropy, but with a variation. The energy levels are the diagonal elements of the Hamiltonian H, in terms of which the partition function can be written as

$$\mathcal{Z} = \mathbf{Tr}\exp[-\beta(H-\mu N)] \tag{29.51}$$

where the trace $\mathbf{Tr}M$ means the sum of the diagonal entries of a matrix M. Introducing the temporary notation $\alpha = \beta\mu$, then

$$\langle N\rangle = \frac{\mathbf{Tr}\,N\exp[-\beta(H-\mu N)]}{\mathbf{Tr}\exp[-\beta(H-\mu N)]} = \beta^{-1}\frac{\partial_\mu \mathcal{Z}}{\mathcal{Z}} = \partial_\alpha \ln \mathcal{Z} \tag{29.52}$$

It then follows

$$\partial_\alpha\langle N\rangle = \frac{1}{\mathcal{Z}}\partial_\alpha^2\mathcal{Z} - \left(\frac{\partial_\alpha \mathcal{Z}}{\mathcal{Z}}\right)^2 = \langle N^2\rangle - \langle N\rangle^2 = \langle \Delta N^2\rangle \tag{29.53}$$

Using the last two equations in Eq. (29.50) gives

$$\frac{\Delta N^2}{N} = \frac{\partial_\alpha\langle N\rangle}{\langle N\rangle} = \frac{N_s\partial_\alpha\theta}{N_s\theta} = \partial_\alpha \ln\theta \tag{29.54}$$

where N_s is the number of available sites and θ is the coverage (fraction of sites that are occupied). The Darken equation is then

$$\frac{D}{D^*} = \frac{\partial(\beta\mu)}{\partial\ln\theta} \tag{29.55}$$

for constant temperature T.

A "Langmuir gas" refers to the absence of interactions between the adatoms themselves or between the adatoms and the metal surface, for which the dependence of μ on the coverage θ arises from both the interaction U term and the entropy term TS in the free energy F of Section 5.3. Regarding the former, the energy will go as the product of ε with the number of pairs of interacting adatoms, with a factor of 2 to prevent over-counting, or $\varepsilon N_p/2 = (\varepsilon/2)\gamma\theta N$. Regarding the later, with S being the entropy, and with $\mu = (\partial F/\partial N)_{T,A}$ (constant area A is used instead of volume V for a surface layer) [403], the non-entropy parts of F give the usual chemical potential μ: it is the entropy part TS that gives the organizational dependence. Finding that term is a problem in combinatorics much like Section 4.4. The configurational entropy is determined by the number of ways to place N indistinguishable adatoms on N_s available sites. From the defining Eq. (4.23), or recalling Eq. (4.26),

$$TS = \beta^{-1}\ln\left(\frac{N_s!}{N!(N_s-N)!}\right) \tag{29.56}$$

$$= \beta^{-1}(-N\ln(N) - (N_s-N)\ln(N_s-N) + N_s\ln(N_s)) \tag{29.57}$$

where the second line uses Sterling's approximation (Eq. (A2.4)) on $n! = n\Gamma(n)$ (but neglects the $\ln(2\pi n)$ part as small compared to n). Writing the chemical potential μ as the sum of the usual contribution from the non-S part of F (call it μ_o) and the part from TS gives

$$\beta\mu(\theta) = \beta\mu_o - \beta\partial_N(TS) = \beta\mu_o + \ln\left(\frac{N}{N_s-N}\right) \tag{29.58}$$

Invoking the definition of coverage, or $N = N_s\theta$, results in

$$\beta\mu(\theta) = \beta\mu_o - \beta\partial_N(TS) = \beta\mu_o + \ln\left(\frac{\theta}{1-\theta}\right) \tag{29.59}$$

where $\beta\mu_o = \beta\gamma\varepsilon\theta^2/2$ from the U term, and so the Darken equation for the (non-interacting) Langmuir gas becomes

$$D = D^*\frac{\partial(\beta\mu(\theta))}{\partial\ln\theta} = \frac{D^*}{1-\theta} \tag{29.60}$$

[16] Bruce Hornsby, lyrics to *The Way It Is*, 1986.

where the $(1-\theta)$ factor accounts for sites which are occupied and therefore excluded from being "jump-to" sites. For the Langmuir gas, $D(\theta) = \Gamma(\theta)\lambda^2(\partial(\beta\mu)/\partial \ln\theta)$ simply recovers $D = \Gamma(0)\lambda^2$ as in the discussion following Eq. (29.32).

If that were all, the effort expended to confirm what was known already would hardly have warranted the investment. But as the caution following Eq. (29.47) makes clear, adatoms *do* interact with each other and with the underlying metal surface. The energy of interaction with the metal surface (call it υ_o) and the pair-wise energy of interaction with each other (call it ε) are inescapably related to the coverage θ. This has two consequences: the interactions affect the diffusion (hopping) rate, but also affect the evaporation rate. The way to relate coverage to these interaction energies is to repeat the above analysis but attend to where those energies enter. It means, though, that the effort to find their contribution is harder than that for the Langmuir gas, although the Langmuir gas provides a useful limiting case.

29.2.7 Adatom interactions

To relate D to D^* when the interaction energies (adatom with surface, or υ_o; adatom with adatom, or ε) are included, assume that the adatom interaction energy is short-ranged and occurs only for nearest neighbors, that is, when the atoms are on adjacent sites, but drops to zero for any separation further away [403, 412]. If there are N_p pairs of such atoms, then a contribution of $-N_p\varepsilon$ to the energy occurs. Assume also that the sites are on a square lattice, so that there are at most $\gamma = 4$ nearest neighbors at the north, east, south, and west squares, as suggested by Figure 29.17.

Each adatom on the lattice contributes an energy υ_o; each adjacent pair of adatoms contributes an energy $-\varepsilon$. The question becomes for such conditions, what is the most probable value of the coverage factor θ [412]? This simple model constitutes the "quasi-chemical" or Bethe–Peierls approximation for the statistics of a lattice gas, and represents a tidy bit of statistical reasoning. It matters for understanding the evaporation rate in Section 29.3 and the work function in Section 31.8. The energy of the lattice in Figure 29.17 is

$$E = \upsilon_o \sum_{j=1}^{\gamma} s_j - \varepsilon s_0 \sum_{j=1}^{\gamma} s_j \tag{29.61}$$

where $s_j = 0$ or 1 for unoccupied or occupied sites, respectively. Let the probability that the center site is in a state $s_0 \to s$ (ignore the subscript; if there isn't one, then the center site is being referred to) and has n nearest neighbors be $P(s,n)$. The sum over all values of s and n must be unity, or

$$\sum_{n=1}^{\gamma} P(s,n) = 1 \tag{29.62}$$

Now comes the interesting approach: the probability of the center site being occupied is related to the energy associated with its occupancy and the number of nearest neighbors it has; the former is proportional to $\exp(-\beta s\varepsilon)$ and the latter is a combinatorics problem familiar from Section 4.4: if there are n adjacent atoms, there are γ ways to place the first atom, $\gamma - 1$ ways to place the second, and so on down to $\gamma + 1 - n$ ways to place the nth. And finally, for every atom on the lattice grid, associate a value $z = e^{\beta\upsilon_o}$, the so-called "fugacity approach" given its similarity to treatments of the grand canonical distribution in statistical mechanics [413] where υ_o here takes the role of μ there, so

$$P(s,n) \propto z^s \frac{\gamma!}{n!(\gamma-n)!} z^n e^{-\beta ns} \tag{29.63}$$

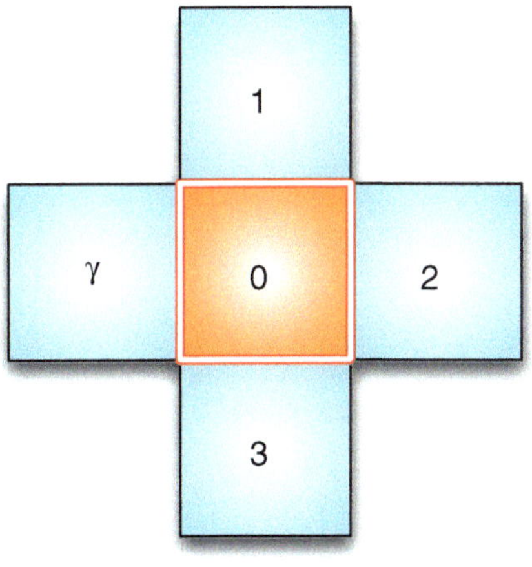

Figure 29.17 A central site labeled 0 and its γ nearest neighbors. The occupancy of the sites will be given by $s_j = (0,1)$ for unoccupied or occupied, respectively, where $j \in \{0,\gamma\}$ denotes the site. $\gamma = 4$ for a square grid of sites.

where the z^s factor is due to the center site, the factorial ratio is due to the combinatorics, the z^n factor is due to the occupancy of the surrounding lattice sites, and the exponential term is due to the interactions of the occupied surrounding sites with the center site.

Consider the $s = 1$ case first. From the binomial theorem, the normalization of $P(1, n)$ is easily deduced, and so

$$P(1, n) = \frac{\gamma!}{n!(\gamma - n)!} \frac{z^n e^{-n\beta\varepsilon}}{(1 + ze^{-\beta\varepsilon})^\gamma} \tag{29.64}$$

The same kind of deductions lead to the $s = 0$ case, or

$$P(0, n) = \frac{\gamma!}{n!(\gamma - n)!} \frac{e^{-n\beta\varepsilon}}{(1 + e^{-\beta\varepsilon})^\gamma} \tag{29.65}$$

Now two questions can be answered quickly, and motion towards a third begun, to give a measure of how adatom motion depends on coverage. First, for non-zero v_o and ε, what is the average occupation $\langle\theta_s\rangle$ given the center site is in a state s, and, second, how does the jump rate $\Gamma(\theta)$ change? The third question, what is $\mu(\theta)$, is sketched with a fuller account in ref. [403]. For the first question, consider the $s = 0$ condition first, for which

$$\langle\theta_0\rangle = \frac{1}{\gamma}\sum_{n=0}^{\gamma} nP(0, n) = \frac{1}{\gamma} z\partial_z \sum_{n=0}^{\gamma} P(0, n) = \frac{z}{1 + z} \tag{29.66}$$

and similarly, for $\langle\theta_1\rangle$, $z \to ze^{-\beta\varepsilon} = z\eta$, where $\eta \equiv e^{-\beta\varepsilon}$, so that

$$\langle\theta_1\rangle = \frac{z\eta}{1 + z\eta} \tag{29.67}$$

The average occupation is the average of the $s = 0$ and $s = 1$ cases, or

$$\langle\theta_s\rangle = \theta\langle\theta_1\rangle + (1 - \theta)\langle\theta_0\rangle \tag{29.68}$$

but because the center site is arbitrary, $\langle\theta_s\rangle = \theta$ and so collecting terms,

$$\frac{\theta}{1 - \theta} = \frac{z(1 + z\eta)}{1 + z} \tag{29.69}$$

which allows v_o to be determined if θ and ε are known. It also has implications for the evaporation rate by solving for z by finding the correct root of the quadratic equation $\eta(1 - \theta)z^2 + (1 - 2\theta)z - \theta = 0$: the correct root corresponds to $z \to 0$ if $\theta \to 0$. Therefore,

$$z(\theta) = \frac{-(1 - 2\theta) + [1 - 4(1 - \eta)\theta(1 - \theta)]^{1/2}}{2\eta(1 - \theta)} \equiv \frac{b(\theta) - 1 + 2\theta}{2\eta(1 - \theta)} \tag{29.70}$$

where $b(\theta) = b(1 - \theta)$ is the radical term. The symmetry in $b(\theta)$ is interpretable as putting an adatom on being equivalent to taking a hole off, what Gomer calls a "particle-hole" symmetry.

Diffusion is affected by the hopping parameter $\Gamma(\theta)$, and that, too, depends on θ and the energies v_o and ε. It will depend on the number of open sites $(\gamma - n)$, the probability of the center site being surrounded by n neighbors, and the jump rate associated with n given by $\Gamma_n = \Gamma(0)/\eta^n$, or

$$\begin{aligned}\Gamma(\theta) &= \frac{1}{\gamma}\sum_{n=0}^{\gamma}(\gamma - n)P(1, n)\Gamma_n = \frac{\Gamma(0)}{(1 + z\eta)^\gamma}\sum_{n=0}^{\gamma}\frac{\gamma - n}{\gamma}\frac{\gamma!}{n!(\gamma - n)!}z^n \\ &= \Gamma(0)\frac{(1 + z)^{\gamma-1}}{(1 + z\eta)^\gamma} = \Gamma(0)\frac{(1 + z)^3}{(1 + z\eta)^4}\end{aligned} \tag{29.71}$$

where the last form puts in $\gamma = 4$ directly.

The last consideration, that of $\mu(\theta)$, is quickly dealt with by observing that μ relates to changes in energy from putting one more adatom on, which is the same as taking away one vacancy. This implies that the change to μ is by a term that is proportional to $\ln(z(\theta)/z(1 - \theta))$, or explicitly [403]

$$\beta\mu(\theta) = \beta\mu_o + \ln\left(\frac{\theta}{1 - \theta}\right) + \frac{z(\theta)}{2}\ln\left(\frac{(b(\theta) - 1 + 2\theta)(1 - \theta)}{(b(\theta) + 1 - 2\theta)\theta}\right) \tag{29.72}$$

so that the modification to the Langmuir form of the Darken equation becomes

$$\frac{\partial(\beta\mu(\theta))}{\partial \ln\theta} = \frac{1}{1-\theta}\left\{1 + \frac{z(\theta)}{2}\left(\frac{1-b(\theta)}{b(\theta)}\right)\right\} \tag{29.73}$$

Look closely: when the interaction energy $\varepsilon = 0$, then $b(\theta) = 1$, and the Langmuir gas results; also, the value $\theta = 1/2$ holds a special place. But when there is interaction, the hopping rate changes as does v_o.

In short, the rapidity with which diffusion occurs depends on the fractional coverage of the surface. When these observations are incorporated into the lattice gas model and studied using Monte Carlo methods, the diffusion constants and hopping rates change in non-trivial ways [403]. The purpose here, however, is to show that the presence of coatings changes how a submonolayer coating of adatoms diffuses, how they evaporate, and (when we return to the subject in Section 31.8) how large the work function $\Phi(\theta)$ is as a consequence.

29.3 Evaporation of coatings

> *The rate of evaporation of atoms v_a increases at any given temperature extremely rapidly as θ increases. Thus at 1000°K. an increase of θ from 0.1 to 0.9 causes a 10^{11}-fold increase in v_a. This indicates very strong repulsive forces between the adatoms, in agreement with the fact that these adatoms tend to be positively charged as indicated by the effect of the film in increasing the electron emission ...A single cesium adatom on a tungsten surface may be regarded as an ion held to the metal by its image force.*
>
> – Irving Langmuir[17] [414]

The mobility and evaporation of cesium atoms on tungsten filaments figured prominently in the development of surface chemistry for which Langmuir received the Nobel Prize in 1932. Although cesium (and barium, for that matter) hopped nicely across the surface, its tendency to wantonly abandon its outer electron to the tungsten metal meant that it wandered the surface more like an ion than an atom. Because it was in turn attracted to its image charge force, the barrier to its evaporation (motion perpendicular to the surface) was much higher than to its migration (motion along the surface). As the surface concentration θ increases, the repulsion amongst the cesium atoms acts to weaken the potential holding it to the surface, and so a strong relationship exists between the evaporation rate, surface coverage, and temperature. The size of cesium compared to the underlying tungsten atoms meant that not much was required for it to get in the way of its neighbors, as suggested in Figure 29.18 using covalent radii from Figure 29.19.

Langmuir [414] (and others afterwards [366, 415]) deduced that the rate of evaporation would be dependent upon the number of nearest neighbors of cesium (for a monolayer of cesium on tungsten, $\gamma = 4$, using the notation of Figure 29.17, but for submonolayer coverage, γ will be smaller) and therefore that the evaporation rate, which is proportional to $d\theta/dt$, should behave as

$$\frac{d\theta}{dt} = -p\theta^{\gamma+1} \tag{29.74}$$

an approximation often used in the modeling of the evaporation of coatings for dispenser cathodes. Integrating Eq. (29.74) results in

$$\theta(t) = \frac{1}{(1+p\gamma t)^{1/\gamma}} \tag{29.75}$$

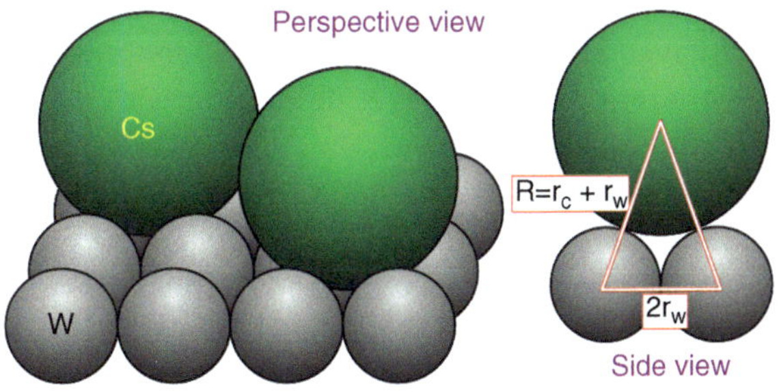

Figure 29.18 (*left*) Schematic of cesium (large green) atoms atop a tungsten surface (small gray); (*right*) Cs and W atoms represented by spheres (r_{Cs} = 253 pm, r_W = 130 pm, as per Figure 29.19. The factor R will reappear in Section 31.8.

[17] Irving Langmuir, Nobel Prize talk, 1932.

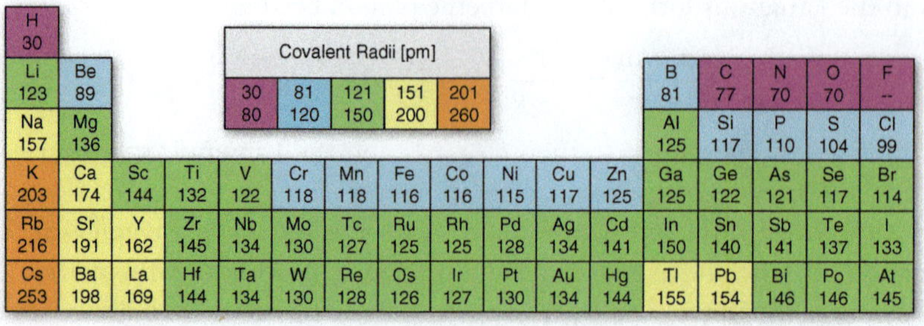

Figure 29.19 Covalent radii of the elements in picometers, excluding the noble gases and the lanthanum and actinium series. The inset shows the ranges associated with the color scheme.

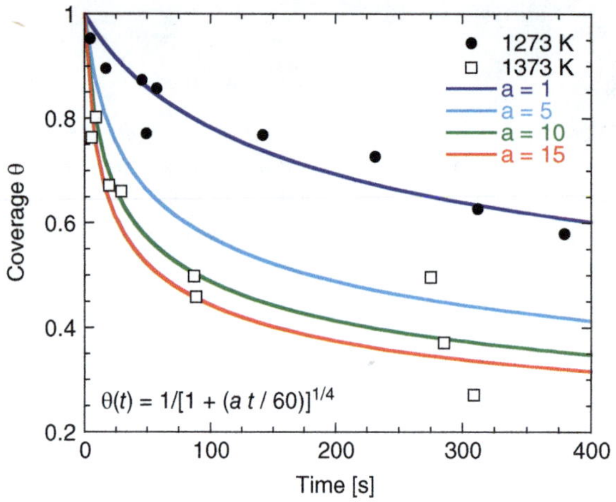

Figure 29.20 Evaporation rate versus coverage θ for barium on tungsten at two different temperatures ($T = 1273$ K for filled circles, $T = 1373$ K for open squares), compared to Eq. (29.75) with $\gamma = 4$ and $a = 60p\gamma$ [1/s], based on Figure 8 of Forman [416].

How does this fare for other coatings, like barium? Consider Eq. (29.75) compared to the data of Forman [416], as in Figure 29.20,[18] which shows a fairly good representation. Such a relation is found to adequately describe other coatings, such as barium and strontium, but Moore and Allison [415] go further, showing that

$$\frac{t}{2\pi\hbar\beta}e^{-\beta\varepsilon} = \mathrm{Erf}(-4\beta\upsilon_o\theta) - \mathrm{Erf}(-4\beta\upsilon_o) \tag{29.76}$$

where Erf(x) is the error function of Section A2.4, and with $\varepsilon = 3.68$ eV and $\upsilon_o = 0.13$ eV for barium. The agreement between their formulation and the data they show is very good when the values of ε and υ_o are adjusted.

As the temperature increases, neutral atom desorption gives way to the desorption of ions [393, 417]. The dynamics of the evaporation process were examined by Shih *et al.* [376] using temperature-programmed desorption studies to see how barium departs from the surface when more than a monolayer is initially present and the temperature is sufficiently high that when barium does evaporate, it does so in a charged state (Ba^+). Three distinctive peak intensities separated by time in Figure 29.21 are interpreted as multilayer desorption (greater than a monolayer of barium is on the surface), followed by monolayer desorption and then a transition to monolayer desorption with a different binding energy representing barium interacting with oxygen on the tungsten surface.

Analogous to the hopping model of Section 29.2, evaporation can be treated as the escape of an adatom oscillating in a Lennard–Jones potential (Eq. (29.25)) where the evaporation rate E_{evap} is the product of a probability of overcoming the barrier P, a surface number density factor σ, a coverage fraction θ, and an inverse time between collisions with the confining potential $(1/\tau)$, or

$$E_{evap} = \frac{1}{\tau}P\sigma\theta \tag{29.77}$$

[18] Figure 8 in ref. [416] labels time in minutes, but that is incompatible with the description in the text, suggesting $\theta = 0.3$ after 5–10 minutes at $T = 1373$ K, indicating that t is in seconds.

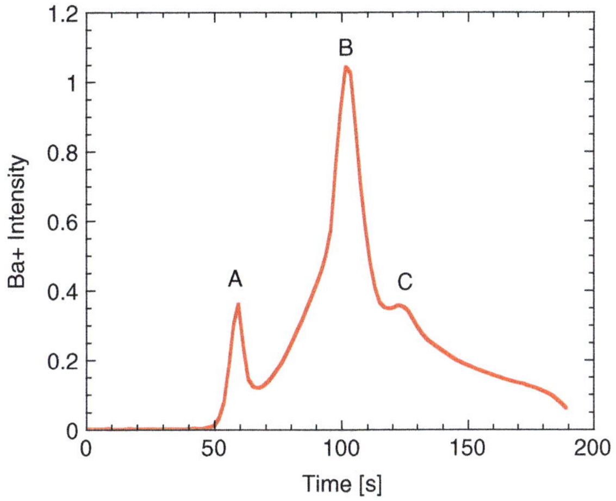

Figure 29.21 Evaporation of Ba^+ showing three distinct evaporation constituents: A, mulitlayer evaporation; B, monolayer Ba evaporation; C, evaporation of Ba and O. Based on Figure 2 of Shih *et al.* [376].

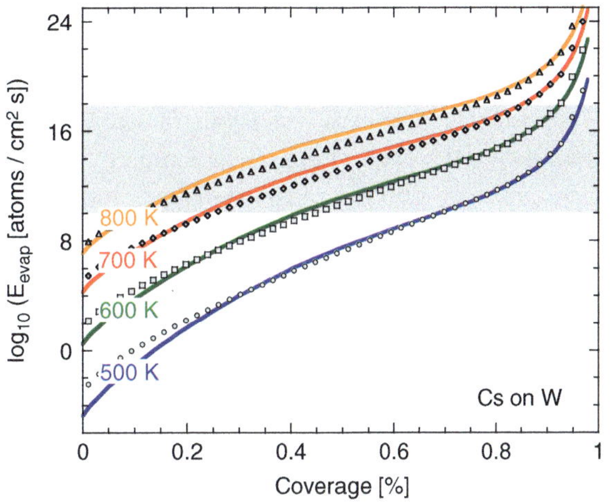

Figure 29.22 Comparison of the theory and simulation of Pan *et al.* [380] (lines) to the data reported by Taylor and Langmuir [418] (symbols; the gray bar represents the region over which experimental data was taken by Taylor and Langmuir on which their projections were based). Based on Figure 9 of ref. [380].

The probability factor P is analogous to Eq. (29.29). The potential well and barrier are a consequence of the charged nature of the adatoms, the θ-dependent v_o and ε familiar from the treatment of Eq. (29.55), and the dependence of the collision frequency on the coverage factor θ, a set of relations that must be solved self-consistently because of the collective dependence on coverage. Pan *et al.* [380] performed the self-consistent calculation numerically and found good agreement between the predictions of the model and the extrapolations of Taylor and Langmuir [418], as shown in Figure 29.22.

29.4 Knudsen flow through pores

This new theory ...states that a gas molecule, when impinging upon a solid wall, is reflected in a direction totally independent of the direction in which it approaches the wall, and that a multitude of molecules impacting on a section of a solid surface is sent out in accordance with the Cos-law as known for the radiation of glowing solid bodies.

– Martin Knudsen [419][19]

[19]The translation is from ref. [420].

The small, numerous, and labyrinthian tubes that make up the pores of a dispenser cathode are responsible for bringing up the coating material (typically barium in thermionic cathodes [366, 372, 378], but also cesium when a dispenser cathode is used as a photocathode [381, 382]) from the depths of the sintered porous tungsten. Although gaseous diffusion is involved, atoms constantly strike and stick to the pore walls, whereupon they evaporate again. The starting amount of impregnant (barium or cesium) in the porous matrix or reservoir, along with the rapidity with which it is transported through the pores, dictates the dispenser cathode operational lifetime [366, 378, 421]. In contrast to the more familiar hydrodynamic flow of fluid mechanics, Knudsen flow deals with the physics of gas transport through tubes under low pressure such that the discrete nature of the atoms (or molecules) matters, and the tracking of their individual trajectories is followed. By such means, Knudsen in 1908 was able to explain anomalies that became apparent when the dimensions of the tube became small compared to the mean free path of the gas atoms. The concepts he introduced eventually found use in understanding gaseous diffusion processes [422] as used, for example, to separate uranium isotopes U-235 and U-238 in the 1940s [423] in the quest for atomic weapons. Knudsen flow has maintained its topicality up to the present day [424] and is even more relevant for understanding of the lifetime of the *controlled* porosity dispenser cathodes of Section 29.5.

Of particular importance is the Knudsen number K_n, which gives a dimensionless measure of what physics matters and when. K_n is a measure of the ratio of two length scales: first, the mean free path $l = \langle v \rangle \tau$ where τ is the mean time between collisions and $\langle v \rangle$ is the average velocity, familiar from Section 3.2; and second, a characteristic dimension d of the enclosing chamber or volume, that is,

$$K_n = \frac{l}{d} \tag{29.78}$$

It can be estimated by considering sequential collisions between two walls [425]. Imagine that an atom has just collided with one wall, has recoiled off the opposing wall, and is just about to strike the first wall again. An uncomplicated visualization is that of two walls as parallel plates, each of area A separated by a length L, but a more imaginative view realizes that the description is more general. The number of collisions is 1, but the time between collisions is twice the time it takes the atom to traverse the length L, or $\delta t = 2L/v\cos\theta$, where $v = \langle v \rangle$ and θ is with respect to the normal to the wall. For $N = \rho V$ atoms, there are $\rho V v \cos\theta / 2L = (\rho A v/2)\cos\theta$ collisions. Averaged over all angles, or $\langle \cos\theta \rangle = 1/2$, this gives the number of collisions per second as $\rho A v/4$. Comparing this to the total number of collisions per second as Nv/d results in

$$d = \frac{4V}{A} \tag{29.79}$$

EXAMPLE: Find d for a cube, a sphere, a cylinder, and concentric spheres.

SOLUTION: For

- a cube of side a, $V = a^3$ and $A = 6a^2$, so $d = 2a/3$
- a sphere of radius r, $V = (4/3)\pi r^3$ and $A = 4\pi r^2$, so $d = 4r/3$
- a cylindrical tube of length L and end radius r, $V = L\pi r^2$ and $A = 2\pi r^2 + 2\pi r L$, so $d = 2rL/(r+L)$
- two concentric spheres of radius a and $b < a$, the volume is $V = (4\pi/3)(a^3 - b^3)$ and $A = (4\pi/3)a^2 + (4\pi/3)b^2$, so $d = 4(a^3 - b^3)/3(a^2 + b^2)$.

The mean free path l takes more work by comparison: although the concept can be put on firm ground by considering Maxwellian velocity distributions and the probabilities of collision [51], a reasonable estimate is given by the following argument. For one kind of gas atom (cesium or barium) of radius r, two neutral atoms will collide if on closest approach they pass within $2r$ of each other. The cross-sectional area is therefore $\sigma = \pi(2r)^2 = 4\pi r^2$. The probability of an atom being struck by one other atom in a volume $V = Al$, where l is understood as the mean free path $l = v\tau$, is the ratio of σ to A. Therefore, the probability of being struck by one of N other atoms is $N\sigma/A$. Setting that probability to unity is a proxy for mean free path, with one small caveat: both atoms are generally moving, rather than one stationary and one moving. When both atoms are moving, their *relative* velocity $\vec{v}_r$ is used in the definition of l, which must be related to the velocities $\vec{v}_1$ and $\vec{v}_2$ of the two atoms. The average velocity of both atoms is the same. Therefore,

$$\begin{aligned} \langle v_r^2 \rangle &= \langle (\vec{v}_1 - \vec{v}_2)^2 \rangle = \langle v_1^2 \rangle + \langle v_2^2 \rangle - 2\langle \vec{v}_1 \cdot \vec{v}_2 \rangle \\ &= \langle v_1^2 \rangle + \langle v_2^2 \rangle = 2\langle v^2 \rangle \end{aligned} \tag{29.80}$$

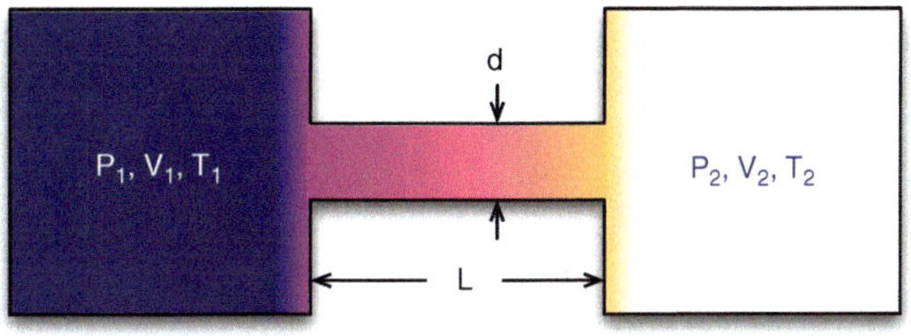

Figure 29.23 Two reservoirs of different pressure, volume, and temperature (P, V, T) connected by a tube of length L and diameter d; the second reservoir, for instance, can be a vacuum. The flow of gas from one to the other is characterized as Knudsen flow.

where because the velocities are randomly directed, the average $\langle \vec{v}_1 \cdot \vec{v}_2 \rangle = 0$, and $\langle v^2 \rangle$ is the average of the velocity squared of one of the atoms, or $\langle v_r^2 \rangle = 2\langle v^2 \rangle$. This will introduce a $\sqrt{2}$ in the velocity defining l. So, the probability of intersection is unity when

$$N\sigma = A = \frac{V}{l\sqrt{2}} \tag{29.81}$$

Letting $N = \rho V$, $\sigma = 4\pi r^2$, and (from the ideal gas law) $\rho = P/k_B T$, then

$$l = \frac{k_B T}{4\sqrt{2}\pi r^2 P} \tag{29.82}$$

Knudsen flow refers to the transport of molecules (or atoms) through a tube connecting two reservoirs, as in Figure 29.23. One of those reservoirs, say the one on the right, will be at vacuum for a thermal cathode, but keep the discussion general and simply assume $P_2 \ll P_1 \equiv P$. Let reservoir 1 contain liquid cesium or barium that has an associated vapor pressure that changes with temperature. A common representation of the vapor pressure from which $P(T)$ is obtained is the *Antoine equation*

$$\log_{10}(P) = A - \frac{B}{T + C} \tag{29.83}$$

whre the values of A, B, and C are available in various databases.[20] For cesium,

$$\log_{10}(P_{Cs}) = 6.5709 - \frac{3453.1}{T + 26.829} \tag{29.84}$$

for P in [Torr] and T in [K], whereas for barium,

$$\log_{10}(P_{Ba}) = 6.9570 - \frac{7599.4}{T - 45.737} \tag{29.85}$$

EXAMPLE: Let a tube fashioned after a dispenser cathode pore be characterized by a diameter of 0.4 μm and a length $L \gg r$. Let it be connected to a reservoir vessel, the pressure in which is characterized by the vapor pressure of cesium, as in Eq. (29.84). The radius of a cesium atom is 0.263 nm. Let the temperature $T = 500$ K. Find K_n.

SOLUTION: Because $L \gg r$, $d = 2R$. Use the ideal gas equation, $\rho = \beta P$ with $\beta = 23.209$ [1/eV]. Finally, observe that 1 Torr is 8.3212×10^{-7} [eV/nm³]. Therefore,

$$K_n = \frac{l}{d} = (16\sqrt{2}\pi R r^2 \beta P(T))^{-1} = 50.703 \tag{29.86}$$

Thus, for the temperature and pressure chosen, the mean free path exceeds the diameter of the pore by over an order of magnitude. Were the cesium to be replaced by barium, a comparable K_n value would be obtained for a much higher T of 1200 K.

The mass flow rate Q through the tube is given by the product of the mass of the gas atom and a flux through an area A, as suggested in Figures 29.24 and 29.25. If the length of the tube is long compared to its diameter *and* to the mean free

[20] See, for example, the NIST Chemistry WebBook, http://webbook.nist.gov, which gives A for P [bar]; here, units are converted so that A gives P [Torr], where 1 bar = 750.062 Torr.

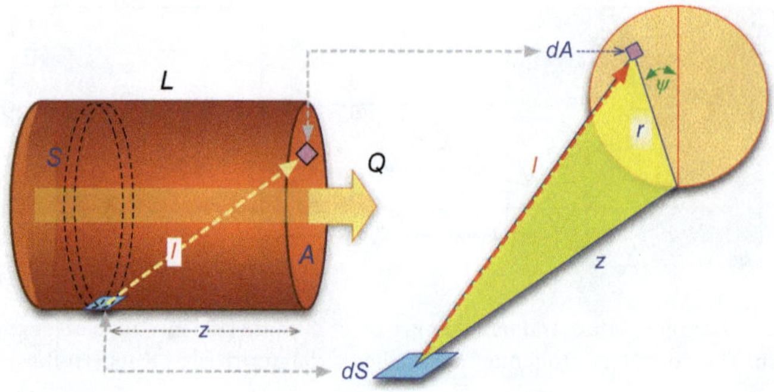

Figure 29.24 The center tube of Figure 29.23, through which gas flows from Reservoir 1 to Reservoir 2. The mass flow rate is Q. The tube is of length L and cross-sectional area A. Cesium atoms that evaporate off a small patch dS on the cylindrical inner surface S a distance z from Reservoir 2 travel to dA and contribute to the mass flow rate.

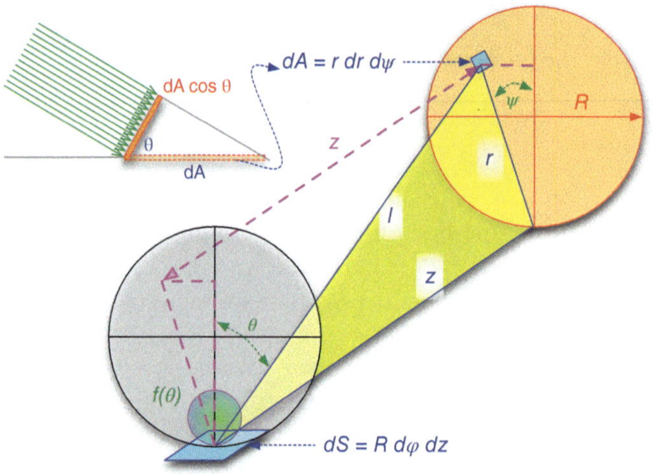

Figure 29.25 Closer view of the relation of dS to dA in Figure 29.25. The $\hat{z}$ axis is along the tube axis. The area normal to dS defines the vertical $\hat{y}$ axis. The function $f(\theta)$ describes the evaporation distribution. R is the radius of the tube. θ is the angle between the vertical axis and the line l linking dS to dA. ψ is the angle between the vertical and the direction r to dA in the x–y plane at the entrance to Reservoir 2 such that $dA = rd\psi dr$; φ is also an angle in the x–y plane such that $dS = Rd\varphi dz$. The solid angle $d\Omega$ originates at dS, encloses the line l, and ends at dA. Inset in upper left: green arrows are the direction of in-flowing atoms; dA is the shadow cast by $dA\cos\theta = (z/l)dA$ onto the plane A.

path l, then the likelihood that the atom had journeyed from Reservoir 1 to Reservoir 2 without encountering anything, be it sticking to the tube wall or colliding with one of its fellow travelers and getting knocked into the direction of a wall, is rather small indeed. Therefore, it may be assumed that all the atoms passing through the end of the tube through the window area marked A have done so after being evaporated off the tube wall. Calculating mass flow is then tantamount to adding up all the atoms that leave the surface element dS and pass through the window element dA.

But there are matters requiring care. First is that the atoms do not approach dA head on, but at an angle: dA is analogous to a shadow of the window that the atoms pass through, the angle between the window and its projection dA onto the window area A, as in the upper left inset of Figure 29.25. Second, the total flux $\mathcal{F}$ *into* the surface dS has to be the same as the total flux coming off the surface, but, as Knudsen pointed out, the atom must be absorbed onto the surface and stay there long enough to lose all memory of its incoming direction, so that the distribution $f(\theta,\phi)$ is normalized to unity. Azimuthal symmetry of emission is reasonable, and Knudsen argued that a cosine-law applied to $f(\theta)$, as was known to be the case for diffuse radiators in optics and which goes by the name of Lambert's cosine law: it results in the same flux per *projected* area as a function of angle θ with respect to the surface normal (see the sub-figure in Figure 29.25). Therefore, setting $f(\theta) = N_c\cos\theta$

$$\mathcal{F} = \mathcal{F}\int_0^{2\pi} d\phi \int_0^{\pi} f(\theta)\sin\theta d\theta = \pi N_c\mathcal{F} \tag{29.87}$$

or $f(\theta) = (1/\pi)\cos\theta$. Such a distribution resembles the small blue-green sphere in Figure 29.25 hovering above dS.

Lastly, although the flux $\mathcal{F}$ into the wall looks like a constant, a moment's reflection shows this cannot be true if the reservoir pressures are different: near Reservoir 1, the pressure is higher, and so the flux must be higher. If the gradients are not large,

$$\mathcal{F}(z) = \mathcal{F}(0) + z(\partial_z \mathcal{F}) \tag{29.88}$$

But remember that flux is already known, or

$$\mathcal{F}(z) = \rho\langle v_x^+\rangle = \frac{\beta P(z)}{(2\pi M\beta)^{1/2}} = \frac{P(z)}{(2\pi M k_B T)^{1/2}} \tag{29.89}$$

where v_x^+ is the velocity component into dS, and $\langle v_x^+\rangle = \langle v\rangle/4$ for a Maxwell–Boltzmann distribution in velocities, as can be deduced in a variant to Eq. (3.16).

EXAMPLE: Prove $\langle v_x^+\rangle = \langle v\rangle/4$ for a Maxwell–Boltzmann distribution, where the + superscript denotes "in the forward direction".

SOLUTION: There is no loss in generality in working in units for which the mass M is equal to 2. Thus $f_{MB}(v) = e^{-v^2}$ where $v = |v|$ and $v^2 = v_x^2 + v_y^2 + v_z^2$. It follows

$$\langle v_x^+\rangle = \frac{\int_0^\infty v_x e^{-v_x^2} dv_x}{\int_{-\infty}^\infty v_x e^{-v_x^2} dv_x} = \frac{1}{2\sqrt{\pi}}; \ \langle v\rangle = \frac{\int_0^\infty v^3 e^{-v^2} dv}{\int_0^\infty v^2 e^{-v^2} dv} = \frac{2}{\sqrt{\pi}}$$

where the differing representations of the integrals involved is a consequence of the peculiar features of a Maxwell–Boltzmann distribution expressed alternately in Cartesian and then polar coordinates. Thus it follows that $\langle v_x^+\rangle = \langle v\rangle/4$.

Now the mass flow rate Q can be evaluated: it is the integral over the flux $\mathcal{F}(z)f(\theta)$ coming off of dS and directed through the solid angle $d\Omega$, or

$$\begin{aligned} Q &= M \iiint dS d\Omega f(\theta)\, \mathcal{F}(z) \\ &= M \int_{-\infty}^{\infty} 2\pi R dz \int_{-\pi/2}^{\pi/2} d\psi \int_0^{2R\cos\psi} \frac{r dr}{l^2} \frac{r\cos\psi}{\pi l} \frac{z}{l} \mathcal{F}(z) \end{aligned} \tag{29.90}$$

where use has been made of $l^2 = r^2 + z^2$ and $d\vec{\Omega}\cdot\hat{n} = z/l$. But observe that the integral over the first term in Eq. (29.88) (a constant) vanishes because the integrand is antisymmetric in z, therefore only the $z\partial_z\mathcal{F}$ term survives. For the second term use the linear approximation

$$\partial_z\mathcal{F}(0) \approx \frac{\mathcal{F}_2 - \mathcal{F}_1}{z_2 - z_1} = \frac{P_2\sqrt{\beta_2} - P_1\sqrt{\beta_1}}{L\sqrt{2\pi M}} \tag{29.91}$$

The remaining integrations are straightforward, and the general Knudsen equation for Q becomes

$$Q = \frac{4}{3}\sqrt{2\pi M}\frac{R^3}{L}\left\{P_2\beta_2^{1/2} - P_1\beta_1^{1/2}\right\} \tag{29.92}$$

For the special case in which the temperatures of the two reservoirs are equal, it simplifies to

$$Q|_{T_1=T_2} = \frac{4}{3}\sqrt{2\pi M\beta}\frac{R^3}{L}\Delta P \tag{29.93}$$

where ΔP is the pressure difference between the reservoirs. Below, making the assumption of equal reservoir temperatures immeasurably eases the discussion, and so Eq. (29.93) will be used for Q. Observe that the infinite tube, or $L \gg 1$, limit was used to eliminate the $\mathcal{F}(0)$ integral. When the tubes are of finite length, then the calculations are more difficult [426].

An expression for Q using a dimensionless tube cross-section parameter W often proves convenient, the relation between W and Q being given by

$$Q = M\langle v_x^+\rangle A(P_2 - P_1)W = \frac{M}{4}\langle v\rangle A(P_2 - P_1)W \tag{29.94}$$

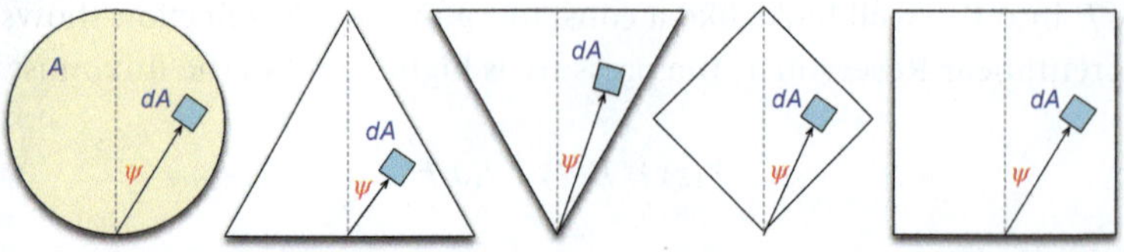

Figure 29.26 Example tube cross-sections for the evaluation of W by Eq. (29.96), such that the height of each is unity. In addition to the circle (left), they are termed triangle, inverted triangle, diamond, and square in Table 29.98, for which $\langle \cos\psi \rangle = 0.849, 0.721, 0.864, 0.916$, and 0.790, respectively.

Cast in this way, the orientation and shape of the tube, which matters in how the gas moves through it, is contained in the behavior of W alone, and it, being dimensionless, allows for convenient comparisons. W is seen to be defined by

$$W \equiv \frac{P_A}{2L}\frac{1}{A}\int dA\cos\psi \equiv \frac{P_A}{2L}\langle \cos\psi \rangle \tag{29.95}$$

where P_A is the *perimeter* of the tube cross-section, $A = \int dA$ is the total cross-sectional area, and ψ is the angle directed to the location of dA in Figure 29.25. Observe that $\langle \cos\psi \rangle$ depends only on the *shape* of the tube, not its total area. An example helps dispel ambiguity, so consider the circular cross-section, or the first of the shapes shown in Figure 29.26, for which $P_A = 2\pi R$, $A = \pi R^2$, and

$$\langle \cos\psi \rangle = \frac{\int_{-\pi/2}^{\pi/2} \cos\psi d\psi \int_0^{2R\cos\psi} rdr}{\int_{-\pi/2}^{\pi/2} d\psi \int_0^{2R\cos\psi} rdr} = \frac{8}{3\pi} = 0.84883 \tag{29.96}$$

where the limits of the r integration specify the rim of the tube. Thus, for a circular pore [427],

$$W_\circ = \frac{8R}{3L} \tag{29.97}$$

The circular pore, however, is unique: all the points along the rim for the evaluation of $\langle \cos\psi \rangle$ are equivalent. When the pore has a non-circular shape, however, the orientation of the pore matters, and where the origin is results in a different value of $\langle \cos\psi \rangle$. For example, the remaining pore shapes shown in Figure 29.26 can be put into integrals of the form

$$\langle \cos\psi \rangle = A^{-1}\int_{x_a}^{x_b} dx \int_{y_a}^{y_b} \frac{ydy}{\sqrt{x^2+y^2}} \tag{29.98}$$

where x_a, x_b, y_a and y_b depend on the pore shape and orientation and are given in Table 29.1. The numerical values of the integral of Eq. (29.98) are shown in the figure caption of Figure 29.26, thereby demonstrating that the value of $\langle \cos\psi \rangle$ depends on where the origin of the integration is. The shapes of Figure 29.26, though, were chosen with care to be the limiting (largest and smallest) cases, so the average over all origin choices should lie in the middle, say near the average of those limiting cases. The averages do not differ greatly from the circular case, and so for the purposes of a life-time model, using the circular result of Eq. (29.97) is convenient and approximately correct. Even so, given that the perimeter P_{pore} and cross-sectional area A_{pore} of the pore need not be circular in general, it is better to call those quantities out specifically, and finally represent Q as

$$Q \approx \frac{2}{3\pi^{3/2}}\sqrt{2M\beta}\frac{A_{pore}P_{pore}}{L}\Delta P(T) \tag{29.99}$$

where $\Delta P(T)$ is the temperature-dependent pressure difference between Reservoir 1 and Reservoir 2 (taken as vacuum).

Table 29.1 Pore integral limits for Eq. (29.98) (recall that the value of $\langle \cos\psi \rangle$ is 0.84883 for a circular pore).

Pore	A^{-1}	x_a	x_b	y_a	y_a	$\langle \cos\psi \rangle$
Triangle	$2\sqrt{3}$	0	$1/\sqrt{3}$	0	$1-\sqrt{3}x$	0.721
Inverted triangle	$(6/5)\sqrt{3}$	0	$1/\sqrt{3}$	0	$x/\sqrt{3}$	0.864
Diamond	4	0	1/2	x	$1-x$	0.916
Square	2	0	1/2	0	1	0.790

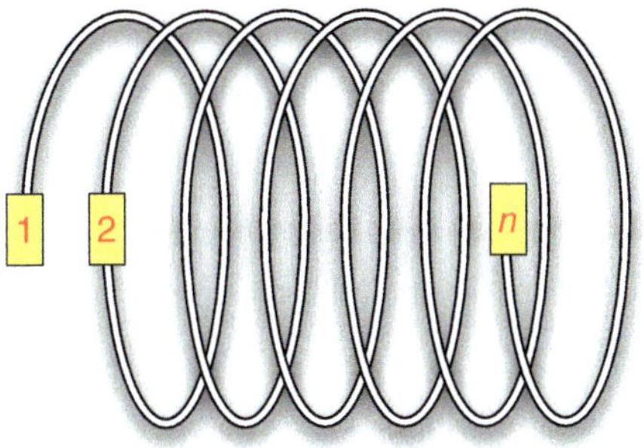

Figure 29.27 A coiled or cork-screwed tube of length L, having a number of turns n. The pitch λ is the ratio of the distance between adjacent threads l/n to the radius r.

In a conventional dispenser cathode, the tubes are hardly straight (as they are in the tubes of the controlled porosity dispenser cathode considered in Section 29.5). As long as the pore diameter is small with respect to the curvature of the tube, characterizing tubes by a tortuosity factor [428] is possible. A simple one-parameter model is suggested by the spring-like shape in Figure 29.27. The length of the tube is given by

$$L = \int_0^{2\pi n} \left\{ r^2 + \left(\frac{l}{2\pi n} \right)^2 \right\}^{1/2} d\theta = l\sqrt{1 + \left(\frac{2\pi}{\lambda} \right)^2} \qquad (29.100)$$

where $\lambda = l/nr$ is the pitch. As the number of turns $n \to 0$, then $L \to l$, but for a very tortuous tube, L can become much greater than l.

29.5 Lifetime of a sintered wire controlled porosity dispenser cathode

The oldest hath borne most; we that are young
Shall never see so much, nor live so long.

– William Shakespeare[21]

Thermionic cathodes operated SCL are at a temperature such that more electrons are emitted than can be pulled across the AK gap. The excess are turned back to the cathode. As a result, the electron cloud in front of the cathode smooths emission non-uniformity that may exist on the cathode surface. The random placement of the pores and the non-uniformity it entails is therefore not necessarily problematic. However, when dispenser cathodes are used for applications which are *not* run SCL, as for gyrotrons and gyro-amplifiers, [12, 13, 221, 429] or for photocathodes for particle accelerators and free electron lasers [380–382, 384], emission non-uniformity is undesirable so that the pore and grain structure may entail problems. Although uniformity can be induced by running the cathodes at a higher temperature to achieve faster barium diffusion and surface coverage, that comes at a cost of lifetime. By the very nature of the sintering process on tungsten grains, and the damage to the pores that results from machining the surface, a uniform placement of clear pores was not possible, for a while.

Controlled porosity dispenser (CPD) cathodes developed by the Naval Research Laboratory [369, 370] showed proof of concept, with laser-drilled or etched holes approximately 5 μm in diameter making the pores. The CPD cathodes demonstrated promising performance in terms of uniformity of emission, although the pore size resulted in a more rapid depletion of the barium than desirable, and costs associated with making the pores was high.

An innovative solution by Calabazas Creek Research (CCR) [392] wound tungsten wires about a quarter the thickness of a human hair[22] on a spool then cut cross-sections of a desired thickness and area to give the densely and uniformly placed pores shown in Figure 29.28. The simple and cost-effective technique was usable with reservoirs that can be made large, which provides practical advantages, a notable one being the absence of a tortuosity factor for tubes of a uniform length. Thus, the Q per tube is the same and a relatively simple estimate of estimate cathode lifetime can be made.

[21] Ref. [37]: *King Lear*, I.v.326–327.

[22] Human hairs vary, so what should have been said is a straight human hair from a caucasian adult (East Asian hair has twice the diameter) that is black (80–100 μm) rather than red (50 μm) or flaxen (20–50 μm). But as tungsten wire thickness varies from 4 μm to 20 μm, a range in hair thicknesses to give the same ratio is permissible.

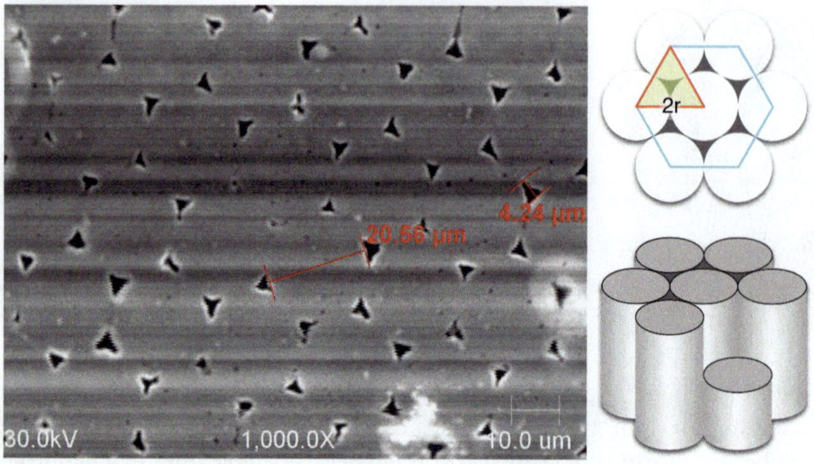

Figure 29.28 Left, SEM of a CCR controlled porosity dispenser cathode manufactured using 20 μm diameter tungsten wires sintered at 2348 K for 1.25 hours, giving rise to approximately triangular pores 4 μm in extent. Image courtesy of R.L. Ives (Calabazas Creek Research), similar to Figure 9 of R.L. Ives, L.R. Falce, S. Schwartzkopf, and R. Witherspoon, *IEEE Trans. El. Dev.*, **52**, 2800, 2005 [392] and used with the permission of IEEE TED. Right, schematic details of the surface (top) and wires (bottom).

Cesium in the reservoir is activated by heating: for dispenser cathodes, the reservoir is filled with cesium or barium producing compounds that liberate cesium or barium when heated. The vapor pressure relation is therefore not that of pure cesium or barium, but something less fulsome (a factor E_d added to the vapor pressure relation will account for it). To evaluate the lifetime then, several factors must be known: the amount of cesium M_o in the reservoir, its flow rate through *one* pore Q, the length L and area A_{pore} of that pore, the number of pores N in the area of the cathode, and the operating temperature T and vapor pressure $P(T)$ in the reservoir. The lifetime Δt is then specified by the relation [380]

$$M_o = NQ\Delta t \tag{29.101}$$

The factor N is determined by first relating the number of pores in the light-blue hexagon, or 6, to the ratio of the area of the hexagon $A_{hex} = 6\sqrt{3}r^2$ to the area of the red equilateral triangle $A_\triangle = \sqrt{3}r^2$ for r being the radius of the wire, as suggested by the upper right inset diagram of Figure 29.28. Assuming that the area of the cathode is much larger than the the cross-sectional area of the tungsten wire, then $N \approx A_o/A_\triangle$, where A_o is the entire area of the cathode.

The area of the pore A_{pore} is affected by the sintering temperature and duration, but take it as $A_\triangle$ with three wedges ("pizza slices") of area $\pi r^2/6$ removed. Likewise the perimeter of the pore is three arc-lengths of $\pi r/3$, or[23]

$$A_{pore} = \left(\sqrt{3} - \frac{\pi}{2}\right) r^2 \tag{29.102}$$

$$P_{pore} = \pi r \tag{29.103}$$

Inserting all the factors into the expression for Δt results in

$$\Delta t = \frac{3M_o}{4\left(\sqrt{3} - (\pi/2)\right)} \frac{L}{A_o r \Delta P(T)} \left(\frac{6\pi k_B T}{M}\right)^{1/2} \tag{29.104}$$

for the CCR sintered wire controlled porosity dispenser cathode. Two features are important. First, although $Q \propto r^3$, the number of pores $N \propto r^{-2}$ so that lifetime is only inversely proportional to r. Second, if the pressure of the second reservoir P_2 is taken to be 0, or the reservoir is at vacuum, then it would seem that $\Delta P(T) = P_{Cs}(T)$ from Eq. (29.84), but in light of the "cesium-producing compounds" nature of how cesium is generated in the first reservoir, a factor to account for a greater difficulty cesium may have in evaporating from other materials rather than itself is accounted for by modifying $\Delta P(T)$ to become

$$\Delta P(T) \to P_{Cs}(T) e^{-E_d/k_B T} \tag{29.105}$$

where E_d serves the role as an activation energy that requires determination by comparison with data.

---

[23] The specification of a tube area and perimeter for non-circular pores has advantages over constructing an "equivalent" pore diameter, but the outcomes of the present approach to ref. [380] are comparable.

EXAMPLE: Estimate the value of E_d for a sintered wire CCR cesium-controlled porosity dispenser cathode using the measured lifetime of 5–10 hours for the following parameters: cathode area $A_o = \pi(0.17\ \text{cm})^2$, thickness $L = 0.762$ mm, radius of tungsten wire $r = 23$ μm, mass of cesium in reservoir $M_o = 7.7$ mg, and temperature $T = 773$ K = 500 °C (measured values courtesy of Eric Montgomery, UMD).

SOLUTION: First take $E_d = 0$ eV. Recall that $M = 132.91$ grams/mole for cesium. Eq. (29.104) then gives $\Delta t = 5.2192$ seconds. Therefore, using the midpoint of the experimentally measured lifetime, $e^{E_d/k_BT} = (7.5\ \text{hours})/(5.2192\ \text{seconds})$, or $E_d = 0.57$ eV.

EXAMPLE: Using E_d from the previous example, estimate the lifetime of a cesium sintered wire CPD cathode for a temperature of 250 °C.

SOLUTION: For $T = 473$ K, $P_{Cs} = 1.9525$ Torr and $e^{-E_d/k_BT} = 3.2432 \times 10^{-6}$. Using the cathode and material properties of the previous example, then $\Delta t = 33.768$ kHrs, or a little under four years.

Extending such an analysis to conventional barium dispenser cathodes requires an estimate of the tortuosity factor as well as E_d. Values of λ of order unity and $E_d \sim 1.4$ eV gave lifetime estimates that were comparable to conventional barium dispenser cathodes of 30 kHrs at 1273 K [380].

CHAPTER 30
Field emitters

Field emitters are conventionally needle shaped in order to provide the required value of the electric field, i.e., $10^7 < F < 10^8$ volts/cm, with convenient values of the applied potential. The emitter tip is usually a single crystal, approximately hemispherical with a radius of the order of 10^{-6} cm. Thus both ϕ and F vary appreciably over the emitting surface ...

– Walter P. Dyke, J. Ken Trolan, Winthrop W. Dolan, and George Barnes [5]

The analysis of field emission experimental data concerns matching current to applied voltage (I-V) rather than current density to *surface* field $J(F)$, as in Section 13.3, where the emphasis on surface is not acknowledged nearly as often as it should be. Applying Fowler–Nordheim theory is nuanced [430], but as before, the simple approximate relations between current and current density of $I = JA$, where A is the area of emission,[1] and between applied voltage V and field of $F = \beta_g V$, where β_g is a field enhancement factor, are oft-used relations that allow the Fowler–Nordheim equation (see Eq. (13.25)) to be morphed into the visually similar but nevertheless very different empirical equation of Eq. (13.30) – recall that it is

$$I(V) = AV^2 \exp\left(-\frac{B}{V}\right) \tag{30.1}$$

Although there are serious issues with making the $I \to JA$ and $V \to V/\beta_g$ replacements [431] so as to recover $J_{FN}(F)$, issues that shall be re-visited in due course, what is interesting here is why these approximations need to be made at all. Visualize holding two metal plates apart with with a dielectric material (in other words, a capacitor) and try to get to fields across the plates that are characteristic of field emission before suffering dielectric breakdown. The dielectric fails first, for example quartz glass (amorphous SiO_2) has a dielectric strength of roughly 160 MV/cm = 0.016 GV/nm [432], which is well below the fields needed in J_{FN} to make it interesting. To get the field at the emission site up in the range of GV/m, emitters are often made into sharp points where the field is strongest at the tip. As a result, the area refers to the apex area (or some fraction of it), and the field enhancement refers to whatever processes cause fields about sharp points to be large. Therefore, it is the multidimensional nature of field emitters that really sets them apart, spawning the myriad technological gambits to make edges and points, and which causes the electron beams from field emission sources to be of a rather different nature than the planar thermal, photo, or secondary emitters they compete with. But more to the point here, it also makes the correct relation for I-V much more complicated than Eq. (30.1) would imply, makes field emitters very susceptible to space charge, and causes their performance to be undermined when groups of them are too close to each other.

A few comments are in order. The methods to treat field enhancement have application to an interesting hybrid known as the Schottky emitter, wherein high temperatures are used for emission, but large fields are used to lower the emission barrier [154]. The physics behind hot needle-like emitters therefore combines thermal-field emission behavior with field enhancement effects, so that though they are not explicitly discussed further here, the physics behind their operation is implicit in what is. Also, not all cold cathodes are pointy: some are flat and rely on an internal field emission model [150, 433–435]. Lastly, the effects of temperature always lurk in the background as when heat is unwanted, its dissipation through the narrow structures of many field emitter sources is problematic [230, 436–441]. In short, what is treated herein is not all that there is, but should be helpful for that which is left out.

30.1 Field enhancement

"Look aloft!" cried Starbuck. "The corpusants! the corpusants!" All the yard-arms were tipped with a pallid fire; and touched at each tri-pointed lightning-rod-end with three tapering white flames, each of the three tall masts was silently burning in that sulphurous air, like three gigantic wax tapers before an alter.

– Herman Mellville[2]

[1] Observe that the relation *should* be $I = \int JdA \approx J_{apex}A_o$, where A_o is the "notional area", but this gets ahead of the narrative.
[2] Herman Melville, *Moby Dick*. New York: Bantam Books, 1986, Chapter 119.

Introduction to the Physics of Electron Emission, First Edition. Kevin L. Jensen.

The typhoons in the Japanese seas that threatened the whaling ship *Pequod* in Melville's epic gave rise to conditions which caused corona discharges at the tips of the masts, in which strong electric fields ionized the air and created St. Elmo's fire (what the coffee-loving first mate Starbuck and others of his time called *corposants* although Melville used a different spelling). Given that the potential difference between cumulonimbus clouds and "ground" are several hundred megavolts, and that the clouds are within 2 km of the ocean surface, gradients on the order of 0.1 eV/μm result, whereas because the dielectric strength of air at normal temperature and pressure is about 3 kV/mm ($F = 3$ eV/μm), the fields at the mast tips must exceed that.

The higher fields on the masts compared to those in the wide open are a result of field enhancement, whereby sharpened structures like the tips of masts, or lightning rods, or needle cathodes cause equipotential lines to bend in the vicinity of a sharpened point. A ship on the ocean distorts the equipotential lines in its vicinity that exist due to the voltage difference between the clouds in the upper atmosphere and sea level, similar to Figure 30.1. Electric fields are a measure of how many equipotential lines are crammed into how small a region: in the figure, the potential difference between adjacent lines (dashed blue or solid red) is 0.2 units (the unit being $F_o a$ as shown below). Near the ship mast, three red lines are crammed into the same vertical space as one blue line in the absence of the ship (blue dashed lines). In other words, the gradient is 0.6 energy units over one length unit for the red lines, but only 0.2 energy units when the ship is absent. Very crudely then, the field enhancement factor, or the ratio between the gradient in the presence of the protrusion compared to the field when the protrusion is absent is three times greater (and gets greater as one zooms in to the top of the mast itself as there are field enhancements atop field enhancements, as suggested by the Schottky hypothesis in Section 30.4).

The equipotential lines of Figure 30.1 seem *ad hoc*, but there is a mathematical model behind them. The law of superposition entails that, as formulated by Eyges ([99], p. 62),

> *Charge and charge alone can produce a potential. If in some problem we pinpoint all the different sources of charge and write an expression for the potential produced by each source, the sum of these partial potentials is the total potential that satisfies the boundary conditions.*

A total potential is therefore the sum of potential distributions due to individual charge distributions, a simple yet flexible model is useful from which to make estimates. The equipotential lines of Figure 30.1 are generated using two point charges of opposite sign placed in a uniform electric field at the points $(z, \rho) = (\pm z_p, 0)$ (the subscript p being for "point"). The reason for choosing two opposite charges rather than one is to insure that all points equidistant between the charges are at zero potential, which fits in nicely with the concept of the presumed ground, or in this case sea, level. The two charges form a *dipole* for which the effects on a background potential die out quickly with distance, but unlike usual dipoles a finite distance $2z_p$ separates the charges themselves. In this most simple of models, the equipotential lines are equivalent to the contour lines of $V(z, \rho)$ given by, in cylindrical coordinates,

$$V(z, \rho) = -F_o z - \frac{4\delta Q}{\sqrt{\rho^2 + (z + z_p)^2}} + \frac{4\delta Q}{\sqrt{\rho^2 + (z - z_p)^2}} \tag{30.2}$$

where the gradient between cloud and sea is represented by the first term, and the dimensionless δ represents the distortion caused by the ship or, more generally, the protrusion. If the relevant length scale of the protrusion is a (e.g., a is the radius of the crude hemisphere that incloses the ship in Figure 30.1), and if the ratio of the energy scale

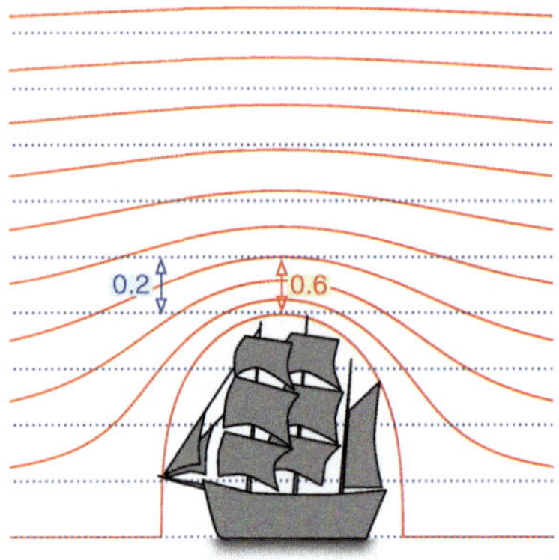

Figure 30.1 The potential difference between adjacent lines is 0.2 units: near the ship mast, three units (red lines) are in the same vertical space as one (blue dashed line) unit in the absence of the ship. The potential gradient, that is, the electric field, therefore has increased approximately by a factor of 0.6/0.2 = 3.

$4\delta Q/a$ to $F_o a$ is denoted by

$$\lambda \equiv \frac{4\delta Q/a}{F_o a} = \frac{4\delta Q}{F_o a^2} \tag{30.3}$$

then

$$\frac{V(z,\rho)}{F_o a} = u(z,\rho) = -\frac{z}{a} - \lambda\left(\frac{a}{r_+} - \frac{a}{r_-}\right) \tag{30.4}$$

$$r_\pm \equiv \sqrt{\rho^2 + (z \pm z_o)^2} \tag{30.5}$$

Importantly, the u function is dimensionless. This interesting equation suggests that problems which scale such that the product $F_o a$ is constant have a commonality. Electric fields are the gradient of potentials, and field enhancement is the ratio of an electric field in the presence of a protrusion compared to that which would exist in the absence of the protrusion. Defined thus, field enhancement is a dimensionless relationship that is a function of the geometry alone, and it has the same value for different fields and length scales so long as λ is the same for both conditions. On the axis, the field enhancement factor $\beta = F/F_o$ is a dimensionless ratio of fields, reflecting a convention to make the terminology *field enhancement* and *β factor* interchangeable. Note, though, that other conventions can be found. In the extended theoretical treatments of field enhancement for a variety of geometrical configurations by Edgcombe *et al.* [442–444] and Forbes *et al.* [445], a distinction is explicitly drawn between the dimensioned factor multiplying the anode potential, as done by Dyke and Dolan [155], and the dimensionless factor multiplying the background field as done here, and they prefer to use $\gamma = F/F_o$ to represent the latter. The reader should beware of both the distinction and the nomenclature. Here,

$$\beta \equiv \frac{F}{F_o} = -\frac{1}{F_o}\partial_z V(z,\rho)|_{z=ha,\rho=0} = -a\partial_z u(z,\rho)|_{z=ha,\rho=0} \tag{30.6}$$

where ha is the height of the protrusion (h is dimensionless), and $\partial_z u(z,\rho) = 1$ when $\delta = 0$ (the protrusion-free case).

For metal-like protrusions, the electric fields are *normal* to the surface, and so the gradients from which the field enhancement factors are determined should be calculated in the direction of the normal as well. Later, this complication will be respected, but here and now it is enough to observe that for the rotationally symmetric problem, the normal at the apex is along $\hat{z}$, so that $\hat{n} \cdot \nabla V = \partial_z V$, a simplification that is too useful to ignore.

EXAMPLE: For the $u(z,\rho)$ function shown in Figure 30.1 (where z is the vertical and the units are such that $a = 1$), find the β factor knowing that $z_o = 0.5$ and $\delta = 0.3$.

SOLUTION: The field enhancement factor is the value of $\partial_z u(z,\rho)$ evaluated on the axis ($\rho = 0$) at the apex ($z = h$). Therefore first find h and then evaluate $\partial_z u(z,\rho)$ at $(z,\rho) = (h,0)$. Finding h requires solving

$$u(h,0) = 0 \rightarrow -h^3 + z_o^2 h + 2\delta z_o = 0$$

for which $h = 0.792739$ is the real root (alternately, using the methods of Section A3.7.2, iterating $h_{j+1} = (2\delta z_o + z_o^2 h_j)^{1/3}$ starting with $h_0 = 0$ and converging to 0.792739 after eight iterations). Using this and $h^2 - z_o^2 = 2\delta z_o/h$ from $u(h,0) = 0$ gives

$$-\partial_x u(z,\rho)|_{z=h,\rho=0} = 1 + \frac{h^3}{\delta z_o} = 4.32123$$

The field F_{tip} at the apex of the hemispherical shape in Figure 30.1 is therefore 4.32 times larger than the background field F_o, or $F_{tip} \approx 4.32 F_o$.

Setting up and solving electrostatic problems in this manner is an example of a *summation problem* [99], where the potential energy everywhere for an electron of charge $-q$ can be evaluated by

$$V(\vec{r}) = 4Q\int \frac{\rho_e(\vec{r}')}{|\vec{r} - \vec{r}'|}d\vec{r}' \tag{30.7}$$

The form of Eq. (30.7) holds true even if the number density $\rho_e(\vec{r})$ is a sum over delta functions $\sum_j \lambda_j \delta(\vec{r} - \vec{r}_j)$ corresponding to point charges, a viewpoint that becomes invaluable in Section 30.3. Contrast this to the separation of variables approach, where a sum over elementary solutions subject to boundary conditions gives the potential, a particularly simple example of which is the case of the hemisphere on plane considered next.

30.2 Hemispheres and notional emission area

Our life is frittered away by detail ...Simplicity, simplicity, simplicity!

– Henry David Thoreau[3]

Just as the point charges of the previous section represented a particularly simple method of introducing field enhancement, the simplest of the expansion methods to do the same involves a hemisphere of radius a on a plane such that far away from the hemisphere the field F_o is constant and in the $\hat{z}$ direction [446]. This elegant little problem is quickly resolved by expressing the potential everywhere in terms of a Legendre polynomial expansion [100], as is often useful for problems with spherical symmetry (remember the treatment of the hydrogen atom in Section 9.1). The hemisphere on a plane of Figure 30.2 has rotational symmetry about the $\hat{z}$ axis, so in spherical coordinates the solution everywhere can be expressed in terms of the Legendre polynomials of Section A2.5 as

$$V(r,\theta) = \sum_{l=0}^{\infty} \left(A_l r^l + \frac{B_l}{r^{l+1}} \right) P_l(\cos\theta) \tag{30.8}$$

Far away from the hemisphere, the potential is that associated with a constant field, and so for $r \to \infty$, $V(r,\theta) \to -F_o z = -F_o r\cos\theta$. As a result, all A_l vanish, save for $A_1 = -F_o$ because all higher P_l involve higher powers of $\cos\theta$. But the hemisphere, too, is held at ground, and so at $r = a$ the potential again vanishes. Because all A_l for $l \neq 1$ are zero so too must B_l be zero for all $l \neq 1$ because there is nothing to cancel them otherwise. But, at $r = a$, it follows that $B_1 = -F_o a^3$ to cancel the A_1 term, and so for a boss (another albeit less exact name for a hemisphere) for which $P_1(\cos\theta) = \cos\theta$

$$V_{boss}(r,\theta) = -F_o r\cos\theta \left\{ 1 - \left(\frac{a}{r}\right)^3 \right\} \tag{30.9}$$

where terms have been gathered up. The normal to the hemisphere is radially outward ($\hat{n} = \hat{r}$), and so the field enhancement factor along the surface of the hemisphere is given by

$$\beta(\theta) = -\frac{1}{F_o}\partial_r V(r,\theta)|_{r=a} = 3\ \cos\theta \tag{30.10}$$

The field enhancement at the apex ($\theta = 0$) is therefore 3. One can verify independently that the same result at the apex is obtained in cylindrical coordinates for which $r = \sqrt{\rho^2 + z^2}$ by finding $\partial_z V$ and evaluating it at $\rho = 0$. But here, the point is more general: *the field declines (as* $\cos\theta$ *for a hemisphere) as one moves off-axis from the apex of a rotationally symmetric shape.* The rapidity with which the field declines is a measure of the curvature of that shape.

The last point deserves dwelling on because of its potential for confusion (pardon the pun). Field enhancement is frequently and implicitly treated as being the ratio of the *on axis* field on the protrusion to the background field [286, 431, 445, 447] (often with the background field being that due to a parallel plate, for which $F_o = V_{anode}/D$, where D is the anode-cathode (AK) gap), but doing so can be possibly misleading [448]. It is not simply the apex of the emitter which figures in field emission: the shape of the protrusion matters, and for that reason understanding that $\beta(\theta)$ *varies over the emission site*, or tip, has pedagogical value.

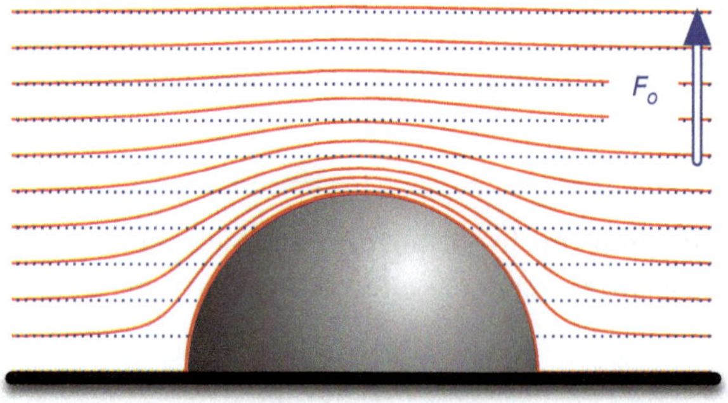

Figure 30.2 A conducting hemisphere on a plane subject to a background field $\vec{F} = F_o\hat{z}$. Both the hemisphere and the plane are held at ground. Red lines are the equipotential lines; blue dashed lines would be the equipotential lines if the hemisphere were absent.

[3]Henry David Thoreau, *Walden, Or, Life in the Woods*, Dover Thrift Editions. New York: Dover Publications, 1995, p. 59.

The shape of the protrusion matters because it affects how much of the surface is apparently emitting. Often the emission area Ω is crudely estimated as the ratio of the total current I_o from the protrusion with the apex current density $J(F_{tip})$, or $\Omega = I_o/J(F_{tip})$, but this overlooks the fact that the current density varies over the tip. That crude approximation can be put on better footing: the *notional emission area* ("notional" being something that only exists in theory or imagination) [449] should be something like

$$\Omega_n \sim \frac{1}{J(F_{tip})} \int_{boss} J(F) d\Omega \tag{30.11}$$

where the integration is over the surface of the hemisphere and the subscript "n" for notional emphasizes that it is not the true area of emission, even though it and the true area scale the same with field. Thus, the so-called crude estimate and the notional area are alike. The active emitting area, though, is larger. As an estimate of it, suppose the active area is defined as the area from which 90% of the emission comes, and further assume that $J(F)$ falls off as a Gaussian in the axial coordinate ρ (which is a better approximation than it appears if $\rho = a \sin\theta$). Then the active area is defined by an axial coordinate ρ_a such that

$$\frac{\int_0^{\rho_a} \exp(-\alpha\rho^2) 2\pi\rho \, d\rho}{\int_0^{\infty} \exp(-\alpha\rho^2) 2\pi\rho \, d\rho} = 0.9 \tag{30.12}$$

or $\rho_a = \sqrt{\ln(10)/\alpha}$. It follows that $\Omega = \ln(10)\Omega_n = 2.3\ \Omega_n$, where $\Omega_n = \pi/\alpha$ in this approximation. Therefore, the active radius is some 52% larger than the notional radius, but both scale as $1/\alpha$.

Armed with that understanding, revisit the hemisphere and pin down what its notional emission area is. As with ρ_a above, a "notional angle" θ_n can be defined such that [346, 450]

$$\int_0^{\theta_n} 2\pi a^2 \sin\theta d\theta = \int_0^{\pi/2} 2\pi a^2 P(\theta) \sin\theta d\theta \tag{30.13}$$

$$P(\theta) = \frac{J(F(\theta))}{J(F(0))} = (\cos\theta)^{2-\nu} \exp\left(-\frac{B_o \Phi^{3/2}(1-\cos\theta)}{3F_o \cos\theta}\right) \tag{30.14}$$

where the field emission current $J(F)$ is given by Eq. (13.25) and $P(\theta)$ acts as a weighting factor. Making a change of variables $x = (1-\cos\theta)/\cos\theta$ and letting $\eta = \cos\theta_n$, then

$$1 - \eta = \int_0^{\infty} (1+x)^{\nu-4} e^{-bx} dx \tag{30.15}$$

$$b \equiv B_o \frac{\Phi^{3/2}}{3F_o} \tag{30.16}$$

Because b is large, the approximation $(1+x)^{4-\nu} \approx e^{(4-\nu)x}$ can be made for small x, and so for a hemisphere,

$$\eta = \cos\theta_n \approx \frac{b+3-\nu}{b+4-\nu} \tag{30.17}$$

The area of the hemisphere is $2\pi a^2$. Therefore, letting $g(F_{tip})$ denote the fraction of the hemisphere that would emit if the current density across it were $J(F_{tip})$ with $F_{tip} = 3F_o$, the notional area is then $2\pi a^2 g(F_{tip})$ or

$$I_{boss}(F_{tip}) = \int J(F(\Omega)) d\Omega = 2\pi a^2 g(F_{tip}) J(F_{tip}) \tag{30.18}$$

$$g(F) = \int_0^{\infty} (1+x)^{4-\nu} e^{-bx} dx \approx \frac{1}{b+4-\nu} \tag{30.19}$$

where the behavior of $g(F)$ is shown in Figure 30.3 for various values of Φ. For low fields and using the small angle approximation $\cos\theta \approx \theta^2/2$, then $\theta_n \approx \sqrt{2g(F)}$.

EXAMPLE: Find the notional angle $\eta = \cos\theta_n$ and the fraction of surface emitting factor $g(F_{tip})$ for a hemisphere using $\Phi = 4.5$ eV and a background field of 2 eV/nm.

SOLUTION: The field at the apex is $F_{tip} = 3F_o = 6$ eV/nm. Using Table 2.3, the factor $b = B_o\Phi^{3/2}/F_{tip} = 10.868$ for $\Phi = 4.5$ eV, and $\nu = (8Q/9\hbar)\sqrt{2m/\Phi} = 0.77281$. Therefore $\eta = 0.92905$ and $\theta_n = 10.9°$. Lastly, $g(3F_o) = 1/(b+4-\nu) = 0.070947$.

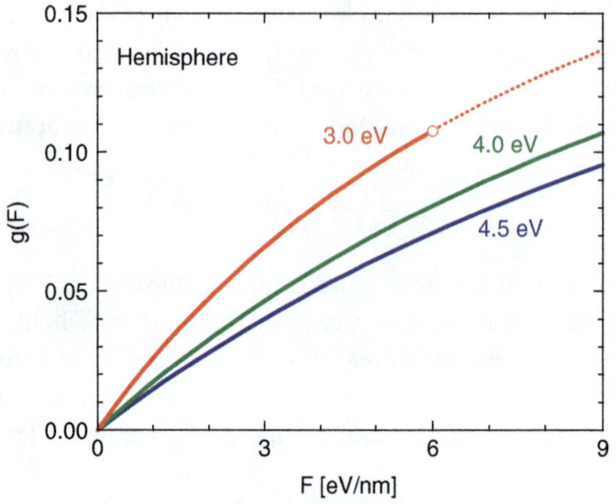

Figure 30.3 The behavior of $g(F)$ for a hemisphere for various work functions assuming that $J(F)$ is given by Eq. (13.25). Because J_{FN} is not defined past $F = \Phi^2/4Q$, the lowest work function $\Phi = 3.0$ eV line is dashed after the maximum field of $F_{max} = 6.25$ eV/nm ($F_{max} = 11.11$ eV/nm and 14.06 eV/nm for $\Phi = 4.0$ eV and 4.5 eV, respectively, and so are not visible in the range shown).

The prototypical manner in which current from a protrusion is wrested from the current density equations of electron emission is encapsulated in Eqs (30.18) and (30.19): current density from differential emission elements has to be summed over the surface of the entire emitting structure. Although differing shapes will give rise to differing $g(F)$, the algorithm of creating an area factor apart from the apex current density is the means by which Eq. (30.11) is realized.

Figuring prominently – or at least not unobtrusively – in the equation for total current from the hemisphere in Eq. (30.18) is the radius a of the hemisphere, which will become the radius of the emission site in due course when quirkier shapes than hemispheres are considered. Backing out the radius of curvature from the equipotential lines of $V(z, \rho)$ is therefore necessary. If $z_s(\rho)$ is the equation of an equipotential such that $V(z_s(\rho), \rho) =$ constant, then a Taylor expansion of $V(z, \rho)$ near the apex ($z = a, \rho = 0$) can be used to find a from the behavior of V. The surface of the hemisphere is defined by the equation $z^2 + \rho^2 = a^2$, or $z_s(\rho) = \sqrt{a^2 - \rho^2}$, that is, $z_s(\rho)$ is a function of ρ^2 itself. Inserting Eq. (30.18) into the Taylor expansion

$$V(z, \rho) = V(a, 0) + (z - a)\,\partial_z V(a, 0) + \frac{1}{2}\rho^2 \partial_\rho^2 V(a, 0) \tag{30.20}$$

where $V(z, \rho) = V(a, 0)$ along the surface and $\partial_\rho V(a, 0)$ vanishes at the apex of the emitter, results in the relation $z_s = a - (\rho^2/2a)$, which is recognized as the leading order expansion of $z = \sqrt{a^2 - \rho^2}$. That relation is a direct consequence of $\partial_z V(a, 0) = -3$ (familiar from the field enhancement calculations) and $\partial_\rho^2 V(a, 0) = -3/a$. As a result, the radius of the hemisphere is related to the gradients of $V(z, \rho)$ evaluated at the apex of the emitter $z_s(0) \equiv z_o$ by

$$a \equiv \frac{\partial_z V(z_o, 0)}{\partial_\rho^2 V(z_o, 0)} \tag{30.21}$$

Although this very important relationship is trivially satisfied for the hemisphere, it will be far more consequential when the shape of the emitter is more engaging. More importantly, it will allow $I_{boss}(F_{tip})$ to be applicable to general emitters, that is, if the field enhancement and the radius of curvature of the apex of a field emitter is known, then Eq. (30.18) can be used to approximate the current from it as long as a suitable $g(F)$ can be found.

That the treatment of notional emission area has been firmly rooted in the usage of the Fowler–Nordheim equation for $J(F)$ is worth keeping in mind. For cold emission, if approximately 11% of the surface field emits, then $g(F) = 0.11$, which implies $\theta_n \approx 30°$. A hot hemisphere is more difficult [451]: the entire surface of the hemisphere can contribute to thermal (or even photo) emission, so that $J_{GTF}(F, T)$ of Eq. (17.61) should be used in Eq. (30.18) under such circumstances. But that is for another time.

30.3 Point charge model

We are what we pretend to be, so we must be careful about what we pretend to be.

– Kurt Vonnegut[4]

[4]Kurt Vonnegut, *Mother Night*. New York: Dell Publishing, 1966, p. v.

Whatever shape the field emitters take on when pressed into the service of devices, it is not the shape of two charges in the configuration of a dipole, or a hemisphere on a plane. Those models are learning tools: they are analytically tractable so that methods can be judged against them. Although real emitters are far more devious, those methods can be brought to bear to reveal hidden features. The advantage of the dipole problem is that the shapes it can treat are malleable and, once chosen, the field enhancement factor is dictated as in Eq. (30.4) by $u(z, \rho)$ regardless of $F_o = V_a/D$, so that $I(V_a)$ curves can be obtained. The advantage of the hemisphere problem is that notional area factors can be readily obtained. The task now is to combine those advantages. To do this, pretend that the protrusion is composed of a distribution of point charges cleverly arranged.

A straight up extension of the dipole problem places additional charges above and below the $z = 0$ plane. That is, the $u(z, \rho)$ of Eq. (30.4) is generalized to n charges and their images. Usage of the images ("dipole model") is better for treating Spindt-type field emitters which have a more conical shape; using only one sign of charge ("monopole" model) results in emitters resembling Taylor cones [230]. To keep the narrative simple, because conical field emitters for $r > 0.35$ are better described by the dipole model, and because a generalization to a line charge model for wire-like emitters follows naturally from the dipole model, fix attention to it, although there are some quirks associated with it for which one must be on guard, for example for too small a charge the equipotential surfaces begin to deform into a floating sphere model which looks more like a balloon tethered to the ground plane: although seemingly unphysical, it resembles the "floating sphere" model of field emitters [452].

There is the matter of what a now represents. When $n = 1$, it was comparable to the radius of the hemisphere-like enclosure. To retain a as a measure of the apex radius, however, there is utility in subscripting the radius of the $n = 1$ point charge, and so refer to it as a_0. Therefore, let

$$u_n(z, \rho) = -\frac{z}{a_0} - a_0 \sum_{j=1}^{n} \lambda_j \left(\frac{1}{r_j^+} - \frac{1}{r_j^-} \right) \tag{30.22}$$

$$r_j^{\pm} \equiv \sqrt{\rho^2 + (z \pm z_j)^2} \tag{30.23}$$

so that the $u(z, \rho)$ of Eq. (30.4) is really $u_1(z, \rho)$. There are many ways to choose the λ_j and z_j, but only some will result in emitter shapes that bear any resemblance to conical Spindt-like field emitters [175] (or, for that matter, melted Taylor cones [31]). The aesthetics of their choice affects the possibility of analytical solutions and is the essence of the point charge model [230, 346, 439, 453, 454].

The parameter governing the magnitude of the jth charge, or λ_j, appears to be independent of the location z_j, but the two are in fact intimately entwined. A thought experiment shows why. Let n point charges plus their images across the $z = 0$ plane be used to model an emitter: if the last point charge λ_n is placed at z_n too far away from z_{n-1}, then its contribution to the zero-equipotential surface is an unphysical disembodied sphere floating above the other point charges. In order to get shapes that correspond to physical emitters, the z_j must be chosen strategically *and* the λ_j then adjusted so that the zero equipotential mimics physical surfaces. Their joint specification can be suitably chosen by appealing to "self-similarity". This is accomplished in two steps:

- First, the $(j + 1)$th point charge location bears the same relation to the jth charge location as the jth charge does to the $(j - 1)$th, enforced by demanding that

$$\frac{z_{j+1} - z_j}{z_j - z_{j-1}} \equiv r \tag{30.24}$$

for all j, with $z_0 \equiv 0$, where the dimensionless r should not be confused with $r_j^{\pm}$.

- Second, all λ_j are set so that z_{n+1} is on the zero equipotential, or

$$u_n(z_{n+1}, 0) = 0 \tag{30.25}$$

These two requirements are sufficient to determine all z_j and λ_j in terms of the number of charges n and the "self-similarity" factor r. Superficially, it seems that $r < 1$, but in fact r is *not* so constrained: it shall be seen that r can take on any value (and in fact some interesting shapes result by having it large). But it is true that for conical or ellipsoidal emitters nominally meant to resemble Spindt-type field emitters that taper down to a point, $r < 1$ is a better choice.

The first requirement is easy: z_1 corresponds to the radius scale of the emitter, or $z_1 = a_0$. For $j = 2$, then $z_2 = z_1 + a_0 r = a_0(1 + r)$. By induction, it follows that

$$z_j = a_0 \sum_{k=0}^{j-1} r^k \rightarrow \frac{z_j}{a_0} = \frac{1 - r^j}{1 - r} \equiv S_j(r) \tag{30.26}$$

where Eq. (A1.4) is invoked and the dimensionless term $S_j(r)$ is introduced and defined by this relation. For convenience, the r argument of S_j is suppressed below. Observe that the nth charge therefore sits at the apex of the $(n - 1)$th line,

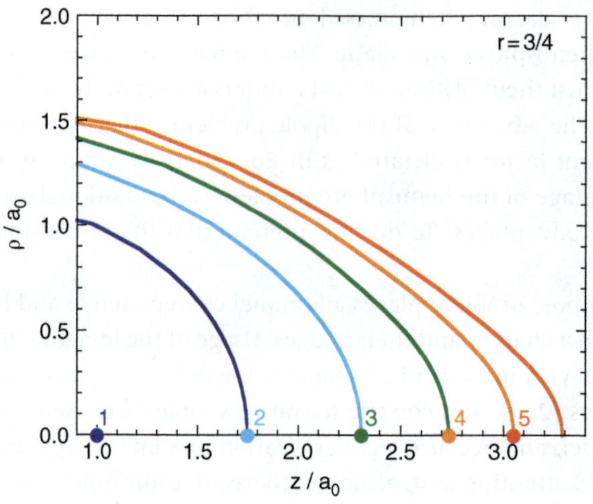

Figure 30.4 Placement of the first five point charges, and the lines they create, with the lines and the point charges of the same color, for $r = 0.75$. The nth charge sits at the apex of the $(n-1)$th line. The z axis is along the horizontal and the ρ axis is vertical, so that the figure is to scale.

as in Figure 30.4. The second requirement is a bit more tricky. It is tantamount to requiring

$$S_{n+1} = \sum_{j=0}^{n} \lambda_j \left(\frac{1}{S_{n+1} + S_j} - \frac{1}{S_{n+1} - S_j} \right) = \sum_{j=0}^{n} \frac{2\lambda_j S_j}{S_{n+1}^2 - S_j^2} \tag{30.27}$$

If these equations are numerically solved for small r, it is rapidly found that λ_j becomes exponentially small as $j \to n$, which unpleasantly complicates the numerical arts. Rather than endure an ugly determination of λ_j, it is better to coax out the small nature explicitly, and so let $\lambda_j(r) \equiv r^j P_j(r)$ where now P_j is to be determined and λ_j calculated from it.

But now for the artistry: a rather pleasant and straightforward substitution using Eq. (30.26) shows that the following relation holds

$$r^j\, S_{n+1-j} = S_{n+1} - S_j \tag{30.28}$$

which eliminates the numerically problematic part of the denominator of Eq. (30.27). Lastly, observe that n can take on any value, in fact S_n can simply be the nth component of a larger vector $\mathbf{S}$ of length N (and similarly P_n with $\mathbf{P}$), with N greater than any value of n likely to be encountered in practice ($N = 24$ works well). This means that a matrix equation of the form

$$\mathbb{M} \cdot \mathbf{P} = \mathbf{S} \tag{30.29}$$

$$M_{n,j} \equiv [\mathbb{M}]_{n,j} = \frac{2S_j}{S_{n+1-j}(S_j + S_{n+1})} \theta_{n,j} \tag{30.30}$$

where $\theta_{n,j} = 1$ if $j \leq n$ and 0 otherwise, is to be solved. Observe that $\mathbb{M}$ is a *lower triangular matrix* $\mathbb{L}$, making the matrix equation to be solved a subset of the LU-decomposition method (where LU denotes lower triangular–upper triangular and refers to when $\mathbb{M} = \mathbb{LU}$) [91, 236], and therefore easily implemented using numerical packages.

An iterative approach, however, is possible from the boundary condition equation $u(z_{n+1}, 0) = 0$, which can be written

$$S_{n+1} = 2 \sum_{j=1}^{n} \frac{P_j S_j}{S_{n+1-j}(S_j + S_{n+1})} \tag{30.31}$$

Taking a cue from knowledge that the matrix solution involving $\mathbb{M}$ involves a lower triangular matrix, it is evident that solving Eq. (30.31) sequentially for P_n is possible. Isolating P_n gives

$$P_n = \frac{1}{2} \left(1 + \frac{S_{n+1}}{S_n} \right) \left[S_{n+1} - 2 \sum_{j=1}^{n-1} \frac{P_j S_j}{S_{n+1-j}(S_j + S_{n+1})} \right] \tag{30.32}$$

from which it is straightforward to show

$$P_1 = \frac{1}{2}(1+r)(2+r) \tag{30.33}$$

$$P_2 = r\frac{(r^3+2r^2+4r+2)(r^2+2r+2)}{2(r+1)(r^2+r+2)} \tag{30.34}$$

after which recourse to numerical evaluation of the sums for $n > 2$ is preferable.

The definitions of field enhancement and apex radius for the point charge model now have to be updated. The derivatives are straightforward but involved, and so only the final form for the dipole case is provided (the monopole case being in refs [453, 454]). For field enhancement,

$$\begin{aligned} \beta_n &= -a_0 \partial_z u_n(z,\rho)|_{z=z_{n+1},\rho=0} \\ &= 1 + \sum_{j=1}^{n} \frac{4P_j S_j S_{n+1}}{r^j [S_{n+1-j}(S_{n+1}+S_j)]^2} \end{aligned} \tag{30.35}$$

Likewise, the apex radius is

$$\frac{a_n}{a_0} = -\frac{1}{a_0}\frac{\partial_z u_n}{\partial_\rho^2 u_n}\bigg|_{z=z_{n+1},\rho=0} = 2\frac{\sum_{j=1}^{n} \frac{P_j}{r^j}\frac{S_j S_{n+1}}{[S_{n+1-j}(S_{n+1}+S_j)]^2}}{\sum_{j=1}^{n} \frac{P_j}{r^{2j}}\frac{S_j(S_j^2+3S_{n+1}^2)}{[S_{n+1-j}(S_{n+1}+S_j)]^2}} \tag{30.36}$$

A numerical evaluation of the various terms is given in Table 30.1 (see also Section A3.25) and shown in Figure 30.5 of the case of $r = 0.75$. Strikingly, even though the height of the emitter never exceeds $z_\infty = 4a_0$ (a series which pays homage to Eq (6.13)), the field enhancement keeps growing as the sharpness of the protrusion increases (a_n decreases).

The trend that β_n exponentially increases, and a_n decreases, with n is even more pronounced for smaller values of r, as shown in Figure 30.6, behavior that figures markedly in Section 30.4 and in studies relating electron emission to heating through changes in $\beta(r)$ (e.g., Figure 2 of ref. [439]). On the other hand, as $r \to 1$, that behavior is broken so that β_n increases more slowly and a_n actually increases itself, with n, behavior that leads to the line charge model of Section 30.6 (again, recall the special status of $x = 1$ in Eq. (6.13), alluding to the special status of $r = 1$ as a transition value). The manner in which the point charges build up – or not – reflects the geometrical structure that they are tasked

Table 30.1 Point charge model evaluation of S_n, P_n, β_n and a_n for the dipole point charge model for the value of $r = 0.75$ and shown in Figure 30.5. Observe that $S_n(r) \to 1/(1-r)$ as $n \to \infty$.

n	$S_n(r)$	$P_n(r)$	$\beta_n(r)$	a_n/a_0
1	1.00000	2.40625	3.96970	0.708589
2	1.75000	1.72054	4.88324	0.629186
3	2.31250	1.53834	6.04215	0.506569
4	2.73438	1.39196	7.40333	0.397150
5	3.05078	1.25858	8.99928	0.307404
6	3.28809	1.13530	10.8780	0.236114
7	3.46606	1.02189	13.0979	0.180418
8	3.59955	0.918287	15.7287	0.137344
9	3.69966	0.824194	18.8528	0.104258
10	3.77475	0.739115	22.5685	7.89674×10^{-2}
11	3.83106	0.662430	26.9925	5.97072×10^{-2}
12	3.87329	0.593466	32.2639	4.50808×10^{-2}
13	3.90497	0.531542	38.5484	3.39983×10^{-2}
14	3.92873	0.475995	46.0437	2.56161×10^{-2}
15	3.94655	0.426207	54.9855	1.92855×10^{-2}
16	3.95991	0.381600	65.6549	1.45101×10^{-2}
17	3.96993	0.341648	78.3876	1.09114×10^{-2}
18	3.97745	0.305872	93.5839	8.20161×10^{-3}
19	3.98309	0.273840	111.722	6.16251×10^{-3}
20	3.98731	0.245162	133.372	4.62896×10^{-3}

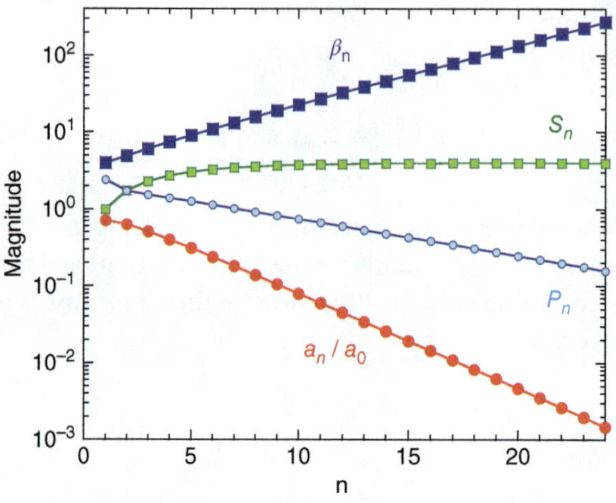

Figure 30.5 The parameters of Table 30.1 shown graphically for $r = 0.75$. The exponential behavior visible after $n > 3$ for β_n and a_n is characteristic of $r < 1$ behavior.

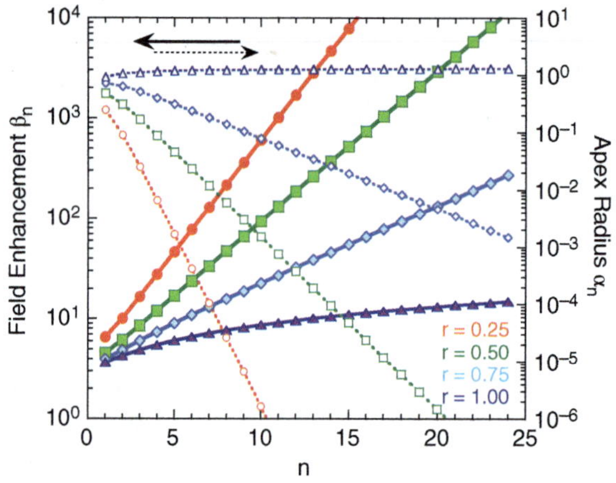

Figure 30.6 Field enhancement β_n (thick solid lines) and apex radius a_n/a_0 (thin dashed lines) as a function of n for $r = 0.25, 0.50, 0.75$ and 1.00 (red circles, green squares, light blue diamonds, dark blue triangles, respectively). With the exception of the $r = 1$ line, all show exponential behavior.

with imitating. But points are not the only distributions of charge that can be pressed into such service. For example, rings of charge of increasing diameter with differing charge can more faithfully model conical emitters [455], showing the power of the technique when harnessed to numerical methods to determine the optimum charge distribution more complex than points but which share the rotational symmetry of the protrusion. In the same spirit (although predating the introduction of the point charge model) modeling emitter structures by charged *spheres* has application to open and closed carbon nanotubes [456]. Such analytical and semi-numerical methods allow for the evaluation of the field enhancement factor, particularly where numerical methods alone cannot capture the very sharp rise in field enhancement, such as when a very small protrusion sits atop a larger one, as considered next.

30.4 Schottky's conjecture

Great fleas have little fleas upon their backs to bite 'em,
And little fleas have lesser fleas, and so ad infinitum.
And the great fleas themselves, in turn, have greater fleas to go on,
While these again have greater still, and greater still, and so on.

– Augustus De Morgan[5]

[5] Augustus De Morgan, *A Budget of Paradoxes*. Waxkeep Publishing, 2015, p. 377 (first published in 1872). De Morgan was a British mathematician and logician. He contributed a section on "mathematical induction" (a term he coined) to *The Penny Cyclopædia*, a source used by Herman Melville while composing *Moby Dick*.

In *Über kalte und warme Elektronenentladungen*[6] [138], Schottky imagined a hemisphere much like Figure 30.2, on top of which another smaller hemisphere sits. Consideration of the equipotential lines led him to conclude that " …wir erhalten dann eine nochmalige Verdoppelung des Potentialgradienten"[7] or, as synopsized by Stern *et al.* [285],

> *If the emitting area lies on a second hemispherical boss lying on the first and is small compared with it, then* $\beta = 3^2$, and so on. [emphasis added]

where Stern refers to the "extra factor due to surface irregularities", or *field enhancement*, as β. In other words, the field enhancement of the top hemisphere (a factor of 3) multiplies the field enhancement of the lower hemisphere (a factor of 3). In short, as long as the topmost protrusion is small by comparison to the bump on which it rests, *field enhancement factors are multiplicative,* a speculation noted and sometimes dubbed Schottky's conjecture in later studies [32, 457]. Mathematically, if β_a and β_b are the field enhancement factors of protrusions a and b on otherwise flat planes, where a is small by comparison to b,[8] then

$$\beta_{total} = \beta_a \beta_b \tag{30.37}$$

But why stop there? As with De Morgan's fleas, there can be bumps on bumps, and the interest is in how they stack up. The point charge model allows for an investigation of the effect. For larger r, understanding the behavior of $P_n(r)$ is required. A numerical evaluation shows that an exponential behavior of $P_n \approx P_o e^{-\alpha n}$ (where α is defined by this relation and is *not* the fine structure constant) seen in Figure 30.5 is more generally apparent, as in Figure 30.7. Clearly, α is a function of r, as implied in Figure 30.8 for the combination $1/(re^{\alpha})$, a combination that will prove useful below. To see how this helps demonstrate Schottky's conjecture, consider the behavior of Eq. (30.35) in greater detail.

The first bump to which all other bumps cling is characterized by $n = 1$: let this primal bump, as it were, be designated by β_o. After explicit expansion and collection of terms, and using Eq. (30.33) to start the iteration, it is

$$\beta_o(r) = 1 + 4P_1(r)\frac{1+r}{r(2+r)^2} = 3 + \frac{2}{r(2+r)} \tag{30.38}$$

Next, insert the approximation $P_j(r) = P_o e^{-\alpha j}$ into Eq. (30.35) along with $S_n(r)$ from Eq. (30.26) to obtain (where the arguments of the S_j and P_j have been suppressed, and where the o subscript on P_o is *not* the $n = 0$ version of $P_n(r)$ but rather a fitting coefficient)

$$\beta_n(r) = 1 + 4P_o \sum_{j=1}^{n} \frac{(1-r)^2(1-r^{n+1})(1-r^j)e^{-\alpha j}}{(1-r^{n+1-j})^2(2-r^j-r^{n+1})^2 r^j} \tag{30.39}$$

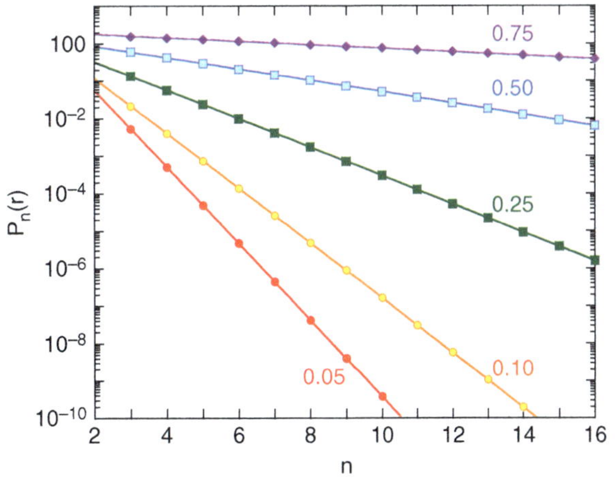

Figure 30.7 Symbols: numerical evaluation of $P_n(r)$ for various r as labeled (0.05, 0.10, 0.25, 0.50, and 0.75). Lines: least squares exponential fits for $P_n(r) \approx P_o e^{-\alpha n}$.

[6] *On cold and warm electron discharges*

[7] " …we then obtain a further doubling of the potential gradient"

[8] Miller *et al.* argue that when "the half-width of the macroprotrusion is less than the height of the microprotrusion" [457] then deviations from Schottky's conjecture occur, so small relates to that.

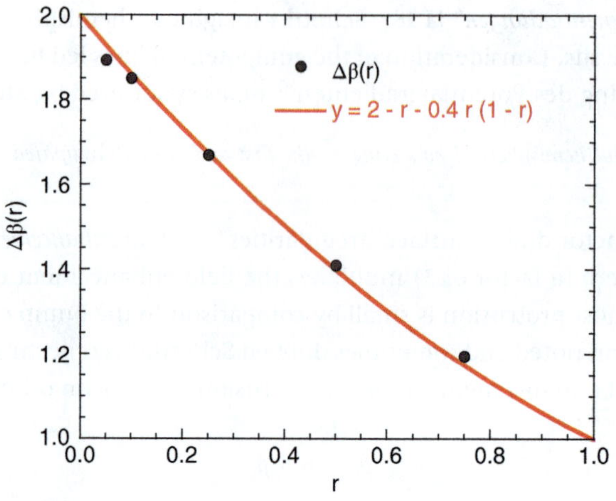

Figure 30.8 The combination $\Delta\beta \equiv 1/(re^{\alpha})$ for α determined by fitting the data of Figure 30.7 to $P_n(r) \approx P_o e^{-n\alpha}$. Also shown is the approximation $\Delta\beta \approx 2 - r - 0.4r(1 - r)$.

Such an equation is a discomforting mess and calls for approximations. As the small r is the situation of greatest interest, observe that the $1/r^j$ term is largest for j approaching n, but under such circumstances $r^j \ll 1$ when n is appreciable.[9] Also observe the combination $e^{-\alpha j}/r^j$ that naturally occurs. Therefore, make the replacements $r^{n+1} \to 0$, $1 - r^j \to 1$, and $2 - r^j \to 2$, and make use of

$$\Delta\beta(r) \equiv \frac{1}{re^{\alpha(r)}} \tag{30.40}$$

to find

$$\beta_n(r) = 1 + P_o(1-r)^2 \sum_{j=1}^{n} \frac{\Delta\beta^j}{(1 - r^{n+1-j})^2} \tag{30.41}$$

which is more accessible. The summation part can likewise be manipulated to give

$$\sum_{j=1}^{n} \frac{\Delta\beta^j}{(1 - r^{n+1-j})^2} = \Delta\beta^n \sum_{j=1}^{n} \frac{\Delta\beta^{1-j}}{(1 - r^j)^2} = \frac{\Delta\beta^n}{(1-r)^2} C_n(r) \tag{30.42}$$

$$C_n(r) \equiv \sum_{j=1}^{n} \Delta\beta^{1-j}\left(\frac{1-r}{1-r^j}\right)^2 \tag{30.43}$$

Now recall the fit of $\Delta\beta(r)$ shown in Figure 30.8. For $n = 1$, $C_1(r) = 1$, whereas for $n \to \infty$, then

$$C_\infty(0) = \sum_{j=1}^{\infty} 2^{1-j} = 2 \tag{30.44}$$

$$C_\infty(1) = \sum_{j=1}^{\infty} j^{-2} = \zeta(2) = \frac{\pi}{6} \tag{30.45}$$

where L'Hôpital's rule has been used for $C_\infty(1)$, and Rieman's zeta function of Eq. (A1.7) makes yet another appearance. The point is, approximating $C_n(r)$ as independent of n really is not a bad approximation. Supposing one is bold enough to do so, what happens? One finds

$$\beta_n(r) \approx 1 + P_o C(r)\Delta\beta(r)^n \approx P_o C(r)\Delta\beta(r)^n \tag{30.46}$$

But that means

$$\beta_n(r) \approx \Delta\beta\ \beta_{n-1}(r) \tag{30.47}$$

---

[9] Why the caveat about "appreciable" n? Remember, the interest is in *bumps on bumps* and the assumption of a perfectly flat base plane is different than that, therefore what is of interest is a few bumps out, say $n = 3 - 4$, where $(0.25)^3 = 0.015625$ and $(0.5)^4 = 0.0625$, thereby approaching what is meant by "negligible." Naturally, departures are expected when r approaches unity, but let's see how far the analysis can be pushed.

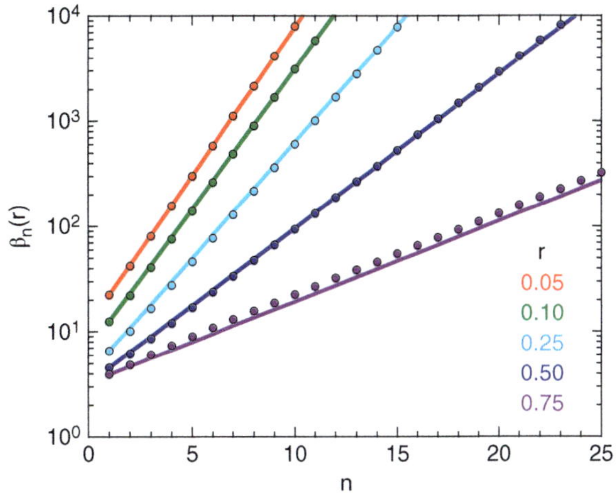

Figure 30.9 Symbols: Numerical evaluation of $\beta_n(r)$ (circles) using Eq. (30.35) and $P_n(r)$ using Eq. (30.32). Lines: Eq. (30.48) using the parameters of Table 30.2. Schottky's conjecture works well throughout, although discrepancies are becoming visible by $r = 0.75$. The algorithms of Section A3.25 were used.

Table 30.2 Parameters for $P_n(r)$ and $\beta_n(r)$ approximations. Least squares estimates of α where $P_n(r) \approx P_o e^{-\alpha n}$. The factors $\Delta\beta \equiv 1/(re^{\alpha})$ and β_o from Eq. (30.38) for using Eq. (30.48) are also given.

r	$\alpha(r)$	$P_o(r)$	$\Delta\beta(r)$	$\beta_o(r)$
0.05	2.3461	6.2397	1.9148	22.5122
0.10	1.6892	3.3736	1.8466	12.5238
0.25	0.8789	1.8943	1.6610	6.5556
0.50	0.3549	1.7033	1.4025	4.6000
0.75	0.1110	2.1839	1.1932	3.9697

is a good approximation. Setting $\beta_1(r) = \beta_o$ of Eq. (30.38) for the primary protrusion, then

$$\beta_{k+1}(r) \approx \Delta\beta^k \beta_o \tag{30.48}$$

where $k = n - 1$ is the number of additional bumps after the primary one. The Schottky conjecture (as Stern *et al.* [285] phrased it) has now been demonstrated: each additional bump past the primary one adds another factor of $\Delta\beta$. But notice that the primary bump has a different genesis than all the other bumps,[10] a sort of nuanced modification of what Schottky said.

Does Eq. (30.48) work? Actually, using the algorithms of Section A3.25 it works surprisingly well, as shown in Figure 30.9.[11] In fact, although departures are visible by $r = 0.75$, the approximation nevertheless remains good up to that relatively large value, although much beyond that, and certainly by the time $r \to 1$, the approximation becomes poor. As a result, for the point charge model Schottky's conjecture works for values of r larger than might be thought applicable.

30.5 Assessment of the tip current models

Nothing is too wonderful to be true if it be consistent with the laws of nature and in such things as these experiment is the best test of such consistency.

– Michael Faraday[12]

[10]The bump *it* sits on is perfectly flat, after all.

[11]Recall, though, from Section 30.3 that for $r < 0.3$ in the dipole model under strong field, the emitter shape is more like the floating sphere shape than a conical emitter shape; this does not change the multiplicative field enhancement, but does affect the visualization of what it represents.

[12]Bence Jones. *The Life and Letters of Faraday (V.2)*. University of California Libraries, 1870. p253 (Entry: March 19, 1849).

The hemisphere model maintains that the current from a field emitter can be known if the field enhancement at its apex and its shape (or, more accurately, its radius) is known. The audacity of that proposition motivates testing how well such a model describes actual field emission data before embarking on models based on line charges, ellipsoids, and boundary element methods. Consider then the data of Figure 2 of Schwoebel *et al.* (ref. [458], but see also refs [459, 460]), which subjects two micro-fabricated emitters to pulsed current processing, where the heating induces blunting of the emitters. That the tips of the emitters are complex (showing several crystal planes of molybdenum subject to "events" corresponding to migration of contaminants, and so on) is evident from field emission electron microscope (FEEM) images [461]. What matters here, though, is that changes to the apex radius (particularly when heated) occur when the sharpness of the tip reflects a balance between surface tensions, which act to blunt the tip, and high fields, which tend to sharpen it [462]. During "field forming" the tips were conditioned (blunted by "thermally activated field assisted self-diffusion") to putatively bring the emitters to the same radius that balanced surface field and surface tension. The surface migration drove emitters to emit uniformly, desirable because emitters of the same radius and material perform comparably, and uniformity of emission over an array, wherein the current load is shared evenly amongst the emitters, contributes to operational reliability. Individual emitters of this type have, remarkably enough, generated a total current of up to 3 mA from a single emitter; at a more modest 300 μA/tip a triangular array of such emitters separated by 4 μm would achieve a current density of 2165 A/cm^2.

Can the current as a function of voltage for these emitters be predicted? This is a different question than projecting performance based on line fitting of I-V data and extracting parameters from it. The present question concerns *using one I-V data point to find the radius* for the data shown in Figure 30.10 and then using the calibrated model to predict the remainder of the I-V curve. The correspondence of the I-V curves with the experimental data, differing only by the value of the apex radius describing differing emitters, gives a measure of the model's fidelity.

A complication is that the emitters are *gated*, in which a close proximity gate (a layer coplanar with the emitter tip having a gate hole through which the tip protrudes [15, 175, 447, 463–465], as in Figure 13.1: the emitter shown there is comparable to the ones generating the data of Figure 30.10) is used to generate the high apex fields using a reasonable and achievable gate potential. A relation between the apex field and the gate potential is therefore required. Such a relation is available from boundary element simulations (Section 30.8.2) and given in Eq. (30.124) (after Eq. (2) of ref. [466]). From it, the apex field F_{tip} can be related to the gate voltage V_g by the relation $F_{tip} = \beta_g V_g$, where β_g has units of [1/length] and is a function of apex radius a_s, gate radius a_g, and cone angle θ_c.

The theory lines of Figure 30.10 may then be determined using Eqs (30.18) and (30.19). Parenthetically, Eq. (30.19) should undergo modifications when the prolate spheroidal models of Section 30.7 are used, as the form of $g(F)$ will alter, although the alteration is weak by comparison to the other terms of I_{boss} when a_s is changed. Calibrate the theory by finding the value of a_s using *one* I-V data point, the other parameters of gate radius a_g and tip cone angle θ_c being determined from examination of the field emitter (e.g., Figure 13.1) or known, as with the work function $\Phi = 4.3$ eV of molybdenum. Although different crystal faces of molybdenum exhibit different Φ and a tip exhibits several faces, the

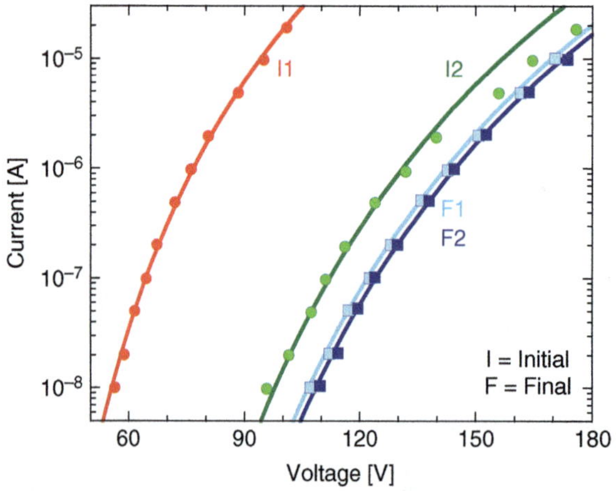

Figure 30.10 I-V data from two as-fabricated (labeled "I" for initial) and after processing (labeled "F" for final) single Spindt-type field emitters similar to those shown in Figure 13.1. Symbols correspond to experimental data; lines correspond to the theoretical model of tip current. Data points digitally extracted from Figure 2 of P.R. Schwoebel, C.A. Spindt, and C. Holland, *J. Vac. Sci. Tech.*, **B21**, 433, 2003 (ref. [458]) as in Table A3.2.

value of $\Phi = 4.3$ eV for molybdenum is neither the lowest nor the highest of the values listed in ref. [135] and is the recommended value of ref. [467]. A method for determining a_s from a single data point is presented in the algorithm of Section A3.26. The results of such an analysis are seen in Figure 30.10: the agreement is rather notable. Note that the agreement for the I2 line would be improved by using a larger work function Φ, as would be the case for a different crystal face or contaminant, but that has not been done.

The ability to predict the remainder of the I-V curve from given parameters and the extrapolated apex radius for a *single* data point bears favorably on the usage of the field enhancement plus emission area factor method of estimating the current of a tip. Although β and $g(F)$ will change, the methodology of integrating current density over differential surface elements is the manner by which total current from a multidimensional field emission structure can be predicted.

The effects of field enhancement, geometric variation, and notional area can start affecting how Eq. (30.1) for $I(V)$ begins to emerge from Eq. (30.18) for the current over a rounded emitter apex. Couple that with the changes wrought by transmission probability effects (thermal, Kemble, and so on) leading to the general thermal–field equation of Eq. (17.61) and it is clear that more prefactors start to aggregate on the basic form of the Fowler–Nordheim equation $J_{FN}(F)$ (or even $I(V)$). The prefactors are not insignificant: they result in variations of 0.1× to 10× the simple form [186]. Forcing all such changes to be carried by field enhancement and apex radius (or energy slope factor β_F) is likely asking too much, so understanding the origins of the variation by recourse to simple models is very useful.

30.6 Line charge models

It is the grand object of all theory to make these irreducible elements as simple and as few in number as possible, without having to renounce the adequate representation of any empirical content whatsoever.

– Albert Einstein[13]

Reflecting the scientific penchant for keeping matters as simple as possible – be they suppositions or models – goes by the name of the *principle of ontological economy*, or, in Latin, *entia non sunt multiplicanda praeter necessitatem* ("entities are not to be multiplied beyond necessity"), a phrasing that is erroneously referred to as *Ockham's razor*.[14] After a point of charge, the next simplest arrangement is a line of charge. But there are different kinds of lines: the simplest one is a line of constant charge per unit length. Such a model has much in common with the prolate spheroidal models of Section 30.7 in that are alternate means of modeling conical-like emitters. A modification of almost comparable simplicity is to let the charge density of the line be varied linearly over its length (tapered). The tapered line charge models become useful for modeling wire-like emitters (carbon fibers or nanotubes). Each is considered here.

30.6.1 Simple line charge model

A line of constant charge per unit length $q\lambda$ in a background field produces a potential everywhere given by [99, 173][15]

$$V_L(z,\rho) = 4Q\lambda \int_{-L}^{L} \frac{dz'}{\sqrt{\rho^2 + (z-z')^2}} = 4Q\lambda Q_0\left[\frac{L}{u(z,\rho)}\right] \tag{30.49}$$

where $Q_0(x)$ is a Legendre function of the second kind as in Eq. (A2.21), and the ellipsoidal functions u and w are related by

$$u(z,\rho) \pm w(z,\rho) \equiv \sqrt{\rho^2 + (z \pm L)^2} \tag{30.50}$$

$$u(z,\rho)w(z,\rho) = Lz \tag{30.51}$$

[13] The Herbert Spencer lecture, delivered at Oxford, June 10, 1933, reproduced in Albert Einstein, *Ideas and Opinions*. New York: Crown Publishers, 1954, p. 272.

[14] As Anthony Flew observes, "These actual words are not in fact to be found in the extant works of William of Ockham." Antony Flew, *A Dictionary of Philosophy*. London: Macmillan, 1979, p. 253.

[15] The notation follows ref. [173], but unfortunately some of the symbols and function names have appeared previously with different meaning. *Cave, lector*.

and where $w(z, \rho)$ will not reappear until Section 30.6.2. As with the point charge model, at the apex of the emitter $z \equiv z_o$, the total potential vanishes. The potential everywhere for a line of charge in a constant background field is then

$$V(z, \rho) = F_o z_o \left\{ -\frac{z}{z_o} + \frac{Q_0[L/u(z, \rho)]}{Q_0[L/u(z_o, 0)]} \right\} \equiv F_o z_o U_c(z, \rho) \tag{30.52}$$

where $U_c(z, \rho)$ (the c subscript is for "constant") is the line charge analog of $u_n(z, \rho)$ in the point charge model: do not confuse $u_n(z, \rho)$ in the point charge model with the analog of $u(z, \rho)$ here, as they mean very different things. Observe that, as with the point charge model, an overall energy scale $F_o z_o$ exists, multiplied by a dimensionless function that will govern the nature of the field enhancement factor β, but in the place of n and r of the point charge model the simple line charge model is now specified by z_o and δ defined by the relation

$$z_o \equiv L(1 + \delta) \tag{30.53}$$

where δ acts like the fraction of the emitter above the extent of L and therefore is expected to bear a very simple relation to the apex radius a when δ is small. Examples of typical choices of $1/Q_0(z_o, 0)$ are shown in Figure 30.11. Observe that in terms of δ,

$$Q_0 \left[\frac{L}{u(z_o, 0)} \right] = \frac{1}{2} \ln \left(\frac{2 + \delta}{\delta} \right) \tag{30.54}$$

The evaluation of β requires knowing $\partial_z u(z, \rho)$ evaluated at $\rho = 0$ and $z = z_o$. But $u(z > L, 0) = z$, and so $\partial_z u(z, 0) = 1$. As a result, for $\delta \ll 1$,

$$\begin{aligned} F_{tip} &= -F_o z_o \partial_z U_c(z_o, 0) = -F_o z_o \partial_z \left(-\frac{z}{z_o} + \frac{Q_0(L/z)}{Q_0(L/z_o)} \right) \Bigg|_{z=z_o} \\ &= F_o + \frac{F_o L z_o}{(z_o^2 - L^2) Q_0(L/z_o)} \\ &\approx F_o \left(1 + \frac{1}{\delta \ln(2/\delta)} \right) \end{aligned} \tag{30.55}$$

a finding that will bear great similarity to the field enhancement for a prolate spheroidal geometry. It is often suggested that field enhancement scales as the inverse tip radius, and the present result shows that this is mostly true, but not quite: it is a little worse than inverse tip radius because of the factor $\ln(2/\delta)$ in the denominator. In the opposite limit, when $\delta \to \infty$, where $Q_0(1/(1 + \delta)) \to 1/\delta$, then $\beta = F_{tip}/F_o \to 2$, which is recognized as the point charge model composed of a single charge at the origin; a single charge has a smaller β than the hemispherical result of $F_{tip} = 3F_o$, a result of the equipotential lines having a softer bend near a point charge in a background field in contrast to a hemisphere.

The apex radius is more involved because of the need to consider $\partial_\rho^2 u(z, \rho)$. It is wise to proceed in stages in evaluating

$$a = \frac{\partial_z U_c(z, \rho)}{\partial_\rho^2 U_c(z, \rho)} \Bigg|_{z=z_o, \rho=0} \tag{30.56}$$

The numerator is straightforward from the experience in evaluating F_{tip}. The denominator takes care. Using

$$\partial_z U_c(z, \rho)|_{z=z_o, \rho=0} = -\frac{1}{z_o} - \frac{L}{(z_o^2 - L^2) Q_0(L/z_o)} \tag{30.57}$$

$$\partial_\rho^2 U_c(z, \rho)|_{z=z_o, \rho=0} = -\frac{L z_o}{(z^2 - L^2)^2 Q_0(L/z_o)} \tag{30.58}$$

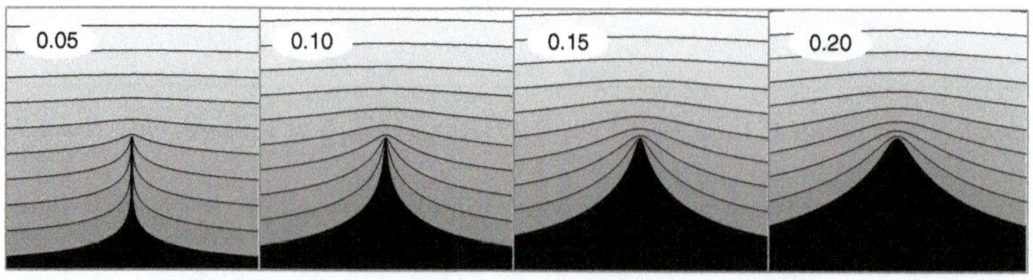

Figure 30.11 Equipotential surfaces of the constant line charge model for various $1/Q_0(z_o, 0)$. The size of the view region is $2L \times 2L$.

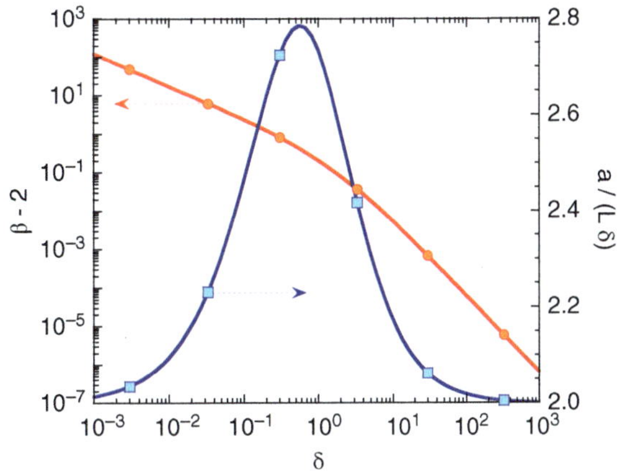

Figure 30.12 Field F_{tip} and radius a as a function of $\delta = (z_o/L) - 1$, represented as $[(F_{tip}/F_o) - 2]$ (left axis, red circles) and $(a/L\delta)$ (right axis, blue squares).

where in the second line use has been made of the following relations, valid when $\rho \to 0$ and $z \to z_o$ after the differentiations:

$$\partial_\rho u(z,\rho) \to 0 \tag{30.59}$$

$$\partial_\rho^2 u(z,\rho) \to \frac{z_o}{z_o^2 - L^2} \tag{30.60}$$

Assembling the pieces, and substituting $z_o = L(1+\delta)$, the apex radius a is then

$$\frac{a}{L} = \frac{\delta(2+\delta)}{1+\delta} + \left[\frac{\delta(2+\delta)}{1+\delta}\right]^2 Q_0\left(\frac{1}{1+\delta}\right) \tag{30.61}$$

Interestingly, $a/L \to 2\delta$ for both $\delta \to 0$ and $\delta \to \infty$. The behavior of $a/(L\delta)$ and F_{tip}/F_o as a function of δ is shown in Figure 30.12.

EXAMPLE: Although $a(\delta)$ can be evaluated once δ is specified, consider the reverse problem: find δ given that $a = 0.01L, 0.1L$, and L.

SOLUTION: The iteration method of Section A3.7.2 works well. Reconfiguring Eq. (30.61), define the $(n+1)$th iteration as

$$\delta_{n+1} = \left(\frac{a}{L}\right)\frac{(1+\delta_n)^2}{(2+\delta_n)(1+\delta_n+\delta_n(2+\delta_n)Q_0(1/(1+\delta_n)))}$$

Starting with $\delta_0 = 0$, by $n = 4$ the iteration has converged to $\delta_5 = 0.0048697, 0.043843$, and 0.36290 for $a/L = 0.01, 0.1$, and 1.0, respectively.

30.6.2 Tapered line charge model

How sharper than a serpent's tooth it is...

– William Shakespeare[16]

The line charge model of Section 30.6.1 was akin, both in methodology and in the shapes of the equipotential surfaces to which it gave rise, to the "monopole" point charge model. For the point charge model, the Taylor-cone-like shapes of the monopole model gave way to the conical emitters of the dipole model, so that the analog for the changes to be wrought on the line charge model are to make the charge below the boundary at $z = 0$ of opposite sign to the line

[16] Ref. [37]: *King Lear*, I.4.193.

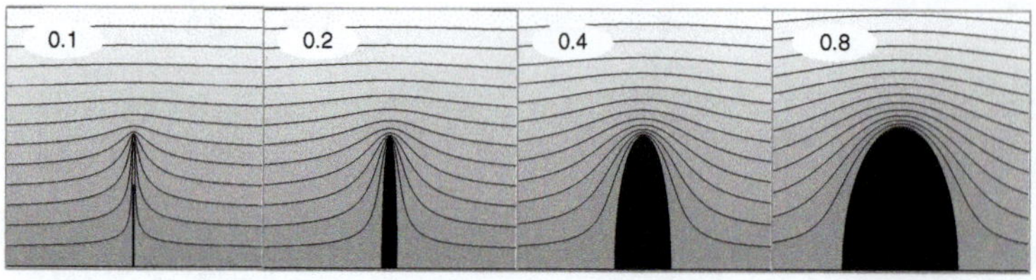

Figure 30.13 Equipotential surfaces of the tapered line charge model for various $1/(z_o Q_0(L/z_o) - L)$. THe size of the view region is $2L \times 2L$.

charge above $z = 0$. But if the purpose is to model wire-like emitters (and it is the purpose to model such emitters), then something more is required: the charge density on the line charge has to be tapered, from a maximum value at the apex of the line charge to a vanishing value at $z = 0$. This will avoid spread-out and odd-shaped bodies in favor of slender and long equipotential surfaces like those of Figure 30.13. The cost is a minor increase in complexity.

For a wire of charge for which the charge per unit length $\lambda \to \lambda_o z/L$ varies linearly over the length of the wire (a "tapered dipole" model), then the potential $V(z, \rho)$ in cylindrical coordinates for a single wire is

$$V(z,\rho) = \frac{4Q\lambda_o}{L}\int_{-L}^{L}\frac{z'dz'}{\sqrt{\rho^2 + (z - z')^2}} - F_o z \tag{30.62}$$

where now $V(z_o, \rho) = 0$ identifies the apex and defines z_o. In terms of the $u(z, \rho)$ and $w(z, \rho)$ functions of Eq. (30.50), this integral is

$$\frac{V(\rho,z)}{F_o z_o} = -\frac{z}{z_o} + \frac{zQ_0(L/u(\rho,z)) - w(\rho,z)}{z_o Q_0(L/z_o) - L} \tag{30.63}$$

As with the point charge model, the factor $F_o z_o$ sets the energy scale. As with the constant line charge treatment, again let $z_o = L(1+\delta)$. Then

$$\beta = \frac{1}{F_o}\partial_z V(z_o, 0) = \frac{1}{\delta(2+\delta)\left[(\delta+1)Q_0\left(\frac{1}{1+\delta}\right) - 1\right]} \tag{30.64}$$

Similarly, for the apex radius, the derivatives of Eq. (30.56) take a bit more effort, but gratifyingly, after simplification,

$$a \to L\delta\left(\frac{2+\delta}{1+\delta}\right) \tag{30.65}$$

which is seen to be the first term in the constant line charge radius of Eq. (30.61).

EXAMPLE: For the constant line charge model, $\beta \to 2$ as $L/a \to 0$, reflecting a single point charge. Find the analogous result for the tapered line charge model.

SOLUTION: The presence of the additional (-1) in the denominator of Eq. (30.65) complicates matters for the $\delta \to \infty$ limit, as it requires more terms to be evaluated in the Taylor expansion of $Q_0[1/(1+\delta)]$. For large δ, using

$$Q_0[1/(1+\delta)] = \frac{1}{2}\ln(1 + 2\delta^{-1}) \approx \delta^{-1} - \delta^{-2} + \frac{4}{3}\delta^{-3} - 2\delta^{-4}$$

where terms up to the fourth order in $(1/\delta)$ must be retained, then asymptotically, $\beta \approx 3 + 30\delta^{-2} \to 3$, or the hemispherical result. This limit is used in the vertical axis of Figure 30.14.

30.7 Prolate spheroidal representation

The point is that the coordinate system should be chosen to fit the problem, to exploit any constraint or symmetry present in it. Then, hopefully, it will be more readily soluble than if we had forced it into a cartesian framework. Quite often "more readily soluble" will mean that we have a partial differential equation that can be split into separate ordinary differential equations, often in 'standard form" in the new coordinate system.

– George Arfken [60], p. 72

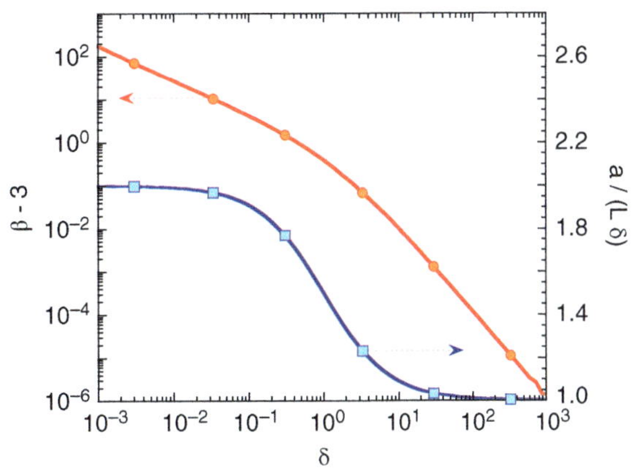

Figure 30.14 Field F_{tip} and radius a as a function of $\delta = (z_o/L) - 1$, represented as $[(F_{tip}/F_o) - 3]$ (left axis, red circles) and $(a/L\delta)$ (right axis, blue squares) when considering the difference between β and its large δ limit.

Poisson's equation is easiest to treat when a coordinate system is such that separation of variables can be used, relying as it does on the Laplacian ∇^2. There are a fair number of coordinate systems to chose from.[17] Of them, the *prolate spheroidal coordinates* are the most useful for treating emitters shaped like needles (ellipsoids of revolution) or cones (hyperbolas of revolution), where, because rotational symmetry is presumed for now, only two of the orthogonal coordinates matter in the discussion. The Laplacian and gradient operator in the prolate spheroidal system is determined in Section A1.3.1.2.

Why pursue yet another representation of a field emitter? Simply, because in orthogonal coordinate systems the emitter surfaces are taken to correspond to when one of the coordinates is held constant, making finding the surface trivial. The field lines, in turn, are along the other coordinate and can be found with little effort. To emphasize the point, *if one coordinate corresponds to the equipotential surface,*[18] *the other coordinate corresponds to the field lines* and this can be extremely useful. Knowing the *surface* and the *field* means that an exploration of the notional emission area factor $g(F)$ becomes far easier, so that an understanding of the relationship between the large-scale geometry and the fraction of the tip emitting can be explored.

The advantages of the prolate spheroidal approach to field emission structures has been often considered [197, 468–470], with various ways to represent the coordinate system; here, specifying the coordinates by angles (η, υ) is preferred,[19] and their relations to the cylindrical coordinates (z, ρ) are

$$\rho = L\sinh(\eta)\sin(\upsilon) \tag{30.66}$$

$$z = L\cosh(\eta)\cos(\upsilon) \tag{30.67}$$

where L is a length scale (the more general case is given by Eq. (A1.64)). Hyperbolic surfaces are defined by constant υ, and ellipsoidal by constant η, both of which are shown in Figure 30.15. The hyperbolic representation allows for a close-spaced anode along $z = 0 \leftrightarrow \upsilon = \pi/2$, whereas the ellipsoidal representation allows putting in a constant background field F_o, but this apparent freedom is rather inflexible: the hyperbolic model is exact only for a diode geometry with the anode a fixed distance away and related to the cone angle, whereas elliptical models are exact for an ellipsoidal emitter all by itself in the presence of a uniform field. The prolate spheroidal treatment therefore does not describe arrays, but it does reveal how the fields along the surface $F(\eta, \upsilon)$, and particularly the apex field F_{tip}, depend on geometry.

Using the separation of variables methods that were so useful for treating the hydrogen atom of Section 9.1 can be used here. Poisson's equation can be recast in the form of

$$\frac{1}{V(\eta,\upsilon)}\nabla^2 V(\eta,\upsilon) = \frac{1}{U(\eta)}\nabla_\eta^2 U(\eta) + \frac{1}{W(\upsilon)}\nabla_\upsilon^2 W(\upsilon) = 0 \tag{30.68}$$

where $V(\eta, \upsilon) \equiv U(\eta)W(\upsilon)$. As the sum of the U and W terms must be 0, they must each equal the same constant but of different signs. Let that constant be $n(n+1)$, where the fact that n is an integer follows from analogous considerations

[17] The second edition of Arfken [60] contains many, but later editions treat fewer.

[18] Or, more precisely, "if the surface of the conductor corresponds to holding one coordinate fixed ...".

[19] In the cited sources, the angles are sometimes (α, β), but as these symbols have been terribly overworked already, a small change in notation is deemed helpful.

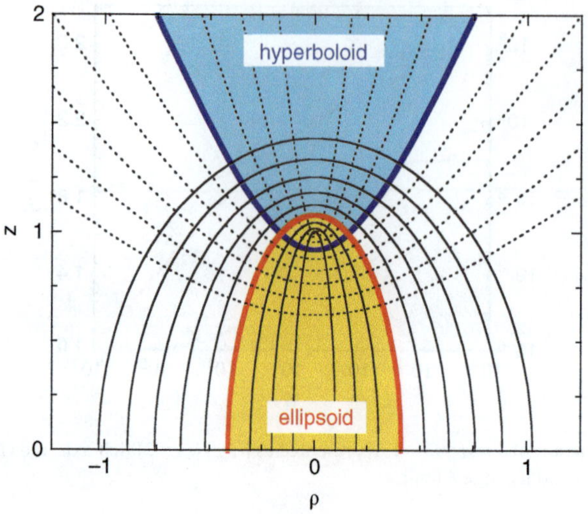

Figure 30.15 The prolate coordinate system: grid lines are in gray. A hyperbolic surface (constant υ) is blue and pointing down: its "anode" would be the $z = 0$ plane, and field lines would correspond to the solid gray lines. An ellipsoidal surface (constant η) is red and pointing up: field lines (in the absence of a background field F_o) would correspond to the dashed gray lines.

that determined l in $l(l+1)$ in the hydrogen atom was likewise an integer. The value of n is determined by the boundary conditions one wishes to impose.

30.7.1 Hyperbolic emitter

For a close-spaced planar anode at $\upsilon = \pi/2$ and a conical emitter defined by $\upsilon = \upsilon_o$, then for the $n = 0$ case and using Eq. (A1.72),

$$(\sin\upsilon\, \partial_\upsilon^2 + \cos\upsilon\, \partial_\upsilon) W_0(\upsilon) = 0 \tag{30.69}$$

This is true for $W(\upsilon)$ equal to a constant, but it is also true for $W(\upsilon) \propto Q_0(\cos\upsilon)$, as can be shown by substitution and using $\partial_\upsilon Q_0(\cos\upsilon) = -1/\sin\upsilon$. The most general solution then is the sum of the constant solution and the solution proportional to $Q_0(\cos\upsilon)$ set to satisfy the boundary conditions that the zero equipotential is at $\upsilon = \upsilon_o$, and so

$$W_0(\upsilon) = V_o\left(1 - \frac{Q_0(\cos\upsilon)}{Q_0(\cos\upsilon_o)}\right) \tag{30.70}$$

The result is independent of η as along the equipotential lines V is constant, and η marks where along the equipotential line the point of interest is situated. The η-dependence of the field along the surface arises because of the gradient operator. From Eq. (A1.74),

$$\hat{\upsilon}\cdot\vec{F} = F_\upsilon(\eta,\upsilon) = -\frac{1}{L(\sinh^2\eta + \sin^2\upsilon)}\partial_\upsilon W_0(\upsilon) \tag{30.71}$$

Recall that, as per the methods of Section 30.2, the apex field $F_{tip} = F(0, \upsilon_o)$ is given by

$$F(\eta,\upsilon_o) = \frac{\sin\upsilon_o}{\sqrt{\sin^2\upsilon_o + \sinh^2\eta}} F_{tip} \tag{30.72}$$

$$F_{tip} = \frac{V_o}{LQ_0(\cos\upsilon_o)\sin^2\upsilon_o} \tag{30.73}$$

Similarly, Eq 30.56 in prolate spheroidal coordinates gives the apex radius

$$a = L\frac{\sin^2\upsilon_o}{\cos\upsilon_o} = D\tan^2\upsilon_o \tag{30.74}$$

where the tip-to-anode distance is $D = L\cos\upsilon_o$. In the limit that $a \ll D$, or $\tan^2\upsilon_o \ll 1$, make use of

$$Q_o(\cos\upsilon_o) = \frac{1}{2}\ln\left(\frac{1+\cos\upsilon_o}{1-\cos\upsilon_o}\right) = \frac{1}{2}\ln\left(\frac{(1+\sec\upsilon_o)^2}{\tan^2\upsilon_o}\right) \tag{30.75}$$

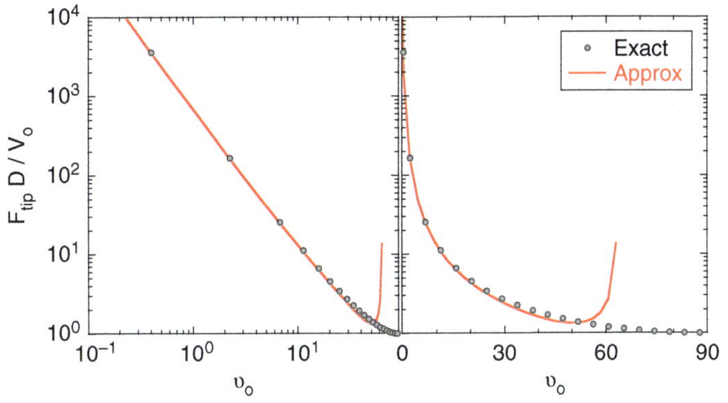

Figure 30.16 The factor $F_{tip}D/V_o$ for the hyperbolic diode configuration, where D is the anode–cathode separation (tip-to-anode) and V_o is the anode potential. υ_o is the cone angle in degrees. Symbols use Eq. (30.73) and the red line corresponds to the asymptotic limit of Eq. (30.76).

and $\cos \upsilon_o \approx 1$ to find [197, 470, 471]

$$F_{tip} \approx \frac{2V_o}{a \ln(4Da)} \tag{30.76}$$

the performance of which is shown in Figure 30.16. The field enhancement factor is therefore comparable to the commonly used form of $\beta = 1/ka$, where k is called the macroscopic component of β [472], with *macroscopic* being with reference to a by the ratio D/a in the logarithm term: macroscopic features do not affect field enhancement with anywhere near the impact that tip radius does [473], a consequence of the equipotential lines being more dependent on the charge distributions closest to them compared to the ones further away.

The impact of field variation as a function of η now allows the notional emission area factor $g(F)$ to be evaluated for the hyperbolic diode emitter. Compared to Eq. (30.19), more work is required: the current density $J(F(\eta, \upsilon_o))$ over a ribbon of area dA must be integrated over the surface. For the hemisphere, $dA = 2\pi r^2 \sin\theta d\theta$, a well-known result, but a generalization to it is needed for surfaces that are rotationally symmetric but for which all points on the surface are not equidistant from a center (hence $rd\theta$ is no longer the width of the ribbon making dA, its length being $2\pi r \sin\theta$ – recall Figure 21.3 in Section 21.2). The problem is to obtain the area of a ribbon along the edge of the disk, such that the width of the ribbon is at an angle to the plane of the disk. If the disk is of radius ρ and has a width dl, then $dA = 2\pi\rho dl$. The element dl, in turn, is the hypotenuse of the amount the ribbon changes in the vertical dz and radial $d\rho$, or

$$dl = \sqrt{dz^2 + d\rho^2} \tag{30.77}$$

and so

$$dA = 2\pi\rho \left(\frac{dl}{d\eta} \right) d\eta = 2\pi L^2 \sin\beta \ \sinh\eta \sqrt{\sin^2\upsilon_o + \sinh^2\eta} \ d\eta \tag{30.78}$$

The integration of current density over this differential element gives the hyperbolic diode $g(F)$ factor as

$$g_h(F_{tip}) = \frac{1}{2\pi a^2 J(F_{tip})} \int J[F(\eta, \upsilon_o)] dA \tag{30.79}$$

$$= \cos^2\upsilon_o \int_0^\infty \frac{(1+x)^\nu e^{-bx}}{\sqrt{1 + x(x+2)\sin^2\upsilon_o}} dx \tag{30.80}$$

where $F_{tip} = F(0, \upsilon_o)$ and $(1+x)^2 = 1 + (\sinh\eta / \sin\upsilon_o)^2$. For small x,

$$g_h(F) \approx \frac{\cos^2\upsilon_o}{b - \nu + \sin^2\upsilon_o} \tag{30.81}$$

is a reasonably good approximation. Compared to the curvature of a hemisphere, the curvature of the hyperbolic cone is more gentle and this is reflected in the 4 in the denominator of Eq. (30.19) being replaced by $4 \to \sin^2\upsilon_o$. This is not a huge change when compared to the magnitude of $b = B_o/F_{tip}$, but it is non-trivial, and it shows the extent to which the macro-scale geometric features intrude upon emission at the apex. An ellipsoid has a sharper apex than a hyperboloid, but not so sharp as the hemispherical boss, creating the expectation that its area factor must lie between these cases.

30.7.2 Ellipsoidal emitter

For a conical emitter in a background field F_o (like the hemispherical boss or the constant line charge density model), such that the ellipsoid is defined by $\eta = \eta_o$, and asymptotically, the potential is that of a constant background field, or

$$\lim_{\eta\to\infty} V(\eta, \upsilon) = -F_o z = -F_o L \cosh\eta \cos\upsilon \tag{30.82}$$

For Poisson's equation $\nabla^2 V = 0$, use separation of variables and let $V(\eta, \upsilon) = F_o L\, U(\eta) W(\upsilon)$ for which

$$\frac{1}{U}\left\{\frac{1}{\sinh\eta}\frac{\partial}{\partial\eta}\left(\sinh\eta \frac{\partial U}{\partial\eta}\right)\right\} = -\frac{1}{W}\left\{\frac{1}{\sin\upsilon}\frac{\partial W}{\partial\upsilon}\left(\sin\upsilon \frac{\partial}{\partial\upsilon}\right)\right\} \tag{30.83}$$

From the asymptotic behavior, $W = \cos\upsilon$, so that

$$-\frac{1}{\cos\upsilon}\left\{\frac{1}{\sin\upsilon}\frac{\partial}{\partial\upsilon}\left(\sin\upsilon \frac{\partial}{\partial\upsilon}\cos\upsilon\right)\right\} = 2 = n(n+1) \tag{30.84}$$

or $n = 1$. As a result, $U(\eta)$ is expected to be of the form

$$U(\eta) = -\cosh\eta + A_1 Q_1(\cosh\eta) \tag{30.85}$$

where $Q_1(x)$ is defined in Eq. (A2.22) but for $x > 1$, or

$$Q_1(\cosh\eta) = \frac{\cosh\eta}{2}\ln\left(\frac{\cosh\eta + 1}{\cosh\eta - 1}\right) - 1 \tag{30.86}$$

and A_1 is determined by the boundary condition that on the surface of the ellipsoid ($\eta = \eta_o$), the potential vanishes. This amounts to

$$U(\eta) = -\cosh\eta\left[1 - \frac{\cosh\eta_o\, Q_1(\cosh\eta)}{\cosh\eta\, Q_1(\cosh\eta_o)}\right] \tag{30.87}$$

and so, just as with the methods of Section 30.2 (compare Eq. (30.9))

$$V(\eta, \upsilon) = -F_o L\cosh\eta_o \cos\upsilon\left(\frac{\cosh\eta}{\cosh\eta_o} - \frac{Q_1(\cosh\eta)}{Q_1(\cosh\eta_o)}\right) \tag{30.88}$$

This is more than coincidence: a sphere is an ellipsoid for which the major and minor axes are equal. Therefore, the ellipsoid result should morph into the hemispherical result under the right conditions: when η becomes large, then $\cosh\eta \approx \sinh\eta \approx e^\eta/2$, so that $z = r\cos\upsilon$ and $\rho = r\sin\upsilon$ if $r = Le^\eta/2$. Thus, consider the large η limit of Eq. (30.88) using $Q_1(x) \approx 1/3x^2$, so that

$$\begin{aligned} V(\eta \gg 1, \upsilon) &= -F_o L\cosh\eta \cos\upsilon + F_o L\cos\upsilon\,\frac{\cosh^3\eta_o}{\cosh^2\eta} \\ &= F_o r\cos\upsilon\left[1 - \left(\frac{a}{r}\right)^3\right] \end{aligned} \tag{30.89}$$

if $a = L\cosh\eta_o$. As a result, the hemispherical model arises from the large η limit of the prolate spheroidal ($n = 1$ ellipsoidal) model.

Flush with the success of having treated the hyperboloid, the ellipsoid evaluations are analogous. The field along the surface is found to be, again using Eq. (A1.74),

$$F(\eta, \upsilon_o) = \frac{\sinh\eta_o \cos\upsilon}{\sqrt{\sin^2\upsilon + \sinh^2\eta_o}} F_{tip} \tag{30.90}$$

$$F_{tip} = \frac{F_o}{Q_1(\cosh\eta_o)\sinh^2\eta_o} \tag{30.91}$$

The apex radius can be found as before, but consider another approach. The apex of the emitter is at $z_o \equiv L\cosh\eta_o$. From the definitions of z and ρ in Eqs (30.66) and (30.67), it follows that along the surface defined by $z = z_s(\rho)$ where

$$1 = \frac{\rho^2}{L^2\sinh^2\eta_o} + \frac{z_s(\rho)^2}{L^2\cosh^2\eta_o} \tag{30.92}$$

Solving for $z_s(\rho)$ and demanding that it correspond to the parabolic relation $z_s(\rho) \approx z_o - (\rho^2/a)$ in the small ρ limit, it is found

$$a = L\frac{\sinh^2\eta_o}{\cosh\eta_o} = z_o\tanh^2\eta_o \tag{30.93}$$

As a final exercise, consider a characterization of the apex field F_{tip} in terms of the dimensions of the ellipsoid itself. Were it a circle, its base radius and height would be the same, but for an ellipsoid the radius of the major radius (or height of the ellipsoid) to the minor radius (or radius of the base), which shall be called R, is given by

$$R = \frac{L\cosh\eta_o\ \cos 0}{L\sinh\eta_o\ \sin\frac{\pi}{2}} = \coth\eta_o \tag{30.94}$$

Now consider the needle-like ($R \to \infty$) and sphere-like ($R \to 1$) limits. Expressing F_{tip} in terms of R results in

$$F_{tip}(R \to 1) = \frac{15F_o}{7 - 2R^2} \tag{30.95}$$

$$F_{tip}(R \gg 1) = \frac{2R^2 - 3}{2(\ln(2R) - 1)}F_o = \frac{2z_o - 3a}{a(\ln(4z_o/a) - 2)}F_o \tag{30.96}$$

Whereas the $R \to 1$ result reproduces the hemispherical result of $F_{tip}/F_o = 3$, the $R \to \infty$ result of $\beta \approx 2z_o/a \ \ln(4z_o/a)$ when $a \ll z_o$ mimics the hyperbolic relation of Eq. (30.76) but with z_o taking the place of D, a result that might have been intuited if one observes that the distance over which the equipotentials take to be free of the bending caused by the protrusion is on the order of the protrusion's height, as in Figure 30.13. Under such conditions, the first reasonably flat equipotential functions much like a virtual anode and is easy to view as such.

Finding the area factor proceeds as before. For the ellipsoid, the area element is

$$dA = 2\pi\rho\left(\frac{dl}{dv}\right)dv = 2\pi L^2\sin\beta\ \sinh\eta\sqrt{\sin^2 v + \sinh^2\eta_o}\ dv \tag{30.97}$$

which, compared to the hyperboloid dA of Eq. (30.78), differs only by the exchange of $d\eta$ with dv and the specification of which angular term is the fixed one. Thus, the ellipsoid in a background field $g(F)$ factor is[20]

$$\begin{aligned} g_e(F_{tip}) &= \frac{1}{2\pi a^2 J(F_{tip})}\int J[F(\eta, v_o)]dA \\ &= \cosh^4\eta_o\int_0^\infty (1+x)^{\nu-1}e^{-bx}\frac{1+x}{[1+(1+x)^2\sinh^2\eta_o]^2}dx \\ &\approx \int_0^\infty (1+x)^{\nu-1}e^{-bx}dx \approx \frac{1}{b-\nu+1} \end{aligned} \tag{30.98}$$

where $(1+x)^2 = (\sinh^2 v + \sin^2\eta_o)/(\cos^2 v\sinh^2\eta_o)$ and, as before, $b = B_o\Phi^{3/2}/F_{tip}$. As anticipated, the ellipsoid $g_e(F)$ is between the hemispherical $g(F)$ and the hyperbolic $g_h(F)$.

The hemispherical, hyperbolic, and ellipsoidal models therefore reveal that the parameter on which tip current most depends is the apex radius due to the field enhancement factor being inversely proportional to it. But conditions further away from the apex, namely, the shape of the emitter, make themselves known through a logarithmic factor in the denominator of the β factor, and in a small change to the notional emission area factor $g(F)$. A comparison of the hemispherical, hyperbolic, and ellipsoidal $g(F)$ factors is shown in Figure 30.17.

30.8 A hybrid analytic-numerical model

> *One ring to rule them all, one ring to find them,*
> *One ring to bring them all and in the darkness bind them.*
>
> – J.R.R. Tolkien[21]

[20] The upper limit corresponds to $v = \pi/2$, but since the dummy variable x has a factor of $\cos v$ in its denominator, its upper limit is ∞.

[21] J.R.R. Tolkien, *The Fellowship of the Ring, The Lord of the Rings*. Boston: Houghton Mifflin, 1987.

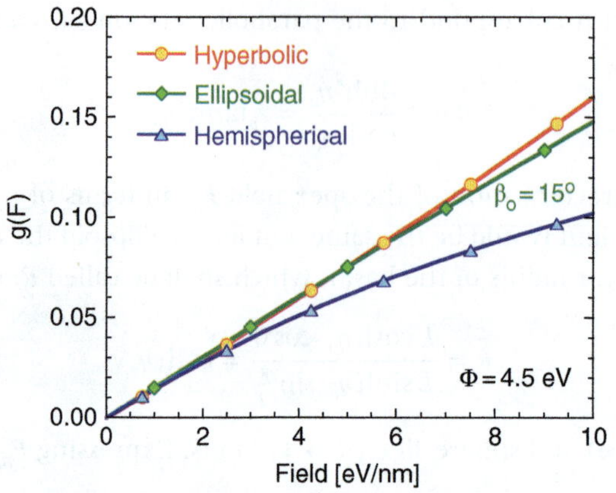

Figure 30.17 Comparison of the notional emission area factors $g(F)$ for the hemispherical (blue triangles, Eq. (30.19)), ellisoidal (green diamonds, Eq. (30.98)) and hyperbolic (red circles, Eq. (30.81) for a cone angle of 15°) emitters. The work function is $\Phi = 4.5$ eV, $v = 0.77281$, and $b = B_o\Phi^{3/2}/F$.

Even with field enhancements of $\beta \approx 500$, to get to the ~ 5 GV/m ($F = 5$ eV/nm) fields required for field emission from refractory metal and semiconductor field emission arrays one still must start with background fields on the order of 50 MV/m, and that is difficult. By way of comparison, dc photoinjectors demand very high fields on the surface of a photocathode with values comparable to 10 MV/m. Alternately, the accelerating grids above the surface of thermionic dispenser cathodes produce surface fields less than that: an accelerating grid 1 cm from the surface of a cathode producing 2.334 A/cm^2 current under space-charge limited conditions requires a grid potential of 10 kV, implying a field of 1 MV/m. Macroscopic fields or distant anodes are not the way to achieve the requisite high fields, as they bring with them the problem of high potentials.[22]

What is needed is to bring the effects of the anode closer to the emitter tip. This can be done by excising a hole in the sheet constituting the anode and bringing it down to the plane of the emitter: electrons shot through such a hole will have to be collected by another conducting surface further away, and *that* surface becomes the true anode. In such a way, the potential of the conducting sheet with the hole in it can be held in the hundreds of volts and still give rise to apex fields that cause field emission. Because such structures can be switched quickly, they can be used to turn on (generate sufficient fields to cause emission) or off (reduced in potential such that sufficient fields do not exist to cause emission) the emitter, and so they are called *gates*. In their physical manifestation, therefore, field emitters sit in small wells with their pointy ends approximately coplanar with the gate plane and centered in the gate holes, just as shown in Figure 13.1. In the spirit of the point and line charge models of Sections 30.3 and 30.6, the simplest conceivable representation of a gated field emitter is therefore a charged sphere in a background field F_o surrounded by a ring of charge representing the gate, such a ring ruling whether or not emission occurs, as it were [474]. A floating sphere, gated or not, is far from a good representation of a field emitter plus extraction grid, but what it does is point the way to an analytic model parameterized by numerical simulations.

30.8.1 Analytic: the saturn model

In memory of the Golden Age, when (Saturn) reigned in Italy, the great feast of the Saturnalia was held every year during the winter. The idea of it was that the Golden Age returned to the earth during the days it lasted. No war could be then declared; slaves and masters ate at the same table; executions were postponed; it was a season for giving presents; it kept alive in men's minds the idea of equality, of a time when all were on the same level.

– Edith Hamilton[23]

[22]To get a feel, the huge MIT Van de Graaff generator, operational around 1933 but now housed in Boston's Museum of Science, was capable of reaching a potential of 5×10^6 V, so that a cathode would have to be placed 10 cm from it to create a field of 50 MV/m, under vacuum conditions.

[23]Edith Hamilton,*Mythology*. Boston: Little, Brown and Company, 1942, p. 45. Although the Romans sought to equate their gods with the Grecian deities by identifying Jupiter with Zeus and Saturn with Chronos (Zeu's father), that was imperfect: Saturn was an agriculture deity; Chronos ate his children.

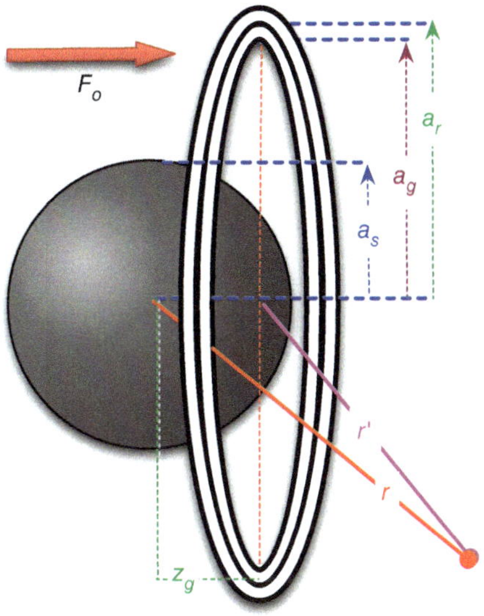

Figure 30.18 The tip is represented by a sphere of radius a_s and the gate by a ring of charge of radius a_r. $\hat{z}$ is normal to the plane of the ring and in the direction of $\vec{F}_o$. The distance between the center of the sphere and the center of the ring is z_g. The gate potential is defined at $V(z_g, a_g) \equiv V_g$. The observation point is a distance r from the center of the sphere, and a distance r' from the center of the ring. The polar angle θ is defined by $\hat{z} \cdot \vec{r} = r\cos\theta$, and the angle α by $\hat{z} \cdot \vec{r}' \equiv r' \cos\alpha$. For purposes of illustration, the sphere representing the tip is grotesquely enlarged.

If the simplest emitter shape is a sphere, then the simplest *gated* emitter is a sphere with a ring about it: as in Figure 30.18, let the sphere be of radius a_s and the ring of charge of radius a_r, with r measuring the distance between an observation point r and the center of the sphere, and r' the distance between the observation point and the gate ring, with the $\hat{z}$ axis normal to the plane of the gate ring. Such a Saturn model is a simple and convenient representation of a gated field emitter [475, 476], but unlike the planet that lent its name, the plane of the gate ring need not pass through the center of the sphere, that is, the origin of the sphere and the origin of the ring need not coincide. Rather, the top of the planet and the ring plane are, as it were, on the same level.

Define the distance between the ring and sphere centers as z_g such that $\vec{r} - \vec{r}' = z_g\hat{z}$, where $\hat{z}$ is in the direction of both the background field $\vec{F}_o = F_o\hat{z}$ and the normal to the plane of the ring. Let θ be the angle $\vec{r}$ makes with $\hat{z}$, and likewise let α be the angle $\vec{r}'$ makes with $\hat{z}$. The gate potential V_g is not defined *at* the ring location, but rather a small distance from it, comparable to, say, half the gate film thickness (on the order of a μm). If a_r is the radius of the ring representing the gate, then $V_g = V(z_g, a_g)$, where $V(z, \rho)$ is the total potential energy.

The potential energy $V(z, \rho)$ is then the sum of three contributions:[24] that due to the background field $V_o = -F_o z$, that due to the ring of charge V_r, and that due to the sphere V_s, or

$$V_o = -F_o z = -F_o r \cos\theta \tag{30.99}$$

$$V_r = \frac{Q_g}{2\pi a_r} \int_0^{2\pi} \frac{dl}{|\vec{r} - \vec{r}'|} \tag{30.100}$$

$$V_s = \sum_{j=0}^{\infty} A_l r^{-l} P_l(\cos\theta) \tag{30.101}$$

where dl is a differential length along the gate ring and $r_g^2 = z_g^2 + a_r^2$ (mind the subscripts) and where if $n_r = q_r/q$ with q_r being the total charge on the gate ring, then $Q_g = 4Qn_r = \lambda/4\pi\varepsilon_0$ in SI units, with $\lambda = q_r/2\pi a_r$ being a charge per unit length. Usage of Q_g is easier, though.

Consider the gate ring first. If ϕ is the azimuthal angle, then $dl = a_r d\phi$ is the differential element, and so [197, 475]

$$V_r(r, \theta) = \frac{Q_g}{2\pi} \int_0^{2\pi} \frac{d\phi}{\sqrt{\rho^2 + (z - z_g)^2 + a_r^2 - 2\rho a_r \cos\phi}} \tag{30.102}$$

$$= \frac{2Q_g}{\pi} \frac{K(p)}{\gamma} \tag{30.103}$$

[24] Observe that the co-mingling of spherical and cylindrical coordinates is done depending on which is most convenient at the time: (r, θ) with α are the spherical coordinates, and (z, ρ) with a_g and a_r are the cylindrical ones.

where $p = \sqrt{4\rho a_r}/\gamma$, $\gamma = [(\rho + a_r)^2 + (z - z_g)^2]^{1/2}$, and the function $K(m)$ is a *complete elliptical integral of the first kind* [60] defined by

$$K(m) = \int_0^{\pi/2} \frac{d\phi}{\sqrt{1 - m^2 \sin^2\phi}} \tag{30.104}$$

Take note of the limits of the integral: apart from a factor of 2 because the integrand is symmetrical, in going from Eq. (30.102) to (30.103) a half-angle formula of the form $\cos(2\phi) = 1 - 2\sin^2\phi$ has been used. Because elliptical integrals can be expanded in a series involving Legendre polynomials, V_r can be rewritten as [99]

$$\frac{2}{\pi}\frac{K(p)}{\gamma} = \frac{1}{r_g}\sum_{l=0}^{\infty}\left(\frac{r}{r_g}\right)^{l+1} P_l(\cos\alpha)P_l(\cos\theta) \tag{30.105}$$

where $\cos\alpha = z_g/r_g = z_g/\sqrt{z_g^2 + a_r^2}$. The value of Q_g is found by demanding that the potential be V_g at a distance $t = a_r - a_g$ from the gate ring (near the gate ring, the equipotential lines, to a good approximation, are toroidal):

$$V_g = \frac{2}{\pi}Q_g\frac{K(p)}{\gamma} + F_o z_g\left[1 - \left(\frac{a_s}{r'_g}\right)^3\right] - \frac{Q_g}{r_g}\sum_{l=0}^{\infty}\left(\frac{a_s}{r'_g}\right)^{l+1} P_l\left(\frac{z_g}{r_g}\right)P_l\left(\frac{z_g}{r'_g}\right) \tag{30.106}$$

where $r'_g = \sqrt{z_g^2 + r_g^2}$. Assuming that $a_s \ll a_g$, then to leading order,

$$Q_g \approx (V_g - F_a z_g)\left[\frac{2K(p)}{\pi\gamma} - \frac{a_s}{r_g r'_g}\right]^{-1} \to \frac{\pi\gamma}{2K(p)}V_g \tag{30.107}$$

where the second form is for when the apex of the emitter is approximately coplanar with the gate (z_g is small) in addition to small apex radius a_s. Finally, the charge on the sphere is given by A_0, or $Q_s = Q_g(a_s/r_g)$, a result that is reminiscent of a charge outside a conducting sphere [99].

For the determination of $F_{tip} = \beta_g V_g$, concentrate on the surface of the sphere, for which $V_o + V_r + V_s = 0$. This requirement specifies the A_l to be

$$A_l = \frac{Q_g}{r_g} a_s^{l+1}\left(\frac{a_s}{r_g}\right)^l P_l(\cos\alpha) - F_o a_s^3\,\delta_{1,l} \tag{30.108}$$

and so the field along the surface $F(\theta)$ is

$$F(\theta) = -3F_o\cos\theta - \frac{Q_g}{a_s r_g}\sum_{l=0}^{\infty}(2l+1)\left(\frac{a_s}{r_g}\right)^l P_l(\cos\alpha)P_l(\cos\theta) \tag{30.109}$$

which suggests that to leading order, $F_{tip} \approx Q_g/(a_s r_g)$. Ferreting out higher order terms requires an approximation to $K(p)$ in Eq. (30.107). To leading order, [477]

$$K(m) \approx \frac{1}{2}\ln\left(\frac{16}{1 - m^2}\right) + (1 - m^2)\left[\frac{1}{8}\ln\left(\frac{16}{1 - m^2}\right) - \frac{1}{4}\right] \tag{30.110}$$

for which the first term dominates as $m \to 1$. When $z_g \approx 0$, it follows that $\gamma = a_g + a_r$ and $p = \sqrt{4a_g a_r}/(a_g + a_r)$. Letting $t = a_r - a_g$, then [475]

$$F_{tip} \approx \frac{\pi}{a_s\ln(8a_g t)}V_g \tag{30.111}$$

There are similarities *and differences* with the other analytic forms for the prolate spheroidal models of Eqs (30.76) and (30.96), where in the logarithm term, a_g has usurped the role of D and z_o and t the role of a_s. This suggests that in a "hybrid" formula to treat physical gated conical emitters, the logarithm term might morph into something like $\ln(k\, a_g/a_s)$, where finding the term $k = k(a_s, a_g)$ requires more than the analytical models brought forth so far to treat emitters (ref. [476] contains a discussion of such log-terms, in addition to an unusual model of a gated field emitter: the "bowling pin" model). Finding k requires numerical input.

30.8.2 Numerical: boundary element methods

The boundary element method has many attractive features which make it especially well suited to numerical solution of these potential field problems[25] when compared to other techniques ...such as finite differences or finite elements ...Other features include high accuracy, the ability to handle an infinite volume without inversion, easier incorporation of meshes which vary dramatically in step size, simpler mesh generation, and change of design with partial rather than complete remeshing.

– Robert L. Hartman, William A. Mackie, and Paul R. Davis [478]

The actual characterization of field emitter performance in an array in order judge their utility for devices which seek to use them [166, 173, 464, 466, 479–483] must take into account the actual physical geometry of the emitter and where it sits in relation to the gates or extraction grids that are used to generate the fields at the apex of the emitters. Although the simple models are intuitively appealing, the locations and potentials of close-proximity electrodes shifts the dependence of emitted current to the *known* and adjustable parameters of gate radii, cone angles, gate and anode potentials, and base-to-gate thicknesses of dielectrics to understand the spread of the beam [484]. Finite difference and finite element analyses, often the workhorses of such simulations, are profoundly undercut by the orders of magnitude differences in size between the emission sites and the micron-scale geometry of the gates and tips. A methodology that avoids fine discretization where it is not really needed and concentrates it where it is (right near the emission site) for rotationally symmetric[26] field emission sources is the boundary element method [475, 478, 485–488]. The boundary element method is perfectly capable of calculating the emission current by itself without recourse to the analytic models of previous sections, but the development of a semi-analytic model (in which the behavior of the field enhancement term, in particular the $k(a_s, a_g)$ term suggested after Eq. (30.111), is inferred) is useful: numerical solutions provide answers, but analytic ones provide insight.

If points, lines, and rings can replace the influence of the gate and model the conical emitter, then a ribbon to model the gate might be better, and going further, if a ribbon of charge can model the edge of a conducting surface then *ribbons of charge can model every conducting surface:* once the boundary conditions are matched by the placement of charge, then the solution $V(z, \rho)$ in the interior is a matter of numerically solving Poisson's equation and the problem is finished (unless space charge comes into play [485]). The trick, as it were, is then knowing how to assign the charge on every ribbon that represents a conducting surface.

The boundary element technique discretizes the boundary into sections. In a rotationally symmetric model, these sections are annular ribbons that can be taken quite small in the vicinity of edges or corners or – most importantly – nanoscale protrusions where field emission occurs. Where the potential does not vary markedly, then the width of the annular ribbons can become quite large. The factors that determine the size of the annular ribbons are tip radius, tip height, gate hole radius, base–gate distance, gate–anode distance, gate thickness, and shape of the emitter, such as those suggested in Figure 30.19. Unlike finite element or finite difference techniques, the boundary element method does *not* require gridding or discretization in the interior region away from the conducting surfaces.

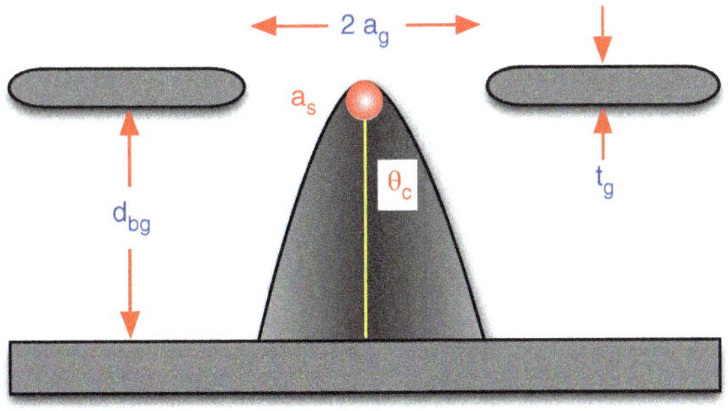

Figure 30.19 Parameters important for the simulation of a field emitter using boundary element methods: a_s = apex radius, a_g = gate radius, d_{bg} = base-to-gate separation, θ_c = cone angle, t_g = gate thickness.

[25]The authors refer to emission from areas vastly smaller than the size of the simulation region.
[26]In principle, the method can be generalized to three dimensions, but in practice the size of the matrices required for inversion explodes and become far more difficult for inversion.

The starting point is rewriting Eq. (30.7) for general number density distributions $\rho_e(\vec{r}')$ to more specific ribbons with sheet charge density $\sigma(\vec{r}')$, or

$$V(\vec{r}) = 4Q\int_\Omega \frac{\sigma(\vec{r}')}{|\vec{r}-\vec{r}'|}d\Omega \tag{30.112}$$

where $d\Omega$ is the differential surface element and σ is the number of charges per unit area. Represent the boundaries as ribbons of charge, each with a constant surface charge density [489] (a faster approach than assuming the charge density varies linearly across the ribbon, which, although more accurate, is less important if the ribbons are sufficiently thin). This causes Eq. (30.112) to become

$$V(\vec{r}) = 4Q\sum_{j=1}^{\infty} 4\sigma_j\sqrt{1+s_j^2}\int_{\rho_j}^{\rho_{j+1}} \rho'\frac{K(p)}{\gamma}d\rho' \tag{30.113}$$

for which the Saturn model has been good preparation. The various components are defined by

$$s_j \equiv \frac{z_{j+1}-z_j}{\rho_{j+1}-\rho_j} \tag{30.114}$$

$$p = \frac{4\rho\rho'}{\gamma} \tag{30.115}$$

$$\gamma = \{(\rho+\rho')^2+(z-z')^2\}^{1/2} \tag{30.116}$$

the potential is specified at the middle of each ribbon, or

$$V(z,\rho) \to V_j \equiv V(z_{j+1/2},\rho_{j+1/2}) \tag{30.117}$$

where $z_{j+1/2} \equiv (z_j+z_{j+1})/2$ and $\rho_{j+1/2} \equiv (\rho_j+\rho_{j+1})/2$. The values of V_j are known: they are simply given by the potentials of the various conducting surfaces (emitter cone, gate plane, or anode plane). What is unknown are the values of the charge densities $\sigma_j = \sigma(z_{j+1/2},\rho_{j+1/2})$. But observe that with the introduction of a matrix defined by [197, 487]

$$[\mathbb{M}]_{ij} = 16Q\sqrt{1+s_j^2}\int_{\rho_j}^{\rho_{j+1}} \rho'\frac{K(p)}{\gamma}d\rho' \tag{30.118}$$

where ρ_i and z_i are buried in the definition of p and γ, then the boundary element method amounts to a solution of the matrix equation

$$\vec{V} = \mathbb{M}\cdot\vec{\sigma} \to \vec{\sigma} = \mathbb{M}^{-1}\cdot\vec{V} \tag{30.119}$$

The inversion of the matrix equation therefore solves for the charge densities, and once *those* are known the fields along the ribbons are quickly deduced by

$$F_j = \frac{q^2}{\varepsilon_0}\sigma_j = 16\pi Q\sigma_j \tag{30.120}$$

There is only one small wrinkle: Eq. (30.118) is easy to numerically calculate *as long as* $(z_{i+1/2},\rho_{i+1/2})$ *does not lie in the center of one of the ribbons* but that happens every time $i=j$, and when it does, the integral is hard to evaluate because $K(p)$ contains a logarithmic singularity of the form $(1/2)\ln(16/(1-p^2))$ as $p\to 1$. Fortunately, the singular portion can be integrated to give

$$\int_{\rho_i}^{\rho_{i+1}} \frac{\rho'}{2\gamma}\ln\left(\frac{16}{1-p^2}\right)d\rho'$$
$$= \frac{\epsilon}{(24\rho)^2}\left\{[96\rho^2-(s^2+2)\epsilon^2]\left[3\ln\left(\frac{16\rho}{\epsilon\sqrt{s^2+1}}\right)+1\right]+192\rho^2\right\} \tag{30.121}$$

where for this equation

$$
\begin{aligned}
\epsilon &= \rho_{i+1} - \rho_i \\
s &= \frac{z_{i+1} - z_i}{\rho_{i+1} - \rho_i} \\
z' &= z_{i+1/2} + s(\rho' - \rho_{i+1/2}) \\
p &= \frac{\sqrt{4\rho' \rho_{i+1/2}}}{\gamma} \\
\gamma &= [(\rho' + \rho_{i+1/2})^2 + (z' - z_{i+1/2})^2]
\end{aligned}
$$

With the singular part of the integral removed, the remainder integral can be readily evaluated using numerical integration means such as those of Section A1.2.3.

30.8.3 Semi-analytic model

These two efforts always go together, as they mutually act and react on each other; they are integral parts of one and the same view, and we have only to ascertain what effect is produced when one or the other has the predominance.

– Carl von Clausewitz[27]

Numerical and analytic techniques both have a role to play, but they need not be kept distinct, and indeed, joining their virtues provides a model with advantages over either. The boundary element method provides a full solution to the problem of field emission from a gated emitter by evaluating the field on each annular ribbon specifying the conducting surfaces, and in particular those ribbons constituting the apex of the field emitter. There is an artistry to the selection of the ribbon widths: they should be very narrow in regions of high curvature (such as near the apex of the emitter), but can be rather wide in regions of low curvature (such as along the shank of the ellipsoid, or the flat regions of the gate and the anode). Not many annular ribbons are required, so that the inversion of Eq. (30.119) is computationally rapid.

Still, having to recalculate the ribbon charges whenever some small change is made is time-consuming, and doing so does not allow for fast evaluations over a range of gate potentials and anode fields. It would be convenient, not to mention enlightening, to have an *analtyical* model of what $F_{tip} = \beta_g V_g$ is. One such method to provide that relation makes use of an approximation to the behavior of the potential $V(z, \rho)$ on axis ($\rho = 0$), with $\bar{z} = z - a_s$ measured from the apex of the emitter (and z from the center of the sphere representing the apex). For small $\bar{z}$, $V(z, 0) \approx F_{tip}\bar{z}$, whereas for large $\bar{z}$, $V(z, 0) \approx F_o z$, where F_o is the field between gate and anode. If $D \gg a_g$ is the anode–gate separation then $F_o = (V_a - V_g)/D$, and $F_o a_s$ is negligible. An expression which captures this behavior is

$$
V(z, 0) \approx \frac{F_{tip}\bar{z}}{V_g + F_{tip}\bar{z}} \left(1 + \frac{F_o \bar{z}}{V_g} \right) V_g \tag{30.122}
$$

where the first fraction captures the small z behavior ($V_g \gg F_{tip}\bar{z}$), and the term in parentheses the large z behavior ($V_g \ll F_o z$). Although this "derivation," such as it is, lacks rigor, it is based on the actual variation compared to boundary element evaluations and therefore captures reasonably well the values of $V(z, 0)$ and its derivatives at $\bar{z} = 0$ and $\bar{z} = D$. Moreover, the hybrid approach it enables gives a reasonable account of the behavior of a single Spindt-type gated emitter [490, 491]. Therefore, return to Eq. (30.111) and the discussion following it to investigate how they can be bent to present needs.

The first step will be to replace V_g in Eq. (30.111) with $V(z_g, 0)$, where z_g is the height above the emitter where $V(z_g, 0) \approx V_g$. This substitution indicates

$$
1 = \frac{\pi z_g V_g}{a_s(V_g + F_{tip} z_g \ln(k\frac{a_g}{a_s}))} \tag{30.123}
$$

[27] Carl von Clausewitz (trans. J.J. Graham), *On War*, new and revised edition, with introduction and notes by Colonel F.N. Maude. London: Kegan Paul, Trench, Trïbner & Co., 1908, Chapter II: End and Means in War, https://www.gutenberg.org/catalog/.

From the hyperbolic model, Eq. (30.74) with $D \to z_g$ suggests that $z_g = a_s/\tan^2 \upsilon_o$, where υ_o the the cone half-angle. Inserting it and solving for F_{tip},

$$F_{tip} \approx \left(\frac{\pi}{\ln(ka_g/a_s)} - \tan^2 \upsilon_o \right) \frac{V_g}{a_s} \equiv \beta_g V_g \tag{30.124}$$

The final component is the approximation for k, for which there is no analytical model. Therefore, boundary element simulations for several gate geometries and cone angles modeled after SRI-like and MIT-like field emitter arrays,[28] as shown in Figure 30.20, are used to develop an approximation to $k(a_s, a_g)$. The geometrical factors that were used in the boundary element method simulations are summarized in Table 30.3. These simulations gave rise to the best fit representation

$$k \approx \frac{1}{54}\left(86 + \frac{a_g}{a_s}\right) \cot \upsilon_o \tag{30.125}$$

and the results of the simulations are shown in Figure 30.20.

Modeling current from a tip is one thing, but describing the distribution of charge that is emitted from a single Spint-type emitter is a more demanding test of the theory. To that end, the theoretical model (a semi-analytic model with parameters determined from boundary element simulations and a trajectory analysis) was used to analyze the spread of a beam of electrons from a single Spindt-type field emitter as experimentally measured using a microfabricated detector system with a Faraday cup anode. The Faraday cup collected current as a function of radial distance from the

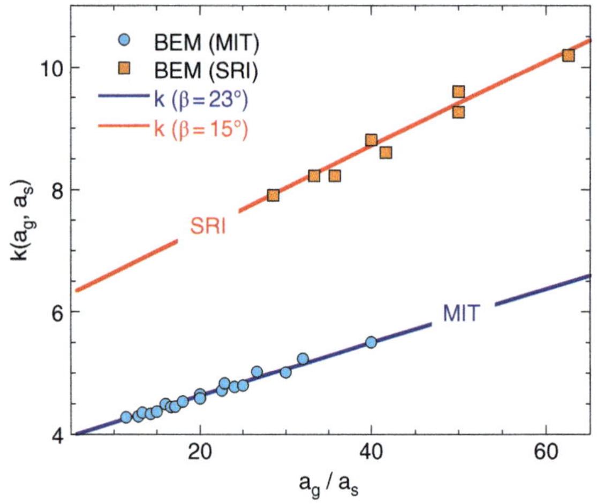

Figure 30.20 Evaluation of k in Eq. (30.124) by boundary element simulations (symbols) for field emitter arrays having the geometries of those fabricated by SRI and MIT, and $k(a_g, a_s)$ as given by Eq. (30.125). Based on Figure 1 of ref. [466].

Table 30.3 Values of gated field emitter parameters as used by boundary element simulations to determine k (Eq. (30.125)) for MIT and SRI parameters. The resulting k values are shown in Figure 30.20.

Parameter	Symbol	Unit	SRI-like	MIT-like
Base-to-gate	d_{bg}	μm	1.00	0.19
Gate thickness	t_g	μm	0.25	0.05
Cone angle	υ_o	Degrees	15	23
Gate radii	a_g	μm	0.2, 0.25	0.08, 0.09, 0.10, 0.12, 0.16
Tip radii	a_s	nm	4, 5, 6, 7	4, 5, 6 7

[28] The physical arrays were developed as part of programs to develop inductive output amplifiers (IOAs) operating at 10 GHz, specifically the ARPA/NRL klystrode and the NRL twystrode (see ref. [492] for a description). Images of the arrays and their characteristics are given in ref. [480]. A 100 W cold cathode TWT was demonstrated by Whaley *et al.* [483] in 2009.

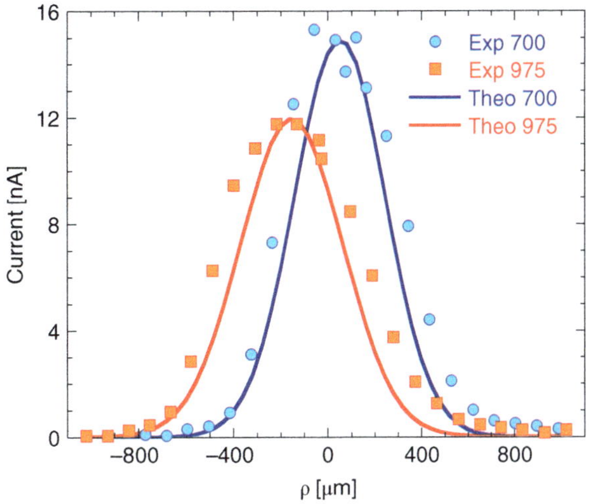

Figure 30.21 Comparison of experimental measurements of emission from a single Spindt-type emitter (SRI geometry) using a microfabricated detector to a theoretical model plus boundary element simulations. Numbers refer to the gate–anode separation in microns. Based on Figure 8 of ref. [490].

axis of symmetry: the microdetector was capable of moving in steps of 3–4 nm [490, 493] and was therefore capable of resolving the spread in the beam. Although the theoretical model used in the analysis presumed a constant k value, the agreement with the experimental characterization of the Gaussian-like spread of the beam was quite good, as seen in Figure 30.21. The varying $k(a_s, a_g)$ model was subsequently developed and successfully used to investigate space charge effects from an array of emitters [491].

30.9 Shielding

To be in company, even with the best, is soon wearisome and dissipating.

– Henry David Thoreau[29]

Take two ellipsoids like the "0.2" case shown in Figure 30.13 and start bringing them together as in Figure 30.22. As they approach each other, the equipotential lines between them will start deforming in that the equipotential lines between the ellipsoids rise. Additionally, the number of lines compressed into the small space above the emitters declines, that is, *the field reduces*. As a consequence, the presence of a nearby emitter causes the β factor of either emitter to become smaller. This effect, known as *screening* or *shielding*, is well-known, particularly for wire-like emitters such as carbon nanotubes [494–497] and carbon fibers [498]. Shielding is even more pronounced in arrays, particularly for ungated emitters.

The modeling of wire-like emitters in particular suggests a parameterized form of accounting for the effects of screening given by

$$\frac{\beta}{\beta_o} = 1 - \exp\left[-a(b/h)\right] \tag{30.126}$$

where β_o is the field enhancement for a single emitter, and β is the enhancement factor of a line of emitters of height h and emitter-to-emitter separation b, where a is a fit parameter suggested to be $a \approx 2.3172$ [499, 500] as in Figure 30.23. Extending the analysis from a one-dimensional line of wires to two-dimensional arrays (where the emitters in the unit cell of an array were on the corners of either a square or an equilateral triangle) using the line charge model [501] tended to confirm Eq. (30.126) but generalized it to become

$$\frac{\beta}{\beta_o} = 1 - \exp\left[-a(b/h)^c\right] \tag{30.127}$$

where a and c are found by fitting (a similar formulation was found by Read and Bowling [496]). Applying Eq. (30.127) to the data digitally extracted from ref. [499] would give $a \approx 2.3668$ and $c \approx 1.176$ for $h = 2$ μm.

[29] Henry David Thoreau, *Walden and Other Writings*, The Modern Library Classics. New York: Modern Library, 2000, p. 128.

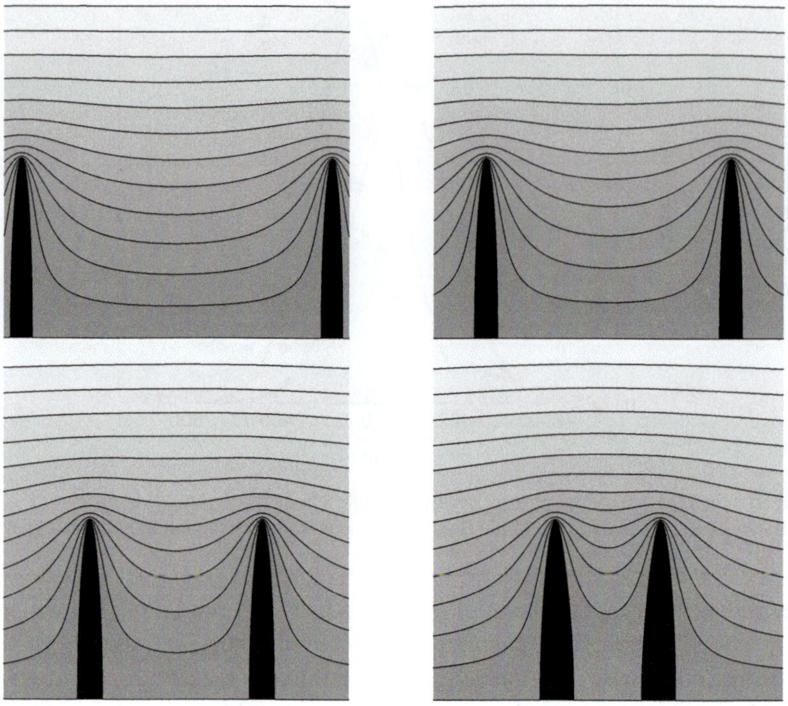

Figure 30.22 Two 0.2 emitters from Figure 30.11 brought progressively closer together. Shielding manifests as a raising of the equipotential lines between them, as well as a softening of the curvature about the apexes. The equipotential lines for all figures are the same. Distance between the emitters starts as $1.8L$ and decreases in increments of $0.6L$ from top left to bottom right.

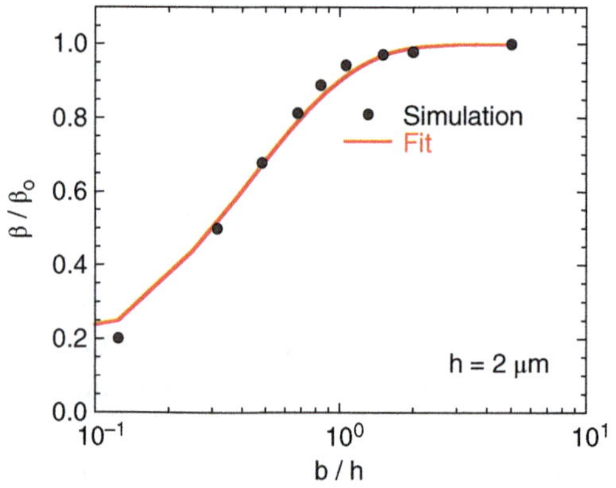

Figure 30.23 Simulations of the ratio of β for a line of emitters separated by a distance b to an isolated emitter β_o, compared to Eq. (30.126) using the parameters $a = 2.3172$ and $h = 2$ μm, based on Figure 4a of Bonard *et al.* [499].

The tapered line charge model of Section 30.6.2 makes for a straightforward analysis of the effect when applied to idealized geometries [501–504]. Let an infinite array of emitters be characterized by their unit cells, or the smallest repeating unit of the array. Two types in particular that reflect how actual arrays are fabricated are the square unit cell (where an emitter is situated at each corner of a square) or a triangular unit cell (where an emitter is situated at each vertex of an equilateral triangle), as in the lower right-hand corners of Figure 30.24. The factor c in Eq. (30.127) differs depending on which unit cell is under consideration. Assuming an infinite array simplifies the calculations for field enhancement factors, enabling the values of a and c to be found and shown in Table 30.4 for the square and triangular array types. Although Eq. (30.126) and its improvement as Eq. (30.127) are based on a fitting of simulation results, there is a way to demonstrate the reasonableness of the parameterization of β/β_o by using arguments derived from the point charge model treatment of Section 30.3, without having to go through the time-consuming methods for the numerical approach of refs [501–505], which involves manually adjusting parameters until the apex radius acquired a desired value.

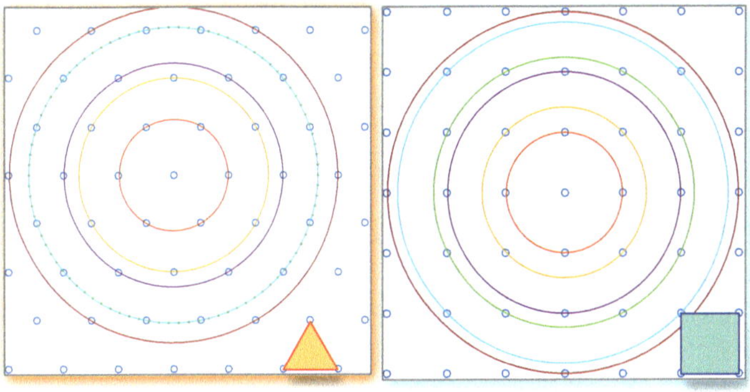

Figure 30.24 Top-down view of an array of wires on a triangular (left) and square (right) grid. The distance between adjacent emitters is b. The radius of the two smallest rings is b and $1.7321b$ with six wires per ring for the triangular array, and b and $1.4142b$ with four wires per ring for the square array. As the rings get larger, the number of wires per ring and the radius of the ring change as shown.

Table 30.4 Values of c and a in Eq. (30.127) for the triangular and square arrays of tapered line charges in the configurations of of Figure 30.24. Also given is a_s, as in the line charge model the nature of the emitter depends on it.

Parameter	Triangular (△)	Square (□)
a	1.265	1.45
c	1.09	1.00
h [μm]	1500	1500
a_s [μm]	1.5	1.5

Most of the charge in the wire for the line charge model aggregates at the ends near the apex. The simplest approximation is to push *all* of the charge to the end: this constitutes a point charge model with only one point charge, plus its image, for each emitter site on the array. A problem, though, is that when the point charges are sufficiently close, the ellipsoidal shape of the equipotential lines for the point charges and their images are undermined, that is, the point charge model fails to account for the overall geometry, particularly near the base. Although such defects can be fixed by the clever placement of yet other charges, it is the qualitative behavior of shielding as emitters very far apart are brought closer together that matters, and for that the qualitative behavior of the simple point charge model is sufficient to reveal what unfolds. Therefore, let point charges and their images occupy the points on the square and triangular arrays. A few points are worth explicit mention. First, all of the charges are in the same plane, as are all of the image charges: as a result and perhaps confusingly until one thinks about it, for an infinite array there is no net force on a charge due to other charges in the same plane, so that a charge is only affected by the image charges. Second, although the tip-to-tip separation is the same for both triangular and square arrays, the triangular array has a higher packing density (number of tips per unit area) than the square array does, as suggested by the number of emitters on the first circle shown in Figure 30.24: the square array has four, but the triangular array has six.

EXAMPLE: Let σ be the number of emitters per unit area. Find $\sigma_{\triangle}$ and $\sigma_{\square}$ for the triangular and square arrays, respectively, where the tip-to-tip separation is b.

SOLUTION: Consider the unit cells shown in Figure 30.24 and epitomized by the color unit cells in the lower right-hand corner.

- In the triangular element, there is $N_{\triangle} = 1/2$ emitters: each vertex has $1/6$ of an emitter, and there are 3 vertices, so $(1/2) \times (1/6) = 1/2$. The area of the triangle is

$$A_{\triangle} = \frac{1}{2} b^2 \sin\left(\frac{\pi}{3}\right) = (\sqrt{3}\, b^2/4) = 0.43301 b^2 \tag{30.128}$$

  Therefore $\sigma_{\triangle} = N_{\triangle}/A_{\triangle} = (2/\sqrt{3})/b^2 = 1.1547/b^2$.
- In the square element, there is $N_{\square} = 1$ emitters: each vertex has $1/4$ of an emitter, and there are 4 vertices. The area of the square is $A_{\square} = b^2$. Therefore, $\sigma_{\square} = N_{\square}/A_{\square} = (1)/(b^2) = 1/b^2$.

When each tapered line charge is replaced by a point charge and its image, the potential energy for the jth charge is

$$V_j(x, y, z) = 4Q_s\left(\frac{1}{r_j^-} - \frac{1}{r_j^+}\right) \quad (30.129)$$

$$r_j^{\pm} = ((x - x_j)^2 + (y - y_j)^2 + (z \pm L)^2)^{1/2} \quad (30.130)$$

where $Q_s = qq_s/16\pi\epsilon_0 = (q_s/q)Q$ and q_s is the magnitude of the point charge. For an infinite array, by symmetry arguments the field at the apex of each emitter is solely in the $\hat{z}$ direction, and so evaluating $F = -\partial_z V$ is sufficient to find the field enhancement β. In units for which all lengths are scaled by L, then the apexes of the emitters reside at $h/L = 1 + \delta$, and so the ratio β/β_o is [505]

$$\frac{\beta}{\beta_o} = \frac{\sum_{j=1}^{N} n_j\{\delta(\rho_j^2 + \delta^2)^{-3/2} - (2+\delta)(\rho_j^2 + (2+\delta)^2)^{-3/2}\}}{\delta^{-2} - (2+\delta)^{-2}} \quad (30.131)$$

where ρ_j is the radius of the jth ring on which the point charge sits, and where n_j is the number of point charges on that ring (e.g., $n_j(\triangle) = 6$ and $n_j(\square) = 4$ for $j = 1$ and 2). Clearly, the larger ρ_j, the less of an influence the jth ring has: as a practical matter, those rings for which $\rho_j > 3b$ can be ignored. Simple geometry shows that for the first six rings,

$$\{(\rho_j/b)^2\}_\square = \{1, 2, 4, 5, 8, 9\} \quad (30.132)$$

$$\{(\rho_j/b)^2\}_\triangle = \{1, 3, 4, 7, 7, 9\} \quad (30.133)$$

(notice that the $j = 4$ and $j = 5$ of the rings for ($\triangle$) have the same radius) and n_j for each of those rings can be easily counted from Figure 30.24, making the evaluation of Eq. (30.131) straightforward.

By way of example, consider the case for $\delta/h = 0.1$ and 0.2 shown in Figure 30.25, with the triangular and square evaluation points compared to fits calculated according to Eq. (30.127) with the parameters of Table 30.5. The parameters tend to have different sizes than those associated with the line charge studies, and that has to do with how charge is distributed: coplanar charges do not give rise to a $\hat{z}$ component in force, whereas the line charge models do. More importantly, as the factor δ/h changes, the best-fit parameters also change, underscoring the interpretation of these parameters as those which capture the effects of geometry and placement. Therefore, although there are differences in magnitude between the values of a and c between the tapered line charge and point charge models, they share qualitatively analogous behavior[30] that has been traced to the placement and distribution of the charges making up the emitter surface.

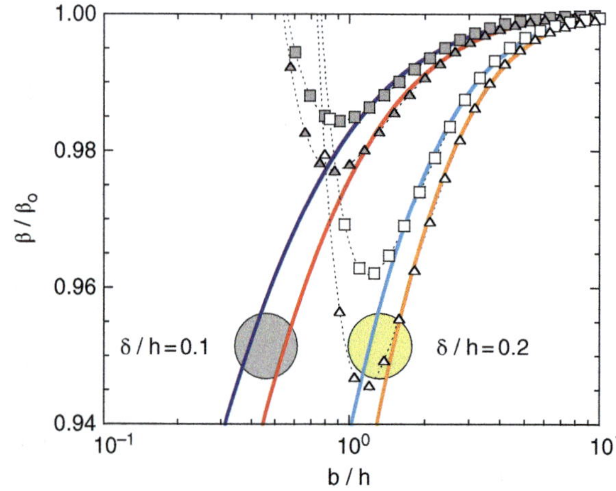

Figure 30.25 Evaluation of Eq. (30.127) for $\delta/h = 0.1$ (dark blue and red lines; filled symbols) and $\delta/h = 0.2$ (light blue and red lines; open symbols) for the point charge model. The best-fit lines use the parameterization of $y(x) = 1 - \exp(-ax^c)$ with values given in Table 30.5.

[30] The upturn in behavior of the point charge model when $b/h \approx 1$ does not have an analog in the tapered line charge model. It coincides with the equipotential lines no longer resembling bumps. The evaluations for $b/h < 1$ is only shown for completeness, not because that behavior imitates the tapered line charge model (it does not): $\beta(b) \to 0$ as $b \to 0$ for the line charge model [501].

Table 30.5 Best fit values of c and a in Eq. (30.127) for the triangular and square arrays of point charges in the configurations of Figure 30.24 as described by Eq. (30.131) and shown in Figure 30.25 for the two cases of $\delta/h = 0.1$ and 0.2.

Parameter	δ/h	△	□
a	0.1	3.7195	4.0509
c	0.1	0.3457	0.3133
a	0.2	2.4897	2.7887
c	0.2	0.4711	0.4290

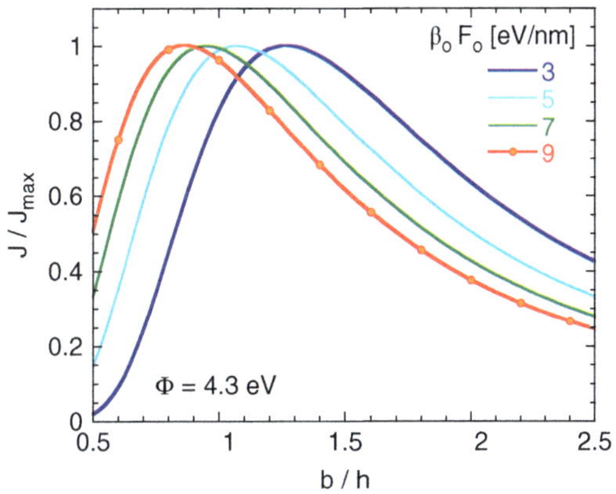

Figure 30.26 The array current density J_{array} normalized to its maximum value as a function of the ratio of the tip-to-tip separation b with the wire height h for *ad hoc* metal-like parameters in a triangular array ($\Phi = 4.3$ eV, $h = 2$ μm, $a_s = 7$ nm, and $\beta_o = 1000$), as evaluated using Eq. (30.134). The values of b_{opt} and J_{max} are shown in Figure 30.27.

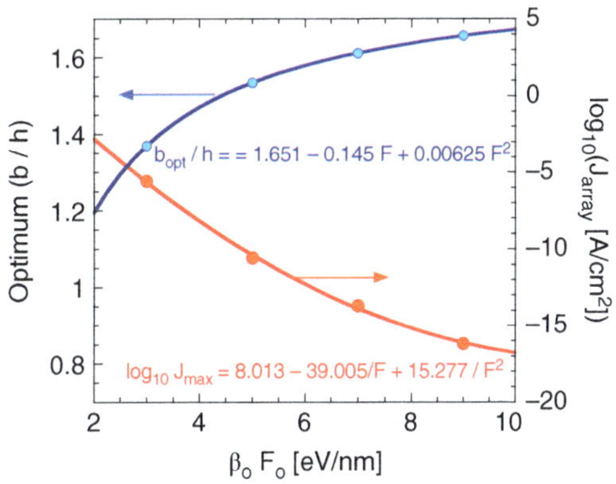

Figure 30.27 The values of b_{opt} and J_{max}, as evaluated using Eq. (30.134) for metal-like parameters ($\Phi = 4.3$ eV, $h = 2$ μm, $a_s = 7$ nm, and $\beta_o = 1000$), as shown in Figure 30.26.

The importance of the behavior of $\beta(b)/\beta_o$ is because it demonstrates that as $b \to 0$, then the field emission current density over the array, $J_{array}(\beta F_o)$, is negligible when $\beta/\beta_o \ll 1$. Conversely, as $b \to \infty$, then J_{array} likewise declines as $1/b^2$. As a result, J_{array} maximizes for an intermediate value of b. That maximum can be found by evaluating the current density from an array of emitters by [501]

$$J_{array}(b,p) = \sigma_p(b) 2\pi a_s^2 g[\beta(b)F_o] J_{FN}[\beta(b)F_o] \tag{30.134}$$

for an array $p = \square$ or $\triangle$, where $\sigma_\square = 1/b^2$ and $\sigma_\triangle = 4/\sqrt{3}b^2$ (see Eq. (30.128)). For a given apex field $\beta_o F_o$ (the field an isolated emitter would experience at its apex), the behavior of J_{array} can be evaluated, as in Figure 30.26. The analysis clearly shows that *the optimal spacing depends on the background field*, with lower fields favoring closer spacing and higher fields favoring greater separation. That the optimal b/h for intermediate field values is near unity (the tips are as far apart from each other as they are high) matches conventional expectations borne of anecdotal data or simulations.

30.10 Statistical variation

Besides the possible ramifications of the resulting spatial nonuniformity, a large variation in the IV characteristics between tips in an array decreases the average current per tip that can be extracted without inducing voltage breakdown events. Thus, for a given tip packing density, the total current density that can be extracted from an array is not proportional to the array size, which is clearly undesirable.
– Paul R. Schwoebel, Charles (Capp) A. Spindt, Chris E. Holland, and John A. Panitz [460]

In the estimation of array current in Eq. (30.134), the implicit assumption is that all of the emitters are identical, but insofar as emission is from nanoscale structures at the tips of the emitters, it stretches credulity to assume that all such protrusions are alike, regardless of the similarity of the larger structures on which they sit. Given the acute dependence of field enhancement β on tip radius a_s over all other parameters, variation in the tip currents of individual emitters must likewise be expected. The question then becomes, how is such statistical variation manifested in current–voltage relations? Or, put another way, why does current density *not* scale with the number of emitters (or the emission area)?

Take a very simple model to start the analysis. Assume that there are N emitters, and that the distribution in radii is such that there are n_i emitters characterized by a radius a_i, so that

$$N = \sum_{i=1}^{\infty} n_i \tag{30.135}$$

$$I_{array} = \sum_{i=1}^{\infty} n_i \, I_{tip}(a_i) \tag{30.136}$$

where clearly not all n_i are non-zero, and where I_{tip} is given by Eq. (30.18). The average current per tip is then

$$I_{ave} = \langle I_i \rangle = \frac{1}{N} \sum_{i=1}^{\infty} n_i \, I_{tip}(a_i) \tag{30.137}$$

If N is sufficiently large, then the summation can be approximated by an integration over tip radii. Let the minimum radius a_j for which n_j is non-zero be a_o (not to be confused with the Bohr radius), and likewise let the maximum radius for which n_j is non-zero be $a_o(1 + \Delta)$. The array current is then

$$I_{array} = \frac{N}{a_o \Delta} \int_{a_o}^{a_o(1+\Delta)} 2\pi a^2 \, g[F(a)] J_{FN}[F(a)] \, da \tag{30.138}$$

A similar formulation would result if, instead of radius, the work function varied [388], but for simplicity take the only variation to be caused by geometry. A simple model then approximates $F(a) \approx F_{tip}(a_o/a)$, so that F_{tip} is the apex field on the *sharpest* emitter characterized by a_o. The simplest model further takes $\nu \approx 1$ for *ellipsoidal* emitters (recall Table (2.3)): those two approximations result in a wonderful cancellation of inconvenient terms (try it), resulting in

$$I_{array} = N \frac{(1 - e^{-b\Delta})}{b\Delta} I_{tip}(F_{tip}) \equiv N_{eff} I_{tip} \tag{30.139}$$

where $b = B_o \Phi^{3/2} / F_{tip}$. Were all the emitters very uniform in apex radius, then Δ would be small, and if $b\Delta$ were small, then the coefficient of $N I_{tip}$ would be approximately unity: under such conditions arrays of emitters *would* scale with area or number: that they do not is an indication of what changes are wrought by Δ. In short, the factor $(1 - e^{-b\Delta})/b\Delta$ acts much like *the fraction of tips that are emitting* if all the tips were identical and had the smallest radius (much like the notional area acts like the fraction of the surface that is emitting if the current density across the surface were the same as the apex value; both the notional area and the fraction of tips emitting factor under-represent how many tips or how much of the surface is actually contributing).

EXAMPLE: Calculate the *fraction of tips emitting* factor of Eq. (30.139) for the copper-like parameter of $\Phi = 4.5$ eV and a tip field of $F_{tip} = 6$ eV/nm for a smallest tip radius of $a_s = 5$ nm and a spread of $\Delta = 2$.
SOLUTION: Using B_o from Table 2.3,

$$b = \frac{B_o \Phi^{3/2}}{F_{tip}} = \left(\frac{6.8309}{\text{eV}^{1/2}\text{nm}} \right) \frac{(4.5\ \text{eV})^{3/2}}{6\ \text{eV/nm}} = 10.868$$

or $b\Delta = 21.736$, the size of which is so large that Eq. (30.139) is well approximated by $N_{eff}/N \approx 1/b\Delta = 0.076678$. Thus, effectively 7.7% of the emitters are contributing. Because b depends on F_{tip}, this percentage increases as the field increases.

As simple and convenient as the uniform radius model is, are there reasons for considering it representative of actual distributions of emitter radii? In spite of its obvious pedagogical value, sadly, the answer is "not quite":[31] microfabrication processes produce distributions in apex radii that are better modeled by Gaussians than a uniform distributions. However, there is a limit to how sharp emitters may become: they must be, at least, greater than zero. To get a distribution that has the semblance of a Gaussian but which forbids negative arguments is the closely related log-normal (LN) distribution, which better describes the radii of actual emitters [388]. Data for carbon nanotubes (Figure 10 of ref. [506]) and explicitly measured for silicon field emitters [507, 508] can be matched by a log-normal distribution (even when matched by a Gaussian, the distribution shows underlying hints of log-normal behavior, as in Figure 3 of ref. [509]). For example, the data of Guerrera *et al.* is reproduced in Figure 30.28 alongside the best log-normal fit.

The log-normal distribution $L(x)$ is defined by the relation in Eq. (29.1), where σ now governs the spread and μ is the median value of x (not to be confused with the chemical potential) where x now refers to apex radii. For $L(x)$, and unlike the Gaussian distribution to which it is related, the mean $\langle x \rangle$ and the median μ are different (recall Section 4.1 for an example of how median and mean differ): the median μ is such that

$$\int_0^{\mu} L(x)dx = \int_{\mu}^{\infty} L(x)dx = \frac{1}{2} \tag{30.140}$$

whereas the mean is $\langle x \rangle = \mu \exp(\sigma^2/2)$. For a discrete set of apex radii a_i with $N \geq i \geq 1$, μ and σ are evaluated by

$$\ln(\mu) = \frac{1}{N}\sum_{i=1}^{N} \ln(a_i) \tag{30.141}$$

$$\sigma^2 = \frac{1}{N}\sum_{i=1}^{N} \left[\ln\left(\frac{a_i}{\mu}\right)\right]^2 \tag{30.142}$$

in terms of which the array current is given by

$$I_{array} \approx N \int_0^{\infty} L(a) I_{tip}(a) da \tag{30.143}$$

Return, now, to Figure 30.28: the larger σ is, then the more likely it is that a small number of very sharp emitters are present. Because of the sensitivity of tip current to apex field, these emitters rapidly dominate their neighbors. An indication of the behavior is given in Figure 30.29, calculated using the algorithms of Section A3.27: only for distributions where σ is small does it appear that most of the emitters "turn on." A second way to see the dominance of the sharpest emitters is shown in Figure 30.30, where for larger σ the sharpest emitters are seen to be responsible for most of the current. For the integrated current, for the larger $\sigma = 0.16$, almost all of the current put out by the array is due to the half of the emitters for which $a_i < a_o$, and so the integrated current curve has taken a hard shift to the left by comparison to the tight distribution $\sigma = 0.02$, for which the number and current distributions are more aligned.

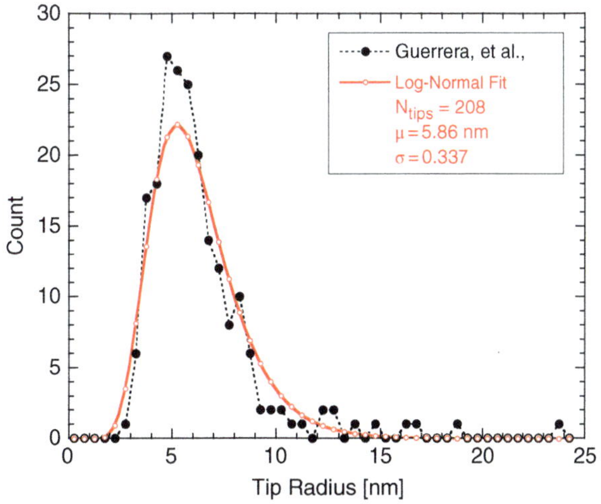

Figure 30.28 The distribution of high aspect ratio silicon field emitters . Red line/open circles is the best-fit log-normal representation based on Eqs (30.141) and (30.142).(data courtesy of S. Guerrera (MIT)): see S.A. Guerrera, L.F. Velasquez-Garcia, and A.I. Akinwande, *IEEE Trans. El. Dev.*, **59**, 2524, 2012, particularly Figure 10, for a description)

[31]The "top hat" distribution may, however, be useful for modeling work function variation due to crystal face or coverage; see ref. [388].

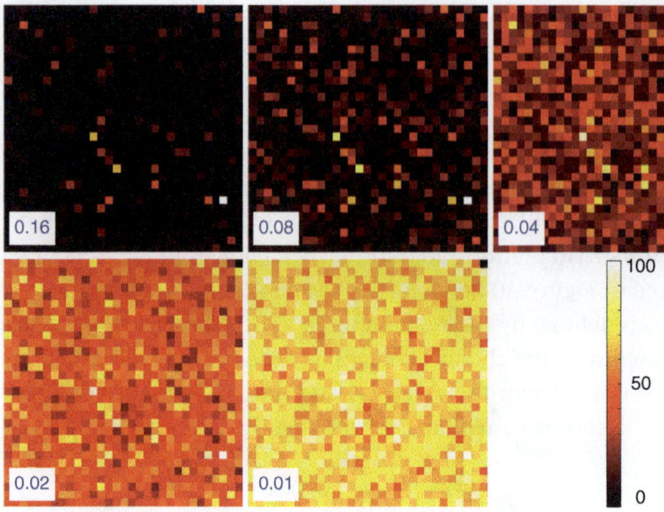

Figure 30.29 A distribution of field emitters tips in a 32 × 32 array with radii a_i log-normal distributed with median $\mu = 6$ nm and σ as shown in the lower left of each panel. A work function of $\Phi = 4.5$ eV is assumed. The apex field is assumed to be $F(a) = F(\mu)(\mu/a)$, with $F(\mu) = 5$ eV/nm. When σ is large, then only a few emitters dominate the emission current, as their contribution becomes well in excess of their neighbors.

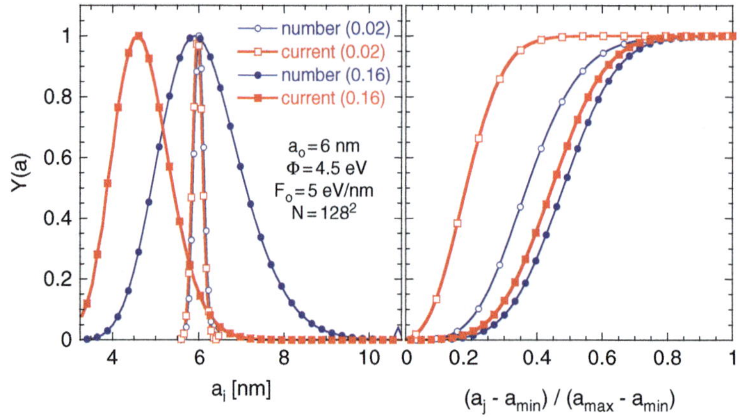

Figure 30.30 Left, the distribution of N^2 tips for $N = 128$: circles correspond to $\sigma = 0.02$, squares to $\sigma = 0.16$. "rn" and "rj" are number n_j and current density J_j normalized to their maximum value (i.e., for density, $Y(a) = n(a)/n_{max}$ and similarly for current). The number of bins $\{n_j\}$ is 64. When σ is large enough, a small number of very sharp emitters exist, which quickly dominate the current that the array produces. Right, the normalized integrated current as a function of the normalized tip radius $(a_i - a_{min})/(a_{max} - a_{min})$, where the maximum and minimum values are determined by the non-zero n_i.

A third and final way to see the impact of the sharpest emitters compared to the median radius emitters is through a representation of the current from the array on a Fowler–Nordheim-like plot, where $\ln(I_{array}/F_{tip}^2)$ is plotted against $1/F_{tip}$, where F_{tip} is the field on the median emitter $a_o = \mu$ and where Eq. (30.137) is used, but with the n_i being log-normal distributed according to Eq. (A3.24). Alongside the Fowler–Nordheim curve for all the emitters are two additional curves: in the first curve, only the output of the first non-zero n_i set of emitters characterizing the sharpest radii is shown. In fact, for the simulation shown, $n_1 = 1$. That it approaches the array current at small field indicates that the array current *is dominated by a very small number of sharpest emitters*. As the field rises, more emitters with radii closer to the median begin to contribute. The second curve is therefore for an array composed solely of identical emitters with tip radii $a_i = a_o = \mu$: the Fowler–Nordheim characteristics of the actual array are seen to converge with this hypothetical set as the field increases, indicating that the mean emitters begin to dominate.

Such a finding bodes poorly for the common practice of inferring apex radii, or even a single field enhancement factor β characteristic of all the emitters, from data on a Fowler–Nordheim plot. Doing so is a rather severe oversimplification, as can be appreciated on general grounds [431] or even careful measurements [510]. Inferring emission area and field enhancement from data is complex and dependent upon assumptions [511–513]. The motivation for doing so may seem to be related to a desire to have an "effective emitter" with which to characterize improvements in emitters or perform simulations. In simulation, however, observe that all the current of the array originating from a few hot spots produces a very different beam than uniform emission over an array, especially if space charge and emittance effects are part of the investigation.

But were one to try to do so anyway, of what must one be careful? The whole premise of "fitting to a Fowler–Nordheim curve" is predicated on a linear fit like $y(x) = C_0 + C_1 x$, where $y(x) = \ln(J/F^2)$ and $x = 1/F$. But supposing one did wish to eke out an effective area and field enhancement anyway, and use the Fowler–Nordheim equation to do so. The assumed linearity of $\ln(I/V^2)$ with $1/V$ plus the approximations $F = \beta V/D$ and $I = A_o J$, where β is the field enhancement, and A_o is an emission area, suggest that the slope and intercept on a Fowler–Nordheim plot is enough to determine field enhancement and emission area. If the tips are distributed in field enhancement (e.g., by being distributed in radius), however, the slope and intercept are no longer unique. The causes are several. First, the underlying Fowler–Nordheim equation of Eq. (13.25) indicates that $\ln(J/F^{2-\nu})$ is linear in $1/F$ rather than $\ln(J/F^2)$. Second, the area factor $g(F)$ is, to leading order, linear in field (recall Figure 30.17), so that $\ln(J/F^{3-\nu})$ should be used. Lastly, the changing slope of Figure 30.31 suggests that the effects of the distribution add their own modification. Isolating the contribution of a

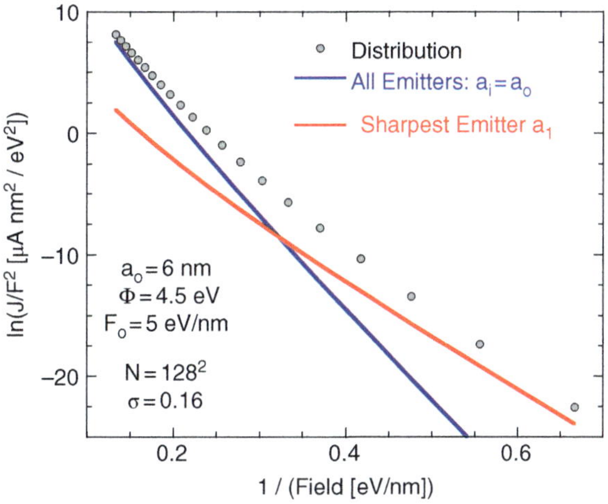

Figure 30.31 The current from the log-normal distribution of emitters of Figure 30.30 for $\sigma = 0.16$, $\mu = 6$ nm $\Phi = 4.5$ eV, and $N^2 = 128^2 = 16384$ (circles) compared to, first, the current only from the $n_1 = 1$ emitters with the smallest apex radius (red line) and, second, an array of N identical emitters all of radius $a_o = \mu = 6$ nm. The sharpest emitters are seen to dominate at low field; the median emitters by virtue of their number dominate at high fields.

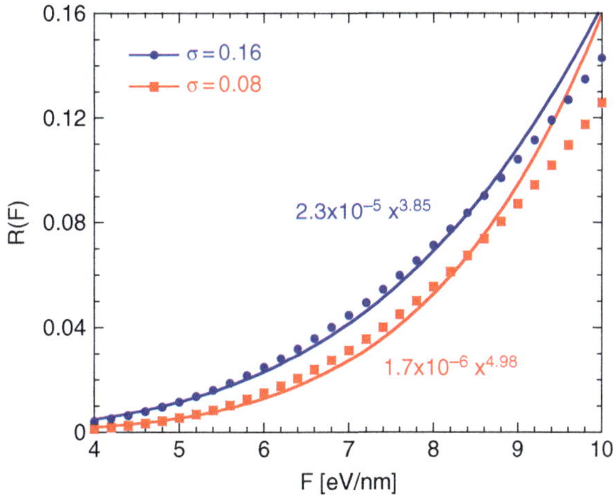

Figure 30.32 $R(F)$ as evaluated using Eq. (30.146) for $\sigma = 0.16$ (blue circles) and 0.08 (red squares), as a function of the field on the sharpest emitter characterized by a_{min}. Best-fit power law fits are overlaid and labeled alongside the curve they fit.

log-normal distribution separate from the area factor and current density is a challenge, but one way is to write

$$N \int_0^\infty L(a) I_{tip}[F(a)] da \equiv N R(F) I_{tip}(F) \tag{30.144}$$

where $F = F_{tip} a_o / a_{min}$ is the field on the sharpest emitter, and where a_o is the median radius and a_{min} the radius of the sharpest emitter present in the array, or

$$1 = N \int_0^{a_{min}} L(a) da \tag{30.145}$$

Using Eq. (13.25) for $J(F)$, Eq. (30.18) for the current and Eq. (30.98) for the area factor $g[F(a)]$, then $R(F)$ is approximated by

$$R(F) = a_{min} \int_{a_{min}}^\infty s^{1+\nu} L(s a_{min}) \exp\left[\frac{B_o \Phi^{3/2} a_{min}}{a_o}(1-s)\right] ds \tag{30.146}$$

Defined in this manner, as the field decreases $R(F) \rightarrow 1/N$ when the sharpest emitter dominates, but $R(F)$ then increases as F increases and more of the blunter emitters contribute. Performing the integration numerically for $\Phi = 4.5$ eV and $a_o = 6$ nm produces the behavior shown in Figure 30.32. Using a power law fit of the form $R(F) \approx R_o F^p$ suggests that $\ln(J/F^{3+p-\nu})$ is linear in $1/F$. As a result, the slope and intercept methods very much depend on where the slopes and intercepts are taken if the $\ln(J/F^2)$ form is used as a consequence of the weak convexity that exists on a standard Fowler–Nordheim plot [197]. As a result, when the actual distribution n_i of emitters is known or can be modeled, and if I_{tip} can be reasonably represented, resorting to Eq. (30.136) is better for modeling the array current.

CHAPTER 31

Photoemitters

...accidental observations have played an important role in the development of electron emitting materials in general and of photoemitters in particular...It is not surprising, therefore, that luck, combined with trial and error, has been so essential in the development of photoemitters.

– Alfred H. Sommer [134]

Among the most challenging applications of photocathodes are particle accelerators [514–516] and light sources such as free electron lasers (FELs) [517–519], both of which demand a great deal in terms of current density and ability to shape the electron pulse. For these applications in particular, the need to overcome the limitations on beam brightness imposed by thermionic cathodes led to the development of high-performance RF photoinjectors [520–523]. As the operational wavelengths of FELs in particular are made smaller and smaller with the goal of an X-ray FEL [524], five metrics of photocathode performance emerge as particularly important:

- *quantum efficiency* governs how many electrons are emitted compared to how many photons are absorbed, with high quantum efficiency being desirable
- *emittance* describes the tendency of an electron beam to spread out as the beam propagates, with low emittance beams being desirable
- the operational *lifetime* of the photocathode before replacement or rejuvenation
- the *ruggedness* or survivability of the photocathode, particularly in the messy environment of an RF photoinjector
- the *response time* of the photocathode, which affects how well pulse shaping of the electron beam can be accomplished.

All high quantum efficiency photocathodes involve cesium, and the highest quantum efficiency photocathodes tend to be semiconductors. Not surprisingly, therefore, electron transport through the bulk material of the photocathode and the conditions of the surface when the photo-excited electrons get there are particularly important to the understanding of photocathode operation. This will lead to a focus on scattering during transport, and the relationship between cesium coverage and low work function (a similar concern of thermionic cathodes, although there barium is used because of its lower evaporation rate), to understand how the processes of emission affect the metrics. Of the five metrics, quantum efficiency and emittance often dominate attention: generally, they increase together [181]. Lifetime appears to be distinct, but a low quantum efficiency requires that high peak/average current applications must increase the intensity of the drive laser power on the photocathode, causing the cathode to heat. Ruggedness refers to how long photocathodes can endure in the injector environment: RF photoinjectors have worse vacuums than DC photoinjectors, and the ions produced erode coatings and damage crystal lattices. Response time refers to the matching of the electron pulse shape to the shape of the drive laser [525]: yield from photocathodes like GaAs or GaAsP has a thermalized electron component which, in the case of negative electron affinity (NEA), can continue to be emitted even after the laser pulse has stopped. Thermal tails are seen experimentally [526] and in Monte Carlo simulations of related photocathodes like Cs_2Te, where the response time is lengthened (approximately 0.4 ps) compared to the fast response for metals [527], as in Figure 31.1.

As long as most of the pulse shape of the electrons mimics the drive laser pulse shape (i.e., the pulse length is long compared to the response time), some simple estimates may guide the theory most useful in understanding photocathode operation. Return to Eq. (14.3) as a starting point. For FELs, rather generic parameters for what is desired are considered in the next example, although they, and the pulse repetition frequency, depend on what the FEL is designed for [528].

EXAMPLE: Develop a relation to estimate the required drive laser power as a function of wavelength λ and quantum efficiency (QE). Ignore reflectivity.

SOLUTION: The QE is the ratio of the number of electrons out compared to the number of photons in, or

$$QE = \left(\frac{J_e}{q}\right) \div \left(\frac{I_\lambda}{\hbar\omega}\right) = \frac{2\pi\hbar c J_e}{q\lambda I_\lambda} = 1239.8\frac{J_e}{\lambda I_\lambda} \tag{31.1}$$

for λ measured in [nm] and I_λ in [Watt/cm^2].

Introduction to the Physics of Electron Emission, First Edition. Kevin L. Jensen.

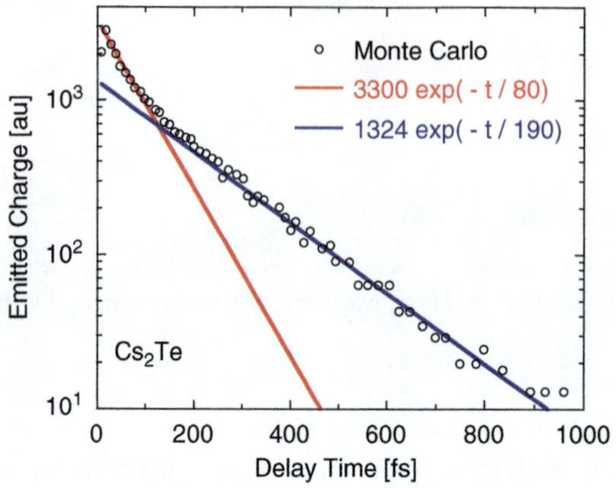

Figure 31.1 Time delay after photo-excitation for emission to occur in a Monte Carlo simulation of Cs_2Te. Monte Carlo data digitally extracted from Figure 5 of ref. [527]. The simulation by Ferrini *et al.* fitted the yield data for a Cs_2Te layer on a molybdenum substrate over a range of photon energies between 3.7 and 5.5 eV.

EXAMPLE: For a charge per bunch of $\Delta Q = 1$ nC, a pulse length of $\delta t = 50$ ps with $\Delta t = 5$ ns between pulses, and an emission area of $A = 0.2\ \text{cm}^2$, find the peak and average currents.
SOLUTION: The peak current $J_e = \Delta Q/\delta t A$ is

$$J_e = \frac{1\ \text{nC}}{(50\ \text{ps})(0.2\ \text{cm}^2)} = 100\frac{\text{A}}{\text{cm}^2}$$

and the average current is smaller by $J_{ave} = J_e(\delta t/\Delta t) = 1\ \text{A/cm}^2$. Using the relation between J_e and I_λ of the previous example, to pull 100 A/cm² from the photocathode for a drive laser with a wavelength of 266 nm and a quantum efficiency of 0.1, the drive laser would have to provide a peak power of 4.66 kW/cm² of incident laser power and an average power of 46.6 W/cm² (the example is for illustration only).

The illustrative example above can be used to develop a table of values relating what is desired in terms of current to what is demanded from the drive laser. A few other considerations matter. The drive laser is usually from a readily available laser (such as an Nd:YAG with an output wavelength of 1064 nm) that is then frequency doubled (to 532 nm), tripled (to 355 nm) or quadrupled (to 266 nm), an amount characterized as the *harmonic* and labeled $n = 2, 3$, or 4, so that $\lambda = \lambda_o/n$. Other drive lasers with other wavelengths, like 800 nm, can be used, of course, but keep the discussion simple. Such conversion generates waste heat in the doubling crystals, the conversion efficiency crudely characterized by 50% for $n = 2$, 30% for $n = 3$, and 10% for $n = 4$. Also, metals are quite rugged, so their lifetimes tend to be long, but good photocathodes like CsK_2Sb which perform well at lower current levels can be taxed so hard that their lifetimes shorten to hours [529]. Similarly, semiconductor photocathodes scattering physics leads to longer response time than metal photocathodes, which affects pulse-shaping demands. A table of laser parameters, as in Table 31.1, current densities, and photocathode candidates can be constructed from such considerations.

Table 31.1 Relation of photocurrent to drive laser. "Harmonic" is the change of the $\lambda_o = 1.06$ μm wavelength to a shorter wavelength using frequency doubling crystals such that wavelength $\lambda = \lambda_o/n$. "Efficiency" is the conversion efficiency of shortening the wavelength. "Temp. Resp." is the temporal response. "Ave P at cath." is the average laser power deposited on the cathode. "Ave P laser" is the average power of the primary 1064 nm Nd:YAG before harmonic conversion. 8760 hours is one year, but metal photocathodes last longer than that.

Property	Units	K_2CsSb	Cs_2Te	GaAs	Cu	Mg
Harmonic	–	2	4	2	4	4
Wavelength	nm	532	266	532	266	266
Efficiency	%	50	10	50	10	10
Starting QE	%	8	5	5	1.4E-2	6.2E-2
Lifetime	hours	4	> 100	58	> 8760	> 8760
Temp. Resp.	–	Prompt	Prompt	< 40	Prompt	Prompt
Vac. tol.	–	Poor	Very good	Poor	Excellent	Excellent
Ave P at cath.	W/cm²	29.13	93.22	46.61	33293	7518
Ave P laser	W/cm²	58.26	932.20	93.22	332932	75178

One consequence of low quantum efficiency evident from Table 31.1 is the substantial power levels deposited on the photocathode: power not leading to photoemission leads to heating of the cathode. Metals subjected to moderate power levels already get hot, an example being the tightly coiled tungsten filament in a standard 60 W incandescent lightbulb, heated by the passage of current through the tungsten wire. How hot can be crudely estimated, as shown in the next example.

EXAMPLE: Estimate the temperature of a tungsten filament knowing that the length of the wire is 2 m and its diameter is 1/100 of an inch. Assume the emissivity of tungsten is $\varepsilon = 0.35$ (it is not a perfect black body emitter). Assume that the length of the coiled tungsten wire is 2.5 cm.

SOLUTION: The temperature of a hot body was treated in Section 3.3. Assume that the tungsten filament is packed into a cylinder half the density of a tungsten wire (the empty space being due to the coil). The radius of the cylinder of tungsten coil is then approximately given by

$$\pi\left(\frac{1}{100}\frac{2.54\text{ cm}}{2}\right)^2(2\text{ m}) = \frac{1}{2}(2.5\text{ cm})\pi r^2$$

or $r \approx 0.4612$ mm. The power $P_o = 60$ W is radiated from this cylinder, and so based on σT^4 law of Section 3.3,

$$\frac{P_o}{2\pi rL} = \varepsilon\sigma_{sb}T^4 \rightarrow T = \left(\frac{P_o}{2\pi\varepsilon\sigma_{sb}Lr}\right)^{1/4} \tag{31.2}$$

or $T \approx 2541$ K after using $P_o = 60$ W and $\sigma_{sb} = 5.67047 \times 10^{-7}$ W/m^2 K^4.

In spite of the crudeness of the previous example, it nevertheless suggests that a metal photocathode will experience temperatures beyond its melting point if driven as hard as the parameters of Table 31.1 would suggest are required, given the vastly larger amounts of laser power (compared to 60 W) that are indicated as needed if a metal photocathode were employed [83, 85, 206]. Semiconductors exhibit greater quantum efficiency and therefore relax drive laser demands. Understanding why involves examining the transport characteristics and the nature of the emitter surface.

Although scandate and barium dispenser cathodes give interesting performance levels when used as photocathodes [23, 530, 531], in general it is true, as Sommer observed [134], that all high quantum efficiency photocathodes involve the use of cesium, either as a coating or integrated into the stoichiometry of the bulk material. On metals, cesium provides a low work function coating. For semiconductors, multi-alkali antimonides and tellurides give much higher quantum efficiencies. Various photocathodes are compared in Figure 31.2, in which quantum efficiency is compared to lifetime (measured by the $1/e$ parameter τ or the amount of time taken for quantum efficiency to degrade by a factor

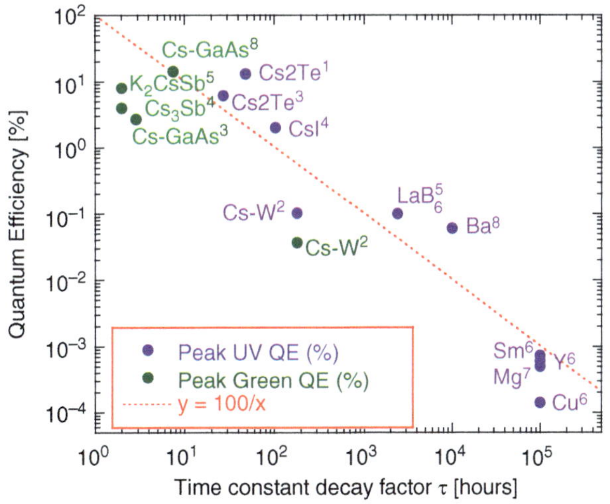

Figure 31.2 The quantum efficiency of various photocathodes or materials (based on Figure 1 of ref. [532]). The red dashed line is used only as a guide to the eye and corresponds to $QE \times \tau = 100$, reinforcing that quantum efficiency scales inversely with the $1/e$ decay factor τ. The superscripts on the chemical formulae identify the reference: 1, Suberlucq [533]; 2, Moody *et al.* [534]; 3, Monaco *et al.* [535]; 4, Siggins *et al.* [536]; 5, Kong *et al.* [537]; 6, Rao *et al.* [538]; 7, Wang *et al.* [539]; 8, Van Loy *et al.* [540]; 9, Calabrese *et al.* [541].

of $1/e$). A pattern emerges: not only is quantum efficiency improved by better transport through the bulk when scattering does not undercut emission, but it is also related to the surface coating and fades as that coating is damaged or disappears.

31.1 Scattering consequences

Nowhere to run ain't got nowhere to go

– Bruce Springsteen[1]

Scattering in metals is a bit different than semiconductors as far as quantum efficiency is concerned. Some hint as to why is revealed by considering a collision between two equal mass particles labeled a and b; the problem is easier working in the reference frame of the b particle, taking the incident a particle to be traveling in the $\hat{x}$ direction, and letting the plane of the collision products be in the x–y plane. Before the collision, there is no $\hat{y}$ component of momentum $\vec{p}$, and after the collision must be the same: therefore, $\hat{y} \cdot \vec{p}_a = mv_y$ and particle b will have $\hat{y} \cdot \vec{p}_b = -mv_y$. This will ease notation in that the initial velocity of particle a can be labeled v_a and the final velocities in the $\hat{x}$ direction can be written v'_a and v'_b. As a result, and after eliminating the mass term, which is the same for both particles, conservation of momentum and energy can be written as

$$v_a = v'_a + v'_b \tag{31.3}$$

$$v_a^2 = v_a'^2 + v_b'^2 + 2v_y^2 \tag{31.4}$$

for which solutions are

$$2v'_a = \sqrt{v_a^2 - 4v_y^2} - v_a \tag{31.5}$$

$$2v'_b = \sqrt{v_a^2 - 4v_y^2} + v_a \tag{31.6}$$

These equations will be treated in greater detail in Section 32.1; here, it is enough to observe that the primary electron shares its energy with the secondary (primed) electrons, and on average the energy is divided evenly amongst the primed electrons. Already that reduction in energy can be fatal to emission if the photon energy does not much exceed the work function. Even more than that, the initial state of the b electron has to be occupied, but the final states of the primed electrons must be *unoccupied*: if not, then the electrons have nowhere to go.

Now the crucial difference between metals and semiconductors can be understood. The b electron is typically near the Fermi level μ in a metal whereas the a electron has an energy typically near $\mu + \hbar\omega$: sharing their energy (if shared evenly) would leave the primed electrons with both their energies above μ in generally unoccupied states if $\hbar\omega > k_B T$, as dictated by the Fermi–Dirac distribution $f_{FD}(E)$ of Section 6.2. Compare that to semiconductors, for which the photon energy must exceed the band gap E_g to excite an electron from the valence band into the conduction band. Were that excited electron to scatter with another valence band electron, *both* of the electrons would have to have an energy greater than E_g or the transition is forbidden, and this does not happen until $\hbar\omega$ is sufficiently large, that is, in the words of Spicer and Herrera-Gomez [180], a "magic window" exists in which there is no electron–electron (*e–e*) scattering for semiconductors, as schematically suggested in Figure 31.3. Electron–phonon (*e–p*) scattering does not involve such a large energy transfer, and so the photoexcited electron keeps its energy longer and therefore has greater opportunity to be emitted.

When scattering of photoexcited electrons does occur in semiconductors, therefore, it is by way of *e–p* scattering, which bleeds the energy off the primary electron far slower than *e–e* scattering. Photoexcited electrons will therefore tend to travel further, and if they do scatter they may nevertheless retain sufficient energy to surmount the barrier if the semiconductor is a positive electron affinity (PEA) material. For NEA materials, for which the vacuum level is either always below the conduction band minimum ("true" NEA) or as a result of band bending when a field is present, even those electrons which have thermalized with the lattice nevertheless may be emitted, as shown in Figure 31.4.

[1]Bruce Springsteen, lyrics to *Born in the USA*, 1984.

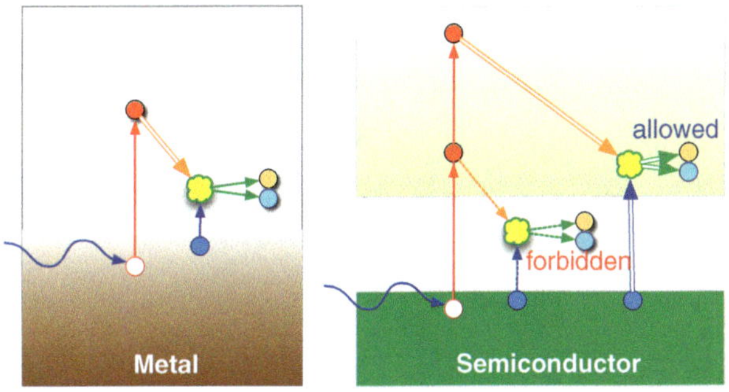

Figure 31.3 Left, e–e scattering in metals: photoexcited electron scatters with an electron at the Fermi level μ, with both final states above μ, so that the scattering is allowed. Right, e–e scattering in semiconductors: photoexcited electron must have sufficient energy so that when it scatters with a valence electron both final states are allowed (thick arrows); if the final states are in the band gap, the scattering is not allowed (thin dashed arrows). Thus, there is a "magic window" in energy where e–e scattering is forbidden, for which the quantum efficiency is larger.

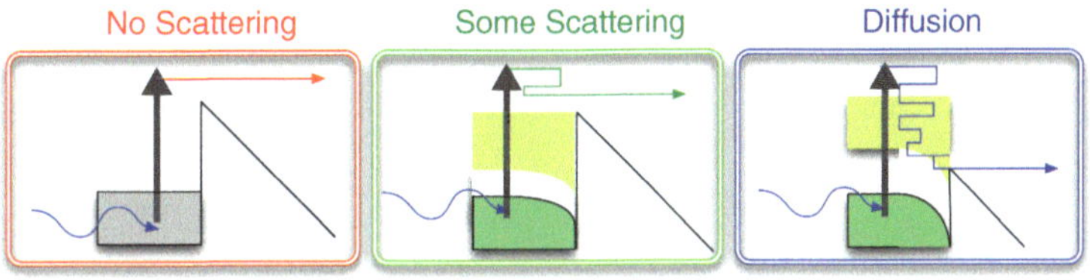

Figure 31.4 Three types of photocathode material characterized by differences in scattering. Left, metal-like: scattering is fatal to emission and primarily unscattered electrons surmount the work function barrier. Middle, semiconductor-like: a few scatterings do not impede emission, so that scattered and unscattered electrons can contribute. Right, drift-diffusion-like: electrons can experience many scattering events and thermalize to the bottom of the conduction band, yet still be emitted.

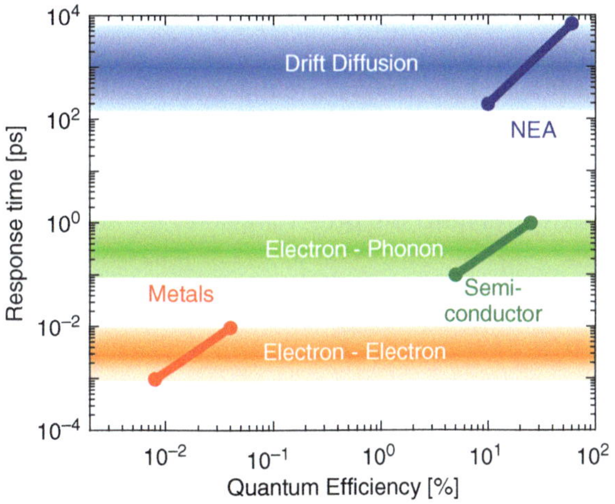

Figure 31.5 The three emission mechanisms of Figure 31.4 characterized in terms of quantum efficiency versus response time (based on Figure 14 in ref. [180]). Metals are fast responders but have low quantum efficiency; conversely, drift-diffusion semiconductors like GaAs have high quantum efficiency but long tails in their emitted pulses. The bands show the dominant scattering mechanism: e–e for metals, e–p for semiconductors, and drift-diffusion for NEA photocathodes.

Scattering, however, slows down electron transport. As a result, a difference in response time exists between the various photocathode material types, with metals being the fastest responders and drift-diffusion materials like GaAs having longer response times. The tracking of the quantum efficiency with the response time is indicated in Figure 31.5. Correlated with scattering by virtue of drive laser requirements, there is finally the question of laser power deposited on the photocathode and photocathode sensitivity to the environment. The relations are compactly shown in Figure 31.6, based on a survey of photocathode metrics and performance by Dowell *et al.* [523], where the vacuum tolerance of the various photocathodes (metal, antinomies, and cesiated GaAs) is shown in conjunction with their average current versus quantum efficiency regimes behind curves of constant current for various wavelengths.

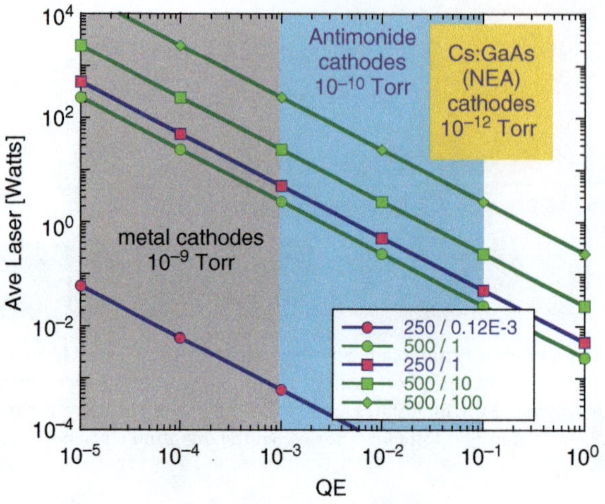

Figure 31.6 Average laser power versus quantum efficiency required to produce average currents. Number labels are "wavelength [nm]/current [mA]". Lines are of the form $I_e = (q\lambda/2\pi\hbar c)\,QE\,I_\lambda$. Ranges for the general cathode types, and their vacuum requirements, are also shown. Based on Figure 1 of ref. [523].

31.2 Basic theory

Much in the letters, nothing in the praise.

– William Shakespeare[2]

Finding the number of electrons to be emitted therefore requires knowing the probability that electrons with an energy E absorb a photon and get to an energy state where photoemission is allowed. The time-dependent Schrödinger's equation is therefore of the form

$$(i\hbar\partial_t - H)|\psi\rangle = 0 \tag{31.7}$$

where the Hamiltonian is the sum of $H = H_o + \delta H$, and where the solutions to

$$(i\hbar\partial_t - H_o)|\psi_o\rangle = 0 \tag{31.8}$$

are known. Introducing the retarded Green's function, $G(t - t')$, where

$$(i\hbar\partial_t - H_o)G(t - t') = \delta(t - t') \tag{31.9}$$

and letting $\theta(t - t')$ be the Heaviside step function (1 for positive argument, 0 for negative), then formally G is given by

$$G^{+}(t - t') = -i\theta(t - t')\exp[i(H_o/\hbar)(t - t')] \tag{31.10}$$

and so

$$|\psi^{+}\rangle = |\psi_o\rangle + \int_{-\infty}^{\infty} G^{+}(t - t')\delta H|\psi^{+}\rangle dt' \tag{31.11}$$

Letting $|\psi\rangle = |a\rangle e^{-i\omega_a t}$, where $\hbar\omega_a = E_a$ and

$$\begin{aligned} 0 &= (H - E_a)|a\rangle \\ 0 &= (H_o - E_a)|a_o\rangle \end{aligned} \tag{31.12}$$

and introducing a small parameter η to make the integrations behave properly, then Eq. (31.11) involves an integral of the form

$$-i\int_{-\infty}^{0} \exp\{i[(H_o - E_a)/\hbar - i\eta]t'\}dt' = \frac{e^{i\eta}}{(E_a - H_o)/\hbar + i\eta} \tag{31.13}$$

[2]Ref. [37]: *Love's Labor's Lost*, V.ii.40.

and so, letting $e^{i\eta} \to 1$, the Lippmann–Schwinger equation

$$|\psi\rangle = |\psi_o\rangle + \frac{1}{(E_a - H_o)/\hbar + i\eta}\delta H|\psi\rangle \equiv |\psi_o\rangle + G_o^+(E_a)\delta H|\psi\rangle \tag{31.14}$$

results. Transitions $W_{a,b}$ from state $|\psi_a\rangle$ to $|\psi_b\rangle$ are given, to leading order, by Fermi's golden rule

$$W_{a,b} = 2\pi\left|\langle\psi_b|\delta H|\psi_a\rangle\right|^2\delta(E_a - E_b) \tag{31.15}$$

Now, when $\delta H \propto \vec{p}\cdot\vec{A}$, where $\vec{A}$ is the radiation field of light [542, 543], then the yield, or number of electrons excited per photon, can be laboriously calculated once the Hamiltonian H_o and the basis states $|\psi_o\rangle$ are specified. But even if they are, transport past the surface barrier constitutes the next non-trivial difficulty. Those who have even made it this far can be excused for experiencing a cold sense of foreboding that an equation of the simplicity of the other canonical emission equations has seemed to pass well beyond reach. Many-body physics is *hard*, even as it is powerful, but if the desire is for a quick computational tool, yearning for a more phenomenological treatment is inevitable. There is hope. It is next.

31.3 Three-step model

Wan waves and wet winds labour,
Weak ships and spirits steer;
They drive adrift, and whither
They wot not who make thither...

– Algernon Charles Swinburne[3]

A phenomenological model of photoexcited electrons migrating in all directions and diminished in numbers due to scattering proved an efficacious description of the yield of photocathodes. The three-step model (TSM) [168, 179, 180, 544, 545] provides a way past the hardship of a proper model,[4] the utility of which continues unabated to describe experimental quantum efficiency and emittance [181, 546]. In the TSM, the photoemission process is separated into the three distinct processes of absorption, transport, and escape. Models for each of them are then developed, from which an estimate of photoemission current can be made. The power of the method and its heuristic appeal lie in the ready visualization of the components, and then in the development of yet more models to treat the parameters on which those components depend. Compared to how the TSM is developed by its progenitors, the derivation here is more improvised so as to introduce concepts before undertaking the moments model in Section 31.4. Five simplifications [179] allow for a tractable TSM:

- the photoexcited electrons are isotropically distributed
- the only scattering events treated are taken to be inelastic
- inelastic scattering is treated in the context of a mean free path that depends only on the photoexcited electron's energy
- the scattering is isotropic
- the normal energy of the photoexcited electron (the energy component into the barrier) must exceed the barrier height for escape, and when it does the probability of escape is unity.

Although such approximations entail conceptual difficulties [547], they allow the evaluation of yield to be separated into the three distinct processes:

1. the penetration into the bulk by photons followed by their *absorption*
2. the *transport* of electrons back to the surface with their numbers whittled away by scatterings with other electrons and phonons
3. the *escape* over (or through) the surface barrier into vacuum.

[3]Algernon Charles Swinburne, *The Garden of Proserpine*, in M.L. Rosenthal, *Poetry in English: An Anthology*. New York: Oxford University Press, 1987, p. 744.

[4]Berglund and Spicer were not so blasé: they spoke of "extreme difficulties ..." instead of "hardship.'.

This demarcation means that the physics of each process needs to be developed. For laser absorption, reflection at the surface and the depth to which the light penetrates mean the reflectivity of the surface and the dielectric constant of the material require examination. For electron current, the transport to the surface and emission into vacuum means scattering rates and energy loss as well as emission probability require examination. The examination of emission can be dismissed by assuming a step-function transmission probability in lieu of Chapter 18.

The fraction of photons absorbed is 1 minus the fraction reflected, or $(1 - R)$, where $R(\omega)$ is the reflectivity as a function of frequency (or wavelength). Of those photons absorbed, only electrons with a starting energy $E > \mu - \hbar\omega$ stand a chance of absorbing them, as the energy levels to which they aspire are otherwise filled as per the Fermi–Dirac distribution, for which the zero temperature limit is a good approximation. Quantum efficiency is then a ratio of the current of electrons that can pass the barrier with the current of all excited electrons. Restrictions on the ability to surmount the barrier introduces the *escape cone*: define θ_m as the maximum angle of a photoexcited electron for which the "normal energy" exceeds the barrier height $(E + \hbar\omega)\cos^2\theta_m = \mu + \phi$ or

$$\cos\theta_m(k) \equiv \left(\frac{\mu + \phi}{E(k) + \hbar\omega}\right)^{1/2} \tag{31.16}$$

Not all electrons make it to the surface, though. The ones that scatter can lose enough energy that they no longer are able to surmount the surface barrier, and so are lost. For metals, a single collision between two electrons can be enough for interfere with emission, and so the *fatal approximation* is often made,[5] whereby any collision thwarts emission [58, 85 179]. If electrons are photoexcited at a depth x into the bulk material and are then traveling at an angle θ with respect to the surface normal, then they must travel a distance $x/\cos\theta$, as in Figure 31.7. A fraction $f_\lambda(E, \cos\theta)$ will therefore survive the journey to the surface. To evaluate that fraction, the number of electrons that are generated at a depth x must be known. Let the optical penetration depth be $\delta(\omega)$. The probability of photoexcitation at a depth x is then proportional to $\exp(-x/\delta)$. The probability an excited electron will scatter and therefore be lost decreases exponentially with distance, with the length scale governed by the mean free path l (distance between scattering events on average) according to $\exp(-r/l)$ where r is the distance traveled (Section 32.2.1). With $x = r\cos\theta$,

$$f_\lambda(\cos\theta, p(E)) = \frac{\int_0^\infty \exp\left(-\frac{x}{\delta} - \frac{x}{l(E)\cos\theta}\right) dx}{\int_0^\infty \exp(-x\delta)\, dx} = \frac{\cos\theta}{\cos\theta + p(E)} \tag{31.17}$$

where

$$p(E) = \frac{\delta(\omega)}{l(E)} = \frac{m\delta(\omega)}{\hbar k(E)\tau(E)} \tag{31.18}$$

is the ratio between the laser penetration depth $\delta(\omega)$ and the mean free path $l(E) = \hbar k\tau/m$ as a function of the *total* electron energy E, and where x is the distance the electron must travel to freedom.

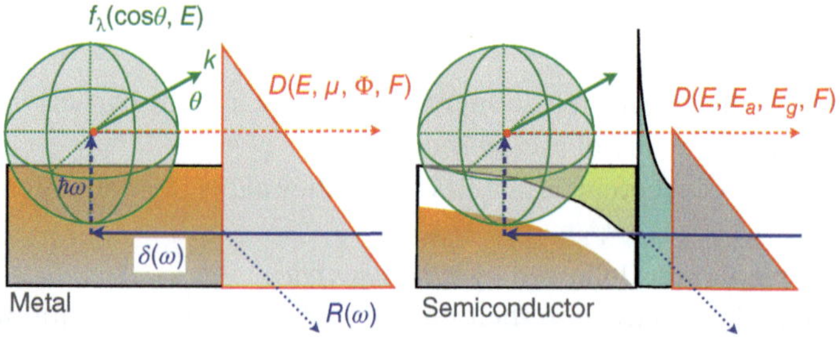

Figure 31.7 Components of the TSM for metals (left) and semiconductors (right): (*i*) absorption and reflection of light $R(\omega)$, (*ii*) tránsport and losses to the surface $f_\lambda(E, \cos\theta)$, and (*iii*) transmission into vacuum with the transmission probability $D(E_x)$ that depends on work function Φ (for metals) or electron affinity E_a (for semiconductors).

[5] But not always: Berglund and Spicer [179], for example, assess the effects of electrons that have scattered but once, and a phenomenological model of the contribution of scattered electrons will be considered in Section 31.7.

EXAMPLE: *Metal-like:* Evaluate $p(E)$ for an electron excited from the Fermi level $\mu = 7$ eV by a photon with wavelength 266 nm ($\hbar\omega = 5$ eV). Assume the laser penetration depth in copper is 12.6 nm. For a hot electron of $E = \mu + \hbar\omega \approx 12$ eV, assume the *e–e* relaxation time is 3.4 fs [85, 548].

SOLUTION: From the energy of the electron of $E = \mu + \hbar\omega = 7$ eV, therefore $k(E) = \sqrt{2mE}/\hbar = 17.747$ 1/nm. Therefore

$$p(E)|_{metal} = \frac{m\delta}{\hbar k(E)\tau} = \frac{(5.6856\ \mathrm{eV}/c^2)\,(12.6\ \mathrm{nm})}{(0.65821\ \mathrm{eV\text{-}fs})\,(17.747\ \mathrm{nm}^{-1})(3.4\ \mathrm{fs})}$$

or $p(E) = 1.804$.

EXAMPLE: *Semiconductor-like:* Evaluate $p(E)$ for an electron excited from the valence band for cesium antimonide (Cs_3Sb) for which [308]: $m_n/m = 0.1176$, band gap $E_g = 1.6$ eV, and laser penetration depth $\delta = 27$ nm and $\tau = 31$ fs.

SOLUTION: The energy of the electron is $E = \hbar\omega - E_g = 0.73053$ eV. Therefore, $k = 0.43788$ 1/nm, and so

$$p(E)|_{semi} = \frac{m\delta}{\hbar k(E)\tau} = \frac{(0.66862\ \mathrm{eV}/c^2)\,(27\ \mathrm{nm})}{(0.65821\ \mathrm{eV\text{-}fs})\,(4.3788\ \mathrm{nm}^{-1})\,(31\ \mathrm{fs})}$$

or $p(E) = 0.2021$.

Taking a moment to appreciate the differences between the Fowler–DuBridge model of Chapter 14 and the TSM is useful. In the modified Fowler–DuBridge model, the number of electrons that survive the odyssey to the surface, as in Eq. (14.6), is separated out in a manner that encapsulates the three steps of the emission process by $(1 - R) \rightarrow$ absorption, $F_\lambda \rightarrow$ transport, and $P(\hbar\omega) \rightarrow$ emission. Where F_λ acts as an overall multiplicative factor to the average probability of emission, the scattering factor f_λ depends on the energy of the scattered electron. A summation over the product of the energy-dependent loss factor $f_\lambda(\cos\theta, p(E))$ with the energy-dependent emission probability is therefore required if the emission probability is not a step function, which forebodes differences arising between the treatment of metals compared to semiconductors as a consequence of the emission barrier (think image charge/Schottky barrier compared to a triangular barrier) in interesting ways, as will be undertaken in the moments model.

31.4 Moments model

Knowledge, the known, and the knower
are the three things that motivate action:
instrument, action, and agent
are the three components of action.

– *Bhagavad Gita*[6]

The reflectivity and laser penetration depth (related to the optical factors) and the relaxation times (related to the *e–e* and *e–p* scattering rates) depend on the photon energy, the former directly, the latter because of how it affects the electron energy. Thus, the factor F_λ of Eq. (14.6) really should not be part of the energy integration. If the TSM is recast in the form of an integration over the possible energies of the emitted electrons and the language of discussion shifts over to a distribution function outlook, then the moments approach results: quantum efficiency QE is a ratio of currents, and that current is the first "moment" of the *emitted* distribution function as in Eq. (7.8), or

$$QE = \{1 - R(\omega)\}\,\frac{M_1(k_z)}{2M_1(k)|_{D=1, f_\lambda=1}} \tag{31.19}$$

where $D(E)$ refers to the transmission probability and f_λ to the scattering factor. The denominator accounts for the current of *all* electrons, so it includes those going in the $(+\hat{z})$ direction as well as the $(-\hat{z})$ direction, thereby giving a factor of 2×. Additionally, *all* excited electrons are considered regardless of what direction they are headed, thereby using k rather than

[6] Stephen Mitchell, *Bhagavad Gita: a new translation*. New York: Harmony Books, 2000, Verse 18.18, p. 185.

k_z in the argument of the moment function in the denominator. For metals, the energy $E(k)$ of the electron is constrained by the initial energy range and the availability of final states: in the zero-temperature limit of the Fermi–Dirac distributions that govern both,

$$f_{FD}(E)(1 - f_{FD}(E + \hbar\omega)) \to \Theta(\mu - E)\Theta(E - \mu + \hbar\omega) \tag{31.20}$$

where $\Theta(x)$ is the Heaviside step function (1 for positive argument, 0 for negative). For semiconductors, in contrast, the photoexcited electron must have an energy above the band gap, for which the energy is constrained by a factor of $\Theta(\hbar\omega + E - E_g)$. For metals then, absorption, transport, and emission combine to describe the current of all electrons making their way to the surface such that their normal energy component exceeds the barrier height and is evaluated by

$$QE \approx (1 - R)\frac{\int_{k_m}^{k_F} k^2 dk \int_0^{\theta_m} (\hbar k \cos\theta/m) f_\lambda(\cos\theta, p) 2\pi \sin\theta d\theta}{2\int_{\sqrt{k_F^2 - k_\omega^2}}^{k_F} k^2 dk \int_0^{\pi/2} (\hbar k/m) 2\pi \sin\theta d\theta} \tag{31.21}$$

where the lower integration limit in the denominator is zero if $\hbar\omega > \mu$ (a situation outside present interest), and where $\hbar^2 k_F^2/2m = \mu$, $\hbar^2 k_m^2/2m = \mu - (\hbar\omega - \phi)$, $\hbar^2 k_\omega^2/2m = \hbar\omega$, and the equation is approximate because scattered electrons are neglected. Relations involving $\theta_m(k)$ in Eq. (31.16) that will prove useful are

$$\sin^2\theta_m(k) = \frac{E(k) - \mu + \hbar\omega - \phi}{E(k) + \hbar\omega} \tag{31.22}$$

$$\sin^2\theta_m(k_F) = \frac{\hbar\omega - \phi}{\mu + \hbar\omega} \tag{31.23}$$

where the second, evaluated for energy at the Fermi level $E(k_F) = \mu$, will be particularly useful, as it is often a small parameter for metals.

31.4.1 Metals

In the case of all things which have several parts and in which the totality is not, as it were, a mere heap, but the whole is something besides the parts…

– Aristotle[7]

Because the small θ region dominates the integrals, the approximation $f_\lambda(\cos\theta, p) \approx \cos\theta/(1 + p)$ is reasonable and convenient given the size of p for metals: with a partially filled conduction band, many final states for a photoexcited electron to scatter into, and the sheer number of roaming electrons that a metal contains with which the photoexcited electron may scatter, the mean free path is small (Section 32.3.3). The approximation results in

$$QE \approx \frac{2(1 - R)}{3(1 + p)k_\omega^2(2k_F^2 - k_\omega^2)} \int_{k_m}^{k_F} k^3 [1 - \cos^3\theta_m(k)]\, dk \tag{31.24}$$

Making the substitution $\cos\theta_m(k) = k_o/\sqrt{k^2 + k_\omega^2}$ results in an integral that can be evaluated analytically to give

$$QE = \frac{2(1 - R)}{3(1 + p)k_\omega^2(2k_F^2 - k_\omega^2)} \times \tag{31.25}$$

$$\left\{ k_F^4 - k_m^4 + 3k_o^2(k_o^2 + 2k_\omega^2) - \frac{4k_o^3(k_F^2 + 2k_\omega^2)}{\sqrt{k_F^2 + k_\omega^2}} \right\} \tag{31.26}$$

where $\hbar^2 k_o^2/2m = \mu + \phi$. When $(\hbar\omega - \phi)/(\mu + \phi)$ is small, it is possible to show[8]

$$QE = \frac{(1 - R)}{(1 + p)} \frac{\mu(\hbar\omega - \phi)^2}{4\hbar\omega(\mu + \hbar\omega)(2\mu - \hbar\omega)} \tag{31.27}$$

[7] Aristotle, *Metaphysica Book VIII* (trans. W.D. Ross), *Great Books of the Western World*, ed. Robert Maynard Hutchins, Vol. 8, Aristotle I. Chicago: Encyclopaedia Britannica, 1952, p. 569.

[8] Showing it using Eq. (31.26) takes work. See the discussion around Eqs (31.99) to (31.103) for an alternate way to the same destination.

Such an equation should give an approximate account of the quantum efficiency of metals up to an overall multiplicative factor C taking into account multiple crystal faces, adsorbates on the surface, effective mass variations, and other factors which may cause only a fraction of the surface to emit.

EXAMPLE: Find the factor C given that $QE = 2.049 \times 10^{-4}$ at $\hbar\omega = 5$ eV based on Figure 3.6 in Section 3.3 and using the following parameters and assumptions: $\mu = 7$ eV, $\Phi = 4.615$ eV, and $F = 50$ eV/μm. Take the reflectivity $R = 0.4$, which is typical for copper at a photon energy of 5.0 eV and treat it as constant (even though it is not). Take the relaxation time in copper to be $\tau = 3.4$ fs (even though it is energy dependent), and take the laser penetration depth to be $\delta = 12.6$ nm.
SOLUTION: First, $k = \sqrt{2m(\mu + \phi)}/\hbar = 17.257$ [1/nm]. Second, the penetration depth to mean free path ratio $p = m\delta/\hbar k\tau = 1.8549$. As a result of the field, $\phi = \Phi - \sqrt{4QF} = 4.3468$ eV. The ratio between the experimental QE and Eq. (31.27) is therefore $C = 0.70504$.

Using Eq. (31.27), the data of Figure 3.6 can be re-examined using the value of $C = 0.67383$ determined from the previous example. The comparison is shown in Figure 31.8, along with the Fowler–DuBridge hypothesis that $QE(\lambda) \propto (\hbar\omega - \phi)^2$ (dashed gray line). The better agreement over the Fowler–DuBridge relation is evident, even though the relaxation time is a function of the photoexcited electron energy $\tau(E + \hbar\omega)$ and the reflectivity is a function of frequency $R(\omega)$ and therefore both are affected by $\hbar\omega$.

As a final observation, the moments-based approach cleaves to the k-based approach of calculating current density, whereas an account based on Spicer's TSM [179] works with energy, where $E = \hbar^2k^2/2m$ is often assumed (the application of the TSM to the present copper data analysis is encapsulated very well in Dowell *et al.* [58]; the energy dependence of $\tau(E)$ will be revisited in Sections 32.3.3 and 32.3.4, the former for *e–e* scattering, as in Figure 32.15, being particularly important for metals). As a result, although the equations for QE for each approach differ in appearance, they are kindred methods.

31.4.2 Semiconductors

In desperation I asked Fermi whether he was not impressed by the agreement between our calculated numbers and his measured numbers. He replied, "How many arbitrary parameters did you use for your calculations?" I thought for a moment about our cut-off procedures and said, "Four." He said, "I remember my friend Johnny von Neumann used to say, with four parameters I can fit an elephant, and with five I can make him wiggle his trunk."

– Freeman Dyson [549]

Semiconductors add complication to the analysis. There are all the considerations of Chapter 22, but in addition, first, the barrier at the surface has an increased triangular shape that affects the transmission probability ($D(E) \rightarrow D_{\triangle}(E)$),

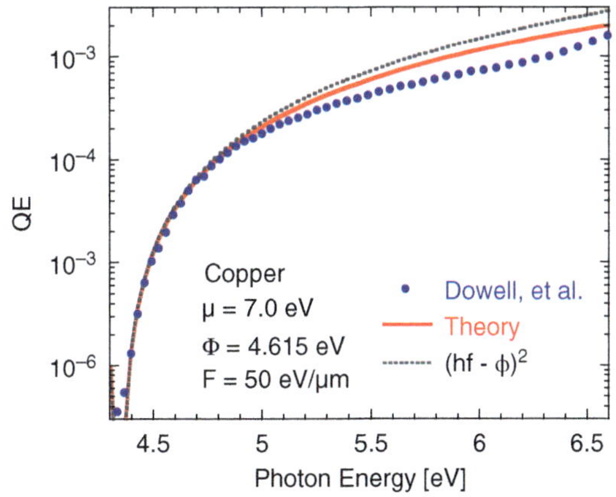

Figure 31.8 Experimental data for the quantum efficiency data of copper (courtesy of David Dowell (SLAC); see also ref. [58]) compared to Eq. (31.27) scaled by $C = 0.70504$. The dashed gray line corresponds to $QE = QE_o[(\hbar\omega - \phi)/(\hbar\omega_o - \phi)]^2$ where QE_o is obtained by setting $\hbar\omega \rightarrow \phi$ in Eq. (31.27) except for the $(\hbar\omega - \phi)^2$ factor.

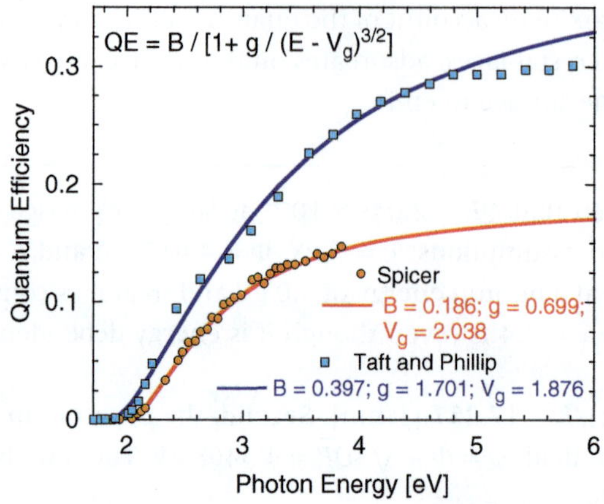

Figure 31.9 The quantum efficiency of Cs_3Sb as measured by Spicer [550] (red dots and line) and by Taft and Phillip [551] (blue squares and line). The fit parameters were based on data digitally extracted from Figure 1 of ref. [308].

and, second, for the "magic window" of semiconductors, scattering effects are not nearly so fatal as for metals. Most importantly, the behavior of QE as a function of photon energy $\hbar\omega$ takes on a different appearance than the quadratic behavior exhibited by metals in Eq. (31.27), as shown by the measurements of Spicer [550] and Taft and Philipp [551] in Figure 31.9 compared to the fit suggested by Spicer to be[9]

$$QE(\hbar\omega) \approx \frac{B}{1 + \dfrac{g}{(\hbar\omega - V_g)^{\nu}}} \tag{31.28}$$

where V_g should be the sum of the band gap plus the electron affinity $E_g + E_a$, and where Spicer found $\nu \approx 3/2$ (not to be confused with the ν of field emission). When $\hbar\omega$ is rather close to V_g, then $QE \propto (\hbar\omega - V_g)^{\nu}$, rather reminiscent of the QE of metals. Here, the parameters B and g are being treated as adjustable, and V_g and ν are perilously close to being treated likewise. Why these parameters are what they are takes some inquiry to constrain just how far they may be adjusted, as well as to what they likely refer.

Return, therefore, to the integral from which QE is calculated, but to avoid having to discuss the value of m_n, the substitution $k = \sqrt{2m_n E}/\hbar$ transforms the integral to one over energy. For semiconductors then,

$$QE = (1 - R(\omega))\frac{\int_{E_a}^{E_m} EdE \int_{x_m}^{1} xdx D_{\Delta}[Ex^2] f_{\lambda}(x, p(E))}{2\int_0^{E_m} EdE \int_0^1 dx} \tag{31.29}$$

where the upper limit of the energy integration is $E_m \equiv \hbar\omega - E_g$ and the lower limit of the $x = \cos\theta$ integration is $x_m = \sqrt{E_a/E}$. Additionally, E is the energy of the photoexcited electron, $x = \cos\theta$, and although some electrons can tunnel through the triangular barrier, the contribution of electrons below the barrier height (characterized by the electron affinity E_a) is negligible. Observe another subtlety: for the metal treatment in Eq. (31.24), $k(E)$ referred to the electron *prior* to the absorption of $\hbar\omega$, whereas now E in Eq. (31.29) refers to the energy *after* absorption, a convenience due to energy being measured with respect to the bottom of the conduction band. Either convention is acceptable as long as the arguments of $p(E)$, $\tau(E)$, and $D(E)$ take into account the photon. The energy of the photoexcited electron is constrained by the difference between the photon energy $\hbar\omega$ and the band gap E_g on the upper limit of integration, and the barrier height E_a on the lower limit.

Suppose that the transmission probability $D_{\Delta}(Ex^2)$ could be treated as a step function, as in $\Theta(Ex^2 - E_a)$, as for metals in Section 31.4.1. The integration over the scattering term f_{λ} in the case of the metal treatment made use of the approximation $f_{\lambda} \approx \cos\theta/(1 + p)$ in Eq. (31.29), but how good is that approximation? From either Eq. (31.21) or Eq. (31.29), the approximation affects the evaluation of

$$F_{\lambda}(x_m, p) \equiv \int_{x_m}^{1} x f_{\lambda}(x, p(E))\, dx \tag{31.30}$$

[9] Although this equation bears scant similarity to Eq. (6) of Spicer [550], the two equations are in fact identical for the approximations Spicer uses.

where $x_m = \sqrt{E_a/E}$ for semiconductors or $\cos\theta_m$ for metals, and the notation of F_λ is meant to evoke fond memories of the modified Fowler–DuBridge equation of Eq. (14.5). The approximation of f_λ used in Section 31.4.1 for metals results in

$$F_\lambda(x_m, p) \approx \int_{x_m}^{1} \frac{x^2}{1+p} dx = \frac{1 - x_m^3}{1+p} \tag{31.31}$$

However, the integration is, in fact, analytic, and gives [308]

$$F_\lambda(x_m, p) = \frac{1}{2}(1 - x_m)(1 + x_m - 2p) - p^2 \ln\left(\frac{x_m + p}{1+p}\right) \tag{31.32}$$

In the limit that $x_m \to 1$, and in terms of the small quantity $\epsilon = (1 - x_m)/(1 + p)$, both approach the same limiting value (the approximate equation behaves as $F_\lambda \approx \epsilon - (1 + p)\epsilon^2$, whereas the exact equation behaves as $F_\lambda \approx \epsilon - (p + (1/2))\epsilon^2$ for small ϵ) and therefore the approximate form of F_λ is adequate. However, when the wavelength becomes small, the differences between the approximate and analytic solutions become more pronounced: in the short wavelength limit $x_m \to 0$, the approximate solution approaches $1/[3(1 + p)]$ whereas the analytic solution approaches

$$F_\lambda(x_m = 0, p) = \frac{1}{2}(1 - 2p) - p^2 \ln\left(\frac{p}{1+p}\right) \tag{31.33}$$

or $(1/2) - p$ when $p \ll 1$ and $1/3p$ when $p \gg 1$. Thus, for the metal-like values where p is large, the approximation is adequate, but for semiconductor parameters where p is small, a noticeable difference accrues, as in Figure 31.10.

There are two problems with using $F_\lambda(p)$ and settling for a modified Fowler–DuBridge equation. The first is that, as a function of energy, $F_\lambda(p)$ is not very friendly and makes for a more difficult integration in Eq. (31.29) so that an analytical solution comparable to that of Eq. (31.27) for metals is going to give way to numerical evaluations. The second problem is that the transmission probability for a triangular barrier *is not a step function*, although the fields associated with photoemission are often such that Eq. (18.46) is a useful limiting approximation, for which

$$D_\triangle(E) \approx \frac{4[E(E - E_a)]^{1/2}}{(E^{1/2} + (E - E_a)^{1/2})^2} \tag{31.34}$$

so that the modified Fowler–DuBridge model of Eq. (14.5) is at best only qualitatively correct for a triangular barrier.

The impact of $D_\triangle(E)$ can be accounted for by using constant but reasonable values for $R(\omega)$ (set to 0.2, although it varies between 0.1 and 0.35 depending on frequency) and p (as labeled) with the band gap E_g and the electron affinity E_a modeled after Cs_3Sb parameters [308]. The consequences are shown in Figure 31.11, showing reasonable agreement, and so V_g and v can be circumscribed by considering the small ($s^2 \equiv (\hbar\omega - E_g - E_a)/E_a \ll 1$) behavior.

To develop an analytic equation with the simplicity of Spicer's Eq. (31.28) is frustrated by the complexity of the integrand, but given that the integrand vanishes at the lower bounds of both integration regions, but rises quickly near the lower bound ($E \approx E_a$) only to become progressively more linear near the upper bound ($E \approx E_m$) as s increases, means that

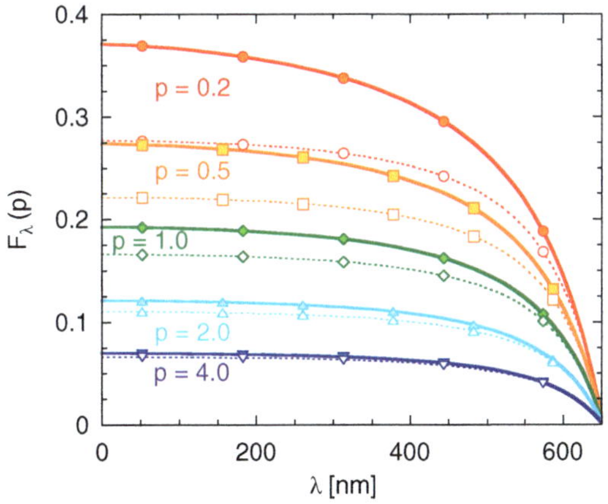

Figure 31.10 Behavior of F_λ as a function of wavelength $\lambda = 2\pi\hbar c/(E + E_g)$ and $x_m = \sqrt{E_a/E}$ for various values of p, as defined by Eq. (31.33). For metal-like (large p) conditions, the approximate form holds, but for semiconductor-like (small p), the approximation degrades for smaller wavelengths.

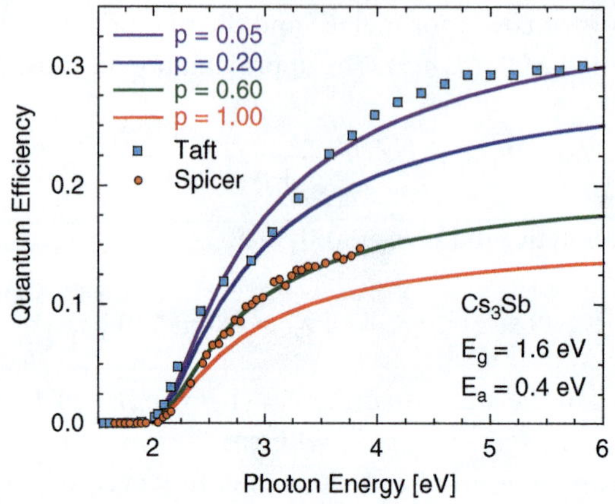

Figure 31.11 A comparison of Eq. (31.29) using constant values of $R(\omega) = 0.2$ and the constant values of p as indicated for the assumed values of $E_g = 1.6$ eV, $E_a = 0.4$ eV (fashioned after Cs_3Sb parameters in ref. [308]). In addition, the data of Figure 31.9 for Taft and Philipp [551] and Spicer [550] are shown.

the integration overall is bounded from below by the trapezoidal approximation of Eq. (A1.9) and from above by twice that approximation (put another way, for a given E and in terms of $s = (x - x_m)/(1 - x_m)$ with $x_m = \sqrt{E_a/E}$, the integrand for the x integration lies above the line $y = s$ and below the line $y = 1$, and the area under the second curve is simply twice the first). Using the trapezoidal approximation, therefore,

$$\int_{x_m}^{1} xD(Ex^2)\frac{x}{x+p}dx \approx \frac{\sqrt{E}-\sqrt{E_a}}{2(1+p)\sqrt{E}}D(E) \tag{31.35}$$

and so using the trapezoidal approximation on the resulting E integration,

$$QE \approx \frac{1-R}{2(1+p)E_m^{3/2}}(E_m - E_a)(\sqrt{E_m} - \sqrt{E_a})D(E_m) \tag{31.36}$$

or, in terms of $s^2 = (\hbar\omega - E_g - E_a)/E_a$, the lower limit is

$$QE > \frac{2(1-R)s^5}{(1+p)(1+s^2)(1+\sqrt{1+s^2})(s+\sqrt{1+s^2})} \tag{31.37}$$

and the upper limit is twice that, as shown in Figure 31.12 In both cases, the dependence on s in the numerator shows that $V_g = E_g + E_a$ as intuited, but that v starts off as 5/2 and approaches unity as s becomes large: the behavior in the intermediate regime therefore hovers about $v \approx 3/2$, as Spicer argued (corresponding to the s^3 line of Figure 31.12).

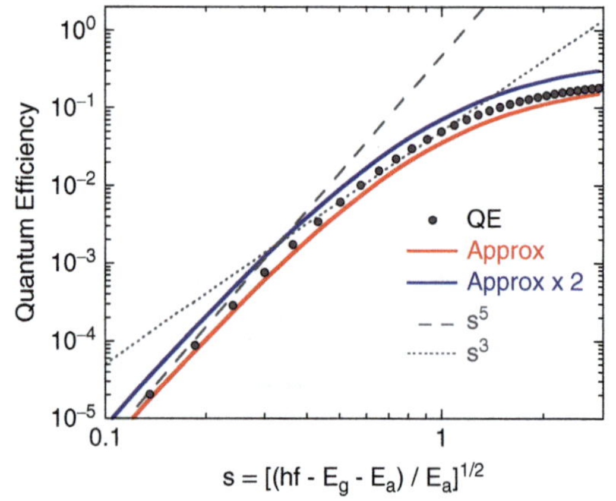

Figure 31.12 A comparison of quantum efficiency numerically evaluated using Eq. (31.29) compared to the lower bound estimate of Eq. (31.37) and twice the lower bound for parameters fashioned after Cs_3Sb ($E_g = 1.6$ eV, $E_a = 0.4$ eV). Gray dashed lines correspond to $QE \propto s^5$ and s^3, the later in keeping with Spicer's finding.

Two of the adjustable parameters ($\nu \sim 3/2$ and $V_g = E_a + E_g$) have now been locked down, but to be usefully predictive equations such as Eq. (31.29) (as well as Eq. (31.27) for that matter) have to circumscribe the other two parameters R and p. Both require knowing the optical properties of the material, and p requires knowing the scattering rates of electrons with other electrons and with the bulk lattice. That will take some work. The optical properties will be considered forthwith. The treatment of scattering is deferred to Chapter 32, in particular the two important cases of electron–electron scattering in Section 32.3.3 and electron–phonon (acoustic) in Section 32.3.4 after Monte Carlo methods motivate their treatment.

31.5 Reflectivity and penetration factors

Obviously then it would be better to assume a finite number of principles. They should, in fact, be as few as possible, consistently with proving what has to be proved.

– Aristotle[10]

Models are needed for reflectivity $R(\omega)$ and the laser penetration depth $\delta(\omega)$ that appears in $p = \delta(\omega)/l(E)$ of $f_\lambda(\cos\theta, E)$ in Eq. (31.18) (the relaxation time $\tau(E)$ that is also a part of p is deferred to Section 32.3). These factors govern how materials and light interact, and can be briefly considered, recognizing that extended treatments [52, 53, 85, 552] and supplementary accounts [55, 80, 553, 554–555] are available. Maxwell's equations for the electromagnetic fields concern the vector relationships between the electric $\vec{\mathcal{E}}$, magnetic $\vec{H}$ fields, and current $\vec{J}$ with regards to their spatial and temporal behavior. Of the four that are normally experienced on first encounters [99, 100, 556, 557], the two that matter here are[11]

$$\vec{\nabla} \times \vec{\mathcal{E}} + \mu_o \partial_t \vec{H} = 0 \tag{31.38}$$

$$\vec{\nabla} \times \vec{H} - \varepsilon_o \partial_t \vec{\mathcal{E}} = \vec{J} \tag{31.39}$$

where $\varepsilon_0 = 0.0552635$ q^2/eV-nm is the permittivity of free space, and $\mu_0 = 1/\varepsilon_0 c^2 = 0.000201335$ eV fs^2/nm q^2 is the permeability of free space in the [nefq] units of Tables 2.1 and 2.2. Consider, then, the quantity $\vec{\nabla} \times (\vec{\nabla} \times \vec{\mathcal{E}}) = \vec{\nabla}(\vec{\nabla} \cdot \vec{\mathcal{E}}) - \nabla^2 \vec{\mathcal{E}}$ and let there be no free charges loitering about so that $\vec{\nabla} \cdot \vec{\mathcal{E}} = 0$ from Poisson's equation. Then the two equations for $\vec{\mathcal{E}}$ and $\vec{H}$ entail[12]

$$\nabla^2 - \frac{1}{c^2}\partial_t^2 \vec{\mathcal{E}} = \frac{16\pi Q}{q^2 c^2}\sigma \partial_t \tilde{\mathcal{E}} \tag{31.40}$$

where Eq. (3.8) has been used to relate current to electric field by reintroducing conductivity σ. For electromagnetic waves characterized by $\vec{\mathcal{E}} = \vec{\mathcal{E}}_o \exp(i\vec{\kappa} \cdot \vec{r} - i\omega t)$, where $\vec{\kappa}$ is the propagation constant (κ is used so as not to confuse matters with k when it is taken to be a damping constant) and ω is the frequency, then

$$-\kappa^2 + \left(\frac{\omega}{c}\right)^2 = -i\frac{16\pi Q\sigma\omega}{q^2 c^2} \rightarrow \kappa = \left(1 + i\frac{16\pi Q\sigma}{\omega q^2}\right)^{1/2}\left(\frac{\omega}{c}\right) \tag{31.41}$$

The coefficient of ω/c is termed the (complex) *index of refraction*, and to emphasize that it is in fact a complex quantity, a small hat as in $\hat{n}$ usually attends the symbol for it, so that

$$\hat{n} = \left(1 + i\frac{16\pi Q\sigma}{\omega q^2}\right)^{1/2} \equiv n - ik \tag{31.42}$$

where n is the usual index of refraction, and k is called the *damping constant*. The presence of ω indicates that both n and k depend on the frequency of incident light. The term σ will fade from view for a while, but returns when the distribution function approach to finding dielectric parameters is undertaken.

Let $\vec{\mathcal{E}} = \mathcal{E}\hat{x}$ so that the magnitude of the force on an electron is $F = q\mathcal{E}$ and that unpleasant notational complication with $\vec{\kappa} \cdot \vec{r}$ can be replaced by κx. When the electron is accelerated over a distance l by a force F, it acquires an energy $\Delta E = Fl$. If

---

[10] Aristotle, *On The Heavens, Book III* (trans. J.L. Stocks), *Great Books of the Western World*, ed. Robert Maynard Hutchins, Vol. 8, Aristotle I. Chicago: Encyclopaedia Britannica, 1952, p. 394.

[11] The electric field notation $\vec{\mathcal{E}}$ reinforces that *light*, not electrons, is under discussion.

[12] The notation for permittivity and permeability are being jettisoned, using $\varepsilon_0 = q^2/16\pi Q$ to enable easy computations using familiar terms with transparent units.

$l = vdt$ where the electron velocity is v, then the rate of change of the energy (a power) is $dE/dt = Fv$. Transferring those ideas to a number density ρ, then the power density transferring over to the electron gas from the wave, or, equivalently, the power that is being extracted from the wave, goes as $\rho Fv = J\mathcal{E} = \sigma\mathcal{E}^2$ (remember that, apart from sign, $J = qv\rho$). The intensity of light (measured in W/cm^2) goes as $|\mathcal{E}|^2$, and its empirical decline as light passes through thin films of material serves to characterize the decay parameters as a function of wavelength.

Because of the imaginary part of $\hat{n}$, the electric field is decaying as the wave propagates into the bulk material. If $y(x)$ is an exponentially decaying function, then the length scale of the exponential decay is $-[\partial_x \ln(y(x))]^{-1}$ (e.g., trying that on $y(x) = y_o e^{-x/a}$ returns a, as it should). The length scale $\delta(\omega)$ governs the exponential decay of the power density or intensity into the bulk material as a function of distance x. Intensity is proportional to $|\mathcal{E}(x)|^2$ so that

$$\delta(\omega) = \{-\partial_x \ln[|\mathcal{E}(z)|^2]\}^{-1} = \frac{c}{2k\omega} \tag{31.43}$$

Experimentally, to measure the absorption coefficient is to measure the transmission through varying thicknesses of material, in which case T goes as [558]

$$T = (1 - R)^2 e^{-d/\delta}/(1 - R^2 e^{-2d/\delta}) \tag{31.44}$$

where R is the reflectivity. In the thick ($d \gg \delta$) limit, the attenuation found above in Eq. (31.43) is obtained.

So what of reflection, then? If the electric field is $\vec{\mathcal{E}} = \hat{x}\mathcal{E}_o \exp(i\kappa x - i\omega t)$, then the associated magnetic field $\vec{H} = \hat{y}H_o \exp(i\kappa x - i\omega t)$ from Eq. (31.38), where $H_o = (c\kappa/\omega)\mathcal{E}_o$. These incident waves are crossing a boundary with transmitted and reflected waves, making for a problem extraordinarily reminiscent of the tunneling problems of Section 18.1. In fact, if $\mathcal{E}$ with the subscripts i, r and t refer to the incident, reflected, and transmitted waves, respectively, then the matching of the waves and their first derivatives at the boundary $x = 0$ results in a matrix equation of the form (where $\kappa_i = \kappa_r$ because both are in the same medium, for example vacuum, and $\hat{\kappa}_t$ is the complex parameter in the bulk material)

$$\begin{pmatrix} 1 & -1 \\ \kappa_i & \kappa_i \end{pmatrix}\begin{pmatrix} \mathcal{E}_o^i \\ \mathcal{E}_o^r \end{pmatrix} = \begin{pmatrix} 1 & -1 \\ \hat{\kappa}_t & \hat{\kappa}_t \end{pmatrix}\begin{pmatrix} \mathcal{E}_o^t \\ 0 \end{pmatrix} \tag{31.45}$$

where the little hats on κ (i.e., $\hat{\kappa}$) denote a complex quantity, and where the top row of the matrix equation refers to matching the waves, and the bottom to matching their derivatives, at $x = 0$. But after experiencing the evaluation of transmission probabilities using the transfer matrix approach of Section 19.1, the present task is straightforward. Solving for $\mathcal{E}_o^r$ in terms of $\mathcal{E}_o^i$ results in

$$E_o^r = \frac{\hat{\kappa}_t - \kappa_i}{\hat{\kappa}_t + \kappa_i} E_o^i \tag{31.46}$$

Defining the reflectivity R as the ratio of the absolute square of the magnitudes of the reflected with the incident wave results in

$$R(\omega) = \frac{|\hat{\kappa}_t - \kappa_i|^2}{|\hat{\kappa}_t + \kappa_i|^2} = \frac{|\hat{n} - 1|^2}{|\hat{n} + 1|^2} = \frac{(n-1)^2 - k^2}{(n+1)^2 - k^2} \tag{31.47}$$

Values of n and k can be empirically found and tabulated, as in refs [55] or [309], but simple models shed light as to what causes them to take the forms they do. Familiarity with the Drude model of conductivity in Section 3.2 and with the dielectric constant in Section 22.6 will help.

31.6 Lorentz–Drude model of the dielectric constant

A theory based on the frequency-dependent dielectric constant in the random phase approximation ... predicts and the experiments confirm essentially free electron-like behavior before the onset of d-band excitations and a plasma frequency modified from that of free electrons due to oscillator strength coupling between valence and d bands and d-band screening effects.

– Herbert R. Philipp and Henry Ehrenreich [553]

The classical theory of absorption and dispersion is due mainly to Lorentz and Drude. The Lorentz model is applicable to insulators... The Drude model is applicable to free-electron metals... Both the Lorentz and Drude models are largely ad hoc, but still useful as starting points and for developing a feeling for optical properties. We shall see that many features of these classical models have quantum mechanical counterparts which are easily understood as generalizations of their classical analogs.

– Frederick Wooten [552]

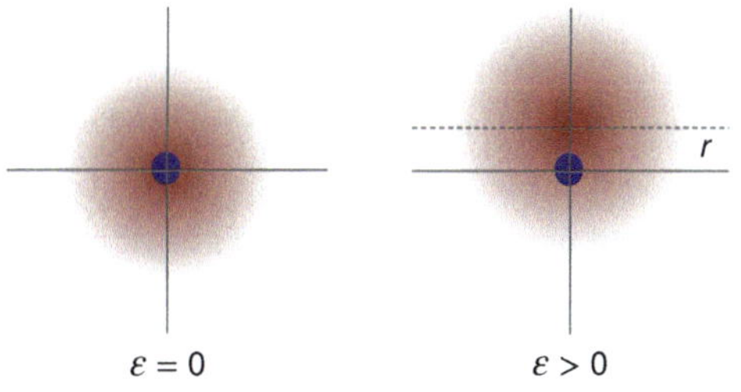

Figure 31.13 The application of $\vec{\mathcal{E}}$ pulls the distribution of negative charge (red fuzzy circle) away from the center of positive charge (small blue circle) in atoms or molecules. The dipoles that form give rise to the polarization vector $\vec{P}$, which is proportional to $\vec{\mathcal{E}}$.

Conductors shield out external fields quite effectively, as roaming electrons in them pile up near the surface so as to completely block external fields from penetrating into the interior (or, for that matter, aggregating around charged inclusions as in Section 22.6). For semiconductors or insulators, however, the electrons may not have the same freedom: the atoms from which they would have otherwise escaped retain a hold on them. In a simple visualization, such atoms in an electric field thereby become *polarized* whereby the center of positive charge is off-set from the center of negative charge and so a dipole is formed, as in Figure 31.13. The dipole moment $\vec{p}$ is proportional to the electric field $\vec{\mathcal{E}}$. Many such little dipoles conspire to produce a macroscopic $\vec{P}$, and so for static conditions[13]

$$\vec{P} = \varepsilon_0 \chi \vec{\mathcal{E}} \tag{31.48}$$

When dielectrics are present, therefore, Maxwell's equations are retained in form by introducing $\vec{D}$ to take over for $\vec{\mathcal{E}}$, where (in conventional notation)

$$\vec{D} = \varepsilon_0 \vec{\mathcal{E}} + \vec{P} = \varepsilon_0 (1 + \chi) \vec{\mathcal{E}} \equiv \varepsilon \vec{\mathcal{E}} \tag{31.49}$$

for which $\vec{\nabla} \cdot \vec{D} = q^2 \rho$ with ρ as the number density of *free* (not bound) charge. When conditions are not static, then the separation of charge that makes the little dipoles and gives rise to $\vec{P}(t)$ is a process that lags behind $\vec{\mathcal{E}}$, and so χ takes over the role of a *response function* whereby

$$\vec{P}(t) = \varepsilon_0 \int_{-\infty}^{t} \chi(t - t') \vec{\mathcal{E}}(t') dt' \tag{31.50}$$

where $\chi(t)$ is a real function. Using the convolution theorem (as done in going from Eq. (22.33) to Eq. (22.35)) gives

$$\vec{P}(\omega) = \varepsilon_0 \hat{\chi}(\omega) \vec{\mathcal{E}}(\omega) \tag{31.51}$$

where the hat notation on $\hat{\chi}$ indicates that it is complex. The relationship between the complex index of refraction and susceptibility is then

$$\hat{n}(\omega)^2 = 1 + \hat{\chi}(\omega) \tag{31.52}$$

Finding $\hat{\chi}(\omega)$ is involved but two kinds of arguments, the first treating the electron as a damped harmonic oscillator and the second using a distribution function approach (a quasi-classical Boltzmann transport equation), are inordinately useful. Such simple models are often used to understand the frequency dependence of absorption and reflection for metals, but also illuminate the same dependence in semiconductors [553]. The latter, though, require a bit more caution (especially for good photoemitters like GaAs) because the carrier density of electrons is less, the semiconductors have a band gap, and the relaxation time is no longer constant as for low frequencies but becomes frequency dependent when the photon energies are sufficiently large (around infrared). Frequency dependence in turn changes the dependence of the absorption coefficient from λ^2 as in the Drude model to λ^3 or λ^4 in a quantum mechanical theory [558–560]. The aim, however, is not to derive that dependence, but rather to show why the reflectivity $R(\omega)$ and the penetration depth $\delta(\omega)$ are frequency dependent in the first place by finding out their relation to $\hat{\chi}(\omega)$ via Eq. (31.52).

[13] There is more to the microscopic to macroscopic transition than just this [99, 556], but the description is adequate for simple models. Still, if the dipole is proportional to the field, as in $\vec{p} = \hat{\alpha}\vec{\mathcal{E}}$, then the macroscopic polarization $\vec{P} = N\hat{\alpha}\vec{\mathcal{E}}$, where $\hat{\alpha}(\omega)$ is the polarizability for a one-electron atom. For those who crave more, see Wootan [552].

31.6.1 Harmonic model

Non-violent resistance to evil does not mean absence of any resistance whatsoever but it means not to resist evil with evil but with good.
– Mahatma Gandhi[14]

Resistance is a fundamental feature of bound particles, manifested as a dampening to forced motion. Consider a simple model first. If the position of a bound electron to the dipoles of Figure 31.13 with a mass m oscillates in response to the time-varying electric field and experiences dampening while doing so, then its acceleration can be related to the forces acting on it by [561]

$$-\vec{F} = m\partial_t^2\vec{r}(t) + \frac{m}{\tau}\partial_t\vec{r}(t) + m\omega_o^2\vec{r}(t) \tag{31.53}$$

where the left-hand side is the force $F = q\vec{\mathcal{E}}$ due to the electric field, while the factors on the right-hand side are the acceleration, the dissipation (or dampening) term, and a restoring force, respectively. Substituting the Fourier transform of $\vec{r}(t)$ into Eq. (31.53) results in

$$\vec{r}(\omega) = \frac{\vec{F}}{m}\left(\omega^2 - \omega_o^2 - i\frac{\omega}{\tau}\right)^{-1} \tag{31.54}$$

where the relaxation time τ is a constant [311] (an assumption of the Drude model). Such an equation is for only one bound electron, but there are many of them doing much the same thing. If the number of electrons per unit volume is ρ_o, then all the little dipoles acting in concert result in

$$\chi(\omega) = \frac{16\pi Q\rho_o}{m}\left(\omega^2 - \omega_o^2 - i\frac{\omega}{\tau}\right)^{-1} \equiv \frac{\omega_p^2}{(\omega^2 - \omega_o^2 - i\frac{\omega}{\tau})} \tag{31.55}$$

where the plasma frequency $\omega_p^2 = 16\pi Q\rho_o/m = q^2\rho_o/\varepsilon_0 m$ has been introduced. Lastly, Eq. (31.55) suggests that the static dielectric constant $\varepsilon(0)$ is given by

$$\varepsilon(0) = \varepsilon_0(1 + \chi(0)) = \varepsilon_0\left(1 + \frac{\omega_p^2}{\omega_o^2}\right) \equiv K_0\varepsilon_0 \tag{31.56}$$

where $K_0\varepsilon_0$ is the static dielectric constant. Going further, and introducing the high frequency limit $\varepsilon_\infty \equiv K_\infty\varepsilon_0$, then

$$K(\omega) = \frac{\varepsilon(\omega)}{\varepsilon_0} = K_\infty + (K_0 - K_\infty)\frac{\omega_0^2}{\omega_0^2 - \omega^2 + i\frac{\omega}{\tau}} \tag{31.57}$$

EXAMPLE: Find the wavelength associated with the plasma frequency ω_p for copper-like parameters ($\mu = 7$ eV), assuming m is the rest mass of the electron.
SOLUTION: From Eq. (6.24), $\mu = \hbar^2k_F^2/2m$ and $\rho_o = k_F^3/(3\pi^2)$, and observing that $\lambda_p = 2\pi c/\omega_p$ along with $R_y = 2Q/a_o = \alpha\hbar c/2a_o$, then

$$\frac{\lambda_p}{a_o} = \frac{\sqrt{3}\pi^{3/2}}{\alpha}\left(\frac{R_y}{\mu}\right)^{3/4} = 1321.7\left(\frac{13.606}{7}\right)^{3/4} = 2175.6$$

or $\lambda_p = 115.13$ nm, a wavelength associated with ultraviolet light.

EXAMPLE: Approximate the static dielectric constant for a semiconductor with $\rho_o = 5 \times 10^{22}$ cm^{-3} assuming one electron per atom and ω_o corresponding to an optical wavelength of 500 nm. Assume $m_n = m$. Contrast to the static dielectric constant of silicon $K_s = 11.9$.
SOLUTION: Numerical evaluation of $\omega_p = \sqrt{16\pi Q\rho_o/m} = 12.615$ rad/fs. For $\lambda = 500$ nm, then $\omega_o = 2\pi c/\lambda = 3.7673$ rad/fs. Consequently,

$$\frac{\varepsilon(0)}{\varepsilon_0} = 1 + \left(\frac{12.615}{3.7673}\right)^2 = 12.212$$

or about the same as silicon.

[14] As quoted in N. Iyer Raghavan, *The Moral and Political Thought of Mahatma Gandhi*. New York: Oxford University Press, 1973, p. 274.

For metals, there is no restoring force (the electrons roam freely) so that $\omega_o = 0$, taking the Lorentz model to the Drude model. When the dielectric function can be separated into bound (Lorentz) and free (Drude) contributions, the optical properties can be related to the band structure and the obtained from reflectance data using Kramers–Kronig relations [552, 553, 561]. Starting from $(n + ik)^2 = 1 + \hat{\chi}$ and then equating the real and imaginary parts of the left- and right-hand sides, Drude theory predicts

$$n^2 - k^2 = 1 - \frac{\omega_p^2 \tau^2}{1 + \omega^2 \tau^2} = \frac{\varepsilon_r}{\varepsilon_0} \tag{31.58}$$

$$2nk = \frac{\omega_p^2 \tau}{\omega(1 + \omega^2 \tau^2)} = \frac{\varepsilon_i}{\varepsilon_0} \tag{31.59}$$

where $\hat{\varepsilon} = \varepsilon_r + i\varepsilon_i$. With a bit of effort, n and k can be expressed in terms of ε_r and ε_i, or

$$2\varepsilon_0 n^2 = \sqrt{\varepsilon_r^2 + \varepsilon_i^2} + \varepsilon_r \tag{31.60}$$

$$2\varepsilon_0 k^2 = \sqrt{\varepsilon_r^2 + \varepsilon_i^2} - \varepsilon_r \tag{31.61}$$

from which $n(\omega)$ and $k(\omega)$ are readily isolated, and $R(\omega)$ evaluated in turn.

31.6.2 Quasi-classical Boltzmann model

So distribution should undo excess,
And each man have enough.

– William Shakespeare[15]

Now that the harmonic model has presaged the impact of resistance (inverse conductivity σ) on the optical parameters, undertake a harder but more insightful reconsideration using a distribution approach, familiar in the transition from Eqs (7.16) to (7.18), along with the perturbation methods in the treatment of Eq. (21.61). That is, let the distribution function f be the sum of a distribution f_0 (such as the Fermi–Dirac distribution) that exists in the absence of fields plus a modification f_1 that accounts for the turn-on of the field, or $f = f_0 + \lambda f_1$ (where here λ is taken from 0 to 1 as the field is turned on, and is not wavelength). For the equilibrium case, Eq. (7.18) becomes (where $\hbar k$ resumes its meaning as a momentum rather than k as a damping term)

$$\partial_t f_0 + \frac{\hbar k}{m} \partial_x f_0 = 0 \tag{31.62}$$

as usual, but it is the next equation of order λ that matters, and is

$$\partial_t f_1 + \frac{\hbar k}{m} \partial_x f_1 - \frac{F}{\hbar} \partial_k f_0 = -\frac{1}{\tau} f_1 \tag{31.63}$$

The difference between f and f_0 can be expressed in a Taylor-like expansion

$$f - f_1 \approx \varphi(x, k, t)(-\partial_E f_0) \tag{31.64}$$

where φ accounts for the variations with x and k. If the force F has spatial and temporal variation, as it will if the $\mathcal{E}$ field arises from light, then

$$F(x, k, t) = q\mathcal{E}_o \exp(i\kappa x - i\omega t) \tag{31.65}$$

It is to be expected, then, that

$$\varphi(x, k, t) = \varphi_o(k) \exp(i\kappa x - i\omega t) \tag{31.66}$$

but if that is the case, and using $\partial_k f = \partial_k E \, \partial_E f$, then Eq. (7.18) becomes

$$\left[\left(-i\omega + i\kappa \frac{\hbar k}{m}\right) \varphi_o(k) - \frac{F_o}{\hbar} \partial_k E\right] (-\partial_E f_o) = -\frac{1}{\tau} \varphi_o(k)(-\partial_E f_o) \tag{31.67}$$

[15] Ref. [37]: *King Lear*, IV.i.70–71.

Upon making the substitutions $\partial_k E = \hbar^2 k/m$ (the parabolic approximation) and $\kappa = \hbar(\omega/c)$ (from Eq. (31.41)) then from equating the coefficients of $(-\partial_E f_o)$ on the left- and right-hand sides,

$$\varphi_o(k) = \frac{\tau F_o}{1 - i\omega\tau}\left(\frac{\hbar k}{m}\right) \tag{31.68}$$

where it has been assumed that the velocity $\hbar k/m$ is much smaller than the speed of light so that $\hbar k/mc \ll 1$.

A parenthetical comment following Eq. (31.42) suggested conductivity σ would resurface, and that time has come. The DC conductivity σ of Section 3.2.1 is changed by the presence of the time-varying field. With an approximation for f_1 now in hand from Eq. (31.64), then

$$\sigma(\omega) = \frac{q}{F}J = \frac{q^2}{2\pi F_o}\int_{-\infty}^{\infty}\frac{\tau F_o}{1 - i\omega\tau}\left(\frac{\hbar k}{m}\right)^2\left(\frac{m}{\pi\hbar^2}\right)\Theta(k_F^2 - k^2)dk \tag{31.69}$$

where the zero-temperature limit of the supply function $f(k)$ (Section 11.1) resulting in the Θ function, is convenient to take. The integration over k results in

$$\sigma(\omega) = \frac{\tau q^2 k_F^3}{3\pi^2 m}(1 - i\omega\tau)^{-1} = \sigma(0)\frac{1 + i\omega\tau}{1 + \omega^2\tau^2} \tag{31.70}$$

where Eqs (3.9) and (6.24) were used to define the DC conductivity $\sigma(0)$, and the τ-dependent fraction rewritten to make the separation into real and imaginary parts easier. The Drude model gives for metals, and in terms of the conductivity,

$$\frac{\varepsilon(\omega)}{\varepsilon_0} = 1 - \frac{\omega_p^2}{\omega(\omega - (i/\tau))} = 1 - \frac{\omega_p^2}{\omega^2 - i\omega\varepsilon_0\omega_p^2\sigma(0)} \tag{31.71}$$

where the equivalence

$$\tau = 16\pi Q\frac{\sigma(0)}{q^2\omega_p^2} = \frac{\sigma(0)}{\varepsilon_0\omega_p^2} \tag{31.72}$$

is invoked (harkening back to Eq. (3.9)).

The small $\omega\tau$ limit for $R(\omega)$ gives the Hagen–Rubens relation: from Eqs (31.58) and (31.59) it follows that as $\omega\tau$ vanishes, then $\varepsilon_r(\omega)$ becomes small and $\varepsilon_i(\omega)$ becomes large. Letting $\varepsilon_r \to 0$ in Eqs. (31.60) and (31.61), then subsequently neglecting terms of order $\sqrt{2\varepsilon_0}$ and $\varepsilon_0/\sqrt{\varepsilon_i(\omega)}$ compared to $\varepsilon_i(\omega)$, it is shown that

$$R(\omega \to 0) \approx 1 - \frac{2}{\omega_p}\sqrt{\frac{2\omega}{\tau}} = 1 - 2\sqrt{\frac{2\varepsilon_0\omega}{\sigma(0)}} \tag{31.73}$$

This suggests that when $\omega\tau \approx (\omega_p\tau)^2/8$, then the reflectivity becomes small: at high enough frequencies (into the UV), metals become increasingly transparent.

The best metal on which to test the prediction of $R(\omega)$ using Eq. (31.47), with $n(\omega)$ and $k(\omega)$ given by Eqs (31.60) and (31.61) using Eq. (31.71), is aluminum as it most closely matches conditions behind the free electron theory in the Drude model [561, 562]. The data can be compared to tabulated values ([309] or [563]) for aluminum if an estimate of τ is available.

EXAMPLE: Find an estimate for the relaxation time τ required for predicting the reflectivity $R(\omega)$ for aluminum using $\hbar\omega_p = 14.98$ eV [564] and $\sigma(0) = 3.5 \times 10^5$ [Ohm-cm]$^{-1}$ [563].

SOLUTION: From Eq. (31.72), and knowing that q^2 Ohm cm = 1602.2 eV fs nm,

$$\tau = 16\pi Q\frac{\sigma(0)}{q^2\omega_p^2} = 3.5 \times 10^5\frac{16\pi Q}{\omega_p^2(1602.2 \text{ eV fs nm})} = 7.6318 \text{ fs}$$

Usage of $\tau = 7.6318$ fs in Eq. (31.71) results in the Theory 1 line of Figure 31.14, which shows that below $\hbar\omega_p$ the reflectivity is high. Since ω_p for aluminum corresponds to $\lambda_p = 83$ nm, aluminum (and metals in general) are mirror-like to wavelengths in the visible: they reflect most of the light incident upon them. The relaxation time, though, was evaluated using the DC conductivity $\sigma(0)$, which begs the question of what if τ were evaluated for frequencies closer to ω_p by some other means? That invites the next example.

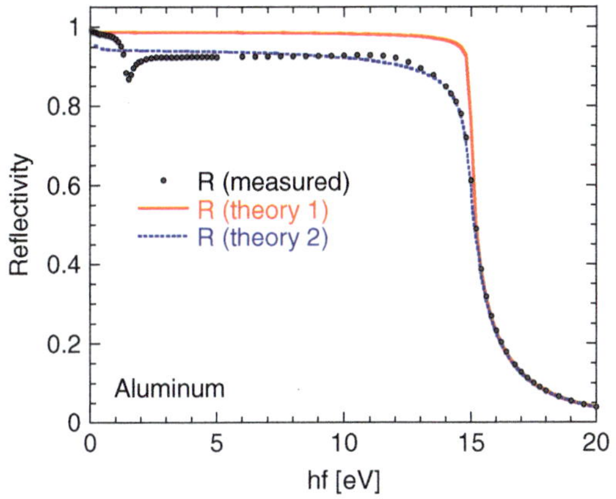

Figure 31.14 Reflectivity of aluminum data [309] compared to Eq. (31.47) via Eqs (31.58) and (31.59) with τ evaluated using $\sigma(0)$ (Theory 1) or $R(14\ \text{eV}/\hbar)$ (Theory 2).

EXAMPLE: Using $\hbar\omega_p = 14.98$ eV as before, but now letting $R(\omega) = 0.8486$ for $\hbar\omega = 14$ eV [309, 563], find an alternate value of τ.

SOLUTION: Using a search algorithm much like that of Section A3.7.1, it is found that a value of $\tau = 1.4967$ results in the given value of $R = 0.8486$ for $\omega = 14\ \text{eV}/\hbar = 22.759$ rad/fs.

Usage of $\tau = 1.4967$ fs in Eq. (31.71) results in the Theory 2 line of Figure 31.14. There is improvement in the intermediate values of ω, although $R(0)$ is under-predicted, even though for ω small ($\hbar\omega < 1$ eV), the prediction is a bit off. Clearly, although the one-parameter (τ) Drude theory behind Eq. (31.71) is surprisingly good, the differences in τ as calculated using the two different approaches suggests strongly that there is more to the physics involved. It takes a quantum mechanical approach to ferret out the right behavior by effectively adding more Lorentzians to ε.

31.6.3 Quantum model

Sean: *If I asked you about war you could refer me to a bevy of fictional and nonfictional material, but you've never been in one…There's nothing you can tell me that I can't read somewhere else.*

– *Good Will Hunting*[16]

Knowledge of how the additional Lorentzians terms to be added to the Drude equations come about is different than knowing how to use those Lorentzians for developing a better model of the reflectivity $R(\omega)$ and decay parameter $\delta(\omega)$. The availability of good and detailed treatments [52, 552, 553, 561] allows the present discussion to focus on how the Lorentzians can be used to obtain reflectivity and decay relations, without becoming mired in arcana.[17] Nevertheless, a demonstration of why the Lorentzian form is plausible instills confidence when using it.

An oscillating electric field $\mathcal{E}(t)$, assumed to be in the $\hat{x}$ direction, is represented by a force term of the form $F_o \cos\omega t$. If the initial state of the electron in an atom is designated $|\phi_o\rangle$ then as a result of the perturbation by an electric field the perturbed wave function takes the form

$$|\psi(t)\rangle = c_o|\phi_o\rangle e^{-i\omega_o t} + \sum_{j\neq o} c_j(t)|\phi_j\rangle e^{-i\omega_j t} \tag{31.74}$$

where the energy of the jth level is characterized by $\hbar\omega_j$ and where the c_j are assumed to be small except for $j = o$ for which c_o is approximately unity,[18] hence its explicit isolation in Eq. (31.74). By the orthogonality of the basis states,

[16] Matt Damon and Ben Affleck, original screenplay of *Good Will Hunting*, 1997.

[17] Examples of such subtleties are what are the different behaviors of the *free* versus *bound* electrons and what is the value of the electric field the harmonic oscillators are oscillating in? Regarding the latter, it is often taken to be that which is at the nucleus of the atom, but the electric field in a material varies locally in and between atoms, and the question becomes rather complex.

[18] The initial state is sometimes labeled by m and sometimes i, but avoiding confusion with electron mass m and $i = \sqrt{-1}$ has virtue.

Schrödinger's equation becomes

$$i\hbar\, \partial_t c_j(t) = \langle \phi_j | F(t)\hat{x} | \phi_o \rangle e^{i(\omega_j - \omega_o)t} \tag{31.75}$$

where the $|\phi_j\rangle$ are solutions to the Hamiltonian without the $F(t)\hat{x}$ contribution. It is convenient to introduce the notation $\omega_{jo} \equiv \omega_j - \omega_o$. Then

$$\begin{aligned} c_j(t) &= \frac{F_o}{i\hbar}\langle \phi_j|\hat{x}|\phi_o\rangle \int_0^t \cos\omega t\, e^{i\omega_{jo}t}\, dt \\ &= \frac{F_o}{2\hbar}\langle \phi_j|\hat{x}|\phi_o\rangle \left\{ \frac{1 - e^{i(\omega_{jo}+\omega)t}}{\omega_{jo}+\omega} + \frac{1 - e^{i(\omega_{jo}-\omega)t}}{\omega_{jo}-\omega} \right\} \end{aligned} \tag{31.76}$$

where the explicit time dependence of $F(t)$ remains in the integrand. Recall that the dielectric constant is associated with the polarization of the atoms (Eq. (31.55) and its discussion), and that the little dipoles go as $\langle q\hat{x}\rangle$. With the $c_j(t)$ in hand so that the wave function is given by Eq. (31.74), and the assumption that all of the $c_j(t)$ save $j = o$ are small, the little dipole can be approximated to leading order (i.e., neglecting quadratic terms in $c_{j\neq o}$), using the orthogonality of $|\phi_j\rangle$, and taking a time average, so as to neglect those terms that oscillate at a higher frequency than the driving force, to give[19]

$$\langle \psi(t)|q\hat{x}|\psi(t)\rangle \approx \frac{2qF_x}{\hbar}\sum_j |x_{jo}|^2 \left(\frac{1}{\omega_{jo}-\omega} + \frac{1}{\omega_{jo}+\omega} \right) \cos(\omega t) \tag{31.77}$$

$$|x_{jo}|^2 \equiv |\langle \phi_j|\hat{x}|\phi_o\rangle|^2 \tag{31.78}$$

The dipole moment is therefore oscillating with the same frequency as the electric field, and so the coefficient of $\cos\omega t$ is what contributes to the definition of the dielectric function. Using $(\omega_{jo}-\omega)^{-1} + (\omega_{jo}+\omega)^{-1} = 2\omega_{jo}/(\omega_{jo}^2 - \omega^2)$ and introducing an oscillator strength factor

$$f_j = \left(\frac{2m}{\hbar^2}\right) \hbar\omega_j |x_{jo}|^2 \tag{31.79}$$

then the atomic polarizability (recall the discussion surrounding Eq. (31.48)) is proportional to $(q^2/m)\sum_j f_j/(\omega_{jo}^2 - \omega^2)$, which, when all the little contributions of the polarizations are summed up, gives the quantum extension to Eq. (31.55). However, there is one complication: no mechanism has been included to account for the transition from one state to another, which causes the initial state to decay as $\exp(-t/\tau_o)$ and the state $|\phi_j\rangle$ to which the electron transitions to grow as $\exp(t/\tau_j)$. That mechanism may be inserted by replacing [552, 553]

$$\hbar\omega_{jo} \to \hbar\omega_{jo} + \frac{i\hbar}{2\tau_j} \tag{31.80}$$

Following the addition of how the imaginary part ratchets through the calculation requires care, but in the end Eq. (31.78) becomes

$$\langle \psi(t)|q\hat{x}|\psi(t)\rangle \to \frac{2qF_x}{\hbar}\sum_j |x_{jo}|^2 \left\{ \frac{2\omega_{jo}e^{i\omega t}}{\omega_{jo}^2 - \omega^2 + i(t/\tau_j)} + \frac{2\omega_{jo}e^{-i\omega t}}{\omega_{jo}^2 - \omega^2 - i(t/\tau_j)} \right\} \tag{31.81}$$

where $(\omega_{jo}\tau_j)^2 \gg 1$ is assumed so as to neglect factors of τ_j^{-2} in the denominator. The term in parentheses can be rewritten as

$$\{\cdots\} = \frac{\omega_{jo}(\omega_{jo}^2 - \omega^2)\cos\omega t}{(\omega_{jo}^2 - \omega^2)^2 + (t/\tau_j)^2} + \frac{(\omega_{jo}/\tau_j)\omega\sin\omega t}{(\omega_{jo}^2 - \omega^2)^2 + (t/\tau_j)^2} \tag{31.82}$$

The first term oscillates with the phase of the electric field, and the second out of phase. The out-of-phase part leads to absorption and transition, and gives rise to the imaginary part of the dielectric function. To simplify matters, drop the o subscript on $\omega_{jo} \to \omega_j$, and then let the oscillator strength factor f_j be defined by

$$f_j = \frac{2m\hbar^2}{\hbar}\omega_j |x_{jo}|^2 \tag{31.83}$$

[19]The $j = o$ term is constant, and therefore taken out of consideration for the time-dependent analysis.

Lastly, using the relation $(a - ib)/(a^2 + b^2) = 1/(a + ib)$ for general a and b results in the quantum version of Eq. (31.55) of

$$\chi(\omega) = \sum_j \frac{f_j \omega_p^2}{\omega_j^2 - \omega^2 + i(\omega/\tau_j)} \tag{31.84}$$

Such is the synopsis of the Lorentz–Drude model of the dielectric function, but to know how to use it is required. The form of the complex dielectric $\hat{\varepsilon}$ is separated into intraband (a *free-electron-like* or Drude part for metals) and interband (a *bound-electron-like* or Lorentz part for insulators) components that look like [564]

$$\hat{\varepsilon}(\omega) = \hat{\varepsilon}_f(\omega) + \hat{\varepsilon}_b(\omega) \tag{31.85}$$

$$\hat{\varepsilon}_f(\omega) = 1 - \frac{f_0 \omega_p^2}{\omega(\omega - i\Gamma_0)} \tag{31.86}$$

$$\hat{\varepsilon}_b(\omega) = \sum_{j=1}^{n} \frac{f_j \omega_p^2}{(\omega_j^2 - \omega^2 + i\omega\Gamma_j)} \tag{31.87}$$

where $\Gamma_j = 1/\tau_j$.[20] For a number of metals, typically the upper limit of the summation n need only be 4 or 5. The values of f_j, ω_j, and Γ_j for j ranging from 0 to 5 (where $\omega_0 \to \omega_p$ for simplicity in the table) is given in Table 31.2. With $\hat{\varepsilon}(\omega)$ in hand, the calculation of n and k via Eqs (31.60) and (31.61) is straightforward, from whence $R(\omega)$ and $\delta(\omega)$ can be obtained from Eqs (31.47) and (31.43). For the values of Table 31.2, the results of such a calculation are shown in Figure 31.15.

How good the model is can be assessed by comparison to copper data[21] to measured values of $n(\omega)$, $k(\omega)$, and $R(\omega)$ as shown in Figure 31.16, the data being chosen from refs. [55] and [309]. Clearly, the Lorentz–Drude model with the table values performs reasonably well in reconstructing the measured data. Apart from the thrill of knowing how to approximately calculate the optical parameters, the shown variation makes it clear that relying on a *single* value of reflectivity or laser penetration depth in the approximation of quantum efficiency (or emittance, for that matter) as a function of wavelength is not desirable: the variation in the optical region of the spectrum is generally too great, a comment that also applies to absorption (which goes as $1 - R$) for the alkali-anitmony compounds and other semiconductors that constitute the favored higher quantum efficiency semiconductors [168, 550].

Table 31.2 Lorentz–Drude model parameters: the oscillator strength factor f_j, the restoring energy $\hbar\omega_j$, the plasma frequency ω_p, and the scattering factor τ_j for various metals of interest to photocathodes. Observe that f_j is dimensionless, but $\hbar\omega_j$ and $\hbar\Gamma_j$ have dimensions of [eV]. Data compiled from Tables 1 and 2 of ref. [564].

Term	Ag	Au	Cu	Al	Ni	W
f_0	0.845	0.760	0.575	0.523	0.096	0.206
$\hbar\Gamma_0$	0.048	0.053	0.030	0.047	0.048	0.064
$\hbar\omega_p$	9.01	9.03	10.83	14.98	15.92	13.22
f_1	0.065	0.024	0.061	0.227	0.100	0.054
$\hbar\Gamma_1$	3.886	0.241	0.378	0.333	4.511	0.530
$\hbar\omega_1$	0.816	0.415	0.291	0.162	0.174	1.004
f_2	0.124	0.010	0.104	0.050	0.135	0.166
$\hbar\Gamma_2$	0.452	0.345	1.056	0.312	1.334	1.281
$\hbar\omega_2$	4.481	0.830	2.957	1.544	0.582	1.917
f_3	0.011	0.071	0.723	0.166	0.106	0.706
$\hbar\Gamma_3$	0.065	0.870	3.213	1.351	2.178	3.332
$\hbar\omega_3$	8.185	2.969	5.300	1.808	1.597	3.580
f_4	0.840	0.601	0.638	0.030	0.729	2.590
$\hbar\Gamma_4$	0.916	2.494	4.305	3.382	6.292	5.836
$\hbar\omega_4$	9.083	4.304	11.18	3.473	6.089	7.498
f_5	5.646	4.384	0.0	0.0	0.0	0.0
$\hbar\Gamma_5$	2.419	2.214	0.0	0.0	0.0	0.0
$\hbar\omega_5$	20.29	13.32	0.0	0.0	0.0	0.0

[20] Whether one uses $\hat{n} = n - ik$ or $\hat{n} = n + ik$ will determine whether one has found $\hat{n}$ or $\hat{n}^*$. However, since one will deal with magnitudes, the distinction is not worth agonizing over. Both conventions are used in the literature. Beware.
[21] See Rakić *et al.* [564] for comparisons to a modification to the model that performs better.

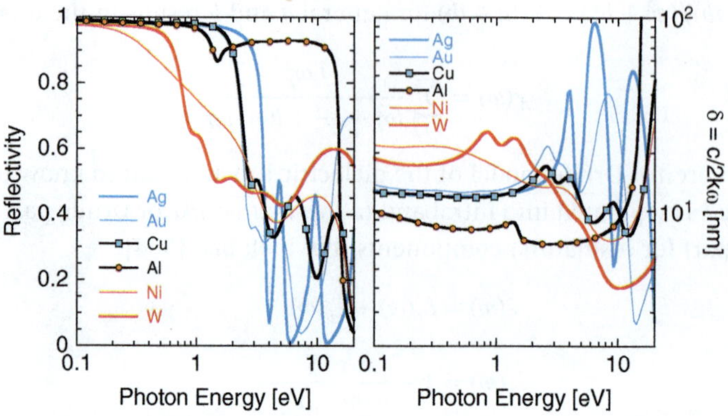

Figure 31.15 The reflectivity $R(\omega)$ (Eq. (31.47)) and the laser penetration depth $\delta(\omega)$ (Eq. (31.43)) calculated using the Lorentz–Drude form of $\hat{\epsilon}(\omega)$ (Eqs (31.86) and (31.87)) and using the parameters of Table 31.2 for typical metals.

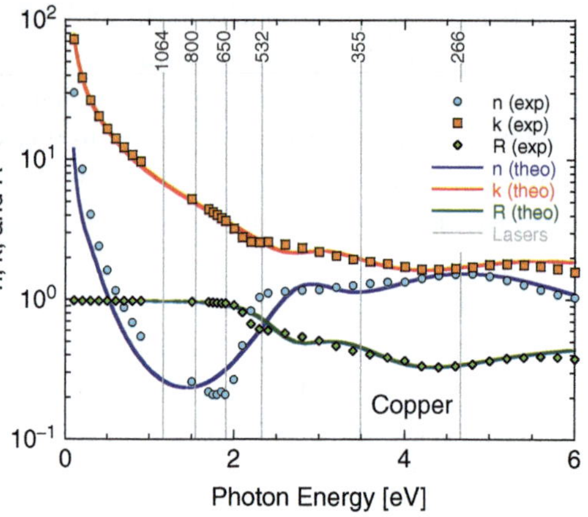

Figure 31.16 Values of n, k, and R obtained from refs. [55] and [309] for copper compared to the Lorentz–Drude model of Eqs (31.86) and (31.87) applied to Eqs (31.60), (31.61), and (31.47) for copper. Gray vertical lines show common laser wavelengths.

31.7 Scattering contributions

From too much love of living,
From hope and fear set free,
We thank with brief thanksgiving
Whatever gods may be
That no life lives for ever;
That dead men rise up never;
That even the weariest river
Winds somewhere safe to sea.

– Algernon Charles Swinburne[22]

To account for the "fatal" removal of electrons by scattering, revisit the equation for quantum efficiency from metals in Eq. (31.21), rendered more useful by fixing k_m and k_o to be

$$k_m \equiv \frac{1}{\hbar}\sqrt{2m[\mu - (\hbar\omega - \phi)]} \tag{31.88}$$

$$k_o \equiv \frac{1}{\hbar}\sqrt{2m(\mu - \hbar\omega)} \tag{31.89}$$

[22]Algernon Charles Swinburne, *The Garden of Proserpine*, in M.L. Rosenthal, *Poetry in English: An Anthology*. New York: Oxford University Press, 1987, p. 744.

Then the QE for metals can be recast as

$$QE = (1 - R)\frac{\int_{k_m}^{k_F} k^2 dk \int_{\varphi(k)}^{1} [kxf_\lambda(x,p)]dx}{\int_{k_o}^{k_F} k^2 dk \int_{-1}^{1} [k]dx} \tag{31.90}$$

The square brackets are intentional: it is convenient to introduce an operator-like notation that denotes the integrals. Therefore, define two operators

$$\hat{O}(\cdots) \equiv \int_{k_m}^{k_F} k^2 dk \int_{\varphi(k)}^{1} [\cdots]dx \tag{31.91}$$

$$\hat{O}_o(\cdots) \equiv \int_{k_o}^{k_F} k^2 dk \int_{-1}^{1} [\cdots]dx \tag{31.92}$$

where the o subscript denotes that the lower limits of the integrals in $\hat{O}$ are different (essentially corresponding to the absence of the impact of ϕ) and where $\varphi(k) = \cos\theta_m(k)$. As a consequence,

$$QE = (1 - R(\omega))\frac{\hat{O}[kx^2/(x+p)]}{\hat{O}_o[k]} \tag{31.93}$$

The denominator $\hat{O}_o$ considers the photoexcited electrons going in all directions, whereas the numerator $\hat{O}$ considers only those within the emission cone towards the surface. The similarity of this notation to $\langle k_z \rangle$, for which $\hbar\langle k_z \rangle$ denotes the average forward momentum of the emitted electrons, is worth noting:

$$\langle k_z \rangle = \frac{\hat{O}[kx^2/(x+p)]}{\hat{O}_o[1]} \tag{31.94}$$

EXAMPLE: Find the relation between QE and $k_{ave} \equiv \langle k_z \rangle$, and show that its variation is small for $\lambda > 266$ nm and copper-like parameters.

SOLUTION: Using Eqs (31.93), (31.94), and (30.26), consider

$$r = \frac{\hat{O}_o[k]}{k_F \hat{O}_o[1]} = \frac{3}{4k_F}\left(\frac{k_F^4 - k_o^4}{k_F^3 - k_o^3}\right) = \frac{3}{4}\left\{\frac{S_4(k_o/k_F)}{S_3(k_o/k_F)}\right\}$$

where $\eta = k_o/k_F = \sqrt{1 - (\hbar\omega/\mu)}$. In the long wavelength limit ($\lambda \to \infty$), $\eta \to 1$, and $S_n(1) = n$ so that $r \to 1$. For $\lambda =$ 266 nm and $\mu = 7$ eV, $\eta = 0.578$. Therefore, the ratio of the two limits is $r(266)/r(\infty) = 1.101$, or a 10% variation over the range of interest.

31.7.1 Dowell–Schmerge relation

It is clearly desirable to maximize the QE and minimize the thermal emittance but due to their interdependence this is difficult.

– David H. Dowell and John F. Schmerge [181]

For unscattered electrons, a relationship between quantum efficiency and emittance can be developed, so that departures from it serve as an indication of the contributions of scattered photoexcited electrons. The formulation of Eq. (31.94) is useful in relating quantum efficiency to emittance $\varepsilon_{n,rms}$ (see also Section 33.3 and Eq. (33.40)) by placing them on comparable footing to draw out similarities between them [565], given that emittance is often represented as

$$\varepsilon_{n,rms} = \frac{\hbar\rho_c}{\sqrt{2mc}}\sqrt{\langle k_\perp^2 \rangle} = \rho_c\sqrt{\frac{MTE}{4mc^2}} \tag{31.95}$$

where $\hbar^2\langle k_\perp^2 \rangle/2m$ is the *mean transverse energy* (MTE) for uniformly emitting rotationally symmetric areas of radius ρ_c. Because $QE \propto \langle k_z \rangle$ (more or less) and $\varepsilon_{n,rms} \propto \sqrt{\langle k_\perp^2 \rangle}$, they should bear a relationship to each other that will have impact on the general desire to maximize quantum efficiency but minimize emittance, something that is thwarted by the tendency of both to increase with $(\hbar\omega - \phi)$. Showing that can be accomplished without necessarily having an analytic form for each by general arguments about how $\hat{O}$ (and $\hat{O}_o$) behave.

The integrals behind $\hat{O}_o[k^n]$ for $n = (0, 1)$ (or any integer for that matter) means that it can be evaluated to give

$$\hat{O}_o(k^n) = \frac{2}{n+1}(k_F^{n+1} - k_o^{n+1}) \tag{31.96}$$

Similarly, the angular integration in $\hat{O}[kx^2/(x+p)]$ is analytic, and gives rise to a function $G(s)$ for $s \to \varphi(k) = \cos\theta_m(k) = k_o/\sqrt{k^2 + k_\omega^2}$, so that

$$G(s) \equiv \frac{1}{2}(1-s)(1-2p+s) + p^2 \ln\left(\frac{1+p}{s+p}\right) \tag{31.97}$$

$$QE = 2(1-R)\frac{\int_{k_m}^{k_F} k^3 G[\varphi(k)]dk}{k_F^4 - k_o^4} \tag{31.98}$$

leaving the k integration in QE to other means. Proceed now in three steps for the determination of the dependence of QE on $\varepsilon_{n,rms}$: (*i*) develop insofar as practical a reasonable approximation to Eq. (31.98) on the assumption that $\hbar\omega - \phi$ is a small quantity, (*ii*) infer the same dependence using more general means, and (*iii*) apply the techniques of the second step to find the dependence of $\varepsilon_{n,rms}$ on $\hbar\omega - \phi$ as well. Having a power law relation for both QE and $\varepsilon_{n,rms}$ in terms of $(\hbar\omega - \phi)$ means that there is a power law relation between them, the determination of which results in the Dowell–Schmerge relation.

For the first step, observe that when the argument of $G(s)$ is close to unity $(s \approx 1)$, then

$$G(s) \approx \frac{1-s}{1+p} - \frac{(2p+1)(1-s)^2}{2(p+1)^2} \tag{31.99}$$

where the first term was used in much of Section 31.4.1. Writing the first term as $G(s) \approx (1-s^2)/[(1+p)(1+s)]$ for $s \to \varphi(k)$ shows that $G(\varphi(k))$ is approximately linear in $(k^2 - k_m^2)$, which vanishes at $k = k_m$ (the energy of the photon is insufficient to cause photoemission even when the electron is directed to the surface): this allows for the approximation

$$G[\varphi(k)] \approx G[\varphi(k_F)]\left(\frac{k^2 - k_m^2}{k_F^2 - k_m^2}\right) \tag{31.100}$$

to be used in Eq. (31.98), and which results in

$$QE \approx (1-R)\frac{2k_F^6 - 3k_m^2 k_F^4 + k_m^6}{6(k_F^4 - k_o^4)(k_F^2 - k_o^2)} G[\varphi(k_F)] \tag{31.101}$$

Using the first term of Eq. (31.99) and the relation of Eq. (31.23) gives rise to the approximation

$$G[\varphi(k_f)] \approx \frac{k_F^2 - k_m^2}{(1+p)(2k_F^2 - k_o^2)} \tag{31.102}$$

Finally, taking $k_F^2 - k_m^2 = (2m/\hbar^2)(\hbar\omega - \phi)$ to be small, then

$$\begin{aligned} QE &\approx \left(\frac{1-R}{1+p}\right)\frac{2k_F^6 - 3k_m^2 k_F^4 + k_m^6}{(k_F^4 - k_o^4)(2k_F^2 - k_o^2)} \\ &\approx \left(\frac{1-R}{1+p}\right)\frac{\mu(\hbar\omega - \phi)^2}{4\hbar\omega(\mu + \hbar\omega)(2\mu - \hbar\omega)} \end{aligned} \tag{31.103}$$

to leading order in $(\hbar\omega - \phi)$, as in Eq. (31.27). The desired relation, therefore, is that $QE \sim (\hbar\omega - \phi)^2$ (albeit other $\hbar\omega$-dependent factors attend it).

For the second step, reproduce Eq. (31.103) using less exact but more intuitive means, as the same means shall be employed to consider the dependence of $\varepsilon_{n,rms}$ on the small factor $(\hbar\omega - \phi)$. The integrand of the numerator of Eq. (31.98) vanishes at the lower limit, enabling an approximation that makes use of Eq. (A1.9) to give[23]

$$\int_{k_m}^{k_F} k^3 G[\varphi(k)]dk \approx \frac{k_F^3}{2} G[\varphi(k_F)](k_F - k_m) \tag{31.104}$$

making use of, first, $(k_F - k_m) = (k_F^2 - k_m^2)/(k_F + k_m) \approx (k_F^2 - k_m^2)/(2k_F)$ and, second, using Eq. (31.102) shows that for $k_F^2 - k_m^2 = (2m/\hbar^2)(\hbar\omega - \phi)$ small, then

$$QE \propto (k_F^2 - k_m^2)^2 = \left(\frac{2m}{\hbar^2}\right)^2 (\hbar\omega - \phi)^2 \tag{31.105}$$

as the first, and longer, approach showed.

Finally, for the third and last step, apply the methods of the second step to the consideration of emittance. In terms of the MTE, emittance $\varepsilon_{n,rms}$, should also depend on $(\hbar\omega - \phi)$, so that a correlation between emittance and quantum efficiency can be formulated. There are a few points to pay attention to. First, $\hbar^2 k_\perp^2/2m = (E + \hbar\omega)\sin^2\theta$ and, second, the averaging $\langle \cdots \rangle$ is calculated with respect to *emitted* electrons, and so the denominator must also include a factor of f_λ. That is,

$$\langle k_\perp^2 \rangle \equiv \frac{\hat{O}[(k^2 + k_\omega^2)(1 - x^2)x/(x + p)]}{\hat{O}[x/(x + p)]} \tag{31.106}$$

where $\hat{O}$ (not $\hat{O}_o$) appears in the denominator. Now applying the same trapazoidal-like rule on both the numerator and the denominator, as done for quantum efficiency previously, shows that

$$\langle k_\perp^2 \rangle \approx \frac{(k_\omega^2 + k_F^2)(1 - \varphi(k_F)^2)}{2k_F} \tag{31.107}$$

but $(1 - \varphi(k_F)^2) = (k_F^2 - k_m^2)/(2k_F^2 - k_o^2) = (\hbar\omega - \phi)/(\hbar\omega + \mu)$, and so from Eq. (31.95) it follows

$$\varepsilon_{n,rms} \propto \sqrt{\hbar\omega - \phi} \tag{31.108}$$

The combined results of Eqs (31.105) and (31.108) taken together therefore gives rise to the Dowell–Schmerge relation [181, 565] expressing the relationship between quantum efficiency and emittance when the fatal approximation holds as

$$QE \propto \varepsilon_{n,rms}^4 \tag{31.109}$$

Compare its predictions to measurements of quantum efficiency and ε as a function of photon energy by Vecchione *et al.* [566] on oxidized antimony, incorporated in Figure 31.17 and recast in a manner to bring out the behavior as a function of $(\hbar\omega - \phi)^n$ for $n = 2$ for QE and $n = 1/2$ for $\varepsilon_{n,rms}$ in Figure 31.18 using fixed values of $R = 1/2$ and $p = 1$. Taking ratios for constant R eliminate its influence, and the variation of the ratio with p is not great. It is seen that the simple theory gives a fair account of the behavior for small $(\hbar\omega - \phi)$ in Figure 31.19, and therefore supports the Dowell–Schmerge relation of Eq. (31.109), but it is also seen that departures occur, which may be due to departures from free-electron-like behavior, the wavelength dependence of both $R(\omega)$ and $p(\omega) = \delta(\omega)/l(E)$, the contribution of scattered electrons, temperature effects (particularly for small $\hbar\omega - \phi$) or even some combination of all of the above. Nevertheless, the presumed constancy of R and p gives a reference point from which to judge the influence of the scattered electron contribution when it matters.

Why the focus on how QE and ε vary with $(\hbar\omega - \phi)$? Insofar as Eq. (31.109) holds, it leads to Dowell and Schmerge's rueful observation introducing the present section: the best that can be obtained is to maximize the quantum efficiency insofar as the emittance levels can be endured. This motivates returning to the derivations (particularly Eq. (31.93) for QE and Eqs (31.95) and (31.106) for $\varepsilon_{n,rms}$) and first extending the result so that $(\hbar\omega - \phi)$ need not be *that* small, and then assessing the impact of scattered electrons (which tend to have lower energy) to the relation.

[23]Knowing that the integrand $f(x)$ vanishes at the lower limit and also behaves approximately like a power law, or $f(x) \sim (x - a)^\alpha$, means one can do better: in place of the trapezoidal rule, a better approximation is $\int_a^b f(x)dx \approx (\alpha + 1)^{-1} f(b)(a - b) = (a - b)^{\alpha+1}/(\alpha + 1)$: the important part here is not the coefficient $1/(\alpha + 1)$ (which does not contribute to ratios) but rather $f(b)(b - a)$.

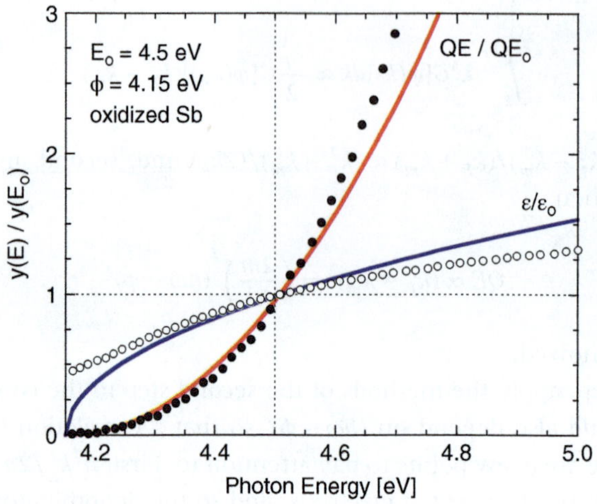

Figure 31.17 The data of Vecchione [566] (data digitally extracted) as a ratio with the value at a photon energy of 4.5 eV (an arbitrary reference value) alongside guide curves of the form $y(E) = y(E_o)[(E - \phi)/(E_o - \phi)]^n$ with $n = 2$ (QE) and $n = 1/2$ (ε) as suggested by theory.

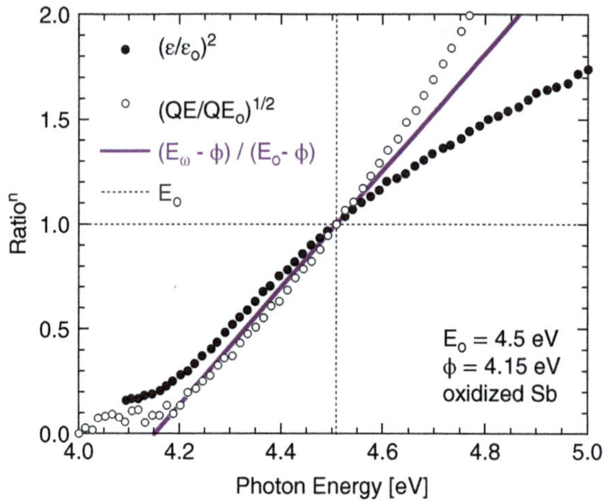

Figure 31.18 The data of Figure 31.17 recast to emphasize its linear behavior as a function of $(\hbar\omega - \phi)^n$ and reveal departures from that linear behavior.

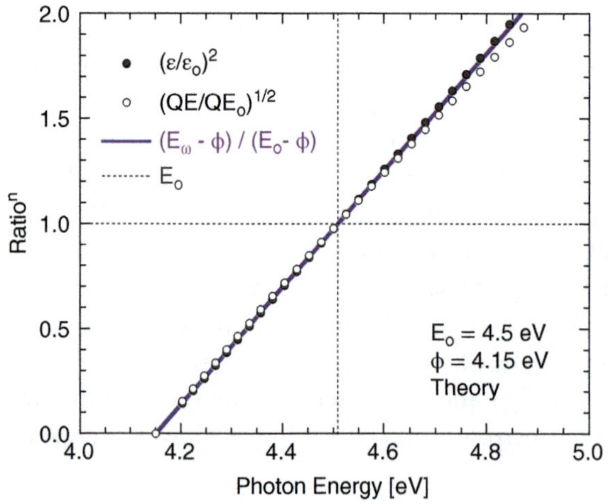

Figure 31.19 The predictions of Eqs (31.105) and (31.108) in the same representation as that of Figure 31.18 for assumed values of $p = 1$ (the variation with p is not large).

31.7.2 Scattered electron contribution

O Joy! that in our embers
Is something that doth live,
That nature yet remembers
What was so fugitive!

– William Wordsworth[24]

Including scattering is not easy. For metals, the colliding electrons tend to share what they have on average, so that if the primary electron has an energy of $\mu + \hbar\omega$ and the target an energy of μ, then both have $(\mu + \hbar\omega/2)$ on average after the collision. That can be a rather substantial change. The situation with semiconductors is a bit better, as changes in energy are associated with phonons, for which the energy losses are not as large (ΔE is on the order of 0.1 eV), but many more scattering events are allowed before so much energy has been lost that emission is no longer possible, and that is a difficult problem often requiring Monte Carlo (Chapter 32). But keeping the discussion just to metals is enough to get the point. Let $f_\lambda(E)$ be the number of electrons that do not scatter before emission, and $\bar{f}_\lambda$ be the number that do, so that all the electrons are accounted for by the equation

$$1 = f_\lambda + \bar{f}_\lambda \tag{31.110}$$

A little reflection reveals this analysis may be iterated, as the scattered electrons follow a similar behavior, and so, using the notation $f_n = f_\lambda(E_n)$

$$1 = f_0 + \bar{f}_0\{f_1 + \bar{f}_1[f_2 + \bar{f}_2(\cdots(f_n + \bar{f}_n)\cdots)]\} \tag{31.111}$$

$$= f_0 + \bar{f}_0 f_1 + \bar{f}_0\bar{f}_1 f_2 + \cdots \left(\prod_{j=0}^{n-1} \bar{f}_j\right) f_n \tag{31.112}$$

with the first term corresponding to no scatterings, the second to one scattering, and so on until n scatterings, where n corresponds to the number of scatterings required before the electron energy has dropped below the barrier height.

Two approximations are useful. First, the photoexcited electron (the primary) colliding with an electron at the Fermi level results in two electrons with, on average, an energy of $\mu + \hbar\omega/2$, but also on average only one travels in the direction of the surface. Second, observe that

$$1 = \frac{\cos\theta}{\cos\theta + p} + \frac{p}{\cos\theta + p} \approx f_\lambda + \frac{p}{1 + p} \tag{31.113}$$

and so $\bar{f}_n \approx p_n/(1 + p_n)$. The second approximation means that the generalization to $\hat{O}[A]$ for some argument A becomes

$$\hat{O}[A] \to \sum_{j=0}^{n} \left(\prod_{l=0}^{j-1} \bar{f}_l\right) \hat{O}[A]_j \equiv \sum_{j=0}^{n} P_j \hat{O}[A]_j \tag{31.114}$$

where the index j on $\hat{O}[A]_j$ means that the energy of the electron before scattering, previously taken to be $\mu + \hbar\omega$ if it is excited from the Fermi level, is now evaluated at E_j. For electron–electron collisions, mathematical induction shows that E_j may be approximated by $E_j \approx \mu + \hbar\omega/2^j$ after the jth scattering but before the $(j + 1)$th, whereas when the energy is removed in small units (as with electron–phonon collisions in semiconductors, where Δ is a small fraction of an electronvolt), then $E_j \approx \hbar\omega - E_g - j\Delta$. For metals, where scattering aggressively removes excess energy beyond μ, the first scattering term dominates and so

$$\frac{\hat{O}[A]}{\hat{O}[B]} \approx \frac{\hat{O}[A]_0 + P_1\hat{O}[A]_1}{\hat{O}[B]_0 + P_1\hat{O}[B]_1} \tag{31.115}$$

with an analogous equation for $\hat{O}_o[A]$ in the evaluation of QE and ε. Because the approximation affects the $\hat{O}$ and $\hat{O}_o$ terms differently by affecting the limits of their integrals in Eqs (31.91) and (31.92) differently, deviations from the Dowell–Schmerge relation between quantum efficiency and emittance arise when the scattered electron contribution is noticeable. An evaluation of the effect, including the dependence of $p(E_j)$ on the pre-collision electron's photoexcited energy and the dependence of the penetration depth $\delta(\omega)$ on frequency (but still neglecting the variation of the reflectivity $R(\omega)$), can show that departure [565]. The results for cesiated copper are shown in Figure 31.20.

[24] William Wordsworth, *Ode: Intimations of Imortality*, in M.L. Rosenthal, *Poetry in English: An Anthology*. New York: Oxford University Press, 1987, p. 521.

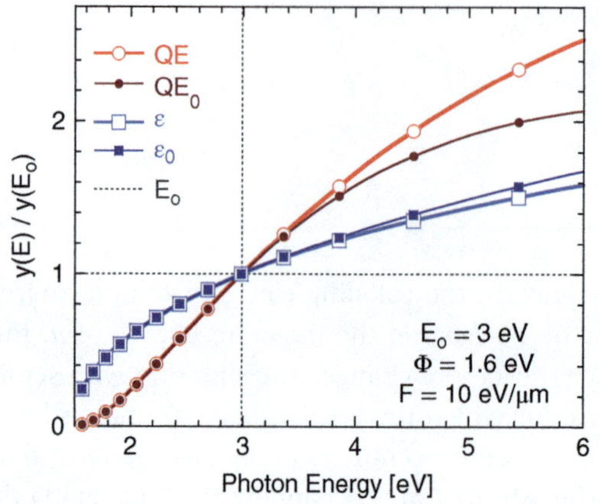

Figure 31.20 Theoretical quantum efficiency for cesiated copper ($\Phi = 1.6$ eV) without (QE_0 and ε_0) and with (QE and ε) scattered electrons as per Eq. (31.115). Based on Figure 1 of ref. [565].

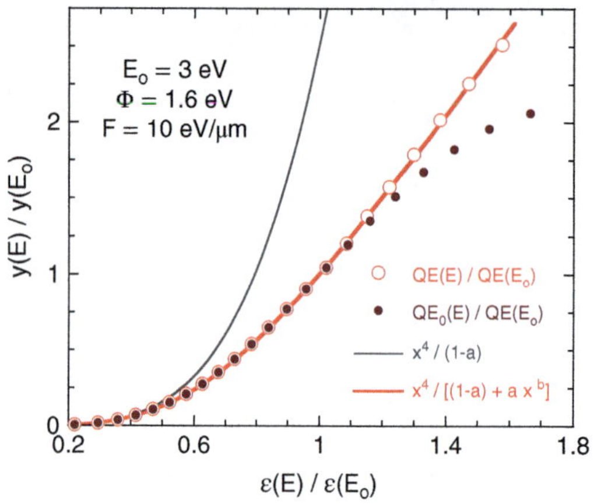

Figure 31.21 The points of Figure 31.20 displayed for QE as a function of ε, both of which depend on E/E_o, where E is the photon energy $\hbar\omega$, based on Figure 2 of K.L. Jensen, *J. Appl. Phys.*, **113**, 056101, 2013. Also shown are fits with $a = 0.6029$ and $b = 2.67$. For small x the Dowell–Schmerge relation is good (the ratio is proportional to x^4); as x increases, the power law shifts so that for large x it is closer to $x^{4-b} \approx x^{4/3}$.

The modifications to the Dowell–Schmerge relation of Eq. (31.109) are intuited by observing that as the photon energy $\hbar\omega$ (denoted E in the figures) increases, the power law relation appears to change that relates QE to $\varepsilon_{n,rms}$ as E transitions from below E_o to above. In terms of $x = E/E_o$, a function which can do this is $x^4/(1 + ax^b)$, where a and b are to be determined, although a better form is $x^4/(\bar{a} + ax^b)$, where $\bar{a} = 1 - a$, as that function has the virtue of being equal to unity at $x = 1$ and therefore is in keeping with the ratios approach to the figures. The range over which the Dowell–Schmerge relation holds, and where it departs, is then visible in Figure 31.21. The question of how the scattered electrons contribute to the current and explain the relation between Figures 31.4 and 31.5 then arises.

Scattered electrons lose energy, thereby delaying their arrival time to the surface as they fall behind the electrons that do not scatter (or scatter a fewer number of times). Let the delay of their arrival to the surface be characterized by a parameter τ, so that a sharp pulse of incident light results in electrons dribbling out with an exponentially decreasing tail. If the laser intensity has a shape $I_\lambda(t)$, then the emitted current should then follow[25]

$$J_e(t) \approx \left(\frac{q}{\hbar\omega}\right)\frac{QE}{\tau}\int_{-\infty}^{t} I_\lambda(s)e^{-(t-s)/\tau}ds \equiv \frac{q}{\hbar\omega}QE\ \mathcal{L}[I_\lambda(t)] \tag{31.116}$$

where the $L[\cdots]$ operator is introduced. Pulse profiles of $I_\lambda(t)$ for which a simple analytic solution results are flat-top and sinusoidal, for which a Fourier representation of $I_\lambda(t)$ is

$$I_\lambda(t) = \left[I_o + \sum_j I_j\cos(\omega_j t + \phi_j)\right]\Theta(t)\Theta(T - t) \tag{31.117}$$

where I_o, I_j and ϕ_j (a phase factor) are constants, $\Theta(x)$ is the Heaviside step function, and $\omega_j = 2\pi j/T$.

[25]The formulation here follows an approach suggested by J.W. Lewellen (LANL) (unpublished).

EXAMPLE: Find $L[I_\lambda(t)]$ assuming only I_o is non-zero (the flat-top).
SOLUTION: For $t < T$, the integral can be rewritten as

$$I_o\mathcal{L}[1] = \frac{I_o}{\tau}e^{-t/\tau}\int_0^t e^{s/\tau}ds = I_o(1 - e^{-t/\tau}) \tag{31.118}$$

For $t > T$, it is instead

$$I_o\mathcal{L}[1] = \frac{I_o}{\tau}e^{-t/\tau}\int_0^T e^{s/\tau}ds = I_o(e^{T/\tau} - 1)e^{-t/\tau} \tag{31.119}$$

EXAMPLE: Find $L[I_\lambda(t)]$ assuming all the coefficients save one I_n is non-zero and all $\phi_j = 0$ (a sinusoidal profile).
SOLUTION: Using $\cos x = (e^{ix} + e^{-ix})/2$ with $x = \omega_n t$, the integration gives for $t < T$

$$I_n\mathcal{L}[\cos\omega_n t] = \frac{I_n(\sin\omega_n t + \omega_n\tau(e^{-t/\tau} - \cos\omega_n t))}{1 + (\omega_n\tau)^2} \tag{31.120}$$

For $t > T$, then $\omega_n T = 2\pi n$ simplifies the sin and cos terms at the integration limits, giving

$$I_n\mathcal{L}[\cos\omega_n t] = \frac{I_n\omega_n\tau}{1 + (\omega_n\tau)^2}(1 - e^{T/t})e^{-t/\tau} \tag{31.121}$$

Although flat-top pulses (with some jitter or shape to them) are possible, laser pulses are commonly Gaussian in shape, for which the I_n are the Fourier transform coefficients (where $e^{i\omega t} = \cos\omega t + i\sin\omega t$ can be used because only the cosine coefficients survive):

$$I_j(\text{Gaussian}) \propto \int_{-\infty}^{\infty} e^{-(t/t_o)^2}e^{i\omega_j t}dt = \sqrt{\pi}t_o\exp[-(\omega_j t_o/2)^2] \tag{31.122}$$

The understanding of the effects of delayed emission are revealed simply by taking all but one of the I_n as non-zero, a useful stratagem as it also provides insight on the effects of laser jitter (random fluctuations in the laser pulse shape) that often accompany pulses and are exaggerated by using frequency doubling crystals to reduce the wavelength of the drive laser [281]. Consider, therefore, a flat-top pulse (non-zero I_o) with a single ripple (non-zero I_n for $j = n$ and zero for all $j \neq n$, or $I_j = I_n\delta_{jn}$), for example let $n = 8$ to provide a reasonably rapid ripple, and so

$$I_\lambda(t) = I_o + I_8\cos(16\pi t/T) \tag{31.123}$$

The current $J_e(t) = (q/\hbar\omega)\ QE\ L[I_\lambda(t)]$ then adopts the shape shown in Figure 31.22 for various values of τ/T: not only does a longer delay time τ result in a softening of the laser jitter, it also elongates the pulse by increasing the emission

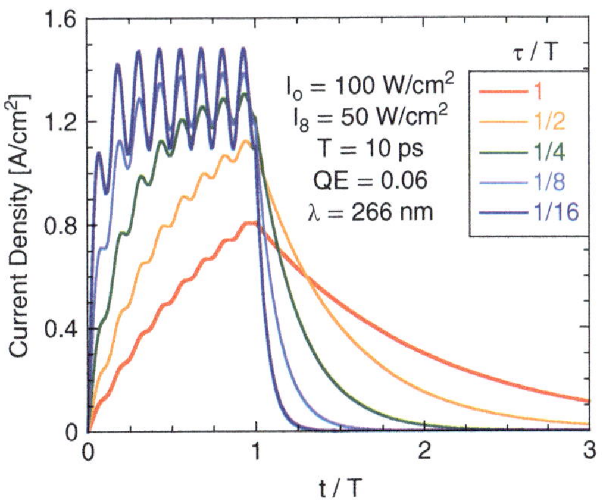

Figure 31.22 Current density as calculated using Eqs (31.116) and (31.117) with $I_o = 2I_n = 100$ W/cm^2, $n = 8$, and $x = t/T$. The other parameters are $\lambda = 266$ nm, $QE = 0.06$, and $T = 10$ ps. The lines are labeled by the values of τ/T.

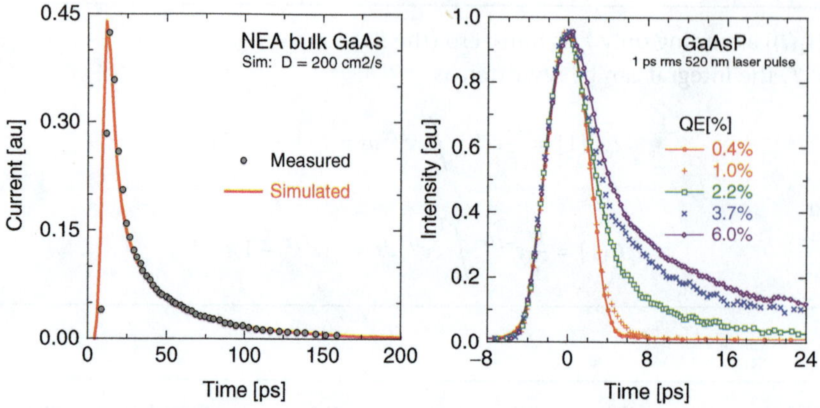

Figure 31.23 Left, photoemission from NEA bulk gallium arsenide: symbols are measured, line is simulated using a diffusion model with $D = 200$ cm^2/s. Based on Figure 3 of P. Hartmann, *et al.*, J. Appl. Phys. **86**, 2245 (1999) (Ref. [567]). Right, measured intensity as a function of time for different GaAsP photocathodes in terms of quantum efficiency using a 520 nm laser pulse of 1 ps root mean square duration: larger QE values are associated with the presence of a more significant emission tail. Based on Figure 9 of I.V. Bazarov, *et al.*, Phys. Rev. ST Accel. Beams **11**, 040702 (2008) (Ref. [526]). All data digitally extracted.

tail. The presence of $e^{-t/\tau}$ in both the $(t < T)$ and $(t > T)$ regimes means that the lag of the pulse rise is of the same shape as the tail of the pulse decline.

Intuitively, τ will depend on how far electrons are excited into the bulk of the material, making it correlate with the laser penetration depth. Therefore, an expectation that metals will be fast responders (short laser penetration depths coupled with the fatal effects of scattering) compared to semiconductors (deep laser penetration depths and not-so-fatal effects of scattering) is in fact borne out: the quantum efficiency of metals is typically below that of semiconductors. The contribution of the tails (which contain scattered and, in the case of NEA photocathodes [23, 516, 523, 526], thermalized electrons) manifests itself in both theoretical/simulation studies and measurements [526, 567], as reproduced in Figure 31.23 for NEA and high quantum efficiency semiconductor photocathodes.

31.7.3 Shell and sphere models

And tomorrow we might not be together
I'm no prophet and I don't know nature's ways

– Carley Simon[26]

The long tails in Figure 31.23 contain electrons that have undergone scattering events and thereby fall behind due to a loss of energy and the randomization of their direction through scattering, compared to the heads for which scattering has not occurred at all. For NEA photocathodes, even electrons that have lost most of their excess energy are still emitted. The nature of such drift-diffusion electrons differs from the single-scattering event model of above. Prophesying how they part ways is undertaken in the algorithm of Section A3.38 for comparing the behavior of drift-diffusion (scattered or "sphere") electrons to the ballistic (unscattered or "shell") contribution using a simple mode that randomizes the direction and reduces the energy of the electrons after every scattering event [568]. The scattering events are taken to occur for $t = n\tau$ with n an integer, and so a random walk is being modeled, with results as shown in Figure 31.24. Although all blue (unscattered) particles have traveled the same distance from the origin, the projection makes it appear they are inside the blue sphere when they are not, in contrast to the red (scattered) drift-diffusion electrons for which the red circle is the root mean square (rms) radius.

Because of the random walk nature of the scattered, or sphere, electrons, the rms radius $(r_{rms})^2 = \left(\sum r_j^2\right)/N$ is expected to increase as $\sqrt{t}$ [51], and so it does as shown in Figure 31.25. Consequently, scattering causes electrons to dawdle in reaching the surface (bottom of figure) and thereby be emitted with lower energy and at a later time than the unscattered electrons. In a semiconductor under "magic window" conditions, the energy loss would not be as extreme, and the

[26]Carley Simon, lyrics to *Anticipation*, 1971.

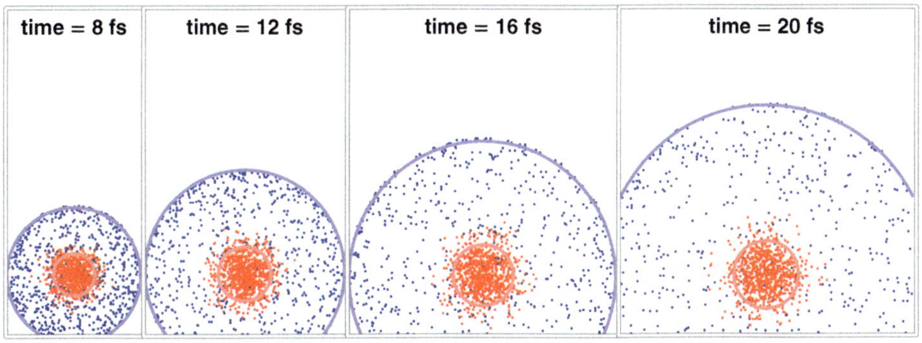

Figure 31.24 Unscattered (shell, blue) and scattered (sphere, red) electrons at $t = 8$, 12, 16, and 20 fs (left to right) for Cu parameters, $\tau = 1$ fs, and $\hbar\omega = 2.33$ eV $\leftrightarrow \lambda = 532$ nm, using the algorithm of Section A3.28. After the jth scattering event, the scattered electron has an energy of $\mu + \hbar\omega/2^j$. Faint red and blue lines are spheres with the root mean squared radius. The surface is at the bottom of the figure.

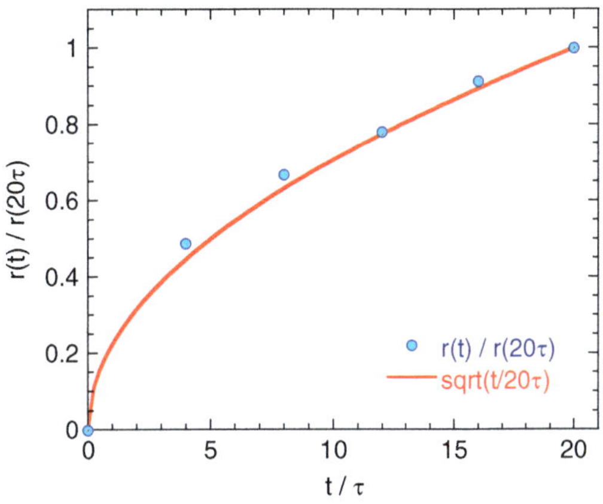

Figure 31.25 The rms radii of the sphere (or scattered) electrons in Figure 31.24. The symbols represent $r(t)/r(t_{max})$ and the line $\sqrt{t/t_{max}}$ for $t_{max} = 20\tau$, ratios being taken to eliminate common factors. By comparison, the shell electrons would have a radius of $v_o t$ and therefore show a linear relation.

expansion would be affected by effective mass variations and band bending, but an analogous evolution occurs, and the resulting behavior helps explain the difference between the initial spike and long tails apparent in Figure 31.23.

When both the shell and the sphere get to the surface, the contribution of each to the emitted current has a different behavior. A manner to estimate it is to evaluate the amount of charge $Q(t) = Q_{shell}(t) + Q_{sphere}(t)$ passing into vacuum for each process, and from that find the time derivative $I(t) = dQ/dt$. For the shell electrons, this is rather easy, as it amounts to finding the portion of a shell past $z = \delta$, or [568]

$$Q_{shell}(t) = \frac{Q_s}{4\pi} \int_0^{2\pi} d\varphi \int_0^{\theta_m} \sin\theta d\theta = Q_s \frac{t}{2(t + t_o)} \tag{31.124}$$

where $\cos\theta_m = R(0)/R(t)$, $\delta = v_o t_o$ is the depth at which the electrons were photoexcited, Q_s is the total charge on the shell,[27] and $R(t) = v_o(t + t_o)$, with $v_o^2 = 2(\mu + \hbar\omega)/m$ as the initial charge and $t_o = \delta/v_o$ the amount of time taken for the shell to first make contact with the surface.

Calculating the contribution from the sphere is a bit more challenging. A diffusively expanding sphere has much in common with the point source models and treatment of the diffusion equation in Section 29.2.1. Compared to the one-dimensional solution to Eq. (29.10), three dimensions requires the product of three such factors, and so the distribution of charge expands like

$$\rho(t) = \frac{Q_d}{(4\pi Dt)^{3/2}} \exp\left(-\frac{r^2}{4Dt}\right) \tag{31.125}$$

[27] As luck would have it, "shell" and "sphere" both begin with "s", so let the sphere charge Q_d be designated with a d subscript to indicate "diffusion", leaving the s subscript to indicate "shell".

and the integral over all space gives $\int \rho(t)dV = Q_d$, as it must. If the velocity of the diffusing electrons is v_d, then the analogous equation to Eq. (31.124) becomes

$$Q_{sphere}(t) = Q_d \frac{\int_{\delta}^{\infty} e^{-z^2/4Dt} dz}{\int_{-\infty}^{\infty} e^{-z^2/4Dt} dz} = \frac{1}{2} Q_d \left\{ 1 - Erf\left(\sqrt{\frac{t_d}{t}} \right) \right\} \tag{31.126}$$

A few convenient tricks are hidden in obtaining Eq. (31.126): (*i*) $r^2 = x^2 + y^2 + z^2$, so that the integrations over x and y coordinates in both numerator and denominator factor out, (*ii*) by forming a ratio of integrals for the purposes of calculation, all the other factors in the coefficient of $\rho(t)$ likewise factor out as well, (*iii*) the origin of the integration has been quietly relocated to where the photoexcited electrons are created, and (*iv*) $t_d = \delta/\sqrt{4D}$ is introduced and $Erf(x)$ is the error function of Section A2.4. Taking the derivative to find the current results in

$$I(t) = Q_s \frac{t_o}{2(t + t_o)^2} + Q_d \left[\frac{t_d}{4\pi t^3} \right]^{1/2} e^{-t_d/t} \tag{31.127}$$

whereas the maximum of the shell electrons occurs right away at ($t = 0$), the maximum of the diffusive electrons occurs at a later time where $\partial_t[t^{-3/2}e^{-t_d/t}] = 0$, or at $t = 2t_d/3$. The sphere contribution to current therefore trails the shell contribution. Such is the quintessential element of one diffusion site: a string of such sites then represents the contribution of a distribution of photoexcited electrons. As pedagogically useful as a shell/sphere distinction is, it represents a difficult way to keep track of the total contribution to emitted current. A Monte Carlo approach is simpler, and illuminates nature's ways by mimicking unfathomable random processes.

31.7.4 A scattering coefficient to quantum efficiency

the black horse of the future,
comes to earth but has no tongue.

– Joan Baez[28]

The Monte Carlo methods behind Figures 31.24 and 31.25 are rather minimalist: after scattering, only the direction is chosen randomly. Predicting the future is a great deal more difficult than that. Monte Carlo can and does account for much more, such as the energy-dependence of the scattering rates, the random time between collisions, different scattering processes weighted according to material properties, a varying number of electrons that are created as a function of depth, the consequences of final state occupation probabilities on the directions the scattered electrons can take, the possibility that the photoexcited electrons impart enough energy to the scattered electron from the Fermi level that one or more of them may still contribute to the photoemission current, and quite a few more complications associated with electron transport [108]. Not all photoexcited electrons and the electrons they excite (the "secondaries") need to be followed, though: once an electron loses enough energy through scattering that it can no longer, if favorably directed, surmount the surface barrier then there is no need to monitor its future evolution (unless needed for band bending and space-charge effects) and it can be dropped from a simulation. Evaluating the ratio R of only the unscattered electrons with all the electrons that surmount the surface barrier thereby gives a measure of how QE can be modified to include scattering. That is, use Monte Carlo to estimate $QE = R(\hbar\omega)QE_o$, where QE_o is the quantum efficiency due to only unscattered electrons (the *fatal approximation*), as in Eq. (31.19). The Monte Carlo methods to do so will treat transport and scattering, but a description of the methodology is deferred to Section 32.1.

Two configurations will be contrasted: the first considers a work function Φ that is comparable to the photon energy $\hbar\omega$, characteristic of bare metals under UV illumination; the second considers a work function brought low through the addition of a cesium surface layer (Section 31.8) so that the photoexcited electron can undergo one to several collisions before its chances of escape vanish. Photoexcited electrons and electrons they excite through scattering are followed until such time that the energetic electrons either loose too much energy to be emitted or pass beyond the surface barrier. Electrons striking the surface barrier with a normal energy that is below the barrier height are reflected even if the total energy exceeds the barrier height: in the language of the escape cone in Eq. (31.16), electrons with total energy $E >$

[28] Joan Baez, lyrics to *Three Horses*, 1971.

$\mu + \phi$ are reflected if $E\cos^2\theta < \mu + \phi$, and such electrons are followed so long as $E > \mu + \phi$. When all of the photoexcited electrons are either lost through an erosion of their kinetic energy through scattering or emitted into vacuum, then a ratio of the number of electrons emitted with the number of *unscattered* electrons emitted is made and defines the factor $R(\hbar\omega)$. Letting N_n refer to the number of electrons emitted after undergoing n scattering events, with N_0 referring to unscattered electrons, then

$$R = \frac{1}{N_0}\sum_{j=0}^{\infty} N_j = 1 + \sum_{j=1}^{\infty} R_j \tag{31.128}$$

where R_j is the ratio of the number of electrons that scatter j times prior to emission compared to the number emitted that do not scatter at all. A few expectations immediately arise. For example, when the photon energy is small, then only x_1 should be non-zero, and it likely scales linearly with $(\hbar\omega - \phi)$. When the photon energy increases, then at some point the secondary electrons may be able to share enough of their energy so that what they scatter with can also be emitted. Therefore, while R may be linear in $(\hbar\omega - \phi)$ for small values, as $(\hbar\omega - \phi)$ increases higher powers should become manifest.

A Monte Carlo simulation counts the number N of electrons that have been emitted and keeps track of the number of times they scatter prior to emission [568]. Contrast two cases, the first (Cu) where the photon energy $\hbar\omega = 4.661$ eV (λ = 266 nm) is close to the work function $\Phi = 4.5$ eV, and the second (Cs-Cu) where the difference is greater because $\Phi = 1.6$ eV. A field of 10 MV/m at the surface causes a further reduction of $\sqrt{4QF} = 0.12$ eV. For Cu, of some 10^6 photoexcited electrons launched, only 1035 are emitted, of which 953 are unscattered and 82 have scattered prior to emission. Compare this to Cs-Cu, for which 39670 electrons are emitted, with 17811 unscattered and 21859 scattered. Therefore, while the fatal approximation is good for simple Cu, for Cs-Cu about half the emitted electrons had undergone some scattering. The contribution of scattered electrons is summarized in Figure 31.26, and based on the study of ref. [568].

A dimensionless term which captures the presumably linear behavior of N with small $(\hbar\omega - \phi)$ yet which should also capture the large photon energy behavior of N increasing rapidly is $x = (\hbar\omega - \phi)/\hbar\omega$. When the ratio of total to unscattered electrons (that is $R = N/N_0$) is shown as a function of x for the cases of Cu and Cs-Cu, then the relations are as in Figure 31.26. In fact, evaluations over a range of wavelengths (with fixed work function) or work function (with fixed wavelength) demonstrate that a parametric fitting defined by

$$R(x) \approx C_0 + C_1 x + C_\nu x^\nu \tag{31.129}$$

performs well, as in Figure 31.27. Although the values of the C_n and ν are determined by simulation, they surprisingly show a weak (but not negligible) indifference to the bulk material. Once found, $R(x)$ may be appended to QE_0 of the fatal approximation to account for the scattered electron contribution via $QE = R\,QE_0$, as demonstrated after the nature of the surface coating is explored (Section 31.8).

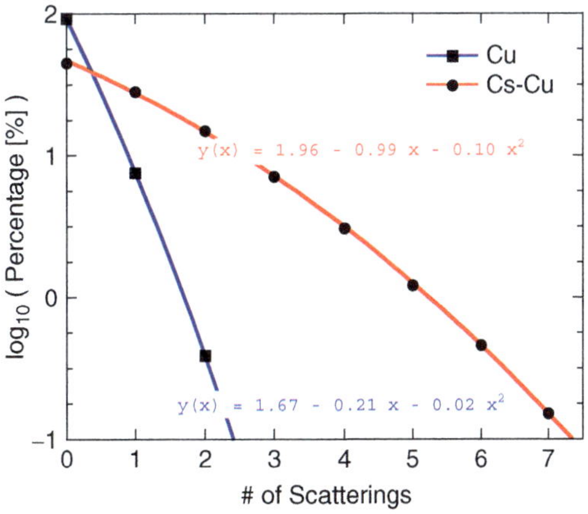

Figure 31.26 The percentage of emitted electrons as a function of the number of scatterings they endured before emission for the cases of copper (Cu: $\Phi = 4.5$ eV) and cesiated copper (Cs-Cu: $\Phi = 1.6$ eV). The photon energy was $\hbar\omega = 4.661$ eV ($\lambda = 266$ nm) and a field of 10 eV/μm caused a Schottky barrier lowering factor of $\sqrt{4QF} = 0.12$ eV. The red and blue lines to guide the eye are from the $y(x)$ equations alongside them.

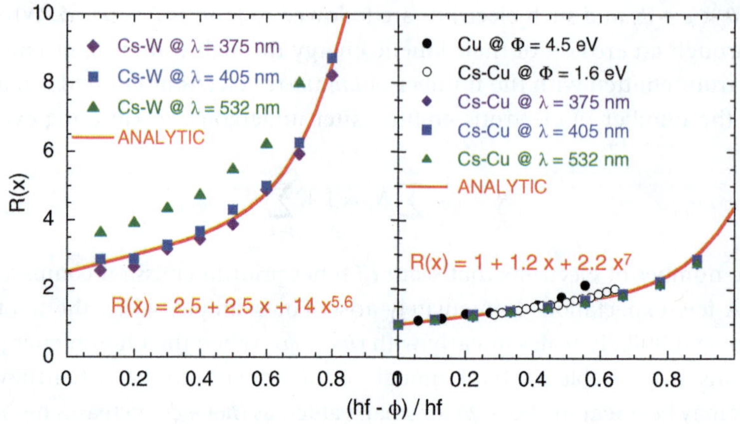

Figure 31.27 Comparison of Eq. (31.129) to Monte Carlo simulations (based on Figures 17 and 21 of ref. [568]) numerically evaluating $R(x)$ with $x = (\hbar\omega - \phi)/\hbar\omega$ for copper and tungsten surfaces with partial coatings of cesium, which reduces the effective work function. Symbols labeled by wavelengths (in [nm]) vary Φ for the fixed wavelength shown; those labeled by energies (in [eV]) vary photon energy for the fixed work function shown.

31.8 Low work function coatings

When a refractory metal is partially covered ($\theta < 1$) by a metallic film, the surface is heterogeneous, the heterogeneity being only quasi-static. The effective work function, which from the experimental standpoint represents the entire surface, is an appropriate average over the work function of sites covered by the film and others that are uncovered. Due to cooperative phenomena between sites of different types, each site cannot be considered independently. Thus, the task of deriving the appropriate averaging procedure is beyond detailed analysis.

Elias P. Gyftopoulos and Jules D. Levine [394]

In the crude capacitor model of Section 21.5, a submonolayer coverage of cesium atoms gave rise to a nanoscale capacitor that caused a work function reduction. The magnitude of that reduction, in Eq. (21.56), was related to the degree to which charge separation occurred between the cesium atoms of the surface layer (which are loose with their outermost electrons) and the tungsten atoms of the bulk (which attract those electrons) as measured by their respective electronegativities (Figure 21.9). Smoluchowski [85, 364] used such concepts to give an account of work function variation amongst the different crystal faces exposed on a metal surface, illustrated in Figure 31.29. Tabulations of work functions, as in Figure 31.28, often obscure the variation with crystal face: the arrangement of the last layer of atoms and the distribution of charge on them makes a difference, as shown in Figure 31.30, where "differences" amount to fractions of an electron volt.

The capacitor model is simple and trivially calculated, but phenomenological. In utter contrast, density functional theory exploits a theorem by Hohenberg and Kohn that a system of interacting electrons in the ground state can be described by a potential determined from the electron charge density [104]. The Kohn–Sham equations can then give one-particle approximations that determine the properties of the system at a fundamental level; its extension to non-uniform electron gases makes use of a local density approximation, which Lang and Kohn used to great effect in describing the image

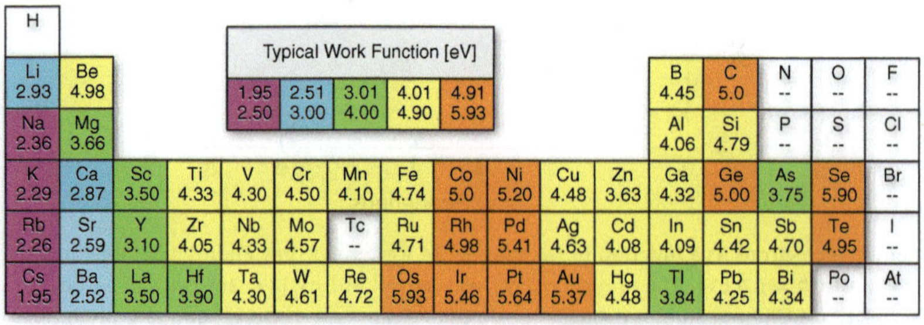

Figure 31.28 Typical work function associated with the elements. For crystals, typically the ⟨100⟩ or ⟨110⟩ face is shown (see Figure 31.30). The inset shows the ranges associated with the color scheme. Compiled from ref. [309] and other sources.

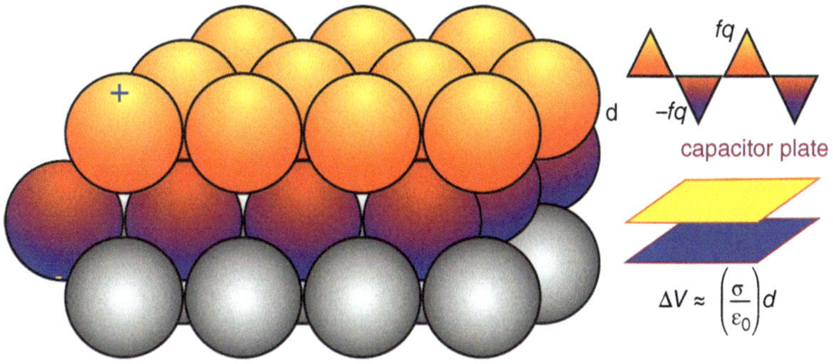

Figure 31.29 Relation of charge redistribution in the atoms on a surface that gives rise to a contribution to the work function. The simple capacitor model is in the inset. The triangular figure to the upper right is a representation suggested by Smoluchowski [364].

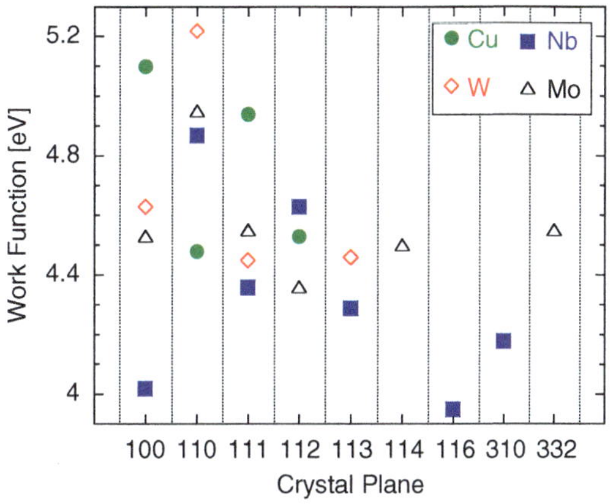

Figure 31.30 Variation of work function Φ with crystal face. Data from ref. [309].

charge potential [85, 340, 344, 348] (Chapter 27). Properties such as effective masses, optical constants, and material parameters are additional benefits [569], but such first principle calculations are difficult and time-consuming. When the need is for something more accurate than the simple models, but less demanding than density functional theory (DFT) or molecular dynamics simulations (e.g., the needs of beam optics codes for modeling electron beams launched into photoinjectors and particle accelerators), something of a more intermediate nature is preferable for predicting $\Phi(\theta)$ from metal and semiconductor surfaces as a function of coverage θ by low work function coatings [84, 308]. Of a number of models that may satisfy such purposes [135], a pedagogically useful one is due to Gyftopoulos and Levine [394, 409], which improves upon an earlier formulation due to Topping [570].

Consider, therefore, a group of measurements of various coatings on the surface of various metals versus coverage, as in Figure 31.31. Schmidt and Gomer [571] observe

> *The $\overline{\phi}$ vs $\overline{n}$ curves of...refractory substrates are generally quite similar: the work function decreases linearly at first, then more slowly, eventually passes through a minimum, and finally approaches the work function of the adsorbate at concentrations corresponding to two to three atomic layers.*

Schmidt and Gomer actually observe a great deal more – they consider the Fowler–Nordheim equation and how it is used to determine work function values as in the experimental field emission investigations of Swanson and Strayer [572] for cesium on various metals (the images of the crystal faces as the coverage changes are particularly edifying), that is, the relations shown in Figure 31.31 all dip through a minimum and asymptotically approach the bulk work function of the coating material. It is that behavior that Gyftopoulos–Levine (GL) theory captures by identifying various relevant parameters and their effects, although a few of those parameters are somewhat squishy, and not all questions of interpretation are resolved with the same clairvoyance or rigor as computational materials physics methods. A modified version (MGL) of that theory, based largely on refs. [85 135, 206, 394] and using more contemporary parameters, was motivated by the performance of dispenser photocathodes [220, 281, 383–386, 530, 573] and used to understand the relationship between quantum efficiency and coverage [206, 381, 382, 532, 534].

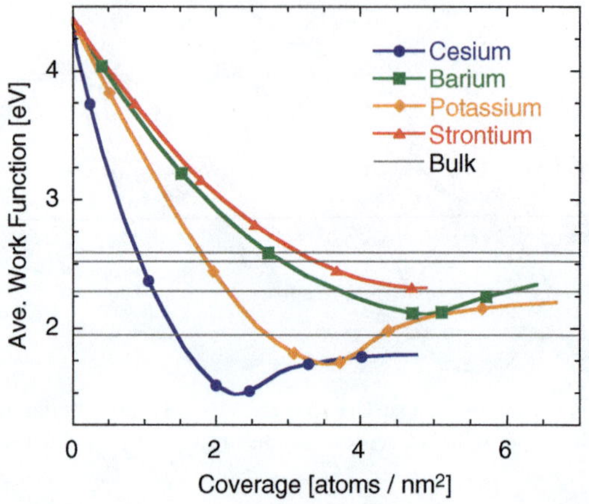

Figure 31.31 Work function of coated tungsten for different coatings as a function of coverage. Gray lines denote bulk values for (bottom to top) Cs, Ba, K, and Sr (based on Figure 22 of L. Schmidt, R. Gomer, Journal of Chemical Physics **42**, 3573 (1965) with the permission of *American Institute of Physics*).

That relation can be understood as follows. A coating atom such as cesium lets go of its electron to atoms such as tungsten, but not completely. How much the electron is "let go" is governed by the electronegativity differences between the atoms (Section 21.5), but also on how close neighboring cesium atoms are: neighbors that are too close get in the way. Gyftopoulos and Levine therefore suggested that the work function should be formed of two parts, the first part $W(\theta)$ governed by the electronegativity differences and the second part $d(\theta)$ by a dipole effect. That is, for a coverage θ representing the fraction of a monolayer,

$$\Phi(\theta) = W(\theta) + d(\theta) \tag{31.130}$$

Mulligan proposed that electronegativity is the mean of the ionization potential and the electron affinity of an atom [102]. The GL hypothesis therefore conforms to intuition in that in the absence of a coating, the work function and electronegativity behave similarly as one moves through the Periodic Table, as in Figure 31.32. A difficulty is that electronegativity is in Pauling units [PU] and work function is in [eV]. In Pauling units, the electronegativity of fluorine (100.45 kJ/mole = 10.411 eV) is 3.98 PU. A rough formula relating Φ to W constructed by Gordy and Thomas [574] was suggested as

$$\Phi[\text{eV}] = 2.27W[\text{PU}] + 0.341 \text{ eV} \tag{31.131}$$

the performance of which is shown in Figure 31.33. The intercept factor of 0.341 eV is argued by Gyftopoulos and Levine to be the energy required to overcome image charge forces, and asserted to be the same for all metals. As a result, if a relationship between electronegativity and coverage, or $W(\theta)$, can be found, then $\Phi(\theta)$ can be found.

A polynomial relationship would be useful, but the determination of the coefficients requires boundary conditions and knowledge of the behavior. Two boundary conditions are readily apparent: when there are no coverage atoms then

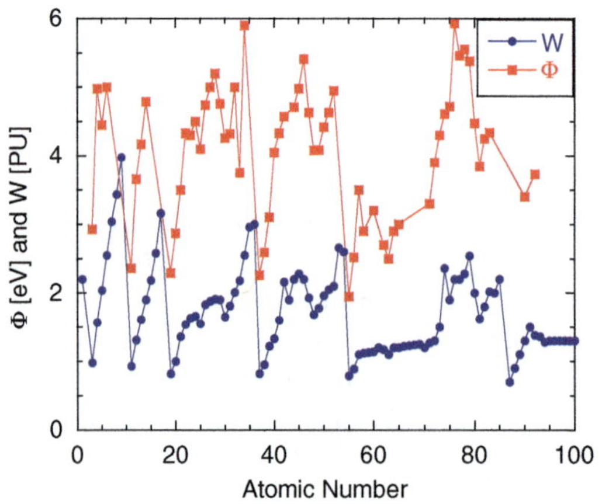

Figure 31.32 The electronegativity W in [PU] and the work function Φ in units of [eV] for various elements as a function of atomic number.

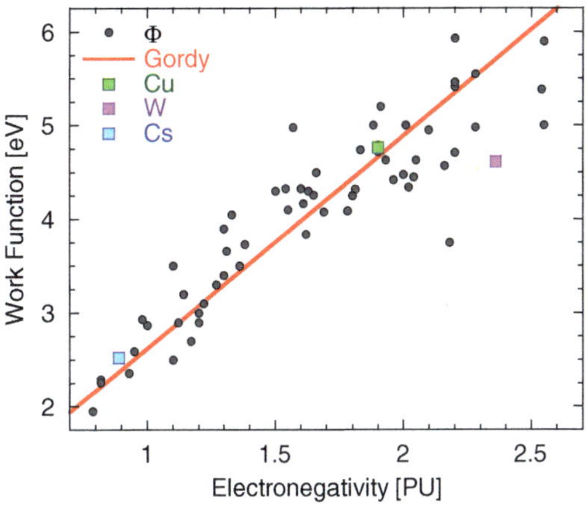

Figure 31.33 The points of Figure 31.32 compared to the equation of Gordy and Thomas in Eq. (31.131). The color squares correspond to copper (Cu), tungsten (W), and cesium (Cs).

the work function $\Phi(\theta)$ should be that of the bulk material or $\Phi(0) = \Phi_w$, and conversely, when there is a monolayer coverage, the monolayer work function should be obtained or $\Phi(1) = \Phi_c$. Moreover, the addition of a few coverage atoms in the former case, or the removal of a few in the latter case, should not make a difference, which is another way of saying the first derivatives $W'(0)$ and $W'(1)$ vanish. Therefore, four equations (two for $W(\theta)$ and two for $W'(\theta)$ at $\theta = 0$ and 1) can be used to determine the coefficients of the polynomial. To summarize, they are[29]

$$\begin{aligned} W(0) &= \Phi_w; W'(0) = 0 \\ W(1) &= \Phi_c; W'(1) = 0 \end{aligned} \tag{31.132}$$

With four equations, the simplest polynomial is a cubic for which $W(\theta) = \sum_{j=0}^{3} C_j \theta^j$. The four equations leads to a matrix relation of the form

$$\begin{pmatrix} 1 & 0 & 0 & 0 \\ 1 & 1 & 1 & 1 \\ 0 & 1 & 0 & 0 \\ 0 & 1 & 2 & 3 \end{pmatrix} \begin{pmatrix} C_0 \\ C_1 \\ C_2 \\ C_3 \end{pmatrix} = \begin{pmatrix} \Phi_w \\ \Phi_c \\ 0 \\ 0 \end{pmatrix} \tag{31.133}$$

with a solution of

$$\begin{pmatrix} C_0 \\ C_1 \\ C_2 \\ C_3 \end{pmatrix} = \begin{pmatrix} \Phi_w \\ 0 \\ 3\Phi_c - 3\Phi_w \\ 2\Phi_w - 2\Phi_c \end{pmatrix} \tag{31.134}$$

A bit of algebra shows that this means

$$W(\theta) = \Phi_c + (\Phi_w - \Phi_c)H(\theta) \tag{31.135}$$

$$H(\theta) \equiv (1 + 2\theta)(1 - \theta)^2 \tag{31.136}$$

with $H(0) = 1$ and $H(1) = H'(1) = H'(0) = 0$ as required.

Evaluating the dipole term $d(\theta)$ remains, and it can be obtained from the electronegativity barrier.[30] A dipole represents a charge separation, and so the dipole term between two atoms should reflect how strongly each atom tugs on the

---

[29]The f and m subscripts of ref. [394] are c and w here, respectively, to suggest "cesium" and "tungsten"; if generalized, they could mean "coating" and "*w*hat is underneath" instead.

[30]In Section 21.5, the electronegativity was designated E_p, but the necessity to discriminate between subscript terms suggests replacing it with something else. Ref. [394] uses $x(\theta)$, but here $X(\theta)$ will be used.

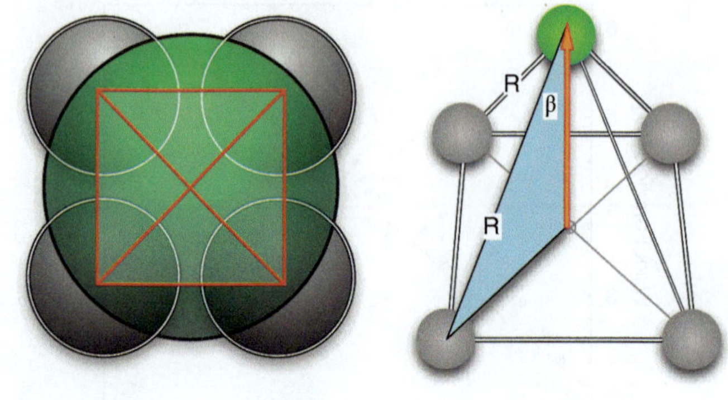

Figure 31.34 Configuration of the coating (green) and substrate (gray) atoms to show the hypotenuse $R = r_c + r_w$ and the angle β. Left, top view; right, perspective view with the locations of the coating (green) and substrate (gray) shown. The orange arrow indicates the direction of M_{cw}.

outermost electrons, that is, it should be proportional to the difference in the electronegativity X_c and X_w of those atoms. Gyftopoulos and Levine argued that a similar conclusion holds when one atom rests upon several atoms. Representing those atoms as hard spheres with a radius r specified by their covalent radius eases the calculation a bit: considering Figure 29.18 suggests that the vertical component of each dipole interaction contributes to the total (the horizontal components cancel) and so for cesium on tungsten, with the distance between adjacent cesium and tungsten atoms being designated $R = r_c + r_w$ and M_{cw} being the dipole of that pair, the sum of the four dipole components is

$$M_o \equiv 4M_{cw} \cos\beta \tag{31.137}$$

where β is the angle R makes with the normal to the surface, as in Figure 31.34, and where

$$M_{cw} = \frac{k}{2.27}(\Phi_w - \Phi_c) \tag{31.138}$$

where 2.27 is from Eq. (31.131) and $k/\varepsilon_0 = 43.256\ \mathrm{nm}^2$ is deduced from electronegativities and molecular dipole moments.[31] Let the length $(k/2.27\varepsilon_0)^{1/2} = (7.0485\pi Qk/q^2)^{1/2} \equiv r_o = 0.43653$ nm, and so

$$M(\theta) = \frac{q^2 r_o^2}{4\pi Q}\cos\beta(\Phi_w - \Phi_c)H(\theta) \tag{31.139}$$

where the dependence of $M(\theta)$ on coverage is reasonably expected to go as $H(\theta)$ (no coverage should result in no dipole, full coverage in the full monolayer dipole, and the addition of a few, or removal of a few, atoms at each extreme does not change matters). The triangle of Figure 29.18 has a base of $2r_w$ and a side of $r_w + r_c$, but that is just a special case when the bulk atoms are touching. If they are pushed apart a small bit, then r_w would have to take that into account. Therefore, let $w/(2r_w)^2$ be the number of atoms per unit area. A generalization to β is then

$$\sin^2\beta = \frac{2}{w}\left(\frac{r_w}{R}\right)^2 \tag{31.140}$$

which finishes the consideration of a dipole by itself.

Now consider the effect of dipoles on each other: they will give rise to fields that act to "depolarize" their neighbors. The "effective" dipole moment $M_e(\theta)$ is therefore the difference between the dipole moment $M(\theta)$ and the depolarizing field $\mathcal{E}(\theta)$, but the depolarizing field is proportional to $M_e(\theta)$ by

$$\mathcal{E}(\theta) = \frac{9}{4\pi\varepsilon_0}\left[\frac{f}{(2r_c)^2}\theta\right]^{3/2} M_e(\theta) = \frac{36Q}{q^2}\left[\frac{f}{(2r_c)^2}\theta\right]^{3/2} M_e(\theta) \tag{31.141}$$

where the new dimensionless factor f is the number of coating atoms per unit cell at monolayer coverage[32] and the coverage θ appears when there is not a monolayer on the surface. The effective dipole $M_e(\theta)$ is therefore the difference

[31] Gyftopoulos and Levine specify $k = 3.83 \times 10^{-30}$ Coulomb-m/V, which is the same as $k/\varepsilon_0 = 43.256\ \mathrm{nm}^2$ here, but without the befuddling exponents.

[32] Calling f by the letter c would have been confusing with the velocity of light. Practicality always requires concessions.

between the dipole $M(\theta)$ with the product of the polarizability factor $4\pi\varepsilon_0 n r_c^3$ (more about which in a moment) and the depolarizing field $\mathcal{E}(\theta)$, or

$$M_e(\theta) = M(\theta) - 4\pi\varepsilon_0 n r_c^3 \mathcal{E}(\theta) = M(\theta) - \frac{nq^2}{4Q} r_c^3 \mathcal{E}(\theta) \tag{31.142}$$

from which $M(\theta)$ can be found. Several things were slid into that equation: first, there is the covalent radius of the coating atoms r_c and, second, there is a new factor n. An atom such as cesium in the alkali metals column of the Periodic Table has only one outer electron so $n = 1$. The alkali earth metals in column two, however, have two atoms, but because they interfere with each other (or "shield" each other) from the atom's nucleus, n is not quite 2 but closer to 1.65 for them. The product of the effective dipole moment $M_e(\theta)$, the surface density of coating atoms $f\theta/(2r_c)^2$, and a factor of $1/\varepsilon_0$ then specify $d(\theta)$, or

$$d(\theta) = -M_e(\theta)\frac{f\theta}{4\varepsilon_0 r_c^2} = -M_e(\theta)\frac{4\pi Q}{q^2}\left(\frac{f\theta}{r_c^2}\right) \tag{31.143}$$

The final piece is the evaluation of $\cos\beta$ based on the geometrical relationships of Figure 31.34. If the area of the base square of the pyramid is $l^2 = w(2r_w)^2$, then with $\sin^2\beta = l^2/2R^2$, it follows

$$\cos\beta = \left\{1 - \frac{2}{w}\left(\frac{r_w}{R}\right)^2\right\}^{1/2} \tag{31.144}$$

Now all the pieces are in place. Putting them together, and defining

$$G(\theta) \equiv \frac{\left(\frac{r_o}{r_c}\right)^2 f\theta\cos\beta}{\left(1 + n\left(\frac{r_c}{R}\right)^3\right)\left(1 + \frac{9n}{8}(f\theta)^{3/2}\right)} \tag{31.145}$$

while remembering that r_c is the covalent radius of the coating atom, r_w of the substrate atom, $R = r_w + r_c$, $r_o = 0.43653$ nm is from the discussion following Eq. (31.138), f and w are dimensionless number of atoms per unit cell factors, and n takes on the value of 1 for coatings from the first column of the Periodic Table (like cesium), but 1.65 for coatings from the second columm (like barium), then the work function Φ as a function of coverage θ (with $\theta = 1$ being a monolayer) is

$$\Phi(\theta) = \Phi_c - (\Phi_c - \Phi_w)H(\theta)[1 - G(\theta)] \tag{31.146}$$

A moment's thought reveals that just as differing crystal faces alter the value of the work function even though the surfaces are composed of atoms of the same material, they will also mess with the value of f and w because different faces position the surface atoms at different relative locations, and the coating atoms are affected by the placement of the atoms under them: as a result, the values f and w take on are not independent. Distinguish between three kinds of crystal faces: the [100] face, the [110] face, and let all others be designated as $[B]$ for "bumpy". Introduce a factor N_o that depends on crystal face, defined such that

$$N_o[100] = 1;\, N_o[110] = 2; \quad N_o[B] = 3 \tag{31.147}$$

The ratios $f/\sqrt{N_o}$ and $w/\sqrt{N_o}$ show approximate consistency from one crystal face to another. The relationship between f and w is then dictated by

$$\begin{aligned} \frac{w}{f}\left(\frac{r_c}{r_w}\right)^2 &= 4(\text{Cs on W, Mo, Ta, ...}) \\ &= 2(\text{Ba on Sr, Th, W, ...}) \end{aligned} \tag{31.148}$$

For generic parameters, the variation of $\Phi(\theta)$ is shown in Figure 31.35.

Putting Eq. (31.146) into the service of comparing theoretical $\Phi(\theta)$ values to experimental measurements for metal surfaces with a submonolayer coverage of coatings encounters difficulties in wresting f from a hard sphere model (on typical and especially sintered surfaces, multiple crystal faces are present), mediating between θ and what is experimentally measured, and even the possibility of contaminants that change the work function (e.g., oxygen on Cs-W or Ba-W tends to *lower* the work function [575, 577]). For example, consider the comparison of various experiments in the literature for Cs and Ba on tungsten shown in Figure 31.36 where the coverage is usually (but not always) given in fractions of a

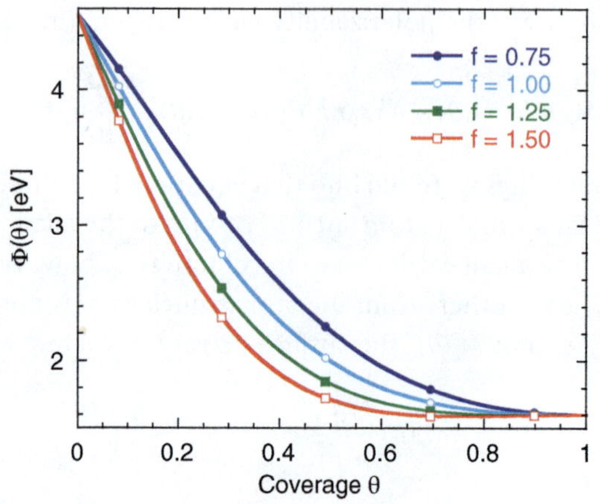

Figure 31.35 Work function Φ as a function of coverage θ for Cs-on-W-like parameters (the default parameters of the Gyftopoulos–Levine algorithm of Section A3.29: $r_w = 0.130$ nm, $r_c = 0.253$ nm, $n = 1.00$, $R_{fw} = 4.0$, $\Phi_c = 1.6$ eV, $\Phi_w = 4.55$ eV.

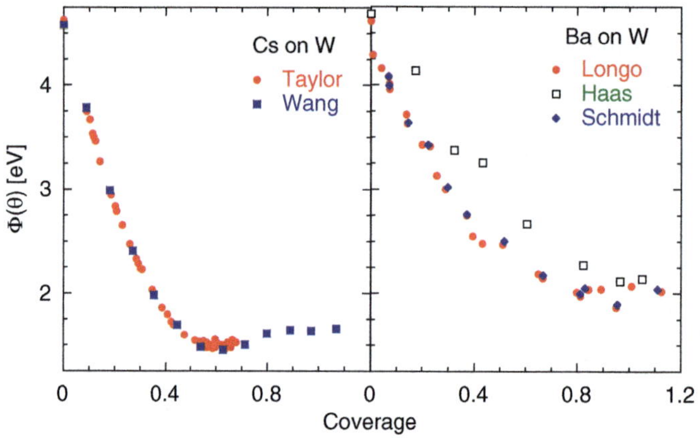

Figure 31.36 Work function Φ as a function of experimentally measured coverage. For Cs: Wang is from Figure 1 of ref. [575], Taylor is from Figure 14 of ref. [418]. For Ba: Longo is from Figure 2 of ref. [576], Haas is from Figure 1 of ref. [577], Schmidt is from Figure 6 of ref. [578]. Data digitally extracted from sources.

monolayer or expressed in terms of deposition time that has to be converted to an equivalent coverage factor. Although the data on the Cs-W side tends to be in reasonable agreement with each other, the data for Ba-W on the right shows variation between the data of Haas *et al.* and that of Longo *et al.* or Schmidt (which tend to be in agreement). Clearly, then, the possibility exists that reported data give coverage up to an overall scale factor, which represents a parameter that must be constrained or determined.

If both f and the scale factor (call it *scale*) that connects θ to the empirical coverage are treated as adjustable parameters, then a constraint must be imposed to find either f or *scale* from the data. Consider then the optimistic assumption that the experimentally reported data is accurate so that *scale* = 1. How might f then be determined? One possible approach is to minimize the rms error. That is, let the empirical data be characterized by N pairs of (θ_j, Φ_j) for $j \in \{1, 2, \cdots N\}$. Then a variation over f such that $\Delta(f)$ defined by the rms error term

$$\Delta(f)^2 = \frac{1}{N}\sum_{j=1}^{N}[\Phi_j - \Phi(\theta_j)]^2 \qquad (31.149)$$

is minimized is sufficient to identify a unique f, where $\Phi(\theta_j)$ is determined using Eq. (31.146).[33] The results of just such a calculation are shown in Figure 31.37, for which the agreement is reasonable, but moreover, the data for Haas alone

[33]One could have just as easily gone the other way by specifying a fixed f and using Eq. (31.149) to identify *scale*, but that is another story. Which one is held fixed depends on the confidence held in the models from which it is constructed.

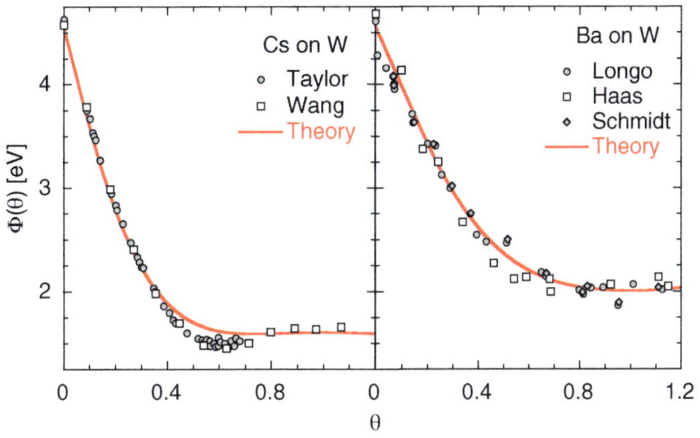

Figure 31.37 The data of Figure 31.36 compared to Eq. (31.146) with f determined such that $\theta = scale \times coverage$. For Wang, Taylor, Longo and Schmidt, f is determined such that $scale = 1$ (Cs: $f = 1.5336$; Ba: $f = 0.800$). For Haas, the value $f = 0.800$ and $scale$ is found by minimizing the least squares difference of Eq. (31.149) between theoretical and experimental values of $\Phi(\theta)$ ($scale = 0.56$).

is brought into rather good agreement with the other data through the determination of a scale factor while otherwise using the f value that was determined for the Longo and Schmidt data lines. Even better agreement can be achieved by adjusting the value of *scale* away from unity, but at that point the effort is more curve fitting than illuminating. What is clear is that GL theory does a respectable job in capturing the variation witnessed for a submonolayer coating atop a bulk metal surface with few adjustable (and constrained) parameters.

31.9 Quantum efficiency of a cesiated surface

Everything in this world is a matter of calculation. Advance then with caution, the balance in your hand. Put into one scale the pleasures which any object may offer; but put fairly into the other the pains which are to follow, and see which preponderates.

– Thomas Jefferson[34]

The GL model for $\Phi(\theta)$ can be combined with the moments model for QE to predict the quantum efficiency $QE(\omega)$ of a metal surface with a varying submonolayer coverage of cesium. Those predictions may then be compared to experimental measurements for cesium deposition onto sintered tungsten surfaces [84, 206, 381, 382, 579 580] much like those of the dispenser cathodes shown in Figure 29.1. In such measurements, cesium deposition occurs simultaneously with QE measurements at various laser wavelengths as the coverage is increased.

The evaluation of QE for different wavelengths as a function of coverage requires an integration over energy and angle in Eq. (31.21), and generally numerical integration is required. In performing the computation, several dependent factors must be kept in mind or evaluated. They are:

- the escape probability (for cesiated metals, approximating the transmission probability by a step function is adequate)
- the optical parameters n and k from which to find the reflectivity $R(\omega)$ in Eq. (31.47) and laser penetration depth $\delta(\omega)$ in Eq. (31.43) (for tungsten, a spline fit to measured data [309] is useful)
- the material constants, bulk and monolayer work functions used in the GL model in Eqs (31.145) and (31.146) [84, 206, 568)
- an approximation to the scattered electron contribution (e.g., Eq. (31.129))
- the relaxation time dependence on photoexcited electron energy (Section 32.3).

A number of these parameters are temperature dependent: if laser intensities are sufficiently low, then a constant temperature approximation is sufficient, but high laser intensities over prolonged times or fast repetition rates cause temperature excursions. If the laser intensity is very high (even if short in duration), a decoupling between the electron and the lattice temperature can occur [83, 85, 206, 581, 582]. For the low laser intensities used in the experiments to which the photoemission model will be measured against [381, 382], the temperature can be treated as constant.

[34] Quoted in Jonathan Haidt, *The Righteous Mind: Why Good People Are Divided By Politics and Religion*. New York: Pantheon Books, 2012, p. 35.

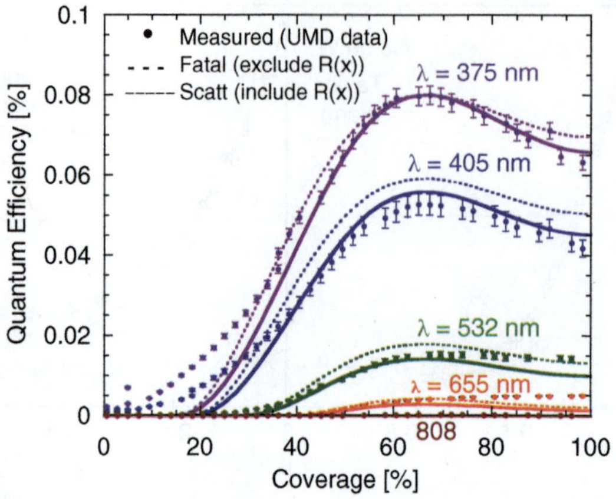

Figure 31.38 Numerical evaluation of $QE(\omega)$ without the scattering factor $R(x)$ (Fatal) and with (Scatt), compared to measured UMD data for cesium evaporated onto a polycrystalline tungsten surface (based on Figure 22 of ref. [568]experimental data courtesy of E. Montgomery (UMD)). All theory lines are calibrated to match the relative peak location and height of the $\lambda = 532$ nm line.

The behavior of the relaxation time that is part of the mean free path between collisions $l(E) = \hbar k(E)\tau(E)/m$ and is a part of the scattering term $p(E)$ of Eq. (31.18) is more difficult. The treatment of relaxation times will be undertaken in Section 32.3; for now, in metals, the relaxation time $\tau(E)$ drops rapidly as the energy of the electron increases over the Fermi energy, and is primarily due to electron–electron collisions.

The predictions from such a model can be compared directly with experiments, for example those undertaken at the University of Maryland in which QE was monitored as cesium was deposited onto a sintered tungsten surface, where rapid and as close to simultaneous measurements from the same spot on the cathode surface necessitate innovative approaches [579, 580]: cesium is mobile and can evaporate quickly, necessitating that QE measurements at different wavelengths be rapid. To make such measurements, a photocathode with a heater coil for measurements at different temperatures (or even gradually increasing temperatures) was used. Direct cesium deposition was performed by using an external source and measuring the coverage by monitoring the mass deposited (e.g., Cs, Sb, K) on the surface using a quartz crystal balance. The configuration enabled the surface to be heated or argon ion cleaned to remove contaminants. The rapid sequential measurement of QE was enabled by using five low-power diode lasers fixed to a rapidly moving stage, so that the QE measurements for various wavelengths were taken from the same spot in close to the same time. A field of 5–30 kV/m was sufficient to prevent space-charge effects. Ion cleaning between runs insured an atomically clean surface, although it roughened the surface as well.

A comparison to the University of Maryland (UMD) data to both the fatal approximation (excluding $R(x)$) or including scattering (using $R(x)$, as in Eq. (31.129)), is shown in Figure 31.38 after the theory is normalized to the $\lambda = 532$ nm reference line [568]. Because of the polycrystalline nature of the sintered tungsten surface, the possibility of contaminants, and other experimental considerations, photoemission cannot be presumed to occur uniformly across the surface, and the normalization reflects that uncertainty. Regardless, once one of the lines is normalized, the theory curves for the fatal approximation account reasonably well for the observed variation in quantum efficiency as a function of wavelength, but when the scattering contribution signified by $R(x)$ is included, the shapes of each line in addition to the relative differences between them is much improved. As a result, the GL model for work function $\Phi(\theta)$ applied to the moments model for quantum efficiency matches the photoemission measurements very well, enough to supply emission models needed by the beam optics codes [583] described in Chapter 33. Hidden in the calculation, however, is the evaluation of the relaxation time $\tau(E)$, the treatment of which must wait until Section 32.3.

CHAPTER 32

Secondary emission cathodes

It appears here …that calculations based on the free electron model of Sommerfeld can lead to a considerably improved understanding of secondary electron emission from metals, so that this phenomenon should have its place along with the many others which have been illuminated by this simple picture.

– Eugene M. Baroody [35]

Those who utilize high brightness photoinjectors often rue that quantum efficiency and photocathode ruggedness do not correlate (Table 31.1), or that what eases demands on the photocathode often increases demands on the drive laser (and vice versa). Recognizing that the measured high secondary yields from diamond in reflection mode (over 100×) observed by Yater and Shih [207, 584] (as shown in Figure 32.1), Ben-Zvi *et al.* [143] envisioned using a thin diamond layer as a secondary electron source to provide copious electrons while protecting the photocathode and easing demands on the drive laser. Studies to investigate the feasibility have been increasingly reported [23, 144, 209, 322, 324, 585–588]. Here the physics is explored.

Consider the secondary yield from diamond in both reflection and transmission mode, the differences of which are suggested in Figure 32.2. Whether secondaries are measured on the same side or on the opposite side from which the primary beam is incident affects what is emitted. In transmission mode, the electrons have further to migrate through a diamond sample, and so higher primary beam energies are often required. When the secondaries are collected and shown as a function of energy, the behavior shown in Figure 32.3 is characteristic. Though perhaps not readily apparent, the lines for transmission mode would for the most part lie atop one another if each were normalized by its maximum value, with only the 10 keV and 8.5 keV lines slightly shifted. The greatest difference between the reflection and transmission lines is the presence of electrons in the higher energy tail for the reflection case, but an absence of such a tail in the transmission case. If one is coming fresh from Section 31.7 this may be no surprise: the secondaries are borne close to the surface in the reflection case, and a fraction of the secondaries can make it to the surface and be emitted without having thermalized; in the case of the transmission measurements, the diamond layer – even at 8.3 μm – is too thick for electrons to pass through without having their excess energy whittled away due to scattering. That scattering has such an effect first motivates a means to treat it, and second has implications for the kinds of beams and the response time associated with a cathode made by using a diamond thin film as a secondary emission source.

32.1 Diamond amplifier concept

…a diamond amplifier functions as follows. Primary electrons with energy of a few keV enter a high purity single-crystal diamond through a metal coating. These electrons excite electron-hole pairs, the number of which typically is about 2 orders of magnitude more than the number of primary electrons, depending on the latter's energy. A fraction of the electrons is lost to recombination near the diamond-metal interface and by diffusion into the metal surface.

– Erdong Wang, Ilan Ben-Zvi, Trivini Rao *et al.* [589]

I saw a highway of diamonds with nobody on it.

– Bob Dylan[1]

Of the various configurations envisioned for a cathode producing bunches of electrons by exploiting the high secondary yield of a hydrogenated diamond surface [209, 322, 585, 586, 588, 589], take a configuration based on refs [144, 322] and illustrated in Figure 32.4 as representative. In it, a high-energy primary beam is incident on the back side of a

[1] Bob Dylan, lyrics to *A Hard Rain's A-Gonna Fall*, 1963.

Introduction to the Physics of Electron Emission, First Edition. Kevin L. Jensen.

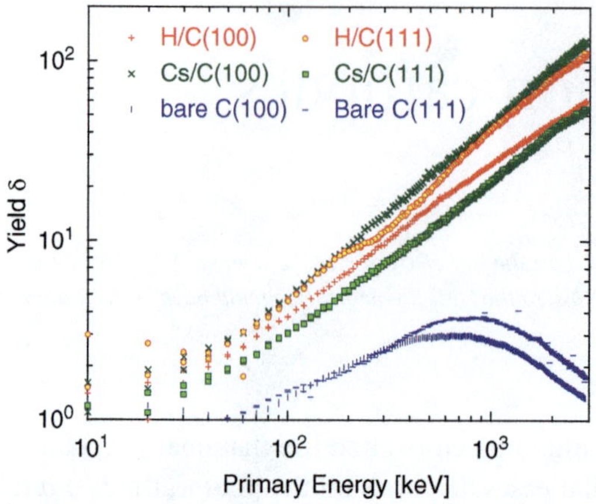

Figure 32.1 Secondary yield from the C100 and C111 faces of hydrogen-terminated and cesium-coated diamond crystals. The high yields indicate that low-energy electrons transport efficiently through diamond regardless of crystal face. The yields have not reached their maximum value for the range of primary electron energies shown. Based on Figures 2 and 5 of ref. [207], data courtesy of J.E. Yater (NRL).

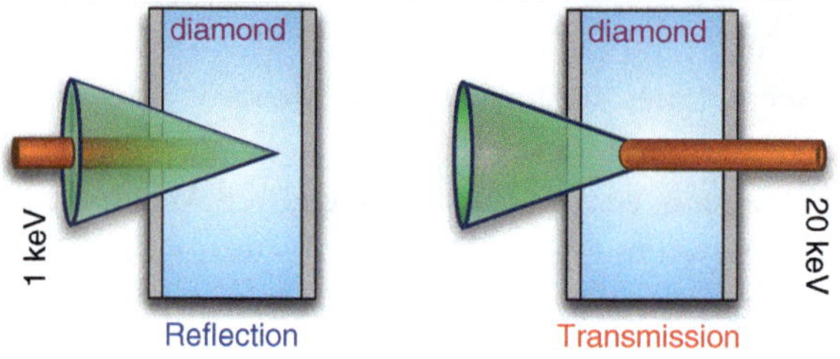

Figure 32.2 Left, in reflection mode, a 1 keV primary beam (red cylinder) incident from the left results in measurement of secondaries emitted to the left (green cone). Right, in transmission mode, a 20 keV primary beam generates secondaries that pass through the thin diamond layer and are collected on the opposite side. After ref. [144].

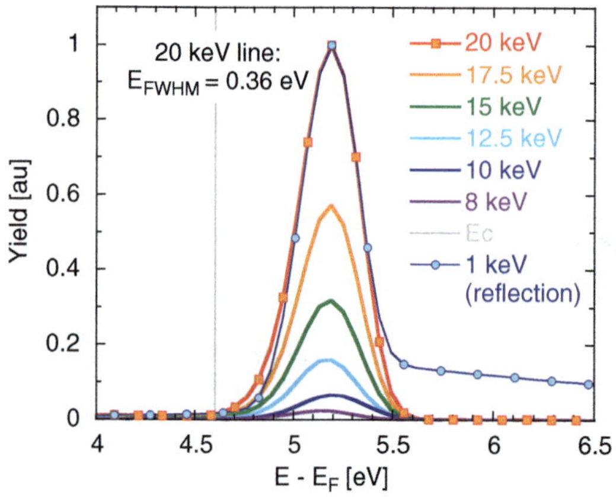

Figure 32.3 The yield of secondary electrons in reflection mode (blue circles) for 1 keV primaries, and in transmission mode (red squares and lines without symbols) for various primary energies. The single-crystal chemical vapor deposition (CVD) diamond thickness was 8.3 μm. *Ec* is the conduction band minimum. Figure based on ref. [144], data courtesy of J. Yater (NRL).

thin diamond flake much like that shown in Figure 15.1; the arrangement of the measurement system is schematically illustrated in Figure 32.5. A potential is maintained across the flake, and it serves to pull the secondaries to the vacuum side, whereupon the hydrogen-terminated (and therefore negative electron affinity, NEA) diamond surface allows their passage into vacuum. Two questions arise: first, for the simple representation, what should be the primary energy and, second, how thick should the flake be? The first question is answered by Baroody's equation of Eq. (15.8): choose $E_o = E_m$ to maximize the yield $\delta(E_o)$. The second question depends on how bunches of electrons migrate through the diamond

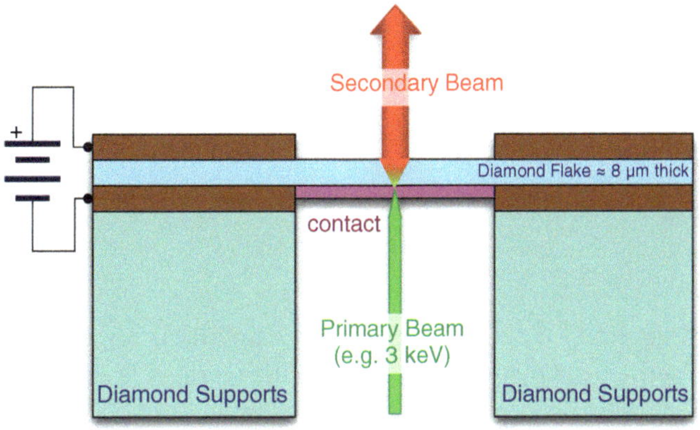

Figure 32.4 The operation and layout of a thin film diamond current amplifier. A high-energy beam of primaries is directed to the back side, generating secondaries. Contacts on the diamond flake have a potential difference which pulls the secondary electrons through the thin flake whereupon a hydrogen-terminated NEA diamond surface allows for their escape. Based on Figure 1 of ref. [322] and Figure 1 of ref. [144].

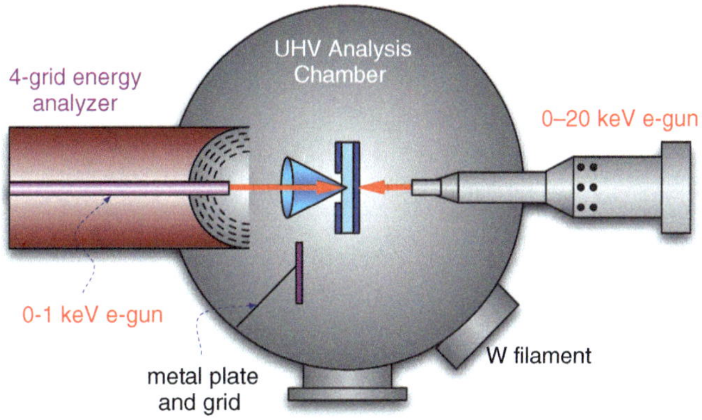

Figure 32.5 Schematic of the UHV analysis chamber used in making the reflection and transmission measurements of secondary emission in thin diamond flakes. Based on Figure 2 of ref. [590].

flake, and therefore depends on what scattering does to that diffusing cluster: why that might matter depends on the uses advocated for a diamond amplifier.

The diamond amplifier was conceived to augment a photocathode for a free electron laser (FEL) [143]. High-performance and X-ray FELs make strong demands on beam quality and emittance [591,592] that are challenging to photocathodes in terms of quantum efficiency, emittance, and lifetime. Existing candidates, as in Table 31.1, can entail unsavory requirements on the drive lasers as well. Therefore, augmenting the "effective" quantum efficiency by enhancing the yield without demanding higher laser intensity eases demands on both photocathode and drive laser, while offering protection to the photocathode from the harsh injector environment as an added benefit.

Other capabilities also matter. An ability to pulse-shape very short bunches entails rapid rise and fall times associated with emission, particularly if a flat (or "top hat") profile is desired for emittance reduction [593, 594]. For example, bunches for the European X-ray FEL are characterized by a trapezoidal pulse 20 ps long with a 4 ps rise/fall time [595], and similar or shorter pulse length requirements exist for other FELs. Although metal and some semiconductor photocathodes have fast response times by comparison, a pulse of electrons forced to diffuse through a thick layer of diamond with an NEA surface might face challenges [209], and so finding the relationship between rise/fall time and diamond thickness matters, particularly if the field across the layer is affected by residual doping levels [322].

The investigation to be undertaken, therefore, is two-fold: first, the physics of how the secondaries are generated by the primary and how deep into the diamond they are made and, second, how those generated secondaries migrate while being accelerated by field and retarded by scattering through the diamond, and how that migration affects how rapidly such pulses rise and fall. A third consideration, namely, how emittance may be affected by the transport of the secondaries through the film, is deferred to Section 33.3.7.

32.1.1 Bethe approximation

Four other Oysters followed them,
And yet another four;
And thick and fast they came at last,
And more, and more, and more–
All hopping through the frothy waves,
And scrambling to the shore.

– Lewis Carroll[2]

The development of Baroody's equation (Eq. (15.8)) used approximations for Eq. (15.1) that relied on a model due to Bethe concerning how high-energy primary electrons shed their energy. A high-energy (primary) electron has more than enough energy to knock core electrons (secondaries) from atoms in the bulk material into the conduction band, where they can roam until such time as they find the surface and are emitted or lost to recombination. Bethe developed a model describing how that shedding of energy occurred (also called a stopping power (SP) formula) [141, 209, 596, 597], so that the number of secondaries per primary could be estimated. Although good for high primary electron energies (higher than several hundred electronvolts), at lower energies it becomes less reliable [598] but nevertheless useful. The energy loss per unit length is given by[3]

$$\frac{dE}{dx} = -\frac{192\pi NQ^2}{E}\sum_{j=1}^{2} Z_{jl} \ln\left(\frac{2E}{BE_j}\right) \tag{32.1}$$

where $Z_{jl} = 2^j$, $BE_j = (2Q/a_o)(Z/j)^2$, $Z = 6$, and $N = \rho/A = (3.512\ \text{g/cm}^3)/(12.0107\ \text{g/mole}) = 0.29241\ \text{mole/cm}^3$ for carbon (of which diamond is composed). If the sum is expanded and the terms collected, an easier form results and is

$$\frac{dE}{dx} = -\frac{192\pi NQ^2}{E}\ln\left(\frac{2^{4/3}Ea_o}{QZ^2}\right) \tag{32.2}$$

EXAMPLE: At what minimum energy does Eq. (32.2) become unphysical (positive energy gain per unit length)? Assume carbon parameters ($Z = 6$).
SOLUTION: Find E such that the logarithm argument equals 1, or

$$E = \frac{QZ^2}{2^{4/3}a_o} = \frac{2^{4/3}(0.3[\text{eV nm}]6^2)}{0.052918[\text{nm}]} = 97.19\ \text{eV}$$

When E drops below ≈ 97 eV, dE/dx changes sign, an unrealistic consequence of the inadequacy of the approximations when the energy transfer is small [596]. The logarithmic sign change is removed, yet the large energy limit retained, by including a factor of 1 in the logarithm, or

$$\frac{dE}{dx} \rightarrow -\frac{192\pi NQ^2}{E}\ln\left(1+\frac{2^{4/3}Ea_o}{QZ^2}\right) \tag{32.3}$$

The inserted factor of unity in Eq. (32.3) need not have been 1: when it is instead a parameter (say d_2) and the factor $2^{4/3}Ea_o/QZ^2$ is augmented by a factor $1/d_1$, with the requirement that the values of d_1 and d_2 must be fitted in conformity with data, then rather good agreement with experiment can be obtained [599]. In the interests of simplicity, though, setting $d_1 = d_2 = 1$ (Eq. (32.3)) will be used. Re-evaluation of the range $R(E_o)$ from Eq. (15.3) now gives

$$R(E_o) = \int_{E_o}^{0}\left(\frac{dE}{dx}\right)^{-1} dE = -\frac{E_o^2}{192\pi NQ^2}\int_0^1 \frac{dx}{\ln\left[1+a(E_o)x\right]} \tag{32.4}$$

[2]Lewis Carroll, *Alice in Wonderland and Through the Looking Glass*. New York: Grosset & Dunlap, 1946, Chapter IV: Tweedledum and Tweedledee.
[3]The present treatment uses a notation dependent on Q instead of α_{fs} as in ref. [209], but the formulae resulting are equivalent. Also, though not treating the Bethe model *per se*, Young [211] successfully compared the electron range energy in Al_2O_3 to the SP formula.

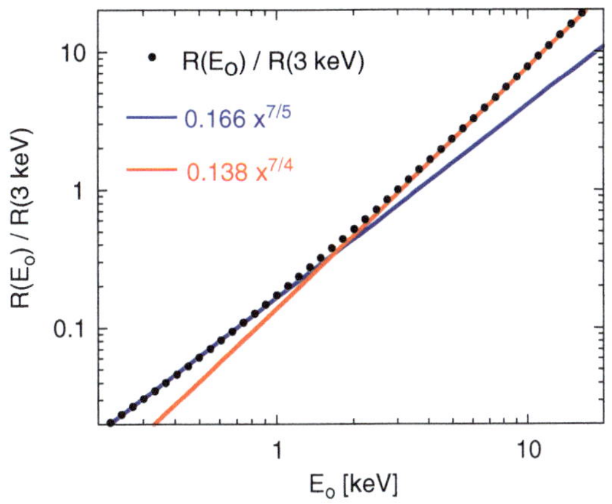

Figure 32.6 Comparison of Eq. (32.4) to the power-law representation of Eq. (15.5) for $n = 0.4$ (blue) and $n = 3/4$ (red), where $R(3\text{ keV}) = 113.64$ nm. A best fit over the range shown (Figure 1 of ref. [209]) finds $R[\text{nm}] = 18.949(E[\text{eV}])^{1.679}$.

where $a(E_o) \equiv 2^{4/3} E_o a_o / QZ^2 = E_o/(97.19\text{ eV})$ for carbon. This is a different sort of relation than that implied by Eq. (15.5), but as Figure 32.6 demonstrates, a power-law representation is not bad, with n in $R(E_o) = A_o E_o^n$ suggested to be between 1.4 and 1.75. For intermediate values of E_o, $R(E_o)[\text{nm}] \approx 18.949(E_o[\text{eV}])^{1.679}$ (e.g., the power-law approximation of ref. [209]). Compare this to the treatment of Kanaya and Okayama [600], who find $R(E_o)[\text{nm}] \approx 19.218(E_o[\text{eV}])^{1.666}$, which is quite close. For the purposes of what follows,[4] it is therefore convenient to let $n = 5/3$ and $nA_o = 18.949$ nm keV$^{-5/3}$.

Pulling a power-law fit out of Eq. (32.4) seems to be extracting an inferior representation from a more accurate and perfectly fine numerical model. Why do it? The power-law form is simply too convenient to pass up because it allows for a friendlier method of calculating $\delta(E_o)$. Used in Eq. (15.1), the power-law form gives

$$\frac{dE}{dx} \approx \frac{E_o}{nR(E_o)} \left[1 - \frac{x}{R(E_o)}\right]^{-(n-1)/n} \tag{32.5}$$

(compare Eq. (15.4)) in terms of which

$$\delta(E_o) = \frac{BE_o}{[\gamma R(E_o)]^{1/n}} e^{-\gamma R(E_o)} F_n[\gamma R(E_o)] \tag{32.6}$$

$$F_n(y) \equiv \int_0^{y^{1/n}} e^{s^n} ds \tag{32.7}$$

in contrast to Eq. (15.6).

EXAMPLE: Find the steps necessary to obtain Young's approximation [141, 601] discussed in Chapter 15 from Eq. (32.6). Let $n = 5/3$.

SOLUTION: Make the substitution $s = y(1 - x)$ in $F_n(y)$ of Eq. (32.7) and then use the exponential approximation

$$(1 - x)^{2/5} \approx e^{-2x/5}$$

The integration is now analytic, and is

$$\delta_e(E_o) = \frac{B}{n} \left(\frac{E_o}{\gamma R(E_o) - (2/5)}\right) \{1 - e^{-\gamma R(E_o) + (2/5)}\} \tag{32.8}$$

If $(n - 1)/n = 2/5$ is small compared to γR and $B' = B/n$, then Young's approximation is obtained.

[4] There are differences between the present treatment and ref. [209], on which it is based, such as n here is $n + 1$ there, and Eq. (15.5) here calls out an n in the denominator, but there n is absorbed into A_o with changes to γ as well as a result.

Might it not be easier to use the simpler exponential approximation $\delta_e(E_o)$ of Eq. (32.8) to compare to measurements rather than enduring the numerical integrals demanded by $F_n(y)$? Bluntly, "No", but why is instructive. In the limit that there are no losses during transport ($\gamma \rightarrow 0$) and the probability of emission is unity ($B \rightarrow 1/\Delta E$, or $p \rightarrow 1$), then the yield $\delta(E_o)$ should simply be just the number of secondaries created. Either by being simple and solving Eq. (15.1) directly with $\gamma = 0$ or by being fancy by using $F_n(y) \approx y^{1/n}$ and taking the limit $\gamma \rightarrow 0$,

$$\lim_{\gamma \to 0} \delta(E_o) = \frac{E_o}{\Delta E} \tag{32.9}$$

when $p = 1$, which just makes sense: each secondary takes an energy of about ΔE to be created, so the number of secondaries is just the number of times that can be done by a primary of initial energy E_o. But that is not the $\gamma \rightarrow 0$ limit of $\delta_e(E_o)$, which is instead (for $n = 5/3$)

$$\lim_{\gamma \to 0} \delta_e(E_o) = \frac{3E_o}{2\Delta E}\left(e^{2/5} - 1\right) = 0.7377\frac{E_o}{\Delta E} \tag{32.10}$$

A temptation might then be to use the easier form of $\delta_e(E_o)$, but appending a coefficient of 1.356 to its overall magnitude so that the number of generated secondaries is brought back into line would result in E_m and δ_m being too large as a consequence. Further machinations beyond that make the approximation become of comparable difficulty to the actual numerical evaluation, and so evaluating $F_n(y)$ numerically from the start is just better.

Compare now the $\delta(E_o)$ of Eq. (32.6) to, for example, the H/C(111) shown in Figure 32.7, one of the sets of data from which high secondary yields from diamond were demonstrated and which helped motivate the diamond amplifier concept. To make the comparison (*i*) evaluate ΔE using the relation $\Delta E = 2.5E_g$ [209, 602], which is compatible with the data of Yater and Shih [207], taking the band gap of diamond to be $E_g = 5.5$ eV, then $\Delta E = 13.75$ eV, (*ii*) from Section 15.1, $B = p/\Delta E$, and (*iii*) the parameters p and γ are all that is left for which to match data to $\delta(E_o)$.

Finding p and γ has some art to it. Actual data is noisy, but taking a polynomial fit of the form $\delta_a(E_o) \approx C_1 E_o - C_2 E_o^2$ smooths out the effects of noise. Fits give $C_1[1/\text{keV}] = 43.588$ and $C_2[1/\text{keV}^2] = 1.8028$. Algorithms such as bisection (Section A3.7.1) can then be used to find γ by requiring $\delta(2\text{ keV})/\delta(1\text{ keV}) = 1.9137$, which occurs for $\gamma = 1.7295$ [1/µm], after which $\delta(1\text{ keV}) = 41.785$ gives $p = 0.58641$, entailing $B = 42.648\text{ keV}^{-1}$. With these values, the theory line of Figure 32.7 is constructed, from which $\delta_m = 187$ for $E_m = 8.6$ keV.

32.1.2 Penetration range and initial conditions

At high primary energies, the high-velocity electrons have a relatively short time to interact with the lattice electrons, and the internal yield per unit length is low. As the primary electrons lose energy, the interaction time increases and so does the yield. The combined effect is that as the primary-electron energy increases, the internal secondary electrons originate deeper beneath the surface.

– Arnold Shih, Joan Yater, Charles Hor, and Richard Abrams [584]

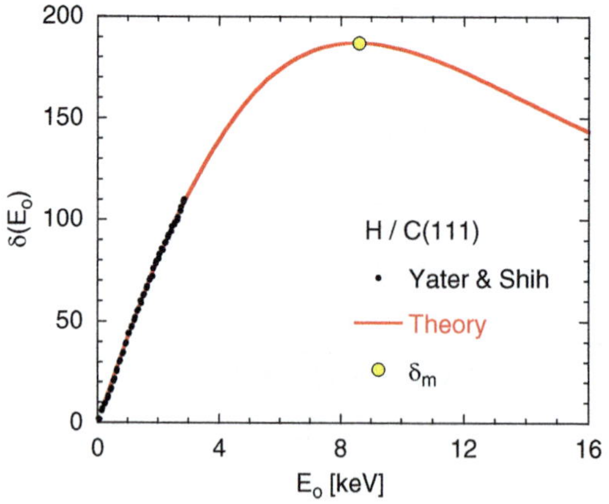

Figure 32.7 Comparison of the yield $\delta(E_o)$ evaluated using Eq. (32.6) to the measured yield from the hydrogen-terminated [111] crystal face of diamond (curve H/C(111) in Figure 7 of Yater and Shih [207]) for the value $p = 0.58641$ and $\gamma = 1.7295$ [1/µm]. The maximum yield occurs at $E_m = 8.6$ keV and is $\delta(E_m) = 187$.

In preparation for following the evolution of secondaries as they migrate through the bulk material, the initial position of the secondaries and their initial direction must be obtained to specify the staring conditions in a Monte Carlo analysis. The direction is relatively straightforward: the secondaries can be assumed to point perpendicularly to the direction of the incident primary beam, although randomly distributed about the azimuthal angle [157, 596].

But how many electrons are deposited as a function of distance into the bulk material? If one assumes that a secondary is generated every time the primary loses ΔE in energy, then the positions where this occurs can be obtained by integrating, or if the positions occur at x_j, then $x_{j+1} \approx x_j + \Delta E(dx/dE)$. But there is an easier way. Recall that $R(E_o)$ is how far a primary beam of energy E_o penetrates into bulk. A primary beam of energy $E_o - \Delta E$ would not penetrate quite as far, but the conception can be inverted:

$$x_j = R(E_o) - R(E_j) \tag{32.11}$$

with $E_j \equiv E_o - j\Delta E$. The number of secondaries (and that would be one) created between x_{j+1} and x_j is then

$$\frac{d}{dx}N(x) = -\frac{1}{\Delta E}\left(\frac{E_j - E_{j-1}}{x_j - x_{j-1}}\right) = \frac{1 \text{ electron}}{R(E_j) - R(E_{j-1})} \tag{32.12}$$

Although Eq. (32.4) should be used for $R(E)$, a quick appreciation of the behavior is given by using Eq. (15.5), for which Figure 32.8 results and matches independent findings [584]. What is evident is that the rate at which the secondaries are created spikes as the primary grinds to a stop. That this is a general feature of particle penetration allows for targeting the depth to which particles are deposited by adjusting the primary beam energy.

EXAMPLE: Estimate how long is required for the primary electron to loose its initial energy E_o. Assume no field exists across the diamond flake, and let $E_o = 220\Delta E = 3.025$ keV. Take $n = 5/3$. Lastly, assume $m_n = 0.57m$ for diamond.

SOLUTION: Although the velocity $v(x)$ of the primary changes by discrete amounts after every collision which generates a secondary, treat $v(x)$ as continuous. From Eq. (32.5), it follows $E(x) = E_o[1 - (x/R(E_o))]^{3/5}$. Introduce $v(x) = \sqrt{2E(x)/m_n}$. As a result,

$$\Delta t = \int_0^{R(E_o)} \frac{dx}{v(x)} = \frac{10R(E_o)}{7v_o} \tag{32.13}$$

where $v_o = \sqrt{2E_o/m_n}$. As a result of the factor (10/7), the time for the primary to come to a stop is approximately 43% longer than the transit time of the primary across $R(E_o)$ if it did not suffer events leading to secondaries. Inserting numbers, $v_o = 32.62$ nm/fs, and so $\Delta t \approx 4$ fs. If a field existed across the diamond flake, then Δt would be smaller because the primary would accelerate between collisions.

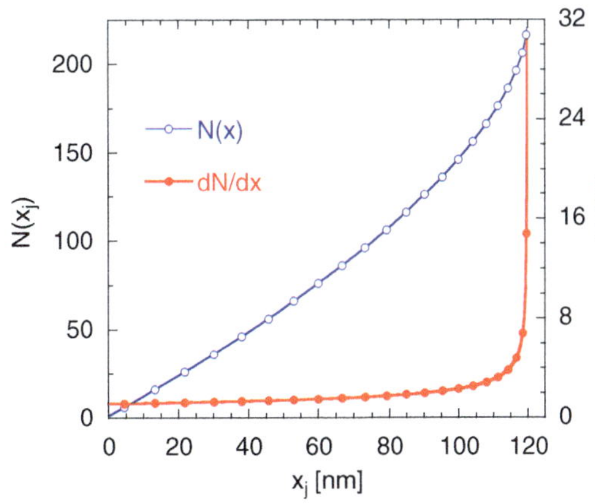

Figure 32.8 Number of secondary electrons generated between x_j and the surface ($N(x)$) and number generated per unit length $dN/dx = 1e^-/(R(E_{j+1}) - R(E_j))$, where 50% of the secondaries (110) are generated in the last 33% of the path of the primary.

Approximating the initial conditions of the secondary electrons is next. The time taken for the primary to stop is orders of magnitude smaller than a picosecond-scale pulse width, and so the secondaries are to a good approximation created all at once with one electron at each x_j, each of which is borne with an initial energy ΔE and directed randomly in a direction perpendicular to the primary electron's path, which bores through the bulk in a straight line. With the initial positions and velocities of the secondaries, their trajectories can be followed, at least until they scatter.

32.2 Monte Carlo methods

If you play a Game of Chance know before you begin
If you are benevolent you will never win

– William Blake[5]

Prior encounters with Monte Carlo, in Section 29.2.4 for choosing how atoms hop or in Section A1.2.5 for evaluating otherwise difficult integrals, are but small examples of how probabilistic methods can be used to solve otherwise difficult or intractable problems. For modeling semiconductor devices and electron transport through materials, the Monte Carlo method is – with no exaggeration – magnificent when time scales are so fast (comparable to the scattering times) or the length scales so small that drift-diffusion and hydrodynamic equations are not appropriate. Good accounts are in the literature [107, 596, 603–605] and in texts [81, 606]; ref. [108] is particularly useful for containing FORTRAN code for various algorithms. The method itself is treated in refs [63, 105, 106]. The diamond amplifier analysis benefits from treatments of Monte Carlo in refs [324, 586, 588, 589, 607]. Other electron sources that exhibit complex physics, such as the Monte Carlo modeling of transport through layers in flat cold cathode structures where space-charge effects lead to current self-quenching in a manner reminiscent of thermionic cathodes [150] are additionally useful.

The unfolding of the Monte Carlo method has general features captured by the flow chart of Figure 32.9. Its essential attribute is the *random* choices made at key decision points that affect particle trajectories, or in the concise description by Peter J. Price [107],

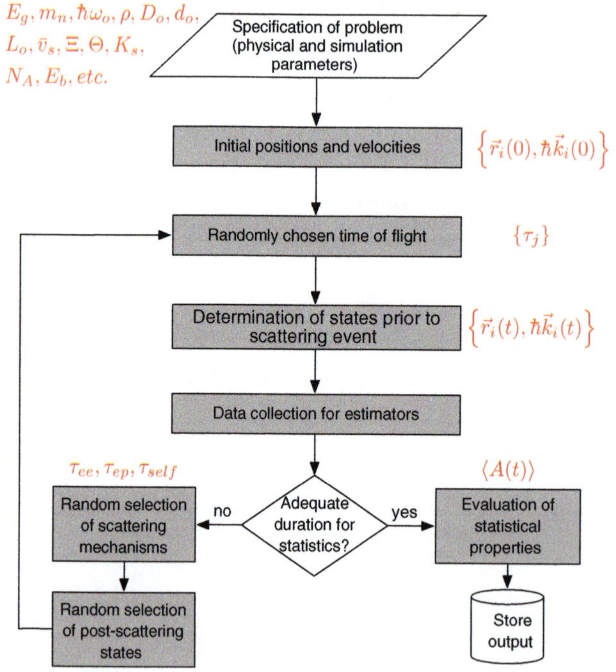

Figure 32.9 The main characteristics of a Monte Carlo program for the simulation of carrier (electron or hole or both) transport in semiconductors, based on a flow chart developed by Jacoboni and Reggiani [603]. Red equations suggest parameters associated with that element of the program.

[5]William Blake, *Blakes Selected Poems.* New York: Dover Publications, 1995, p. 58.

The motion of a mobile electron in a solid, in the quasi-classical representation, is a sequence of trajectories governed by the external fields (and by any internal macroscopic structure as in a junction) alternating with scatterings determined by interaction with lattice motion and other microscopic deviations from crystal perfection. Both the duration of a trajectory ("path") before termination by a scattering, and the electron state resulting from the scattering, are to be treated as random variables with given distributions. These random variables are produced in a determinate way as functions of one or more values of a standard random variable.

as reflected in the prominent features of the flow chart of Figure 32.9 identifying where reliance on random numbers and the decision point associated with scattering events occur. When transport can be treated using a semiclassical Boltzmann transport equation, a pedagogical approach suffices, but when device structures become much smaller so that quantum mechanical effects (discrete energy levels and wave packet interactions with scattering centers) must be considered, then the semiclassical methods lose their validity and what Monte Carlo entails as a consequence becomes much more involved [608]. The simple models are sufficient to treat the temporal shape of emitted secondaries.

32.2.1 Poisson statistics

Now I'm winding up staring at an empty glass
Ah ah ah ah ah ah ah ah.
Cause it's so easy to say that you'll forget your past.

– Greg Kihn and Steve Wright[6]

Electron emission, radioactive decay, line queueing, and yes, scattering, are kinds of random processes in which the occurrence of a previous event (emission of an electron, scattering with something) does not affect the likelihood of the next event, in contrast with events that are very much affected by the outcome of the prior event (having grandchildren, Russian roulette). That is, the processes are said to have no memory of what has come before. The noise associated with individual or discrete electron emission events, for example, is called "shot noise" and is of great interest for thermal [148, 221, 609, 610] and field emission cathodes [150, 175, 611–615]. Such a process is characterized by a single time scale τ used to parameterize the passage of time between subsequent events, a rather remarkable state of affairs. Why the simple assumption of processes which "forget their past" results in an exponential relation in time is worth investigating.

So what is the question? The one posed by Hamlet is a good start: existence of a state or its absence pretty much covers all cases, and if the probability of occurrence is p, then Hamlet's formulation is equivalent to $1 = p + q$, which is the same as $q = 1 - p$ after rearranging.[7] The self-evident simplicity of the single event is rapidly complicated when more than one event is considered. For the harder question of a maximum of n events occurring in a measure of time, it is still the case that all possible outcomes are accounted for by

$$1 = (p+q)^n = \sum_{j=0}^{n} \frac{n!}{k!(n-k)!} p^k q^{n-k} \tag{32.14}$$

where the binomial theorem in Eq. (5.16) is met once again. Observe that the sum is just the probability that an event did not occur at all ($k = 0$) plus the probability that it occurred only once ($k = 1$) and so on till the last instance where every chance an event could have occurred for n times in fact occurred. The combinatoric factor $n!/k!(n-k)!$ (Section 4.4) out front allows for the order of the events not mattering, only that they occurred in the span of time $t = n\,\delta t$. For example, the probability an event occurs three times for $n = 4$, given $1 = p^4 + q^4 + 4pq^3 + 6p^2q^2 + 4p^3q$, is then $4p^3(1-p)$.

For a given time duration, if the unit of time were, say, reduced (that is to say, n increased), then the probability p that an event occurring in that amount of time would have to be smaller, even as np stayed the same. This makes intuitive sense: a cold thermionic cathode, for example, may emit electrons so infrequently (the probability of emission being related to $J_{RLD}(T)$) that changing the measurement of the time between events in units of seconds to milliseconds should not affect the measured estimate of number of electrons emitted per hour by much, for example at 350 K and $\Phi = 2$ eV a barium dispenser cathode of area 1 cm^2 emits five electrons an hour on average. A Poisson process, then, conceptually takes the number of time units $n \to \infty$, even as $np \to \lambda$ is held fixed. Conceivably, however, two of those aforementioned six electrons could be emitted within picoseconds of each other, even though the average time between emission events is far longer: being uncorrelated events (no memory), the duration of time to the next event could be anything.

---

[6] Greg Kihn and Steve Wright, lyrics to *The Breakup Song (They Don't Write 'Em)*, 1981.

[7] Hamlet's formulation *"To be, or not to be – that is the question"* (ref. [37], *Hamlet* III.i.56) is then rephrased as "p or not-p", as it were.

Now the question is: what is the probability that an event (like the emission of an electron or its scattering) has occurred k times in the length of time under consideration, given that the number of time units n making up that interval becomes very large?[8] That is, what is the large n limit of Eq. (32.14) for a fixed λ?

EXAMPLE: Assuming $p = \lambda/n$, find

$$P_k(\lambda) \equiv \lim_{n\to\infty} \frac{n!}{k!(n-k)!} p^k q^{n-k} \tag{32.15}$$

SOLUTION: Use Stirling's approximation (Eq. (A2.4)) to find

$$\frac{n!}{k!(n-k)!} p^k q^{n-k} \approx \left(1 - \frac{\lambda}{n}\right)^{n-k} \left(1 - \frac{k}{n}\right)^{k-n-1/2} \frac{\lambda^k}{k!} e^{-k}$$

For small x, use $(1+x)^n = e^{n \ln(1+x)} \approx e^{nx}$ to show

$$\left(1 - \frac{\lambda}{n}\right)^{n-k} \approx \left(1 - \frac{\lambda}{n}\right)^{-k} \exp\left(-\lambda - \frac{\lambda^2}{2n}\right)$$

$$\left(1 - \frac{k}{n}\right)^{k-n-1/2} \approx \exp\left(k + \frac{k(1-k)}{2n}\right)$$

As a result

$$P_k(\lambda) \approx \lim_{n\to\infty} \frac{\lambda^k}{k!} \left(1 - \frac{\lambda}{n}\right)^{-k} \exp\left(-\lambda + \frac{k(1-k) - \lambda^2}{2n}\right)$$

The $n \to \infty$ limit can be taken forthwith, leaving

$$P_k(\lambda) = \frac{\lambda^k}{k!} e^{-\lambda} \tag{32.16}$$

Generating random times between events and then measuring how many events occur before the time between the start of the clock and last event exceeds λ shows just how good P_k is as an approximation to the distribution, as suggested in Figure 32.10 using the algorithm of Section A3.30.1. The $k = 0$ term has a special place: it is the probability that *no* events occur in the time $t = 1$ unit, for which $P_0 \propto \exp(-\lambda)$. For an arbitrary number of units, this becomes $P_0(t) \propto e^{-\lambda t}$, or the probability that an event has *not* occurred decreases exponentially with time. Taking into account that integrating over all time must result in a probability of unity, then

$$P_0(t)dt = e^{-\lambda t} \lambda dt \tag{32.17}$$

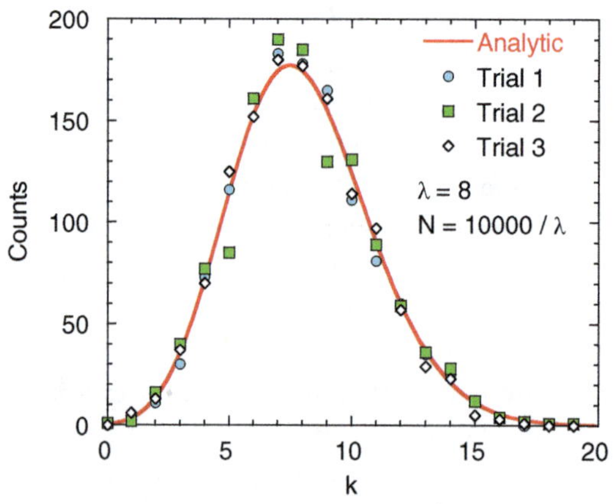

Figure 32.10 Distribution of events for sequentially emitted particles Eq. (32.16) (red line) in three simulations or trials based on the algorithm of Section A3.30.1. Events randomly occur until the time separation between first and last exceeds λ. The number of events is $N/\lambda = 1250$ (based on ref. [209]).

[8] The time unit must be small so that the time between successive events can be resolved regardless of how close they are, which accounts for the $n \to \infty$ limit.

32.2.2 Random number generation

...Get your facts first, and ...then you can distort 'em as much as you please.

– Mark Twain[9]

Deep in Section A3.30.1 is the relation $x_{j+1} = x_j - \ln(r)$, where x_j is represented by `xt` in the algorithm, and the random number $0 \leq r \leq 1$ is represented by `rand()` in the programming language of the example. Why $-\ln(r)$? There is a reason for that, and it is based on how distributions of random numbers are generated.

Let $P(x)$ be the probability distribution such that $P(x)dx$ is the probability that an "event" has occurred. It is useful to have a method to construct $P(x)$ from the uniform distribution $\mathcal{U}(0,1)$ of random numbers $0 \leq r \leq 1$.[10] The acceptance–rejection (AR) method of von Neumann [63] then provides a means to generate random numbers distributed according to $P(x)$ from a uniformly generated random number r. The algorithm is

1. Generate two random number r_1 and r_2.
2. Compare r_1 to $P(r_2)$.
3. If $r_1 < P(r_2)$ then accept r_1; if not, try again.

Although such a method works for any $P(x)$, if $P(x)$ has regions where it is sharply peaked and other regions where it is small, then employing the AR scheme can become inefficient, meaning more time than perhaps desired is spent in regions where $P(x)$ is small.

Some forms of $P(x)$ enable a more elegant method. The examples chosen here will be when $P(x)$ is linear and when it is an exponential (like $P_0(t)$ in Eq. (32.17)). $P(x)$ is normalized so that

$$\int_0^L P(x)dx = 1 \tag{32.18}$$

where L is the range of x beyond which the distribution $P(x)$ is zero (although L can be infinite, as in the case of $P_0(t)$). Separate the range of integration into regions that contain the same number of events. If there are n such regions, then

$$\int_{x_j}^{x_{j+1}} P(x)dx = \frac{1}{n} \tag{32.19}$$

for $j = 0, 1, \ldots n-1$ with n large, which serves to define the x_j. Observe that if $P(x)$ were constant (as it would be for a uniformly distributed random variable), then $1/n = \Delta x$ is the width of the region. Now consider the two cases:

- For the linear case with $L = 1$, then $P(x) = 2x$. Eq. (32.19) now entails that $x_j^2 - x_{j-1}^2 = 1/n$, or

$$x_j = \left(\frac{j}{n}\right)^{1/2} \tag{32.20}$$

  If n is large, then $(j/n) \approx r$ where r is the uniformly distributed random number. As a result, a random number x that is linearly distributed can be generated from a uniformly distributed r by $x(r) = \sqrt{r}$.
- For the exponential case where $P(x) = e^{-x}$ (λ is taken as 1 and $L \to \infty$), then Eq. (32.19) entails $\exp(-x_{j-1}) - \exp(-x_j) = 1/n$, or

$$x_j = -\ln\left(1 - \frac{j}{n}\right) \tag{32.21}$$

  Again, letting $(j/n) \to r$ identifies $x(r) = -\ln(1-r)$. However, if $0 \leq r \leq 1$, then so does $0 \leq (1-r) \leq 1$, and we can just as well make the replacement $r \to 1-r$ as it will not change the distribution of r. Therefore

$$x(r) = -\ln(r) \tag{32.22}$$

  a result corresponding to the $m = 0$ (no "occurrences") case of an algorithm described by Knuth (ref. [411], p. 132) for the treatment of Poisson distributions.

There are alternate methods to generate an exponential-like distributions to circumvent the computational burden of low probability events by grouping them into classes from which selections are made [450, 616], but the two options here are enough to explore. An algorithm to do so, which is used to generate Figure 32.11, is given in Section A3.30.2.

---

[9] Quoted in Rudyard Kipling, *From Sea to Sea: Letters of Travel* (1899). Project Gutenberg, EBook #32977, Contents of Part II, XXXVII, 2010, p. 167.
[10] Programming languages often have calls that generate a random number. In the language of MATLAB, it is `rand()`; in FORTRAN, it is `rnunf()`.

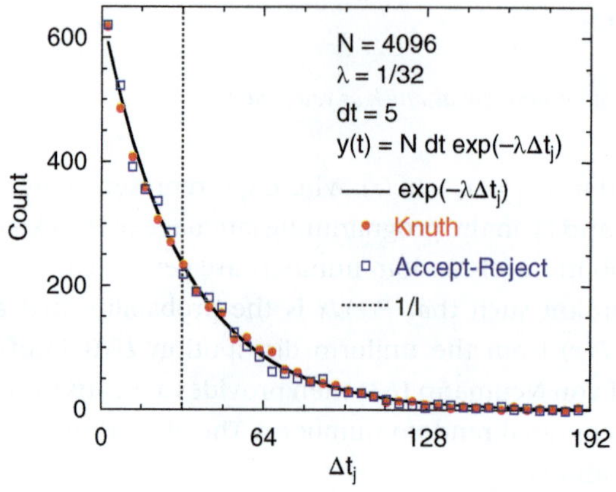

Figure 32.11 A comparison of Eq. (32.17) ($\exp(-\lambda \Delta t_j)$) with Eq. (32.22) (Knuth) and the AR algorithm. Based on Figure 2 of ref. [450].

32.2.3 Scattering and self-scattering

What fates impose, that men must needs abide;
It boots not to resist both wind and tide.

– William Shakespeare[11]

Now that a random number generation method has been developed to determine the time to the next scattering event $\Delta t = -\tau \ln(r)$, with τ being the average scattering duration, for a secondary electron based on Eq. (32.22), a means of determining the scattering mechanism must be chosen by a second random process. There are five or so scattering mechanisms that will form the core of those considered here, but choosing amongst them is not simply choosing them randomly. There is more art than that to the process.

The relaxation time τ has been cropping up regularly, from when Eq. (3.18) first made mention of it in introducing conductivity in the Drude model of Section 3.2 right up through its appearance in the optical parameters in the Drude–Lorentz model of Section 31.6. How τ appears in Eq. (7.18) (see also Eq. (31.63)) is of most interest here: it is a measure of how long a distribution f takes to "relax" back to its equilibrium distribution. Clearly, collisions are involved, wherein electrons scatter with their brethren, but also with lattice atoms, imperfections, and other impediments to their flow; it is also intuitive that each one of those processes takes their own particular toll on how departures from the equilibrium distribution evolve. One might, for example, write the impact of those various processes in Eq. (31.63) as (for simplicity, consider time independent and spatially uniform conditions)

$$-\frac{F}{\hbar}\partial_k f_0 = -\frac{1}{\tau_1} f_1 - \frac{1}{\tau_2} f_1 - \cdots = -\sum_{j=1}^{n} \frac{1}{\tau_j} f_1 \tag{32.23}$$

where each scattering process $\tau_j(k)$, of which there are n, has its own energy dependence. If scattering is characterized by an overall relaxation time τ, then this line of thought suggests *Matthiessen's rule* of

$$\frac{1}{\tau} = \sum_{j=1}^{n} \frac{1}{\tau_j} \tag{32.24}$$

Actually, Matthiessen's rule dealt with partial resistivities [52] where $\rho = \sum_j \rho_j$, but insofar as resistivity is the inverse of conductivity, or $\rho_j = 1/\sigma_j$, and $\sigma_j \propto \tau_j$, speaking of τ amounts to much the same thing. The probability that a scattering event characterized by τ occurs is then related to the sum of the probabilities for each scattering process.

Consider the overall probability for the occurrence of a scattering event in a short time interval δt, and represent it as $P\delta t$. The probability that a scattering *will not* occur in that same time interval is $1 - P\delta t$. If there are two time intervals, then the probability that an event will occur in the second interval is the product of the probability that nothing will happen in the first interval and something will in the second, or $W_2 = (1 - P\delta t_1)P\delta t_2$ (such language is reminiscent of

[11] Ref. [37]: *The Third Part of King Henry the Sixth,* IV.iii.58–59.

$W_n(k)$ in Eq. (4.27) of Section 4.4). That easily generalizes into the probability that something happens in the $(n+1)$th interval is (assuming all the intervals are the same δt)

$$W_{n+1} = (1 - P\delta t)^n = \exp\left[n \ln(1 - P\delta t)\right] \tag{32.25}$$

but if $\delta t \to dt$ becomes infinitesimally small, then the ln term may be expanded, or

$$W_{n+1} \propto \exp\left(-\sum_{j=1}^{n} P\delta t\right) Pdt \to W(t)dt = P\exp\left[-\int_0^t Pdt'\right] dt \tag{32.26}$$

Although it was seemingly implied that P was the same for each interval, observe that the treatment does not require this: P can change, for example, as the electron energy changes, and so

$$W(t)dt = P[\vec{k}(t)] \exp\left\{-\int_0^t P[\vec{k}(t')]dt'\right\} dt \tag{32.27}$$

Were P constant, however, then $W(t) \to W_o(t)$ could be written as

$$W_o(t)dt = \frac{1}{\tau}\exp(-t/\tau)dt \tag{32.28}$$

which would be very nice indeed: the total relaxation time τ could be thought composed of individual relaxation time components τ_j, from which one chosen at random (in proportion to how much τ depended upon it according to Eq. (32.24)) would dictate the nature of the scattering event. Nice, but not possible: a huge difference exists between Eqs. (32.27) and (32.28).

A trick to bridge the gap is to introduce another, or *self-scattering* term τ_0 which simply does not do anything: if the electron scatters according to it, the electron simply goes on its way unmolested. But the introduction of such a scattering term does bring Eq. (32.27) into Eq. (32.28), at the price of making the self-scattering τ_0 long enough so that τ takes on its maximum value (i.e., $1/\tau$ takes on the minimum value of $\sum_{j=1}^{n}(1/\tau_j)$ – mind the upper limit) which serves to define τ_{n+1} by

$$\frac{1}{\tau} = \sum_{j=1}^{n+1}\frac{1}{\tau_j} = \frac{1}{\tau_{n+1}} + \sum_{j=1}^{n}\frac{1}{\tau_j} \equiv \min\left(\sum_{j=1}^{n}\frac{1}{\tau_j}\right) \tag{32.29}$$

with $\tau_{n+1}(k)$ defined to make this so. That will make for many self-scattering events that just do not do anything. The price, however, is worth it: the calculation of the free-flight times becomes tractable [603].

In practice, then, for the secondary emission simulation, the choice of which scattering event occurs is found by creating a vector P, the elements of which are [209]

$$P_k = \frac{\sum_{j=1}^{k}\tau_j^{-1}}{\sum_{j=1}^{6}\tau_j^{-1}} \equiv \tau_{\min}\sum_{j=1}^{k}\frac{1}{\tau_j} \tag{32.30}$$

where the five scattering mechanisms of Section 32.3 are considered. For each electron, a random number $0 \le r \le 1$ is generated, from which the electron is allowed to travel for $t = -\tau_{\min}\ln(r)$, after which a scattering process characterized by τ_k is chosen from the five candidates, with τ_k determined by finding a k that satisfies $P_k \le r < P_{k+1}$. If none is selected because $r \ge P_5$, then a self-scattering event has occurred.

32.3 Relaxation time

Relaxation, then, is not an end; for it is taken for the sake of activity.

– Aristotle[12]

Understanding how to calculate the initial and final states of a particle (such as an electron) undergoing a scattering event based on the properties of the scattering potential, be it due to other similar particles, oscillations of the lattice

[12] Aristotle, *Nicomachean Ethics Book X* (trans. W.D. Ross), *Great Books of the Western World*, ed. Robert Maynard Hutchins, Vol. 9, Aristotle II. Chicago: Encyclopaedia Britannica, 1952, Chapter 6, p. 431.

(phonons), impurities charged or otherwise, and imperfections in the crystal structure is complex [80–82, 97, 294, 617, 618]. The specific components of most use to the scattering of secondaries in diamond will be the focus [85, 209, 308].

A potential $\hat{U}$ will change the dynamics of an incident electron, characterized by the unperturbed wave function $|\psi_{\mathbf{k}}(t)\rangle \equiv |\mathbf{k}\rangle e^{-i\omega_{\mathbf{k}} t}$ by scattering it into other states.[13] As in Sections 8.2 and 21.6, the final state is expressed as a sum over the unperturbed states with weights $c_{\mathbf{k}'}$. If the interaction potential is weak, then the original wave function will dominate or $c_{\mathbf{k}} \gg c_{\mathbf{k}'}$ for $\mathbf{k} \neq \mathbf{k}'$. This entails

$$|\psi(t)\rangle = \sum_{\mathbf{k}} c_{\mathbf{k}}(t)\,|\mathbf{k}\rangle\, e^{-i\omega_{\mathbf{k}} t} \tag{32.31}$$

where the unperturbed wave functions are the solutions to

$$\hat{H}_0\,|\mathbf{k}\rangle = E(\mathbf{k})\,|\mathbf{k}\rangle = \hbar\omega_{\mathbf{k}}\,|\mathbf{k}\rangle \tag{32.32}$$

Schrödinger's equation takes the form

$$i\hbar\partial_t\,|\psi(t)\rangle = \left(\hat{H}_0 + \hat{U}\right)|\psi(t)\rangle \tag{32.33}$$

which, in light of Eq. (32.31), becomes

$$i\hbar\partial_t c_{\mathbf{k}'} = \sum_{\mathbf{k}} c_{\mathbf{k}}\left\langle \mathbf{k}'\right|\hat{U}(t)\left|\mathbf{k}\right\rangle \exp\left[i\left(\omega_{k'} - \omega_k\right)t\right] \tag{32.34}$$

where $\mathbf{k}'$ is the final state (after scattering) and $\mathbf{k}$ is the initial state (prior to scattering). Letting $\hat{U}(t) = \hat{U}_o^{\mp} e^{\pm i\omega t}$ where the sign $\pm$ will come to denote the emission or absorption of phonons, then

$$i\hbar c_{\mathbf{k}'}(t) = \sum_{k} c_{\mathbf{k}}\langle \mathbf{k}'|\hat{U}_o^{\mp}|\mathbf{k}\rangle \frac{\exp\left[i\left(\omega_{\mathbf{k}'} - \omega_{\mathbf{k}} \pm \omega\right)t\right] - 1}{\left(\omega_{\mathbf{k}'} - \omega_{\mathbf{k}} \pm \omega\right)} \tag{32.35}$$

Now invoke the fact that the interaction potential is "weak" so that one state dominates (the *Born approximation*) in that $c_{\mathbf{k}'} = \delta_{\mathbf{k}\,\mathbf{k}'}$, so that

$$c_{\mathbf{k}'}(t) \approx \left\langle \mathbf{k}'\right|\hat{U}_o^{\mp}\left|\mathbf{k}\right\rangle \left\{\frac{\exp\left\{i\left(\omega_{\mathbf{k}'} - \omega_{\mathbf{k}} \pm \omega\right)t\right\} - 1}{i\hbar\left(\omega_{\mathbf{k}'} - \omega_{\mathbf{k}} \pm \omega\right)}\right\} \tag{32.36}$$

The probability that the scattered electron is in the state $|\mathbf{k}'\rangle$ at time t is $|c_{\mathbf{k}'}(t)|^2 dt$: the probability that it *ends up* in that state is then the integration of $|c_{\mathbf{k}'}(t)|^2 dt$ over all time. That appears daunting, but observe that the time dependence is wrapped up in the term in brackets, and if the substitution $\Omega \equiv \left(\omega_{\mathbf{k}'} - \omega_{\mathbf{k}} \pm \omega\right)$, then the time-dependent part of $c_{\mathbf{k}'}(t)$ looks like the far less threatening

$$\frac{e^{i\Omega t} - 1}{i\Omega t} = e^{i\Omega t/2}\frac{\sin(\Omega t/2)}{(\Omega t/2)} \tag{32.37}$$

which is a rather sharply peaked function. The integration over all time therefore contains an integration that looks like

$$\int_{-\infty}^{\infty}\left(\frac{\sin x}{x}\right)^2 dx = \pi \tag{32.38}$$

The integrand is sharply peaked about $x = 0$: knowing that the function $\phi_n(x) = (n\pi x^2)^{-1}(\sin nx)^2$ is a delta sequence [400], or a function that in the limit of $n \to \infty$ behaves as a Dirac δ function, then applied to the present circumstances,

$$\left|c_{\mathbf{k}'}(t)\right|^2 \approx \left|\left\langle \mathbf{k}'\right|\hat{U}_o^{\mp}\left|\mathbf{k}\right\rangle\right|^2\left(\frac{2\pi t}{\hbar}\right)\delta\left(E(\mathbf{k}') - E(\mathbf{k}) \pm \hbar\omega\right) \tag{32.39}$$

as long as the time between collisions is much longer than how long the collision takes to occur [80]. The *rate* at which transitions take place, or $|c_{\mathbf{k}'}(t)|^2/t \equiv S(\mathbf{k}, \mathbf{k}')$, is then

$$S(\mathbf{k}, \mathbf{k}') = S^{+}(\mathbf{k}, \mathbf{k}') + S^{-}(\mathbf{k}, \mathbf{k}') \tag{32.40}$$

$$S^{\pm}(\mathbf{k}, \mathbf{k}') = \left(\frac{2\pi}{\hbar}\right)\left|\left\langle \mathbf{k}'\right|\hat{U}_o^{\mp}\left|\mathbf{k}\right\rangle\right|^2\delta\left(E(\mathbf{k}') - E(\mathbf{k}) \pm \hbar\omega\right) \tag{32.41}$$

[13] In the present section, let the vector $\vec{k}$ be instead denoted by $\mathbf{k}$ to avoid clutter, with arrows above kets, subscripts, and primed terms.

As long as something does not interfere with the scattering occurring, then the *relaxation time* $\tau(\mathbf{k})$ is defined by

$$\frac{1}{\tau(\mathbf{k})} = \sum_{\mathbf{k}'} S(\mathbf{k}, \mathbf{k}') \tag{32.42}$$

If something does interfere, like a high concentration of carriers or the occupation of the final state (as occurs for metals and degenerate semiconductors), then accommodations do have to be made. Now Eq. (32.42) governs how often collisions occur, but knowledge of how long either momentum or energy takes to randomize through collisions is also of interest, necessitating the introduction of a *momentum* and *energy* relaxation time [80, 81]. When the change in momentum matters, and if the initial direction is defined as the $\hat{z}$ direction, then evaluation of the momentum relaxation time τ_m introduces a factor

$$\frac{\hat{z} \cdot (\mathbf{k} - \mathbf{k}')}{\hat{z} \cdot \mathbf{k}} = 1 - \frac{(\mathbf{k} \cdot \mathbf{k}')}{k^2} = 1 - \frac{k'}{k} \cos\theta \tag{32.43}$$

that is appended to $S(\mathbf{k},\mathbf{k}')$ in Eq. (32.40); for energy, the insertion of

$$1 - \frac{E(\mathbf{k}')}{E(\mathbf{k})} = 1 - \left(\frac{k'}{k}\right)^2 \tag{32.44}$$

defines the energy relaxation time τ_E when appended to Eq. (32.40). Observe that for *elastic scattering* where there is no change in energy (i.e., $k' = k$) then $1/\tau_E \to 0$, meaning there is no relaxation in energy.

Specifying the initial and final states, the nature of the potentials, the effects of band structure, and much else enormously complicates the evaluation of relaxation times for various processes that matter, more important ones being: (*i*) scattering off other electrons,(*ii*) acoustic phonon scattering, (*iii*) optical and polar optical scattering, (*iv*) ionized and neutral impurity scattering, and (*v*) others like piezoelectric scattering. The evaluation of the related relaxation times is sufficiently complex to have generated a large literature [80–82, 84, 85, 209, 294, 548, 559, 560, 617, 619, 620]. For describing how a bunch of scattering electrons behaves in transport through a diamond layer, though, the exact form is less important than the qualitative effect, and that affords some flexibility, here expressed as allowing two simple models in particular to stand in for the remainder in such a way that electron–electron and the electron–phonon scattering factors feel plausible when they are introduced rather than derived. The former have much in common with scattering off an ionized potential, while the latter are more like scattering off sinusoidal potentials. Although the discussion of the scattering times is motivated by considerations of secondary emission, the discussion is kept more general so as to be relevant to photoemission as well, particularly with regards to electron–electron scattering.

32.3.1 Ionized center

...somehow the arc of the moral universe is long but it bends toward justice.

– Martin Luther King, Jr.[14]

A bare Coulomb potential ($4Q/r$) can, allegedly like justice, bend the trajectory of a passing charged particle no matter how far away it is. In the parlance of scattering as a cross-section, the bare Coulomb potential therefore has an infinite cross-section [90]. For scattering as a relaxation time, τ can be related to cross-section σ and a volume V by the velocity v as

$$\frac{1}{\tau} = v\sigma V \tag{32.45}$$

An infinite cross-section σ therefore implies a vanishing relaxation time: a charged particle is *always* getting knocked off course. In materials, that will change due to the presence of screening.

Modifying the Coulomb potential by adding a screening factor $e^{-\kappa r}$ results in the Yukawa potential of Eq. (21.89): were the ionized core to be multiply charged, then $Q \to ZQ$ in the Coulomb part (e.g., a doubly ionized potential would have $Z = 2$), but for simplicity, ignore that complication. The operator $\hat{U}_o$ is

$$\hat{U}_o = \frac{4Q}{K_s \hat{r}} \exp(-\kappa|\hat{r}|) \tag{32.46}$$

[14]Martin Luther King, Jr., Sermon at the Temple Israel of Hollywood, 26 February 1965, http://www.americanrhetoric.com/speeches/mlktempleisraelhollywood.htm.

Recalling the methods of Sections 8.2.1 and 8.2.4 extended to three dimensions, with familiarity of Section 22.6 and in particular Eq. (22.36), then in some detail[15]

$$\begin{aligned}\langle \mathbf{k}|\hat{U}_o|\mathbf{k}'\rangle &= \frac{1}{V^2}\int d\mathbf{r}d\mathbf{r}'\,\langle \mathbf{k}|\mathbf{r}\rangle\langle \mathbf{r}|\hat{U}_o|\mathbf{r}'\rangle\langle \mathbf{r}'|\mathbf{k}'\rangle \\ &= \frac{1}{V^2}\int d\mathbf{r}d\mathbf{r}'\frac{4Q}{K_s r}e^{-\kappa r}\delta(\mathbf{r}-\mathbf{r}')\exp(i\mathbf{k}\cdot\mathbf{r}-i\mathbf{k}'\cdot\mathbf{r}') \\ &= \frac{1}{V}\int d\mathbf{r}\frac{4Q}{K_s r}e^{-\kappa r}\exp(i\mathbf{p}\cdot\mathbf{r})\end{aligned} \tag{32.47}$$

where V is volume and $\mathbf{p}=\mathbf{k}-\mathbf{k}'$. Writing out the integral explicitly, using $d\mathbf{r}=r^2\sin\theta\, dr d\theta d\phi$, then

$$\begin{aligned}\langle \mathbf{k}|\hat{U}_o|\mathbf{k}'\rangle &= \frac{2\pi}{V}\left(\frac{4Q}{K_s}\right)\int_0^\infty re^{-\kappa r}dr\int_{-1}^{1}e^{iprx}dx \\ &= \frac{1}{V}\left(\frac{4Q}{K_s}\right)\frac{4\pi}{\kappa^2+p^2}\end{aligned} \tag{32.48}$$

where $x=\cos\theta$ and the integrals are straightforward. As a result, using the continuum version of Eq. (32.42), and observing that $p=|\mathbf{k}-\mathbf{k}'|$ entails

$$p=\sqrt{k^2+k'^2-2kk'\cos\theta}\rightarrow\sqrt{2k^2-2k^2\cos\theta}=2k\sin(\theta/2) \tag{32.49}$$

because the energy δ function demands $k=k'$, then

$$\frac{1}{\tau(k)}=\frac{8}{V}\left(\frac{2\pi}{\hbar}\right)\left(\frac{4Q}{K_s}\right)^2\int_0^\infty\frac{k^2dk\delta[E(\mathbf{k})-E(\mathbf{k}')]}{\left[\kappa^2+4k^2\sin^2(\theta/2)\right]^2} \tag{32.50}$$

There are four modifications that now must be made. First, the relation is for only one scattering center. In a crystal, there are generally $N_i=V\rho_i$ scattering centers, where ρ_i is the number density of them, so that $1/\tau\rightarrow N_i/\tau\equiv 1/\tau_{ii}$. Second, to perform the integration over dk the relation

$$\delta[E(\mathbf{k})-E(\mathbf{k}')]=\delta\left[\frac{\hbar^2}{2m_n}(k^2-k'^2)\right]=\frac{m_n}{\hbar^2}\delta(k-k') \tag{32.51}$$

must be employed. Third, to get the momentum relaxation time, recall that the replacement $(1-\cos\theta)$ must be included in the integration. Lastly, introduce $\varsigma=2k/\kappa$, whereupon

$$\frac{1}{\tau_{ii}(k)}=\frac{\pi m_n}{2(\hbar k)^3}\left(\frac{4Q}{K_s}\right)^2\int_0^\infty\frac{\sin\theta(1-\cos\theta)d\theta}{\left[\varsigma^{-2}+\sin^2(\theta/2)\right]^2} \tag{32.52}$$

The integral is analytic, as can be shown by the substitution $x=\cos\theta$ and undoing the half-angle formula for $\sin^2(\theta/2)$. It gives

$$\int_0^\infty\frac{\sin\theta(1-\cos\theta)d\theta}{\left[\varsigma^{-2}+\sin^2(\theta/2)\right]^2}=4\ln(1+\varsigma^2)-\frac{4\varsigma^2}{1+\varsigma^2} \tag{32.53}$$

The result is the *Brooks–Herring* (BH) relation, for which

$$\frac{1}{\tau_{ii}(k)}=\frac{16\pi m_n\rho_i}{(\hbar\kappa)^3}\left(\frac{4Q}{K_s}\right)^2\frac{1}{\varsigma^3}\left(\ln(1+\varsigma^2)-\frac{\varsigma^2}{1+\varsigma^2}\right)\equiv\frac{1}{\tau_o G(\varsigma)} \tag{32.54}$$

where the replacement $k=\varsigma\kappa/2$ has been made so that the final result can be expressed in terms of the dimensionless ς in the function $G(\varsigma)$ defined by this equation.

Much hinges on the value that κ takes on. Recall that the screening factor κ was introduced to prevent infinities that otherwise attend scattering from a Coulomb potential: were there no screening ($\kappa\rightarrow 0$ or $\varsigma\rightarrow\infty$) then the relaxation time would go to infinity as $\varsigma^3/(2\ln(\varsigma))$ as in Figure 32.12. Alternately, in the *Conwell–Weisskopf* (CW) approximation, the infinities are avoided by truncating the offending parts of the integration at a lower scattering angle [80, 621]. But

[15]Here, it is easier to use the bold face notation for $\vec{r}\rightarrow\mathbf{r}$ to indicate vectors.

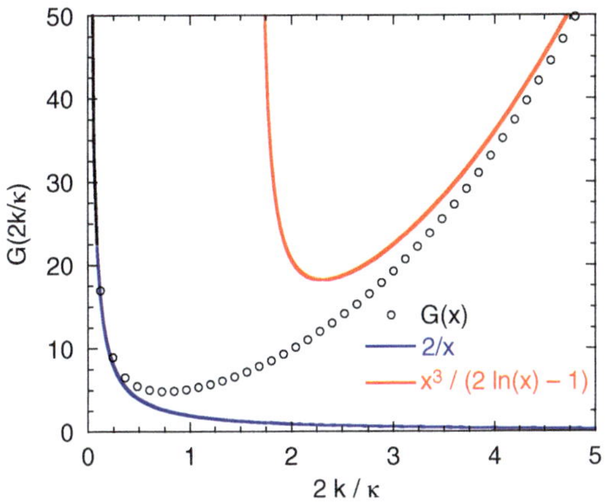

Figure 32.12 Behavior of $G(2k/\kappa)$ of Eq. (32.54), alongside its asymptotes.

here, analogous to the BH model using κ, then if the screening of the ionized impurities by electrons is governed by Debye's screening constant of Eq. (21.88), then

$$\varsigma = k\left(\frac{k_B T K_s}{4\pi Q\rho_o}\right)^{1/2} \tag{32.55}$$

The transition between the BH and CW approximations is governed by the length scale associated with the average separation of the charged scattering centers: if $1/(2\rho_i^{1/3}) > \kappa$, then use of the BH approximation is good [81, 621].

EXAMPLE: For parameters loosely suggested by diamond, or $m_n = 0.57m$, $T = 300$ K, $K_s = 5.7$, and assuming that the electron density and charged ion densities are $\rho_o = 10^{17}$ cm^{-3} and $\rho_i = \rho_e/10$, find $\tau_{ii}(k)$ when $\varsigma = 1$.
SOLUTION: In [nefq] units of Table 2.2, Eq. (32.55) becomes

$$\kappa = \left(\frac{K_s k_B T}{16\pi Q\rho_o}\right)^{1/2} = \left(\frac{5.7\cdot(300/11604.5)}{16\pi\cdot 0.35999\cdot 0.0001}\right)^{1/2} = 9.0241\ \text{nm}^{-1}$$

Setting $\varsigma = 1$ entails that $k = \kappa/2$ and

$$G(1) = \frac{1}{\ln(2) - (1/2)} = 5.1774$$

From

$$\tau_o = \frac{(\hbar\kappa)^3 (K_s/Q)^2}{256\pi m_n \rho_i} = \frac{209.56\cdot 250.71}{256\pi\cdot 3.2408\cdot 10^{-5}} = 2.0157\times 10^6\ \text{fs}$$

then $\tau_{ii}(k = \kappa/2) = \tau_o G(1) = 10.436$ ns.

32.3.2 Sinusoidal potential

Trying to understand the way nature works involves a most terrible test of human reasoning ability. It involves subtle trickery, beautiful tightropes of logic on which one has to walk in order not to make a mistake in predicting what will happen.

– Richard P. Feynman[16]

In contrast to the screened charge potential of Eq. (32.46), consider instead a potential that is oscillatory and extends over all space (it could, for example, be just one component of the Fourier representation of a real potential). The potential takes the form of

$$\hat{U}_o = 2\alpha_{\mathbf{q}} \cos(\mathbf{q}\cdot\hat{\mathbf{r}}) \equiv \hat{U}_o^+ + \hat{U}_o^- \tag{32.56}$$

[16] Richard P. Feynman, *The Meaning of it All: Thoughts of a Citizen Scientist.* Reading, Mass.: Perseus Books, 1998, p. 15.

which joins the usual $e^{i\mathbf{q}\cdot\hat{\mathbf{r}}}$ of Fourier transforms with its complex conjugate so as to render the potential $\langle \mathbf{r}|\hat{U}|\mathbf{r}\prime\rangle$ real. It is quick work to show

$$\begin{aligned}\langle \mathbf{k}|\hat{U}_o^{\pm}|\mathbf{k}'\rangle &= \frac{\alpha_{\mathbf{q}}}{V^2}\int d\mathbf{r}\, d\mathbf{r}'\, \langle \mathbf{k}|\mathbf{r}\rangle \langle \mathbf{r}|e^{\pm i\mathbf{q}\cdot\hat{\mathbf{r}}}|\mathbf{r}'\rangle \langle \mathbf{r}'|\mathbf{k}'\rangle \\ &= \frac{\alpha_{\mathbf{q}}}{V^2}\int d\mathbf{r}\, d\mathbf{r}' e^{i\mathbf{k}\cdot\mathbf{r}} e^{\pm i\mathbf{q}\cdot\mathbf{r}} e^{-i\mathbf{k}\cdot\mathbf{r}} \delta(\mathbf{r}-\mathbf{r}') \\ &= \alpha_{\mathbf{q}}\delta(\mathbf{k}-\mathbf{k}'\pm\mathbf{q})\end{aligned} \tag{32.57}$$

As a result,

$$S^{\pm}(\mathbf{k},\mathbf{k}') = \left(\frac{2\pi}{\hbar}\right)|\alpha_{\mathbf{q}}|^2\delta\left[E(\mathbf{k})-E(\mathbf{k}')\pm\hbar\omega\right]\delta\left(\mathbf{k}-\mathbf{k}'\pm\mathbf{q}\right) \tag{32.58}$$

and $S(\mathbf{k},\mathbf{k}') = S^{+}(\mathbf{k},\mathbf{k}') + S^{-}(\mathbf{k},\mathbf{k}')$ as in Eq. (32.40).

How $\omega(\mathbf{q})$ behaves is given by the dispersion relation. There are two types to consider, one being where $\omega(\mathbf{q}) = \omega_o$ is a constant and the other being where $\omega(\mathbf{q}) = v_s q$, where v_s is the sound velocity. Though not apparent yet, the second type occurs for small values of q and is associated with the collective motion of lattice atoms, and so is called the *acoustic* mode. The first type is correlated with lattice atoms moving out of phase with each other and because such motion interacts strongly with light, it is called the *optical* mode [53, 292]. Insofar as lattice motion can be treated in the language of particle interactions, the vibrations are termed "phonons". The literature [80–82, 294, 618] is extensive, whereas acquiring a feeling for what is involved will be more useful here (following ref. [85]).

Consider a linear chain of atoms, but of two types, and let the harmonic forces holding them in place be visualized as little springs, much like in Figure 32.13. If the masses were equal, or $M_A = M_B = M$, then the force on the nth atom connected to its nearest neighbors by springs with a spring constant g would be

$$M\ddot{u}_n = g\left[(u_{n+1}-u_n)-(u_n-u_{n-1})\right] = g\left(u_{n+1}+u_{n-1}-2u_n\right) \tag{32.59}$$

On the other hand, if the masses were different, then there would be two equations, one for atom n and the the other for its neighbor $n+1$, given by

$$\begin{aligned}M_A\ddot{u}_n &= g\left(u_{n+1}+u_{n-1}-2u_n\right) \\ M_B\ddot{u}_{n+1} &= g\left(u_{n+2}+u_n-2u_{n+1}\right)\end{aligned} \tag{32.60}$$

If solutions of the form $u_j = C_j \exp(i\omega t - ijqa)$ for $j = n$ or $n+1$, and where ja is the position of the jth atom, then Eq. (32.60) becomes

$$\begin{aligned}M_A\omega^2 C_A &= 2g\left(C_A - C_B\cos(qa)\right) \\ M_B\omega^2 C_B &= 2g\left(C_B - C_A\cos(qa)\right)\end{aligned} \tag{32.61}$$

Two equations for two unknowns C_A and C_B suggests a matrix solution of the form

$$\begin{bmatrix} M_A\omega^2 - 2g & 2g\cos(qa) \\ 2g\cos(qa) & M_B\omega^2 - 2g \end{bmatrix}\begin{pmatrix} C_A \\ C_B \end{pmatrix} = 0 \tag{32.62}$$

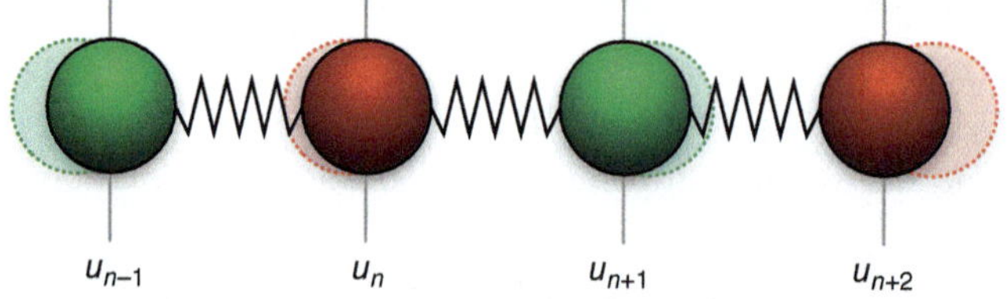

Figure 32.13 An arrangement of two types of atoms (A = green, B = red) on a linear chain characterized by different masses M_A and M_B, and connected by springs with a spring constant g. The positions of the nth mass is denoted $u_n(t)$: the solid spheres show the equilibrium position, and the dashed faded spheres their positions at some time t.

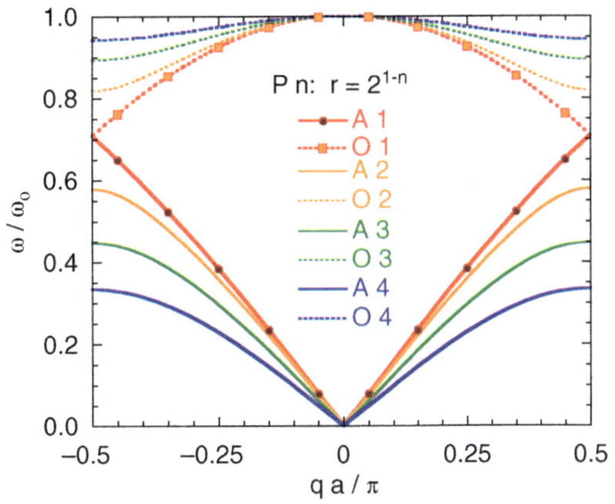

Figure 32.14 The behavior of Eq. (32.64) for various values of r. The lines are labeled by $[P, n]$, where P is either acoustic (A) or optical (O), and n refers to the power of $r = 2^{1-n} = 1$, 1/2, 1/4, and 1/8 for the curves shown. The acoustic branch behaves as $v_s q$, and the optical branch as ω_o, for q small.

For a solution, the determinant of the 2×2 matrix must equal zero, and so after a bit of manipulation,

$$M_A M_B \omega^4 - 2g\left(M_A + M_B\right)\omega^2 + 4g^2 \sin^2(qa) = 0 \tag{32.63}$$

for which the quadratic equation yields the solutions. With the introduction of the ratio $r = M_A/M_B$, the "reduced mass" $M = M_A M_B/(M_A + M_B)$, and the frequency $\omega_g \equiv \sqrt{2g/M}$, the solutions take on the more compact form of

$$\left(\frac{\omega_\pm}{\omega_g}\right)^2 = \frac{1}{2}\left[1 \pm \sqrt{1 - \frac{4r}{(1+r)^2}\sin^2(qa)}\right] \tag{32.64}$$

The behavior of $\omega_\pm/\omega_g$ given in Eq. (32.64) is shown in Figure 32.14, from which it is seen that the small q values for r small to $r \approx 1$ are visually similar.

Only because it is particularly simple to do so, consider the behavior of $\omega_\pm/\omega_g$ for $r = 1$ rather than general r, for which

$$\left(\frac{\omega_\pm}{\omega_g}\right)^2 = \frac{1}{2}\left(1 \pm \cos(qa)\right) \tag{32.65}$$

Finding the behavior of $\omega_\pm$ to order (qa) amounts to the replacements $\cos(qa) \to 1$ and $\sin(qa) \to qa$. For ω_+, it is easily seen to be

$$\omega_+ \to \omega_o \sqrt{(1/2)(1+1)} = \omega_o \tag{32.66}$$

The same replacements for ω_- take a bit more finesse as the replacement appears to make the right-hand side vanish. Consider instead the equivalent expression

$$\omega_- = \omega_o \frac{\sin(qa)}{\sqrt{2[1+\cos(qa)]}} \to \omega_o \frac{qa}{\sqrt{2[1+1]}} = \frac{qa\omega_o}{2} \tag{32.67}$$

which is easier than Taylor expanding the original relation to first order.

The two different behaviors have now been identified, with ω_- being the *acoustic* phonon mode and ω_+ being the *optical* phonon mode.

- **Acoustic:** the dispersion relation $\omega_- = v_s q$ gives the sound velocity as

$$v_s = \frac{1}{2}\omega_o a = a\sqrt{\frac{g}{2M}} \tag{32.68}$$

  where the A and B atoms are mostly moving in unison. In the spring metaphor, such coordinated motion does not overly compress the springs, so the energy of such oscillations is small.
- **Optical:** The frequency $\omega_+ \approx \omega_o$ is approximately constant for small q. It corresponds to when the A and B atoms are moving out of phase, that is, in opposite directions. As a result, the springs can be thought of as experiencing substantial compression (or elongation), so the energy of such oscillations is large.

Table 32.1 Sound velocity compared to the Bohm–Staver (BS) relation of Eq. (32.72) for longitudinal (long) waves. The ratio is measured over the BS values of v_S. Sound velocities similar to ref. [55]. Fermi energies based on ref. [86], p. 150, except for tungsten.

Element	v_s (long) [m/s]	μ [eV]	M [g/mole]	v_s (BS) v [m/s]	Ratio (m/BS)
Berylium	12890	14.14	9.012	10046.1	1.28
Magnesium	5770	7.13	24.305	4343.9	1.33
Aluminum	6420	11.63	26.982	5265.5	1.22
Iron	4994	11.10	55.845	3575.6	1.40
Copper	4760	7.00	63.546	2661.9	1.79
Zinc	4210	9.38	65.380	3037.8	1.39
Silver	3650	5.48	107.868	1807.7	2.02
Tungsten	5220	11.00	183.840	1961.8	2.66
Gold	3240	5.51	196.967	1341.4	2.42
Lead	2160	9.37	207.200	1705.5	1.27

The spring constant g can be related to the deformation potentials Ξ for various crystals [80, 85, 292]. Because the motion of the ions tends to drag the much lighter electrons with them, there is interaction between lattice vibrations and electrons, and such interaction can be described in the language of electron–phonon scattering. A quick understanding of what the sound velocity depends on can be understood from a statistical mechanical argument.

A method to approximate the sound velocity in metals is, curiously enough, to treat the positive ions as a plasma and talk of their "plasma frequency" [622], where ω_p was discussed in the context of Eq. (31.55). The electrons are light and will therefore adhere to the ions as they vibrate. A conceptual shortcut is to run through the discussion as though electrons were being discussed, but swap the electron mass m for the ion mass M (equivalently, think of the ions as just fat electrons). Recalling that the plasma frequency is then

$$\omega_p = \left(\frac{q^2\rho_o}{K_s\varepsilon_0 M}\right)^{1/2} = \left(\frac{16\pi Q\rho_o}{K_s M}\right)^{1/2} \tag{32.69}$$

where the first form is in the conventional [MKSA] units and the later in [nefq] units (Table 2.2), M represents the mass of the ion and K_s is the static dielectric constant. Recall that the dielectric constant gives rise to a shielded Coulomb potential $V(\vec{r}) = 4Qe^{-k_{TF}r}/r$ where k_{TF} is the Thomas–Fermi wavenumber of Eq. (21.85). From the Fourier transforms $V(\vec{k})$ and $V_o(\vec{k})$, evaluated much like Eq. (32.48), the relation $V(\vec{k}) = V_o(\vec{k})/K_s(k)$ results in

$$K_s(k) = \frac{k^2 + k_{TF}^2}{k^2} \tag{32.70}$$

which therefore entails that

$$\omega_p^2 = \frac{16\pi Q\rho_o k^2}{M(k^2 + k_{TF}^2)} \tag{32.71}$$

Taking the sound velocity as $v_s = d\omega_p/dk$ in the limit $k \to 0$ then gives

$$v_s^2 = \frac{16\pi Q\rho_o}{Mk_{TF}^2} = \frac{4a_o Qk_F^2}{3M} = \frac{2\mu}{3M} \tag{32.72}$$

where the relations $a_o = \hbar/\alpha mc$, $Q = \alpha\hbar c/4$, $\rho_o = k_F^3/(3\pi^3)$ (remember Eq. (6.24)) and $\mu = \hbar^2 k_F^2/2m$. Eq. (32.72) is known as the *Bohm–Staver* (BS) relation, and for the simple metals it gives a reasonable account of the sound velocity, as shown in Table 32.1.

32.3.3 Electron–electron scattering

Even if suffering makes you gentler and wiser, it is still bad for you. It is still a bad thing that this was the way you managed to become gentler and wiser, rather than some other way.

– Terry Eagleton[17]

[17] Terry Eagleton, *On Evil*. New Haven: Yale University Press, 2010, pp. 134–135.

What causes energetic and otherwise hot-headed electrons to calm down and equilibrate with their environment (or *thermalize*) is their mutually harsh interactions with each other. The scattering between carriers (particularly electrons) is quite important in photoemission models for metals as in Section 31.7, or in highly doped semiconductors, but is almost ignorable for scattering mechanisms operative in high purity diamond. Its importance though recommends τ_{ee} be treated in any discussion of scattering, and so it is.

Conceptually, the scattering between electrons bears similarity to the scattering of an electron off an ionized impurity, with the difference being that the ions are much more massive. The evaluation of the scattering time τ_{ee} is arduous [80, 85, 97, 548, 617, 623]. When undertaken, τ_{ee} is found to be[18]

$$\tau_{ee}(k) = \frac{2\hbar a_o}{\pi Q}\left(\frac{K_s E}{k_B T}\right)^2 \left\{\left[1+\left(\frac{E-\mu}{\pi k_B T}\right)^2\right]\gamma\left(\frac{2k}{q_o}\right)\right\}^{-1} \tag{32.73}$$

where μ is the chemical potential, and the other terms have their meanings as in Table 2.3. The function $\gamma(x)$ is defined by

$$\gamma(x) \equiv \frac{x^3}{4}\left(\frac{x}{1+x^2} - \frac{\arccos[1/(x^2+1)]}{\sqrt{x^2+2}} + \arccos\left[1/\sqrt{x^2+1}\right]\right) \tag{32.74}$$

and the parameter q_o is from the Thomas–Fermi approximation to k_{TF} (Eq. (21.85)) now modified by an inserted factor of K_s (5.2 corresponds to copper) to account for *d*-band contributions, or

$$q_o = \left(\frac{3\rho_o}{2K_s\varepsilon_0\mu}\right)^{1/2} = \left(\frac{4k_F}{\pi K_s a_o}\right)^{1/2} \tag{32.75}$$

These equations are busy, but can be simplified. For metals, $\gamma(x)$ is reasonably well approximated, when $(2k/q_o)^2$ is large, by

$$\gamma\left(\frac{2k}{q_o}\right) \approx \left(\frac{E}{\mu}\right)^{3/2}\gamma\left(\frac{2k_F}{q_o}\right) \tag{32.76}$$

Consequently, if τ_o is defined by

$$\tau_o \equiv \frac{2\hbar a_o}{\pi Q}\frac{(K_s\beta\mu)^2}{\gamma\left(2k_F/q_o\right)} \tag{32.77}$$

then (recalling that $E = \hbar^2k^2/2m$ and $\mu = \hbar^2k_F^2/2m$)

$$\tau_{ee}(k) \approx \tau_o(T)\left(\frac{k}{k_F}\right)\left\{1+\left[\frac{\beta(E-\mu)}{\pi}\right]^2\right\}^{-1} \tag{32.78}$$

The performance of $\tau_{ee}(k)$ (Eq. (32.73)) compared to the approximation of Eq. (32.78) is shown in Figure 32.15, for which the agreement is good.

Two limiting cases in the behavior of $\tau_{ee}(k)$ in terms of $\beta(E-\mu)$ arise. First, when $(E-\mu) \gg \pi k_B T$, then the $(1/k_BT)^2$ factor in the curly brackets of Eq. (32.78) cancels the same factor in τ_o of Eq. (32.77), rendering τ_{ee} approximately temperature independent. In the opposite limit $(E-\mu) \ll \pi k_B T$, when E hugs the Fermi energy μ, causes a different behavior to come into play. The average relaxation time $\langle\tau_{ee}(k)\rangle$ due to *e–e* scattering is more germane: the averaging is done over the probability that the initial scattering state is occupied (governed by $f_{FD}(E)$ and the final scattering state is unoccupied (governed by $1-f_{FD}(E)$). Because $f_{FD}(E)(1-f_{FD}(E)$ acts like a Delta function $\delta(E-\mu)$ as in Eq. (7.25), for low and room temperature conditions, then $\langle\tau_{ee}(k)\rangle \propto \tau_o$, and the temperature dependence of τ_o then suggests

$$\langle\tau_{ee}(k)\rangle \propto \frac{2\hbar a_o K_s^2}{\pi Q\gamma(2k_F/q_o)}\left(\frac{\mu}{k_BT}\right)^2 \rightarrow \frac{1}{A_oT^2} \tag{32.79}$$

The form $1/\langle\tau_{ee}\rangle \rightarrow 1/\tau_{ee} = A_oT^2$ is commonly encountered [86, 292] and is useful in considering intense laser heating of metals [206, 624, 625] where the electron population heats faster than the background lattice.

---

[18] The cited references can differ depending on the approximations made. The present text follows refs [85, 548, 617].

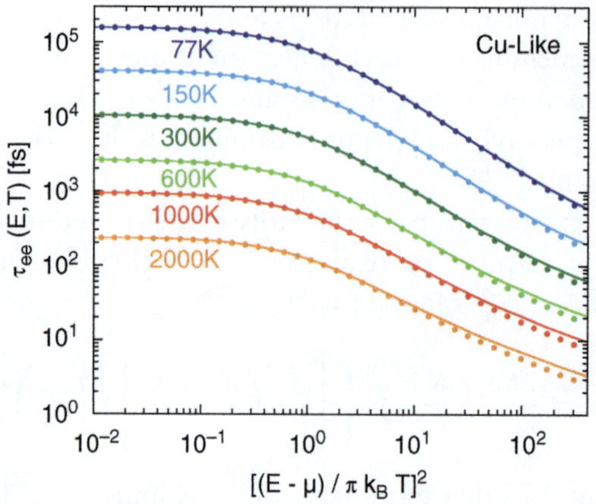

Figure 32.15 Comparison of τ_{ee} (Eq. (32.73)) with its approximation (Eq. (32.78)) as a function of $z = \beta(E - \mu)/\pi$ for values typical of copper ($\mu = 7$ eV, $m_n = m$, and $K_s = 5.2$.) as performed in Section A3.31. The approximate form works well over the range shown but begins to depart when $E - \mu$ becomes large with respect to $k_B T$.

32.3.4 Acoustic phonon scattering

Good, good, good, good vibrations
She's giving me excitations

– The Beach Boys[19]

Phonons are lattice vibrations, so that a proper account of them should quantize the energy levels of coupled harmonic oscillators (governed by a Bose–Einstein distribution $n(\hbar\omega_{\mathbf{q}})$) and explore how oscillations in the surrounding electron density (governed by a Fermi–Dirac distribution) varies with their motion. Such a program is difficult [80, 82, 97, 560, 617], but much can nevertheless be uncovered by starting more or less in the middle. If one is prepared to accept that[20]

$$\frac{1}{\tau_{ep}\left(\mathbf{k}_2\right)} = \frac{1}{(2\pi)^2\hbar}\int d\mathbf{q}\left|\alpha_{\mathbf{q}}\right|^2\left\{\left[f_+ + n\right]\delta_+ + \left[1 + n - f_-\right]\delta_-\right\} \tag{32.80}$$

where notational simplicity is achieved by dropping $\mathbf{q}$ and FD subscripts and $f_{FD}(E_2 \pm \hbar\omega_{\mathbf{q}}) \to f_\pm$, $n(\hbar\omega_{\mathbf{q}}) \to n$, and $\delta_\pm = \delta(E_2 - E_1 \pm \hbar\omega_{\mathbf{q}})$. Looking at the δ-function arguments (the vector $\mathbf{q}$ and its magnitude q do not refer to the unit of charge also called q – the context should make the distinction apparent)

$$E_2 - E_1(\mathbf{q}) \pm \hbar\omega_{\mathbf{q}} = \frac{\hbar^2}{2m}\left(q^2 + 2\mathbf{k}\cdot\mathbf{q}\right) \pm \hbar\omega_{\mathbf{q}} = \hbar\omega\left(\frac{v_k}{v_s}\cos\theta \pm 1\right) \tag{32.81}$$

where $v_k = \hbar k/m$ and v_s is the familiar sound velocity from the dispersion relation $\omega_{\mathbf{q}} = v_s q$, and the inconvenient $\mathbf{q}$ subscripts on $\omega_{\mathbf{q}}$ are abandoned. But this means that the δ functions only affect integrations over angle associated with $\mathbf{q}$, not its magnitude q. When those integrations are performed [85, 623], what is left resembles

$$\frac{1}{\tau_{ep}\left(\mathbf{k}\right)} = \frac{2\lambda\pi}{\left(2k_F v_s\right)^2}\int_0^{\omega_D} d\omega\,\omega^2 \mathcal{F}\left[\beta(E-\mu), \beta\hbar\omega\right]\Theta\left(v_k - v_s\right) \tag{32.82}$$

$$\mathcal{F}(x,y) = \frac{(\cosh x + 1)(\cosh y + 1)}{(\cosh x + \cosh y)\sinh y} \tag{32.83}$$

where λ is a constant (not a wavelength) born of $|\alpha_{\mathbf{q}}|^2$: it relates the dimensionless λ to the deformation potentials that govern lattice vibrations [80, 85]. The factor $\hbar\omega_D \equiv k_B T_D$ uses the Debye temperature T_D of Section 7.4.3: the frequency ω_D is specified by the maximum number of modes that can exist for the vibrations of the lattice. In a one-dimensional chain of N atoms, for example, there could not be more than N modes, and similar arguments apply to three dimensions.

[19] Brian Wilson and Mike Love, lyrics to *Good Vibrations*, 1966.
[20] If one is not, see ref. [85], Section III.F for its origin.

The thrust has been to obtain the function $\mathcal{F}(x, y)$ because near the Fermi level $E \approx \mu$, and so

$$\mathcal{F}\left[\beta(E-\mu), \beta\hbar\omega\right] \approx \frac{2}{\sinh(\beta\hbar\omega)} \tag{32.84}$$

Working with it gives

$$\frac{1}{\tau_{ep}\left(k_F\right)} = \frac{\lambda\pi}{\hbar\left(\hbar k_F v_s\right)^2}\left(k_B T\right)^3 \int_0^{x_D} \frac{x^2}{\sinh(x)} dx \tag{32.85}$$

where $x_D = T_D/T$ and $x = \beta\hbar\omega$. If x_D is small, as would be the case in a high-temperature limit, then

$$\tau_{ep}(k \approx k_F) \approx \frac{\hbar}{2\pi\lambda}\left(\frac{1}{k_B T}\right) \to \frac{1}{B_o T} \tag{32.86}$$

as often encountered in discussions of conductivity (e.g., the Drude model) and laser heating of metals that distinguish between the temperature of the lattice and that of the electrons over short times, as in Eq. (32.79).

The temperature behavior of acoustic phonon scattering is a bit more complicated, though. In switching over to the momentum relaxation time as per Eq. (32.43) and performing the integrations, the net effect is to introduce yet another factor of ω^2 into the integrand with $\mathcal{F}$ [97]. Recalling the evaluation of conductivity, which made use of $\langle\tau\rangle$ for which τ_{ep} is a dominant contribution, then the form gives rise to the *Bloch–Grüneisen* relation [561, 619, 626, 627] for which the integral form is

$$\frac{1}{\left\langle\tau_{ep}\right\rangle} \approx \frac{8\pi\lambda k_B T_D}{\hbar}\left(\frac{T}{T_D}\right)^5 \int_0^{T_D/T} \frac{x^5}{(e^x-1)(1-e^{-x})} dx \tag{32.87}$$

an equation which hearkens back to Eq. (7.44) in form. The form of the integral is a specific case of a more general one encountered for particles obeying Bose–Einstein and Fermi–Dirac statistics of the form [84]

$$W_\pm(n, x) = \int_0^x \frac{s^n}{(e^s \pm 1)(1 \pm e^{-s})} ds \tag{32.88}$$

so that the Bloch–Grüneisen function is equivalent to $W_-(5, x)$. Approximating that function is pedagogically useful.

Consider the high-temperature limit first, for which T_D/T becomes small: the small x behavior of $W_-(5, x)$ is found by Taylor expanding the denominator, or

$$W_-(5, x \ll 1) \int_0^x \frac{12 s^5}{s^2(s^2+12)} ds \approx 6x^2 - 72\ \ln\left(1 + \frac{x^2}{12}\right) \tag{32.89}$$

for which the leading order behavior is $(T_D/T)^4/4$. When coupled with the coefficient of Eq. (32.87), the linear behavior of $1/\left\langle\tau_{ep}\right\rangle \propto T$ for sufficiently high temperature results, as expected.

For low temperature, T_D/T becomes large. Rewrite the defining integral as $W_-(5, x) = W_-(5, \infty) - \delta W$ where

$$\delta W(x) = \int_x^\infty \frac{s^5}{(e^s-1)(1-e^{-s})} ds \tag{32.90}$$

Consider $W_-(5, \infty)$ first. Treating e^{-x} as an expansion parameter, then

$$W_-(5, \infty) = \sum_{n=1}^{\infty} n \int_0^\infty x^5 e^{-nx} dx = \Gamma(6) \sum_{n=1}^{\infty} n^{-6} = \Gamma(6)\zeta(5) \tag{32.91}$$

where the Riemann ζ function of Section 6.3 makes yet another appearance, $\zeta(6) = 1.03693$, and where $\Gamma(6) = 5! = 120$, so that $W_-(5, \infty) = 124.431$. As for $\delta W(x)$ it can be transformed into

$$\delta W(x) = -\int_0^{e^{-x}} \frac{(\ln(s))^5}{(1-s)^2} ds \tag{32.92}$$

An approximation may be had by noting s^5 can get rather large in Eq. (32.90) before e^{-s} dominates it (the maximum of $s^n e^{-s}$ occurs at $s = n$ and $e^5 = 148.413$), so that the approximation $e^{-s} \ll 1$ in the denominator allows for the consideration of the simpler integral

$$\begin{aligned}\delta W(x) &\approx \int_x^\infty s^5 e^{-s} ds \\ &= -\left(x^5 + 5x^4 + 20x^3 + 60x^2 + 120x + 120\right) e^{-x}\end{aligned} \tag{32.93}$$

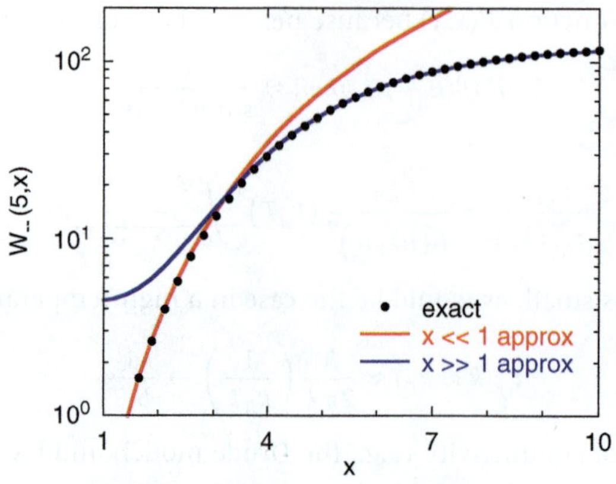

Figure 32.16 The Bloch–Grüneisen function $W\_(5,x)$ compared to its small and large approximations given by Eqs (32.89) and (32.93), respectively. An algorithm for its evaluation is given in Section A3.32.

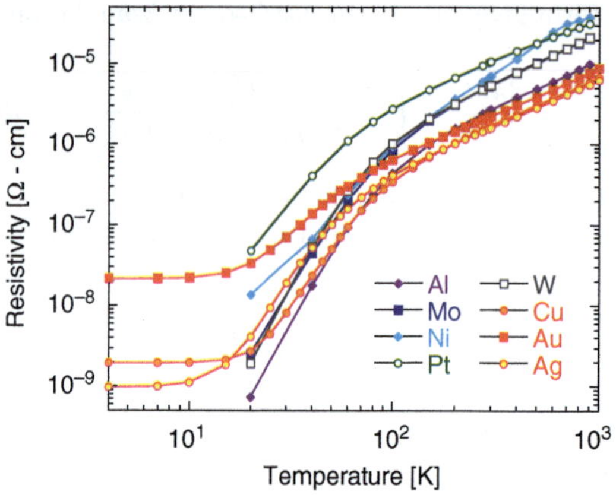

Figure 32.17 The resistivity of various metals. Although dominated by acoustic phonon scattering at most temperatures, at low temperatures other mechanisms come into play.

The limiting case approximations of Eqs (32.89) and (32.93) are compared with the numerical evaluation of $W\_(5,x)$ in Figure 32.16.

Conductivity $\sigma(T)$ is the inverse of resistivity $\rho(T)$, which for metals is rather varied, as in Figure 32.17. That variation seems to suggest an underlying simplification may be out of reach, but then Eq. (7.29) can be written as

$$\frac{\rho(T)}{\rho_o(T_o)} = \frac{\langle \tau(T_o) \rangle}{\langle \tau(T) \rangle} \tag{32.94}$$

where $\rho_o(T_o)$ is the resistivity at a reference temperature T_o. Insofar as acoustic phonon scattering is the primary contribution to resistivity, the ratio of the relaxation times in Eq. (32.94) can be approximated by the *acoustic* relaxation times (except for low temperature, where other relaxation times contribute via Eq. (32.24)). It is possible to use $W\_(5,x)$ to ascertain the Debye temperature first encountered near Eqs. (7.43) and (7.44). If so, and using $W\_(5,1) = 0.236616$, then

$$\frac{\rho(T)}{\rho(T_D)} = 4.22626\left(\frac{T}{T_D}\right)^5 W\_\left(5, \frac{T_D}{T}\right) \tag{32.95}$$

which is a substantial simplification: one ratio (T/T_D) controls the shape of the resistivity. The value of T_D must be found: a least squares algorithm can be used. Let the measured values of resistivity ρ_j be at the temperatures T_j (i.e., $\rho(T_j) = \rho_j$) and let the data be ordered by $T_j > T_{j-1}$. Then the quantity S defined by

$$S(T) = \sum_{j=1} N \left\{ \ln \left[\frac{\rho_N}{\rho_j} \left(\frac{T_j}{T_N}\right)^5 \frac{W\_(5,T/T_j)}{W\_(5,T/T_N)} \right] \right\}^2 \tag{32.96}$$

Table 32.2 The Debye temperatures T_D and the corresponding resistivity for various metals, as determined using the algorithms of Section A3.32. Values of T_D from ref. [86] (Kittel) are also shown.

Metal	T_D (fit)	ρ_D [$\mu\Omega$-cm]	T_D [86]
Ag	238.2	1.258	225
Au	184.6	1.339	165
Cu	349.0	2.056	343
W	385.8	7.492	400
Pt	188.8	6.318	240
Ni	519.2	19.197	450
Al	416.9	4.060	428
Mo	391.6	7.811	450

is minimized when $T = T_D$. An algorithm to do so, given in Section A3.32 for data culled from Table (9d-2a) of the AIP Handbook [55] for various metals, and from Matula [296] for copper, gold, and silver, is used to find the entries of T_D shown in Table 32.2. When the resistivity data is plotted in the coordinates of ρ_j/ρ_D vs T_j/T_D (where $\rho_D = \rho(T_D)$), then, as Figure 32.18 demonstrates, the resistivities of metals follow the behavior suggested by Eq. (32.95). As Table 32.2 shows in its comparisons to T_D from the literature, acoustic phonons catch most, but not all, of the resistivity behavior for typical temperatures. In some cases, the form $1/\tau \approx A_o T^2 + B_o T$ (Eqs (32.79) and (32.86)) does well in capturing what is of interest [206, 625], the coefficients being related to material parameters.

The cold realization that diamond is quite far from being a metal can no longer be contained. But diamond is a crystal, as are the metals, and it is electron migration through the lattice that gives rise to acoustic scattering. The acoustic phonon relaxation time can be transformed from its previous metal-friendly form [85] to one more appropriate [80, 209, 561] for examining the diamond amplifier. That form is

$$\tau_{ac} = \frac{\pi\hbar^3 \rho v_s^2}{4m_n \Xi^2\, k(E)\, k_B T_D} \left[\left(\frac{T}{T_D}\right)^5 W_{-}(5, T_D/T) \right]^{-1} \tag{32.97}$$

where $\rho = 3.51525$ g/cm^3 is the mass density, Ξ is the *acoustic deformation potential* (variously estimated at 8.8 eV [628], 8.7 eV [603], and even 5.5 eV [629]), and v_s is the sound velocity (18000 m/s) for diamond. Compared to metals, the Debye temperature for diamond is substantial, weighing in at 1860 K (or $k_B T = 0.1603$ eV). Therefore, to leading order, $\tau_{ac} \approx \pi\hbar^3 \rho v_s^2 / 4m\Xi^2 k_B T k(E)$ is good enough at room temperature. The hyperbolic relation (Eq. (23.27))

$$\frac{\hbar^2 k(E)^2}{2m_n} = E'\left(1 - \frac{E'}{E_g}\right) = E\left(1 + \frac{E}{E_g}\right) \tag{32.98}$$

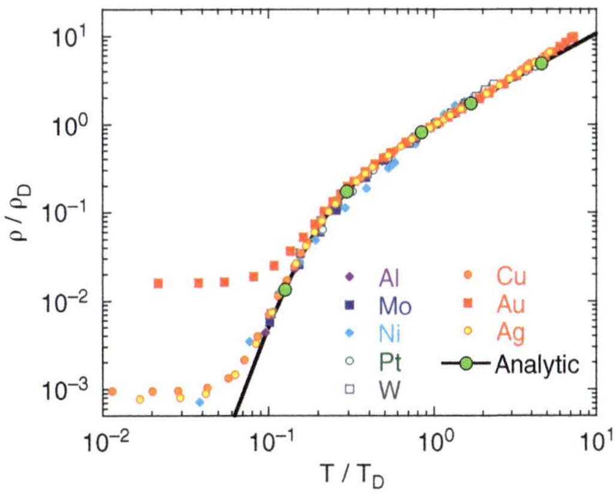

Figure 32.18 The resistivity values of Figure 32.17 but as $\rho(T)/\rho_D$ vs T/T_D, as in Eq. (32.95), for the Debye temperatures of Table 32.2.

relates k to $E = E' - E_g$, but the size of diamond's band gap ($E_g = 5.5$ eV) lessens the hyperbolic modification to the parabolic relation for electrons not much above the conduction band. For typical values of secondary electron energy, τ_{ac} starts in the range of 10–100 fs.

EXAMPLE: Find τ_{ac} in diamond for a secondary electron for two cases: first, for the secondary energy at its creation of 13.75 eV and, second, after it thermalizes to $k_B T$ at room temperature. Use the values $m_n = 0.57m$ and $\Xi = 8.7$ eV.
SOLUTION: In [nefq] units, $k = \sqrt{14.961 \times 13.75 \times 19.25} = 26.833$ nm^{-1}. Next, with $\rho v_s^2 = 7108.7$ eV/nm^3, then $\pi\hbar^3 \rho v_s^2 / 4k_B T_D m_n \Xi^2 = 40.495$ fs/nm and $k_B T_D = 0.16028$ eV. When $E = 13.75$ eV, $k(E) = 26.833$ [1/nm], whereas for $E = k_B T = 0.025852$ eV, it is $k(E) = 0.62336$ [1/nm]. With $T/T_D = 0.16129$, then $W\_(5, T_D/T) = 74.678$. So, $\tau_{ac} = 185.14$ fs for $E = 13.75$ eV and $\tau_{ac} = 7.9693$ ps for $E = 0.025852$ eV.

32.3.5 Optical phonon and ionized impurity scattering

Well something's lost, but something's gained
In living every day.

– Joni Mitchell[21]

The last two scattering processes to be considered are optical phonon and ionized impurity, although there are others [80–82, 306]. Non-polar optical phonon (NOP) relaxation times involve the emission or absorption of an optical phonon $\hbar\omega_o$ and are given by [294]

$$\tau_{NOP,e} = \frac{\sqrt{2}\pi\hbar^2\rho}{D_o^2 m_n^{3/2}} \frac{\hbar\omega_o}{\left(1 + n\left(\beta\hbar\omega_o\right)\right)\left(E - \hbar\omega_o\right)^{1/2}} \tag{32.99}$$

$$\tau_{NOP,a} = \frac{\sqrt{2}\pi\hbar^2\rho}{D_o^2 m_n^{3/2}} \frac{\hbar\omega_o}{n\left(\beta\hbar\omega_o\right)\left(E + \hbar\omega_o\right)^{1/2}} \tag{32.100}$$

where e and a subscripts correspond to "emission" and "absorption" of a phonon with energy $\hbar\omega_o$, and $n(x) = 1/(e^x - 1)$ is the Bose–Einstein distribution governing phonons, with $\beta\hbar\omega_o = \hbar\omega_o/k_B T = 6.3051$ at $T = 300$ K for diamond. The factor D_o is the energy per atomic displacement [80]. When the energy is high, emission dominates, and secondary electrons that are just born lose most of their excess energy via optical phonon emission. If there is an internal field in the diamond, then energy acquired through acceleration in the field can likewise be lost this way. The secondaries are initially borne with a rather large energy of $2.5E_g = 13.75$ eV: each non-polar optical scattering event associated with "emission" of a phonon releases $\hbar\omega_o = 0.163$ eV to the lattice (whereas absorption would have the electron acquire that energy). The dependence of the relaxation time on $(E - \hbar\omega)$ means that although NOP is very effective in reducing the electron energy while it is high, as the electron energy drops there comes a point where NOP fades in importance and acoustic phonon scattering dominates, as in Figure 32.21.

Although the diamond is grown as pure as possible, boron doping invariably occurs, albeit at small levels.[22] Nevertheless, it serves as a source of neutral and ionized impurity scattering. The neutral relaxation time is

$$\tau_{ni} = \frac{Q m_n c}{5\hbar^3 K_o N_n} \tag{32.101}$$

The ionized impurity rate is more involved: an asymptotic form [209, 308] valid when the phonon energy is small compared to the band gap is

$$\frac{1}{\tau_{ii}(E)} \approx \frac{\pi\hbar^3}{Q m_n^2}\left(\frac{N_i}{K_o^2}\right)\beta \ln(2) \frac{E_g^2}{\sqrt{E\left(E + E_g\right)}} \tag{32.102}$$

[21] Joni Mitchell, lyrics to *Both Sides Now*, 1969.
[22] A doping of 10^{14} atoms/cm^3 is considered small. It corresponds to about 100 atoms per cubic micron or, put another way, a boron atom every 0.215 μm.

In the absence of fields, the ionized impurity density N_i is obtained from the neutral impurity density N_n by Eq. (22.17), with $E_b = 0.365$ eV. At room temperature, $N_i/N_b < 2 \times 10^{-7}$. However, when fields are present the impurities are ionized within the depletion region (Section 24.4). These ionized boron atoms can shield out an externally applied field over a length (the depletion width) given by $L_D = F/(8\pi Q N_i)$, where F is the vacuum field (using the vacuum field bypasses a formula requiring the dielectric constant K_s). For example, if $F = 50$ eV/μm (corresponding to $\mathcal{E} = 50$ MV/m), then $L_D \approx 55.26$ μm. As diamond films thinner than a depletion width will be seen to be desirable, then the boron is assumed to be ionized throughout [209].

32.3.6 Scattering angles

I shall be telling this with a sigh
Somewhere ages and ages hence:
Two roads diverged in a wood, and I –
I took the one less traveled by,
And that has made all the difference.

– Robert Frost[23]

In a Monte Carlo simulation, knowing that a scattering event has occurred is only part of the problem. The other part is knowing down what road the scattered electron heads, and with what energy. Consider first a collision between equal mass particles (representing electrons in a metal) to understand how the energy and momentum are constrained by a possible lack of available final states, and then return to the problem of diamond which forbids final states being within its band-gap. What is allowed makes all the difference.

The collision of a primary electron (a) with a target (b) electron is most easily understood by shifting from the lab frame to a frame where the $\hat{z}$ axis is in the direction of the incident particle and the target particle is at rest, as in Figure 32.19. The notational convention is as follows: the velocity of the A electron (primary) in the rest frame of the B electron is designated $\vec{u}$. After the collisions, the daughter electrons A' and B' have velocities $\vec{u}'$ and $\vec{v}'$, respectively, in the B frame (primes therefore indicate post-collision electrons). Given that the masses of the particles are equal, the equations for conservation of energy and momentum become simple: in going from the lab frame to the rest frame, they are

$$E_a + E_b = E_a' + E_b' \Rightarrow u^2 = u'^2 + v'^2 \tag{32.103}$$

$$\vec{v}_a + \vec{v}_b = \vec{v}_a' + \vec{v}_b' \Rightarrow \vec{u} = \vec{u}' + \vec{v}' \tag{32.104}$$

To satisfy both equations is to ensure that $|\vec{u}' + \vec{v}'|^2 = u'^2 + v'^2$, in other words the vectors $\vec{u}'$ and $\vec{v}'$ must be orthogonal, or $\vec{u}' \cdot \vec{v}' = 0$. In terms of the rotation matrices (Section 8.2.1)

$$R_x(\theta) \equiv \begin{bmatrix} 1 & 0 & 0 \\ 0 & \cos\theta & \sin\theta \\ 0 & -\sin\theta & \cos\theta \end{bmatrix} \tag{32.105}$$

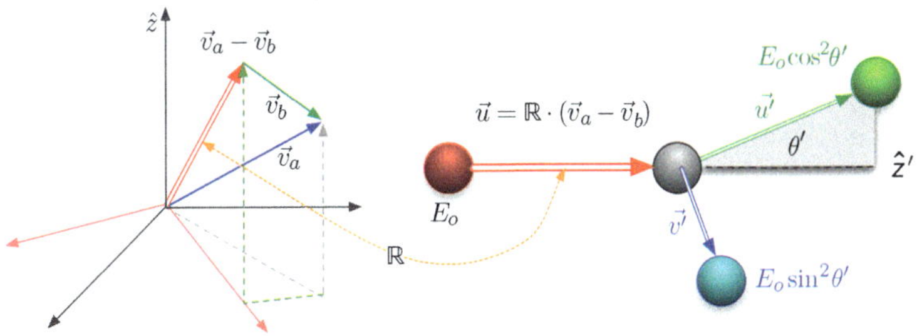

Figure 32.19 The frame in which to analyze the collision is best accomplished in the rest frame of the b electron. The coordinate system is rotated by the rotation matrix $\mathbb{R}$. In this rest frame, the daughter electrons emerge at 90° to each other, so the Pythagorean theorem can be used to find their energy.

[23] Robert Frost, *The Road Not Taken*, 1916.

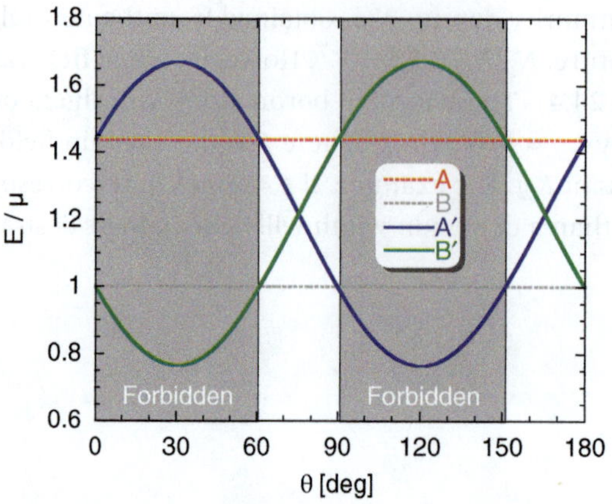

Figure 32.20 The final states of electrons A' and B', with energies E'_a and E'_b. The regions where one of them falls below the Fermi energy μ are grayed out: such final states are forbidden in a metal, as those states are already occupied.

$$R_y(\phi) \equiv \begin{bmatrix} \cos\phi & 0 & \sin\phi \\ 0 & 1 & 0 \\ -\sin\phi & 0 & \cos\phi \end{bmatrix} \tag{32.106}$$

$$R_z(\psi) \equiv \begin{bmatrix} \cos\psi & \sin\psi & 0 \\ -\sin\psi & \cos\psi & 0 \\ 0 & 0 & 1 \end{bmatrix} \tag{32.107}$$

then the rotation of the coordinate system corresponds to

$$\begin{aligned} \mathbb{R}\left(\varphi, \theta, \varphi'\right) &\equiv R_z(-\varphi)R_y(\theta)R_z(\varphi') \\ \vec{v}'_a &= \mathbb{R} \cdot \vec{u}' + \vec{v}_b \\ \vec{v}'_b &= \mathbb{R} \cdot \vec{v}' + \vec{v}_b \end{aligned} \tag{32.108}$$

Velocities parallel to $\vec{u}$ are not affected by a rotation $R_z(\varphi')$, as no components along $\hat{x}'$ or $\hat{y}'$ are present. Hence, the angle φ' can be randomly chosen as a rotation about the polar axis. That leaves θ': on it falls the burden of having to be compatible with conservation of energy and momentum for the final states or, to ensure that $\vec{u}' \cdot \vec{v}' = 0$ and $u'^2 + v'^2 = u^2$, then

$$\begin{aligned} \vec{u}' &= |\vec{u}| \cos\theta' \begin{pmatrix} \sin\theta' \\ 0 \\ \cos\theta' \end{pmatrix} \\ \vec{v}' &= |\vec{u}| \sin\theta' \begin{pmatrix} -\cos\theta' \\ 0 \\ \sin\theta' \end{pmatrix} \end{aligned} \tag{32.109}$$

The values that $|\vec{u}'|$ and $|\vec{v}'|$ can take are shown in the lines of Figure 32.20.

32.3.7 Allowed final states

The reason why they could never answer the question "How could it possibly happen?" is that it's the wrong question ... the question is "Why doesn't it happen more often?"

–Woody Allen[24]

[24] Woody Allen, *Hannah and Her Sisters*. New York: Vintage Books, 1987, p. 101.

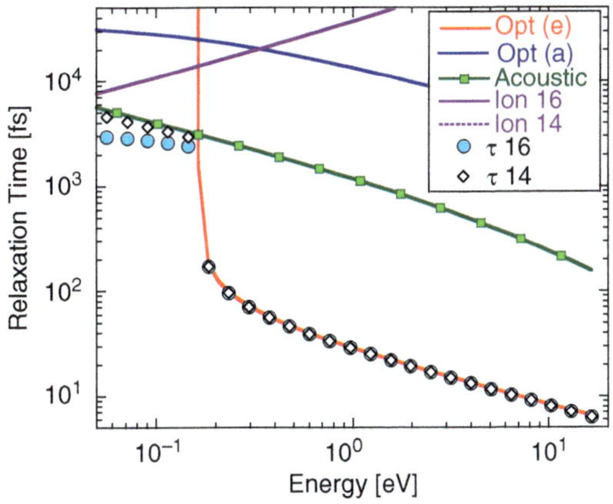

Figure 32.21 Relaxation time components and the sum τ via Eq. (32.24) for the dominant scattering mechanisms for diamond. Suffixes of "14" and "16" are $\log_{10} N_i$ [atoms/cm^3]. The ionized impurity relaxation time τ_{ii} for $N_i = 10^{14}$ atoms/cm^3 is not visible in the range shown. Based on Figure 4 of ref. [209] but using the values in Table 32.3.

Given the large number of electrons in a metal, it might seem that they should scatter incessantly, but their scattering is constrained to only be allowed if the final states are allowed; a similar constraint exists for semiconductors (recall Section 7.1). For metals, it is a reasonably good approximation that all states with energy below the Fermi energy are filled, and therefore neither of the scattered electrons (the primary one and its target) can have them as final states. As a consequence, not all of the solutions of Eq. (32.108) are possible. Rather, the "allowed final states" are – for metals – those for which $E_b' > \mu$: the least energetic of the daughter electrons still must possess an energy higher than the Fermi energy. Those constraints result in the gray bands of Figure 32.20.

For a semiconductor with a band gap E_g, an analogous situation holds: final states within the band gap are forbidden, but with an additional wrinkle in that the target electron must originate in the valence band, but both daughters must be in the conduction band. Because the photon energy must exceed a threshold for this to occur, a "magic window" develops in which electron–electron scattering is precluded, so that phonon scattering represents the only energy loss mechanism.

Using this understanding reveals that the final direction of the post-scattered daughter electron is not uniformly distributed in the polar angle θ (although it is in the azimuthal angle φ). Rather, the final state θ is obtained by generating a random number r and evaluating θ according to the *Conwell–Weisskopf* approach of [603]

$$\cos\theta_{CW} = \frac{rE^2 - E_{ii}^2}{rE^2 + E_{ii}^2} \tag{32.110}$$

or the *Brooks–Herring* approach of

$$\cos\theta_{BH} = \frac{4rE + (2r-1)E_{ii}}{4rE + E_{ii}} \tag{32.111}$$

where E_{ii} is evaluated according to

$$E_{ii} = \frac{4Q}{K_s}\left(\frac{4\pi}{3}N_A^-\right)^{1/3} \tag{32.112}$$

where N_A^- is the number density of the ionized boron acceptors. For diamond with $K_s = 5.7$, a value of $N_A^- \approx 10^{14}$ cm^{-3} entails that $E_{ii} = 1.89$ meV, so that most scattering events will not change the direction of the primary electron very much. Although the Brooks–Herring approach is favored at high energy conditions when the unintentional boron doping is small, both Brooks–Herring and Conwell–Weisskopf lead to preferential forward-directed scattering and the Conwell–Weisskopf approach has simplicity.

With the relaxation times in place, the effects of temperature, doping, and flake thickness can be investigated. For simplicity, ignore the band bending that boron doping entails (Section 24.4): although it makes a rather substantial difference in losses and transit times [209, 322], the constant-internal-field approximation ($F = F_{vac}/K_s$) behind the simulations to be discussed below represents, as it were, a "best-case scenario" in that transport through the diamond will be at its most rapid and expansion of the secondary bunch will be reduced.

32.4 Monte Carlo and diamond amplifier response time

...then the Father balanced his golden scales, and in them he set two fateful portions of death, which lays men prostrate, one for Achilleus, and one for Hektor, breaker of horses, and balanced it by the middle; and Hecktor's death-day was heavier and dragged downward toward death, and Phoibos Apollo forsook him.

– Homer[25]

In consequential matters, Monte Carlo methods can provide practical answers. One such matter is the the characteristics of electron bunches created from a diamond flake after a high-energy primary beam is incident on it, particularly with regards to its temporal shape. Let the primary beam of electrons be characterized by an energy E_o. Roughly $E_o/\Delta E$ secondaries are generated, each with an initial energy of about $2.5E_g = 13.75$ eV (e.g., for $E_o = 3$ keV, then 218 secondaries are generated). At such high energies, the internal field across the flake will have little sway over the directions the electrons are inclined to go, so that unless the field is particularly strong and the electron penetration range $R(E_o)$ rather deep, many of the secondaries will be lost to the back contact before they lose sufficient energy so that the field can direct their migration to the vacuum side of the diamond flake in earnest. If doping substantially reduces the field at the back contact, the losses can be up to 50%, another reason that the thinness of the flake is a design goal (recombination effects, whereby electrons join with holes and are lost, can also reduce the number of electrons [209, 589, 630]). The back contact can sweep the holes out of the simulation as well, and more complete Monte Carlo simulations which include them [588] show this quite well, an example being shown in Figure 32.22, demonstrating that the hole and electron populations in a diamond amplifier simulation can separate in a thin diamond layer relatively rapidly (in that figure, it occurs in under 3 ps). As a result, for the simple Monte Carlo to be considered below for analyzing how an electron bunch exits a thin diamond flake, and thereby estimating its rise/fall time, attention need only be focused on the electron population.

A constant field approximation is adequate when the boron concentration is $< 10^{14}$ cm^{-3} and the flake is under 10 μm. With the initial conditions and scattering mechanisms in place, Monte Carlo enables following the electrons' tortuous path through a thin diamond layer. Not surprisingly, and as previewed in Figure 32.22, the shape of the bunch expands

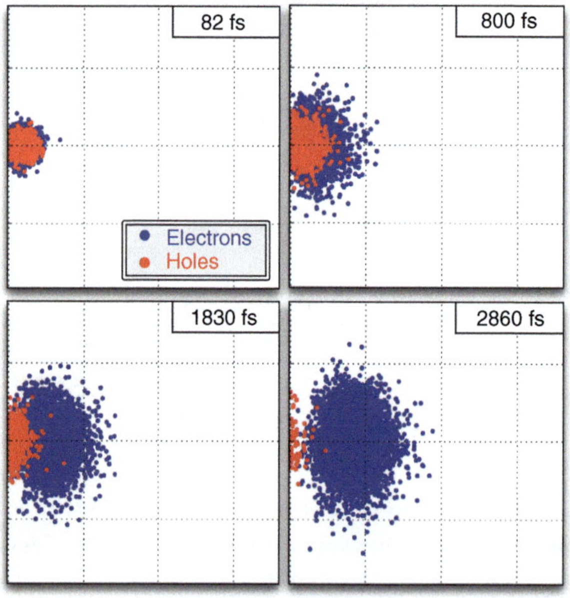

Figure 32.22 Electron (blue circles) and holes (red circles) generated by a 2.7 keV primary electron beam incident on a diamond surface from the left. A 3 MV/m applied field induces the electron cloud to drift away from the surface so that after 3 ps the holes have left the diamond region. Grid lines are every 0.5 μm. (Images courtesy of D. Dimitrov (Tech-X), and based on Figure 6 from D.A. Dimitrov, R. Busby *et al.* Multiscale Three-Dimensional Simulations of Charge Gain and Transport in Diamond, *J. Appl. Phys.*, **108**, 073712, 2010, with the permission of AIP Publishing).

[25] Homer (trans. Richmond Lattimore), *The Illiad*. Chicago: University of Chicago Press, 1951, Book 22, lines 209–213.

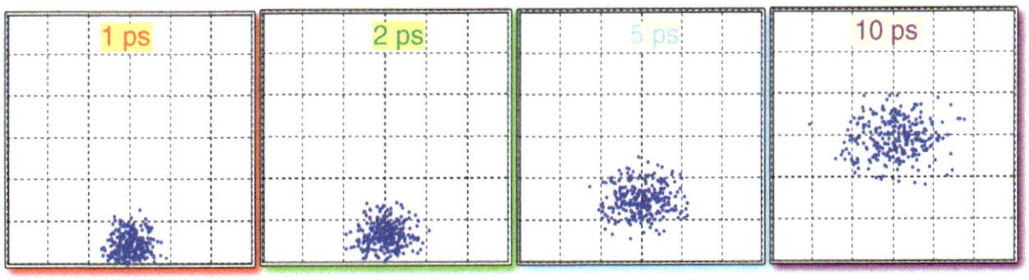

Figure 32.23 Side view of the transport of secondaries. Primaries are incident from the bottom and the field pulls them up to the top (vacuum side). Observe the apparent expansion of the bunch as it migrates. The numbers at the top give the time of the snapshot (1 ps, 2 ps, 3 ps, and 10 ps from left to right). Based on Figure 5(a) of ref. [322].

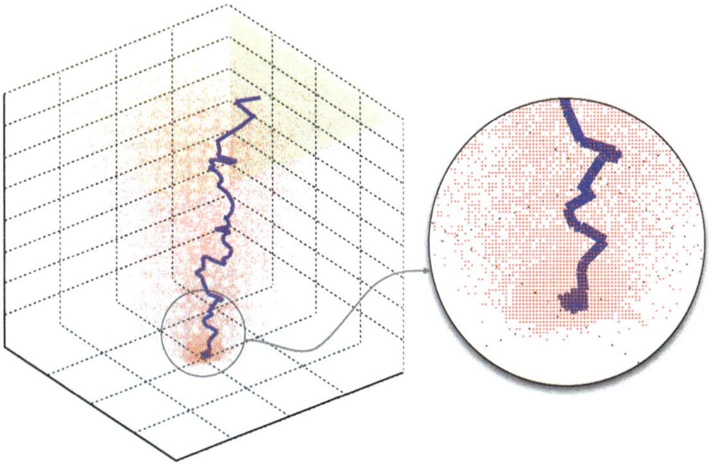

Figure 32.24 Sites of scattering (red dots) and the path of one electron (blue line). The vertical $\hat{z}$ scale is 8 μ, whereas the $\hat{x}$ and $\hat{y}$ axes are 4 μm. The blue hue base is the side from which the primaries are incident; the yellow hue vacuum side is at the top of the figure. Many (red) scattering events occur near where the secondaries are generated (as in inset) as they lose their excess energy to optical phonon (emission) scattering. The blue line indicates the trajectory quickly becomes diffusive and scattering ever present, which causes the bunch radius to expand. Based on Figure 3(a) of ref. [322].

as it is pulled across the flake by the internal field F, further illustrated in Figure 32.23. A measure of the number of scattering events, of which there are many, is indicated in two ways in Figure 32.24: first, by representing the location of every scattering event by a red dot and, second, by following the path of one electron as it migrates. The distribution of red dots reveals a very tight concentration of events near the injection site, where optical phonon emission processes rapidly sap the excess energy of the secondaries. When the energy of the electrons has dropped to the point where the internal field can begin to exert an inexorable draw to the vacuum side, acoustic phonon scattering dominates and the electron bunch acquires a diffusive character, the sphere of electrons expanding as it moves. Even a bunch excited instantaneously (a delta-function primary pulse) will expand, and as that diffusive sphere of electrons crosses into vacuum, the size of the bunch will affect the apparent rise/fall time of the emission process. A simple model can be made to account for it. Monitoring the number of electrons in the simulation is therefore an effective way of estimating losses to the back contact, quantifying when the bunch has started its migration across the flake in earnest, and – most importantly – getting a feel for how much time the bunch requires to exit the flake. An example is shown in Figure 32.25, wherein 364 electrons generated by one primary at an energy $E_o = 5$ keV are followed as they migrate through a 4 μm thick diamond flake subject to an internal field of $F = 34$ eV/μm (or $\mathcal{E} = 34$ MV/m). The regions naturally divide into (gray) losses to the back contact, (yellow) transport through diamond flake under field to the surface, and (blue) exit from the flake and the characterization of the rize/fall time.

Imagine that a uniform sphere of charge moves past a boundary at L (the width of the diamond flake), and let the center of the sphere be given by $z(t)$ and its radius by $r(t)$. How much of the sphere has passed the boundary $z = L$ is indicated by

$$s \equiv \frac{L - z(t)}{r(t)} \tag{32.113}$$

Electrons to the right of s have moved past L; those to the left are still in the diamond flake. When $s = 1$, the sphere has just touched the right boundary, and when $s = -1$, the entire sphere has passed $z = L$. The fraction of the sphere that has

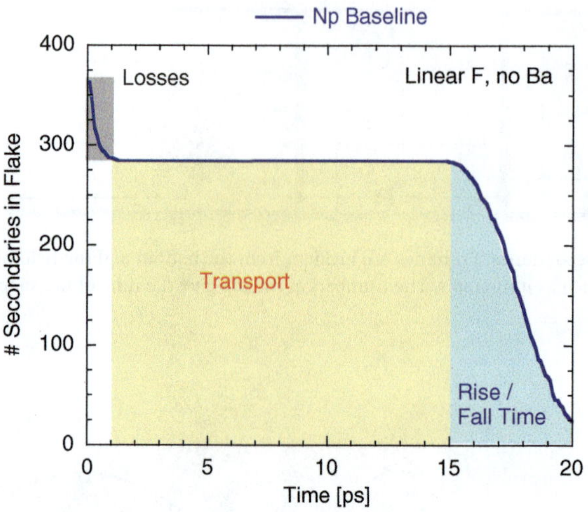

Figure 32.25 Characterization of the three regions of secondary generation, transport, and emission for $E_o = 5$ keV and $L = 4$ μm in a Monte Carlo simulation. Diamond is assumed free of boron impurities. Losses to the back contact result in a decline of the initial number of secondaries (gray). After they thermalize, they are pulled across the flake by the internal field (hellow). On encountering the surface, they are emitted (blue). How fast their numbers decline characterizes the rise/fall time.

passed into vacuum is then a summation over disks of area $\pi\rho^2$ with $\rho^2 = r^2 - z'^2$ and differential thickness dz'. If the total charge in the sphere is Q_o, then the uniform charge density is $Q_o/(4\pi r^3/3)$. The summation over all disks for which $z' > L - z$ is then

$$\Delta Q(s) = \frac{3Q_o}{4\pi r^3}\int_{L-z}^{r} \pi(r^2 - z'^2)dz' = \frac{1}{4}(2+s)(1-s)^2 Q_o \tag{32.114}$$

which has the proper behavior: at $s = 1$, no charge has escaped ($Q(s < -1) = 0$), when $s = 0$ half of the charge has escaped $Q(0) = Q_o/2$, and when $s = -1$, the entire sphere has just emerged ($Q(s > 1) = Q_o$).

Such a model is limited from the outset: a bunch of secondaries are better modeled by a diffusive sphere with a density that decreases with radius much like Eq. (31.125). Still, a uniform sphere model provides a rapid and qualitatively useful representation, and makes understanding the the speed of the sphere perhaps more intuitive. In a uniform field, how fast the sphere moves is related to the *saturation velocity* v_o of the semiconductor.[26] Why the electrons only accelerate to a limiting velocity value is heuristically understood to be when an equilibrium has set in between processes whereby the electron gains energy (acceleration in a field and optical phonon absorption) and processes whereby energy is lost (optical phonon emission). More concretely, an electron will accelerate in the internal field in diamond until it acquires enough energy to create a phonon. The mean free path for this to occur will be $v_o\tau_{e,a}$, where the subscripts on τ denote the different processes of absorption and emission. The rate at which the accelerated electrons shed their excess energy is therefore the difference between the absorption and emission rates, or

$$Fv_o \sim \hbar\omega_o\left(\frac{1}{\tau_a} - \frac{1}{\tau_e}\right) \tag{32.115}$$

On the other hand, the force on an electron is related to its change in momentum, which is the time between collisions (i.e., $F = dp/dt \approx \Delta p/\tau$). For electrons at a mean energy, the change in momentum after an emission is close to the negative of the change after an absorption, and so

$$F \sim \frac{\Delta p_a}{\tau_a} - \frac{\Delta p_e}{\tau_e} \approx m_n v_o\left(\frac{1}{\tau_e} + \frac{1}{\tau_a}\right) \tag{32.116}$$

Taking the ratio of these equations, and observing that in Eq. (32.99), because of the behavior of the Bose–Einstein factor $n(\beta\hbar\omega_o)$, the emission and absorption τ are separated in magnitude by an exponential factor e^x with $x = \beta\hbar\omega_o$,

$$v_o \sim \frac{\hbar\omega_o}{m_n v_o}\left(\frac{e^x - 1}{e^x + 1}\right) \tag{32.117}$$

26 The saturation velocity is often designated v_s in the sources, but will be called v_o here so as not to conflate it unintentionally with the sound velocity.

Table 32.3 Parameters used in the calculation of the relaxation times for high purity diamond (after Table 1 of ref. [209]).

Definition	Symbol	Value	Unit
Common			
Effective mass	m_n/m	0.57	–
Internal field ($q\mathcal{E}$)	F	10	eV/μm
Primary energy (E_o)	F	10	eV/μm
Band gap	E_g	3	keV
Secondary	ΔE	$2.5E_g$	eV
Static dielectric	K_s	5.7	–
Number density	N	0.2924	moles/cm^3
Mass density	ρ	3.51525	g/cm^3
Impurity density	N_A	$10^{14} - 10^{16}$	cm^{-3}
Boron activation energy	E_b	0.365	eV
Acoustic phonon			
Sound velocity	v_s	0.018	nm/fs
Deformation potential	Ξ	8.4	eV
Debye temperature	T_D	1860	K
Sound velocity	$\tilde{v}_s$	0.018	nm/fs
Non-polar optical			
Phonon energy	$\hbar\omega_o$	0.163	eV
Optical deformation potential	D_o	$\sqrt{3/2}(d_o/L_o)$	eV/nm
Deformation term	d_o	61.2	eV
Lattice constant	L_o	0.357	nm

or, solving for v_o,

$$v_o^2 \sim \frac{\hbar\omega_o}{m_n}\left(\frac{e^x-1}{e^x+1}\right) = \frac{\hbar\omega_o}{m_n}\tanh(x/2) \tag{32.118}$$

Showing a greater concern regarding the coefficients, the final form becomes, on replacing $x \to \hbar\omega_o/k_BT = \beta\hbar\omega_o$ [294],

$$v_o = \left[\frac{3\hbar\omega_o}{4m_n}\tanh\left(\frac{1}{2}\beta\hbar\omega_o\right)\right]^{1/2} \tag{32.119}$$

Making use of Table 32.3 parameters, $v_o = 1.9387$ nm/fs. The charge center of the sphere can therefore be approximated by the saturation velocity, or $z(t) \approx v_o t$. The current passing the surface boundary is then given by taking the time derivative acting on $z(t)$ (not $r(t)$, as even though the spherical bunch is expanding, its center of charge remains at the center), or

$$\Delta I(s) = -\frac{3v_o}{4r}Q_o\left(1-s^2\right)\Theta\left(1-s^2\right) \tag{32.120}$$

where $v_o/r(t)$ is a characteristic time that increases the longer the bunch spends in the flake and $\Theta(x)$ is the Heaviside step function (unity for positive argument, zero for negative), which insures that the current does not start until the sphere reaches the vacuum interface and stops when the sphere has passed into vacuum.

The bunches of a Monte Carlo simulation of a cluster of frenetically scattering electrons does not resemble a fixed density sphere so much as it resembles a diffusively expanding one. Rather than the hard boundary of Figure 32.26, the fuzzy sphere of Figure 32.23 is more in keeping with the probable behavior. The behavior of Eq. (32.114) bears a similarity to a hyperbolic tangent function, that is

$$\frac{1}{4}(2+s)(1-s)^2 \sim \frac{1}{1+e^{-y}} \tag{32.121}$$

for $y = 3s + O(s^3)$. Thus, there is reason to expect that

$$Q(t) \approx \frac{Q_o}{1+\exp[-(t-t_o)/\delta t]} \equiv Q_o h(t-t_o) \tag{32.122}$$

and in fact, monitoring the time dependence of particles passing into vacuum using a Monte Carlo simulation [209] shows the approximation works well, as in Figure 32.27: the surface of hydrogen-terminated diamond is NEA, so that

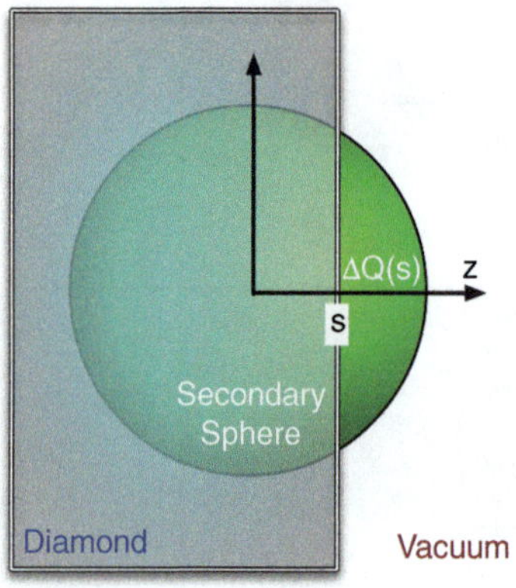

Figure 32.26 Side view of a sphere with uniformly distributed charge crossing the surface of a diamond flake into vacuum. $\Delta Q(s)$ is the portion of the sphere that has moved into vacuum, and s is the dimensionless measure of the position of the sphere with respect to the thickness L of the diamond flake. After on Figure 13 in ref. [209].

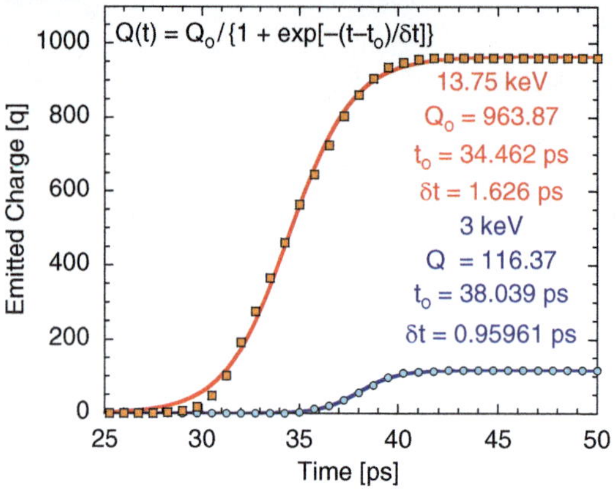

Figure 32.27 Fits to charge emitted into vacuum ($Q(s)$) for diamond-like parameters for an internal field $F = 10$ eV/μm and a flake thickness of 8 μm for two different primary energies of 3 and 13.75 keV. For optical phonons, $\hbar\omega_o = 0.163$ eV. Q_o, t_o, and δt are as occur in Eq. (32.122). The 3 keV line corresponds to Figure 14 of ref. [209].

there is no barrier to emission and what passes the vacuum boundary therefore contributes to $Q(t)$. The constant internal field approximation allows more of the initial secondaries for the higher primary energy $E_o = 13.75$ keV to survive than for $E_o = 3$ keV, as the penetration depth for the former is greater than for the latter, making the barrier to the back contact harder to surmount.

The parameters t_o and δt are dependent upon the primary energy E_o. For $E_o = 3$ keV and a flake of thickness $L = 8$ μm, then $Q_o \approx 116$, $t_o = 38$ ps, and $\delta t = 0.966$ ps, whereas for E_o of 13.75 keV, $Q_o = 963$, $t_o = 34.5$ ps, and $\delta t = 1.62$ ps. That is, a higher beam energy means fewer electrons are lost to the back contact and the greater penetration of the primaries results in a greater spread that affects δt. The rapidity with which $h(t)$ changes from near zero to near unity is the meaning of the rise/fall time, and δt is the parameter which sets it.

Consider now a string of bunches like pearls on a chain. The charge for a single bunch is as found above, namely, $\Delta Q(t) = Q_o h(t - t_o)$. The total current is then the sum over the individual bunch currents

$$\Delta I(t) = Q_o \frac{d}{dt} h(t - t_o) = \frac{Q_o}{\delta t} h(t - t_o) \left[1 - h(t - t_o)\right] \tag{32.123}$$

which, because of the similarity of the $h(t)$ functions in form to the Fermi–Dirac functions, means the form of Eq. (32.123) is similar to that of Eq. (7.24) in its peakiness. The total current is the summation over all the spherical elements distributed in time, or $t_o \rightarrow t'$ and

$$I(t) = \frac{Q_{tot}}{\delta t} \int_{-\infty}^{\infty} P(t') h(t - t') \left[1 - h(t - t')\right] dt' \tag{32.124}$$

where the distribution $P(t')$ is normalized. That is, the integration over all time should give the total charge Q_{tot} emitted, as easily shown:

$$\int_{-\infty}^{\infty} I(t)dt = \frac{Q_{tot}}{\delta t}\int_{-\infty}^{\infty} P(t')dt' \int_{-\infty}^{\infty} h(t-t')\left[1-h(t-t')\right] dt$$
$$= Q_{tot}\int_{-\infty}^{\infty} P(t')dt' = Q_{tot} \tag{32.125}$$

The diffusive expansion of a bunch of electrons in the diamond flake has consequences for how fast current can be turned on (rise to its maximum value) and turned off (reduced to a negligible level) in comparison to the duration Δt of the primary beam. In the simplest case, let $P(t)$ be a top-hat distribution in which the beam of primaries turns on at $t = -\Delta t/2$ and turns off at $t = \Delta t/2$. $P(t)$ can be represented as a product of two Heaviside step functions, or $P(t) = \Theta(\Delta t/2 + t)\Theta(\Delta t/2 - t)/\Delta t$, where the Δt in the denominator is from the normalization requirement. When inserted into Eq. (32.124) it follows

$$I(t) = \frac{Q_{tot}}{\delta t \Delta t}\int_{-\Delta t/2}^{\Delta t/2} h(t-s)\left[1-h(t-s)\right] ds$$
$$= \frac{Q_{tot}}{\Delta t}\left[h(t+\Delta t/2) - h(t-\Delta t/2)\right] \tag{32.126}$$

Because one h function is rising as the other falls, the choice of the integration region insures that the midpoint (at $t = 0$) corresponds to the maximum. Letting t_a and t_b be such that

$$p = \frac{I(t_a)}{I(0)}; \qquad 1 - p = \frac{I(t_b)}{I(0)} \tag{32.127}$$

then the rise/fall time corresponds to $t_a - t_b$ depending on the value p takes on (e.g., $p = 1\%$). For small p, a rewriting of Eq. (32.126) is useful, using the notation $u = \exp(t/\delta t)$ and $C = \exp(\Delta t/2\delta t)$, to render it as

$$\frac{I(t)}{I(0)} = \frac{u(C+1)^2}{(u+1)(C^2u+1)} \tag{32.128}$$

from which the limiting cases of pulse length large compared to rise/fall time ($\Delta t \gg \delta t$) can be contrasted with the reverse when the pulse length is smaller than the rise/fall time ($\Delta t \ll \delta t$). When $C \to 1$, the primary beam of electrons is deposited all at once ($\Delta t \to 0$), but when $C \to \infty$, the primary beam is very long compared to the processes that give rise to δt.

EXAMPLE: Find the rise/fall time for the $\Delta/\delta t = 0$ line of Figure 32.28, an *ad hoc* definition being the time separation between where $I(t)/I(0)$ changes from 99% to 1% ($\epsilon = 0.01$).

SOLUTION: The center of $I(t)$, or $t = 0$, is close to where $I(t_b)/I(0) = 1 - \epsilon \approx 1$. Therefore, $t_b/\delta t \approx 0$. For t_a, inserting $C = 1$ into Eq. (32.128) results in, where $a = \exp(t_a/\delta t)$,

$$\epsilon = \frac{4a}{(a+1)^2} \to a \approx \frac{4}{\epsilon}$$

Thus, $t_a - t_b \approx \delta t \ \ln(4/\epsilon) = 5.9915\ \delta t$.

EXAMPLE: Find the rise/fall time for the $\Delta/\delta t = 32$ line of Figure 32.28, an *ad hoc* definition being the time separation between where $I(t)/I(0)$ changes from 99% to 1% ($\epsilon = 0.01$).

SOLUTION: Introduce $a = \exp(t_a/\delta t)$, $b = \exp(t_b/\delta t)$, and $C = \exp[\Delta t/(2\delta t)]$. For t_b, $C \gg 1$, and from Eq. (32.128),

$$\varepsilon \approx \frac{uC^2}{(a+1)C^2u} = \frac{1}{(a+1)} \to t_a \approx \delta t \ \ln\left(\frac{1-\varepsilon}{\varepsilon}\right)$$

Analogously, for t_a,

$$1 - \varepsilon \approx \frac{1}{(b+1)} \to t_b \approx \delta t \ \ln\left(\frac{\varepsilon}{1-\varepsilon}\right)$$

and so $t_a - t_b \approx 2\delta t \ \ln[(1-\varepsilon)/\varepsilon] = 9.1902\ \delta t$.

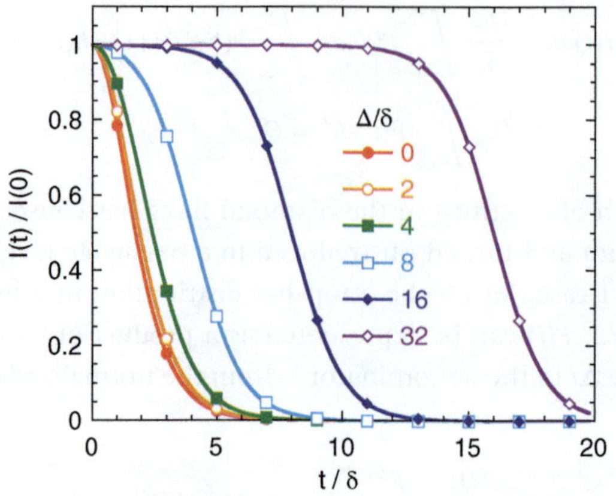

Figure 32.28 Response characteristics of $I(t)/I(0)$ for $\Delta = n\delta$ with n indicated in the labels and $I(t)$ as defined in Eq. (32.126). The time is shifted such that the center of the pulse is at $t = 0$.

Therefore, regardless of whether the primary pulse is much longer than the characteristic time ($\Delta t > \delta t$) or much shorter ($\Delta t < \delta t$), the rise/fall time is nevertheless constrained to be several multiples of δt. As an example, the rise/fall time for $E_o = 3$ keV with $L = 8$ μm (as in the example following Eq. (32.122)), then $t_a - t_b \approx 2\ \ln(99) \times 0.966$ ps $= 8.877$ ps.

A second analytically tractable model is the Gaussian distribution, $P(t) \propto \exp(-t^2/2\Delta t^2)$. When the characteristic time δt is much larger than the characteristic time Δt of the distribution, then because Eq. (32.126) can be well-approximated by a Gaussian itself, their joint integration in $I(t)$ is possible. First attend to the Gaussian parameterization. Let $f(x) = \ln[h(x)(1 - h(x))] = \ln[h(x)h(-x)]$. Performing a Taylor expansion about $x = 0$ gives

$$\begin{aligned} f(x) &\approx f(0) - \frac{x^2}{\delta t^2}\left[h(0)^2 - 2h(0) + 1\right] \\ &= -\ln(4) - \frac{x^2}{4\delta t^2} \end{aligned} \tag{32.129}$$

When inserted into the defining equation for $I(t)$ if Eq. (32.124), then

$$I(t) = \frac{Q_{tot}}{\delta t \Delta t}\int_{-\infty}^{\infty} \frac{1}{4}\exp\left[-\frac{s^2}{4\delta t^2}\right]\exp\left[-\frac{(s+t)^2}{2\Delta t^2}\right] ds \tag{32.130}$$

Using the general integral relation

$$\int_{-\infty}^{\infty} e^{-ax^2 \pm bx} dx = \sqrt{\frac{\pi}{a}}\exp\left(\frac{b^2}{4a}\right) \tag{32.131}$$

results in

$$I(t) = \frac{\sqrt{\pi}Q_{tot}}{2\sqrt{2\delta t^2 + \Delta t^2}}\exp\left(-\frac{t^2}{4\delta t^2 + 2\Delta t^2}\right) \tag{32.132}$$

As a result, a small pulse length Δt of the primary beam increases the width (in time) of the resulting Gaussian pulse. Put another way, even if the primary beam of electrons were deposited all at once with no spread in time, the bunch that would emerge would have a Gaussian shape as a result of diffusive expansion during transport through the flake.

Intuitively, δt is a measure of how long the diffusion is allowed to take place in the diamond flake and so is governed by both the internal field in the diamond, as well as the width of the flake itself. The simple Monte Carlo methods of Section 32.2 are enough to make an estimate of the center of charge position $z(t)$ of the bunch as it travels through a layer of thickness L, and of its root mean squared (rms) radius $r(t)$ about its center of charge, as the discussion surrounding Eq. (32.122) suggests. For a linear field across the flake, a constant saturation velocity and a diffusive expansion imply that $z(t) \sim t$ and $r(t) \sim \sqrt{t}$, behaviors which simulation readily confirm as in Figure 32.29 for two different fields. If the current from the flake is to be delivered in short pulses, then the size of $r(t)$ matters, and therefore how long the bunch dwells in the flake matters. Therefore, if short sculpted pulses are desired, fast rise/fall times entail that the flake should be as thin as practical.

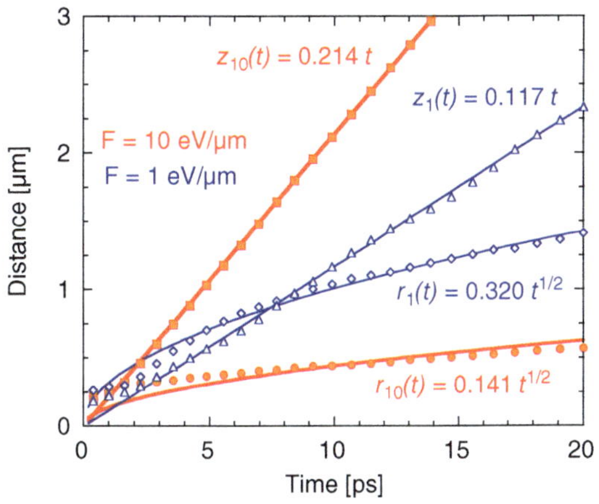

Figure 32.29 Symbols, Monte Carlo evaluation of charge center $z(t)$ and rms radius $r(t)$ of a bunch of secondaries transporting through a thin flake under field for diamond-like parameters. Lines, equations show least squares fits to the forms of $z(t) \approx v_o t$ and $r(t) = a\sqrt{t}$, based on a constant drift velocity (for z) or diffusion (for r). Red, $F = 10$ eV/μm; blue, $F = 1$ eV/μm. Based on Figure 10 in ref. [209].

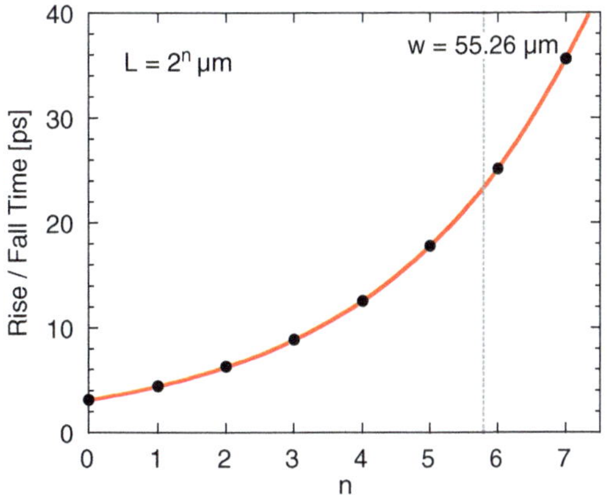

Figure 32.30 Rise/fall time t_{ab} variation with flake thickness (calibrated by $t_o = 8.877$ ps when $L_o = 8$ μm) for $L = 2^n$ μm, as per Eq. (32.133). Also shown is a depletion width for boron concentrations of 10^{14} cm^{-3} and typical fields. Based on Figure 16 in ref. [209].

The rise/fall time $t_{ab} = t_a - t_b$ can now be related to the thickness of the diamond flake itself. If the radius of the bunch does not change appreciably as the sphere passes into vacuum, then t_{ab} should behave something like $t_{ab} \propto r(t)/v_o$. The time for the sphere to cross the thickness L of the flake is $t \approx L/v_o$, so that $t_{ab} \propto \sqrt{L}$. As a result,

$$t_{ab}(L) \approx t_o \left(\frac{L}{L_o} \right)^{1/2} = 3.1385\sqrt{L} \tag{32.133}$$

for $t_o = 8.877$ ps and $L_o = 8$ μm. The relationship between rise/fall time and flake thickness is indicated in Figure 32.30, where it is compared to a typical depletion width $w = 55.26$ μm for a boron concentration of 10^{14} cm^{-3} (recall Eq. (16.4)). The boron concentration matters more than has been let on: the internal fields in the diamond flake are not linear as so far assumed: rather, the charged impurities tend to shield the back contact and retard the acceleration of the newly minted secondaries, contributing to losses and spreading the beam further [322]. Still, if performance is below par in the linear field model, matters are only going to get worse when doping (intentional or not) is considered, so the models are a best-case scenario.

An obsession over rise/fall time compared to other features of emitted bunches of electrons is seemingly disproportionate to its importance, but there are reasons for the interest. The diamond amplifier is an innovative solution to hard-to-meet demands placed on the electron sources for FELs and advanced light sources. The performance of an FEL depends on the charge within a bunch, and how well its spreading can be constrained, so rise/fall times matter. Spreading is related to emittance: high current densities from the cathode are demanded, and the very strong fields at the surface of a photocathode are to counter space-charge forces that exist in the bunch in an effort to keep spatial features of a character to meet stringent FEL requirements. Bunch shape therefore matters, with a top-hat profile in particular being preferable to Gaussian or rounded shapes for emittance reduction [593, 631, 632]. The European X-ray FEL, for example, uses a trapezoidal pulse 20 ps long with a 4 ps rise/fall time [595], with other FELs seeking similar shapes and rise/fall times [633]. Therefore, usage of a diamond amplifier to provide a secondary electron emission boost to the performance of a photocathode, for example, entails more than just raw current or bunch charge needs to be practical.

CHAPTER 33

Electron beam physics

His greatness weighed, his will is not his own.
For he himself is subject to his birth.

– William Shakespeare[1]

Liberating electrons from their bondage is not the end of the tale: the variety of electron sources speaks to the many technologies that use them, and so the nature of the beams produced becomes a deciding factor in their utility. The applications of radar, communications, and electronic countermeasures (Chapter 1) rely on the production of electromagnetic radiation with wavelengths in the range of meters to millimeters [429] and shorter (terahertz and beyond) [13, 634]. Others such as scanning and tunneling electron microscopy [162], electron beam lithography [16], high power microwave [635] and directed energy [591], energy conversion and optoelectronics [636], and particle accelerator/advanced light source and X-ray free electron lasers (FELs) [523] are no less exacting in the properties of the electron beam even as the needs of each application are phenomenally different in terms of bunch charge, pulse duration, response time, and beam spread – all of which will be recast in the language of current density, emittance, and brightness. How the beam is born figures strongly in what the beam does, and describing that is a hugely complex undertaking [148, 637, 638].

A few simple considerations (such as the beam has cylindrical symmetry, and the magnetic field is static) do not undercut an appreciation of what awaits. Feral electrons are typically not well behaved: they flee in directions often not normal to the surface (leading to emittance), and they generally try to get away from each other (leading to space-charge effects). To corral the disbanding electrons, emission is often into a magnetic field region so that the transverse velocity components of the electron to the direction of the magnetic field $\vec{B}$ simply causes them to spiral about the field lines. Thus, if an electron with a velocity $\vec{v}$ is traveling through a region with a magnetic field $\vec{B} = B\hat{z}$, then the force equation $m\ddot{\vec{r}} = \vec{F} = -q\vec{v} \times \vec{B}$ will describe their subsequent motion. By considering motion in a region without an electric field, the energy of the electron is not changing. By conservation of angular momentum, the angular velocity $v_\theta = \rho\dot{\theta} = \rho\omega_o$ is likewise constant. So, setting the centrifugal force to the magnetic force,

$$m\frac{v_\theta^2}{\rho} = m\rho\omega_o^2 = q\rho\omega_o B \tag{33.1}$$

from which the *cyclotron resonance frequency* ω_o is defined by

$$\omega_o = \frac{qB}{m} \tag{33.2}$$

which depends only on the magnitude of the magnetic field B, not the kinetic energy of the electron. As electrons are borne from the surface with a distribution of transverse velocities v_θ,[2] this means that all of them gyrate about the magnetic field lines in orbits with the same frequency ω_o, the radii of which are also distributed because $v_\theta = \rho\omega_o$.

EXAMPLE: Find the radius ρ_o of an electron for which the transverse kinetic energy $(1/2)m(v_x^2 + v_y^2) = mv_\theta^2$ is equal to the average thermal energy $k_B T$ (Eq. (3.14)). Let the magnetic field $B = 0.7$ Tesla (a representative value for FELs [639]), and let the temperature of the thermionic emitter be $T = 1400$ K.

[1] Ref. [37]: *Hamlet*, I.iii.17–18.

[2] An angular velocity is not associated with a change in energy whereas a radial velocity is, so when electrons are not being accelerated use v_θ instead of $v_\perp$.

Introduction to the Physics of Electron Emission, First Edition. Kevin L. Jensen.

SOLUTION: Evaluate $qcB = 0.20985$ eV/nm and $k_BT = 0.12064$ eV. Relating ρ_o to the cyclotron resonance frequency ω_o from Eq. (33.2), then

$$\rho_o = \frac{mv_\theta}{qB} = \frac{\sqrt{mc^2 k_B T}}{qcB} = 1.1832 \ \mu\text{m}$$

As the last example shows, the radii of the orbits compared to the radius of a standard thermionic dispenser cathode, with a diameter from millimeters to centimeters, is several orders of magnitude smaller. In simulation, the launch of a "cold beam" that neglects the initial transverse velocity components of the electrons [640] is easier and often sufficiently accurate for linear beam devices with large beam tunnels. When the emission area is very small, or when the beam tunnel is very small, neglecting the traverse velocity components during the launching of the electron beam can undermine the simulation. When high current density inside beam tunnels is desired, the problem is compounded. So as to not limit thermal cathode lifetimes by operating them at excessive temperatures, a substantial compression of the beam from the cross-section of the cathode to the cross-section of the beam tunnel is performed, for example using a Pierce gun [12, 638, 641]. Indeed, the high current density of field emitter arrays coupled with a rapid modulation capability motivated their consideration [480, 492, 642, 643] and demonstration [483] in inductive output amplifiers (IOAs). As the transverse velocity components of the beam do not go away, simulations of very small beam tunnels see their effects [644]. Why transverse velocity components matter will lead to the development of a beam envelope equation and a consideration of the effects of emittance.

33.1 Electron orbits and cathode area

"What is a Caucus-race?" said Alice; not that she wanted much to know ... "Why," said the DoDo, "the best way to explain it is to do it." ... First, it marked out a race-course, in a sort of circle ("the exact shape doesn't matter," it said), and then all the party were placed on the course, here and there. There was no "One, two, three, and away," but they began running when they liked, and left off when they liked, so that it was not easy to know when the race was over.

– Lewis Carroll[3]

As a result of the fact that the angular frequency of the electron's orbit about the field line depends only on the magnetic field, all the electrons return to their starting position in the x–y plane at the same time (but not along the $\hat{z}$ axis). If the electrons are emitted from the surface of a flat cathode, then those starting positions are distributed across the emitting area, with the initial transverse velocities randomly directed: when the electrons move away from the cathode surface, some will spiral to the center, but others will spiral away. Letting the radius of the electron orbit be ρ_o, and the radius of the cathode area be R, then those orbits can be examined for various values of $C = \rho_o/R$.

If the emitted electrons are Maxwell–Boltzmann (MB) distributed, then a rather elegant method for calculating the starting transverse velocity components v_x and v_y all at once is possible. Recall from Eq. (3.12) (or Eq. (5.2), which is a variation) that for MB distributions in velocity,

$$f_{MB}(v_\perp) = f_{MB}(v_x) f_{MB}(v_y) \tag{33.3}$$

where $\hat{z}$ is reserved for the direction of the $\vec{B}$ field. This allows for the creation of an MB distribution of transverse velocities in a manner different than the algorithm of Section 5.1 when an MB distribution for one component of the velocity, v_x, was generated. First, distributions in v^2 are required, for which an MB distribution is of the form $e^{-\beta E}$ for $E = (1/2)mv^2$, meaning that something like Eq. (32.22) can be employed. The other requirement is that $v_x^2 + v_y^2 = v_o^2$ (where v_o stands in for either $v_\perp$ or v_θ), and therefore appending trigonometric functions to the random v_x and v_y once v_o is obtained satisfies that condition. Thus, to generate v_x and v_y, a pair of numbers n_1 and n_2 is generated using two random numbers r_1 and r_2, from which

$$\begin{aligned} n_1 &= \sqrt{-2\ \ln(r_1)}\ \cos(2\pi r_2) \\ n_2 &= \sqrt{-2\ \ln(r_1)}\ \sin(2\pi r_2) \end{aligned} \tag{33.4}$$

[3] Lewis Carroll, *Alice's Adventures in Wonderland*. Racine, Wis.: Whitman Publishing Company, 1945, Chapter III.

If v_o is a characteristic velocity (say, $v_o = \sqrt{k_B T/m}$, where the usual factor of 2 is missing because $v_o^2 = v_x^2 + v_y^2$), then $v_x = v_o n_1$ and $v_y = v_o n_2$. To take the next step and make v_o correspond to v_θ takes a bit more. Because $x = \rho\cos\theta$ and $y = \rho\sin\theta$ then (letting the dot notation denote time derivative)

$$v_x = \dot{x} = v_\rho \cos\theta - v_\theta \sin\theta$$

$$v_y = \dot{y} = v_\rho \sin\theta + v_\theta \cos\theta \tag{33.5}$$

where $\dot{\rho} = v_\rho$ and $\rho\dot{\theta} = v_\theta$. Knowing $\vec{v} = v_x\hat{x} + v_y\hat{y}$ and $\hat{\theta} = -\hat{x}\sin\theta + \hat{y}\cos\theta$ gives (ignoring any consideration of v_z for the moment)

$$v_\theta = \hat{\theta}\cdot\vec{v} = -v_x \sin\theta + v_y \cos\theta \tag{33.6}$$

In order for v_o of Eq. (33.4) to correspond to v_θ then the identification $\theta \equiv 2\pi n_2 + (\pi/2)$ is required to insure $\hat{\theta}\cdot\vec{v}_o = 0$.[4] In short, then, for two random numbers r_1 and r_2,

$$v_x = \left(\frac{k_B T}{m}\right)^{1/2} \sqrt{-2\,\ln(r_1)}\,\cos\left(2\pi r_2 + \frac{\pi}{2}\right) \tag{33.7}$$

$$v_y = \left(\frac{k_B T}{m}\right)^{1/2} \sqrt{-2\,\ln(r_1)}\,\sin\left(2\pi r_2 + \frac{\pi}{2}\right) \tag{33.8}$$

Assuming that the launch sites on a cathode of radius R are randomly placed, and the transverse velocity v_θ is MB distributed, then the spiraling electrons overlaid on the cathode area resemble Figure 33.1 for various values of $C = \rho_o/R$, as calculated using the algorithm of Section A3.33: the starting points relative to the cathode area are the same in each sub-figure. The radius of the orbits need not be that much smaller than the cathode radius before significant excursions by spiraling electrons are noticeable beyond the radius of the cathode

The starting points characterized by small black dots in Figure 33.2 are special: after a time $t = 2\pi/\omega_o$ all of the electrons regardless of their velocity v_θ return to those dots. As a result, the size of the circle enclosing all the trajectories oscillates periodically in time with a frequency ω_o. If the radius of the cathode is now the initial radius of the beam $R \to R(t=0)$, then the radius of the beam follows a harmonic oscillation. It is often desirable to bring the energy of the electrons in a beam up to their maximum energy as rapidly as possible to minimize longitudinal spread: when done, $v_z/c \equiv \beta$ is not negligible. The location of the electrons along z is a measure of how much time has passed, and so the time derivative of the harmonic oscillations that $R(t)$ undergoes can be replaced instead by derivatives with respect to z via $d/dt = v_z d/dz$. Because the total velocity is dominated by v_z the *paraxial equation* [638] can be obtained. In the absence of effects that bedevil the final form of that equation, the harmonic oscillations of the radius of the beam that are made apparent by

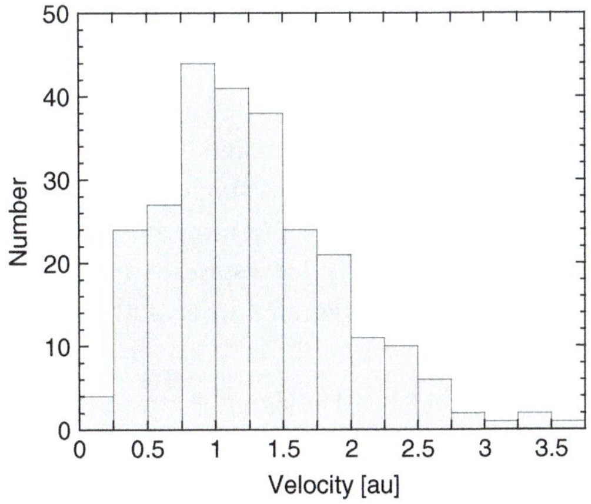

Figure 33.1 An MB of transverse velocities v_θ evaluated using the algorithm of Section A3.33. The velocity is in units of $\sqrt{k_B T/m}$. The total number of particles is 256. From these velocities, the orbital radii of Figure 33.1 are calculated according to $v_\theta = \rho\omega_o$.

[4]Were v_o identified with v_ρ, then $\theta = 2\pi n_2$ directly, as can be confirmed by $\hat{\rho}\cdot\vec{v}$.

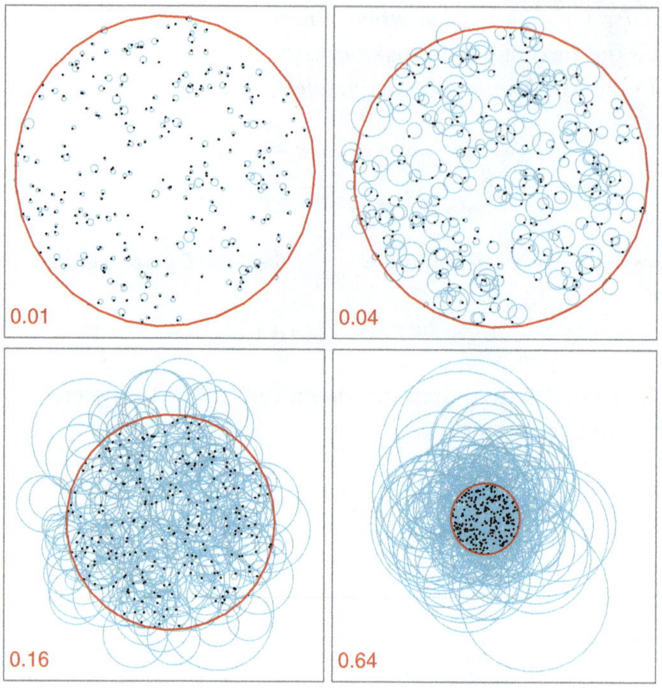

Figure 33.2 Starting points (black dots) of 256 emitted electrons from a cathode area given by the red circle. Magnetic field is normal to the plane of the figure. Placement is random but the starting points are the same in each quadrant. The blue circles represent the trajectories the electrons take over one cycle. The figure is generated using the algorithm of Section A3.33 and labeled by ρ_o/R (lower left corner), where ρ_o is the initial radius for the mean velocity (Figure 33.1) and R is the radius of the emitting surface.

spiraling orbits shown in Figure 33.2 are described by

$$\frac{d^2}{dz^2}R + \left(\frac{qB}{2\beta\gamma mc}\right)^2 R = 0 \tag{33.9}$$

where B is magnetic field and ω_o from Eq. 33.2 has been dressed with the relativistic factors $\beta = v_z/c$ and $\gamma = 1/\sqrt{1-\beta^2}$. Electrons accelerated rapidly to high energies make v_z a fraction, and sometimes a large fraction, of the speed of light, for which the closeness of v_z to the overall velocity magnitude v makes using v in evaluating β and γ expedient.

33.2 Beam envelope equation

Fain would I run into the crowd for shelter and warmth; but cannot prevail with myself to mix with such deformity. I call upon others to join me, in order to make a company apart; but no one will hearken me.

– David Hume[5]

As soon as the beam is born, however, the electrons inside it agitate to get away from each other due to their mutual repulsion, and so a space-charge modification must be added. A single particle gives rise to a $1/r^2$ force, as is well known; almost as well-known is that a line of charge gives rise to a $1/\rho$ force [99]. A fast way to such conclusions is to use the Gaussian pillbox argument [99] familiar from Eq. (16.35): for a sphere of radius r surrounding a point charge of charge q, then $\vec{F}$ is in the direction of $\hat{r}$, and so equating the integrated force dotted to the surface normal over the area of the sphere, or $\int \vec{F}\cdot d\vec{S} = q^2/\varepsilon_0$, gives

$$4\pi r^2 F = 16\pi Q \rightarrow F = \frac{4Q}{r^2} \tag{33.10}$$

For a beam, the surface is a cylinder of length l (where λ is the charge per unit length) and radius ρ which contains a charge λl, and so

$$2\pi\rho l F = 16\pi Q\frac{\lambda l}{q} \rightarrow F = \frac{8Q\lambda}{q\rho} \tag{33.11}$$

[5]David Hume, *A Treatise of Human Nature*, 2nd edn. Oxford: Oxford University Press, 1978, Book I, Part IV, Section VII, p. 264.

The charge in the beam, however, is moving, and so relativistic factors enter it, too. A careful derivation [638] results in the ratio of the beam current I_a to the Alfvén–Lawson current defined by

$$I_o = \frac{4\pi\varepsilon_0 mc^3}{q} = qc\left(\frac{mc^2}{4Q}\right) = \frac{qc}{\alpha^2 a_o} \tag{33.12}$$

where the second form conjoins the fundamental time unit ($c/a_o = 1.76515 \times 10^{-4}$ fs), with a ratio of the energy scales associated with the electron (ratio of the potential energy $4Q/a_o$ of an electron in the hydrogen atom with the electron rest energy mc^2, or $4Q/a_o mc^2 = 1/\alpha^2$) as in Section 2.3.

EXAMPLE: Evaluate the Alfvén-Lawson current I_o
SOLUTION: Using the second of the forms of Eq. (33.12)

$$I_o = qc\frac{511000 \text{ eV}/c^2}{4(0.35999 \text{ eV-nm})} = 1.0639 \times 10^8 \frac{q}{\text{fs}} = 17045 \text{ A}$$

where 1 A = 6241.5 q/fs from Table 2.2. Using the third of the forms of Eq. (33.12) is perhaps easier and involves the characteristic charge (q) and time ($\delta t = a_o/c = 1.7651 \times 10^{-4}$ fs) of the hydrogen atom (Section 2.2.1), or $q/\alpha^2\delta t =$ 17.045 kA.

Electrons are ejected with transverse velocity as soon as they exit the surface, which cannot be corrected and that leads to the intrinsic emittance of the cathode: no amount of focusing or beam manipulation (*aka* beam optics) after emission will undo what the statistical nature of emission does to mess up a beam, in contrast to other causes of oscillations in the beam radius from which recovery is possible [645]. Emittance causes the cross-sectional area of the beam to seem to swell in the beam tunnel, as in Figure 33.2. More (much more) will be said or can be read [638] on emittance ε, but for now it is sufficient to note that incorporating the effects of emittance in the beam envelope equation along with space charge results in [646]–[649]

$$\frac{d^2}{dz^2}R + \left(\frac{qB}{2\beta\gamma mc}\right)^2 R - \frac{2}{(\gamma\beta)^3}\left(\frac{I_a}{I_o}\right)\frac{1}{R} - \frac{\varepsilon^2}{R^3} = 0 \tag{33.13}$$

where the factors after the gradient term d^2R/dz^2 are the magnetic term, the space-charge term, and the emittance term, respectively: that they keep company thusly indicates their linked nature, as in Figure 33.3, in affecting the overall beam quality and therefore device operation.

For electrons sprinting along at a good fraction of the speed of light (and even electrons at 1000 eV do so, with $v/c = 0.063$), the transverse velocity components are overshadowed (an electron at k_BT for $T = 1400$ K has $v/c = 6.9 \times 10^{-4}$). It is therefore convenient to take $v_z \approx v$. The relativistic factors $\beta = v/c$ and $\gamma = 1/\sqrt{1-\beta^2}$ are then related to the acceleration potential $V_b = q\varphi_b$ that brings the beam up to speed by

$$\gamma mc^2 = mc^2 + V_b \tag{33.14}$$

Big as V_b may be, it is generally less than mc^2, and so

$$(\beta\gamma)^2 = \frac{V_b}{mc^2}\left(2 + \frac{V_b}{mc^2}\right) \approx \frac{2V_b}{mc^2} \tag{33.15}$$

As a result, $\beta\gamma \approx \sqrt{2V_b/mc^2}$.

Now return to Eq. (33.13): were there no space-charge or emittance effects, then the equation is of the form $\partial_z^2 R + \omega^2 R = 0$, indicating that R oscillates harmonically. On the other hand, as the combination of space-charge forces and emittance effects increases, at some point those terms match and cancel the ω^2 term, resulting in $\partial_z^2 R = 0$. Such a matching is known as *Brillouin flow* and results in a smooth tapered beam entering the beam tunnel (although other physical effects rapidly complicate such ideal conditions [650]). As the interaction between the electrons in the beam and the circuit, be it a traveling wave tube (TWT), klystron, FEL, or other linear beam device, is strongest for electrons closest to the beam tunnel walls, this is good: the absence of oscillations keeps the electrons where they should be if they can be brought and kept there in the first place. In some devices, in fact, annular (or hollow) beams are advantageous [651, 652]. Electron beams are affected by the countervailing presence of ions [641] but that complication is ignored. Setting the sum of

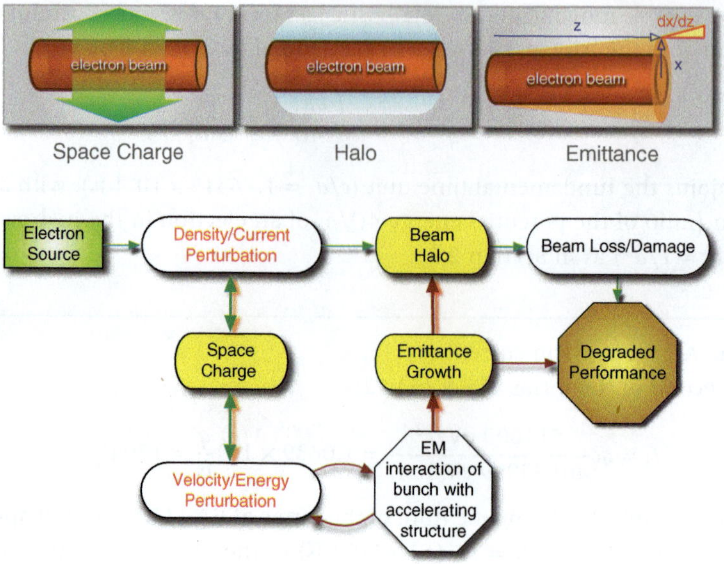

Figure 33.3 Factors that affect electron beams and how they interact: Top, (*i*) space charge changes velocity and causes distortion, (*ii*) halo is made of electrons outside the core part of the beam which can hit beam tunnel walls and cause damage, and (*iii*) emittance is (in part) the change dx/dz of the beam due to transverse velocity components of electrons and causes spreading. Bottom, how space charge, halo, and emittance interact and affect beam quality, with double arrows indicating feedback (based on a figure by P.G. O'Shea).

the magnetic, space-charge, and emittance terms in Eq. (33.13) to zero, the factors can be regrouped and rearranged to give [451]

$$\frac{I_a}{\pi R^2} = \frac{I_o}{8\pi}\left(\frac{2V_b}{mc^2}\right)^{1/2}\left(\frac{qB}{mc}\right)^2\left\{1 - \frac{8mV_b\varepsilon^2}{q^2B^2R^4}\right\} \tag{33.16}$$

where the $\beta\gamma \approx \sqrt{2V_b/mc^2}$ is used. The left-hand side is the current in the beam tunnel over the area of the beam, that is, the current density of the beam. The factor in curly brackets is therefore *the degree to which emittance reduces the current density of the beam* for Brillouin flow conditions, that is,

$$J_{beam}(\varepsilon) = J_{beam}(0)\{1 - \delta(\varepsilon)\} \tag{33.17}$$

where the fractional reduction in emittance is characterized by the factor

$$\delta(\varepsilon) \equiv \frac{8mV_b\varepsilon^2}{q^2B^2R^4} \tag{33.18}$$

For concreteness, consider a particular class of vacuum electronic devices used for microwave amplification and what constrains the push of the technology into the terahertz regime. Generally, the frequency of the device scales with the inverse radius of the beam tunnel. The quest for higher frequency therefore entails both smaller beam tunnels and higher current densities in those tunnels (Chapter 1 and Table 33.1): as the frequency range of the devices is pushed higher, the

Table 33.1 Maximum tolerable emittance (in units of [mm-mrad]) such that $\delta \le 0.01$ in Eq. (33.20). Pierce-type and magnetron injection are gun types described in Appendices A1 and A2 of ref. [638]. Others are for generic cases in the literature.

Type	*I* [A]	*R* [cm]	V_b [q kV]	*B* [T]	*f* [GHz]	Max. ε
Pierce-type [638]	0.25	0.6	5.0	0.01	1–4	75.3
Magnetron [638]	480.0	0.6	500	0.08	2–4	49.5
Gyrotron	0.05	0.344	15.0	9.7	263	13796
S-band MBK	32.0	1.3	0.53	0.22	2–4	23940
Generic TWT	0.288	0.1	10.0	0.09	1–4	13.2
Generic TWT	1.138	0.128	25.0	0.09	1–4	13.6
CCTWT [656]	0.5	0.0559	19.2	0.8	27–40	26.5
Serpentine TWT [655]	0.12	0.019	11.7	0.3	220	1.48

maximum tolerable emittance goes from well in excess of what thermionic cathodes actually produce to a range far more challenging. If the maximum magnetic B field in a compact configuration achievable with permanent magnets is 1 Tesla, and V_b is kept comparable to 20 kV to keep the device small while avoiding problems associated with voltage insulation [13], then the maximum achievable current density in the beam tunnel is limited by the emittance of the beam for a given beam tunnel radius. Noting that (qcB) has units of [eV/nm], let

$$R_o \equiv \frac{\sqrt{8mV_b}}{qB} \tag{33.19}$$

so that

$$\delta = \left(\frac{\varepsilon_0 R_o}{R^2}\right)^2 \tag{33.20}$$

which reinforces that ε has units of length. The actual units of emittance are [radians × length]: sometimes [mm-mrad] with mrad denoting milli-radians are used, sometimes [μm] is used. In the early literature, a factor of π would appear to reinforce the radians unit. The convention employed is not always obvious. Here, the unit is chosen to be [mm-mrad]. Clearly, then, emittance is a problem when $\varepsilon \approx (R^2/R_o)\sqrt{\delta}$ for a particular allowable δ. For example purposes, taking δ to be approximately 1% is agreeable to the human penchant for choosing pleasing magnitudes bereft of significance (it makes calculations easy), but actual emittance limits are more circumspect: given the conjunction of emittance with space charge in ejecting electrons from the core part of the beam (halo), and the damage that halo causes [653, 654], acceptable ranges are often much more discriminating. The beam tunnel diameters of the last row of Table 33.1, described in ref. [655], are comparable to 150 μm, with tunnels less than 100 μm anticipated in the future. For tunnels of comparable dimensions, emittance matters.

EXAMPLE: Find R_o assuming $B = 1$ Tesla and $V_b = 20$ keV. From it, determine the maximum tolerable emittance if $J_{beam}(\varepsilon)$ is 99% of the emittance-free $J_{beam}(0)$ ($\delta = 0.01$) for a beam required to be $R = 0.1$ cm.
SOLUTION: Using $qcB = 0.29979$ eV/nm in Eq. (33.19) gives

$$R_o = \frac{\left[(2 \times 10^4 \text{ eV})(5.11 \times 10^6 \text{ eV})\right]^{1/2}}{0.29979 \text{ eV/nm}} = 0.095378 \text{ cm}$$

Consequently, $\varepsilon \leq \sqrt{\delta}R^2/R_o = 104.85$ mm-mrad is required.

EXAMPLE: Consider a high-frequency serpentine TWT [655] or an extended interaction Klystron [644], for which the beam current is on the order of 100 mA, the beam tunnel current density is desired to be 100–225 A/cm^2, the acceleration potential is 25 kV, the magnetic field is 0.3 Tesla, and the beam tunnel radius is 130 μm [451]. Find the maximum emittance if $\delta < 1\%$.
SOLUTION: Using $Bqc = 0.089938$ eV/nm, then

$$\varepsilon^2 < \delta R^4 \frac{(qcB)^2}{8V_b mc^2} = 0.01(130\mu\text{m})^2 \frac{(0.089938 \text{ eV/nm})^2}{(25 \text{ keV})(0.511 \text{ MeV})}$$

or $\varepsilon < 0.47545$ μm.

For FElLs [657], the laser beam optical mode in the wiggler constrains the electron beam to fit within it for the laser threshold to be reached. If the radius of the mode is w, the divergence angle is θ, and the wavelength of the FEL is λ, then the waist-divergence product has to be such that $w\theta = \lambda/\pi$. Because emittance describes the spread of the beam, then [633]

$$\varepsilon \approx \frac{\lambda}{4\pi} \tag{33.21}$$

The factor of 4 has to do with how emittance has been historically defined, and the factor of π has to do with how emittance is defined in phase space (Section 33.3 and refs [658, 659]). Sources with emittances on the order of a micron are objectives, and in the case of an X-ray FEL [20, 524, 660], even smaller [523, 661].

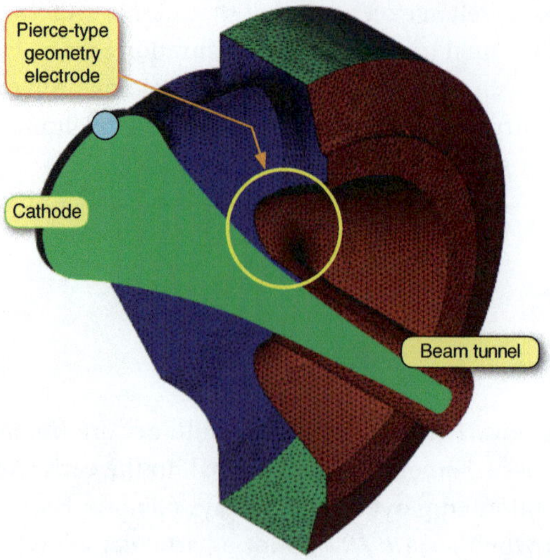

Figure 33.4 Modeling the compression of an electron beam from a curved thermal emitter into a beam tunnel for a Pierce-type diode using the PIC-code `MICHELLE`, similar to that considered in Ref. [583]. Electron trajectories are colored bright green: the apparent solidity of the beam is a consequence of many of them together. Sky-blue dot indicates the region of Fig. 33.5. Image and simulation courtesy of J. J. Petillo (*Leidos*).

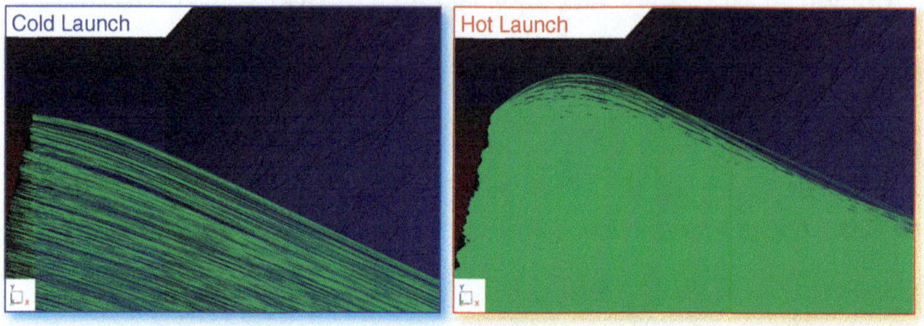

Figure 33.5 Launch sites at the edge of the cathode (sky-blue circle region of Figure 33.4). Left, cold launch in which trajectories are normal to the surface of the cathode (no transverse velocity). Right, hot launch in which a distribution of transverse velocities is applied to the emitted electron at launch. Images courtesy of J. Petillo (Leidos).

A recitation of numbers and equations, though, is not nearly as effective as conveying the consequences of emittance as a numerical simulation of an electron beam. Launching electrons from a cathode, compressing the beam to the diameter of the beam tunnel, and following the bunches down the length of the device is the ambition of many codes, for example `MAGIC`, `PARMELA`, `elegant`, `IMPACT`, `WARP`, `Neptune`, `Vorpal` and `MICHELLE`, and others [588, 640, 656, 662–668] are the basis of studies of beam dynamics. Particle-in-cell (PIC) codes are a particular kind useful to model amplifiers and accelerators. Consider, therefore, a simulation to investigate the consequences of including intrinsic emittance using `MICHELLE` [640] for a Pierce-type diode employing a thermal cathode [583] in which the beam is compressed to the size of the beam tunnel, as in Figure 33.4.

The simulation takes the gun to be at 5 kV with a thermal dispenser cathode (analogous to that shown in Figure 12.1) with a radius of 0.25 mm and producing 100 mA at a temperature of 1500 K. In a "cold launch," electrons are born with no transverse velocity components, so that their trajectories are normal to the cathode surface, whereas in a "hot launch," electrons are randomly given a transverse velocity component comparable to $(1/2)mv_\perp^2 \sim k_B T \approx 0.086$ eV, but otherwise the simulation methodologies of the cold and hot launches are the same. The electrons are followed to the entrance of the beam tunnel: the effects of transverse velocity are clearly seen in the larger diameter of the beam, where the beam tunnels have been oriented so as to face each other with the cold launch on the left and the hot launch on the right of Figure 33.6. *Ad hoc* parameters of $B = 0.7$ Tesla, $\varepsilon = 45$ mm-mrad, and $V_b = 5\,q$ kV $= 5$ keV, then $\delta(\varepsilon) = 0.24$ are used. The increase in the beam radius is visually apparent.

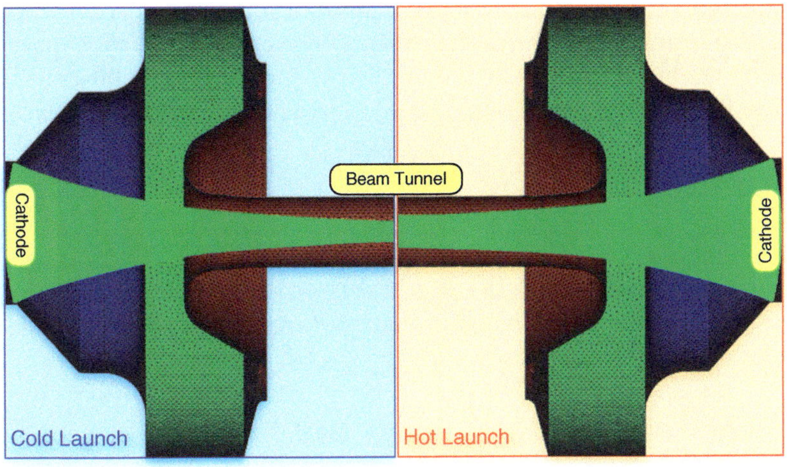

Figure 33.6 Trajectories of the electrons from the cathode surface for the cold (left) and hot (right) conditions of Figure 33.5. Observe the larger beam diameter for the hot launch conditions. Images courtesy of J. Petillo (Leidos).

33.3 Emittance for flat and uniform surfaces

> *Each point on the surface of the source is emitting particles with different initial magnitude and direction of the velocity vector. This intrinsic* thermal velocity spread *remains present in the beam at any distance* downstream *from the source … The* emittance *provides a quantitative basis, or a* figure of merit, *for describing the quality of the beam.*
>
> – Martin Reiser [638], p. 56

> *…but as for fate, I think that no man yet has escaped it once it has taken its first form, neither brave man nor coward.*
>
> – Homer[6]

Emittance was previously characterized (Chapter 31) as a tendency of a beam to spread as it propagates. Although beams can be focused, the initial emittance of the beam, or the *intrinsic* emittance, cannot be undone in the same way that beam optics can undo spreading, and so the emittance generated at the cathode becomes important because it can be limiting. In thinking of emittance, keeping an optical analogy in mind is useful. The difference between the emittance of a flashlight and that of a laser pointer is instructive: flashlight beams spread out, but laser beams stay tightly collimated. Flashlights are said to have a large emittance and lasers a small one.

To be precise, spreading is how the transverse extent of the beam changes as the beam moves along the $\hat{z}$ axis. The transverse axis can be either $\hat{x}$ or $\hat{y}$, though the emittance of each is the same if the beam is rotationally symmetric so that mean values of x^2 and y^2 or k_x^2 and k_y^2, are particularly simple. For ease of discussion, then, assume a rotationally symmetric beam and speak of the $\hat{x}$ emittance in place of both. Emittance should give a measure of $x' \equiv dx/dz$ in addition to x for each electron in the beam. The empirical definition of emittance is then the area of the shape that encloses all of the (x, x') pairs for a beam.

Difficulties arise almost immediately in such a definition. For electrons accelerated in fields, for example, two problems are easy to see. First, all the points are *distributed* and therefore the area is difficult to quantify: should it enclose *most* of the (x, x') points, or include the odd outliers? Second, the points can be so distributed that two beams may seem to have comparable areas enclosing the (x, x') points, but the one exhibiting more painful contortions to enclose the points is difficult to work with. That is, as expressed by J. Buon [669],

> *The usual definition of emittance as the area of a limiting contour suffers not only from the arbitrariness of the chosen contour, but also from its distortion when time is running. It may become so twisted and bent that its area is no more representative of the spread of the particles.*

---

[6]Homer, *The Iliad* (trans. Richmond Lattimore). Chicago: University of Chicago Press, 1962, Book 6, lines 488–489.

A working definition of emittance should account for whatever torture of the shape enclosing the (x, x') points is in store. Lastly, as the electrons become relativistic, their spread in velocity is such that emittance appears to decrease markedly as $v \to c$ for all the electrons in the bunch: an emittance that changes thus is not as useful a metric of beam quality as one that describes the beam over its evolution consistently, that is, a metric that remains *invariant* as the beam accelerates.

33.3.1 Liouville's theorem and normalized rms emittance

> *...Epaminondas demonstrated that the phalanx system could be adapted to achieve decisive tactical manoeuvre in the face of the enemy. At Leuctra, outnumbered 11,000 to 6000, he quadrupled the strength of his left wing and, masking his weakness on the right, led his massed column in a charge ... the Spartans failed to reinforce the threatened section in time and were broken, for considerable loss to themselves and almost none to the Thebans.*
>
> – John Keegan[7]

Phalanx warfare in 371 BC, when the battle of Leuctra was fought, normally proceeded by spreading the phalanx evenly across the boundary of engagement. By deceptively massing his outnumbered troops on one side, Epaminondas was able to overwhelmed the Spartans on the left wing and thereby carry the battle, as the losses inflicted on the side that broke and ran in phalanx engagements tended to be disproportionately heavy. The effective deployment of men on a battlefield, or arrangement of electrons in phase space, affects their utility, particularly if the manner of their grouping makes the phase space area seem more foreboding than it is.

The grouping of electrons, as it were, has the added complication of keeping track of the velocity, and because visualization is perplexing on first encounter, consider a rather trivial and therefore artificial example, but one which nevertheless catches the spirit of the difficulties. Suppose that the distribution of (x, x') points is such that the initial distribution of phase space points is enclosed by a square region of size LL'. Consider linear forces first, so that the x coordinate $x(t)$ of an electron can be described by

$$x(t) = x_o + v_o t + \frac{F}{2m}t^2 \tag{33.22}$$

letting $a = F/m$ be only in the $\hat{x}$ direction. If there is no acceleration in the $\hat{z}$ direction, then $x' = dx/dz = (dx/dt)/(dz/dt) = v_x(t)/v_z$ with $v_x(t) = v_o + at$ (this is a highly artificial condition simply to illustrate a point; in real linear beam devices, the acceleration that matters *is* in the $\hat{z}$ direction, but that is for later).

In fact, suppose that there is an $N \times N$ array of electrons as in Figure 33.7, positioned such that all in the same row have the same velocity, and all in the same column have the same starting position, as in Figure 33.8: the total area is then found by summing over all the small phase space rhombus areas dA where

$$dA(t) = \Delta x(t)\Delta x'(t) \tag{33.23}$$

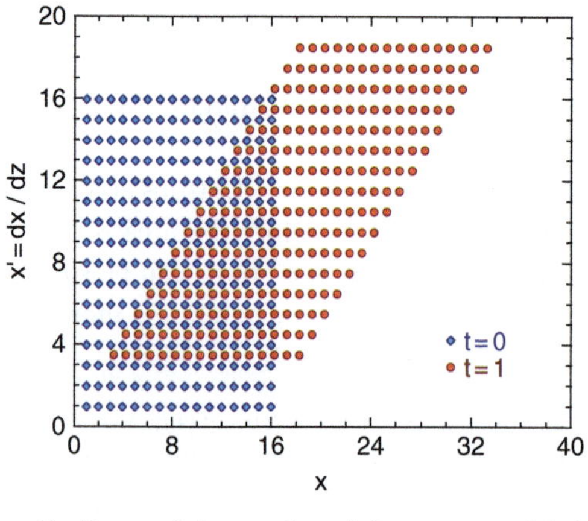

Figure 33.7 Initial positions (blue diamonds) of an $N \times N$ array of electrons in x–x' phase space, and their position at a time t later, assuming motion in a constant force $\vec{F} = F\hat{x}$ environment: there is no acceleration in the $\hat{z}$ direction, so that $dx/dz = v_x(t)/v_z$. Units are such that the initial x and x' positions are integers, and $a = 2.5$. The area of the red parallelogram is the same as the area of the blue rectangle.

---

[7] John Keegan, *A History of Warfare*. New York: Vintage Books, 1994, p. 258.

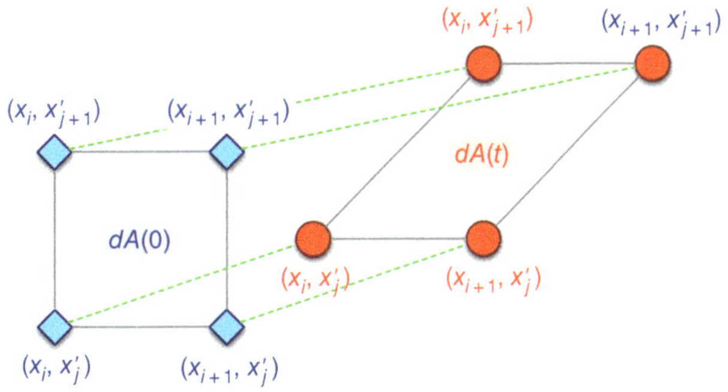

Figure 33.8 Evolution of phase space points from Figure 33.7 in close up: blue are at time = 0, red are at time = t, with the index i denoting row element and j denoting column element. The area defined by the four points is $dA(0)$ and $dA(t)$ for blue and red, respectively.

For the (i, j)th point, with i denoting position in row and j denoting position in column, and all the initial velocities in the jth row equal to $v_j(0)$,

$$x_i(t) = x_i(0) + v_j(0)t + \frac{1}{2}at^2 \tag{33.24}$$

$$x'_j(t) = \frac{1}{v_z}\left(v_j(0) + at\right) \tag{33.25}$$

and so

$$\begin{aligned} dA_{ij}(t) &= \left[x_{i+1}(t) - x_i(t)\right]\left[x'_{j+1}(t) - x'_j(t)\right] \\ &= \left[x_{i+1}(0) - x_i(0)\right]\left[x'_{j+1}(0) - x'_j(0)\right] = dA_{ij}(0) \end{aligned} \tag{33.26}$$

because of the cancelation of the t-dependent terms. In other words, the area of the phase space element is unchanging (recall Eq. (7.15)), and so the sum of all the small dA areas to get the area of the beam in phase space does not change either. Another way of saying this is that as time progresses, the phase space distribution $f(x, x', 0)$ maps on to a new distribution $f(x, x', t)$ and the mapping, being governed by a Hamiltonian, occurs by equations of motion that leave volumes in phase space invariant. Such is the essence of *Liouville's theorem*.

What about when the distribution of phase space points gets, in the words of Buon, "twisted and bent," and how does that complicate the empirical definition of emittance as an area that encloses the phase space points? Consider just one row of Figure 33.7, but let the force acting on the particles be something more complicated, for example one in which the force is a function of position. The twisting that may result might resemble something like Figure 33.9. Although an S-like enclosing area can be found for the red points at a later time t compared to the straight line on which the blue points lie at time 0 (for which the area of the enclosing ellipse is vanishingly small), as a practical consideration as far as the utility of the resulting beam is concerned, the enclosing shape is better represented as a fat ellipse: the effective area

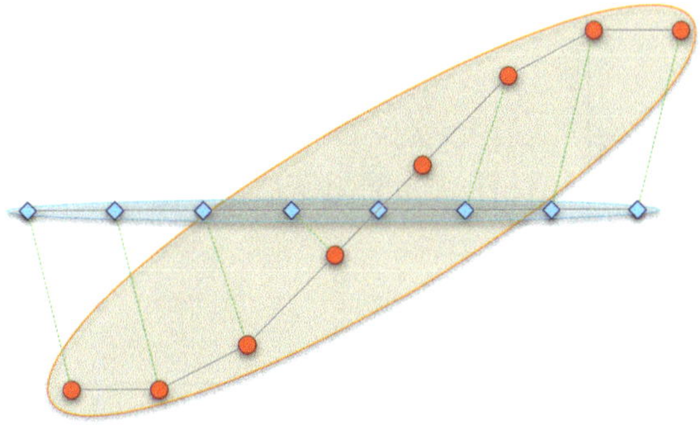

Figure 33.9 Evolution of a line of phase space points with very little phase space area to a contorted shape that requires a larger ellipse to enclose the points, with dashed green lines showing the mapping. Even though a boundary that hugs the shape of the S-like curve would have the same area, the ellipse better entails the difficulties of dealing with the resulting beam.

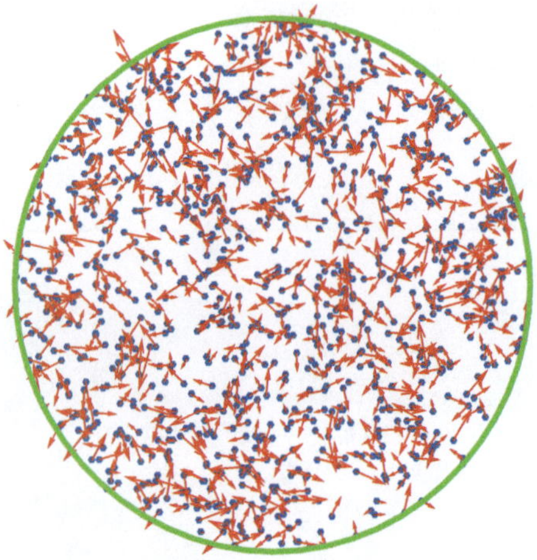

Figure 33.10 Position (blue dots) and transverse velocity (red arrows) of electrons. The length of the red arrows corresponds to the magnitude $|\vec{v}_\perp|$ and the arrow indicates the direction. The data for the representation is recast in Figure 33.11. The green circle shows the boundary of the beam.

in phase space has increased so that emittance (like the lop-sided phalanx of Epaminondas) can appear as though it were larger than it is. A good definition of emittance should take that behavior into account.

A definition of emittance that computationally captures the impact of twisted and bent distributions is to use rms values of x and x'. It is often represented as [638, 646, 658, 670, 671]

$$\varepsilon = \sqrt{\langle x^2 \rangle \langle x'^2 \rangle - \langle xx' \rangle^2} \tag{33.27}$$

Sometimes the definition has a 4π out front, sometimes just a 4, depending on the ulterior motives of the author. Such factors appear in the *trace space* definition of emittance, but the factor of 4 is eliminated when considering the rms form of Eq. (33.27) (the treatment by Reiser [638] of trace space, rms factors, and invariance is invaluable). Although this captures the effects of twisted distributions, it does not (yet) account for the shrinking of emittance as the beam is accelerated in the $\hat{z}$ direction. Therefore, return to the small black dots of Figure 33.2 representing the starting points of the cyclical electron motion on top of a circle the size of which gives an indication of the magnitude of the transverse velocity $\vec{v}_\perp$. Now let the black dots show the starting point and a red vector indicate the direction and magnitude of $\vec{v}_\perp$, as in Figure 33.10: such a representation captures four dimensions (x, y, v_x, v_y) of the six-dimensional distribution function $f(\vec{r}, \vec{v})$ encountered in Section 5.1. A representation more in keeping with (x, x') is to use one axis for position and another for velocity: such slices of phase space are shown in Figure 33.11. The shape and size of the area enclosing the points representing the particles in the beam therefore gives a measure of emittance. Still, v_x and x' differ.

Taking a cue from Eq. (33.25) relating x' to v_x, make the replacement of

$$x' = \frac{\Delta x}{\Delta z} = \frac{v_x \Delta t}{v_z \Delta t} = \frac{v_x}{v_z} \tag{33.28}$$

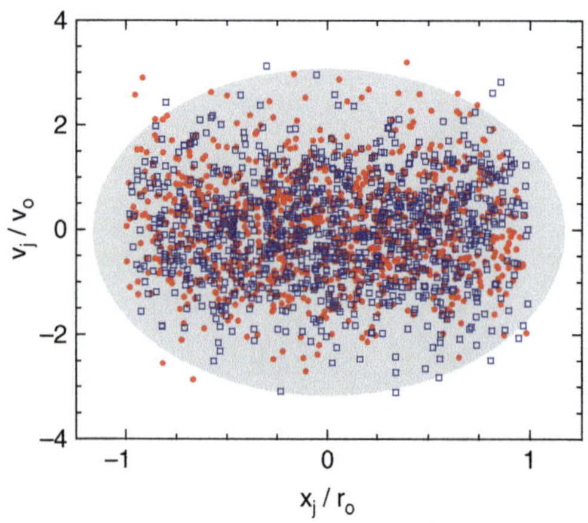

Figure 33.11 Position and velocity pairs, red corresponding to the pair (x, v_x) and blue to the pair (y, v_y), where x_j denotes x or y and similarly with v_j for the $\vec{r}$ and $\vec{v}_\perp$ of Figure 33.10. The factors r_o and v_o are the radius of the emission area and the average transverse velocity. The gray ellipse is an area in phase space that encloses most of the dots, and will be related to ε.

For electron beam devices, the electrons are aggressively accelerated in the $\hat{z}$ direction to avoid emittance growth and space-charge issues. Consequently, the difference between v_z and the total velocity magnitude v is not all that great, and so making the replacement $v_z \to v$ is useful in the relativistic factors $\beta = v/c$ and $\gamma = 1/\sqrt{1-\beta^2}$. Additionally, it is not the *velocity* that matters in the phase space descriptions so much as it is the *momentum* $m|\vec{v}| = \hbar|\vec{k}|$. As the beam becomes relativistic,

$$\hbar k_z = m v_z \to (\gamma m)(\beta c) = \beta\gamma mc \tag{33.29}$$

where the behavior of $\beta\gamma$ in the denominator of Eq. (33.28) is what causes the emittance to decrease as the kinetic energy increases. Therefore, the *normalized* emittance takes out the $\beta\gamma$ factor and genuflects to the usage of $\hbar k$ over x', and so [672]

$$\beta\gamma\varepsilon = \varepsilon_{n,rms} \equiv \frac{\hbar}{mc}\sqrt{\langle x^2\rangle \left\langle k_x^2\right\rangle - \left\langle xk_x\right\rangle^2} \tag{33.30}$$

Emittance can be brought to very low values simply by reducing the current in the beam: space-charge effects fade, small cathode areas can be used, and so on. But low current beams are often contrary to, or at least not only, what is desired. As a result, a related figure of merit gets both current *and* emittance together, and is called *brightness B*. Brightness relates the amount of current I to the area in which it can be focused A and how it diverges into a solid angle Ω by [658]

$$B = \frac{I}{A\Omega} \propto \frac{I}{\varepsilon^2} \tag{33.31}$$

where the constant of proportionality depends on the distribution of the particles in the bunch, for example the proportionality constant for a hollow shell (the so-called Kapchinskij and Vladimirskij distribution) is 2 [659]. Using the normalized rms emittance in Eq. (33.31) results in the normalized brightness.

33.3.2 Thermal emittance

The way things are does not determine the way they ought to be.

– Michael J. Sandel[8]

For symmetrical distributions in (x, k_x) space, $\left\langle xk_x\right\rangle$ vanishes, as Figure 33.11 makes plausible: for every point on the right side $(x > 0)$ a point on the left side $(x < 0)$ matches it (or would for a sufficiently dense distribution of points), and so for beams from uniformly emitting rotationally symmetric surfaces, the factor $\left\langle xk_x\right\rangle \to 0$. Additionally, for such beams $\left\langle x^2\right\rangle = \left\langle \rho^2\right\rangle/2$ and $\left\langle k_x^2\right\rangle = \left\langle k_\rho^2\right\rangle/2$ as is straightforward to show.

EXAMPLE: For a rotationally symmetric and uniformly emitting cathode of radius R, first show that $\left\langle x^2\right\rangle = R^2/2$ where R is the rms radius of the cathode, and second show that $\hbar^2\left\langle k_x^2\right\rangle/2m = E_\perp/2$.

SOLUTION: Define $\rho^2 = x^2 + y^2$ and $k_\perp^2 = k_x^2 + k_y^2$. Then

$$\left\langle x^2\right\rangle = \frac{1}{\pi\rho_c^2}\int_{-\infty}^{\infty} x^2 \int_{-\infty}^{\infty} \Theta(\rho_c^2 - x^2 - y^2)dxdy = \left\langle y^2\right\rangle$$

where the last equality follows from the symmetrical nature of x and y. As a result,

$$\left\langle x^2 + y^2\right\rangle = \left\langle \rho^2\right\rangle \to \left\langle x^2\right\rangle = \left\langle y^2\right\rangle = \frac{\left\langle \rho^2\right\rangle}{2}$$

For $R = \sqrt{\langle\rho^2\rangle}$ then the result follows. Applying the same logic to $\left\langle k_x^2\right\rangle$ and letting $E_x = \hbar^2k_x^2/2m$ and $E_\perp = E_x + E_y$, then

$$\left\langle k_x^2 + k_y^2\right\rangle = \left\langle k_\perp^2\right\rangle \to \frac{\hbar^2}{2m}\left\langle k_x^2\right\rangle = \frac{\hbar^2}{2m}\left\langle k_y^2\right\rangle = \frac{\langle E_\perp\rangle}{2}$$

[8] Michael J. Sandel, *Justice: What's the Right Thing to Do*. New York: Farrar, Straus and Giroux, 2009, p. 165.

For uniformly emitting and rotationally symmetric emission areas, it is then common to define the *mean transverse energy (MTE)* as $\langle E_\perp \rangle$ and so write the normalized rms emittance as

$$\varepsilon_{n,rms} = \frac{R}{2}\sqrt{MTE} \tag{33.32}$$

For thermal emission, the emitted energy distribution is MB. In cylindrical coordinates k_ρ takes over for the transverse $k_\perp$, and so[9]

$$\langle k_\rho^2 \rangle = \frac{\int_{-\infty}^{\infty} k_\rho^2 \exp(-E_\rho/k_B T) 2\pi k_\rho dk_\rho}{\left[\int_{-\infty}^{\infty} \exp(-E_x/k_B T) dk_x\right]^2} \tag{33.33}$$

The integrations can be performed analytically to give

$$\langle k_\rho^2 \rangle = \frac{4\pi(mk_BT/\hbar^2)^2}{2\pi mk_BT/\hbar^2} = \frac{2mk_BT}{\hbar^2} \tag{33.34}$$

which might have been expected from the equipartition theorem of Eq. (5.13). The result identifies $MTE = k_BT$, and so in the case of thermal emission,

$$\varepsilon_{n,rms}(\text{thermal}) = \frac{R}{2}\sqrt{\frac{k_BT}{mc^2}} \tag{33.35}$$

so that emittance is often identified by a temperature for thermal emitters.

33.3.3 Photoemission emittance

The application of the mean transverse energy approach to calculating the emittance of a photocathode is a bit more nuanced [181, 673], as the evaluation of the moments in Eq. (31.21) require a consideration of the photon energy's contribution to the emitted electron energy. For a flat surface (one-dimensional emission barrier), continuity of the electron wave function and its first derivative require that the transverse momentum component $k_\perp$ be conserved across the barrier. The total energy of the electron prior to encountering that barrier, however, is $E + \hbar\omega$. This means for metals,

$$\frac{\hbar^2 k_\perp^2}{2m} = (E + \hbar\omega)\sin^2\theta \tag{33.36}$$

As a result, and very much like the calculation of quantum efficiency in Eq. (31.21) using the moments approach, emittance for photoemission becomes

$$\varepsilon_{n,rms} = \frac{R}{2}\left(\frac{\hbar}{mc}\right)\sqrt{\frac{M_2(k_\perp)}{2M_0(k_\perp)}} \tag{33.37}$$

where, recall, the subscript of M_n specifies the power to which its argument is taken in the integrand as per Eq. (7.8). Substituting the various approximations behind the integrand terms into Eq. (31.21) as applied to $M_n(k_\perp)$ gives [673]

$$\begin{aligned} M_n(k_\perp) \approx & \frac{1}{8\pi^2}\left(\frac{2m}{\hbar^2}\right)^{(n+3)/2} \frac{1}{(n+2)(1+p_o)} \\ & \times \int_{\phi-\hbar\omega}^{0} \frac{\sqrt{\mu+s}(s+\hbar\omega-\phi)^{(n+2)/2}}{\mu+s+\hbar\omega} ds \end{aligned} \tag{33.38}$$

For photon energy $\hbar\omega$ close to the Schottky-reduced work function ϕ, the integral may be approximated by

$$M_n(k_\perp) \approx \frac{\sqrt{2m\mu/\hbar^2}\left[2m(\hbar\omega-\phi)/\hbar^2\right]^{(n+2)/2}}{4\pi^2(\mu+\hbar\omega)(1+p_o)(n+4)(n+2)} \tag{33.39}$$

Much of what dresses this approximation to M_n is stuff independent of n: as a result, ratios of M_2 with M_0 will eliminate most of it. What is left is

$$\varepsilon_{n,rms}(\text{photo}) \approx \frac{R}{2}\sqrt{\frac{\hbar\omega-\phi}{3mc^2}} \tag{33.40}$$

[9] One may wonder what became of the k_y and k_z integrations in the denominator. In cylindrical coordinates for the MB distribution, all of the k_j integrals are equal (hence the square of the term in the denominator). Further, the integrals containing k_z appear the same in both the numerator and denominator, and so cancel.

which is the *Dowell–Schmerge* formula for photocathode emittance [181].[10] Like the thermal emittance of Eq. (33.35), Eq. (33.40) is the product of a characteristic radius with the square root of a characteristic energy factor ($k_B T$ for thermal, $(\hbar\omega - \phi)/3$ for photo) in ratio with the electron rest energy mc^2, an intuitively pleasing result.

33.3.4 Field emission emittance

And ere a man hath power to say "Behold!"
The jaws of darkness do devour it up:
So quick bright things come to confusion.

– William Shakespeare[11]

The evaluation of emittance from a planar surface for field emission is similar to the moments approach for photoemission but without the addition of a photon energy and using the transmission probability associated with tunneling rather than a step function for emission over the barrier. Recall that in the Fowler–Nordheim model, the transmission probability can be obtained (e.g., Eqs (13.11) or (17.11)) as

$$-\ln D(k_z) = \frac{4}{3\hbar F}\sqrt{2m\Phi^3}\, v(y) - \frac{2}{\hbar F}\sqrt{2m\Phi}\, t(y)\left(E_z - \mu\right) \tag{33.41}$$

Although the full behavior of $v(y)$ captured in Eq. (13.23) will be used, the variation of $t(y)$ is weak and so approximating it by $t(y) \to t(y_o) = 1 + (1/6e) = 1.06131$ is convenient. When used in the moments approach, the MTE for field emission becomes

$$\left\langle k_\rho^2 \right\rangle = \frac{\int_0^{k_F} dk_z D\left(k_z\right) \int_0^{\sqrt{k_F^2 - k_z^2}} k_\rho^3 dk_\rho}{\int_0^{k_F} dk_z D\left(k_z\right) \int_0^{\sqrt{k_F^2 - k_z^2}} k_\rho dk_\rho} \tag{33.42}$$

The integrations over k_ρ are easily dispatched, giving

$$\left\langle k_\rho^2 \right\rangle = \frac{1}{2}\frac{\int_0^{k_F} dk_z D\left(k_z\right)\left(k_F^2 - k_z^2\right)^2}{\int_0^{k_F} dk_z D\left(k_z\right)\left(k_F^2 - k_z^2\right)} \approx \frac{2k_F^2}{\left[4\mu\hbar F(2m\Phi)^{1/2} t(y) - 1\right]} \tag{33.43}$$

and so the emittance associated with a one-dimensional field emission barrier becomes

$$\varepsilon_{n,rms}(\text{field}) = R\frac{\hbar k_F}{mc}\left[\frac{4\mu}{\hbar F}(2m\Phi)^{1/2} t(y) - 1\right]^{-1/2} \tag{33.44}$$

Field emitters have always been described as exceptionally bright sources, but their brightness comes not from the MTE values but rather their smaller emission areas, as will be seen in Section 33.4.

33.3.5 Comparisons of the one-dimensional emittance equations

…since statistical data are rarely uniform, they may be grouped or "interpreted" in different ways. Do not hesitate to use only that part of the statistical data which suits your case, especially if you do not expect to be challenged.

– Nicholas Capaldi[12]

Insofar as the photo, field, and thermal emission equations for current density and emittance can be related to a moments-based model, they can be compared. Secondary emission (from diamond) is a breed apart, and so will be taken up separately. By virtue of the approximation of a planar emission area, comparisons of $\varepsilon_{n,rms}$ amount to comparisons between $\sqrt{MTE}$ for the three emission mechanisms: ratios between ε will simply factor out the radius of the circular emitting area. This is just as well: comparing the emittances of different electron sources with different emission areas

[10] The mistaken neglect of the photon energy $\hbar\omega$ contribution to $k_\perp^2$ in ref. [674] introduced a factor of $\sqrt{\mu/(\mu + \hbar\omega)}$ that was identified and corrected in ref. [181]. Science is, as Carl Sagan approvingly observed, "self-correcting" (Carl Sagan, *Cosmos*. New York: Random House, 1980, p. 333).

[11] William Shakespeare, *A Midsummer Night's Dream*, I.1.147–149.

[12] Nicholas Capaldi, *The Art of Deception: An Introduction to Critical Thinking*. Buffalo, NY: Prometheus Books, 1987, p. 58.

and operating conditions is cherry-picking of data in the worst way. Removing the dependence on cathode area by first considering equal emission area conditions allows for greater appreciation for which cathodes are bright and which are not when emission area is re-introduced.

Each emission mechanism has a characteristic parameter on which ε depends: thermal emission involves $k_B T$, photoemission of course involves $\hbar\omega$, and field emission most certainly will squeeze in $B_o \Phi^{3/2}/F$. Once the characteristic parameter is identified based on typical currents drawn from typical applications, the variation in emittance as parameters change can be addressed. Representative cases for the various processes are considered in the next few examples.

EXAMPLE: Find $\sqrt{MTE}$ for standard thermal emission conditions.

SOLUTION: A typical current density as used in a generic TWT (Table 33.1) is given by

$$J = \frac{0.288 \text{ A}}{\pi(0.1 \text{ cm})^2} = 9.16732 \frac{\text{A}}{\text{cm}^2}$$

For typical parameters $\Phi = 2$ eV and $F = 0.1$ eV/μm, then $J_{RLD}(T)$ as per Eq. (12.3) will occur for $T = 1357.09$ K. As a result,

$$\sqrt{MTE(\text{thermal})} = \sqrt{\frac{0.116945 \text{ eV}}{510999 \text{ eV}}} = 0.000478388$$

For a cathode of radius 0.1 cm, then $\varepsilon_{n,rms}(\text{thermal}) = 0.239194$ mm-mrad.

EXAMPLE: Find $\sqrt{MTE}$ for standard photoemission conditions.

SOLUTION: Consider a metal photoemitter of radius $r = 0.5$ cm with $\Phi = 4.5$ eV and $\mu = 7$ eV that is $R = 50\%$ reflective with a penetration depth of $\delta = 12.8$ nm at a wavelength of $\lambda = 266$ nm. Let the relaxation time be $\tau = 4$ fs, typical of e–e collisions, as in Figure 32.15. Let the maximum field correspond to $F = 115$ eV/μm. It follows that $\phi = \Phi - \sqrt{4QF} = 4.09307$ eV, $k_o = \sqrt{2m(\mu + \phi)}/\hbar = 17.0634$ nm^{-1}, and so $p = m\delta/\hbar k_o \tau = 1.61994$. When such values are used in Eq. (31.27), then $QE = 2.12268 \times 10^{-4}$. Let $J(\hbar\omega)$ correspond to a charge of 0.01 nC emitted in 50 ps from a circular area with a diameter of 1 cm. Collectively, such values, nominally suggested by refs [675] and [676], entail that the laser intensity be $I_\lambda = 5.59165$ kW/cm^2 as per $J = (q/\hbar\omega)\, QE\, I_\lambda$. Then, using Eq. (33.40),

$$\sqrt{MTE(\text{photo})} = \left\{ \frac{(4.66106 - 4.09307) \text{ eV}}{3 \times 510999 \text{ eV}} \right\}^{1/2} = 0.000608697$$

EXAMPLE: Find $\sqrt{MTE}$ for standard field emission conditions.

SOLUTION: For a field emitter with $\Phi = 4.5$ eV and $\mu = 7$ eV subject to an apex field of $F = 6$ eV/nm for an apex radius such that the the notional emission area is $\pi r^2 = \pi(20 \text{ nm})^2$, then $J(F)\pi r^2 = 52.28$ μA, a high but not atypical value [460]. Expressing emittance in terms of $B = (4/3\hbar)\sqrt{2m\Phi^3} = B_o \Phi^{3/2} = 63.0458$ eV/nm, where B_o is as per Table 2.3 and $t(y_o)$ from Eq. (13.24), then

$$\sqrt{MTE(\text{field})} = \left\{ \frac{8\mu}{mc^2} \left[3t(y_o) \frac{\mu}{\Phi} \left(\frac{B}{F} \right) - 1 \right]^{-1} \right\}^{1/2}$$
$$= \left(\frac{8 \times 7}{510999[53.8264 - 1]} \right)^{1/2} = 0.00144032$$

Each of the MTE relations can be put in the form of $MTE/MTE_o = x/x_o$, where x is given by $k_B T$ (thermal: $k_B T_o = 0.116945$ eV), $F\Phi/B$ (field: $F_o \Phi/B = 0.414064$ eV), or Δ/Δ_o with $\Delta = \hbar\omega - \phi$ (photo: $\Delta_o = \hbar\omega_o - \phi = 0.567995$ eV).

Observe that in the examples, the values of $\sqrt{MTE}$ for (thermal, photo, field) are in the proportions $(1, 1.272, 3.011)$, that is, MTE(thermal) is the smallest of them, and MTE(field) is the largest for the chosen values of T_o, F_o, and ω_o (which are themselves somewhat arbitrarily chosen). This is not in contradiction to the expectation that field emission sources have unparalleled brightness: what has not been included in the analysis – yet – is the consequences of emission area. By increasing the laser power and substantially reducing the emission area, the emittance of photocathodes can be made better than thermal sources. Field emission, with its nanoscale emission areas, is breathtakingly small, on the order of $\varepsilon_{n,rms}$(field) = 14 nm-mrad for $r = 20$ nm and other parameters of the field emission example above.

However – and this is a rather consequential "however" – field emitters are sharp little structures from which electrons are borne with a rather substantial transverse energy kick. Of course, thermal and photocathodes also have structure, and such surface "roughness" is implicated in their emittance values exceeding the theoretical limits, particularly for photocathodes [677–679], but the sharpness of field emitters is particularly legendary, and how curvature of the surface contributes to the emittance, especially for an array of emission sites, merits its own treatment in Section 33.4. Transverse emittance, though, is not the only concern. Longitudinal energy spread leads to longitudinal emittance, and it can interact with space charge in particularly mischievous ways for the high current density associated with photocathodes [514, 680–682]. All told, thermal emitters have advocates [683], and field emitters have opportunity [20, 24] apart from what one-dimensional models imply.

33.3.6 Space charge and emittance

...men do not make events, but events make men.

– Clarence Darrow[13]

What electrons do depends on what is going on around them. Although transverse emittance has received most of the attention, a gentle familiarization with some issues surrounding longitudinal emittance merit a look because of how space charge affects it. Longitudinal emittance is caused by differences in forward energy. The position of a non-relativistic electron in a bunch is given by the kinematic equation Eq. (33.24) with $a = F/m$. For a pulse length of δt, the separation between an electron at the head of a bunch to one in the tail is then simply dictated by the difference in launch times, where the tail electron is emitted at a time δt after the head electron, or

$$x(t) - x(t - \delta t) = v_o \delta t + \frac{F\delta t}{2m}(2t - \delta t) \tag{33.45}$$

where v_o can be the maximum energy of a photoexcited electron $v_o = \sqrt{2(\hbar\omega - \phi)/m}$ or perhaps the thermal energy $v_o = \sqrt{k_B T/m}$. Although the head and tail electrons can have different velocities (e.g., if the tail electron has a larger transverse energy on emission so that its forward energy is close to zero) and therefore induce a larger separation $x(t) - x(t - \delta t)$, at surface fields in the MV/m range, such distinctions get trifling, and it is simpler to take v_o as the same for both head and tail (e.g., in a 1 MV/m field, an electron is accelerated to the 2.6× the room temperature thermal velocity in 1 ps).

Compare such separations with the radius R of the emission area. For representative parameters of photoinjectors [684, 685], take $F = 1$ eV/nm and $R = 0.1$ cm for a $\delta t = 10$ ps pulse, and let the electrons emerge with a thermal velocity for room temperature conditions ($v_o = \sqrt{k_B T/m} = 67.431$ nm/ps). Under such conditions, $[x(t) - x(t - \delta t)]/R = 0.062233$: the thickness of the disk of charge compared to its radius is such that the bunch is a pancake (sometimes described as "all ends").

Demands on the photoinjectors to reduce emittance make matters more interesting. A reduction in area decreases transverse emittance, and the acceleration of the beam to near-light velocities as rapidly as possible has all the electrons in the bunch traveling at nearly the same speed so that further elongation of the pulse does not continue. Space charge, however, creates issues. Keeping the visualization of the pancake of charge in mind, if the charge density of the disk is σ, then a field on each side of the disk of $\delta F = 8\pi Q\sigma = q^2\sigma/2\varepsilon_0$ is created (Section 16.1 and the Gaussian pillbox argument of Eq. (16.35), and recall that σ is a *number* charge density). Electrons at the head of the pulse therefore feel a force pushing them forward, whereas those in the tail feel a force pushing them back, relative to a charge in the center of the pulse.

[13]Quoted in Irving Stone, *Clarence Darrow for the Defense*. New York: Signet, 1971, p. 133.

EXAMPLE: Find the space-charge field associated with a disk of radius $r = 0.05$ cm containing $\Delta q = 0.1$ nC of charge. Compare its magnitude to a background field corresponding to $F = 50$ eV/μm.

SOLUTION: Using $\delta F = 8\pi Q\sigma$ with σ being the number charge per unit area, then

$$\delta F = 8\pi Q\left(\frac{\Delta q}{q}\right)\frac{1}{\pi r^2} = 0.00719\frac{\text{eV}}{\text{nm}}$$

or 14.38% of F.

The field δF is *added* to the force accelerating the electrons in the head, and *taken away* from that for the electrons in the tail, causing Eq. (33.45) to become

$$x(t) - x(t-\delta t) = v_o\delta t + \frac{F-\delta F}{2m}(2t-\delta t)\delta t + \frac{\delta F}{m}t^2 \tag{33.46}$$

Now consider the slightly more interesting *ad hoc* example of a $\Delta q = 1$ nC bunch emitted in 4 ps, still with $R = 0.1$ cm, where $F = 50$ eV/μm (corresponding to a field of 50 MV/m). In the absence of space-charge forces, after a time $t = 4\delta t$, then $x(t) - x(t-\delta t) = 0.49274$, but when the space-charge contributions are included, then it becomes 1.4913: *right after emission, space-charge forces elongate the pulse.*

That is not all space charge does. For the expected circumstance of when the electrons have different velocities on emission as a consequence of how far the photon energy $\hbar\omega$ exceeds the barrier height ϕ (if a metal) or E_a (if a semiconductor), those that emerge at a sprint separate themselves from those that dawdle (either because of a different normal energy component, or a smaller total energy), and that difference is exacerbated by the space charge such that the energy differences can be made much more consequential, increase the longitudinal emittance, and put complicated structure on the beam [686]. Such changes greatly complicate how the beam is accelerated and couples in with non-uniformity of emission and its relation to both longitudinal and transverse emittance [687]. As a result, and particularly for the treatment of RF photoinjectors for particle accelerators and FELs where the bunch charges can be large, the accelerating fields intense, and the emission areas by no means uniform, the implicit assumptions behind the gentle theory here (or elsewhere) are not adequate. Understanding the evolution of the beams then requires the intervention of beam optics simulation codes to make headway.

33.3.7 Secondary emission and emittance

All right then. I'll go to hell…

– Mark Twain[14]

At the moral climax of Twain's masterpiece novel, Huck agonizes over acting in accordance with prevailing social norms of his time and place, but resolves instead to do what he naïvely thinks is egregiously wrong (but ethically exemplary) by choosing not to betray his raft-mate Jim to the authorities. Perhaps no other literary example distills so well that sometimes what seems to be the worst decision actually represents the right choice when viewed beyond a narrow purview. In a way, secondary emission provides another example.

The treatment of photoemission (Chapter 31) looked askance at scattering because of its effects on diminishing quantum yield (particularly for metals) while at the same time elongating response time (particularly for semiconductors). It seems suspicious that scattering can be put to good use after it has been decried, but with emittance, scattering represents an opportunity for a short and simple reason: for NEA surfaces such as hydrogen-terminated diamond, electrons that have thermalized with the lattice can nevertheless be emitted, and that improves the transverse emittance, as well as the longitudinal emittance. This is no surprise in the literature: the virtues of thermalization of hot electrons for high quantum efficiency NEA DC photocathodes (like GaAs) are appreciated [526, 529, 688, 689]. Bringing the subject up in the context of secondary emission, though, is because the secondaries in diamond are created with such a *large* energy above the conduction band minimum (about 13.75 eV or so), so that reducing such a magnificent energy distribution to a thermal spread is rather dramatic and offers advantages for diamond amplifiers [324, 529, 586, 587, 590, 690, 690] beyond simply allowing for a narrative with more panache.

[14]Mark Twain, *Greenwich Unabridged Library Classics: The Adventures of Huckleberry Finn*. New York: Chatham River Press, 1982, p. 282.

The consequences of scattering on secondary emission are clearly evident by comparing secondary emission in reflection mode with those in transmission mode (Section 32.1). The behavior of the electron energy distribution curves (EDCs) show the non-thermalized emitted electrons by the high-energy tail of the reflected EDC compared to the transmission EDC where it is absent, as was also apparent in Figure 32.3: such tails are counter to the demand for narrow energy spreads needed by some applications, but they do contribute to high yields in reflection mode operation. To enable a thermalized beam, the diamond flake must be thick enough to eliminate the non-thermal component, yet not so thick that the secondary yield is reduced too far. Relying on transmission current rather than reflected has one other advantage: it protects the NEA surface of the hydrogen-terminated diamond flake, or allows the diamond flake to protect in turn a more sensitive electron source such as a photocathode from a damaging environment such as an RF photoinjector. A diamond flake thinner than the depletion width L_D (Section 32.4) allows a field to be maintained across it, which further improves the yield. For such reasons, the range over which thermalization takes place needs to be compared to the penetration depth of the primary beam as a function of its energy E_o in order to maximize the yield.

Measurements of the EDCs in transmission and reflection mode of high purity and very thin (< 5 μm) boron-doped diamond films using chemical vapor deposition, when compared to Monte Carlo simulations, determine the minimum thickness diamond flakes that will allow for complete thermalization of the secondaries before they are emitted [144].[15] In that study, two fixed-thickness nanocrystalline diamond flakes, one of 2.3 μm thickness and one of 0.65 μm thickness, were subjected to a range of primary electron beam energies. The penetration depth $R(E_o)$ decreased when E_o was small, and at some point the high-energy tail of the EDC was eliminated, leaving a narrow distribution with a full width at half maximum (FWHM) of about 0.35 eV composed of secondaries that had thermalized with the diamond lattice.

For the thicker (2.3 μm) flake, no transmission occurred until the primary beam energy reached $E_o = 13.5$ keV, and the transmission that did occur did not have a high energy tail. When the beam energy is increased to $E_o = 17$ keV, the first signs of a non-thermal tail became evident. For $E_o > 17$ keV, the total transmission continued to increase, but the relative proportion of emitted electrons in the tail stayed the same. The measurements are shown in Figure 33.12, where the beam is normalized to its maximum value. More accurately, because experimental measurement has uncertainty, the beam is normalized rather to $y = [I(E) - I_{min}]/[I_{max} - I_{min}]$ where I_{max} and I_{min} are the maximum intensity for the former and the minimum (or representative minimum because of the noise behavior) of the latter. The thermalized electrons all show the same behavior at the large peak near $E = 6.5$ eV. The non-thermalized electrons make up the tail, and are progressively eliminated as E_o decreases due to both decreased penetration of the diamond flake as per Eq. (32.4) and the fewer number of collisions needed to thermalize the primary. To underscore this interpretation, the difference between reflection mode for a much lower primary energy of $E_o = 1$ keV and the $E_o = 16$ keV transmission mode line of Figure 33.12 is shown in Figure 33.13. The thermalized component is the same for both, but the non-thermalized tail apparent in the reflection data is absent in the transmission data, understandable because the reflection mode secondaries are borne very close to the surface and are able to escape before they are brought to heel by scattering.

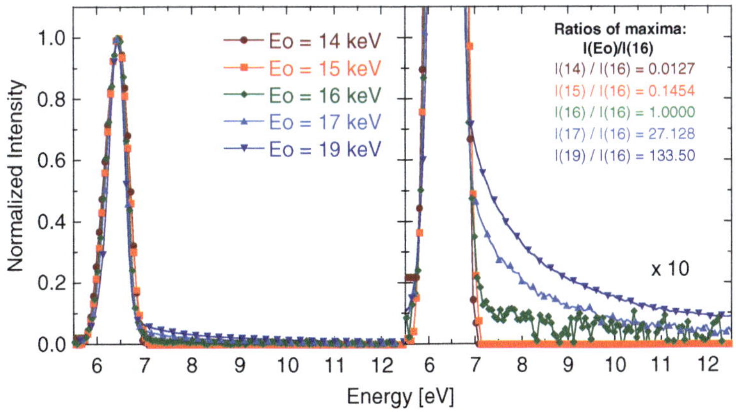

Figure 33.12 Emission intensity as a function of energy for various primary beam energies E_o normalized so that the maximum value of all lines coincide. As E_o declines, the tail of non-thermalized electrons is progressively eliminated. Based on Figure 3 of J.E. Yater, K.L. Jensen, T.I. Feygelson, and B.B. Pate, *MRS Advances*, 2016. Data courtesy of J. Yater (NRL).

[15]The CASINO Monte Carlo program [691] was used by Yater *et al.* in contrast to the Monte Carlo methods of Section 32.2.

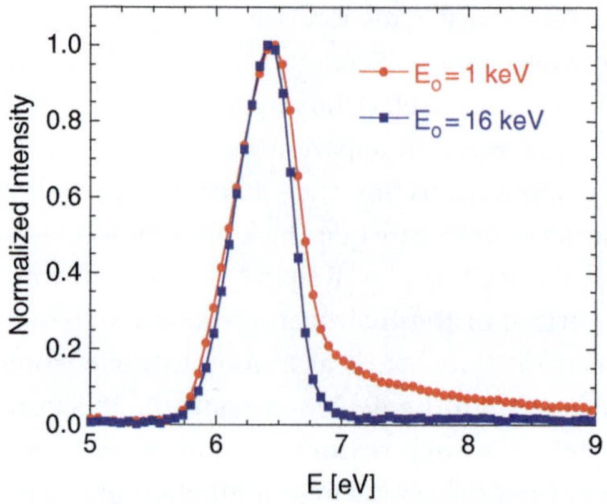

Figure 33.13 Emission intensity as a function of energy for various primary beam energies E_o normalized so that the maximum value of all lines coincide. As E_o declines, the tail of non-thermalized electrons is progressively eliminated. Based on Figure 3 of J.E. Yater, K.L. Jensen, T.I. Feygelson, and B.B. Pate, *MRS Advances*, 2016. Data courtesy of J. Yater (NRL).

The manner in which the secondaries loose their energy and thermalize, however, dictates that the diamond flake thickness is an example of the Goldilocks principle:[16] in weighing between the extremes of thermalization and losses, there is a region that is "just right" where both the emission and thermalization are jointly maximized, and Monte Carlo, by allowing the secondaries to be followed and their energy loss monitored, shows how. Matching the experimental arrangement, different primary energies can be, via simulation, injected into a flake region and contours of their energy loss monitored: when the losses have dropped enough, say to just past where 5% of the initial energy contour is, then that is the Goldilocks thickness. Two such examples are given in Figure 33.14, which shows the cross-section of a diamond flake for which high-energy primary electrons are incident from the bottom and generate a line of secondaries each with an initial energy of ≈ 13.75 eV. The secondaries then diffuse and lose energy, and the contour lines mark the extent of their loss as a percentage of their initial value. The broccoli-like contour shapes show the fanning out of the secondaries as they drop in energy, with the largest contour being that for which the energy has dropped to 5% of its initial value of 13.75 eV, the average energy at which secondaries are created by the primary. The injection sites of the primary are offset so as to enable easier comparison of the contours. If the thickness of the diamond flake is made to coincide with the largest contour (the 5% line), then the electrons will have thermalized just as they are emitted into vacuum (top boundary of

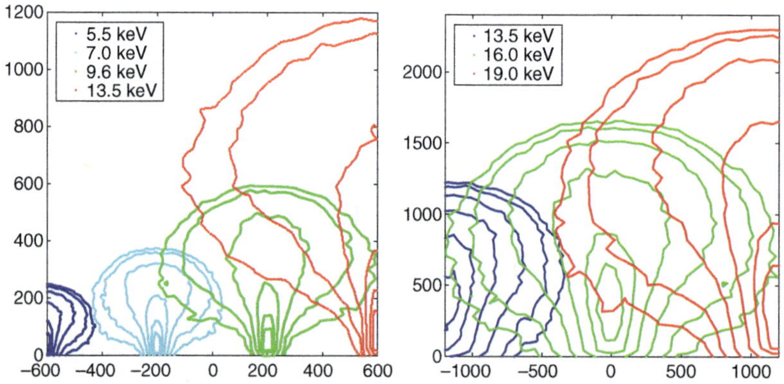

Figure 33.14 Energy contours for various primary energies, each color associated with a different E_o. Contours are drawn at the values of 5%, 10%, 25%, 50%, 75% and 90% for largest to smallest contours of a given color, respectively, with the various cases spread out along the x axis. The left-hand figure is for a box of size 1200 nm × 1200 nm, whereas the right is 2400 nm × 2400 nm. The left shows the primary energies for E_o = 5.5, 7.0, 9.6 and 13.5 keV, whereas the right shows 13.5, 16.0, and 19.0 keV because a greater area is being shown.

[16]The "Goldilocks principle", named after the children's story of a larcenous young girl resorting to breaking and entering to obtain edibles and bedding of intermediate temperature and firmness, respectively, has wider application: it is used to describe the just right conditions of certain planets like Earth that support liquid water and enable life, up to the minute balancing of the physical constants of Nature that have to be just so if life in the Universe can develop - the later discussion invariably being associated with the philosophically fecund "Anthropic Principle". See John D. Barrow, *The Constants of Nature* (Vintage Books, New York, 2004) or Brian Greene, *The Hidden Reality* (Knopf, New York, 2011).

figure). Differences with respect to field across the flake and the thickness of the simulation region occur with respect to the left and right images, just as variation in flake thickness is associated with differences in grain boundaries and band bending in the nanaocrystalline diamond. Thus, for example, if the flake is 650 nm thick, then the most energetic primary to be used is 9.6 keV, as the 5% contour is just at the 600 nm mark. Such studies conclude, therefore, that for primary beam energies of 5 keV, a minimum flake thickness of 1 μm is required to insure thermalization of the beam (although a flake twice that thickness may be preferable as it adds mechanical strength) [590].

33.4 Emittance for a bump

Once more unto the breach, dear friends, once more ...

– William Shakespeare[17]

As the Bard tells it, the exhortations of the English king Henry V to his men to yet again storm a break in the walls of the French port of Harfleur enabled the taking of the town in the year 1415, the opening move in Henry V's ambitions to press his claims on the French throne. Incensed, the French overtook the English at Agincourt, intent on exterminating the invaders. Astonishingly, the wearied English force defeated the substantially larger French force in the resulting battle, an outcome that owed much to the use of the longbow and a preponderance of archers, whose task, in a description by John Keegan,[18] was to

> *...provoke the French into attacking, and it was therefore essential that their arrows should "group" as closely as possible on target. To translate their purpose into modern artillery language, they had to achieve a very narrow 100° zone* (i.e., *that belt of territory into which* all *missiles fell) and a Time on Target effect* (i.e., *all their missiles had to arrive simultaneously).*

The arrow barrage precipitated its desired effect. Longbow arrows could punch through armor for ranges closer than 100 m, but at the much longer distances of the first volley that provoked the French to charge, the plate steel making up the helmets and armor of men afforded greater protection. The effect on the less protected and therefore wounded and panicking horses of the French cavalry was an entirely different matter. The descriptions of a volley of arrows in a small area descending at a sharp angle and arriving on a plane at a set time is very reminiscent of the description of "emittance". Where the arrows landed and the angle at which they protruded from the ground (or whatever they lodged in) brings to mind x and dx/dz in a tightly clustered pin cushion of phase space rendered in the relatively antiseptic Figure 33.10 compared to the carnage of the battlefield, and invites rumination on what impact on the trajectories of arrows – or electrons – that being launched from an uneven surface causes.[19] So, once more unto the breach.

In the absence of wind resistance (drag is neglected), arrows follow a parabolic path making the kinematics of their ascent symmetric to their descent, and so for the sheer convenience of it, consider just the descent part of the trajectory. From apogee to ground, the path is equivalent to shooting a projectile horizontally from a height h equivalent to the apogee. Setting $t = 0$ to be when the arrows are at apogee, then the initial velocity is v_o, the initial height is found from $z(t) = 0 = h - (1/2)gt^2$ (where $g = 9.8$ m/s is the gravitational constant), and the range from $x(t) = v_o t$. The velocity in the $\hat{z}$ direction is simply $v_z(t) = -gt$. Thus, it is easy to show

$$\frac{dx}{dz} = \frac{v_o}{\sqrt{2g(h-z)}} = \frac{v_o}{v_z(t)} \tag{33.47}$$

As a result, $v_z(dx/dz) = v_o$, which hearkens back to Eq. 33.30 for the capturing of the behavior of a property that does not change as the "beam" of arrows accelerates, just as if the radius of the cathode is R and the rms velocity is v_o, then $\beta\gamma\varepsilon = Rv_o/c$, that is, it is *invariant*. So far, so good.

Arrows released horizontally with a velocity v_o from an elevated height h follow the same trajectory, but launching from a height suggests variations. If the arrows are launched normal to the surface from which they are fired (say, straight out of a window), the change of the surface from flat to curved means that many of the arrows come out in different directions. Variations in the strength of the archer means that there are also variations in the value of v_o for each projectile. If the stronger archers cluster, then the initial velocity and the number of particles changes across the

[17] Ref. [37]: *The Life of King Henry the Fifth*, III.i.1.

[18] John Keegan, *The Face of Battle*. New York: Viking Press, 1976, Chapter 2: Archers versus Infantry and Cavalry.

[19] It is to be hoped that such speculation would follow, rather than lead, deep pondering about whether pressed claims over territory and an asserted right to rule justify four-digit body counts in a single battle, a fraction of which were captured prisoners.

Figure 33.15 Electron emission from a carbon fiber field emitter as simulated using the beam optics code MICHELLE. The diameter of the carbon fiber is 35 μm and the length of the fiber (not all of which is shown) is 1.5 mm, comparable to experimentally tested carbon fibers. Light blue lines denote the equipotentials. For the trajectories, the deep blue lines contain the most current, green are intermediate amounts, and red contain the least current. Based on Figure 19 of ref. [451]. Image courtesy of J.J. Petillo (Liedos).

surface. All of these are readily understandable in how they affect the "time on target", but less analogous modifications, such as gravity varying in magnitude and direction near the surface, the projectiles repelling each other in flight, and their spiraling about the Earth's magnetic field, really complicate matters: by the time these particles get to ground, they will arrive at various times (x) and stick into the ground at various angles (dx/dz). The sophomoric understanding of emittance as a cluster of phase space points seemingly considered at one point in time in one plane gets convoluted.

Following the trajectory of particles subject to such variations is the task of beam optics codes, and such codes have become remarkably versatile. An example of electrons emitted from a wire-like surface subject to many of the aforementioned complications above is shown in Figure 33.15 for the simulation of field emission from a 35 μm diameter and 1.5 mm long carbon fiber [448, 451], with even more modifications: the effects of the adjacent carbon fiber off to the right, not seen in the view shown, are inferred from the unsymmetric blue equipotential lines hugging the surface representing the fiber (the lines on the right do not drop as quickly, signaling the presence of the neighbor). Such codes can follow the growth in emittance from all the varied complications through the life of the beam or electron bunch.

Nevertheless, what happens near the surface of the cathode, the so-called initial conditions of the beam or bunch's subsequent trajectory, is tough, particularly when the emission sites are on the nanoscale and the small geometric shapes making up the surface of the emitter in Figure 33.15 are μm sized and typically larger. The contributions to emittance due to unruly features and conditions at the surface of the cathode are termed *intrinsic emittance*: no amount of beam optics will undo the damage to phase space that the cathode puts on the distribution, and so intrinsic emittance is an irreducible part of ε and, sometimes, the dominant part. Simple models relating to how it manifests itself are therefore still useful, first for simple understanding and second for figuring out how the initial conditions (the cathode's "virtual anode") in beam optics codes can be constructed.

It seems obvious that care must be taken with each complication. The paths the electrons take to the (virtual) anode and their time of arrival means that a method of calculating trajectories in the high fields near the apex of the emitter is needed if emittance is to be addressed. The changing of the current density over the surface and the launch direction (the normal to the surface) affect the transverse velocity components that survive to the anode. Lastly, the presence of more than one emitter bodes ominously for how $\langle x^2 \rangle$ is to be calculated. Begin systematically and start with the trajectories.

33.4.1 Trajectory evaluation

Prevent trouble before it arises.
Put things in order before they exist...
The giant pine tree
grows from a tiny sprout.
The journey of a thousand miles starts from beneath your feet.

– Lao Tzu[20]

[20]Lao Tzu, *Tao Te Ching: By Lao Tzu* (trans. Stephen Mitchell). London: Francis Lincoln Limited, 2013, p. 64.

The trajectory equations above, namely, $z(t) = h - (1/2)gt^2$, $v_z(t) = -gt$, and $x(t) = v_o t$, are the simple solutions to Newton's equation for constant force expressed as

$$m\frac{d^2}{dt^2}\vec{r}(t) = \vec{F} \tag{33.48}$$

with $\vec{r}(t) = x(t)\hat{x} + z(t)\hat{z}$, $\vec{F} = -mg\hat{z}$, and where h, v_o, and g are constant. Because gravity is constant and directed along one of the cartesian coordinate axes, the equation separates into separate equations for $x(t)$ and $z(t)$, and each is solved separately. The methods to be discussed apply to either $x(t)$ or $z(t)$, but for simplicity pick $x(t)$ and $dx/dt = v(t)$. The task is then to numerically integrate the equations: naïvely, it may seem that the algorithms of Section A1.4 may be useful, but alas, in those algorithms *both* boundaries $x(t = 0)$ and $x(t = t_{max})$ are known, whereas in the present case only the initial conditions $x(t = 0)$ and $v(t = 0)$ are known, and that makes a difference. Still, the finite difference methods there are useful to understand before undertaking the approaches described here.

Accurate methods such as the Runge–Kutta (RK) approach [236] are useful for numerically solving ordinary differential equations (ODEs). A commonly used form of RK is fourth-order accurate. Nevertheless, consider the leap-frog and average value methods here for calculating the trajectories even though they are only second-order accurate: their simplicity makes their implementation easy, and accuracy can always be improved simply by reducing the time step size.

The leap-frog method is easier to introduce. The first step is to recognize that there are really two differential equations inside Eq.(33.48), one for $x(t)$ and one for $v(t)$, and they satisfy

$$\frac{d}{dt}x(t) = v(t) \tag{33.49}$$

$$\frac{d}{dt}v(t) = \frac{F}{m} \tag{33.50}$$

where $F = \vec{F}\cdot\hat{x}$. Of course, the force may be position and time dependent. Now the artistry comes in: by the central difference approximation methods of Section A1.3.2, the derivatives are

$$x(t + \Delta t) - x(t) = v\left(t + \frac{\Delta t}{2}\right)\Delta t \tag{33.51}$$

$$v\left(t + \frac{\Delta t}{2}\right) - v\left(t - \frac{\Delta t}{2}\right) = \frac{F[x(t)]\Delta t}{m} \tag{33.52}$$

The notation is already distasteful, but a method to bring it to elegant simplicity is to let the time coordinate, and therefore the position and velocity terms, be discretized such that

$$\begin{aligned} t &\rightarrow t_j \equiv j\Delta t \\ x(t) &\rightarrow x(t_j) \equiv x_j \\ v(t - \Delta t/2) &\rightarrow v(t_{j-1/2}) \equiv v_j \end{aligned} \tag{33.53}$$

It is essential to observe that whereas position is defined on the time steps t_j, the velocity is defined on the halfway point between the time steps, or at $t_{j-1/2}$. Using Eq. (33.53) in Eq. (33.51) gives

$$x_{j+1} - x_j = \Delta t\, v_j \tag{33.54}$$

$$v_{j+1} - v_j = \Delta t\, a_j \tag{33.55}$$

where the acceleration $a_j \equiv F(x_j)/m$ is introduced. Is it good? An nth-order method is exact for a polynomial for which the largest power present is n. Motion in a constant gravitational field is parabolic (second-order in time), and so the leap-frog and average value methods are exact under such circumstances. Consider how this is shown for the leap-frog approach. From Eq. (33.54), $x_{j+1} = x_j + \Delta t\, v_j$ but now recall that v_j is defined at the time step between the x_j time steps, that is,

$$x[(j + 1)\Delta t] = x(j\Delta t) + v[(j + 1/2)\Delta t]\, \Delta t \tag{33.56}$$

From Eq. (33.55), though, $v[(j + 1/2)\Delta t] = v(j\Delta t) - (g/2)\Delta t$ and so

$$\begin{aligned} x[(j + 1)\Delta t] &= x(j\Delta t) + \Delta t\left[v(j\Delta t) - \frac{1}{2}g\Delta t\right] \\ &= x(j\Delta t) + v(j\Delta t)\,\Delta t - \frac{1}{2}g\Delta t^2 \end{aligned} \tag{33.57}$$

which is the solution $x(t+\Delta t) = x(t) + v(t)\Delta t - (g/2)\Delta t^2$ to motion in a gravitational field and confirms the leap-frog approach is accurate to second order. The final step is to express the leap-frog algorithm in a manner suitable for repeated evaluations by formulating it as a matrix equation. When put in a form that will accommodate the second (or average value) approach as well, it is

$$\begin{pmatrix} 1 & -\Delta t \\ 0 & 1 \end{pmatrix} \begin{pmatrix} x_{j+1} \\ v_{j+1} \end{pmatrix} = \begin{pmatrix} x_j \\ v_j \end{pmatrix} + \begin{pmatrix} 0 & 0 \\ \Delta t & 0 \end{pmatrix} \begin{pmatrix} a_j \\ a_{j+1} \end{pmatrix} \tag{33.58}$$

for which the solution is

$$\begin{pmatrix} x_{j+1} \\ v_{j+1} \end{pmatrix} = \begin{pmatrix} 1 & \Delta t \\ 0 & 1 \end{pmatrix} \begin{pmatrix} x_j \\ v_j \end{pmatrix} + \begin{pmatrix} \Delta t^2 & 0 \\ \Delta t & 0 \end{pmatrix} \begin{pmatrix} a_j \\ a_{j+1} \end{pmatrix} \tag{33.59}$$

Unfortunately the leap-frog approach has complexity below the surface that may be confusing: the indices on x and v *refer to different times*, and so a facile reading of the equations Eq. (33.54) and (33.55) that result leads to trouble for the unaware. That, plus having to specify what $v_0 = v(-\Delta t/2)$ is requires one to be careful in setting up the calculation of a trajectory. Such demerits, as it were, should be weighed against the advantage of only needing the acceleration term $a(t)$ evaluated at the time t (and, as shall be seen below, fewer computations).

The second approach conveniently defines $x(t)$ and $v(t)$, and by extension $a(t) = F(t)/m$, at the same time steps $t \to t_j \equiv j\Delta t$, or $x_j \equiv x(j\Delta t)$, $v_j \equiv v(j\Delta t)$ and $a_j \equiv a(j\Delta t)$. Doing so, however, means that the central difference approximation of Eq. (A1.85) has lost its midpoint (why $v[(j+1/2)\Delta t]$ was introduced in the first place). Observing that the midpoint can be approximated by the average of the end points, as in Eq. (A1.9), then a reasonable approximation is $s_{j+1/2} = (s_j + s_{j+1})/2$, where s stands in for x, v and a. The defining relations are now

$$x_{j+1} = x_j + \frac{1}{2}(v_j + v_{j+1})\Delta t \tag{33.60}$$

$$v_{j+1} = v_j + \frac{1}{2}(a_j + a_{j+1})\Delta t \tag{33.61}$$

To show that the method is (like leap-frog) second-order accurate, use the gravitational example for which $a_j = a_{j+1} = -g$. Then it immediately follows that $v_{j+1} = v_j - g\Delta t$ and so $x_{j+1} = x_j + v_j\Delta t - (g/2)\Delta t^2$ simply and naturally. These are, of course, the usual elementary equations for motion in a constant gravitational field that bedevil undergrads, just written differently. Lastly, the solution of the associated matrix equation is

$$\begin{pmatrix} x_{j+1} \\ v_{j+1} \end{pmatrix} = \begin{pmatrix} 1 & \Delta t \\ 0 & 1 \end{pmatrix} \begin{pmatrix} x_j \\ v_j \end{pmatrix} + \frac{\Delta t}{4} \begin{pmatrix} \Delta t & \Delta t \\ 2 & 2 \end{pmatrix} \begin{pmatrix} a_j \\ a_{j+1} \end{pmatrix} \tag{33.62}$$

The advantage of having x and v defined at the same values of time has come at the cost of needing to know a_{j+1} in the average value approach, something the leap-frog algorithm did not require. For constant acceleration, that is not a concern, but it is for more general problems involving forces that change with position (such as space charge or fields near a protrusion). Approximating $a_{j+1} \approx a_j$ renders the entire simulation only first order in accuracy because the equation for v_{j+1} becomes only first-order accurate, independent of the fact that the equation for x_{j+1} is second order: accuracy is always set by the worst performer, much like a chain being only as strong as its weakest link.

The weakest link can be fixed. Let the value of x_{j+1} that comes about from using the first-order approximation $a_{j+1} \to a^* = a_j$ in Eq. (33.62) be denoted x^*, that is, x^* is the *first order approximation* to x_{j+1}. The second-order approximation to a_{j+1} is then $a_{j+1} = a(x^*) = F(x^*)/m$, and that value is used when Eq. (33.62) is evaluated a second time. In such a way, a good bit of the second-order accuracy is recovered. The price of having the same time points used for $x(t)$ and $v(t)$ is at the cost of having to perform the matrix evaluation twice in order to achieve second-order accuracy. The result of an average value method approach compared to a leap-frog approach is compared to the analytic result in Figure 33.16 for the harmonic oscillator, evaluated using the code of Section A3.34: both are seen to be good. The results would be better (the small drift away from the analytic result over time) if an additional constraint like conservation of energy were imposed at each time step, a method often employed in particle simulations [606].

The accuracy of the leap-frog and average value methods can be assessed by applying them to a quintessential problem, namely the harmonic oscillator of Section 9.3, which is an excellent problem on which to test them: it has a known analytical solution and it is not monotonically increasing or decreasing, and therefore tends to flush out how errors like to propagate. As a practical matter, the importance of having *an analytically exact solution to a problem on which numerical solutions are applied* to test the quality of the numerical approaches cannot be overemphasized. Insert the known solution

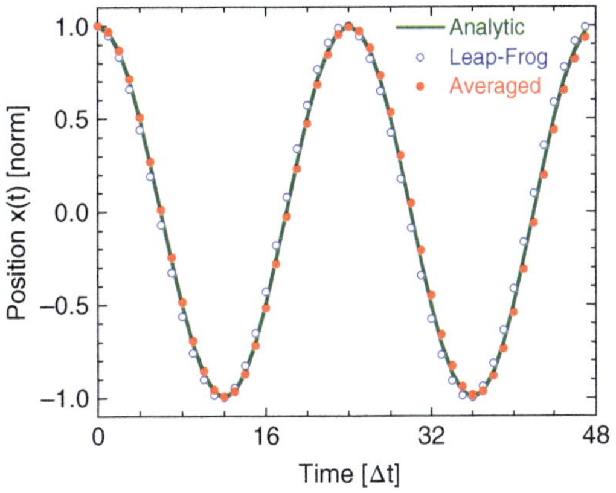

Figure 33.16 Comparison of the leap-frog and average value methods for evaluating the trajectory $x(t)$ of a harmonic oscillator for units of $x(0) = 1$ and $\Delta t = 1$ using the algorithm of Section A3.34 for $N = 48$ and $\omega\Delta t = 4\pi/N$.

$x(t) = A\cos(\omega t + \phi)$ into the discrete equations Eqs (33.54) and (33.60) for the purposes of discerning what the error terms are. It follows that $v(t) = -\omega A\sin(\omega t + \phi)$ and $a(t) = -\omega^2 x(t)$ with $t = n\Delta t$. Let $\phi = 0$ and $A = 1$ for convenience. For the leap-frog method, Eq. (33.54) becomes[21]

$$\frac{1}{48}\omega^4\Delta t^3\cos(n\Delta t) - \frac{1}{24}\omega^3\Delta t^2\sin(n\Delta t) \approx 0 \tag{33.63}$$

$$\frac{1}{24}\omega^4\Delta t^2\cos(n\Delta t) \approx 0 \tag{33.64}$$

where the first equation derives from x_{j+1} and the second from v_{j+1}. For the average value method, Eq. (33.60) becomes

$$\frac{1}{12}\omega^4\Delta t^3\cos(n\Delta t) + \frac{1}{12}\omega^3\Delta t^2\sin(n\Delta t) \approx 0 \tag{33.65}$$

$$\frac{1}{12}\omega^4\Delta t^2\cos(n\Delta t) - \frac{1}{12}\omega^5\Delta t^3\sin(n\Delta t) \approx 0 \tag{33.66}$$

where use is made of the small angle relations $\sin\theta \approx \theta - \theta^3/6$ and $\cos\theta \approx 1 - \theta^2/2$ only for those terms not containing n, as $t = n\Delta t$ can be large. Therefore, both methods are of order Δt^2 accurate as already known, but the leap-frog approach has the smaller error, perhaps expected due to the roughness of approximating a midpoint by an average value of the endpoints of a time step.

Comparing the output of Figure 33.16 suggests that the average value method is performing somewhat better than the leap frog method, and this may appear perplexing. There is a reason for it: the average value method is a somewhat ham-handed effort to claim the benefits of *implicit* methods over *explicit* methods in solving a differential equation [236, 242]. Suppose that instead of solving for one trajectory $x(t)$, one required trajectories for N particles, and therefore had an N-component vector $|x\rangle$ where $x_n(t)$ is the nth component of the vector representing the history of the nth particle. Paranthetically, the ket notation $|x\rangle$ is used only as a convenient notation of a vector quantity on which matrices will act without implying a relationship to anything quantum mechanical, even though the implicit methods discussed below are essential, for example, in time-evolving quantum phase space distributions [126, 264, 325] or modeling the heating of an electron gas in photoemission [85]. The distinction between implicit and explicit methods concerns the language of matrices. Let $\mathbb{M}$ be a matrix that in some way scrambles the values of a vector. For example, in the solution of the diffusion equation, where $\partial_t u(x,t) = K\partial_x^2 u(x,t)$, then $\mathbb{M}$ would be the finite difference representation of $K\partial_x^2$. Letting $|u(t)\rangle$ be the vector of $u(x_n, t)$ values, where n denotes the individual particles, then the difference between the schemes is that

- *explicit* methods rely only on past values to calculate future values, so that

$$|u(t + \Delta t)\rangle = |u(t)\rangle + \mathbb{M}\,|u(t)\rangle \tag{33.67}$$

[21] This can be shown by a straightforward and routine albeit tedious analysis using the relations of Eq. (A2.1), the details of which are omitted as an act of kindness.

- *implicit* methods use future values in the solution and solve

$$|u(t+\Delta t)\rangle = |u(t)\rangle + \frac{1}{2}\mathbb{M}\{|u(t+\Delta t)\rangle + |u(t)\rangle\} \tag{33.68}$$

The implicit scheme looks much like the average value method in that the average of a future $|u(t+\Delta t)\rangle$ and past $|u(t)\rangle$ are used to create the vector on which $\mathbb{M}$ acts, but it demands that a much harder matrix equation must be solved. After rearranging, Eq. (33.68) becomes

$$|u(t+\Delta t)\rangle = \left\{\mathbb{I} - \frac{1}{2}\mathbb{M}\right\}^{-1}\left\{\mathbb{I} + \frac{1}{2}\mathbb{M}\right\}|u(t)\rangle \tag{33.69}$$

where $\mathbb{I}$ is the identity matrix. As an aside, if Schrdinger's equation was being solved rather than the heat diffusion equation, then the matrix $\mathbb{M} \to -i\mathbb{M}$ in Eq. (33.69). The right-hand side is then just the second-order accurate approximation to

$$|\psi(t+\Delta t)\rangle = \exp(-i\Delta t\ \mathbb{M})\,|\psi(t)\rangle \tag{33.70}$$

and the propagator representation of Eq. (8.22) is recovered.

Matrix inversions can be hard and time-consuming, and are not usually considered unless there is good cause. The good cause is that much larger time steps are enabled and stability is achieved at such a level that the additional effort is well worth the computational expense (e.g., finding out how an electron gas heats during photoemission [85, 206]). The average value method, then, is seen to try to filch the benefits of implicit schemes by taking a guess as to what the future value is and using the guess in the calculation of x_{j+1}, a disposition it shares with RK methods [236].

The actual evaluation of the trajectories is on hold until, first, the initial velocity of the electron is specified and, second, the analytical *impulse* approximation is introduced. Both will naturally emerge in a consideration of the emittance and how it is to be calculated for a curved surface.

33.4.2 Hemispheres and emittance

> ...*in their endeavor to show that nature does nothing in vain, i.e., nothing which is useless to man, they only seem to have demonstrated that nature, the gods, and men are all mad together.*
>
> – Benedictus de Spinoza[22]

The tranquil and elegant derivations of emittance for planar surfaces are shattered when the primary feature of surfaces, namely, their *roughness*, are considered, and any understanding of a useful cathode must deal with it. Roughness comes in two flavors, as shown in Figure 33.17: the *physical* roughness is due to undulations and protrusions on the surface, and *chemical* roughness is caused by work function variation with crystal faces that cause fluctuations in equipotential lines near the cathode surface [321, 692]. In fact, both occur at the same time, as images from a field emitter tip quickly demonstrate in Figure 33.18: on the left, a field emission electron microscope (FEEM) image shows regions of high

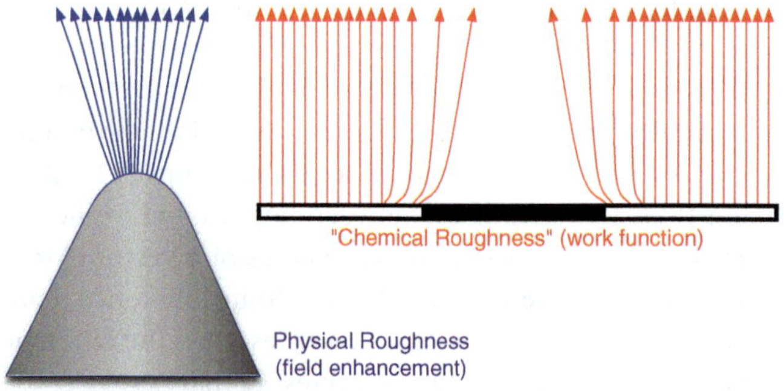

Figure 33.17 Emittance due to trajectory alterations by physical variations in surface field due to physical structure (left) and chemical variations in crystal face work function (right).

[22] Benedictus de Spinoza and Dagobert D. Runes, *Spinoza: Dictionary*. New York: Philosophical Library, 1951, pp. 100–101. This source is all the more interesting for having its foreward authored by Albert Einstein.

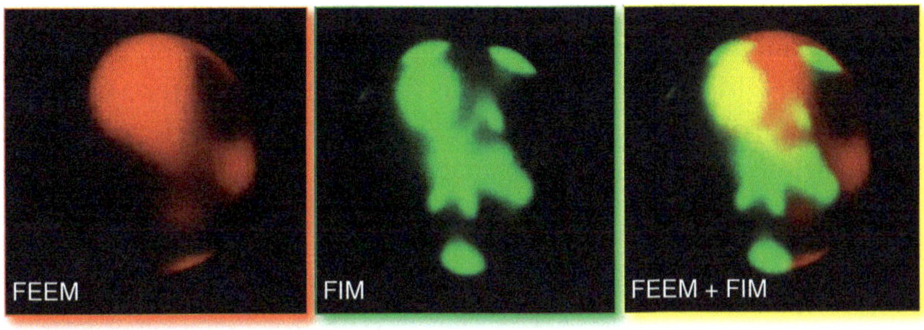

Figure 33.18 Images from a Spindt-type field emitter tip comparable to that shown in Figure 13.1. Left, field emission electron microscope image (FEEM). Middle, field ion microscope image (FIM). Right, FEEM plus FIM. Images for the analysis courtesy of Capp A. Spindt and Paul R. Schwoebel (SRI International)

geometric field enhancement and low work function sites where electron emission is likeliest, in the middle, a field ion microscope (FIM) image shows regions of high geometric field enhancement and high work function where ion emission is likeliest, and lastly, on the right, the combined FEEM plus FIM images show where conditions are favorable to both (yellow), namely, where high field enhancement and moderate work function overlap. Where both regions are dark is where the work function is high, thwarting emission regardless of the field.

Finding field variations for work function changes is in essence a numerical boundary value problem for which numerical packages are available, but a discussion about it is less enlightening for that reason. Field variation due to geometric structure, on the other hand, is easier to take in on first encounter because the idea of an electron being emitted normal to an undulating surface suggests right away the idea of a transverse velocity. Not only that, but the models which can be brought to bear to treat emittance from a bump for examining the effects of bending trajectories have analogs to the curved equipotential lines near work function patches, so the trajectory methods to be described below are similar in both cases. Therefore, choosing to treat the model with a well-defined surface is simultaneously more enlightening and easier.

To compare the emittance from a surface with structure for the various emission mechanisms of thermal, field, and photoemission, having a good analytical model on which to test all three is desirable. A useful one is the hemispherical model of Section 30.2, and using it will follow ref. [451].[23] As found in Eq. (30.9), the potential outside the sphere is, in spherical coordinates,

$$V_f(r,\theta) = -F_o r\cos\theta\left[1-\left(\frac{a}{r}\right)^3\right] \tag{33.71}$$

The relations $z = r\cos\theta$ and $\rho = r\sin\theta$ take the spherical coordinates over into the cylindrical coordinates. Finding the fields in the $\hat{\rho}$ and $\hat{z}$ directions is straightforward

$$F_z = F_o\left\{1+\frac{a^3(2z^2-\rho^2)}{(\rho^2+z^2)^{5/2}}\right\} = F_o\left[1+\left(\frac{a}{r}\right)^3(3\cos^2\theta-1)\right] \tag{33.72}$$

$$F_\rho = \frac{3F_o a^3\rho z}{(\rho^2+z^2)^{5/2}} = 3F_o\left(\frac{a}{r}\right)^3\cos\theta\sin\theta \tag{33.73}$$

where the first relation is in cylindrical and the second in spherical notation. At the surface, $F_z = 3F_o\cos^2\theta$ and $F_\rho = 3F_o\cos\theta\sin\theta$, so that the force is radially outward with a magnitude $F_r = 3F_o\cos\theta$ as per Eq. (30.10). Electrons ejected from the surface and accelerated by such forces will follow trajectories that can be evaluated using the methods of Section 33.4.1. Recalling that emittance for a rotationally symmetric problem goes as $\varepsilon_{n,rms} = (\hbar/2mc)\sqrt{\langle\rho^2\rangle\langle k_\rho^2\rangle}$ because $\langle x^2\rangle = \langle\rho^2\rangle/2$ and $\langle k_x^2\rangle = \langle k_\rho^2\rangle/2$ in Eq. (33.30) (the cross term vanishes), dealing with ρ and k_ρ is a bit simpler as F_ρ can be used.

In a beam optics code, all the particles in a bunch have known positions and velocities, and so the evaluation of the emittance of the bunch is well defined. In contrast, treating emission as continuous makes the evaluation of emittance

[23] Saying "mostly follow" would have been more accurate. First, ref. [451] introduces the impulse approximation late in the development, but here, it is introduced from the outset. Second $\sigma(\theta)$ has changed to Eq. (33.77), which has consequences albeit small ones. Third, Eq. (33.115) is mildly different compared to its progenitor.

less well-defined. Take a cue from Agincourt (Section 33.3.1), but instead of evaluating the position and angle of arrows after they have stuck into the ground (or, regrettably, something standing on the ground), consider instead the group of arrows that have the same time of flight or, here, the electrons that have traveled for a time t from the cathode. Such a definition will differ from electrons that have crossed a plane (such as a phosphor screen) because of small differences in arrival time, but at least the definition gives an approximate measure that can be used to compare disparate sources with very different surfaces and operating conditions.

33.4.3 Impulse approximation

If we cannot bear too much reality, we might at least avoid too much fantasy.

– John Dominic Crossan[24]

Finding the trajectories of electrons near surfaces with geometric structure is very difficult and is the bane of simulation, particularly when that structure is nanoscale, but whatever strategy concocted to do so should not be too fantastical. Far away from the surface, $\vec{F} \to F_o\hat{z}$ and the familiar ballistic equations analogous to the gravitational problem take over. How far out is suggested by Figure 33.19, where the magnitude of the force components F_z and F_ρ are evaluated for the hemispherical potential of Eq. 33.71: the later becomes negligible and the former transitions over to the background field F_o over a length scale of na where a is the radius of the hemisphere and n is between 2 and 3: the modifications to the constant field are proportional to $(a/r)^3$, for which $(a/r)^3 = 1/n^3$ or between $1/8$ and $1/27$ for $2 < n < 3$. Consequently, the influence of the hemisphere is only at the initial part of the trajectory, suggesting that the initial conditions encapsulated in v_o might be profitably modified. But how to modify v_o is ambiguous. The initial velocity v_o in the ballistic equations appears to have been given a boost, or *impulse*, to account for acceleration near the hemisphere. Therefore, a simple approximation is then to make the replacement $v_o \to \sigma_o v_o$, where σ_o is a dimensionless factor that dictates the size of the impulse. Fantasy, to be sure, but not an egregious one.

The estimation of σ_o can be made by setting the change in kinetic energy to the change in potential energy as an electron moves from the enhanced field near the surface of the emitter to a region where the potential has more or less become that due to the constant background field. That is, using Eq. (33.71), let

$$\frac{1}{2}mv_o^2\left[\sigma(\theta)^2 - 1\right] = V_f(a,\theta) - V_f(na,\theta) \tag{33.74}$$

and so

$$\sigma(\theta)^2 = 1 + \left(\frac{2F_o a}{mv_o^2}\right)\left(\frac{n^3 - 1}{n^2}\right)\cos(\theta) \tag{33.75}$$

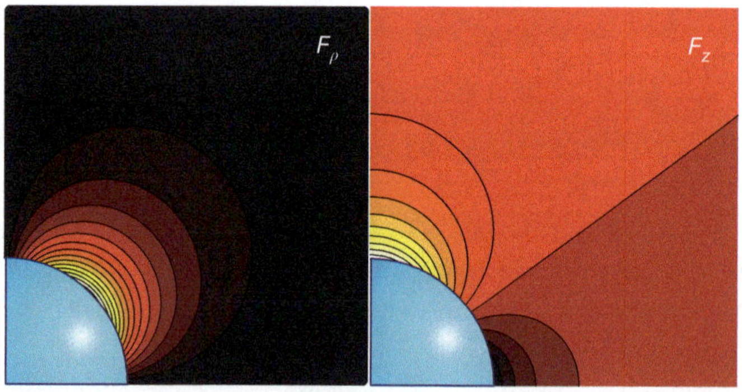

Figure 33.19 The field components associated with the hemispherical potential shown in Figure 30.2. The blue region represents the hemispherical emitter. The size of the region shown is $3a$ where a is the radius of the hemisphere. Contours are equispaced between the maximum ($3F_o$ for F_z, $3F_o/2$ for F_ρ, white) and minimum (0, black). Based on Figures 10 and 11 of ref. [451].

[24] John Dominic Crossan, *God & Empire*. New York: HarperCollins, 2007, p. 202.

which makes sense: emission from the apex gets a good impulse ($\sigma(0) \equiv \sigma_o$), but at the base of the hemisphere, where the fields vanish, then the impulse should not occur ($\sigma(\pi/2) \to 1$). Observe the natural introduction of a dimensionless factor κ defined as the ratio of the energy scale associated with the field and that associated with the initial kinetic energy

$$\kappa \equiv \frac{F_o a}{m v_o^2} \tag{33.76}$$

which, coupled with an estimate drawn from Figure 33.19 suggesting $n = 2$, gives

$$\sigma(\theta) = \left[1 + \left(\frac{7\kappa}{2}\right)\cos\theta\right]^{1/2} \tag{33.77}$$

As a result, replace $v_o \to \sigma(\theta)v_o$ in the ballistic trajectory equations, which thereby accounts for field enhancement effects near the bump, and so

$$\begin{aligned} \rho(t) &= a\sin\theta + \sigma(\theta)v_o t\sin\theta \\ z(t) &= a\cos\theta + \sigma(\theta)v_o t\cos\theta + \frac{F_o}{2m}t^2 \end{aligned} \tag{33.78}$$

The impulse approximation compared to the numerical trajectory evaluation of Eq. (33.60) treated in Section 33.4.1 is shown in Figure 33.20.[25] Letting $\tau = mv_o/F$ be a characteristic time scale, then define $t \equiv p\tau$. In terms of the dimensionless factors κ and p, then Eq. (33.78) becomes

$$\frac{\rho}{a} = (1 + \sigma(\theta)p)\sin\theta \tag{33.79}$$

$$\frac{z}{a} = (1 + \sigma(\theta)p)\cos\theta + \frac{1}{2}\kappa p^2 \tag{33.80}$$

With both $x(t) = \rho(t)\cos\varphi$ and $z(t)$ known, then it is finally possible to obtain the much sought factor $x'(t)$, now readily seen to be

$$x'(t) = \frac{v_x(t)}{v_z(t)} = \frac{\sigma(\theta)v_o\sin\theta\cos\varphi}{\sigma(\theta)v_o\cos\theta + (F_o t/m)} \equiv \frac{\sin\theta\cos\varphi}{\cos\theta + (p/\sigma(\theta))} \tag{33.81}$$

Observe that if $y'(t)$ were under consideration, the relation would be much the same but for $\cos\varphi \to \sin\varphi$, where φ is the azimuthal angle such that $x = \rho\cos\varphi$ and $y = \rho\sin\varphi$.

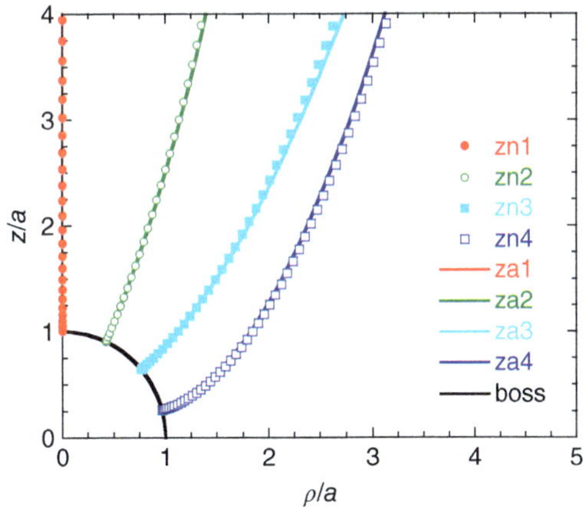

Figure 33.20 Numerical evaluation of trajectories using Eq. (33.60) compared to the analytic impulse model of Eq. (33.78), where *zn* refers to numerical (circles and squares) and *za* to the impulse model (lines). The algorithm is given in Section A3.35 for the input parameters $N_p = 4$ (four trajectories), $\theta_{max} = 75°$, and $\kappa = 50$.

[25]The prior version of the impulse approximation (Eq. (68) and (69) of ref. [451]) made use of the *ad hoc* form $\sigma \approx \sigma_o\cos\theta$, suggested by the intuition that the impulse factor σ must decline as θ increases. The present treatment shows that $\cos\theta$ by itself is a tad too aggressive, and that Eq. (33.77), being based on Eq. (33.75), gives a better account.

33.4.4 Single hemisphere emittance parameters

A critical factor in the operation of coherent radiation sources such as free electron lasers (FEL), gyrotrons, etc., is the quality of the electron beams …the problems which are cathode-related such as patchiness, nonuniform emission, roughness, and nonconstant work function, etc., may in many cases place a fundamental limit on the brightness of an electron beam.

– Yue-Ying Lau [693]

When comparing the thermal, photo, and field emission mechanisms, the development of parameters depends on scales characteristic of each process so that direct comparisons between them can be made. If distances are given in units of a and time in units of τ such that $t = p\tau$, then $\beta_z = v_z/c = (p+1)v_o/c$. Next, emission shall vary down the side of the hemisphere to a degree specified by the emission mechanism being considered. The notional angle $\cos\theta_n \equiv \eta$ suggested by Eq. (30.17) and defined by

$$1 - \eta = \int_0^{\theta_n} \sin\theta d\theta = \int_0^{\pi/2} P(\theta)\sin\theta d\theta \tag{33.82}$$

becomes useful to develop scaling relations using the methods of Section 30.2. Because each emission mechanism (thermal, field, or photoemission) differs, so will the forms of $P(\theta) \equiv J[F(\theta)]/J(F_{tip})$, η, and v_o differ. From $P(\theta)$, $\Delta(\theta)$, defined after the integral of Eq. (30.13), is

$$\Delta(\theta) \equiv \frac{I(\theta)}{I_{boss}} = \frac{1}{1-\eta}\int_0^{\theta} P(\theta')\sin\theta' d\theta' \tag{33.83}$$

where I_{boss} is the total integrated current over the surface of the hemisphere. $\Delta(\theta)$ therefore gives a measure of what fraction of the current is emitted by the surface of the hemisphere inside the polar angle θ.

The velocity v_o needs a bit more effort: using methods motivated by Eq. (3.5), but in a way closer to the spirit of Eq. (7.9), of $J = q\rho\langle v\rangle$, then, from Eq. (11.1)

$$v_o \equiv \langle v\rangle \equiv \frac{\int_0^{\infty}(\hbar k/m)D(k)f(k)dk}{\int_0^{\infty} D(k)f(k)dk} \tag{33.84}$$

where $D(k)$ is the transmission probability, $f(k)$ is the supply function, and $\hbar k/m$ is a velocity. Likewise, the evaluation of $\langle v^2\rangle$ would involve a factor of $(\hbar k/m)^2$ in the numerator, so that, for example, the rms velocity is

$$v_{rms} = \sqrt{\langle v^2\rangle} = (\hbar/m)\sqrt{\langle k^2\rangle} \tag{33.85}$$

The velocities v_o and v_{rms} are *not* the same, although they can be close: the former is required to launch the electrons in a trajectory calculation, but the latter is required in the evaluation of emittance right at the surface.

In what follows, the calculation of the notional angle factor η, the integrated current fraction $\Delta(\theta)$, the launch velocity v_o, and the rms velocity v_{rms} is undertaken. Add to that list a dimensionless characteristic energy ratio parameter going by the designation b: it is crafted from the field- and/or temperature- and/or frequency-dependent characteristic energy parameters of each mechanism on which the other terms will depend, and will be found on a case-by-case basis. The evaluations require terms drawn from Tables 2.3 and 33.2.

Table 33.2 Emission parameters used in the evaluation of emittance (see also Tables 2.2, 2.3, and 33.3).

Symbol	Definition	Relation	Unit
k_F	Fermi wavenumber	$\sqrt{2m\mu}/\hbar$	1/nm
β	Inverse temperature	$1/k_BT$	eV^{-1}
ϕ	Schottky loweredΦ	$\Phi - \sqrt{4QF}$	eV
a	Hemisphere radius	–	nm
$\sigma(\theta)$	Impulse factor	Eq. (33.77)	nm
v_o	Launch velocity	Eq. (33.84)	–
v_{rms}	rms velocity	Eq. (33.85)	–
η	Notional cosine angle	Eq. (33.88)	–
$\Delta(\theta)$	Integrated current fraction	Eq. (33.83)	–
$b(F,T)$	Characteristic energy ratio	Process dependent	–

33.4.4.1 Thermal emittance

Begin by considering the current density J of the Richardson–Laue–Dushman equation given by Eq. (12.3). The field dependence resides in the Schottky lowered work function $\phi = \Phi - \sqrt{4QF}$. Knowing that $F(\theta)/F_o = 3\cos\theta$ for the field along the surface of a hemisphere (Eq. (30.10)), then the field- and temperature-dependent energy ratio parameter $b(F,T)$ is

$$b(F,T) = \frac{\sqrt{4QF}}{k_B T} = \beta\sqrt{4QF} \tag{33.86}$$

in terms of which the emission probability factor $P(\theta)$ is given by

$$P(\theta) = \exp\left[b\left(\sqrt{\cos\theta} - 1\right)\right] \tag{33.87}$$

The notional cosine angle is defined by

$$\begin{aligned}\eta &= 1 - \int_0^{\pi/2} e^{b(\sqrt{\cos\theta}-1)} \sin\theta d\theta \\ &= 1 - \frac{2}{b}\int_{-b}^{0}\left(1+\frac{u}{b}\right)e^u du = 1 - \frac{2}{b} + \frac{2}{b^2}\left(1-e^{-b}\right)\end{aligned} \tag{33.88}$$

where the second line has used a change of integration variable to $u = b(\sqrt{\cos\theta} - 1)$. The integrated current fraction $\Delta\theta$ follows a similar evaluation, and results in

$$\Delta(\theta) = 1 - \frac{1 + \left(b\sqrt{\cos\theta} - 1\right)e^{b\sqrt{\cos\theta}}}{1 + (b-1)e^b} \tag{33.89}$$

Using methods already considered in the evaluation of Eq. (3.14), the launch velocity and rms velocity are given by

$$v_o = \left(2k_B T/\pi m\right)^{1/2} \tag{33.90}$$

$$v_{rms} = \left(k_B T/m\right)^{1/2} \tag{33.91}$$

EXAMPLE: Evaluate b, η, v_o, and v_{rms} for typical parameters associated with a thermal emitter, namely, an electric field of $\mathcal{E} = 333.33$ kV/m (or $3F_o = 1$ eV/μm) and $T = 1200$ K (or $k_B T = 0.103408$ eV).

SOLUTION: The evaluations follow sequentially:

- The field at the apex of the hemisphere is $F = 3F_o$. Therefore, $b = \sqrt{4QF}/(k_B T) = (0.0379469\text{ eV})/(0.103408\text{ eV}) = 0.366962$.
- By Eq. (33.88), $\eta = 0.111875$, or $\theta_n = 83.56°$.
- $v_o = (0.206816/\pi 510999)^{1/2}(299.792\text{ nm/fs}) = 0.134862$ nm/fs.
- $v_{rms} = (0.103408/510999)^{1/2}(299.792\text{ nm/fs}) = 0.190723$ nm/fs.

33.4.4.2 Field emittance

Next consider the current density J of the Fowler–Nordheim equation given by Eq. (13.25), a form which shows the field dependence explicitly. As before $F(\theta)/F_o = 3\cos\theta$ for the field along the surface of a hemisphere (Eq. (30.10)). The energy ratio parameter $b(F)$ depends only on field and is

$$b(F) = \frac{B_o \Phi^{3/2}}{F} \tag{33.92}$$

in terms of which the emission probability factor $P(\theta)$ is given by

$$P(\theta) = (\cos\theta)^{2-\nu}\exp\left[b(1-\sec\theta)\right] \tag{33.93}$$

The notional cosine angle is as already encountered in Eq. (30.17) and is

$$\eta = \frac{b+3-\nu}{b+4-\nu} \tag{33.94}$$

The integrated current fraction $\Delta(\theta)$ makes use of the same approximations encountered in evaluating Eq. (30.15) and is

$$\Delta(\theta) = 1 - \exp[-(b + 4 - \nu)(\sec\theta - 1)] \tag{33.95}$$

Lastly, the launch velocity and rms velocity are best evaluated by assuming that the chemical potential μ is large and emission is concentrated about $k \approx k_F$ so that

$$v_o \approx \frac{\hbar k_F}{m}\left(1 - \frac{\Phi}{3t_o b\mu}\right)^2 \tag{33.96}$$

$$v_{rms} \approx \frac{\hbar k_F}{m}\left(\frac{3t_o b\mu - \Phi}{3t_o b\mu + \Phi}\right)^2 \tag{33.97}$$

where t_o is familiar from Eq. (13.24) and the discussion following it.

EXAMPLE: Evaluate b, η, v_o, and v_{rms} for typical parameters associated with a field emitter, namely, an apex field of $F = 8$ eV/nm and copper-like parameters.

SOLUTION: The evaluations follow sequentially. With $B_o\Phi^{3/2} = 65.2073$ eV/nm and $\nu = 2B_oQ/3\sqrt{\Phi} = 0.772808$ then

- $b = (65.2073/8) = 8.15091$
- $\eta = 0.912112$, or $\theta_n = 24.2°$
- using $3bt_o = 25.952$, then $v_o = 1.49241$ nm/fs
- $v_{rms} = 1.42113$ nm/fs.

33.4.4.3 Photo emittance

Lastly, consider the current density J based on the low temperature limit of the Fowler–DuBridge equation of Eq. (14.16). The field dependence resides in the Schottky lowered work function $\phi = \Phi - \sqrt{4QF}$. The field- and frequency-dependent energy ratio parameter $b(F, \omega)$ is

$$b(F,\omega) = \frac{\hbar\omega - \phi}{\sqrt{4QF}} \tag{33.98}$$

in terms of which the emission probability factor $P(\theta)$ is given by

$$P(\theta) = b^{-2}\left(b - 1 + \sqrt{\cos\theta}\right)^2 \tag{33.99}$$

The notional cosine angle is defined by

$$\eta = \frac{4b - 1}{6b^2} \tag{33.100}$$

The integrated current fraction $\Delta(\theta)$ is long-winded to evaluate but relies only on simple integrations, and is

$$\Delta(\theta) = 1 - \cos\theta\left[\frac{6(b-1)^2 + 8(b-1)\sqrt{\cos\theta} + 3\cos\theta}{6(b-1)^2 + 8(b-1) + 3}\right] \tag{33.101}$$

Lastly, the evaluation of v_o and v_{rms} contains a feature that thermal and field evaluations do not: the photoemitted electron has absorbed the photon's energy, and it is the excited energy that matters. The approximations behind the Fowler–DuBridge approach are adequate to deal with that complication: the difference between the vacuum level and the conduction band minimum is the height of the barrier $\mu + \phi$, and it is removed from the energy of the photoexcited electron by the means of $E' = E + \hbar\omega - (\mu + \phi)$, where the prime indicates the vacuum level energy. The remaining integrals are a challenge to evaluate, and so under the approximation that $(\hbar\omega - \phi)/\mu$ is a small parameter, the integrands in the velocity equations can be expanded to leading order to find

$$v_o' = \frac{\hbar k_F}{m}\left[\frac{\hbar\omega - \phi}{3\mu}\left(1 - \frac{\hbar\omega - \phi}{8\mu}\right)\right]^{1/2} \tag{33.102}$$

$$v_{rms}' = \frac{\hbar k_F}{m}\left[\frac{\hbar\omega - \phi}{3\mu}\left(1 - \frac{\hbar\omega - \phi}{12\mu}\right)\right]^{1/2} \tag{33.103}$$

where the primes are to remind one that the quantities are defined on the vacuum side, not the material side. For expediency, the field dependence of ϕ is not shown, but evaluation requires knowing that $\phi = \Phi - \sqrt{4QF}$ is the Schottky lowered work function.

EXAMPLE: Evaluate b, η, v_o, and v_{rms} for typical parameters associated with photoemission, namely, $F = 75$ eV/μm and $\lambda = 250$ nm.
SOLUTION: The required factors are the Schottky factor $\sqrt{4QF} = 0.328629$ eV, $\hbar\omega = 4.95937$ eV, and $\hbar\omega - \phi = 0.787997$ eV. Then

- $b = (4.95937 - 4.17137)/(0.328629) = 2.39783$
- $\eta = 0.249042$, or $\theta_n = 75.58°$
- from $(\hbar\omega - \phi)/\mu = 0.112571$, then $v_o = 0.301821$ nm/fs
- $v_{rms} = 0.302539$ nm/fs.

33.4.4.4 Parameter comparisons

Each of the emission mechanisms (thermal, field, photo) are employed with characteristic sizes, fields, temperatures, and conditions. For the purposes of comparison, each technology should be evaluated under the conditions in which it operates. Therefore, consider the following representative values:

- **Thermal:** Dispenser cathodes used in high-frequency microwave amplifiers operate space-charge limited using an extraction grid that plays the role of the "anode" in Child's law (Eq. (16.18)). For $J_{CL} = 2$ A/cm^2 and an anode–cathode gap of $D = 250$μm, then $V_a/D = 0.2638$ eV/μm [492]. A background field of $F_o = 0.333$ eV/μm is therefore in excess of the space-charge limit but comparable to extraction grid technology. Machining of the cathode surface results in grooves and bumps with feature sizes of a few microns [281]. Let $a = 2$ μm.
- **Field:** A molybdenum field emitter array used in a 100 W TWT generated 100 mA using 50,000 emitters [483]. Assume that 40% of the emitters produce all the current, and that they have a characteristic apex radius of $a = 5$ nm. Apex fields of 8 eV/nm are therefore required, so let $F_o = (8/3)$ eV/nm.
- **Photo:** A multi-alkali antimonide photocathode was used in the first high average power photoinjector [520]. The drive laser had a wavelength of 527 nm, and the surface field was 26 MV/m, corresponding to $F_o = 26$ eV/μm. Such fields are comparable to more recent X-ray FEL fields [676]. Recent metal photocathode characterizations [695] suggest a surface roughness equivalent to $a = 0.2$ μm.

With these values, the characteristic parameters can be obtained, as in Table 33.3. Additionally, the characteristic integrated current fraction $\Delta(\theta)$ (Eq. (33.83)) can be compared, as in Figure 33.21.

Table 33.3 Emittance parameters and values for the evaluation of emittance based on copper. When "–" appears in the equation and/or units column, that parameter is not needed for the emittance parameters for that emission mechanism. See also Table 33.2.

Term	Equation and/or unit	Thermal	Field	Photo
μ	[eV]	–	7	7
Φ	[eV]	–	4.5	4.5
a	[nm]	2000	5	200
F	$3F_o$ [eV/nm]	0.001	8	0.075
T	[K]	1200	–	–
λ	[nm]	–	–	250
ϕ	$\Phi - \sqrt{4Q(3F_o)}$ [eV]	4.4621	1.1059	4.1714
v_o	33.90, 33.96, 33.102, [nm/fs]	0.1076	1.4924	0.30182
τ	mv_o/F [fs]	1835.4	3.1820	68.642
κ	$F_o a/mv_o^2$	30.381	3.1587	72.402
σ_o	$\sqrt{1+(7\kappa/2)}$	10.360	3.4721	15.950
b	33.86, 33.92, 33.98	0.36696	8.1509	2.3978
η	33.88, 33.94, 33.100	0.11187	0.91211	0.24904
θ_n	arccos(η) [degrees]	83.577°	24.201°	75.579°
$\Delta(\theta_n)$	33.89, 33.95, 33.101	90.524%	66.591%	82.647%

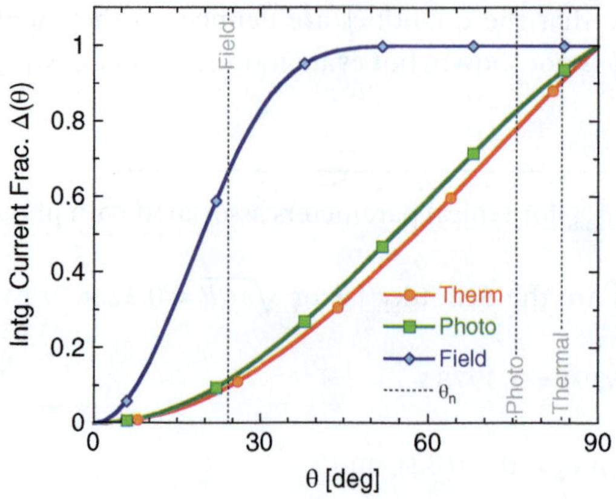

Figure 33.21 Evaluation of $\Delta(\theta)$ using Eqs (33.89), (33.95), and (33.101), and the parameters of Table 33.3. Dashed vertical labeled lines indicate the location of θ_n. Based on Figure 4 of ref. [451].

33.4.5 A notional emittance for a hemisphere

Consider whence each thing is come, and of what it consists, and into what it changes, and what kind of a thing it will be when it has changed, and that it will sustain no harm.

– Marcus Aurelius[26]

If a factor $\mathcal{O}(\theta,\varphi)$ is specified over the surface of the hemisphere, then the calculation of its mean value proceeds according to

$$\langle \mathcal{O} \rangle = \frac{\int_0^{\pi} \sin\theta d\theta \int_0^{2\pi} \mathcal{O}(\theta,\varphi) P(\theta,\varphi) d\varphi}{\int_0^{\pi} \sin\theta d\theta \int_0^{2\pi} P(\theta,\varphi) d\varphi} \tag{33.104}$$

where $P(\theta,\varphi)$ is the weighting factor. For the rotationally symmetric hemisphere, field enhancement varies only with θ, and so $P(\theta,\varphi) \to P(\theta) = J[F(\theta)]/J[F(0)]$, where J is the current density. After experiencing what goes behind the evaluation of field enhancement, and the varied forms the canonical equations of emission take, finding $\langle \mathcal{O} \rangle$ is promising to be an ordeal. The same was said of the evaluation of emission area in Section 30.2. Then, as now, the consideration of the *notional* quantity served as a convenient stand-in to the quantity itself: in the case of area, although the notional area is always smaller than the actual emission area, both scale the same way as the field (or other governing parameter) changes. In the same way, the otherwise frightful complexity of the evaluation of $\langle x(t)^2 \rangle$ and the other terms of $\varepsilon_{n,rms}$ can be sidestepped by considering instead their *notional* values and observing that these values serve as lower bound estimates, but which will also scale in the same manner as the more accurate representation involving $P(\theta)$. If θ_n is the notional angle, then the notional average quantities will be defined by

$$\langle \mathcal{O} \rangle = \frac{1}{4\pi} \int_0^{\theta_n} \sin\theta d\theta \int_0^{2\pi} \mathcal{O}(\theta,\varphi) d\varphi \tag{33.105}$$

a relation that provides a profoundly easier analytical representation.

Letting $\mathcal{O}(\theta,\varphi) = x(t)^2$ is a perfect place to start. From Eq. (33.78), $x(t) = a(1+\sigma p)\sin\theta\cos\varphi$ but this can be approximated by $x(t) \approx a\sigma p \sin\theta\cos\varphi$ because in general $\sigma p \gg 1$. Therefore, the corrections due to the launch site on the hemisphere being a fraction of a radius a off-axis is largely irrelevant, and so it can be ignored. Inserting the approximate form into Eq. (33.105) results in

$$\begin{aligned} \langle x(t)^2 \rangle &= \frac{(v_o t)^2}{4\pi} \int_0^{\theta_n} \sigma(\theta)^2 \sin^3\theta d\theta \int_0^{2\pi} \cos^2\varphi d\varphi \\ &\approx \frac{(p\sigma_o v_o \tau)^2}{12} (2+\eta)(1-\eta)^2 \end{aligned} \tag{33.106}$$

where the second line follows after using $\eta = \cos\theta_n$ and replacing $t \to p\tau$.

[26] Marcus Aurelius, *The Meditations of Marcus Aurelius, The Harvard Classics: Plato, Epictetus, Marcus Aurelius*, ed. Charles W. Eliot, Collector's Edition. Danbury, CT: Grolier Enterprises Inc., 1980, p. 289.

The next term corresponds to $x'(t)^2 = (v_x/v_z)^2$, which can use Eq. (33.81). The θ dependence of $\sigma(\theta)$ is an impediment to analytic integration of what remains. Observe, however, that as θ increases, so does the denominator, and so making the swap $\sigma(\theta) \to \sigma_o$ gives the lower bound estimate to $\langle x'(t)^2 \rangle$, not to mention its virtue in enabling an analytic solution to the integral. Letting $s = \cos\theta$ then

$$\left\langle x'(t)^2 \right\rangle = \frac{1}{4} S\left(\frac{p}{\sigma_o}, \eta\right) \tag{33.107}$$

where $S(\bar{p}, \eta)$ is defined by (with $\bar{p} = p/\sigma_o$)

$$S(\bar{p}, \eta) = \int_{\eta}^{1} \frac{1-s^2}{(\bar{p}+s)^2} ds = 2\bar{p} \ln\left(\frac{1+\bar{p}}{\eta+\bar{p}}\right) + \frac{(1-\eta)(1-\eta-2\bar{p})}{\eta+\bar{p}} \tag{33.108}$$

The cross term $\langle x(t)x'(t)\rangle$ is the final piece, and the same substitutions yields

$$\left\langle x(t)x'(t) \right\rangle = \frac{p v_o \tau}{4} \int_0^{\theta_n} \frac{\sigma(\theta)\sin^3\theta}{\sigma(\theta)\cos\theta + p} d\theta \approx \frac{p\sigma_o v_o \tau}{4} R\left(\frac{p}{\sigma_o}, \eta\right) \tag{33.109}$$

where

$$R(\bar{p}, \eta) = (1-\bar{p}^2) \ln\left(\frac{1+\bar{p}}{\eta+\bar{p}}\right) + \frac{1}{2}\left[(\bar{p}-\eta)^2 - \left(1-\bar{p}^2\right)^2\right] \tag{33.110}$$

and so, when the velocity of the electrons in the current is non-relativistic, the emittance for a solitary hemispherical bump is

$$\varepsilon_x > \frac{1}{4} \tau p \sigma_o v_o \sqrt{\frac{1}{3}(1-\eta)^2(2+\eta)^2 S(\eta) - R(\eta)^2} \tag{33.111}$$

which is satisfying until one actually tries to use it. The two groups of terms in the radical are very closely equal to each other so that ferreting out the scaling behavior is not easy, the invariance of $\beta\gamma\varepsilon_x$ is not discernible at all in the large p limit, and thoughts (*contra* Marcus Aurelius) turn unkind. But before indulging dark musings, it is worth reflecting on the consequences of such cancelation: that the "direct" term $\langle x^2 \rangle \langle x'^2 \rangle$ is largely canceled by the "cross" term $\langle xx' \rangle^2$ is not just a peculiarity of hemispheres, as the cancellation likewise occurs for needle-like field emitters modeled by ellipsoids where the electron trajectories follow field lines in a prolate spheroidal geometry [661]). The emittance of single protrusions is quite small as a result of the cancelation, and therefore the brightness (Eq. (33.31)) of needle-like cathodes, particularly field and Schottky emitters [154], is significant even though the transverse velocity off the apex of the emitter can be comparatively large when the current is high. There is evidently more to the story.

A pleasing method to coax out the remnants after the cancellation is by treating the integrals as a kind of distribution function in conjunction with another term for which the moments are taken. Consider the following definitions

$$\left\langle x^2 \right\rangle = \frac{1}{4} N\left(\sigma_o v_o p \tau\right)^2 \int_0^{\theta_o} g(\theta)\, d\theta \tag{33.112}$$

$$\left\langle x'^2 \right\rangle = \frac{N}{4p^2}\left(\sigma_o v_o p \tau\right)^2 \int_0^{\theta_o} f(\theta)^2 g(\theta)\, d\theta \tag{33.113}$$

$$\left\langle xx' \right\rangle = \frac{1}{4} N \sigma_o v_o p \tau \int_0^{\theta_o} f(\theta) g(\theta)\, d\theta \tag{33.114}$$

where the functions $f(\theta)$, $g(\theta)$ and normalization N are defined by

$$f(\theta) = \frac{\sigma_o}{\left(p + \sigma_o \cos\theta\right)} \tag{33.115}$$

$$g(\theta) = \frac{1}{N} \sin^3\theta \tag{33.116}$$

$$N = \frac{1}{3}(1-\eta)^2(2+\eta) \tag{33.117}$$

where N is so defined that $g(\theta)$ is normalized to unity, and $g(\theta)$ takes on the role of a distribution for which $f(\theta)$ becomes the term for which moments are taken. Why do that? Because with $\langle f^n \rangle$ defined by

$$f_n \equiv \langle f^n \rangle = \int_0^{\theta_n} [f(\theta)]^n g(\theta) d\theta \tag{33.118}$$

then use can be made of (in a manner completely synonymous with Eq. (8.65))

$$\Delta f_2 \equiv \left\langle \Delta f^2 \right\rangle = \int_0^{\theta_n} (f(\theta) - \langle f \rangle)^n g(\theta) d\theta$$
$$= \left\langle f^2 - 2f\langle f \rangle + \langle f \rangle^2 \right\rangle = f_2 - 2f_1^2 + f_1^2 = f_2 - f_1^2 \tag{33.119}$$

and so

$$\varepsilon_x = \frac{N}{4}\left(p\sigma_o v_o \tau\right)\sqrt{\Delta f_2} \tag{33.120}$$

Because Δf_2 is well behaved (it is not the difference of two very nearly equal terms), the scaling behavior of emittance can be found. To obtain an approximate analytical solution in terms of $s = \cos\theta$ and $\eta = \cos\theta_n$, where θ_n is the notional angle (Eq. (30.13)) and $f(\cos\theta)$ is defined by Eq. (33.115)

$$f(s) = \frac{1}{s + (p/\sigma_o)} \tag{33.121}$$

then because $f(1) < f_1 < f(\eta)$, a middle value $s_o = \cos\theta_o$ exists where $f_1 = f(s_o)$, which in turn gives

$$f(s) - f_1 = f(s) - f(s_o) = \frac{s_o - s}{[s_o + (p/\sigma_o)][s + (p/\sigma_o)]} \tag{33.122}$$

The evaluation of Δf_2 will therefore encounter an integral of the form

$$\Delta f_2 = \frac{1}{N\left[s_o + (p/\sigma_o)\right]^2}\int_\eta^1 \frac{(1-s^2)(s-s_o)^2}{\left[s + (p/s_o)\right]^2} ds \tag{33.123}$$

Such an integral undoubtedly has a closed-form solution, but that is not the point: how it depends on $(1-\eta)$ is, and so consider approximations that are more or less sufficient to capture the flavor even while introducing a smidgen of error. Two approximations that are reasonable for doing so are first, the location of s_o is between 1 and η so approximate it by $s_o \approx (1+\eta)/2$ and, second, evaluate the denominator of the integrand at $s = s_o$ and pull it out, leaving an integral that is trivial to evaluate:

$$\int_\eta^1 \frac{(1-s^2)(s-s_o)^2}{\left[s + (p/\sigma_o)\right]^2} ds \approx \frac{(2\eta + 3)(1-\eta)^4}{15\left[1 + \eta + (2p/\sigma_o\right]^2} \tag{33.124}$$

As crude as the approximations are, they capture most of the behavior as seen in Figure 33.22. When put into the equation for emittance, and considering the long time limit ($p \gg 1$), then

$$\varepsilon_{n,rms} \approx C_o(p)\sqrt{(2+\eta)(2\eta+3)(1-\eta)^3} \tag{33.125}$$

$$C_o(p) = \frac{\sqrt{5}\,\tau\sigma_o^3 p^2 v_o^2}{30c\left[\sigma_o(1+\eta) + 2p\right]^2} \rightarrow \frac{\sqrt{5}\,\tau\sigma_o^3 v_o^2}{120c} \tag{33.126}$$

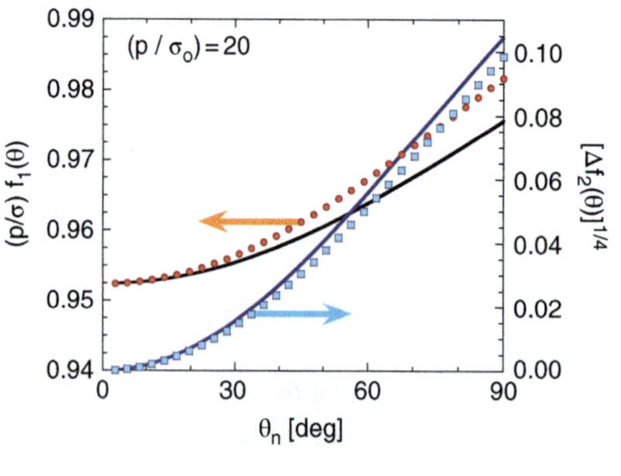

Figure 33.22 The values of $(p/\sigma_0)f_1 = (p/\sigma_0)\langle f(\theta)\rangle$ and $[\Delta f_2]^{1/4}$ as a function of notional angle ($\eta = \cos\theta_n$), evaluated numerically (symbols) and using the approximations behind Eq. (33.124) (lines).

where the large time ($p \to \infty$) limit of $C_o(p)$ is shown.[27] The factor $C_o(\infty)$ serves as a characteristic emittance coefficient for each process of thermal, field, and photoemission.

Emittance for a hemisphere has several distinct features. First, the emittance depends on the characteristic time constant $\tau = mv_o/F$, which in turn depends on the launch velocity v_o and the typical field associated with that emission mechanism. Second, emittance depends on the behavior of the notional emission angle θ_n through the factor $(1-\eta)^3$, and that is a strong function of how rapidly emission falls off down the side of the bump. As field emitters have small θ_n, the emittance of a single field emitter is small and field emitters are therefore quite bright sources, particularly when the substantial current density on apex is taken into account. Third, for a hemisphere, the emittance is *independent of hemisphere radius a*. Lastly, the emittance caused by electrons being launched at an angle to the beam direction due to surface roughness is governed by v_o, whereas the intrinsic one-dimensional emittance due to the emission process itself is governed by v_{rms}, and these quantities are comparable. In treatments in the literature, these two differing kinds of emittance contributions are sometimes termed '"slope" and "field distortion" or "tilt" effects [677, 692]. Treating field distortion is generally difficult due to an absence of knowledge about how the field lines behave except in special circumstances in which they can be exactly determined in theoretical models (e.g., the *tour de force* by Lau [693]). Additionally, how the field away from the hemisphere (or any protrusion) induces further variations complicate how the electrons are accelerated, leading to, for example, an $\varepsilon^2_{\text{field}}$ term [696]. An account of the total emittance from a rough cathode surface must include them all, and a method to do so is

$$\varepsilon^2_{total} = \varepsilon^2_{1D} + \varepsilon^2_{bump} + \varepsilon^2_{field} + \cdots \tag{33.127}$$

where "1D" is the emittance due to transverse energy contributions in the one-dimensional model, "bump"' is due to the effects of a transversal launch from a rough (or hemisphere-like) surface, "field" is for field non-uniformity effects during acceleration close to the cathode surface, and "$\cdots$" is for whatever else Nature chooses to bedevil emittance with, be it near the cathode or acquired elsewhere during the beam's propagation.

Knowing that $\eta < 1$ for thermal, field and photoemission sources, a comparison of $C_o(\infty)$ for them using the parameters of Table 33.3 results in $C_o(\infty)$ [nm] = 1.4688, 0.018439, and 1.5771 for thermal, field, and photoemission, respectively. If only emission was strictly normal to the surface (it is not), then it would seem that field emission sources are far and away the brightest emitters of the bunch because of *both* C_o and $(1-\eta)^3$, and for single emitters, they are. But for most electron sources, emission is not from a single bump, but rather (and particularly for field emitters) from an array of sites, and that small observation has consequences for the $(1-\eta)$ dependence by removing the ability of the cross term $\langle xx' \rangle^2$ to offset $\langle x^2 \rangle \langle x'^2 \rangle$.

33.4.6 An array of emitters

> *...if their actions be directed according to their particular judgements, and particular appetites, they can expect thereby no defense ...(f)or being distracted in opinions concerning the best use and application of their strength, they do not help, but hinder one another....*
>
> – Thomas Hobbes[28]

Emitter surfaces are not atomically flat. They have structure and variation that can be significant, as images of various thermal, field, and photoemitter surfaces show [176, 382, 480, 677–679, 697–704]. The cumulative effect of all those busybody protrusions or emission sites is to complicate making a good beam. There is nothing simple in how they do so, but simple models nevertheless give a feel.

Realistic thermionic and photoemission surfaces have varying surface feature size and random placement. In contrast, arrays of field emitters are made in a triangular array rather than a square lattice so as to increase packing density: for a tip-to-tip spacing, or pitch, of l, a square array has (1 emitter)$/l^2$, whereas a triangular array has (1 emitter)$/[l^2 \sin(\pi/3)] =$ 1.1547 emitters $/l^2$. Be that as it may, nevertheless model the emitters as sites on a square lattice that are uniformly sized with the number of sites being M^2 and the area of the array being L^2, such that

$$L^2 = [(M-1)l]^2 \tag{33.128}$$

[27]The form of this equation is similar to that of ref. [451], but differs in that here $f_o = 1/[s_o + (p/\sigma_o)]$ is retained and $s_o \approx (1+\eta)/2$ is not used. The crudeness of the estimate $s_o \approx (1+\eta)/2$ makes Eq. (33.125) differ by about a factor of 2 compared to a more careful treatment (the value of s_o really does matter), but tedious Taylor expansions are avoided. If the point is to recover the $(1-\eta)^3$ dependence, the price is a bargain.

[28]Thomas Hobbes and C.B. Macpherson, *Leviathan*. Harmondsworth: Penguin, 1968, "Of Common-wealth", Part II, Chapter 17, p. 224.

Lastly, let the trajectories of a single boss be unaffected by its neighbors,[29] which takes l on the order of several hemisphere radii a at least. By way of comparison: (*i*) the microstructure on scandate dispenser cathodes can have feature sizes from 5 to 20 μm [702], (*ii*) pitches of Spindt-type field emitters are a few micrometers [176], (*iii*) atomic force microscopy images of SLAC's linac coherent light source (LCLS) photocathode [696] have undulations with a period of about 10 μm, (*iv*) roughness parameters assumed in a theoretical treatment [705] took the periodic bumps to be tens of nanometers apart on a photocathode, (*v*) sputter-deposited and solid lead photocathodes [706] have feature sizes on the order of 1 μm, and (*vi*) aligned carbon nanotube fibers can be placed 10 μm apart [701]. Lastly, assume that the M^2 protrusions are uniformly sized and emitting, the latter a condition which can be relaxed afterwards.

Because knowledge of the emitter's location in an array is now required, consider a redefinition of $x(t)$ to account for it [451, 661, 707]: let the location of the emitter in the ith row and jth column be designated by appending the location of the center to $x(t)$, and the same with the location $y(t)$, and so

$$\begin{aligned} x(t) &\rightarrow X_{ij}(t) = x_{ij} + x(t) \\ y(t) &\rightarrow Y_{ij}(t) = y_{ij} + y(t) \end{aligned} \tag{33.129}$$

$$\begin{aligned} x_{ij} &= (2j - M - 1)\,l \\ y_{ij} &= (2i - M - 1)\,l \end{aligned} \tag{33.130}$$

where, as expected, $X_{ij}(t)$ is independent of the row (i) and $Y_{ij}(t)$ is independent of the column (j). Now, $\langle X_{ij}(t)^2 \rangle$ takes the place of $\langle x(t)^2 \rangle$, where

$$\langle X_{ij}(t)^2 \rangle = \frac{1}{4\pi M^2} \sum_{i=1}^{M} \sum_{j=1}^{M} \int X_{ij}(t)^2 d\Omega \tag{33.131}$$

where $d\Omega/4\pi$ includes all the factors which appear in Eq. 33.104 apart from M^2, which is needed for turning the summations into averages. For a macroscopic array, then $L^2 \gg \langle x(t)^2 \rangle$, leading to the independence of X_{ij} on j. As a consequence of $\langle x_{ij} \rangle = 0$ (in an infinite array, emitters on one side of the emitter under examination are balanced by emitters on the other side) then

$$\langle X_{ij}(t)^2 \rangle = \langle x_{ij}^2 \rangle + \langle x(t)^2 \rangle = \frac{1}{3}\left(M^2 - 1\right) l^2 + \langle x(t)^2 \rangle \tag{33.132}$$

EXAMPLE: Show $\langle x_{ij}^2 \rangle = (1/3)\left(M^2 - 1\right) l^2$ as in Eq. (33.132).
SOLUTION: From the definition of averaging and using Eq. (A1.2),

$$\begin{aligned} \langle x_{ij}^2 \rangle &\equiv \frac{l^2}{M^2} \sum_{i=1}^{M} \sum_{j=1}^{M} (2j - M - 1)^2 \\ &= \frac{l^2}{M} \sum_{j=1}^{M} \left[4j^2 - 4j(M+1) + (M+1)^2\right] \\ &= l^2 \left(\frac{2}{3}(M+1)(2M+1) - (M+1)^2\right) = \frac{l^2}{3}\left(M^2 - 1\right) \end{aligned}$$

Convert over to L by $(M^2 - 1)l^2/3 = (M+1)L^2/12(M-1)$ and then taking the large M limit to get $L^2/12$. Now L^2 is big, certainly far more so than $\langle x(t)^2 \rangle$, which is evaluated over one emission site. Therefore, $\langle X_{ij}(t)^2 \rangle$ is of the size $L^2/12$. Now, the factor $\langle x'(t)^2 \rangle$ is not affected by the array and so Eq. (33.107) can be used. That leaves the cross-term $\langle X_{ij}(t)x'(t) \rangle = \langle x(t)x'(t) \rangle$, which therefore does not change in size and is therefore negligible compared to $L^2 \langle x'(t)^2 \rangle / 12$. The important thing is that $\langle x(t)^2 \rangle \propto N$ whereas $\langle X_{ij}(t)^2 \rangle \propto L^2/12$ so that a factor of $N \propto (1-\eta)^2$ (seen in Eq. (33.112)) seems to *have gone missing*, meaning in short that

$$\varepsilon_{n,rms} \propto (1 - \eta)\frac{L\sigma_o v_o}{12c} \tag{33.133}$$

[29]This approximation is not as bad as it sounds: apart from being reasonable for widely spaced bumps, for the extreme case of field emitters on a (perfect) regular lattice, apart from shielding, the neighbor emitters are symmetrically placed and have a reduced effect on trajectories.

instead of the $(1-\eta)^3$ as before: the effect of the array has been to enlarge the emittance due to an emission area factor L^2 via a weaker dependence on $(1-\eta)$, an effect particularly consequential for field emission as η for it is near unity.

33.5 Emittance and realistic surfaces

33.5.1 Emittance and undulating surfaces

> *...we are now reaching the thermal emittance as the limit to increasing the beam brightness. However, the effective thermal, or "intrinsic", emittance depends on several effects, including the crystallinity, surface roughness, surface impurities, and QE non-uniformity. Thus it is very challenging to measure and combine all these phenomena into a complete and useful physical model.*
>
> – David H. Dowell *et al.* [523]

Machining or processing of a cathode surface can leave grooves that have a spacing of approximately 10 μm for example the dispenser cathode surfaces of refs [281] and [282], but such structure has analogs in the aforementioned LCLS photocathode of ref. [696]. The impact of grooves and undulations is a recurrent topic of theoretical treatments [693, 705, 708] and is particularly of interest when limits on emittance are essential [679].

Consideration of a sinusoidally undulating surface by Dowell *et al.* [696] assessing several contributions to emittance gives a feel for the relative weight of each to the final emittance. The thermal component (intrinsic ε_i) is as already encountered in Eq. (33.40). Including the cosine-modulated (tilt ε_t) of the surface then amends the thermal component by

$$\varepsilon_{i+t} = \frac{R}{2}\sqrt{\frac{\hbar\omega - \phi}{3mc^2}\left[1 + \frac{1}{2}(kz_o)^2\right]} \tag{33.134}$$

where the surface height is modeled by $z(x) = z_o \cos(kx)$. The impact of the electron accelerating through the curved fields caused by F near the ridges [692] is approximated by

$$\varepsilon_f = \frac{R}{2}\sqrt{\frac{\pi k z_o^2 F}{4mc^2}} \tag{33.135}$$

and so

$$\varepsilon_{i+t+f} = \frac{R}{2}\sqrt{\frac{\hbar\omega - \phi}{3mc^2}\left[1 + \frac{1}{2}(kz_o)^2\right] + \frac{\pi k z_o^2 F}{4mc^2}} \tag{33.136}$$

EXAMPLE: Evaluate ε_i using the values of ref. [696] of $\lambda = 255$ nm, $F = 57.5$ eV/μm, and $\Phi = 4.8$ eV. Let $R = 2$ mm.
SOLUTION: The photon energy is $\hbar\omega = \hbar(2\pi/\lambda) = 4.8621$ eV. The Schottky factor is $\sqrt{4QF} = \sqrt{1.44 \times 0.0575} = 0.28775$ eV, so that $\phi = 4.5123$ eV. Therefore,

$$\varepsilon_i = (1\text{ mm})\sqrt{\frac{4.8621 - 4.5123}{3(511000)}} = 0.47773 \text{ mm-mrad}$$

EXAMPLE: Evaluate the tilt correction $(kz_o)^2/2$ using the values of ref. [696] of $z_o = 17$ nm and $k = (\pi/5)$ μm^{-1}. Let $R = 2$ mm.
SOLUTION: Straight substitution yields

$$\frac{1}{2}(kz_o)^2 = \frac{1}{2}\left(\frac{17\pi}{5000}\right)^2 = 5.7046 \times 10^{-5}$$

making the correction very small so that ε_t is less than 1% of ε_i.

EXAMPLE: Evaluate the field emittance ε_f using the values of ref. [696] given in the previous examples. Find the total emittance. Let $R = 2$ mm.

SOLUTION: Using kz_o from before and $Fz_o = 0.9775$ eV, then

$$\varepsilon_f = (1\text{ mm})\sqrt{\frac{\pi \times 0.9775 \times 0.010681}{4 \times 511000}} = 0.12668\text{ mm-mrad}$$

and so

$$\varepsilon_{i+t+f} = \sqrt{\varepsilon_i^2 + \varepsilon_t^2 + \varepsilon_f^2} = 0.49426\text{ mm-mrad}$$

The thermal emittance ε_i dominates.

33.5.2 Intrinsic emittance and space charge

...beam dynamics in an accelerator can be very sensitive to the details of the 6D distribution at the source. Even if a beam appears to be well behaved in configuration space, distortions in velocity space can reappear as density structures later downstream. Note that this fine structure persists for a long time.

– Rami A. Kishek *et al.* [666]

With such an understanding in hand, the linkages of space-charge, halo, and emittance in Figure 33.3 can be better understood. The evolution of a bunch of charge represents the interplay between the initial spatial variations or non-uniformities of emission at the cathode, the fluctuations of current density induced by space-charge interactions, and the spatial and energy spreads caused by emittance. Particle-in-cell codes can capture that interplay for a presumed value of emittance at the cathode. Could not, then, the emittance of the cathode be "predicted" if the initial distribution can be specified and monitored as it propagates? That, in fact, was the goal of an experiment performed at the University of Maryland's Electron Ring (UMER) designed to understand the waves and collective oscillations carried by a space-charge-dominated beam [666].

By passing the electron beam through a mask containing five holes, very clear structure was put on the beam in the shape of a *quincunx* (four items in the corners of a square or rectangle with the fifth one in the center). That beam was then allowed to propagate a distance and imaged on a phosphor plane, as shown in Figure 33.23. The starting emittance of the beam affects its subsequent evolution through its interplay with space-charge forces, and by assuming various

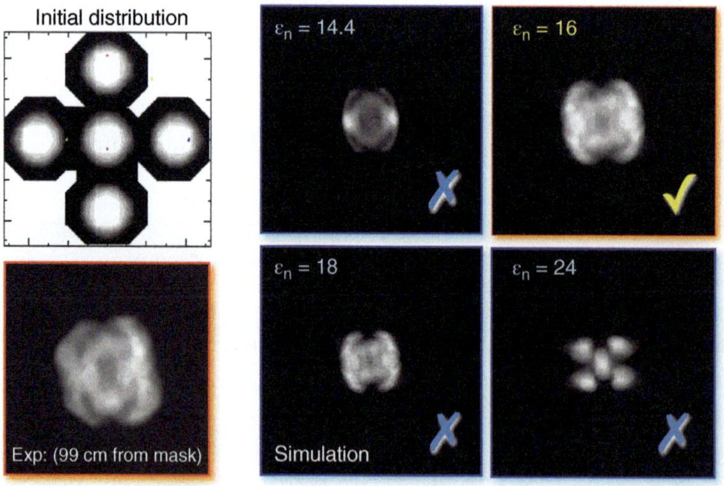

Figure 33.23 Left, the experimental launching and mixing of five beamlets (top image) over a length of 104 cm to a phosphor screen (bottom image). Right, simulations using PIC code WARP with different starting emittances but otherwise modeled after the experimental arrangement. Of the four cases considered, only the one marked with a check replicated the experimental behavior, showing the beam preserved details of the density structure and the emittance affected the evolution [666]. Images courtesy of R. Kishek (UMD). Right-hand side based on Figure 2 of R.A. Kishek, S. Bernal, C.L. Bohn, D. Grote, I. Haber, H. Li, P.G. O'Shea, M. Reiser, and M. Walter, Phys. Plasmas, **10**, 2016, 2003.

starting values of that emittance and comparing the final simulated shape with the experimentally measured one, a fair match of the starting emittance was thereby discerned. The experimental/simulation comparison therefore provides a rather striking example of the realization of the underlying physics behind many of the simple models behind the treatments herein, and serves as fair warning for the consequences of foolishly taking the initial conditions of a beam for granted. What is more tantalizing is that such "chaotic mixing" characterizes more than just electron beams: gravitational interactions also exhibit a $1/r^2$ force, and so, amazingly, beam experiments such as those at UMER can address hypotheses coming from as unexpected directions as galactic dynamics, with a twist: gravity is attractive, so the dynamics of the beam interaction have to be played in *reverse*.

PART VI
Appendices

It is also reported that Ptolemy once asked Euclid if there was not a shorter road to geometry than through the Elements, and Euclid replied that there was no royal road to geometry.

– Proclus[1]

The practical implementation of formulae requires numerical computation, and what Euclid allegedly opined about geometry is true of mathematics in general: there is no Royal Road.[2] There are, however, ways to ease the road one is on. These appendices provide assistance to do so in the form of numerical methods, algorithms, and means to evaluate functions of particular interest for emission physics.

[1]Proclus, *A Commentary on the First Book of Euclids Elements* (trans. Glenn R. Morrow). Princeton, NJ: Princeton University Press, 1970, p. 57.
[2]The Royal Road built by Darius I in the 5th century BC allowed rapid travel across a section of the Persian Empire, almost from the Aegean Sea to the Persian Gulf. Herodotus (Herodotus, *The History of Herodotus* (trans. George Rawlinson, *Great Books of the Western World*, ed. Robert Maynard Hutchins, Vol. 6, Herodotus, Thucydides. Chicago: Encyclopaedia Britannica, 1952, Book 8, passage 98, p. 277)) speaks of the unparalleled speed of the Persian messengers along the Royal Road, observing "these men will not be hindered from accomplishing at their best speed the distance which they have to go, either by snow, or rain, or heat, or by the darkness of night", from which the inscription on the James Farley Post Office (New York City) drew its inspiration and became the creed associated with the US Post Office.

Appendices

APPENDIX 1

Summation, integration, and differentiation

For all the wisdom of a lifetime, no one can ever tell.

– Rod Stewart[1]

A1.1 Series

The following series are useful.

$$\sum_{j=1}^{N} j = \frac{1}{2}N(N+1) \tag{A1.1}$$

$$\sum_{j=1}^{N} j^2 = \frac{1}{6}N(N+1)(2N+1) \tag{A1.2}$$

The following Taylor series are encountered with regularity. Derivatives of each side give further series relations of use.

$$\frac{1}{1 \pm x} = \sum_{j=0}^{\infty} (\mp x)^j \tag{A1.3}$$

$$\frac{1 - x^N}{1 - x} = \sum_{j=0}^{N-1} x^j \tag{A1.4}$$

$$\ln(1 \pm x) = -\sum_{j=1}^{\infty} \frac{(\mp x)^j}{j} \tag{A1.5}$$

$$e^{\pm x} = \sum_{j=0}^{\infty} \frac{(\pm x)^j}{j!} \tag{A1.6}$$

The following series are associated with Riemann's ζ function:

$$\sum_{j=1}^{N} \frac{1}{j^n} = \zeta(n) \tag{A1.7}$$

$$\sum_{j=1}^{N} \frac{(-1)^{j+1}}{j^n} = (1 - 2^{1-n})\,\zeta(n) \tag{A1.8}$$

A1.2 Integration

A1.2.1 Series summation

The trapezoidal approximation [236] to the numerical integration of a function $f(x)$ between x_0 and x_1, and perhaps the simplest of the methods usually considered, amounts to approximating $f(x) \approx ax + b$ and finding

$$I_1 = \int_{x_0}^{x_1} f(x)dx \approx \frac{1}{2}(f(x_1) + f(x_0))(x_1 - x_0) \tag{A1.9}$$

a formula often said to be exact for polynomials up to and including degree 1 (a precise way of saying "linear").

[1] Jim Cregan, Kevin Savigar, Bob Dylan, and Rod Stewart, lyrics to *Forever Young*, 1988.

Extending the integration over N regions bounded by x_j that are equally spaced such that $x_{j+1} - x_j = \Delta$ for $0 \le j \le N$ results in the composite summation formula to approximate the integral I_N by (where the notation $f(x_j) \equiv f_j$ is used)

$$I_N = \int_{x_0}^{x_N} f(x)dx \approx S_N \equiv \Delta \sum_{j=0}^{N} f_j - \frac{\Delta}{2}(f_0 + f_N) \tag{A1.10}$$

where the designation S_N reinforces that a simple summation of the function $f(x_j)$ is involved.

Simpson's rule is a modification of the trapezoidal rule by evaluating $f(x)$ in Eq. (A1.9) at the midpoint: a curve that passes through three points is a quadratic, so that Simpson's rule is expected to be exact up to degree 2, and is given by

$$I_2 = \int_{x_0}^{x_2} f(x)dx \approx Q_2 = \frac{\Delta}{3}(f_2 + 4f_1 + f_0) \tag{A1.11}$$

In fact, letting $f(x) = x^\alpha$ shows that Eq. (A1.11) is exact for

$$\frac{2^{\alpha+1}}{\alpha + 1} = \frac{1}{3}(2^\alpha + 4) \tag{A1.12}$$

that is, for $\alpha = 2$ and 3: the approximation is therefore good up to degree 3. Using Simpson's rule to generalize Eq. (A1.10) results in a summation for which the odd j and even j terms have coefficients of 2 or 1, respectively, and give rise to the widely used composite Simpson's rule to estimate I_N [236]. The series formula using Simpson's rule will be designated Q_N and is defined by

$$Q_N = \frac{\Delta}{3}\left\{ -f_0 + f_N + 2\sum_{j=0}^{(N/2)-1} \left(2f_{2j+1} + f_{2j}\right) \right\} \tag{A1.13}$$

which evidently is a bit more involved to code. Higher-order polynomials result in inserting more intermediate points (e.g., Bode's rule uses five evaluations and is accurate up to degree 5, but the resulting generalization of Eq. (A1.10) results in sums with more complex coefficients). All involve algorithms that are increasingly of greater difficulty than S_N. The higher-order approaches all achieve their accuracy by including more points in the integration region, as it is known that a polynomial of order N requires $N + 1$ points to specify it [243].

It would be convenient to have a summation formula close to the simplicity of Eq. (A1.10) with an accuracy closer to the higher degree approximations, but without the alternating coefficients. The method to do this is to make use of the derivatives of $f(x)$, and leads to the Euler–MacLauren series method. This would seem to make the development of a simple summation formula harder, not easier, because now in addition to evaluations of $f(x)$, evaluations of $df/dx \equiv f'(x)$ at x_j are also required, but this expectation will be shown to be subverted. It is found that

$$I_1 \approx M_1 = \frac{\Delta}{2}(f_1 + f_0) + \frac{\Delta^2}{2}(f_1' - f_0') \tag{A1.14}$$

$$I_2 \approx M_2 = \frac{\Delta}{2}(f_2 + 2f_1 + f_0) + \frac{\Delta^2}{2}(f_2' - f_0') \tag{A1.15}$$

where $f(x_j) \equiv f_j$ and $df/dx \equiv f'$. It is seen that for $f(x) = x^\alpha$, then explicit evaluation of I_2 gives (compare Eq. (A1.12))

$$\frac{2^{\alpha+1}}{\alpha + 1} = 1 + \frac{2^\alpha}{24}(12 - \alpha) \tag{A1.16}$$

which is likewise satisfied exactly for $\alpha = 2$ or 3. For higher values of α, the relative error characteristic of Simpson's rule compared to the modified trapezoidal rule of Eq. (A1.15) is shown in Figure A1.1, and so the modified trapezoidal rule outperforms Simpson's rule.

Observing that Eq. (A1.15) is not dependent on f_1', the composite modified trapezoidal rule can be created, in which the prime terms in Eq. (A1.15) are opposite in sign. Cancellation thereby occurs, resulting in only the derivatives at the integration limits requiring evaluation. Thus, I_N is approximated by M_N defined by

$$M_N = S_N + \frac{\Delta^2}{12}(f_0' - f_N') \tag{A1.17}$$

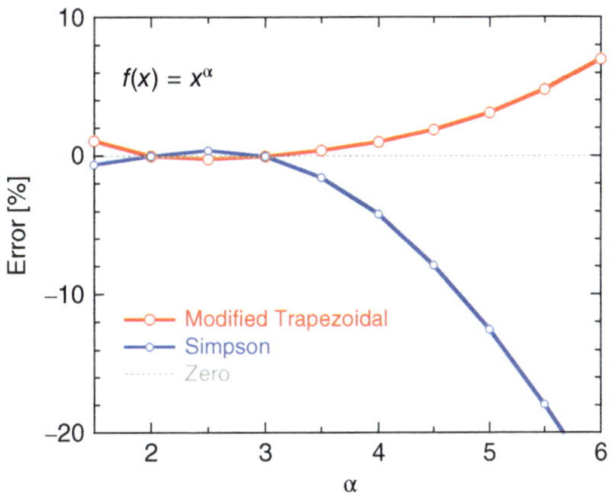

Figure A1.1 Trapezoidal approximation applied to integrals. Comparison of the errors associated with the modified trapezoidal approximation of Eq. (A1.15) and Simpson's rule of Eq. (A1.11) for $f(x) = x^\alpha$.

and is exact up to degree 3, as shown by explicit evaluation with $f(x) = x^\alpha$, yet the formula is no more difficult to evaluate than the trapezoidal rule S_N augmented by two additional derivative terms, and is therefore simpler to implement than Q_N and generally more accurate.

It is worth pointing out that often the trapezoidal rule S_N is treated as a lower-order approximation than the composite Simpson's rule Q_N, but this comes from neglecting the endpoint derivatives in the factor $\frac{\Delta^2}{12}(f_0' - f_N')$ of M_N: S_N is good to degree 3 apart from the boundary terms. Two examples are chosen so that $f_0' = f_2' = 0$, and so $M_N \to S_N$, to compare S_N to Q_N and show that the former is closer to the actual integral.

- For $f(x) = \cos^2(x)$ for $\Delta = \pi/2$. Explicit evaluation shows that $I_2 = S_2 = \pi/2$ but $Q_2 = \pi/3$. Similarly, for $\Delta = \pi/N$ and $N = 8$, $I_8 = S_8 = \pi/2$.
- For $f(x) = [x(1-x)]^2$ and let $N\Delta = 1$. Explicit evaluation shows that $I_2 = 16/15$, whereas $S_2 = 1$ and $Q_2 = 4/3$. Also, $I_8 = I_2 = 1.0667$ but $S_8 = 273/256 = 1.0664$.

EXAMPLE: Compare S_8, M_8, and Q_8 for $\Delta = 3/8$, $x_0 = 0$, and $f(x) = x^4$ to the analytical result I_8.
SOLUTION: Direct integration shows $I_8 = 243/5 = 48.6$. Series summation shows that $S_8 = 408483/8192 = 49.8636$, $M_8 = 398115/8192 = 48.5980$, and $Q_8 = 99549/2084 = 48.6079$. It is seen that M_8 is the most accurate approximation.

A1.2.2 Series expansion

If an integrand can be Taylor expanded in terms of quantities that can be term-by-term integrated, another method for the evaluation of integrals is possible. Examples based on series expansions of $\ln(1+x)$ and $1/(1+x)$, as found in Chapter 17, are a case in point. Let a function $g(x)$ be such that

$$g(x) = \sum_{j=0}^{\infty} f_j(x) \tag{A1.18}$$

Let G be the integral of $g(x)$ and $F_j(x) = \int f_j(x)dx$. Explicitly,

$$G = \int_a^b g(x)dx = \sum_{j=0}^{\infty}(F_j(b) - F_j(a)) \tag{A1.19}$$

which is a seemingly obvious finding. The utility arises if the series involving F_n is summable, and especially if F_j at one of the limits vanishes. Examples give the flavor.

EXAMPLE: Let $g(x) = 1/(1-x)$ with $a = 0$ and $b < 1$. Find G.
SOLUTION: The Taylor expansion of $g(x)$ reveals $f_j(x) = x^j$ so that $F_j(x) = x^{j+1}/(j+1)$. Therefore

$$G = \int_0^b \frac{dx}{1-x} = -\sum_{j=1}^{\infty} \frac{b^j}{j} = -\ln(1-b) \tag{A1.20}$$

where the summation has been re-indexed.

EXAMPLE: Using the previous example, find G for $g(x) = x^3/(1-x)$ with $a = 0$ and $b < 1$.
SOLUTION: Although the integral can be evaluated by repeated integration by parts, the series method is faster. From $f_j(x) = x^{j+3}$ and $F_j(x) = x^{j+4}/(j+4)$ it follows

$$G = \int_0^b \frac{x^3 dx}{1-x} = -\sum_{j=4}^{\infty} \frac{b^j}{j} = -\ln(1-b) - \sum_{j=1}^{3} \frac{b^j}{j} \tag{A1.21}$$

where Eq. (A1.20) has been used.

EXAMPLE: Find the integral of $x/(e^x - 1)$ with $a = 0$ and $b \to \infty$.
SOLUTION: To use the series method, the integrand must be recast in terms of a $g(x)$ that allows for a convergent series expansion. Observe that

$$\frac{1}{e^x - 1} = \sum_{j=1}^{\infty} e^{-jx} \tag{A1.22}$$

The term $e^{-jx} < 1$ for the range of integration so a convergent series expansion is straightforward. With $f_j(x) = xe^{-jx}$ then $F_j(x) = -(1+jx)e^{-jx}/j^2$ and so

$$\int_0^b \frac{x}{e^x - 1} dx = \sum_{j=1}^{\infty} \frac{1}{j^2}\left(1 - (1+jb)e^{-jb}\right) \tag{A1.23}$$

which, in the large b limit, is $\sum_{j=1}^{\infty} j^{-2} = \zeta(2)$.

Observe a lesson from the last example. Consider the integral $\int_0^b xdx/(1-e^{-x})$: as $b \to \infty$ the integral diverges. Situations arise in which the large, possibly even divergent, part of the integral may cancel the divergent part of another integral, leaving finite remainders that are of interest. For the integral in question, if over most of the integration region the denominator is well approximated by unity, then remove unity from the integrand. Thus, for $b \gg 1$,

$$\int_0^b \frac{xdx}{1-e^{-x}} \approx \int_0^b xdx + \int_0^{\infty} \frac{xdx}{e^x - 1} = \frac{1}{2}b^2 + \zeta(2) \tag{A1.24}$$

a method that is useful in the development of the Fowler function of Eq. (14.9), albeit with a different integrand.

Lastly, the choice of $(1-x)$ and $(1-e^{-x})$ in the denominator of the example integrals was for convenience: the more topical cases $(1+x)$ and $(1+e^{-x})$ represents the inclusion of a factor $(-1)^{j+1}$ in $f_j(x)$, the consequences of which are left to the pleasure of self-discovery.

A1.2.3 Gaussian quadrature

A smooth function $f(x)$ can be expanded in a series of orthogonal polynomials where the locations at which $f(x)$ are evaluated, x_j, and the weights given to those points, w_j, are to be determined. It is useful to first rescale x in Eq. (A1.9) and consider instead the equation

$$G = \int_{-1}^{1} f(x)dx \approx G_n = \sum_{j=1}^{n} w_j f(x_j) \tag{A1.25}$$

for a given n. There are $2n$ unknowns (n apiece for the w_j and x_j), and to determine them, in the case of the Legendre polynomials (for which the x_j are the zeros) let $f(x)$ be represented by a power series of $2n$ terms, or

$$f(x) = \sum_{j=0}^{2n-1} a_j x^j \tag{A1.26}$$

The values of a_j are arbitrary, so n equations have to be satisfied. The equation associated with each a_j is of the form

$$0 = \sum_{k=1}^{n} w_k x_k^j \quad (j = \text{odd}) \tag{A1.27}$$

$$\frac{2}{j+1} = \sum_{k=1}^{n} w_k x_k^j \quad (j = \text{even}) \tag{A1.28}$$

Consider the lowest order cases $n = 2$ and $n = 3$ below (the case $n = 1$ corresponds to the trapezoidal approximation).

EXAMPLE: Find the n roots x_j and the weights w_j for $n = 2$.
SOLUTION: For j odd and even, Eqs (A1.27) and (A1.28) are concisely represented as (left and right, respectively):

$$\begin{bmatrix} x_1 & x_2 \\ x_1^3 & x_2^3 \end{bmatrix}\begin{bmatrix} w_1 \\ w_2 \end{bmatrix} = \begin{bmatrix} 0 \\ 0 \end{bmatrix}$$

$$\begin{bmatrix} 1 & 1 \\ x_1^2 & x_2^2 \end{bmatrix}\begin{bmatrix} w_1 \\ w_2 \end{bmatrix} = \begin{bmatrix} 2 \\ 2/3 \end{bmatrix}$$

The top (odd) matrix equation is satisfied if the determinant of the 2×2 matrix vanishes, or $x_1 x_2 (x_2^2 - x_1^2) = 0$, therefore $x_1 = -x_2$. The top row of the same matrix then demands $w_1 = w_2$. Making these substitutions into the bottom (even) matrix shows that $2w_1 = 2$ and $2x_1^2 w_1 = 2/3$, which sets $w_1 = 1$ and $x_1 = 1/\sqrt{3}$. Observe, therefore, that for $n = 2$, there are two different roots $x_1 = -x_2 = 1/\sqrt{3}$ and one weight $w_1 = w_2 = 1$.

EXAMPLE: Find the n roots x_j and the weights w_j for $n = 3$.
SOLUTION: For j odd, Eqs (A1.27) and (A1.28) are again concisely represented as 3×3 matricies (top and bottom, respectively):

$$\begin{bmatrix} x_1 & x_2 & x_3 \\ x_1^3 & x_2^3 & x_3^3 \\ x_1^5 & x_2^5 & x_3^5 \end{bmatrix}\begin{bmatrix} w_1 \\ w_2 \\ w_3 \end{bmatrix} = \begin{bmatrix} 0 \\ 0 \\ 0 \end{bmatrix}$$

$$\begin{bmatrix} 1 & 1 & 1 \\ x_1^2 & x_2^2 & x_3^2 \\ x_1^4 & x_2^4 & x_3^4 \end{bmatrix}\begin{bmatrix} w_1 \\ w_2 \\ w_3 \end{bmatrix} = \begin{bmatrix} 2 \\ 2/3 \\ 2/5 \end{bmatrix}$$

Setting the determinant of the odd (top) matrix to 0, it is found $x_1 = 0$ and $x_2 = -x_3$. The topmost row of the odd (top) matrix then entails that $w_2 = w_3$. Inserting these values into the bottom (even) matrix results in three equations: $w_1 + 2w_2 = 2$, $2x_2^2 w_2 = 2/3$, and $2x_2^4 w_2 = 2/5$. The ratio of the second and third entail that $x_2 = \sqrt{3/5}$, which, when inserted into the second equation, sets $w_2 = 5/9$ and lastly $w_1 = 8/9$. Observe, therefore, that for $n = 3$ there are three different roots, $x_1 = 0$, $x_2 = -x_3 = \sqrt{3/5}$, and two different weights, $w_2 = w_3 = 5/9$ and $w_1 = 8/9$.

Higher-order equations are more involved, although induction holds. First, the (odd) matrix determines that the roots come in $\pm$ pairs for n even and the unpaired root for n odd is zero, and that the weights are likewise paired. Second, the (even) matrix then allows for the determination of the independent roots and weights. By $n = 4$, the procedure for their numerical determination has become non-trivial, and so consulting the authoritative Abramowitz and Stegun tables [103] for $n \geq 4$ provides roots and weights for higher-order Gaussian quadratures. An algorithm for their calculation is given in Press *et al.* [236].

That such induction is valid is reinforced by observing that the x_j are roots of the Legendre polynomials that form the basis of the Gaussian quadrature method (others exist depending on the limits of integration). The Legendre polynomial $P_n(x)$ is defined by the recurrence relationship [103]

$$(n+1)P_{n+1}(x) = (2n+1)xP_n(x) - nP_{n-1}(x) \tag{A1.29}$$

with $P_0(x) = 1$ and $P_1(x) = x$. Thus, $P_2(x) = (3x^2 - 1)/2$ and $P_3(x) = x(5x^2 - 3)/2$, from which the roots can readily be inferred as given in the examples above, which shows that the $\pm$ root characteristic is a consequence of P_{2n} being a polynomial in x^2 and P_{2n+1} the product of x with a polynomial in x^2. In the case of $n = 4$, $P_4(x) = (35x^4 - 30x^2 + 3)/8$ has the roots $x_1 = -x_2 = (2\sqrt{30} - 15)/35$ and $x_3 = -x_4 = (2\sqrt{30} + 15)/35$. The reduced 2×2 matrix entailed by Eq. (A1.28) is then trivially inverted to show $w_1 = w_2 = (18 + \sqrt{30})/36 = 0.65214515$ and $w_3 = w_4 = (18 - \sqrt{30})/36 = 0.34785485$.

As with the composite trapezoidal method, so too can a composite Gaussian quadrature method be used to evaluate subregions of a larger integral, and so achieve numerical results of high accuracy for smoothly varying $f(x)$.

A1.2.4 Sharply peaked integrands

The Dirac delta function $\delta(x)$ operates according to the relation

$$f(0) = \int_{-\infty}^{\infty} f(x)\delta(x)dx \tag{A1.30}$$

Certain functions, such as $\delta_l(x) = 1/(e^x + 1)(e^{-x} + 1)$, when sharply peaked by comparison to other functions in the integrand, act much like a Dirac delta function, for example Eq. (7.24) in the treatment of electrical conductivity and leading to Eq. (7.27). The method of evaluating the slowly varying function where the integrand is maximum so that it can be removed from the integral is a generally useful technique, an example being in the treatment of $N(n, s)$ function in Section (17.5). In short,

$$f(x_m) \approx \int_{-\infty}^{\infty} f(x)\delta_l(x)dx \tag{A1.31}$$

where x_m is the location of the maximum of the integrand, which is close to but nevertheless different than 0. The statement is less profound than it may appear, as it is simply the consequence of Taylor expanding $f(x)$ about x_m and only keeping the first (constant) term, but at the (sometimes unbearable) cost that x_m must be found. Numerical evaluation can assess the quality of the approximation, but the test case example here is analytically tractable.

Consider the case $f(x) = e^{ax}$ for a small and $\delta_l(x) = 1/(e^x + 1)(e^{-x} + 1)$, so that the integral in question is

$$F_l(a) = \int_{-\infty}^{\infty} f(x)\delta_l(x)dx = \int_{-\infty}^{\infty} \frac{e^{ax}}{(e^x + 1)(e^{-x} + 1)}dx \tag{A1.32}$$

where the assertion to be shown is then that $F_l(a) \approx e^{ax_m} = f(x_m)$, where $x_m(a)$ is the location of the maximum of the integrand $f(x)\delta_l(x)$. Using the following integral relations, derived from the observation that $\delta_l(x) = -\partial_x(e^x + 1)^{-1}$, integration by parts, and Reimann's zeta function (Eq. (A2.6))

$$\int_{-\infty}^{\infty} \delta_l(x) = 1 \tag{A1.33}$$

$$\int_{-\infty}^{\infty} x^2\delta_l(x)dx = \frac{\pi^2}{3} = 2\zeta(2) \tag{A1.34}$$

$$\int_{-\infty}^{\infty} x^4\delta_l(x)dx = \frac{7\pi^4}{15} = 42\zeta(4) \tag{A1.35}$$

(integrals with odd powers of x are identically zero because $\delta_l(x)$ is an even function) along with a Taylor expansion $e^{ax} \approx 1 + ax + (ax)^2/2$ show that to leading order for small a,

$$F_l(a) \approx 1 + \frac{1}{6}(\pi a)^2 + \frac{7}{360}(\pi a)^4 \tag{A1.36}$$

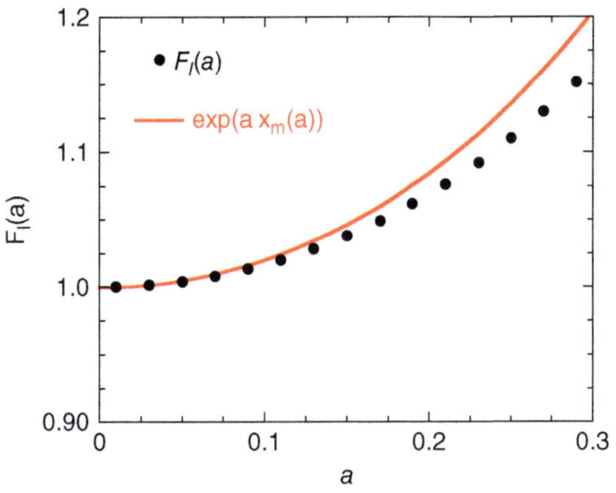

Figure A1.2 Demonstration that the integration of a slowly varying function $f(x) = e^{ax}$ for a small with the comparatively sharp function $\delta_l(x)$ is approximately $f(x_m)$, where x_m is the location of the maximum of $f(x)\delta_l(x)$.

Compare this to the approximation of Eq. (A1.31). The maximum of the integrand $f(x)\delta_l(x)$ is found by determining where its derivative vanishes. This occurs when

$$2a + (1 + a)e^{-x_m} - (1 - a)e^{x_m} = 0 \tag{A1.37}$$

which has solutions

$$x_m(a) = \ln\left(\frac{1 + a}{1 - a}\right) \tag{A1.38}$$

an analytical form for x_m being the motive for the choice of $f(x)$. As Figure A1.2 demonstrates, $F_l(a) \approx f(x_m(a))$ is a good approximation for small a.

A1.2.5 Monte Carlo integration

For some integrals of sufficient complexity, or which possess features that make their solution using the usual numerical methods foreboding or problematic, the use of randomly generated numbers offers a solution. The ability of random numbers or observations to make an estimation to quantities of interest has long been appreciated, perhaps one of the more intriguing examples being the Buffon needle problem [106], a captivating example of an "experimental" Monte Carlo calulation in which a needle of length L thrown onto a square grid: the needle will cross the grid lines, and the expectation value of the number of times n the needle crosses them is $\langle n \rangle = 4L/\pi$. Thus, the value of π can be empirically determined.

First encounters with Monte Carlo, however, are often towards the evaluation of integrals. For low-dimensional integrals, numerical methods surpass Monte Carlo for accuracy, but for multidimensional integrals, or ones with particularly intensive integrands, Monte Carlo methods can be preferable [63, 105, 106]. The illustration here, however, will be for one-dimensional integrals easily compared to analytical results.

First review expectation values and random numbers in a treatment quite similar to but simpler than Section (4.1). The expectation value of a randomly occurring quantity X_j that is observed N times is

$$\langle X \rangle = \frac{1}{N}\sum_{j=1}^{N} X_j \tag{A1.39}$$

If the values of X_j can take on only one of n values Y_k (n being the number of "bins" and k taking values of 1 to n, for example), and the number of times that Y_k occurs in N events is Np_k, with p_k being the fraction of the times, the expectation can be re-expressed as

$$\langle X \rangle = \sum_{k=1}^{n} Y_j p_j \tag{A1.40}$$

For example, using dice throws, if the random event is the value of the die face, then $n = 6$. In the limit of large N, the fraction of throws showing a face k is $p_k = 1/6$ for all k and $Y_k = k$, so that $\langle X \rangle$ converges to $7/2$.[2] Or, say, n could equal 3, with p_1 being the fraction of times the face is 1, p_2 being the fraction of times the face is a 2 or a 3, and p_3 the fraction of times it is a 4, 5, or 6. Thus, the expectation value can be evaluated using a variety of distribution functions of the random events and they should give the same result.

Use these ideas to examine integrals. Let the integral of a function $0 \le f(x) \le 1$ defined over the range $0 \le x \le 1$ be desired: there is no loss of generality for doing so, as x and $f(x)$ can be so normalized, and the explanation becomes easier. That is

$$\mathcal{F} \equiv \int_0^1 f(x)dx \tag{A1.41}$$

Let there also be a normalized probability distribution function $p(x)$ (e.g., uniform, linear, and so on) such that $p(x) \ge 0$ for all $0 \le x \le 1$. Some desirable distributions are defined over all x, but by remapping the integration variable, such as $y(x) = x/(1 - x^2)^{1/2}$, then the range of y is from $-\infty$ to ∞ and, for example, the Gaussian distribution is $p_g[y(x)] = \exp(-y^2)/\sqrt{\pi}$. The expectation value of some quantity $\mathcal{O}$ is then

$$\langle \mathcal{O} \rangle = \int_0^1 \mathcal{O}(x)p(x)dx \approx \sum_{j=1}^{n} \mathcal{O}(x_k)p(x_k) = \frac{1}{N}\sum_{j=1}^{N} \mathcal{O}_j \tag{A1.42}$$

If the quantity in question is $\mathcal{O}(x) = f(x)/p(x)$, then the expectation value is $\mathcal{F}$, but this means that $f(x)/p(x)$ is to be taken as the random event with a probability distribution of $p(x)$. In other words,

$$\mathcal{F} \approx \frac{1}{N}\sum_{j=1}^{N}\left(\frac{f_j}{p_j}\right) \tag{A1.43}$$

In programming, one random number distribution stands out, and that is the uniformly distributed random number r with the associated *uniform distribution* $p_u(r) = 1$, so that

$$\mathcal{F}_u \approx \frac{1}{N}\sum_{j=1}^{N} f(r_j) \tag{A1.44}$$

where r_j is the jth generated random number. This relation is by far the easiest to implement, but it is not always the most accurate. Alternatives are then a question of calculating other useful distributions from the uniform distribution (in addition to Section 32.2.2, see refs. [63, 105]). Let the random number h be associated with the distribution $p(x)$. Then a relation exists

$$\int_0^h p(x)\, dx = \int_0^r 1\, dr = r \tag{A1.45}$$

where $p_u(r) = 1$ is used in the second integral. For choices of $p(x)$ which are simple and integrable, then $h(r_j) = h_j$ can be found from a generated r_j. Why would this ever be done? Simply, because *the more* $p(x)$ *looks like* $f(x)$, *the more accurate Eq.* (A1.43) *becomes*.

If $p(x)$ is not simply integrable, then it may nevertheless be obtained using the acceptance–rejection algorithm of Section (32.2.2), which can be contorted until, here, it takes the form of generating two random numbers r and r' and adding 1 to a running sum if $r < f(r')$. Such an approach is effectively evaluating an integral by bounding it in a unit area, taking a scattering of random locations in that area, and only counting up those points that lie below the $f(x)$ curve. The ratio of points below $f(x)$ to the total number of points is then the ratio of the area under the curve to the unit area, which is the integral.

For an integrable case, consider the following example, where $p(x) = (1 + \nu)x^\nu$, where ν is an arbitrary power. It is straightforward to show $\int_0^1 p(x)dx = 1$. Therefore, the inversion of Eq. (A1.45) gives

$$h(r) = r^{1/(\nu+1)} \tag{A1.46}$$

[2]This is not a particularly interesting example: it would be more interesting to use *two* dice and show the expectation value for the *sum* of the faces is closer to 6, but that is a longer example. The one die case is, however, an almost uniquely simple example, and that is the point.

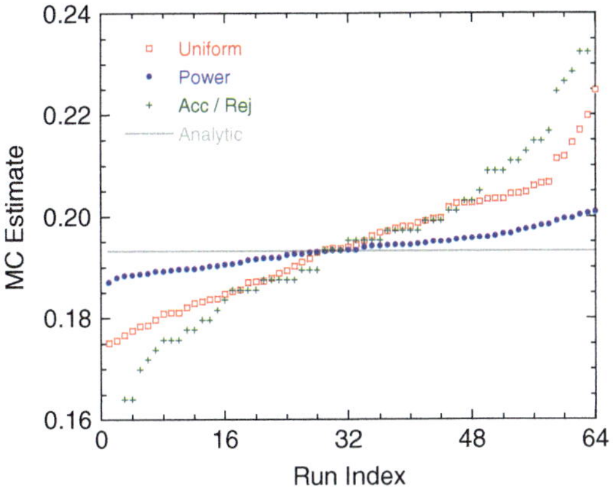

Figure A1.3 Comparison of the uniform distribution (Eq. (A1.44)) and the power law distribution (Eq. (A1.47) for $p(x) = 3x^2$) for the estimation of the integral of $f(x) = (e^{ax} - 1)/(e^a - 1)$ with $a = 5$ in the code of Section A3.36. Also shown is the acceptance–rejection approach discussed in text and a gray line for the analytical result of $\mathcal{F} = 0.193216$.

Then, defining $h(r_j) = h_j$ the approximation becomes

$$\mathcal{F} \approx \frac{1}{N} \sum_{j=1}^{N} \left(\frac{f(h_j)}{p(h_j)} \right) \tag{A1.47}$$

A comparison of the uniform, the power law, and the acceptance rejection methods is demonstrated for the function $f(x) = (e^{ax} - 1)/(e^a - 1)$, for which $\mathcal{F} = a^{-1} - (e^a - 1)^{-1}$. For large a, this is a rapidly rising function and so choosing $p(x) = 3x^2$ is a useful distribution. The results are shown in Figure A1.3 for $a = 5$, in which $N_j = 64$ separate evaluations of $\mathcal{F}$ are made then sorted from lowest to highest and compared to the analytical result. The spread of the N_j estimates about the analytic value gives a feel for the accuracy of the method, and because $p(x)$ mimics the shape of $f(x)$, it is seen to be the most accurate in general. The algorithm is implemented in Section A3.36.

A1.3 Differentiation

A1.3.1 Orthogonal coordinates

A method to find the gradient operator $\vec{\nabla}$ and the Laplacian ∇^2 in other orthogonal coordinate systems, particularly cylindrical and spherical, but also prolate spheroidal coordinate systems [60], starting with the Cartesian system, is given.

A1.3.1.1 Radial coordinates (r, θ, φ)

The relation between cartesian (x, y, z) and spherical (r, θ, φ) coordinates is

$$x = r \sin\theta \cos\varphi \tag{A1.48}$$

$$y = r \sin\theta \sin\varphi \tag{A1.49}$$

$$z = r \cos\theta \tag{A1.50}$$

The differential changes associated with them can be rendered in matrix notation to relate $\vec{dr}$ in cartesian coordinates ($\vec{dr}_c$) to the same quantity in spherical coordinates ($\vec{dr}_s$) in the form $\vec{dr}_c = \hat{R}(r, \theta, \varphi) \cdot \vec{dr}_s$. Explicitly,

$$\begin{pmatrix} dx \\ dy \\ dz \end{pmatrix} = \begin{pmatrix} \sin\theta\cos\varphi & r\cos\theta\cos\varphi & -r\sin\theta\sin\varphi \\ \sin\theta\sin\varphi & r\cos\theta\sin\varphi & r\sin\theta\cos\varphi \\ \cos\theta & -r\sin\theta & 0 \end{pmatrix} \begin{pmatrix} dr \\ d\theta \\ d\varphi \end{pmatrix} \tag{A1.51}$$

Finding the spherical differential elements $\vec{dr}_s$ in terms of the cartesian $\vec{dr}_c$ then involves inverting $\hat{R}(r, \theta, \varphi)$, the inversion of which requires finding its determinant. Although ominous, the inversion can be less painfully accomplished using a cofactor expansion method [91] to show $|\hat{R}(r,\theta,\varphi)| = r\sin\theta$. Thus,

$$\begin{pmatrix} dr \\ d\theta \\ d\varphi \end{pmatrix} = \begin{pmatrix} \sin\theta\cos\varphi & \sin\theta\sin\varphi & \cos\theta \\ r^{-1}\cos\theta\cos\varphi & r^{-1}\cos\theta\sin\varphi & -r^{-1}\sin\theta \\ -r^{-1}\csc\theta\sin\varphi & r^{-1}\csc\theta\cos\varphi & 0 \end{pmatrix} \begin{pmatrix} dx \\ dy \\ dz \end{pmatrix} \tag{A1.52}$$

Finding partials when shifting from one coordinate system to another can be read off the matrix elements. For example,

$$\begin{aligned} \partial_x &= (\partial_x r)\partial_r + (\partial_x\theta)\partial_\theta + (\partial_x\varphi)\partial_\varphi \\ &= \sin\theta\cos\varphi\ \partial_r + \frac{1}{r}\cos\theta\cos\varphi\ \partial_\theta - \frac{1}{r}\frac{\sin\varphi}{\sin\theta}\ \partial_\varphi \end{aligned} \tag{A1.53}$$

where the coefficients of each ∂ term are read off the corresponding row (e.g., the coefficient $(\partial_x r)$ of ∂_r is read off the dr row). The same procedure yields

$$\partial_y = \sin\theta\sin\varphi\partial_r + \frac{1}{r}\cos\theta\sin\varphi\partial_\theta + \frac{1}{r}\frac{\cos\varphi}{\sin\theta}\partial_\varphi \tag{A1.54}$$

$$\partial_z = \cos\theta\partial_r - \frac{1}{r}\sin\theta\partial_\theta \tag{A1.55}$$

In cartesian coordinates, the gradient and Laplacian operators are expressed as

$$\vec{\nabla} = \hat{x}\partial_x + \hat{y}\partial_y + \hat{z}\partial_z \tag{A1.56}$$

$$\nabla^2 = \partial_x^2 + \partial_y^2 + \partial_z^2 \tag{A1.57}$$

Finding their forms in spherical coordinates is aided with the introduction of the h_α, h_β, and h_γ terms, where a general orthogonal coordinate system is specified by (α,β,γ). Define

$$h_\alpha \equiv \sqrt{(\partial_\alpha x)^2 + (\partial_\alpha y)^2 + (\partial_\alpha z)^2} \tag{A1.58}$$

with analogous equations for h_β and h_γ. A compact formula for the gradient is

$$\vec{\nabla} = \frac{1}{h_\alpha}\frac{\partial}{\partial\alpha} + \frac{1}{h_\beta}\frac{\partial}{\partial\beta} + \frac{1}{h_\gamma}\frac{\partial}{\partial\gamma} \tag{A1.59}$$

The Laplacian is a bit more involved, being

$$\nabla^2 = \frac{1}{h_\alpha h_\beta h_\gamma}\left\{\frac{\partial}{\partial\alpha}\left(\frac{h_\beta h_\gamma}{h_\alpha}\frac{\partial}{\partial\alpha}\right) + \frac{\partial}{\partial\beta}\left(\frac{h_\alpha h_\gamma}{h_\beta}\frac{\partial}{\partial\beta}\right) + \frac{\partial}{\partial\gamma}\left(\frac{h_\alpha h_\beta}{h_\gamma}\frac{\partial}{\partial\gamma}\right)\right\} \tag{A1.60}$$

but it is seen that the $(h_\alpha h_\beta/h_\gamma)$ fractions cycle through the subscripts, and so the algorithm is easy to remember.

Spherical coordinates are specified by $\alpha = r$, $\beta = \theta$, and $\gamma = \varphi$. It is straightforward to show, using Eqs (A1.48)–(A1.50), that

$$h_r = 1,\ h_\theta = \frac{1}{r},\ h_\varphi = \frac{1}{r\sin\theta} \tag{A1.61}$$

Therefore,

$$\vec{\nabla} = \hat{r}\frac{\partial}{\partial r} + \hat{\theta}\frac{1}{r}\frac{\partial}{\partial\theta} + \hat{\varphi}\frac{1}{r\sin\theta}\frac{\partial}{\partial\varphi} \tag{A1.62}$$

and, with a bit more work, the Laplacian is

$$\nabla^2 = \frac{1}{r^2}\frac{\partial}{\partial r}\left(r^2\frac{\partial}{\partial r}\right) + \frac{1}{r^2\sin\theta}\frac{\partial}{\partial\theta}\left(\sin\theta\frac{\partial}{\partial\theta}\right) + \frac{1}{r^2\sin^2\theta}\frac{\partial^2}{\partial\varphi^2} \tag{A1.63}$$

A1.3.1.2 Prolate spheroidal coordinates (η, υ, γ)

The relation between cartesian (x, y, z) and the (angular representation) prolate spheroidal (η, υ, γ) coordinates is

$$x = L\sinh(\eta)\sin(\upsilon)\cos(\gamma) \tag{A1.64}$$

$$y = L\sinh(\eta)\sin(\upsilon)\sin(\gamma) \tag{A1.65}$$

$$z = L\cosh(\eta)\cos(\upsilon) \tag{A1.66}$$

where L sets the length scale. The h_η factors are

$$h_\eta = L\,(\cosh(\eta)^2 - \cos(\upsilon)^2) = L\,(\sinh(\eta)^2 + \sin(\upsilon)^2) \tag{A1.67}$$

$$h_\upsilon = L\,(\cosh(\eta)^2 - \cos(\upsilon)^2) = L\,(\sinh(\eta)^2 + \sin(\upsilon)^2) \tag{A1.68}$$

$$h_\gamma = L\ \sinh(\eta)\sin(\upsilon) \tag{A1.69}$$

Observe that $h_\eta = h_\upsilon$. For rotationally symmetric geometries such as the conical/Spindt-like and ellipsoidal/wire-like emitters of Figure 30.15, there is no γ dependence, for which $\rho = (x^2 + y^2)^{1/2}$ and z as the cylindrical coordinates, and (η, υ) as the prolate spheroidal coordinates. The components of the Laplacian of Eq. (A1.60) are then constructed from

$$h_\eta h_\upsilon h_\gamma = L^3(\sinh(\eta)^2 + \sin(\upsilon)^2)\sinh(\eta)^2\sin(\upsilon) \tag{A1.70}$$

$$\partial_\eta\left(\frac{h_\upsilon h_\gamma}{h_\eta}\partial_\eta\right) = L\sin(\upsilon)(\sinh(\eta)\partial_\eta^2 + \cosh(\eta)\partial_\eta) \tag{A1.71}$$

$$\partial_\upsilon\left(\frac{h_\gamma h_\eta}{h_\upsilon}\partial_\upsilon\right) = L\sinh(\eta)(\sin(\upsilon)\partial_\upsilon^2 + \cos(\upsilon)\partial_\upsilon) \tag{A1.72}$$

$$\partial_\gamma\left(\frac{h_\eta h_\upsilon}{h_\gamma}\partial_\gamma\right) = L\left(\frac{\sinh(\eta)^2 + \sin(\upsilon)^2}{\sinh(\eta)\sin(\upsilon)}\right)\partial_\gamma^2 \tag{A1.73}$$

Similarly, the gradient components are

$$\vec{\nabla} = \hat{\eta}\frac{1}{h_\eta}\frac{\partial}{\partial_\eta} + \hat{\upsilon}\frac{1}{h_\upsilon}\frac{\partial}{\partial_\upsilon} + \hat{\gamma}\frac{1}{h_\gamma}\frac{\partial}{\partial_\gamma} \tag{A1.74}$$

for which the simple substitutions of Eq. (A1.67) and its brethren are straightforward.

A1.3.2 Finite differences

The approximation of derivatives of a function $y = f(x)$ given that $f(x)$ is known at a finite number of discrete locations x_j, or $y_j = f(x_j)$ is often desired. Finite difference methods can then be used to find solutions to differential equations [242], and are often closely related to methods used in approximation and interpolation [103, 243]. A brief and slightly idiosyncratic account is given. The Taylor expansion of a function is well known to be given by (where primes represent derivatives with respect to argument)

$$\begin{aligned} f(x) &= f(0) + xf'(0) + \frac{1}{2!}x^2 f''(0) + \cdots \\ &\equiv \sum_{n=0}^{\infty}\frac{x^n}{n!}(\partial_x^n f(x)|_{x=0}) \end{aligned} \tag{A1.75}$$

Further, let $y = f(x)$ be specified on evenly spaced locations $x_j = j\Delta x$, where j can take on positive and negative values[3] so that $y_j \equiv f(x_j)$. For notational ease, let

$$\Delta x^n(\partial_x^n f(x)|_{x=0}) \equiv f^n \tag{A1.76}$$

so that $f^0 = f(x = 0)$, and so on. Nothing is inherently special about $x = 0$: the expansion point could be for any x_o, but one can always redefine the coordinate system so that the expansion point is at the origin. Then the Taylor expansion

[3] The points x_j do not have to be evenly spaced, but the resulting formalism is much easier if they are; the extension to non-uniformly spaced points is an exercise left to those with the time to consider it.

formula of Eq. (A1.75) takes the form of

$$y_j = \sum_{n=0}^{\infty} \frac{j^n}{n!} f^n \tag{A1.77}$$

This way of expressing a Taylor expansion makes it suitable for matrix methods. Let $|y\rangle$ and $|f\rangle$ be vectors whose kth entries are y_k and f^k, respectively, and let $\mathbb{M}$ be a matrix whose components are specified by $M_{j,n} = j^n/n!$. Then the inversion

$$|y\rangle = \mathbb{M}|f\rangle \rightarrow |f\rangle = \mathbb{M}^{-1}|y\rangle \tag{A1.78}$$

provides what is sought, as the approximations to the derivatives specified by $|f\rangle$ are provided by the equi-spaced points specified by $|y\rangle$. Observe that because the length N of both vectors is the same, that length specifies the order $N-1$ of the approximation. A few special cases are instructive.

A1.3.2.1 First-order approximations

The first-order approximations involve two points and assume that the underlying polynomial that passes through the y_j is linear (i.e., is a polynomial of order 1). The y_j can be chosen anywhere, but accuracy generally increases if one of the points is at $j = 0$ and Δx is kept small. If the second point is at $j = 1$ the scheme is termed an *upwind differencing scheme* (UDS), and if $j = -1$ then it is a *downwind differencing scheme* (DDS). The matrix equation for downwind is

$$\begin{pmatrix} f^0 \\ f^1 \end{pmatrix} = \begin{pmatrix} 1 & -1 \\ 1 & 0 \end{pmatrix}^{-1} \begin{pmatrix} y_{-1} \\ y_0 \end{pmatrix} = \begin{pmatrix} y_0 \\ y_0 - y_{-1} \end{pmatrix} \tag{A1.79}$$

Unpacking the result, the top row asserts $f^0 = f(x=0) = y_0$, not surprisingly. The second row gives the approximation to the first derivative in the DDS to be (after invoking Eq. (A1.76))

$$\left(\frac{dy}{dx}\right)_{DDS} \approx \frac{1}{\Delta x}(y_0 - y_{-1}) \tag{A1.80}$$

In exactly the same way, the UDS

$$\begin{pmatrix} f^0 \\ f^1 \end{pmatrix} = \begin{pmatrix} 1 & 0 \\ 1 & 1 \end{pmatrix}^{-1} \begin{pmatrix} y_0 \\ y_1 \end{pmatrix} = \begin{pmatrix} y_0 \\ y_1 - y_0 \end{pmatrix} \tag{A1.81}$$

gives

$$\left(\frac{dy}{dx}\right)_{UDS} \approx \frac{1}{\Delta x}(y_1 - y_0) \tag{A1.82}$$

The accuracy of the approximation can be assessed albeit crudely by applying the algorithm to an $(N+1)$th order polynomial, or $f(x) = (x+1)^2$ for the present case, comparing the known result to the approximation

$$\frac{f^1}{\Delta x} = 2 \leftrightarrow \frac{y_1 - y_0}{\Delta x} = 2 + \Delta x \tag{A1.83}$$

so the approximation is said to be accurate to first order (the second term on the right-hand side is the error of the approximation).

A1.3.2.2 Second-order approximations

Adding another y_j increases the accuracy of the interpolating polynomial, as a quadratic function can uniquely pass through three points. In the *central difference scheme* (CDS), the y_j values after and before y_0 are used to give

$$\begin{pmatrix} f^0 \\ f^1 \\ f^2 \end{pmatrix} = \left\{ \frac{1}{2} \begin{pmatrix} 2 & -2 & 1 \\ 1 & 0 & 0 \\ 2 & 2 & 1 \end{pmatrix} \right\}^{-1} \begin{pmatrix} y_{-1} \\ y_0 \\ y_1 \end{pmatrix} = \frac{1}{2} \begin{pmatrix} 2y_0 \\ y_1 - y_{-1} \\ 2y_1 - 4y_0 + 2y_{-1} \end{pmatrix} \tag{A1.84}$$

so that the CDS approximations to the first and second derivative are

$$\left(\frac{dy}{dx}\right)_{CDS} \approx \frac{1}{2\Delta x}(y_1 - y_{-1}) \tag{A1.85}$$

$$\left(\frac{d^2y}{dx^2}\right)_{CDS} \approx \frac{y_1 - 2y_0 + y_{-1}}{\Delta x^2} \tag{A1.86}$$

Two other schemes exist that are higher-order versions of the UDS and DDS. Both can be found from expanding

$$\begin{pmatrix} f^0 \\ f^1 \\ f^2 \end{pmatrix} = \left\{ \frac{1}{2} \begin{pmatrix} 2 & 0 & 0 \\ 2 & \pm 2 & 1 \\ 2 & \pm 4 & 4 \end{pmatrix} \right\}^{-1} \begin{pmatrix} y_0 \\ y_{\pm 1} \\ y_{\pm 2} \end{pmatrix} \tag{A1.87}$$

giving, for the first derivative, the approximations

$$\left(\frac{dy}{dx}\right)_{SDS} \approx \mp \frac{1}{2\Delta x}(3y_0 - 4y_{\pm 1} + y_{\pm 2}) \tag{A1.88}$$

$$\left(\frac{d^2y}{dx^2}\right)_{SDS} \approx \frac{1}{\Delta x^2}(y_0 - 2y_{\pm 1} + y_{\pm 2}) \tag{A1.89}$$

Observe that the second derivative has the same coefficients as the CDS scheme, the only difference being that the indices are shifted up or down by one unit, which is understandable because the scheme is accurate for quadratic polynomials (for which the second derivative is a constant). Testing the algorithms on $f(x) = (x + 1)^3$ gives

$$\left(\frac{f^1}{\Delta x}\right)_{CDS} = 3 \leftrightarrow \frac{y_1 - y_{-1}}{2\Delta x} = 3 + \Delta x^2 \tag{A1.90}$$

$$\left(\frac{f^1}{\Delta x}\right)_{SDS} = 3 \leftrightarrow \frac{-3y_0 + 4y_1 - y_2}{2\Delta x} = 3 - 2\Delta x^2 \tag{A1.91}$$

$$\frac{f^2}{\Delta x^2} = 6 \leftrightarrow \frac{y_0 - 2y_1 + y_2}{\Delta x^2} = 6 + 6\Delta x \tag{A1.92}$$

and similarly for the downwind version. Therefore, the first derivative is accurate to order Δx^2 and the second derivative to order Δx. Observe also that the error associated with the CDS scheme for the first derivative, where the expansion point is in the middle, is smaller than for the up and downwind SDS schemes, where the expansion point is on the edge. This is a general trend that the interior expansion points are more accurate.

It is left as an exercise to show that the fourth-order scheme involves solving

$$\begin{pmatrix} f^0 \\ f^1 \\ f^2 \\ f^3 \\ f^4 \end{pmatrix} = \left\{ \frac{1}{24} \begin{pmatrix} 24 & -48 & 48 & -32 & 16 \\ 24 & -24 & 12 & -4 & 1 \\ 24 & 0 & 0 & 0 & 0 \\ 24 & 24 & 12 & 4 & 1 \\ 24 & 48 & 48 & 32 & 16 \end{pmatrix} \right\}^{-1} \begin{pmatrix} y_{-2} \\ y_{-1} \\ y_0 \\ y_1 \\ y_2 \end{pmatrix} \tag{A1.93}$$

The expansion is a bit more involved to calculate, but doing so gives

$$\begin{pmatrix} f^0 \\ f^1 \\ f^2 \\ f^3 \\ f^4 \end{pmatrix} = \frac{1}{12} \begin{pmatrix} 12y_0 \\ -y_2 + 8y_1 - 8y_{-1} + y_{-2} \\ -y_2 + 16y_1 - 30y_0 + 16y_{-1} - y_{-2} \\ 6y_2 - 12y_1 + 12y_{-1} - 6y_{-2} \\ 12y_2 - 48y_1 + 72y_0 - 48y_{-1} + 12y_{-2} \end{pmatrix} \tag{A1.94}$$

from which the finite differencing operators can be read off. The fourth-order CDS was shown, but the method is trivially extended to UDS and DDS by simply changing the range of j and inverting the resulting matrix.

For example, a test of the algorithm on the polynomial $f(x) = (x+1)^5$ gives, using a compact representation of the derivatives (left-hand side) with their finite difference approximations (right-hand side),

$$\begin{pmatrix} 1 \\ 5 \\ 20 \\ 60 \\ 120 \end{pmatrix} \approx \begin{pmatrix} 1 \\ 5 - 4\Delta x^4 \\ 20 \\ 60 + 30\Delta x^2 \\ 120 \end{pmatrix} \tag{A1.95}$$

A1.4 Numerical solution of an ordinary differential equation

Numerical methods to solve ordinary differential equations (ODEs), such as Runge–Kutta and others [236], are well known and often part of numerical programs. Nevertheless, there is always room at the bottom for a quick and dirty method: human cleverness takes time and computer stupidity is remarkably fast, so that when scheming is too time-consuming simple recourse to a finer discretization to increase accuracy can be effective. Poisson's equation in particular can be cast in a form that requires no memory overhead and is therefore compact.

Consider, therefore, the numerical solution of the differential equation defined by

$$\frac{\partial^n}{\partial x^n} u(x) = v(x) \tag{A1.96}$$

where the cases $n = 1$ are called first order, $n = 2$ called second order (the only two to be considered here), $u(x)$ is to be found and $v(x)$ is known (or at least known on the points $x_j = j\Delta x : j \in \{1, N\}$). If x is scaled so that the boundaries are at 0 and 1, that is, $u(x = 0)$ and $u(x = 1)$ are known, then $\Delta x = 1/(N+1)$. In that case, $u_0 = u(x_0 = 0)$ and $u_{N+1} = u(x_{N+1} = 1)$ are the boundary conditions.

A1.4.1 First order

Using the finite difference scheme of Eq. (A1.85), and doing so for $N = 5$ to keep the results simple but easily generalizable, results in a matrix equation of the form $\mathbb{M}|u\rangle = 2\Delta x|v\rangle + |u_{bc}\rangle$, or explicitly as

$$\begin{bmatrix} 0 & 1 & 0 & 0 & 0 \\ -1 & 0 & 1 & 0 & 0 \\ 0 & -1 & 0 & 1 & 0 \\ 0 & 0 & -1 & 0 & 1 \\ 0 & 0 & 1 & -4 & 3 \end{bmatrix} \begin{pmatrix} u_1 \\ u_2 \\ u_3 \\ u_4 \\ u_5 \end{pmatrix} = 2\Delta x \begin{pmatrix} v_1 \\ v_2 \\ v_3 \\ v_4 \\ v_5 \end{pmatrix} + \begin{pmatrix} u_0 \\ 0 \\ 0 \\ 0 \\ 0 \end{pmatrix} \tag{A1.97}$$

Note a few peculiar features about the form of this equation. First, the discretization element Δx has been brought to the right-hand side. Second, and this is important, u_{N+1} is not a part of the boundary condition vector (the second vector on the right-hand side of Eq. (A1.97)): to make it so would be to over-constrain the problem and that in turn would cause the solution to fail in that the $N \times N$ differencing matrix would not have an inverse. As a consequence, the Nth row of the differencing matrix uses a DDS. This is only one choice: another is to let the first row contain a UDS and let u_{N+1} specify the boundary condition.

For smooth functions, such a cheap solution is surprisingly accurate. By way of demonstration, consider two examples. For the first, let $v(x) = e^{-x}$ with boundary conditions $u(0) = 0$, so that the analytical solution is $u(x) = 1 - e^{-x}$. Then even for $N = 5$ Eq. (A1.97) becomes

$$\begin{pmatrix} u_1 \\ u_2 \\ u_3 \\ u_4 \\ u_5 \end{pmatrix} = \frac{1}{12} \begin{bmatrix} 4 & -4 & 4 & -3 & 1 \\ 4 & 0 & 0 & 0 & 0 \\ 4 & 0 & 4 & -3 & 1 \\ 4 & 0 & 4 & 0 & 0 \\ 4 & 0 & 4 & 1 & 1 \end{bmatrix} \begin{pmatrix} e^{-1/6} \\ e^{-2/6} \\ e^{-3/6} \\ e^{-4/6} \\ e^{-5/6} \end{pmatrix} = \begin{pmatrix} 0.1534 \\ 0.2822 \\ 0.3922 \\ 0.4843 \\ 0.5633 \end{pmatrix} \tag{A1.98}$$

which is good to 1%. Not bad. Observe further that the structure of the inverse differencing matrix has a peculiar repeating pattern that suggests a fast solution, although the presence of the DDS corrupts it a bit, thus suggesting the second example.

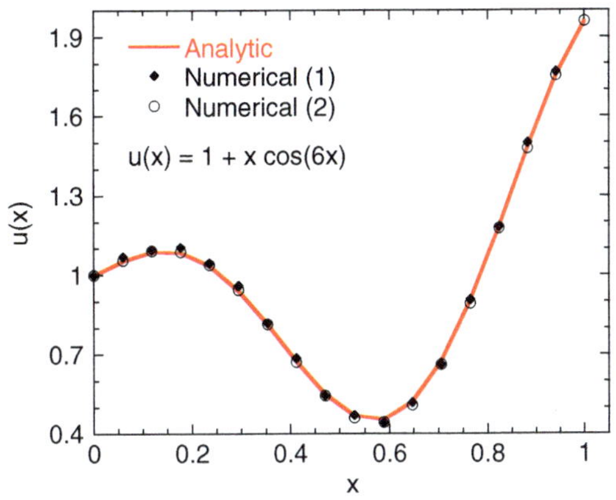

Figure A1.4 A comparison of the analytic form of $u(x) = 1 + x\cos(6x)$ to its determination from numerically solving $\partial_x u(x) = v(x)$ using the code of Section A3.15.1 for first order (1) and Section A3.15.2 for second order (2).

The solution to a generalization of Eq. (A1.98) may be found by using Gauss–Jordan elimination on the matrix $\mathbb{M}$ and simultaneously performing the same operations on the right-hand side vector [91], although it will be easier to deal with N even and so let $N = 2n$. The simplicity of $\mathbb{M}$ results in a clean algorithm that is not entirely unexpected. Using the notation u_j = the jth element of $|u\rangle$, the solution is obtained by first performing the increasing j replacements for j from 2 to n (remember, $N = 2n$)

$$u_2 \leftarrow 2\Delta x v_1 + u_0 \tag{A1.99}$$

$$u_{2j} \leftarrow u_{2j-2} + 2\Delta x v_{2j-1} \tag{A1.100}$$

followed by the decreasing j replacements (again for $j = 1$ to n)

$$u_{2n-1} \leftarrow u_{2n-2} + (3/4)\Delta x v_{2n-1} - (1/4)\Delta x v_{2n} \tag{A1.101}$$

$$u_{2n+1-2j} \leftarrow u_{2n+3-2j} - 2\Delta x v_{2n+2-2j} \tag{A1.102}$$

The algorithm, rendered to code in Section A3.15.1 and applied to the trial function $u(x) = x\cos(6x) + 1$, is shown in Figure A1.4 under the name "Numerical (1)".

Observe that the algorithm of Eq. (A1.99) has elements and features very reminiscent of the CDS of Eq. (A1.85) and a whiff of the trapezoidal integration scheme of Section A1.2.1, but the intriguing partitioning of a downwind versus an upwind recurrent pattern demonstrates the differences brought to bear by using a matrix method to solve an ODE; observe further that it is the downwind differencing operator that connects the down and up portions of the Eq. (A1.99) algorithm.

The application of finite difference methods to solving ODEs has more art to it than the broad brushstrokes here: there is much hidden here that is part of more formal treatments [242], as well as similarities to interpolation and approximation methods [103, 243].

A1.4.2 Second order

The algorithm for solving Eq. (A1.96) for the case $n = 2$ results in a matrix equation of the form $\mathbb{M}|u\rangle = \Delta x^2|v\rangle + |u_{bc}\rangle$ or, for the case $N = 5$, explicitly as

$$\begin{bmatrix} -2 & 1 & 0 & 0 & 0 \\ 1 & -2 & 1 & 0 & 0 \\ 0 & 1 & -2 & 1 & 0 \\ 0 & 0 & 1 & -2 & 1 \\ 0 & 0 & 0 & 1 & -2 \end{bmatrix} \begin{pmatrix} u_1 \\ u_2 \\ u_3 \\ u_4 \\ u_5 \end{pmatrix} = \Delta x^2 \begin{pmatrix} v_1 \\ v_2 \\ v_3 \\ v_4 \\ v_5 \end{pmatrix} - \begin{pmatrix} u_0 \\ 0 \\ 0 \\ 0 \\ u_6 \end{pmatrix} \tag{A1.103}$$

where the ability to use both boundary conditions results in much simpler matrix elements in row $N = 5$. As with the first-order case, the simplicity of $\mathbb{M}$ results in an even more elegant algorithm. Using the notation u_j = the jth element of

$|u\rangle$, then the solution is obtained by first performing (for $j = 2$ to N)

$$u_j \leftarrow (2u_j + u_{j-1})\frac{j}{j+1} \tag{A1.104}$$

followed by (for $j = N - 1$ to 1)

$$u_j \leftarrow u_j + u_{j+1}\frac{j}{j+1} \tag{A1.105}$$

followed by $|u\rangle \rightarrow -(1/2)|u\rangle$. The algorithm, detailed in Section A3.15.2 and applied to the trial function $u(x) = x\cos(6x) + 1$, is shown in Figure A1.4 under the name "Numerical (2)". Observe that very slight differences are discernible between the two numerical methods, as is to be expected, but also note how small N may be to obtain reasonable results compared to the analytical values.

APPENDIX 2
Functions

Very important functions can be performed very wastefully and often are ...It is far easier to cut the function than the waste.
– John Kenneth Galbraith[1]

A2.1 Trigonometric functions

Relations between the sine and cosine functions when their arguments are the sum or difference of two angles are repeatedly employed:

$$\sin(\theta \pm \phi) = \sin\theta\cos\phi \pm \cos\theta\sin\phi$$
$$\cos(\theta \pm \phi) = \cos\theta\cos\phi \pm \sin\theta\sin\phi \tag{A2.1}$$

When $\theta = \phi$ then from $\cos^2\theta + \sin^2\theta = 1$, it follows that

$$\sin(2\theta) = 2\cos\theta\sin\theta$$
$$\cos(2\theta) = 1 - 2\sin^2\theta \tag{A2.2}$$

A2.2 Gamma function

The gamma function can be defined by the relation $\Gamma(p+1) = p\Gamma(p)$ or by Euler's integral formula

$$\Gamma(p) = \int_0^\infty x^{p-1}e^{-x}dx \tag{A2.3}$$

For $p = n =$ an integer, it is related to the factorial function by $\Gamma(n+1) = n!$. For large values of p, Stirling's approximation is useful and is the leading part of the relation

$$\Gamma(p) \approx p^{p-1}e^{-p}\sqrt{2\pi p}\left(1 + \frac{1}{12p}\right) \tag{A2.4}$$

The function has poles at negative integers, but for negative non-integer values, the relation

$$\Gamma(p)\Gamma(1-p) = \frac{\pi}{\sin(\pi p)} \tag{A2.5}$$

is useful, and shows $\Gamma(1/2) = \sqrt{\pi}$.

A2.3 Riemann zeta function

There are many defining equations (both series and integral) that provide a means to numerically evaluate Riemann's zeta function. This section shall instead focus on representations useful in calculations involving particles, and they are

[1]John Kenneth Galbraith, *The Affluent Society*. Boston: Houghton Mifflin, 1998, p. 178.

Introduction to the Physics of Electron Emission, First Edition. Kevin L. Jensen.

given by

$$\zeta(p) = \frac{1}{\Gamma(p)} \int_0^\infty \frac{x^{p-1}}{e^x - 1} dx \tag{A2.6}$$

$$= \frac{1}{(1 - 2^{1-p})\Gamma(p)} \int_0^\infty \frac{x^{p-1}}{e^x + 1} dx \tag{A2.7}$$

Using a series expansion of the denominator of the form $1/(e^x - 1) = \sum_{j=1}^{\infty} e^{-jx}$, each term in the series can be integrated to give the more familiar series representation of $\zeta(p)$ as

$$\zeta(p) = \sum_{j=1}^{\infty} j^{-p} \tag{A2.8}$$

For integer n, $\zeta(2n)$ is easily evaluated by noting the relationship between it and the Fourier transform of $f(x) = x^2$ over the interval $[-\pi, \pi]$, for which the Fourier cosine coefficients are $a_0 = 2\pi^2/3$ and $a_n = 4(-1)^n n^{-2}$. For $x = \pi$, $\cos(nx) = (-1)^n$, then $f(x) = a_0/2 + \sum_1^\infty a_n \cos(nx)$ from which

$$\pi^2 = \frac{\pi^2}{3} + 4\sum_{n=1}^{\infty} n^{-2} = \frac{\pi^2}{3} + 4\zeta(2) \tag{A2.9}$$

and so $\zeta(2) = \pi^2/6$. Similar reasoning using $f(x) = x^{2n}$ shows $\zeta(4) = \pi^4/90$, $\zeta(6) = \pi^6/945$, and $\zeta(8) = \pi^8/9450$.

The value of $\zeta(3) = 1.2020569$ is known as Apéry's constant.

A method for the numerical evaluation of $\zeta(p)$ in MATLAB makes use of the `integral` call. An example is shown for the evaluation of $\zeta(3/2)$ and $\zeta(3)$, with the output compared to the exact values.

```
%-----------------------------------------
% Riemann's Zeta function zeta(p)
%-----------------------------------------
% zeta(3) and zeta(1.5) are calculated and
% compared to exact values, but the method is valid
% for p arbitrary. gamma(p) is the Gamma function
% for which gamma(n+1) = n! and gamma(1/2)=sqrt(pi)
% zeta30ext and zeta15ext are exact values
%-----------------------------------------

fx = @(x,p) (x.^(p-1))./(exp(x) + 1.0);

zeta30ext = 1.202056903159594285399738161511;
zeta15ext = 2.612375348685488343348567567924;

p = 3.0;
zeta30 = integral(@(x)fx(x,p),0.0,Inf)/...
         ((1.0 - 2.0^(1-p))*gamma(p));
zerror30 = 100*(zeta30ext - zeta30)/zeta30ext;

p = 1.5;
zeta15 = integral(@(x)fx(x,p),0.0,Inf)/...
         ((1.0 - 2.0^(1-p))*gamma(p));
zerror15 = 100*(zeta15ext - zeta15)/zeta15ext;
```

The output is -3.6944×10^{-14} and -1.6999×10^{-14} for *zerror*30 and *zerror*15, respectively.

A2.4 Error function

The error function $Erf(x)$ is the integral over a Gaussian, or

$$Erf(x) \equiv \frac{2}{\sqrt{\pi}} \int_0^x e^{-s^2} ds \tag{A2.10}$$

For large argument,

$$Erf(x \gg 1) \approx 1 - \frac{2}{\sqrt{\pi}} e^{-x^2} \left(\frac{1}{2x} - \frac{1}{4x^3} \right) \tag{A2.11}$$

For small argument, a series expansion is

$$Erf(x) = e^{-x^2} \sum_{j=0}^{\infty} \frac{x^{2j+1}}{\Gamma\left(j + \frac{3}{2}\right)} \tag{A2.12}$$

A2.5 Legendre polynomials

Consider the Laplacian in spherical coordinates as given in Eq. (A1.60) for problems with rotational symmetry (no φ dependence). For the equation $\nabla^2 \psi(r,\theta) = 0$, then $\psi(r,\theta)$ can be represented at $\psi(r,\theta) = \sum_l R_l(r) P_l(\theta)$, as in the treatment of the hydrogen atom of Section 2.2.2. The eigenvalue equation for $P_l(\cos\theta)$ is

$$\frac{1}{\sin\theta} \frac{d}{d\theta} \left(\sin\theta \frac{dP_l}{d\theta} \right) + l(l+1) P_l(\cos\theta) = 0 \tag{A2.13}$$

Letting $x = \cos\theta$, then *Rodrigues' formula* for the generation of $P_l(x)$ is

$$P_l(x) = \frac{(-1)^l}{2^l l!} \frac{d^l}{dx^l} (1 - x^2)^l \tag{A2.14}$$

for which the first few are

$$P_0(x) = 1 \tag{A2.15}$$

$$P_1(x) = x \tag{A2.16}$$

$$P_2(x) = \frac{1}{2}(3x^2 - 1) \tag{A2.17}$$

$$P_3(x) = \frac{1}{2}(5x^3 - 3x) \tag{A2.18}$$

A recursion relation to get to higher $P_l(x)$ can be used which makes use of lower l terms, or

$$(l+1)P_{l+1} - (2l+1)xP_l + lP_{l-1} = 0 \tag{A2.19}$$

The normalization is

$$\int_{-1}^{1} P_n(x) P_l(x) dx = \frac{2}{2l+1} \delta_{n,l} \tag{A2.20}$$

Legendre's differential equation also admits non-polynomial solutions called Legendre polynomials of the second kind, and are designated by $Q_n(x)$. The first few of them are

$$Q_0(x) = \frac{1}{2} \ln\left(\frac{1+x}{1-x}\right) \tag{A2.21}$$

$$Q_1(x) = \frac{x}{2} \ln\left(\frac{1+x}{1-x}\right) - 1 \tag{A2.22}$$

$$Q_2(x) = \frac{3x^2 - 1}{4} \ln\left(\frac{1+x}{1-x}\right) - \frac{3x}{2} \tag{A2.23}$$

If $Q_n(x)$ is encountered with $x > 1$, as in Section 30.7.2, then the replacement $(1-x) \rightarrow (x-1)$ in the $\ln(\cdots)$ function is made.

A2.6 Airy functions

An extensive literature on Airy functions is available, the treatment here is after ref. [103]. Airy functions are solutions to the differential equation

$$\partial_x^2 \psi - x\psi = 0 \tag{A2.24}$$

Two real solutions Ai($\pm x$) and Bi($\pm x$) exist, where $x \geq 0$. For positive argument, Bi(x) and Ai(x) are exponentially increasing and decreasing, respectively. For negative argument Ai($-x$) and Bi($-x$) are oscillatory, as in Figure A2.1 and Table A2.1. The combinations of Airy functions highlight the additional asymptotic properties shown in Figure A2.2. From the combinations of Zi(c, x) functions (Eqs 18.22 and (18.24)) in the form Zi(c, x) Zi($-c, x$) it follows

$$\mathrm{Zi}(1, x)\,\mathrm{Zi}(-1, x) = \mathrm{Ai}(x)\mathrm{Bi}(x) = \frac{G_1(x)G_{-1}(x)}{4\pi\sqrt{x}} \tag{A2.25}$$

$$\mathrm{Zi}(i, x)\,\mathrm{Zi}(-i, x) = \mathrm{Ai}(-x)^2 + \mathrm{Bi}(-x)^2 = \frac{|G_i(x)|^2}{4\pi\sqrt{x}} \tag{A2.26}$$

where $x \geq 0$ and the G functions are of order unity. The Wronskian of the Airy functions is defined by

$$\mathrm{Ai}(x)\partial_x\,\mathrm{Bi}(x) - \mathrm{Bi}(x)\partial_x\,\mathrm{Ai}(x) = \frac{1}{\pi} \tag{A2.27}$$

with the same equation holding for $x \to -x$. At the origin $x = 0$ the Airy functions are (where a prime denotes derivative with respect to argument)

$$\mathrm{Ai}(0) = \frac{1}{3^{2/3}\Gamma(2/3)} = 0.355028 \tag{A2.28}$$

$$\mathrm{Ai}'(0) = -\frac{1}{3^{1/3}\Gamma(1/3)} = -0.258819 \tag{A2.29}$$

$$\mathrm{Bi}(0) = \frac{1}{3^{1/6}\Gamma(2/3)} = 0.614927 \tag{A2.30}$$

$$\mathrm{Bi}'(0) = \frac{3^{1/6}}{\Gamma(1/3)} = 0.448288 \tag{A2.31}$$

Series expansions are, for positive argument,

$$\mathrm{Ai}(x) = \frac{1}{2\sqrt{\pi}x^{1/4}}e^{-\zeta}\sum_{k=0}^{\infty}(-1)^k b_k \zeta^{-k} \tag{A2.32}$$

$$\mathrm{Bi}(x) = \frac{1}{\sqrt{\pi}x^{1/4}}e^{\zeta}\sum_{k=0}^{\infty} b_k \zeta^{-k} \tag{A2.33}$$

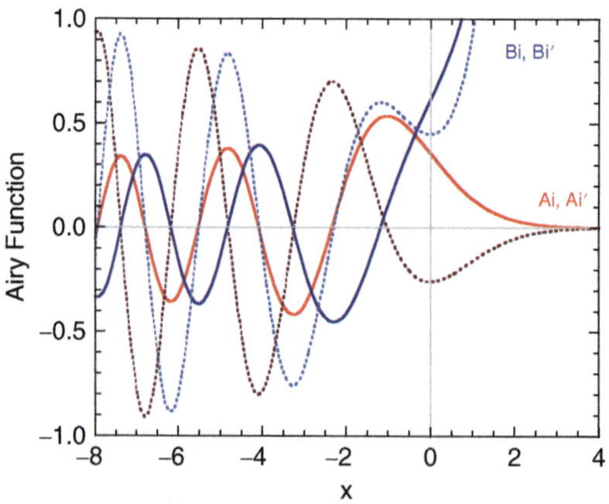

Figure A2.1 The Airy functions Ai(x) (red) and Bi(x) (blue) (solid lines) with their derivatives Ai′(x) and Bi′(x) (dashed lines).

Table A2.1 Table of Airy function values.

x	Ai(x)	Ai′(x)	Bi(x)	Bi′(x)
−4.00	−0.070266	−0.790629	0.392235	−0.116671
−3.75	−0.251613	−0.632454	0.317185	−0.467801
−3.50	−0.375534	−0.343443	0.168940	−0.693116
−3.25	−0.419013	−0.002454	−0.016034	−0.759759
−3.00	−0.378814	0.314584	−0.198290	−0.675611
−2.75	−0.268491	0.551338	−0.344376	−0.478387
−2.50	−0.112325	0.678853	−0.432422	−0.220420
−2.25	0.061599	0.695016	−0.453921	0.045904
−2.00	0.227407	0.618259	−0.412303	0.278795
−1.75	0.365483	0.478652	−0.319503	0.452495
−1.50	0.464257	0.309187	−0.191785	0.557908
−1.25	0.520045	0.139080	−0.045867	0.599814
−1.00	0.535561	−0.010161	0.103997	0.592376
−0.75	0.517773	−0.125991	0.247780	0.554475
−0.50	0.475728	−0.204082	0.380353	0.505934
−0.25	0.418725	−0.246389	0.501400	0.465151
0.00	0.355028	−0.258819	0.614927	0.448288
0.25	0.291164	−0.249062	0.728747	0.469861
0.50	0.231694	−0.224911	0.854277	0.544573
0.75	0.179336	−0.193175	1.006931	0.690300
1.00	0.135292	−0.159147	1.207424	0.932436
1.25	0.099645	−0.126487	1.484388	1.310203
1.50	0.071749	−0.097382	1.878942	1.886212
1.75	0.050570	−0.072854	2.452271	2.761582
2.00	0.034924	−0.053090	3.298095	4.100682
2.25	0.023655	−0.037759	4.563206	6.172558
2.50	0.015726	−0.026251	6.481661	9.421423
2.75	0.010269	−0.017864	9.432379	14.588170
3.00	0.006591	−0.011913	14.037329	22.922215
3.25	0.004160	−0.007793	21.330905	36.554851
3.50	0.002584	−0.005004	33.055507	59.164320
3.75	0.001580	−0.003158	52.183238	97.173147
4.00	0.000952	−0.001959	83.847071	161.92668

where[2] $\zeta \equiv 2x^{3/2}/3$ and the b_k are defined by

$$b_k = \frac{\Gamma\left(3k+\frac{1}{2}\right)}{54^k\Gamma(k+1)\Gamma\left(k+\frac{1}{2}\right)} \tag{A2.34}$$

The series expansions for negative argument, as well as the derivatives, can be inferred from the change $\zeta \rightarrow -i\zeta$ as a consequence of $x \rightarrow -x$ and the behavior of the Zi(c,x) functions of Eq. (18.24), but as the b_k increase as $k!$ for large k, usage of the series requires caution in tunneling calculations in the form of truncating the series [85].

Alternately, using Taylor expansions and Eq. (A2.24), the values of Ai(x) and Bi(x) can be found from tabulated values $\text{Ai}(x_j) = \text{Ai}_j$ and so on by

$$\begin{aligned}\text{Ai}(x_j+\delta) \approx &\left(1+\frac{1}{2}x_j\delta^2+\frac{1}{6}\delta^3+\frac{1}{24}x_j^2\delta^4\right)\text{Ai}_j\\ &+\delta\left(1+\frac{1}{6}x_j\delta^2+\frac{1}{12}\delta^3\right)\text{Ai}'_j\end{aligned} \tag{A2.35}$$

with the same equation for Ai → Bi, as well as with $x \rightarrow -x$. As with the series expansions, care with arguments and the limitations of the expansion should be kept in mind.

[2]The notation follows ref. [103]: ζ should not be confused with the zeta function of Section A2.3.

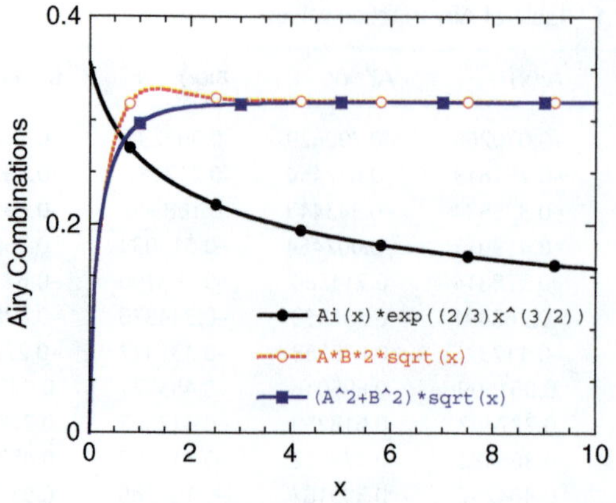

Figure A2.2 Various combinations of the Airy functions Ai($\pm x$) and Bi($\pm x$) showing their interrelated behavior: (*i*) black solid line, solid circles shows Ai(x) exp($2x^{3/2}/3$) to reveal the non-exponential part, (*ii*) red dashed line, open circles shows the combination $\sqrt{x}$Ai(x)Bi(x), and (*iii*) blue solid line, solid squares shows the combination $\sqrt{x}(\mathrm{Ai}(-x)^2 + \mathrm{Bi}(-x)^2)$.

A2.7 Lorentzian functions

Lorentzian functions of the form

$$\delta_n(x) = \frac{n}{\pi}\frac{1}{1+(nx)^2} \tag{A2.36}$$

are useful representations of the Dirac delta function in the limit $n \to \infty$, that is,

$$\lim_{n\to\infty} \int_{-\infty}^{\infty} \delta_n(x)g(x)dx = \int_{-\infty}^{\infty} \delta(x)g(x)dx = g(0) \tag{A2.37}$$

This can be demonstrated as follows [400]. Consider the following three integrals for $a \to \infty$, where $g_m > |g(x)|$ for all x:

$$\int_{-\infty}^{-a} \delta_n(x)g(x)dx < g_m \int_{-\infty}^{-a} \delta_n(x)dx = g_m\left(\frac{1}{2} - \frac{1}{\pi}\arctan(an)\right) \tag{A2.38}$$

$$\int_{a}^{\infty} \delta_n(x)g(x)dx < g_m \int_{a}^{\infty} \delta_n(x)dx = g_m\left(\frac{1}{2} - \frac{1}{\pi}\arctan(an)\right) \tag{A2.39}$$

$$\int_{-a}^{a} \delta_n(x)g(x)dx = g(0)\frac{2}{\pi}\arctan(an) \tag{A2.40}$$

For the case that $a = 1/\sqrt{n}$, as $n \to \infty$, $a \to 0$, but $\arctan(\sqrt{n}) \to \pi/2$. But this demonstrates that Eqs (A2.38) and (A2.39) both vanish, whereas Eq. (A2.40) survives and equals $g(0)$. In other words, $\delta_n(x)$ mimics a Dirac delta function.

Alternately, because $\delta_n(x)$ is symmetric in x, integrals of odd powers of x with it vanish. If $g(x)$ can be well approximated by a Taylor expansion to first order, or $g(x) \approx g(0) + xg'(0)$, over the range where $\delta_n(x)$ is significant, then

$$\int_{-\infty}^{\infty} \delta_n(x)g(x)dx \approx \int_{-\infty}^{\infty} \delta_n(x)\{g(0) + xg'(0)\}dx = g(0) \tag{A2.41}$$

APPENDIX 3

Algorithms

And you may ask yourself
Am I right?...Am I wrong?
And you may say to yourself
My God!...What have I done?!

– Talking Heads[1]

A3.1 Permutation algorithm

An algorithm for generating permutations for $n = 6$ using the matrix methods of Eq. (4.25) by iteratively populating the matrix C is given.

For each n loop, C is overwritten and enlarged (i.e., goes from $C^{n-1} \rightarrow C^n$) by (*i*) inserting C^n on the first block-row in Eq. (4.25), (*ii*) filling the upper triangular part of C^n, (*iii*) filling the lower triangular part of C^n, and finally (*iv*) replacing the diagonals of C^n with $\overline{n}$. The algorithm is

Require: C^n is matrix of integers with $N_{rows} = n!$ and $N_{columns} = n$

Require: On input, $C \rightarrow C^2 = \begin{pmatrix} 2 & 1 \\ 1 & 2 \end{pmatrix}$

```
for n = 3 to 6 do
  n_l = (n − 1)!
  for j = n to 2 do
    for k = 1 to n_l do
      C(k,j) = C(k,j − 1)
    end for
  end for
  for i = 2 to n − 1 do
    for j = 1 to n do
      for k = 1 to n_l do
        C((i − 1)n_l + k, j + 1) = C(k, j + 1)
      end for
    end for
  end for
  for i = 2 to n do
    for j = 1 to i do
      for k = 1 to n_l do
        C((i − 1)n_l + k, j) = C(k, j + 1)
      end for
    end for
  end for
  for i = 1 to n do
    for k = 1 to n_l do
```

[1] David Byrne, Brian Eno, Chris Frantz, Jerry Harrison, and Tina Weymouth, lyrics to *Once in a Lifetime*, 1981.

$C((i-1)n_l+k,i)=n$
end for
end for
end for
Concatenate the columns and sort the resulting $n! \times 1$ array.

A3.2 Birthday algorithm

The function $W(k)$ defined in Eq. (4.27) for a given n is evaluated exactly for k, and then compared to its approximate form of Eq. (4.28). The approximate form, after being shown to be quite good, uses the method of bisection (Section A3.7.1) to find the value of k_a for which $W(k_a, n) = W_o$.

```
%% Birthday Problem
%  Find k such that W(k) = n! / (n-k)! n^k = 1/2

if exist('n','var')    == 1; else n    = 365; end
if exist('k','var')    == 1; else k    =  52; end
if exist('Wo','var')   == 1; else Wo   = 0.5; end
if exist('iout','var') == 1; else iout =  10; end

W = @(x,n) exp(-x).*(1 - (x/n)).^(x - n - 0.5);

Wk = zeros(1,k); Wk(1) = 1; kx = (1:k);
for j = 2:k;
    Wk(j) = (n-j+1)*Wk(j-1)/n;
end

Waprx = W(kx,n);

plot(kx,Wk,'o',kx,Waprx);

% Bisection search for W(k) = Wo
% Require Wk(kmax) < Wo and Wk(kmin) > Wo
kmax = k;kmin = 1;
for j = 1:24;
    ka = 0.5*(kmax + kmin);
    za = W(ka,n);
    if za < Wo;
        kmax = ka;
    else
        kmin = ka;
    end
end

if iout > 0;
    Mout = [kx',Wk',Waprx'];
    dlmwrite('BD.txt',Mout,'delimiter',...
             '\t','precision',6);
end
```

A3.3 Least squares fitting of data

A3.3.1 Theory

A best-fit line through data $\{x_i, y_i\}$ is often needed to identify a parameter, such as the extraction of work function Φ from thermionic emission data. A method to do so is to minimize the sum of the square of the deviations between the experimental data $\{y_i\}$ and the fit $f(\{x_i\})$. This process is known as least squares fitting [709].

Consider perhaps the most common case, where $f(x) = A + Bx$. Define the sum of the square of the differences between y_i and $f(x_i)$ as χ, or

$$\chi = \sum_{i=1}^{N} (y_i - A - Bx_i)^2 \tag{A3.1}$$

The error is minimized by finding A and B such that that $\partial_A \chi = \partial_B \chi = 0$. This results in two equations:

$$\sum_{i=1}^{N} (y_i - A - Bx_i) = 0 \tag{A3.2}$$

$$\sum_{i=1}^{N} (y_i - A - Bx_i)\, x_i = 0 \tag{A3.3}$$

which can be solved for A and B using conventional means. Here it is profitable to reconsider the problem in terms of matrix and vector manipulations to harness the full power of computational programming languages that are widely available and which are, like FORTRAN or MATLAB, well-suited for dealing with matrices.

Using a bra-ket notation, let $|x\rangle$ be the column vector of the independent variable, and $|y\rangle$ the column vector of the dependent variable. It follows that $\langle x|$ and $\langle y|$ are their transposes (row vectors). Similarly, let $|1\rangle$ be a column vector such that all the entries are 1. The following relations then hold

$$\sum_{i=1}^{N} 1 = \langle 1|1\rangle = N \tag{A3.4}$$

$$\sum_{i=1}^{N} x_i = \langle 1|x\rangle = \langle x|1\rangle = N x_o \tag{A3.5}$$

$$\sum_{i=1}^{N} x_i^2 = \langle x|x\rangle \tag{A3.6}$$

The requirement of Eqs (A3.2) and (A3.3) can be concisely represented as

$$\begin{pmatrix} \langle 1|1\rangle & \langle 1|x\rangle \\ \langle x|1\rangle & \langle x|x\rangle \end{pmatrix} \begin{pmatrix} A \\ B \end{pmatrix} = \begin{pmatrix} \langle 1|y\rangle \\ \langle x|y\rangle \end{pmatrix} \tag{A3.7}$$

For a small number of data points N, the evaluation of each matrix element, and the inversion of the 2×2 matrix, is no difficulty, but for large N, the summations can become large as well, causing the determinant of the 2×2 matrix to be the difference of two large and possibly nearly equal numbers. It is profitable to introduce, therefore, the vector

$$|\bar{x}\rangle = |x\rangle - |1\rangle \frac{\langle 1|x\rangle}{\langle 1|1\rangle} = |x\rangle - x_o|1\rangle \tag{A3.8}$$

in terms of which the equations for A and B are compactly written as

$$A = \langle x| \left\{ \frac{|x\rangle\langle 1| - |1\rangle\langle x|}{N\langle \bar{x}|\bar{x}\rangle} \right\} |y\rangle \tag{A3.9}$$

$$B = \frac{\langle \bar{x}|y\rangle}{\langle \bar{x}|\bar{x}\rangle} \tag{A3.10}$$

where the factor in the curly brackets in A is seen to be an $n \times n$ matrix, and the quantity $\langle \overline{x} | \overline{x} \rangle$ is recognized as proportional to the determinant of the 2×2 matrix of Eq. (A3.7) as shown by

$$\begin{aligned} \langle 1 \,|\, 1 \rangle \langle x \,|\, x \rangle - \langle 1 \,|\, x \rangle \langle x \,|\, 1 \rangle &= N \langle x \,|\, x \rangle - N^2 x_o^2 \\ &= N(\langle x | - x_o \langle 1 |)(| x \rangle - x_o | 1 \rangle) \\ &= N \langle \overline{x} \,|\, \overline{x} \rangle \end{aligned}$$

where the equality is a consequence of Eq. (A3.5). The forms of Eqs (A3.9) and (A3.10) are ideally suited to programming languages that can vector multiply.

The extension to equations that are exponential rather than linear is as follows. Let $y \approx \exp(A + Bx)$: the only change in Eqs (A3.9) and (A3.10) is the replacement $|y\rangle \to |\ln(y)\rangle$ (i.e., each y_i is replaced by $\ln(y_i)$). The methodology can also be extended to $y = Af(x) + Bg(x)$ (see ref. [709]).

A3.3.2 Matrix method for linear fit

The implementation of Eqs (A3.9) and (A3.10) in MATLAB is given the following routine, which is designed to fit the temperature versus current density data to an equation of the form $\ln(J/T^2) \approx A + B/k_B T$. The least squares parameter B is then an approximation to the negative of the work function, or $-\Phi$.

```
%-----------------------------------------
% Routine Least Squares Fit exponential
%-----------------------------------------
% Finds best A and B for log(y) = A + B*x
%-----------------------------------------

load T.txt;
load J.txt;
T = T'; J = J';
x = 11604.5./T; y = J./(T.^2);
n = length(T);

% Bx = <x| , Kx = |x> , and so on
BI = ones(1,n);
KI = ones(n,1);
Bx = x;
Kx = x';
Bxb = Bx - mean(x);
Kxb = Bxb';
Bf = log(y);
Kf = Bf';

% Best fit for y = exp(A + B*x)
A = Bx*(Kx*BI-KI*Bx)*Kf/(n*(Bxb*Kxb));
B = (Bxb*Kf)/(Bxb*Kxb);

% Compare raw data to best fit ya(xa)
xa = 1900 + 450*(0:100)/100;
ya = (xa.^2).*exp(A + B*11604.5./xa);

plot(T,J,'o',xa,ya,'r')
    xlabel('Temperature␣[Kelvin]','FontSize',18);
    ylabel('Current␣Density␣[A/cm^2]','FontSize',18);
    title('Table␣II,␣Dushman','FontSize',18)
```

Table A3.1 $J(T)$ data. Data files `T.txt` and `J.txt` containing the temperature [K] and current density [A/cm^2] of Table I of ref. [67], where the emission area is taken to be 0.1825 cm^2 from a tungsten filament.

i	T.txt	J.txt
1	1935.5	0.00051178
2	1986.5	0.0010811
3	2036.0	0.0021737
4	2077.5	0.0037173
5	2086.5	0.0041951
6	2102.0	0.0051304
7	2134.5	0.0074630
8	2134.5	0.0077753
9	2158.0	0.010422
10	2182.0	0.013907
11	2204.0	0.017912
12	2231.0	0.024137
13	2235.0	0.025205
14	2271.5	0.037671
15	2280.0	0.040515
16	2306.0	0.053655

A3.3.3 Input data

The input data required in the MATLAB code is contained in the files `T.txt` and `J.txt`, each of which contains a single column of numbers. The numbers are given in Table A3.1.

A3.4 Monty hall algorithm

The Monty Hall "measurement" of Eq. (8.37) can be applied sequentially (same contestant over N time steps) or thought of as the same contest applied to N different contestants all at once. Either way, the running average of the number of wins for j games models a progression over time. In MATLAB, an algorithm to implement Eq. (8.37) and record the number of wins for the STAY (do not switch curtains when offered) algorithm versus the SWITCH (switch curtains when offered) and graphically represented as in Figure 8.2 is given.

```
%-----------------------------------------
% Routine Monty
%-----------------------------------------
% Finds time evolution of running average of wins for
% the STAY vs. SWITCH strategy in the Monty Hall Pb.
%-----------------------------------------

N = 1000;
prize = zeros(1,N); %initialize prize vector
picks = zeros(1,N); %initialize picks vector

%Hide the prizes, and let the contestants pick curtains
for j = 1:N;
    prize(j) = floor(3*rand())+1;
    picks(j) = floor(3*rand())+1;
end

%initialize the counters
igame = (1:N);      %game counter (time)
```

```
istay = zeros(1,N);%initialize wins for staying vector
iswch = zeros(1,N);%initialize wins for switching vector
nstay = istay;      %running average of istay
nswch = iswch;      %running average of iswch

for j = 1:N;
    if prize(j) == picks(j);
        %it does not matter what the host does: she only
        %wins if her initial pick was right.
        istay(j) = 1;
    else
        %whatever curtain host shows her, she switches
        % to other one:  she only wins if her initial
        % pick was not right.
        iswch(j) = 1;
    end;
    nstay(j) = sum(istay(1:j))/j; %calc running average
    nswch(j) = sum(iswch(1:j))/j; %calc running average
end;
nstay = 100*nstay;             %Convert to a percent
nswch = 100*nswch;             %Convert to a percent

plot(igame,nstay,'r',igame,nswch,'b');

    fid = fopen('Curtain.txt','w');
    fprintf(fid,'%s␣\t␣%s␣\t␣%s␣\n','Game',...
            'Stay', 'Switch');
    for j = 1:N;
        fprintf(fid,'%12.8f␣\t␣%12.8f␣\t␣%12.8f␣\n',...
        igame(j),nstay(j),nswch(j));
    end
    fclose(fid);
    disp('FINISHED:␣Generation␣of␣␣...
␣␣␣␣␣␣␣␣␣␣␣␣␣␣output␣file␣<Curtain.txt>')
```

A3.5 Wave function and density algorithm

Using the notation in Section 9.2.3.1, two calculations are performed. First, the time evolution of a wave packet incident on a step function barrier is plotted over time. Second, the time-independent density profile near the step function barrier is plotted. The output is collected for various parameters and shown in Figures 9.9 and 9.10.

```
%%Program PSIplot
% Set up and solve the time dependent problem of a
% wave packet incident on a potential step, and the
% time independent problem % of the density of
% electrons near a potential step using coordinates
% such that Ek = k^2 and k =(pi/4)*(j/N) for
% simplicity.

%-----------------------------------------
%% Functions
```

```
%-------------------------------------------
%wave vector on barrier side
kappa = @(x,c)  sqrt(-c.^2 + x.^2);

%reflection probability
rkfun = @(x,y)  (x - y)./(x + y);

%transmission probability
tkfun = @(x,y)  2*x./(x + y);

%-------------------------------------------
%%Parameters
%-------------------------------------------
N = 48;Nx = 100;Nt = 24;
L = 40.0;Lt = 40.0;
C  = (1:N);C = (1.0 - cos(pi*C/20))./C;
k  = 0.25*pi*(1:N)/N; E = k.^2;
ko = 1.750*max(k);

%-------------------------------------------
%%Reflection, Transmission, and Wave Vectors
%-------------------------------------------
kp = kappa(k,ko);
rk = rkfun(k,kp);
tk = tkfun(k,kp);

%-------------------------------------------
%% Figure 1:  Time Dependent wave function
%-------------------------------------------
%position vector
xx = -L + 1.5*L*((1:Nx)-1)/(Nx-1);

%time vector
tt = - Lt + 2*Lt*((1:Nt)-1)/(Nt-1);

%Initialize Psi Matrix
Psi = zeros(Nt,Nx);
for it = 1:Nt;
    for ix = 1:Nx;
        for jn = 1:N;
            if xx(ix) < 0.0;
                Psi(it,ix) = Psi(it,ix) + C(jn)*...
                    (exp(1i*k(jn)*xx(ix)) + rk(jn)*...
                    exp(-1i*k(jn)*xx(ix)))...
                    *exp(-1i*E(jn)*tt(it));
            else
            Psi(it,ix) = Psi(it,ix) + C(jn)*tk(jn)...
                    *exp(1i*kp(jn)*xx(ix))*...
                    exp(-1i*E(jn)*tt(it));
            end
        end
    end
```

```
end
Psi2 = (abs(Psi)).^2;          %Density
figure(1);
waterfall(xx,tt,Psi2);
axis tight square; view(0,35);grid off;axis off

%------------------------------------------
%% Figure 2:  Time-independent density
%------------------------------------------
%initialize density
dens = zeros(1,Nx);

%Position vector
xx = -3*pi + 5*pi*((1:Nx)-1)/(Nx-1);

%mimic supply function
C  = ((N+1)^2 - (1:N).^2)/(N+1)^2;

for ix = 1:Nx;
    for jn = 1:N;
        if xx(ix) < 0.0;
            dens(ix) = dens(ix) + abs(C(jn)*...
                (exp(1i*k(jn)*xx(ix)) + ...
                rk(jn)*exp(-1i*k(jn)*xx(ix)))).^2;
        else
            dens(ix) = dens(ix) + ...
            abs(C(jn)*tk(jn)*exp(1i*kp(jn)*xx(ix))).^2;
        end
    end
end
dens = dens/dens(1);          %normalize to boundary value
figure(2);
xzero = [0,0]; yzero = [0, 1.2];
plot(xx/pi,dens,xzero,yzero,'--r'); axis([-3 2 0 1.2]);
```

A3.6 Hydrogen atom algorithms

An acceptance–rejection (AR) algorithm is used to judge whether or not an electron is "measured" at a coordinate (x, y, z) for the case $y = 0$ (i.e., the probability plot is for a slide in the x–z plane). The AR algorithm works by generating a random number r_n and comparing it to the probability evaluated from the wave function $\psi_{nml}(x, y, z)$: if r_n is greater, the electon is "not found" and if smaller the electron is "found". Although the position can be chosen randomly, the algorithm instead considers the three-dimensional space to be composed of n^3 small volumes and uses the AR technique to determine whether or not an electron is in that small volume. Such an approach makes the integration along a coordinate, as in Eq. (9.52), more simply represented as a summation over the occupancy of the n cells along that dimension.

```
%%Program HydrogenSlice:  calculation of probability
% factors associated with the hydrogen atom but along
% a plane

% formula for n = 3, l = 2 hydrogenic wave function
RY = @(x) (x.^2 - 9*x + 13.5).*exp(-x/3);
% no need to put in phase exp(i*phi), as that factor
```

```
% will cancel when the absolute value is taken

% default variables
if exist('n','var')     == 1; else n    =  1000  ;end
if exist('xmax','var') == 1; else xmax =  20   ;end

x = xmax*(2*(1:n) - n - 1)/(n-1); z=x;

% Create the Psi^2 = rho matrix
% calculate for y = 0 (in the x-z plane)
rho = zeros(n,n); %rho = |Psi|^2
for i = 1:n;
    for k = 1:n;
        r = sqrt(x(i)^2 + z(k)^2);
        rho(i,k) = RY(r).^2;
    end
end

% Scatter plot of probability distribution
%  in the x-z plane

xt = zeros(1,n^2); zt = xt;
rho = rho/max(max(rho));
nt = 0;
for i = 1:n;
    for k = 1:n;
        if rand() < rho(i,k);
            nt = nt+1;
            xt(nt) = x(i);
            zt(nt) = z(k);
        end
    end
end
xt = xt(1:nt);zt = zt(1:nt); %trim the vectors
figure(1);
plot(xt,zt,'rs','MarkerSize',2);
axis([x(1) x(n) x(1) x(n)]);
axis square;
```

The second code file performs the actual summation over the y coordinate, and therefore is to be compared to Eq. (9.52). Whether $n = 3, l = 2$ or $n = 2, l = 1$ is considered depends on which formula is commented out.

```
%%Program HydrogenPsi:  calculation of probability
% factors associated with the hydrogen atom

%% choose the orbital
if exist('n','var') == 1; else n = 2 ;end
if n == 3;
    % formula for n = 3, l = 2 hydrogenic wave function
    RY = @(r,z) exp(-r/3).*(3*z.^2 - r.^2); %dzz orbital
else
    % formula for n = 2, l = 1 hydrogenic wave function
    RY = @(r,z) exp(-r/3).*z; %pz orbital
```

```
end
% no need to put in phase exp(i*phi), as that factor
% will not remain when the absolute value is taken
% default variables
if exist('Nx','var') == 1; else Nx = 64;end
if exist('xmax','var') == 1; else xmax = 15 ;end

x = xmax*(2*(1:Nx) - Nx - 1)/(Nx-1); y=x; z=y;

%% Create the Psi^2 = rho matrix
rho = zeros(Nx,Nx,Nx); %rho = |Psi|^2
for i = 1:Nx;
    for j = 1:Nx;
        for k = 1:Nx;
            r = sqrt(x(i)^2 + y(j)^2 + z(k)^2);
            rho(i,j,k) = RY(r,z(k)).^2;
        end
    end
end

% Integrate over x then contour plot dnsx
rhoxz = zeros(Nx,Nx);
for j = 1:Nx;
    for k = 1:Nx;
        for i = 1:Nx;
            rhoxz(k,j) = rhoxz(k,j) + rho(i,j,k);
        end
    end
end
rhoxz = rhoxz/max(max(rhoxz)); %Normalization
figure(2);
contour(x,y,rhoxz,10);axis square;title('rho(x,z)');
grid off;

%% Scatter plot of probability distribution using
%  acceptance / rejection algorithm
xt = zeros(1,Nx^3); yt = xt; zt = xt;
rho = rho/max(max(max(rho)));
nt = 0;
for i = 1:Nx;
    for j = 1:Nx;
        for k = 1:Nx;
            if rand() < rho(i,j,k);
                nt = nt+1;
                xt(nt) = x(i);
                yt(nt) = y(j);
                zt(nt) = z(k);
            end
        end
    end
end
xt = xt(1:nt);yt = yt(1:nt); zt = zt(1:nt);
```

```
figure(3);
plot3(xt,yt,zt,'ro','MarkerSize',2);
axis([-xmax xmax -xmax xmax -xmax xmax]);view(40,20)
axis square;grid on;
```

A3.7 Root-finding Methods

A3.7.1 Method of bisection

The zeros of the function $H_n(\lambda)$, defined in Eq. (15.13), are determined using the method of bisection. Initially $\lambda_{min} = 0$ and λ_{max} set large enough that $H_n(\lambda_{max}) > 0$. At each iteration, the midpoint λ_h value is found and $H_n(\lambda_h)$ evaluated. Depending on whether the result is greater than or less than zero, the maximum or the minimum λ value, respectively, is set to the midpoint value. Because $\lambda_{min} < \lambda_h < \lambda_{max}$, as the separation $\lambda_{max} - \lambda_{min}$ shrinks the value of λ_m is converged upon. After $\lambda_m(n)$ is found, δ/δ_m is evaluated based on Eq. (15.8) and either plotted or output to a data file `Baroody.txt`. In the code, λ is `lam` and $H_n(\lambda)$ of Eq. (15.13) is `Hnlam`. Output is shown in Figures 15.2 and 15.3. The calculated zeros $\lambda_m(n)$ are shown in Figure 15.8.

```
%% Secondary Emission Baroody Function
%-------------------------------------------
% Input Terms:
%   n    = power of E in R(E)  = E^n / A*n
%   lmax = initial guess of how large lam can be
%   lmin = initial guess of how small lam can be
%   N    = number of bisections to zoom in on lam
%   xmax = size of plotting region E/Em for output
%   iout = 0 - plot data; 1 - output data

%%Functions
%-------------------------------------------
fba = @(x,a,p,n)  exp(a.*(x.^n - 1).*p.^n);
dfba = @(x,a,n)  (1.0 + a*n*(x.^n - 1.0)).*...
                  exp(a.*(x.^n - 1));

%%Parameters
%-------------------------------------------
if exist('n','var') == 1; else n = 1.35;end
if exist('lmax','var')  == 1; else lmax = 10.0;end
if exist('lmin','var')  == 1; else lmin =  0.0;end
if exist('N','var')     == 1; else    N =   16;end
if exist('Nx','var')    == 1; else   Nx =  101;end
if exist('xmax','var')  == 1; else xmax =  6.0;end
if exist('iout','var')  == 1; else iout =    1;end

%Step 1:  Test evaluation:  For lam = 2 and p = 1/2,
%         deldelm = 0.8231054
%-------------------------------------------
p = 0.5;
deldelm = p.*integral(@(x)fba(x,2.0,p,n),0.0,1.0)./...
            integral(@(x)fba(x,2.0,1.0,n),0.0,1.0);

%Step 2:  Find lambda(n) using method of bisection
%         such that Hn(lambda(n)) = 0
%-------------------------------------------
```

```
for j = 1:N;
    lam = 0.5*(lmax+lmin);
    Hnlam = integral(@(x) dfba(x,lam,n),0.0,1.0);
    if Hnlam > 0; %
        lmin = lam;
    else
        lmax = lam;
    end
end
%Taylor Correction:  y = y0 + (x-x0)*(y1-y0)/(x1-x0)
%                     Solve for x(y=0)
Hnlam1 = integral(@(x) dfba(x,lmax,n),0.0,1.0);
Hnlam0 = integral(@(x) dfba(x,lmin,n),0.0,1.0);
lam = lmin - (Hnlam0/(Hnlam1-Hnlam0))*(lmax - lmin);
%Check:  does Hnlam = 0?
Hnlam = integral(@(x) dfba(x,lam,n),0.0,1.0);

%Step 3:  Plot the function (del / delm)
pn = (0:Nx-1)*xmax/(Nx-1);
yn = zeros(1,Nx);
for j = 1:Nx;
    yn(j) = pn(j).*...
          integral(@(x)fba(x,lam,pn(j),n),0.0,1.0)./...
          integral(@(x)fba(x,lam,1.0,n),0.0,1.0);
end

if iout > 0;
    format long
    fid = fopen('Baroody.txt','w');
    fprintf(fid,'%s_\t_%s_\n','p','del/delm');
    for j = 1:Nx;
        fprintf(fid,'%12.8f_\t_%12.8f_\n',pn(j),yn(j));
    end
    fclose(fid);
    format short
else
    plot(pn,yn);
end
```

A3.7.2 Method of iteration

When an equation to be solved can be put into the form of $x = f(x)$ (an example being finding the maximum field which satisfies Eq. (17.90)), then iterative methods can under certain conditions allow for a solution. Graphically, the method is simple: the solution is the intersection of the two curves $y_1 = x$ and $y_2 = f(x)$. The method involves taking a reasonable estimate of x, calculating $f(x)$, and feeding that solution back into $f(x)$ as its argument ("iterating") until convergence is achieved. If one were to plot the iteration points, it would be seen that they bounce between the y_1 and y_2 lines until they converge on the intersection point.

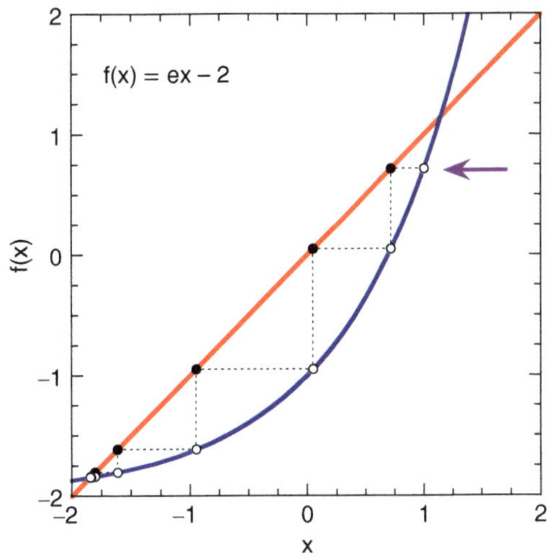

Figure A3.1 A graphical representation of the iteration process used to find a solution to $e^x - x - 2 = 0$ using iteration of $x = f(x)$ with $f(x) = e^x - 2$. Open circles represent $f(x_{j-1})$ and black circles represent x_j. The purple arrow indicates x_1.

> **EXAMPLE:** Find a root of $e^x - x - 2 = 0$ using iteration.
> **SOLUTION:** Isolating x, then $x = f(x)$ implies $f(x) = e^x - 2$. Starting with $x_1 = 1$ and letting $x_{j+1} = f(x_j)$, then x_1 through x_{10} are given by (1.0000, 0.7183, 0.0509, −0.9478, −1.6124, −1.8006, −1.8348, −1.8404, −1.8412, −1.8414), after which the last significant digit no longer changes.

A plot of the iterated points of the previous example, shown in Figure A3.1, shows a bouncing between the lines $y = x$ and $y = f(x)$, with the points coming progressively closer to the unique point where $x = f(x)$. The iterations appear to be "attracted" to the lower root, instead of the upper root near $x = 1$. The iterative process is not as rapid as, for example, Newton's method, where derivatives of the function are used, but then again, in the iterative method, the calculations of derivatives are not needed.

It might be thought that to find the other root, x_1 should be chosen to the right of it, say at $x_1 = 1.2$, but that would be catastrophic: $f(x_1)$ would be larger, not smaller, and subsequent iterations would grow out of control. It is seen that for the iteration process to work, $f'(x_\infty) < 1$, where the prime indicates derivative with respect to argument. Knowing that, a method to find the upper root by iteration rewrites the equation to be solved: isolating the x in e^x gives the equation $x = \ln(x + 2)$ so that $f(x) = \ln(x + 2)$. Then 10 or so iterations show that $x_\infty \approx 1.1462$.

So far, only the case of positive $f'(x_\infty)$ has been considered: an interesting behavior for negative $f'(x_\infty)$ occurs. Consider the example $x = \cos(x)$, with a solution at $x_\infty = 0.7391$. The example is interesting in that $|f'(x)| < 1$, and so iteration will converge on x_∞ regardless of the x_1 chosen. A plot of the iteration sequence, however, shows the iterations spiraling towards the solution, as in Figure A3.2.

A3.7.3 Final correction

Although bisection is rugged and iteration elegant, both are moderate accuracy approaches. When greater accuracy is required, the algorithms must be run longer for convergence to an acceptable error level, but if the functions are smooth, short cuts to the final answer are possible.

An example is the Taylor correction employed in the algorithm of Section A3.7.1. After a number of iterations, the maximum and minimum values are so close that $H_n(\lambda)$ appears linear for $\lambda_{min} \le \lambda \le \lambda_{max}$. For a general function $f(x)$ that is approximately linear between $x_0 \le x \le x_1$, then the Taylor expansion is to leading order

$$f(x) \approx f(x_0) + (x - x_0)f'(x_0) \tag{A3.11}$$

where the prime on f' indicates derivative with respect to argument. Solving for x,

$$x \approx x_0 - \frac{f(x_0)}{f'(x_0)} \tag{A3.12}$$

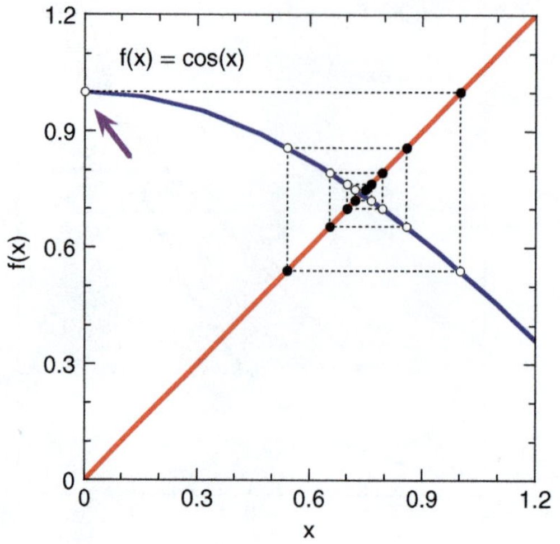

Figure A3.2 A graphical representation of the iteration process used to find a solution to $x = \cos(x)$. Open circles represent $f(x_{j-1})$ and black circles represent x_j. The purple arrow indicates x_1.

In numerical treatments, when x_1 and x_0 are so close that $f(x)$ is essentially linear between them, a more convenient approximation is to use finite difference approximations to the derivative to obtain

$$x \approx x_0 - (x_1 - x_0)\left(\frac{f(x_0)}{f(x_1) - f(x_0)}\right) \tag{A3.13}$$

which has the added advantage of relying only on the evaluation of $f(x)$ rather than its derivative. An iteration or two of this correction, with the evaluated x replacing x_0, hones in on the root nicely.

An even more rapid honing in on the root is possible if $f(x)$ is approximately no worse than quadratic between x_0 and x_1. Let $x_h = (x_1 + x_0)/2$ be the halfway point, and define

$$f_o = f(x_0) \tag{A3.14}$$

$$d_1 = -3f(x_0) + 4f(x_h) - f(x_1) \tag{A3.15}$$

$$d_2 = 4f(x_0) - 8f(x_h) + 4f(x_1) \tag{A3.16}$$

where the forms of d_1 and d_2 are familiar from upwind/downwind approximations to derivatives and second-order derivatives in finite differencing schemes, with $d_1 \approx 2\Delta f'(x_0)$ and $d_2 \approx 4\Delta^2 f''(x_0)$ with $\Delta = x_1 - x_0$. The quadratic approximation to the root is then

$$x = x_0 - (x_1 - x_0)\left(\frac{2f_o}{d_1 + \sqrt{d_1^2 - 2f_o d_2}}\right) \tag{A3.17}$$

The recommended approach to iteration, therefore, is to use it to get to the neighborhood of the root, and then, if the estimated root is close enough that $f(x)$ can be approximated by a linear or quadratic function in that neighborhood, use the Taylor correction method a few times to obtain the final approximation.

A3.8 Thermal–field algorithm

The behavior of the current density integrand $D(E_x)f(E_x)$, where D is the transmission probability and f is the supply function, is investigated using a form of the Gamow factor $\theta(E_x)$ that is linear in energy and found using a Taylor expansion near $E \approx \mu$, as in the development of the Fowler–Nordheim equation (see Chapter 13). The algorithm is used to evaluate the numerical data for the figures of Section 17.2. The code evaluates $\theta(E)$, which is called `theta`. It is evaluated on a mesh grid of equally spaced E_x and F evaluation points. After each curve $D(E_x)f(E_x)$ is evaluated, the curve is normalized to its maximum value, so as to mimic the behavior of Figure 17.4. If the spacing ΔF is such that not too many curves are evaluated, a second plot in waterfall format is generated. Figures based on this algorithm are shown in Figure 17.7 and 17.8. Observe that the term "kb" is equivalent to $1/k_B$. A convenient numerical evaluation of the Gamow factor $\theta(E)$ and $\beta_F(E)$ for the image charge potential may be obtained using Eq. (17.92).

```
%% Gadzuk and Plummer-like Normal Energy Distribution
%  using a linearized Gamow Factor Theta(E)
%-----------------------------------------

%%Functions:  Supply function and Nordheim functions
%             using Forbes & Deane approximation
%-----------------------------------------
fok = @(x,b,m)  1.0./(1.0 + exp(b.*(x - m)));
vx  = @(x)      1.0 + (1.0/3)*(x.^2).*( log(x) - 3.0);
tx  = @(x)      1.0 + (1.0/9)*(x.^2).*(-log(x) + 1.0);

%%Parameters
%-----------------------------------------
if exist('Chem','var') == 1; else Chem =  7.0   ;end
if exist('Phi','var')  == 1; else Phi  =  4.5   ;end
if exist('Temp','var') == 1; else Temp =  1570.0;end
if exist('DelE','var') == 1; else DelE =  0.02  ;end
if exist('DelF','var') == 1; else DelF =  0.20  ;end
if exist('Emin','var') == 1; else Emin =  6.00  ;end
if exist('Emax','var') == 1; else Emax =  10.0  ;end
if exist('Fmin','var') == 1; else Fmin =  2.40  ;end
if exist('Fmax','var') == 1; else Fmax =  3.60  ;end

%Fundamental Constants and needed terms
Q = 0.3599911;m = 5.6856297;hbar = 0.6582119;
kb = 1.0/11604.5;

%Create grid over which values of D(E)fo(E) are
% evaluated as a function of energy Ex and field Fo
[Ex, Fo] = meshgrid(Emin:DelE:Emax , Fmin:DelF:Fmax);

%Energy Slope Factors
betaT = 1.0/(kb*Temp);
betaF = (2.0./(hbar*Fo)).*sqrt(2*m*Phi).*...
        tx(sqrt(4*Q*Fo)./Phi);

%Linearized Gamow Factor
Ec    = Chem + (2*Phi/3).*vx(sqrt(4*Q*Fo)./Phi)./...
        tx(sqrt(4*Q*Fo)./Phi);
theta = betaF.*(Ec - Ex);

%Grid values - calculated then normalized to maximum
dJdE = fok(Ex,betaT,Chem)./(1.0 + exp(theta));
maxJ = max(dJdE');
Ny = length(maxJ);
for j = 1:Ny;
    dJdE(j,:) = dJdE(j,:)/maxJ(j);
end
```

```
%Plotting Section (do not make waterfall plots
%                  if too many lines)
figure(1);
surf(Ex,Fo,dJdE);axis tight square;
shading flat;view(2)
xlabel('Energy_[eV]','FontSize',16);
ylabel('Field_[eV/nm]','FontSize',16)

if length(Fo(:,1)) < 50;
    figure(2);
    waterfall(Ex,Fo,dJdE);axis tight square;
    grid off; axis off;
end
```

A3.9 Gamow factor algorithm

An algorithm for the evaluation of the Gamow factor $\theta(E)$ for the image charge barrier is given. See Sections 11.2, 11.3, and 17.4, and Eq. (17.78).

```
%% Evaluation of the Gamow Factor for Image charge
%  using full integral function
%-----------------------------------------

%%Functions:  Supply function and Nordheim functions
%             using Forbes & Deane approximation
%-----------------------------------------
REx = @(x,a)  0.25*(sin(2*x).^2)./sqrt(a+(sin(x)).^2);
vy  = @(x) 1.0 - (x.^2)*(1.0 - log(x)/3);
ty  = @(x) 1.0 + (x.^2)*(1.0 - log(x))/9;


%%Parameters
%-----------------------------------------
if exist('Chem','var') == 1; else Chem =  7.0   ;end
if exist('Phi','var')  == 1; else Phi  =  4.5   ;end
if exist('Nx','var')   == 1; else Nx   =  21    ;end
if exist('Emin','var') == 1; else Emin =  Chem  ;end
if exist('Fo','var')   == 1; else Fo   =  8.0  ;end

%Fundamental Constants and needed terms
Q = 0.3599911;m=5.6856297;hbar=0.6582119;kb=11604.5;

aa = 4*Q*Fo;
Ex = Emin + (Phi - sqrt(4*Q*Fo))*(0:Nx)/Nx;

Bo = (4.0/(3*hbar))*sqrt(2*m*Phi^3);
Co = (2.0/hbar)*sqrt(2*m*Phi);

xx = Chem + Phi - Ex;

LE = sqrt(xx.^2 - aa)./Fo;
yE = aa./(2*(xx + sqrt(xx.^2 - aa)).*sqrt(xx.^2 - aa));
```

```
Nx = length(Ex);
FREx = zeros(size(Ex)); FNx = FREx; FQx = FREx;
ys = sqrt(aa)/Phi;
Thetao = (Bo*vy(ys)/Fo) + (Co*ty(ys)/Fo)*Chem;
BetaF = Co*ty(ys)/Fo;
BetaQ = (pi/hbar)*sqrt(2*m)*(Q/Fo^3)^0.25;
for i = 1:Nx;
        FREx(i) = integral(@(x)REx(x,yE(i)),0,0.5*pi);
    FNx(i) = Thetao - BetaF*Ex(i);
    FQx(i) = BetaQ*(Chem + Phi - sqrt(4*Q*Fo) - Ex(i));
end

FREx = (4*sqrt(2*m)/hbar)*FREx.*sqrt(Fo.*LE.^3);

FREx = real(FREx);

figure(1);
plot(Ex,FREx,'o',Ex,FNx,'r',Ex,FQx,'b');
axis tight square;
xlabel('Energy_[eV]','FontSize',16);
ylabel('Theta(E)','FontSize',16)

fid = fopen('ThetaE.txt','w');
fprintf(fid,'%s_\t_%s_\t_%s_\t_%s\n','E_[eV]',...
       'Theta(E)','ThetaFN','ThetaQuad');
for j = 1:Nx;
    fprintf(fid,...
        '%12.8f_\t_%12.8f_\t_%12.8f_...\t_%12.8f_\n',...
        Ex(j),FREx(j),FNx(j),FQx(j));
end
fclose(fid);
disp('FINISHED:_Generation_of_data_file_<ThetaE.txt>')
```

A3.10 Triangular barrier $D(E)$

An algorithm for the evaluation of the Gamow factor $\theta(E)$ for the triangular barrier (the $Q \rightarrow 0$ of the image charge barrier) is given. See Section 18.2.3.

```
%% Evaluation of the Transmission probability D(k)
%  for a triangular barrier
%  Output:  col 1 = E, col 2 = D(E), col 3 = Step D(E)
%-------------------------------------------

% Airy Function combination formulae:
% ABx = Ai^2 + Bi^2; ABp = Ai'^2 + Bi'^2
ABx = @(x) airy(0,x).^2 + airy(2,x).^2;
ABp = @(x) airy(1,x).^2 + airy(3,x).^2;

%Fundamental Constants and needed terms
m=5.6856297;hbar=0.6582119;
```

```
%% Parameters
%----------------------------------------
if exist('N','var')    == 1; else N    = 50    ;end
if exist('F','var')    == 1; else F    = 6.0   ;end
if exist('mu','var')   == 1; else mu   = 7.0   ;end
if exist('Phi','var')  == 1; else Phi  = 4.5   ;end
if exist('Emax','var') == 1; else Emax = 4.0  ;end

%% Vectors:  Ex is energy relative to mu + Phi;
%            ko = k at barrier maximum
%            kxp and kxm are k above and below ko
Ex = (1:N)*Emax/N;
f = 2*m*F/hbar^2;f1 = f^(1.0/3); f2 = f^(2.0/3);
ko = sqrt(2*m*(mu+Phi))/hbar;
kxp = sqrt(2*m*(mu+Phi+2*Ex))/hbar;
kxm = sqrt(2*m*(mu+Phi-Ex))/hbar;
zp = (kxp.^2 - ko^2)/f2;  %For above barrier
zm = (ko^2 - kxm.^2)/f2;  %For below barrier

Dkp = 4*f1*kxp./(pi*((kxp.^2).*ABx(-zp) +...
      f2*ABp(-zp) + (2 * f1*kxp/pi)));
Dkm = 4*f1*kxm./(pi*((kxm.^2).*ABx( zm) +...
      f2*ABp( zm) + (2 * f1*kxm/pi)));

kxm = kxm(N:-1:1); %reverse order the kxm
Dkm = Dkm(N:-1:1); %and reverse order the Dkm

kx = [kxm,kxp]; Dk = [Dkm,Dkp]; %stitch vectors together
kappa = sqrt(max(kx.^2 - ko.^2,0.0));
Dkstep = (4*kx.*kappa./(kx+kappa).^2); %zero field case

Ex = ((hbar*kx).^2)/(2*m);
plot(Ex,Dk,Ex,Dkstep);

%% Clear junk and output data
clear kxm kxp Dkm Dkp f f1 f2 ko zp zm m hbar

Hout = [Ex',Dk', Dkstep'];
filenam= ['Airy',num2str(F),'.txt'];
dlmwrite(filenam,Hout,'delimiter','\t','precision',6);
disp(['FINISHED:␣Generation␣of␣data␣file:␣',filenam])
```

A3.11 Evaluation of $H_c(u)$

An algorithm for the evaluation of $H_c(u)$ in Figure 18.8 and Table 18.1 is given.

```
%% Evaluation of the Hc(u) function
%  using TDMA method of solving tridiagonal matrix
%----------------------------------------

Uxn = @(x,n,y) 12*(-2*y*(n+1.0) + 3*x.*(x+1.0));
Dxn = @(x,n,y) 12*( 2*y*(n+1.0) + 3*x.*(x-1.0));
```

```
Cxn = @(x)      5.0 - 72*x.^2;
bo  = @(x)      0.25*((x+3.0)*(x^2+1.0) +...
                2*sqrt(2)*(x-1.0)*(x^2-1.0));

%%Parameters
%-----------------------------------------
if exist('N','var') == 1; else N =   25   ;end
if exist('c','var') == 1; else c =   1.0   ;end

% Position vector:  uj
% up, center, and down diagonals:  vUj, vCj, and vDj
% Boundary conditions:  Hc(0) = bo(c); Hc(1) = Hcone
% Hcone is related to the Airy functions
% evaluated at u = 1: they can be evaluated
% using MATLAB routines as follows

if c == 1i;
    Zione = airy(0,c^2) - 1i*airy(2,c^2);
elseif c == -1i;
    Zione = airy(0,c^2) + 1i*airy(2,c^2);
elseif c == -1.0;
    Zione = airy(0,c^2);
else
    Zione = airy(2,c^2);
end
Hcone = 2*sqrt(pi)*Zione*exp(-2*c/3);

% set up position and boundary vector and
% tridiagonal matrix; Hx corresponds to (N+1)*Hc(u)
% here to make implementation easier.
% We will rescale u later to fix this
uj = (1:N); Hx = uj;
vDj = Dxn(uj,N,c);
vCj = Cxn(uj);
vUj = Uxn(uj,N,c);
vBj = zeros(1,N);
vBj(1) = -Dxn(1,N,c)*bo(c);
vBj(N) = -Uxn(N,N,c)*Hcone;

% Start tridiagonal matrix solution algorithm without
% make sure vUj is not zero (it never is here)
% and exploiting fact that vBj=0 except at j=1 and N
% Downward elimination part:
vUj(1) = vUj(1)/vCj(1);
vBj(1) = vBj(1)/vCj(1);
for i = 2:N-1;
    temp = vCj(i) - vDj(i)*vUj(i-1);
    vUj(i) = vUj(i)/temp;
    vBj(i) = - vDj(i) * vBj(i-1)/temp;
end
vBj(N)= (vBj(N) - vDj(N)*vBj(N-1))/...
      (vCj(N) - vDj(N)*vUj(N-1));
```

```
% Upward part
Hx(N) = vBj(N);
for i = N-1:-1:1;
    Hx(i) = vBj(i) - vUj(i)*Hx(i+1);
end

uj = uj/(N+1);  %time to rescale u so 0 < u < 1

% All finished
%---------------------------------------
%% Compare to MATLAB routines to see accuracy

Hexact = zeros(1,N);
zj = uj.^(-2.0/3);
for i = 1:N;
    if c == 1i;
        Hexact(i) = airy(0,-1.0/uj(i).^(2.0/3))...
            - 1i*airy(2,-1.0/uj(i).^(2.0/3));
    elseif c == -1i;
        Hexact(i) = airy(0,-1.0/uj(i).^(2.0/3))...
            + 1i*airy(2,-1.0/uj(i).^(2.0/3));
    elseif c == -1.0;
        Hexact(i) = airy(0,1.0/uj(i).^(2.0/3));
    else
        Hexact(i) = airy(2,1.0/uj(i).^(2.0/3));
    end
end
Hexact = Hexact * 2*sqrt(pi)./...
       ((uj.^(1.0/6)).*exp(2*c./(3*uj)));

Hexact = [bo(c),Hexact,Hcone]; %add on boundary terms
Hx     = [bo(c),Hx,Hcone];     %add on boundary terms
uj     = [0,uj,1.0];           %add on boundary terms

Hout = [uj',real(Hexact'),imag(Hexact'), ...
     real(Hx'),imag(Hx')];
dlmwrite('Hx.txt',Hout,'delimiter','\t','precision',5)
```

A3.12 Transfer matrix algorithm

There are features of the implementation of the transfer matrix algorithm (TMA) evaluation that require particular attention:

- It is assumed that at the left and right boundaries, $V(x) = 0$. If the right boundary were at a different potential, then $D(k) \to (k_r/k)|t(k)|^2$, where $\hbar k_r = \sqrt{2m(E - V_r}$ and V_r is the potential of the right-hand boundary.
- $V(j)$ corresponds to the potential to the *right* of $x(j)$.
- The quantity `rm` is the ratio of the effective mass of the electron in bulk material with the rest mass, or $r = m^*/m$. For example, for GaAs, $r = 0.067$ but for metals, $r = 1.0$. Typical semiconductor values are often around 0.2, and such values make more visually interesting $D(k)$.
- $x(1)$ is a point prior to the onset of the potential barrier. $x(N_v)$ is the point where the barrier ends. Thus, $V(1) = V(N_v) = 0.0$.

- The very odd choice of $k_{max} \leftarrow 3.1416k_{max}(1 : N)/(N\pi)$ (where the notation $(1 : N)$ creates a vector of integers from 1 to N) is simply to make the values of k not rational because $(3.1416/\pi) \approx 1$ is not rational (cannot be expressed as a ratio of integers). As a result, k_j would only be 0 by an amazing coincidence: if it were 0, the calculation of $\mathbb{S}_n$ would fail. The prevention of the $k_j = 0$ circumstance is better avoided through the use of an `IF-THEN` construct to explicitly exclude the $k_j = 0$ occurrence, but the simple "cheat" here is easier, results in faster code, and for the most part always works.
- In the code, `k(i)` refers to the incident momentum values k and is the argument of $D(k)$. `ko(j)` refers to k_j and is defined by $k_o(j) = \sqrt{2m(E - V_j)}/\hbar$. The notation could be better, but is not.
- The output to a file called `DkTMA.txt` is in the form of two columns spaced by a tab character using MATLAB-specific calls. The first column is the values of k. The second column is the values of $D(k)$.

A3.12.1 Transfer matrix evaluation

Implemented in MATLAB, the calculation of Eq. (19.3) may be accomplished as follows (where `1i` $= \sqrt{-1} = i$):

```
function [Spqx] = FUNSmat(k,kp,x)
% Calculation of Transfer matrix component

Sx = zeros(2);

Sx(1,1) = (k+kp).*exp( 1i*(kp-k).*x);
Sx(1,2) = (k-kp).*exp(-1i*(k+kp).*x);
Sx(2,1) = (k-kp).*exp( 1i*(k+kp).*x);
Sx(2,2) = (k+kp).*exp(-1i*(kp-k).*x);

Spqx = Sx./(2*k);
```

A3.12.2 Matrix multiplication algorithm

Once the matrices defined by Eq. (19.3) are available, from which $\mathbb{S}_n$ is constructed as in Eq. (19.1), then the multiplication of the matrices in the proper order is needed. Implemented in MATLAB, the procedure may be accomplished as follows:

```
%% Program QBarrier
%  Transmission probability calculated using a
%  transfer matrix approach with plane waves
%-----------------------------------------
% Requires:   FUNSmat.m
% Can use:    VxRTD, VxQUAD, or equivalent

%Fundamental Constants and needed terms
m=5.6856297;hbar=0.6582119;

%% Parameters

if exist('iout','var') == 1; else iout  = 0    ;end
if exist('rm','var')   == 1; else rm    = 0.3  ;end
if exist('N','var')    == 1; else N     = 100  ;end
if exist('Vx','var')   == 1;
    Nv = length(Vx)-1;
else
    % Default case if no Vx, xj available:  RTD-like
    % k(j) is the value of kappa to right of x(j)
    % Potential structure beings at x(2)
```

```
    Nv   = 4;
    Vx = 0.2*[ 0.0, 1.0, 0.0, 1.0, 0.0];
    xj =      [-1.0, 0.0, 2.0, 6.0, 8.0];
end
ko = sqrt(2*rm*m*Vx)/hbar;
if exist('kmax','var') == 1; else kmax=2.0*max(ko) ;end
if exist('xj','var') == 1; else xj =(0:Nv)*8.0/Nv ;end

%% Solve for Dk
% next line puts k on non-rational basis
k = 3.1416*kmax*(1:N)/(N*pi);
Dk = zeros(1,N);
for i = 1:N;
    Sx = [1,0;0,1];
    for j = Nv:-1:1;
        kj  = sqrt(k(i)^2 - ko( j)^2);
        kjp = sqrt(k(i)^2 - ko(j+1)^2);
        Sx = FUNSmat(kj,kjp,xj(j+1))*Sx;
    end
    Dk(i) = 1.0./(abs(Sx(1,1))^2);
end

if iout > 0;
    Hout = [k',Dk'];
    filenam= ['DkTMA.txt'];
    dlmwrite(filenam,Hout,'delimiter',...
                 '\t','precision',6);
end;
```

A3.12.3 Quadratic barrier $V(x)$ algorithm

The following algorithm, implemented in MATLAB, creates V_j and x_j for use by the algorithms of Sections A3.12.1 and A3.12.2 for the case of a quadratic barrier potential profile. Setting $N_b = 2$ results in a single rectangular barrier with the same height and width as the $N_b = 2$ case of Section A3.12.4. The vectors x_p and V_p are for visualization purposes (they serve no function in the $\mathbb{S}_n$ matrix).

```
%% Program VxQuad
%  Potential Barrier associated with Nb-barrier RTDs

%% Input terms
if exist('iout','var') == 1; else iout = 0   ;end
if exist('L','var')   == 1; else L   =  5.0  ;end
if exist('Vo','var')  == 1; else Vo  =  0.3  ;end
if exist('rm','var')  == 1; else rm  =  0.25 ;end
if exist('Nb','var')  == 1; else Nb  =  5    ;end

%% Set up
Nb = max(Nb,2); %Avoids Nx < 2 catastropies
xj = (1:Nb); xj = L*(2*xj - Nb - 1)/(Nb-1);
Vx = Vo*(1.0 - (xj/L).^2);
Nv = length(Vx);
```

```
%% Plot the potential
Nv = length(Vx);
for j = 1:Nv-1;
    xp(2*j-1) = xj(j);
    Vp(2*j-1) = Vx(j);
    xp(2*j)   = xj(j+1);
    Vp(2*j)   = Vx(j);
end
xp(2*Nv-1) = xj(Nv);
Vp(2*Nv-1) = Vx(Nv);
xp(2*Nv)   = xj(Nv) - xj(1);
Vp(2*Nv)   = Vx(Nv);

plot(xp,Vp,'-_o','MarkerSize',8);
hold on;
plot(xj,Vx,'r_.','MarkerSize',30);
hold off;


if iout > 0;
    Hout = [xp',Vp'];
    filenam= ['Q_Vxp.txt'];
    dlmwrite(filenam,Hout,'delimiter','\t', ...
             'precision',6);
    Hout = [xj',Vx'];
    filenam= ['Q_Vx.txt'];
    dlmwrite(filenam,Hout,'delimiter','\t', ...
             'precision',6);
    clear Hout filenam
end
```

A3.12.4 Resonant tunneling diode/superlattice $V(x)$ algorithm

The following algorithm, implemented in MATLAB, creates V_j and x_j for use by the algorithm of Section A3.12.1 for the case of a resonant tunneling diode (RTD) potential profile. Setting $N_b = 2$ results in a single rectangular barrier with the same height and width as the $N_b = 2$ case of Section A3.12.3. Setting $N_b = 3$ results in the prototypical RTD [183]. The vectors x_p and V_p are for visualization purposes (they serve no function in the $\mathbb{S}_n$ matrix).

```
%% Program VxRTD
%  Potential Barrier associated with Nb-barrier RTDs

%% Input terms
if exist('Lbar','var')  == 1; else Lbar  =  5.0   ;end
if exist('Lwel','var')  == 1; else Lwel  =  5.0   ;end
if exist('Vo','var')    == 1; else Vo    =  0.3   ;end
if exist('rm','var')    == 1; else rm    =  0.25  ;end
if exist('Nb','var')    == 1; else Nb    =  1     ;end
if exist('iout','var')    == 1; else iout = 0     ;end

%% Set up first Barrier-Well-Barrier
xj(1) = -Lwel;
Vx(1) = 0.0;
xj(2) = 0.0;
```

```
Vx(2)  = Vo;
xj(3)  = xj(2) + Lbar;
Vx(3)  = 0.0;

Nv = length(Vx);

%% Add more
xjo = xj - xj(1); %xjo is unit RTD structure (position)
Vxo = Vx; %Vxo is the unit RTD structure (potential)
for j =2:Nb;
    Nv = length(Vx);
    xj  = [xj,xj(Nv) + xjo];
    Vx  = [Vx,Vxo];
end

%% Plot the potential
Nv = length(Vx);
for j = 1:Nv-1;
    xp(2*j-1)  = xj(j);
    Vp(2*j-1)  = Vx(j);
    xp(2*j)    = xj(j+1);
    Vp(2*j)    = Vx(j);
end
xp(2*Nv-1)  = xj(Nv);
Vp(2*Nv-1)  = Vx(Nv);
xp(2*Nv)    = xj(Nv) - xj(1);
Vp(2*Nv)    = Vx(Nv);

plot(xp,Vp,'-␣o','MarkerSize',8);
hold on;
plot(xj,Vx,'r␣.','MarkerSize',30);
hold off;

if iout > 0;
    Hout = [xp',Vp'];
    filenam= ['RTD_Vxp.txt'];
    dlmwrite(filenam,Hout,'delimiter',...
          '\t','precision',6);
    Hout = [xj',Vx'];
    filenam= ['RTD_Vx.txt'];
    dlmwrite(filenam,Hout,'delimiter',...
          '\t','precision',6);
    clear Hout filenam
end
```

A3.12.5 Screened ("Yukawa") coulomb potentials

The following algorithm, implemented in MATLAB, compares the screened Coulomb potentials for the copper-like and silicon-like parameters of Section 21.7. For Cu, κ is given by the Thomas–Fermi formula of Eq. (21.85). For Si, κ is given by the Debye formula of Eq. (21.88). The values of R are twice the covalent radius of the atom, and the lattice is assumed to be cubic.

```
%% Program Vxions
% Thomas-Fermi and Debye shielded Coulomb Barriers

%Fundamental Constants and needed terms:  units
%  are in eV, nm, fs, and q = 1
m=5.6856299; hbar=0.65821193; Qo = 0.35999112;
kb = 1.0/11604.519; ao = 0.052917721;

% Vxkap = V(x) / (4Q/R) and is dimensionless
Vxkap = @(x,c) -exp(-c*abs(x))./abs(x);

if exist(   'Nx','var') == 1; else     Nx =      64    ;end
if exist(    'N','var') == 1; else      N =      16    ;end
if exist( 'Vmax','var') == 1; else Vmax   =     1.0    ;end
if exist( 'Vmin','var') == 1; else Vmin   =    -8.0    ;end
if exist('Vmin2','var') == 1; else Vmin2  =    Vmin    ;end
if exist( 'xmax','var') == 1; else xmax   =     1.2    ;end

%% kappa factors

% Thomas Fermi Screening Factor (copper)
RCu = 2*0.138; ChemCu = 7.0;
kf = sqrt(2*m*ChemCu)/hbar;
kapCu = sqrt(4*kf/(pi*ao)); alfcu = kapCu*RCu;

%Debye Screening factor (silicon)
RSi = 2*0.111; Ks = 11.9; T = 300;
Q = (Ks-1.0)*Qo/(Ks + 1.0);
rho = 1.0e18 /(1e7)^3;
kapSi = sqrt(16*pi*Q*rho/(kb*T)); alfsi = kapSi*RSi;

%% Single Ion
x = -xmax + 2*xmax*((1:Nx)-1)/(Nx-1);

VSio  = Vxkap(x,alfsi);
VCuo  = Vxkap(x,alfcu);

%% cubic array of ions (2N+1)^3 atoms
VSi = zeros(1,Nx); VCu = zeros(1,Nx);
for ix = 1:Nx;
    for j = -N:N;
        for k = -N:N;
            for l = -N:N;
                rs = sqrt((j - x(ix)).^2 + k.^2 + l.^2);
                VSi(ix) = VSi(ix) + Vxkap(rs,alfsi);
                VCu(ix) = VCu(ix) + Vxkap(rs,alfcu);
            end
        end
    end
end
VSi = VSi - max(VSi);
VCu = VCu - max(VCu);
```

```
figure(1); plot(x,VSio,x,VCuo);
axis([x(1) x(Nx) Vmin Vmax])
figure(2); plot(x,VSi,x,VCu);
axis([x(1) x(Nx) Vmin Vmax])

Vout = [x',VSio',VSi',VCuo',VCu'];
dlmwrite('VxionsOut.txt',Vout,'delimiter',...
             '\t','precision',6)
```

A3.13 Semiconductors and doping density

The following algorithm, implemented in MATLAB, finds the temperature dependence of the chemical potential $\mu(T)$ and the carrier concentration (taken to be electrons) $\rho_e(T)$ under the assumption that the acceptor doping concentration $N_A = 0$. The algorithm makes use of the approximation of Eq. (22.22). The algorithm is the basis of Figures 22.5, 22.6, and 22.8. It relies on a bisection search algorithm to find $\mu(T)$, knowing that $-E_g/2 \leq \mu(T) \leq \mu_o$.

```
%% PROGRAM Doping.m
%  Evaluation of the Fermi level for a
%  given donor doping level
%  Units are eV, nm, fs, q = 1

% Fundamental constants:
% [kb] = eV/K; [mo] = eV/c^2; [hbar] = eV-fs
kb = 1.0/11604.5192; mo = 5.685629853;
hbar = 0.6582119282;
Nc = sqrt(2*mo^3)/((pi^2)*(hbar^3));

% FD formula
fxe = @(x)  1.0./(1.0 + 2*exp(x));

if exist('Nd','var')   == 1; else Nd   =  1.e-6 ;end
if exist('Nt','var')   == 1; else Nt   =  51    ;end
if exist('Nb','var')   == 1; else Nb   =  24    ;end
if exist('Tmin','var') == 1; else Tmin =  6.0   ;end
if exist('Tmax','var') == 1; else Tmax = 1006.0 ;end
if exist('Eg','var')   == 1; else Eg   =  1.17  ;end
if exist('Ed','var')   == 1; else Ed   = -0.045 ;end
if exist('C0mn','var') == 1; else C0mn = -Eg/2  ;end
if exist('C0mx','var') == 1; else C0mx =  10.5  ;end


T = Tmin + (Tmax - Tmin)*((1:Nt)-1)/(Nt-1);
Mu = zeros(size(T)); rho = Mu;

MRX = zeros(Nt,5);


for j = 1:Nt;
    beta = 1.0/(kb*T(j));
    cNv = 0.5*sqrt(pi)*(Nc/beta^1.5);
    Cmax = C0mx;
```

```
    Cmin = C0mn;
    for k = 1:Nb;
        Chem = 0.5*(Cmax+Cmin);
        be =  beta*Chem;
        bh = -beta*(Chem + Eg);
        rhoe = cNv * exp( beta*Chem);
        rhoh = cNv * exp(-beta*(Chem+Eg));
        Ndp  = Nd*fxe(beta*(Chem - Ed));
        if((rhoh - rhoe) + Ndp < 0.0);
            Cmax = Chem;
        else
            Cmin = Chem;
        end
    end
    Mu(j) = 0.5*(Cmax+Cmin);
    rho(j) = cNv * exp(beta*Mu(j))/Nd;
end
MRX(:,1) = T; MRX(:,2) = Mu; MRX(:,4) = rho;


for j = 1:Nt;
    beta = 1.0/(kb*T(j));
    cNv = (Nc/beta^1.5);
    Cmax = C0mx;
    Cmin = C0mn;
    for k = 1:Nb;
        Chem = 0.5*(Cmax+Cmin);
        be =  beta*Chem;
        bh = -beta*(Chem + Eg);
        rhoe = cNv * FUNfp( beta*Chem);
        rhoh = cNv * FUNfp(-beta*(Chem+Eg));
        Ndp  = Nd*fxe(beta*(Chem - Ed));
        if((rhoh - rhoe) + Ndp < 0.0);
            Cmax = Chem;
        else
            Cmin = Chem;
        end
    end
    Mu(j) = 0.5*(Cmax+Cmin);
    rho(j) = cNv * exp(beta*Mu(j))/Nd;
end
MRX(:,3) = Mu; MRX(:,5) = rho;


Edx = Ed*ones(1,Nt); Egx = -0.5*Eg*ones(1,Nt);
figure(1);plot(T,Mu,T,Edx,T,Egx);
xlabel('Temp_[K]','FontSize',18);
ylabel('\mu_[eV]','FontSize',18);
Txt = ['log(Nd)_=_', num2str(log10(Nd))];
title(Txt,'FontSize',18);

Edx = ones(1,Nt);
```

```
figure(2);plot(T,rho,T,Edx);
xlabel('Temp_[K]','FontSize',18);
ylabel('\rho_/N_d','FontSize',18);
Txt = ['log(Nd)_=_', num2str(log10(Nd))];
title(Txt,'FontSize',18);
axis([0 Tmax 0 3])

filnam = ['DopeNd' int2str(-int8(log10(Nd))) '.txt']

dlmwrite(filnam,MRX,'delimiter','\t','precision',8)
```

A3.14 Band bending: accumulation layer

The following algorithm, implemented in MATLAB, finds $\phi(x)$ in an accumulation layer at the surface of a semiconductor exposed to a vacuum field F_{vac} using silicon-like parameters and using the approximation of Eq. (24.32). Quantities such as F_β, evaluated in the examples of Section 24.3, are also obtained. Output is shown in the graphs of Figure 24.6.

```
%% Evaluation of Phi(x) for an accumulation region
%  Silicon-like parameters for semiconductor surface
%  under a vacuum field
%-----------------------------------------

%%Functions:  Analytic F(phi)
%-----------------------------------------
Fphi = @(x)  sqrt(exp(x) - 1 - x);

%%Parameters
%-----------------------------------------
if exist('Chemo','var') == 1; else Chemo   =   -0.20 ;end
if exist('Nx','var')    == 1; else Nx      =   40    ;end
if exist('T','var')     == 1; else T       =   300.0 ;end
if exist('Ks','var')    == 1; else Ks      =    11.9 ;end
if exist('Phis','var')  == 1; else Phis    =   0.020 ;end
if exist('idebug','var')== 1; else idebug  =    0          ;end

%Fundamental Constants and needed terms
Q = 0.3599911;m=5.6856297;hbar=0.6582119;kb=1.0/11604.5;

Nc   = sqrt(2*m^3)/((pi^2)*(hbar^3));
beta = 1.0/(kb*T);
Fb   = sqrt((8*pi*Q/Ks)*(Nc/beta^2.5));
Fo   = Fb*sqrt(0.5*sqrt(pi)*exp(beta*Chemo));

Phi = Phis*(Nx + 1 - (1:Nx))/Nx;
u   = beta*Phi;
Fld = Fo*Fphi(u);
Fvac = Ks*Fo*Fphi(beta*Phis);
rho = Nc*sqrt(pi)*exp(beta*Chemo)/beta^1.5;
xj  = zeros(size(Phi));

if Chemo < 0.0;
    Lamda = sqrt(2)/(Fo*beta);
```

```
else
    Lamda = sqrt(Ks/(4*sqrt(Chemo)*pi*Q*Nc));
end

for j = 2:Nx;
    xj(j) = xj(j-1) + 2*(Phi(j) - Phi(j-1))/...
            (Fld(j) + Fld(j-1));
end

PhiD = Phis*exp(xj/Lamda);

plot(xj,Phi,'o',xj,PhiD);
```

A3.15 Simple ODE solvers

A3.15.1 First order

An algorithm is given for the simple numerical solution of a one-dimensional equation of the form

$$\partial_x u(x) = v(x) \tag{A3.18}$$

and applied to the trial function $u(x) = x\cos(6x) + 1$. The number of points is $N = 2n$ with a default value of $n = 8$. The algorithm is designed so as not to require memory for the creation of a coefficients matrix $\mathbb{M}$ or necessitate its inversion, nor require memory space to create the $|v\rangle$ vector because the $|u\rangle$ is overwritten (see discussion in Section A1.4.1). The algorithm was used in Figure A1.4.

```
%% Program GradientM
%  demo of a simple ODE solution for du/dx = v(x)

% Create a vector of length N = 2n
if exist('n','var') == 1; else n = 8 ;end

% Sample definition of u(x) and du/dx for a smoothly
% varying but integrable function
ux = @(x)  x.*cos(6*x)+1.0;
du = @(x)  cos(6*x)-6*x.*sin(6*x);

dx = 1.0/(2*n+1);
x = (1:2*n)*dx;

v = 2*dx*du(x); uo = ux(x);

% initialize the u vector
u = zeros(1,2*n);

% upward solution for even j
u(2) = v(1) + ux(0);
for j =2:n;
    u(2*j) = u(2*j-2) + v(2*j-1);
end

% downward solution for odd j
u(2*n-1) = u(2*n-2) + 0.75*v(2*n-1) - 0.25*v(2*n);
```

```
for j = 2:n;
    u(2*n + 1 - 2*j) = u(2*n+3-2*j) - v(2*n+2-2*j);
end

% plot output
plot(x,u,'r',x,uo,'k.')

% % Output matrix
MRX = zeros(2*n,3);
MRX(:,1) = x'; MRX(:,2) = u'; MRX(:,3) = ux(x)';
dlmwrite('ode1.txt',MRX,'delimiter','\t','precision',8)
```

A3.15.2 Second order

An algorithm is given for the simple numerical solution of a one-dimensional equation of the form

$$\partial_x^2 u(x) = v(x) \tag{A3.19}$$

such as can be used in the solution of Poisson's equation (Eq. (24.48)). The algorithm is designed so as not to require memory for the creation of a coefficients matrix $\mathbb{M}$ or necessitate its inversion, nor require memory space to create the $|v\rangle$ vector as the $|u\rangle$ is overwritten (see discussion in Section A1.4.2). The algorithm is again (as in Section A3.15.1) applied to the trial function $u(x) = x\cos(6x) + 1$. As before, the number of points is $N = 2n$ with a default value of $n = 8$. The algorithm was used in Figure A1.4. The two components are:

- The preparation of the initial vectors is done by `Program Poisson`
- The matrix solution for $\mathbb{M} \cdot \vec{u} = \vec{v}$ is done by `FUNode2`

```
%% Program Poisson
%  demonstration of a simple ODE solution for
%       (d/dx)^2 u(x) = v(x)
%  The boundary conditions are uo = u(0), uN = u(1)
%      where x is normalized to between 0 and 1
%      that is, x(j) = j/(N+1), so that x = 0
%      corresponds to j = 0, and x = 1 corresponds
%      to j = N+1

% Create a vector of length N
if exist('n','var') == 1; else N = 16 ;end

% Sample definition of u(x) and du/dx for a
% smoothly varying but integrable function
ux = @(x)  x.*cos(6*x)+1.0;
du = @(x)  -12*(3*x.*cos(6*x) + sin(6*x));

% create normalized x vector
dx = 1.0/(N+1);
x = (1:N)*dx;

% create v vector, then call ODE2 solver
v = (dx^2)*du(x);
u = FUNode2(v,ux(0),ux(1));

% plot output, then export output to tab-delimited file
plot(x,u,'r',x,ux(x),'ko')
```

```
MRX = zeros(N,3);
MRX(:,1) = x'; MRX(:,2) = u'; MRX(:,3) = ux(x)';
dlmwrite('ode2.txt',MRX,'delimiter','\t','precision',8)
```

```
function ux = FUNode2(u,uo,uN);
%% Solution to the equation (d/dx)^2 u(x) = v(x)
% On input, u is the v vector
% On completion, u is overwritten with the solution

%% Initialization
N = length(u);
u(1) = u(1) - uo;  % left boundary condition
u(N) = u(N) - uN;  %right boundary condition

%% Solution (Matrix inversion)
% increasing j
for j =2:N;
    u(j) = j*(2*u(j) + u(j-1))/(j+1);
end

% decreasing j
for j = N-1:-1:1;
    u(j) = u(j) + j*u(j+1)/(j+1);
end

%% Final factor multiplication and output solution
ux = -0.5*u;
```

An example of the usage of FUNode2 for the solution of Poisson's equation for doping in a thin diamond film was encountered in Section 24.4 for the generation of Figure 24.14. The code for that figure is:

```
%% Program DopePhi
%  finding potential in diamond for constant doping of
%  nitrogen and boron.  Boron is throughout;  Nitrogen
%  extends only to xn.  Width of diamond is L

% Create a vector of length N and a subregion of nc
if exist('N','var')  == 1; else N  =  24 ;end
if exist('nc','var') == 1; else nc =  6 ;end
% Set up doping rho_b for boron, and rho_n for nitrogen
%   and let rho_b = nb*rho0, rho_n = nn*rho0
%   where rho0 = 10^16 atoms/cm^3.
if exist('nb','var') == 1; else nb =  0.01 ;end
if exist('nn','var') == 1; else nn =  2.0  ;end
if exist('R','var')  == 1; else  R =  0.03 ;end

% Create dimensionless s vector (call it x temporarily)
dx = 1.0/(N+1);
x = (1:N)*dx;
sn = (nc + 0.5)*dx;

% Doping Profiles (signs are explicitly included)
boron = -nb*ones(1,N);
```

```
nitro =   nn*[ones(1,nc) zeros(1,N-nc)];
v = nitro + boron;

%% Analytic solution:  create each s in region as needed
s   = x(1:nc);
ya  = 0.5*(nn - nb)*s.^2 + (nb - nn*sn - R)*s;
s   = x(nc+1:N);
yb  = -0.5*nb*(s.^2) + (nb - R)*s - 0.5* nn *sn^2;
s = x; y = [ya, yb];

%% Numerical Solution:  create v vector, then call ODE2
%  Observe the usage of the analytic solution to find
%  the right boundary uN for the purposes of comparing
%  to the analytic solution (this would normally just
%  be chosen to a desired value).
v = (dx^2)*v;
uo = 0.0; uN = -0.5* nb + (nb - R) - 0.5* nn *sn^2;
u = FUNode2(v,uo,uN);

% plot output, then export output to tab-delimited file
% The numerical solution is u; the analytic is y
plot(s,y,'r',s,u,'ko');

MRX = zeros(N,3);
MRX(:,1) = x'; MRX(:,2) = u'; MRX(:,3) = y';
dlmwrite('dope.txt',MRX,'delimiter','\t','precision',8)

clear v ya yb MRX x y
```

A3.16 Current through a metal–insulator–metal diode

The current density through the barrier created by an insulator between two metal boundaries, as schematically shown in Figure 24.1, is found through the numerical implementation of Eq. (24.12) in the following algorithm. For convenience, a parameter p is introduced and defined by the relation $p = f^{1/3}$. Output is written to a file labeled by the bias value.

```
%% Program DkMIM
%  Solution of simple rectangular barrier under bias
%  Note:  standard theory used f^(1/3).  Let p = f^(1/3)

%% Needed terms

if exist('re','var')     == 1; else re    =  0.5   ;end
if exist('Phib','var')   == 1; else Phib  =  1.0   ;end
if exist('Vb','var')     == 1; else Vb    =  3.0   ;end
if exist('L','var')      == 1; else L     =  5.0   ;end
if exist('N','var')      == 1; else N     =  200   ;end
m=re*5.6856297;hbar=0.6582119;

%%  Main part of Program
Field = Vb/L;
p = (2*m*Field/hbar^2)^(1./3);
```

```
ko = sqrt(2*m*Phib)/hbar;
kb = sqrt(2*m*Vb)/hbar;

if exist('kmax','var') == 1; else kmax = 5.0  ;end
if exist('kmin','var') == 1; else kmin = 1.0  ;end
if N > 1;
    k = kmin + (kmax-kmin)*((1:N)-1)/(N-1);
else
    k = kmin;
end

%% Begin matrix solution
% Airy(k,z):  k = 0 -- Ai(z), k = 1 -- Ai'(z)
%             k = 2 -- Bi(z), k = 3 -- Bi'(z)

Dk = zeros(1,N);

for ik = 1:N;
    kp = sqrt(k(ik)^2 + kb^2);
    zb  = (ko^2 - k(ik)^2 - L*p^3)/p^2;
    zo = (ko^2 - k(ik)^2)/(p^2);

    % right boundary
    Sgl = [1.0, 1.0; 1i*kp, -1i*kp];
    Trans = [-p*airy(1,zb), -airy(0,zb); ...
                p*airy(3,zb), airy(2,zb)];
    Sgl = Trans*Sgl;

    % left boundary
    Trans = [airy(2,zo), airy(0,zo); ...
                -p*airy(3,zo), -p*airy(1,zo)];
    Sgl = Trans*Sgl;
    Trans = [k(ik), -1i; k(ik), 1i];
    Sgl = (pi/(2*p*k(ik)))* Trans*Sgl;

    Dk(ik) = (kp/k(ik))/abs(Sgl(1,1))^2;
end

Dkfn = (4.0*k/ko).* ...
            exp(-(4.0/(3*p^3))*(max(ko^2 - k.^2,0.0).^1.5));
Dkfn = min(1,Dkfn);

%% Plots and Output

figure(1); plot(k,Dkfn,k,Dk);
xlabel('k_[1/nm]','FontSize',16);
ylabel('D(k)','FontSize',16)
legend('FN','Dk');
figure(2); semilogy(k,Dkfn,k,Dk);
xlabel('k_[1/nm]','FontSize',16);
ylabel('D(k)','FontSize',16)
legend('FN','Dk');
```

```
Hout = [k',Dk',Dkfn'];
filenam = ['Dk',num2str(10*Vb),'.txt'];
dlmwrite(filenam,Hout,'delimiter','\t','precision',6);
```

A3.17 Field emission from semiconductors

The evaluation of field emission from semiconductors requires the determination of the band bending associated with a given applied vacuum field, and then its incorporation into a modified Fowler–Nordheim-like equation. The algorithm is written in two parts. In the first, the vacuum field at the surface is related to the extent of the band bending factor ϕ. Because F_{vac} is the desired parameter for showing results on a Fowler–Nordheim plot, a method is then given to find the maximum ϕ_{max} and minimum ϕ_{min} associated with the largest and smallest F_{max} and F_{min}. This can be done quickly using the methods of Section A3.7.2 applied to Eq. (24.27) (indeed, the example in that section is quite similar). The ϕ associated with a particular F can be found by

$$\beta\phi_{j+1} = \ln\left\{\beta\phi_j + 1 + \left(\frac{F_{max}}{F_o}\right)^2\right\} \tag{A3.20}$$

and analogously for F_{min}, where

$$F_o = \frac{K_s\sqrt{4\pi^3}N_cQe^{\beta\mu_o}}{K_s\beta^{5/2}} \tag{A3.21}$$

with $\phi_0 = 1$, and j ranging from 1 to approximately 10 (for good convergence). Observe also that the work function Φ is replaced by $\chi - \phi$, although the code treats Φ as χ so as to keep the correspondence to the usual implementation clear.

The method works for finding both the maximum and minimum associated ϕ. Letting the ϕ be uniformly spaced gives a range of fields for the subsequent evaluations of J_F. The algorithm to set up these evaluations is:

```
%% Parameters
%fundamental Constants in units of eV, fs, nm, q
    rkb = 1.0/11604.50529;
    rhbarc = 197.3269631; rhbar = 0.658211899;
    rmo = 5.685629657; afs = 1.0/137.0359997;
%end fundamental constants

gx = @(x)  sqrt(exp(x) - 1.0 - x);

if exist('T','var')     == 1; else T    = 300.0;end
if exist('Phi','var')   == 1; else Phi  = 4.0   ;end
if exist('Nf','var')    == 1; else Nf   = 100    ;end
if exist('re','var')    == 1; else re   = 0.2   ;end
if exist('Ks','var')    == 1; else Ks   = 8.0   ;end
if exist('mu','var')    == 1; else mu   = -0.1   ;end
if exist('Fmax','var')   == 1; else Fmax   = 10.0 ;end
if exist('Fmin','var')   == 1; else Fmin   = 1.0 ;end
if exist('Nloop','var')   == 1; else Nloop   =  10 ;end

%% F(phi) plots
Nc = sqrt(2*(re*rmo)^3)/(rhbar*(pi*rhbar)^2);
beta = 1.0/(rkb*T);
Q = 0.25*afs*rhbarc;
Cof = sqrt(4*sqrt(pi^3)*Nc*Q*exp(beta*mu)/ ...
        (Ks*beta^2.5));
```

```
% Use iteration to find the proper pmax and pmin
Fo = Ks*Cof;
pmax = 1.0;pmin = 1.0;
for j = 1:Nloop;
    pmax = log(beta*pmax + 1 + (Fmax/Fo)^2)/beta;
    pmin = log(beta*pmin + 1 + (Fmin/Fo)^2)/beta;
end
Fmax = Fo*gx(beta*pmax);
Fmin = Fo*gx(beta*pmin);

p = pmin + (pmax-pmin)*(0:Nf)/Nf;
Fp = Fo*gx(beta*p);
figure(2); plot(p,Fp);
xlabel('p [eV]'); ylabel('F(p)_[eV/nm]');

%% Standard Fowler Nordheim Plots
Jf = zeros(size(Fp)); Js = Jf;
for j = 0:Nf;
    Jf(j+1) = FUNfns(Fp(j+1),4.5,T,1.0,1.e10);
    Js(j+1) = FUNfns(Fp(j+1),Phi-p(j+1),T,re,Ks);
end

x = 1.0./Fp;
FNf = log(Jf./Fp.^2);
FNs = log(Js./Fp.^2);

figure(1); plot(x,FNf,x,FNs,'o');
xlabel('1/F'); ylabel('ln(J/F^2)');

Vout = [p',Fp',x',FNf',FNs'];
dlmwrite('FNsemicon.txt',Vout,'delimiter',...
            '\t','precision',6)


%% clean up
clear rkb rhbarc rhbar rmo afs Acm2 Nc Q Cof j
```

The calls to the semiconductor field emission current calculation method are made using the algorithm

```
%% The Fowler-Nordheim Current Density
%  modified by effective mass re and dielectric Ks
%  This function call is self-contained:  all
%     constansts and functions
%       required are embedded.
%  Input units:  eV fs nm q=1
%  Outut units:  Amp/cm^2
%  Arguments:  F, Phi, T, re, Ks
%           for re = 1, Ks > 1, gives metal FN equation
function Jfns = FUNfns(F,Phi,T,re,Ks)

%fundamental Constants in units of eV, fs, nm, q
    rkb = 1.0/11604.5192; rhbarc = 197.3269719;
```

```
    rhbar = 0.6582119282; rmo = 5.685629853;
    afs = 1.0/137.03600; Acm2 = 1.602176565e10;
%end fundamental constants

% Semiconductor-modified FN-like terms
eta = (Ks - 1.0)/(Ks + 1.0);
Q = 0.25*afs*rhbarc;
ys = sqrt(4*eta*Q*F)/Phi;

% Rule out barrier being pulled down past Fermi level;
% Do so in way that plotting does not show the bad data.
if ys > 1.0;
    Jfns = NaN;
    return
end

% Forbes functions:  a little faster to code because of
% changes to Q and Phi
vys = 1.0 + (ys.^2).*(log(ys) - 3.0)/3;
tys = 1.0 + (ys.^2).*(1.0 - log(ys))/9;

% FN terms
Afn = (re^1.5)/(16*(pi^2)*rhbar*Phi*tys.^2);
Bfn = (4.0/(3*rhbar))*sqrt(2*re*rmo*Phi^3)*vys;

x   = (2*rkb*T/rhbar)*sqrt(2*re*Phi)*tys/F;

% Temperature Effects
sig = ((1.0 + x.^2)./(1.0 - x.^2)) - ...
          (2.0 - ((pi^2)/6))*x.^2 + ...
      0.25*(((7*pi^4)/90)-8)*x.^4;

% The final equation
Jfns = Afn*(F.^2).*exp(-Bfn./F)*sig*Acm2;
```

These algorithms were used for the preparation of Figures 24.7 (relation between F and ϕ) and 24.8 for the representation of J_F in Fowler–Nordheim coordinates.

A3.18 Roots of the quadratic image charge barrier

A method of rapidly finding the roots of the equation $V_o - Fx - (Q/x) + \gamma x^2 = 0$ is given, based on the initial choices of Eq. (24.63) used in the recursion relation of Eq. (24.62) as per the iterative prescription of Section A3.7.2.

```
%% Numerical Evaluation of the Roots of
%  V(x) = Vo - Fx - Q/x + gx^2

Xn = @(x,V,F,Q,g)  (g*x.^3 - V*x + 2*Q).*x...
                   /(Q - F*x.^2 + 2*g*x.^3);

Q = 0.3599911214;

if exist('Vo','var') == 1; else Vo = 4.0   ;end
if exist('F','var')  == 1; else F  = 2.0   ;end
```

```
if exist('g','var')  == 1; else g  = 0.22   ;end
if exist('Ks','var') == 1; else Ks = 8.0   ;end
if exist('Nx','var') == 1; else Nx = 12     ;end

Qs = Q/Ks;

Ra = zeros(Nx,3);

% Zeroth order approximations
Ra(1,1)  = 2*Qs/(Vo + sqrt(Vo^2 - 4*Qs*F));
Ra(1,3)  = (F^3 + Qs*g^2)/(g*(F^2 + g*Vo));
Ra(1,2)  = Qs/(g*Ra(1,1)*Ra(1,3));

% Iterative Solution
for j = 2:Nx;
    for ix = 1:3;
        Ra(j,ix)  = Xn(Ra(j-1,ix),Vo,F,Qs,g);
    end
end

% Output
dlmwrite('VxRoots.txt',Ra,'delimiter',...
       '\t','precision',6)
x = [x',y'];
dlmwrite('vx.txt',x,'delimiter','\t','precision',6)
```

A3.19 Zeros of the airy function

The following algorithm starts with Eq. (26.22) to produce an approximate value of w_n (called y in the code) for which $Ai(w_n) = 0$, and improves it using Newtonian iteration via Eq. (26.23) to provide the values used in Table 26.1. The algorithm then implements Eq. (26.28) to make Figure 26.5.

```
%% Create a table of the first 20 Airy Function zeros
clear; clc
%Control flags
iout = 1;

% Defining functions
wn    = @(x)  -(x.^(2./3)).*(1 + (5./(48*x.^2)));

if exist('N','var') == 1; else N = 32;end

n = (1:N);
M = zeros(N,3);

M(:,1)  = n;
x = 3*pi*(4*n-1)/8;

% reasonable estimate of the zeros
y = wn(x);
M(:,2)  = y;
```

```
% High accuracy: iterative improvement
% by Newtonian Iteration
y = y - (airy(0,y)./airy(1,y));
y = y - (airy(0,y)./airy(1,y));
M(:,3) = y;

if iout > 0;
    filenam= 'AiZeros.txt';
    dlmwrite(filenam,M,'delimiter','\t','precision',10);
end

clear M n filenam x

%% unscaled Graph
Nx = 100;
s = -2.0*((1:Nx)-1)/(Nx-1);
Na = [2, 4, 8, 16];
Srho = zeros(4,Nx);
M = zeros(Nx,5);
M(:,1) = s;

for j = 1:4;
    for k = 1:Na(j);
        for i = 1:Nx;
            Srho(j,i) = Srho(j,i) + (Na(j)^2 - k^2)*...
                (airy(0,y(k) - s(i))./airy(1,y(k))).^2;
        end
    end
    Srho(j,:) = Srho(j,:)/Na(j)^2;
    M(:,1+j) = Srho(j,:);
end

if iout > 0;
    filenam= 'UnscaledAiS.txt';
    dlmwrite(filenam,M,'delimiter','\t','precision',10);
end
figure(1); plot(s,Srho)

%% scaled Graph
Nx = 200;
s = -2.0*(((1:Nx)-1)/(Nx-1)).^2;
Srho = zeros(4,Nx);
M = zeros(Nx,5);
M(:,1) = s;

for j = 1:4;
    for k = 1:Na(j);
        for i = 1:Nx;
          Srho(j,i) = Srho(j,i) + (Na(j)^2 - k^2)*...
          (airy(0,y(k) - Na(j)*s(i))./airy(1,y(k))).^2;
        end
    end
```

```
    Srho(j,:) = Srho(j,:)/Na(j)^2;
    M(:,1+j) = Srho(j,:);
end

if iout > 0;
    filenam= 'ScaledAiS.txt';
    dlmwrite(filenam,M,'delimiter','\t','precision',10);
end
figure(2); plot(s,Srho)
```

A3.20 Atomic sphere radius r_s

The following algorithm first calculates the data points used to construct Figure 27.3, and then finds the minimum of Σ_ε as a function of r_s for a range of r_i, which is used to construct Figure (27.4). Observe that Ekxco is given by Eq. (27.22) and DEkxco is its derivative with respect to r_s (represented by x in the algorithm). Similarly, EIcor is the sum of $\varepsilon_{ee} + \varepsilon_{ei}$, and DEIcor is its derivative with respect to r_s. Lastly, after using the MATLAB routine MIN to find the approximate location of the minimum, four iterations using finite different approximations (Section A1.3.2) are used to solve the quadratic equation

$$y(x_o + \delta) = 0 = y(x_o) + \delta\, y'(x_o) + \frac{1}{2}\delta^2\, y''(x_o) \tag{A3.22}$$

to find the better estimate $x_o + \delta$ that is used to find corrections to x_o. The toggle $C_i = 0$ turns off the inclusion of $\varepsilon_{ee} + \varepsilon_{ei}$; $C_i = 1$ includes it.

```
%% Program AtomicR
%  Evaluation of minimum of Sigma-eps: a method of
%  determining rs if the ionic radius of a metal
%  (valency = 1) is given

% Formulae
%  kinetic, exchange, and correlation function
%  Ci = 0 turns off the Ground state energy
%     = 1 turns on  the Ground state energy
Ci = 0;

% Kinetic + Exchange + Correlation & Deriviative
Ekxco = @(x) (2.21./x.^2) - (0.9163./x) -...
             0.874./(2.946 + x + 3.362*sqrt(x));
DEkxco = @(x) (-4.42./x.^3) + (0.9163./x.^2) +...
             0.874*(1.0 + 1.681./sqrt(x))./...
             ((2.946 + x + 3.362*sqrt(x)).^2);
%  Ion core & Derivative
EIcor  =@(x,a) 3*(-3*x.^2 + 5*a.^2)./(5*x.^3);
DEIcor =@(x,a) 9*(   x.^2 - 5*a.^2)./(5*x.^4);

N = 50; s = (0:N-1)/(N-1);
x = -0.01+exp(log(100)*s);
Na = 5;
a= (0:Na-1);
y = zeros(N,Na);
for i = 1:N;
    for j = 1:Na;
        y(i,j) = Ekxco(x(i)) + Ci*EIcor(x(i),a(j));
```

```
    end
end

Mx = [x',y];
dlmwrite('IonCore.txt',Mx,'delimiter',...
            '\t','precision',6)
clear x y Mx i j Na N s
%% Find rs (here called x) given ri (here called a)

Del = 0.004;      % Used to make finite difference...
                      %... approximations
Na = 41;          % Sets size of the ionic radii vector a
a = 6*(((1:Na)-1)/(Na-1)).^1.5;  % Gives better spacing
rs = zeros(1,Na);
iout = 0;         % turn on to monitor error assessment

for i = 1:Na;
    % find minimum using methods based in MIN function
    x = 0.1 + 8.0*(0:100)/100;
    y = Ekxco(x) + Ci*EIcor(x,a(i));
    [~, io] = min(y); xo = x(io);
    for j = 1:4;
      fp = DEkxco(xo + Del) + Ci*DEIcor(xo + Del,a(i));
      fm = DEkxco(xo - Del) + Ci*DEIcor(xo - Del,a(i));
      fo = DEkxco(xo)       + Ci*DEIcor(xo,a(i));
      xo = xo-4*Del*fo/(fp - fm +...
           sqrt(abs((fp-fm)^2+8*fo*(2*fo-fm-fp))));
    end
    rs(i) = xo;
    if iout > 0;
        disp(['dEkxc/dr␣=␣',...
           num2str(DEkxco(xo) + Ci*DEIcor(xo,a(i)))]);
    end;
end
clear Del fm fo fp i j io Na x y xo yo

% Compare to data Kittel (Table 9 p 78 & Table 1 p 126)
%    of "Introduction to Solid State Physics, 7th Ed."
% aex is converted to ri by dividing by Bohr radius
%      [    Li,      Na,       K,      Rb,...
%           Cs,      Cu,      Ag,      Au]
aex = [1.2850, 1.8330, 2.5132, 2.7967,...
       3.1557, 2.5510, 2.3810, 2.5888];
rex = [3.2500, 3.9300, 4.8600, 5.2000,...
       5.6300, 2.6700, 3.0200, 3.0100];

plot(a,rs,aex,rex,'o'); axis([0 4 1 6]);

Mx = [a',rs'];
dlmwrite('rvsa.txt',Mx,'delimiter','\t','precision',6);
```

A3.21 Sodium exchange-correlation potential

The following algorithm evaluates the electron density $\rho(x)$ for the infinite barrier using Eq. (26.5), forms a hyperbolic tangent approximation $\rho_a(x)$ to it using Eqs (26.52) and (26.18), then uses $\rho_a(x)$ to construct the exchange correlation potential using only ε_{ex} (Section 27.2) and ε_{cor} (Section 27.3).

```
%% Program ExCor
%  Evaluation of the potential profile near a surface
%  based on simple exchange-correlation calculation
%  Hardwired to parameters for Sodium and neglect of
%     ion core term

% Hyperbolic Density approximation
Rhoax  =@(x,a,b) a./(1.0 + exp(b*x));

%lambda from Friedel model
lam = (64*sqrt(2)/(9*pi^4))*(16 + 12*pi - 3*pi^2);
mo = 5.6856299; hbar = 0.65821193; ao = 0.052918;

% Sodium Parameters
chem = (hbar*kf).^2/(2*mo);  % Chemical Potential of Na
kf = sqrt(2*mo*chem)/hbar; % fermi wavenumber
rs = (3.0/(4*pi*rho*ao^3))^(1.0/3); %Wigner-Seitz radius
Phi = 2.29;                    % Work function of Na
xi = - (3*pi)/(8*kf);          % positive background
F = 2.0;                       % ad hoc field

Nx = 201;
z = -10*pi + 24*pi*(0:Nx-1)/(Nx-1);
x = (z/(2*kf)) - xi;

Vf = F*max(x,0.0);

rhox = zeros(1,Nx); Vimg = rhox; Vxc = rhox;
rhoa = Rhoax(kf*x,rho,lam);
for i = 1:Nx;
    rhox(i) = rho*FUNXic(z(i));
    Vimg(i) = Vimage(x(i),chem+Phi,F);
    Vxc(i) = FUNxcor(rho) - FUNxcor(rhoa(i)) - Vf(i);
end

Mx = zeros(Nx,4);
Mx(:,1) = rhox; Mx(:,2) = rhoa;
Mx(:,3) = Vxc; Mx(:,4) = Vimg;
figure(2);plot(x,Mx);

Mx = [x',Mx];
dlmwrite('NaXCor.txt',Mx,'delimiter',...
             '\t','precision',6);
```

A3.22 Field-dependent work function

The following algorithm evaluates the current density using Eq. (13.25) and compares the cases where the work function is field dependent ($\Phi \rightarrow \Phi_o + 2Fx_o$) and where it is not ($\Phi_o$). The resulting curve is fitted to a quadratic equation.

```
%% Program Animag
%  Comparison of Analytic Image Charge to standard
%  Fowler Nordheim equation

% constants
hbar = 0.658212;
m = 5.68563;
Q = 0.359991;
Phio = 4.5;
chem = 7.0;
Nf = 41;

% Field Emission: Jfo = straight FN, Jf = analytic image
% Method:  make a vector of F values.  Evaluate xo using
%
F = 2.0 + 8*(0:Nf-1)/(Nf-1);
xo = hbar./sqrt(2*m*(chem + Phio - sqrt(4*Q*F)));
Phi = Phio + 2*F.*xo;

Jfo = zeros(1,Nf); Jf = Jfo;
for i = 1:Nf;
    Jfo(i) = FUNjfn(F(i),Phio);
    Jf(i)  = FUNjfn(F(i),Phi(i));
end
y = Jf./Jfo;
P = polyfit(F,y,2); ya = polyval(P,F);
figure(2); plot(F,y,'o',F,ya);

Mx = [F',Phi',Jfo',y',ya'];
dlmwrite('AnFN.txt',Mx,'delimiter','\t','precision',6);
```

A3.23 Digitizing an image file

The following algorithm takes a gray-scale PNG image file (specifically, the ones shown in Figure 29.9) and returns a matrix of numerical values that can be processed using MATLAB. In the present circumstance, to give Figure 29.10, the maximum values were cut off at z_{max} to allow the regions of low intensity to be seen in greater contrast, and a floor cut off to bring the low intensity regions into greater contrast trough the usage of z_{min}. A vertical white line can be imposed on the data by setting the elements along that line to z_{max}.

Because digitized data from images is noisy (which makes the line plots derived from them, such as Figure 29.11, rather spiky), a smoothing algorithm was employed by the name of `SmoothZ`, the effects of which are to replace each element of the image matrix Z_{ij} by a weighted average. If Z_i is a vector of values, then it amounts to

$$Z_i' = \frac{Z_{i+1} + R_o Z_i + Z_{i-1}}{R_o + 2} \quad \text{(A3.23)}$$

and the extension to a matrix Z_{ij} is straightforward. The parameter `nloop` controls the number of times the smoothing process is implemented. The effect is to mildly blur the image.

```
clc;clear;
Ro = 2; nloop = 0; zmax = 60; zmin = 5; Nx = 512;
io = 205;

%% Raw first
Z1=double(imread('18.png'));max(max(Z1))
figure(1); surf(Z1); axis tight square; shading flat;
axis([1 Nx 1 Nx 0 100]); view(2); colormap gray;
axis off; grid off;
%% processed first
Z=Z1; SmoothZ; Z1 = Z;
Z1 = max(Z1,zmin) - zmin; Z1 = min(Z1,zmax);
za = Z1(:,io);
Z1(:,io) = zmax;
figure(2); surf(Z1); axis tight square; shading flat
axis([1 Nx 1 Nx 0 (zmax)]); view(2); colormap hot;
axis off; grid off;

%% Raw second
Z2=double(imread('23.png'));max(max(Z2))
figure(3); surf(Z2); axis tight square; shading flat;
axis([1 Nx 1 Nx 0 100]); view(2); colormap gray;
axis off; grid off;
%% processed second
Z=Z2; SmoothZ; Z2 = Z;
Z2 = max(Z2,zmin) - zmin; Z2 = min(Z2,zmax);
zb = Z2(:,io);
Z2(:,io) = zmax;
figure(4); surf(Z2); axis tight square; shading flat
axis([1 Nx 1 Nx 0 (zmax)]); view(2); colormap hot;
axis off; grid off;

%% line plot
figure(5); xa = (1:Nx); xb = xa+16;
plot(xa,za,'r',xb,zb,'b');
axis([10 500 0 60]);
legend('18','23');
Mx = zeros(Nx,4);
Mx(:,1) = xa; Mx(:,2) = za; Mx(:,3) = xb; Mx(:,4) = zb;
dlmwrite('CsW.txt',Mx,'delimiter','\t','precision',6)
```

A3.24 Lattice gas algorithm

A two-dimensional grid of sites is randomly populated with adatoms so that they preferentially cluster near the center. At each time step, the atoms are allowed an opportunity to jump to an adjacent site if one is available; the order the atoms are chosen is based on a *shuffle* algorithm, modeled after a version given by Knuth ("Algorithm P: (shuffling)", Ref. [411], p. 139), which shuffles the order of the indices of a vector. This algorithm is used to generate Figure 29.15. The shuffling algorithm is called `shuffle(Nc)`.

```
%% Program HoppingCs
% A simple algorithm to allow point particles
% to hop from site to site on a square lattice.
```

```
% The length scale is set by the radius
% of the initial distribution.

% Pore distribution function
Pore = @(x,a) 1.0./(1.0 + exp(a*(x.^2 - 1)));

% Set up Initial State;
% ro controls the box size; a controls sharpness of
% the initial distribution; po controls probabiity
% of a jump.
% N is tne number of coordinates on a box side
% Nt is the number of time recordings
% Ns is the number of time steps between recordings
% Nr is the number of r coordinates to plot

if exist('ro','var') == 1; else ro =  2 ;end
if exist( 'a','var') == 1; else a  = 16 ;end
if exist('po','var') == 1; else po =0.2 ;end
if exist( 'N','var') == 1; else N  =128 ;end
if exist('Nt','var') == 1; else Nt =  6 ;end
if exist('Ns','var') == 1; else Ns =250 ;end
if exist('Nr','var') == 1; else Nr = 16 ;end

x = (2*(1:N)-N-1)*ro/(N-1);y = x;
z = Pore(x,a);
figure(1); plot(x,z); clear z

% Initial distribution
Mx = zeros(N); Mxo = Mx;
R = rand(N);ic = zeros(1,N^2); jc = ic;
Nc = 0;
for i = 1:N;
  for j = 1:N;
    if R(i,j) < Pore(sqrt(x(i)^2 + y(j)^2),a);
      Nc = Nc+1; ic(Nc) = i; jc(Nc) = j;
    end;
  end;
end;
ic = ic(1:Nc);
jc = jc(1:Nc);
xc  = (2*ic -N-1)*ro/(N-1);
yc  = (2*jc -N-1)*ro/(N-1);

% rho is the radial distribution;
% vecxy is the matrix of occupied sites
rho = zeros(Nt,Nr); vecxy = zeros(2*Nt,Nc);
vecxy(1,:) = xc; vecxy(2,:) = yc;
for i = 1:Nc; Mxo(ic(i), jc(i)) = 1; end;
rho(1,:) = FUNrhosum(x,y,Mxo,ro,Nr);

ico = ic; jco = jc;
```

```
for k = 2:Nt;
  for l = 1:Ns;
    % set up hopping directions
    ihop = floor(4*rand(1,Nc));
    % see if they hop, and if the hop-to
    % space is empty, move them
    % Hop directions:
    % 0 = +x; 1 = +y; 2 = -x; 3 = -y
    % shuffle the vectors:
    ix = shuffle(Nc);
    ic = ic(ix); jc = jc(ix);
    for i = 1:Nc; Mx(ic(i), jc(i)) = 1; end;
    for i = 1:Nc;
      if rand() < po;
        switch ihop(i);
          case 0
            n = ic(i)+1; if n > N;n = 1 ;end;
            if Mx(n,jc(i)) < 1;
              Mx(n,jc(i)) = 1; Mx(ic(i),jc(i)) = 0;
              ic(i) = n;
            end;
          case 1
            n = jc(i)+1; if n > N;n = 1 ;end;
            if Mx(ic(i),n) < 1;
              Mx(ic(i),n) = 1; Mx(ic(i),jc(i)) = 0;
              jc(i) = n;
            end;
          case 2
            n = ic(i)-1; if n<1;n = N ;end;
            if Mx(n,jc(i)) < 1;
              Mx(n,jc(i)) = 1; Mx(ic(i),jc(i)) = 0;
              ic(i) = n;
            end;
          otherwise
            n = jc(i)-1; if n<1;n = N ;end;
            if Mx(ic(i),n) < 1;
              Mx(ic(i),n) = 1; Mx(ic(i),jc(i)) = 0;
              jc(i) = n;
            end;
        end;
      end;
    end;
  end;
  rho(k,:) = FUNrhosum(x,y,Mx,ro,Nr);
  vecxy(2*k-1,:) = (2*ic -N-1)*ro/(N-1);
  vecxy(2*k,:)   = (2*jc -N-1)*ro/(N-1);
end;

clear i j R

% Two ways to plot Mx that are equivalent:
xco = (2*ico-N-1)*ro/(N-1); yco=(2*jco-N-1)*ro/(N-1);
```

```
xc  = (2*ic -N-1)*ro/(N-1); yc =(2*jc -N-1)*ro/(N-1);
figure(2); plot(xco,yco,'bo',xc,yc,'r.');
axis square tight; axis([-ro ro -ro ro]);
figure(3); plot(rho'); axis square tight;

dlmwrite('Rhot.txt',rho','delimiter',...
            '\t','precision',6)
dlmwrite('vecxy.txt',vecxy','delimiter',...
            '\t','precision',6)

function ix = shuffle(Nc)
% returns array of indicies that will shuffle vector
vx = (1:Nc);
for i = Nc:-1:1;
    io = floor(i*rand())+1;
    vo      = vx(i);
    vx(i)   = vx(io);
    vx(io)  = vo;
end
ix = vx;

function rx = FUNrhosum(x,y,Mx,ro,Nr)
N = length(x);
rx = zeros(1,Nr);
for i = 1:N;
    for j = 1:N;
        n = floor((Nr+1)*sqrt(x(i)^2 + y(j)^2)/ro)+1;
        if Nr+1 > n;rx(n) = rx(n) + Mx(i,j) ;end;
    end;
end;
nx = (1:Nr)-0.5; rx = rx./nx;
```

A3.25 Evaluation of the point charge model functions

The following algorithm evaluates $S_n(r)$, constructs $P_n(r)$ from the iteration algorithm of Eq. (30.32), and then evaluates $\beta_n(r)$ using Eq. (30.35). Lastly, $\beta_n(r)$ is approximated by $\Delta\beta^{n-1}\beta_o$ using Eqs (30.38) and (30.48) (Schottky's conjecture). The output was used in Figure 30.9.

```
%% Program PnrCalc
%  calculation of Pn(r) using a recursion relation

if exist('r','var') == 1; else r = 0.75;end
if exist('Np','var') == 1; else Np = 25;end

betao = @(x) 1.0 + ((2*(1 + x)^2)./(x.*(2+x)));

S = zeros(1,Np); P = S;
S = [S,(1.0 - r.^(Np+1))/(1.0 - r)];  %S longer than P

%% Create Sn(r):
S(1) = 1.0;
if r < 0.99;
    for j = 2:Np; S(j) = (1.0 - r^j)/(1-r); end;
```

```
else
    for j = 2:Np; S(j) = 1.0 + r*S(j-1); end;
end

%% Create Pn(r):
P(1) = 0.5*(1.0 + r)*(2.0 + r);
for n = 2:Np;
    P(n) = 0.5*S(n+1);
    for j = 1:n-1;
        P(n) = P(n) -  ...
                   (P(j)*S(j)/(S(n+1-j)*(S(j) + S(n+1))));
    end
    P(n) = P(n)*(1.0 + (S(n+1)/S(n)));
end

%% Create Betan(r):

beta = ones(1,Np);
for n = 1:Np;
    for j = 1:n;
        beta(n) = beta(n) + (4*P(j)*S(j)*S(n+1)./...
            ((r^j)*(S(n+1-j)*(S(n+1) + S(j)))^2));
    end;
end;

%% Approximate beta
x = (1:Np);
% ao = -(d/dx) ln(Px(r));
ao = polyfit(x,log(P),1); ao = -ao(1);
bo = betao(r);
delb = 1.0/(r*exp(ao));
betaA = zeros(1,Np);
betaA(1) = bo;
for j = 2:Np;
    betaA(j) = delb*betaA(j-1);
end;

semilogy(x,beta,'o',x,betaA);
xlabel('n','FontSize',24);
ylabel('\beta_n(r)','FontSize',24)
legend('exact','approx')

S = S(1:Np); %make S, P, and beta the same size again;

%% Send< to Output file
M = zeros(Np,5);
M(:,1) = x';M(:,2) = S';M(:,3) = P';
M(:,4) = beta';M(:,5) = betaA';
dlmwrite('Pnr.txt',M,'delimiter','\t','precision',6)
```

A3.26 Modeling of Field emitter *I*(*V*) data

The algorithm takes the experimental $I(V)$ data of Figure 30.10 (based on Figure 2 of ref. [458] and reproduced in Table A3.2 using the field enhancement and area factor models of Eqs (30.18) and (30.19) based on the hemisphere model) and analyzes it using field enhancement and notional emission area concepts. Refinements of the hemisphere model using a prolate spheroidal model will change $g(F)$, but will not dramatically change the theoretical $I(V)$ curve. The algorithm used a single data point labeled by $i_o = 3$ to make an estimate of the tip radius a_s using one of the lower current data points, after which Eq. (30.124) provides the apex field value and Eq. (30.18) provides the estimate of the tip current. The default parameters of $a_g = 375$ nm, $\theta_c = 15°$, and $\Phi = 4.3$ eV are assumed based on fabricated geometry and accepted values. The method of bisection (Section A3.7.1) is used to find a_s. The input data of Table 30.10 is presumed to exist as comma-delimited numerical data in a file called `Figdat.txt`.

```
%% Program SchwoebelFig
%  Take digitally extracted data from
%  P. Schwoebel, et al., JVSTB, Fig 2 and analyze it
%  using the field enhancement + notional area model

% Load Data

load Figdat.txt;
Siv = Figdat; clear Figdat;

%% Assign data:

V1 = Siv(:,1); I1 = Siv(:,2);
V2 = Siv(:,3); I2 = Siv(:,4);
V3 = Siv(:,5); I3 = Siv(:,6);
V4 = Siv(:,7); I4 = Siv(:,8);

%% Analyze Data

betag = @(x,g,b) ((pi/log(((86 + (g/x))/...
    (54*tan(b)))*(g/x))) - (tan(b)^2))/x;

% Input values:  Mo work function of 4.3 eV
% recommended by Samsonov
```

Table A3.2 $I(V)$ digitally extracted data from Figure 2 of ref. [458] and shown in Figure 30.10 for the initial and final voltage (V) and current (I).

V1	I1	V2	I2	V3	F1	V4	F2
56.30	1.014×10^{-8}	95.79	9.864×10^{9}	107.1	1.014×10^{8}	109.6	1.052×10^{8}
58.78	2.026×10^{8}	101.6	2.008×10^{8}	112.1	2.063×10^{8}	114.3	2.101×10^{8}
61.55	5.037×10^{8}	107.4	4.947×10^{8}	117.1	5.129×10^{8}	119.5	5.320×10^{8}
64.58	9.881×10^{8}	111.1	9.797×10^{8}	122.6	1.006×10^{7}	124.0	1.025×10^{7}
67.35	2.010×10^{7}	116.1	1.940×10^{7}	128.1	2.010×10^{7}	129.8	2.010×10^{7}
72.04	4.907×10^{7}	124.0	4.914×10^{7}	135.8	5.182×10^{7}	137.8	5.089×10^{7}
76.18	9.803×10^{7}	131.8	9.468×10^{7}	142.7	9.626×10^{7}	144.4	9.983×10^{7}
80.60	1.958×10^{6}	139.7	1.927×10^{6}	150.7	1.994×10^{6}	152.7	2.031×10^{6}
88.33	4.868×10^{6}	155.9	4.880×10^{6}	161.5	4.868×10^{6}	163.7	4.957×10^{6}
94.96	9.725×10^{6}	164.6	9.664×10^{6}	170.4	1.009×10^{5}	173.7	9.904×10^{6}
101.0	1.908×10^{5}	175.8	1.862×10^{5}	180.9	2.015×10^{5}	182.8	1.943×10^{5}

```
if exist('N','var') == 1;    else N     = 76; end
if exist('io','var') == 1;   else io    = 3; end
if exist('ag','var') == 1;   else ag    = 375.0; end
if exist('bc','var') == 1;   else bc    = 15*pi/180; end
if exist('Phi','var') == 1;  else Phi   = 4.3;  end

% Fowler Nordheim parameters (numerical values
% for constants used)
Bo = 6.8308898*sqrt(Phi^3);
nu = 1.6393731/sqrt(Phi);
Ao = (1.368477e-6 / Phi)*(280.16579*Phi^2)^nu;

an = zeros(1,4);

for k = 1:4;
    amax = 50.0;
    amin = 0.5;
    Io = Siv(io,2*k);
    Vo = Siv(io,2*k-1);
    for j = 1:24;
        ao = (amax + amin)/2;
        F = betag(ao,ag,bc)*Vo;
        Itest = Ao*(F^(2-nu))*...
                exp(-Bo/F)*2*pi*(ao^2) *...
            F/(Bo + F*(4 - nu));
        if Itest > Io; amin = ao; else amax = ao;end
    end
    an(k) = (amax + amin)/2;
end

Th = zeros(N,4);
Vx = 50 + 150*(0:N-1)/(N-1);
for j = 1:N;
    F = betag(an(1),ag,bc)*Vx(j);
    Th(j,1) = Ao*(F^(2-nu))*...
            exp(-Bo/F)*2*pi*(an(1)^2) *...
        F/(Bo + F*(4 - nu));
    F = betag(an(2),ag,bc)*Vx(j);
    Th(j,2) = Ao*(F^(2-nu))*...
            exp(-Bo/F)*2*pi*(an(2)^2) *...
        F/(Bo + F*(4 - nu));
    F = betag(an(3),ag,bc)*Vx(j);
    Th(j,3) = Ao*(F^(2-nu))*...
            exp(-Bo/F)*2*pi*(an(3)^2) *...
        F/(Bo + F*(4 - nu));
    F = betag(an(4),ag,bc)*Vx(j);
    Th(j,4) = Ao*(F^(2-nu))*...
            exp(-Bo/F)*2*pi*(an(4)^2) *...
        F/(Bo + F*(4 - nu));
end;
```

```
semilogy(V1,I1,'o',V2,I2,'s',V3,I3,'d',V4,I4,'v',Vx,Th);
xlabel('Voltage_[V]','FontSize',24);
ylabel('Current_[A]','FontSize',24);
axis([50 190 5e-9 5e-5]);
```

A3.27 Modeling a log-normal distribution of field emitters

In modeling an array of field emitters for which the apex radius $\{a_i\}$ is log-normal distributed, a conventional approach might be to choose the radius with a probability based on Eq. (29.1), but here a much faster algorithm is presented. First, observe that the number n_i of emitters with a radius a_i is given by

$$n_i = N \int_{a_{i-1}}^{a_i} L(x)dx = \frac{1}{2}(Erf(A_i) - Erf(A_{i-1})) \tag{A3.24}$$

where $Erf(x)$ is the error function of Eq. (A2.10), N is the total number of emitters, and

$$A(a) = \frac{1}{\sqrt{2}\sigma} \ln\left(\frac{a}{\mu}\right) \tag{A3.25}$$

The emitters are then placed sequentially on the array and their positions are "shuffled" by the `Shuffle(Nc)` algorithm of Section A3.24. As a result, the array has randomly placed emitters that are log-normal distributed.

A3.27.1 Driver

The `Driver` routine loads the parameters, finds the maximum and minimum a_j values, whereby $n_j = 0$ for larger or smaller a_j, and plots the results. Finding the maximum and minimum a_j values is accomplished by `zoomas`, a relatively cheap method of redistributing the a_j values over a region where the n_j are non-zero.

```
%% Program Driver:
%  This program populates values needed and runs StatTip

if exist('Fo','var')   == 1; else Fo    = 5.0; end
if exist('sig','var')  == 1; else sig   = 0.08;end
if exist('ao','var')   == 1; else ao    = 16;  end
if exist('amin','var') == 1; else amin  = 0;   end
if exist('amax','var') == 1; else amax  = 50;  end
if exist('nj','var')   == 1; else nj    = 32;  end
if exist('N','var')    == 1; else N     = 128; end

% Cheap method to find the max/min values of as
for j = 1:5
    StatTip;
    zoomas;
end

% Mo plots the sorted Jfo; M plots random placement Jo
M = zeros(N);Mo = zeros(N);
Jmax = max(Jf);
for j = 1:N;
    Mo(:,j)  = Jfo((j-1)*N+1:j*N)/Jmax;
    M(:,j)   = Jf((j-1)*N+1:j*N)/Jmax;
end
```

```
% Set at least one tip to zero current so surface plot
% better shows the distribution of current
Mo(N-1,N-1) = 0.0;
M(N-1,N-1) = 0.0;

%% Plot results
figure(1);
plot(as,ntips/max(ntips),as,jtips/max(jtips));
figure(2); surf(Mo); axis square tight;
    shading flat; view(2); colormap hot
figure(3); surf(M); axis square tight;
    shading flat; view(2); colormap hot

%% Send to Output file
Mout = zeros(nj,3);
Mout(:,1) = as';Mout(:,2) = ntips';Mout(:,3) = jtips';
filn = ['sig',num2str(1000*sig),'.txt'];
dlmwrite(filn,Mout,'delimiter','\t','precision',6);
clear Mout filn aso kmin Ao Bo j nu Phi F
```

A3.27.2 StatTip

Each value in the vector `Jf` is the total tip current for an ellipsoidal emitter with apex field $F(a_j) = F_{tip}a_o/a_j$, where $a_o \leftrightarrow \mu$ of the log-normal distribution and F_{tip} is the field on the median tip. The vector `Jf` is shuffled and mapped over to the array matrix $[\mathbb{M}]_{ij} = M(i,j)$. When $\mathbb{M}$ is plotted as a surface map, each pixel mimics the output of the tip. Figure 30.29 was generated in this way for various values of σ.

```
%% Program StatTip
%  Create I(F) data for a statistical distribution

if exist('nj','var') == 1;   else nj    = 24;  end
if exist('N','var') == 1;    else N     = 64;  end
if exist('ao','var')   == 1; else ao    = 8;   end
if exist('amin','var') == 1; else amin  = 2;   end
if exist('amax','var') == 1; else amax  = 24;  end
if exist('Phi','var') == 1;  else Phi   = 4.5; end
if exist('Fo','var') == 1;   else Fo    = 3.0; end
if exist('sig','var') == 1;  else sig   = 0.4; end
if exist('ish','var') == 1;  else ish=shuffle(N^2);end

% Fowler Nordheim parameters (numerical)
Bo = 6.8308898*sqrt(Phi^3);
nu = 1.6393731/sqrt(Phi);
Ao = (1.368477e-6 / Phi)*(280.16579*Phi^2)^nu;

% Tip radii:  aso prevents maximum field from
% exceeding 12 eV/nm by "blunting" the tip to
% aso it does
aso = max(Fo*ao/12.0,amin);
as = aso + (amax-aso)*(0:nj-1)/(nj-1);
F = Fo*ao./as;
```

```
Jf = zeros(1,N^2); jtips = zeros(size(as));
ntips = floor(N^2*GenLN(as,ao,sig));

% eliminate the larger as that do not have tips
% absorb the left-over tips into the largest tip cell
ntips(nj) = N^2 - sum(ntips(1:nj-1));
for j = 1:nj;
    if ntips(nj-j) < 1;
        ntips(nj-j) = ntips(nj-j+1);
        ntips(nj-j+1) = 0;
    else
        break;
    end
end

% Calculate the tip current for each tip
kmin = 0;
for j = 1:nj;
    if ntips(j) > 0;
        Jtemp = Ao*(F(j)^(2-nu))* ...
                exp(-Bo/F(j))*2*pi*(as(j)^2) * ...
            F(j)/(Bo + F(j)*(1 - nu));
        Jf(kmin+1:kmin+ntips(j)) = Jtemp;
        jtips(j) = ntips(j)*Jtemp;
        kmin = kmin + ntips(j);
    end;
end

% Jfo is sorted; Jf is randomized placement
Jfo = Jf*160.2176487;        %convert to microamps
Jf = Jfo(ish)*160.2176487;   %convert to microamps
```

A3.27.3 zoomas

This routine resets a_{max} and a_{min} based on the number of emitters, so that the resulting distribution of a_j contains nj bins for which the number of emitters in the *j*th bin is n_j.

```
for j = 1:nj;
    if ntips(j) < 1;
        amin = as(j);
    else
        break
    end
end
for j = 1:nj;
    if ntips(nj-j+1) < 1;
        amax = as(nj-j+1);
    else
        break
    end
end
```

A3.27.4 GenLN

This routine performs the calculation of the number of emitters n_j for $a_j \le a < a_{j+1}$ for emitters that are log-normal distributed.

```
function Lx = GenLN(as,mu,sig)
% Function GenLN
%  Generate the ntips parameter
%  according to a log-normal distribution

exps = @(x) exp(-x.^2)/sqrt(pi);

nj = length(as);

if exist('sig','var') == 1;  else sig   =  0.2;   end
if exist('mu','var')  == 1;  else mu    =  6;   end
if exist('xmax','var') == 1; else xmax  =  3*mu;   end

xj = log(as/mu)/(sig*sqrt(2));

yj = zeros(1,nj);
yj(1)  = integral(exps,-Inf,xj(1));
for i = 2:nj-1;
    yj(i) = integral(exps,xj(i-1),xj(i));
end
yj(nj) = integral(exps,xj(nj),Inf);

Lx = yj;

end
```

A3.28 Simple shell and sphere algorithm

Two populations of electrons are created at a depth of $\delta = 12.8$ nm into copper: the first population (blue = "shell") is electrons that travel without scattering, and the second population (red = "sphere") is electrons that travel for a time $\tau = 1$ fs then scatter to a random direction with a new energy $E_j = \mu + \hbar\omega/2^j$. The initial velocity of all electrons was set to $v_o = \sqrt{(2/m)(\mu + \hbar\omega)}$ for $\lambda = 532$ nm. All shell electrons are at the same distance $r = (x^2 + y^2 + z^2)^{1/2}$ from the origin, although the plots only show the (x, z) coordinates, as in Figure 31.24. The terms r_a and r_o are the rms radii, and are used to represent the red and blue lines. Lastly, in choosing the random direction, it must be ensured that the choice of *any* direction is equally probable. Although $\phi = 2\pi r$ with r a random number chooses the azimuthal angle, the polar angle requires greater care [108] as the integration over a surface element $d\sigma$ on a sphere of radius a involves $d\sigma = a^2 \sin\theta d\theta d\phi$. Therefore, it is $\cos\theta$ that must be chosen randomly, and so for two random numbers r_1 and r_2,

$$\phi = 2\pi r_1 \tag{A3.26}$$

$$\cos\theta = 1 - 2r_2 \tag{A3.27}$$

```
%-----------------------------------------
% Script ShellSphere
%-----------------------------------------
% Expanding Shell (unscattered) is BLUE
% Sphere (scattered) is RED
```

```
% Scattered electron algorithm:  e- travel for
%   a time tau and then scatter:  when scattered,
%   the electron's energy is reduced as though it
%   had scattered with an electron from the Fermi
%   level (that is, its energy goes to the average
%   energy of the two electrons)
%-----------------------------------------

m = 5.685629853; hbar = 0.6582119282;
c = 299.792458; chem = 7.0;

if exist('N','var')      == 1; else      N = 800;   end
if exist('Nt','var')     == 1; else     Nt = 8;     end
if exist('tau','var')    == 1; else    tau = 1.0;   end
if exist('zinit','var')  == 1; else  zinit = 12.8;  end
if exist('zmin','var')   == 1; else   zmin = 0.0;   end
if exist('lambda','var') == 1; else lambda = 532.0; end

% Assume all e- initially excited to chem + hbar*omega
hf = 2*pi*hbar*c/lambda;
vmax = sqrt((2/m)*(chem + hf));

% circle coordinates
cirx = cos(2*pi*(0:90)/90);
ciry = sin(2*pi*(0:90)/90);

phi = 2*pi*rand(1,N);
costheta = 1.0 - 2*rand(1,N);
sintheta = sqrt(1.0 -costheta.^2);
epx = sintheta.*cos(phi);
epy = sintheta.*sin(phi);
epz = costheta;
vx = vmax*epx; vy =vmax*epy; vz = vmax*epz;
vax=vx;vay=vy;vaz=vz;

xp = zeros(1,N); yp = xp; zp = xp + zinit;
xa = xp;
ya = yp;
za = zp;

% set window size
if exist('zmax','var') == 1; else zmax = 5*zinit; end
xmax = zmax/2; xmin = - xmax;

for j = 1:Nt;
    vo = sqrt((2.0/m)*(chem + (hf/2^j)));
    for i = 1:N
        xp(i) = xp(i) + vx(i)*tau;
        yp(i) = yp(i) + vy(i)*tau;
        zp(i) = zp(i) + vz(i)*tau;
        xa(i) = xa(i) + vax(i)*tau;
        ya(i) = ya(i) + vay(i)*tau;
```

```
        za(i) = za(i) + vaz(i)*tau;
        phi = 2*pi*rand();
        costheta = 1.0 - 2*rand();
        sintheta = sqrt(1.0 - costheta.^2);
        epx = sintheta.*cos(phi);
        epy = sintheta.*sin(phi);
        epz = costheta;
        vax(i) = vo*epx;vay(i)=vo*epy;vaz(i)=vo*epz;
    end

    scatter(xp,zp,8,'filled','b');hold on;
    xcen = sum(xa)/N;
    ycen = sum(ya)/N;
    zcen = sum(za)/N;
    rp = sqrt(sum((xp-xcen).^2...
            +(yp-ycen).^2+(zp-zcen).^2)/N);
    zcen = sum(zp)/N;
    plot(rp*cirx,rp*ciry+zcen,'LineWidth',4,...
            'Color',[0.6 0.6 1]);
    scatter(xa,za,8,'filled','r');
    xcen = sum(xa)/N;
    ycen = sum(ya)/N;
    zcen = sum(za)/N;
    ra = sqrt(sum((xa-xcen).^2 +...
            (ya-ycen).^2 + (za-zcen).^2)/N);
    plot(ra*cirx,ra*ciry+zcen,'LineWidth',4,...
            'Color',[1 0.6 0.6]);
    hold off;
    axis([xmin xmax zmin zmax]);
    axis square; grid off;
    title(['time_=_',num2str(j*tau),'_fs'],...
            'FontSize',24);
    axis off;
end
```

A3.29 Gyftopoulos–Levine work function algorithm

An algorithm for the evaluation of the work function $\Phi(\theta)$ as a function of coverage θ, where $\theta = 1$ is a monolayer, as specified by Eq. (31.146), is given. It relies on the function call `FUNglx` that requires seven input parameters to evaluate $\Phi(\theta)$ on output: $[r_w, r_c, f, n, R_{wf}, \Phi_w, \Phi_c]$. Observe that f is called `fac` and R_{wf} called `rmf` in the body of the algorithm. The default parameters are nominally modeled after cesium on tungsten. Code is inserted for plotting the results and an `io` controls which parameter of f, n, or R_{wf} is changed.

```
%-----------------------------------------
%% Program Gyftopoulos-Levine
%-----------------------------------------
%  Calculation of work function Phi as a
%  function of surface coverage theta

% Parameters
if exist('N','var')    == 1; else N    = 50;    end
```

```
if exist('rw','var')   == 1; else rw   = 0.130;end
if exist('rc','var')   == 1; else rc   = 0.253;end
if exist('fac','var')  == 1; else fac  = 1.0;  end
if exist('n','var')    == 1; else n    = 1.0;  end
if exist('rmf','var')  == 1; else rmf  = 4.0;  end
if exist('Phic','var') == 1; else Phic = 1.6;  end
if exist('Phiw','var') == 1; else Phiw = 4.55; end
if exist('io','var')   == 1; else io   = 1;    end

pxo = [rw, rc, fac, n, rmf, Phiw, Phic];

theta = ((1:N)-1)/(N-1);

switch io
    case 1
        % f variation:  Examine Phi for various fac
        Phix = zeros(N,4);
        px = pxo;
        for j = 1:4;
            px(3) = 0.5 + j*0.25;
            Phix(:,j) = FUNglx(px,theta);
        end

        figure(1);
        plot(theta,Phix)
        xlabel('\theta','FontSize',18);
        ylabel('\Phi(\theta)','FontSize',18);
        legend('f␣=␣0.75','f␣=␣1.00',...
             'f␣=␣1.25','f␣=␣1.50');
        title('Variation␣with␣f','FontSize',24);
        axis([0 1 1.5 4.6]);axis square

        Mx = [theta',Phix];
        dlmwrite('fac.txt',Mx,'delimiter',...
                    '\t','precision',5);

    case 2
        % n variation  Examine Phi for two values of n

        Phix = zeros(N,2);
        px = pxo;
        filn = 'nfac';
        for j = 1:2;
            px(4) = 1 + (j-1)*0.65;
            Phix(:,j) = FUNglx(px,theta);
        end

        figure(2)
        plot(theta,Phix)
        xlabel('\theta','FontSize',18);

        ylabel('\Phi(\theta)','FontSize',18);
```

```
        legend('n␣=␣1.00','n␣=␣1.65');
        title('Variation␣with␣n','FontSize',24);
        axis([0 1 1.5 4.6]);axis square

        Mx = [theta',Phix];
        dlmwrite('nfac.txt',Mx,'delimiter',...
                      '\t','precision',5);

    case 3
        % rmf variation Examine Phi for 2 values of rmf

        Phix = zeros(N,2);
        px = pxo;
        for j = 1:2;
            px(5) = 2*j;
            Phix(:,j) = FUNglx(px,theta);
        end

        figure(3)
        plot(theta,Phix)
        xlabel('\theta','FontSize',18);
        ylabel('\Phi(\theta)','FontSize',18);
        legend('rmf␣=␣2','rmf␣=␣4');
        title('Variation␣with␣rmf','FontSize',24);
        axis([0 1 1.5 4.6]);axis square

        Mx = [theta',Phix];
        dlmwrite('rmf.txt',Mx,'delimiter',...
                      '\t','precision',5);
end

%-----------------------------------------
function G = FUNglx(px,theta)
%-----------------------------------------

Hx = @(x) (2*x+1).*(1 - x).^2;

rw = px(1); rc = px(2); fac = px(3);
n = px(4); rmf = px(5);
Phiw = px(6); Phic = px(7);
ro=0.43652798610891956;
R = rc + rw;
wac = rmf*fac*(rw/rc)^2;
cosb = sqrt(1.0 - (2.0/wac)*(rw/R)^2);
numr = fac*theta*cosb*(ro/rc)^2;
deno = (1.0 + n*(rc/R)^3)*(1+(9*n/8)*(fac*theta).^1.5);
Gx = (numr./deno);
G = Phic + (Phiw-Phic).*Hx(theta).*(1.0 - Gx);
return
```

A3.30 Poisson distributions

A3.30.1 Generation

The Poisson distribution of Eq. (32.16) is generated in one of two ways: first, by simulating random events, where the duration after the last random event is given by $\Delta t_j = -\ln r$, with r a random number and k determined by

$$\sum_{j=1}^{k} \Delta t_j \leq \lambda \tag{A3.28}$$

for three separate times, and, second, by simply plotting Eq. (32.16) as a continuous function of k. The output is the basis of Figure 32.10.

```
%% PoisMaker
%  Given lambda, make a Poisson distribution four
%    different times (four simulations) and compare
%    to the analytical form
%  Terms:
%    np = maximum number of points to be shown
%    Ntotal = a big enough number so that simulation
%              will likely end before ix > Ntotal/lambda
%    iout = a flag to control the generation of outupt
%    lambda = the Poisson parameter
%  Analytic form
%    Pk(t) = (lambda^k / k!) exp(-lambda)

if exist('np','var')      == 1; else np     = 160  ;end
if exist('Ntotal','var')  == 1; else Ntotal = 10000;end
if exist('iout','var')    == 1; else iout   = 1    ;end
if exist('lambda','var')  == 1; else lambda = 8.0  ;end

%% set up maximum values and zeroed out matricies
%  it is unlikely that k will ever exceed 20 * lambda
%  so restrict size of matrix holdng the data to Nkmax
Nkmax = round(lambda*20);
Pk = zeros(3,Nkmax);
Nt = 0;
Nx = round(Ntotal/lambda);

%% Here is the main (numerical) part.
%  j is the simulation number
%  k is the number of events occuring
%  xt keeps adding events until they exceed lambda
%  -log(rand()) is the random time to the next event

for j=1:3;
    for l = 1:Nx;
        xt = 0.0;
        k = 0;
        while xt < lambda;
            xt = xt - log(rand());
            k = k+1;
        end
        Pk(j,k) = Pk(j,k) + 1;
```

```
        Nt = max(Nt,k);
    end
end

%% Here is the generation of the analytical form
%  It makes use of n! = n * Gamma(n),
%  and gammaln = ln(Gamma(n))
%  However many entries Pk contains,
%  just view Nt of them
Nt = max(Nt,round(2.3*lambda));
np = max(Nt,np);

Pk = Pk(:,1:Nt);
kk = (0:Nt-1);

xk = (0:np)*Nt/np;
yk = Nx*exp(-lambda+xk*log(lambda)-gammaln(xk+1.0));

%% Send to Output file
if iout > 0;
    Mout = [xk',yk'];
    dlmwrite('APk.txt',Mout,'delimiter',...
                 '\t','precision',6);
    Mout = [kk',Pk'];
    dlmwrite('NPk.txt',Mout,'delimiter',...
                 '\t','precision',6);
end
```

A3.30.2 Test of acceptance–rejection method

A test of the assignment $\Delta t_j = -\ln r$ as suggested by ref. [411] compares its results for $p_0(\Delta t)$ to that of the acceptance–rejection method. Both are compared to $Np_0(t) = (N\lambda)\exp(-\lambda\Delta t)$. The output is the basis for Figure 32.11.

```
% PoisTest
% Test of Poisson distribution algorithm
% Notes:  Del  is the mean time beween events
%             Delj is the individual time between events
%                  from a Poisson distribution algorithm.
% If r is a random number, then Delj = - Del*ln(r) is
%    amount of time that passes until the "next" event

if exist('np','var') == 1; else np = 12;end
if exist('nb','var') == 1; else nb = 6;end
if exist('Del','var') == 1; else Del = 32.0;end
if exist('iout','var') == 1; else iout = 0;end

Nb = 2^nb;
Np = 2^np;

% METHOD 1: Straightforward
Delj = -Del*log(rand(1,Np));
xdel = ((1:Nb)-0.5)*10*Del/Nb;
```

```
yx(1,:) = (Np*10/Nb)*exp(-xdel/Del);
yx(2,:) = hist(Delj,xdel);

% METHOD 2:  Acceptance / Rejection

p = exp(-1.0/Del);
Delx = zeros(1,Np);

for j = 1:Np;
    for k = 1:20000;
        if rand() > p;
            Delx(j) = k;
            break;
        end;
    end;
end
yx(3,:) = hist(Delx,xdel);
%%
plot(xdel,yx,'o');
legend('Exp','Knuth','AR');

if iout > 0;
    Mout = [xdel',yx'];
    dlmwrite('POIS.txt',Mout,'delimiter',...
             '\t','precision',6);
    disp('FINISHED: Headers = x, Exp, Knuth, AR')
end
```

A3.31 Electron–electron relaxation time

An algorithm is given for the evaluation of the electron–electron relaxation time $\tau_{ee}(k)$ where $E = \hbar^2 k^2/2m$, as in Eq. (32.73) and contained in the matrix $\mathbb{M}$, and compared to its approximation by Eq. (32.78) (see also Eq. (6) of ref. [548] but note that their ε_0 should rather be ε_0^2) and contained in the matrix $\mathbb{M}_a$. Default values are copper-like for the parameters of T [K] = 77, 150, 300, 600, 1000, and 2000. As a numerical feature, observe that $z = \beta(E - \mu)/\pi$: the points chosen are so as to be uniformly spaced on a log–log plot of $\tau_{ee}(k)$ vs z by setting

$$z_j = \exp\left(-5 + 11\frac{(j-1)}{(N-1)}\right) \tag{A3.29}$$

so that $z_1 = 0.00674 < 0.01$ and $z_N = 403 > 300$, where the second numbers are the limits of z in Figure 32.15.

```
%% Tauee
%  Calculation of Lugovskoy-Bray relaxation time

%  fundamental constants and input parameters
kb = 1.0/11604.52; m = 5.68563; hbar = 0.6582119;
Q = 0.3599911; ao = 0.05291772;

if exist('Nx','var')   == 1; else Nx   = 40 ;end
if exist('mu','var')   == 1; else mu   = 7.0;end
if exist('Ks','var')   == 1; else Ks   = 5.2; end
if exist('iout','var')== 1; else iout  = 0;end
```

```
%% set up primary quantities
kf = sqrt(2*m*mu)/hbar;
qo = sqrt(4*kf/(pi*Ks*ao));
xo = 2*kf/qo;
yo = 1.0 /sqrt(1.0 + xo^2);
Gammo = (0.25*xo^3)*(xo*yo^2 + acos(yo) -...
        yo*acos(yo^2)/sqrt(1.0 + yo^2));
T = [77,150,300,600,1000,2000]; Nt = length(T);
z = exp(-5 + 11*((1:Nx) - 1)/(Nx - 1));

%% Initialize Output Matricies

M  = zeros(Nx,Nt);
Ma = zeros(Nx,Nt);

%% Loop through the temperatures
for j = 1:Nt;
    beta = 1.0/(kb*T(j));
    E = mu + (pi/beta)*z;
    k = sqrt(2*m*E)/hbar;
    x  = 2*k./qo;
    y  = 1.0./sqrt(1.0 + x.^2);
    Gamma = (0.25*x.^3).*(x.*y.^2 + acos(y) -...
            y.*acos(y.^2)./sqrt(1.0 + y.^2));
    tauo = (2*hbar*ao/(pi*Q*Gammo))*(Ks*beta*mu)^2;

    M(:, j) = Gammo*tauo*((E/mu).^2)./((z + 1).*Gamma);
    Ma(:,j) = tauo* sqrt(E/mu)./ (z + 1);
end

%% Send to Output file
if iout > 0;
    Mout = [z',M];
    dlmwrite('tauee.txt',Mout,'delimiter',...
                 '\t','precision',6);
    Mout = [z',Ma];
    dlmwrite('taueeA.txt',Mout,'delimiter',...
                 '\t','precision',6);
end
```

A3.32 Resistivity and the Debye temperature

If discrete data $\{x_i, y_i\}$ can be modeled by a relation of the form $y_\lambda(x)$, where λ is a parameter, then a means of finding λ is to minimize the *least squares difference* between the set of $\{y_i\}$ and the values of $y_\lambda(x_i)$ with $i \in \{1, N\}$. That is, λ is adjusted until $S(\lambda)$ defined by

$$S(\lambda) \equiv \sum_{j=1}^{N} (y_i - y_\lambda(x_i))^2 \tag{A3.30}$$

takes on its smallest value at $\lambda = \lambda_o$. In some circumstances, if $y(x)$ grows exponentially large (as in Eq. (32.96)), it can prove more convenient to consider $\ln(y_i)$ instead, in which case,

$$S(\lambda) \to \sum_{j=1}^{N} \left[\ln \left(\frac{y_i}{y_\lambda(x_i)} \right) \right]^2 \tag{A3.31}$$

Near $\lambda \approx \lambda_o$, $S(\lambda)$ is approximately parabolic. Letting $S(\lambda) = C_1\lambda^2 + C_2\lambda + C_3$, then an approximation to λ_o is $\lambda_o \approx -C_2/2C_1$. Therefore first a polynomial fitting to a quadratic occurs, and then λ_o is estimated in the algorithm, or rather, by evaluating $S(\lambda)$ over n points $\lambda_j = p_j\lambda$ and using these points to construct a quadratic polynomial fit to $S(\lambda)$ from which the coefficients C_j are found. Generally, the first guess of T_D is the largest value of T in the data sets (culled from Tables A3.3 and A3.4), but occasionally this is a poor choice, and so an option is included to specify the initial value of T_D from which the iterations begin if needed. Generally, algorithms that require such intervention on the part of the user are not robust, but

Table A3.3 Resistivity of copper (Cu), silver (Ag), and gold (Au) in units of [$\mu\Omega$-cm] from Matula [296].

T [K]	Cu	Ag	Au
1.0000	0.0020000	0.0010000	0.022000
4.0000	0.0020000	0.0010000	0.022000
7.0000	0.0020000	0.0010300	0.022100
10.000	0.0020200	0.0011500	0.022600
15.000	0.0021800	0.0018900	0.025800
20.000	0.0028000	0.0042200	0.034600
25.000	0.0044900	0.0095500	0.050200
30.000	0.0082800	0.019400	0.072500
35.000	0.014700	0.034100	0.10180
40.000	0.023900	0.053900	0.14100
45.000	0.035800	0.077300	0.18100
50.000	0.051800	0.10400	0.22100
55.000	0.072700	0.13200	0.27000
60.000	0.097100	0.16200	0.30800
70.000	0.15400	0.22500	0.39500
80.000	0.21500	0.28900	0.48100
90.000	0.28100	0.35400	0.56600
100.00	0.34800	0.41800	0.65000
125.00	0.52200	0.57300	0.85700
150.00	0.69900	0.72600	1.0610
175.00	0.87400	0.87800	1.2620
200.00	1.0460	1.0290	1.4620
225.00	1.2170	1.1790	1.6620
250.00	1.3870	1.3290	1.8640
273.15	1.5430	1.4670	2.0510
293.00	1.6780	1.5870	2.2140
300.00	1.7250	1.6290	2.2710
350.00	2.0630	1.9320	2.6850
400.00	2.4020	2.2410	3.1070
500.00	3.0900	2.8750	3.9140
600.00	3.7920	3.5310	4.8750
700.00	4.5140	4.2090	5.8160
800.00	5.2620	4.9120	6.8080
900.00	6.0410	5.6380	7.8620
1000.0	6.8580	6.3960	8.9860
1100.0	7.7170	7.2150	10.191
1200.0	8.6260	8.0890	11.486
1300.0	9.5920		12.854
1235.1		8.4150	
1337.6			13.388
1357.6	10.171		

Table A3.4 Resistivity of select metals (aluminum, iron, molybdenum, nickel, platinum, and tungsten) in units of [$\mu\Omega$-cm] from the AIP Handbook, Table (9d-2a) [55].

T [K]	Al	Fe	Mo	Ni	Pt	W
20	0.000755	–	0.00261	0.014	0.0484	0.00196
40	0.0181	0.0758	0.0457	0.068	0.409	0.0544
60	0.0959	0.271	0.206	0.242	1.107	0.266
80	0.245	0.693	0.482	0.545	1.922	0.606
100	0.442	1.28	0.858	0.96	2.755	1.02
150	1.006	3.15	1.99	2.21	4.76	2.09
200	1.587	5.20	3.13	3.67	6.77	3.18
273	2.417	8.57	4.85	6.16	9.60	4.82
293	2.650	9.61	5.34	6.93	10.5	5.28
298	2.709	9.87	5.47	7.12	10.7	5.39
300	2.733	9.98	5.52	7.20	10.8	5.44
400	3.87	16.1	8.02	11.8	14.6	7.83
500	4.99	23.7	10.6	17.7	18.3	10.3
600	6.13	32.9	13.1	25.5	21.9	13.0
700	7.35	44.0	15.8	32.1	25.4	15.7
800	8.70	57.1	18.4	35.5	28.7	18.6
900	10.18	–	21.2	38.6	32.0	21.5

calculations such as these are not something to be repeated, and so it is left as interactive. The interactive lines are under the heading "Driver" with (in MATLAB) the main algorithm `ExtractBD` invoked by entering its name and a return, with each line corresponding to a different metal.

The data sets are assumed to be text files labeled `rhoXX.txt`, where `XX` is a stand-in for the element to be examined, for example for copper, the file is `rhoCu.txt`. Especially with the data from ref. [296], the first few ρ_j, and sometimes the last few, are being affected by processes other than acoustic phonon, and so those data points are excluded using the parameters `ilo` and `ihi`.

```
%% Driver:  calling arguments for Program ExtractBD
clear;          iout = 0; ilo=5; ihi=3; flno='Ag';
ExtractBD;
clear;          iout = 0; ilo=7; ihi=3; flno='Au';
ExtractBD;
clear;          iout = 0; ilo=7; ihi=0; flno='Cu';
ExtractBD;
clear;          iout = 0; ilo=0; ihi=0; flno= 'W';
ExtractBD;
clear; TD=230; iout = 0; ilo=0; ihi=0; flno='Pt';
ExtractBD;
clear; TD=450; iout = 0; ilo=2; ihi=4; flno='Ni';
ExtractBD;
clear;          iout = 0; ilo=0; ihi=0; flno='Al';
ExtractBD;
clear;          iout = 0; ilo=0; ihi=0; flno='Mo';
ExtractBD;
clear;          iout = 0; ilo=0; ihi=0; flno='Pt';
ExtractBD;


%-----------------------------------------
% Program ExtractBD
%-----------------------------------------
% Iterative Search for Debye Temperature from rho(T)
```

```
% data Perform a least-squares search to find what value
% of TD provides the best correspondence to experimental
% rho(T) data.  The input data file (default case) is
% for copper, so load <rhoCu.txt>:  change this if
% considering different conductivity data for other
% metals.
%
% Units of R are assumed to be microOhm-cm, so convert
% it first to Ohm-cm which are the units of this program

if exist('flno','var') == 1; else flno = 'Cu'   ;end
if exist( 'ilo','var') == 1; else ilo  = 5  ;end
if exist( 'ihi','var') == 1; else ihi  = 0  ;end
if exist( 'NL','var')  == 1; else NL   = 8  ;end
fln = ['rho',flno,'.txt'];
Rtmp = load(fln);
T = Rtmp(:,1); R = 1e-6*Rtmp(:,2); clear Rtmp
Nt = length(T);
T = T(ilo+1:Nt-ihi); R = R(ilo+1:Nt-ihi);

% Number of Data points
Nplot = 50;

%% Fit high temp resistivity data to Bloch Grun fit
% Find the Debye Temperature
% least squares search algorithm
Nt = length(T);
if exist('TD','var') == 1; else TD = 200;end
Npoly = 5; pj = zeros(1,Npoly); yj = pj;
pj = 1.0 + 0.01*(2*(1:Npoly) - Npoly - 1)/(Npoly - 1);
for jd = 1:NL;
    yj = zeros(1,Npoly);
    TDj = pj*TD;
    for j = 1:Npoly;
        for k = 1:Nt;
          yj(j) = yj(j) + ...
              (log(((T(k)/T(Nt))^5)*(R(Nt)/R(k))*...
          (FUNbg(TDj(j)/T(k))/FUNbg(TDj(j)/T(Nt))))).^2;
        end
    end
    Cpoly = polyfit(pj,yj,2);
    TD = -TD*Cpoly(2)/(2*Cpoly(1)); TD = max(TD,1);
end
disp([flno,'␣␣TD␣=␣',num2str(TD)])
clear Cpoly ihi ilo j jd k NL Npoly TDj
Rtmp = load(fln);
T = Rtmp(:,1); R = 1e-6*Rtmp(:,2); clear Rtmp
Nt = length(T);

for j = Nt-1:-1:1;
    if TD > T(j);
        rhoD = R(j) + ((R(j+1) - R(j))*...
```

```
                        (TD - T(j))/(T(j+1)-T(j)));
            break
        end
end
disp([flno,'__rhoD_=_',num2str(rhoD)]);

Ra = ((T/TD).^5).*FUNbg(TD./T)/FUNbg(1);

figure(1); loglog(T/TD,R/rhoD,'+',T/TD,Ra,'g',1,1,'ro');
xlabel('T/T_D','FontSize',18);
ylabel('\rho/\rho_D','FontSize',18);
title([flno,num2str(round(TD))],'FontSize',20);
% axis([0.01 10 0.0005 20]);

%% Send to Output file
if exist( 'iout','var') == 1; else iout = 1 ;end
if iout > 0;
    Mout = [T,R,T/TD,R/rhoD,Ra];
    dlmwrite(['TD-',flno,'.txt'],Mout,...
                    'delimiter','\t','precision',6);
end
```

The algorithm uses $W\_(5,x)$, evaluated by `FUNbg(x)`, which in turn uses Eqs (32.88) and (32.92), depending on the size of the argument. The code is:

```
function yx = FUNbg(y)

fx = @(x) (x.^5)./((exp(x) - 1.0).*(1.0 - exp(-x)));
gx = @(x) -(log(x).^5)./((1.0 - x).^2);

zo = 124.4313306172022;
Nx = length(y);
z = exp(-y);
s = zeros(Nx,1);
for j = 1:Nx
    if y(j) < 20.0;
        s(j) = integral(@(x)fx(x),0.0,y(j));
    else
        s(j) = zo - integral(@(x)gx(x),0.0,z(j));
    end
end

yx = s;
```

A3.33 Orbits in a magnetic field

Using the methods of Section 33.1 a Maxwell–Boltzmann distribution of transverse velocities is created according to Eqs (33.7) and (33.8). The output is used to create the data behind Figures 33.1 and 33.2: the `OrbitsFig` file is used in the former, and the `MDvo.txt` file in the later.

```
%%Program Orbits
% find the initial position of N electrons and plot
% their circular orbits about the magnetic field lines
```

```
if exist( 'N','var')      == 1; else N     = 256 ;end
if exist( 'Nt','var')     == 1; else Nt    = 32  ;end
if exist( 'cof','var')    == 1; else cof   = 0.1 ;end
if exist( 'iseed','var') == 1; else iseed = 5   ;end

% Generate Maxwell Boltzmann Distribution of velocities
ra = rand(1,N); rb = rand(1,N);
vo = sqrt(-2*log(ra));
theta = 2*pi*rb + 0.5*pi;
vx = vo.*cos(theta);
vy = vo.*sin(theta);

% Create random uniform distribution of launch sites
ntotal =  0;
rx = zeros(1,N); ry = zeros(1,N);

rng(iseed);
while ntotal < N;
    a = 2*rand()-1; b = 2*rand()-1;
    if sqrt(a^2 + b^2) < 1;
        ntotal = ntotal + 1;
        rx(ntotal) = a;
        ry(ntotal) = b;
    end
end
% subtract off the starting points
rx = rx - cof*vx;
ry = ry - cof*vy;

% Create phi for drawing circles
phi = 2*pi*(0:Nt-1)/(Nt-1);

% Create the matricies xp and yp that contain orbits
xp = zeros(N,Nt); yp = xp;
for i = 1:N;
    xp(i,:) = rx(i) + cof*vo(i)*cos(phi + theta(i));
    yp(i,:) = ry(i) + cof*vo(i)*sin(phi + theta(i));
end;

ymax =  max(max(yp)); xmax = max(max(xp));
ymax = max(ymax,xmax);
ymin = -min(min(yp)); xmin = -min(min(xp));
ymin = max(ymin,xmin);
L = max(ymax,ymin);

% Plot the initial starting points and the orbits
plot(xp(1,:),yp(1,:),'c');
axis([-L L -L L]); axis square
hold on;
for j = 2:N;
    plot(xp(j,:), yp(j,:),'c');
```

```
end
plot(xp(:,1),yp(:,1),'k.');
plot(cos(phi),sin(phi),'r','LineWidth',2)
hold off;
axis off;

clear a b i j ntotal xmax xmin ymin
clear ra rb rx ry iseed vx vy

%% Output to files

print -dpdf OrbitsFig

% Send to Output file
dlmwrite(['MDvo.txt'],vo','delimiter',...
              '\t','precision',6);
```

A3.34 Trajectory of a harmonic oscillator

Using the discrete methods of Section 33.4.1, the evaluation of the trajectory of a one-dimensional harmonic oscillator for which $x(t) = cos(\omega t)$ is compared to its evaluation using a leap-frog (Eq. (33.59)) and an average value (Eq. (33.62)) method. The output is the basis for Figure 33.16.

```
%-------------------------------------------
% PROGRAM TrajectoryHO
%-------------------------------------------
% plotting of trajectories for the harmonic oscillator
% x = cos(w*t);  v = -w*sin(w*t);
% . . . . . . . . . . . . . . . . . . . . . . . .
% Input terms:
% Nt = Number of time steps
%  n = to evaluate frequency by w = 2*pi*n/Nt
% Note that Delta-t = 1 and x(0) = 1 for convenience
% . . . . . . . . . . . . . . . . . . . . . . . .


% Set up needed term
if exist('Nt','var')  == 1; else Nt   = 48;end
if exist('n','var')   == 1; else n    = 2 ;end
if exist('iout','var')== 1; else iout = 0 ;end
w = 2*pi*n/Nt;

t = (1:Nt) -1;
xj = zeros(1,Nt); xj(1) = 1;
vj = zeros(1,Nt);

%% Leap frog algorithm
Px = [1; -0.5*w*sin(-0.5*w)];
Mo = [1, 1; 0, 1];
for j = 1:Nt-1
    Ax = - (w^2)*[xj(j); xj(j)];
    Px = Mo*Px + Ax;
```

```
    xj(j+1) = Px(1);
end
xlf = xj;

%% Average algorithm
Px = [1; 0];
Mo = [1, 1; 0, 1]; Do = [0.25, 0.25; 0.5, 0.5];
for j = 1:Nt-1
    Ax = - (w^2)*[xj(j); xj(j)]; %Prediction
    Ps = Mo*Px + Do*Ax;
    Ax = - (w^2)*[xj(j); Ps(1)]; %Correction
    Px = Mo*Px + Do*Ax;
    xj(j+1) = Px(1);
end
xav = xj;

%% Plotting and output section
% first show the trajectory x(t), then> show how
% both methods compare to the analytical result
%    x(t) = cos(w*t)
% by looking at the differences

xt = cos(w*t);
figure(1);plot(t,xt,'k',t,xlf,'ro',t,xav,'bs');
legend('Analytic','Leapfrog','Average');
title('Trajectory','FontSize',20);
xlabel('time_[\Delta_t]','FontSize',18);
ylabel('position_x(t)','FontSize',18);

figure(2);plot(t,xt-xlf,'-ro',t, xt - xav,'-bs');
legend('Leapfrog','Average');
title('Difference_with_Analytic','FontSize',20);
xlabel('time_[\Delta_t]','FontSize',18);
ylabel('position_x(t)','FontSize',18);

if iout > 0;
    Mxt = [t',xt',xlf',xav'];
    filenam= 'xvt.txt';
    dlmwrite(filenam,Mxt,'delimiter',...
                '\t','precision',8);
end
```

A3.35 Trajectories for emission from a hemisphere

The numerical evaluation of electron trajectories from a hemisphere is done by using Eq. (33.62) and makes use of the average method of Eq. (33.60). The results are compared alongside the analytical evaluation using Eq. (33.78). The output data ($x = \rho/a$ vs $y = z/a$) is shown in Figure 33.20.

```
%-----------------------------------------
% PROGRAM TRAJECTORIES
%-----------------------------------------
% Dimensionless Representation:
```

```
%     (tau = time, a = boss radius
% x = rho/a;  u = vrho / vo;
% y =   z/a;  w =   vz / vo;
% t = tau s;
% . . . . . . . . . . . . . . . . . . . . . . . . . .
% Input terms:
% Np = [N]umber of [p]articles
% Nt = [N]umber of [t]ime steps
% ds = dimensionless time increment
% kappa = Fo * a / m * vo^2 where
%        Fo = Field [eV/nm]
%         a = boss radius [nm],
%         m = electron mass [eV/c^2],
%        vo = initial velocity [nm/fs]
%-----------------------------------------
%% Dimensionless Frho Fz in terms of rho = a*x, z = a*y
% fx = @(x,y) 1.0 - (x^2 - 2*y^2)/(x^2+y^2)^2.5;
% fy = @(x,y) 3*x*y/(x^2+y^2)^2.5;
% . . . . . . . . . . . . . . . . . . . . . . . . . .
fy = @(x,y) 1.0 +...
  ((-x.^2 + 2*y.^2)./((x.^2+y.^2).^2.5));
fx = @(x,y) 3*x.*y./((x.^2+y.^2).^2.5);

% Set up needed term

if exist('Np','var') == 1; else Np = 8;end
if exist('Nt','var') == 1; else Nt = 24;end
if exist('ds','var') == 1; else ds = 0.05;end
if exist('thetamax','var') == 1;...
   else thetamax = 45.0;end
if exist('kappa','var') == 1; else kappa = 15.0;end
if exist('nsig','var') == 1; else nsig = 2;end

theta = thetamax*pi*(0:Np-1)/(180*(Np-1));
sigma = sqrt(1.0 +...
  (2*(nsig^3 - 1)*kappa/nsig^2)*cos(theta));

%% Initial conditions and Integration.
% x = rho/a, y = z/a, u = vrho/vo, w = vz/vo
% zero out all the vectors and allocate their space
% for computation

x = zeros(Nt,Np); y = x; u = x; w = x;
xu = x; yu = x; xc = x; yc = x;

% Initial launch sites and velocities
x(1,:) = sin(theta); y(1,:) = cos(theta);
u(1,:) = sin(theta); w(1,:) = cos(theta);

%% Analytic trajectories:  set up time
ts = (0:Nt-1)*ds;
for ip = 1:Np;
```

```
    xu(:,ip) = sin(theta(ip))*(1.0 + ts);
    xc(:,ip) = sin(theta(ip))*(1.0 + ts*sigma(ip));
    yu(:,ip) = cos(theta(ip))*(1.0 + ts)...
               + 0.5*kappa*ts.^2;
    yc(:,ip) = cos(theta(ip))*(1.0 + ts*sigma(ip))...
               + 0.5*kappa*ts.^2;
end

%% Numerical trajectories
for j = 1:Nt-1;
    xs = x(j,:) + ds*u(j,:) +...
             0.5*kappa*(ds^2)*fx(x(j,:),y(j,:));
    ys = y(j,:) + ds*w(j,:) +...
             0.5*kappa*(ds^2)*fy(x(j,:),y(j,:));
    u(j+1,:) = u(j,:) +...
               0.5*kappa*ds*(fx(xs,ys)+fx(x(j,:),y(j,:)));
    w(j+1,:) = w(j,:) +...
               0.5*kappa*ds*(fy(xs,ys)+fy(x(j,:),y(j,:)));
    x(j+1,:) = x(j,:) + 0.5*ds*(u(j,:)+u(j+1,:));
    y(j+1,:) = y(j,:) + 0.5*ds*(w(j,:)+w(j+1,:));
end

% For the j-th trajectory, plot x(j,:) vs. y(j,:)
```

A3.36 Monte Carlo and integration

The numerical algorithm to generate Figure A1.3 in Section A1.2.5 is given, which calculates Eq. (A1.47) for the uniform ($p(x) = 1$), the power law ($p(x) = (1 + \nu)x^{\nu}$ with $\nu = 2$), and the acceptance–rejection method. The code has been explicitly made to be compact, so that each estimate of the integral is performed using a sum over vector elements. The acceptance–rejection method employs a trick of $\max(x, 0)/x = 1$ if $x > 0$ and 0 otherwise.

```
%% Program Monte Carlo
%  Compare three different methods of evaluating a
%  definite integral using Monte Carlo techniques:
%   1. Acceptance / Rejection        A/R
%   2. Uniform p(x) distribution     Uni
%   3. Power Law p(x) distribution   Pow
%  Do this calculation Nj times and compare the output:
%   The more accurate method(s) will cluster around ya

clear;clc; %Clear the work space

%%Parameters
Nx = 512; Nj =  64;
a  =  5; p=2;
iseed = 5;
rng(iseed);  % initialize the random number generator

% create the normalized function f(x) and its
% analytcal integrated value (from 0 to 1) ya
ya = (1/a) - 1/(exp(a) -1);
Fx = @(x,a) (exp(a*x) - 1)/(exp(a)-1);
```

```
% Initialize the vectors
Iuni = zeros(1,Nj); Ipow = zeros(1,Nj);
Iaj  = zeros(1,Nj); Io = ya*ones(1,Nj);

%% Do all the Monte Carlo calculations in the same loop
for j = 1:Nj;
    rj = rand(1,Nx); % create same random vector for all

    %% Method 1:  Uniform random variable
    hj = rj; pj = ones(1,Nx);
    Iuni(j) = sum(Fx(hj,a)./pj)/Nx;

    %% Method 2:  Power Law random variable
    hj = rj.^(1/(p+1)); pj = (1+p)*hj.^p;
    Ipow(j) = sum(Fx(hj,a)./pj)/Nx;

    %% Method 3:  Acceptance - Rejection
    hj = Fx(rj,a); rp = rand(1,Nx);
    Iaj(j) = sum(max(hj - rp, 0.0)./(hj - rp))/Nx;

end

%% Create index (x) and sort the data:  shows better
%  how well the data clusters around the actual value ya
x = (1:Nj)';
Itot = [sort(Iuni)',sort(Ipow)',sort(Iaj)',Io'];
figure(4); plot(x,Itot);
legend('Uni','Pow','A/R','Analytic');
title('Monte_Carlo_Comparison','FontSize',24);
xlabel('Run_Index','FontSize',18);
ylabel('MC_Estimate','FontSize',18);

%% Output the data
Mx = [x,Itot];
dlmwrite('Monte.txt',Mx,'delimiter','\t','precision',6);
```

References

1. James, R. (1989) A history of radar. *IEE Review*, **35** (9), 343–349.
2. Kevles, D.J. (1987) *The Physicists: the History of a Scientific Community in Modern America*, Harvard University Press, Cambridge, MA.
3. Skolnik, M. (2002) Role of radar in microwaves. *IEEE Trans. Microwave Theory Tech.*, **50** (3), 625–632.
4. Walker, J.S. (1997) *Prompt and utter destruction: Truman and the use of atomic bombs against Japan*, University of North Carolina Press, Chapel Hill.
5. Dyke, W.P., Trolan, J.K., Dolan, W.W., and Barnes, G. (1953) The field emitter: Fabrication, electron microscopy, and electric field calculations. *J. Appl. Phys.*, **24** (5), 570–576.
6. Martin, E., Trolan, J., and Dyke, W. (1960) Stable, high density field emission cold cathode. *J. Appl. Phys.*, **31**, 782.
7. Charbonnier, F.M., Dyke, W.P., Garrett, L., and Barbour, J.P. (1963) Basic and applied studies of field emission at microwave frequencies. *Proc. IEEE*, **51** (7), 991–1004.
8. Gold, S. and Nusinovich, G. (1997) Review of high-power microwave source research. *Rev. Sci. Instrum.*, **68** (11), 3945–3974.
9. Shiffler, D., Luginsland, J., Ruebush, M., Lacour, M., Golby, K., Cartwright, K., Haworth, M., and Spencer, T. (2004) Emission uniformity and shot-to-shot variation in cold field emission cathodes. *IEEE Trans. Plas. Sci.*, **32** (3), 1262–1266.
10. Granatstein, V.L. and Armstrong, C. (1999) Special issue on new vistas for vacuum electronics. *Proc. IEEE*, **87** (5), 699–701.
11. Parker, R.K., Abrams, Richard H., J., Danly, B.G., and Levush, B. (2002) Vacuum electronics. *IEEE Trans. Microwave Theory Tech.*, **50** (3), 835–845.
12. Gilmour, A. (1986) Microwave tubes, in *The Artech House microwave library*, Artech House, Dedham, MA.
13. Booske, J. (2008) Plasma physics and related challenges of millimeter-wave-to-terahertz and high power microwave generation. *Phys. Plasmas*, **15** (5), 055 502.
14. Spindt, C.A., Holland, C., Brodie, I., Mooney, J., and Westerberg, E. (1989) Field-emitter arrays applied to vacuum fluorescent display. *IEEE Trans. Electron Devices*, **36** (1), 225–228.
15. Temple, D. (1999) Recent progress in field emitter array development for high performance applications. *Materials Science and Engineering R-reports*, **24** (5), 185–239.
16. Jensen, W.D., Gerhard Jr., F., and Koffman, D. (2005) Electron beam lithography, in *Van Nostrand's Scientific Encyclopedia* (eds G.D. Considine and P.H. Kulik), John Wiley and Sons Inc, pp. 1771–1774.
17. Pawley, J. (1997) The development of field-emission scanning electron microscopy for imaging biological surfaces. *Scanning*, **19**, 324–336.
18. Lewellen, J.W. and Noonan, J. (2005) Field-emission cathode gating for RF electron guns. *Phys. Rev. Spec. Top. Accel. Beams*, **8** (3), 033 502.
19. Brau, C.A. (1997) High-brightness photoelectric field-emission cathodes for free-electron laser applications. *NucL Instr. and Meth. in Phys. Res. A*, **393** (1-3), 426–429.
20. Patterson, B.D., Abela, R., Braun, H.H., Flechsig, U., Ganter, R., Kim, Y., Kirk, E., Oppelt, A., Pedrozzi, M., Reiche, S., Rivkin, L., Schmidt, T., Schmitt, B., Strocov, V.N., Tsujino, S., and Wrulich, A.F. (2010) Coherent science at the swissFEL x-ray laser. *New Journal of Physics*, **12**.
21. Neil, G.R., Behre, C., Benson, S.V., Bevins, M., Biallas, G., Boyce, J., Coleman, J., Dillon-Townes, L., Douglas, D., Dylla, H., Evans, R., Grippo, A., Gruber, D., Gubeli, J., Hardy, D., Hernandez-Garcia, C., Jordan, K., Kelley, M., Merminga, L., Mammosser, J., Moore, W., Nishimori, N., Pozdeyev, E., Preble, J., Rimmer, R., Shinn, M., Siggins, T., Tennant, C., Walker, R., Williams, G., and Zhang, S. (2006) The JLAB high power ERL light source. *NucL Instr. and Meth. in Phys. Res. A*, **557** (1), 9–15.
22. Sinclair, C.K. (2006) DC photoemission electron guns as ERL sources. *NucL Instr. and Meth. in Phys. Res. A*, **557** (1), 69–74.
23. Rao, T., Burrill, A., Chang, X., Smedley, J., Nishitani, T., Garcia, C., Poelker, M., Seddon, E., Hannon, F., Sinclair, C., Lewellen, J., and Feldman, D. (2006) Photocathodes for the energy recovery linacs. *NucL Instr. and Meth. in Phys. Res. A*, **557** (1), 124–130.
24. Tsujino, S., le Pimpec, F., Raabe, J., Buess, M., Dehler, M., Kirk, E., Gobrecht, J., and Wrulich, A. (2009) Static and optical field enhancement in metallic nanotips studied by two-photon photoemission microscopy and spectroscopy excited by picosecond laser pulses. *Appl. Phys. Lett.*, **94** (9).
25. Marrese, C., Polk, J.E., Jensen, K.L., Gallimore, A.D., Spindt, C.A., Fink, R.L., and Palmer, W.D. (2000) Chapter 11: Performance of field emission cathodes in xenon electric propulsion systems environments, in *Micropropulsion for small spacecraft* (eds M.M. Micci and A.D. Ketsdever), American Institute of Aeronautics and Astronautics, Reston, Va.
26. Johnson, L., Gilchrist, B., Estes, R., and Lorenzini, E. (1999) Overview of future NASA tether applications. *Adv. Space Res.*, **24** (8), 1055–1063.
27. Gilchrist, B., Jensen, K.L., Gallimore, A., and Severns, J. (2000) Space based applications for FEA cathodes (FEAC), in *Mat. Res. Soc. Symp. Proc.*, vol. 621, Materials Research Society, pp. R481–R487.

Introduction to the Physics of Electron Emission, First Edition. Kevin L. Jensen.

28. Morris, D., Gilchrist, B., and Gallimore, A. (2001) The effect of plasma density and emitter geometry on space charge limits for field emitter array electron charge emission into a space plasma, in *2001 Space Technology and Applications International Forum - 2001 Aip Conf. Proc.*, vol. **552**(1) (eds M.S. El-Genk and M.J. Bragg), AIP, Albuquerque, New Mexico, pp. 467–472.
29. O'Shea, P.G., Bender, S., Carlsten, B., Early, J., Feldman, D.W., Feldman, R., Johnson, W., Lumpkin, A., Sheffield, R., Springer, R., Stein, W., and Young, L. (1991) Performance of the photoinjector accelerator for the Los Alamos free-electron laser, in *Proceedings of the IEEE Particle Accelerator Conference*, pp. 2754–2756.
30. Wilson, P.B. (2003) Gradient limitation in accelerating structures imposed by surface melting. *Report No. SLAC-PUB-9953*.
31. Wilson, P.B. (2007) Formation of taylor cones on a molten metal surface followed by ion injection into the vacuum, in *High Gradient Workshop*, Stanford, CA.
32. Norem, J., Wu, V., Moretti, A., Popovic, M., Qian, Z., Ducas, L., Torun, Y., and Solomey, N. (2003) Dark current, breakdown, and magnetic field effects in a multicell, 805 MHz cavity. *Phys. Rev. Spec. Top. Accel. Beams*, **6** (7), 072 001.
33. Dushman, S. (1930) Thermionic emission. *Rev. Mod. Phys.*, **2** (4), 0381–0476.
34. DuBridge, L. and Roehr, W. (1932) The thermionic and photoelectric work - functions of molybdenum. *Phys. Rev.*, **42** (1), 52–57.
35. Baroody, E.M. (1950) A theory of secondary electron emission from metals. *Phys. Rev.*, **78** (6), 780.
36. Fowler, R.H. and Nordheim, L. (1928) Electron emission in intense electric fields. *Philos. Trans. R. Soc. London, Ser. A*, **119** (781), 173–181.
37. Shakespeare, W. and Harbage, A. (1977) *William Shakespeare: The Complete Works*, The Viking Press, New York, NY.
38. Bohr, N. (1913) On the constitution of atoms and molecules, part i. *Philos. Mag.*, **26 (151)** (9), 1–24.
39. de Broglie, L. (1929) The wave nature of the electron. *Nobel Lecture, December 12*, http://nobelprize.org/physics/laureates/1929.
40. Dirac, P.A.M. (1958) *The Principles of Quantum Mechanics.*, Clarendon Press, Oxford.
41. Dicke, R.H. and Wittke, J.P. (1960) *Introduction to quantum mechanics*, Addison-Wesley Pub. Co, Reading, MA.
42. Schiff, L.I. (1968) *Quantum Mechanics*, McGraw-Hill, New York.
43. Gasiorowicz, S. (1974) *Quantum physics.*, Wiley, New York.
44. Shankar, R. (1980) *Principles of Quantum Mechanics*, Plenum Press, New York.
45. Houston, E.J. (1884) Notes on phenomena in incandescent lamps. *Transactions of the American Institute of Electrical Engineers*, **1**, 11–18.
46. Allison, D.K. (1981) *New eye for the Navy: the origin of radar at the Naval Research Laboratory*, vol. 8466, Naval Research Laboratory (U.S.). Technical Information Division.
47. Thomson, J. (2010) Cathode rays. *Philos. Mag.*, **90** (supl), 25–29.
48. Drude, P. (1900) On the electron theory of metals. *Ann. Phys.-Berlin*, **1** (3), 566–613.
49. Millikan, R.A. and Eyring, C.F. (1926) Laws governing the pulling of electrons out of metals by intense electrical fields. *Phys. Rev.*, **27** (1), 51.
50. Millikan, R.A. and Lauritsen, C. (1928) Relations of field currents to thermionic currents. *Proc. Natl. Acad. Sci. USA*, **14**, 45–49.
51. Reif, F. (1965) *Fundamentals of Statistical and Thermal Physics*, McGraw-Hill, New York.
52. Hummel, R.E. (2011) *Electronic properties of materials*, Springer, New York.
53. Ziman, J.M. (1985) *Principles of the Theory of Solids*, Academic Press, Cambridge.
54. Corak, W.S., Garfunkel, M.P., Satterthwaite, C.B., and Wexler, A. (1955) Atomic heats of copper, silver, and gold from 1 k to 5 k. *Phys. Rev.*, **98** (6), 1699–1707.
55. Gray, D.E. (1972) *American Institute of Physics Handbook. Section editors: Bruce H. Billings [and others] Coordinating editor: Dwight E. Gray.*, McGraw-Hill, New York.
56. Wright, E., Mather, J., Bennett, C., Cheng, E., Shafer, R., Fixsen, D., Eplee, R., Isaacman, R., Read, S., Boggess, N., Gulkis, S., Hauser, M., Janssen, M., Kelsall, T., Lubin, P., Meyer, S., Moseley, S., Murdock, T., Silverberg, R., Smoot, G., Weiss, R., and Wilkinson, D. (1991) A preliminary measurement of the cosmic microwave background spectrum by the cosmic background explorer (cobe) l satellite. *Astrophys. J.*, **381** (1, 1), 200–209.
57. Einstein, A. (1905) Concerning an heuristic point of view toward the emission and transformation of light. *Ann. Phys.-Berlin*, **17** (6), 132–148.
58. Dowell, D.H., King, F.K., Kirby, R.E., Schmerge, J.F., and Smedley, J.M. (2006) In situ cleaning of metal cathodes using a hydrogen ion beam. *Phys. Rev. Spec. Top. Accel. Beams*, **9** (6), 063 502.
59. Woodward, P.M. (1980) *Probability and Information Theory With Applications To Radar (Radar Library)*, Artech Print on Demand.
60. Arfken, G.B. (1970) *Mathematical Methods for Physicists*, Academic Press, New York.
61. Kubo, R. (1965) *Statistical mechanics, an advanced course with problems and solutions*, North-Holland Pub. Co. Interscience Publishers, Amsterdam New York.
62. Askerov, B.M. and Figarova, S.R. (2010) *Thermodynamics, Gibbs Method And Statistical Physics Of Electron Gases*, Springer, Heidelberg, New York.
63. Rubinstein, R.Y. (1981) *Simulation and the Monte Carlo method*, Wiley series in probability and mathematical statistics, Wiley, New York.
64. Feynman, R.P. (1972) *Statistical Mechanics: a Set of Lectures*, Frontiers in Physics, W.A. Benjamin, Reading, MA.
65. Reichl, L.E. (1987) *A Modern Course in Statistical Physics*, Dover, New York.
66. Lawden, D.F. (2005) *Principles of thermodynamics and statistical mechanics*, Dover Publications, Mineola, NY.
67. Dushman, S. (1923) Electron emission from metals as a function of temperature. *Phys. Rev.*, **21** (6), 0623–0636.
68. Fowler, R.H. (1928) The photo-electric threshold frequency and the thermionic work function. *Philos. Trans. R. Soc. London, Ser. A*, **118** (779), 229–232.
69. Becker, J. and Brattain, W. (1934) The thermionic work function and the slope and intercept of richardson plots. *Phys. Rev.*, **45** (10), 0694–0705.

70. Birrell, N.D. and Davies, P.C.W. (1982) *Quantum Fields in Curved Space*, Cambridge Monographs on Mathematical Physics, Cambridge University Press, Cambridge, New York.
71. Morris, R. (1997) *Achilles in the quantum universe: the definitive history of infinity*, Henry Holt, New York.
72. Blakemore, J.S. (1987) *Semiconductor Statistics*, Dover, New York.
73. Pierret, R.F. (1988) *Semiconductor Fundamentals*, Modular series on solid state devices, Vol. 1, Addison-Wesley Pub. Co., Reading, MA.
74. Chandrasekhar, S. (1945) Ralph Howard Fowler, 1889–1944. *The Astrophysical Journal*, **101**, 1.
75. Smith, E.V. P. and Jacobs, K.C. (1973) *Introductory Astronomy And Astrophysics*, Saunders, Philadelphia.
76. Asimov, I. (1972) *Asimov's Guide to science*, Basic Books, New York.
77. Keegan, J. (1993) *A History of Warfare*, Alfred A. Knopf, distributed by Random House, Inc., New York.
78. Quéré, Y. (1998) *Physics of Materials*, Gordon and Breach Science Publishers, Amsterdam.
79. Fowler, R.H. (1928) The restored electron theory of metals and thermionic formulae. *Philos. Trans. R. Soc. London, Ser. A*, **117** (778), 549–552.
80. Ridley, B.K. (1999) *Quantum Processes in Semiconductors*, Clarendon Press, Oxford.
81. Lundstrom, M. (2000) *Fundamentals of Carrier Transport*, Cambridge University Press, Cambridge, New York.
82. Ziman, J.M. (2001) *Electrons and Phonons: the Theory of Transport Phenomena in Solids*, Oxford Classic Texts in the Physical Sciences, Clarendon Press, Oxford University Press, Oxford, New York.
83. Bechtel, J.H. (1975) Heating of solid targets with laser pulses. *J. Appl. Phys.*, **46** (4), 1585–1593.
84. Jensen, K.L., Moody, N.A., Feldman, D.W., Montgomery, E.J., and O'Shea, P.G. (2007) Photoemission from metals and cesiated surfaces. *J. Appl. Phys.*, **102** (7), 074 902–074 919.
85. Jensen, K.L. (2007) *Electron Emission Physics*, Advances in Imaging Electron Physics 149, Academic Press, San Diego, CA.
86. Kittel, C. (1996) *Introduction to Solid State Physics*, Wiley, New York, 7th edn..
87. Poundstone, W. (1992) *Prisoner's Dilemma*, Doubleday, New York.
88. Axelrod, R.M. (1990) *The Evolution of Cooperation*, Penguin Books, London, New York.
89. Farmelo, G. (2003) *It Must Be Beautiful: Great Equations of Modern Science*, Granta UK.
90. Goldstein, H. (1980) *Classical Mechanics*, Addison-Wesley Pub. Co, Reading, MA.
91. Anton, H. (1977) *Elementary Linear Algebra*, Wiley, New York.
92. Flitney, A.P. and Abbott, D. (2002) Quantum version of the Monty Hall problem. *Phys. Rev. A*, **65** (6), 062 318.
93. Heisenberg, W. (1950) *The physical principles of the quantum theory*, Dover Publications, New York.
94. Powers, T. (1993) *Heisenberg's War: The Secret History of the German Bomb*, Knopf, distributed by Random House, New York.
95. Gomer, R. (1993) *Field Emission and Field Ionization*, American Vacuum Society Classics, American Institute of Physics, New York.
96. Gadzuk, J.W. and Plummer, E.W. (1971) Energy distributions for thermal field emission. *Phys. Rev. B*, **3** (7), 2125.
97. Rammer, J. (2004) *Quantum Transport Theory*, Frontiers in Physics 99, Westview, Boulder, CO.
98. Aitchison, I.J.R. (1982) *An informal introduction to gauge field theories*, Cambridge University Press, Cambridge, New York.
99. Eyges, L. (1972) *The Classical Electromagnetic Field*, Addison-Wesley Pub. Co, Reading, MA.
100. Lorrain, P., Corson, D.R., and Lorrain, F. (1987) *Electromagnetic Fields and Waves*, W.H. Freeman & Company.
101. Elliott, J.P. and Dawber, P.G. (1979) *Symmetry in Physics (Vol. 1 & 2)*, Oxford University Press, New York.
102. Gray, H.B. (1964) *Electrons and Chemical Bonding*, W.A. Benjamin, New York.
103. Abramowitz, M. and Stegun, I.A. (1965) *Handbook of mathematical functions, with formulas, graphs, and mathematical tables*, Dover Publications, New York.
104. Sutton, A.P. (1993) *Electronic Structure of Materials*, Oxford Science Publications, Clarendon Press, Oxford University Press, Oxford, New York.
105. Sobol, I.M. (1974) *The Monte Carlo method*, University of Chicago Press, Chicago.
106. Hammersley, J.M. and Handscomb, D.C. (1964) *Monte Carlo methods*, Methuen's Monographs on Applied Probability and Statistics, Methuen Wiley, London, New York.
107. Price, P.J. (1979) Monte Carlo calculation of electron transport in solids, in *Semiconductors and Semimetals Volume 14* (eds R.K. Willardson and A.C. Beer), Academic Press, New York, pp. 249–308.
108. Tomizawa, K. (1993) *Numerical Simulation of Submicron Semiconductor Devices*, The Artech House Materials Science Library, Artech House, Boston.
109. Oppenheimer, J.R. (1928) Three notes on the quantum theory of aperiodic effects. *Phys. Rev.*, **31** (1), 66.
110. Murphy, E.L. and Good, R.H. (1956) Thermionic emission, field emission, and the transition region. *Phys. Rev.*, **102** (6), 1464–1473.
111. Modinos, A. (1984) *Field, Thermionic, and Secondary Electron Emission Spectroscopy*, Plenum Press, New York.
112. Kemble, E. (1935) A contribution to the theory of the BWK method. *Phys. Rev.*, **48** (6), 549–561.
113. Kane, E. (1961) Theory of tunneling. *J. Appl. Phys.*, **32** (1), 83–91.
114. Jeffrey, A. (2004) *Handbook of Mathematical Formulas and Integrals*, Elsevier Academic Press, Amsterdam, Boston.
115. Wigner, E. (1932) On the quantum correction for thermodynamic equilibrium. *Phys. Rev.*, **40** (5), 749.
116. Feynman, R.P. (1987) Negative probability, in *Quantum implications: essays in honour of David Bohm* (eds B.J. Hiley and F.D. Peat), published in the USA by Routledge and Kegan Paul in association with Methuen, New York, p. 235.
117. Hillery, M., O'Connell, R., Scully, M., and Wigner, E. (1984) Distribution functions in physics – fundamentals. *Phys. Rep.*, **106** (3), 121–167.
118. Kim, Y.S. and Noz, M.E. (1991) *Phase Space Picture of Quantum Mechanics: Group Theoretical Approach*, World Scientific, Singapore, Teaneck, NJ.
119. Ferry, D. and Zhou, J.R. (1993) Form of the quantum potential for use in hydrodynamic equations for semiconductor device modeling. *Phys. Rev. B*, **48** (11), 7944–7950.

120. Lee, H.W. (1995) Theory and application of the quantum phase-space distribution functions. *Phys. Rep.*, **259** (3), 147–211.
121. Bohm, D. and Hiley, B. (1985) Unbroken quantum realism, from microscopic to macroscopic levels. *Phys. Rev. Lett.*, **55** (23), 2511–2514.
122. Dewdney, C. and Hiley, B. (1982) A quantum potential description of one-dimensional time-dependent scattering from square barriers and square wells. *Foundations of Physics*, **12** (1), 27–48.
123. Vigier, J.P., Dewdney, C., Holland, P.R., and Kyprianidis, A. (1987) Causal particle trajectories and the interpretation of quantum mechanics, in *Quantum implications: essays in honour of David Bohm* (eds B.J. Hiley and F.D. Peat), Routledge, New York, pp. 169–204.
124. Na, K. and Wyatt, R. (2001) Quantum trajectories for resonant scattering. *International Journal of Quantum Chemistry*, **81** (3), 206–213.
125. Jensen, K.L. and Buot, F.A. (1989) Particle trajectory tunneling – a novel approach to quantum transport, in *Conference Series (99), Institute of Physics*, pp. 137–140.
126. Frensley, W.R. (1990) Boundary conditions for open quantum systems driven far from equilibrium. *Rev. Mod. Phys.*, **62** (3), 745.
127. Jensen, K.L. and Buot, F.A. (1991) Numerical simulation of intrinsic bistability and high-frequency current oscillations in resonant tunneling structures. *Phys. Rev. Lett.*, **66** (8), 1078–1081.
128. Hsu, Y. and Wu, G.Y. (1992) Wigner trajectories in resonant tunneling diodes under transverse magnetic field. *J. Appl. Phys.*, **71** (1), 304–307.
129. Razavy, M. (1996) Wigner trajectories in quantum tunneling. *Phys. Lett. A*, **212** (3), 119–122.
130. Jensen, K.L. and Ganguly, A. (1992) A comparison of the transmission coefficient and the Wigner function approaches to field emission. *Compel – the International Journal for Computation and Mathematics in Electrical and Electronic Engineering*, **11** (4), 457–470.
131. Jensen, K.L. and Ganguly, A. (1994) Time-dependent, self-consistent simulations of field emission from silicon using the Wigner distribution function. *J. Vac. Sci. Technol. B*, **12** (2), 770–775.
132. Langmuir, I. (1913) The effect of space charge and residual gases on thermionic currents in high vacuum. *Phys. Rev.*, **2** (6), 450–486.
133. Richardson, O. (1913) The emission of electrons from tungsten at high temperatures – an experimental proof that the electric current in metals is carried by electrons. *Science*, **38** (1), 57–61.
134. Sommer, A.H. (1983) The element of luck in research – photocathodes 1930 to 1980. *J. Vac. Sci. Technol. A*, **1**, 119–124.
135. Haas, G.A. and Thomas, R.E. (1968) Chapter 2: Thermionic emission and work function, in *Techniques of Metals Research Vol. VI, Part 1* (ed. R.F. Bunshah), Interscience Publishers, New York, pp. 91–262.
136. Lilienfeld, J. (1922) The auto-electronic discharge and its application to the construction of a new form of x-ray tube. *Am. J. Roentgenol.*, **9**, 172–179.
137. Kleint, C. (1993) On the early history of field emission including attempts of tunneling spectroscopy. *Prog. Surf. Sci.*, **42** (1-4), 101–115.
138. Schottky, W. (1923) Über kalte und warme elektronenentladungen (cold and warm electron discharge). *Zeitschrift Für Physik*, **14** (1), 63–106.
139. Bonzel, H. and Kleint, C. (1995) On the history of photoemission. *Prog. Surf. Sci.*, **49** (2), 107–153.
140. DuBridge, L.A. and Hergenrother, R. (1933) The effect of temperature on the energy distribution of photoelectrons. I. Normal energies. *Phys. Rev.*, **44** (11), 0861–0865.
141. Dekker, A. (1958) Secondary electron emission, in *Solid state physics: Advances in Research and Applications Vol. (6)* (eds F. Seitz and D. Turnbull), Academic Press, New York.
142. Wooldridge, D. (1939) Theory of secondary emission. *Phys. Rev.*, **56** (6), 562–578.
143. Ben-Zvi, I., Rao, T., Burrill, A., Chang, X., Grimes, J., Rank, J., Segalov, Z., and Smedley, J. (2007) Diamond secondary emitter. *Int. J. Mod. Phys. A*, **22** (22), 3759–3775.
144. Yater, J.E., Shaw, J., Jensen, K.L., Feygelson, T., Myers, R., Pate, B., and Butler, J. (2011) Secondary electron amplification using single-crystal CVD diamond film. *Diam Relat Mater*, **20** (5-6), 798–802.
145. Luginsland, J., Lau, Y., Umstattd, R., and Watrous, J. (2002) Beyond the Child–Langmuir law: A review of recent results on multidimensional space-charge-limited flow. *Phys. Plasmas*, **9** (5), 2371–2376.
146. Langmuir, I. and Blodgett, K. (1923) Currents limited by space charge coaxial cylinders. *Phys. Rev.*, **22** (4), 0347–0356.
147. Langmuir, I. and Blodgett, K. (1924) Currents limited by space charge between concentric spheres. *Phys. Rev.*, **24** (1), 49–59.
148. Birdsall, C.K. and Bridges, W.B. (1966) *Electron dynamics of diode regions*, Academic Press, New York.
149. Ang, L.K. and Zhang, P. (2007) Ultrashort-pulse child-langmuir law in the quantum and relativistic regimes. *Phys. Rev. Lett.*, **98** (16), 164 802–164 804.
150. Mumford, P. and Cahay, M. (1998) Space-charge effects and current self-quenching in a metal/CdS/LaS cold cathode. *J. Appl. Phys.*, **84** (5), 2754–2767.
151. De Boer, J.H. (1935) *Electron Emission and Adsorption Phenomena*, Cambridge Series of Physical Chemistry, Cambridge University Press, Cambridge.
152. van der Ziel, A. (1968) *Solid State Physical Electronics*, Prentice-Hall, Englewood Cliffs, N.J.
153. Dobretsov, L. and Gomoyunova, M. (1971) *Emission electronics*, Israel Program for Scientific Translations, Jerusalem.
154. Fransen, M., Van Rooy, T., Tiemeijer, P., Overwijk, M., Faber, J., and Kruit, P. (1999) On the electron-optical properties of the ZrO/W Schottky electron emitter, in *Adv. Imag. Elect. Phys.*, vol. III (ed. P. Hawkes), Academic Press, p. 91.
155. Dyke, W. and Dolan, W.W. (1956) Field emission, in *Adv. Electron El. Phys. 8* (eds G. Metson and L. Marton), Academic Press Inc., New York, pp. 90–187.
156. Swanson, L. and Bell, A. (1973) Recent advances in field electron microscopy of metals, in *Adv. Imag. Elect. Phys. 32* (eds M. L. and M. Claire), Academic Press, pp. 193–309.
157. Murata, K. and Kyser, D. (1987) Monte-Carlo methods and microlithography simulation for electron and x-ray-beams. *Adv. Imag. Elect. Phys.*, **69**, 175–259.

158. Hawkes, P.W. (1996) Electron optics, in *Adv. Imag. Elect. Phys. 96* (ed. M. Tom), Elsevier, pp. 405–412.
159. Binh, V.T., Garcia, N., and Purcell, S. (1996) Electron field emission from atom-sources: Fabrication, properties, and applications of nanotips, in *Adv. Imag. Elect. Phys. 95* (ed. W.H. Peter), Elsevier, pp. 63–153.
160. Fransen, M., van Rooy, T., and Kruit, P. (1999) Field emission energy distributions from individual multiwalled carbon nanotubes. *Appl. Surf. Sci.*, **146** (1-4), 312–327.
161. Binh, V.T. and Semet, V. (2007) Planar cold cathodes, in *Advances in Imaging and Electron Physics*, vol. **148** (ed. P. Hawkes), Academic Press, pp.1–73.
162. Swanson, L. and Schwind, G. (2009) A review of the cold-field electron cathode, in *Adv. Imag. Elect. Phys. 159* (ed. P. Hawkes), Academic Press, pp.63–100.
163. Becker, J. (1935) Thermionic electron emission and adsorption. Part I. thermionic emission. *Rev. Mod. Phys.*, **7** (2), 0095–0128.
164. Herring, C. and Nichols, M. (1949) Thermionic emission. *Rev. Mod. Phys.*, **21** (2), 185–270.
165. Gadzuk, J.W. and Plummer, E.W. (1973) Field emission energy distribution (feed). *Rev. Mod. Phys.*, **45** (3), 487–548.
166. Zhu, W. (2001) *Vacuum Microelectronics*, Wiley, New York.
167. Görlich, P. (1959) Recent advances in photoemission, in *Advances in Electronics and Electron Physics*, vol. 11 (ed. L. Marton), Academic Press, New York, pp. 1–30.
168. Sommer, A. and Spicer, W. (1965) Photoelectric emission, in *Photoelectronic Materials and Devices* (ed. S. Larach), Van Nostrand, Princeton, NJ, p. 175.
169. Rougeot, H. and Baud, C. (1979) Negative electron affinity photoemitters, in *Advances in Electronics and Electron Physics*, vol. **48** (ed. L. Marton), Academic Press, New York, pp. 1–36.
170. McKay, K.G. (1948) Secondary electron emission, in *Advances in Electronics and Electron Physics*, vol. 11 (ed. L. Marton), Academic Press, New York, pp. 66–130.
171. Hacehnberg, O. and Brauer, W. (1959) Secondary electron emission from solids, in *Advances in Electronics and Electron Physics*, vol. 11 (ed. L. Marton), Academic Press, New York, pp. 413–498.
172. Christov, S. (1967) Unified theory of thermionic and field emission from semiconductors. *Phys. Status Solidi*, **21** (1), 159.
173. Jensen, K.L. (2003) Electron emission theory and its application: Fowler–Nordheim equation and beyond. *J. Vac. Sci. Technol. B*, **21** (4), 1528–1544.
174. Christov, S. (1966) General theory of electron emission from metals. *Phys. Status Solidi*, **17** (1), 11–26.
175. Spindt, C.A., Brodie, I., Humphrey, L., and Westerberg, E.R. (1976) Physical properties of thin-film field emission cathodes with molybdenum cones. *J. Appl. Phys.*, **47** (12), 5248–5263.
176. Schwoebel, P.R. and Brodie, I. (1995) Surface-science aspects of vacuum microelectronics. *J. Vac. Sci. Technol. B*, **13** (4), 1391–1410.
177. Forbes, R.G. (2003) Field electron and ion emission from charged surfaces: a strategic historical review of theoretical concepts. *Ultramicroscopy*, **95** (1-4), 1–18.
178. Dubridge, L.A. (1933) Theory of the energy distribution of photoelectrons. *Phys. Rev.*, **43** (9), 0727–0741.
179. Berglund, C. and Spicer, W. (1964) I – Photoemission studies of copper and silver: Theory. *Phys. Rev.*, **136** (4A), A1030.
180. Spicer, W.E. and Herrera-Gomez, A. (1993) Modern theory and applications of photocathodes. *Proc. SPIE Int. Soc. Opt. Eng.*, **2022** (1), 18–35.
181. Dowell, D.H. and Schmerge, J.F. (2009) Quantum efficiency and thermal emittance of metal photocathodes. *Phys. Rev. Spec. Top. Accel. Beams*, **12** (7), 074 201–11.
182. Barbour, J.P., Dolan, W., Trolan, J., Martin, E., and Dyke, W. (1953) Space-charge effects in field emission. *Phys. Rev.*, **92** (1), 45–51.
183. Tsu, R. and Esaki, L. (1973) Tunneling in a finite superlattice. *Appl. Phys. Lett.*, **22** (11), 562–564.
184. Segré, E. (1977) *Nuclei and particles: an introduction to nuclear and subnuclear physics*, W.A. Benjamin, Reading, MA.
185. Berry, M. and Mount, K. (1972) Semiclassical approximations in wave mechanics. *Rep. Prog. Phys.*, **35** (4), 315–397.
186. Forbes, R. (2008) Physics of generalized Fowler–Nordheim-type equations. *J. Vac. Sci. Technol. B*, **26**, 788.
187. Miller, S.C. and Good, R.H. (1953) A WKB-type approximation to the Schrodinger equation. *Phys. Rev.*, **91** (1), 174.
188. Jensen, K.L. (2012) A quantum dipole-modified work function for a simplified electron emission barrier. *J. Appl. Phys.*, **111**, 054 916–054 910.
189. Shelton, H. (1957) Thermionic emission from a planar tantalum crystal. *Phys. Rev.*, **107** (6), 1553–1557.
190. Guth, E. and Mullin, C.J. (1942) Electron emission of metals in electric fields III. The transition from thermionic to cold emission. *Phys. Rev.*, **61** (5-6), 339.
191. Sprangle, P., Peñano, J., Hafizi, B., Gordon, D., Gold, S., Ting, A., and Mitchell, C. (2011) High average current electron guns for high-power free electron lasers. *Phys. Rev. Spec. Top. Accel. Beams*, **14**, 020 702.
192. Dolan, W.W. and Dyke, W.P. (1954) Temperature-and-field emission of electrons from metals. *Phys. Rev.*, **95** (2), 327.
193. Forbes, R.G. and Deane, J.H.B. (2007) Reformulation of the standard theory of Fowler–Nordheim tunnelling and cold field electron emission. *Proc. R. Soc. London, Ser. A*, **463** (2087), 2907–2927.
194. Nordheim, L.W. (1928) The effect of the image force on the emission and reflexion of electrons by metals. *Proc. R. Soc. London A*, **121** (788), 626–639.
195. Burgess, R.E., Kroemer, H., and Houston, J.M. (1953) Corrected values of Fowler–Nordheim field emission functions v(y) and s(y). *Phys. Rev.*, **90** (4), 515.
196. Jensen, K.L. and Ganguly, A. (1993) Numerical simulation of field emission and tunneling – a comparison of the Wigner function and transmission coefficient approaches. *J. Appl. Phys.*, **73** (9), 4409–4427.
197. Jensen, K.L. (2001) Theory of field emission, in *Vacuum Microelectronics* (ed. W. Zhu), Wiley, New York, p. 33.

198. Forbes, R.G. (1999) Use of a spreadsheet for Fowler–Nordheim equation calculations. *J. Vac. Sci. Technol. B*, **17** (2), 534–541.
199. Jensen, K.L. (2003) On the application of quantum transport theory to electron sources. *Ultramicroscopy*, **95** (1-4), 29–48.
200. Forbes, R.G. (2006) Simple good approximations for the special elliptic functions in standard Fowler–Nordheim tunneling theory for a Schottky–Nordheim barrier. *Appl. Phys. Lett.*, **89** (11), 113 122.
201. Deane, Jonathan, H.B. and Forbes, Richard, G. (2008) The formal derivation of an exact series expansion for the principal Schottky–Nordheim barrier function v, using the Gauss hypergeometric differential equation. *J. Phys. A*, **41** (39), 395 301.
202. Jensen, K.L. (2007) General formulation of thermal, field, and photoinduced electron emission. *J. Appl. Phys.*, **102** (2), 024 911–024 911.
203. Fowler, R.H. (1931) The analysis of photoelectric sensitivity curves for clean metals at various temperatures. *Phys. Rev.*, **38** (1), 45–56.
204. DuBridge, L.A. (1932) A further experimental test of Fowler's theory of photoelectric emission. *Phys. Rev.*, **39** (1), 108–118.
205. Bechtel, J.H., Smith, W.L., and Bloembergen, N. (1977) Two-photon photoemission from metals induced by picosecond laser pulses. *Phys. Rev. B*, **15** (10), 4557–4563.
206. Jensen, K.L., Feldman, D.W., Moody, N.A., and O'Shea, P.G. (2006) A photoemission model for low work function coated metal surfaces and its experimental validation. *J. Appl. Phys.*, **99** (12), 124 905–19.
207. Yater, J.E. and Shih, A. (2000) Secondary electron emission characteristics of single-crystal and polycrystalline diamond. *J. Appl. Phys.*, **87** (11), 8103–8112.
208. Lye, R. and Dekker, A. (1957) Theory of secondary emission. *Phys. Rev.*, **107** (4), 977–981.
209. Jensen, K.L., Yater, J.E., Shaw, J.L., Myers, R.E., Pate, B.B., Butler, J.E., and Feygelson, T. (2010) Bunch characteristics of an electron beam generated by a diamond secondary emitter amplifier. *J. Appl. Phys.*, **108** (4), 044 509–044 512.
210. Yang, J., Wang, C.F., Hu, E.L., and Butler, J.E. (2006) Lift-off process to get free-standing high quality single crystal diamond films and suspended single crystal diamond devices. *MRS Proceedings*, **956**.
211. Young, J. (1956) Penetration of electrons and ions in aluminum. *J. Appl. Phys.*, **27** (1), 1–4.
212. Phillips, M. (1991) Electric charge, in *Encyclopedia of Physics* (eds R.G. Lerner and G.L. Trigg), 2nd edn, VCH Publishers, Inc., New York, p. 207.
213. Mönch, W. (1995) *Semiconductor Surfaces and Interfaces*, Springer Series in Surface Sciences 26, Springer-Verlag, New York.
214. Rokhlenko, A., Jensen, K.L., and Lebowitz, J.L. (2010) Space charge effects in field emission: One dimensional theory. *J. Appl. Phys.*, **107** (1), 014 904–014 910.
215. Read, F. and Bowring, N. (2004) Simulation of thermionic cathodes. *NucL Instr. and Meth. in Phys. Res. A*, **531** (3), 407–415.
216. Rokhlenko, A. and Lebowitz, J. (2004) Space charge limited flow of a thin electron beam confined by a strong magnetic field. *Phys. Plasmas*, **11** (10), 4559–4563.
217. Rokhlenko, A. and Lebowitz, J.L. (2012) Space charge effects on externally injected current in planar diodes: Existence of multiple stationary states. *J. Appl. Phys.*, **111** (8), 083 306.
218. Caflisch, R.E. and Rosin, M.S. (2012) Beyond the Child–Langmuir limit. *Phys. Rev. E*, **85** (5), 056 408.
219. Adler, E. and Longo, R. (1986) Effect of nonuniform work function on space-charge-limited current. *J. Appl. Phys.*, **59** (4), 1022–1027.
220. Valfells, A., Feldman, D.W., Virgo, M., O'Shea, P.G., and Lau, Y.Y. (2002) Effects of pulse-length and emitter area on virtual cathode formation in electron guns. *Phys. Plasmas*, **9** (5), 2377–2382.
221. Lau, Y., Jensen, K.L., and Levush, B. (2000) A comparison of flicker noise and shot noise on a hot cathode. *IEEE Trans. Plasma Sci.*, **28** (3), 794–797.
222. Umstattd, R., Carr, C., Frenzen, C., Luginsland, J., and Lau, Y. (2005) A simple physical derivation of Child–Langmuir space-charge-limited emission using vacuum capacitance. *Am. J. Phys.*, **73** (2), 160–163.
223. Zhu, Y.B., Zhang, P., Valfells, A., Ang, L.K., and Lau, Y.Y. (2013) Novel scaling laws for the Langmuir–Blodgett solutions in cylindrical and spherical diodes. *Phys. Rev. Lett.*, **110** (26).
224. Koh, W.S., Ang, L.K., and Kwan, T.J.T. (2006) Multidimensional short-pulse space-charge-limited flow. *Phys. Plasmas*, **13** (6), 063 102.
225. Pedersen, A., Manolescu, A., and Valfells, A. (2010) Space-charge modulation in vacuum microdiodes at thz frequencies. *Phys. Rev. Lett.*, **104** (17), 175 002.
226. Forbes, R.G. (2008) Call for experimental test of a revised mathematical form for empirical field emission current-voltage characteristics. *Appl. Phys. Lett.*, **92** (19), 193 105–193 103.
227. Griswold, M.E., Fisch, N.J., and Wurtele, J.S. (2010) An upper bound to time-averaged space-charge limited diode currents. *Phys. Plasmas*, **17** (11).
228. Geittner, P., Gartner, G., and Raasch, D. (2000) Low temperature properties of Ba-dispenser cathodes. *J. Vac. Sci. Technol. B*, **18** (2), 997–999.
229. Lumpkin, A. (1991) in *Observations on Field-emission Electrons From the Los Alamos FEL Photoinjector, Vol. 3*, Proceedings of the IEEE Particle Accelerator Conference, San Francisco, CA, pp. 1967–1969.
230. Jensen, K.L., Lau, Y.Y., Feldman, D.W., and O'Shea, P.G. (2008) Electron emission contributions to dark current and its relation to microscopic field enhancement and heating in accelerator structures. *Phys. Rev. Spec. Top. Accel. Beams*, **11** (8), 081 001.
231. Young, R.D. (1959) Theoretical total-energy distribution of field-emitted electrons. *Phys. Rev.*, **113** (1), 110–114.
232. Forbes, R.G. (2004) Use of energy-space diagrams in free-electron models of field electron emission. *Surface and Interface Analysis*, **36** (5-6), 395–401.
233. Cutler, P. and Nagy, D. (1965) Use of a new potential barrier model in Fowler–Nordheim theory of field emission. *Surf. Sci.*, **3** (1), 71–94.
234. Guth, E. and Mullin, C.J. (1941) Electron emission of metals in electric fields II. Field dependence of the surface photo-effect. *Phys. Rev.*, **59** (11), 867.
235. Jensen, K.L., Feldman, D.W., and O'Shea, P.G. (2005) Time dependent models of field-assisted photoemission. *J. Vac. Sci. Technol. B*, **23** (2), 621–631.

236. Press, W.H., Flannery, B.P., Teukolsky, S.A., and Vetterling, W.T. (1992) *Numerical Recipes in Fortran: the Art of Scientific Computing*, Cambridge University Press, Cambridge, New York.
237. Chapman, S.J. (1998) *Fortran 90/95 for scientists and engineers*, WCB/McGraw-Hill, Boston.
238. Jensen, K.L., Feldman, D.W., and O'Shea, P.G. (2003) Advanced photocathode simulation and theory. *Nucl. Instr. and Meth. in Phys. Res. A*, **507**, 238–241.
239. Watson, G.N. (1995) *A treatise on the theory of Bessel functions*, Cambridge Mathematical Library, Cambridge University Press, Cambridge, New York.
240. Forbes, R.G. and Deane, J. (2011) Transmission coefficients for the exact triangular barrier: an exact general analytical theory that can replace Fowler and Nordheim's 1928 theory. *Philos. Trans. R. Soc. London, Ser. A*, **471** (online).
241. Hochstadt, H. (1986) *The functions of mathematical physics*, Dover Books on Physics and Chemistry, Dover Publications, New York.
242. Smith, G.D. (1985) *Numerical Solution of Partial Differential Equations: Finite Difference Methods*, Oxford Applied Mathematics and Computing Science Series, Clarendon Press, Oxford University Press, Oxford, New York.
243. Davis, P.J. (1975) *Interpolation and approximation*, Dover Publications,New York.
244. Fröman, N. and Fröman, P.O. (1965) *JWKB approximation: contributions to the theory*, North-Holland Pub. Co.; [sole distributors for U.S.A.: Interscience Publishers, New York], Amsterdam.
245. Vassell, M., Lee, J., and Lockwood, H. (1983) Multibarrier tunneling in Ga1-xAlxAs/GaAs heterostructures. *J. Appl. Phys.*, **54** (9), 5206–5213.
246. Ando, Y. and Itoh, T. (1987) Calculation of transmission tunneling current across arbitrary potential barriers. *J. Appl. Phys.*, **61** (4), 1497–1502.
247. Jonsson, B. and Eng, S. (1990) Solving the Schrodinger equation in arbitrary quantum-well potential profiles using the transfer-matrix method. *IEEE J. Quantum Electron.*, **26** (11), 2025–2035.
248. Mayer, A. and Vigneron, J. (1999) Transfer-matrix quantum-mechanical theory of electronic field emission from nanotips. *J. Vac. Sci. Technol. B*, **17** (2), 506–514.
249. Lui, W.W. and Fukuma, M. (1986) Exact solution of the Schrodinger equation across an arbitrary one-dimensional piecewise-linear potential barrier. *J. Appl. Phys.*, **60** (5), 1555–1559.
250. Brennan, K. and Summers, C. (1987) Theory of resonant tunneling in a variably spaced multiquantum well structure – an Airy function-approach. *J. Appl. Phys.*, **61** (2), 614–623.
251. Chung, M., Cutler, P., Miskovsky, N., and Kumar, N. (2000) Calculations of field emission from AlxGa1-xN as a function of stoichiometry. *J. Vac. Sci. Technol. B*, **18** (2), 919–922.
252. Furno, M., Bonani, F., and Ghione, G. (2007) Transfer matrix method modelling of inhomogeneous Schottky barrier diodes on silicon carbide. *Solid-State Electron.*, **51** (3), 466–474.
253. Rangaswamy, K., Cahay, M., and Jensen, K.L. (2005) Shot noise power spectrum of planar field emitters. *J. Vac. Sci. Technol. B*, **23** (2), 380–388.
254. Esaki, L. (1973) Long journey into tunneling. *Nobel Lecture, December 12*. URL http://nobelprize.org/physics/laureates/1973/esaki-lecture .pdf.
255. Price, P.J. (1988) Theory of resonant tunneling in heterostructures. *Phys. Rev. B*, **38** (3), 1994–1998.
256. Price, P.J. (1989) Simple theory of double-barrier tunneling. *IEEE Trans. Electron Devices*, **36** (10), 2340–2343.
257. Gadzuk, J.W. (1970) Resonance-tunneling spectroscopy of atoms adsorbed on metal surfaces: Theory. *Phys. Rev. B*, **1** (5), 2110.
258. Binh, V.T., Purcell, S.T., Garcia, N., and Doglioni, J. (1992) Field-emission electron spectroscopy of single-atom tips. *Phys. Rev. Lett.*, **69** (17), 2527–2530.
259. Litovchenko, V., Evtukh, A., Kryuchenko, Y., Goncharuk, N., Yilmazoglu, O., Mutamba, K., Hartnagel, H., and Pavlidis, D. (2004) Quantum-size resonance tunneling in the field emission phenomenon. *J. Appl. Phys.*, **96** (1), 867–877.
260. Feynman, R. (1948) Space-time approach to non-relativistic quantum mechanics. *Rev. Mod. Phys.*, **20** (2), 367–387.
261. Feynman, R. (1951) An operator calculus having applications in quantum electrodynamics. *Phys. Rev.*, **84**, 108–128.
262. Forbes, R. (2008) On the need for a tunneling pre-factor in Fowler–Nordheim tunneling theory. *J. Appl. Phys.*, **103** (11), 114 911.
263. Cutler, P. and Gibbons, J. (1958) Model for the surface potential barrier and the periodic deviations in the Schottky effect. *Phys. Rev.*, **111** (2), 394–402.
264. Buot, F.A. and Jensen, K.L. (1990) Lattice Weyl–Wigner formulation of exact many-body quantum-transport theory and applications to novel solid-state quantum-based devices. *Phys. Rev. B*, **42** (15), 9429.
265. Jensen, K.L. and Buot, F.A. (1990) Numerical aspects on the simulation of IV characteristics and switching times of resonant tunneling diodes. *J. Appl. Phys.*, **67** (4), 2153–2155.
266. Tsuchiya, H., Ogawa, M., and Miyoshi, T. (1991) Simulation of quantum transport in quantum devices with spatially varying effective mass. *IEEE Trans. Electron Devices*, **38** (6), 1246–1252.
267. Miyamoto, K. and Yamamoto, H. (1998) Resonant tunneling in asymmetrical double-barrier structures under an applied electric field. *J. Appl. Phys.*, **84** (1), 311–318.
268. Zhao, P., Cui, H., Woolard, D., Jensen, K.L., and Buot, F.A. (2000) Simulation of resonant tunneling structures: Origin of the I-V hysteresis and plateau-like structure. *J. Appl. Phys.*, **87** (3), 1337–1349.
269. Buot, F.A., Zhao, P., Cui, H., Woolard, D., Jensen, K.L., and Krowne, C. (2000) Emitter quantization and double hysteresis in resonant-tunneling structures: a nonlinear model of charge oscillation and current bistability. *Phys. Rev. B*, **61** (8), 5644–5665.
270. Price, P.J. (2003) Electron tunneling from channel to gate. *Appl. Phys. Lett.*, **82** (13), 2080–2081.
271. Price, P.J. (2004) A tunnelling formalism. *Semiconductor Science and Technology*, **19** (4), S241–S243.

272. Testè, P. and Chabrerie, J.P. (1996) Some improvements concerning the modelling of the cathodic zone of an electric arc (ion incidence on electron emission and the 'cooling effect'). *J. Phys. D*, **29** (3), 697.
273. Paschen, F. (1889) Ueber die zum funkenübergang in luft, wasserstoff und kohlensäure bei verschiedenen drucken erforderliche potentialdifferenz. *Ann. Phys.-Berlin*, **273** (5), 69–96.
274. Schonhuber, M. (1969) Breakdown of gases below paschen minimum: Basic design data of high-voltage equipment. *Power Apparatus and Systems, IEEE Transactions on*, **PAS-88** (2), 100–107.
275. Go, D.B. and Pohlman, D.A. (2010) A mathematical model of the modified Paschenäôs curve for breakdown in microscale gaps. *J. Appl. Phys.*, **107** (10), 103 303.
276. Torres, J. and Dhariwal, R. (1999) Electric field breakdown at micrometre separations. *Nanotechnology*, **10** (1), 102–107.
277. Tirumala, R. and Go, D.B. (2010) An analytical formulation for the modified Paschenäôs curve. *Appl. Phys. Lett.*, **97** (15), 151 502.
278. Rumbach, P. and Go, D.B. (2012) Fundamental properties of field emission-driven direct current microdischarges. *J. Appl. Phys.*, **112** (10), 103 302.
279. Go, D.B. (2013) Theoretical analysis of ion-enhanced thermionic emission for low-temperature, non-equilibrium gas discharges. *J. Phys. D*, **46** (3), 035 202.
280. Li, Y., Tirumala, R., Rumbach, P., and Go, D. (2013) The coupling of ion-enhanced field emission and the discharge during microscale breakdown at moderately high pressures. *IEEE Trans. Plasma Sci.*, **41** (1), 24–35.
281. Jensen, K.L., Feldman, D.W., Virgo, M., and O'Shea, P.G. (2003) Measurement and analysis of thermal photoemission from a dispenser cathode. *Phys. Rev. Spec. Top. Accel. Beams*, **6** (8), 083 501.
282. Jensen, K.L., Lau, Y., and Jordan, N. (2006) Emission nonuniformity due to profilimetry variation in thermionic cathodes. *Appl. Phys. Lett.*, **88** (16), 164 105.
283. Wilson, P. (2007) A theory for the RF surface field for various metals at the destructive breakdown limit. *Report No. SLAC-PUB-12354*, **877**.
284. Padamsee, H. and Knobloch, J. (1999) The nature of field emission from microparticles and the ensuing voltage breakdown, in *High Energy Density Microwaves High Energy Density Microwaves Aip Conf. Proc.*, vol. **474**(1) (eds R.M. Phillips and K. McMillen), AIP, Pajaro Dunes, CA, pp. 212–248.
285. Stern, T., Gossling, B., and Fowler, R. (1929) Further studies in the emission of electrons from cold metals. *Philos. Trans. R. Soc. London, Ser. A*, **124** (795), 699–723.
286. Lewis, T. (1955) High field electron emission from irregular cathode surfaces. *J. Appl. Phys.*, **26** (12), 1405–1410.
287. Radmilovic-Radjenovic, M. and Radjenovic, B. (2008) An analytical relation describing the dramatic reduction of the breakdown voltage for the microgap devices. *Europhys. Lett.*, **83** (2), 25 001.
288. Jensen, K.L. (2001) Electron transmission through modified schottky barriers, in *Mat. Res. Soc. Symp. Proc.*, vol. 685, Materials Research Society, pp. 365–370.
289. Kapur, P. and Peierls, R. (1937) Penetration into potential barriers in several dimensions. *Philos. Trans. R. Soc. London, Ser. A*, **163** (A915), 606–610.
290. Dirac, P. (1929) Quantum mechanics of many-electron systems. *Philos. Trans. R. Soc. London, Ser. A – Containing Papers of a Mathematical and Physical Character*, **123** (792), 714–733.
291. Kronig, R. and Penney, W. (1931) Quantum mechanics of electrons on crystal lattices. *Philos. Trans. R. Soc. London, Ser. A*, **130** (814), 499–513.
292. Ashcroft, N.W. and Mermin, N.D. (1976) *Solid state physics*, Holt, Rinehart and Winston, New York.
293. Faraday, M. (1952) *Experimental Researches in Electricity, Great Books of the Western World*, vol. **45**, Encyclopaedia Britannica, Inc. (ed. Robert Maynard Hutchins.), Chicago.
294. Ferry, D.K. (2000) *Semiconductor Transport*, Taylor and Francis, London, New York.
295. Hoffmann, A., Reuschel, K., and Rupprecht, H. (1959) Measurement of the Hall effect and conductivity of super-pure silicon. *J. Phys. Chem. Solids*, **11**, 284–287.
296. Matula, R.A. (1979) Electrical resistivity of copper, gold, palladium, and silver. *J. Phys. Chem. Ref. Data*, **8** (4), 1147–1298.
297. Dynes, R. and Lee, P. (1984) Localization, interactions, and the metal-insulator transition. *Science*, **223** (4634), 355–360.
298. Cutler, P., Miskovsky, N., Lerner, P., and Chung, M. (1999) The use of internal field emission to inject electronic charge carriers into the conduction band of diamond films: a review. *Appl. Surf. Sci.*, **146** (1-4), 126–133.
299. Ghahramani, E. and Sipe, J.E. (1989) Pressure dependence of the band gaps of semiconductors. *Phys. Rev. B*, **40** (18), 12 516.
300. O'Donnell, K. and Chen, X. (1991) Temperature-dependence of semiconductor band-gaps. *Appl. Phys. Lett.*, **58** (25), 2924–2926.
301. Bludau, W., Onton, A., and Heinke, W. (1974) Temperature-dependence of band-gap of silicon. *J. Appl. Phys.*, **45** (4), 1846–1848.
302. Panish, M. and Casey, H. (1969) Temperature dependence of energy gap in GaAs and GaP. *J. Appl. Phys.*, **40** (1), 163.
303. Clark, C., Dean, P., and Harris, P. (1964) Intrinsic edge absorption in diamond. *Proc. R. Soc. London, Ser. A*, **277** (1370), 312–329.
304. Pierret, R.F. (1987) *Advanced semiconductor fundamentals*, Modular series on solid state devices, Vol 6, Addison-Wesley Pub. Co., Reading, MA.
305. Stratton, R. (1962) Theory of field emission from semiconductors. *Phys. Rev.*, **125** (1), 67–82.
306. Jensen, B.L. (1973) Quantum-theory of free carrier absorption in polar semiconductors. *Ann. Phys.-New York*, **80** (2), 284–360.
307. Atwater, H.A. (1974) *Introduction to general relativity*, Pergamon Press, Oxford, New York.
308. Jensen, K.L., Jensen, B., Montgomery, E.J., Feldman, D.W., O'Shea, P.G., and Moody, N.A. (2008) Theory of photoemission from cesium antimonide using an alpha-semiconductor model. *J. Appl. Phys.*, **104** (4), 044 907–044 910.
309. Weaver, J. and Frederikse, H. (2007) Chapter 12: Optical properties of selected elements, in *CRC Handbook of Chemistry and Physics, 88th Edition* (ed. D.R. Lide), CRC, p. 12.126.

310. Berger, L. (1997) *Semiconductor Materials*, CRC, Boca Raton, FL.
311. Jensen, B.L. and Jensen, K.L. (2009) The quantum mechanical extension of the Drude Zener theory and the optical constants of an alpha semiconductor. *Journal of Computational and Theoretical Nanoscience*, **6**, 1770–1788.
312. Stratton, R. (1955) Field emission from semiconductors. *Proceedings of the Physical Society. Section B*, **68** (10), 746–757.
313. Baskin, L., Lvov, O., and Fursey, G. (1971) General features of field emission from semiconductors. *Phys. Status Solidi B*, **47** (1), 49–62.
314. Simmons, J.G. (1963) Generalized formula for the electric tunnel effect between similar electrodes separated by a thin insulating film. *J. Appl. Phys.*, **34** (6), 1793–1803.
315. Rhoderick, E. (1982) Metal-semiconductor contacts. *IEE Proc.*, **129** (1), 1.
316. Rhoderick, E.H. (1978) *Metal-semiconductor contacts*, Clarendon Press, Oxford.
317. Stratton, R. (1964) Electron tunneling with diffuse boundary conditions. *Phys. Rev. a*, **136** (3A), A837.
318. Jensen, K.L. (1995) Improved Fowler–Nordheim equation for field emission from semiconductors. *J. Vac. Sci. Technol. B*, **13** (2), 516–521.
319. Forbes, R.G. (2012) Extraction of emission parameters for large-area field emitters, using a technically complete Fowler–Nordheim-type equation. *Nanotechnology*, **23** (9), 095 706–12.
320. Lerner, P., Cutler, P.H., and Miskovsky, N. (1997) Theoretical analysis of field emission from a metal diamond cold cathode emitter. *J. Vac. Sci. Technol. B*, **15** (2), 337–342.
321. Robertson, J. (1999) Mechanisms of electron field emission from diamond, diamond-like carbon, and nanostructured carbon. *J. Vac. Sci. Technol. B*, **17** (2), 659–665.
322. Jensen, K.L., Shaw, J.L., Yater, J.E., and Pate, B.B. (2015) Enhancing secondary yield of a diamond amplifier using a nitrogen layer. *J. Appl. Phys.*, **117** (21), 214 501.
323. Brillson, L. (1994) Metal-semiconductor interfaces. *Surf. Sci.*, **299/300**, 909–927.
324. Dimitrov, D.A., Smithe, D., Cary, J.R., Ben-Zvi, I., Rao, T., Smedley, J., and Wang, E. (2015) Modeling electron emission and surface effects from diamond cathodes. *J. Appl. Phys.*, **117**, 055 708.
325. Jensen, K.L. and Buot, F.A. (1991) The methodology of simulating particle trajectories through tunneling structures using a Wigner distribution approach. *IEEE Trans. Electron Devices*, **38** (10), 2337–2347.
326. Stern, F. (1973) Quantum properties of surface space-charge layers. *CRC Critical Reviews in Solid State Sciences*, **4** (1-4), 499–514.
327. Christov, S. (1969) Injected electron currents through insulators. *Phys. Status Solidi*, **32** (2), 509–519.
328. Waters, R. and Van Zeghbroeck, B. (1998) Fowler–Nordheim tunneling of holes through thermally grown SiO2 on p+ 6H-SiC. *Appl. Phys. Lett.*, **73** (25), 3692–3694.
329. Lenzlinger, M. and Snow, E.H. (1969) Fowler–Nordheim tunneling into thermally grown SiO2. *J. Appl. Phys.*, **40** (1), 278–283.
330. Lopez-Villanueva, J., Jimenez Tejada, J., Cartujo, P., Bausells, J., and Carceller, J. (1991) Analysis of the effects of constant current Fowler–Nordheim tunneling injection with charge trapping inside the potential barrier. *J. Appl. Phys.*, **70** (7), 3712–3720.
331. Hiraiwa, A., Saito, T., Daicho, A., and Kawarada, H. (2013) Accuracy assessment of sheet-charge approximation for Fowler–Nordheim tunneling into charged insulators. *J. Appl. Phys.*, **114** (13), 134 501.
332. Frenkel, J. (1938) On pre-breakdown phenomena in insulators and electronic semi-conductors. *Phys. Rev.*, **54** (8), 647–648.
333. Karpus, V. and Perel', V. (1986) Multiphonon ionization of deep centers in semiconductors in an electric field. *Sov. Phys. JETP*, **64** (6), 1376.
334. Ganichev, S., Ziemann, E., Prettl, W., Yassievich, I., Istratov, A., and Weber, E. (2000) Distinction between the Poole–Frenkel and tunneling models of electric-field-stimulated carrier emission from deep levels in semiconductors. *Phys. Rev. B*, **61** (15), 10 361–10 365.
335. Göhl, A., Günther, B., Habermann, T., Müller, G., Schreck, M., Thürer, K., and Stritzker, B. (2000) Field emission mechanism from undoped chemical vapor deposition diamond films. *J. Vac. Sci. Technol. B*, **18** (2), 1031–1034.
336. Shaw, J. and Itoh, J. (2001) Silicon field emitter arrays, in *Vacuum Microelectronics* (ed. W. Zhu), Wiley, New York, p. 187.
337. Kiejna, A. and Wojciechowski, K. (1981) Work function of metals – relation between theory and experiment. *Prog. Surf. Sci.*, **11** (4), 293–338.
338. Jensen, K.L. (1999) Exchange-correlation, dipole, and image charge potentials for electron sources: Temperature and field variation of the barrier height. *J. Appl. Phys.*, **85** (5), 2667–2680.
339. Forbes, R.G. (1999) The electrical surface as Centroid of the surface-induced charge. *Ultramicroscopy*, **79** (1-4), 25–34.
340. Lang, N. and Kohn, W. (1970) Theory of metal surfaces: Charge density and surface energy. *Phys. Rev. B*, **1** (12), 4555–4568.
341. Kiejna, A. (1991) Image potential matched self-consistently to an effective potential for simple-metal surfaces. *Phys. Rev. B*, **43** (18), 14 695–14 698.
342. Jones, W. and March, N.H. (1985) *Theoretical Solid State Physics*, vol. 2, Dover Publications, New York.
343. Prutton, M. (1994) *Introduction to Surface Physics*, Clarendon Press, Oxford University Press, Oxford, New York.
344. Lang, N. and Kohn, W. (1973) Theory of metal-surfaces – induced surface charge and image potential. *Phys. Rev. B*, **7** (8), 3541–3550.
345. Forbes, R.G. (1997) On the problem of establishing the location of the electrical surface (image plane) for a tungsten field emitter. *Zeitschrift Fur Physikalische Chemie-international Journal of Research in Physical Chemistry and Chemical Physics*, **202**, 139–161.
346. Jensen, K.L. (2014) Field emission: Fundamental theory to usage, in *Wiley Encyclopedia of Electrical and Electronics Engineering* (ed. J.G. Webster), John Wiley & Sons, Inc.
347. Jensen, K.L. (1999) Semianalytical model of electron source potential barriers. *J. Vac. Sci. Technol. B*, **17** (2), 515–519.
348. Lang, N. and Kohn, W. (1971) Theory of metal surfaces – work function. *Phys. Rev. B*, **3** (4), 1215–1223.
349. Kiejna, A. (1981) Simple nonlocal calculation of Jellium metal-surface properties. *Phys. Status Solidi B*, **105** (1), 147–153.
350. Smith, J.R. (1969) Self-consistent many-electron theory of electron work functions and surface potential characteristics for selected metals. *Phys. Rev.*, **181** (2), 522–529.

351. Appelbaum, J. and Hamann, D. (1972) Variational calculation of the image potential near a metal surface. *Phys. Rev. B*, **6** (4), 1122.
352. Kiejna, A. and Wojciechowski, K. (1996) *Metal Surface Electron Physics*, Elsevier Science, Oxford, New York.
353. Fetter, A.L. and Walecka, J.D. (1971) *Quantum theory of many-particle systems*, McGraw-Hill, San Francisco.
354. Gell-Mann, M. and Brueckner, K.A. (1957) Correlation energy of an electron gas at high density. *Phys. Rev.*, **106** (2), 364.
355. Pines, D. (1962) *The many-body problem: a lecture note and reprint volume.*, Frontiers in Physics, W.A. Benjamin, New York.
356. Ceperley, D.M. and Alder, B. (1980) Ground-state of the electron-gas by a stochastic method. *Phys. Rev. Lett.*, **45** (7), 566–569.
357. Jones, W. and March, N.H. (1985) *Theoretical Solid State Physics*, vol. 1, Dover Publications, New York.
358. Nicolaou, N. and Modinos, A. (1975) Band-structure effects in field-emission energy distributions in tungsten. *Phys. Rev. B*, **11** (10), 3687.
359. Joachain, C.J. (1975) *Quantum collision theory*, North-Holland Pub. Co., American Elsevier Pub. Co., Amsterdam, New York.
360. Kohn, W. and Sham, L. (1965) Self-consistent equations including exchange and correlation effects. *Phys. Rev.*, **140** (4A), 1133.
361. Kiejna, A. (1995) Response of a stabilized-Jellium surface to a static electric field. *Surf. Sci.*, **333**, 1167–1171.
362. Perdew, J. (1995) Simple theories for simple metals – face-dependent surface energies and work-functions. *Prog. Surf. Sci.*, **48** (1-4), 245–259.
363. Kiejna, A. (1999) Stabilized Jellium – simple model for simple-metal surfaces. *Prog. Surf. Sci.*, **61** (5-6), 85–125.
364. Smoluchowski, R. (1941) Anisotropy of the electronic work function of metals. *Phys. Rev.*, **60** (9), 661–674.
365. Brodie, I. (1995) Uncertainty, topography, and work function. *Phys. Rev. B*, **51** (19), 13 660–13 668.
366. Rittner, E., Ahlert, R., and Rutledge, W. (1957) Studies on the mechanism of operation of the L cathode. 1. *J. Appl. Phys.*, **28** (2), 156–166.
367. Blevis, E.H. (1959) *The Temperature Dependence of the Work Functions of Single Crystal Surfaces of Copper*, Ph.D. thesis, McGill University.
368. Il'chenko, L. and Kryuchenko, Y. (1995) External-field penetration effect on current-field characteristics of metal emitters. *J. Vac. Sci. Technol. B*, **13** (2), 566–570.
369. Thomas, R., Gibson, J., Haas, G., and Abrams, R. (1990) Thermionic sources for high-brightness electron beams. *IEEE Trans. Electron Devices*, **37** (3), 850–861.
370. Falce, L. and Thomas, R. (1978) Controlled porosity dispenser cathode: Iridium-barium oxide, in *Technical Digest of the IEEE Int. El. Devices Meet.*, vol. **24**, IEEE, pp. 156–159.
371. Longo, R. (1978) Long life, high current density cathodes, in *Technical Digest of the IEEE Int. El. Devices Meet.*, vol. **24**, IEEE, pp. 152–155.
372. Green, M. (1980) The m-type cathode - no longer magic?, in *Technical Digest of the IEEE Int. El. Devices Meet.*, vol. **26**, IEEE, pp. 471–474.
373. Thomas, R.E. (1985) Diffusion and desorption of barium and oxygen on tungsten dispenser cathode surfaces. *Appl. Surf. Sci.*, **24** (3-4), 538–556.
374. Shih, A., Berry, A., Marrian, C., and Haas, G. (1987) Os-coated cathode for very high emission-density applications. *IEEE Trans. Electron Devices*, **34** (5), 1193–1200.
375. Gartner, G., Geittner, P., Raasch, D., and Wiechert, D. (1999) Supply and loss mechanisms of ba dispenser cathodes. *Appl. Surf. Sci.*, **146** (1-4), 22–30.
376. Shih, A., Yater, J.E., and Abrams, R. (1999) Thermal desorption of Ba from tungsten. *Appl. Surf. Sci.*, **146**, 1–6.
377. Jensen, K.L., Lau, Y., and Levush, B. (2000) Migration and escape of barium atoms in a thermionic cathode. *IEEE Trans. Plasma Sci.*, **28** (3), 772–781.
378. Longo, R. (2003) Physics of thermionic dispenser cathode aging. *J. Appl. Phys.*, **94** (10), 6966–6975.
379. Yamamoto, S. (2006) Fundamental physics of vacuum electron sources. *Rep. Prog. Phys.*, **69** (1), 181–232.
380. Pan, Z., Jensen, K.L., and Montgomery, E.J. (2013) Modeling the resupply, diffusion, and evaporation of cesium on the surface of controlled porosity dispenser photocathodes. *J. Appl. Phys.*, **114** (10), 104 906.
381. Montgomery, E.J., Feldman, D.W., O'shea, P., Pan, Z., Sennett, N., Jensen, K.L., and Moody, N.A. (2008) Advances in cesium dispenser photocathodes: Modeling and experiment. *Journal of Directed Energy*, **3**, 66–79.
382. Moody, N.A., Jensen, K.L., Feldman, D.W., Montgomery, E.J., and O'Shea, P.G. (2007) Factors affecting performance of dispenser photocathodes. *J. Appl. Phys.*, **102** (10), 104 901–104 909.
383. Travier, C., Devanz, G., Leblond, B., and Mouton, B. (1997) Experimental characterization of candela photo-injector. *NucL Instr. and Meth. in Phys. Res. A*, **393** (1-3), 451–454.
384. Leblond, B., Kuznetsov, G., and Batazova, M. (1996) Short pulse photoemission from lab6-based materials. *NucL Instr. and Meth. in Phys. Res. A*, **372** (3), 562–566.
385. Travier, C., Leblond, B., Bernard, M., Cayla, J., Thomas, P., and Georges, P. (1995) Candela photo-injector experimental results with a dispenser photocathode. *Proc IEEE Part Accel Conf*, **2**, 945–947 vol.2.
386. Septier, A., Sabary, F., Dudek, J., and Boumiz, A. (1992) An alkali dispenser photocathode (Al-Li)-Ag-O-Li. *Comptes Rendus De L Academie Des Sciences Serie II*, **314** (6), 569–574.
387. Olkin, I., Gleser, L.J., and Derman, C. (1980) *Probability Models And Applications*, Macmillan, New York.
388. Jensen, K.L. and Marrese-Reading, C. (2003) Emission statistics and the characterization of array current. *J. Vac. Sci. Technol. B*, **21** (1), 412–417.
389. Feinleib, M. and Green, M. (1984) High current density cathodes – an update, in *Technical Digest of the IEEE Int. El. Devices Meet.*, vol. **30**, IEEE, San Francisco, CA, pp. 314–317.
390. Vaughan, R. (1986) A synthesis of the Longo and Eng cathode emission models. *IEEE Trans. Electron Devices*, **33** (11), 1925–1927.
391. Jensen, K.L., Lebowitz, J., Lau, Y.Y., and Luginsland, J. (2012) Space charge and quantum effects on electron emission. *J. Appl. Phys.*, **111**, 054 917–054 919.
392. Ives, R., Falce, L., Schwartzkopf, S., and Witherspoon, R. (2005) Controlled porosity cathodes from sintered tungsten wires. *IEEE Trans. Electron Devices*, **52** (12), 2800–2805.

393. Langmuir, I. and Taylor, J.B. (1932) The mobility of caesium atoms adsorbed on tungsten. *Phys. Rev.*, **40**, 463–464.

394. Gyftopoulos, E.P. and Levine, J.D. (1962) Work function variation of metals coated by metallic films. *J. Appl. Phys.*, **33** (1), 67.

395. Jones, G.L. and Grant, J.T. (1983) Determination of barium and calcium evaporation rates from impregnated tungsten dispenser cathodes. *Appl. Surf. Sci.*, **16** (1-2), 25–39.

396. Wahnstrom, G. and Zhdanov, V. (1991) Dynamics and statistics of surface-diffusion at finite coverages. *Surf. Sci.*, **247** (1), 74–82.

397. Swanson, L.W., Strayer, R.W., Charbonnier, F.M., and Cooper, E.C. (1964) Field-emission investigation of thermal desorption and surface diffusion of cesium on tungsten. *NASA Conference Report*, **B054357**, 43.

398. Kleint, C. (1971) Surface diffusion model of adsorption-induced field emission flicker noise. 1. Theory. *Surf. Sci.*, **25** (2), 394–410.

399. Gomer, R. (1990) Diffusion of adsorbates on metal-surfaces. *Rep. Prog. Phys.*, **53** (7), 917–1002.

400. Butkov, E. (1968) *Mathematical Physics*, Addison-Wesley Pub. Co, Reading, MA.

401. Antczak, G. and Ehrlich, G. (2005) The beginnings of surface diffusion studies. *Surf. Sci.*, **589** (1-3), 52–66.

402. Ghatee, M., H. and Shams-Abadi, H. (2001) Linear exp-6 isotherm for compressed molten cesium over the whole liquid range including metal–nonmetal transition and Tc. *The Journal of Physical Chemistry B*, **105** (3), 702–710.

403. Reed, D.A. and Ehrlich, G. (1981) Surface diffusion, atomic jump rates and thermodynamics. *Surf. Sci.*, **102** (2-3), 588–609.

404. Murch, G. and Thorn, R. (1979) Isothermal transport in the lattice gas. *Philos. Mag. A*, **40** (4), 477–484.

405. Kutner, R. (1981) Chemical diffusion in the lattice gas of non-interacting particles. *Phys. Lett. A*, **81** (4), 239–240.

406. Reed, D. and Ehrlich, G. (1981) Surface diffusivity and the time correlation of concentration fluctuations. *Surf. Sci.*, **105** (2-3), 603–628.

407. Ala-nissila, T. and Ying, S. (1992) Theory of classical surface-diffusion. *Prog. Surf. Sci.*, **39** (3), 227–323.

408. Krylov, S.Y., Beenakker, J.J.M., and Tringides, M.C. (1999) On the theory of surface diffusion: kinetic versus lattice gas approach. *Surf. Sci.*, **420** (2-3), 233–249.

409. Levine, J.D. and Gyftopoulos, E.P. (1964) Adsorption physics of metals partially covered by metalic particles. II desoprtion rates of atoms and ions. *Surf. Sci.*, **1**, 225–241.

410. Mandel, J. (1984) *The Statistical Analysis Of Experimental Data*, Dover, New York.

411. Knuth, D.E. (1981) *Seminumerical Algorithms, The Art of Computer Programming*, vol. 2, Addison-Wesley Pub. Co., Reading, MA.

412. Peierls, R. (1936) Statistical theory of adsorption with interaction between the adsorbed atoms. *Proceedings of the Cambridge Philosophical Society*, **32**, 471–476.

413. Thompson, C.J. (1971) *Mathematical statistical mechanics*, Macmillan, New York.

414. Langmuir, I. (1933) Surface chemistry. *Chemical Reviews*, **13** (2), 147–191.

415. Moore, G.E. and Allison, H.W. (1955) Adsorption of strontium and of barium on tungsten. *J. Chem. Phys.*, **23** (9), 1609–1621.

416. Forman, R. (1976) Surface studies of barium and barium oxide on tungsten and its application to understanding the mechanism of operation of an impregnated tungsten cathode. *J. Appl. Phys.*, **47** (12), 5272–5279.

417. Becker, J. (1926) Thermionic and adsorption characteristics of caesium on tungsten and oxidized tungsten. *Phys. Rev.*, **28** (2), 0343–0361.

418. Taylor, J.B. and Langmuir, I. (1933) The evaporation of atoms, ions and electrons from caesium films on tungsten. *Phys. Rev.*, **44** (6), 423–458.

419. Knudsen, M. (1909) Die gesetze der molekularströmung und der inneren reibungsströmung der gase durch röhren. *Ann. Phys.-Berlin*, **28**, 75–130.

420. Knudsen, M. (1995) The laws of molecular flow and of inner friction flow of gases through tubes. *Journal of Membrane Science*, **100** (1), 23–25.

421. Roquais, J.M., Poret, F., le Doze, R., Ricaud, J.L., Monterrin, A., and Steinbrunn, A. (2003) Barium depletion study on impregnated cathodes and lifetime prediction. *Appl. Surf. Sci.*, **215** (1-4), 5–17.

422. DeMarcus, W. (1956) The problem of Knudsen flow Part i. General theory. *United States Atomic Energy Commission Report K-1302*, **Part 1**, 1–18.

423. Rhodes, R. (1986) *The making of the atomic bomb*, Simon & Schuster, New York.

424. Steckelmacher, W. (1986) Knudsen-flow 75 years on – the current state-of-the-art for flow of rarefied-gases in tubes and systems. *Reports on Progress in Physics*, **49** (10), 1083–1107.

425. Reich, G. (1990) On the significance of mean free-path to vacuum physics and technology. *Vacuum*, **41** (7-9), 2041–2044.

426. Pao, Y. (1970) Knudsen flow through a long circular tube. *Phys. Fluids*, **13** (2), 527–528.

427. Steckelmacher, W. (1978) Effect of cross-sectional shape on molecular flow in long tubes. *Vacuum*, **28** (6-7), 269–275.

428. DeMarcus, W. (1957) The problem of Knudsen flow Part VI. Tortuosity. *United States Atomic Energy Commission Report K-1302*, **Part VI**, 1–14.

429. Granatstein, V.L., Parker, R., and Armstrong, C. (1999) Vacuum electronics at the dawn of the twenty-first century, in *Proc. IEEE*, vol. **87**(5), pp. 702–716.

430. Forbes, R.G. (1999) Refining the application of Fowler–Nordheim theory. *Ultramicroscopy*, **79** (1-4), 11–23.

431. Nicolaescu, D. (1993) Physical basis for applying the Fowler–Nordheim J-E relationship to experimental IV data. *J. Vac. Sci. Technol. B*, **11** (2), 392–395.

432. Lohrmann, D.R. and Wu, D.C. (1999) Measurement of the dielectric strength of quartz crystalline material, *Tech. Rep. NRL/FR/5740-99-9919*, Naval Research Laboratory.

433. Binh, V., Dupin, J., Adessi, C., and Semet, V. (2001) Solid-state field-controlled emitters: a thin-film technology solution for industrial cathodes. *Solid-State Electronics*, **45** (6), 1025–1031.

434. Semet, V., Binh, V., Zhang, J., Yang, J., Khan, M., and Tsu, R. (2004) Electron emission through a multilayer planar nanostructured solid-state field-controlled emitter. *Appl. Phys. Lett.*, **84** (11), 1937–1939.

435. Semet, V., Binh, V., Zhang, J., Yang, J., Khan, M., and Tsu, R. (2005) Composite-layered solid-state field controlled emitter for a better control of the cathode surface barrier. *J. Vac. Sci. Technol. B*, **23** (2), 824–830.

436. Ancona, M. (1995) Thermomechanical analysis of failure of metal field emitters. *J. Vac. Sci. Technol. B*, **13** (6), 2206–2214.

437. Charbonnier, F. (1998) Arcing and voltage breakdown in vacuum microelectronics microwave devices using field emitter arrays: Causes, possible solutions, and recent progress. *J. Vac. Sci. Technol. B*, **16** (2), 880–887.

438. Purcell, S.T., Vincent, P., Journet, C., and Binh, V.T. (2002) Hot nanotubes: Stable heating of individual multiwall carbon nanotubes to 2000 K induced by the field-emission current. *Phys. Rev. Lett.*, **88** (10), 105 502.

439. Keser, A.C., Antonsen, T.M., Nusinovich, G.S., Kashyn, D.G., and Jensen, K.L. (2013) Heating of micro-protrusions in accelerating structures. *Phys. Rev. Spec. Top. Accel. Beams*, **16**, 092 001.

440. Fairchild, S.B., Boeckl, J., Back, T.C., Ferguson, J.B., Koerner, H., Murray, P.T., Maruyama, B., Lange, M.A., Cahay, M.M., Behabtu, N., Young, C.C., Pasquali, M., Lockwood, N.P., Averett, K.L., Gruen, G., and Tsentalovich, D.E. (2015) Morphology dependent field emission of acid-spun carbon nanotube fibers. *Nanotechnology*, **26** (10), 105 706. URL http://stacks.iop.org/0957-4484/26/i=10/a=105706.

441. Cahay, M., Zhu, W., Fairchild, S., Murray, P., Back, T., and Gruen, G. (2016) Multiscale model of heat dissipation mechanisms during field emission from carbon nanotube fibers. *Appl. Phys. Lett.*, **108** (3), 033 110.

442. Edgcombe, C. and Valdre, U. (2002) Experimental and computational study of field emission characteristics from amorphous carbon single nanotips grown by carbon contamination I. Experiments and computation. *Philos. Mag. B*, **82** (9), 987–1007.

443. Edgcombe, C. (2002) Experimental and computational study of field emission characteristics from amorphous carbon single nanotips grown by carbon contamination II. Theory. *Philos. Mag. B*, **82** (9), 1009–1026.

444. Edgcombe, C.J. (2010) New dimensions for field emission: Effects of structure in the emitting surface, in *Adv. Imag. Elect. Phys. 162* (ed. P. Hawkes), Academic Press, pp. 77–127.

445. Forbes, R.G., Edgcombe, C., and Valdre, U. (2003) Some comments on models for field enhancement. *Ultramicroscopy*, **95** (1-4), 57–65.

446. Martinelli, R.U. (1973) Effects of cathode bumpiness on the spatial resolution of proximity focused image tubes. *Appl. Optics*, **12** (8), 1841–1845.

447. Hunt, C., Trujillo, J., and Orvis, W. (1991) Structure and electrical characteristics of silicon field emission microelectronic devices. *IEEE Trans. Electron Devices*, **38** (10), 2309–2313.

448. Shiffler, D.A., Tang, W., Jensen, K.L., Golby, K., LaCour, M., Petillo, J.J., and Harris, J.R. (2015) Effective field enhancement factor and the influence of emitted space charge. *J. Appl. Phys.*, **118** (8), 083 302–6.

449. Forbes, R.G. and Jensen, K.L. (2001) New results in the theory of Fowler–Nordheim plots and the modelling of hemi-ellipsoidal emitters. *Ultramicroscopy*, **89** (1-3), 17–22.

450. Jensen, K.L., Shiffler, D.A., Rittersdorf, I.M., Lebowitz, J.L., Harris, J.R., Lau, Y.Y., Petillo, J.J., Tang, W., and Luginsland, J.W. (2015) Discrete space charge affected field emission: Flat and hemisphere emitters. *J. Appl. Phys.*, **117** (19), 194 902–17.

451. Jensen, K.L., Shiffler, D.A., Petillo, J.J., Pan, Z., and Luginsland, J.W. (2014) Emittance, surface structure, and electron emission. *Phys. Rev. Spec. Top. Accel. Beams*, **17** (4), 043 402–043 419.

452. Utsumi, T. (1991) Vacuum microelectronics: what's new and exciting. *IEEE Trans. Electron Devices*, **38** (10), 2276–2283.

453. Jensen, K.L. (2010) Space charge effects in field emission: Three dimensional theory. *J. Appl. Phys.*, **107** (1), 014 905–014 909.

454. Jensen, K.L. (2011) Space charge, emittance, trajectories, and the modeling of field emitter arrays. *J. Vac. Sci. Technol. B*, **29** (2), 02B101–7.

455. Filip, L.D., Carey, J.D., and Silva, S.R.P. (2011) Exact equipotential profile mapping: A self-validating method. *J. Appl. Phys.*, **109** (8).

456. Kokkorakis, G., Modinos, A., and Xanthakis, J. (2002) Local electric field at the emitting surface of a carbon nanotube. *J. Appl. Phys.*, **91** (7), 4580–4584.

457. Miller, R., Lau, Y., and Booske, J. (2009) Schottky's conjecture on multiplication of field enhancement factors. *J. Appl. Phys.*, **106** (10), 104 903.

458. Schwoebel, P.R., Spindt, C.A., and Holland, C. (2003) Spindt cathode tip processing to enhance emission stability and high-current performance. *J. Vac. Sci. Technol. B*, **21** (1), 433–435.

459. Schwoebel, P.R., Olson, R., Panitz, J., and Brodie, A. (2000) In situ cleaning of microfabricated field emitter cathodes. *J. Vac. Sci. Technol. B*, **18** (5), 2579–2582.

460. Schwoebel, P.R., Spindt, C.A., Holland, C., and Panitz, J. (2001) Emission uniformity enhancement between microfabricated tips in cold cathode arrays. *J. Vac. Sci. Technol. B*, **19** (2), 582–584.

461. Schwoebel, P.R., Spindt, C.A., Holland, C., and Panitz, J. (2001) Field emission current cleaning and annealing of microfabricated cold cathodes. *J. Vac. Sci. Technol. B*, **19** (3), 980–987.

462. Dyke, W.P., Charbonnier, F.M., Strayer, R.W., Floyd, R.I., Barbour, J.P., and Trolan, J. (1960) Electrical stability and life of the heated field emission cathode. *J. Appl. Phys.*, **31** (5), 790–805.

463. Temple, D., Palmer, W., Yadon, L., Mancusi, J., Vellenga, D., and McGuire, G. (1998) Silicon field emitter cathodes: Fabrication, performance, and applications. *J. Vac. Sci. Technol. A*, **16** (3), 1980–1990.

464. Parameswaran, L., Harris, C., Graves, C., Murphy, R., and Hollis, M. (1999) Resistive arc protection for field-emitter-array cold cathodes used in x-band inductive output amplifiers. *J. Vac. Sci. Technol. B*, **17** (2), 773–777.

465. Makishima, H., Miyano, S., Imura, H., Matsuoka, J., Takemura, H., and Okamoto, A. (1999) Design and performance of traveling-wave tubes using field emitter array cathodes. *Appl. Surf. Sci.*, **146** (1-4), 230–233.

466. Jensen, K.L. (1998) Analytical model of an emission-gated twystrode using a field emitter array. *J. Appl. Phys.*, **83** (12), 7982–7992.

467. Fomenko, V.S. and Samsonov, G.V. (1966) *Handbook of thermionic properties; electronic work functions and Richardson constants of elements and compounds*, Plenum Press Data Division [Consultants Bureau Enterprises], New York.

468. Kosmahl, H. (1991) Analytic evaluation of field emission enhancement factors for ellipsoidal cones and elliptic cross-section wedges. *IEEE Trans. Electron Devices*, **38** (6), 1534–1537.

469. He, J., Cutler, P., Miskovsky, N., Feuchtwang, T., Sullivan, T., and Chung, M. (1991) Derivation of the image interaction for nonplanar pointed emitter geometries – application to field emission IV characteristics. *Surf. Sci.*, **246** (1-3), 348–364.

470. Zuber, J., Jensen, K.L., and Sullivan, T. (2002) An analytical solution for microtip field emission current and effective emission area. *J. Appl. Phys.*, **91** (11), 9379–9384.

471. Martin, E., Charbonnier, F.M., Dolan, W.W., Dyke, W.P., Pitman, H.W., and Trolan, J.K. (1960) Research on field emission cathodes, *Tech. Rep. 59-20*, Wright Air Development Division, Wright Patterson AFB, OH.

472. Gotoh, Y., Kawamura, Y., Niiya, T., Ishibashi, T., Nicolaescu, D., Tsuji, H., Ishikawa, J., Hosono, A., Nakata, S., and Okuda, S. (2007) Derivation of length of carbon nanotube responsible for electron emission from field emission characteristics. *Appl. Phys. Lett.*, **90** (20), 203 107.

473. Bonard, J., Salvetat, J., Stockli, T., Forro, L., and Chatelain, A. (1999) Field emission from carbon nanotubes: Perspectives for applications and clues to the emission mechanism. *Appl Phys A*, **69** (3), 245–254.

474. Jensen, K.L., Phillips, P., Nguyen, K., Malsawma, L., and Hor, C. (1996) Electron emission from a single spindt-type field emitter structure: Correlation of theory and experiment. *Appl. Phys. Lett.*, **68** (20), 2807–2809.

475. Jensen, K.L., Zaidman, E.G., Kodis, M., Goplen, B., and Smithe, D. (1996) Analytical and seminumerical models for gated field emitter arrays. 1. Theory. *J. Vac. Sci. Technol. B*, **14** (3), 1942–1946.

476. Dvorson, L., Ding, M., and Akinwande, A. (2001) Analytical electrostatic model of silicon conical field emitters. Part I. *IEEE Trans. Electron Devices*, **48** (1), 134–143.

477. Dwight, H.B. (1961) *Tables of integrals and other mathematical data.*, Macmillan, New York.

478. Hartman, R., Mackie, W., and Davis, P. (1994) Use of boundary-element methods in-field emission computations. *J. Vac. Sci. Technol. B*, **12** (2), 754–758.

479. Calame, J., Gray, H., and Shaw, J.L. (1993) Analysis and design of microwave amplifiers employing field-emitter arrays. *J. Appl. Phys.*, **73** (3), 1485–1504.

480. Jensen, K.L., Abrams, R., and Parker, R. (1998) Field emitter array development for high frequency applications. *J. Vac. Sci. Technol. B*, **16** (2), 749–753.

481. Spindt, C.A., Holland, C., Schwoebel, P.R., and Brodie, I. (1998) Field emitter array development for microwave applications. II. *J. Vac. Sci. Technol. B*, **16** (2), 758–761.

482. Jensen, K.L. (1999) Field emitter arrays for plasma and microwave source applications. *Phys. Plasmas*, **6** (5), 2241–2253.

483. Whaley, D., Duggal, R., Armstrong, C., Bellew, C., Holland, C., and Spindt, C. (2009) 100 W operation of a cold cathode TWT. *IEEE Trans. Electron Devices*, **56** (5), 896–905.

484. Tang, C.M., Swyden, T.A., and Ting, A.C. (1995) Planar lenses for field emitter arrays. *J. Vac. Sci. Technol. B*, **13** (2), 571–575.

485. Hartman, R., Mackie, W., and Davis, P. (1996) Three-dimensional axisymmetric space charge simulation via boundary elements and emitted particles. *J. Vac. Sci. Technol. B*, **14** (3), 1952–1957.

486. Jensen, K.L., Zaidman, E.G., Kodis, M., Goplen, B., and Smithe, D. (1996) Analytical and seminumerical models for gated field emitter arrays. 2. Comparison of theory to experiment. *J. Vac. Sci. Technol. B*, **14** (3), 1947–1951.

487. Zaidman, E.G., Jensen, K.L., and Kodis, M. (1996) A, B, and C characterization of gated field emission arrays for radio frequency device performance. *J. Vac. Sci. Technol. B*, **14** (3), 1994–1999.

488. Watcharotone, S., Ruoff, R., and Read, F. (2008) Possibilities for graphene for field emission: modeling studies using the BEM. *Physics Procedia*, **1** (1), 71–75.

489. Renau, A., Read, F., and Brunt, J. (1982) The charge-density method of solving electrostatic problems with and without the inclusion of space-charge. *J. Phys. E*, **15** (3), 347–354.

490. Jensen, K.L., Mukhopadhyay-Phillips, P., Zaidman, E.G., Nguyen, K., Kodis, M., Malsawma, L., and Hor, C. (1997) Electron emission from a single Spindt-type field emitter: Comparison of theory with experiment. *Appl. Surf. Sci.*, **111**, 204–212.

491. Jensen, K.L., Kodis, M., Murphy, R., and Zaidman, E.G. (1997) Space charge effects on the current-voltage characteristics of gated field emitter arrays. *J. Appl. Phys.*, **82** (2), 845–854.

492. Murphy, R. and Kodis, M. (2001) *Cold cathode microwave devices*, in *Vacuum Microelectronics* (ed. W. Zhu), Wiley, New York, p. 349.

493. Phillips, P., Hor, C., Malsawma, L., Jensen, K.L., and Zaidman, E.G. (1996) Design and construction of apparatus for characterization of gated field emitter array electron emission. *Rev. Sci. Instrum.*, **67** (6), 2387–2393.

494. Groning, O., Kuttel, O., Emmenegger, C., Groning, P., and Schlapbach, L. (2000) Field emission properties of carbon nanotubes. *J. Vac. Sci. Technol. B*, **18** (2), 665–678.

495. Nilsson, L., Groening, O., Emmenegger, C., Kuettel, O., Schaller, E., Schlapbach, L., Kind, H., Bonard, J.M., and Kern, K. (2000) Scanning field emission from patterned carbon nanotube films. *Appl. Phys. Lett.*, **76** (15), 2071–2073.

496. Read, F. and Bowring, N. (2004) Field enhancement factors of one-dimensional and two-dimensional arrays of nanotubes. *Microelectronic Engineering*, **73-74**, 679–685.

497. Chen, G., Wang, W., Peng, J., He, C., Deng, S., Xu, N., and Li, Z. (2007) Screening effects on field emission from arrays of (5,5) carbon nanotubes: Quantum mechanical simulations. *Phys. Rev. B*, **76** (19), 195 412.

498. Tang, W., Golby, K., LaCour, M., and Knowles, T. (2014) Experimental study of electric field screening of four carbon fiber cathodes in a linear array configuration. *IEEE Trans. Plas. Sci.*, **42** (10), 2580–2581.

499. Bonard, J., Weiss, N., Kind, H., Stockli, T., Forro, L., Kern, K., and Chatelain, A. (2001) Tuning the field emission properties of patterned carbon nanotube films. *Adv. Matter.*, **13** (3), 184–188.

500. Zhu, Y., Yu, T., Cheong, F., Xu, X., Lim, C., Tan, V., Thong, J., and Sow, C. (2005) Large-scale synthesis and field emission properties of vertically oriented CuO nanowire films. *Nanotechnology*, **16** (1), 88.

501. Harris, J.R., Jensen, K.L., and Shiffler, D.A. (2015) Dependence of optimal spacing on applied field in ungated field emitter arrays. *AIP Advances*, **5** (8), 0871 982–7.

502. Harris, J., Jensen, K., and Shiffler, D. (2015) Modelling field emitter arrays using line charge distributions. *J. Phys. D*, **48** (38), 385 203.

503. Harris, J.R., Jensen, K.L., Shiffler, D.A., and Petillo, J.J. (2015) Shielding in ungated field emitter arrays. *Appl. Phys. Lett.*, **106** (20), 201 603–4.

504. Harris, J.R., Jensen, K.L., and Shiffler, D.A. (2015) Modelling field emitter arrays using line charge distributions. *J. Phys. D*, **48** (38).

505. Harris, J.R., Jensen, K.L., and Shiffler, D.A. (2016) A control of bulk and edge screening effects in two-dimensional arrays of ungated field emitters. *J. Vac. Sci. Technol. B*, **34** (4), 041 215–10.

506. Nilsson, L., Gröning, O., Gröning, P., and Schlapbach, L. (2001) Scanning anode field emission microscopy on carbon nanotube thin film emitters, in *Cold Cathodes: Proceedings of the First International Symposium*, vol. 2000-28 (eds M. Cahay, K.L. Jensen, P.D. Mumford, J. Yater, R.A. Murphy, D. Temple, and V.J. Kapoor), Electrochemical Society, Pennington, NJ, vol. **2000-28**, p. 52.

507. Hong, C. and Akinwande, A. (2003) Silicon metal-oxide-semiconductor field effect transistor/field emission array fabricated using chemical mechanical polishing. *J. Vac. Sci. Technol. B*, **21** (1), 500–505.

508. Guerrera, S., Velasquez-Garcia, L., and Akinwande, A. (2012) Scaling of high-aspect-ratio current limiters for the individual ballasting of large arrays of field emitters. *Electron Devices, IEEE Transactions on*, **59** (9), 2524–2530.

509. Cole, M.T., Teo, K.B.K., Groening, O., Gangloff, L., Legagneux, P., and Milne, W.I. (2014) Deterministic cold cathode electron emission from carbon nanofibre arrays. *Scientific Reports*, **4**, 4840.

510. Nilsson, L., Groening, O., Kuettel, O., Groening, P., and Schlapbach, L. (2002) Microscopic characterization of electron field emission. *J. Vac. Sci. Technol. B*, **20** (1), 326–337.

511. Mackie, W.A., Xie, T.B., and Davis, P. (1999) Transition metal carbide field emitters for field-emitter array devices and high current applications. *J. Vac. Sci. Technol. B*, **17** (2), 613–619.

512. Mackie, W.A., Southall, L.A., Xie, T.B., Cabe, G.L., Charbonnier, F.M., and McClelland, P.H. (2003) Emission fluctuation and slope-intercept plot characterization of pt and transition metal carbide field-emission cathodes in limited current regimes. *J. Vac. Sci. Technol. B*, **21** (4), 1574–1580.

513. Forbes, R.G., Deane, J., Hamid, N., and Sim, H. (2004) Extraction of emission area from Fowler–Nordheim plots. *J. Vac. Sci. Technol. B*, **22** (3), 1222–1226.

514. Rosenzweig, J., Barov, N., Hartman, S., Hogan, M., Park, S., Pellegrini, C., Travish, G., Zhang, R., Davis, P., Hairapetian, G., and Joshi, C. (1994) Initial measurements of the UCLA rf photoinjector. *NucL Instr. and Meth. in Phys. Res. A*, **341** (1-3), 379–385.

515. Lewellen, J. (2005) Future directions in electron sources, in *Proc. IEEE Particle Acc. Conf.*, Knoxville, TN, pp. 563–567.

516. Hernandez-Garcia, C., O'Shea, P.G., and Stutzman, M.L. (2008) Electron sources for accelerators. *Physics Today*, **61** (2), 44–49.

517. Fraser, J.S. and Sheffield, R.L. (1987) High-brightness injectors for rf-driven free-electron lasers. *IEEE J. Quantum Electron.*, **23** (9), 1489–1496.

518. Feldman, D.W., Bender, S., Carlsten, B., Early, J., Feldman, R., Johnson, W., Lumpkin, A., O'Shea, P.G., Stein, W., Sheffield, R., and Mckenna, K. (1991) Performance of the Los Alamos HIBAF accelerator at 17 mev. *NucL Instr. and Meth. in Phys. Res. A*, **304** (1-3), 224–227.

519. Travier, C. (1991) Rf guns – bright injectors for FEL. *NucL Instr. and Meth. in Phys. Res. A*, **304** (1-3), 285–296.

520. Dowell, D.H., Davis, K.J., Friddell, K.D., Tyson, E.L., Lancaster, C.a., Milliman, L., Rodenburg, R.E., Aas, T., Bemes, M., Bethel, S.Z., Johnson, P.E., Murphy, K., Whelen, C., Busch, G.E., and Remelius, D.K. (1993) First operation of a photocathode radio frequency gun injector A. High duty factor. *Appl. Phys. Lett.*, **63** (15), 2035–2037.

521. O'Shea, P.G., Bender, S.C., Byrd, D.A., Early, J.W., Feldman, D.W., Fortgang, C.M., Goldstein, J.C., Newnam, B.E., Sheffield, R.L., Warren, R.W., and Zaugg, T.J. (1993) Ultraviolet free-electron laser driven B. A high-brightness 45-mev electron beam. *Phys. Rev. Lett.*, **71** (22), 3661.

522. Pagani, C., Michelato, P., and Serafini, L. (1994) High brightness, long pulse, electron beam production with SC photo-injectors. *NucL Instr. and Meth. in Phys. Res. A*, **340** (1), 17–25.

523. Dowell, D., Bazarov, I., Dunham, B., Harkay, K., Hernandez-Garcia, C., Legg, R., Padmore, H., Rao, T., Smedley, J., and Wan, W. (2010) Cathode RandD for future light sources. *NucL Instr. and Meth. in Phys. Res. A*, **622**, 685–697.

524. Bostedt, C., Bozek, J.D., Bucksbaum, P.H., Coffee, R.N., Hastings, J.B., Huang, Z., Lee, R.W., Schorb, S., Corlett, J.N., Denes, P., Emma, P., Falcone, R.W., Schoenlein, R., Doumy, G., Kanter, E.P., Kraessig, B., Southworth, S., Young, L., Fang, L., Hoener, M., Berrah, N., Roedig, C., and DiMauro, L.F. (2013) Ultra-fast and ultra-intense x-ray sciences: first results from the linac coherent light source free-electron laser. *J. Phys. B*, **46** (16, SI).

525. Bazarov, I.V., Kim, A., Lakshmanan, M.N., and Maxson, J.M. (2011) Comparison of dc and superconducting rf photoemission guns for high brightness high average current beam production. *Phys. Rev. Spec. Top. Accel. Beams*, **14** (7), 072 001–072 012.

526. Bazarov, I.V., Ouzounov, D.G., Dunham, B.M., Belomestnykh, S.A., Li, Y., Liu, X., Meller, R.E., Sikora, J., Sinclair, C.K., Wise, F.W., and Miyajima, T. (2008) Efficient temporal shaping of electron distributions for high-brightness photoemission electron guns. *Phys. Rev. Spec. Top. Accel. Beams*, **11** (4), 040 702.

527. Ferrini, G., Michelato, P., and Parmigiani, F. (1998) A Monte Carlo simulation of low energy photoelectron scattering in Cs2Te. *Solid State Commun.*, **106** (1), 21–26.

528. Colson, W. (2001) Short wavelength free electron lasers in 2000. *NucL Instr. and Meth. in Phys. Res. A*, **475** (1-3), 397–400.

529. Mammei, R.R., Suleiman, R., Feingold, J., Adderley, P.A., Clark, J., Covert, S., Grames, J., Hansknecht, J., Machie, D., Poelker, M., Rao, T., Smedley, J., Walsh, J., McCarter, J.L., and Ruiz-Osés, M. (2013) Charge lifetime measurements at high average current using a K_2CsSb photocathode inside a dc high voltage photogun. *Phys. Rev. Spec. Top. Accel. Beams*, **16** (3), 033 401.
530. Leblond, B. (1992) Short pulse photoemission from a dispenser cathode under the 2nd, 3rd and 4th harmonics of a picosecond Nd-yag laser. *Nucl. Instr. and Meth. in Phys. Res. A*, **317** (1-2), 365–372.
531. Jensen, K.L., Feldman, D.W., Virgo, M., and O'Shea, P.G. (2003) Infrared photoelectron emission from scandate dispenser cathodes. *Appl. Phys. Lett.*, **83** (6), 1269–1271.
532. Jensen, K.L. and Montgomery, E.J. (2009) Photoemission theory and the development of high performance photocathodes. *Journal of Computational and Theoretical Nanoscience*, **6**, 1754–1769.
533. Suberlucq, G. (1996) Development and production of photocathodes for the CLIC test facility, in *Proc. 18th Int. FEL Conf.*, Rome, Italy, pp. 20–30.
534. Moody, N.A., Jensen, K.L., Feldman, D.W., O'Shea, P.G., and Montgomery, E.J. (2007) Prototype dispenser photocathode: Demonstration and comparison to theory. *Appl. Phys. Lett.*, **90** (11), 114 108–114 103.
535. Monaco, L., Michelato, P., Pagani, C., Sertore, D., Bähr, J.W., Han, J.H., Krasilnikov, M., Lederer, S., Schreiber, S., and Stephan, F. (2007) High QE photocathodes performance during operation at flash/pitz photoinjectors, in *Proc. IEEE Particle Acc. Conf.*, Albuquerque, NM.
536. Siggins, T., Sinclair, C., Bohn, C., Bullard, D., Douglas, D., Grippo, A., Gubeli, J., Krafft, G.A., and Yunn, B. (2001) Performance of a DC GaAs photocathode gun for the Jefferson lab FEL. *Nucl. Instr. and Meth. in Phys. Res. A*, **475** (1-3), 549–553.
537. Kong, S.H., Kinross-Wright, J., Nguyen, D.C., and Sheffield, R.L. (1995) Photocathodes for free electron lasers. *Nucl. Instr. and Meth. in Phys. Res. A*, **358** (1-3), 272–275.
538. Srinivasan-Rao, T., Fischer, J., and Tsang, T. (1991) Photoemission studies on metals using picosecond ultraviolet laser pulses. *J. Appl. Phys.*, **69** (5), 3291–3296.
539. Wang, X., Rao, T., Batchelor, K., Benzvi, I., and Fischer, J. (1995) Measurements on photoelectrons from a magnesium cathode in a microwave electron-gun. *Nucl. Instr. and Meth. in Phys. Res. A*, **356** (2-3), 159–166.
540. Van Loy, M.D., Young, A.T., and Leung, K.N. (1993) Measurements of barium photocathode quantum yields at four excimer laser wavelengths. *Appl. Phys. Lett.*, **63** (4), 476–478.
541. Calabrese, R., Ciullo, G., Dellamea, G., Egeni, G., Guidi, V., Lamanna, G., Lenisa, P., Maciga, B., Rigato, V., Rudello, V., Tecchio, L., Yang, B., and Zandolin, S. (1994) Performance of a GaAs electron source. *Nucl. Instr. and Meth. in Phys. Res. A*, **340** (1), 109–113.
542. Kane, E. (1962) Theory of photoelectric emission from semiconductors. *Phys. Rev.*, **127** (1), 131–141.
543. Mahan, G.D. (1970) Theory of photoemission in simple metals. *Phys. Rev. B*, **2** (11), 4334.
544. Krolikowski, W. and Spicer, W. (1969) Photoemission studies of the noble metals. I. Copper. *Phys. Rev.*, **185** (3), 882–900.
545. Powell, R., Spicer, W., Fisher, G., and Gregory, P. (1973) Photoemission studies of cesium telluride. *Phys. Rev. B*, **8**, 3987.
546. Maldonado, J.R., Pianetta, P., Dowell, D.H., Corbett, J., Park, S., Schmerge, J., Trautwein, A., and Clay, W. (2012) Experimental verification of the 3-step model of photoemission for energy spread and emittance measurements of copper and CsBr-coated copper photocathodes suitable for free electron laser applications. *Appl. Phys. Lett.*, **101** (23), 231 103–231 104.
547. Courths, R. and Hufner, S. (1984) Photoemission experiments on copper. *Physics Reports – Review Section of Physics Letters*, **112** (2), 53–171.
548. Lugovskoy, A.V. and Bray, I. (1998) Electron relaxation and excitation processes in metals. *J. Phys. D*, **31** (23), L78–L84.
549. Dyson, F. (2004) A meeting with Enrico Fermi. *Nature*, **427** (6972), 297.
550. Spicer, W. (1958) Photoemissive, photoconductive, and optical absorption studies of alkali-antimony compounds. *Phys. Rev.*, **112**, 114–122.
551. Taft, E. and Philipp, H. (1959) Structure in the energy distribution of photoelectrons from K_3Sb and Cs_3Sb. *Phys. Rev.*, **115** (6), 1583–1586.
552. Wooten, F. (1972) *Optical Properties Of Solids*, Academic Press, New York.
553. Philipp, H.R. and Ehrenreich, H. (1963) Optical properties of semiconductors. *Phys. Rev.*, **129** (4), 1550–1560.
554. Calkin, M.G. and Nicholson, P.J. (1967) Electrodynamics of a semiclassical free-electron gas. *Rev. Mod. Phys.*, **39** (2), 361.
555. Berger, L. (2007) Optical properties of selected inorganic and organic solids, in *CRC Handbook of Chemistry and Physics, 88th Edition* (ed. D.R. Lide), CRC, p. 12.141.
556. Marion, J.B. and Heald, M.A. (1980) *Classical electromagnetic radiation*, Academic Press, New York.
557. Brau, C.A. (2004) *Modern problems in classical electrodynamics*, Oxford University Press, New York.
558. Jensen, B. (1982) Quantum theory of the complex dielectric constant of free carriers in polar semiconductors. *IEEE J. Quantum Electron.*, **18** (9), 1361–1370.
559. Jensen, B.L. (1975) Quantum-mechanical extension of Boltzmann transport equation and optical absorption in semiconductors. *Ann. Phys.-New York*, **95** (2), 229–266.
560. Jensen, B.L. (1983) Review of the theory of infrared and far-infrared free-carrier behavior in semiconductors, in *Reviews of infrared and millimeter waves Vol. 8* (ed. K.J. Button), Academic Press, New York, pp. 127–171.
561. Ibach, H. and Lüth, H. (2003) *Solid-state Physics: An Introduction to Principles of Materials Science*, Advanced Texts in Physics, Springer, Berlin, New York.
562. Levinson, H.J., Greuter, F., and Plummer, E. (1983) Experimental band structure of aluminum. *Phys. Rev. B*, **27** (2), 727.
563. Rakic, A. (1995) Algorithm for the determination of intrinsic optical constants of metal-films – application to aluminum. *Appl. Optics*, **34** (22), 4755–4767.
564. Rakic, A., Djurisic, A., Elazar, J., and Majewski, M. (1998) Optical properties of metallic films for vertical cavity optoelectronic devices. *Appl. Optics*, **37** (22), 5271–5283.
565. Jensen, K.L. (2013) Scattering and the relationship between quantum efficiency and emittance. *J. Appl. Phys.*, **113** (5), 056 101.

566. Vecchione, T. (2012) QE and emittance from free electron metals, in *Photocathode Physics for Photoinjectors Workshop (P3)*, Cornell University, Ithaca, NY. URL http://edms.classe.cornell.edu/agenda/conferenceOtherViews.py?confId=15.
567. Hartmann, P., Bermuth, J., Von Harrach, D., Hoffmann, J., Kobis, S., Reichert, E., Aulenbacher, K., Schuler, J., and Steigerwald, M. (1999) A diffusion model for picosecond electron bunches from negative electron affinity GaAs photocathodes. *J. Appl. Phys.*, **86** (4), 2245–2249.
568. Jensen, K.L., Montgomery, E., Feldman, D.W., O'Shea, P.G., Harris, J., Lewellen, J., and Moody, N. (2011) Multiple scattering effects on quantum efficiency and response time for cesiated metal photocathodes. *J. Appl. Phys.*, **110** (3), 034 504–034 513.
569. Shabaev, A., Mehl, M., and Efros, A.L. (2015) Energy band structure of CuInS 2 and optical spectra of CuInS 2 nanocrystals. *Phys. Rev. B*, **92** (3), 035 431.
570. Topping, J. (1927) On the mutual potential energy of a plane network of doublets. *Philos. Trans. R. Soc. London, Ser. A*, **114** (766), 67–72.
571. Schmidt, L. and Gomer, R. (1965) Adsorption of potassium on tungsten. *J. Chem. Phys.*, **42** (10), 3573–3598.
572. Swanson, L.W. and Strayer, R.W. (1968) Field electron microscopy studies of cesium layers on various refractory metals: Work function change. *J. Chem. Phys.*, **48** (6), 2421–2442.
573. Septier, A., Sabary, F., Dudek, J., Bergeret, H., and Leblond, B. (1991) A binary Al/Li alloy A. A new material for the realization of high-intensity pulsed photocathodes. *NucL Instr. and Meth. in Phys. Res. A*, **304** (1-3), 392–395.
574. Gordy, W. and Thomas, W.J.O. (1956) Electronegativities of the elements. *J. Chem. Phys.*, **24** (2), 439–444.
575. Wang, C. (1977) High photoemission efficiency of submonolayer cesium-covered surfaces. *J. Appl. Phys.*, **48** (4), 1477–1479.
576. Longo, R., Adler, E., and Falce, L. (1984) Dispenser cathode life prediction model, in *Technical Digest of the IEEE Int. El. Devices Meet.*, vol. **30**, pp. 318–321.
577. Haas, G.A., Shih, A., and Marrian, C.R.K. (1983) Interatomic Auger analysis of the oxidation of thin Ba films II. Applications to impregnated cathodes. *Appl. Surf. Sci.*, **16** (1-2), 139–162.
578. Schmidt, L.D. (1967) Adsorption of barium on tungsten: Measurements on individual crystal planes. *J. Chem. Phys.*, **46** (10), 3830–3841.
579. Moody, N.A., Feldman, D.W., O'Shea, P.G., Jensen, K.L., and Balter, A. (2005) Measurement of low work function cesiated metals for use in dispenser photocathodes, in *Proc. 27th FEL Conf.*, Stanford, CA, pp. 168–171.
580. Montgomery, E.J., Pan, Z., Leung, J., Feldman, D.W., O'Shea, P.G., and Jensen, K.L. (2009) Towards a robust, efficient dispenser photocathode: the effect of recesiation on quantum efficiency. *Advanced Accelerator Concepts*, **1086**, 599–603.
581. Girardeau-Montaut, J., Girardeau-Montaut, C., and Moustaizis, S. (1994) Femtosecond nonlinear single-photon photoelectric-emission from tungsten a. 248 nm. *J. Phys. D*, **27** (4), 848–851.
582. Girardeau-Montaut, J. and Girardeau-Montaut, C. (1995) Theory of ultrashort nonlinear multiphoton photoelectric-emission from metals. *Phys. Rev. B*, **51** (19), 13 560–13 567.
583. Petillo, J.J., Nelson, E., Deford, J., Dionne, N., and Levush, B. (2005) Recent developments to the Michelle 2-d/3-d electron gun and collector modeling code. *IEEE Trans. Electron Devices*, **52** (5), 742–748.
584. Shih, A., Yater, J., Hor, C., and Abrams, R. (1997) Secondary electron emission studies. *Appl. Surf. Sci.*, **111**, 251–258.
585. Smedley, J., Ben-Zvi, I., Burrill, A., Chang, X., Grimes, J., Rao, T., Segalov, Z., and Wu, Q. (2006) *Electron amplification in diamond*, in *12th Adv. Acc. Concepts Workshop*, vol. **877**(1) (eds M. Conde and C. Eyberger), AIP Conf. Proc., Lake Geneva, Wisconsin (USA), pp. 672–679.
586. Dimitrov, D.A., Bushy, R., Bruhwiler, D.L., Cary, J.R., Ben-Zvi, I., Rao, T., Chang, X., Smedley, J., and Wu, Q. (2007) 3D simulations of secondary electron generation and transport in a diamond amplifier for photocathodes, in *Proc IEEE Part Accel Conf*, Albuquerque, NM, p. THPAS020.
587. Chang, X., Ben-Zvi, L., Burrill, A., Grimes, J., Rao, T., Segalov, Z., Smedley, J., and Wu, Q. (2007) Recent progress on the diamond amplified photo-cathode experiment, in *Recent progress on the diamond amplified photo-cathode experiment*, Brookhaven Natl Lab, New York, pp. 3788–3790.
588. Dimitrov, D., Busby, R., Cary, J., Ben-Zvi, I., Rao, T., Smedley, J., Chang, X., Keister, J., Wu, Q., and Muller, E. (2010) Multiscale three-dimensional simulations of charge gain and transport in diamond. *J. Appl. Phys.*, **108** (7), 073 712.
589. Wang, E., Ben-Zvi, I., Rao, T., Dimitrov, D.A., Chang, X., Wu, Q., and Xin, T. (2011) Secondary-electron emission from hydrogen-terminated diamond: Experiments and model. *Phys. Rev. Spec. Top. Accel. Beams*, **14**, 111 301.
590. Yater, J.E., Jensen, K.L., Feygelson, T.I., and Pate, B.B. (2016) Secondary electron transmission studies of the electron diffusion and thermalization processes in thin cvd diamond films. *MRS Advances*, **1** (16), 1081–1086.
591. Committee on a Scientific Assessment of Free-Electron Laser Technology for Naval Applications, N.R.C. (ed.) (2009) *Scientific Assessment of High-power Free-electron Laser Technology*, National Academies Press, Washington, DC.
592. Lewellen, J. (2004) High-brightness electron guns for linac-based light sources. *Proceedings of the SPIE - International Society for Optical Engineering*, **5534**, 22–36.
593. Parazzoli, C. and Koltenbah, B. (2000) End-to-end simulation of a visible 1 Kw FEL. *NucL Instr. and Meth. in Phys. Res. A*, **449** (1-2), 1–15.
594. Loos, H., Dowell, D., Sheehy, B., Shen, Y., Tsang, T., Wang, X., Serafini, L., Boscolo, M., Ferrario, M., Petrarca, M., and Vicario, C. (2005) Experimental studies of temporal electron beam shaping at the DUV-FEL accelerator, in *Proc. 27th FEL Conf.*, Stanford, CA, p. THPP066.
595. Altarelli, M., Brinkmann, R., Chergui, M., Decking, W., Dobson, B., Düsterer, S., Grübel, G., Graeff, W., Graafsma, H., Hajdu, J., Marangos, J., Pflüger, J., Redlin, H., Riley, D., Robinson, I., Rossbach, J., Schwarz, A., Tiedtke, K., Tschentscher, T., Vartaniants, I., Wabnitz, H., Weise, H., Wichmann, R., Witte, K., Wolf, A., Wulff, M., and Yurkov, M. (2007) The European X-ray free-electron laser technical design report, *Tech. Rep. 2006-097*, DESY, Hamburg, Germany.
596. Murata, K., Kyser, D., and Ting, C. (1981) Monte-Carlo simulation of fast secondary-electron production in electron-beam resists. *J. Appl. Phys.*, **52** (7), 4396–4405.
597. Pandey, L. and Rustgi, M. (1989) A comparative study of electron transport phenomenon in the Kev range. *J. Appl. Phys.*, **66** (12), 6059–6064.

598. Ziaja, B., London, R., and Hajdu, J. (2006) Ionization by impact electrons in solids: Electron mean free path fitted over a wide energy range. *J. Appl. Phys.*, **99** (3), 033 514.

599. Jablonski, A. and Powell, C. (2006) A predictive formula for the electron stopping power. *Journal of Surface Analysis*, **13** (2), 170–177.

600. Kanaya, K. and Okayama, S. (1972) Penetration and energy-loss theory of electrons in solid targets. *J. Phys. D*, **5** (1), 43–58.

601. Young, J. (1957) Dissipation of energy by 2.5-10 kev electrons in al2o3. *J. Appl. Phys.*, **28** (5), 524–525.

602. Klein, C.a. (1968) Bandgap dependence and related features of radiation ionization energies in semiconductors. *J. Appl. Phys.*, **39** (4), 2029–2038.

603. Jacoboni, C. and Reggiani, L. (1983) The Monte-Carlo method for the solution of charge transport in semiconductors with applications to covalent materials. *Rev. Mod. Phys.*, **55** (3), 645–705.

604. Fischetti, M.V. and Laux, S.E. (1988) Monte Carlo analysis of electron transport in small semiconductor devices including band-structure and space-charge effects. *Phys. Rev. B*, **38** (14), 9721.

605. Fischetti, M. (1991) Monte-carlo simulation of transport in technologically significant semiconductors of the diamond and zincblende structures.1. homogeneous transport. *IEEE Trans. Electron Devices*, **38** (3), 634–649.

606. Hockney, R.W. and Eastwood, J.W. (1981) *Computer simulation using particles*, McGraw-Hill International Book Co., New York.

607. Watanabe, T., Teraji, T., Ito, T., Kamakura, Y., and Taniguchi, K. (2004) Monte Carlo simulations of electron transport properties of diamond in high electric fields using full band structure. *J. Appl. Phys.*, **95** (9), 4866–4874.

608. Sverdlov, V., Ungersboeck, E., Kosina, H., and Selberherr, S. (2008) Current transport models for nanoscale semiconductor devices. *Mat. Sci. Eng. R*, **58** (6), 228–270.

609. Smullin, L.D. and Haus, H.A. (1959) *Noise in electron devices*, Technology Press Books in Science and Engineering, published jointly by the Technology Press of Massachusetts Institute of Technology and Wiley, Cambridge, New York.

610. Brodie, I. (1961) Emission fluctuations of tungsten-based barium dispenser cathodes. *J. Appl. Phys.*, **32**, 2039.

611. Cahay, M., Jensen, K.L., and von Allmen, P. (2002) Noise issues in cold cathodes for vacuum microelectronic applications, in *Noise and fluctuations control in electronic devices* (ed. A.A. Balandin), American Scientific Publishers, Stevenson Ranch, CA.

612. Rangaswamy, K., Cahay, M., and Jensen, K.L. (2004) Influence of image force potential on the shot noise properties of field emitters. *Appl. Phys. Lett.*, **85** (17), 3763–3765.

613. de Jonge, N., Allioux, M., Oostveen, J.T., Teo, K.B.K., and Milne, W.I. (2005) Low noise and stable emission from carbon nanotube electron sources. *Appl. Phys. Lett.*, **87** (13), 133 118–133 113.

614. Charbonnier, F.M., Southall, L.A., and Mackie, W.A. (2005) Noise and emission characteristics of NbC/Nb field emitters. *J. Vac. Sci. Technol. B*, **23** (2), 723–730.

615. Sun, S. and Ang, L.K. (2012) Shot noise of low energy electron field emission due to Klein tunneling. *J. Appl. Phys.*, **112** (1).

616. Bortz, A., Kalos, M., and Lebowitz, J. (1975) A new algorithm for Monte Carlo simulation of Ising spin systems. *J. Comput. Phys.*, **17** (1), 10–18.

617. Wagner, D.K. and Bowers, R. (1978) Radio-frequency size effect – tool for investigation of conduction electron-scattering in metals. *Adv. Phys.*, **27** (5), 651–746.

618. Nag, B.R. (1980) *Electron Transport in Compound Semiconductors*, Springer-Verlag,Berlin, New York.

619. Wilson, A.H. (1938) The electrical conductivity of the transition metals. *Philos. Trans. R. Soc. London, Ser. A*, **167** (A931), 0580–0593.

620. Kane, E. (1966) Simple model for collision effects in photoemission. *Phys. Rev.*, **147** (1), 335–339.

621. Ridley, B. (1977) Reconciliation of Conwell–Weisskopf and Brooks–Herring formulas for charged-impurity scattering in semiconductors – 3rd-body interference. *J. Phys. C*, **10** (10), 1589–1593.

622. Bardeen, J. and Pines, D. (1955) Electron–phonon interaction in metals. *Phys. Rev.*, **99** (4), 1140–1150.

623. Gasparov, V.A. and Huguenin, R. (1993) Electron–phonon, electron–electron and electron surface scattering in metals from ballistic effects. *Adv. Phys.*, **42** (4), 393–521.

624. Girardeau-Montaut, J., Girardeau-Montaut, C., Moustaizis, S., and Fotakis, C. (1993) High-current density produced by femtosecond nonlinear single-photon photoelectric-emission from gold. *Appl. Phys. Lett.*, **62** (4), 426–428.

625. Papadogiannis, N., Moustaizis, S., and Girardeau-Montaut, J. (1997) Electron relaxation phenomena on a copper surface via nonlinear ultrashort single-photon photoelectric emission. *J. Phys. D*, **30** (17), 2389–2396.

626. Westlake, D. and Alfred, L. (1968) Determination of Debye characteristic temperature of vanadium from bloch-gruneisen relation. *Journal of Physics and Chemistry of Solids*, **29** (11), 1931–1934.

627. White, G.K. and Woods, S.B. (1959) Electrical and thermal resistivity of the transition elements at low temperatures. *Philos. Trans. R. Soc. London, Ser. A*, **251** (995), 273–302.

628. Bardeen, J. and Shockley, W. (1950) Deformation potentials and mobilities in non-polar crystals. *Phys. Rev.*, **80** (1), 72–80.

629. Reggiani, L., Bosi, S., Canali, C., Nava, F., and Kozlov, S. (1981) Hole-drift velocity in natural diamond. *Phys. Rev. B*, **23** (6), 3050–3057.

630. Yater, J.E., Shih, A., Butler, J., and Pehrsson, P. (2002) Electron transmission studies of diamond films. *Appl. Surf. Sci.*, **191** (1-4), 52–60.

631. Serafini, L. and Rosenzweig, J.B. (1997) Envelope analysis of intense relativistic quasilaminar beams in rf photoinjectors: a theory of emittance compensation. *Phys. Rev. E*, **55** (6), 7565.

632. Brachmann, A., Coffee, R., Dowell, D., Edstrom, S., Emma, P., Gilevich, S., Hays, G., Hering, P., Huang, Z., Miahnahri, a., Nuhn, H., Ratner, D., Reis, D., White, W., Wu, J., and Xiang, D. (2009) LCLS drive laser shaping experiments, in *Proc 31st FEL Conf.*, Liverpool, UK, p. WEOA03.

633. Neil, G.R. and Merminga, L. (2002) Technical approaches for high-average-power free-electron lasers. *Rev. Mod. Phys.*, **74** (3), 685–701.

634. Booske, J.H., Dobbs, R.J., Joye, C.D., Kory, C.L., Neil, G.R., Park, G.S., Park, J., and Temkin, R.J. (2011) Vacuum electronic high power terahertz sources. *IEEE Tranansactions on Terahertz Science and Technology*, **1** (1), 54–75.

635. Shiffler, D., Ruebush, M., Zagar, D., LaCour, M., Golby, K., Clark, M., Haworth, M., and Umstattd, R. (2002) Emission uniformity and emittance of explosive field-emission cathodes. *IEEE Trans. Plas. Sci.*, **30** (4), 1592–1596.
636. Nojeh, A. (2014) Carbon nanotube electron sources: From electron beams to energy conversion and optophononics. *ISRN Nanomaterials*, **2014**, 23.
637. Humphries, S. (1986) *Principles of Charged Particle Acceleration*, Wiley, New York.
638. Reiser, M. (1994) *Theory and Design of Charged Particle Beams*, Wiley Series in Beam Physics and Accelerator Technology, Wiley, New York.
639. Nguyen, D.C., Sheffield, R.L., Fortgang, C.M., Goldstein, J.C., Kinross-Wright, J.M., and Ebrahim, N.A. (1998) Self-amplified spontaneous emission driven by a high-brightness electron beam. *Phys. Rev. Lett.*, **81** (4), 810.
640. Petillo, J., Panagos, D., Ovtchinnikov, S., Burke, A., Kostas, C., Jensen, K.L., Levush, B., Held, B., DeFord, J., and Nelson, E. (2010) 19.2: Modeling emission processes in the finite-element michelle gun and collector simulation code, in *IEEE International Vacuum Electronics Conference (IVEC)*, Monterey, CA, pp. 441–442.
641. Gilmour, A., Thelen, D., Macmullen, A., Kodis, M., Reynolds, R., and Emerson, R. (2004) Ion instabilities initiated in the gun of a linear beam microwave tube. *IEEE Trans. Plasma Sci.*, **32** (3), 1251–1261.
642. Zaidman, E.G. and Kodis, M. (1991) Emission gated device issues. *IEEE Trans. Electron Devices*, **38** (10), 2221–2228.
643. Whaley, D., Gannon, B., Smith, C., Armstrong, C., and Spindt, C. (2000) Application of field emitter arrays to microwave power amplifiers. *IEEE Trans. Plasma Sci.*, **28** (3), 727–747.
644. Dobbs, R., Roitman, A., Horoyski, P., Hyttinen, M., Sweeney, D., Steer, B., Khanh, N., Wright, E., Chernin, D., Burke, A., Calame, J., Levush, B., Barker, N.S., Booske, J., and Blank, M. (2010) Design and fabrication of terahertz extended interaction klystrons, in *35th Int. Conf. IRMMW-THZ*, IEEE, pp. 978–1–4244–6657–3.
645. Carlsten, B.E. (1998) Emittance growth due to radial density variations of an emittance-dominated electron beam in a channel with continuous acceleration and focusing. *Phys. Plasmas*, **5** (4), 1148–1161.
646. Lapostolle, P.M. (1971) Possible emittance increase through filamentation due to space charge in continuous beams. *IEEE Trans. Nucl. Sci.*, **18** (3), 1101.
647. Sacherer, F. (1971) RMS envelope equations with space charge. *IEEE Trans. Nucl. Sci.*, **18** (3), 1105.
648. Sun, A. and Sun, B. (1999) Focusing characteristics of intense beam solenoid lens. *Rev. Sci. Instrum.*, **70** (8), 3303–3307.
649. Whaley, D.R. (2014) Practical design of emittance dominated linear beams for RF amplifiers. *IEEE Trans. Electron Devices*, **61** (6), 1726–1734.
650. Christenson, P., Chernin, D., Garner, A., and Lau, Y. (1996) Resistive destabilization of cycloidal electron flow and universality of (near-) Brillouin flow in a crossed-field gap. *Phys. Plasmas*, **3** (12), 4455–4462.
651. Lau, Y. (1987) Some design considerations on using modulated intense annular electron beams for particle acceleration. *J. Appl. Phys.*, **62** (2), 351–356.
652. Kodis, M., Jensen, K.L., Zaidman, E.G., Goplen, B., and Smithe, D. (1996) Optimization of field emission arrays for inductive output amplifiers. *J. Vac. Sci. Technol. B*, **14** (3), 1990–1993.
653. Bohn, C.L. (1993) Transverse phase-space dynamics of mismatched charged-particle beams. *Phys. Rev. Lett.*, **70** (7), 932.
654. Bohn, C.L. and Sideris, I.V. (2003) Fluctuations do matter: Large noise-enhanced halos in charged-particle beams. *Phys. Rev. Lett.*, **91** (26), 264 801.
655. Joye, C.D., Calame, J.P., Nguyen, K.T., and Garven, M. (2012) Microfabrication of fine electron beam tunnels using UV-LIGA and embedded polymer monofilaments for vacuum electron devices. *Journal of Micromechanics and Microengineering*, **22** (1), 015 010.
656. Cooke, S., Chernyavskiy, I., Stanchev, G., Levush, B., and Antonsen, T. (2012) GPU-accelerated large-signal device simulation using the 3D particle-in-cell code 'neptune', in *39th Int. Conf. Plas. Sci.*, IEEE, Edinburgh, UK.
657. O'Shea, P.G. and Freund, H.P. (2001) Laser technology – free-electron lasers: Status and applications. *Science*, **292** (5523), 1853–1858.
658. Brau, C. (1990) *Free-Electron Lasers, Adv. Electron El. Phys. Suppl.*, vol. 22, Academic Press, Boston.
659. Brau, C. (2004) What brightness means, in *Physics and Applications of High Brightness Electron Beams: Proceedings of the ICFA Workshop Chia Laguna, Sardinia, Italy 1-6 July 2002* (eds J. Rosenzweig, G. Travish, and L. Serafini), World Scientific Publishing Company, p. 20.
660. Feldhaus, J., Krikunova, M., Meyer, M., Möller, T., Moshammer, R., Rudenko, A., Tschentscher, T., and Ullrich, J. (2013) AMO science at the FLASH and European XFEL free-electron laser facilities. *J. Phys. B*, **46** (16), 164 002.
661. Jensen, K.L., O'Shea, P.G., Feldman, D.W., and Shaw, J.L. (2010) Emittance of a field emission electron source. *J. Appl. Phys.*, **107** (1), 014 903–014 914.
662. Kodis, M., Jensen, K.L., Zaidman, E.G., Goplen, B., and Smithe, D. (1996) Operation and optimization of gated field emission arrays in inductive output amplifiers. *IEEE Trans. Plasma Sci.*, **24** (3), 970–981.
663. Hernandez, M., Fisher, A., Meyerhofer, D., Miller, R., Palmer, D., Park, S., Reis, D., Schmerge, J., Weaver, J., Wiedermann, H., Winick, H., and Yeremian, D. (1997) Emittance measurements for the SLAC gun test facility, in *Proc. IEEE Particle Acc. Conf.*, IEEE, pp. 2840–2842.
664. Borland, M., Chae, Y., Emma, P., Lewellen, J., Bharadwaj, V., Fawley, W., Krejcik, P., Limborg, C., Milton, S., Nuhn, H., Soliday, R., and Woodley, M. (2002) Start-to-end simulation of self-amplified spontaneous emission free electron lasers from the gun through the undulator. *Nucl. Instrum. Methods A*, **483** (1-2), 268–272.
665. Haber, I., Bieniosek, F., Celata, B., Friedman, A., Grote, D., Henestroza, E., Vay, J., Bernal, S., Kishek, R., O'Shea, P.G., Reiser, M., and Herrmannsfeldt, W. (2002) End-to-end simulation: the front end. *Laser and Particle Beams*, **20** (3), 431–433.
666. Kishek, R., Bernal, S., Bohn, C., Grote, D., Haber, I., Li, H., O'Shea, P.G., Reiser, M., and Walter, M. (2003) Simulations and experiments with space-charge-dominated beams. *Phys. Plasmas*, **10** (5), 2016–2021.
667. Nelson, E. and Petillo, J.J. (2004) An analysis of the basic space-charge-limited emission algorithm in a finite-element electrostatic gun code. *IEEE Trans. Plasma Sci.*, **32** (3), 1223–1235.

668. Neumann, J.G., Fiorito, R.B., O'Shea, P.G., Loos, H., Sheehy, B., Shen, Y., and Wu, Z. (2009) Terahertz laser modulation of electron beams. *J. Appl. Phys.*, **105** (5).

669. Buon, J. (1992) Beam phase space and emittance, in *CERN Accelerator School: 5th General Accelerator Physics Course*, LAL-RT-90-15-REV, p. 27.

670. Lawson, J.D., Lapostolle, P.M., and Gluckstern, R.L. (1973) Emittance, entropy and information. *Part. Accel.*, **5**, 61–65.

671. Brau, C.A. (1998) High-brightness electron beams – small free-electron lasers. *Nucl. Instr. and Meth. A*, **407** (1-3), 1–7.

672. O'Shea, P.G. (1998) Reversible and irreversible emittance growth. *Phys. Rev. E*, **57**, 1081–1087.

673. Jensen, K.L., O'Shea, P.G., and Feldman, D.W. (2010) Emittance of a photocathode: Effects of temperature and field. *Phys. Rev. Spec. Top. Accel. Beams*, **13** (8), 080 704–080 708.

674. Jensen, K.L., O'Shea, P.G., Feldman, D.W., and Moody, N.A. (2006) Theoretical model of the intrinsic emittance of a photocathode. *Appl. Phys. Lett.*, **89** (22), 224 103.

675. Schmerge, J.F., Bolton, P., Clendenin, J., Decker, F.J., Dowell, D., Gierman, S., Limborg, C., and Murphy, B. (2001) Transverse emittance measurements on an S-band photocathode rf electron gun. *Report No. SLAC-PUB-8963*.

676. Sannibale, F., Filippetto, D., Papadopoulos, C.F., Staples, J., Wells, R., Bailey, B., Baptiste, K., Corlett, J., Cork, C., De, Santis, S., Dimaggio, S., Doolittle, L., Doyle, J., Feng, J., Quintas, D.G., Huang, G., Huang, H., Kramasz, T., Kwiatkowski, S., Lellinger, R., Moroz, V., Norum, W.E., Padmore, H., Pappas, C., Portmann, G., Vecchione, T., Vinco, M., Zolotorev, M., and Zucca, F. (2012) Advanced photoinjector experiment photogun commissioning results. *Phys. Rev. Spec. Top. Accel. Beams*, **15** (10).

677. Karkare, S. and Bazarov, I. (2011) Effect of nanoscale surface roughness on transverse energy spread from GaAs photocathodes. *Appl. Phys. Lett.*, **98** (9).

678. Vecchione, T., Feng, J., Wan, W., Padmore, H.A., Ben-Zvi, I., Liang, X., Ruiz-Oses, M., Rao, T., Smedley, J., and Dowell, D. (2012) Effect of roughness on emittance of potassium cesium antimonide photocathodes, in *Int. Particle Accel. Conf.*, New Orleans, LA, p. MOPPP041.

679. Schubert, S., Ruiz-Oses, M., Ben-Zvi, I., Kamps, T., Liang, X., Muller, E., Muller, K., Padmore, H., Rao, T., Tong, X., Vecchione, T., and Smedley, J. (2013) Bi-alkali antimonide photocathodes for high brightness accelerators. *APL Materials*, **1** (3), 032 119–032 116.

680. Dowell, D.H., Joly, S., and Loulergue, A. (1997) The space charge limits of longitudinal emittance in RF photoinjectors, in *AIP Conference Proceedings*, vol. 398 (eds S. Chattopadhyay, J. McCullough, and P. Dahl), pp. 793–802.

681. Harris, J.R., Lewellen, J.W., and Poole, B.R. (2012) Transverse-longitudinal correlations in electron guns. *J. Appl. Phys.*, **112** (2), 023 304–5.

682. Graves, W.S., Kaertner, F.X., Moncton, D.E., and Piot, P. (2012) Intense superradiant x rays from a compact source using a nanocathode array and emittance exchange. *Phys. Rev. Lett.*, **108** (26), 263 904–5.

683. Sprangle, P., Penano, J., Hafizi, B., Gordon, D., Gold, S., Ting, A., and Mitchell, C. (2009) *High average current electron guns for high-power FELs, Tech. Rep. NRL/MR/6790-09-9230*, Naval Research Laboratory, Washington, DC.

684. Srinivasan-Rao, T., Ben-Zvi, I., Burrill, A., Citver, G., Hershcovitch, A., Pate, D., Reuter, A., Scaduto, J., Zhao, Q., Zhao, Y., Delayen, J., and Kneisel, P. (2003) Design, construction and status of all niobium superconducting photoinjector at BNL, in *Proc. Particle Accel. Conf.*, IEEE, MOPB010, pp. 92–94.

685. Todd, A. (2006) State-of-the-art electron guns and injector designs for energy recovery linacs (ERL). *NucL Instr. and Meth. in Phys. Res. A*, **557** (1), 36–44.

686. Dowell, D.H., Joly, S., Loulergue, A., de Brion, J.P., and Haouat, G. (1997) Observation of space-charge driven beam instabilities in a radio frequency photoinjector. *Phys. Plasmas*, **4** (9), 3369–3379.

687. Zhou, F., Ben-Zvi, I., Babzien, M., Chang, X.Y., Doyuran, A., Malone, R., Wang, X.J., and Yakimenko, V. (2002) Experimental characterization of emittance growth induced by the nonuniform transverse laser distribution in a photoinjector. *Phys. Rev. Spec. Top. Accel. Beams*, **5** (9), 094 203.

688. Sinclair, C.K. (2003) Very high voltage photoemission electron guns, in *Proc Particle Accel. Conf.*, IEEE, MOPB006, pp. 76–80.

689. Bazarov, I.V., Dunham, B.M., Hannon, F., Li, Y., Liu, X., Miyajima, T., Ouzounov, D.G., and Sinclair, C.K. (2007) Thermal emittance measurements from negative electron affinity photocathodes, in *Proc. Particle Acc. Conf.*, IEEE, Albuquerque, NM, TUPMS020.

690. Wu, Q., Ben-Zvi, I., Burrill, A., Chang, X., Kayran, D., Rao, T., and Smedley, J. (2007) Thermal emittance measurement design for diamond secondary emission, in *Proc. Particle Acc. Conf.*, IEEE, Albuquerque, NM, TUPMS089.

691. Drouin, D., Couture, A.R., Joly, D., Tastet, X., Aimez, V., and Gauvin, R. (2007) CASINO V2.42 – a fast and easy-to-use modeling tool for scanning electron microscopy and microanalysis users. *Scanning*, **29** (3), 92–101.

692. Bradley, D., Allenson, M., and Holeman, B. (1977) The transverse energy of electrons emitted from GaAs photocathodes. *J. Phys. D*, **10** (1), 111.

693. Lau, Y. (1987) Effects of cathode surface-roughness on the quality of electron-beams. *J. Appl. Phys.*, **61** (1), 36–44.

694. Jensen, K.L. and Cahay, M. (2006) General thermal-field emission equation. *Appl. Phys. Lett.*, **88** (15), 154 105.

695. Maldonado, J., Pianetta, P., Dowell, D., Smedley, J., and Kneisel, P. (2010) Performance of a CsBr-coated Nb photocathode at room temperature. *J. Appl. Phys.*, **107** (1), 013 106.

696. Dowell, D., Padmore, H., Vecchione, T., Schmerge, J., and Wan, W. (2012) Models for cathode roughness, in *Photocathode Physics for Photoinjectors Workshop (P3)*, Cornell University, Ithaca, NY.

697. Higuchi, T., Nakamura, O., Matsumoto, S., and Uda, E. (1999) Pore geometry of dispenser cathode surface vs. emission characteristics, and Ba recovery characteristics after ion bombardment. *Appl. Surf. Sci.*, **146** (1-4), 51–61.

698. Bonard, J., Dean, K., Coll, B., and Klinke, C. (2002) Field emission of individual carbon nanotubes in the scanning electron microscope. *Phys. Rev. Lett.*, **89** (19), 197 602–4.

699. Smedley, J., Rao, T., and Zhao, Q. (2005) Photoemission studies on niobium for superconducting photoinjectors. *J. Appl. Phys.*, **98** (4), 043 111.

700. Smedley, J., Rao, T., Sekutowicz, J., Kneisel, P., Langner, J., Strzyzewski, P., Lefferts, R., and Lipski, A. (2005) Progress on lead photocathodes for superconducting injectors, in *Progress on lead photocathodes for superconducting injectors*, IEEE, Knoxville, TN, pp. 4099–4101.
701. Milne, W.I., Teo, K.B.K., Minoux, E., Groening, O., Gangloff, L., Hudanski, L., Schnell, J.P., Dieumegard, D., Peauger, F., Bu, I.Y.Y., Bell, M.S., Legagneux, P., Hasko, G., and Amaratunga, G.A.J. (2006) Aligned carbon nanotubes/fibers for applications in vacuum microwave amplifiers. *J. Vac. Sci. Technol. B*, **24** (1), 345–348.
702. Wang, Y., Wang, J., Liu, W., Zhang, K., and Li, J. (2007) Development of high current-density cathodes with scandia-doped tungsten powders. *IEEE Trans. Electron Devices*, **54** (5), 1061–1070.
703. Barik, R., Tanwar, A., Baek, I., Kihoon, E., Bera, A., Sattorov, M., Min, S., Kwon, O., and Park, G.S. (2012) Effect of surface roughness of high current density nano-particle cathode for terahertz vacuum devices application, in *Intl. Vac. El. Conf / Intl. Vac. El. Sources Conf.*, IEEE, Monterey, CA, pp. 369–370.
704. Spindt, C., Holland, C., and Schwoebel, P.R. (2015) Thermal field forming of spindt cathode arrays. *J. Vac. Sci. Technol. B*, **33** (3).
705. Krasilnikov, M. (2006) THPPH013: Impact of the cathode roughness on the emittance of an electron beam, in *Proceedings of FEL Conference*, BESSY, Berlin, Germany, pp. 583–586.
706. Smedley, J., Rao, T., Sekutowicz, J., Kneisel, P., Langner, J., Strzyzewski, P., Lefferts, R., and Lipski, A. (2005) Progress on lead photocathodes for superconducting injectors, in *Proc. IEEE Particle Acc. Conf.*, IEEE, Knoxville, TN, pp. 4099–4101.
707. Rhee, M.J. and Boulais, K.A. (1991) Root-mean-square emittance of multiple beam systems. *Phys. Fluids B*, **3** (7), 1781–1786.
708. Finn, J., Antonsen, T.M., J., and Manheimer, W. (1988) Space-charge-limited and temperature-limited electron flow in the vicinity of edges and conical points. *IEEE Trans. Plas. Sci.*, **16** (2), 281–289.
709. Taylor, J.R. (1982) *An introduction to error analysis: the study of uncertainties in physical measurements*, University Science Books, Mill Valley, CA.

Index

Page numbers followed by *f* and *t* indicate figures and tables, respectively. Numbers followed by n indicate footnotes.

Introduction to the Physics of Electron Emission, First Edition. Kevin L. Jensen.

Printed in the USA/Agawam, MA
September 16, 2026

910661.207